Philosophy of Science

Second Edition

The frontispiece (engraved by Francesco Curti, 1603–70) is from *Almagestum Novum* (Bologna, 1651), by the Jesuit astronomer Giambattista Riccioli (1598–1671). In the decades following the condemnation of Galileo, Riccioli was an ardent critic of the Copernican theory. He conceded that Galileo's discovery of the phases of Venus had refuted the Ptolemaic system but insisted that Tycho Brahe's system, in which the earth does not move, captured all the observational and mathematical advantages of the Copernican theory with none of its physical and theological disadvantages. Riccioli's book (whose title is a deliberate reference to the "old" *Almagest* of Ptolemy, now discredited) gives an exhaustive survey of arguments for and against the Copernican theory, and concludes that Tycho Brahe's system (modified slightly by Riccioli) is more plausible.

Thus, Riccioli's frontispiece shows his own version of the Tychonic system weighing more heavily in the scales of evidence than its Copernican rival. In Riccioli's variant, Mercury, Venus, and Mars are satellites of the sun, but, unlike Brahe's original scheme, Jupiter and Saturn are centered on the earth. The figure holding the scales and the armillary sphere combines features of Urania (the muse of astronomy) and Astraea (the goddess of justice). On the left is hundred-eyed Argus, observing the sun through a telescope held to an eye on his knee. His words allude to Psalm 8, verse 3: "When I consider thy heavens, the work of thy fingers. . . ." On the right is a quotation from Psalm 104, verse 5: ". . . it [the earth] should not be removed for ever." At the bottom lies Ptolemy with his discarded system. Ptolemy rests his hand on the coat of arms of the prince of Monaco (to whom the *Almagestum Novum* was dedicated) magnanimously acknowledging the correction of his errors. At the top are depicted recent astronomical discoveries of the seventeenth century: Mercury and Venus displaying crescent phases; Saturn with two "handles"—this was prior to Huygens's ring hypothesis; Jupiter with four moons and two bands parallel to its equator (a feature first noted by Riccioli); a heavily cratered moon; and a comet soaring through the heavens like a spotted cannonball. In the center at the top is the Hebrew word *Yah-Veh* and a reference to the Wisdom of Solomon 11, verse 20: "But thou hast ordered all things by measure and number and weight." On the left and right are quotations from Psalm 19, verse 2: "Day unto day uttereth speech, and night unto night sheweth knowledge."

Although Riccioli's book had no effect on the debate over the Copernican theory—by the middle of the seventeenth century, almost all scientists and astronomers were Copernicans—it illustrates one of the most important contests between rival theories in the history of science.

Philosophy of Science

THE CENTRAL ISSUES

SECOND EDITION

Martin Curd | J. A. Cover | Christopher Pincock

W. W. Norton & Company

NEW YORK | LONDON

Copyright © 2013, 1998 by W. W. Norton & Company, Inc.

Manufacturing by: LSC Communications Crawfordsville
Production manager: Ashley Horna

Library of Congress Cataloging-in-Publishing Data

   Philosophy of science : the central issues / [edited by] Martin Curd, J.A. Cover,
Chris Pincock. — 2nd ed.
     p. cm.
   Includes bibliographical references and indexes.
   ISBN 978-0-393-91903-5 (pbk.)
   1. Science—Philosophy. I. Curd, Martin. II. Cover, J. A. (Jan A.), 1958–
III. Pincock, Chris.
   Q175. P5129 2013
   501—dc23

                                         2012017371

ISBN: 978-0-393-91903-5 (pbk.)

W. W. Norton & Company, Inc., 500 Fifth Avenue, New York, N.Y. 10110
www.wwnorton.com

W. W. Norton & Company Ltd., 15 Carlisle Street, London W1D 3BS

3 4 5 6 7 8 9 0

To the memory of
Willard Van Orman Quine (1908–2000)
Wesley C. Salmon (1925–2001)
Peter Lipton (1954–2007)
Ernan McMullin (1924–2011)

Contents

3 | The Duhem-Quine Thesis
and Underdetermination

4 | Induction, Prediction, and Evidence

5 | Confirmation and Relevance: Bayesian Approaches

6 | Models of Explanation

7 | Laws of Nature

8 | Intertheoretic Reduction

9 | Empiricism and Scientific Realism

PREFACE

Philosophy of science has grown considerably in the fourteen years since the first edition of our anthology was published. Not the least of these changes has been the appearance of several fine anthologies, single-author texts, and online resources in a field where such instructional material used to be scarce. Many people have written to us—teachers, students, scientists, and engineers—to express their appreciation for our selection of articles and the book's main distinguishing feature, namely, the commentaries accompanying each chapter. This response has persuaded us that *Philosophy of Science* will continue to play a useful role alongside more conventional texts, both inside and outside the classroom. Consequently, the format of articles followed by commentaries has been retained in this new edition. Six articles have been deleted and nine added in light of suggestions from teachers, readers, and reviewers.

It is a pleasure to acknowledge the help we received in preparing this revised edition of our book. Special thanks go to our reviewers: Marthe Chandler, Len Clark, Keith J. Cooper, Richard Creath, Evan Fales, Stuart S. Glennan, Paul Humphreys, David Magnus, Deborah G. Mayo, and Bonnie Paller (first edition); Donald Scott Carson, Mathias Frisch, Jill North, and Janet Stemwedel (second edition). We are also grateful for the (often detailed) comments we received from Prasanta Bandyopadhyay, Rod Bertolet, Hasok Chang, Patricia Kenig Curd, Paul Draper, Hanspeter Fetz, Stuart Glennan, Adolf Grünbaum, Don Howard, Kenneth Long, Eric Scerri, and Ted Ulrich. Many of the changes in the present volume are the direct result of their criticisms and advice. Working with the staff at Norton has been a pleasure. That this volume exists at all is due largely to the persistence, patience, and skill of our two editors: Allen Clawson (first edition) and Peter Simon (second edition). We would also like to acknowledge the sterling work of Betty Pessagno (for manuscript editing), Megan Jackson (for obtaining the permissions for production), and Debbie Masi at Westchester Book Group (who guided the entire production process). Paul Coppock (first edition) and Enid Zafran (second edition) prepared the indexes. This book also owes much to Rosanne Galloway for her gracious hospitality and beautiful garden in Bermuda where most of the revisions for this edition were made. Last but certainly not least there are our wives. Thank you: Patricia, Karen, and Tansel.

Martin Curd
Jan Cover
Christopher Pincock

GENERAL INTRODUCTION

The philosophy of science is at least as old as Aristotle, but it has risen to special prominence since the beginning of the twentieth century. As scientists have made tremendous advances in fields as diverse as genetics, geology, and quantum mechanics, increasing numbers of philosophers have made science their focus of study. In its broadest terms, the philosophy of science is the investigation of philosophical questions that arise from reflecting on science. What makes these questions philosophical is their generality, their fundamental character, and their resistance to solution by empirical disciplines such as history, sociology, and psychology.

Contrasting different sorts of questions can bring out the difference between the philosophy of science and other disciplines that study science. For example, "When was the planet Neptune discovered?" is primarily a question for historians, not for philosophers.[1] Similarly, "Why did Soviet biologists under Stalin reject Mendelian genetics?" or "Why did James Watson underrate the contributions of Rosalind Franklin to the work that led to the discovery of the double helix structure of DNA?" fall within the domains of sociology, political science, and psychology. Contrast these questions with the following: "When is a theory confirmed by its predictions?" "Should we be realists about all aspects of well-established theories?" "What is a law of nature?" These questions are philosophical. They cannot be answered simply by finding out what has happened in the past or what people now believe.

For similar reasons, philosophical questions about science cannot be answered by the sciences themselves (although being able to answer these questions often depends on having a good understanding of scientific theories). A geneticist at the National Cancer Institute, for example, might ask whether certain people are born with a natural immunity to AIDS and set out to answer this question through empirical research. But if our geneticist asked "What is a law of nature?" or "What is science?" or "When is a theory confirmed?" she would not discover the answer by doing more science.

The central questions in the philosophy of science do not belong to science as such; they are *about* science, but not *part of it*. Of course, scientists can be (and sometimes have been) philosophers of science. The point is that when people are doing philosophy of science, they are not (usually) doing science per se, and most philosophers of science (at least since the twentieth century) have not been practicing scientists. Thus, the philosophy of science is not a branch of science but belongs to philosophy, and it intersects with other areas of philosophy, such as epistemology, metaphysics, and the philosophy of language.

The aim of *Philosophy of Science: The Central Issues* is to introduce the reader to the main currents in modern philosophy of science. It is primarily intended for use in introductory courses at both the undergraduate and graduate levels. In order to keep the book within manageable bounds, some difficult decisions had to be made about what to include and what to exclude. In making these decisions we were guided by our own experience in teaching at Purdue and the recommendations of our reviewers who contributed significantly to the book's development.

The first and in some ways the easiest decision was to exclude almost entirely the social sciences and concentrate on the natural sciences.* In this we followed the lead of other texts. The philosophical questions raised by disciplines such as history, psychology, sociology, and anthropology are fascinating and important. But they are so different from the questions one encounters in physics, biology, and chemistry that they would require another volume, comparable in length to this one, in order to address them adequately.

A second decision, which we made at the outset, was to avoid foundational questions about the concepts, structure, and content of particular theories and to focus instead on general issues that arise across scientific disciplines. Thus, this volume is organized around wide-ranging philosophical topics and problems, not individual theories or sciences. Details of particular sciences are introduced rarely and only when necessary for evaluating a philosophical position or argument (as, for example, in the chapter on reduction). In this way we hope to avoid the trap of turning a philosophy of science course into a minicourse in science and to keep the focus on the philosophy in the philosophy of science. It also has the advantage of making courses based on this book accessible to students (even those at the graduate level) whose background in the sciences may be slight or nonexistent. For the same reason of accessibility, we have confined our selections to readings that use no more than a bare minimum of logical or mathematical notation.

Our approach, then, is focused on philosophical topics and problems, not on particular sciences and theories. A consequence of this topics-and-problems approach is that the chapters of *Philosophy of Science: The Central Issues* pay little attention to tracing the historical development of the philosophy of science. Although we devote some time to filling in some of the essential historical background in the commentaries, this volume is not historical in the way that it treats ideas, arguments, or philosophers. What connects the readings (and the discussions of them in the commentaries) is their focus on common themes, arguments, and criticisms, regardless of whether the authors share the same nationality, are writing in the same decade, or belong to the same school. Our approach is not antihistorical, but it is largely ahistorical.

Because of the many sharp disagreements within the philosophy of science and the unresolved character of nearly all the fundamental questions

---

* One notable exception is our inclusion in chapter 8 of Jerry Fodor's influential article defending the autonomy of the special sciences (which include sociology, psychology, and economics).

that philosophers ask about science, an anthology seemed to us to be the only sane choice for a book intended for use in the classroom. But the anthology format brings with it a problem that just about every teacher of philosophy of science has had to confront. Hardly any of the readings, whether old classics or brand-new articles, were written with students in mind. Rather, they were published in books and professional journals, addressed primarily to fellow professionals. Thus, they often presuppose an awareness of issues, positions, and arguments, both in the philosophy of science and in philosophy more generally, that most students lack. Consequently, even the brightest students can find it hard to understand the material they are being asked to read, discuss, and evaluate. The most common complaint voiced by the teachers we spoke with in the several years that went into planning and writing this book is that many of the readings in the existing anthologies are too sophisticated—they make too many references to the history of science and allude too frequently to philosophical ideas and arguments for the beginning student to get much out of them. What was needed, and what we have tried to provide here, is a serious, comprehensive guide that will really help students in their first encounter with the readings. Thus, in addition to short introductions to each chapter, we have written extended and often detailed commentaries on the readings. Getting the tone and level of detail right in these commentaries has been the hardest and most rewarding part of the book's development. Much of the fine tuning and, in some cases, the inclusion and deletion of entire sections, was guided by our reviewers. We have strived to make each commentary and the sections within them self-contained so that each can be used independently of the rest. And in order to maximize the pedagogical usefulness of *Philosophy of Science: The Central Issues*, each reading is linked explicitly with one or more of the sections into which the commentaries are divided. In this way, where one should look in the commentaries for discussion, explanation, background, and analysis of any of the fifty-two separate readings in the book should be clear.

At the end of this volume there is a glossary, a bibliography, and indexes of names and subjects. The glossary is comprehensive: it covers most of the terms that may be new to the reader or that are being used in an unfamiliar way. The bibliography is divided into nine sections, one for each chapter. Inevitably, this involves some repetition of titles of books and articles, but our aim was to provide the reader with suggestions for further reading, at an appropriate level, about the issues discussed in each chapter's commentary. Consequently, not everything cited in the commentaries appears in the bibliography, some items appear in the bibliography more than once, and there are some entries in the bibliography that are not mentioned in the commentaries.

The difference between an anthology and a heap of articles lies in their organization. But any system of division will be, to some extent, artificial and misleading: artificial because of the interrelated character of the issues in the philosophy of science and misleading because it might suggest that the readings in one chapter are not connected with those in another. Thus, as with any collection of this kind, the reader or teacher needs to bear in mind that not everything pertinent to, say, the topic of laws will be found in the chapter devoted to laws and that relevant readings and commentaries might also appear in the chapters on induction, confirmation, and reduction (as indeed, in this case, they do). Moreover, this is a collection of readings on related topics, not an extended narrative with a beginning, a middle, and an end. Users of the book should not feel constrained by the order of the chapters or even, in most cases, by the order of the readings within those chapters, when deciding what to read first, what to read second, and so on. Obviously, we have arranged the material in an order that makes sense to us, trying wherever possible to juxtapose readings that speak to the same or closely related issues, but many different arrangements are possible and may be preferable, depending on one's interests and teaching goals.

In articles referring to material reprinted in this volume, and in all footnotes, internal page numbers are enclosed in square brackets. In the commentaries, internal page numbers appear in parentheses.

■ | Note

1. This is not to deny that the question might raise philosophical issues concerning the concept of discovery. For example, while Galileo was observing Jupiter and its moons with his telescope, he saw Neptune and recorded its position on at least two occasions. Up to that time, no one had seen Neptune (because it is not visible to the naked eye). Galileo apparently regarded what he saw as a fixed star, not a planet (though he did note that the object appeared to have moved). Did Galileo discover Neptune some 234 years before its official discovery in 1846? See Stillman Drake and Charles T. Kowal, "Galileo's Sighting of Neptune," *Scientific American* 243 (1980): 52-59. For further discussion of the concept of discovery and its application to theoretical entities such as the electron, see Peter Achinstein, "Who Really Discovered the Electron?" in Jed Buchwald and Andrew Warwick, eds., *Histories of the Electron: The Birth of Microphysics* (Cambridge, Mass.: MIT Press, 2001), reprinted in Peter Achinstein, *Evidence, Explanation, and Realism* (New York: Oxford University Press, 2010), and Theodore Arabatzis, *Representing Electrons: A Biographical Approach to Theoretical Entities* (Chicago: University of Chicago Press, 2006).

Philosophy of Science

1 |
Science and
Pseudoscience

INTRODUCTION

Parapsychology is defined by its practitioners as the study of extrasensory perception (ESP) and paranormal powers such as telekinesis. ESP includes such alleged psychic phenomena as telepathy, clairvoyance, and precognition. Shunned for decades by the scientific establishment, parapsychologists received official recognition in 1969 when the American Association for the Advancement of Science (the AAAS) admitted the Parapsychological Association as an affiliate member. Many scientists are unhappy with this decision, since they regard parapsychology as a pseudoscience. In 1979, the renowned physicist John A. Wheeler wrote a blistering letter to the president of the AAAS urging that the parapsychologists be expelled from the association. Wheeler wrote, "We have enough charlatanism in this country today without needing a scientific organization to prostitute itself to it. The AAAS has to make up its mind whether it is seeking popularity or whether it is strictly a scientific organization."[1]

The debate about the nature of science—about its scope, methods, and aims—is as old as science itself. But this debate becomes especially heated when one group of practitioners accuses another group of practicing pseudoscience. Many individuals, groups, and theories have been accused of being pseudoscientific, including Freud and psychoanalysis, astrology, believers in the paranormal, Immanuel Velikovsky and Erich von Daniken (whose best-selling books *Worlds in Collision* and *Chariots of the Gods* excited the wrath of Carl Sagan and the scientific establishment), and, most recently, the self-styled advocates of creation-science. The proponents of astrology, the paranormal, psychoanalysis, and creation-science engage in research, write books, and publish articles, but their work is typically found in popular magazines and bookstores rather than refereed journals and science libraries. They are seldom funded by the National Science Foundation or elected to the National Academy of Sciences. They are outside of the scientific establishment and are kept out by those who regard themselves as real scientists.

If our only concern were to label certain people "pseudoscientists," we might simply check where their work is published and how their theories have been received by the scientific community. But we are concerned with the reasons certain doctrines are considered pseudoscientific; it is those reasons that interest philosophers of science.

Some philosophers have proposed necessary conditions for genuine science. That is, they have offered characteristics that any discipline or field of study must possess in order to qualify as genuine science. These characteristics are often called *demarcation criteria* because they can be used to differentiate science from its counterfeit: if a discipline fails to meet one of these conditions, then it is judged to be nonscientific.

Philosophers of science have often disagreed about demarcation criteria. In this chapter Karl Popper, Thomas Kuhn, Imre Lakatos, and Paul Thagard each defend a different set of necessary conditions for genuine science. Popper's view, that a scientific theory must be open to refutation by making testable predictions, has been very influential, especially among working scientists. Kuhn, Lakatos, and Thagard all reject Popper's claim that falsifiability is the hallmark of genuine science but disagree about what should replace it. All three address whether a theory or discipline's claim to scientific legitimacy depends on historical considerations, such as how theories have developed over time.

The chapter ends with an exchange of views between Michael Ruse and Larry Laudan about the credentials of creation-science. Ruse, a prominent philosopher of biology, served as an expert witness in a trial concerning the constitutionality of an Arkansas law requiring public school biology teachers to present creationism as a viable scientific alternative to evolutionary theory. Under Ruse's guidance, the judge in the case drew up a list of five criteria for genuine science and concluded that creation-science failed on all five counts. Laudan not only criticizes the items on this list (which includes Popper's falsifiability) but also doubts whether there are any demarcation criteria that all scientific theories must satisfy.

■ | Note

1. Quoted in Jack W. Grove, *In Defence of Science* (Toronto: University of Toronto Press, 1989), 137. See also Martin Gardner, *Science: Good, Bad and Bogus* (Buffalo, N.Y.: Prometheus Books, 1981), 185–206, and Richard Wiseman, *Paranormality: Why We See What Isn't There* (London: Macmillan, 2011). The Parapsychological Association is still a member of the AAAS.

KARL POPPER

Science: Conjectures and Refutations

Mr. Turnbull had predicted evil consequences, . . . and was now doing the best in his power to bring about the verification of his own prophecies.
—Anthony Trollope [*Phineas Finn*, ch. 25]

■ | I

When I received the list of participants in this course and realized that I had been asked to speak to philosophical colleagues* I thought, after some hesitation and consultation, that you would probably prefer me to speak about those problems which interest me most, and about those developments with which I am most intimately acquainted. I therefore decided to do what I have never done before: to give you a report on my own work in the philosophy of science, since the autumn of 1919 when I first began to grapple with the problem, '*When should a theory be ranked as scientific?*' or '*Is there a criterion for the scientific character or status of a theory?*'

The problem which troubled me at the time was neither, 'When is a theory true?' nor, 'When is a theory acceptable?' My problem was different. I *wished to distinguish between science and pseudo-science*; knowing very well that science often errs, and that pseudo-science may happen to stumble on the truth.

I knew, of course, the most widely accepted answer to my problem: that science is distinguished from pseudo-science—or from 'metaphysics'—by its *empirical method*, which is essentially *inductive*, proceeding from observation

FROM Karl Popper, *Conjectures and Refutations* (London: Routledge & Kegan Paul, 1963), 33–39.

* This essay was originally presented as a lecture at Peterhouse College at Cambridge University in the summer of 1953 as part of a course on developments and trends in contemporary British philosophy, organized by the British Council. It was originally published as "Philosophy of Science: A Personal Report," in *British Philosophy in Mid-Century*, ed. C. A. Mace (London: Allen & Unwin, 1957).

or experiment. But this did not satisfy me. On the contrary, I often formulated my problem as one of distinguishing between a genuinely empirical method and a non-empirical or even a pseudo-empirical method—that is to say, a method which, although it appeals to observation and experiment, nevertheless does not come up to scientific standards. The latter method may be exemplified by astrology, with its stupendous mass of empirical evidence based on observation—on horoscopes and on biographies.

But as it was not the example of astrology which led me to my problem I should perhaps briefly describe the atmosphere in which my problem arose and the examples by which it was stimulated. After the collapse of the Austrian Empire there had been a revolution in Austria: the air was full of revolutionary slogans and ideas, and new and often wild theories. Among the theories which interested me Einstein's theory of relativity was no doubt by far the most important. Three others were Marx's theory of history, Freud's psycho-analysis, and Alfred Adler's so-called 'individual psychology'.*

There was a lot of popular nonsense talked about these theories, and especially about relativity (as still happens even today), but I was fortunate in those who introduced me to the study of this theory. We all—the small circle of students to which I belonged—were thrilled with the result of Eddington's eclipse observations which in 1919 brought the first important confirmation of Einstein's theory of gravitation. It was a great experience for us, and one which had a lasting influence on my intellectual development.†

The three other theories I have mentioned were also widely discussed among students at that time. I myself happened to come into personal contact with Alfred Adler, and even to co-operate with him in his social work among the children and young people in the working-class districts of Vienna where he had established social guidance clinics.

It was during the summer of 1919 that I began to feel more and more dissatisfied with these three theories—the Marxist theory of history, psycho-analysis, and individual psychology; and I began to feel dubious about their claims to scientific status. My problem perhaps first took the simple form, 'What is wrong with Marxism, psycho-analysis, and individual psychology?

* For a fascinating autobiographical account of Popper's youthful flirtation and painful disenchantment with Marxism, see "A Crucial Year: Marxism; Science and Pseudoscience," in *The Philosophy of Karl Popper*, ed. Paul A. Schilpp (La Salle, Ill.: Open Court, 1974), 1:23–29. There is also an extended criticism of Freud in Karl R. Popper, *Realism and the Aim of Science* (New York: Routledge, 1983), 163–74.

† Einstein's general theory of relativity entails that light rays must bend in a gravitational field. Organized by Sir Arthur Eddington, two Royal Astronomical Society expeditions were dispatched to observe the solar eclipse of 1919, and verified that starlight was indeed deflected by the sun by the amount that Einstein had predicted. The *Times* of London reported this success as the most remarkable scientific event since the discovery of the planet Neptune. The light-bending test of relativity theory is discussed in "Popper's Demarcation Criterion," in the commentary on chapter 1, and in "Two Arguments for Explanationism," in the commentary on chapter 4.

Why are they so different from physical theories, from Newton's theory, and especially from the theory of relativity?'

To make this contrast clear I should explain that few of us at the time would have said that we believed in the *truth* of Einstein's theory of gravitation. This shows that it was not my doubting the *truth* of those other three theories which bothered me, but something else. Yet neither was it that I merely felt mathematical physics to be more *exact* than the sociological or psychological type of theory. Thus what worried me was neither the problem of truth, at that stage at least, nor the problem of exactness or measurability. It was rather that I felt that these other three theories, though posing as sciences, had in fact more in common with primitive myths than with science; that they resembled astrology rather than astronomy.

I found that those of my friends who were admirers of Marx, Freud, and Adler, were impressed by a number of points common to these theories, and especially by their apparent *explanatory power*. These theories appeared to be able to explain practically everything that happened within the fields to which they referred. The study of any of them seemed to have the effect of an intellectual conversion or revelation, opening your eyes to a new truth hidden from those not yet initiated. Once your eyes were thus opened you saw confirming instances everywhere: the world was full of *verifications* of the theory. Whatever happened always confirmed it. Thus its truth appeared manifest; and unbelievers were clearly people who did not want to see the manifest truth; who refused to see it, either because it was against their class interest, or because of their repressions which were still 'un-analysed' and crying aloud for treatment.

The most characteristic element in this situation seemed to me the incessant stream of confirmations, of observations which 'verified' the theories in question; and this point was constantly emphasized by their adherents. A Marxist could not open a newspaper without finding on every page confirming evidence for his interpretation of history; not only in the news, but also in its presentation—which revealed the class bias of the paper—and especially of course in what the paper did *not* say. The Freudian analysts emphasized that their theories were constantly verified by their 'clinical observations'. As for Adler, I was much impressed by a personal experience. Once, in 1919, I reported to him a case which to me did not seem particularly Adlerian, but which he found no difficulty in analysing in terms of his theory of inferiority feelings, although he had not even seen the child. Slightly shocked, I asked him how he could be so sure. 'Because of my thousandfold experience,' he replied; whereupon I could not help saying: 'And with this new case, I suppose, your experience has become thousand-and-one-fold.'

What I had in mind was that his previous observations may not have been much sounder than this new one; that each in its turn had been interpreted in the light of 'previous experience', and at the same time counted as additional confirmation. What, I asked myself, did it confirm? No more than that a case could be interpreted in the light of the theory. But this meant very little, I reflected, since every conceivable case could be interpreted in the light of Adler's theory, or equally of Freud's. I may illustrate this by two very different examples of human behaviour: that of a man who pushes a child into the water with the intention of drowning it; and that of a man who sacrifices his life in an attempt to save the child. Each of these two cases can be explained with equal ease in Freudian and in Adlerian terms. According to Freud the first man suffered from repression (say, of some component of his Oedipus complex), while the second man had achieved sublimation. According to Adler the first man suffered from feelings of inferiority (producing perhaps the need to prove to himself that he dared to commit some crime), and so did the second man (whose need was to prove to himself that he dared to rescue the child). I could not think of any human behaviour which could not be interpreted in terms of either theory. It was precisely this fact—that they always fitted, that they were always confirmed— which in the eyes of their admirers constituted the strongest argument in favour of these theories. It began to dawn on me that this apparent strength was in fact their weakness.

With Einstein's theory the situation was strikingly different. Take one typical instance—Einstein's prediction, just then confirmed by the findings of Eddington's expedition. Einstein's gravitational theory had led to the result that light must be attracted by heavy bodies (such as the sun), precisely as material bodies were attracted. As a consequence it could be calculated that light from a distant fixed star whose apparent position was close to the sun would reach the earth from such a direction that the star would seem to be slightly shifted away from the sun; or, in other words, that stars close to the sun would look as if they had moved a little away from the sun, and from one another. This is a thing which cannot normally be observed since such stars are rendered invisible in daytime by the sun's overwhelming brightness; but during an eclipse it is possible to take photographs of them. If the same constellation is photographed at night one can measure the distances on the two photographs, and check the predicted effect.

Now the impressive thing about this case is the *risk* involved in a prediction of this kind. If observation shows that the predicted effect is definitely absent, then the theory is simply refuted. The theory is *incompatible with certain possible results of observation*—in fact with results which everybody before Einstein would have expected.[1] This is quite different from the situation I have previously described, when it turned out that the theories in question were compatible with the most divergent human behaviour, so that it was practically impossible to describe any human behaviour that might not be claimed to be a verification of these theories.

These considerations led me in the winter of 1919–20 to conclusions which I may now reformulate as follows.

1 It is easy to obtain confirmations, or verifications, for nearly every theory—if we look for confirmations.

2 Confirmations should count only if they are the result of *risky predictions*; that is to say, if, unenlightened by the theory in question, we should have expected an event which was incompatible with the theory—an event which would have refuted the theory.

3 Every 'good' scientific theory is a prohibition: it forbids certain things to happen. The more a theory forbids, the better it is.

4 A theory which is not refutable by any conceivable event is non-scientific. Irrefutability is not a virtue of a theory (as people often think) but a vice.

5 Every genuine *test* of a theory is an attempt to falsify it, or to refute it. Testability is falsifiability; but there are degrees of testability: some theories are more testable, more exposed to refutation, than others; they take, as it were, greater risks.

6 Confirming evidence should not count *except when it is the result of a genuine test of the theory*; and this means that it can be presented as a serious but unsuccessful attempt to falsify the theory. (I now speak in such cases of 'corroborating evidence'.)

7 Some genuinely testable theories, when found to be false, are still upheld by their admirers—for example by introducing *ad hoc* some auxiliary assumption, or by re-interpreting the theory *ad hoc* in such a way that it escapes refutation. Such a procedure is always possible, but it rescues the theory from refutation only at the price of destroying, or at least lowering, its scientific status. (I later described such a rescuing operation as a '*conventionalist twist*' or a '*conventionalist stratagem*'.)

One can sum up all this by saying that *the criterion of the scientific status of a theory is its falsifiability, or refutability, or testability.*

▪ | II

I may perhaps exemplify this with the help of the various theories so far mentioned. Einstein's theory of gravitation clearly satisfied the criterion of falsifiability. Even if our measuring instruments at the time did not allow us to pronounce on the results of the tests with complete assurance, there was clearly a possibility of refuting the theory.

Astrology did not pass the test. Astrologers were greatly impressed, and misled, by what they believed to be confirming evidence—so much so that they were quite unimpressed by any unfavourable evidence. Moreover, by making their interpretations and prophecies sufficiently vague they were able to explain away anything that might have been a refutation of the theory had the theory and the prophecies been more precise. In order to escape falsification they destroyed the testability of their theory. It is a typical soothsayer's trick to predict things so vaguely that the predictions can hardly fail: that they become irrefutable.

The Marxist theory of history, in spite of the serious efforts of some of its founders and followers, ultimately adopted this soothsaying practice. In some of its earlier formulations (for example in Marx's analysis of the character of the 'coming social revolution') their predictions were testable, and in fact falsified.[2] Yet instead of accepting the refutations the followers of Marx re-interpreted both the theory and the evidence in order to make them agree. In this way they rescued the theory from refutation; but they did so at the price of adopting a device which made it irrefutable. They thus gave a 'conventionalist twist' to the theory; and by this stratagem they destroyed its much advertised claim to scientific status.

The two psycho-analytic theories were in a different class. They were simply non-testable, irrefutable. There was no conceivable human behaviour which could contradict them. This does not mean that Freud and Adler were not seeing certain things correctly: I personally do not doubt that much of what they say is of considerable importance, and may well play its part one day in a psychological science which is testable. But it does mean that those 'clinical observations' which analysts naïvely believe confirm their theory cannot do this any more than the daily confirmations which astrologers find in their practice.[3] And as for Freud's epic of the Ego, the Super-ego, and the Id, no substantially stronger claim to scientific status can be made for it than for Homer's collected stories from Olympus. These theories describe some facts, but in the manner of myths. They contain most interesting psychological suggestions, but not in a testable form.

At the same time I realized that such myths may be developed, and become testable; that historically speaking all—or very nearly all—scientific theories originate from myths, and that a myth may contain important anticipations of scientific theories. Examples are Empedocles' theory of evolution by trial and error, or Parmenides' myth of the unchanging block universe in which nothing ever happens and which, if we add another dimension, becomes Einstein's block universe (in which, too, nothing ever happens, since everything is, four-dimensionally speaking, determined and laid down from the beginning). I thus felt that if a theory is found to be non-scientific, or 'metaphysical' (as we might say), it is not thereby found to be unimportant, or insignificant, or 'meaningless', or 'nonsensical'.[4] But it cannot claim to be backed by empirical evidence in the scientific sense—although it may easily be, in some genetic sense, the 'result of observation'.

(There were a great many other theories of this pre-scientific or pseudo-scientific character, some of them, unfortunately, as influential as the Marxist interpretation of history; for example, the racialist interpretation of history—

another of those impressive and all-explanatory theories which act upon weak minds like revelations.)

Thus the problem which I tried to solve by proposing the criterion of falsifiability was neither a problem of meaningfulness or significance, nor a problem of truth or acceptability. It was the problem of drawing a line (as well as this can be done) between the statements, or systems of statements, of the empirical sciences, and all other statements—whether they are of a religious or of a metaphysical character, or simply pseudo-scientific. Years later—it must have been in 1928 or 1929—I called this first problem of mine the *'prob-lem of demarcation'*. The criterion of falsifiability is a solution to this problem of demarcation, for it says that statements or systems of statements, in order to be ranked as scientific, must be capable of conflicting with possible, or conceivable, observations. . . .

▪ | Notes

1. This is a slight oversimplification, for about half of the Einstein effect may be derived from the classical theory, provided we assume a ballistic theory of light.

2. See, for example, my *Open Society and Its Enemies* [Routledge & Kegan Paul, 1945], ch. 15, section iii, and notes 13–14.

3. 'Clinical observations', like all other observations, are *interpretations in the light of theories* . . . ; and for this reason alone they are apt to seem to support those theories in the light of which they were interpreted. But real support can be obtained only from observations undertaken as tests (by 'attempted refutations'); and for this purpose *criteria of refutation* have to be laid down beforehand: it must be agreed which observable situations, if actually observed, mean that the theory is refuted. But what kind of clinical responses would refute to the satisfaction of the analyst not merely a particular analytic diagnosis but psycho-analysis itself? And have such criteria ever been discussed or agreed upon by analysts? Is there not, on the contrary, a whole family of analytic concepts, such as 'ambivalence' (I do not suggest that there is no such thing as ambivalence), which would make it difficult, if not impossible, to agree upon such criteria? Moreover, how much headway has been made in investigating the question of the extent to which the (conscious or unconscious) expectations and theories held by the analyst influence the 'clinical responses' of the patient? (To say nothing about the conscious attempts to influence the patient by proposing interpretations to him, etc.) Years ago I introduced the term *'Oedipus effect'* to describe the influence of a theory or expectation or prediction *upon the event which it predicts* or describes: it will be remembered that the causal chain leading to Oedipus' parricide was started by the oracle's prediction of this event. This is a characteristic and recurrent theme of such myths, but one which seems to have failed to attract the interest of the analysts, perhaps not accidentally. (The problem of confirmatory dreams suggested by the analyst is discussed by Freud, for example in *Gesammelte Schriften* [Complete works], III, 1925, where he says on p. 314: 'If anybody asserts that most of the dreams which can be utilized in an analysis . . . owe their origin to [the analyst's] suggestion, then no objection can be made from the

point of view of analytic theory. Yet there is nothing in this fact', he surprisingly adds, 'which would detract from the reliability of our results.')

4. The case of astrology, nowadays a typical pseudo-science, may illustrate this point. It was attacked, by Aristotelians and other rationalists, down to Newton's day, for the wrong reason—for its now accepted assertion that the planets had an 'influence' upon terrestrial ('sublunar') events. In fact Newton's theory of gravity, and especially the lunar theory of the tides, was historically speaking an offspring of astrological lore. Newton, it seems, was most reluctant to adopt a theory which came from the same stable as for example the theory that 'influenza' epidemics are due to an astral 'influence'. And Galileo, no doubt for the same reason, actually rejected the lunar theory of the tides; and his misgivings about Kepler may easily be explained by his misgivings about astrology.

Thomas S. Kuhn

Logic of Discovery or Psychology of Research?

Among the most fundamental issues on which Sir Karl [Popper] and I agree is our insistence that an analysis of the development of scientific knowledge must take account of the way science has actually been practiced. That being so, a few of his recurrent generalizations startle me. One of these provides the opening sentences of the first chapter of the *Logic of Scientific Discovery*: 'A scientist', writes Sir Karl, 'whether theorist or experimenter, puts forward statements, or systems of statements, and tests them step by step. In the field of the empirical sciences, more particularly, he constructs hypotheses, or systems of theories, and tests them against experience by observation and experiment.[1] The statement is virtually a cliché, yet in application it presents three problems. It is ambiguous in its failure to specify which of two sorts of 'statements' or 'theories' are being tested. That ambiguity can, it is true, be eliminated by reference to other passages in Sir Karl's writings, but the generalization that results is historically mistaken. Furthermore, the mistake proves important, for the unambiguous form of the description misses just that characteristic of scientific practice which most nearly distinguishes the sciences from other creative pursuits.

There is one sort of 'statement' or 'hypothesis' that scientists do repeatedly subject to systematic test. I have in mind statements of an individual's best guesses about the proper way to connect his own research problem with the corpus of accepted scientific knowledge. He may, for example, conjecture that a given chemical unknown contains the salt of a rare earth, that the obesity of his experimental rats is due to a specified component in their diet, or that a newly discovered spectral pattern is to be understood as an effect of nuclear spin. In each case, the next steps in his research are intended to try out or test the conjecture or hypothesis. If it passes enough or stringent enough tests, the scientist has made a discovery or has at least resolved the puzzle he had been set. If not, he must either abandon the

From Imre Lakatos and Alan Musgrave, eds., *Criticism and the Growth of Knowledge* (Cambridge: Cambridge University Press, 1970), 4–10.

puzzle entirely or attempt to solve it with the aid of some other hypothesis. Many research problems, though by no means all, take this form. Tests of this sort are a standard component of what I have elsewhere labelled 'normal science' or 'normal research', an enterprise which accounts for the overwhelming majority of the work done in basic science. In no usual sense, however, are such tests directed to current theory. On the contrary, when engaged with a normal research problem, the scientist must *premise* current theory as the rules of his game. His object is to solve a puzzle, preferably one at which others have failed, and current theory is required to define that puzzle and to guarantee that, given sufficient brilliance, it can be solved.[2] Of course the practitioner of such an enterprise must often test the conjectural puzzle solution that his ingenuity suggests. But only his personal conjecture is tested. If it fails the test, only his own ability not the corpus of current science is impugned. In short, though tests occur frequently in normal science, these tests are of a peculiar sort, for in the final analysis it is the individual scientist rather than current theory which is tested.

This is not, however, the sort of test Sir Karl has in mind. He is above all concerned with the procedures through which science grows, and he is convinced that 'growth' occurs not primarily by accretion but by the revolutionary overthrow of an accepted theory and its replacement by a better one.[3] (The subsumption under 'growth' of 'repeated overthrow' is itself a linguistic oddity whose *raison d'être* may become more visible as we proceed.) Taking this view, the tests which Sir Karl emphasizes are those which were performed to explore the limitations of accepted theory or to subject a current theory to maximum strain. Among his favourite examples, all of them startling and destructive in their outcome, are Lavoisier's experiments on calcination,* the eclipse expedition of 1919,† and the recent experiments on parity conserva-

---

* Calcination occurs when a metal is burned in air, forming a calx or oxide. According to the phlogiston theory, metals (and all other combustible substances) are compounds of an earthy calx and the fiery element, phlogiston. When a metal burns, the phlogiston is released, leaving the calx as a residue. Because metals gain weight when they are calcined, some proponents of the phlogiston theory conjectured that phlogiston must have negative weight. Others inferred that some other substance must combine with the metal when the phlogiston is released. By careful experiments in the 1770s, Antoine Lavoisier (1743–94) showed that the weight gained during calcination is entirely due to the metal combining with a gas in the air, which he named oxygen. Lavoisier's oxygen theory of calcination (and, more generally, of combustion) overthrew the phlogiston theory and gave rise to a revolution in chemistry. See James B. Conant, ed., *The Overthrow of the Phlogiston Theory: The Chemical Revolution of 1775–1789* (Cambridge, Mass.: Harvard University Press, 1950); reprinted in *Harvard Case Histories in Experimental Science*, ed. J. B. Conant and J. K. Nash (Cambridge, Mass.: Harvard University Press, 1966). See also Alan Musgrave, "Why Did Oxygen Supplant Phlogiston? Research Programmes in the Chemical Revolution," in *Method and Appraisal in the Physical Sciences*, ed. C. Howson (Cambridge: Cambridge University Press, 1976), 181–209.

† For information about the eclipse expedition of 1919 and its role in confirming Einstein's general theory of relativity, see the preceding reading by Karl Popper,

tion.*[4] All, of course, are classic tests, but in using them to characterize scientific activity Sir Karl misses something terribly important about them. Episodes like these are very rare in the development of science. When they occur, they are generally called forth either by a prior crisis in the relevant field (Lavoisier's experiments or Lee and Yang's[5]) or by the existence of a theory which competes with the existing canons of research (Einstein's general relativity). These are, however, aspects of or occasions for what I have elsewhere called 'extraordinary research', an enterprise in which scientists do display very many of the characteristics Sir Karl emphasizes, but one which, at least in the past, has arisen only intermittently and under quite special circumstances in any scientific speciality.[6]

I suggest then that Sir Karl has characterized the entire scientific enterprise in terms that apply only to its occasional revolutionary parts. His emphasis is natural and common: the exploits of a Copernicus or Einstein make better reading than those of a Brahe or Lorentz;† Sir Karl would not be the first if he mistook what I call normal science for an intrinsically uninteresting enterprise. Nevertheless, neither science nor the development of knowledge

---

"Science: Conjectures and Refutations." Further discussion can be found in "Popper's Demarcation Criterion," in the commentary on chapter 1, and in "Two Arguments for Explanationism," in the commentary on chapter 4.

* Kuhn is referring to the experiments performed by Chien-Shiung Wu and her associates in 1956–57, which verified the conjecture of Tsung Dao Lee and Chen Ning Yang that parity is not conserved in weak interactions. Wu's results were soon confirmed by other groups and Lee and Yang received the Nobel prize in physics in 1957 for their discovery of parity violation. For a description of Wu's experiment and an explanation of its revolutionary significance, see Eugene Wigner, "Violations of Symmetry in Physics," *Scientific American* 213 (1965): 28–36 and Martin Gardner, *The New Ambidextrous Universe*, 3d rev. ed. (New York: W. H. Freeman, 1990).

† For Kuhn, Tycho Brahe (1546–1601) and H. A. Lorentz (1853–1928) exemplify the conservative scientist practicing normal science. Brahe objected to Copernicus's revolutionary theory of a heliocentric universe on physical, astronomical, and religious grounds, proposing in its place his own version of a geostatic system. Like Ptolemy, Brahe had the sun moving around the earth, but unlike Ptolemy, he made the other planets orbit round the sun. In this way, Brahe was able to capture many of the explanatory features of Copernicus's theory without having to attribute any motion to the earth. Lorentz, like most physicists of his day, believed that light and other electromagnetic radiation propagates in an aether that is at rest with respect to absolute space. In order to account for the null result of the Michelson-Morley experiment, Lorentz (and, independently, Fitzgerald) postulated the famous Lorentz-Fitzgerald contraction according to which all physical objects contract in their direction of motion. Lorentz later introduced time dilation, thus obtaining the Lorentz transformations that lie at the heart of Einstein's special theory of relativity; but unlike Einstein, Lorentz worked within a classical framework of absolute space and time. The theories of Lorentz and Einstein are compared in Jonathan Powers, *Philosophy and the New Physics* (New York: Methuen, 1982), ch. 3; Michel Janssen, "Reconsidering a Scientific Revolution: The Case of Einstein *versus* Lorentz," *Physics in Perspective* 4 (2002): 421–46; and Robert DiSalle, *Understanding Space-Time: The Philosophical Development of Physics from Newton to Einstein* (Cambridge: Cambridge University Press, 2006), ch. 4.

is likely to be understood if research is viewed exclusively through the revolutions it occasionally produces. For example, though testing of basic commitments occurs only in extraordinary science, it is normal science that discloses both the points to test and the manner of testing. Or again, it is for the normal, not the extraordinary practice of science that professionals are trained; if they are nevertheless eminently successful in displacing and replacing the theories on which normal practice depends, that is an oddity which must be explained. Finally, and this is for now my main point, a careful look at the scientific enterprise suggests that it is normal science, in which Sir Karl's sort of testing does not occur, rather than extraordinary science which most nearly distinguishes science from other enterprises. If a demarcation criterion exists (we must not, I think, seek a sharp or decisive one), it may lie just in that part of science which Sir Karl ignores.

In one of his most evocative essays, Sir Karl traces the origin of 'the tradition of critical discussion [which] represents the only practicable way of expanding our knowledge' to the Greek philosophers between Thales and Plato, the men who, as he sees it, encouraged critical discussion both between schools and within individual schools.[7] The accompanying description of Presocratic discourse is most apt, but what is described does not at all resemble science. Rather it is the tradition of claims, counterclaims, and debates over fundamentals which, except perhaps during the Middle Ages, have characterized philosophy and much of social science ever since. Already by the Hellenistic period mathematics, astronomy, statics and the geometric parts of optics had abandoned this mode of discourse in favour of puzzle solving. Other sciences, in increasing numbers, have undergone the same transition since. In a sense, to turn Sir Karl's view on its head, it is precisely the abandonment of critical discourse that marks the transition to a science. Once a field has made that transition, critical discourse recurs only at moments of crisis when the bases of the field are again in jeopardy.[8] Only when they must choose between competing theories do scientists behave like philosophers. That, I think, is why Sir Karl's brilliant description of the reasons for the choice between metaphysical systems so closely resembles my description of the reasons for choosing between scientific theories.[9] In neither choice, as I shall shortly try to show, can testing play a quite decisive role.

There is, however, good reason why testing has seemed to do so, and in exploring it Sir Karl's duck may at last become my rabbit.* No puzzle-solving enterprise can exist unless its practitioners share criteria which, for that

---

* The duck-rabbit is a visually ambiguous drawing, made popular among philosophers by Ludwig Wittgenstein in his *Philosophical Investigations* (1953). It can be seen either as a duck's head with a long beak or as a rabbit's head with long ears, but it cannot be seen as both at the same time. It is a favorite with philosophers of science (such as Kuhn, Hanson, and Feyerabend) wishing to emphasize the theory-ladenness of observation.

group and for that time, determine when a particular puzzle has been solved. The same criteria necessarily determine failure to achieve a solution, and anyone who chooses may view that failure as the failure of a theory to pass a test. Normally, as I have already insisted, it is not viewed that way. Only the practitioner is blamed, not his tools. But under the special circumstances which induce a crisis in the profession (e.g. gross failure, or repeated failure by the most brilliant professionals) the group's opinion may change. A failure that had previously been personal may then come to seem the failure of a theory under test. Thereafter, because the test arose from a puzzle and thus carried settled criteria of solution, it proves both more severe and harder to evade than the tests available within a tradition whose normal mode is critical discourse rather than puzzle solving.

In a sense, therefore, severity of test-criteria is simply one side of the coin whose other face is a puzzle-solving tradition. That is why Sir Karl's line of demarcation and my own so frequently coincide. That coincidence is, however, only in their *outcome*; the *process* of applying them is very different, and it isolates distinct aspects of the activity about which the decision—science or non-science—is to be made. Examining the vexing cases, for example, psychoanalysis or Marxist historiography, for which Sir Karl tells us his criterion was initially designed,[10] I concur that they cannot now properly be labelled 'science'. But I reach that conclusion by a route far surer and more direct than his. One brief example may suggest that of the two criteria, testing and puzzle solving, the latter is at once the less equivocal and the more fundamental.

To avoid irrelevant contemporary controversies, I consider astrology rather than, say, psychoanalysis. Astrology is Sir Karl's most frequently cited example of a 'pseudo-science'.[11] He says: 'By making their interpretations and prophecies sufficiently vague they [astrologers] were able to explain away anything that might have been a refutation of the theory had the theory and the prophecies been more precise. In order to escape falsification they destroyed the testability of their theory.'[12] Those generalizations catch something of the spirit of the astrological enterprise. But taken at all literally, as they must be if they are to provide a demarcation criterion, they are impossible to support. The history of astrology during the centuries when it was intellectually reputable records many predictions that categorically failed.[13] Not even astrology's most convinced and vehement exponents doubted the recurrence of such failures. Astrology cannot be barred from the sciences because of the form in which its predictions were cast.

Nor can it be barred because of the way its practitioners explained failure. Astrologers pointed out, for example, that, unlike general predictions about, say, an individual's propensities or a natural calamity, the forecast of an individual's future was an immensely complex task, demanding the utmost skill, and extremely sensitive to minor errors in relevant data. The configuration of the stars and eight planets was constantly changing; the astronomical tables used to compute the configuration at an individual's birth were notoriously imperfect; few men knew the instant of their birth with the requisite precision.[14] No wonder, then, that forecasts often failed. Only after astrology itself became implausible did these arguments come to seem question-begging.[15] Similar arguments are regularly used today when explaining, for example, failures in medicine or meteorology. In times of trouble they are also deployed in the exact sciences, fields like physics, chemistry, and astronomy.[16] There was nothing unscientific about the astrologer's explanation of failure.

Nevertheless, astrology was not a science. Instead it was a craft, one of the practical arts, with close resemblances to engineering, meteorology, and medicine as these fields were practised until little more than a century ago. The parallels to an older medicine and to contemporary psychoanalysis are, I think, particularly close. In each of these fields shared theory was adequate only to establish the plausibility of the discipline and to provide a rationale for the various craft-rules which governed practice. These rules had proved their use in the past, but no practitioner supposed they were sufficient to prevent recurrent failure. A more articulated theory and more powerful rules were desired, but it would have been absurd to abandon a plausible and badly needed discipline with a tradition of limited success simply because these desiderata were not yet at hand. In their absence, however, neither the astrologer nor the doctor could do research. Though they had rules to apply, they had no puzzles to solve and therefore no science to practise.[17]

Compare the situations of the astronomer and the astrologer. If an astronomer's prediction failed and his calculations checked, he could hope to set the situation right. Perhaps the data were at fault: old observations could be re-examined and new measurements made, tasks which posed a host of calculational and instrumental puzzles. Or perhaps theory needed adjustment, either by the manipulation of epicycles, eccentrics, equants, etc., or by more fundamental reforms of astronomical technique. For more than a millennium these were the theoretical and mathematical puzzles around which, together with their instrumental counterparts, the astronomical research tradition was constituted. The astrologer, by contrast, had no such puzzles. The occurrence of failures could be explained, but particular failures did not give rise to research puzzles, for no man, however skilled, could make use of them in a constructive attempt to revise the astrological tradition. There were too many possible sources of difficulty, most of them beyond the astrologer's knowledge, control, or responsibility. Individual failures were correspondingly uninformative, and they did not reflect on the competence of the prognosticator in the eyes of his professional compeers.[18] Though astronomy and astrology were regularly practised by the same people, including Ptolemy, Kepler, and Tycho Brahe, there was never an astrological equivalent of the puzzle-solving astronomical tradition. And without puzzles, able first to challenge and then to attest the ingenuity of the individual practitioner, astrology could not have become a science even if the stars had, in fact, controlled human destiny.

In short, though astrologers made testable predictions and recognized that these predictions sometimes failed, they did not and could not engage in the sorts of activities that normally characterize all recognized sciences. Sir Karl is right to exclude astrology from the sciences, but his over-concentration on science's occasional revolutions prevents his seeing the surest reason for doing so.

That fact, in turn, may explain another oddity of Sir Karl's historiography. Though he repeatedly underlines the role of tests in the replacement of scientific theories, he is also constrained to recognize that many theories, for example the Ptolemaic, were replaced before they had in fact been tested.[19] On some occasions, at least, tests are not requisite to the revolutions through which science advances. But that is not true of puzzles. Though the theories Sir Karl cites had not been put to the test before their displacement, none of these was replaced before it had ceased adequately to support a puzzle-solving tradition. The state of astronomy was a scandal in the early sixteenth century. Most astronomers nevertheless felt that normal adjustments of a basically Ptolemaic model would set the situation right. In this sense the theory had not failed a test. But a few astronomers, Copernicus among them, felt that the difficulties must lie in the Ptolemaic approach itself rather than in the particular versions of Ptolemaic theory so far developed, and the results of that conviction are already recorded. The situation is typical.[20] With or without tests, a puzzle-solving tradition can prepare the way for its own displacement. To rely on testing as the mark of a science is to miss what scientists mostly do and, with it, the most characteristic feature of their enterprise. . . .

■ | **Notes**

1. Popper [1959], p. 27.

2. For an extended discussion of normal science, the activity which practitioners are trained to carry on, see my [1962], pp. 23–42, and 135–42. It is important to notice that when I describe the scientist as a puzzle solver and Sir Karl describes him as a problem solver (e.g. in his [1963], pp. 67, 222), the similarity of our terms disguises a fundamental divergence. Sir Karl writes (the italics are his), 'Admittedly, our expectations, and thus our theories, may precede, historically, even our problems. Yet *science starts only with problems*. Problems crop up especially when we are disappointed in our expectations, or when our theories involve us in difficulties, in contradictions'. I use the term 'puzzle' in order to emphasize that the difficulties which *ordinarily* confront even the very best scientists are, like crossword puzzles or chess puzzles, challenges only to his ingenuity. *He* is in difficulty, not current theory. My point is almost the converse of Sir Karl's.

3. Cf. Popper [1963], pp. 129, 215 and 221, for particularly forceful statements of this position.

4. For example, Popper [1963], p. 220.

5. For the work on calcination see, Guerlac [1961]. For the background of the parity experiments see, Hafner and Presswood [1965].

6. The point is argued at length in my [1962], pp. 52–97.

7. Popper [1963], chapter 5, especially pp. 148–52.

8. Though I was not then seeking a demarcation criterion, just these points are argued at length in my [1962], pp. 10–22 and 87–90.

9. Cf. Popper [1963], pp. 192–200, with my [1962], pp. 143–58.

10. Popper [1963], p. 34 [see this volume, pp. 4–5].

11. The index to Popper [1963] has eight entries under the heading 'astrology as a typical pseudo-science'.

12. Popper [1963], p. 37 [see this volume, pp. 7–8].

13. For examples see, Thorndike [1923–58], 5, pp. 225 ff.; 6, pp. 71, 101, 114.

14. For reiterated explanations of failure see, *ibid.* 1, pp. 11 and 514 f.; 4, p. 368; 5, p. 279.

15. A perceptive account of some reasons for astrology's loss of plausibility is included in Stahlman [1956]. For an explanation of astrology's previous appeal see, Thorndike [1955].

16. Cf. my [1962], pp. 66–76.

17. This formulation suggests that Sir Karl's criterion of demarcation might be saved by a minor restatement entirely in keeping with his apparent intent. For a field to be a science its conclusions must be *logically derivable* from *shared premises*. On this view astrology is to be barred not because its forecasts were not testable but because only the most general and least testable ones could be derived from accepted theory. Since any field that did satisfy this condition *might* support a puzzle solving tradition, the suggestion is clearly helpful. It comes close to supplying a sufficient condition for a field's being a science. But in this form, at least, it is not even quite a sufficient condition, and it is surely not a necessary one. It would, for example, admit surveying and navigation as sciences, and it would bar taxonomy, historical geology, and the theory of evolution. The conclusions of a science may be both precise and binding without being fully derivable by logic from accepted premises. Cf. my [1962], pp. 35–51. . . .

18. This is not to suggest that astrologers did not criticize each other. On the contrary, like practitioners of philosophy and some social sciences, they belonged to a variety of different schools, and the inter-school strife was sometimes bitter. But these debates ordinarily revolved about the *implausibility* of the particular theory employed by one or another school. Failures of individual predictions played very little role. Compare Thorndike [1923–58], 5, p. 233.

19. Cf. Popper [1963], p. 246.

20. Cf. my [1962], pp. 77–87.

■ | References

Braithwaite [1953]: *Scientific Explanation*, 1953.

Guerlac [1961]: *Lavoisier—The Crucial Year*, 1961.

Hafner and Presswood [1965]: 'Strong Interference and Weak Interactions', *Science*, 149, pp. 503–10.

Hawkins [1963]: Review of Kuhn's 'Structure of Scientific Revolutions', *American Journal of Physics*, 31.

Hempel [1965]: *Aspects of Scientific Explanation*, 1965.

Kuhn [1958]: 'The Role of Measurement in the Development of Physical Science', *Isis*, 49, pp. 161–93.

Kuhn [1962]: *The Structure of Scientific Revolutions*, 1962.

Lakatos [1963–4]: 'Proofs and Refutations', *The British Journal for the Philosophy of Science*, 14, pp. 1–25, 120–39, 221–43, 296–342.

Popper [1935]: *Logik der Forschung*, 1935.*

Popper [1945]: *The Open Society and its Enemies*, 2 vols, 1945.

Popper [1957]: *The Poverty of Historicism*, 1957.

Popper [1959]: *Logic of Scientific Discovery*, 1959.

Popper [1963]: *Conjectures and Refutations*, 1963.

Stahlman [1956]: 'Astrology in Colonial America: An Extended Query', *William and Mary Quarterly*, 13, pp. 551–63.

Thorndike [1923–58]: *A History of Magic and Experimental Science*, 8 vols, 1923–58.

Thorndike [1955]: 'The True Place of Astrology in the History of Science', *Isis*, 46, pp. 273–8.

* Although "1935" appears on its title page, Popper's *Logik der Forschung* was published in Vienna, by Julius Springer Verlag, in November 1934. Most authors follow Popper in listing its publication date as 1934.

Imre Lakatos

Science and
Pseudoscience

Man's respect for knowledge is one of his most peculiar characteristics. Knowledge in Latin is *scientia*, and science came to be the name of the most respectable kind of knowledge. But what distinguishes knowledge from superstition, ideology or pseudoscience? The Catholic Church excommunicated Copernicans, the Communist Party persecuted Mendelians on the ground that their doctrines were pseudoscientific. The demarcation between science and pseudoscience is not merely a problem of armchair philosophy: it is of vital social and political relevance.

Many philosophers have tried to solve the problem of demarcation in the following terms: a statement constitutes knowledge if sufficiently many people believe it sufficiently strongly. But the history of thought shows us that many people were totally committed to absurd beliefs. If the strength of beliefs were a hallmark of knowledge, we should have to rank some tales about demons, angels, devils, and of heaven and hell as knowledge. Scientists, on the other hand, are very sceptical even of their best theories. Newton's is the most powerful theory science has yet produced, but Newton himself never believed that bodies attract each other at a distance. So no degree of commitment to beliefs makes them knowledge. Indeed, the hallmark of scientific behaviour is a certain scepticism even towards one's most cherished theories. Blind commitment to a theory is not an intellectual virtue: it is an intellectual crime.

Thus a statement may be pseudoscientific even if it is eminently 'plausible' and everybody believes in it, and it may be scientifically valuable even if it is unbelievable and nobody believes in it. A theory may even be of supreme scientific value even if no one understands it, let alone believes it.

The cognitive value of a theory has nothing to do with its psychological influence on people's minds. Belief, commitment, understanding are states

From Imre Lakatos, *Philosophical Papers*, vol. 1 (Cambridge: Cambridge University Press, 1977), 1–7. Written in early 1973, this was originally presented as a radio lecture broadcast by the Open University (30 June 1973).

of the human mind. But the objective, scientific value of a theory is independent of the human mind which creates it or understands it. Its scientific value depends only on what objective support these conjectures have in facts. As Hume said:

> If we take in our hand any volume; of divinity, or school metaphysics, for instance; let us ask, does it contain any abstract reasoning concerning quantity or number? No. Does it contain any experimental reasoning concerning matter of fact and existence? No. Commit it then to the flames. For it can contain nothing but sophistry and illusion.*

But what is 'experimental' reasoning? If we look at the vast seventeenth-century literature on witchcraft, it is full of reports of careful observations and sworn evidence—even of experiments. Glanvill, the house philosopher of the early Royal Society, regarded witchcraft as the paradigm of experimental reasoning. We have to define experimental reasoning before we start Humean book burning.

In scientific reasoning, theories are confronted with facts; and one of the central conditions of scientific reasoning is that theories must be supported by facts. Now how exactly can facts support theory?

Several different answers have been proposed. Newton himself thought that he proved his laws from facts. He was proud of not uttering mere hypotheses: he only published theories proven from facts. In particular, he claimed that he deduced his laws from the 'phenomena' provided by Kepler. But his boast was nonsense, since according to Kepler, planets move in ellipses, but according to Newton's theory, planets would move in ellipses only if the planets did not disturb each other in their motion. But they do. This is why Newton had to devise a perturbation theory from which it follows that no planet moves in an ellipse.

One can today easily demonstrate that there can be no valid derivation of a law of nature from any finite number of facts; but we still keep reading about scientific theories being proved from facts. Why this stubborn resistance to elementary logic?

There is a very plausible explanation. Scientists want to make their theories respectable, deserving of the title 'science', that is, genuine knowledge. Now the most relevant knowledge in the seventeenth century, when science was born, concerned God, the Devil, Heaven and Hell. If one got one's conjectures about matters of divinity wrong, the consequence of one's mistake was eternal damnation. Theological knowledge cannot be fallible: it must be beyond doubt. Now the Enlightenment thought that we were fallible and ignorant about matters theological. There is no scientific theology

---

* These famous lines are from the final paragraph of David Hume's *An Enquiry Concerning Human Understanding*, first published in 1748 (under the title *Philosophical Essays Concerning Human Understanding*).

and, therefore, no theological knowledge. Knowledge can only be about Nature, but this new type of knowledge had to be judged by the standards they took over straight from theology: it had to be proven beyond doubt. Science had to achieve the very certainty which had escaped theology. A scientist, worthy of the name, was not allowed to guess: he had to prove each sentence he uttered from facts. This was the criterion of scientific honesty. Theories unproven from facts were regarded as sinful pseudoscience, heresy in the scientific community.

It was only the downfall of Newtonian theory in this century which made scientists realize that their standards of honesty had been utopian. Before Einstein most scientists thought that Newton had deciphered God's ultimate laws by proving them from the facts. Ampère, in the early nineteenth century, felt he had to call his book on his speculations concerning electromagnetism: *Mathematical Theory of Electrodynamic Phenomena Unequivocally Deduced from Experiment*. But at the end of the volume he casually confesses that some of the experiments were never performed and even that the necessary instruments had not been constructed!

If all scientific theories are equally unprovable, what distinguishes scientific knowledge from ignorance, science from pseudoscience?

One answer to this question was provided in the twentieth century by 'inductive logicians'. Inductive logic set out to define the probabilities of different theories according to the available total evidence. If the mathematical probability of a theory is high, it qualifies as scientific; if it is low or even zero, it is not scientific. Thus the hallmark of scientific honesty would be never to say anything that is not at least highly probable. Probabilism has an attractive feature: instead of simply providing a black-and-white distinction between science and pseudoscience, it provides a continuous scale from poor theories with low probability to good theories with high probability. But, in 1934, Karl Popper, one of the most influential philosophers of our time, argued that the mathematical probability of all theories, scientific or pseudoscientific, given *any* amount of evidence is zero.* If Popper is right, scientific theories are not only equally unprovable but also equally improbable. A new demarcation criterion was needed and Popper proposed a rather stunning one. A theory may be scientific even if there is not a shred of evidence in its favour, and it may be pseudoscientific even if all the available evidence is in its favour. That is, the scientific or non-scientific character of a theory can be determined independently of the facts. A theory is 'scientific' if one is prepared to specify in advance a crucial experiment (or observation) which can falsify it, and it is pseudoscientific if one refuses to specify such a 'potential falsifier'. But if so, we do not demarcate scientific theories from pseudoscientific ones, but rather scientific method from non-scientific method. Marxism, for a Popperian, is scientific if the Marxists are prepared

---

* Popper's argument for this claim can be found in Appendix *vii of *The Logic of Scientific Discovery* (New York: Basic Books, 1959), 363–67.

to specify facts which, if observed, make them give up Marxism. If they refuse to do so, Marxism becomes a pseudoscience. It is always interesting to ask a Marxist, what conceivable event would make him abandon his Marxism. If he is committed to Marxism, he is bound to find it immoral to specify a state of affairs which can falsify it. Thus a proposition may petrify into pseudoscientific dogma or become genuine knowledge, depending on whether we are prepared to state observable conditions which would refute it.

Is, then, Popper's falsifiability criterion the solution to the problem of demarcating science from pseudoscience? No. For Popper's criterion ignores the remarkable tenacity of scientific theories. Scientists have thick skins. They do not abandon a theory merely because facts contradict it. They normally either invent some rescue hypothesis to explain what they then call a mere anomaly or, if they cannot explain the anomaly, they ignore it, and direct their attention to other problems. Note that scientists talk about anomalies, recalcitrant instances, not refutations. History of science, of course, is full of accounts of how crucial experiments allegedly killed theories. But such accounts are fabricated long after the theory had been abandoned. Had Popper ever asked a Newtonian scientist under what experimental conditions he would abandon Newtonian theory, some Newtonian scientists would have been exactly as nonplussed as are some Marxists.

What, then, is the hallmark of science? Do we have to capitulate and agree that a scientific revolution is just an irrational change in commitment, that it is a religious conversion? Tom Kuhn, a distinguished American philosopher of science, arrived at this conclusion after discovering the naivety of Popper's falsificationism. But if Kuhn is right, then there is no explicit demarcation between science and pseudoscience, no distinction between scientific progress and intellectual decay, there is no objective standard of honesty. But what criteria can he then offer to demarcate scientific progress from intellectual degeneration?

In the last few years I have been advocating a methodology of scientific research programmes, which solves some of the problems which both Popper and Kuhn failed to solve.

First, I claim that the typical descriptive unit of great scientific achievements is not an isolated hypothesis but rather a research programme. Science is not simply trial and error, a series of conjectures and refutations. 'All swans are white' may be falsified by the discovery of one black swan. But such trivial trial and error does not rank as science. Newtonian science, for instance, is not simply a set of four conjectures—the three laws of mechanics and the law of gravitation. These four laws constitute only the 'hard core' of the Newtonian programme. But this hard core is tenaciously protected from refutation by a vast 'protective belt' of auxiliary hypotheses. And, even more importantly, the research programme also has a 'heuristic', that is, a powerful problem-solving machinery, which, with the help of sophisticated mathematical techniques, digests anomalies and even turns them into positive evidence. For instance, if a planet does not move exactly as it should, the Newtonian scientist checks his conjectures concerning atmospheric refraction, concerning propagation of light in magnetic storms, and hundreds of other conjectures which are all part of the programme. He may even invent a hitherto unknown planet and calculate its position, mass and velocity in order to explain the anomaly.

Now, Newton's theory of gravitation, Einstein's relativity theory, quantum mechanics, Marxism, Freudianism, are all research programmes, each with a characteristic hard core stubbornly defended, each with its more flexible protective belt and each with its elaborate problem-solving machinery. Each of them, at any stage of its development, has unsolved problems and undigested anomalies. All theories, in this sense, are born refuted and die refuted. But are they equally good? Until now I have been describing what research programmes are like. But how can one distinguish a scientific or progressive programme from a pseudoscientific or degenerating one?

Contrary to Popper, the difference cannot be that some are still unrefuted, while others are already refuted. When Newton published his *Principia*, it was common knowledge that it could not properly explain even the motion of the moon; in fact, lunar motion refuted Newton. Kaufmann, a distinguished physicist, refuted Einstein's relativity theory in the very year it was published.* But all the research programmes I admire have one characteristic in common. They all predict novel facts, facts which had been either undreamt of, or have indeed been contradicted by previous or rival programmes. In 1686, when Newton published his theory of gravitation, there were, for instance, two current theories concerning comets. The more popular one regarded comets as a signal from an angry God warning that He will strike and bring disaster. A little known theory of Kepler's held that comets were celestial bodies moving along straight lines. Now according to Newtonian theory, some of them moved in hyperbolas or parabolas never to return; others moved in ordinary ellipses. Halley, working in Newton's programme, calculated on the basis of observing a brief stretch of a comet's path that it would return in seventy-two years' time; he calculated to the minute when it would be seen again at a well-defined point of the sky. This was incredible. But seventy-two years later, when both Newton and Halley were long dead, Halley's comet returned exactly as Halley predicted. Similarly, Newtonian scientists predicted the existence and exact motion of small planets which had never been observed before. Or let us take Einstein's programme. This programme made the stunning prediction that if one measures the

---

* Here, as elsewhere in this reading, Lakatos is using the word *refuted* rather loosely. For Lakatos, a refutation is any apparently well-founded result that seems to be inconsistent with a theory. In the two cases he mentions—Newton and the moon, Einstein and Kaufmann's experiments on beta rays—the "refutations" were later shown to be spurious: the moon's motion is not actually inconsistent with Newton's theory, and Kaufmann's results were due to experimental error. For an account of Kaufmann's experiments and Einstein's reaction to them, see Arthur I. Miller, *Albert Einstein's Special Theory of Relativity* (Reading, Mass.: Addison-Wesley, 1981).

distance between two stars in the night and if one measures the distance between them during the day (when they are visible during an eclipse of the sun), the two measurements will be different. Nobody had thought to make such an observation before Einstein's programme. Thus, in a progressive research programme, theory leads to the discovery of hitherto unknown novel facts. In degenerating programmes, however, theories are fabricated only in order to accommodate known facts. Has, for instance, Marxism ever predicted a stunning novel fact successfully? Never! It has some famous unsuccessful predictions. It predicted the absolute impoverishment of the working class. It predicted that the first socialist revolution would take place in the industrially most developed society. It predicted that socialist societies would be free of revolutions. It predicted that there will be no conflict of interests between socialist countries. Thus the early predictions of Marxism were bold and stunning but they failed. Marxists explained all their failures: they explained the rising living standards of the working class by devising a theory of imperialism; they even explained why the first socialist revolution occurred in industrially backward Russia. They 'explained' Berlin 1953, Budapest 1956, Prague 1968. They 'explained' the Russian–Chinese conflict. But their auxiliary hypotheses were all cooked up after the event to protect Marxian theory from the facts. The Newtonian programme led to novel facts; the Marxian lagged behind the facts and has been running fast to catch up with them.

To sum up. The hallmark of empirical progress is not trivial verifications: Popper is right that there are millions of them. It is no success for Newtonian theory that stones, when dropped, fall towards the earth, no matter how often this is repeated. But so-called 'refutations' are not the hallmark of empirical failure, as Popper has preached, since all programmes grow in a permanent ocean of anomalies. What really count are dramatic, unexpected, stunning predictions: a few of them are enough to tilt the balance; where theory lags behind the facts, we are dealing with miserable degenerating research programmes.

Now, how do scientific revolutions come about? If we have two rival research programmes, and one is progressing while the other is degenerating, scientists tend to join the progressive programme. This is the rationale of scientific revolutions. But while it is a matter of intellectual honesty to keep the record public, it is not dishonest to stick to a degenerating programme and try to turn it into a progressive one.

As opposed to Popper the methodology of scientific research programmes does not offer instant rationality. One must treat budding programmes leniently: programmes may take decades before they get off the ground and become empirically progressive. Criticism is not a Popperian quick kill, by refutation. Important criticism is always constructive: there is no refutation without a better theory. Kuhn is wrong in thinking that scientific revolutions are sudden, irrational changes in vision. The history of science refutes both Popper and Kuhn: on close inspection both Popperian crucial experiments and Kuhnian revolutions turn out to be myths: what normally happens is that progressive research programmes replace degenerating ones.

The problem of demarcation between science and pseudoscience has grave implications also for the institutionalization of criticism. Copernicus's theory was banned by the Catholic Church in 1616 because it was said to be pseudoscientific.* It was taken off the index in 1820 because by that time the Church deemed that facts had proved it and therefore it became scientific. The Central Committee of the Soviet Communist Party in 1949 declared Mendelian genetics pseudoscientific and had its advocates, like Academician Vavilov, killed in concentration camps; after Vavilov's murder Mendelian genetics was rehabilitated; but the Party's right to decide what is science and publishable and what is pseudoscience and punishable was upheld. The new liberal Establishment of the West also exercises the right to deny freedom of speech to what it regards as pseudoscience, as we have seen in the case of the debate concerning race and intelligence. All these judgments were inevitably based on some sort of demarcation criterion. This is why the problem of demarcation between science and pseudoscience is not a pseudo-problem of armchair philosophers: it has grave ethical and political implications.

* Lakatos's account of why the Catholic Church placed Copernicus's book on the index in 1616 is idiosyncratic and anachronistic. A book by Paolo Foscarini (defending the view that the earth's motion is consistent with the Bible) was condemned as false and banned. Copernicus's *De Revolutionibus* was suspended pending the addition of certain "corrections" emphasizing the purely hypothetical nature of the Copernican astronomical theory. The Church's position seems to have been: (1) a realistic interpretation of the Copernican hypothesis is inconsistent with the Church's interpretation of the Bible; (2) the Copernican hypothesis is incapable of being proven true by any set of observations; consequently, given the presumed authority of the Church's interpretation of the Bible, (3) the Copernican hypothesis should be regarded merely as a convenient instrument for representing astronomical data. In short, the issue for the Congregation of the Index was not whether Copernicus's theory was pseudoscientific but whether it should be interpreted realistically. See Maurice A. Finocchiaro, *Defending Copernicus and Galileo: Boston Studies in the Philosophy of Science*, Vol. 280 (Dordrecht: Springer, 2010).

PAUL R. THAGARD

Why Astrology
Is a Pseudoscience

Most philosophers and historians of science agree that astrology is a pseudoscience, but there is little agreement on *why* it is a pseudoscience. Answers
range from matters of verifiability and falsifiability, to questions of progress
and Kuhnian normal science, to the different sorts of objections raised by a
large panel of scientists recently organized by *The Humanist* magazine. Of
course there are also Feyerabendian anarchists* and others who say that no
demarcation of science from pseudoscience is possible. However, I shall
propose a complex criterion for distinguishing disciplines as pseudoscientific; this criterion is unlike verificationist and falsificationist attempts in
that it introduces social and historical features as well as logical ones

I begin with a brief description of astrology. It would be most unfair to
evaluate astrology by reference to the daily horoscopes found in newspapers
and popular magazines. These horoscopes deal only with sun signs, whereas
a full horoscope makes reference to the "influences" also of the moon and
the planets, while also discussing the ascendant sign and other matters.

Astrology divides the sky into twelve regions, represented by the familiar signs of the Zodiac: Aquarius, Libra and so on. The sun sign represents
the part of the sky occupied by the sun at the time of birth. For example,
anyone born between September 23 and October 22 is a Libran. The ascendant sign, often assumed to be at least as important as the sun sign, represents
the part of the sky rising on the eastern horizon at the time of birth, and
therefore changes every two hours. To determine this sign, accurate knowledge of the time and place of birth is essential. The moon and the planets

FROM P. Asquith and I. Hacking, eds., *Proceedings of the Philosophy of Science
Association* Vol. 1 (East Lansing, Mich.: Philosophy of Science Association, 1978),
223–34.

* Paul Feyerabend (1924–94) used the term *epistemological anarchism* in his
Against Method (London: New Left Books, 1975), arguing that there is no rational
method in science and that the only principle consistent with scientific progress is
"anything goes."

(of which there are five or eight depending on whether Uranus, Neptune and Pluto are taken into account) are also located by means of charts on one of the parts of the Zodiac. Each planet is said to exercise an influence in a special sphere of human activity; for example, Mars governs drive, courage and daring, while Venus governs love and artistic endeavor. The immense number of combinations of sun, ascendant, moon and planetary influences allegedly determines human personality, behavior and fate.

Astrology is an ancient practice, and appears to have its origins in Chaldea, thousands of years B.C. By 700 B.C., the Zodiac was established, and a few centuries later the signs of the Zodiac were very similar to current ones. The conquests of Alexander the Great brought astrology to Greece, and the Romans were exposed in turn. Astrology was very popular during the fall of the Republic, with many notables such as Julius Caesar having their horoscopes cast. However, there was opposition from such men as Lucretius and Cicero.

Astrology underwent a gradual codification culminating in Ptolemy's *Tetrabiblos* [20], written in the second century A.D. This work describes in great detail the powers of the sun, moon and planets, and their significance in people's lives. It is still recognized as a fundamental textbook of astrology. Ptolemy took astrology as seriously as he took his famous work in geography and astronomy; this is evident from the introduction to the *Tetrabiblos*, where he discusses two available means of making predictions based on the heavens. The first and admittedly more effective of these concerns the relative movements of the sun, moon and planets, which Ptolemy had already treated in his celebrated *Almagest* [19]. The secondary but still legitimate means of prediction is that in which we use the "natural character" of the aspects of movement of heavenly bodies to "investigate the changes which they bring about in that which they surround" ([20], p. 3). He argues that this method of prediction is possible because of the manifest effects of the sun, moon and planets on the earth, for example on weather and the tides.

The European Renaissance is heralded for the rise of modern science, but occult arts such as astrology and alchemy flourished as well. Arthur Koestler has described Kepler's interest in astrology: not only did astrology provide Kepler with a livelihood, he also pursued it as a serious interest, although he was skeptical of the particular analyses of previous astrologers ([13], pp. 244–248). Astrology was popular both among intellectuals and the general public through the seventeenth century. However, astrology lost most of this popularity in the eighteenth century, when it was attacked by such figures of the Enlightenment as Swift [24] and Voltaire [29]. Only since the 1930's has astrology again gained a huge audience: most people today know at least their sun signs, and a great many believe that the stars and planets exercise an important influence on their lives.

In an attempt to reverse this trend, Bart Bok, Lawrence Jerome and Paul Kurtz drafted in 1975 a statement attacking astrology; the statement was signed by 192 leading scientists, including 19 Nobel prize winners. The statement raises three main issues: astrology originated as part of a magical world view, the planets are too distant for there to be any physical foundation for astrology, and people believe it merely out of longing for comfort ([2], pp. 9f.). None of these objections is ground for condemning astrology as pseudoscience. To show this, I shall briefly discuss articles written by Bok [1] and Jerome [12] in support of the statement.

According to Bok, to work on statistical tests of astrological predictions is a waste of time unless it is demonstrated that astrology has some sort of physical foundation ([1], p. 31). He uses the smallness of gravitational and radiative effects of the stars and planets to suggest that there is no such foundation. He also discusses the psychology of belief in astrology, which is the result of individuals' desperation in seeking solutions to their serious personal problems. Jerome devotes most of his article to the origins of astrology in the magical principle of correspondences. He claims that astrology is a system of magic rather than science, and that it fails "not because of any inherent inaccuracies due to precession or lack of exact knowledge concerning time of birth or conception, but rather because its interpretations and predictions are grounded in the ancients' magical world view" ([12], p. 46). He does however discuss some statistical tests of astrology, which I shall return to below.

These objections do not show that astrology is a pseudoscience. First, origins are irrelevant to scientific status. The alchemical origins of chemistry ([11], pp. 10–18) and the occult beginnings of medicine [8] are as magical as those of astrology, and historians have detected mystical influences in the work of many great scientists, including Newton and Einstein. Hence astrology cannot be condemned simply for the magical origins of its principles. Similarly, the psychology of popular belief is also in itself irrelevant to the status of astrology: people often believe even good theories for illegitimate reasons, and even if most people believe astrology for personal, irrational reasons, good reasons may be available.[1] Finally the lack of a physical foundation hardly marks a theory as unscientific ([22], p. 2). Examples: when Wegener [31] proposed continental drift, no mechanism was known, and a link between smoking and cancer has been established statistically [28] though the details of carcinogenesis remain to be discovered. Hence the objections of Bok, Jerome and Kurtz fail to mark astrology as pseudoscience.

Now we must consider the application of the criteria of verifiability and falsifiability to astrology. Roughly, a theory is said to be verifiable if it is possible to deduce observation statements from it. Then in principle, observations can be used to confirm or disconfirm the theory. A theory is scientific only if it is verifiable. The vicissitudes of the verification principle are too well known to recount here ([9], ch. 4). Attempts by A. J. Ayer to articulate the principle failed either by ruling out most of science as unscientific, or by ruling out nothing. Moreover, the theory/observation distinction has increasingly come into question. All that remains is a vague sense that testability somehow is a mark of scientific theories ([9], ch. 4; [10], pp. 30–32).

Well, astrology *is* vaguely testable. Because of the multitude of influences resting on tendencies rather than laws, astrology is incapable of making precise predictions. Nevertheless, attempts have been made to test the reality of these alleged tendencies, using large scale surveys and statistical evaluation. The pioneer in this area was Michel Gauquelin, who examined the careers and times of birth of 25,000 Frenchmen. Astrology suggests that people born under certain signs or planets are likely to adopt certain occupations: for example, the influence of the warlike planet Mars tends to produce soldiers or athletes, while Venus has an artistic influence. Notably, Gauquelin found *no significant correlation* between careers and either sun sign, moon sign, or ascendant sign. However, he did find some statistically interesting correlations between certain occupations of people and the position of certain planets at the time of their birth ([5], ch. 11, [6]). For example, just as astrology would suggest, there is a greater than chance association of athletes and Mars, and a greater than chance association of scientists and Saturn, where the planet is rising or at its zenith at the moment of the individual's birth.

These findings and their interpretation are highly controversial, as are subsequent studies in a similar vein [7]. Even if correct, they hardly verify astrology, especially considering the negative results found for the most important astrological categories. I have mentioned Gauquelin in order to suggest that through the use of statistical techniques astrology is at least *verifiable*. Hence the verification principle does not mark astrology as pseudoscience.

Because the predictions of astrologers are generally vague, a Popperian would assert that the real problem with astrology is that it is not falsifiable: astrologers cannot make predictions which if unfulfilled would lead them to give up their theory. Hence because it is unfalsifiable, astrology is unscientific.

But the doctrine of falsifiability faces serious problems as described by Duhem [4], Quine [21], and Lakatos [15]. Popper himself noticed early that no observation ever guarantees falsification: a theory can always be retained by introducing or modifying auxiliary hypotheses, and even observation statements are not incorrigible ([17], p. 50). Methodological decisions about what can be tampered with are required to block the escape from falsification. However, Lakatos has persuasively argued that making such decisions in advance of tests is arbitrary and may often lead to overhasty rejection of a sound theory which *ought* to be be saved by anti-falsificationist stratagems ([15], pp. 112 ff.). Falsification only occurs when a better theory comes along. Then falsifiability is only a matter of replaceability by another theory, and since astrology is in principle replaceable by another theory, falsifiability provides no criterion for rejecting astrology as pseudoscientific. We saw in the discussion of Gauquelin that astrology can be used to make predictions about statistical regularities, but the non-existence of these regularities does not falsify astrology; but here astrology does not appear worse than the best of scientific theories, which also resist falsification until alternative theories arise.[2]

Astrology cannot be condemned as pseudoscientific on the grounds proposed by verificationists, falsificationists, or Bok and Jerome. But undoubtedly astrology today faces a great many unsolved problems ([32], ch. 5). One is the negative result found by Gauquelin concerning careers and signs. Another is the problem of the precession of the equinoxes, which astrologers generally take into account when heralding the "Age of Aquarius" but totally neglect when figuring their charts. Astrologers do not always agree on the significance of the three planets, Neptune, Uranus and Pluto, that were discovered since Ptolemy. Studies of twins do not show similarities of personality and fate that astrology would suggest. Nor does astrology make sense of mass disasters, where numerous individuals with very different horoscopes come to similar ends.

But problems such as these do not in themselves show that astrology is either false or pseudoscientific. Even the best theories face unsolved problems throughout their history. To get a criterion demarcating astrology from science, we need to consider it in a wider historical and social context.

A demarcation criterion requires a matrix of three elements: theory, community, historical context. Under the first heading, "theory", fall familiar matters of structure, prediction, explanation and problem solving. We might also include the issue raised by Bok and Jerome about whether the theory has a physical foundation. Previous demarcationists have concentrated on this theoretical element, evident in the concern of the verification and falsification principles with prediction. But we have seen that this approach is not sufficient for characterizing astrology as pseudoscientific.

We must also consider the *community* of advocates of the theory, in this case the community of practitioners of astrology. Several questions are important here. First, are the practitioners in agreement on the principles of the theory and on how to go about solving problems which the theory faces? Second, do they care, that is, are they concerned about explaining anomalies and comparing the success of their theory to the record of other theories? Third, are the practitioners actively involved in attempts at confirming and disconfirming their theory?

The question about comparing the success of a theory with that of other theories introduces the third element of the matrix, historical context. The historical work of Kuhn and others has shown that in general a theory is rejected only when (1) it has faced anomalies over a long period of time and (2) it has been challenged by another theory. Hence under the heading of historical context we must consider two factors relevant to demarcation: the record of a theory over time in explaining new facts and dealing with anomalies, and the availability of alternative theories.

We can now propose the following principle of demarcation:

A theory or discipline which purports to be scientific is *pseudoscientific* if and only if:

1 it has been less progressive than alternative theories over a long period of time, and faces many unsolved problems; but

2 the community of practitioners makes little attempt to develop the theory towards solutions of the problems, shows no concern for attempts to evaluate the theory in relation to others, and is selective in considering confirmations and disconfirmations.

Progressiveness is a matter of the success of the theory in adding to its set of facts explained and problems solved ([15], p. 118; cf. [26], p. 83).

This principle captures, I believe, what is most importantly unscientific about astrology. First, astrology is dramatically unprogressive, in that it has changed little and has added nothing to its explanatory power since the time of Ptolemy. Second, problems such as the precession of equinoxes are outstanding. Third, there are alternative theories of personality and behavior available: one need not be an uncritical advocate of behaviorist, Freudian, *or* Gestalt theories to see that since the nineteenth century psychological theories have been expanding to deal with many of the phenomena which astrology explains in terms of heavenly influences. The important point is not that any of these psychological theories is established or true, only that they are growing alternatives to a long-static astrology. Fourth and finally, the community of astrologers is generally unconcerned with advancing astrology to deal with outstanding problems or with evaluating the theory in relation to others.[3] For these reasons, my criterion marks astrology as pseudoscientific.*

This demarcation criterion differs from those implicit in Lakatos and Kuhn. Lakatos has said that what makes a series of theories constituting a research program scientific is that it is progressive: each theory in the series has greater corroborated content than its predecessor ([15], p. 118). While I agree with Lakatos that progressiveness is a central notion here, it is not sufficient to distinguish science from pseudoscience. We should not brand a nonprogressive discipline as pseudoscientific unless it is being maintained against more progressive alternatives. Kuhn's discussion of astrology focuses on a different aspect of my criterion. He says that what makes astrology unscientific is the absence of the paradigm-dominated puzzle solving activity characteristic of what he calls normal science ([14], p. 9). But as Watkins has suggested, astrologers are in some respects model normal scientists: they concern themselves with solving puzzles at the level of individual horoscopes, unconcerned with the foundations of their general theory or para-

---

* Ten years later, Thagard offered a revised account of pseudoscience in chapter 9 of his book *Computational Philosophy of Science* (Cambridge, Mass.: MIT Press, 1988). This revised account is discussed in our commentary on chapter 1.

digm ([30], p. 32). Hence that feature of normal science does not distinguish science from pseudoscience. What makes astrology pseudoscientific is not that it lacks periods of Kuhnian normal science, but that its proponents adopt uncritical attitudes of "normal" scientists despite the existence of more progressive alternative theories. (Note that I am not agreeing with Popper [18] that Kuhn's normal scientists are unscientific; they can become unscientific only when an alternative paradigm has been developed.) However, if one looks not at the puzzle solving at the level of particular astrological predictions, but at the level of theoretical problems such as the precession of the equinoxes, there is some agreement between my criterion and Kuhn's; astrologers do not have a paradigm-induced confidence about solving theoretical problems.

Of course, the criterion is intended to have applications beyond astrology. I think that discussion would show that the criterion marks as pseudoscientific such practices as witchcraft and pyramidology, while leaving contemporary physics, chemistry and biology unthreatened. The current fad of biorhythms, implausibly based like astrology on date of birth, cannot be branded as pseudoscientific because we lack alternative theories giving more detailed accounts of cyclical variations in human beings, although much research is in progress.[4]

One interesting consequence of the above criterion is that a theory can be scientific at one time but pseudoscientific at another. In the time of Ptolemy or even Kepler, astrology had few alternatives in the explanation of human personality and behavior. Existing alternatives were scarcely more sophisticated or corroborated than astrology. Hence astrology should be judged as not pseudoscientific in classical or Renaissance times, even though it is pseudoscientific today. Astrology was not simply a perverse sideline of Ptolemy and Kepler, but part of their scientific activity, even if a physicist involved with astrology today should be looked at askance. Only when the historical and social aspects of science are neglected does it become plausible that pseudoscience is an unchanging category. Rationality is not a property of ideas eternally: ideas, like actions, can be rational at one time but irrational at others. Hence relativizing the science/pseudoscience distinction to historical periods is a desirable result.

But there remains a challenging historical problem. According to my criterion, astrology only became pseudoscientific with the rise of modern psychology in the nineteenth century. But astrology was already virtually excised from scientific circles by the beginning of the eighteenth. How could this be? The simple answer is that a theory can take on the appearance of an unpromising project well before it deserves the label of pseudoscience. The Copernican revolution and the mechanism of Newton, Descartes and Hobbes undermined the plausibility of astrology.[5] Lynn Thorndike [27] has described how the Newtonian theory pushed aside what had been accepted as a universal natural law, that inferiors such as inhabitants of earth are ruled and governed by superiors such as the stars and the planets. William Stahlman [23] has described how the immense growth of science in the seventeenth century contrasted with the stagnation of astrology. These developments provided good reasons for discarding astrology as a promising pursuit, but they were not yet enough to brand it as pseudoscientific, or even to refute it.

Because of its social aspect, my criterion might suggest a kind of cultural relativism. Suppose there is an isolated group of astrologers in the jungles of South America, practicing their art with no awareness of alternatives. Are we to say that astrology is *for them* scientific? Or, going in the other direction, should we count as alternative theories ones which are available to extraterrestrial beings, or which someday will be conceived? This wide construal of "alternative" would have the result that our best current theories are probably pseudoscientific. These two questions employ, respectively, a too narrow and a too broad view of alternatives. By an alternative theory I mean one generally available in the world. This assumes first that there is some kind of communication network to which a community has, or should have, access. Second, it assumes that the onus is on individuals and communities to find out about alternatives. I would argue (perhaps against Kuhn) that this second assumption is a general feature of rationality; it is at least sufficient to preclude ostrichism as a defense against being judged pseudoscientific.

In conclusion, I would like to say why I think the question of what constitutes a pseudoscience is important. Unlike the logical positivists, I am not grinding an anti-metaphysical ax, and unlike Popper, I am not grinding an anti-Freudian or anti-Marxian one.[6] My concern is social: society faces the twin problems of lack of public concern with the advancement of science, and lack of public concern with the important ethical issues now arising in science and technology, for example around the topic of genetic engineering. One reason for this dual lack of concern is the wide popularity of pseudoscience and the occult among the general public. Elucidation of how science differs from pseudoscience is the philosophical side of an attempt to overcome public neglect of genuine science.[7]

■ | Notes

1. However, astrology would doubtlessly have many fewer supporters if horoscopes tended less toward compliments and pleasant predictions and more toward the kind of analysis included in the following satirical horoscope from the December, 1977, issue of *Mother Jones*: VIRGO (Aug. 23–Sept. 22). You are the logical type and hate disorder. This nitpicking is sickening to your friends. You are cold and unemotional and sometimes fall asleep while making love. Virgos make good bus drivers.

2. For an account of the comparative evaluation of theories, see [26].

3. There appear to be a few exceptions; see [32].

4. The fad of biorhythms, now assuming a place beside astrology in the popular press, must be distinguished from the very interesting work of Frank Brown and others on biological rhythms. For a survey, see [5].

5. Plausibility is in part a matter of a hypothesis being of an appropriate *kind*, and is relevant even to the acceptance of a theory. See [26], p. 90, and [25].

6. On psychoanalysis see [3]. I would argue that Cioffi neglects the question of alternatives to psychoanalysis and the question of its progressiveness.

7. I am grateful to Dan Hausman and Elias Baumgarten for comments.

▪ | References

[1] Bok, Bart J. "A Critical Look at Astrology." In [2]. Pages 21–33.

[2] ———, Jerome, Lawrence E., and Kurtz, Paul. *Objections to Astrology.* Buffalo: Prometheus Books, 1975.

[3] Cioffi, Frank. "Freud and the Idea of a Pseudoscience." In *Explanation in the Behavioral Sciences.* Edited by R. Borger and F. Cioffi. Cambridge: Cambridge University Press, 1970. Pages 471–499.

[4] Duhem, P. *The Aim and Structure of Physical Theory.* (trans.) P. Wiener. New York: Atheneum, 1954. (Translated from 2nd edition of *La Théorie physique: son objet et sa structure.* Paris: Marcel Rivière & Cie, 1914.)

[5] Gauquelin, Michel. *The Cosmic Clocks.* Chicago: Henry Regnery, 1967.

[6] ———. *The Scientific Basis of Astrology.* New York: Stein and Day, 1969.

[7] ———. "The Zelen Test of the Mars Effect." *The Humanist* 37 (1977): 30–35.

[8] Haggard, Howard W. *Mystery, Magic, and Medicine.* Garden City: Doubleday, Doran & Company, 1933.

[9] Hempel, Carl. *Aspects of Scientific Explanation.* New York: The Free Press, 1965.

[10] ———. *Philosophy of Natural Science.* Englewood Cliffs: Prentice-Hall, 1966.

[11] Ihde, Aaron J. *The Development of Modern Chemistry.* New York: Harper and Row, 1964.

[12] Jerome, Lawrence E. "Astrology: Magic or Science?" In [2]. Pages 37–62.

[13] Koestler, Arthur. *The Sleepwalkers.* Harmondsworth: Penguin, 1964.

[14] Kuhn, T. S. "Logic of Discovery or Psychology of Research?" In [16]. Pages 1–23.

[15] Lakatos, Imre. "Falsification and the Methodology of Scientific Research Programmes." In [16]. Pages 91–195.

[16] ——— and Musgrave, Alan. (eds.). *Criticism and the Growth of Knowledge.* Cambridge: Cambridge University Press, 1970.

[17] Popper, Karl. *The Logic of Scientific Discovery.* London: Hutchinson, 1959. (Originally published as *Logik der Forschung.* Vienna: J. Springer, 1935.)

[18] ———. "Normal Science and its Dangers." In [16]. Pages 51–58.

[19] Ptolemy. *The Almagest (The Mathematical Composition).* (As printed in Hutchins, Robert Maynard (ed.). *Great Books of the Western World*, Volume 16. Chicago: Encyclopedia Britannica, Inc., 1952. Pages 1–478.)

[20] ———. *Tetrabiblos*. Edited and translated by F. E. Robbins. Cambridge: Harvard University Press, 1940.

[21] Quine, W. V. O. "Two Dogmas of Empiricism." In *From a Logical Point of View*. New York: Harper & Row, 1963. Pages 20–46. (Originally published in *The Philosophical Review* 60 (1951): 20–43.)

[22] Sagan, Carl. "Letter." *The Humanist* 36 (1976): 2.

[23] Stahlman, William D. "Astrology in Colonial America: An Extended Query." *William and Mary Quarterly* 13 (1956): 551–563.

[24] Swift, Jonathan. "The Partridge Papers." In *The Prose Works of Jonathan Swift*, Volume 2. Oxford: Basil Blackwell, 1940–1968. Pages 139–170.

[25] Thagard, Paul R. "The Autonomy of a Logic of Discovery." In L. W. Sumner *et al.*, eds., *Pragmatism and Purpose: Essays Presented to Thomas A. Goudge* (Toronto: University of Toronto Press, 1981), 248–60.

[26] ———. "The Best Explanation: Criteria for Theory Choice." *Journal of Philosophy* 75 (1978): 76–92.

[27] Thorndike, Lynn. "The True Place of Astrology in the History of Science." *Isis* 46 (1955): 273–278.

[28] U.S. Department of Health, Education and Welfare. *Smoking and Health: Report of the Advisory Committee to the Surgeon General of the Public Health Service*. Washington, D.C.: U.S. Government Printing Office, 1964.

[29] Voltaire. "Astrologie" and "Astronomie." *Dictionnaire Philosophique*. In *Oeuvres Complètes de Voltaire*, Volume XVII. Paris: Garnier Frères, 1878–1885. Pages 446–453.

[30] Watkins, J. W. N. "Against 'Normal Science'." In [16]. Pages 25–37.

[31] Wegener, Alfred. "Die Entstehung der Kontinente." *Petermanns Geographische Mitteilung* 58 (1912): 185–195, 253–256, 305–309.

[32] West, J. A. and Toonder, J. G. *The Case for Astrology*. Harmondsworth: Penguin, 1973.

MICHAEL RUSE

Creation-Science Is Not Science

In December 1981 I appeared as an expert witness for the plaintiffs and the American Civil Liberties Union (ACLU) in their successful challenge of Arkansas Act 590, which demanded that teachers give "balanced treatment" to "creation-science" and evolutionary ideas.[1] My presence occasioned some surprise, for I am an historian and philosopher of science. In this essay, I do not intend to apologize for either my existence or my calling, nor do I intend to relive past victories[2]; rather, I want to explain why a philosopher and historian of science finds the teaching of "creation-science" in science classrooms offensive.

Obviously, the crux of the issue—the center of the plaintiffs' case—is the status of creation-science. Its advocates claim that it is genuine science and may, therefore, be legitimately and properly taught in the public schools. Its detractors claim that it is not genuine science but a form of religion—dogmatic Biblical literalism by another name. Which is it, and who is to decide?

It is somewhat easier to describe who should participate in decisions on this issue. On the one hand, one naturally appeals to the authority of religious people and theologians. Does creation-science fit the accepted definitions of a religion? (In Arkansas, the ACLU produced theologians who said that indeed it did.) One also appeals to the authority of scientists. Does creation-science fit current definitions of science? (In Arkansas, the ACLU produced scientists who said that indeed it did not.)[3]

Having, as it were, appealed to the practitioners—theologians and scientists—a link still seems to be missing. Someone is needed to talk at a more theoretical level about the nature of science—any science—and then show that creation-science simply does not fit the part. As a philosopher and an historian, it is my job to look at science, and to ask precisely those questions about defining characteristics.

FROM *Science, Technology, and Human Values* 7 No. 40 (Summer 1982): 72–78.

■ | What Is Science?

It is simply not possible to give a neat definition—specifying necessary and sufficient characteristics—which separates all and only those things that have ever been called "science." The concept "science" is not as easily definable as, for example, the concept "triangle." Science is a phenomenon that has developed through the ages—dragging itself apart from religion, philosophy, superstition, and other bodies of human opinion and belief.[4]

What we call "science" today is a reasonably striking and distinctive set of claims, which have a number of characteristic features. As with most things in life, some items fall on the borderline between science and non-science (e.g., perhaps Freudian psychoanalytic theory). But it is possible to state positively that, for example, physics and chemistry are sciences, and Plato's Theory of Forms and Swedenborgian theology are not.[5]

In looking for defining features, the obvious place to start is with science's most striking aspect—it is an empirical enterprise about the real world of sensation. This is not to say that science refers only to observable entities. Every mature science contains unobservables, like electrons and genes, but ultimately, they refer to the world around us. Science attempts to understand this empirical world. What is the basis for this understanding? Surveying science and the history of science today, one thing stands out: science involves a search for order. More specifically, science looks for unbroken, blind, natural regularities (laws). Things in the world do not happen in just any old way. They follow set paths, and science tries to capture this fact. Bodies of science, therefore, known variously as "theories" or "paradigms" or "sets of models," are collections of laws.[6]

Thus, in Newtonian physics we find Newton's three laws of motion, the law of gravitational attraction, Kepler's laws of planetary motion, and so forth. Similarly, for instance, in population genetics we find the Hardy-Weinberg law. However, when we turn to something like philosophy, we do not find the same appeal to empirical law. Plato's Theory of Forms only indirectly refers to this world. Analogously, religion does not insist on unbroken law. Indeed, religious beliefs frequently allow or suppose events outside law or else events that violate law (miracles). Jesus feeding the 5,000 with the loaves and fishes was one such event. This is not to say that religion is false, but it does say that religion is not science. When the loaves and fishes multiplied to a sufficiency to feed so many people, things happened that did not obey natural law, and hence the feeding of the 5,000 is an event beyond the ken of science.[7]

A major part of the scientific enterprise involves the use of law to effect explanation. One tries to show why things are as they are—and how they fall beneath or follow from law (together perhaps with certain specified initial conditions). Why, for example, does a cannon ball go in a parabola and not in a circle? Because of the constraints of Newton's laws. Why do two blue-eyed parents always have blue-eyed children? Because this trait obeys Mendel's first law, given the particular way in which the genes control eye-color.

A scientific explanation must appeal to law and must show that what is being explained had to occur. The explanation excludes those things that did not happen.[8]

The other side of explanation is *prediction*. The laws indicate what is going to happen: that the ball will go in a parabola, that the child will be blue-eyed. In science, as well as in futurology, one can also, as it were, predict backwards. Using laws, one infers that a particular, hitherto-unknown phenomenon or event took place in the past. Thus, for instance, one might use the laws of physics to infer back to some eclipse of the sun reported in ancient writings.

Closely connected with the twin notions of explanation and prediction comes *testability*. A genuine scientific theory lays itself open to check against the real world: the scientist can see if the inferences made in explanation and prediction actually obtain in nature. Does the chemical reaction proceed as suspected? In Young's double slit experiment, does one find the bands of light and dark predicted by the wave theory? Do the continents show the expected after-effects of drift?

Testability is a two-way process. The researcher looks for some positive evidence, for *confirmation*. No one will take seriously a scientific theory that has no empirical support (although obviously a younger theory is liable to be less well-supported than an older theory). Conversely, a theory must be open to possible refutation. If the facts speak against a theory, then it must go. A body of science must be *falsifiable*. For example, Kepler's laws could have been false: if a planet were discovered going in squares, then the laws would have been shown to be incorrect. However, in distinguishing science from nonscience, no amount of empirical evidence can disprove, for example, the Kantian philosophical claim that one ought to treat people as ends rather than merely as a means. Similarly, Catholic religious claims about transubstantiation (the changing of the bread and wine into the body and blood of Christ) are unfalsifiable.[9]

Science is *tentative*. Ultimately, a scientist must be prepared to reject his theory. Unfortunately, not all scientists are prepared to do in practice what they promise to do in theory; but the weaknesses of individuals are counterbalanced by the fact that, as a group, scientists do give up theories that fail to answer to new or reconsidered evidence. In the last 30 years, for example, geologists have reversed their strong convictions that the continents never move.

Scientists do not, of course, immediately throw their theories away as soon as any counter-evidence arrives. If a theory is powerful and successful, then some problems will be tolerated, but scientists must be prepared to change their minds in the face of the empirical evidence. In this regard, the scientists differ from both the philosophers and the theologians. Nothing in the real world would make the Kantian change his mind, and the Catholic is equally dogmatic, despite any empirical evidence about the stability of bread and wine. Such evidence is simply considered irrelevant.[10]

Some other features of science should also be mentioned, for instance, the urge for simplicity and unification; however, I have now listed the major characteristics. Good science—like good philosophy and good religion—presupposes an attitude that one might describe as professional *integrity*. A scientist should not cheat or falsify data or quote out of context or do any other thing that is intellectually dishonest. Of course, as always, some individuals fail; but science as a whole disapproves of such actions. Indeed, when transgressors are detected, they are usually expelled from the community. Science depends on honesty in the realm of ideas. One may cheat on one's taxes; one may not fiddle the data.[11]

■ | Creation-Science Considered

How does creation-science fit the criteria of science listed in the previous section? By "creation-science" in this context, I refer not just to the definition given in Act 590, but to the whole body of literature which goes by that name. The doctrine includes the claims that the universe is very young (6,000 to 20,000 years), that everything started instantaneously, that human beings had ancestry separate from apes, and that a monstrous flood once engulfed the entire earth.[12]

LAWS—NATURAL REGULARITIES

Science is about unbroken, natural regularity. It does not admit miracles. It is clear, therefore, that again and again, creation-science invokes happenings and causes outside of law. For instance, the only reasonable inference from Act 590 (certainly the inference that was accepted in the Arkansas court) is that for creation-science the origin of the universe and life in it is not bound by law. Whereas the definition of creation-science includes the unqualified phrase "sudden creation of the universe, energy and life from nothing," the definition of evolution specifically includes the qualification that its view of origins is "naturalistic." Because "naturalistic" means "subject to empirical law," the deliberate omission of such a term in the characterization of creation-science means that no laws were involved.

In confirmation of this inference, we can find identical claims in the writings of creation scientists: for instance, the following passage from Duane T. Gish's popular work *Evolution—The Fossils Say No!*

CREATION. By creation we mean the bringing into being of the basic kinds of plants and animals by the process of sudden, or fiat, creation described in the first two chapters of Genesis. Here we find the creation by God of the plants and animals, each commanded to reproduce after its own kind using processes which were essentially instantaneous.

We do not know how God created, what processes He used, *for God used processes which are not now operating anywhere in the natural universe.* This is why we refer to divine creation as special creation. We cannot discover by scientific investigations anything about the creative processes used by God.[13]

By Gish's own admission, we are not dealing with science. Similar sentiments can be found in *The Genesis Flood* by John Whitcomb, Jr., and Henry M. Morris:

> But during the period of Creation, God was introducing order and organization and energization into the universe in a very high degree, even to life itself! *It is thus quite plain that the processes used by God in creation were utterly different from the processes which now operate in the universe!* The Creation was a unique period, entirely incommensurate with this present world. This is plainly emphasized and reemphasized in the divine revelation which God has given us concerning Creation, which concludes with these words: 'And the heavens and the earth were *finished,* and *all* the host of them. And on the seventh day God *finished* His work which He had made; and He *rested* on the seventh day from *all His work* which He had made. And God blessed the seventh day, and hallowed it; because that in it He *rested* from *all* his work which God had created and made.' In view of these strong and repeated assertions, is it not the height of presumption for man to attempt to study Creation in terms of present processes?[14]

Creation scientists generally acknowledge this work to be *the* seminal contribution that led to the growth of the creation-science movement. Morris, in particular, is the father figure of creation-science and Gish his chief lieutenant.

Creation scientists also break with law in many other instances. The creationists believe that the Flood, for example, could not have just occurred through blind regularities. As Whitcomb and Morris make very clear, certain supernatural interventions were necessary to bring about the Flood.[15] Similarly, in order to ensure the survival of at least some organisms, God had to busy himself and break through law.

EXPLANATION AND PREDICTION

Given the crucial role that law plays for the scientist in these processes, neither explanation nor prediction is possible where no law exists. Thus, explanation and prediction simply cannot even be attempted when one deals with creation-science accounts either of origins or of the Flood.

Even against the broader vistas of biology, creation-science is inadequate. Scientific explanation/prediction must lead to the thing being explained/predicted, showing why that thing obtains and not other things. Why does the ball go in a parabola? Why does it not describe a circle? Take an important and pervasive biological phenomenon, namely, "homologies," the isomorphisms between the bones of different animals. These similarities were recognized as pervasive facets of nature even before Darwin published *On the Origin of Species*. Why are the bones in the forelimbs of men, horses, whales, and birds all so similar, even though the functions are quite different? Evolutionists explain homologies naturally and easily, as a result of common descent. Creationists can give no explanation, and make no predictions. All they can offer is the disingenuous comment that homology signifies nothing, because classification is all man-made and arbitrary anyway. Is it arbitrary that man is not classified with the birds?[16] Why are Darwin's finches distributed in the way that we find on the Galapagos? Why are there 14 separate species of this little bird, scattered over a small group of islands in the Pacific on the equator? On those rare occasions when Darwin's finches do fly into the pages of creation-science, it is claimed either that they are all the same species (false), or that they are a case of degeneration from one "kind" created back at the beginning of life.[17] Apart from the fact that "kind" is a term of classification to be found only in Genesis, this is no explanation. How could such a division of the finches have occurred, given the short span that the creationists allow since the Creation? And, in any case, Darwin's finches are anything but degenerates. Different species of finch have entirely different sorts of beaks, adapted for different foodstuffs—evolution of the most sophisticated type.[18]

TESTABILITY, CONFIRMATION, AND FALSIFIABILITY

Testability, confirmation, and falsifiability are no better treated by creation-science. A scientific theory must provide more than just after-the-fact explanations of things that one already knows. One must push out into the frontiers of new knowledge, trying to predict new facts, and risking the theory against the discovery of possible falsifying information. One cannot simply work at a secondary level, constantly protecting one's views against threat: forever inventing *ad hoc* hypotheses to save one's core assumptions.

Creation scientists do little or nothing by way of genuine test. Indeed, the most striking thing about the whole body of creation-science literature is the virtual absence of *any* experimental or observational work by creation scientists. Almost invariably, the creationists work exclusively with the discoveries and claims of evolutionists, twisting the conclusions to their own ends. Argument proceeds by showing evolution (specifically Darwinism) wrong, rather than by showing Creationism right.

However, this way of proceeding—what the creationists refer to as the "two model approach"—is simply a fallacious form of argument. The views of people like Fred Hoyle and N. C. Wickramasinghe, who believe that life comes from outer space, are neither creationist nor truly evolutionist.[19] Denying evolution in no way proves Creationism. And, even if a more straightforward either/or between evolution and Creationism existed, the perpetually negative approach is just not the way that science proceeds. One must find one's own evidence in favor of one's position, just as physicists, chemists, and biologists do.

Do creation scientists ever actually expose their theories and ideas to test? Even if they do, when new counter-empirical evidence is discovered, creation scientists appear to pull back, refusing to allow their position to be falsified.

Consider, for instance, the classic case of the "missing link"—namely, that between man and his ancestors. The creationists say that there are no plausible bridging organisms whatsoever. Thus, this super-gap between man and all other animals (alive or dead) supposedly underlines the creationists' contention that man and apes have separate ancestry. But what about the australopithecines, organisms that paleontologists have, for most of this century, claimed are plausible human ancestors? With respect, argue the creationists, australopithecines are not links, because they had apelike brains, they *walked* like apes, and they used their knuckles for support, just like gorillas. Hence, the gap remains.[20]

However, such a conclusion can be maintained only by blatant disregard of the empirical evidence. *Australopithecus afarensis* was a creature with a brain the size of that of an ape which walked upright.[21] Yet the creationists do not concede defeat. They then argue that the *Australopithecus afarensis* is like an orangutan.[22] In short, nothing apparently makes the creationists change their minds, or allows their views to be tested, lest they be falsified.

Tentativeness

Creation-science is not science because there is absolutely no way in which creationists will budge from their position. Indeed, the leading organization of creation-science, The Creation Research Society (with 500 full members, all of whom must have an advanced degree in a scientific/technological area), demands that its members sign a statement affirming that they take the Bible as literally true.[23] Unfortunately, an organization cannot require such a condition of membership, and then claim to be a scientific organization. Science must be open to change, however confident one may feel at present. Fanatical dogmatism is just not acceptable.

Integrity

Creation scientists use any fallacy in the logic books to achieve their ends. Most particularly, apart from grossly distorting evolutionists' positions, the creation-scientists frequently use inappropriate or incomplete quotations. They take the words of some eminent evolutionist, and attempt to make him or her say exactly the opposite to that intended. For instance, in *Creation: The Facts of Life*, author Gary E. Parker constantly refers to "noted Harvard geneticist" Richard Lewontin as claiming that the hand and the eye are the best evidence of God's design.[24] Can this reference really be true? Has the author of *The Genetic Basis of Evolutionary Change*[25] really foresworn Darwin for Moses? In fact, when one looks at Lewontin's writings, one finds that he says that *before Darwin*, people believed the hand and the eye to be the effect of direct design. Today, scientists believe that such features were produced by the natural process of evolution through natural selection; but, a reader learns nothing of this from Parker's book.

What are the essential features of science? Does creation-science have any, all, or none of these features? My answer to this is none. By every mark of what constitutes science, creation-science fails. And, although it has not been my direct purpose to show its true nature, it is surely there for all to see. Miracles brought about by an intervening supervising force speak of only one thing. Creation "science" is actually dogmatic religious Fundamentalism. To regard it as otherwise is an insult to the scientist, as well as to the believer who sees creation-science as a blasphemous distortion of God-given reason. I believe that creation-science should not be taught in the public schools because creation-science is not science.

■ | Notes

1. In fact, Act 590 demanded that *if* one teach[es] evolution, *then* one must also teach creation-science. Presumably a teacher could have stayed away from origins entirely—albeit with large gaps in some courses.

2. For a brief personal account of my experiences, see Michael Ruse, "A Philosopher at the Monkey Trial," *New Scientist* (1982): 317–319.

3. Judge William Overton's ruling on the constitutionality (or, rather, unconstitutionality) of Act 590 gives a fair and full account of the various claims made by theologians (including historians and sociologists of religion) and scientists.

4. In my book, *The Darwinian Revolution: Science Red in Tooth and Claw* (Chicago, IL: University of Chicago Press, 1979), I look at the way science was breaking apart from religion in the 19th century.

5. What follows is drawn from a number of basic books in the philosophy of science, including R. B. Braithwaite, *Scientific Explanation* (Cambridge, England: Cambridge University Press, 1953); Karl R. Popper, *The Logic of Scientific Discovery* (London: Hutchinson, 1959); E. Nagel, *The Structure of Science* (London: Routledge and Kegan Paul, 1961); Thomas S. Kuhn, *The Structure of Scientific Revolutions* (Chicago, IL: University of Chicago Press, 1962); and C. G. Hempel, *Philosophy of Natural Science* (Englewood Cliffs, NJ: Prentice-Hall, 1966). The discussion is the same as what I provided for the plaintiffs in a number of position papers. It also formed the basis of my testimony in court, and, as can be seen from Judge Overton's ruling, was accepted by the court virtually verbatim.

6. One sometimes sees a distinction drawn between "theory" and "model." At the level of this discussion, it is not necessary to discuss specific details. I consider vari-

ous uses of these terms in my book, *Darwinism Defended: A Guide to the Evolution Controversies* (Reading, MA: Addison-Wesley, 1982).

7. For more on science and miracles, especially with respect to evolutionary questions, see my *Darwinian Revolution, op. cit.*

8. The exact relationship between laws and what they explain has been a matter of much debate. Today, I think most would agree that the connection must be fairly tight—the thing being explained should follow. For more on explanation in biology see Michael Ruse, *The Philosophy of Biology* (London: Hutchinson, 1973); and David L. Hull, *Philosophy of Biological Science* (Englewood Cliffs, NJ: Prentice-Hall, 1974). A popular thesis is that explanation of laws involves deduction from other laws. A theory is a body of laws bound in this way: a so-called "hypothetico-deductive" system.

9. Falsifiability today has a high profile in the philosophical and scientific literature. Many scientists, especially, agree with Karl Popper, who has argued that falsifiability is *the* criterion demarcating science from non-science (see especially his *Logic of Scientific Discovery*). My position is that falsifiability is an important part, but only one part, of a spectrum of features required to demarcate science from non-science. For more on this point, see my *Is Science Sexist? And Other Problems in the Biomedical Sciences* (Dordrecht, Holland: D. Reidel Publishing Company, 1981).

10. At the Arkansas trial, in talking of the tentativeness of science, I drew an analogy in testimony between science and the law. In a criminal trial, one tries to establish guilt "beyond a reasonable doubt." If this can be done, then the criminal is convicted. But, if new evidence is ever discovered that might prove the convicted person innocent, cases can always be reopened. In science, too, scientists make decisions less formally but just as strongly—and get on with business, but cases (theories) can be reopened.

11. Of course, the scientist as citizen may run into problems here!

12. The key definitions in Arkansas Act 590, requiring "balanced treatment" in the public schools, are found in Section 4 [of the Act]. Section 4(a) does not specify exactly how old the earth is supposed to be, but in court a span of 6,000 to 20,000 years emerged in testimony.

The fullest account of the creation-science position is given in Henry M. Morris, ed., *Scientific Creationism* (San Diego, CA: Creation-Life Publishers, 1974).

13. Duane T. Gish, *Evolution—The Fossils Say No!* (San Diego, CA: Creation-Life Publishers, 1973), pp. 22–25, his italics.

14. John Whitcomb, Jr., and Henry M. Morris, *The Genesis Flood* (Philadelphia, PA: Presbyterian and Reformed Publishing Company, 1961), pp. 223–224, their italics.

15. *Ibid*, p. 76.

16. See Morris, *op. cit.*, pp. 71–72, and my discussion in *Darwinism Defended, op. cit.*

17. For instance, in John N. Moore and H. S. Slusher, *Biology: A Search for Order in Complexity* (Grand Rapids, MI: Zondervan, 1977).

18. D. Lack, *Darwin's Finches* (Cambridge, England: Cambridge University Press, 1947).

19. Fred Hoyle and N. C. Wickramasinghe, *Evolution from Space* (London: Dent, 1981).

20. Morris, *op. cit.*, p. 173.

21. Donald Johanson and M. Edey, *Lucy: The Beginnings of Humankind* (New York, NY: Simon and Schuster, 1981).

22. Gary E. Parker, *Creation: The Facts of Life* (San Diego, CA: Creation-Life Publishers, 1979), p. 113.

23. For details of these statements, see [footnote] 7 in Judge Overton's ruling.

24. Parker, *op. cit.* See, for instance, pp. 55 and 144. The latter passage is worth quoting in full:

> Then there's 'the marvelous fit of organisms to the environment,' the special adaptations of cleaner fish, woodpeckers, bombardier beetles, etc., etc.,—what Darwin called 'Difficulties with the Theory,' and what Harvard's Lewontin (1978) called 'the chief evidence of a Supreme Designer.' Because of their 'perfection of structure,' he says, organisms 'appear to have been carefully and artfully designed.'

The pertinent article by Richard Lewontin is "Adaptation," *Scientific American* (September 1978).

25. Richard C. Lewontin, *The Genetic Basis of Evolutionary Change* (New York, NY: Columbia University Press, 1974).

Larry Laudan

Commentary: Science at the Bar—Causes for Concern

In the wake of the decision in the Arkansas Creationism trial *(McLean v. Arkansas)*[1] the friends of science are apt to be relishing the outcome. The creationists quite clearly made a botch of their case and there can be little doubt that the Arkansas decision may, at least for a time, blunt legislative pressure to enact similar laws in other states. Once the dust has settled, however, the trial in general and Judge William R. Overton's ruling in particular may come back to haunt us; for, although the verdict itself is probably to be commended, it was reached for all the wrong reasons and by a chain of argument which is hopelessly suspect. Indeed, the ruling rests on a host of misrepresentations of what science is and how it works.

The heart of Judge Overton's Opinion is a formulation of "the essential characteristics of science." These characteristics serve as touchstones for contrasting evolutionary theory with Creationism; they lead Judge Overton ultimately to the claim, specious in its own right, that since Creationism is not "science," it must be religion. The Opinion offers five essential properties that demarcate scientific knowledge from other things: "(1) It is guided by natural law; (2) it has to be explanatory by reference to natural law; (3) it is testable against the empirical world; (4) its conclusions are tentative, i.e., are not necessarily the final word; and (5) it is falsifiable."

These fall naturally into two families: properties (1) and (2) have to do with lawlikeness and explanatory ability; the other three properties have to do with the fallibility and testability of scientific claims. I shall deal with the second set of issues first, because it is there that the most egregious errors of fact and judgment are to be found.

At various key points in the Opinion, Creationism is charged with being untestable, dogmatic (and thus non-tentative), and unfalsifiable. All three charges are of dubious merit. For instance, to make the inter-linked claims that Creationism is neither falsifiable nor testable is to assert that Creationism

From *Science, Technology, and Human Values* 7 No. 41 (Fall 1982): 16–19.

makes no empirical assertions whatever. That is surely false. Creationists make a wide range of testable assertions about empirical matters of fact. Thus, as Judge Overton himself grants (apparently without seeing its implications), the creationists say that the earth is of very recent origin (say 6,000 to 20,000 years old); they argue that most of the geological features of the earth's surface are diluvial in character (i.e., products of the postulated worldwide Noachian deluge); they are committed to a large number of factual historical claims with which the Old Testament is replete; they assert the limited variability of species. They are committed to the view that, since animals and man were created at the same time, the human fossil record must be paleontologically co-extensive with the record of lower animals. It is fair to say that no one has shown how to reconcile such claims with the available evidence—evidence which speaks persuasively to a long earth history, among other things.

In brief, these claims are testable, they have been tested, and they have failed those tests. Unfortunately, the logic of the Opinion's analysis precludes saying any of the above. By arguing that the tenets of Creationism are neither testable nor falsifiable, Judge Overton (like those scientists who similarly charge Creationism with being untestable) deprives science of its strongest argument against Creationism. Indeed, if any doctrine in the history of science has ever been falsified, it is the set of claims associated with "creation-science." Asserting that Creationism makes no empirical claims plays directly, if inadvertently, into the hands of the creationists by immunizing their ideology from empirical confrontation. The correct way to combat Creationism is to confute the empirical claims it does make, not to pretend that it makes no such claims at all.

It is true, of course, that some tenets of Creationism are not testable in isolation (e.g., the claim that man emerged by a direct supernatural act of creation). But that scarcely makes Creationism "unscientific." It is now widely acknowledged that many scientific claims are not testable in isolation, but only when embedded in a larger system of statements, some of whose consequences can be submitted to test.

Judge Overton's third worry about Creationism centers on the issue of revisability. Over and over again, he finds Creationism and its advocates "unscientific" because they have "refuse[d] to change it regardless of the evidence developed during the course of the[ir] investigation." In point of fact, the charge is mistaken. If the claims of modern-day creationists are compared with those of their nineteenth-century counterparts, significant shifts in orientation and assertion are evident. One of the most visible opponents of Creationism, Stephen Gould, concedes that creationists have modified their views about the amount of variability allowed at the level of species change. Creationists do, in short, change their minds from time to time. Doubtless they would credit these shifts to their efforts to adjust their views to newly emerging evidence, in what they imagine to be a scientifically respectable way.

Perhaps what Judge Overton had in mind was the fact that some of Creationism's core assumptions (e.g., that there was a Noachian flood, that man did not evolve from lower animals, or that God created the world) seem closed off from any serious modification. But historical and sociological researches on science strongly suggest that the scientists of any epoch likewise regard some of their beliefs as so fundamental as not to be open to repudiation or negotiation. Would Newton, for instance, have been tentative about the claim that there were forces in the world? Are quantum mechanicians willing to contemplate giving up the uncertainty relation? Are physicists willing to specify circumstances under which they would give up energy conservation? Numerous historians and philosophers of science (e.g., Kuhn, Mitroff, Feyerabend, Lakatos) have documented the existence of a certain degree of dogmatism about core commitments in scientific research and have argued that such dogmatism plays a constructive role in promoting the aims of science. I am not denying that there may be subtle but important differences between the dogmatism of scientists and that exhibited by many creationists; but one does not even begin to get at those differences by pretending that science is characterized by an uncompromising open-mindedness.

Even worse, the *ad hominem* charge of dogmatism against Creationism egregiously confuses doctrines with the proponents of those doctrines. Since no law mandates that creationists should be invited into the classroom, it is quite irrelevant whether they themselves are close-minded. The Arkansas statute proposed that Creationism be taught, not that creationists should teach it. What counts is the epistemic status of Creationism, not the cognitive idiosyncrasies of the creationists. Because many of the theses of Creationism are testable, the mind set of creationists has no bearing in law or in fact on the merits of Creationism.

What about the other pair of essential characteristics which the *McLean* Opinion cites, namely, that science is a matter of natural law and explainable by natural law? I find the formulation in the Opinion to be rather fuzzy; but the general idea appears to be that it is inappropriate and unscientific to postulate the existence of any process or fact which cannot be explained in terms of some known scientific laws—for instance, the creationists' assertion that there are outer limits to the change of species "cannot be explained by natural law." Earlier in the Opinion, Judge Overton also writes "there is no scientific explanation for these limits which is guided by natural law," and thus concludes that such limits are unscientific. Still later, remarking on the hypothesis of the Noachian flood, he says: "A worldwide flood as an explanation of the world's geology is not the product of natural law, nor can its occurrence be explained by natural law." Quite how Judge Overton knows that a worldwide flood "cannot" be explained by the laws of science is left opaque; and even if we did not know how to reduce a universal flood to the familiar laws of physics, this requirement is an altogether inappropriate standard for ascertaining whether a claim is scientific. For centuries scientists

have recognized a difference between establishing the existence of a phenomenon and explaining that phenomenon in a lawlike way. Our ultimate
goal, no doubt, is to do both. But to suggest, as the *McLean* Opinion does
repeatedly, that an existence claim (e.g., there was a worldwide flood) is
unscientific until we have found the laws on which the alleged phenomenon depends is simply outrageous. Galileo and Newton took themselves to
have established the existence of gravitational phenomena, long before anyone was able to give a causal or explanatory account of gravitation. Darwin
took himself to have established the existence of natural selection almost a
half-century before geneticists were able to lay out the laws of heredity on
which natural selection depended. If we took the *McLean* Opinion criterion
seriously, we should have to say that Newton and Darwin were unscientific;
and, to take an example from our own time, it would follow that plate tectonics is unscientific because we have not yet identified the laws of physics and
chemistry which account for the dynamics of crustal motion.

The real objection to such creationist claims as that of the (relative)
invariability of species is not that such invariability has not been explained
by scientific laws, but rather that the evidence for invariability is less robust
than the evidence for its contrary, variability. But to say as much requires
renunciation of the Opinion's other charge—to wit, that Creationism is not
testable.

I could continue with this tale of woeful fallacies in the Arkansas ruling, but that is hardly necessary. What is worrisome is that the Opinion's line
of reasoning—which neatly coincides with the predominant tactic among
scientists who have entered the public fray on this issue—leaves many loop-
holes for the creationists to exploit. As numerous authors have shown, the
requirements of testability, revisability, and falsifiability are exceedingly *weak*
requirements. Leaving aside the fact that (as I pointed out above) it can be
argued that Creationism already satisfies these requirements, it would be
easy for a creationist to say the following: "I will abandon my views if we find
a living specimen of a species intermediate between man and apes." It is, of
course, extremely unlikely that such an individual will be discovered. But, in
that statement the creationist would satisfy, in one fell swoop, all the formal
requirements of testability, falsifiability, and revisability. If we set very weak
standards for scientific status—and, let there be no mistake, I believe that all
of the Opinion's last three criteria fall in this category—then it will be quite
simple for Creationism to qualify as "scientific."

Rather than taking on the creationists obliquely and in wholesale fashion
by suggesting that what they are doing is "unscientific" *tout court* (which is
doubly silly because few authors can even agree on what makes an activity
scientific), we should confront their claims directly and in piecemeal fashion
by asking what evidence and arguments can be marshalled for and against
each of them. The core issue is not whether Creationism satisfies some unde-
manding and highly controversial definitions of what is scientific; the real
question is whether the existing evidence provides stronger arguments for

evolutionary theory than for Creationism. Once that question is settled, we will know what belongs in the classroom and what does not. Debating the scientific status of Creationism (especially when "science" is construed in such an unfortunate manner) is a red herring that diverts attention away from the issues that should concern us.

Some defenders of the scientific orthodoxy will probably say that my reservations are just nitpicking ones, and that—at least to a first order of approximation—Judge Overton has correctly identified what is fishy about Creationism. The apologists for science, such as the editor of *The Skeptical Inquirer*, have already objected to those who criticize this whitewash of science "on arcane, semantic grounds . . . [drawn] from the most remote reaches of the academic philosophy of science."[2] But let us be clear about what is at stake. In setting out in the *McLean* Opinion to characterize the "essential" nature of science, Judge Overton was explicitly venturing into philosophical terrain. His *obiter dicta* are about as remote from well-founded opinion in the philosophy of science as Creationism is from respectable geology. It simply will not do for the defenders of science to invoke philosophy of science when it suits them (e.g., their much-loved principle of falsifiability comes directly from the philosopher Karl Popper) and to dismiss it as "arcane" and "remote" when it does not. However noble the motivation, bad philosophy makes for bad law.

The victory in the Arkansas case was hollow, for it was achieved only at the expense of perpetuating and canonizing a false stereotype of what science is and how it works. If it goes unchallenged by the scientific community, it will raise grave doubts about that community's intellectual integrity. No one familiar with the issues can really believe that anything important was settled through anachronistic efforts to revive a variety of discredited criteria for distinguishing between the scientific and the non-scientific. Fifty years ago, Clarence Darrow asked, *à propos* the Scopes trial, "Isn't it difficult to realize that a trial of this kind is possible in the twentieth century in the United States of America?" We can raise that question anew, with the added irony that, this time, the pro-science forces are defending a philosophy of science which is, in its way, every bit as outmoded as the "science" of the creationists.

▪ | Notes

1. *McLean v. Arkansas Board of Education*, 529 F. Supp. 1255 (E.D. Ark. 1982). For the text of the law, the decision, and essays by participants in the trial, see *Science, Technology, and Human Values* 7 No. 40 (Summer 1982), and also *Creationism, Science, and the Law [:The Arkansas Case*, ed. Marcel C. La Follette (Cambridge, Mass.: MIT Press, 1983).]

2. "The Creationist Threat: Science Finally Awakens," *The Skeptical Inquirer* 3 (Spring 1982): 2–5.

1 | COMMENTARY

1 | COMMENTARY

1.1 | Popper's Demarcation Criterion

In "Science: Conjectures and Refutations," Karl Popper explains how he came to formulate his falsifiability criterion for the scientific status of a theory. He recognized that it was not enough to use the so-called empirical (or inductive) method of generalizing from observation and experience, for by this standard astrology might well qualify as genuine science. So why, Popper wondered, were Freudian psychoanalysis, Adlerian "individual psychology," and the Marxist theory of history more like astrology than astronomy, more like myth than science?

His answer came from noting that, while proponents of these disciplines found confirming evidence for their theories at every turn, they made no predictions that could be disconfirmed by evidence. With deliberate irony, Popper describes "the incessant stream of confirmations, of observations which 'verified' the theories in question" (5). Moreover, it seemed to Popper as though just about anything, even apparent counterevidence, could be explained in Freudian or Adlerian or Marxist terms. In marked contrast to this were certain features characterizing one of the most important physical theories of this century. Popper recounts how impressed he was by the bold prediction of the bending of starlight near the surface of the sun made by Einstein's general theory of relativity. This prediction was verified by two astronomical expeditions to observe the total solar eclipse of 29 May 1919— one to Brazil, the other to the west coast of Africa—organized by the British cosmologist Sir Arthur Eddington. Photographic plates produced during these expeditions revealed that starlight was indeed deflected by the sun by an amount very close to Einstein's prediction of 1.75 seconds of arc. This crucial observation led to the overthrow of Newton's theory of gravity by Einstein's general theory of relativity.[1]

Unlike Marx's theory of history and Adler's theory of the inferiority complex, Einstein's theory ran a serious risk of refutation by predicting the result of an observational test *before* the test was made. Popper sees this possibility of refutation by observation and experiment as the hallmark of genuine science. Agreement with known facts, or the ability to explain known facts, is not enough to make a theory scientific. Whereas the Marxists and Adlerians saw confirmation of their theories everywhere and recognized nothing that their theories could not explain, Einstein's theory is refutable because, by its very nature, it is incompatible with certain possible results of observation—it is open to falsifying tests. Popper insists that in order to be scientific a theory must take a risk by predicting something new. Thus, Popper advocates falsifiability (testability), not verifiability (confirmability), as the demarcation criterion for distinguishing science from pseudoscience.

Falsifiability

Despite its simplicity and initial plausibility, there is much that is unclear and controversial about Popper's demarcation criterion. Part of the unclarity arises because Popper shifts back and forth between two different notions—between *falsifiability* as a logical property of statements (requiring that scientific statements logically imply at least one testable prediction) and *falsifiability* as a term prescribing how scientists should act. According to Popper, scientists should test their theories by trying to refute them; when a prediction disagrees with observation and experiment, they should abandon their theories as refuted. Falsifiability in this second, prescriptive sense implies falsifiability in the first sense, for it is only by making testable predictions that a theory—made up of scientific statements—can be refuted. But the implication does not hold in the other direction. It is perfectly possible for a theory such as Marxism to imply at least one testable prediction (say, that all socialist revolutions will occur among the proletariat of industrialized capitalist nations) and yet, when the prediction turns out to be false (because the Russian and Chinese revolutions occurred in societies that were preindustrial and feudal), the adherents of the theory refuse to regard it as refuted and strive to explain away the anomaly. Thus, a theory that is scientific in Popper's (first) logical sense might be judged pseudoscientific in Popper's (second) methodological sense because of the behavior of its proponents.

Many philosophers have criticized the prescriptive, methodological aspect of Popper's demarcation criterion. They argue that abandoning a theory the instant it makes a false prediction would rule out too much good science. (This criticism, made by Kuhn and Lakatos, among others, will be discussed later.) Some philosophers also object to the first sense of Popper's falsifiability criterion, that falsifiability is a logical property of scientific statements, on the grounds that it is too weak. Take any statement, however implausible or crazy it may sound, and conjoin it with a respectable scientific theory. The crazy statement, *C*, might be the claim that aliens visited the earth during the Pleistocene era and removed all traces of their visit before departing. Although *C* is not a tautology, it makes no testable predictions. The respectable scientific theory, *T*, could be from any field whatever—geology, chemistry, physics, or astronomy. The conjunction, (*T* & *C*), makes lots of testable predictions since its logical consequences include all the predictions made by *T* alone. Thus, (*T* & *C*) satisfies Popper's falsifiability criterion. The moral is clear: having testable consequences is a very weak requirement. At best, perhaps, it is a necessary condition for genuine science, and many statements that satisfy it are not part of science. Thus, presumably, Popper was not claiming that all falsifiable statements are scientific; he was merely claiming that in order to be scientific, a statement must be falsifiable.[2]

POPPER AND THE THEORY OF EVOLUTION

Popper claimed for several decades that the principle of natural selection in Darwin's theory of evolution fails to satisfy his falsifiability criterion (that is, falsifiability as a logical property of statements) and thus, in some important sense, that it is not scientific but "metaphysical." Popper recanted this belief, appropriately enough, when he delivered the first Darwin Lecture at Darwin College, Cambridge University, in 1977.[3] This is important for three reasons. First, it illustrates how difficult it can be to decide whether or not a component of a scientific theory is falsifiable. Second, it illustrates the complexity of Popper's position, since Popper never condemned the whole of Darwin's theory as a pseudoscience even when he judged that an important part of that theory could not be falsified. Third, it sheds some light—if just a little—on the position of creationists who, much to Popper's dismay, have appealed to Popper's (pre-1977) writings for support in their crusade against the theory of evolution.

Before his recantation, Popper expressed reservations about Darwin's theory by saying that "Darwinism is not a testable scientific theory, but a *metaphysical research programme*—a possible framework for testable scientific theories."[4] What Popper meant by this is that Darwin's theory, when expressed in very general terms as a group of claims about heredity, random mutation, and differential survival, does not make any predictions about which species (or indeed whether any species) will evolve. Popper thought that prediction and explanation seem to occur because we forget that adaptation or fitness is implicitly defined in terms of survival. Thus, while it may seem as if we have explained why a particular species now thrives by saying that it adapted to its environment, Popper judged this to be no explanation at all. Rather, he said that the claim that a species now living has adapted to its environment is "almost tautological," that is, true by definition.[5]

Popper's charge that the phrase "the survival of the fittest" is tantamount to a tautology (that to survive is to be fittest) has been repeated by creationists such as Henry Morris, who have then denied that evolutionary theory as whole is either empirical or testable.[6] But even if a theory includes some elements that are true by definition or untestable for some other reason, it hardly follows that the theory as a whole or specific versions of it are untestable. Indeed, Popper regarded Darwinism as similar in this regard to atomism and field theory. In his view, these are all metaphysical generalizations that make no predictions and hence are untestable. Nonetheless, they are of great scientific value because they give rise to specific theories that *are* testable and have been tested. So when Popper judges a proposition to be unfalsifiable and "metaphysical," he is not claiming that the proposition has no scientific value, nor is he asserting that any theory associated with it is pseudoscientific.

But is "the survival of the fittest" a tautology as has been charged? One problem in assessing this accusation is that "the survival of the fittest" is a phrase, not a proposition, and only propositions can be tautologies.[7] What is needed is a precise statement of the allegedly tautologous proposition. Is the proposition in question a definition of *fitness* (or *relative adaptiveness*) in terms of the probability of reproductive success? Or is it the historical claim that the traits in current populations are the result of natural selection (i.e., selection of the fittest ancestral variants)? In his Darwin Lecture, Popper opted for the latter and then noted, correctly, that it is an empirical matter whether natural selection or some other mechanism (such as genetic drift) is responsible for the traits we now find in a population of organisms. Thus, Popper conceded that the principle of natural selection is falsifiable and testable.

Before concluding this section, there is one further small matter concerning Darwin's theory and Popper's criterion of falsifiability. Sometimes the claim is made (often, but not always, by creationists) that Darwin's theory (and, presumably, other sciences such as paleontology, geology, and cosmology) are unscientific because they are, at least in part, historical. Evolutionary theory, we are told, makes claims about historical events, many of which occurred before the advent of any human observers on this planet. Historical events are unique and unrepeatable. Therefore, critics conclude, Darwin's theory cannot be tested or refuted.[8] This argument, as Popper himself has emphasized, is invalid: its conclusion does not follow from its premises.[9] Claims about historical events, even events that occurred millions of years ago can be tested (and thus, in principle, refuted) by using them to make predictions about the evidence we should find now if the historical claims are true: cometary collisions with the earth leave craters and abnormally high concentrations of iridium in the surrounding rocks; animals and plants leave fossils; the "big bang" still resonates in the form of background microwave radiation in space.

1.2 | Kuhn's Criticisms of Popper

One of the many people who have challenged Popper's appeal to falsifiability as a demarcation criterion is Thomas Kuhn. In his book, *The Structure of Scientific Revolutions* (1962), Kuhn insisted that if we are to arrive at an adequate characterization of science, close attention must be paid to its history: a proper philosophy of science should reflect the history of science. On this account, philosophy of science ought to describe the way scientists actually behave and the way that science has evolved over time. Doing so shows that not all scientific activity is of the same kind. Kuhn thinks that scientific activity falls into two distinct types: normal science and extraordinary (or revolutionary) science.

Normal Science and Puzzle Solving

During periods of normal science, scientists take for granted the major theories of their day and content themselves with what Kuhn calls puzzle solving. In some respects, the puzzle-solving aspect of normal science is like trying to do the exercises at the back of a physics or chemistry textbook. The aim of practicing scientists is not to call into question Newtonian mechanics or the laws of thermodynamics, but rather to see whether they can solve problems by using these accepted theories in conjunction with other assumptions and models. Just as failure to get the right answer to an exercise is regarded as a failure of the student, not of the theory, so, too, failure to solve a puzzle during a period of normal science is considered the fault of the scientist using the theory, not the fault of the theory itself. Only very rarely, during periods of extraordinary science, do scientists deliberately question the received theories of their day and attempt to refute them. Typically such periods of extraordinary science arise because of repeated failures to solve puzzles. If a theory is refuted, then it must be replaced by another theory that is at least as general in scope. Science, like nature, abhors a vacuum: scientists will give up a global theory only when they have an even better theory to adopt in its place. When such a replacement occurs, we have a scientific revolution. (For a much fuller discussion of Kuhn's views on scientific revolutions, see chapter 2, "Rationality, Objectivity, and Values in Science," below.)

Kuhn agrees with Popper and many other philosophers of science that astrology is a pseudoscience. In this, as in many other cases, Popper's criterion of demarcation (severity of testing) leads to the same verdict as Kuhn's criterion (puzzle solving). But Kuhn rejects Popper's demarcation criterion and with it Popper's explanation of why astrology is pseudoscientific. Popper insists that by formulating their accounts in suitably vague terms, astrologers are able to "explain away anything that might have been a refutation of the theory" (8). For Popper, this emphasis on confirmation and avoidance of testability or falsification marks the difference between pseudoscience and science. Kuhn's account of why astrology is a pseudoscience is quite different from Popper's. Kuhn points out that astrology was finally abandoned by scientists around the middle of the seventeenth century, mainly as a consequence of the Copernican revolution. But throughout its history, astrology was notoriously unreliable and its predictions often failed. Interestingly, these frequent failures were never given as a reason for thinking that astrology is false until after astrology had been abandoned. During its heyday, astrology was regarded rather as medicine and meteorology once were—as an imprecise study of an enormously complex subject. In any case, astrology can scarcely be reckoned a nonscience simply because it made predictions that turned out to be false. Still, Kuhn insists, astrology never was a science. Although astrologers use rules of thumb to cast horoscopes, astrology has no central theory and no puzzle-solving tradition of a sort characterizing normal science. Though astrologers often made false predictions, they had no way of learning from their mistakes, something that is a hallmark of genuine science. Thus astrology was, and remains, at best a craft and not a science.

Scientific Revolutions

In "Logic of Discovery or Psychology of Research?" Kuhn gives several examples of scientific revolutions: the overthrow of Newton's theory of gravity by Einstein's general theory of relativity; the replacement of the phlogiston theory by Lavoisier's new chemical theory (in which the addition of oxygen, not the release of phlogiston, is responsible for the burning of metals in air); the experimental confirmation of Lee and Yang's theory that the weak interaction—the nuclear process responsible for the release of electrons during radioactive decay—does not conserve parity. One of Kuhn's main criticisms of Popper is that sincere attempts to refute theories are quite rare in science. Such attempts are usually confined to the periods of extraordinary science that immediately precede scientific revolutions. Thus, according to Kuhn, Popper's falsifiability account of science fails to describe normal science. If falsifiability were the criterion marking off science from pseudoscience, then genuine science as it is done most of the time, being normal and not extraordinary, would be improperly classified as pseudoscientific.

As we have seen, Kuhn rejects Popper's falsifiability criterion as an account of normal science; but how well does it fit those episodes of extraordinary science (scientific revolutions) in which large-scale theories are refuted and replaced? According to Kuhn, another flaw in Popper's historically insensitive treatment is that in some scientific revolutions—Kuhn gives the Copernican revolution as an example—the old theory (Ptolemy's geocentric theory) was replaced by the new theory (Copernicus's heliocentric theory) *before* the old theory was refuted. For example, Galileo's telescopic observations of the phases of Venus, the moons of Jupiter, and the motion of sunspots were made at least sixty years after the publication of Copernicus's *De Revolutionibus* (1543) and only after Galileo had become a convinced Copernican. Arguably, Ptolemy's theory (in which the earth is stationary at the center of the universe) was decisively refuted only when Newton's theory of mechanics and gravity was accepted. (Newton's *Principia* was published in 1687.) Newton's theory showed that it was physically impossible for the entire heavens to rotate around the earth's north-south axis. When Copernicus proposed his new theory, most astronomers thought that the Ptolemaic theory could solve all its problems by adjusting a few parameters. Hardly anyone thought that Ptolemy's theory had been severely tested and found irreparably wanting. Here again, Kuhn argues, Popper's account of science does not fit the history of science. There is more to science and being scientific than falsifiability and testing.

1.3 | Lakatos and Scientific Research Programmes

In "Science and Pseudoscience," Imre Lakatos notes that genuine *scientia* (knowledge) cannot be marked off from impostors simply in terms of the number of people who believe it or how strongly they believe it. The worst of pseudoscience has, in the past, commanded dogged assent from large numbers of intellectuals. Nor can we rest a criterion of demarcation on the commonplace assertion that genuine science is supported by the observable facts. For, Lakatos asks, how could this criterion be justified? Like Kuhn and Popper, Lakatos agrees that no scientific theory can be deduced from observational and experimental facts. When scientists such as Newton and Ampère claimed that their theories were not *hypotheses* but proven *truths* because they were deduced from experiments and observations, they were simply wrong.

WHY ALL THEORIES ARE UNPROVABLE

We can appreciate Lakatos's point by considering a single example: Newton's theory of gravitation. Newton's theory says that *every* particle of matter in the universe attracts every other particle with a force according to an inverse square law. Newton's theory is a universal generalization that applies to every particle of matter, anywhere in the universe, at any time. But however numerous they might be, our observations of planets, falling bodies, and projectiles concern only a finite number of bodies during finite amounts of time. So the scope of Newton's theory vastly exceeds the scope of the evidence. It is possible that all our observations are correct, and yet Newton's theory is false because some bodies not yet observed violate the inverse square law. Since "All *F*s are *G*" cannot be deduced from "Some *F*s are *G*," Newton's theory cannot be proven by logically deducing it from the evidence. As Lakatos points out, this prevents us from claiming that scientific theories, unlike pseudoscientific theories, can be proven from observational facts. The truth is that no theory can be deduced from such facts. *All* theories are unprovable, scientific and unscientific alike.

WHY ALL THEORIES ARE IMPROBABLE

While conceding that scientific theories cannot be proven, most people still believe that theories can be made more probable by evidence. Lakatos follows Popper in denying that any theory can be made probable by any amount of evidence. Popper's argument for this controversial claim rests on the analysis of the objective probability of statements given by inductive logicians.[10] Consider a card randomly drawn from a standard deck of fifty-two cards. What is the probability that the card selected is the ten of hearts? Obviously, the answer is 1/52. There are fifty-two possibilities, each of which is equally likely and only one of which would render true the statement "This card is the ten of hearts." Now consider a scientific theory that, like Newton's theory of gravitation, is universal. The number of things to which Newton's theory applies is, presumably, infinite. Imagine that we name each of these things by numbering them 1, 2, 3, . . . , *n*, ... There are infinitely many ways the world could be, each equally probable.

1 obeys Newton's theory, but none of the others do.

1 and 2 obey Newton's theory, but none of the others do.

1, 2, and 3 obey Newton's theory, but none of the others do.

. . . .

All bodies (1, 2, 3, . . . , n, . . .) obey Newton's theory.

Since these possibilities are infinite in number, and each of them has the same probability, the probability of any one of them must be 0.[11] But only one, the last one, represents the way the world would be if Newton's theory were true. So the probability of Newton's theory (and any other universal generalization) must be 0.

Now one might think that, even if the initial probability of a theory must be 0, the probability of the theory when it has been confirmed by evidence will be greater than 0. As it turns out, the probability calculus denies this. Let our theory be T, and let our evidence for T be E. We are interested in $P(T/E)$, the probability of T given our evidence E. Bayes's theorem (which follows logically from the axioms of the probability calculus) tells us that this probability is:

$$P(T/E) = \frac{P(E/T) \times P(T)}{P(E)}.$$

If the initial probability of T—that is, $P(T)$—is 0, then $P(T/E)$ must also be 0.[12] Thus, no theory can increase in objective probability, regardless of the amount of evidence for it. For this reason, Lakatos joins Popper in regarding all theories, whether scientific or not, as equally unprovable and equally improbable.

THE METHODOLOGY OF SCIENTIFIC RESEARCH PROGRAMMES

The failure to specify demarcation criteria along the intuitively attractive lines of "whatever is proved or made probable by evidence" might suggest returning to the Popperian model. But like Kuhn (and Thagard after him), Lakatos rejects Popper's falsifiability criterion as a solution to the demarcation problem. Scientists rarely specify in advance of observation and experiment those results that, if found, would refute their theories. At best, such results would be regarded as anomalous or recalcitrant, not as genuine refutations. Even when they are first proposed, some theories are (or are thought to be) inconsistent with the known data. Newton's gravitational theory is a good example. By his own admission, Newton was unable to reconcile his theory with the known orbit of the earth's moon. (This anomaly was later cleared up by Alexis Clairaut who in 1749 found a mistake in Newton's calculations.[13]) But Newton did not immediately abandon his theory as refuted. Later, after the discovery of the planet Uranus (by William Herschel in 1781), it was noted that Uranus did not move precisely as Newton's theory predicted. Again, scientists did not abandon the inverse square law; rather, they postulated another planet, as yet unobserved, which was perturbing the orbit of Uranus. This hypothetical new planet was eventually discovered and given the name Neptune.[14]

In order to make sense of the ways in which scientists protect their theories from refutation, Lakatos proposes that scientific theories be regarded as having three components: a hard core, a protective belt, and a positive heuristic. The hard core of Newton's theory consists of his three laws of motion plus the inverse square law of gravitational attraction. These are basic postulates that scientists were extremely reluctant to give up. The protective belt consists of many auxiliary hypotheses such as assumptions about the number and the masses of the planets. The positive heuristic tells scientists how to solve problems using the theory and how to respond to anomalies by revising the protective belt. Lakatos proposes that we stop thinking of scientific theories as frozen in time but instead regard theories as historically extended scientific research programmes. The Newtonian research programme covered several centuries. For much of its history it was progressive. Why? Because in dealing with anomalies and other problems, the Newtonian programme continued to predict novel facts.

According to Lakatos, Popper is wrong in thinking that a crucial experiment can (or should) instantly refute a theory. As Kuhn has shown, the actual history of science teaches us otherwise: genuine scientific progress (as opposed to degenerating science or pseudoscience) is not simply a matter of one theory remaining unrefuted while others are falsified. But Lakatos is equally critical of Kuhn for suggesting that scientific revolutions are largely irrational affairs, dependent on a kind of group psychology. Were Kuhn right, there would be no objective way of marking off scientific progress from scientific regress or decay. Instead, Lakatos suggests that scientific change occurs as the result of competition between rival research programmes. If one programme is progressive (because it continues to predict novel facts), and if its rival is degenerating, then most scientists will, rationally, switch their allegiance. In this way, progressive research programmes replace degenerating ones. (For a critical discussion of Lakatos's methodology of scientific research programmes as a normative theory of science, see Colin Howson and Peter Urbach, "The Duhem Problem" in chapter 3 of this volume.)

1.4 | Thagard on Why Astrology Is a Pseudoscience

Paul Thagard takes up Lakatos's notion of scientific theories as research programmes, and develops it into an explicit criterion for demarcating science from pseudoscience. In "Why Astrology Is a Pseudoscience," Thagard surveys several different proposals for a demarcation criterion that would explain why astrology is a pseudoscience and finds each of them deficient. In light of the alchemical and occult origins of chemistry and medicine, one cannot uncritically cite astrology's origin in magic as what makes it a pseudoscience. (Indeed, this merely postpones the question, Why is magic not itself a genuine science?) Nor can the supposed immunity from testing, verification, or falsification be what makes astrology a pseudoscience. As Thagard notes, some astrological claims (about the influence of planetary positions at the time of one's birth on one's personality and future career, for example) are testable. Moreover, Thagard agrees with Kuhn and Lakatos that abandoning a theory the moment one of its predictions failed would be irrational. Many of our best scientific theories have been modified in the light of failed predictions and recalcitrant observations. Hasty rejection would nip too many good theories in the bud, before they had the chance to grow and blossom.

Contrary to Kuhn, Thagard claims that modern astrology does indeed present a number of unsolved problems (such as accommodating the precession of the equinoxes and planets that were discovered many centuries after Ptolemy's death). This undercuts the Kuhnian proposal that astrology fails as a science simply because it lacks a tradition of problem solving. Against Lakatos, Thagard suggests that absence of progress is not by itself a sufficient condition of pseudoscience, since it might be nonprogressive only in periods when it faces no progressive competitors. Despite these differences, however, Thagard agrees with Kuhn that judgments about the scientific status of a theory or discipline must involve both a social and a historical dimension, and he agrees with Lakatos that progress is necessary for genuine science.

Thagard's Definition of Pseudoscience

In light of his criticisms of Kuhn and Lakatos, Thagard proposes two conditions that are necessary and sufficient for a theory or discipline to be pseudoscientific.

A theory or discipline which purports to be scientific is pseudoscientific if and only if:

1 it has been less progressive than alternative theories over a long period of time, and faces many unsolved problems; but

2 the community of practitioners makes little attempt to develop the theory towards solutions of the problems, shows no concern for attempts to evaluate the theory in relation to others, and is selective in considering confirmations and disconfirmations. (32)

According to these conditions, astrology is pseudoscientific in part because it has not changed much since the time of Ptolemy. Unsolved problems have accumulated, and as Thagard notes, we now have (since the nineteenth century) psychological theories that do a better job of explaining and predicting human behavior. Despite this competition from psychology, astrologers have shown little interest in improving their theory or in evaluating it with respect to rivals.

Thagard concludes by isolating a number of interesting logical (and, to some, startling) consequences of his demarcation criterion. One might view the acceptability of these consequences as a measure of the plausibility of his proposal. First, some current fads, such as pyramidology and biorhythms, would not be considered pseudosciences because, at the moment, they lack serious competitors. Second, a theory can be scientific at one time and pseudoscientific at a later time; being scientific is not an unchanging property of a theory. Third, Thagard concludes that astrology used to be a genuine science but became pseudoscientific only when modern psychology arose in the late nineteenth century. If this is correct, then those scientists (the vast majority) who rejected astrology as pseudoscientific in the eighteenth century were being irrational.

THAGARD'S LATER THOUGHTS ABOUT PSEUDOSCIENCE

Because of objections to his demarcation principle for pseudoscience, especially the objection that nothing can be a pseudoscience unless it has competitors, Thagard changed his views. In his book *Computational Philosophy of Science* (1988) he gives up trying to provide necessary and sufficient conditions for pseudoscience. Instead, he offers contrasting profiles of genuine science and pseudoscience. Relative progressiveness and a concern with confirmation and disconfirmation are still presented as hallmarks of science (and their absence is still associated with pseudoscience), but Thagard no longer claims that science must always possess these features or that pseudoscience must necessarily lack them. Thagard also introduces two new criteria for pseudoscience. One of these criteria is that pseudoscientific theories are often highly complex and riddled with ad hoc hypotheses. This provides some grounds for judging a doctrine pseudoscientific on its content, even if it currently has no scientific competitors.

Thagard's second new criterion concerns the sort of reasoning employed by many practitioners of pseudoscience, such as astrologers, namely, reasoning based on resemblances. Instead of testing causal claims by looking for statistical correlations, pseudoscientists are often content to rest their beliefs on superficial analogies. Traditional astrology is full of this sort of "resemblance thinking." For example, the planet Mars often has a reddish appearance, and so astrologers associate it with blood, war, and aggression. From this they conclude that Mars causes (or, at least, has a tendency to cause) aggressive personalities in people born at the appropriate time. In a similar way believers in folk medicine recommend turmeric as a treatment for jaundice and powdered rhinoceros horn as a cure for impotence.

As Thagard recognizes, not all pseudosciences employ resemblance thinking, and some pseudosciences employ reasoning based on statistical correlations that mimics, to some extent, reasoning found in the genuine sciences. Proponents of biorhythms, for example, rest much of their case on alleged correlations as do Velikovsky and von Daniken when they appeal to common elements in ancient myths to support their astronomical theories. Thus, in Thagard's revised account of pseudoscience, none of the elements mentioned—using resemblance thinking, refusing to seek confirmations and disconfirmations, ignoring alternative theories, trafficking in ad hoc hypotheses, sticking with theories that fail to progress—is a necessary feature of pseudoscience, and genuine sciences might, from time to time, share one or two of these features. But, Thagard claims, pseudosciences usually have most of these features and genuine sciences nearly always lack most of them. Thus, the difference between science and pseudoscience is a matter of degree rather than kind, although Thagard remains convinced that the difference of degree is usually large and obvious.

1.5 | Creation-Science and the Arkansas Trial

The search for demarcation criteria is not simply a curiosity to entertain armchair intellectuals or a pastime for students in philosophy of science. Consider, for instance, the 1982 case of *McLean v. Arkansas Board of Education*.[15] At issue in the case was the constitutionality of Arkansas Act 590, which required teachers to give "balanced treatment" to both evolutionary theory and creationism in the biology classes taught in public schools. Act 590 describes "evolution-science" and "creation-science" as competing scientific models of the origin of species and offers the following definition for creation-science:

> "Creation-science" means the scientific evidences [*sic*] for creation and inferences from those scientific evidences. Creation-science includes the scientific evidences and related inferences that indicate: (1) Sudden creation of the universe, energy, and life from nothing; (2) The insufficiency of mutation and natural selection in bringing about development of all living kinds from a single organism; (3) Changes only within fixed limits of originally created kinds of plants and animals; (4) Separate ancestry for man and apes; (5) Explanation of the earth's geology by catastrophism, including the occurrence of a worldwide flood; and (6) A relatively recent inception of the earth and living kinds.[16]

Judge Overton's Opinion

The task of the presiding judge, William Overton, was to decide whether Act 590 violates the Constitution of the United States. He reasoned that Act 590 is consistent with the Constitution only if the act satisfies the Establishment Clause of the First Amendment, which says that "Congress shall make no law respecting an establishment of religion, or prohibiting the free exercise thereof." The Supreme Court of the United States has for many years applied the articles of the Bill of Rights (the first ten amendments to the Constitution) not only to federal legislation but also to the laws passed by individual states. The Supreme Court's interpretation of the Establishment Clause has evolved into a three-part test for the constitutionality of any legislation involving religion. It was this three-part test that Judge Overton applied to Arkansas's Act 590. Failing any one of these three parts is sufficient to render a piece of legislation unconstitutional. Here is the test:

> First, the statute must have a secular purpose; second, its principal or primary effect must be one that neither advances nor inhibits religion . . . ; finally, the statute must not foster "an excessive government entanglement with religion."[17]

Judge Overton thought it clear that Act 590 was passed by the Arkansas General Assembly with the specific intention of advancing religion, and that fact alone—the lack of a secular purpose—would suffice to invalidate the statute. But Judge Overton wanted to show that Act 590 also fails the second and third parts of the three-part test. In order to show that Act 590 fails the second part, it is necessary to show that the statute either advances or inhibits religion as its "principal or primary effect." To accomplish this, Judge Overton thought it necessary to establish that creation-science is not a genuine science. For, as he argued (at the end of part IV(D) of his Opinion), "Since creation-science is not science, the conclusion is inescapable that the only real effect of Act 590 is the advancement of religion."[18]

Thus Judge Overton entered the philosophical debate over the criteria for genuine science. He sought guidance from expert witnesses, especially from a philosopher of biology, Michael Ruse.[19] It was primarily Ruse who developed the five characteristics that Overton lists as essential (necessary conditions) for genuine science:

1 it is guided by natural law,
2 it has to be explanatory by reference to natural law;
3 it is testable against the empirical world;
4 its conclusions are tentative, i.e., they are not necessarily the final word; and
5 it is falsifiable.

RUSE ON THE STATUS OF CREATION-SCIENCE

In "Creation-Science Is Not Science" Ruse defends the five items on Overton's list and argues that creation-science satisfies none of them. Ruse sees an intimate connection between items (1) and (2) on the list: it is only because scientific theories posit natural laws that the theories are able to explain; genuinely to explain something is to show why, given the relevant circumstances, it had to happen, and that requires an appeal to laws. (For more on explanation and laws, see chapter 6, "Models of Explanation," and chapter 7, "Laws of Nature.") Since creation-science posits acts of creation that are miraculous and unlawlike, Ruse concludes that it is not scientific. He also points out that creation-scientists make few if any testable predictions. Most of the time, creationists content themselves with describing the evidence in ways that are consistent with their doctrines. For example, creationists regard the common pattern of bones in the forelimbs of humans, bats, whales, and other mammals as an instance of God's design plan for mammals, but they offer no reason why this particular pattern exists rather than some other pattern or several different patterns. Evolutionists follow Darwin in explaining the pattern as the result of common descent: because all mammals have descended from a common ancestor, they share a common anatomical structure. Ruse concludes his case against creation-science by noting that most creationist research aims at trying to find flaws in evolutionary theory rather than making testable predictions based on the creationists' own theory. Modern creationists are dogmatic (not tentative) about their fundamental beliefs and show little or no interest in trying to falsify them.

Laudan's Criticisms of Ruse

In his "Commentary: Science at the Bar—Causes for Concern," Larry Laudan chastises Ruse for perpetuating a view of science that, he claims, both of them know to be false. Laudan denies that philosophers of science would accept any list of characteristics as capturing the essence of science. Why, for example, should explanation by means of laws be regarded as a necessary condition for a theory to be scientific? Many theories begin by describing a new phenomenon, and only later, if at all, explain the phenomenon in a lawlike way. For example, Galileo discovered that all bodies released near the surface of the earth fall with the same acceleration but offered no explanation for this. Similarly, Newton claimed to have "deduced" the universal law of gravitation "from the phenomena" but accepted action at a distance as ultimately unexplainable.[20] Indeed, if one accepts the deductive-nomological model of explanation, according to which scientific explanations are deductive arguments with at least one statement of a law in their premises (see chapter 6, "Models of Explanation"), then in any such explanation there will remain, at least provisionally, something that is not explained, namely the premises that do the explaining. So Laudan rejects item (2) from the Ruse-Overton list as too strong.

Laudan also criticizes items (3) and (5) as being too weak, since, he argues, they are all too easily satisfied. Any theory, even a theory like creation-science that posits a divine creator, implies something about the observable world. For example, many creationists claim that all living things were created at the same time fewer than 50,000 years ago and that a worldwide flood caused many of the geological features now observed on the earth. As

Laudan sees it, the flaw with creationism is not that such claims are untestable or unfalsifiable but rather that they have been tested and falsified.[21]

The debate about demarcation criteria surfaced again in connection with an alternative to evolution known as intelligent design (ID). Advocates of ID reject many of the "young earth" aspects of creation science, while still criticizing evolution on the grounds that certain features of cells and organisms (such as bacterial flagella, the blood clotting cascade, and the immune system) are too complex and intricate to have arisen through unguided natural selection. Proponents of ID claim that the best explanation of these features is the action of an intelligent designer. Whether this designer was supernatural or extraterrestrial (or even a time-traveling human) is deliberately left unspecified by ID theorists in order to avoid the charge that they are advocating a doctrine that is explicitly religious.[22]

As with creation science, many aspects of intelligent design have been discussed in connection with a trial and a judge's decision. In *Tammy Kitzmiller et al. v. Dover Area School District et al.* (2005) Judge Jones concluded that requiring high school teachers to tell their students that Darwin's theory is "not a fact" but a theory that continues to be tested, that Darwin's theory has "gaps," and that students should consult a particular ID "reference book" (*Of Pandas and People*) for an alternative "explanation of the origin of life" violates the Establishment Clause of the First Amendment. Of particular interest in Judge Jones's lengthy and comprehensive ruling was his determination (in Section 4, "Whether ID Is Science") that apart from its other flaws ID is not science because it allows for supernatural causation.[23] Citing the testimony of Robert Pennock, Kenneth Miller, and other expert witnesses who testified at the Dover trial, Judge Jones declared that the restriction to natural causes (which Jones refers to as "methodological naturalism") is one of the ground rules of modern science. In effect, Judge Jones declared that relying exclusively on natural causes, forces, and explanations is a demarcation criterion for genuine science and that ID fails this test.[24]

1.6 | Summary

In this chapter, we have explored a number of attempts to demarcate science from pseudoscience. But the results have been curiously inconclusive. Most scientists and philosophers of science readily agree that such things as pyramidology and creation-science are not genuine sciences, but there is no consensus on why this is so. Like obscenity, most people are able to recognize pseudoscience when they encounter it but find it much harder to explain why what they have encountered is pseudoscientific. The stress on explanation is important here. What we are seeking, as philosophers of science, is not just a handy way of detecting pseudoscience (on the basis, say, of a majority vote of the National Academy of Sciences) but a philosophically informative account of what makes a discipline genuinely scientific.

Despite the defects of his own demarcation criterion—falsifiability—Popper deserves credit for disposing of one tempting answer to the demarcation problem. No appeal to confirming evidence, by itself, is going to distinguish genuine science from its counterfeit. Inventing an elaborate hypothesis that is consistent with the known facts is just too easy. Popper's fruitful idea was to seek the demarcation between science and pseudoscience, not in confirmation, but in falsification. The hallmark of true science lies in the willingness of scientists to use their theories to make testable predictions. If the predictions fail, then the theory from which they are deduced should be abandoned as false. Unfortunately, Popper's simple idea does not work. As Lakatos and Thagard explain, falsifiability is both too weak and too strong. It is too weak because it would allow as scientific any number of claims that are testable in principle but that are not, by any stretch of the imagination, scientific. It is too strong because it would rule out as unscientific many of the best theories in the history of science. Few scientists give up their theories simply because they have come into conflict with observation and experiment. Instead, they either look for a flaw in the data, or they modify their theories. The rejection of a theory simply because is disagrees with the facts (or what are taken to be facts) is the exception rather than the rule in science.

In differing ways, Kuhn, Lakatos, and Thagard each propose that there is a historical and a social dimension to judgments concerning the scientific status of a theory. All three insist that when we ask of a theory "Is it genuinely scientific?" it is a mistake to look at the theory as if it were a snapshot, caught at an instant of time. Rather, they argue, we have to consider how the theory has developed, especially how the theory has been modified to deal with new problems and recalcitrant data. For Kuhn, this means seeing the theory as part of a larger whole—what Kuhn calls a *paradigm*. (For more on Kuhn's notion of a paradigm see chapter 2, "Rationality, Objectivity, and Values in Science.") Thagard adopts Lakatos's notion of a scientific research programme in order to define a demarcation criterion. On this approach, roughly speaking, a theory is pseudoscientific if the research programme with which it is associated has been less progressive over time than has its rivals. As suggested in our discussion of Thagard's proposal, this comparative-progress definition of pseudoscience has a number of startling consequences. For example, some modern fads, such as pyramidology, might fail to qualify as pseudosciences simply because, at the moment, they lack competitors. Because of these defects, Thagard has avoided giving necessary and sufficient conditions for pseudoscience in his more recent writings.

As the debate between Ruse and Laudan concerning the status of creation-science makes clear, judgments about pseudosciences often depend on detailed considerations of the nature of law, explanation, confirmation, and falsification. It is highly unlikely that any simple-minded, one- or two-sentence definition of science will yield a plausible demarcation criterion that we can use to label and condemn as pseudoscientific those theories (and their advocates) that fail to meet the standards of good science. Ultimately, discriminating between science and its counterfeit depends on a detailed understanding of how science works. Despite the variety and complexity of the many different theories and activities that are, by common consent, genuinely scientific, are there general principles concerning explanation, confirmation, testing, and the like that these theories and activities share? In the rest of our book, some important attempts to answer this question will be explained and evaluated. Thus, what follows can be seen as an attempt to answer the questions left unanswered in this first chapter.

■ | Notes

1. Newton's theory also predicts the bending of starlight if light rays are regarded as a stream of particles traveling at the speed of light. Because inertial mass is exactly equal to gravitational mass, the orbit of any object moving around the sun depends only on the velocity of the moving object, not on its mass. (The same thing is true of bodies near the surface of the earth. If you throw two objects with the same velocity in the same direction, then they will follow the same path regardless of their mass.) Thus, we do not have to know the mass of the light particles in order to calculate how they will move when close to the sun. But, Newton's theory predicts an amount of bending which is only half of that predicted by Einstein. The difference arises because Einstein's theory entails that the gravitational field close to the sun is slightly stronger than in Newton's theory. Thus, it is not that Einstein's theory predicted a kind of effect, the bending of starlight, that Newton's theory did not. Rather, both theories gave competing predictions of its magnitude, and Einstein's prediction was more nearly right. Interestingly, observations made during some later eclipses (1929, 1947) found deviations that were higher than those predicted by Einstein. But more recent observations are in closer agreement with Einstein's theory, and none of the observations agrees with Newton's. (The issue of whether theories such as Newton's theory of gravity can be conclusively refuted is discussed in chapter 3, "The Duhem-Quine Thesis and Underdetermination.") Eddington's role in this episode is controversial because he threw out as biased one set of observations that agreed with Newton's prediction. For details about the difficulties of making the eclipse observations and competing interpretations of Eddington's behavior, see John Earman and Clark Glymour, "Relativity and Eclipses: The British Eclipse Expeditions of 1919 and Their Predecessors," *Historical Studies in the Physical Sciences* 11 (1980): 49–85; Deborah Mayo, "Novel Evidence and Severe Tests," *Philosophy of Science* 58 (1991): 523–52; and Harry Collins and Trevor Pinch, *The Golem: What Everyone Should Know about Science* (Cambridge: Cambridge University Press, 1993). For a defense of the analysis of the eclipse data by Eddington and Dyson (the other expedition leader), see Daniel Kennefick, "Testing Relativity from the 1919 Eclipse—A Question of Bias," *Physics Today* 62:3 (2009): 37–42.

2. It is tempting to try to rule out cases such as (T & C) by requiring not merely that the theory as a whole make testable predictions but that each individual component of the theory also make testable predictions. See chapter 3 for a discussion of whether any significant scientific theory could meet this additional requirement.

3. See Karl R. Popper, "Natural Selection and the Emergence of Mind," *Dialectica* 32 (1978): 339–55.

4. Karl R. Popper, "Autobiography of Karl Popper," in *The Philosophy of Karl Popper*, ed. Paul A. Schilpp (La Salle, Ill.: Open Court, 1974), 1: 134.

5. *The Philosophy of Karl Popper*, 1: 137. For excellent discussions of the tautology problem, see chapter 2 of Elliott Sober, *The Nature of Selection* 2d ed. (Chicago, Ill.: University of Chicago Press, 1993), and chapter 4, "The Structure of the Theory of Natural Selection," in Robert N. Brandon, *Adaptation and Environment* (Princeton, N.J.: Princeton University Press, 1990).

6. Thus, somewhat paradoxically given the Arkansas trial (discussed later in this commentary), one of the themes in creationist literature is that creationism and evolutionary theory are both equally *un*scientific because neither makes testable predictions. Needless to say, this position is hard to reconcile with another creationist theme, namely, that evolutionary theory has been significantly disconfirmed by a variety of evidence.

7. Another problem is that tautologies, strictly speaking, are propositions that are true solely in virtue of their logical form. Presumably, the issue is not whether some biological statement is a tautology but whether it is analytic. Analytic statements, on one characterization of analyticity, are statements that are true solely in virtue of the meanings of the words and symbols used to express them. In chapter 3, there is an extended discussion of Quine's thesis that no line can be drawn, even in principle, between statements that are analytic and those that are not. If Quine is right, then the charge of being tautologous evaporates.

8. See, for example, the authors quoted in Beverly Halstead, "Popper: Good Philosophy, Bad Science?" *New Scientist* (17 July 1980): 215–17.

9. Karl R. Popper, "Letter on Evolution," *New Scientist* (21 August 1980): 611.

10. The argument that follows is a simplified version of the one given in Appendix *vii of Karl R. Popper, *The Logic of Scientific Discovery* (New York: Basic Books, 1959), 363–77.

11. The possible hypotheses enumerated in the text—only the first particle obeys Newton's theory, only the first two particles obey Newton's theory, etc.—are exclusive: if any one of them is true, then all the others must be false. If each hypothesis has the same prior probability, p, and there are n of them, then the probability that at least one of the hypotheses is true is $n \times p$. (See axiom 3, the special addition rule, in "Bayes's Theorem and the Axioms of Probability Theory" in the commentary on chapter 5.) Since $n \times p$ is a probability, it cannot be greater than 1. So, if n is infinite, p cannot be finite. Thus, p must be 0. The derivation of this result depends on assuming that each hypothesis has the same prior probability, something that Bayesians deny. For this and other Bayesian criticisms of Popper's argument, see Colin Howson, "Must the Logical Probability of Laws Be Zero?" *British Journal for the Philosophy of Science* 35 (1973): 153–63, and Alan Chalmers, "The Bayesian Approach," in chapter 5.

12. For a fuller discussion of this and other applications of Bayes's theorem to issues in confirmation, see chapter 5.

13. As Newton realized, the main irregularities in the motion of the moon are due to the attraction of the sun. The force exerted on the moon by the sun is a rather large fraction (1/89 at new and full moon) of the force exerted by the earth. As in all

such three-body problems, no exact solution of Newton's equations is possible. Because the moon is close to the earth, even small perturbations are easily observed. This requires the calculations to be extended down to very small terms. Initially, Clairaut's calculations yielded a rate of precession of the moon's apogee of 20 degrees per year, only half the real amount. At first, Clairaut speculated that Newton's inverse-square law gravitational formula was incorrect for small distances and should be supplemented by an extra term varying as the inverse fourth power of the distance. But on extending his calculations to include higher order terms that had been neglected in his original approximation, Clairaut found that his first result was doubled. So Newton's theory was vindicated. Scholars have since found that Newton himself had obtained the correct result in his unpublished papers. For more on the problem of the moon, see Anton Pannekoek, A History of Astronomy, rpt. (1951; New York: Dover Publications, 1990) ch. 30.

14. See Morton Grosser, The Discovery of Neptune (Cambridge, Mass.: Harvard University Press, 1962). For a lively criticism of the oft-repeated claim (by Popper, Lakatos, and others) that Newton's theory was prima facie falsified by the discovery of perturbations in the orbit of Uranus, see Greg Bamford, "Popper and His Commentators on the Discovery of Neptune: A Close Shave for the Law of Gravitation?" Studies in History and Philosophy of Science 27 (1996): 207–32.

15. Judge Overton's opinion in this case is reprinted in Science 215 (1982): 934–43, in Science, Technology, and Human Values 7 No. 40 (1982): 28–42, and in both Michael Ruse, ed., But Is It Science? 1st ed. (Buffalo, N.Y.: Prometheus Books, 1988) and Michael Ruse and Robert T. Pennock, eds., But Is It Science? 2d ed. (Amherst, N.Y.: Prometheus Books, 2009).

16. Balanced Treatment for Creation-Science and Evolution-Science Act (1981), 73d General Assembly, State of Arkansas, Act 590 sec. 4; reprinted in Science, Technology, and Human Values 7 No. 40 (1982): 11.

17. William R. Overton, "Opinion in McLean v. Arkansas," Science, Technology, and Human Values 7 No. 40 (1982): 29.

18. For a criticism of this inference and other aspects of Overton's opinion, see Philip L. Quinn, "The Philosopher of Science as Expert Witness," in Science and Reality, ed. J. T. Cushing, C. F. Delaney, and G. M. Gutting (Notre Dame, Ind.: University of Notre Dame Press, 1984), 32–53. As Quinn explains in a later article, he agrees with Overton's conclusion that Arkansas Act 590 is unconstitutional because it lacks a secular purpose. What he criticizes is Overton's attempt to show that Act 590 has the advancement of religion as its primary effect because, as it is alleged, creation-science fails each of the five conditions on Ruse's list deemed necessary for genuine science. Like Laudan, Quinn argues that each of Ruse's conditions is either not necessary for genuine science (because some bona fide sciences lack it) or, when properly interpreted, is possessed by creation-science. According to Quinn, the proper thing to say about creation-science is not that we can show that it is not science but that, at best, it is dreadful science. See Philip L. Quinn, "Creationism, Methodology, and Politics," in Michael Ruse, ed., But Is It Science? 1st ed., 395–99.

19. For an entertaining account of Ruse's participation in the Arkansas trial and a transcript of his testimony, see Michael Ruse, ed., But Is It Science? 1st ed., 13–35, 287–306; the transcript is also reprinted in But Is It Science? 2d ed., 253–78.

20. In the General Scholium of the *Principia*, added to the second edition of 1713, Newton wrote: "But hitherto I have not been able to discover the cause of those properties of gravity [i.e., the proportionality of gravitational force to the quantity of matter and its variation with the inverse square of distance], and I frame no hypotheses [*hypotheses non fingo*]; for whatever is not deduced from the phenomena is to be called an hypothesis; and hypotheses, whether metaphysical or physical, whether of occult qualities or mechanical, have no place in experimental philosophy. In this philosophy particular propositions are inferred from the phenomena, and afterwards rendered general by induction." Isaac Newton, *Philosophiae naturalis principia mathematica*, vol. 2, trans. A. Motte, rev. F. Cajori (Berkeley: University of California Press, 1934), 547. Throughout the *Principia*, Newton argues that gravity cannot be explained by any mechanism acting by direct contact, as Descartes and the Cartesians had hypothesized. This seems to leave only two choices, both of which Newton entertained in his writings: either gravity is due to the direct action of God, or it is caused by an aether, itself composed of particles between which forces act at a distance across empty space. Many philosophers of science, notably Duhem and Popper, have criticized Newton's claim that his own theory can be "deduced from the phenomena." See "Duhem's Critique of Inductivism: The Attack on Newtonian Method," in the commentary on chapter 3.

21. The debate between Ruse and Laudan is continued in the pages of *Science, Technology, and Human Values*. See Michael Ruse, "Response to the Commentary: *Pro Judice*," 7 (Fall 1982): 19–23, Larry Laudan, "More on Creationism," 8 (Winter 1983): 36–38, and Michael Ruse, "The Academic as Expert Witness," 11 (Spring 1986): 68–73. These and other relevant articles are conveniently reprinted in Michael Ruse, ed., *But Is It Science?* 1st ed., and in Michael Ruse and Robert T. Pennock, eds., *But Is It Science?* 2d ed. Elsewhere, Laudan has argued that the wide diversity of scientific beliefs and activities and the failure of the attempts by Popper, Thagard, and others to solve the demarcation problem make it unlikely that we will ever find a demarcation criterion in the form of necessary conditions for genuine science. See Larry Laudan, "The Demise of the Demarcation Problem," in *Physics, Philosophy, and Psychoanalysis*, ed. R. S. Cohen and L. Laudan (Dordrecht, Netherlands: D. Reidel, 1983), 111–28; reprinted in Larry Laudan, *Beyond Positivism and Relativism* (Boulder, Co.: Westview Press, 1996) and Michael Ruse and Robert T. Pennock, eds., *But Is It Science?* 2d ed.

22. For a representative sample of writings by ID proponents and their critics see William A. Dembski and Michael Ruse, eds., *Debating Design* (Cambridge: Cambridge University Press, 2004).

23. The entire Dover decision is available online at http://www.pamd.uscourts.gov/kitzmiller/kitzmiller_342.pdf. An excerpt from the decision, along with Robert Pennock's "Expert Report" and excerpts from the expert witness testimony of Michael Behe, are included in Michael Ruse and Robert T. Pennock, eds., *But Is It Science?* 2d ed. The other two flaws criticized by Judge Jones in Section 4 of his decision are (i) the failure of ID criticisms of evolutionary theory to withstand scientific scrutiny and (ii) the "two-model" fallacy of thinking that the ID attacks on evolutionary theory are sufficient to secure a measure of credibility for ID as the only reasonable alternative. The weakness of the ID case against evolutionary theory is detailed in Kenneth R. Miller, *Only a Theory* (New York: Viking Penguin, 2008). Miller's book is based in part on his testimony at the Dover trial. A lively account of the trial and

the personalities involved can be found in Matthew Chapman, *40 Days and 40 Nights* (New York: HarperCollins, 2007).

24. Two opposing philosophical evaluations of naturalism as a demarcation criterion can be found in Bradley Monton, *Seeking God in Science: An Atheist Defends Intelligent Design* (Buffalo, N.Y.: Broadview Press, 2009), ch. 2, and Robert T. Pennock, "Can't Philosophers Tell the Difference Between Science and Religion?: Demarcation Revisited," *Synthese* 178 (2011): 177–206, reprinted in Michael Ruse and Robert T. Pennock, eds., *But Is It Science?* 2d ed. See also the discussion of Darwinism and methodological naturalism in Elliott Sober, *Did Darwin Write the Origin Backwards?* (Amherst, N.Y.: Prometheus Books, 2011).

2 | Rationality, Objectivity, and Values in Science

INTRODUCTION

Much of the recent debate about the role of values in science and the nature of scientific objectivity has been crucially affected by the work of Thomas Kuhn. Clark Glymour (writing in 1980) aptly described Kuhn's *The Structure of Scientific Revolutions* as "very likely the single most influential work on the philosophy of science that has been or will be written in this century."[1] It is certainly the most widely read. Since its publication in 1962, the University of Chicago Press has sold over one million copies of the English-language edition, and the book had been translated into at least thirty-six languages.[2] The frequency with which the term *paradigm* is bandied about in disciplines as diverse as literary criticism, art history, sociology of knowledge, and theology attest to the far-reaching character of Kuhn's work.

What made Kuhn's book so controversial was its rejection of many of the ways of thinking about science that had become standard during the first half of the twentieth century. Instead of giving logical analyses of individual scientific theories or constructing formal models of concepts such as explanation and confirmation, Kuhn turned to psychology, sociology, and history in order to draw a picture of science that, he claimed, was far more faithful to the original than anything that philosophers of science had yet proposed. The first reading in this chapter, "The Nature and Necessity of Scientific Revolutions," is taken from the heart of Kuhn's path-breaking book. In it Kuhn explains why he thinks that reason and evidence can play only a limited role in determining the outcome of scientific revolutions and why we must abandon the traditional assumption that science progresses by getting ever closer to the truth.

Soon after *The Structure of Scientific Revolutions* was published, Kuhn was criticized for portraying science, at least during scientific revolutions, as an irrational affair that is largely "a matter for mob psychology."[3] Many philosophers of science also deplored Kuhn's rejection of objective progress and realism in favor of relativism about truth and instrumentalism about theories. In response to these criticisms, Kuhn added a substantial postscript to *The Structure of Scientific Revolutions* in 1970[4] and delivered a lecture in 1973, "Objectivity, Value Judgment, and Theory Choice," which is reprinted as the second reading in this chapter. In this widely cited paper, Kuhn tried to defuse his critics' accusations of irrationalism and subjectivism by acknowledging that, during scientific revolutions, proponents of rival paradigms often share a number of important cognitive values. To the extent that arguments can be based on these shared values, scientific revolutions are rational. But Kuhn still insisted that there is no set of universal rules for choosing between rival theories, that cognitive values are ultimately a matter of subjective preference that transcends rationality, and that nonrational psychological and social factors must play a vital role in determining which theory wins the allegiance of the scientific community. Moreover, Kuhn continued to reject scientific realism, persisting in his view that scientific theories should be regarded as instruments for solving puzzles rather than as literal descriptions (or would-be descriptions) of reality.

Despite their influence on other academic disciplines, Kuhn's views are primarily about science, and it is on the basis of their adequacy as a philosophically informative account of science that they must be assessed. The next two readings in this chapter (by Ernan McMullin and Larry Laudan) explain and evaluate Kuhn's claims about the role of values in science, especially during those upheavals that constitute scientific revolutions. In "Rationality and Paradigm Change in Science," McMullin takes Kuhn to task for denying that the notions of objective progress and truth are relevant to understanding scientific revolutions, illustrating his criticism of Kuhn's instrumentalism with an analysis of the Copernican revolution in which the sun-centered astronomy of Copernicus replaced the earth-centered astronomy of Ptolemy. Laudan also attacks Kuhn, but from a different angle. In "Kuhn's Critique of Methodology," he argues that Kuhn has failed to establish that standards for appraising theories—Kuhn's cognitive values—are always so ambiguous, imprecise, and paradigm-relative as to preclude rational scientific agreement about which of two competing theories is the better or more strongly warranted by the available evidence. Laudan accuses Kuhn of having seriously exaggerated the extent to which proponents of rival paradigms must disagree about the interpretation of cognitive values and their relative importance when judging theories.

In the remaining two readings, by Helen Longino and Kathleen Okruhlik, respectively, the emphasis shifts from cognitive values in scientific revolutions to contextual values in normal science. In Longino's "Values and Objectivity," the main topic is the role that contextual values (in the form of background beliefs, possibly of a sexist or racist nature) play in everyday scientific assumptions about theories by shaping what scientists recognize as evidence for their theories. Two key questions are whether such beliefs could be eliminated from science, even in principle, and whether the presence of such beliefs prevents science from being objective. Longino defends a contextualist analysis of evidence and locates scientific objectivity, not in rules for choosing between theories, but in a social organization that permits diverse viewpoints and encourages criticism. In "Gender and the Biological Sciences," Okruhlik gives examples of androcentric bias in the life sciences to illustrate the ways in which contextual values can affect scientific judgment. She then compares and contrasts several different feminist critiques of science and defends a version of feminist empiricism.

Since all the readings in this chapter concern the role of values in the practice of science, it is helpful to consider what kinds of things values are and what sorts of values are relevant to the question of scientific objectivity and rationality. The important distinction between cognitive and contextual values is explored in the first section of the commentary that follows the readings. This section also includes a brief discussion of the value-neutrality thesis—that is, the thesis that only cognitive values should play a role in the decisions that scientists make about theories.

■ | Notes

1. Clark Glymour, *Theory and Evidence* (Princeton, N.J.: Princeton University Press, 1980), 94. The quotation comes from the opening sentence of a chapter entitled "New Fuzziness and Old Problems." By attesting to Kuhn's influence on the discipline, Glymour is not endorsing the approach of Kuhn and his successors to the philosophy of science. Rather, he berates the "new fuzziness" for failing to solve the "old problems" of explaining what makes evidence relevant to theory and why evidence varies in its confirming power. These traditional questions of confirmation theory are addressed in chapters 4 and 5 below.

2. In 2012, the University of Chicago Press re-issued *The Structure of Scientific Revolutions* in a 50th-anniversary fourth edition. Although the type was reset, the pagination of the chapters closely matches the second and third editions.

3. Imre Lakatos, "Falsification and the Methodology of Scientific Research Programmes," in *Criticism and the Growth of Knowledge*, ed. I. Lakatos and A. Musgrave (Cambridge: Cambridge University Press, 1970), 178. Kuhn replies to this and similar criticisms in the second reading in chapter 2, "Objectivity, Value Judgment, and Theory Choice."

4. As well as adding the *Postscript* to the second edition of *Structure*, Kuhn also inserted nine lines on page 84 of Chapter VIII ("The Response to Crisis") with the result that the pagination of the second and third editions of *Structure* differs from

that of the first edition. Beginning with the first page (92) of Chapter IX, all pages are numbered one higher than in the first edition (which is the edition that Laudan refers to in his article). The added lines reveal an important concession. Originally Kuhn asserted that all crises end with the emergence of a new paradigm. In 1970 Kuhn retracted this statement by recognizing that crises can also end in two other ways: either by normal science proving able to resolve the crisis without requiring a paradigm change or by scientists concluding that no solution is possible in the present state of the field and setting the problem aside (with the hope that future scientists will be able to solve it). So, what began as a bold and contentious claim—that all crises end with the emergence of a new paradigm—became the weaker but uncontroversial assertion that either crises end with a new paradigm or they do not.

Thomas S. Kuhn

The Nature and Necessity of Scientific Revolutions

... What are scientific revolutions, and what is their function in scientific development? ... Scientific revolutions are here taken to be those non-cumulative developmental episodes in which an older paradigm is replaced in whole or in part by an incompatible new one. There is more to be said, however, and an essential part of it can be introduced by asking one further question. Why should a change of paradigm be called a revolution? In the face of the vast and essential differences between political and scientific development, what parallelism can justify the metaphor that finds revolutions in both?

One aspect of the parallelism must already be apparent. Political revolutions are inaugurated by a growing sense, often restricted to a segment of the political community, that existing institutions have ceased adequately to meet the problems posed by an environment that they have in part created. In much the same way, scientific revolutions are inaugurated by a growing sense, again often restricted to a narrow subdivision of the scientific community, that an existing paradigm has ceased to function adequately in the exploration of an aspect of nature to which that paradigm itself had previously led the way. In both political and scientific development the sense of malfunction that can lead to crisis is prerequisite to revolution. Furthermore, though it admittedly strains the metaphor, that parallelism holds not only for the major paradigm changes, like those attributable to Copernicus and Lavoisier, but also for the far smaller ones associated with the assimilation of a new sort of phenomenon, like oxygen or X-rays. Scientific revolutions ... need seem revolutionary only to those whose paradigms are affected by them. To outsiders they may, like the Balkan revolutions of the early twentieth century, seem normal parts of the developmental process. Astronomers, for example, could accept X-rays as a mere addition to knowledge, for their

From Thomas S. Kuhn, *The Structure of Scientific Revolutions*, 3d ed. (Chicago: University of Chicago Press, 1996), 92–110.

paradigms were unaffected by the existence of the new radiation. But for men like Kelvin, Crookes, and Roentgen, whose research dealt with radiation theory or with cathode ray tubes, the emergence of X-rays necessarily violated one paradigm as it created another. That is why these rays could be discovered only through something's first going wrong with normal research.*

This genetic aspect of the parallel between political and scientific development should no longer be open to doubt. The parallel has, however, a second and more profound aspect upon which the significance of the first depends. Political revolutions aim to change political institutions in ways that those institutions themselves prohibit. Their success therefore necessitates the partial relinquishment of one set of institutions in favor of another, and in the interim, society is not fully governed by institutions at all. Initially it is crisis alone that attenuates the role of political institutions as we have already seen it attenuate the role of paradigms. In increasing numbers individuals become increasingly estranged from political life and behave more and more eccentrically within it. Then, as the crisis deepens, many of these individuals commit themselves to some concrete proposal for the reconstruction of society in a new institutional framework. At that point the society is divided into competing camps or parties, one seeking to defend the old institutional constellation, the others seeking to institute some new one. And, once that polarization has occurred, *political recourse fails.* Because they differ about the institutional matrix within which political change is to be achieved and evaluated, because they acknowledge no supra-institutional framework for the adjudication of revolutionary difference, the parties to a revolutionary conflict must finally resort to the techniques of mass persuasion, often including force. Though revolutions have had a vital role in the evolution of political institutions, that role depends upon their being partially extrapolitical or extrainstitutional events.

---

* Kuhn discusses the discovery of X rays in chapter 6 of *The Structure of Scientific Revolutions* and in his paper, "The Historical Structure of Scientific Discovery," *Science* 136 (1962): 760–64, reprinted in T. S. Kuhn, *The Essential Tension* (Chicago, Ill.: University of Chicago Press, 1977), 165–77. X rays are produced in a high-tension vacuum tube when cathode rays hit the glass wall of the tube. X rays were discovered accidentally by Roentgen in 1895 during his investigation of cathode rays. Roentgen noticed that a radiation detection screen at some distance from his vacuum tube glowed when current was passing through the tube, even though the apparatus was shielded. After seven weeks of intense experimental work, Roentgen ruled out cathode rays as a cause of the glow and announced his discovery of a new form of radiation that could pass easily through matter. His announcement was initially greeted with skepticism and surprise; Kelvin, for example, at first thought it must be a hoax.

Kuhn regards Roentgen's discovery as paradigm breaking because it required physicists to revise the way they performed and interpreted their experiments with cathode ray tubes. It also inspired the search for other new forms of radiation such as gamma rays and the notorious "N rays." (N rays were ultimately proved to be spurious—see Irving J. Langmuir. "Pathological Science," *Physics Today* 42 (1989): 36–48, for a revealing account of how respectable scientists can believe they are detecting and measuring phenomena that turn out to be nonexistent.)

The remainder of this essay aims to demonstrate that the historical study of paradigm change reveals very similar characteristics in the evolution of the sciences.* Like the choice between competing political institutions, that between competing paradigms proves to be a choice between incompatible modes of community life. Because it has that character, the choice is not and cannot be determined merely by the evaluative procedures characteristic of normal science, for these depend in part upon a particular paradigm, and that paradigm is at issue. When paradigms enter, as they must, into a debate about paradigm choice, their role is necessarily circular. Each group uses its own paradigm to argue in that paradigm's defense.

The resulting circularity does not, of course, make the arguments wrong or even ineffectual. The man who premises a paradigm when arguing in its defense can nonetheless provide a clear exhibit of what scientific practice will be like for those who adopt the new view of nature. That exhibit can be immensely persuasive, often compellingly so. Yet, whatever its force, the status of the circular argument is only that of persuasion. It cannot be made logically or even probabilistically compelling for those who refuse to step into the circle. The premises and values shared by the two parties to a debate over paradigms are not sufficiently extensive for that. As in political revolutions, so in paradigm choice—there is no standard higher than the assent of the relevant community. To discover how scientific revolutions are effected, we shall therefore have to examine not only the impact of nature and of logic, but also the techniques of persuasive argumentation effective within the quite special groups that constitute the community of scientists.

To discover why this issue of paradigm choice can never be unequivocally settled by logic and experiment alone, we must shortly examine the nature of the differences that separate the proponents of a traditional paradigm from their revolutionary successors. That examination is the principal object of this section. . . . We have, however, already noted numerous examples of such differences, and no one will doubt that history can supply many others. What is more likely to be doubted than their existence—and what must therefore be considered first—is that such examples provide essential information about the nature of science. Granting that paradigm rejection has been a historic fact, does it illuminate more than human credulity and confusion? Are there intrinsic reasons why the assimilation of either a new sort of phenomenon or a new scientific theory must demand the rejection of an older paradigm?

First notice that if there are such reasons, they do not derive from the logical structure of scientific knowledge. In principle, a new phenomenon might emerge without reflecting destructively upon any part of past scientific practice. Though discovering life on the moon would today be destructive

---

* By "the remainder of this essay," Kuhn means the final five chapters of *The Structure of Scientific Revolutions* of which the present selection (from chapter 9) is a crucial part.

of existing paradigms (these tell us things about the moon that seem incompatible with life's existence there), discovering life in some less well-known part of the galaxy would not. By the same token, a new theory does not have to conflict with any of its predecessors. It might deal exclusively with phenomena not previously known, as the quantum theory deals (but, significantly, not exclusively) with subatomic phenomena unknown before the twentieth century. Or again, the new theory might be simply a higher level theory than those known before, one that linked together a whole group of lower level theories without substantially changing any. Today, the theory of energy conservation provides just such links between dynamics, chemistry, electricity, optics, thermal theory, and so on. Still other compatible relationships between old and new theories can be conceived. Any and all of them might be exemplified by the historical process through which science has developed. If they were, scientific development would be genuinely cumulative. New sorts of phenomena would simply disclose order in an aspect of nature where none had been seen before. In the evolution of science new knowledge would replace ignorance rather than replace knowledge of another and incompatible sort.

Of course, science (or some other enterprise, perhaps less effective) might have developed in that fully cumulative manner. Many people have believed that it did so, and most still seem to suppose that cumulation is at least the ideal that historical development would display if only it had not so often been distorted by human idiosyncrasy. There are important reasons for that belief. . . . Nevertheless, despite the immense plausibility of that ideal image, there is increasing reason to wonder whether it can possibly be an image of *science*. After the pre-paradigm period the assimilation of all new theories and of almost all new sorts of phenomena has in fact demanded the destruction of a prior paradigm and a consequent conflict between competing schools of scientific thought. Cumulative acquisition of unanticipated novelties proves to be an almost non-existent exception to the rule of scientific development. The man who takes historic fact seriously must suspect that science does not tend toward the ideal that our image of its cumulativeness has suggested. Perhaps it is another sort of enterprise.

If, however, resistant facts can carry us that far, then a second look at the ground we have already covered may suggest that cumulative acquisition of novelty is not only rare in fact but improbable in principle. Normal research, which *is* cumulative, owes its success to the ability of scientists regularly to select problems that can be solved with conceptual and instrumental techniques close to those already in existence. (That is why an excessive concern with useful problems, regardless of their relation to existing knowledge and technique, can so easily inhibit scientific development.) The man who is striving to solve a problem defined by existing knowledge and technique is not, however, just looking around. He knows what he wants to achieve, and he designs his instruments and directs his thoughts accordingly. Unanticipated novelty, the new discovery, can emerge only to the extent that his anticipations about nature and his instruments prove wrong. Often the importance of the resulting discovery will itself be proportional to the extent and stubbornness of the anomaly that foreshadowed it. Obviously, then, there must be a conflict between the paradigm that discloses anomaly and the one that later renders the anomaly lawlike. . . . There is no other effective way in which discoveries might be generated.

The same argument applies even more clearly to the invention of new theories. There are, in principle, only three types of phenomena about which a new theory might be developed. The first consists of phenomena already well explained by existing paradigms, and these seldom provide either motive or point of departure for theory construction. When they do, . . . the theories that result are seldom accepted, because nature provides no ground for discrimination. A second class of phenomena consists of those whose nature is indicated by existing paradigms but whose details can be understood only through further theory articulation. These are the phenomena to which scientists direct their research much of the time, but that research aims at the articulation of existing paradigms rather than at the invention of new ones. Only when these attempts at articulation fail do scientists encounter the third type of phenomena, the recognized anomalies whose characteristic feature is their stubborn refusal to be assimilated to existing paradigms. This type alone gives rise to new theories. Paradigms provide all phenomena except anomalies with a theory-determined place in the scientist's field of vision.

But if new theories are called forth to resolve anomalies in the relation of an existing theory to nature, then the successful new theory must somewhere permit predictions that are different from those derived from its predecessor. That difference could not occur if the two were logically compatible. In the process of being assimilated, the second must displace the first. Even a theory like energy conservation, which today seems a logical superstructure that relates to nature only through independently established theories, did not develop historically without paradigm destruction. Instead, it emerged from a crisis in which an essential ingredient was the incompatibility between Newtonian dynamics and some recently formulated consequences of the caloric theory of heat. Only after the caloric theory had been rejected could energy conservation become part of science.[1] And only after it had been part of science for some time could it come to seem a theory of a logically higher type, one not in conflict with its predecessors.* It is hard to see how new

---

* According to the caloric theory, heat is a conserved fluid: its total quantity remains constant in its interactions with matter. So, for example, during the operation of a heat engine, no heat is destroyed when work is produced. This contradicts the energy conservation principle, according to which heat must be converted into an equivalent amount of work when the engine operates. As Kuhn writes, acceptance of the conservation-of-energy principle required the rejection of the caloric theory of heat. This rejection was hard to achieve because of the many well-confirmed results in thermodynamics obtained by Sadi Carnot using the caloric

theories could arise without these destructive changes in beliefs about nature. Though logical inclusiveness remains a permissible view of the relation between successive scientific theories, it is a historical implausibility.

A century ago it would, I think, have been possible to let the case for the necessity of revolutions rest at this point. But today, unfortunately, that cannot be done because the view of the subject developed above cannot be maintained if the most prevalent contemporary interpretation of the nature and function of scientific theory is accepted. That interpretation, closely associated with early logical positivism and not categorically rejected by its successors, would restrict the range and meaning of an accepted theory so that it could not possibly conflict with any later theory that made predictions about some of the same natural phenomena. The best-known and the strongest case for this restricted conception of a scientific theory emerges in discussions of the relation between contemporary Einsteinian dynamics and the older dynamical equations that descend from Newton's *Principia*. From the viewpoint of this essay these two theories are fundamentally incompatible in the sense illustrated by the relation of Copernican to Ptolemaic astronomy: Einstein's theory can be accepted only with the recognition that Newton's was wrong. Today this remains a minority view.[2] We must therefore examine the most prevalent objections to it.

The gist of these objections can be developed as follows. Relativistic dynamics cannot have shown Newtonian dynamics to be wrong, for Newtonian dynamics is still used with great success by most engineers and, in selected applications, by many physicists. Furthermore, the propriety of this use of the older theory can be proved from the very theory that has, in other applications, replaced it. Einstein's theory can be used to show that predictions from Newton's equations will be as good as our measuring instruments in all applications that satisfy a small number of restrictive conditions. For example, if Newtonian theory is to provide a good approximate solution, the relative velocities of the bodies considered must be small compared with the velocity of light. Subject to this condition and a few others, Newtonian theory seems to be derivable from Einsteinian, of which it is therefore a special case.*

But, the objection continues, no theory can possibly conflict with one of its special cases. If Einsteinian science seems to make Newtonian dynamics

---

theory. Thus, accepting the energy conservation principle required not only overturning the prevailing caloric paradigm but also rewriting the foundations of thermodynamics. See D. S. L. Cardwell, *From Watt to Clausius: The Rise of Thermodynamics in the Early Industrial Age* (Ithaca, N.Y.: Cornell University Press, 1971), for a fascinating account of this revolution in physics.

* Attempts to understand scientific change by viewing the relation between earlier theories and their successors as a type of reduction are discussed in chapter 8. A sustained attack on the traditional account of reduction—including the derivability requirement discussed by Kuhn below—can be found in Paul Feyerabend, "How to Be a Good Empiricist" also included in chapter 8.

wrong, that is only because some Newtonians were so incautious as to claim that Newtonian theory yielded entirely precise results or that it was valid at very high relative velocities. Since they could not have had any evidence for such claims, they betrayed the standards of science when they made them. In so far as Newtonian theory was ever a truly scientific theory supported by valid evidence, it still is. Only extravagant claims for the theory—claims that were never properly parts of science—can have been shown by Einstein to be wrong. Purged of these merely human extravagances, Newtonian theory has never been challenged and cannot be.

Some variant of this argument is quite sufficient to make any theory ever used by a significant group of competent scientists immune to attack. The much-maligned phlogiston theory, for example, gave order to a large number of physical and chemical phenomena. It explained why bodies burned—they were rich in phlogiston—and why metals had so many more properties in common than did their ores. The metals were all compounded from different elementary earths combined with phlogiston, and the latter, common to all metals, produced common properties. In addition, the phlogiston theory accounted for a number of reactions in which acids were formed by the combustion of substances like carbon and sulphur. Also, it explained the decrease of volume when combustion occurs in a confined volume of air—the phlogiston released by combustion "spoils" the elasticity of the air that absorbed it, just as fire "spoils" the elasticity of a steel spring.[3] If these were the only phenomena that the phlogiston theorists had claimed for their theory, that theory could never have been challenged. A similar argument will suffice for any theory that has ever been successfully applied to any range of phenomena at all.

But to save theories in this way, their range of application must be restricted to those phenomena and to that precision of observation with which the experimental evidence in hand already deals.[4] Carried just a step further (and the step can scarcely be avoided once the first is taken), such a limitation prohibits the scientist from claiming to speak "scientifically" about any phenomenon not already observed. Even in its present form the restriction forbids the scientist to rely upon a theory in his own research whenever that research enters an area or seeks a degree of precision for which past practice with the theory offers no precedent. These prohibitions are logically unexceptionable. But the result of accepting them would be the end of the research through which science may develop further.

By now that point too is virtually a tautology. Without commitment to a paradigm there could be no normal science. Furthermore, that commitment must extend to areas and to degrees of precision for which there is no full precedent. If it did not, the paradigm could provide no puzzles that had not already been solved. Besides, it is not only normal science that depends upon commitment to a paradigm. If existing theory binds the scientist only with respect to existing applications, then there can be no surprises, anomalies, or crises. But these are just the signposts that point the way to extraordinary science. If positivistic restrictions on the range of a theory's legitimate applicability are taken literally, the mechanism that tells the scientific community what problems may lead to fundamental change must cease to function. And when that occurs, the community will inevitably return to something much like its pre-paradigm state, a condition in which all members practice science but in which their gross product scarcely resembles science at all. Is it really any wonder that the price of significant scientific advance is a commitment that runs the risk of being wrong?

More important, there is a revealing logical lacuna in the positivist's argument, one that will reintroduce us immediately to the nature of revolutionary change. Can Newtonian dynamics really be *derived* from relativistic dynamics? What would such a derivation look like? Imagine a set of statements, E_1, E_2, . . . , E_n, which together embody the laws of relativity theory. These statements contain variables and parameters representing spatial position, time, rest mass, etc. From them, together with the apparatus of logic and mathematics, is deducible a whole set of further statements including some that can be checked by observation. To prove the adequacy of Newtonian dynamics as a special case, we must add to the E_i's additional statements, like $(v/c)^2 \ll 1$, restricting the range of the parameters and variables. This enlarged set of statements is then manipulated to yield a new set, N_1, N_2, . . . , N_m, which is identical in form with Newton's laws of motion, the law of gravity, and so on. Apparently Newtonian dynamics has been derived from Einsteinian, subject to a few limiting conditions.

Yet the derivation is spurious, at least to this point. Though the N_i's are a special case of the laws of relativistic mechanics, they are not Newton's Laws. Or at least they are not unless those laws are reinterpreted in a way that would have been impossible until after Einstein's work. The variables and parameters that in the Einsteinian E_i's represented spatial position, time, mass, etc., still occur in the N_i's; and they there still represent Einsteinian space, time, and mass. But the physical referents of these Einsteinian concepts are by no means identical with those of the Newtonian concepts that bear the same name. (Newtonian mass is conserved; Einsteinian is convertible with energy. Only at low relative velocities may the two be measured in the same way, and even then they must not be conceived to be the same.) Unless we change the definitions of the variables in the N_i's, the statements we have derived are not Newtonian. If we do change them, we cannot properly be said to have *derived* Newton's Laws, at least not in any sense of "derive" now generally recognized. Our argument has, of course, explained why Newton's Laws ever seemed to work. In doing so it has justified, say, an automobile driver in acting as though he lived in a Newtonian universe. An argument of the same type is used to justify teaching earth-centered astronomy to surveyors. But the argument has still not done what it purported to do. It has not, that is, shown Newton's Laws to be a limiting case of Einstein's. For in the passage to the limit it is not only the forms of the laws that have changed. Simultaneously we have had to alter the fundamental structural elements of which the universe to which they apply is composed.

This need to change the meaning of established and familiar concepts is central to the revolutionary impact of Einstein's theory. Though subtler than the changes from geocentrism to heliocentrism, from phlogiston to oxygen, or from corpuscles to waves [as an account of the nature of light], the resulting conceptual transformation is no less decisively destructive of a previously established paradigm. We may even come to see it as a prototype for revolutionary reorientations in the sciences. Just because it did not involve the introduction of additional objects or concepts, the transition from Newtonian to Einsteinian mechanics illustrates with particular clarity the scientific revolution as a displacement of the conceptual network through which scientists view the world.

These remarks should suffice to show what might, in another philosophical climate, have been taken for granted. At least for scientists, most of the apparent differences between a discarded scientific theory and its successor are real. Though an out-of-date theory can always be viewed as a special case of its up-to-date successor, it must be transformed for the purpose. And the transformation is one that can be undertaken only with the advantages of hindsight, the explicit guidance of the more recent theory. Furthermore, even if that transformation were a legitimate device to employ in interpreting the older theory, the result of its application would be a theory so restricted that it could only restate what was already known. Because of its economy, that restatement would have utility, but it could not suffice for the guidance of research.

Let us, therefore, now take it for granted that the differences between successive paradigms are both necessary and irreconcilable. Can we then say more explicitly what sorts of differences these are? The most apparent type has already been illustrated repeatedly. Successive paradigms tell us different things about the population of the universe and about the population's behavior. They differ, that is, about such questions as the existence of subatomic particles, the materiality of light, and the conservation of heat or of energy. These are the substantive differences between successive paradigms, and they require no further illustration. But paradigms differ in more than substance, for they are directed not only to nature but also back upon the science that produced them. They are the source of the methods, problem-field, and standards of solution accepted by any mature scientific community at any given time. As a result, the reception of a new paradigm often necessitates a redefinition of the corresponding science. Some old problems may be relegated to another science or declared entirely "unscientific." Others that were previously non-existent or trivial may, with a new paradigm, become the very archetypes of significant scientific achievement. And as the problems change, so, often, does the standard that distinguishes a real scientific solution from a mere metaphysical speculation, word game, or mathematical play. The normal-scientific tradition that emerges from a scientific revolution is not only incompatible but often actually incommensurable with that which has gone before.

The impact of Newton's work upon the normal seventeenth-century tradition of scientific practice provides a striking example of these subtler effects of paradigm shift. Before Newton was born the "new science" of the century had at last succeeded in rejecting Aristotelian and scholastic explanations expressed in terms of the essences of material bodies. To say that a stone fell because its "nature" drove it toward the center of the universe had been made to look a mere tautological word-play, something it had not previously been. Henceforth the entire flux of sensory appearances, including color, taste, and even weight, was to be explained in terms of the size, shape, position, and motion of the elementary corpuscles of base matter. The attribution of other qualities to the elementary atoms was a resort to the occult and therefore out of bounds for science. Molière caught the new spirit precisely when he ridiculed the doctor who explained opium's efficacy as a soporific by attributing to it a dormitive potency.* During the last half of the seventeenth century many scientists preferred to say that the round shape of the opium particles enabled them to sooth the nerves about which they moved.[5]

In an earlier period explanations in terms of occult qualities had been an integral part of productive scientific work. Nevertheless, the seventeenth century's new commitment to mechanico-corpuscular explanation proved immensely fruitful for a number of sciences, ridding them of problems that had defied generally accepted solution and suggesting others to replace them. In dynamics, for example, Newton's three laws of motion are less a product of novel experiments than of the attempt to reinterpret well-known observations in terms of the motions and interactions of primary neutral corpuscles. Consider just one concrete illustration. Since neutral corpuscles could act on each other only by contact, the mechanico-corpuscular view of nature directed scientific attention to a brand-new subject of study, the alteration of particulate motions by collisions. Descartes announced the problem and provided its first putative solution. Huyghens, Wren, and Wal-

---

* Molière satirizes *virtus dormitiva* (dormitive potency) as an explanation of opium's power to induce sleep in his last play, *Le malade imaginaire* (The imaginary invalid, 1673). As Kuhn suggests (with deliberate irony) in his next sentence, the purported explanations given by the mechanico-corpuscular philosophy could be just as superficial as those offered by the Aristotelians and scholastics. More importantly, as Kuhn notes in the following paragraphs, the *vis inertiae* (force of inertia) and gravitational action-at-a-distance that Newton attributed to matter bear more than a casual resemblance to the essences and powers of the Aristotelians. In this respect, Newton was more of an Aristotelian than his corpuscularian contemporaries. Newton's achievement lay not in banishing essences and powers from science but in discovering the precise mathematical laws according to which they operate and being able to use those laws to make testable predictions. For further thoughts in this direction, see Rudolf Carnap, "The Value of Laws: Explanation and Prediction" in chapter 6 below.

lis carried it still further, partly by experimenting with colliding pendulum bobs, but mostly by applying previously well-known characteristics of motion to the new problem. And Newton embedded their results in his laws of motion. The equal "action" and "reaction" of the third law are the changes in quantity of motion experienced by the two parties to a collision. The same change of motion supplies the definition of dynamical force implicit in the second law. In this case, as in many others during the seventeenth century, the corpuscular paradigm bred both a new problem and a large part of that problem's solution.[6]

Yet, though much of Newton's work was directed to problems and embodied standards derived from the mechanico-corpuscular world view, the effect of the paradigm that resulted from his work was a further and partially destructive change in the problems and standards legitimate for science. Gravity, interpreted as an innate attraction between every pair of particles of matter, was an occult quality in the same sense as the scholastics' "tendency to fall" had been. Therefore, while the standards of corpuscularism remained in effect, the search for a mechanical explanation of gravity was one of the most challenging problems for those who accepted the *Principia* as paradigm. Newton devoted much attention to it and so did many of his eighteenth-century successors. The only apparent option was to reject Newton's theory for its failure to explain gravity, and that alternative, too, was widely adopted. Yet neither of these views ultimately triumphed. Unable either to practice science without the *Principia* or to make that work conform to the corpuscular standards of the seventeenth century, scientists gradually accepted the view that gravity was indeed innate. By the mid-eighteenth century that interpretation had been almost universally accepted, and the result was a genuine reversion (which is not the same as a retrogression) to a scholastic standard. Innate attractions and repulsions joined size, shape, position, and motion as physically irreducible primary properties of matter.[7]

The resulting change in the standards and problem-field of physical science was once again consequential. By the 1740's, for example, electricians could speak of the attractive "virtue" of the electric fluid without thereby inviting the ridicule that had greeted Molière's doctor a century before. As they did so, electrical phenomena increasingly displayed an order different from the one they had shown when viewed as the effects of a mechanical effluvium that could act only by contact. In particular, when electrical action-at-a-distance became a subject for study in its own right, the phenomenon we now call charging by induction could be recognized as one of its effects. Previously, when seen at all, it had been attributed to the direct action of electrical "atmospheres" or to the leakages inevitable in any electrical laboratory. The new view of inductive effects was, in turn, the key to Franklin's analysis of the Leyden jar and thus to the emergence of a new and Newtonian paradigm for electricity. Nor were dynamics and electricity the only scientific fields affected by the legitimization of the search for forces innate to matter. The large body of eighteenth-century literature on chemical affinities and replacement series also derives from this supramechanical aspect of Newtonianism. Chemists who believed in these differential attractions between the various chemical species set up previously unimagined experiments and searched for new sorts of reactions. Without the data and the chemical concepts developed in that process, the later work of Lavoisier and, more particularly, of Dalton would be incomprehensible.[8]* Changes in the standards governing permissible problems, concepts, and explanations can transform a science. . . .

Other examples of these nonsubstantive differences between successive paradigms can be retrieved from the history of any science in almost any period of its development. For the moment let us be content with just two other and far briefer illustrations. Before the chemical revolution, one of the acknowledged tasks of chemistry was to account for the qualities of chemical substances and for the changes these qualities underwent during chemical reactions. With the aid of a small number of elementary "principles"—of which phlogiston was one—the chemist was to explain why some substances are acidic, others metalline, combustible, and so forth. Some success in this direction had been achieved. We have already noted that phlogiston explained why the metals were so much alike, and we could have developed a similar argument for the acids. Lavoisier's reform, however, ultimately did away with chemical "principles," and thus ended by depriving chemistry of some actual and much potential explanatory power. To compensate for this loss, a change in standards was required. During much of the nineteenth century failure to explain the qualities of compounds was no indictment of a chemical theory.[9]

Or again, Clerk Maxwell shared with other nineteenth-century proponents of the wave theory of light the conviction that light waves must be propagated through a material ether. Designing a mechanical medium to support such waves was a standard problem for many of his ablest contemporaries. His own theory, however, the electromagnetic theory of light, gave no account at all of a medium able to support light waves, and it clearly

---

* The oxygen theory of Antoine Lavoisier (1743–94) owed much to experiments on calcination and the isolation of gases by Joseph Priestley, Carl Scheele, and Henry Cavendish, all of whom were proponents of the phlogiston theory. Similarly, the atomic theory of John Dalton (1766–1844) was indebted to the discovery of the law of equivalent proportions (the basis for assigning equivalent weights to chemical elements) and the law of constant proportions (that regardless of how a compound is made, it always contains the same ratio of elements by weight), discoveries made within the Newtonian paradigm referred to by Kuhn. The laws of equivalent and constant proportions led Dalton to formulate the law of multiple proportions (that when two elements can form more than one compound, the weights of one element that combine with a fixed weight of the other are always in a simple numerical ratio) and the confirmation of this law played a central role in Dalton's case for the atomic theory.

made such an account harder to provide than it had seemed before. Initially, Maxwell's theory was widely rejected for those reasons. But, like Newton's theory, Maxwell's proved difficult to dispense with, and as it achieved the status of a paradigm, the community's attitude toward it changed. In the early decades of the twentieth century Maxwell's insistence upon the existence of a mechanical ether looked more and more like lip service, which it emphatically had not been, and the attempts to design such an ethereal medium were abandoned. Scientists no longer thought it unscientific to speak of an electrical "displacement" without specifying what was being displaced.* The result, again, was a new set of problems and standards, one which, in the event, had much to do with the emergence of relativity theory.[10]

These characteristic shifts in the scientific community's conception of its legitimate problems and standards would have less significance to this essay's thesis if one could suppose that they always occurred from some methodologically lower to some higher type. In that case their effects, too, would seem cumulative. No wonder that some historians have argued that the history of science records a continuing increase in the maturity and refinement of man's conception of the nature of science.[11] Yet the case for cumulative development of science's problems and standards is even harder to make than the case for cumulation of theories. The attempt to explain gravity, though fruitfully abandoned by most eighteenth-century scientists, was not directed to an intrinsically illegitimate problem; the objections to innate forces were neither inherently unscientific nor metaphysical in some pejorative sense. There are no external standards to permit a judgment of that sort. What occurred was neither a decline nor a raising of standards, but simply a change demanded by the adoption of a new paradigm. Furthermore,

---

* Kuhn is referring to the difficulty of understanding what the term D represents physically in Maxwell's equations for the electromagnetic field in free space. Maxwell introduced the term D, calling it the displacement current by analogy with the ordinary electric current that flows when a wire is connected to the terminals of a battery. When a state of electric polarization is induced in a dielectric medium, there is a transient change in the electric field. This changing electric field acts just like an electric current in producing a magnetic field. When the medium is a real, physical substance (such as an insulator between the plates of a condensor), we can easily imagine that the displacement current arises because charged particles (electrons) are moved slightly in the direction of the applied electric field. But what happens in free space, where there is no physical substance and no charged particles? According to Maxwell's equations, variations in the displacement current give rise to a changing magnetic field that, in turn, induces an electric field, and so on. In this way, electromagnetic waves (such as radio waves and light rays) can propagate in a vacuum. But what is the medium, the so-called ether, in which these waves propagate? As Kuhn remarks, despite the empirical success of Maxwell's equations, all attempts to construct a consistent mechanical model of the ether met with failure. Eventually, with the successful integration of Maxwell's equations within the special theory of relativity, physicists stopped asking what the displacement current in free space is a displacement of and contented themselves with defining D operationally, in terms of quantities that can be observed and measured.

that change has since been reversed and could be again. In the twentieth century Einstein succeeded in explaining gravitational attractions, and that explanation has returned science to a set of canons and problems that are, in this particular respect, more like those of Newton's predecessors than of his successors. Or again, the development of quantum mechanics has reversed the methodological prohibition that originated in the chemical revolution. Chemists now attempt, and with great success, to explain the color, state of aggregation, and other qualities of the substances used and produced in their laboratories. A similar reversal may even be underway in electromagnetic theory. Space, in contemporary physics, is not the inert and homogeneous substratum employed in both Newton's and Maxwell's theories; some of its new properties are not unlike those once attributed to the ether; we may someday come to know what an electric displacement is.

By shifting emphasis from the cognitive to the normative functions of paradigms, the preceding examples enlarge our understanding of the ways in which paradigms give form to the scientific life. Previously, we had principally examined the paradigm's role as a vehicle for scientific theory. In that role it functions by telling the scientist about the entities that nature does and does not contain and about the ways in which those entities behave. That information provides a map whose details are elucidated by mature scientific research. And since nature is too complex and varied to be explored at random, that map is as essential as observation and experiment to science's continuing development. Through the theories they embody, paradigms prove to be constitutive of the research activity. They are also, however, constitutive of science in other respects, and that is now the point. In particular, our most recent examples show that paradigms provide scientists not only with a map but also with some of the directions essential for map-making. In learning a paradigm the scientist acquires theory, methods, and standards together, usually in an inextricable mixture. Therefore, when paradigms change, there are usually significant shifts in the criteria determining the legitimacy both of problems and of proposed solutions.

That observation returns us to the point from which this section began, for it provides our first explicit indication of why the choice between competing paradigms regularly raises questions that cannot be resolved by the criteria of normal science. To the extent, as significant as it is incomplete, that two scientific schools disagree about what is a problem and what a solution, they will inevitably talk through each other when debating the relative merits of their respective paradigms. In the partially circular arguments that regularly result, each paradigm will be shown to satisfy more or less the criteria that it dictates for itself and to fall short of a few of those dictated by its opponent. There are other reasons, too, for the incompleteness of logical contact that consistently characterizes paradigm debates. For example, since no paradigm ever solves all the problems it defines and since no two paradigms leave all the same problems unsolved, paradigm debates always involve the question: Which problems is it more significant to have solved? Like the issue of competing standards, that question of values can be answered only in terms of criteria that lie outside of normal science altogether, and it is that recourse to external criteria that most obviously makes paradigm debates revolutionary. . . .

■ | Notes

1. Silvanus P. Thompson, *Life of William Thomson Baron Kelvin of Largs* (London, 1910), I, 266–81.

2. See, for example, the remarks by P. P. Wiener in *Philosophy of Science*, XXV (1958), 298.

3. James B. Conant, *Overthrow of the Phlogiston Theory* (Cambridge, 1950), pp. 13–16; and J. R. Partington. *A Short History of Chemistry* (2d ed.; London, 1951), pp. 85–88. The fullest and most sympathetic account of the phlogiston theory's achievements is by H. Metzger, *Newton, Stahl, Boerhaave et la doctrine chimique* (Paris, 1930), Part II.

4. Compare the conclusions reached through a very different sort of analysis by R. B. Braithwaite, *Scientific Explanation* (Cambridge, 1953), pp. 50–87, esp. p. 76.

5. For corpuscularism in general, see Marie Boas, "The Establishment of the Mechanical Philosophy," *Osiris*, X (1952), 412–541. For the effect of particle-shape on taste, see *ibid.*, p. 483.

6. R. Dugas, *La mécanique au XVIIe siècle* (Neuchâtel, 1954), pp. 177–85, 284–98, 345–56.

7. I. B. Cohen, *Franklin and Newton: An Inquiry into Speculative Newtonian Experimental Science and Franklin's Work in Electricity as an Example Thereof* (Philadelphia, 1956), chaps. vi–vii.

8. For electricity, see *ibid.*, chaps. viii–ix. For chemistry, see Metzger, *op. cit.*, Part I.

9. E. Meyerson, *Identity and Reality* (New York, 1930), chap. x.

10. E. T. Whittaker, *A History of the Theories of Aether and Electricity*, II (London, 1953), 28–30.

11. For a brilliant and entirely up-to-date attempt to fit scientific development into this Procrustean bed, see C. C. Gillispie, *The Edge of Objectivity: An Essay in the History of Scientific Ideas* (Princeton, 1960).

Thomas S. Kuhn

Objectivity, Value Judgment, and Theory Choice

In the penultimate chapter of a controversial book first published fifteen years ago, I considered the ways scientists are brought to abandon one time-honored theory or paradigm in favor of another. Such decision problems, I wrote, "cannot be resolved by proof." To discuss their mechanism is, therefore, to talk "about techniques of persuasion, or about argument and counterargument in a situation in which there can be no proof." Under these circumstances, I continued, "lifelong resistance [to a new theory] . . . is not a violation of scientific standards. . . . Though the historian can always find men—Priestley, for instance—who were unreasonable to resist for as long as they did, he will not find a point at which resistance becomes illogical or unscientific."[1] Statements of that sort obviously raise the question of why, in the absence of binding criteria for scientific choice, both the number of solved scientific problems and the precision of individual problem solutions should increase so markedly with the passage of time. Confronting that issue, I sketched in my closing chapter a number of characteristics that scientists share by virtue of the training which licenses their membership in one or another community of specialists. In the absence of criteria able to dictate the choice of each individual, I argued, we do well to trust the collective judgment of scientists trained in this way. "What better criterion could there be," I asked rhetorically, "than the decision of the scientific group?"[2]

A number of philosophers have greeted remarks like these in a way that continues to surprise me. My views, it is said, make of theory choice "a matter for mob psychology."[3] Kuhn believes, I am told, that "the decision of a scientific group to adopt a new paradigm cannot be based on good reasons of any kind, factual or otherwise."[4] The debates surrounding such choices must, my critics claim, be for me "mere persuasive displays without deliberative

FROM Thomas S. Kuhn, *The Essential Tension: Selected Studies in Scientific Tradition and Change* (Chicago: University of Chicago Press, 1977), 320–39. This essay was originally presented as the Machette Lecture, delivered at Furman University, Greenville, S.C., on 30 November 1973.

94

substance."[5] Reports of this sort manifest total misunderstanding, and I have occasionally said as much in papers directed primarily to other ends. But those passing protestations have had negligible effect, and the misunderstandings continue to be important. I conclude that it is past time for me to describe, at greater length and with greater precision, what has been on my mind when I have uttered statements like the ones with which I just began. If I have been reluctant to do so in the past, that is largely because I have preferred to devote attention to areas in which my views diverge more sharply from those currently received than they do with respect to theory choice.

What, I ask to begin with, are the characteristics of a good scientific theory? Among a number of quite usual answers I select five, not because they are exhaustive, but because they are individually important and collectively sufficiently varied to indicate what is at stake. First, a theory should be accurate: within its domain, that is, consequences deducible from a theory should be in demonstrated agreement with the results of existing experiments and observations. Second, a theory should be consistent, not only internally or with itself, but also with other currently accepted theories applicable to related aspects of nature. Third, it should have broad scope: in particular, a theory's consequences should extend far beyond the particular observations, laws, or subtheories it was initially designed to explain. Fourth, and closely related, it should be simple, bringing order to phenomena that in its absence would be individually isolated and, as a set, confused. Fifth—a somewhat less standard item, but one of special importance to actual scientific decisions—a theory should be fruitful of new research findings: it should, that is, disclose new phenomena or previously unnoted relationships among those already known.[6] These five characteristics—accuracy, consistency, scope, simplicity, and fruitfulness—are all standard criteria for evaluating the adequacy of a theory. If they had not been, I would have devoted far more space to them in my book, for I agree entirely with the traditional view that they play a vital role when scientists must choose between an established theory and an upstart competitor. Together with others of much the same sort, they provide *the* shared basis for theory choice.

Nevertheless, two sorts of difficulties are regularly encountered by the men who must use these criteria in choosing, say, between Ptolemy's astronomical theory and Copernicus's, between the oxygen and phlogiston theories of combustion, or between Newtonian mechanics and the quantum theory. Individually the criteria are imprecise: individuals may legitimately differ about their application to concrete cases. In addition, when deployed together, they repeatedly prove to conflict with one another; accuracy may, for example, dictate the choice of one theory, scope the choice of its competitor. Since these difficulties, especially the first, are also relatively familiar, I shall devote little time to their elaboration. Though my argument does demand that I illustrate them briefly, my views will begin to depart from those long current only after I have done so.

Begin with accuracy, which for present purposes I take to include not only quantitative agreement but qualitative as well. Ultimately it proves the most nearly decisive of all the criteria, partly because it is less equivocal than the others but especially because predictive and explanatory powers, which depend on it, are characteristics that scientists are particularly unwilling to give up. Unfortunately, however, theories cannot always be discriminated in terms of accuracy. Copernicus's system, for example, was not more accurate than Ptolemy's until drastically revised by Kepler more than sixty years after Copernicus's death. If Kepler or someone else had not found other reasons to choose heliocentric astronomy, those improvements in accuracy would never have been made, and Copernicus's work might have been forgotten. More typically, of course, accuracy does permit discriminations, but not the sort that lead regularly to unequivocal choice. The oxygen theory, for example, was universally acknowledged to account for observed weight relations in chemical reactions, something the phlogiston theory had previously scarcely attempted to do. But the phlogiston theory, unlike its rival, could account for the metals' being much more alike than the ores from which they were formed. One theory thus matched experience better in one area, the other in another.* To choose between them on the basis of accuracy, a scientist would need to decide the area in which accuracy was more significant. About that matter chemists could and did differ without violating any of the criteria outlined above, or any others yet to be suggested.

However important it may be, therefore, accuracy by itself is seldom or never a sufficient criterion for theory choice. Other criteria must function as well, but they do not eliminate problems. To illustrate I select just two—consistency and simplicity—asking how they functioned in the choice between the heliocentric and geocentric systems. As astronomical theories both Ptolemy's and Copernicus's were internally consistent, but their relation to related theories in other fields was very different. The stationary central

---

* Supporters of the phlogiston theory argued that metals are similar because they all contain phlogiston, which is released as heat and fire when metals burn in air. (This argument is not very impressive, since it fails to explain why carbon and other nonmetallic combustible materials, which are also supposed to be rich in phlogiston, are not at all like metals.) Lavoisier's oxygen theory offered no explanation of why metals resemble each other, but it did predict that all metals become heavier when they burn, and this prediction was confirmed by weighing experiments. Metals gain weight when they burn because burning involves the chemical combination of the metal with the oxygen in the air to form an oxide. The heat associated with oxidation was attributed by Lavoisier to the release of caloric fluid that, according to his theory, surrounds the particles of oxygen. This is a typical example of what has come to be known as a "Kuhn loss": when one paradigm replaces another, not every problem that was solved by the old paradigm can be solved by the new one, even though the new paradigm solves problems that the old one either ignored or could not solve. Modern science has finally succeeded in explaining the similarity among metals: metals are shiny, conduct electricity, etc. because they all contain free electrons that are not bound to individual atoms.

earth was an essential ingredient of received physical theory, a tight-knit body of doctrine which explained, among other things, how stones fall, how water pumps function, and why the clouds move slowly across the skies. Heliocentric astronomy, which required the earth's motion, was inconsistent with the existing scientific explanation of these and other terrestrial phenomena. The consistency criterion, by itself, therefore, spoke unequivocally for the geocentric tradition.

Simplicity, however, favored Copernicus, but only when evaluated in a quite special way. If, on the one hand, the two systems were compared in terms of the actual computational labor required to predict the position of a planet at a particular time, then they proved substantially equivalent. Such computations were what astronomers did, and Copernicus's system offered them no labor-saving techniques; in that sense it was not simpler than Ptolemy's. If, on the other hand, one asked about the amount of mathematical apparatus required to explain, not the detailed quantitative motions of the planets, but merely their gross qualitative features—limited elongation, retrograde motion, and the like—then, as every schoolchild knows, Copernicus required only one circle per planet, Ptolemy two. In that sense the Copernican theory was the simpler, a fact vitally important to the choices made by both Kepler and Galileo and thus essential to the ultimate triumph of Copernicanism. But that sense of simplicity was not the only one available, nor even the one most natural to professional astronomers, men whose task was the actual computation of planetary position.

Because time is short and I have multiplied examples elsewhere, I shall here simply assert that these difficulties in applying standard criteria of choice are typical and that they arise no less forcefully in twentieth-century situations than in the earlier and better-known examples I have just sketched. When scientists must choose between competing theories, two men fully committed to the same list of criteria for choice may nevertheless reach different conclusions. Perhaps they interpret simplicity differently or have different convictions about the range of fields within which the consistency criterion must be met. Or perhaps they agree about these matters but differ about the relative weights to be accorded to these or to other criteria when several are deployed together. With respect to divergences of this sort, no set of choice criteria yet proposed is of any use. One can explain, as the historian characteristically does, why particular men made particular choices at particular times. But for that purpose one must go beyond the list of shared criteria to characteristics of the individuals who make the choice. One must, that is, deal with characteristics which vary from one scientist to another without thereby in the least jeopardizing their adherence to the canons that make science scientific. Though such canons do exist and should be discoverable (doubtless the criteria of choice with which I began are among them), they are not by themselves sufficient to determine the decisions of individual scientists. For that purpose the shared canons must be fleshed out in ways that differ from one individual to another.

Some of the differences I have in mind result from the individual's previous experience as a scientist. In what part of the field was he at work when confronted by the need to choose? How long had he worked there; how successful had he been; and how much of his work depended on concepts and techniques challenged by the new theory? Other factors relevant to choice lie outside the sciences. Kepler's early election of Copernicanism was due in part to his immersion in the Neoplatonic and Hermetic movements of his day; German Romanticism predisposed those it affected toward both recognition and acceptance of energy conservation; nineteenth-century British social thought had a similar influence on the availability and acceptability of Darwin's concept of the struggle for existence. Still other significant differences are functions of personality. Some scientists place more premium than others on originality and are correspondingly more willing to take risks; some scientists prefer comprehensive, unified theories to precise and detailed problem solutions of apparently narrower scope. Differentiating factors like these are described by my critics as subjective and are contrasted with the shared or objective criteria from which I began. Though I shall later question that use of terms, let me for the moment accept it. My point is, then, that every individual choice between competing theories depends on a mixture of objective and subjective factors, or of shared and individual criteria. Since the latter have not ordinarily figured in the philosophy of science, my emphasis upon them has made my belief in the former hard for my critics to see.

What I have said so far is primarily simply descriptive of what goes on in the sciences at times of theory choice. As description, furthermore, it has not been challenged by my critics, who reject instead my claim that these facts of scientific life have philosophic import. Taking up that issue, I shall begin to isolate some, though I think not vast, differences of opinion. Let me begin by asking how philosophers of science can for so long have neglected the subjective elements which, they freely grant, enter regularly into the actual theory choices made by individual scientists? Why have these elements seemed to them an index only of human weakness, not at all of the nature of scientific knowledge?

One answer to that question is, of course, that few philosophers, if any, have claimed to possess either a complete or an entirely well-articulated list of criteria. For some time, therefore, they could reasonably expect that further research would eliminate residual imperfections and produce an algorithm able to dictate rational, unanimous choice. Pending that achievement, scientists would have no alternative but to supply subjectively what the best current list of objective criteria still lacked. That some of them might still do so even with a perfected list at hand would then be an index only of the inevitable imperfection of human nature.

That sort of answer may still prove to be correct, but I think no philosopher still expects that it will. The search for algorithmic decision procedures has continued for some time and produced both powerful and illuminating results. But those results all presuppose that individual criteria of choice can be unambiguously stated and also that, if more than one proves relevant, an appropriate weight function is at hand for their joint application. Unfortunately, where the choice at issue is between scientific theories, little progress has been made toward the first of these desiderata and none toward the second. Most philosophers of science would, therefore, I think, now regard the sort of algorithm which has traditionally been sought as a not quite attainable ideal. I entirely agree and shall henceforth take that much for granted.

Even an ideal, however, if it is to remain credible, requires some demonstrated relevance to the situations in which it is supposed to apply. Claiming that such demonstration requires no recourse to subjective factors, my critics seem to appeal, implicitly or explicitly, to the well-known distinction between the contexts of discovery and of justification.[7]* They concede, that is, that the subjective factors I invoke play a significant role in the discovery or invention of new theories, but they also insist that that inevitably intuitive process lies outside of the bounds of philosophy of science and is irrelevant to the question of scientific objectivity. Objectivity enters science, they continue, through the processes by which theories are tested, justified, or judged. Those processes do not, or at least need not, involve subjective factors at all. They can be governed by a set of (objective) criteria shared by the entire group competent to judge.

I have already argued that that position does not fit observations of scientific life and shall now assume that that much has been conceded. What is now at issue is a different point: whether or not this invocation of the distinction between contexts of discovery and of justification provides even a plausible and useful idealization. I think it does not and can best make my point by suggesting first a likely source of its apparent cogency. I suspect that my critics have been misled by science pedagogy or what I have elsewhere called textbook science. In science teaching, theories are presented together with exemplary applications, and those applications may be viewed as evidence. But that is not their primary pedagogic function (science students are distressingly willing to receive the word from professors and texts). Doubtless *some* of them were *part* of the evidence at the time actual decisions were being made, but they represent only a fraction of the considerations relevant to the decision process. The context of pedagogy differs almost as much from the context of justification as it does from that of discovery.

---

* The "well-known" distinction to which Kuhn refers—between the contexts of discovery and justification—has been endorsed by many philosophers of science. The phrases *context of discovery* and *context of justification* were coined by Hans Reichenbach in *Experience and Prediction* (Chicago: University of Chicago Press, 1938), ch. 1. For further discussion, see "The Problem of Description," in the commentary on chapter 4 below.

Full documentation of that point would require longer argument than is appropriate here, but two aspects of the way in which philosophers ordinarily demonstrate the relevance of choice criteria are worth noting. Like the science textbooks on which they are often modelled, books and articles on the philosophy of science refer again and again to the famous crucial experiments:* Foucault's pendulum, which demonstrates the motion of the earth; Cavendish's demonstration of gravitational attraction; or Fizeau's measurement of the relative speed of sound in water and air. These experiments are paradigms of good reason for scientific choice; they illustrate the most effective of all the sorts of argument which could be available to a scientist uncertain which of two theories to follow; they are vehicles for the transmission of criteria of choice. But they also have another characteristic in common. By the time they were performed no scientist still needed to be convinced of the validity of the theory their outcome is now used to demonstrate. Those decisions had long since been made on the basis of significantly more equivocal evidence. The exemplary crucial experiments to which philosophers again and again refer would have been historically relevant to theory choice only if they had yielded unexpected results. Their use as illustrations provides needed economy to science pedagogy, but they scarcely illuminate the character of the choices that scientists are called upon to make.

Standard philosophical illustrations of scientific choice have another troublesome characteristic. The only arguments discussed are, as I have previously indicated, the ones favorable to the theory that, in fact, ultimately triumphed. Oxygen, we read, could explain weight relations, phlogiston could not; but nothing is said about the phlogiston theory's power or about the oxygen theory's limitations. Comparisons of Ptolemy's theory with Copernicus's proceed in the same way. Perhaps these examples should not be given since they contrast a developed theory with one still in its infancy. But philosophers regularly use them nonetheless. If the only result of their doing so were to simplify the decision situation, one could not object. Even historians do not claim to deal with the full factual complexity of the situations they describe. But these simplifications emasculate by making choice totally unproblematic. They eliminate, that is, one essential element of the decision situations that scientists must resolve if their field is to move ahead. In those situations there are always at least some good reasons for each possible choice. Considerations relevant to the context of discovery are then

---

* A crucial experiment is one that conclusively falsifies one of two rival theories or hypotheses, thereby establishing its rival as well confirmed or true. Thus, for example, Kuhn describes Foucault's pendulum as a crucial experiment because it conclusively refutes the hypothesis that the earth is stationary, thereby "demonstrating" the motion of the earth. For further discussion, see the section, "Why Crucial Experiments Are Impossible in Physics," in the commentary on chapter 3.

relevant to justification as well; scientists who share the concerns and sensi-
bilities of the individual who discovers a new theory are ipso facto likely to
appear disproportionately frequently among that theory's first supporters.
That is why it has been difficult to construct algorithms for theory choice,
and also why such difficulties have seemed so thoroughly worth resolving.
Choices that present problems are the ones philosophers of science need to
understand. Philosophically interesting decision procedures must function
where, in their absence, the decision might still be in doubt.

That much I have said before, if only briefly. Recently, however, I have
recognized another, subtler source for the apparent plausibility of my critics' position. To present it, I shall briefly describe a hypothetical dialogue
with one of them. Both of us agree that each scientist chooses between competing theories by deploying some Bayesian algorithm which permits him
to compute a value for $p(T,E)$, i.e., for the probability of a theory T on the
evidence E available both to him and to the other members of his professional group at a particular period of time. "Evidence," furthermore, we both
interpret broadly to include such considerations as simplicity and fruitfulness. My critic asserts, however, that there is only one such value of p, that
corresponding to objective choice, and he believes that all rational members of the group must arrive at it. I assert, on the other hand, for reasons
previously given, that the factors he calls objective are insufficient to determine in full any algorithm at all. For the sake of the discussion I have conceded that each individual has an algorithm and that all their algorithms
have much in common. Nevertheless, I continue to hold that the algorithms
of individuals are all ultimately different by virtue of the subjective considerations with which each must complete the objective criteria before any
computations can be done. If my hypothetical critic is liberal, he may now
grant that these subjective differences do play a role in determining the hypothetical algorithm on which each individual relies during the early stages of
the competition between rival theories. But he is also likely to claim that, as
evidence increases with the passage of time, the algorithms of different individuals converge to the algorithm of objective choice with which his presentation began. For him the increasing unanimity of individual choices is
evidence for their increasing objectivity and thus for the elimination of
subjective elements from the decision process.

So much for the dialogue, which I have, of course, contrived to disclose the non sequitur underlying an apparently plausible position. What
converges as the evidence changes over time need only be the values of p
that individuals compute from their individual algorithms. Conceivably
those algorithms themselves also become more alike with time, but the
ultimate unanimity of theory choice provides no evidence whatsoever that
they do so. If subjective factors are required to account for the decisions that
initially divide the profession, they may still be present later when the profession agrees. Though I shall not here argue the point, consideration of the

occasions on which a scientific community divides suggests that they actually do so.

My argument has so far been directed to two points. It first provided evidence that the choices scientists make between competing theories depend not only on shared criteria—those my critics call objective—but also on idiosyncratic factors dependent on individual biography and personality. The latter are, in my critics' vocabulary, subjective, and the second part of my argument has attempted to bar some likely ways of denying their philosophic import. Let me now shift to a more positive approach, returning briefly to the list of shared criteria—accuracy, simplicity, and the like— with which I began. The considerable effectiveness of such criteria does not, I now wish to suggest, depend on their being sufficiently articulated to dictate the choice of each individual who subscribes to them. Indeed, if they were articulated to that extent, a behavior mechanism fundamental to scientific advance would cease to function. What the tradition sees as eliminable imperfections in its rules of choice I take to be in part responses to the essential nature of science.

As so often, I begin with the obvious. Criteria that influence decisions without specifying what those decisions must be are familiar in many aspects of human life. Ordinarily, however, they are called, not criteria or rules, but maxims, norms, or values. Consider maxims first. The individual who invokes them when choice is urgent usually finds them frustratingly vague and often also in conflict one with another. Contrast "He who hesitates is lost" with "Look before you leap," or compare "Many hands make light work" with "Too many cooks spoil the broth." Individually maxims dictate different choices, collectively none at all. Yet no one suggests that supplying children with contradictory tags like these is irrelevant to their education. Opposing maxims alter the nature of the decision to be made, highlight the essential issues it presents, and point to those remaining aspects of the decision for which each individual must take responsibility himself. Once invoked, maxims like these alter the nature of the decision process and can thus change its outcome.

Values and norms provide even clearer examples of effective guidance in the presence of conflict and equivocation. Improving the quality of life is a value, and a car in every garage once followed from it as a norm. But quality of life has other aspects, and the old norm has become problematic. Or again, freedom of speech is a value, but so is preservation of life and property. In application, the two often conflict, so that judicial soul-searching, which still continues, has been required to prohibit such behavior as inciting to riot or shouting fire in a crowded theater. Difficulties like these are an appropriate source for frustration, but they rarely result in charges that values have no function or in calls for their abandonment. That response is barred to most of us by an acute consciousness that there are societies with other values and that these value differences result in other ways of life, other decisions about what may and what may not be done.

I am suggesting, of course, that the criteria of choice with which I began function not as rules, which determine choice, but as values, which influence it. Two men deeply committed to the same values may nevertheless, in particular situations, make different choices as, in fact, they do. But that difference in outcome ought not to suggest that the values scientists share are less than critically important either to their decisions or to the development of the enterprise in which they participate. Values like accuracy, consistency, and scope may prove ambiguous in application, both individually and collectively; they may, that is, be an insufficient basis for a *shared* algorithm of choice. But they do specify a great deal: what each scientist must consider in reaching a decision, what he may and may not consider relevant, and what he can legitimately be required to report as the basis for the choice he has made. Change the list, for example by adding social utility as a criterion, and some particular choices will be different, more like those one expects from an engineer. Subtract accuracy of fit to nature from the list, and the enterprise that results may not resemble science at all, but perhaps philosophy instead. Different creative disciplines are characterized, among other things, by different sets of shared values. If philosophy and engineering lie too close to the sciences, think of literature or the plastic arts. Milton's failure to set *Paradise Lost* in a Copernican universe does not indicate that he agreed with Ptolemy but that he had things other than science to do.

Recognizing that criteria of choice can function as values when incomplete as rules has, I think, a number of striking advantages. First, as I have already argued at length, it accounts in detail for aspects of scientific behavior which the tradition has seen as anomalous or even irrational. More important, it allows the standard criteria to function fully in the earliest stages of theory choice, the period when they are most needed but when, on the traditional view, they function badly or not at all. Copernicus was responding to them during the years required to convert heliocentric astronomy from a global conceptual scheme to mathematical machinery for predicting planetary position. Such predictions were what astronomers valued; in their absence, Copernicus would scarcely have been heard, something which had happened to the idea of a moving earth before. That his own version convinced very few is less important than his acknowledgment of the basis on which judgments would have to be reached if heliocentricism were to survive. Though idiosyncrasy must be invoked to explain why Kepler and Galileo were early converts to Copernicus's system, the gaps filled by their efforts to perfect it were specified by shared values alone.

That point has a corollary which may be more important still. Most newly suggested theories do not survive. Usually the difficulties that evoked them are accounted for by more traditional means. Even when this does not occur, much work, both theoretical and experimental, is ordinarily required before the new theory can display sufficient accuracy and scope to generate widespread conviction. In short, before the group accepts it, a new theory has been tested over time by the research of a number of men, some working within it, others within its traditional rival. Such a mode of development, however, *requires* a decision process which permits rational men to disagree, and such disagreement would be barred by the shared algorithm which philosophers have generally sought. If it were at hand, all conforming scientists would make the same decision at the same time. With standards for acceptance set too low, they would move from one attractive global viewpoint to another, never giving traditional theory an opportunity to supply equivalent attractions. With standards set higher, no one satisfying the criterion of rationality would be inclined to try out the new theory, to articulate it in ways which showed its fruitfulness or displayed its accuracy and scope. I doubt that science would survive the change. What from one viewpoint may seem the looseness and imperfection of choice criteria conceived as rules may, when the same criteria are seen as values, appear an indispensable means of spreading the risk which the introduction or support of novelty always entails.

Even those who have followed me this far will want to know how a value-based enterprise of the sort I have described can develop as a science does, repeatedly producing powerful new techniques for prediction and control. To that question, unfortunately, I have no answer at all, but that is only another way of saying that I make no claim to have solved the problem of induction. If science did progress by virtue of some shared and binding algorithm of choice, I would be equally at a loss to explain its success. The lacuna is one I feel acutely, but its presence does not differentiate my position from the tradition.

It is, after all, no accident that my list of the values guiding scientific choice is, as nearly as makes any difference, identical with the tradition's list of rules dictating choice. Given any concrete situation to which the philosopher's rules could be applied, my values would function like his rules, producing the same choice. Any justification of induction, any explanation of why the rules worked, would apply equally to my values. Now consider a situation in which choice by shared rules proves impossible, not because the rules are wrong but because they are, as rules, intrinsically incomplete. Individuals must then still choose and be guided by the rules (now values) when they do so. For that purpose, however, each must first flesh out the rules, and each will do so in a somewhat different way even though the decision dictated by the variously completed rules may prove unanimous. If I now assume, in addition, that the group is large enough so that individual differences distribute on some normal curve, then any argument that justifies the philosopher's choice by rule should be immediately adaptable to my choice by value. A group too small, or a distribution excessively skewed by external historical pressures, would, of course, prevent the argument's transfer.[8] But those are just the circumstances under which scientific progress is itself problematic. The transfer is not then to be expected.

I shall be glad if these references to a normal distribution of individual differences and to the problem of induction make my position appear very close to more traditional views. With respect to theory choice, I have never thought my departures large and have been correspondingly startled by such charges as "mob psychology," quoted at the start. It is worth noting, however, that the positions are not quite identical, and for that purpose an analogy may be helpful. Many properties of liquids and gases can be accounted for on the kinetic theory by supposing that all molecules travel at the same speed. Among such properties are the regularities known as Boyle's and Charles's law. Other characteristics, most obviously evaporation, cannot be explained in so simple a way. To deal with them one must assume that molecular speeds differ, that they are distributed at random, governed by the laws of chance. What I have been suggesting here is that theory choice, too, can be explained only in part by a theory which attributes the same properties to all the scientists who must do the choosing. Essential aspects of the process generally known as verification will be understood only by recourse to the features with respect to which men may differ while still remaining scientists. The tradition takes it for granted that such features are vital to the process of discovery, which it at once and for that reason rules out of philosophical bounds. That they may have significant functions also in the philosophically central problem of justifying theory choice is what philosophers of science have to date categorically denied.

What remains to be said can be grouped in a somewhat miscellaneous epilogue. For the sake of clarity and to avoid writing a book, I have throughout this paper utilized some traditional concepts and locutions about the viability of which I have elsewhere expressed serious doubts. For those who know the work in which I have done so, I close by indicating three aspects of what I have said which would better represent my views if cast in other terms, simultaneously indicating the main directions in which such recasting should proceed. The areas I have in mind are: value invariance, subjectivity, and partial communication. If my views of scientific development are novel—a matter about which there is legitimate room for doubt—it is in areas such as these, rather than theory choice, that my main departures from tradition should be sought.

Throughout this paper I have implicitly assumed that, whatever their initial source, the criteria or values deployed in theory choice are fixed once and for all, unaffected by their participation in transitions from one theory to another. Roughly speaking, but only very roughly, I take that to be the case. If the list of relevant values is kept short (I have mentioned five, not all independent) and if their specification is left vague, then such values as accuracy, scope, and fruitfulness are permanent attributes of science. But little knowledge of history is required to suggest that both the application of these values and, more obviously, the relative weights attached to them have varied markedly with time and also with the field of application. Furthermore, many of these variations in value have been associated with particular changes in scientific theory. Though the experience of scientists provides no philosophical justification for the values they deploy (such justification would solve the problem of induction), those values are in part learned from that experience, and they evolve with it.

The whole subject needs more study (historians have usually taken scientific values, though not scientific methods, for granted), but a few remarks will illustrate the sort of variations I have in mind. Accuracy, as a value, has with time increasingly denoted quantitative or numerical agreement, sometimes at the expense of qualitative. Before early modern times, however, accuracy in that sense was a criterion only for astronomy, the science of the celestial region. Elsewhere it was neither expected nor sought. During the seventeenth century, however, the criterion of numerical agreement was extended to mechanics, during the late eighteenth and early nineteenth centuries to chemistry and such other subjects as electricity and heat, and in this century to many parts of biology. Or think of utility, an item of value not on my initial list. It too has figured significantly in scientific development, but far more strongly and steadily for chemists than for, say, mathematicians and physicists. Or consider scope. It is still an important scientific value, but important scientific advances have repeatedly been achieved at its expense, and the weight attributed to it at times of choice has diminished correspondingly.

What may seem particularly troublesome about changes like these is, of course, that they ordinarily occur in the aftermath of a theory change. One of the objections to Lavoisier's new chemistry was the roadblocks with which it confronted the achievement of what had previously been one of chemistry's traditional goals: the explanation of qualities, such as color and texture, as well as of their changes. With the acceptance of Lavoisier's theory such explanations ceased for some time to be a value for chemists; the ability to explain qualitative variation was no longer a criterion relevant to the evaluation of chemical theory. Clearly, if such value changes had occurred as rapidly or been as complete as the theory changes to which they related, then theory choice would be value choice, and neither could provide justification for the other. But, historically, value change is ordinarily a belated and largely unconscious concomitant of theory choice, and the former's magnitude is regularly smaller than the latter's. For the functions I have here ascribed to values, such relative stability provides a sufficient basis. The existence of a feedback loop through which theory change affects the values which led to that change does not make the decision process circular in any damaging sense.

About a second respect in which my resort to tradition may be misleading. I must be far more tentative. It demands the skills of an ordinary language philosopher, which I do not possess. Still, no very acute ear for language is required to generate discomfort with the ways in which the terms "objectivity" and, more especially, "subjectivity" have functioned in this paper. Let me briefly suggest the respects in which I believe language has gone astray. "Subjective" is a term with several established uses: in one of these it is opposed to "objective," in another to "judgmental." When my critics describe the idiosyncratic features to which I appeal as subjective, they resort, erroneously I think, to the second of these senses. When they complain that I deprive science of objectivity, they conflate that second sense of subjective with the first.

A standard application of the term "subjective" is to matters of taste, and my critics appear to suppose that that is what I have made of theory choice. But they are missing a distinction standard since Kant when they do so. Like sensation reports, which are also subjective in the sense now at issue, matters of taste are undiscussable. Suppose that, leaving a movie theater with a friend after seeing a western, I exclaim: "How I liked that terrible potboiler!" My friend, if he disliked the film, may tell me I have low tastes, a matter about which, in these circumstances, I would readily agree. But, short of saying that I lied, he cannot disagree with my report that I liked the film or try to persuade me that what I said about my reaction was wrong. What is discussable in my remark is not my characterization of my internal state, my exemplification of taste, but rather my *judgment* that the film was a potboiler. Should my friend disagree on that point, we may argue most of the night, each comparing the film with good or great ones we have seen, each revealing, implicitly or explicitly, something about how he *judges* cinematic merit, about his aesthetic. Though one of us may, before retiring, have persuaded the other, he need not have done so to demonstrate that our difference is one of judgment, not taste.

Evaluations or choices of theory have, I think, exactly this character. Not that scientists never say merely, I like such and such a theory, or I do not. After 1926 Einstein said little more than that about his opposition to the quantum theory.* But scientists may always be asked to explain their choices, to exhibit the bases for their judgments. Such judgments are eminently discussable, and the man who refuses to discuss his own cannot expect to be taken seriously. Though there are, very occasionally, leaders of scientific

---

* Presumably Kuhn meant "1936," given that from 1927 to 1936 Einstein and Bohr carried on a debate about quantum mechanics that has been described as "one of the great intellectual disputes in the history of science." The debate culminated in the famous EPR paper of 1936 in which the authors (Einstein, Podolsky, and Rosen) argued that quantum mechanics could not give a complete description of reality. During this period, Einstein was an articulate and relentless critic of the so-called Copenhagen interpretation of quantum mechanics. See Arthur Fine, *The Shaky Game: Einstein, Realism, and the Quantum Theory* 2d ed. (Chicago, Ill.: University of Chicago Press, 1996), and Dugald Murdoch, *Niels Bohr's Philosophy of Physics* (Cambridge: University of Cambridge Press, 1987). The description quoted above is from page 155 of Murdoch's book.

taste, their existence tends to prove the rule. Einstein was one of the few, and his increasing isolation from the scientific community in later life shows how very limited a role taste alone can play in theory choice. Bohr, unlike Einstein, did discuss the bases for his judgment, and he carried the day. If my critics introduce the term "subjective" in a sense that opposes it to judgmental—thus suggesting that I make theory choice undiscussable, a matter of taste—they have seriously mistaken my position.

Turn now to the sense in which "subjectivity" is opposed to "objectivity," and note first that it raises issues quite separate from those just discussed. Whether my taste is low or refined, my report that I liked the film is objective unless I have lied. To my judgment that the film was a potboiler, however, the objective-subjective distinction does not apply at all, at least not obviously and directly. When my critics say I deprive theory choice of objectivity, they must, therefore, have recourse to some very different sense of subjective, presumably the one in which bias and personal likes or dislikes function instead of, or in the face of, the actual facts. But that sense of subjective does not fit the process I have been describing any better than the first. Where factors dependent on individual biography or personality must be introduced to make values applicable, no standards of factuality or actuality are being set aside. Conceivably my discussion of theory choice indicates some limitations of objectivity, but not by isolating elements properly called subjective. Nor am I even quite content with the notion that what I have been displaying are limitations. Objectivity ought to be analyzable in terms of criteria like accuracy and consistency. If these criteria do not supply all the guidance that we have customarily expected of them, then it may be the meaning rather than the limits of objectivity that my argument shows.

Turn, in conclusion, to a third respect, or set of respects, in which this paper needs to be recast. I have assumed throughout that the discussions surrounding theory choice are unproblematic, that the facts appealed to in such discussions are independent of theory, and that the discussions' outcome is appropriately called a choice. Elsewhere I have challenged all three of these assumptions, arguing that communication between proponents of different theories is inevitably partial, that what each takes to be facts depends in part on the theory he espouses, and that an individual's transfer of allegiance from theory to theory is often better described as conversion than as choice. Though all these theses are problematic as well as controversial, my commitment to them is undiminished. I shall not now defend them, but must at least attempt to indicate how what I have said here can be adjusted to conform with these more central aspects of my view of scientific development.

For that purpose I resort to an analogy I have developed in other places. Proponents of different theories are, I have claimed, like native speakers of different languages. Communication between them goes on by translation, and it raises all translation's familiar difficulties. That analogy is, of course, incomplete, for the vocabulary of the two theories may be identical, and most words function in the same ways in both. But some words in the basic as well as in the theoretical vocabularies of the two theories—words like "star" and "planet," "mixture" and "compound," or "force" and "matter"—do function differently. Those differences are unexpected and will be discovered and localized, if at all, only by repeated experience of communication breakdown. Without pursuing the matter further, I simply assert the existence of significant limits to what the proponents of different theories can communicate to one another. The same limits make it difficult or, more likely, impossible for an individual to hold both theories in mind together and compare them point by point with each other and with nature. That sort of comparison is, however, the process on which the appropriateness of any word like "choice" depends.

Nevertheless, despite the incompleteness of their communication, proponents of different theories can exhibit to each other, not always easily, the concrete technical results achievable by those who practice within each theory. Little or no translation is required to apply at least some value criteria to those results. (Accuracy and fruitfulness are most immediately applicable, perhaps followed by scope. Consistency and simplicity are far more problematic.) However incomprehensible the new theory may be to the proponents of tradition, the exhibit of impressive concrete results will persuade at least a few of them that they must discover how such results are achieved. For that purpose they must learn to translate, perhaps by treating already published papers as a Rosetta stone or, often more effective, by visiting the innovator, talking with him, watching him and his students at work. Those exposures may not result in the adoption of the theory; some advocates of the tradition may return home and attempt to adjust the old theory to produce equivalent results. But others, if the new theory is to survive, will find that at some point in the language-learning process they have ceased to translate and begun instead to speak the language like a native. No process quite like choice has occurred, but they are practicing the new theory nonetheless. Furthermore, the factors that have led them to risk the conversion they have undergone are just the ones this paper has underscored in discussing a somewhat different process, one which, following the philosophical tradition, it has labelled theory choice.

■ | Notes

1. *The Structure of Scientific Revolutions*, 2d ed. (Chicago, 1970), pp. 148,151–52, 159. All the passages from which these fragments are taken appeared in the same form in the first edition, published in 1962.

2. Ibid., p. 170.

3. Imre Lakatos, "Falsification and the Methodology of Scientific Research Programmes," in I. Lakatos and A. Musgrave, eds., *Criticism and the Growth of*

Knowledge (Cambridge, 1970), pp. 91–195. The quoted phrase, which appears on p. 178, is italicized in the original.

4. Dudley Shapere, "Meaning and Scientific Change," in R. G. Colodny, ed., *Mind and Cosmos: Essays in Contemporary Science and Philosophy,* University of Pittsburgh Series in the Philosophy of Science, vol. 3 (Pittsburgh, 1966), pp. 41–85. The quotation will be found on p. 67.

5. Israel Scheffler, *Science and Subjectivity* (Indianapolis, 1967), p. 81.

6. The last criterion, fruitfulness, deserves more emphasis than it has yet received. A scientist choosing between two theories ordinarily knows that his decision will have a bearing on his subsequent research career. Of course he is especially attracted by a theory that promises the concrete successes for which scientists are ordinarily rewarded.

7. The least equivocal example of this position is probably the one developed in Scheffler, *Science and Subjectivity,* chap. 4.

8. If the group is small, it is more likely that random fluctuations will result in its members' sharing an atypical set of values and therefore making choices different from those that would be made by a larger and more representative group. External environment—intellectual, ideological, or economic—must systematically affect the value system of much larger groups, and the consequences can include difficulties in introducing the scientific enterprise to societies with inimical values or perhaps even the end of that enterprise within societies where it had once flourished. In this area, however, great caution is required. Changes in the environment where science is practiced can also have fruitful effects on research. Historians often resort, for example, to differences between national environments to explain why particular innovations were initiated and at first disproportionately pursued in particular countries, e.g., Darwinism in Britain, energy conservation in Germany. At present we know substantially nothing about the minimum requisites of the social milieux within which a sciencelike enterprise might flourish.

Ernan McMullin

Rationality and Paradigm Change in Science

As we look back at the first responses of philosophers of science to Thomas Kuhn's classic *The Structure of Scientific Revolutions* [SSR], we are struck by their near unanimity toward the challenge that the book posed to the rationality of science. Kuhn's account of the paradigm changes that for him constituted scientific revolutions was taken by many to undermine the rationality of the scientific process itself. The metaphors of conversion and gestalt switch, the insistence that defenders of rival paradigms must inevitably fail to make contact with each other's viewpoints, struck those philosophical readers whose expectations were formed by later logical empiricism as a deliberate rejection of the basic requirements of effective reason giving in the natural sciences.

Kuhn responded to this reading of SSR in a lengthy Postscript to the second edition of his book in 1970 and in the reflective essay "Objectivity, Value Judgment, and Theory Choice" in 1977.[1] He labored to show that the implications of his new account of scientific change for the *rationality* of that change were far less radical than his critics were taking them to be. But his disavowals were not, in the main, taken as seriously as he had hoped they would be; the echoes of the rhetoric of SSR still lingered in people's minds. It seems worth returning to this ground, familiar though it may seem, in order to assess just what Kuhn *did* have to say about how paradigm change comes about in science. We will see that the radical thrust of his account of science was indeed not directed so much against the rationality of theory choice as against the epistemic, or truthlike, character of the theories so chosen.

From Paul Horwich, ed., *World Changes: Thomas Kuhn and the Nature of Science* (Cambridge, Mass.: MIT Press, 1993), 55–78.

1 | Good Reasons for Paradigm Change

The theme that recurs in Kuhn's discussions of paradigm change is a two-sided one. On one hand, he wanted to emphasize the fundamental role played by "good reasons" in motivating theory change in science. Notable among these is the perception of anomaly, the growing awareness that something is wrong, which makes it possible for alternatives to be seriously viewed *as* alternatives. On the other hand, these reasons are never coercive in their own right in forcing change; the reasons in favor of a new paradigm cannot *compel* assent. There is no precise point at which resistance to the change of paradigm becomes illogical.[2] Proponents of the new paradigm and defenders of the old one may each be able to lay claim to be acting "rationally"; the fact that neither side can persuade the other does not undermine the claim each can make to have good reasons for what they assert. "The point I have been trying to make," Kuhn says in the Postscript to SSR, "is a simple one, long familiar in philosophy of science. Debates over theory-choice cannot be cast in a form that fully resembles logical or mathematical proof. . . . Nothing about that relatively familiar thesis implies either that there are no good reasons for being persuaded or that those reasons are not ultimately decisive for the group. Nor does it even imply that the reasons for choice are different from those usually listed by philosophers of science: accuracy, simplicity, fruitfulness, and the like. What it should suggest, however, is that such reasons function as values and that they can thus be differently applied, individually and collectively, by men who concur in honoring them."[3]

It is with the implications of this thesis that I will be mainly concerned in this essay. The values a good theory is expected to embody enable comparisons to be made, even when the rival theories are incommensurable. Kuhn makes it clear that "incommensurable" for him does not imply "incomparable." SSR, he notes, "includes many explicit examples of comparisons between successive theories. I have never doubted either that they were possible or that they were essential at times of theory choice."[4] What he wanted to emphasize, he says, is that "successive theories are incommensurable (which is not the same as incomparable) in the sense that the referents of some of the terms which occur in both are a function of the theory within which those terms appear," and hence that there is no neutral language available for purposes of comparison. Nonetheless, translation is in principle possible.[5] But to translate another's theory is still not to make it one's own. "For that one must go native, discover that one is thinking and working in, not simply translating out of, a language that was previously foreign."[6] And that transition cannot simply be willed, he maintained, however strong the reasons for it may be. This is what enabled him to maintain his most characteristic claim, even after the qualifiers he inserted in the Postscript: "The conversion experience that I have likened to a gestalt switch remains, therefore, at the heart of the revolutionary process. Good reasons for choice provide motives for conversion and a climate in which it is more likely to occur. Translation may, in addition, provide points of entry for the neural reprogramming that, however inscrutable at this time, must underlie conversion. But neither good reasons nor translation constitute conversion, and it is that process we must explicate in order to understand an essential sort of scientific change."[7]

How is the transition to be explicated? Kuhn has only some hints to offer: "With respect to divergences of this sort, no set of choice criteria yet proposed is of any use. One can explain, as the historian characteristically does, why particular men made particular choices at particular times. But for that purpose one must go beyond the list of shared criteria to characteristics of the individuals who make the choice. One must, that is, deal with characteristics which vary from one scientist to another without thereby in the least jeopardizing their adherence to the canons that make science scientific."[8]

And he mentions such characteristics as previous experience as a scientist, philosophical views, personality differences. In the years since *SSR* appeared, sociologists of science have made much of these factors, often in ways that Kuhn himself would disavow. It was his stress on the role of these factors, he later remarked, that led critics to dub his views "subjectivist." They forgot his stress on the "shared criteria" that guide (but do not dictate) theory choice.[9] I will take him at his word here, assuming that the rationality of theory choice in his account rests on the persistence of these criteria that enable theories to be compared and evaluated, relatively to one another, even when they are incommensurable.

2 | How Deep Do Revolutions Go?

Here we immediately encounter a difficulty. Do these criteria persist? Can they bridge paradigm differences? How deep, in short, do revolutions go? There is an ambiguity in Kuhn's response to this question. In a celebrated paragraph in *SSR*, he describes paradigm change as follows: "Like the choice between competing political institutions, that between competing paradigms proves to be a choice between incompatible modes of community life. Because it has that character, the choice is not and cannot be determined merely by the valuative procedures characteristic of normal science, for these depend in part upon a particular paradigm, and that paradigm is at issue. When paradigms enter, as they must, into a debate about paradigm choice, their role is necessarily circular. Each group uses its own paradigm to argue in that paradigm's defense."[10]

Since the evaluative procedures depend on the paradigm, and the paradigm itself is in question, there can be no agreed-upon way to adjudicate the choice between rival paradigms. Though he goes on to say that the resulting circularity does not *necessarily* undercut the arguments used, he concludes that the status of such arguments can at best be only that of persuasion.

They "cannot be made logically or even probabilistically compelling for those who refuse to step into the circle. The premises and values shared by the two parties to a debate over paradigms are not sufficiently extensive for that."[11]

What prevents the rival parties from agreeing as to which paradigm is the better, then, is in part the fact that the norms in terms of which this debate could be carried on are themselves part of the paradigm, so that there is no neutral methodological ground, or at least not enough to enable agreement to be reached. How important is this sort of "circularity" to Kuhn's account of the inability of either side in a paradigm debate to muster an entirely cogent argument in its own behalf? If a circularity in regard to evaluative procedures were to hold in general in such cases, then scientific revolutions *would* indeed seem to be the irrational, or at least minimally rational, affairs that Kuhn's critics take him to be saying they are. One way to find out is to direct attention to the examples he gives of scientific revolutions and ask what paradigm change amounts to in each of these cases.

When the question is put in this way, it is clear that there is a striking difference in the depth of the different changes classified by Kuhn as "revolutions." At one end of the spectrum is the Copernican revolution, the charting of which led him to the writing of SSR in the first place. At the other end would be, for example, the discovery of X rays. Somewhere in the middle might come the discovery of the oxygen theory of combustion.[12] We have a choice in some cases, it would seem, between saying that only a small part of the paradigm changed and saying that an entire paradigm changed but that the "paradigm" in this case comprised only a fraction of the beliefs, procedures, and so forth, of the scientists involved.

Take the case of X rays. Kuhn insists that their discovery did accomplish a revolution in his sense. Yet he recognizes that at first sight this episode scarcely seems to qualify. After all, no fundamental change of theory occurred. No troublesome anomalies were noted in advance. There was no prior crisis to signal that a revolution might be at hand. Why then, he asks, can we not regard the discovery of X rays as a simple extension of the range of electromagnetic phenomena? Because, he responds, it "violated deeply entrenched expectations . . . implicit in the design and interpretation of established laboratory procedures."[13] The use of a particular apparatus "carries with it the assumption that only certain sorts of circumstances will arise." Roentgen's discovery "denied previously paradigmatic types of instrumentation their right to that title." That was sufficient, in his view, to constitute it a "revolution" in the sense in which he is proposing to use that term.

I will call this a shallow revolution because so much was left untouched by it. Electromagnetic theory was not replaced or even altered in any significant way. There were no challenges to accepted ways of assessing theory or to what counts as proper explanation. The textbooks, the sets of approved problem solutions, did not change much. What changed were the experimental procedures used in working with cathode-ray equipment and the expected outcomes of such work. And, of course, there were some important long-range implications for theory (as we now know). Such "revolutions" ought, it would seem, to be fairly frequent. Much would depend on how literally one should take the criteria Kuhn specifies as being the symptoms of impending revolution: previous awareness of anomaly and a resistance to a threatened change in procedures or categories.[14]

We are much more likely to think in terms of "revolution" in cases where one large-scale theory replaces another. Kuhn's favorite example is the replacement of phlogiston theory by the oxygen theory of combustion.[15] It meant a reformulation of the entire field of chemistry, a new conceptual framework, a new set of problems. Another example he gives of this sort of intermediate revolution, as we might call it, is the discovery of the Leyden jar and the resulting emergence of "the first full paradigm for electricity."[16] Prior to this discovery, Kuhn remarks, no single paradigm governed electrical research. A number of partial theories were applied, none of them entirely successful. The new conceptual framework enabled normal science to get under way, even though one-fluid and two-fluid theories were still in competition.

These changes involved the formulation of a new and more comprehensive theory. But they left more or less unchanged the epistemic principles governing the paradigm debate itself. Both sides would have agreed as to what counts as evidence, as to how claims should be tested. Or more accurately, to the extent that the scientists involved would have disagreed on these issues, their disagreements would not have been paradigm-dependent to any significant extent. So far as we can tell, Priestley and Lavoisier applied the same sorts of criteria to the assessment of theory, though they might not have attached the same weight to each criterion.

In Kuhn's favorite example of a scientific revolution, the Copernican one, this was, of course, not the case. This was a revolution of a much more fundamental sort because it involved a change in what counted as a good theory, in the procedures of justification themselves. It was not abrupt; indeed, it took a century and a half, from Copernicus's *De revolutionibus* to Newton's *Principia*, to consummate. And what made it revolutionary was not just the separation of Newtonian cosmology or Newtonian mechanics from their Aristotelian counterparts but the gradual transformation in the very idea of what constitutes valid evidence for a claim about the natural world, as well as in people's beliefs about how that world is ordered at the most fundamental level.[17]

It can thus be called a *deep revolution*, by contrast with the others described above. The Aristotelians and the Galileans totally disagreed as to how agreement itself should be brought about. So did the Cartesians and the Newtonians. The Galileans made use of idealization, of measurement, of mathematics, in ways the Aristotelians believed were illegitimate. The Newtonians allowed a form of explanation that the Cartesians were quite sure was improper. The shift in paradigm here meant a radical shift in the methodology of paradigm debate itself. Paradigm replacement means something much more thoroughgoing in such a case.

Have there been other deep revolutions in the more recent history of natural science? Newton's success means the success of a methodology which is still roughly the methodology of natural science today. Perhaps only one deep revolution was needed to get us to what Kuhn calls "mature" science. The two major revolutions in the physics of our own century did not run quite so deep. But they *did* involve principles of natural order, that is, shared assumptions as to what count as acceptable ways of articulating physical process at its most basic level. In the quantum revolution, what separated Bohr and Einstein was not just a difference in theoretical perspective but a disagreement as to what counted as good science and why. Quantum theory, in its Copenhagen interpretation, came much closer to a deep paradigm replacement than it would have done in Einstein's way of taking it.

In the Postscript to *SSR*, Kuhn addressed the ambiguity of the notion of paradigm and proposed a new label. A disciplinary matrix is the answer to the question, "What does [a community of specialists] share that accounts for the relative fullness of their professional communication and the relative unanimity of their professional judgments?"[18] Some of its principal components, he says, are symbolic generalizations, models of the underlying ontology of the field under investigation, concrete problem solutions, and the values governing theory appraisal.

It is clear, then, that for there to be a revolution in Kuhn's sense of the term this last component does not have to be at issue. Only in a deep revolution does one side challenge the other in regard to the appropriate methodology of theory assessment. When X rays were discovered, there was no dispute as to how their reality should be tested. When a Kuhnian revolution takes place, it is evidently not necessary that the entire paradigm should change. Only a part of the disciplinary matrix need be affected for there to be a sufficient change in worldview to qualify as "revolutionary." What 'revolutionary' means in practice is a change that falls outside the normal range of puzzle-solving techniques and whose resolution cannot, therefore, be brought about by the ordinary resources of the paradigm.

The implicit contrast is between puzzle solving, with its definitive ways of deciding whether a puzzle really *is* solved, and paradigm debate, where no such means of ready resolution exists. Whether so sharp a contrast is warranted by the actual practice of science may well be questioned. Decision between rival theories is an everyday affair in any active part of science. There may be an accepted general framework within which problems are formulated, but new data constantly pose challenges to older subtheories within that framework. This was the main issue dividing Kuhn and his Popperian critics in the late 1960s. It is clear in retrospect that there was merit on both sides of that dispute but that each was focusing on a particular aspect of scientific change to the exclusion of others.

The appraisal of rival theories within a paradigm is not a simple matter of puzzle solving. The history of high-energy physics over the past thirty years, for example, has seen one theory dispute after another. The notorious divisions at the moment among paleontologists about the causes of the Cretaceous extinction or between planetary physicists about the origin of the moon are only two of the more obvious reminders of the fact that deep-seated disagreement about the merits of alternative theories is a routine feature of science at its most "normal." As we have seen, Kuhn traced the roots of paradigm disagreement to two different sources: an "incommensurability" of a complex sort between two ways of looking at the world and a set of criteria for theory choice that function as values to be maximized rather than as an effective logic of decision. But this latter source of difference characterizes theory disputes generally and not just the more intractable ones that Kuhn terms paradigm disagreements. What we have here, I suspect, is a spectrum of different levels of intractability, not just a sharp dichotomy between revolutions and puzzle solutions. Nevertheless, Kuhn's dichotomy, though rather idealized, did serve to bring out in a forceful and dramatic way how complex, and how far from a simple matter of demonstration, the choice between theoretical alternatives ordinarily is.

3 | The Virtues of a Good Theory

What makes this choice a *rational* one for Kuhn, as we have seen, is the fact that scientists are guided by what they would regard as the virtues of a good theory. And there has been a certain constancy in that regard, according to him, across all but perhaps the deepest of revolutions: "I have implicitly assumed that, whatever their initial source, the criteria or values deployed in theory choice are fixed once and for all, unaffected by their transitions from one theory to another. Roughly speaking, but only roughly speaking, I take that to be the case. If the list of relevant values be kept short (I have mentioned five, not all independent) and if their specification is left vague, then such values as accuracy, scope, and fruitfulness are permanent attributes of science."[19]

This is a strong assertion indeed. Ironically, it is stronger than that now made by some of those who, like Laudan and Shapere, have chided Kuhn in the past for his subjectivism.[20] They argue that the values involved in theory choice are in no sense fixed; Shapere objects to any such claim as an objectionable form of essentialism. According to Laudan and Shapere, these values themselves change gradually as theories change or are replaced. They change for *reasons*, they insist, these reasons functioning as some sort of higher-level arbitration. But there is no limit in principle as to how *much* they might change over time. To put this in a more direct way, there is no constraint on how different the criteria of a good theory might be in the science of the far future from those we rely on today, unlikely though a radical shift might be.[21] In the original text of SSR, Kuhn proposed what sounds like a rather different view:

> [W]hen paradigms change, there are usually significant shifts in the criteria determining the legitimacy both of problems and of proposed solutions. . . . [This is] why the choice between competing paradigms regularly raises questions that cannot be resolved by the criteria of normal science. To the extent, as significant as it is incomplete, that two scientific schools disagree about what is a problem and what a solution, they will inevitably talk through each other when debating the relative merits of their respective paradigms. In the partially circular arguments that regularly result, each paradigm will be shown to satisfy more or less the criteria that it dictates for itself and to fall short of a few of those dictated by its opponent.[22]

The criteria governing theory choice are described here as strongly paradigm-dependent and thus as suffering "significant shifts" from one paradigm to the next. The resulting partial circularity in paradigm assessment leads rival scientists to "talk through each other." This was the theme, of course, that Paul Feyerabend picked up on. One can see how severely it limits the notion that there are "good reasons" for paradigm change. Here, then, is a clear instance of how Kuhn's later construals soften the radical overtones of the earlier work.

Kuhn does not hesitate to speak of the values involved in theory appraisal as "permanent attributes of science." He allows that the manner in which these values are understood and the relative weights attached to them have changed in the past and may change again in the future. But he wants to emphasize that these changes at the metalevel tend to be slower and smaller in scale than the changes that can occur at the level of theory:

> [I]f such value changes had occurred as rapidly or been as complete as the theory changes to which they related, then theory choice would be value choice, and neither could provide justification for the other. But, historically, value change is ordinarily a belated and largely unconscious concomitant of theory choice, and the former's magnitude is regularly smaller than the latter's. For the functions I have here ascribed to values, such relative stability provides a sufficient basis. The existence of a feedback loop through which theory change affects the values which led to that change does not make the decision process circular in any damaging sense.[23]

One would need, however, to know just how and why changes in theory bring about changes at the metalevel of theory assessment in order to judge how large these latter changes might become without undermining the claim that a rational choice is being made. Is the "relative stability" of the criteria governing theory choice a contingent historical finding, or is it a necessary feature of any activity claiming the title of science? There are suggestions of both views in the passage I have just quoted. Historically, these values have in fact been stable, Kuhn remarks. But he adds that if they were not, if one had to choose the criteria of choice themselves in the act of choosing between theories, there would be no fulcrum. The process would lack justification; it would be circular in a way that would be damaging to its claim to qualify as science.

The presumption appears to be that *really* deep revolutions do not occur, that is, revolutions where there is *no* sharing of epistemic values between one paradigm and the other. Kuhn allows that large-scale theory change may involve smaller-scale changes in the values believed to be appropriate to theory appraisal. In such cases, adoption of the new paradigm carries with it adoption of a somewhat different "rationality" at the metalevel. The advantages of the new theory are so marked, in terms of a minimal level of shared values, that a shift in the values themselves is ultimately taken to be warranted. This, it can be argued, is what happened in the seventeenth century as the balance shifted between Aristotelians and Galileans. Galileo set out to undermine Aristotle's physics in its *own* terms first and then to present an alternative that, in terms of consistency, empirical adequacy, and future potential, could claim a definite advantage, even in terms of criteria the Aristotelian might be brought to admit. That, at any rate, would be the grounds, in Kuhn's perspective, for regarding the Scientific Revolution as a "rational" shift in the way in which natural science was carried on.

In a recent essay Kuhn argues that we learn to use the term 'science' in conjunction with a cluster of other terms like 'art', 'medicine', 'philosophy'. To know what science is, is to know how it relates to these other activities.[24] Identifying an activity as scientific is to single out "such dimensions as accuracy, beauty, predictive power, normativeness, generality, and so on. Though a given sample of activity can be referred to under many descriptions, only those cast in this vocabulary of disciplinary characteristics permit its identification as, say, science; for that vocabulary alone can locate the activity close to other scientific disciplines and at a distance from disciplines other than science. That position, in turn, is a necessary property of all referents of the modern term, 'science.'"[25]

He immediately qualifies this last very strong claim by noting that not every activity that qualifies as "scientific" need be predictive, not all need be experimental, and so forth. And there is no sharp line of demarcation between science and nonscience. Nonetheless, there is a well-defined cluster of values whose pursuit marks off scientific from other activities in a relatively unambiguous way and that gives the term 'science' the position it occupies in the "semantic field." This marking off is not a mere matter of convention. The taxonomy of disciplines has developed in an empirical way; a real learning has taken place. If someone were to deny the rationality of learning from experience, we would not know what he or she is trying to say. One cannot, he maintains, further *justify* the norms for rational theory choice. He cites C.G. Hempel to the effect that this inability is a testimony to our continuing failure to solve the classical problem of induction.[26]

Kuhn rests his case, then, both for the rationality of science and for its distinctiveness as a human activity mainly on the values governing theory choice in science. But he does not chronicle their history, disentangle them from one another except in a cursory way, or inquire in any detail into how and why they have changed in the ways they have. Many of these variations, he remarks, "have been associated with particular changes in scientific theory. Though the experience of scientists provides no philosophical justification for the values they deploy (such justification would solve the problem of induction), those values are in part learned from that experience, and they evolve with it."[27]

But what justification other than the experience of scientists is *needed* to justify the values they deploy? Kuhn has, I suspect, altogether too lofty a view of what "philosophical" justification might amount to. And he has too readily allowed himself to be intimidated by that most dire of philosophers' threats: "That *can't* be right: if it were, it would solve the problem of induction." My own guess is that attention to the role of values in theory appraisal might well dissolve the problem Hume bequeathed us about the grounds for inductive inference. But whether that be true or not, the criteria employed by scientists in theory evaluation enjoy whatever sanction is appropriate to something learned in, and tested by, experience.

4 | How Might Epistemic Values Be Validated?

Suppose a scientist were to doubt whether a particular value, say simplicity, is really a desideratum in a practical situation of theory choice facing him or her. The rationality of the choice depends, presumably, on what sort of answer can be given to this kind of question. Two different sorts of answers suggest themselves. One is to look at the track record and decide how good a guide simplicity has proved to be in the past. (There are obvious problems about how the criterion itself is to be understood, but I will bracket these for the moment.) A quite different sort of response would be that simplicity is clearly a desideratum of theory because____, where we fill the blank with a reason why on the face of it, a simple theory is more likely to be a good theory (if indeed one *can* find a convincing reason). Both of these responses would, of course, need further clarification before they could begin to carry any conviction.

First, what does it mean to ask how good a guide simplicity has been in the past? Guide to what? Some kind of ordering of means and ends is clearly needed here. Some of the values we have been talking about seem to function as goals of the scientific enterprise itself: predictive accuracy (empirical adequacy) and explanatory power are the most obvious candidates. One can trace each of these goals back a very long way in human history. In some sense, they may be as old as humanity itself. The story of how they developed in the ancient world, how the skills of prediction came to be prized in many domains, how explanatory accounts of natural process came to be constructed, is a familiar one. Less familiar is the realization that these goals were not linked together in any organic way at the beginning. Indeed, they were long considered antithetical in the domain of astronomy, the most highly developed part of the knowledge of nature in early times. One of the consequences, perhaps the most important consequence, of the Copernican revolution was to show that they *are* compatible, that they can be successfully blended. This was an empirical discovery about the sort of universe we live in. It was something we *learned* and that now we *know*.

Each of these goals has come to be considered valuable in its own right, an end in itself.[28] An activity that gives us accurate knowledge of the world we live in and consequently power over its processes can come to seem worthwhile for all sorts of reasons. An activity that allows us to understand natural process, that allows our imaginations to reach out to realms inaccessible to our senses, holds immediate attraction. What it is to understand will, of course, shift as the principles of natural order themselves shift. So this goal of explaining lacks the definiteness of the goal of predicting; as theory changes, so will the contours of what counts as explaining.

Much more would have to be said about all this, but I am going to press on to make my main point.[29] Other epistemic values serve as *means* to these ends; they help to identify theories more likely to predict well or to explain. Some of these are quite general and would apply to any epistemic activity. Logical consistency (absence of contradiction) and compatibility with other accepted knowledge claims would be among these. They are obviously not goals in themselves; they would not motivate us to carry on an activity in the first place. But we have found that these values are worth taking seriously as *means*. Or should I say, it has always been obvious that we must not neglect them, if it is knowledge we are seeking?

Other values are more specific to science, for example, fertility, unifying power, and coherence (i.e., absence of ad hoc features). Once again, these are clearly not primary goals. They are not so much deliberately aimed at as esteemed when present. And they are esteemed not in themselves but because they have proved to be the marks of a "good" theory, a theory that will serve well in prediction and explanation. A long story could be told about this, beginning with Kepler, Boyle, and Huygens and working through Herschel, Whewell, and a legion of others who have drawn attention to the significance of these three virtues.

Once again, the story is an ambiguous one: it can be told in two quite different ways. According to one way of telling it, these values can be shown to have played a positive historical role in theory choice; we have gradually learned to trust them as clues. According to the other, a series of acute thinkers (some of the most prominent of them listed above) have realized that these values *ought* to serve as indicators of a good theory. These are what one would *expect* a priori from a theory that purported to predict accurately and explain correctly. When Kepler and Boyle drew attention to the importance of such criteria, it was not to point to their efficacy in the earlier history of natural philosophy but to recommend them on general epistemic grounds.[30]

The question of how to validate the values that customarily guide scientific theory choice can now be addressed more directly. The goals of predictive accuracy (empirical adequacy) and explanatory power serve to define the activity of science itself, in part at least. If, as Kuhn notes, one relinquishes the goal of producing an accurate account of natural regularity, the activity one is engaged in may be worthwhile, but it is not science.[31] The notion of epistemic justification does not directly apply to the goals themselves. One might ask, of course, whether the pursuit of these goals is justifiable on *moral* grounds. Or one might ask, as a means of determining whether effort expended on them is worthwhile, whether the goals are in fact attainable. We have learned that in general they *are* attainable. This is something one could not have known a priori. And we have learned much about the *methods* that have to be followed for theory construction to get under way, methods of experiment, of conceptual idealization, of mathematical formulation, and the rest. All of this had to be *learned*, and no doubt there is still much to discover in this regard.

The other values, being instrumental, are justified when it is shown that they serve as means to the ends defined by the primary goals. And this, as we have seen, can be done in two ways: by an appeal to what we have learned from the actual practice of science or by an analysis in epistemological terms of the aims of theory and what, in consequence, the marks of a good theory should be. Ideally, both ways need to be followed, each serving as check for the other. The appeal to historical practice works not so much as a testimony to what values have actually guided scientists in their theory choices but as a finding that reliance on certain values has *in fact* served the primary goals of science. Might it cease to?

This is the Humean echo that seems to worry Kuhn so much. One might respond, as he does, that learning from experience is part of what it is to be rational. We cannot *demonstrate* that experience will continue to serve as a reliable guide. But demonstration is not what is called for. Kuhn has done more than anyone else, perhaps, to show that rational theory choice does not require the cogency of demonstration. We know that the predictive powers of natural science have enormously increased, and we know something of the theory characteristics that have served to promote this expansion. No future development could, so far as I can see, lead us to deny these knowledge claims, which rest not just on a perception of past regularities but on an understanding, partial at least, of why these regularities took the course they did. We can, and almost surely will, learn more about what to look for in a good theory. But no further evidence seems to be needed to show that coherence in a theory is a value to be sought, so that, other things being equal, a more coherent theory is to be preferred to a less coherent one.

5 | Rationality without Realism?

Over the years since SSR appeared, Kuhn has, as we have seen, become more and more explicit about the basic rationality that underlies theory choice in science. It is a complex rationality with many components, allowing much latitude for difference among the defenders of different theories. But it has remained relatively invariant since the deep revolution that brought it into clear focus in the seventeenth century. One might almost speak of a *convergence* here. Kuhn clearly believes that scientists have a pretty good grip on the values that *ought* to guide the appraisal of rival theories, and that this grip has improved as it has been tested against a wider and wider variety of circumstances.

But he has not softened his stance in regard to the truth character of theories in the least. In a well-known passage in the Postscript, he insists that the only sort of progress that science exhibits is in puzzle solving: later theories solve more puzzles than earlier ones, or (to put this in a different idiom) they predict better. But there is, he insists, "no coherent direction of ontological development"; there is no reason to think that successive theories approximate more and more closely to the truth."[32] "The notion of a match between the ontology of a theory and its 'real' counterpart in nature now seems to me illusive in principle."[33] Kuhn thus rejects in a most emphatic way the traditional realist view that the explanatory success of a theory gives reason to believe that entities like those postulated by the theory exist, i.e., that the theory is at least approximately true.

He does not argue for this position in SSR, aside from a remark about Einstein's physics being closer in some respects to Aristotle's than to Newton's. But it is clear what the grounds for it are in his mind: the incommensurability of successive paradigms implies a discontinuity between their ontologies. By separating the issues of comparability and commensurability, he believes he can retain a more or less traditional view in regard to the former while adopting an instrumentalist one in regard to the latter. The radical challenge of SSR is directed not at rationality but at realism. The implications of the familiar Kuhnian themes of holism and paradigm replacement are now seen to be more significant for the debate about realism than for the issue of scientific rationality, on which they had so great an initial impact.

Kuhn's influence on the burgeoning antirealism of the last two decades can scarcely be overestimated. His views on theory change, on problems about the continuity of reference, are reflected in the work of such notable critics of realism as Arthur Fine, Bas van Fraassen, and especially Larry

Laudan.[34] Kuhn's own emphasis on science as a puzzle-solving enterprise would lead one to interpret him in an instrumentalist manner. At this point I am obviously not going to open a full-scale debate on realism versus instrumentalism.[35] But I would like to pull out one thread from that notorious tangle. Kuhn's way of securing scientific rationality by focusing on the values proper to theory choice might well have led him (I argue) to a more sympathetic appreciation of realism. I am not saying that rationality and realism are all of a piece, that to defend one is to commit oneself to the other. Most of the current critics of realism would be emphatic in their defense of the overall rationality of scientific change. But a closer study of the values to which Kuhn so effectively drew attention should, to my mind, raise a serious question about the adequacy of an instrumentalist construal of the puzzle-solving metaphor. If such a construal is adopted, it is hard to make sense of those many episodes in the history of science where values other than mere predictive accuracy played a decisive role in the choice between theories.

To show this, I will focus on a case history from Kuhn's own earlier work, *The Copernican Revolution*. At issue are the relative merits of the Ptolemaic and the Copernican systems prior to Galileo's work. Kuhn points out that there was little to choose between the two on the score of predictive accuracy. "Judged on purely practical grounds," he concludes, "the Copernican system was a failure; it was neither more accurate nor significantly simpler than its Ptolemaic predecessors."[36] Yet it persuaded some of the best astronomers of the time. And it was they who ultimately produced the "simple and accurate" account that carried the day. How *did* it persuade them? In Kuhn's view, "The real appeal of sun-centered astronomy was aesthetic rather than pragmatic. To astronomers the initial choice between Copernicus' system and Ptolemy's could only be a matter of taste, and matters of taste are the most difficult of all to define or debate."[37]

But such matters cannot be regarded as unimportant, he goes on, as the success of the Copernican Revolution itself testifies. Whatever it was that persuaded so many of those most skilled in astronomy to make what we would now regard as the right step obviously must be looked at with care. Those who were equipped "to discern geometric harmonies" obviously found "a new neatness and harmony" in the heliocentric system. What Copernicus offered was "a new and aesthetic harmony" that somehow carried conviction in the right quarters.

But now let us see how Copernicus's own argument went, in the crucial chapter 10 of book 1 of *De revolutionibus*. He points to two different sorts of clues. First, the heliocentric model allows one to specify the order of the planets outward from the central body in an unequivocal way, which Ptolemy's model could not do. Furthermore, the Copernican model has the planetary periods increase as one moves outward from the sun, just as one would expect. What Copernicus claims to discover in the new way of ordering the planets is a "clear bond of harmony," "an admirable symmetry." But why should this carry conviction, especially since (as Kuhn emphasizes)

Copernicus in the end had to retain an inelegant and far from harmonious-seeming tangle of epicycles?

He had stronger arguments. The heliocentric model could *explain*, that is, provide the *cause* of, a whole series of features of the planetary motions that Ptolemy simply had to postulate as given, as inexplicable in their own right. For example, even in ancient times it had been suggested that Venus and Mercury appear to have the sun as their center of rotation, since, unlike the other planets, they accompany the sun in its motion across our sky. Or again, it had long been noted that the superior planets (Mars, Jupiter, Saturn) are at their brightest when in opposition (rising together in the evening or setting together in the morning). Assuming that brightness is a measure of relative distance, this is explained if we are viewing the planetary motions from a body that itself is orbiting the sun as center. This "proves," Copernicus somewhat optimistically concludes, that the center of motion of the superior planets is the same as that of the inferior planets, namely the sun.

Kuhn comments that it does "not actually prove a thing. The Ptolemaic system explains these phenomena as completely as the Copernican," although the latter can be said to be "more natural."[38] Here I must disagree. The Ptolemaic system does not *explain* the phenomena mentioned above at all. Ptolemy is forced to postulate that the center of the epicycle for both Venus and Mercury always lies on the line joining the earth and sun. Kuhn says that in this way Ptolemy "accounts for" this feature of their motions. But this is surely not *accounting for* in the sense of explaining. Kuhn evidently equates prediction and explanation in these passages, not an unusual assumption at the time his book was written.

But he allows that Copernicus gives a "far more natural" account than does Ptolemy. Why? And what does 'natural' mean in the lexicon of an instrumentalist? Ptolemy's restriction on the deferent radii swept out by Venus and Mercury "is an 'extra' device, an *ad hoc* addition,"[39] one that Copernicus can discard. Kuhn is surely on the right track here. But this is *not* an aesthetic argument, an appeal to taste. Copernicus himself makes the genre to which it belongs quite clear. He says that [he] is able to assign the cause of these features of the planetary motions, whereas Ptolemy is not. There is no reason in Ptolemy's system for them, other than the mere need to get the predictions right. They are, as Kuhn himself says, ad hoc.

Copernicus gives another set of arguments based on the retrograde motions. Their relative size and frequency from one planet to another and the lack of any such motions on the part of the sun and moon are exactly what one would be led to expect in a system where we are observing the motions from the third planet and the moon is not a true planet but a satellite of earth. Later, in the *Mysterium cosmographicum*, Kepler developed these arguments more fully and added some of his own, for example, the striking fact that in the Ptolemaic model, the period of rotation for each planet on either the deferent or the epicycle circle is exactly one year, something which seemed like an extraordinary piece of adjustment, especially since Ptolemy took the planets to be dynamically independent of one another. Kepler is clear that the issue here is one of causal explanation; one of the systems can provide such an explanation, the other cannot. He is also clear that the criterion of prediction alone will not be enough to decide in all cases between two rival accounts of the planetary motions and thus that a different genre of argument (he calls it "physical") is needed.[40] This he urged as a refutation of the instrumentalism of his opponent, Ursus.

The competition may have been neutral between Ptolemy and Copernicus where *prediction* of planetary motions was concerned, but the two systems were quite unequal as *explanation*. No better illustration could be found of the distinction between these two concepts, and of the consequent importance of criteria of theory appraisal other than that of predictive or descriptive accuracy. Copernicus's criterion of "naturalness," the elimination of ad hoc features, the virtue that might today be called coherence, is not aesthetic; it is epistemic. He is not just appealing to his reader's taste, or sense of elegance. He is not assuming that the simpler, the more beautiful, models are more likely to be true. He is saying that a theory that makes causal sense of a whole series of features of the planetary motions is more likely to be true than one that leaves these features unexplained.

Copernicus and those who followed him believed that they had good arguments for the reality of the earth's motion around the sun. They sometimes overstated the force of those arguments, to be sure, using terms like 'proof' and 'demonstration'. The natural philosophers of the day were not yet accustomed to the weaker notions of likelihood and probability. Galileo found, to his cost, that he had to speak in terms of demonstration if his claims for the Copernican system were to be taken seriously. He did not have a demonstration, but from our perspective, he called effectively on the criterion of coherence in his critique of the geostatic alternative, just as Copernicus had earlier done.

As we look back on those debates, we are ready to allow that the coherence arguments of Copernicus and Galileo *did* carry force, that they *did* give a motive for accepting the new heliocentric model as true. And their force came from something other than predictive advantage. Kuhn's point in regard to theory assessment, one that became clearer in his successive formulations of it, was that the different theory values were not reducible to one another, and hence that no simple algorithm, no logic of confirmation such as the logical positivists had sought, underlay real-life theory decision. What I have tried to do here is to carry this insight further and to note the special epistemic weight carried by certain of these values. Besides coherence, one could make similar cases for fertility and unifying power. It is hard to make sense of the role played by these values if one adopts the instrumentalist standpoint that Kuhn feels compelled to advocate.

The case for scientific realism rests in large part on these "superempirical" values. That is, when we ask about a particular theory, how likely is it that it is true (correlatively, how likely is it that something like the explanatory entities it postulates actually exist), it is to these virtues that we are inclined to turn. To say that a theory simply "saves the phenomena," though this carries *some* epistemic weight, leaves open the suspicion of its being ad hoc. If a theory be thought of simply as an hypothetico-deductive device, it would seem plausible to suppose that other devices might account as well or better for the phenomena to be explained. It is only when the *temporal* dimension is added, when a theory is evaluated in a historical context, when its success in unifying domains over time or in predicting new sorts of phenomena are taken into account, that conviction begins to emerge. Theories are not assessed simply as predictors; they are not confirmed purely by the enumeration of consequences.

My conclusion is that the diversity of the expectations scientists hold up for their theories argues not only for the tentative character of theory choice, Kuhn's original point, but also for its properly epistemic character. This leaves us, of course, with a problem: how can the difficulties in regard to incommensurability be reconciled with the epistemic force of such arguments as that of Copernicus? Kuhn emphasized the discontinuities of language across theory change so strongly that he left no room for the possibility of convergence, for the possibility that the theories of the paleontologists of today, for example, not only solve more puzzles than those of yesteryear but also tell us, with high degree of likelihood, what actually happened at distant epochs in the earth's past.

The Kuhnian heritage is thus a curiously divided one. Kuhn wanted to maintain the rational character of theory choice in science while denying the epistemic character of the theory chosen. The consequent tensions are, of course, familiar to every reader of current philosophy of science. Thirty years later, *The Structure of Scientific Revolutions* still leaves us with an agenda.

■ | Notes

1. [Hereafter "Objectivity," from] *The Essential Tension (ET)*, pp. 320–339 [reprinted in this volume, pp. 94–110]. In his effort to ward off the charge of subjectivism, Kuhn might also have pointed to "The Function of Measurement in Modern Physical Science" (*Isis* 52 [1961]: 161–190; reprinted in *ET*, pp. 178–224), which appeared before SSR and whose theme was that "measurement can be an immensely powerful weapon in the battle between two theories" (*ET*, p. 211), that "the comparison of numerical predictions . . . has proved particularly successful in bringing scientific controversies to a close" (*ET*, p. 213). Or he could have recalled an even earlier paper, "The Essential Tension" (*The Third University of Utah Research Conference on the Identification of Scientific Talent*, ed. C. W. Taylor [Salt Lake City: University of Utah Press, 1959], 162–174; reprinted in *ET*, pp. 225–239), whose title referred to the opposition between the themes of tradition and innovation in science and which argued that it is the very effort to work within a tightly construed tradition that leads eventually to the recognition of anomalies that in

turn prepares the way for revolution (*ET*, p. 234). One further paper that Kuhn might have called on was "A Function for Thought Experiments" (*L'Aventure de la science, Mélanges Alexandre Koyré* [Paris: Hermann, 1964], vol. 2, pp. 307–334; reprinted in *ET*, pp. 240–265), which describes how failures of expectation induce the crisis that is the usual prelude to paradigm change (*ET*, p. 263).

2. "Objectivity," p. 320 [p. 94].

3. *The Structure of Scientific Revolutions*, 2nd ed. (Chicago: University of Chicago Press, 1970), p. 199.

4. "Metaphor in Science," in *Metaphor and Thought*, ed. Andrew Ortony (Cambridge: Cambridge University Press, 1979), 409–419; see p. 416.

5. In a recent essay Kuhn distinguishes between translation and interpretation and shows how communication can occur even where languages are incommensurable ("Commensurability, Comparability, Communicability," *PSA 1982* [Philosophy of Science Association], 1983: 669–688). In a comment Philip Kitcher remarks that Kuhn, in his later readings of SSR, has progressively weakened the dramatic doctrine of the original work in ways, be it said, of which Kitcher approves ("Implications of Incommensurability," *PSA 1982*, 1983: 689–703).

6. SSR, p. 204.

7. SSR, p. 204.

8. "Objectivity," p. 324 [p. 97].

9. "Objectivity," p. 325 [p. 98].

10. SSR, p. 94 [p. 81].

11. SSR, p. 94 [p. 81].

12. In SSR Kuhn himself distinguishes between "major paradigm changes, like those attributable to Copernicus and Lavoisier," and "the far smaller ones associated with the assimilation of a new sort of phenomenon, like oxygen or X-rays" (p. 92) [p. 79].

13. SSR, p. 58.

14. SSR, p. 62. It is not clear to me that the discovery of X rays satisfies either of these criteria in any other than a minimal way.

15. SSR, p. 199.

16. SSR, p. 62. There might be some question as to whether, in fact, a single theory of electricity did emerge at this time. But that is not to the point of my inquiry.

17. I have worked out this theme in some detail in my "Conceptions of Science in the Scientific Revolution," in *Reappraisals of the Scientific Revolution*, ed. David Lindberg and Robert Westman (Cambridge: Cambridge University Press, 1990).

18. SSR, p. 182.

19. "Objectivity," p. 335 [p. 105].

20. See Larry Laudan, *Science and Values* (Berkeley: University of California Press, 1984) [part of Ch. 4 of *Science and Values* is reprinted on pp. 131–143, this volume]; Dudley Shapere, *Reason and the Search for Knowledge* (Dordrecht: Reidel, 1984). I have discussed the ironies of this particular divergence more fully in "The

Shaping of Scientific Rationality," in *Construction and Constraint*, ed. E. McMullin (Notre Dame: University of Notre Dame Press, 1988), pp. 1–47.

21. Nicholas Rescher defends a somewhat similar position in regard to how different from ours the "science" carried on by the inhabitants of a distant planet might be: "Science *as we have it*—the only 'science' that we ourselves know—is a specifically human artifact that must be expected to reflect in significant degree the particular characteristics of its makers. Consequently, the prospect that an alien 'science'-possessing civilization has a *science* that we could acknowledge (if sufficiently informed) as representing the same general line of inquiry as that in which we ourselves are engaged seems extremely implausible" ("Extraterrestrial Science," *Philosophia Naturalis* 21 [1984]: 400–424; see p. 413).

22. *SSR*, pp. 109–110 [p. 92].

23. "Objectivity," p. 336 [p. 106].

24. "Rationality and Theory Choice," *Journal of Philosophy* 80 (1983): 563–570; see p. 567.

25. Ibid., p. 568.

26. C. G. Hempel, "Valuation and Objectivity in Science," in *Physics, Philosophy, and Psychoanalysis*, ed. R. S. Cohen and L. Laudan (Dordrecht: Reidel, 1983), 73–100.

27. "Objectivity," p. 335 [p. 106].

28. See my "Values in Science," *PSA* 1982, 1983: 3–25.

29. The story sketched so lightly here is told in much more detail in my "Goals of Natural Science," *Proceedings of the American Philosophical Association* 58 (1984): 37–64.

30. For a fuller historical treatment, see my "Conceptions of Science in the Scientific Revolution."

31. "Rationality and Theory Choice," p. 569.

32. *SSR*, p. 206.

33. *SSR*, p. 206.

34. Laudan's much-quoted essay, "A Confutation of Convergent Realism," in *Scientific Realism*, ed. J. Leplin (Berkeley: University of California Press, 1984), 218–249 [an edited version is reprinted on pp. 1108–28, this volume], presents in detail the sort of arguments that Kuhn would need to support his own rejection of convergence.

35. See my "Case for Scientific Realism," in *Scientific Realism*, ed. J. Leplin, 8–40, and "Selective Anti-realism," *Philosophical Studies* 61 (1991): 97–108.

36. *The Copernican Revolution* (New York: Random House, 1957), p. 171.

37. Ibid., p. 172.

38. Ibid., p. 178.

39. Ibid., p. 172.

40. Kepler's clearest treatment of this issue will be found in the *Apologia Tychonis contra Ursum* (1600). See Nicholas Jardine's translation of this work in *The Birth of*

History and Philosophy of Science (Cambridge: Cambridge University Press, 1984). Michael Gardner extracts a "Kepler principle" to the effect that it counts in favor of the realistic acceptance of a theory if it explains facts that competing theories merely postulate. See "Realism and Instrumentalism in Pre-Newtonian Astronomy," in *Testing Scientific Theories*, ed. John Earman (Minneapolis: University of Minnesota Press, 1983), 201–265; p. 256.

LARRY LAUDAN

Kuhn's Critique of Methodology

Several writers (e.g., Quine, Hesse, Goodman) have asserted that the rules or principles of scientific appraisal underdetermine theory choice. For reasons I have tried to spell out elsewhere,[1] such a view is badly flawed. Some authors, for instance, tend to confuse the logical underdetermination of theories by data with the underdetermination of theory choice by methodological rules. Others (e.g., Hesse and Bloor) have mistakenly taken the logical underdetermination of theories to be a license for asserting the causal underdetermination of our theoretical beliefs by the sensory evidence to which we are exposed.[2] But there is a weaker, and much more interesting, version of the thesis of underdetermination, which has been developed most fully in Kuhn's recent writings. Indeed, it is one of the strengths of Kuhn's challenge to traditional philosophy of science that he has "localized" and given flesh to the case for underdetermination, in ways that make it prima facie much more telling. In brief, Kuhn's view is this: if we examine situations where scientists are required to make a choice among the handful of paradigms that confront them at any time, we discover that the relevant evidence and appropriate methodological standards fail to pick out one contender as unequivocally superior to its extant rival(s). I call such situations cases of "local" underdetermination, by way of contrasting them with the more global forms of underdetermination (which say, in effect, that the rules are insufficient to pick out any theory as being uniquely supported by the data). Kuhn offers four distinct arguments for local underdetermination. Each is designed to show that, although methodological rules and standards do constrain and delimit a scientist's choices or options, those rules and standards are never sufficient to compel or unequivocally to warrant the choice of one paradigm over another.

FROM Larry Laudan, *Science and Values: The Aims of Science and Their Role in Scientific Debate* (Berkeley: University of California Press, 1984), 87–102.

1 THE "AMBIGUITY OF SHARED STANDARDS" ARGUMENT

Kuhn's first argument for methodological underdetermination rests on the purported ambiguity of the methodological rules or standards that are shared by advocates of rival paradigms. The argument first appeared in *The Structure of Scientific Revolutions* (1962) and has been extended considerably in his later *The Essential Tension* (1977). As he put it in the earlier work, "lifelong resistance [to a new theory] . . . is not a violation of scientific standards . . . though the historian can always find men—Priestley, for instance—who were unreasonable to resist for as long as they did, he will not find a point at which resistance becomes illogical or unscientific."[3] Many of Kuhn's readers were perplexed by the juxtaposition of claims in such passages as these. On the one hand, we are told that Priestley's continued refusal to accept the theory of Lavoisier was "unreasonable"; but we are also told that Priestley's refusal was neither "illogical" nor "unscientific." To those inclined to think that being "scientific" (at least in the usual sense of that term) required one to be "reasonable" about shaping one's beliefs, Kuhn seemed to be talking gibberish. On a more sympathetic construal, Kuhn seemed to be saying that a scientist could always interpret the applicable standards of appraisal, whatever they might be, so as to "rationalize" his own paradigmatic preferences, whatever they might be. This amounts to claiming that the methodological rules or standards of science never make a real or decisive difference to the outcome of a process of theory choice; if any set of rules can be used to justify any theory whatever, then methodology would seem to amount to just so much window dressing. But that construal, it turns out, is a far cry from what Kuhn intended. As he has made clear in later writings, he wants to bestow a positive, if (compared with the traditional view) much curtailed, role on methodological standards in scientific choice.

What Kuhn apparently has in mind is that the shared criteria, standards, and rules to which scientists explicitly and publicly refer in justifying their choices of theory and paradigm are typically "ambiguous" and "imprecise," so much so that "individuals [who share the same standards] may legitimately differ about their application to concrete cases."[4] Kuhn holds that, although scientists share certain cognitive values "and must do so if science is to survive, they do not all apply them in the same way. Simplicity, scope, fruitfulness, and even accuracy can be judged differently (which is not to say they may be judged arbitrarily) by different people."[5] Because, then, the shared standards are ambiguous, two scientists may subscribe to "exactly the same standard" (say, the rule of simplicity) and yet endorse opposing viewpoints.

Kuhn draws some quite large inferences from the presumed ambiguity of the shared standards or criteria. Specifically, he concludes that every case of theory choice must involve an admixture of objective and subjective factors, since (in Kuhn's view) the shared, and presumably objective, criteria are too amorphous and ambiguous to warrant a particular preference. He puts the point this way: "I continue to hold that the algorithms of individuals are all ultimately different by virtue of the subjective considerations with which each [scientist] must complete the objective criteria before any computations can be done."[6] As this passage makes clear, Kuhn believes that, because the shared criteria are too imprecise to justify a choice, and because—despite that imprecision—scientists do manage to make choices, those choices *must* be grounded in individual and subjective preferences different from those of his fellow scientists. As he says, "every individual choice between competing theories depends on a mixture of objective and subjective factors, or of shared and individual criteria."[7] And, the shared criteria "are not by themselves sufficient to determine the decisions of individual scientists."[8]

This very ambitious claim, if true, would force us to drastically rethink our views of scientific rationality. Among other things, it would drive us to the conclusion that every scientist has different reasons for his theory preferences from those of his fellow scientists. The view entails, among other things, that it is a category mistake to ask (say) why physicists think Einstein's theories are better than Newton's; for, on Kuhn's analysis, there must be as many different answers as there are physicists. We might note in passing that this is quite an ironic conclusion for Kuhn to reach. Far more than most writers on these subjects, he has tended to stress the importance of community and socialization processes in understanding the scientific enterprise. Yet the logic of his own analysis drives him to the radically individualistic position that every scientist has his own set of reasons for theory preferences and that there is no real consensus whatever with respect to the grounds for theory preference, not even among the advocates of the same paradigm. Seen from this perspective, Kuhn tackles . . . the problem of consensus by a maneuver that trivializes the problem: for if we must give a separate and discrete explanation for the theory preferences of each member of the scientific community—which is what Kuhn's view entails—then we are confronted with a gigantic mystery at the collective level, to wit, why the scientists in a given discipline—each supposedly operating with his own individualistic and idiosyncratic criteria, each giving a different "gloss" to the criteria that are shared—are so often able to agree about which theories to bet on. But we can leave it to Kuhn to sort out how he reconciles his commitment to the social psychology of science with his views about the individual vagaries of theory preference. What must concern us is the question whether Kuhn has made a plausible case for thinking that the shared or collective criteria must be supplemented by individual and subjective criteria.

The first point to stress is that Kuhn's thesis purports to apply to all scientific rules or values that are shared by the partisans of rival paradigms, not just to a selected few, notoriously ambiguous ones. We can grant straightaway that some of the rules, standards, and values used by scientists ("simplicity" would be an obvious candidate) exhibit precisely that high degree of ambiguity which Kuhn ascribes to them. But Kuhn's general argument for the impotence of shared rules to settle disagreements between scientists working in different paradigms cannot be established by citing the occasional example. Kuhn must show us, for he claims as much, that there is something in the very nature of those methodological rules that come to be shared among scientists which makes the application of those rules or standards invariably inconclusive. He has not established this result, and there is a good reason why he has not: it is false. To see that it is, one need only produce a methodological rule widely accepted by scientists which can be applied to concrete cases without substantial imprecision or ambiguity. Consider, for instance, one of Kuhn's own examples of a widely shared scientific standard, namely, the requirement that an acceptable theory must be internally consistent and logically consistent with accepted theories in other fields. (One may or may not favor this methodological rule. I refer to it here only because it is commonly regarded, including by Kuhn, as a methodological rule that frequently plays a role in theory evaluation.)

I submit that we have a very clear notion of what it is for a theory to be internally consistent, just as we understand perfectly well what it means for a theory to be consistent with accepted beliefs. Moreover, on at least some occasions we can tell whether a particular theory has violated the standard of (internal or external) consistency. Kuhn himself, in a revealing passage, grants as much; for instance, when comparing the relative merits of geocentric and heliocentric astronomy, Kuhn says that "the consistency criterion, by itself, therefore, spoke unequivocally for the geocentric tradition."[9] (What he has in mind is the fact that heliocentric astronomy, when introduced, was inconsistent with the then reigning terrestrial physics, whereas the assumptions of geocentric astronomy were consistent with that physics.) Note that in this case we have a scientific rule or criterion "speaking unequivocally" in favor of one theory and against its rival. Where are the inevitable imprecision and ambiguity which are supposed by Kuhn to afflict all the shared values of the scientific community? What is ambiguous about the notion of consistency? The point of these rhetorical questions is to drive home the fact that, even by Kuhn's lights, some of the rules or criteria widely accepted in the scientific community do not exhibit that multiplicity of meanings which Kuhn has described as being entirely characteristic of methodological standards.

One could, incidentally, cite several other examples of reasonably clear and unambiguous methodological rules. For instance, the requirements that theories should be deductively closed or that theories should be subjected to controlled experiments have not generated a great deal of confusion or disagreement among scientists about what does and does not constitute closure or a control. Or, consider the rule that theories should lead successfully to the prediction of results unknown to their discoverer; so far as I am aware, scientists have not differed widely in their construal of the meaning of this rule. The significance of the nonambiguity of many methodological concepts and rules is to be found in the fact that such nonambiguity refutes one of Kuhn's central arguments for the incomparability of paradigms and for its corollary, the impotence of methodology as a guide to scientific rationality. There are at least some rules that are sufficiently determinate that one can show that many theories clearly fail to satisfy them. We need not supplement the shared content of these objective concepts with any private notions of our own in order to decide whether a theory satisfies them.

2 THE "COLLECTIVE INCONSISTENCY OF RULES" ARGUMENT

As if the ambiguity of standards was not bad enough, Kuhn goes on to argue that the shared rules and standards, when taken as a collective, "repeatedly prove to conflict with one another."[10] For instance, two scientists may each believe that empirical accuracy and generality are desirable traits in a theory. But, when confronted with a pair of rival (and thus incompatible) theories, one of which is more accurate and the other more general, the judgments of those scientists may well differ about which theory to accept. One scientist may opt for the more general theory; the other, for the more accurate. They evidently share the same standards, says Kuhn, but they end up with conflicting appraisals. Kuhn puts it this way: ". . . in many concrete situations, different values, though all constitutive of good reasons, dictate different conclusions, different choices. In such cases of value-conflict (e.g., one theory is simpler but the other is more accurate) the relative weight placed on different values by different individuals can play a decisive role in individual choice."[11]

Because many methodological standards do pull in different directions, Kuhn thinks that the scientist can pretty well go whichever way he likes. Well, not quite any direction he likes, since—even by Kuhn's very liberal rules—it would be unreasonable for a scientist to prefer a theory (or paradigm) which failed to satisfy any of the constraints. In Kuhn's view, we should expect scientific disagreement or dissensus to emerge specifically in those cases where (a) no available theory satisfied all the constraints and (b) every extant theory satisfied some constraints not satisfied by its rivals. That scientists sometimes find themselves subscribing to contrary standards, I would be the first to grant. Indeed, . . . the discovery of that fact about oneself is often the first prod toward readjusting one's cognitive values. But Kuhn is not merely saying that this happens occasionally; he is asserting that such is the nature of any set of rules or standards which any group of reasonable scientists might accept. As before, our verdict has to be that Kuhn's highly ambitious claim is just that; he never shows us why families of methodological rules should always or even usually be internally inconsistent. He apparently expects us to take his word for it that he is just telling it as it is.[12] I see no reason why we should follow Kuhn in his global extrapolations from the tiny handful of cases he describes. On the contrary, there are good grounds for resisting, since there are plenty of sets of consistent methodological standards. Consider, for instance, one of the most influential documents of nineteenth-century scientific methodology, John Stuart Mill's *System of Logic*. Mill offered there a set of rules or canons for assessing the soundness of causal hypotheses. Nowadays those rules are still called "Mill's methods," and much research in the natural and social sciences utilizes them, often referring to them as the methods of agreement, difference, and concomitant variations. To the best of my knowledge, no one has ever shown that Mill's methods exhibit a latent tendency toward contradiction or conflict of the sort that Kuhn regards as typical of systems of methodological rules. To go back further in history, no one has ever shown that Bacon's or Descartes's or Newton's or Herschel's famous canons of reasoning are internally inconsistent. The fact that numerous methodologies of science may be cited which have never been shown to be inconsistent casts serious doubts on Kuhn's claim that any methodological standards apt to be shared by rival scientists will tend to exhibit mutual inconsistencies.

Kuhn could have strengthened his argument considerably if, instead of focusing on the purported tensions in sets of methodological rules, he had noted, rather, that whenever one has more than one standard in operation, it is conceivable that we will be torn in several directions. And this claim is true, regardless of whether the standards are strictly inconsistent with one another or not (just so long as there is not a complete covariance between their instances). If two scientists agree to judge theories by two standards, then it is trivially true that, depending upon how much weight each gives to the two standards, their judgments about theories may differ. Before we can make sense of how to work with several concurrent standards, we have to ask (as Kuhn never did) about the way in which these standards do (or should) control the selection of a preferred theory. Until we know the answer to that question, we will inevitably find that the standards are of little use in explaining scientific preferences. Kuhn simply assumes that all possible preference structures (i.e., all possible differential weightings of the applicable standards) are equally viable or equally likely to be exemplified in a working scientist's selection procedures. . . .

To sum up the argument to this point: I have shown that Kuhn is wrong in claiming that all methodological rules are inevitably ambiguous and in claiming that scientific methodologies consisting of whole groups of rules always or even usually exhibit a high degree of internal "tension." Since these two claims were the linchpins in Kuhn's argument to the effect that shared criteria "are not by themselves sufficient to determine the decisions of individual scientists,"[13] we are entitled to say that Kuhn's effort to establish a general form of local underdetermination falls flat.

3   The Shifting Standards Argument

Equally important to Kuhn's critique of methodology is a set of arguments having to do with the manner in which standards are supposed to vary from one scientist to another. In treating Kuhn's views on this matter, I follow Gerald Doppelt's excellent and sympathetic explication of Kuhn's position.[14] In general, Kuhn's model of science envisages two quite distinct ways in which disagreements about standards might render scientific debate indeterminate or inconclusive. In the first place, the advocates of different paradigms may subscribe to different methodological rules or evaluative criteria. Indeed, "may" is too weak a term here, for, as we have seen, Kuhn evidently believes that associated with each paradigm is a set of methodological orientations that are (at least partly) at odds with the methodologies of all rival paradigms. Thus, he insists that whenever a "paradigm shift" occurs, this process produces "changes in the standards governing permissible problems, concepts and explanations."[15] This is quite a strong claim. It implies, among other things, that the advocates of different paradigms invariably have different views about what constitutes a scientific explanation and even about what constitutes the relevant facts to be explained (viz., the "permissible problems"). If Kuhn is right about these matters, then debate between the proponents of two rival paradigms will involve appeal to different sets of rules and standards associated respectively with the two paradigms. One party to the dispute may be able to show that his theory is best by his standards, while his opponent may be able to claim superiority by his.

. . . Kuhn is right to say that scientists sometimes subscribe to different methodologies (including different standards for explanation and facticity). But he has never shown, and I believe him to be chronically wrong in claiming, that disagreements about matters of standards and rules neatly coincide with disagreements about substantive matters of scientific ontology. Rival scientists advocating fundamentally different theories or paradigms often have the same standards of assessment (and interpret them identically); on the other hand, adherents to the same paradigm will frequently espouse different standards. In short, methodological disagreements and factual disagreements about basic theories show no striking covariances of the kind required to sustain Kuhn's argument about the intrinsic irresolvability of interparadigmatic debate. It was the thrust of my earlier account of "piecemeal change" to show why Kuhn's claims about irresolvability will not work.

But, of course, a serious issue raised by Kuhn still remains before us. If different scientists sometimes subscribe to different standards of appraisal (and that much is surely correct), then how is it possible for us to speak of the resolution of such disagreements as anything other than an arbitrary closure? To raise that question presupposes a picture of science which I [have] sought to demolish*. . . . Provided there are mechanisms for rationally resolving disagreements about methodological rules and cognitive values . . . ,

---

* Laudan is referring to what, in *Science and Values*, he calls the hierarchical model of scientific rationality. See the commentary following the readings in this chapter for a discussion of this model and Laudan's reasons for preferring his reticulational model.

the fact that scientists often disagree about such rules and values need not, indeed should not, be taken to show that there must be anything arbitrary about the resolution of such disagreements.

4 The Problem-Weighting Argument

As I have said earlier, Kuhn has another argument up his sleeve which he and others think is germane to the issue of the rationality of comparative theory assessment. Specifically, he insists that the advocates of rival paradigms assign differential degrees of importance to the solution of different sorts of problems. Because they do, he says that they will often disagree about which theory is better supported, since one side will argue that it is most important to solve a certain problem, while the other will insist on the centrality of solving a different problem. Kuhn poses the difficulty in these terms: "if there were but one set of scientific problems, one world within which to work on them, and one set of standards for their solution, paradigm competition might be settled more or less routinely by some process like counting the number of problems solved by each. But, in fact, these conditions are never met completely. The proponents of competing paradigms are always at least slightly at cross purposes . . . the proponents will often disagree about the list of problems that any candidate for paradigm must resolve."[16]

In this passage Kuhn runs together two issues which it is well to separate: one concerns the question (just addressed in the preceding section) about whether scientists have different standards of explanation or solution; the other (and the one that concerns us here) is the claim that scientists working in different paradigms want to solve different problems and that, because they do, their appraisals of the merits of theories will typically differ. So we must here deal with the case where scientists have the same standards for what counts as solving a problem but where they disagree about which problems are the most important to solve. As Kuhn puts it, scientific controversies between the advocates of rival paradigms "involve the question: which problems is it more significant to have solved? Like the issue of competing standards, that question of values can be answered only in terms of criteria that lie outside of normal science altogether."[17] Kuhn is surely right to insist that partisans of different global theories or paradigms often disagree about which problems it is most important to solve. But the existence of such disagreement does not establish that interparadigmatic debate about the epistemic support of rival paradigms is inevitably inconclusive or that it must be resolved by factors that lie outside the normal resources of scientific inquiry.

At first glance, Kuhn's argument seems very plausible: the differing weights assigned to the solution of specific problems by the advocates of rival paradigms may apparently lead to a situation in which the advocates of rival paradigms can each assert that their respective paradigms are the best because they solve precisely those problems they respectively believe to be the most important. No form of reasoning, insists Kuhn, could convince either side of the merits of the opposition or of the weakness of its own approach in such circumstances.

To see where Kuhn's argument goes astray in this particular instance, we need to dissect it at a more basic level. Specifically, we need to distinguish two quite distinct senses in which solving a problem may be said to be important. A problem may be important to a scientist just in the sense that he is particularly curious about it. Equally, it may be important because there is some urgent social or economic reason for solving it. Both sorts of considerations may explain why a scientist regards it as urgent to solve the problem. Such concerns are clearly relevant to explaining the motivation of scientists. But these senses of problem importance have no particular epistemic or probative significance. When we are assessing the evidential support for a theory, when we are asking how well supported or well tested that theory is by the available data, we are not asking whether the theory solves problems that are socially or personally important. Importance, in the sense chiefly relevant to this discussion, is what we might call epistemic or probative importance. One problem is of greater epistemic or probative significance than another if the former constitutes a more telling test of our theories than does the latter.

So, if Kuhn's point is to be of any significance for the epistemology of science (or, what amounts to the same thing, if we are asking how beliefworthy a theory is), then we must imagine a situation in which the advocates of different paradigms assign conflicting degrees of epistemic import to the solution of certain problems. Kuhn's thesis about such situations would be, I presume, that there is no rational machinery for deciding who is right about the assignment of epistemic weight to such problems. But that seems wrongheaded, or at least unargued, for philosophers of science have long and plausibly maintained that the primary function of scientific epistemology is precisely to ascertain the (epistemic) importance of any piece of confirming or disconfirming evidence. It is not open to a scientist simply to say that solving an arbitrarily selected problem (however great its subjective significance) is of high probative value. Indeed, it is often true that the epistemically most salient problems are ones with little or no prior practical or even heuristic significance. (Consider that Brownian motion was of decisive epistemic significance in discrediting classical thermodynamics, even though such motion had little intrinsic interest prior to Einstein's showing that such motion was anomalous for thermodynamics.) The whole point of the theory of evidence is to desubjectify the assignment of evidential significance by indicating the kinds of reasons that can legitimately be given for attaching a particular degree of epistemic importance to a confirming or refuting instance. Thus, if one maintains that the ability of a theory to solve a certain problem is much more significant epistemically than its ability to solve another, one must be able to give reasons for that epistemic preference. Put differently, one has to be able to show that the probative significance of the one problem for testing theories of a certain sort is indeed greater than that of the other. He might do so by showing that the former outcome was much more surprising than or more general than the latter. One may thus be able to motivate a claim for the greater importance of the first problem over the second by invoking relevant epistemic and methodological criteria. But if none of these options is open to him, if he can answer the question, "Why is solving this problem more important probatively than solving that one?" only by replying, in effect, "because I am interested in solving this rather than that," then he has surrendered any claim to be shaping his beliefs rationally in light of the available evidence.

We can put the point more generally: the rational assignment of any particular degree of probative significance to a problem must rest on one's being able to show that there are viable methodological and epistemic grounds for assigning that degree of importance rather than another. Once we see this, it becomes clear that the degree of empirical support which a solved problem confers on a paradigm is not simply a matter of how keenly the proponents of that paradigm want to solve the problem.

Let me expand on this point by using an example cited extensively by both Kuhn and Doppelt: the Daltonian "revolution" in chemistry. As Doppelt summarizes the Kuhnian position, ". . . the pre-Daltonian chemistry of the phlogiston theory and the theory of elective affinity achieved reasonable answers to a whole set of questions effectively abandoned by Dalton's new chemistry."[18] Because Dalton's chemistry failed to address many of the questions answered by the older chemical paradigm, Kuhn thinks that the acceptance of Dalton's approach deprived "chemistry of some actual and much potential explanatory power."[19] Indeed, Kuhn is right in holding that, during most of the nineteenth century, Daltonian chemists were unable to explain many things that the older chemical traditions could make sense of. On the other hand, as Kuhn stresses, Daltonian chemistry could explain a great deal that had eluded earlier chemical theories. In short, "the two paradigms seek to explain different kinds of observational data, in response to different agendas of problems."[20] This "loss" of solved problems during transitions from one major theory to another is an important insight of Kuhn's. . . . But this loss of problem-solving ability through paradigm change, although real enough, does not entail, as Kuhn claims, that proponents of old and new paradigms will necessarily be unable to make congruent assessments of how well tested or well supported their respective paradigms are.

What leads Kuhn and Doppelt to think otherwise is their assumption that the centrality of a problem on one's explanatory agenda necessarily entails one's assigning a high degree of epistemic or probative weight to that problem when it comes to determining how well supported a certain theory or paradigm is. But that assumption is usually false. In general, the observations to which a reasonable scientist attaches the most probative or epistemic weight are those instances that test a theory especially "severely" (to use Popper's splendid term). The instances of greatest probative weight in the history of science (e.g., the oblate shape of the "spherical" earth, the Arago disk experiment, the bending of light near the sun, the recession of Mercury's perihelion, the reconstitution of white light from the spectrum) have generally not been instances high on the list of problems that scientists developed their theories to solve. A test instance acquires high probative weight when, for example, it involves testing one of a theory's surprising or counterintuitive predictions, or when it represents a kind of crucial experiment between rival theories. The point is that a problem or instance does not generally acquire great probative strength in testing a theory simply because the advocates of that theory would like to be able to solve the problem. Quite the reverse, many scientists and philosophers would say. After all, it is conventional wisdom that a theory is not very acutely tested if its primary empirical support is drawn from the very sort of situations it was designed to explain. Most theories of experimental design urge—in sharp contrast with Kuhn—that theories should not be given high marks simply because they can solve the problems they were invented to solve. In arguing that the explanatory agenda a scientist sets for himself automatically dictates that scientist's reasoned judgments about well-testedness, Kuhn and Doppelt seem to have profoundly misconstrued the logic of theory appraisal.

Let us return for a moment to Kuhn's Dalton example. If I am right, Dalton might readily have conceded that pre-Daltonian chemistry solved a number of problems that his theory failed to address. Judged as theories about the qualitative properties of chemical reagents, those theories could even be acknowledged as well supported *of their type*. But Dalton's primary interests lie elsewhere, for he presumably regarded those earlier theories as failing to address what he considered to be the central problems of chemistry. But this is not an epistemic judgment; it is a pragmatic one. It amounts to saying: "These older theories are well-tested and reliable theories for explaining certain features of chemical change; but those features happen not to interest me very much." In sum, Kuhn and Doppelt have failed to offer us any grounds for thinking that a scientist's judgment about the degree of evidential support for a paradigm should or does reflect his personal views about the problems he finds most interesting. That, in turn, means that one need not share an enthusiasm for a certain paradigm's explanatory agenda in order to decide whether the theories that make up that paradigm are well tested or ill tested. It appears to me that what the Kuhn-Doppelt point really amounts to is the truism that scientists tend to invest their efforts exploring paradigms that address problems those scientists find interesting. That is a subjective and pragmatic matter which can, and should, be sharply distinguished from the question whether one paradigm or theory is better tested or better supported than its rivals. Neither Kuhn nor Doppelt has made plausible the claim that, because two scientists have different degrees of interest in solving different sorts of problems, it follows that their epistemic judgments of which theories are well tested and which are not will necessarily differ.

We are thus in a position to conclude that the existence of conflicting views among scientists about which problems are interesting apparently entails nothing about the *incompatibility* or *incommensurability* of the epistemic appraisals those scientists will make. That in turn means that these real differences of problem-solving emphasis between advocates of rival paradigms do nothing to undermine the viability of a methodology of comparative theory assessment, insofar as such a methodology is epistemically rather than pragmatically oriented. It seems likely that Kuhn and Doppelt have fallen into this confusion because of their failure to see that acknowledged differences in the motivational appeal of various problems to various scientists constitutes no rationale for asserting the existence of correlative differences in the probative weights properly assigned to those problems by those same scientists.

The appropriate conclusion to draw from the features of scientific life to which Kuhn and Doppelt properly direct our attention is that the pursuit of (and doubtless the recruitment of scientists into) rival paradigms is influenced by pragmatic as well as by epistemic considerations. That is an interesting thesis, and probably a sound one, but it does nothing to undermine the core premise of scientific epistemology: that there are principles of empirical or evidential support which are neither paradigm-specific, hopelessly vague, nor individually idiosyncratic. More important, these principles are sometimes sufficient to guide our preferences unambiguously.[21]

■ | Notes

1. See Laudan [1990].

2. See ibid. for a lengthy treatment of some issues surrounding underdetermination of theories.

3. Kuhn, 1962, p. 159 [p. 94].

4. Kuhn, 1977, p. 322 [p. 95].

5. Ibid., p. 262.

6. Ibid., p. 329 [p. 101].

7. Ibid., p. 325; see also p. 324 [pp. 98, 97].

8. Ibid., p. 325 [p. 97].

9. Ibid., p. 323 [p. 97].

10. Ibid., p. 322 [p. 95].

11. Kuhn, 1970, p. 262.

12. "What I have said so far is primarily simply descriptive of what goes on in the sciences at times of theory choice" (Kuhn, 1977, p. 325 [p. 98]).

13. Kuhn, 1977, p. 325 [p. 97].

14. Doppelt, 1978. Whereas Kuhn's own discussion of these questions in *The Structure of Scientific Revolutions* rambles considerably, Doppelt offers a succinct and perspicacious formulation of what is, or at least what should have been, Kuhn's argument. Although I quarrel with Doppelt's analysis at several important points, my own thoughts about these issues owe a great deal to his writings.

15. Kuhn, 1962, p. 105 [p. 90].

16. Ibid., pp. 147–148.

17. Ibid., p. 110 [pp. 92–93].

18. Doppelt, 1978, p. 42.

19. Kuhn, 1962, p. 107 [p. 90].

20. Ibid., p. 43.

21. Even on the pragmatic level, however, it is not clear that the Doppeltian version of Kuhn's relativistic picture of scientific change will stand up, for Doppelt is at pains to deny that there can be any short-term resolution between the advocates of rival axiologies. If the arguments of [*Science and Values*, ch. 2] have any cogency, it seems entirely possible that pragmatic relativism, every bit as much as its epistemic counterpart, is question begging.

▪ | References

Doppelt, Gerald (1978). "Kuhn's Epistemological Relativism: An Interpretation and Defense," *Inquiry* 21: 33–86.

Kuhn, Thomas (1962). *The Structure of Scientific Revolutions*. Chicago: University of Chicago Press.

——(1977). *The Essential Tension*. Chicago: University of Chicago Press.

Laudan, Larry (1990). "Demystifying Underdetermination." In C. Wade Savage, ed., *Minnesota Studies in the Philosophy of Science*, Volume XIV, *Scientific Theories*. Minneapolis: University of Minnesota Press. [Reprinted in chapter 3, pp. 288–320, of this volume.]

Helen E. Longino

Values and Objectivity

Objectivity is a characteristic ascribed variously to beliefs, individuals, theories, observations, and methods of inquiry. It is generally thought to involve the willingness to let our beliefs be determined by "the facts" or by some impartial and nonarbitrary criteria rather than by our wishes as to how things ought to be. A specification of the precise nature of such involvement is a function of what it is that is said to be objective. In this chapter I will review some common ideas about objectivity and argue that the objectivity of science is secured by the social character of inquiry. This chapter is a first step, therefore, towards socializing cognition.

Some part of the popular reverence for science has its origin in the belief that scientific inquiry, unlike other modes of inquiry, is by its very nature objective. In the modern mythology, the replacement of a mode of comprehension that simply projects human needs and values into the cosmos by a mode that views nature at a distance and dispassionately "puts nature to the question," in the words of Francis Bacon, is seen as a major accomplishment of the maturing human intellect.[1] The development of this second mode of approaching the natural world is identified, according to this view, with the development of science and the scientific method. Science is thought to provide us with a view of the world that is objective in two seemingly quite different senses of that term. In one sense objectivity is bound up with questions about the truth and referential character of scientific theories, that is, with issues of scientific realism. In this sense to attribute objectivity to science is to claim that the view provided by science is an accurate description of the facts of the natural world as they are; it is a correct view of the objects to be found in the world and of their relations with each other. In the second sense objectivity has to do with modes of inquiry. In this sense to attribute objectivity to science is to claim that the view provided by science is one

---

From Helen Longino, *Science as Social Knowledge: Values and Objectivity in Scientific Inquiry* (Princeton, N.J.: Princeton University Press, 1990), pp. 62–82.

achieved by reliance upon nonarbitrary and nonsubjective criteria for developing, accepting, and rejecting the hypotheses and theories that make up the view. The reliance upon and use of such criteria as well as the criteria themselves are what is called scientific method. Common wisdom has it that if science is objective in the first sense it is because it is objective in the second.

At least two things can be intended by the ascription of objectivity to scientific method. Often scientists speak of the objectivity of data. By this they seem to mean that the information upon which their theories and hypotheses rest has been obtained in such a way as to justify their reliance upon it. This involves the assumption or assurance that experiments have been properly performed and that quantitative data have not been skewed by any faults in the design of survey instruments or by systematic but uncharacteristic eccentricities in the behavior of the sample studied. If a given set of data has been objectively obtained in this sense, one is thereby licensed to believe that it provides a reliable view of the world in the first of the two senses of objectivity distinguished above. . . . While objective, that is, reliable, measurement is indeed one crucial aspect of objective scientific method,[2] it is not the only dimension in which questions about the objectivity of methods can arise. In ascribing (or denying) objectivity to a method we can also be concerned about the extent to which it provides means of assessing hypotheses and theories in an unbiased and unprejudiced manner.

In this chapter I will explore more deeply the nature of this second mode of scientific objectivity and its connection with the logic of discourse in the natural sciences. . . . Logical positivists have relied upon formal logic and a priori epistemological requirements as keys to developing the logical analysis of science, while their historically minded wholist critics have insisted upon the primacy of scientific practice as revealed by study of the history of science. According to the former view, science does indeed appear to be, by its very nature, free of subjective preference, whereas according to the latter view, subjectivity plays a major role in theory development and theory choice. Witnesses to the debate seem to be faced with a choice between two unacceptable alternatives: a logical analysis that is historically unsatisfactory and a historical analysis that is logically unsatisfactory. This kind of dilemma suggests a debate whose participants talk past one another rather than addressing common issues. Certainly part of the problem consists in attempts to develop a comprehensive account of science on the basis either of normative logical constraints or of empirical historical considerations. My analysis makes no pretense to totality or completion. It suggests, rather, a framework to be filled-in and developed both by epistemologists whose task is to develop criteria and standards of knowledge, truth, and rational belief and by historians and sociologists whose task is to make visible those historical and institutional features of the practice of science that affect its content. . . . To make way for this interdisciplinary framework, I begin by briefly reviewing the treatment of objectivity and subjectivity in the competing analyses of the logic of science.

■ | Objectivity, Subjectivity, and Individualism

The positivist analysis of confirmation guaranteed the objectivity of science by tying the acceptance of hypotheses and theories to a public world over whose description there can be no disagreement. Positivists allow for a subjective, nonempirical element in scientific inquiry by distinguishing between a context of discovery and a context of justification.[3] The context of discovery for a given hypothesis is constituted by the circumstances surrounding its initial formulation—its origin in dreams, guesses, and other aspects of the mental and emotional life of the individual scientist. Two things should be noted here. First, these nonempirical elements are understood to be features of an individual's psychology. They are treated as randomizing factors that promote novelty rather than as beliefs or attitudes that are systematically related to the culture, social structure, or socioeconomic interests of the context within which an individual scientist works. Secondly, in the context of justification these generative factors are disregarded, and the hypothesis is considered only in relation to its observable consequences, which determine its acceptability. This distinction enables positivists to acknowledge the play of subjective factors in the initial development of hypotheses and theories while guaranteeing that their acceptance remains untainted, determined not by subjective preferences but by observed reality. The subjective elements that taint its origins are purged from scientific inquiry by the methods characteristic of the context of justification: controlled experiments, rigorous deductions, et cetera. When one is urged to be objective or "scientific," it is this reliance on an established and commonly accepted reality that is being recommended. The logical positivist model of confirmation simply makes the standard view of scientific practice more systematic and logically rigorous.

As long as one takes the positivist analysis as providing a model to which any inquiry must conform in order to be objective and rational, then to the degree that actual science departs from the model it fails to be objective and rational. As noted above with respect to evidence and inference, both the historians and philosophers who have attacked the old model and those who have defended it have at times taken this position. The only disagreement with respect to objectivity, then, seems to be over the question of whether actual, historical science does or does not realize the epistemological ideal of objectivity. Defenders of the old model have argued that science ("good science") does realize the ideal. Readers of Kuhn and Feyerabend take their arguments to show that science is not objective, that objectivity has been fetishized by traditionalists. These authors themselves have somewhat more subtle approaches. While Kuhn has emphasized the role of such subjective factors as personality, education, and group commitments in theory choice, he also denies that his is a totally subjectivist view. . . . He suggests that values such as relative simplicity and relative problem-solving ability can and do function as nonarbitrary criteria in theory acceptance.

Such values can be understood as internal to inquiry, especially by those to whom scientific inquiry just is problem solving.[4] Feyerabend, on the other hand, has rejected the relevance to science of canons of rationality or of general criteria of theory acceptance and defends a positive role for subjectivity in science.[5]

. . . How can the contextualist analysis of evidence, with its consequent denial of any logically guaranteed independence from contextual values, be accommodated within a perspective that demands or presupposes the objectivity of scientific inquiry?

As a first step in answering this question it is important to distinguish between objectivity as a characteristic of scientific method and objectivity as a characteristic of individual scientific practitioners or of their attitudes and practices. The standard accounts of scientific method tend to conflate the two, resulting in highly individualistic accounts of knowledge. Both philosophical accounts assume that method, the process by which knowledge is produced, is the application of rules to data. The positivist or traditional empiricist account of objectivity attributes objectivity to the practitioner to the extent that she or he has followed the method. Scientific method, on this view, is something that *can* be practiced by a single individual: sense organs and the capacity to reason are all that are required for conducting controlled experiments or practicing rigorous deduction. For Kuhn and for the contextualist account sketched above rationality and deference to observational data are not sufficient to guarantee the objectivity of individuals. For Kuhn this is because these intellectual activities are carried out in the context of a paradigm assented to by the scientific community. But, although Kuhn emphasizes the communitarian nature of the sciences, the theory of meaning he developed to account for the puzzling aspects of scientific change that first drew his attention reduces that community to a solipsistic monad incapable of recognizing and communicating with other monads/communities. Kuhn's account is, thus, as individualist as the empiricist one. The contextualist account makes the exercise of reason and the interpretation of data similarly dependent on a context of assumptions. Why is it not subject to the same problems?

▪ | Objectivity, Criticism, and Social Knowledge

Two shifts of perspective make it possible to see how scientific method or scientific knowledge is objective even in the contextualist account. One shift is to return to the idea of science as practice. The analysis of evidential relations outlined above was achieved by thinking about science as something that is done, that involves some form of activity on the part of someone, the scientist. Because we think the goal of the scientist's practice is knowledge, it is tempting to follow tradition and seek solutions in abstract or universal rules. Refocussing on science as practice makes possible the second shift, which involves regarding scientific method as something practiced not primarily by individuals but by social groups.

The social nature of scientific practice has long been recognized. In her essay "Perception, Interpretation and the Sciences" Marjorie Grene discusses three aspects of the social character of science.[6] One she sees as the existence of the scientific disciplines as "social enterprises," the individual members of which are dependent on one another for the conditions (ideas, instruments, et cetera) under which they practice. Another related aspect is that initiation into scientific inquiry requires education. One does not simply declare oneself a biologist but learns the traditions, questions, mathematical and observational techniques, "the sense of what to do next," from someone who has herself or himself been through a comparable initiation and then practiced. One "enters into a world" and learns how to live in that world from those who already live there. Finally, as the practitioners of the sciences all together constitute a network of communities embedded in a society, the sciences are also among a society's activities and depend for their survival on that society's valuing what they do. Much of the following can be read as an elaboration of these three points, particularly as regards the outcome, or product, of scientific practices, namely scientific knowledge. What I wish particularly to stress is that the objectivity of scientific inquiry is a consequence of this inquiry's being a social, and not an individual, enterprise.

The application of scientific method, that is, of any subset of the collection of means of supporting scientific theory on the basis of experiential data, requires by its nature the participation of two or more individuals. Even brief reflection on the actual conditions of scientific practice shows that this is so. Scientific knowledge is, after all, the product of many individuals working in (acknowledged or unacknowledged) concert. As noted earlier, scientific inquiry is complex in that it consists of different kinds of activities. It consists not just in producing theories but also in (producing) concrete interactions with, as well as models—mechanical, electrical, and mathematical—of, natural processes. These activities are carried out by different individuals, and in this era of "big science" a single complex experiment may be broken into parts, each of which will be charged to a different individual or group of individuals. The integration and transformation of these activities into a coherent understanding of a given phenomenon are a matter of social negotiations.

One might argue that this is at least in principle the activity of a single individual. But, even if we were to imagine such group efforts as individual efforts, scientific knowledge is not produced by collecting the products of such imagined individuals into one whole. It is instead produced through a process of critical emendation and modification of those individual products by the rest of the scientific community. Experiments get repeated with variations by individuals other than their originators, hypotheses and theories are critically examined, restated, and reformulated before becoming an accepted part of the scientific canon. What are known as scientific break-throughs build, whether this is acknowledged or not, on previous work and rest on a tradition of understandings, even when the effect of the break-through will be to undermine those understandings.[7]

The social character of scientific knowledge is made especially apparent by the organization of late twentieth-century science, in which the production of knowledge is crucially determined by the gatekeeping of peer review. Peer review determines what research gets funded and what research gets published in the journals, that is, what gets to count as knowledge. Recent concern over the breakdown of peer review and over fraudulent research simply supports the point. The most startling study of peer review suggested that scientific papers in at least one discipline were accepted on the basis of the institutional affiliation of the authors rather than the intrinsic worth of the paper.[8] Commentary on the paper suggested that this decision procedure might be more widespread. Presumably the reviewers using the rule assume that someone would not get a job at X institution if that person were not a top-notch investigator, and so her/his experiments must be well-done and the reasoning correct. Apart from the errors in that assumption, both the reviewer and the critic of peer review treat what is a social process as an individual process. The function of peer review is not just to check that the data seem right and the conclusions well-reasoned but to bring to bear another point of view on the phenomena, whose expression might lead the original author(s) to revise the way they think about and present their observations and conclusions. To put this another way, it is to make sure that, among other things, the authors have interpreted the data in a way that is free of their subjective preferences.

The concern over the breakdown of peer review, while directed at a genuine problem, is also exaggerated partly because of an individualist conception of knowledge construction. Peer review prior to publication is not the only filter to which results are subjected. The critical treatment *after* publication is crucial to the refining of new ideas and techniques. While institutional bias may also operate in the postpublication reception of an idea, other factors, such as the attempt to repeat an experiment or to reconcile incompatible claims, can eventually compensate for such misplaced deference. Publication in a journal does not make an idea or result a brick in the edifice of knowledge. Its absorption is a much more complex process, involving such things as subsequent citation, use and modification by others, et cetera. Experimental data and hypotheses are transformed through the conflict and integration of a variety of points of view into what is ultimately accepted as scientific knowledge.[9]

What is called scientific knowledge, then, is produced by a community (ultimately the community of all scientific practitioners) and transcends the contributions of any individual or even of any subcommunity within the larger community.[10] Once propositions, theses, and hypotheses are developed, what will become scientific knowledge is produced collectively through the clashing and meshing of a variety of points of view. The relevance of these features of the sociology of science to objectivity will be apparent shortly.

The social character of hypothesis acceptance underscores the publicity of science. This publicity has both social and logical dimensions. We are accustomed to thinking of science as a public possession or property in that it is produced for the most part by public resources—either through direct funding of research or through financial support of the education of scientists. The social processes described underscore another aspect of its publicity; it is itself a public resource—a common fund of assertions presumably established to a point beyond question. It thereby constitutes a body of putative truths that can be appealed to in defense or criticism of other claims.

From a logical point of view the publicity of science includes several crucial elements. First, theoretical assertions, hypotheses, and background assumptions are all in principle public in the sense of being generally available to and comprehensible to anyone with the appropriate background, education, and interest. Second, the states of affairs to which theoretical explanations are pegged (in evidential and explanatory relationships) are public in the sense that they are intersubjectively ascertainable. . . . This does not require a commitment to a set of theory-free, eternally acceptable observation statements but merely a commitment to the possibility that two or more persons can agree about the descriptions of objects, events, and states of affairs that enter into evidential relationships. Both features are consequences of the facts (1) that we have a common language which we use to describe our experience and within which we reason and (2) that the objects of experience which we describe and about which we reason are purported to exist independently of our seeing and thinking about them.[11]

These two aspects of the logical publicity of science make criticism of scientific hypotheses and theories possible in a way that is not possible, for instance, for descriptions of mystical experience or expressions of feeling or emotion. First, a common language for the description of experience means that we can understand each other, which means in turn that we can accept or reject hypotheses, formulate and respond to objections to them. Second, the presupposition of objects existing independently of our perception of them imposes an acceptance of constraints on what can be said or reasonably believed about them. Such acceptance implies the relevance of reports and judgments other than our own to what we say or believe. There is no way, by contrast, to acquire the authority sufficient to criticize the description of a mystical experience or the expression of a particular feeling or emotion save by having the experience or emotion in question, and these are not had in the requisite sense by more than one person. By contrast, the logical publicity of scientific understanding and subject matter makes them and hence the authority to criticize their articulation accessible to all.[12] It should be said that these constitute necessary but not sufficient conditions for the possibility of criticism, a point I shall return to later. It is the possibility of intersubjective criticism, at any rate, that permits objectivity in spite of the context dependence of evidential reasoning. Before developing this idea further let me outline some of the kinds of criticism to be found in scientific discourse.

There are a number of ways to criticize a hypothesis. For the sake of convenience we can divide these into evidential and conceptual criticism to reflect the distinction between criticism proceeding on the basis of experimental and observational concerns and that proceeding on the basis of theoretical and metatheoretical concerns.[13] Evidential criticism is familiar enough: John Maddox, editor of *Nature*, criticizing Jacques Benveniste's experiments with highly diluted antibody solutions suggesting that immune responses could be triggered in the absence of even one molecule of the appropriate antibody;[14] Richard Lewontin analyzing the statistical data alleged to favor Jensen's hypothesis of the genetic basis of I.Q.,[15] Stephen Gould criticizing the experiments of David Barash purporting to demonstrate punitive responses by male mountain bluebirds to putative adultery on the part of their female mates.[16] Such criticism questions the degree to which a given hypothesis is supported by the evidence adduced for it, questions the accuracy, extent, and conditions of performance of the experiments and observations serving as evidence, and questions their analysis and reporting.[17]

Conceptual criticism, on the other hand, often stigmatized as "metaphysical," has received less attention in a tradition of discourse dominated by empiricist ideals. At least three sorts can be distinguished. The first questions the conceptual soundness of a hypothesis—as Einstein criticized and rejected the discontinuities and uncertainties of the quantum theory;[18] as Kant criticized and rejected, among other things, the Newtonian hypotheses of absolute space and time, a criticism that contributed to the development of field theory.[19] A second sort of criticism questions the consistency of a hypothesis with accepted theory—as traditionalists rejected the heliocentric theory because its consequences seemed inconsistent with the Aristotelian physics of motion still current in the fifteenth and sixteenth centuries;[20] as Millikan rejected Ehrenhaft's hypothesis of subelectrons on the basis not only of Millikan's own measurements but of his commitment to a particulate theory of electricity that implied the existence of an elementary electric charge.[21] A third sort questions the relevance of evidence presented in support of a hypothesis: relativity theorists could deny the relevance of the Michelson-Morley interferometer experiment to the Lorentz-Fitzgerald contraction hypothesis by denying the necessity of the ether;[22] Thelma Rowell and others have questioned the relevance of certain observations of animal populations to claims about dominance hierarchies within those populations by criticizing the assumptions of universal male dominance underlying claims of such relevance;[23] critics of hypotheses about the hazards of exposure to ionizing radiation direct their attention to the dose-response model with which results at high exposures are projected to conditions of low exposures.[24] Thus most of the debate centers not on the data but on the assumptions in light of which the data are interpreted. This last form of criticism, though related to evidential considerations, is grouped with the forms of conceptual criticism because it is concerned not with how accurately the data has been measured and reported but with the assumptions in light of which that data is taken to be evidence for a given hypothesis in the first place. Here it is not the material presented as evidence itself that is challenged but its relevance to a hypothesis.

All three of these types of criticism are central to the development of scientific knowledge and are included among the traditions of scientific discourse into which the novice is initiated. It is the third type of criticism, however, which amounts to questioning the background beliefs or assumptions in light of which states of affairs become evidence, that is crucial for the problem of objectivity. Objectivity in the sense under discussion requires a way to block the influence of subjective preference at the level of background beliefs. While the possibility of criticism does not totally eliminate subjective preference either from an individual's or from a community's practice of science, it does provide a means for checking its influence in the formation of "scientific knowledge." Thus, even though background assumptions may not be supported by the same kinds of data upon which they confer evidential relevance to some hypothesis, other kinds of support can be provided, or at least expected.[25] And in the course of responding to criticism or providing such support one may modify the background assumption in question. Or if the original proponent does not, someone else may do so as a way of entering into the discourse. Criticism is thereby transformative. In response to criticism, empirical support may be forthcoming (subject, of course, to the limitations developed above). At other times the support may be conceptual rather than empirical. Discussions of the nature of human judgment and cognition and whether they can be adequately modelled by computer programs, and of the relation of subjectively experienced psychological phenomena to brain processes, for instance, are essential to theoretical development in cognitive science and neuropsychology respectively. But these discussions involve issues that are metaphysical or conceptual in nature and that, far from being resolvable by empirical means, must be resolved (explicitly or implicitly) in order to generate questions answerable by such means. The contextual analysis of evidential relations shows the limits of purely empirical considerations in scientific inquiry. Where precisely these limits fall will differ in different fields and in different research programs.

As long as background beliefs can be articulated and subjected to criticism from the scientific community, they can be defended, modified, or abandoned in response to such criticism. As long as this kind of response is possible, the incorporation of hypotheses into the canon of scientific knowledge can be independent of any individual's subjective preferences. Their incorporation is, instead, a function in part of the assessment of evidential support. And while the evidential relevance to hypotheses of observations and experiments is a function of background assumptions, the adoption of these assumptions is not arbitrary but is (or rather can be) subject to the kinds of controls just discussed. This solution incorporates as elements both the social character of the production of knowledge and the public accessibility of the material with which this knowledge is constructed.

Sociologically and historically, the molding of what counts as scientific knowledge is an activity requiring many participants. Even if one individual's work is regarded as absolutely authoritative over some period—as for instance, Aristotle's and later Newton's were—it is eventually challenged, questioned, and made to take the role of contributor rather than sole author—as Aristotle's and Newton's have been. From a logical point of view, if scientific knowledge were to be understood as the simple sum of finished products of individual activity, then not only would there be no way to block or mitigate the influence of subjective preference but scientific knowledge itself would be a potpourri of merrily inconsistent theories. Only if the products of inquiry are understood to be formed by the kind of critical discussion that is possible among a plurality of individuals about a commonly accessible phenomenon, can we see how they count as knowledge rather than opinion.

Objectivity, then, is a characteristic of a community's practice of science rather than of an individual's, and the practice of science is understood in a much broader sense than most discussions of the logic of scientific method suggest. These discussions see what is central to scientific method as being the complex of activities that constitute hypothesis testing through comparison with experiential data—in principle, if not always in reality, an activity of individuals. What I have argued here is that scientific method involves as an equally central aspect the subjection of hypotheses and the background assumptions in light of which they seem to be supported by data to varieties of conceptual criticism, which is a social rather than an individual activity.[26]

The respect in which science is objective, on this view, is one that it shares with other modes of inquiry, disciplines such as literary or art criticism and philosophy.[27] The feature that has often been appealed to as the source of the objectivity of science, that its hypotheses and theories are accepted or rejected on the basis of observational, experimental data, is a feature that makes scientific inquiry empirical. In the positivist account, for instance, it was the syntactically and deductively secured relation of hypotheses to a stable set of observational data that guaranteed the objectivity of scientific inquiry. But, as I've argued, most evidential relations in the sciences cannot be given this syntactic interpretation. In the contextual analysis of evidential relations, however, that a method is empirical in the above sense does not mean that it is also objective. A method that involved the appeal to observational or experimental data but included no controls on the kinds of background assumptions in light of which their relevance to hypotheses might be determined, or that permitted a weekly change of assumptions so that a hypothesis accepted in one week on the basis of some bit of evidence *e* would be rejected the next on the same basis, would hardly qualify as objective. Because the relation between hypotheses and evidence is mediated by background assumptions that themselves may not be subject to empirical confirmation or disconfirmation, and that may be infused with metaphysical or normative considerations, it would be a mistake to identify the objectivity of scientific methods with their empirical features alone. The process that can expose such assumptions is what makes possible, even if it cannot guarantee, independence from subjective bias, and hence objectivity. Thus, while rejecting the idea that observational data alone provide external standards of comparison and evaluation of theories, this account does not reject external standards altogether. The formal requirement of demonstrable evidential relevance constitutes a standard of rationality and acceptability independent of and external to any particular research program or scientific theory. The satisfaction of this standard by any program or theory, secured, as has been argued, by intersubjective criticism, is what constitutes its objectivity.

Scientific knowledge is, therefore, social knowledge. It is produced by processes that are intrinsically social, and once a theory, hypothesis, or set of data has been accepted by a community, it becomes a public resource. It is available to use in support of other theories and hypotheses and as a basis of action. Scientific knowledge is social both in the ways it is created and in the uses it serves.

■ | Objectivity by Degrees

I have argued both that criticism from alternative points of view is required for objectivity and that the subjection of hypotheses and evidential reasoning to critical scrutiny is what limits the intrusion of individual subjective preference into scientific knowledge. Are these not two opposing forms of social interaction, one dialogic and the other monologic? Why does critical scrutiny not simply suppress those alternative points of view required to prevent premature allegiance to one perspective? How does this account of objectivity not collapse upon itself? The answer involves seeing dialogic and monologic as poles of a continuum. The maintenance of dialogue is itself a social process and can be more or less fully realized. Objectivity, therefore, turns out to be a matter of degree. A method of inquiry is objective to the degree that it permits *transformative* criticism. Its objectivity consists not just in the inclusion of intersubjective criticism but in the degree to which both its procedures and its results are responsive to the kinds of criticism described. I've argued that method must, therefore, be understood as a collection of social, rather than individual, processes, so the issue is the extent to which a scientific community maintains critical dialogue. Scientific communities will be objective to the degree that they satisfy four criteria necessary for achieving the transformative dimension of critical discourse: (1) there must be recognized avenues for the criticism of evidence, of methods, and of assumptions and reasoning; (2) there must exist shared standards that critics can invoke; (3) the community as a whole must be responsive to such criticism; (4) intellectual authority must be shared equally among qualified practitioners. Each of these criteria requires at least a brief gloss.

Recognized Avenues for Criticism

The avenues for the presentation of criticism include such standard and public forums as journals, conferences, and so forth. Peer review is often pointed to as the standard avenue for such criticism, and indeed it is effective in preventing highly idiosyncratic values from shaping knowledge. At the same time its confidentiality and privacy make it the vehicle for the entrenchment of established views. This criterion also means that critical activities should receive equal or nearly equal weight to "original research" in career advancement. Effective criticism that advances understanding should be as valuable as original research that opens up new domains for understanding; pedestrian, routine criticism should be valued comparably to pedestrian and routine "original research."

Shared Standards

In order for criticism to be relevant to a position it must appeal to something accepted by those who hold the position criticized. Similarly, alternative theories must be perceived to have some bearing on the concerns of a scientific community in order to obtain a hearing. This cannot occur at the whim of individuals but must be a function of public standards or criteria to which members of the scientific community are or feel themselves bound. These standards can include both substantive principles and epistemic, as well as social, values. Different subcommunities will subscribe to different but overlapping subsets of the standards associated with a given community. Among values the standards can include such elements as empirical adequacy, truth, generation of specifiable interactions with the natural or experienced world, the expansion of existing knowledge frameworks, consistency with accepted theories in other domains, comprehensiveness, reliability as a guide to action, relevance to or satisfaction of particular social needs. Only the first of these constitutes a necessary condition that any research program must meet or aspire to meet, and even this requirement may be temporarily waived and is subject to interpretation.

The list shares some elements with the list Thomas Kuhn presents in his essay "Objectivity, Value Judgment, and Theory Choice,"[28] and like the items in his list they can be weighted differently in different scientific communities and they must be more precisely formulated to be applicable. For example, the requirement that theories have some capability to generate specifiable interactions with the natural or experienced world will be applied differently as the sorts of interactions desired in a community differ. The particular weighting and interpretation assigned these standards will vary in different social and historical contexts as a function of cognitive and social needs. Furthermore, they are not necessarily consistent. . . . The goals of truth or accurate representation and expansion of existing knowledge frameworks exist in some tension with each other.

Standards do not provide a deterministic theory of theory choice. Nevertheless, it is the existence of standards that makes the individual members of a scientific community responsible to something besides themselves. It is the open-ended and nonconsistent nature of these standards that allows for pluralism in the sciences and for the continued presence, however subdued, of minority voices. Implicit or explicit appeals to such standards as I've listed underwrite many of the critical arguments named above.

COMMUNITY RESPONSE

This criterion requires that the beliefs of the scientific community as a whole and over time change in response to the critical discussion taking place within it. This responsiveness is measured by such public phenomena as the content of textbooks, the distribution of grants and awards, the flexibility of dominant world views. Satisfaction of this criterion does not require that individuals whose data and assumptions are criticized recant. Indeed, understanding is enhanced if they can defend their work against criticism.[29] What is required is that community members pay attention to the critical discussion taking place and that the assumptions that govern their group activities remain logically sensitive to it.

EQUALITY OF INTELLECTUAL AUTHORITY

This Habermasian criterion is intended to disqualify a community in which a set of assumptions dominates by virtue of the political power of its adherents.[30] An obvious example is the dominance of Lamarckism in the Soviet Union in the 1930s. While there were some good reasons to try experiments under the aegis of a Lamarckian viewpoint, the suppression of alternative points of view was a matter of politics rather than of logic or critical discussion. The bureaucratization of United States science in the twentieth century tends similarly to privilege certain points of view.[31] The exclusion, whether overt or more subtle, of women and members of certain racial minorities from scientific education and the scientific professions has also constituted a violation of this criterion. While assumptions about race and about sex are not imposed on scientists in the United States in the way assumptions about inheritability of acquired traits were in the Soviet Union, . . . assumptions about sex structure a number of research programs in biology and behavioral sciences. Other scholars have documented the role of racial assumptions in the sciences.[32] The long-standing devaluation of women's voices and those of members of racial minorities means that such assumptions have been protected from critical scrutiny.

The above are criteria for assessing the objectivity of communities. The objectivity of individuals in this scheme consists in their participation in the collective give-and-take of critical discussion and not in some special relation (of detachment, hardheadedness) they may bear to their observations. Thus understood, objectivity is dependent upon the depth and scope of the transformative interrogation that occurs in any given scientific community. This communitywide process ensures (or can ensure) that the hypotheses ultimately accepted as supported by some set of data do not reflect a single individual's idiosyncratic assumptions about the natural world. To say that a theory or hypothesis was accepted on the basis of objective methods does not entitle us to say it is true but rather that it reflects the critically achieved consensus of the scientific community. In the absence of some form of privileged access to transempirical (unobservable) phenomena it's not clear we should hope for anything better.

The weight given to criticism in the formation of knowledge represents a social consensus regarding the appropriate balance between accurate representation and knowledge extension. Several conditions can limit the extent of criticism and hence diminish a scientific community's objectivity without resulting in a completely or intentionally closed society (for example, such as characterized Soviet science under Stalin or some areas of Nazi science).

First of all, if scientific inquiry is to have any effect on a society's ability to take advantage of natural processes for the improvement of the quality of its life, criticism of assumptions cannot go on indefinitely. From a logical point of view, of course, criticism of background assumptions, as of any general claim, can go on ad infinitum. The philosophical discussion of inductive reasoning is an example of such unending (though not useless) debate. The utility of scientific knowledge depends on the possibility of finding frameworks of inquiry that remain stable enough to permit systematic interactions with the natural world. When critical discussion becomes repetitive and fixed at a metalevel, or when criticism of one set of assumptions ceases to have or does not eventually develop a connection to an empirical research program, it loses its relevance to the construction of empirical knowledge. It is the intrinsic incapacity of so-called "creation science" to develop a fruitful research program based on its alleged alternative to evolutionary theory that is responsible for the lack of attention given to it by the contemporary United States scientific community. The appeal by its advocates to pluralistic philosophies of science seems misguided, if not disingenuous.

Secondly, these critical activities, however crucial to knowledge building, are de-emphasized in a context that rewards novelty and originality, whether of hypotheses or of experimental design. The commoditization of scientific knowledge—a result of the interaction of the requirements of career advancement and of the commercial value of data—diminishes the attention paid to the criticism of the acquisition, sorting, and assembling of data. It is a commonplace that in contemporary science papers reporting negative results do not get published.

In the third place, some assumptions are not perceived as such by any members of the community. When, for instance, background assumptions are shared by all members of a community, they acquire an invisibility that renders them unavailable for criticism. They do not become visible until individuals who do not share the community's assumptions can provide alternative explanations of the phenomena without those assumptions, as, for example, Einstein could provide an alternative explanation of the Michelson-Morley interferometer experiment. Until such alternatives are available, community assumptions are transparent to their adherents. In addition, the substantive principles determining standards of rationality within a research program or tradition are for the most part immune to criticism by means of those standards.

From all this it follows again that the greater the number of different points of view included in a given community, the more likely it is that its scientific practice will be objective, that is, that it will result in descriptions and explanations of natural processes that are more reliable in the sense of less characterized by idiosyncratic subjective preferences of community members than would otherwise be the case. The smaller the number, the less likely this will be.[33] Because points of view cannot simply be allowed expression but must have an impact on what is ultimately thought to be the case, such diversity is a necessary but not a sufficient condition for objectivity. Finally, these conditions reinforce the point that objectivity is a matter of degree. While the conditions for objectivity are at best imperfectly realized, they are the basis of an ideal by reference to which particular scientific communities can be evaluated. Ascertaining in greater detail the practices and institutional arrangements that facilitate or undermine objectivity in any particular era or current field, and thus the degree to which the ideal of objectivity is realized, requires both historical and sociological investigation. . . .

■ | Conclusion

On the positivist analysis of scientific method it is hard to understand how theories purporting to describe a nonobservable underlying reality, or containing descriptive terms whose meaning is independent of that of so-called observational terms, can be supported. On the antiempiricist wholist account it is just as difficult to understand how the theories that are developed have a bearing on intersubjective reality. Each of these approaches is also unable to account for certain facts about the actual practice of science. The absolute and unambiguous nature of evidential relations presented in the positivist view cannot accommodate the facts of scientific change. The incommensurability of theories in the wholist view cannot do justice to the lively and productive debate that can occur among scientists committed to different theories. Each of these modes of analysis emphasizes one aspect of scientific method at the expense of another, and each produces an individualist logic of scientific method that fails adequately to reflect the social nature of scientific discourse. Furthermore, the emphasis on theories distorts scientific growth and practice. Scientists rarely engage in the construction or evaluation of comprehensive theories. Their constructive, theoretical activity tends to consist much more in the development of individual or interrelated hypotheses (as laws, generalizations, or explanations) from the complex integration of observation and experiment with background assumptions. Success in expanding the scope of an explanatory idea via such complex integration plays as important a role in its acceptance as the survival of falsifying tests. Accounts of validation in the sciences must take account both of the role of background assumptions in evidential reasoning and of the roles of (sometimes) conflicting goals of inquiry with respect to which hypotheses and theories are assessed. The logic that reflects the structure of this activity will have to abandon some of the simplicity of the positivist account, but what it loses in elegance it will surely regain in application.

The analysis conducted in this chapter means that values can enter into theory-constructive reasoning in two major ways—through an individual's values or through community values. The fact that a bit of science can be analyzed as crucially dependent on contextual values or on value-laden background assumptions does not necessarily mean that someone is attempting to impose his/her wishes on the natural world without regard to what it might really be like. More customarily such analysis should be taken as showing the way in which such contextual features have facilitated the use of given data or observations as evidence for some hypothesis by an individual or by a community. Because community values and assumptions determine whether a given bit of reasoning will pass or survive criticism and thus be acceptable, individual values as such will only rarely be at issue in these analyses. When an individual researcher's values enable her or him to make inferences at variance with those of the scientific community, this is less evidence of strongly eccentric individualism than of allegiance to some other social (political or religious) community.[34]

The contextualist view produces a framework within which it is possible to respect the complexity of science, to do justice to the historical facts and to the current practice of science, and to avoid paradox. In addition, it is possible to articulate a standard of comparison independent of and external to any particular theory or research project. In making inter-theoretic comparison possible it offers the basis (an expanded basis) upon which to develop criteria of evaluation. Finally, the social account of objectivity and scientific knowledge to which the contextualist account of evidence leads seems more true to the fact that scientific inquiry is not always as free from subjective preference as we would wish it to be. And even though the resulting picture of objectivity differs from what we are used to, our intuition that scientific inquiry at its best is objective is kept intact by appealing to the spirit of criticism that is its traditional hallmark.[35]

■ | Notes

1. This mythology originates with the founders of modern science—compare Isaac Newton's "Rules of Reasoning in Philosophy" in Newton (1953), pp. 3–5—and has come to be the standard view.

2. It has become a subject of increased concern lately in light of several alleged incidents of data faking. Compare Broad (1981).

3. Hempel (1966), pp. 3–18, and Popper (1962), pp. 42–59.

4. Laudan (1977) does articulate criteria for what counts as progress. These are not necessarily criteria or standards for truth.

5. Feyerabend (1975).

6. Grene (1985).

7. James Watson's account of the discovery of the molecular structure of DNA, read in conjunction with the story of Rosalind Franklin's contributions to that discovery in Sayre (1975), provides a vivid example of this interdependence. See Watson (1968). Participant accounts of recent developments in one or another science usually offer good illustrations of this point. Weinberg (1977) and Feinberg (1978) account for the mid-1970s states of cosmology and microphysics, respectively. Each presents what can be called the current canon in its field, making clear the dependence of its production upon the activity and interaction of many individual researchers.

8. See Peters and Ceci (1982, 1985) and the associated commentary. For additional discussion of peer review see Glazer (1988); Goleman (1987); Cole and Cole (1977); Cole, Cole, and Simons (1981).

9. In what I take to be a similar vein, Bruno Latour (1987) claims that in science a statement made by an individual becomes a fact only as a consequence of what others do with the statement. Latour, however, emphasizes the agonistic as opposed to the cooperative dimension of social relations in the sciences.

10. The precise extension of "scientific community" is here left unspecified. If it includes those interested in and affected by scientific inquiry, then it is much broader than the class of those professionally engaged in scientific research. For a discussion of these issues and some consequences of our current restricted understanding of the scientific community see Addelson (1983).

11. One might say that the language game of science presupposes the independent existence of objects of experience. Contemporary arguments about scientific realism can be understood as arguments about (1) the nature of this presupposition and (2) what categories of objects it covers.

12. To avoid possible confusion about the point being made here, I wish to emphasize that I am contrasting the descriptive statements of science with expressions of

emotion. *Descriptions* of emotion and other subjective states may be as objective as other kinds of description, if the conditions for objectivity can be satisfied. Objectivity as it is being discussed here involves the absence (or control) of subjective *preference* and is not necessarily divorced from our beliefs about our subjective states. Locke (1968) discusses the different ways in which privacy is properly and improperly attributed to subjective states (pp. 5–12).

13. The distinction between the different kinds of concerns relevant to the development and evaluation of theories is discussed for different purposes and with significant differences in detail by Buchdahl in a discussion of criteria choice, by Laudan in a discussion of the problems that give rise to the development of theory and by Schaffner in a discussion of categories for comparative theory evaluation. A more complete categorization of concerns and types of criticism than that offered here requires a more thorough study of past and present scientific practice. See Gerd Buchdahl (1970); Larry Laudan (1977); and Kenneth Schaffner (1974).

14. Maddox, Randi, and Stewart (1988) and Benveniste's reply in Benveniste (1988). The chapter "Laboratories" in Latour (1987) can be read as providing a series of examples of evidential criticism (pp. 63–100).

15. Lewontin (1970, 1974).

16. Gould (1980).

17. The latter two kinds of questions are concerned with the objectivity of data, a notion mentioned above.

18. Bernstein (1973), pp. 137–177.

19. Williams (1966), pp. 32–63. A somewhat different account is presented by Hesse (1965), pp. 170–180.

20. Kuhn (1957), pp. 100–133, 185–192.

21. Holton (1978).

22. Jaffe (1960), pp. 95–103.

23. Rowell (1974).

24. See Longino (1987).

25. Conceptual criticism of this sort is a far cry from the criticism envisaged by Popper. For him metaphysical issues must be decided empirically, if at all. (And if they cannot be so tested, they lack significance.)

26. This is really a distinction between the number of points of view (minds) required. Many individuals (sharing assumptions and points of view) may be involved in testing a hypothesis (and commonly are in contemporary experiments). And though this is much rarer, one individual may be able to criticize her or his own evidential reasoning and background assumptions from other points of view.

27. This is not to deny the importance of distinguishing between different modes of understanding—for instance, between scientific, philosophical, and literary theories—but simply to deny that objectivity can serve as any kind of demarcation criterion.

28. Kuhn (1977) [pp. 94–110].

29. Beatty (1985) makes a similar point.

30. Invocation of this criterion confirms the kinship of this account of objectivity with the account of truth that Jürgen Habermas has developed as part of his theory of communicative competence. . . .

31. See Levins and Lewontin (1985), pp. 197–252, for further discussion of this point.

32. See Gould (1981); Lewontin, Rose, and Kamin (1984); Richardson (1984).

33. This insistence on the variety of points of view required for objectivity is developed on a somewhat different basis for the social sciences by Sandra Harding (1978).

34. This should not be taken to mean that social inequality and marginalization are necessary for objectivity but rather that differences in perspective are. A scientific community existing in a (utopian at this point) society characterized by thoroughgoing inclusivity and equality might indeed encourage the persistence of divergent points of view to ensure against blindness to its own assumptions.

35. Note added in proof. Three books read since completing the manuscript also draw attention in varying degrees to the social character of cognitive processes in science: Peter Galison, *How Experiments End* (Chicago, IL: University of Chicago Press, 1987); David Hull, *Science as a Process* (Chicago, IL: University of Chicago Press, 1988); and Sharon Traweek, *Beamtimes and Lifetimes: The World of High Energy Physicists* (Cambridge, MA: Harvard University Press, 1988).

■ | References

Addelson, Kathryn Pyne. 1983. "The Man of Professional Wisdom." In *Discovering Reality: Feminist Perspectives on Epistemology, Metaphysics, Methodology, and Philosophy of Science*, ed. Sandra Harding and Merrill Hintikka, pp. 165–186. Dordrecht: D. Reidel.

Beatty, John. 1985. "Pluralism and Panselectionism." In *PSA 1984*, ed. Peter Asquith and Philip Kitcher, pp. 25–83. East Lansing, MI: Philosophy of Science Association.

Benveniste, Jacque. 1988. "Reply to Maddox, Randi and Stewart." *Nature* 334: 291.

Bernstein, Jeremy. 1973. *Einstein*. Bungay: William Collins and Son, Ltd.

Broad, William. 1981. "Fraud and the Structure of Science." *Science* 212: 137–141.

Buchdahl, Gerd. 1970. "History of Science and Criteria of Choice." In *Minnesota Studies in the Philosophy of Science*, ed. Roger Steuwer, 5: 205–230. Minneapolis: University of Minnesota Press.

Cole, Stephen, Jonathan R. Cole, and Gary Simons. 1981. "Chance and Consensus in Peer Review." *Science* 214: 881–886.

Cole, Stephen, Leonard Rubin, and Jonathan R. Cole. 1977. "Peer Review and the Support of Science." *Scientific American* 237, no. 4: 34–41.

Feinberg, Gerald. 1978. *What Is the World Made Of?* New York: Anchor Press.

Feyerabend, Paul K. 1975. *Against Method*. London: Verso.

Glazer, Sarah. 1988. "Combating Science Fraud." *Editorial Research Reports* 2: 390–399.

Goleman, Daniel. 1987. "Failing to Recognize Bias in Science." *Technology Review* 90 (November–December): 26–27.

Gould, Stephen J. 1980. "Sociobiology and the Theory of Natural Selection." In *Sociobiology: Beyond Nature/Nurture?* ed. George Barlow and James Silverberg, pp. 257–269. Boulder, CO: Westview Press.

———. 1981. *The Mismeasure of Man.* New York: W.W. Norton and Co.

Grene, Marjorie. 1985. "Perception, Interpretation and the Sciences." In *Evolution at a Crossroads,* ed. David Depew and Bruce Weber, pp. 1–20. Cambridge, MA: MIT Press.

Harding, Sandra. 1978. "Four Contributions Values Can Make to the Objectivity of the Social Sciences." In *Proceedings of the 1978 Beinnial Meeting of the Philosophy of Science Association,* ed. Peter Asquith and Ian Hacking, pp. 199–209. East Lansing, MI: Philosophy of Science Association.

Hempel, Carl Gustav. 1966. *Philosophy of Natural Science.* Englewood Cliffs, NJ: Prentice Hall.

Hesse, Mary. 1965. *Forces and Fields,* Totowa, NJ: Littlefield Adams.

Holton, Gerald. 1978. "Subelectrons, Presuppositions, and the Millikan-Ehrenhaft Dispute." In *The Scientific Imagination,* pp. 25–83. Cambridge: Cambridge University Press.

Jaffe, Bernard. 1960. *Michelson and the Speed of Light.* Garden City, NY: Doubleday and Co.

Kuhn, Thomas. 1957. *The Copernican Revolution.* Cambridge, MA: Harvard University Press.

———. 1977. *The Essential Tension.* Chicago: University of Chicago Press.

Latour, Bruno. 1987. *Science in Action.* Cambridge MA: Harvard University Press.

Laudan, Larry. 1977. *Progress and Its Problems.* Berkeley, CA: University of California Press.

Levins, Richard, and Richard Lewontin. 1985. *The Dialectical Biologist.* Cambridge, MA: Harvard University Press.

Lewontin, Richard. 1970. "Race and Intelligence." *Bulletin of the Atomic Scientists* 26 (March): 2–8.

———. 1974. "The Analysis of Variance and the Analysis of Causes." *American Journal of Human Genetics* 26: 400–411.

Lewontin, Richard, Steven Rose, and Leon Kamin. 1984. *Not in Our Genes: Biology, Ideology and Human Nature.* New York: Pantheon Books.

Locke, Don. 1968. *Myself and Others.* London: Oxford University Press.

Longino, Helen. 1987. "What's Really Wrong with Quantitative Risk Assessment?" In *PSA 1986,* ed. Arthur Fine and Peter Machamer. pp. 376–383. East Lansing: MI: Philosophy of Science Association.

Maddox, John, James Randi, and Walter W. Stewart. 1988. "High Dilution Experiments a Delusion." *Nature* 334: 287–290.

Newton, Isaac. 1953. "Rules of Reasoning in Philosophy." In *Newton's Philosophy of Nature*, ed. H. S. Thayer, pp. 3–5. New York: Hafner.

Peters, Donald, and Stephen Ceci. 1982. "Peer Review Practices of Psychological Journals: The Fate of Published Articles Submitted Again." *Behavioral and Brain Sciences* 5: 187–195.

———. 1985. "Beauty Is in the Eye of the Beholder." *Behavioral and Brain Sciences* 8, no. 4. 747–749.

Popper, Karl. 1959. *The Logic of Scientific Discovery*. London: Hutchinson.

———. 1962. *Conjectures and Refutations*. New York: Basic Books.

Richardson, Robert C. 1984. "Biology and Ideology: The Interpenetration of Science and Values." *Philosophy of Science* 51: 396–420.

Rowell, Thelma. 1974. "The Concept of Dominance." *Behavioral Biology* 11: 131–154.

Sayre, Anne. 1975. *Rosalind Franklin and DNA*. New York: W. W. Norton.

Schaffner, Kenneth. 1974. "Einstein versus Lorentz: Research Programmes and the Logic of Theory Evaluation." *British Journal for the Philosophy of Science*: 45–78.

Watson, James. 1968. *The Double Helix*. New York: Atheneum.

Weinberg, Steven. 1977. *The First Three Minutes*. New York: Basic Books.

Williams, L. Pearce. 1966. *The Origins of Field Theory*. New York: Random House.

Kathleen Okruhlik

Gender and the Biological Sciences

Feminist critiques of science provide fertile ground for any investigation of the ways in which social influences may shape the content of science. Many authors working in this field are from the natural and social sciences; others are philosophers. For philosophers of science, recent work on sexist and androcentric bias in science raises hard questions about the extent to which reigning accounts of scientific rationality can deal successfully with mounting evidence that gender ideology has had deep and extensive effects on the development of many scientific disciplines.

Feminist critiques of biology have been especially important in the political struggle for gender equality because biologically determinist arguments are so often cited to 'explain' women's oppression. They explain why it is 'natural' for women to function in a socially subordinate role, why men are smarter and more aggressive than women, why women are destined to be homebodies, and why men rape. Genes, hormones, and evolutionary processes are cited as determinants of this natural order and ultimately as evidence that interventions to bring about a more egalitarian and just society are either useless or counterproductive.

The critiques of biology are also *epistemically* important because of the position that biology occupies in the usual hierarchy of the sciences—somewhere between physics, on the one hand, and the social sciences, on the other. Very often feminist critiques of the social sciences are dismissed out of hand by philosophers of science on the grounds that the social sciences aren't science anyway; and so the feminist critiques, however devastating, are said to tell us nothing about the nature of real science. It is, however, not quite so easy to dismiss biology as pseudo-science; and so the critiques in this area assume added significance. If we are to infer in light of the feminist critiques anything about the nature of science (its rationality, its objectivity,

FROM *Biology and Society, Canadian Journal of Philosophy*, Supplementary vol. 20 (1994): 21–42.

its degree of insulation from social influences, its character as an individual or collective enterprise), then the biological sciences are perhaps the best place to start. Hence this essay. It has four parts. In section I, several case studies of gender ideology in the biological sciences are reviewed. This review provides a common stock of examples for discussion purposes and the opportunity to indicate very briefly how standard theses in philosophy of science can provide partial illumination of them. In section II, the possible epistemic significance of these case studies (and others like them) is addressed in light of alternative conceptions of science available in the feminist literature. The third part of the essay develops an account of the relation between contexts of discovery and justification that makes room for the sorts of social and cultural influences on science exemplified by gender bias while still allowing room for fairly robust notions of objectivity and rationality. Finally, in section IV, an attempt is made to locate this account vis-à-vis others represented among feminist critiques of science.

I | Some Case Studies

Consider first a 1988 article entitled "The Importance of Feminist Critiques for Contemporary Cell Biology," authored by the Biology and Gender Study Group at Swarthmore College.[1] The article discusses the ways in which contemporary research is still shackled by outmoded models of the relationship between egg and sperm in reproduction. In particular, commitment to the Sleeping Beauty/Prince Charming model of egg and sperm may have blinded researchers and theoreticians to some of the facts about human reproduction. Just as women are seen to be passive and men active, so traditionally have egg and sperm been assigned the traditional feminine and masculine roles. The egg waits passively while the sperm heroically battles upstream, struggles against the hostile uterus, courts the egg, and (if victorious) penetrates by burrowing through, thereby excluding all rival suitors. The egg's only role in this saga is to select which rival will be successful.

The notion that the male semen awakens the slumbering egg appeared as early as 1795 and has been influential ever since. In the last fifteen years, however, some rival accounts have challenged the old narrative by making the egg an energetic partner in fertilization. For example, using electron microscopy it can be shown that the sperm doesn't just burrow through the egg, as previously thought. Instead, the egg directs the growth of small, finger-like projections of the cell surface to *clasp* the sperm and slowly draw it in. This mound of microvilli extending to the sperm was discovered in 1895 when the first photographs of sea urchin fertilization were published; but it has largely been ignored until recently.

What matters for our purposes here is not whether the newer theory is entirely correct (it is still controversial), but that its very existence as a rival to the more established views throws into sharp relief the questionable assumptions of the older model. It shows us how pre-existing theoretical assumptions inform which questions we ask, which hypotheses we investigate, and which data we decide to ignore as evidentially insignificant. These considerations are sometimes relegated to the context of discovery and are said to be epistemically irrelevant to the actual content of science. This is a topic to which we shall return later. In the meantime, let us investigate some cases in which the controverted question is not whether some data are evidentially significant at all, but which interpretation should be placed upon the same data as the result of differing theoretical commitments.

Many feminist criticisms of primatology and sociobiology focus on the fact that male struggle, male competition, and male inventiveness are portrayed as the bases for human evolution. In familiar passages from *The Descent of Man* quoted by Ruth Hubbard and other critics, Darwin attributes evolutionary development in human beings almost exclusively to male activity.

> [Men have had] to defend their females, as well as their young, from enemies of all kinds, and to hunt for their joint subsistence. But to avoid enemies or to attack them with success, to capture wild animals, and to fashion weapons, requires the aid of the higher mental faculties, namely observation, reason, invention, or imagination. These various faculties will thus have been continually put to the test and selected during manhood.

'Thus,' the discussion ends, 'man has ultimately become superior to woman' and it is a good thing that men pass on their characteristics to their daughters as well as to their sons, 'otherwise it is probable that man would have become as superior in mental endowment to woman, as the peacock is in ornamental plumage to the peahen.'[2]

The influence of Darwin's androcentric bias has not been limited to evolutionary biology, since that theory functions as an auxiliary hypothesis in many other disciplines. Consider, for example, anthropology. If one holds the view that man-the-hunter is chiefly responsible for human evolutionary development, one interprets fossil evidence in light of the changing behavior of males. Helen Longino and Ruth Doell, for example, in a very important 1983 article called 'Body, Bias, and Behavior: A Comparative Analysis of Reasoning in Two Areas of Biological Science,'[3] trace the ways in which the androcentric account attributes the development of tool use to male hunting behavior. Some recent work, however, has suggested that up to 80% of the subsistence diet of what used to be called hunter-gatherer societies came from female gatherers. If that is the background theory informing one's interpretation of the evidence, then quite a different account of that same evidence emerges. This is how Longino and Doell summarize the point:

> By contrast [with the androcentric account], the gynecentric story explains the development of tool use as a function of female behavior, viewing it as a

response to the greater nutritional stress experienced by females during pregnancy, and later in the course of feeding their young through lactation and with foods gathered from the surrounding savanna. Whereas man-the-hunter theorists focus on stone tools, woman-the-gatherer theorists see tool use developing much earlier and with organic materials such as sticks and reeds. They portray females as innovators who contributed more than males to the development of such allegedly human characteristics as greater intelligence and flexibility. Women are said to have invented the use of tools to defend against predators while gathering and to have fashioned objects to serve in digging, carrying, and food preparation.

Again, what matters here is not that the gynecentric hypothesis be *true* but rather that it makes obvious the extent to which the standard interpretation of the anthropological evidence has been colored by androcentric bias.

The cases examined so far are instances in which attention to the theory-ladenness of observation or the underdetermination of theory by data shed some light on the way in which pre-existing theoretical commitments regarding sex and gender may influence decisions about which questions get asked, which data must be accounted for and which can safely be ignored, as well as which interpretation among those that are empirically adequate is actually adopted. There are other cases in which attention to the Duhem-Quine thesis is helpful. Even if the body of relevant data has already been strictly delimited with preferred interpretations settled upon, and even if the test hypothesis has been selected, it is still to some extent an open question how one ought to respond to apparently falsifying information. Although one may simply reject the test hypothesis, it is also possible to pin the blame for a failed prediction on one of the background assumptions that was used to generate the failed prediction. The arrow of *modus tollens*, in other words, may be redirected away from the test hypothesis and toward one or more of the auxiliaries. This, of course, can be a perfectly respectable and useful response to failed prediction; but it does raise interesting questions about what factors (social as well as more narrowly 'cognitive') motivate our decisions to protect some hypotheses from falsification. It also draws attention to the important role played in theory assessment by our background assumptions, a role that is particularly crucial in the present discussion since so few of our background assumptions about sex and gender have been subjected to systematic scrutiny.

Certain hypotheses seem to survive one falsification after another, with the blame for failure and the subsequent adjustment always being located elsewhere in the system of beliefs. I have in mind here recent developments in neuroanatomy which are directed to explaining intelligence differences between women and men, particularly as these relate to alleged male superiority with respect to mathematical and spatial ability. Anne Fausto-Sterling, in her book *Myths of Gender*,[4] has surveyed some recent theories; the following examples are taken from her discussion.

It has been suggested that spatial ability is X-linked and therefore exhibited more frequently in males than in females; that high levels of prenatal androgen increase intelligence; that lower levels of estrogen lead to superior male ability at 'restructuring' tasks. Some have held that female brains are more lateralized than male brains and that greater lateralization interferes with spatial functions. Others have argued that female brains are *less* lateralized than male brains and that less lateralization interferes with spatial ability. Some have attempted to save the hypothesis of X-linked spatial ability from refuting evidence by suggesting that the sex-linked spatial gene can be expressed only in the presence of testosterone. Others have argued that males are smarter because they have more uric acid than females.

None of these hypotheses is well-supported by the evidence and most seem to be clearly refuted. What is interesting for our purposes is that for many researchers the one element of the theoretical network they are unwilling to surrender in the face of recalcitrant data is the assumption that there must be predominantly *biological* reasons for inferior intellectual achievement in women.

Some have found this situation reminiscent of nineteenth-century craniometry's well-known attempt to explain inferior female intelligence by appealing to brain size. This is a case also discussed by Fausto-Sterling. The 'bigger is better' hypothesis foundered on the elephant problem (if absolute size were the true measure of intelligence, elephants should be smarter than people). So it was suggested that the true measure of intelligence lay in the proportion of brain mass to body mass; but this proportion favored women, and so the hypothesis was quickly rejected. The proposal that greater intelligence is linked to a lower ratio of facial bones to cranial bones ran afoul of the 'bird problem.' So it was suggested that the frontal lobes are the seat of the intellect, and men have bigger frontal lobes; the parietal lobes are larger in women. This hypothesis was surrendered when newer research pointed to the parietal lobes as the seat of the intellect. So the data were re-evaluated to show that *really* women have smaller parietal lobes . . . and so the saga continued. The one component of the theoretical network that scientists were unwilling to give up in the face of apparent falsification was the underlying assumption that women are *biologically* determined to be less intelligent than men. It is no wonder that feminist critics find the same pattern reinstated in current debates about gender and mathematical ability.

In the preceding cases, appeal has been made to such standard philosophical theses as the theory-ladenness of observation, the underdetermination thesis, and the Duhem-Quine thesis in order to suggest how gender ideology could permeate the biological sciences even on fairly standard accounts of theory appraisal. In these cases, we might want to say that external values have been imported into science; but the values are *implicit* in these cases and often exposed only in light of a rival hypothesis embedding conflicting values. The situation is different in the last set of cases in this rapid review of the literature. In the medical sciences, values or norms are often quite explicit. When one has to judge who is healthy and who is diseased, what body types are desirable and which not, the concepts involved are explicitly normative as well as descriptive. This opens the door for types of gender bias other than those discussed above. In one type, different ideals are set for male and female; these ideals are said to be 'complementary' but really only the male is seen as fully human. Another type of bias occurs when a single norm is adopted for both males and females, but is in actuality a male rather than human norm.

A nice historical example of the complementarity problem is developed in Londa Schiebinger's excellent book, *The Mind Has No Sex? Women in the Origins of Modern Science*.[5] Schiebinger documents the changes that occurred in representations of male and female anatomy as a concerted effort was made in the eighteenth century to ground gender differences in anatomy. If differences between masculinity and femininity could be located in the *bones* of the organism, in its infrastructure, then there would be a modern scientific account of difference, and it would no longer be necessary to rely on the old heat models of Aristotle and Galen to do the job.

Prior to this time, male and female skeletons had been portrayed as similar; they were not sexualized. Sometimes the sex of the skeleton was not identified; sometimes the front view was represented as male, the back view as female. But all this changed in the years between 1730 and 1790.

> The materialism of the age led anatomists to look first to the skeleton, as the hardest part of the body, to provide a "ground plan" for the body and give a "certain and natural" direction to the muscles and other parts of the body attached to it. If sex differences could be found in the skeleton, then sexual identity would no longer depend on differing degrees of heat (as the ancients had thought), nor would it be a matter of sex organs appended to a neutral human body (as Vesalius had thought). Instead, sexuality would be seen as penetrating every muscle, vein, and organ attached to and moulded by the skeleton.[6]

The male and female ideals that emerged were very different from one another. The male skeleton was typified by a big head and strong shoulders; its animal analogue was the horse, which sometimes appeared in the background of male skeletal drawings. The female skeleton had a large pelvis, a long elegant neck, and a smallish head. She had much in common with the ostrich who sometimes decorated her portrait. Those skeletons which approximated most closely to cultural ideals of masculinity and femininity were favored for just that reason over drawings that were in some sense more accurate.

It is worth noting that one way the largeness of the female pelvis and smallness of the head are emphasized is by depicting a very narrow rib cage. Fausto-Sterling points out that there may have been more than just the power of ideology at work here. It may be that some of the corpses on which the drawings were modeled had their rib cages compressed by long-term use of the corset. This reminds us again that Ruth Hubbard[7] and others are correct when they argue that it is wrong to think of the body as a purely biological infrastructure onto which the socio-cultural crud of gender accretes. Although the distinction between 'sex' as biological and 'gender' as socially assigned has in many respects served feminist theorizing well, it has sometimes led to the mistaken assumption that all biological attributes are given in some absolute sense. *Sex* as well as gender is socially constructed, at least in part. Such 'physical givens' as height, bone density, and musculature are to a large extent determined by cultural practice.

The skeletal case is one in which the male and female norms are said to be complementary, but the male is treated as more fully human. In other cases, there is allegedly a single human norm, but on closer inspection it turns out to be masculine. It has been suggested that the treatment of menstruation, pregnancy, and childbirth as diseases or medical emergencies may be traced to the fact that these are not things that happen to the ideal healthy human being who is, of course, male. The ideal healthy lab rat is also male. His body, his hormones, and his behaviors define the norm; so he is used in experiments. Female hormones and their effects are just *nuisance variables* that muck up the works, preventing experimenters from getting at the pure, clean, stripped-down essence of rathood as instantiated by the male model. Insofar as the female of the species is truly a rat (or truly a human being), she is covered by the research on males. Insofar as she is not included in that research, it is because she is not an archetypal member of her own species. The dangerous effects of such research procedures, especially in the biomedical sciences, are just now being documented. For far too long, the assumption underlying these experimental designs (that males are the norm) simply went unchallenged.

Elisabeth Lloyd [has written] a book called [*The Case of the Female Orgasm*], and it includes a lovely example of a male norm masquerading as human. Various sociobiologists, when advancing theories about the evolutionary origins of the female orgasm, have cited detailed statistics about the nature, length, frequency, and repeatability of orgasm in order to support their origin stories. When tracking down their footnotes, Lloyd discovered that these statements—which were being used to explain the origins of the *female* orgasm—were in fact based on data about *male* orgasms. The sleight of hand was typically accomplished by referring to the male subjects as 'individuals' rather than males!

II | Varieties of Feminist Critique

Case studies such as those canvassed above are interesting in their own right, but they leave open the question of what we are to make of them. Two contexts in which this question arises interest me particularly.

1 In the feminist literature the question that has been foremost in the last few years is whether these case studies are examples of 'bad science'

or whether, on the contrary, they show that science is intrinsically and irredeemably androcentric.

2 In philosophy of science the question too often has been: what does this have to do with philosophy of science?

The two questions are related, and I should like to tackle them together. With respect to the first, Sandra Harding's tripartite taxonomy of feminist epistemologies has been extremely influential.[8] In order to deal with the bewildering diversity of feminist critiques of science, Harding proposes dividing them into three categories: feminist empiricism, standpoint epistemologies, and feminist postmodernism.

'Feminist empiricism' diagnoses failures such as those sketched above as failures of science to live up to its own ideals. Androcentric bias has gotten in the way of rigorous application of scientific method; but if the canons of science had been adhered to faithfully, episodes such as those above could have been avoided. For feminist empiricism, the standpoint of the knower is epistemically irrelevant; any bias originating from that standpoint will be eliminated by proper application of objective methods.

This assumption is denied by 'standpoint epistemologists' who argue that the credentials of the knowledge claim depend in part on the situation of the knower. Just as Hegel's slave could know more than the master, so women (or feminists) may enjoy an epistemic advantage over men (or non-feminists). A science based upon the standpoint of women would be an improvement over current science, according to standpoint epistemology. In this sense it is still a 'successor science' project, since its aim is to produce a *better* (epistemically superior) account of the world. A number of problems have been pointed out with this approach, but the most damaging criticism has been the insistence that there is no single feminist standpoint. Just as the standpoint of women differs from that of men, so also the standpoint of poor women differs from that of rich women, the standpoint of black women from that of white women, the standpoint of lesbians from that of heterosexual women, and so on. On what grounds could one of these be privileged over the other as a standpoint from which to describe the world?

This fracturing of identities and hence of standpoints has led some theorists to embrace what Harding calls 'feminist postmodernism' by giving up altogether the endeavor to become more and more objective and by accepting the existence of an irreducible plurality of alternative narratives about the way the world is. The notion of a scientific method that might allow us to transcend the constraints of culture, time, and place is repudiated once and for all by feminist postmodernists. Transtheoretical criteria for rationality and objectivity are dismissed as products of a masculine mythology, and the 'successor science' project is abandoned.

Although Harding's taxonomy has been very helpful in facilitating analyses of the diverse philosophical commitments of feminist critics, I fear that it also tends to obscure a promising possibility, one that takes into account the ways in which social structures (like the institution of gender) affect the very content of science without surrendering altogether the ideal of rational theory choice. In the following section, such a position is sketched.

III | Science and Shared Social Values

Traditionally, philosophy of science has been quite willing to grant that social and psychological factors (including perhaps gender) play a role in science; but that role has been a strictly delimited one, contained entirely within the so-called context of discovery, or alternatively within those episodes called 'bad science' in which the canons of rationality were clearly violated in favor of other interests. (The Lysenko Affair is a standard example here.) In the context of discovery or theory generation, says the traditional story, anything goes: the source of one's hypotheses is epistemically irrelevant; all that matters is the context of justification. If you arrived at your hypothesis by reading tea leaves, it doesn't matter so long as the hypothesis is confirmed or corroborated in the context of justification. You test the hypothesis in the tribunal of nature and if it holds up, then you're justified in holding on to it—whatever its origins. The idea here is that the canons of scientific theory choice supply a sort of filter which removes social, psychological, and political contaminants as a hypothesis passes from one context to the next.

This view made a certain amount of sense in the first half of this century when models of theory evaluation held that hypotheses were compared directly to nature. But this account, which shears the context of discovery or theory generation of all epistemic significance, makes no sense at all given current models of scientific rationality that view theory choice as irreducibly *comparative*. That is, we now recognize that one does not actually compare the test hypothesis to nature directly in the hope of getting a 'yes' or 'no' ('true' or 'false') answer; nor does one compare it to all logically possible rival hypotheses. We can only compare a hypothesis to its extant rivals—that is, to other hypotheses which have actually been articulated to account for phenomena in the same domain and developed to the point of being testable. So the picture underlying current debates regarding theory choice looks something like this:

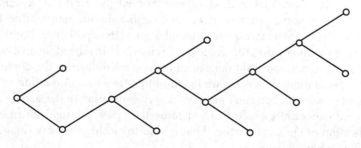

Each of the nodes is meant to represent a decision point at which the scientist must choose among alternative rivals. Methodological objectivists argue that so long as the proper machinery of theory assessment is brought to bear at each of the nodes, the rationality of science is preserved. How the nodes were generated in the first place is irrelevant, so long as the right decisions are made at each juncture. There may be interesting sociological stories to tell about the generation of the various alternative hypotheses, but sociological influences are effectively screened from affecting the content of science by the decision procedure operating at the nodes. This procedure will tell us which theory is preferable to its extant rivals on purely objective grounds.

My point, however, is that even if we grant for the sake of argument that scientific method is itself free of contamination by non-cognitive factors and that the decision procedure operates perfectly at the nodes, nothing in this procedure will insulate the content of science from sociological influences *once we grant that these influences do affect theory generation*. If our choice among rivals is irreducibly comparative, as it is on this model, then scientific methodology cannot guarantee (even on the most optimistic scenario) that the preferred theory is true—only that it is epistemically superior to the other *actually available* contenders. But if all these contenders have been affected by sociological factors, nothing in the appraisal machinery will completely 'purify' the successful theory.

Suppose, for the sake of example, that the graph represents the history of theories about female behavior. These theories may in many respects be quite different from one another; but if they have all been generated by males operating in a deeply sexist culture, then it is likely that all will be contaminated by sexism. Non-sexist rivals will never even be generated. Hence the theory which is selected by the canons of scientific appraisal will simply be the best of the sexist rivals; and the very *content* of science will be sexist, no matter how rigorously we apply objective standards of assessment in the context of justification. In fact, the best of the sexist theories will emerge more and more highly *confirmed* after successive tests.

So, if my account is right, it doesn't necessarily follow that the presence of androcentrism and sexism in science makes rational theory choice impossible, but it *does* follow that scientific method *by itself* as currently understood cannot be counted upon to eliminate sexist or androcentric bias from science. Note that methodological rationalists can still have (approximately) monotonic progress. Every choice among alternatives may be a rational choice. Science can (in principle) get better and better. But this in no way guarantees that the content of science is insulated against social influences. Once you grant that social factors may influence the context of theory generation, then you *have* to admit that they may also influence the content of science. You can't just give theory generation to the social scientists and expect to exclude them at some later date through the rigorous application of epistemic virtue. That is akin to closing the barn door after the horses have escaped.

Let me make the same point in a different way. One of David Bloor's[9] favorite arguments in support of social influences on theory content is based on the well-known underdetermination thesis. This is, of course, the claim that the data cannot pick out a single theory which uniquely accounts for them. There are, in principle, an infinite number of rival contenders that could do the same job. So, Bloor argues that if the data aren't completely determining our theory choices, then something else must be doing the job—and, of course, his favorite candidates for that job are sociological in character.

Larry Laudan's reply[10] is that there is an unfortunate tendency in the recent literature to overestimate underdetermination. Underdetermination, he says, would be a problem if we were actually faced with an infinitude or even a pair of empirically adequate rival theories. But, as a matter of fact, he says, we never encounter such an embarrassment of riches. We're lucky if we get even *two* rivals that are credible contenders for theory acceptance. We're certainly never faced with more than a small handful of competing alternatives. And we can always find (at least in the passage of time) good cognitive reasons for preferring one of these to the others. So, he concludes, although the underdetermination thesis may pose nice problems in principle, these never figure into *actual* theory choice. This is, of course, another way of stating his claim that theory choice is irreducibly comparative in nature—that our choices will always be among a finite class of extant alternatives, not among an infinitude of in-principle rivals. Since there will always be good reasons for preferring one of the extant rivals to the others, he claims that Bloor's invocation of social determinants is effectively undercut.

Notice, however, that there is an important sense in which this argument strategy simply begs the question. We can still ask why just *this* class of contenders was generated, given that others were equally compatible with the data. To say that once the rivals are fully articulated, our choice among them can be rational is to leave untouched the prior question of how our options came to be determined in the particular ways that they are. As long as Laudan concedes (as he does) that non-cognitive factors play a role in the posing of questions, the weighting of problems, and the initial articulation of theory, he cannot be sure that these factors will be eliminated in the context of justification. I stress once again that it is his attempt to maintain the *conjunction* of two views that gets him into trouble. The first of these views is that the context of discovery is normatively insignificant; the second is that theory appraisal is irreducibly comparative in nature. Once the second claim is made we *must* grant that factors affecting theory generation acquire normative significance.

The argument here is *not* that we should abolish the distinction between contexts of discovery and justification, but that we must recognize that on a comparative model factors that influence theory development and theory generation must necessarily influence our confirmation practices and hence the very content of science.

It is important to stress here that this argument about confirmation practices applies not only to test hypotheses but also to the auxiliary assumptions that jointly constitute the relevant background theory. How a particular piece of evidence bears on a hypothesis depends in large measure upon the collateral assumptions that come into play. It is here that the relationship between biology and the social sciences is particularly interesting because the traffic between the two is largely at this level. This is illustrated in some of the examples I cited earlier in this paper. For instance, in the man-the-hunter example, the relevant auxiliary assumptions are imported from evolutionary biology. In particular, it is the assumption that it is the male struggle for survival that drives the human evolutionary process that dictates in large measure what should count as evidence and how it should be interpreted. Conversely, in the Sleeping Beauty/Prince Charming model of the egg-sperm interaction, the biological data are informed by sociological assumptions about appropriate male and female roles. Donna Haraway's work in primatology[11] provides nice examples of how experimental design is influenced by background assumptions. She has traced the development of primatology since 1900, showing how political principles of hierarchy and male dominance have been embedded in that science, re-enforcing a theory of primate social organization in which a large, aggressive male is portrayed as defending a hierarchically organized troop and territory, enjoying first choice in food, sex, and grooming, and deciding troop movements. Consequently, when Carpenter undertook his highly acclaimed work on rhesus monkeys, he removed dominant males to test his organizing hypothesis about the source of social order but undertook no control study in which other members of the group were removed. We can't control for every possible variable in our experimental designs; so which we take into account depends on what our background theory tells us may be relevant. If the components of that background theory are never called into question or challenged by a serious rival, our experimental practices will continue to embody potentially problematic assumptions.

The claim here is *not* that the traffic in auxiliary assumptions makes a pernicious form of holism inevitable or that these auxiliaries are not themselves (potentially) testable,[12] but that they provide points at which gender biases from one discipline are easily transported into another. Furthermore, because of the pervasiveness of gender ideology in our culture, these assumptions generally are not called into question and are sometimes not even noticed. It is usually the case that they come to light only in the presence of a rival hypothesis.

The argument here is not restricted to hypothetico-deductive forms of confirmation and cannot be evaded by an appeal to Clark Glymour's 'bootstrapping' model.[13] Bootstrap confirmation does not make background assumptions dispensable but explicitly recognizes their crucial role: 'Hypotheses are not tested by themselves but only in relation to their fellows within the theory. Confirmation is a three-place relation, not a two-place relation.

Large parts of the theory may be invoked in confirming, from given evidence, any of its hypotheses.'[14]

I have been arguing all along that *even if we grant that the standards of theory assessment are free of contamination by non-cognitive factors*, nonetheless, non-cognitive values may permeate the very content of science. Stating the thesis in this way seemed useful because it avoided the messy controversy regarding the culture-bound evolution of scientific method itself. Even *granting* the transcendence of method, in other words, the scientific product could itself be radically culture-bound.

I should mention in bringing this line of argument to a close, however, that what has been granted for the sake of argument is probably not plausible in the final analysis. Scientific method itself is developed and articulated by culture-bound individuals and so the arguments which applied at the object level of theory content will likely apply at the meta-level of theory evaluation as well. Although we may have *good reasons* for making certain methodological changes, (e.g., for moving from single blind to double blind experiments), our methodological choices will be limited by the range of alternatives already actualized.

Finally, I should touch very briefly on the implications of the preceding argument regarding the scope of models of rationality and its implications for science policy.

These appear to be the two alternatives: (1) We could simply acknowledge the reduced scope of models of rationality and make more modest claims for the objectivity of science; or (2) We could attempt to enlarge our model of rationality so that it takes into account the context of theory generation. That is, if we acknowledge that the context of theory generation has normative significance, then we may want to alter science policy in the light of a new normative account of theory generation.

Once we recognize that the content of science is affected by the social arrangements that govern its practice and production, then those social arrangements acquire *epistemic* significance as do the affirmative action programs and other interventions undertaken to alter those social arrangements. Any adequate philosophy of science will have to take this into account.

IV | Reviewing the Situation[15]

How does the account sketched above fit into Harding's taxonomy of feminist critiques of science? Clearly it shares much in common with so-called 'feminist-empiricism' insofar as it is a successor science project that aims at ever-increasing objectivity and rationality through the use of established scientific methods. It parts company with feminist empiricism, however, in at least two important respects. First, it recognizes that current methodologies simply do not take into account the epistemic significance of the social arrangements that govern the activities scientists undertake and the products they produce. Any adequate methodology will have to control for the biases introduced by these social arrangements just as it has to control for other sources of bias. (It has become fashionable recently to eschew talk of 'bias' on the ground that such talk implies the possibility of science that is entirely free of bias. I don't think the implication holds, and so I continue to speak of gender bias. We aim to eliminate other forms of partiality without thinking we'll ever be entirely successful; the same regulative ideal seems perfectly serviceable in discussions of androcentrism.)

Second, the feminist empiricism described by Harding does not appear to challenge the assumption in much traditional methodology that the rationality of the scientific community is just individual rationality writ large, a simple summation of individual rationalities. In the account sketched above, it is the rationality of the scientific community that is enhanced by inclusion of diverse strategies at the individual level. The kinds of bias discussed above can be systematically addressed only at the community level; no adequate program of *individual* rehabilitation could be prescribed in advance. Only the inclusion of diverse standpoints will bring about the conditions under which change is possible.

Is the current proposal then a kind of standpoint epistemology? Not precisely: epistemic privilege on this analysis does not attach to the individual woman (or feminist) but to the community that includes her standpoint along with others. The individual standpoints on this account are starting points. Furthermore, it is important to stress that on this analysis nothing depends on women having a different psychological make-up from men or different 'ways of knowing.' The distinctive mark of the work of the feminist critics cited above is not that it is holistic, intuitive, subjective, emotional, nurturant, or non-linear. Instead, what gives it focus and distinction is the fact that it is informed by a social and political viewpoint different from that which has dominated science and science studies.

Couldn't men have done exactly the same work? Yes, it is logically possible. But the connection here is not about necessary or sufficient conditions. It is about contingencies: about causal factors that operate, not from a God's eye point of view nor in the infinite long run, but here and now. It is not a logical necessity but also no accident that the advent of certain scientific hypotheses coincided with increased political power for women and increased representation of women in the academy and scientific communities.

Does the position advanced here have any affinities with feminist postmodernism? The overlap is minimal but not non-existent. My position is perfectly compatible with the rejection of metaphysical realism (perhaps that is even required) but not with the wholesale rejection of objectivity and rationality. The important point is that these two (metaphysical realism and objectivity) are separable, a point too often obscured in the postmodern literature. (Indeed, one often gets the impression from postmodern accounts that the logical positivists were metaphysical realists.)

I find feminist postmodernism unattractive for the usual reasons. I believe that feminist theories in science are *superior to* (cognitively preferable to) their sexist rivals, not simply that they provide alternative narratives. And I believe that postmodernism with its emphasis on fractured identities as well as on epistemic relativism provides no adequate basis for the political action feminism requires. There is much of value, however, in postmodernism's emphasis on the requirement of *local* problem-solving. Gender bias manifests itself in different ways in different sciences. There is no single 'feminist method' that will reveal and eliminate that bias. There is no 'feminist paradigm' that can be imposed from above and no reason to believe (as many postmodems appear to believe) that gender bias in physics, for example, will be of the same kind or degree as that in biology. Real change in science will occur only when specific rival theories are developed by scientists who have both a thorough grounding in their own disciplines and a commitment to questioning biases introduced by social arrangements of science.

I believe, therefore, that it is possible to do justice to the range and depth of gender bias in the biological sciences without sacrificing altogether the traditional ideals of objectivity and rationality; but doing so will require that we take into account the social structure of science. Case studies of the sort summarized in the first part of this paper show the necessity of coming to grips with the ways in which social factors can influence the development of science, and they demonstrate the extent to which some standard philosophical tools can partially illuminate the origins and diversity of ideological biases in science. These tools, however, are inadequate to the task at hand so long as they are embedded within an outmoded and indefensible conception of the scientific process that limits the influence of social factors to the context of discovery. Mainstream philosophy of science continues to ignore feminist critiques of science at its own peril.[16]

■ | Notes

1. The Biology and Gender Study Group, 'Importance of Feminist Critiques for Contemporary Cell Biology,' in *Feminism and Science*, Nancy Tuana, ed. (Bloomington: Indiana University Press 1989), 172–87.

2. Charles Darwin, *The Descent of Man* (1871). Cited by Ruth Hubbard in 'Have Only Men Evolved?' in *Biological Woman: The Convenient Myth*, Ruth Hubbard, Mary Sue Henifin, and Barbara Fried, eds. (Cambridge, MA: Schenkman 1982), 17–45.

3. Helen Longino and Ruth Doell, 'Body, Bias, and Behavior: A Comparative Analysis of Reasoning in Two Areas of Biological Science,' *Signs* 9 (1983), 206–27.

4. Anne Fausto-Sterling, *Myths of Gender: Biological Theories About Women and Men* (New York: Basic Books 1985).

5. Londa Schiebinger, *The Mind Has No Sex? Women in the Origins of Modern Science* (Cambridge, MA: Harvard University Press 1989), 189–213.

6. Ibid., 191.

7. See *The Politics of Women's Biology* (New Brunswick: Rutgers University Press 1990).

8. Sandra Harding, *The Science Question in Feminism* (Ithaca, NY: Cornell University Press 1986). See also S. Harding, *Whose Science? Whose Knowledge? Thinking From Women's Lives* (Ithaca, NY: Cornell University Press 1991).

9. David Bloor, *Knowledge and Social Imagery* (London: Routledge and Kegan Paul 1976).

10. Larry Laudan, 'The Pseudo-Science of Science?' *Philosophy of the Social Sciences* 11 (1981), 195–97.

11. See, for example, 'Primatology Is Politics by Other Means,' *PSA 1984*, vol. 2 (East Lansing, MI: Philosophy of Science Association 1985).

12. See Helen Longino, *Science as Social Knowledge* (Princeton: Princeton University Press 1990) [a portion of which is excerpted above, pp. 144–64].

13. Clark Glymour, *Theory and Evidence* (Princeton: Princeton University Press 1980).

14. Ibid., 151.

15. The argument in this section was also sketched in my essay, 'Birth of a New Physics or Death of Nature?' in *Women and Reason*, E. Harvey and K. Okruhlik, eds. (Ann Arbor, MI: University of Michigan Press 1991).

16. Work on this project was supported by a grant from the Social Science and Humanities Research Council for which I am grateful. I would also like to thank J. R. Brown and Alison Wylie for useful discussions during the project's early days and my sister Peggy Okruhlik for making its completion possible. This paper results in part from amalgamating two earlier typescripts that circulated widely. The first was called 'A Locus of Values in Science' and dates from 1984; the second, 'Gender Ideology and Science,' was first drafted in 1988.

2 | COMMENTARY

2.1 | Values in Science

Contrary to what you might expect, none of the readings in chapter 2 concern moral values. There has been an unfortunate tendency, especially in public debates about education, to assume that all values are moral values. But there are many different kinds of values that people approve of and wish to promote. They range from aesthetic to religious, from cognitive to non-cognitive. Moral values—ideas about what things are morally good and which actions morally right—are just one sort of value among many. Of course, moral values are relevant to the practice of science. Scientists are people, after all, and, as such, their behavior should conform to the norms of morality. For example, there are moral constraints on what kinds of experiments can be performed (especially those involving human subjects) and how they should be conducted; there are rules concerning the use of other scientists' ideas and prohibitions against faking or fudging experimental data; there can also be serious moral concerns about the ways one's research might be used.[1] Clearly, then, there is a moral dimension to science just as there is with any human activity. But when the authors in this chapter talk about values in science, they are not referring to moral values; instead, their focus is on either cognitive or contextual values.

COGNITIVE VALUES

Cognitive values (sometimes called *epistemic values*) have to do with cognition or knowledge. Now, one might think that the paramount epistemic value in science is truth: scientists have as their principal aim the acceptance of true claims about the natural world (and the rejection of false ones). But on closer inspection, this assumption proves flawed. For one thing, truth by itself cannot be sufficient as a characterization of the goal of science. Why not? Because so many of the true statements we could make about the natural world have little or no scientific value. Imagine, for example, that a biologist wants to increase our store of scientific knowledge by counting the precise number of hairs on individual dogs at various times on various days, not to test a theory or experiment with a drug to prevent hair loss but simply to know the canine hair-count for its own sake. Even if the information that the biologist collects is true, it has negligible scientific value. Scientists are interested not merely in discovering truths about the world, but in discovering interesting truths, usually in the form of general theories and laws with predictive power. These criteria of scientific excellence—generality and predictive power—and many others besides (such as explanatory power and simplicity) are among the cognitive values of science. They are not the same as truth, and despite their relation to truth, they do not seem to be reducible to truth in any obvious way.

Cognitive values are, in an important way, partly constitutive of what science is. They help to define the nature of science and to distinguish it from other truth-seeking enterprises, such as genealogy, logic, and theology (to name but three). They underlie many of our judgments about what constitutes acceptable scientific method and practice. For this reason, philosopher of science Helen Longino refers to them as *constitutive values*. In the readings for chapter 2, there is much discussion of whether the existence of several different, sometimes competing, cognitive (constitutive) values compromises the objectivity of science or undermines the rationality of scientific change.

A second problem with taking truth as the sole epistemic value is that scientists cannot tell whether a theory is true simply by inspecting it; they need to perform experiments and make observations. Interesting scientific theories involve empirical generalizations of universal scope and are often about entities and processes that cannot be directly observed no matter how sharp one's vision. Thus in order to reach judgments about the truth of a theory, scientists must rely on other criteria that, they hope, have some connection with truth. For example, commonly a theory is believed more likely to be true if it gives rise to correct predictions. (See chapters 4 and 5 for a detailed discussion of the relationship between prediction and confirmation.) Similarly, but more controversially, a theory is believed more likely to be true if it explains a wide variety of phenomena. More controversially still, a theory is believed more likely to be true if, other things being equal, it is simpler than its rivals. The moral is clear: even if truth is the primary cognitive value, we need other cognitive values to help us decide which of our theories are true. When choosing between competing theories, various cognitive values, such as predictive power, explanatory scope, and simplicity, may not always point in the same direction, so reasonable disagreements might arise about whether (or to what extent) these values are indicators of truth. Thus, even within the realm of cognitive values, debates about the relative importance of particular values and their merits as indicators of truth are bound to arise. So even if we accept that truth is the primary virtue of a scientific theory, value debates within science are inevitable.

Contextual Values

What are contextual values? Roughly speaking, they are norms, preferences, beliefs, and interests that are unrelated to the cognitive aims of science. For example, in the 1930s the official German scientific establishment became so infected with Nazi ideology that it rejected Einstein's theory of relativity on the grounds that its author was Jewish. Limiting one's teaching and research solely to theories whose authors are Aryan would be to adopt an obnoxious (and stupid) contextual value in the practice of science. More subtle examples of contextual values include the preference for theories (especially in the social and biological sciences) that conform to widespread, unexamined, and usually false beliefs about gender and race: in short, racial and sexual bias.

Helen Longino uses the term *contextual values* because these values vary with time and across cultures (and even from one scientist to another). Contextual values show a much greater degree of dependence on society and culture than do their cognitive counterparts because they are not tied in any close and essential way to the fundamental aims of science, such as the search for truth. Rejecting the constitutive (cognitive) values of science is impossible without ceasing to do science, but scientists can (and do) have different sets of contextual values that can (and do) affect the way they do science.

THE VALUE-NEUTRALITY THESIS AND ITS CRITICS

Until quite recently, most philosophers of science accepted the standard view that, insofar as science is value laden, the values in question are exclusively cognitive ones. Contextual values have no place in science; or, at least, *should* have no place. Because contextual values are unrelated to the fundamental goals of science, they should play no role in scientific decision making. In that sense, science should be value-free (i.e., free of contextual values, not free of values of any kind). This view is also called the *value-neutrality thesis*. According to the value-neutrality thesis, when contextual values intrude into science, their influence is invariably pernicious. "Good" science is objective and thus free from subjective values, whether they are the subjective preferences of a single individual or the cultural biases of an entire society. Contextual values distort the judgment of individual scientists and diminish scientific objectivity. Although it is an empirical issue (and thus not one that philosophers have any particular competence to address), proponents of the standard view assume that episodes of contextually infected "bad" science have been historically rare. Indeed, those holding the standard view assume that the constitutive values of science and the institutional mechanisms based on them are a prophylaxis against such infection and an effective cure for it when it occasionally occurs.

As we shall see in our later discussion, Helen Longino rejects several key aspects of the standard view and the value-neutrality thesis. She defends the view—referred to as *contextualism*—that all scientific judgments depend, to a greater or lesser degree, on contextual values. But, she argues, this value-ladenness of science is not incompatible with scientific objectivity. Longino focuses primarily on the appraisal of theories and hypotheses after they have been formulated and examines the process by which scientists decide which are confirmed by the available evidence. She argues that a proper analysis of the relation between evidence and theory shows that all judgments about when and to what extent a piece of evidence confirms a theory depend on background beliefs. Since these background beliefs can vary widely, different scientists can reach different judgments about the significance of an observation or experiment. Longino calls this the *contextualist analysis of evidence*.

If the contextualist analysis of evidence is correct and all judgments about confirmation and evidence are relative to the subjective beliefs of the individual researcher, how can science be objective? The answer, Longino thinks, is to realize that scientific knowledge is the product of a community of scientists. Scientific knowledge is, she claims, social knowledge. As such, it emerges through a process of criticism (such as peer review) based on shared community standards (not individual ones). The social nature of objectivity does not guarantee that science will be free from all subjective preference, but if certain criteria are met (she lists four: recognized avenues for criticisms, shared standards, community response, and equality of intellectual authority), to that extent science will be objective. Thus, in several respects, Longino challenges the value-neutrality thesis: she denies that scientists are free from contextual values when they decide which theories are confirmed by the evidence; she rejects the contention that scientists ought to be free from all contextual values as an unrealizable ideal resulting from a faulty understanding of the relationship between theory and evidence; and she insists that the neutrality thesis is mistaken in assuming that contextual values necessarily compromise the objectivity of science.

Okruhlik's article complements Longino's by describing several cases of sexist bias in modern biological research. Unlike some feminist critics of science (such as Sandra Harding), Okruhlik does not think that we should abandon rationality and objectivity as self-serving male myths. Rather, Okruhlik argues that once we understand how science works, especially in coming up with hypotheses in the first place, we can retain the notion of scientific rationality while recognizing that the scientific method cannot, by itself, be relied upon to eliminate sexist bias. One of Okruhlik's key arguments is that, even if the process of choosing among rival theories were fully objective and rational, if all the theories under consideration are sexist so, too, will be the theory selected as the best. Thus, even though theory appraisal may be free from bias, the contamination of theory generation by contextual factors will inevitably affect the content of science.

2.2 | Kuhn's Analysis of Science

Paradigms and Normal Science

At the heart of Kuhn's characterization of science is his notion of a paradigm. An area of inquiry attains the status of a genuine science when it acquires its first paradigm. At least in *The Structure of Scientific Revolutions*, Kuhn portrays this transition as occurring only once in the history of a scientific field, and he refers to immature science ("something less than science") as being preparadigmatic.[2] To use terminology introduced in chapter 1 of this book, Kuhn is treating the possession of a paradigm as a *demarcation criterion*, something that essentially distinguishes genuine science from things that are not science (or which might merely pretend to be science).

Once they have acquired the status of being genuine sciences, Kuhn portrays scientific disciplines as going through cycles of normal science and revolutionary science. Normal science is the way science is usually done (except for those relatively rare periods of extraordinary or revolutionary science). The hallmark of normal science is that scientists work under the aegis of a paradigm that they all accept. Typical activities of normal science are making more accurate measurements of constants, looking for entities and processes that the paradigm tells us must exist, extending the paradigm to new areas and types of phenomena, reconciling the paradigm with recalcitrant data, and removing the conceptual difficulties that afflict even the most successful paradigms. Kuhn refers to all these paradigm-based activities as *puzzle solving*. To call something a puzzle (whether it is a crossword puzzle, a jigsaw puzzle, or a scientific puzzle) is to imply that it has a solution. The failure of a scientist to solve a scientific puzzle reflects on the competence of the researcher, not on the soundness of the paradigm. Thus, although paradigms typically float in a sea of anomalies, normal scientific activity is directed towards removing those anomalies, not towards refuting the paradigm.

Kuhn portrays normal science as dogmatic: students are taught the paradigm without being encouraged to question it or seriously consider alternatives; debates about fundamentals are discouraged or suppressed; the main scientific activity—puzzle solving—takes for granted that the prevailing paradigm is basically correct. (For more on Kuhn's account of normal science, see his "Logic of Discovery or Psychology of Research?" in chapter 1 of this book and the section "Kuhn's Criticisms of Popper" in the commentary on that chapter.)

Although the concept of a paradigm is fundamental to Kuhn's analysis of science, he does not define it clearly and precisely in *The Structure of Scientific Revolutions*. In his *Postscript*, Kuhn acknowledges his critics' complaint that he had used the term *paradigm* ambiguously. In response to this criticism, Kuhn introduces two new terms to stand for two of the most important senses of *paradigm*: *exemplars* and *disciplinary matrices*. Exemplars include the solutions to problems used in the education of scientists: typically these solutions appear at the end of chapters in science texts, in laboratory exercises, and on examinations. More advanced technical solutions are published in journals. Kuhn's point about these exemplars is that they are crucial to teaching students how to use theories to solve problems; merely memorizing a few equations is not enough. Students become scientists by recognizing and embracing the tacit knowledge implicit in the exemplars.

Kuhn uses the phrase *disciplinary matrix* to stand for his second sense of *paradigm*. Disciplinary matrices include exemplars, but they also contain, among their main components, symbolic generalizations, metaphysical commitments, heuristic models, and values. It is this second, more inclusive sense of *paradigm* that Kuhn employs in many of his more challenging and controversial claims about science. Henceforth, when we speak of paradigms, it will be in terms of Kuhn's notion of disciplinary matrices.

Kuhn makes several claims about normal science that have elicited much criticism. These claims are important because they underlie many of the assertions that Kuhn makes about scientific revolutions. Here we will simply list these claims with brief comments, reserving criticisms for later.

■ NORMAL SCIENCE IS BASED ON A PARADIGM.

The fundamental entity that drives normal science is not an individual theory but a paradigm. Paradigms are collections of diverse elements (ontological, epistemological, methodological, and axiological) that tell scientists what experiments to perform, which observations to make, how to modify their theories, how to make choices between competing theories and hypotheses, and so on.

■ NORMAL SCIENCE IS DOGMATIC.

Normal science is a relatively dogmatic affair, consisting largely of what Kuhn calls puzzle solving. In Kuhn's words, normal science is a "mopping-up operation," "an attempt to force nature into the preformed and relatively inflexible box that the paradigm supplies" (*The Structure of Scientific Revolutions*, 24).

■ NORMAL SCIENCE IS OBJECTIVELY PROGRESSIVE.

Kuhn equates progress with cumulativeness. Since (when things are going well) scientists accumulate solutions to scientific puzzles, normal science must be progressive. Thus, at least with respect to normal science, Kuhn agrees that science is progressive in an objective sense. When we turn to scientific revolutions, things are radically different.

Because of their importance in providing philosophical support for Kuhn's claims about both normal and revolutionary science, we list the doctrines underlying Kuhn's arguments. Later, we shall examine these doctrines and the arguments based upon them in greater depth.

■ THE HOLISTIC CHARACTER OF PARADIGMS

Paradigms contain a variety of elements—from explicit claims about the kinds of entities and processes that populate the world to implicit commitments to various methodological values, from precisely defined techniques to loosely articulated heuristic models. Nonetheless, Kuhn presupposes that paradigms are integrated wholes. Because the elements of a paradigm are inextricably intertwined, scientists must accept or reject the paradigm as whole; they cannot accept some elements and reject others.

■ THE THEORY-LADENNESS OF OBSERVATION

The theories that scientists accept (and the paradigms in which they are embedded) significantly affect what scientists observe. In *The Structure of Scientific Revolutions*, Kuhn subscribes to the thesis of the theory-ladenness of observation, according to which what scientists observe depends on what they believe.

■ THE THEORY-DEPENDENCE OF MEANING

The theories that scientists accept (and the paradigms in which they are embedded) significantly affect what scientists mean by the theoretical terms they employ. In particular, theoretical terms such as *planet, electron, oxygen, gene*, and *mass* derive their meaning, at least in part, from the theories in which they occur.

SCIENTIFIC REVOLUTIONS

In "The Nature and Necessity of Scientific Revolutions," Kuhn tells us that "scientific revolutions are here taken to be those non-cumulative developmental episodes in which an older paradigm is replaced in whole or in part by an incompatible new one" (79). In brief, scientific revolutions occur when scientists switch from one paradigm to another.

It is clear from *The Structure of Scientific Revolutions* that Kuhn does not intend simply to define scientific revolutions as noncumulative and leave it at that. Nor does his claim rest merely on history, on scientific revolutions past. Rather, while he frequently illustrates his claims by appealing to history, he also argues that scientific revolutions must be noncumulative because of certain presumed truths about the nature of paradigms, observation, and meaning.[3] Thus, while Kuhn claims that his view of science accurately reflects the way science in fact works, he is not simply describing selected episodes from the history of science and generalizing from them. Like many of the philosophers of science that he criticizes, Kuhn, too, draws conclusions about the way science must develop on the basis of philosophical arguments. Let us begin with Kuhn's conclusions, and then examine his arguments.

Below is a summary of the six main claims that Kuhn makes about scientific revolutions in "The Nature and Necessity of Scientific Revolutions" and elsewhere in *The Structure of Scientific Revolutions*.

■ SCIENTIFIC REVOLUTIONS ARE PARADIGM SHIFTS.

Paradigms are as vital to our understanding of scientific revolutions as they are to our understanding of normal science. Scientific revolutions involve not merely the replacement of one theory by another but a wholesale shift from one network of scientific commitments, beliefs, and values to another.

■ SCIENTIFIC REVOLUTIONS ARE TOTAL.

Scientific revolutions are always total: "Though logical inclusiveness remains a permissible view of the relation between successive scientific theories, it is a historical implausibility" (*The Structure of Scientific Revolutions*, 98). Paradigms are mutually exclusive: one cannot consistently hold both; commitment to one paradigm logically precludes commitment to its rival.

■ SCIENTIFIC REVOLUTIONS ARE RELATIVELY SUDDEN AND
  UNSTRUCTURED EVENTS.

Kuhn compares scientific revolutions with gestalt switches, flashes of intuition, and religious conversions. Like some conversion experiences, scientific revolutions are preceded by a period of debate. Indeed, Kuhn argues that, given the dogmatism of normal science, a period of extraordinary or revolutionary science is required to relax the normal suppression of debate about fundamentals. Kuhn describes this period of debate as a crisis brought about by the accumulation of anomalies under the prevailing paradigm. But when the revolution finally occurs, it is rapid and unstructured. Hence, Kuhn's book appears to be mistitled, since, according to Kuhn, there is no structure of scientific revolutions. But the appearance of paradox is removed by distinguishing between revolutionary science (the structured process leading up to the paradigm switch) and a scientific revolution (the unstructured event of paradigm switch). We make much the same distinction between, say, the process of dying and the event of death.

■ EXTRAORDINARY (REVOLUTIONARY) SCIENCE IS NONDOGMATIC.

This is clear from the previous claim. Because normal science is so dogmatic, the process leading up to a paradigm switch must involve a much freer kind of scientific debate, in which dissent and questioning are tolerated. Otherwise, scientific revolutions would be impossible.

■ SCIENTIFIC REVOLUTIONS CANNOT BE DECIDED BY
  RATIONAL DEBATE.

Kuhn insists that scientific revolutions necessarily have a nonrational character. The choice between one paradigm and its rival cannot be decided using only the values and methods of normal science. Arguments in favor of a paradigm are bound to be circular, taking for granted doctrines that proponents of a rival paradigm reject. So paradigm acceptance via revolution is essentially a sociological, not a rational phenomenon—a matter of persuasion and conversion, not of rational argument. "As in political revolutions, so in paradigm choice—there is no standard higher than the assent of the relevant community" (*The Structure of Scientific Revolutions*, 94).

■ SCIENTIFIC REVOLUTIONS ARE NOT OBJECTIVELY
   PROGRESSIVE.

Time and again Kuhn proclaims that there is no genuine, objective progress through scientific revolutions; changes of paradigms cannot be said to bring us closer to the truth. Once a scientific revolution has occurred, the writers of the new textbooks and histories of science make the development of science *look* cumulative and linear, but that is simply window dressing. It has no more objective significance than did the histories of political development written by Marxist ideologues in the former Soviet Union. In science, as in politics, history is rewritten by the winners.

All six of Kuhn's general claims about scientific revolutions have drawn fire from critics. But particular attention has been focused on the last two theses because they deny that science as a whole is either objectively rational or objectively progressive—a position that challenges the traditional image of science. According to Kuhn, judgments about scientific rationality and progress can only be made relative to paradigms. With good reason, then, Kuhn's overall position is aptly described as a version of relativism.

2.3 | Six Kuhnian Arguments for Relativism

Kuhn has at least six different arguments for his conclusions about the lack of rationality and objective progress in scientific revolutions. These concern the theory-ladenness of observation, meaning variance, problem weighting, shifting standards, the ambiguity of shared standards, and the collective inconsistency of rules. We shall briefly describe each of these arguments and criticize the first two; criticisms of the last four can be found in the discussions of McMullin and Laudan.

THEORY-LADENNESS OF OBSERVATION

In accounts of confirmation and disconfirmation, the fate of theories is often decided, at least in part, by observations. (For a more detailed consideration of confirmation, see chapters 4 and 5.) Typically, a theory makes a prediction concerning something we should be able to observe. Sometimes, the observation is just a matter of looking and seeing; more often, the observation requires the use of instruments or special apparatus, or the performance of an experiment. When two rival theories make incompatible predictions about the outcome of an experiment, we should be able to decide between them (at least in principle) by performing a crucial experiment. Indeed, the empirical component provided by observation usually is supposed to confer on science its objectivity and rationality: choices about scientific theories are considered objective and rational largely because acceptable theories must agree with observation. Science is considered progressive because current theories are in closer agreement with observations than were the theories that they replaced. Whether or not empirical evidence (such as that provided by observations) can ever determine theory choice is considered in chapter 3; for the moment the issue is whether observation can serve as a rational basis, however partial, for accepting or rejecting individual theories or for deciding between rival theories. Kuhn's argument regarding the theory-ladenness of observation denies that observation can play this role.

The key premise in Kuhn's argument is that observation is theory laden. What does this mean? Roughly, it means that what scientists observe depends on the theories they accept. In order for this to support Kuhn's denial that observation can serve as a rational basis for making decisions about theories, the thesis must imply that scientists embedded in competing paradigms interpret the world so differently that they "see different things" when making their observations, and what those scientists observe can never disconfirm or undermine their confidence in their own theory in favor of the theory of a competing paradigm.

When put in this stark fashion, the theory-ladenness of observation thesis seems wildly implausible. Imagine that Kepler and Ptolemy are watching the sun climb over the horizon at sunrise.[4] Kepler accepts the Copernican theory that the earth rotates on its axis once a day and revolves around the sun once a year. Ptolemy accepts the Ptolemaic theory that the earth is completely stationary and that it is the sun that makes both a daily rotation and an annual revolution around the earth. For Kepler, the earth is a planet and the sun is stationary; for Ptolemy, the sun is a planet and earth is at rest. Now, do Kepler and Ptolemy "see the same thing" at dawn when they observe the rising sun? The proponent of the theory-ladenness thesis might argue thus: Kepler's sun is at rest and Ptolemy's sun is moving; it is impossible for the same object to be both at rest and in motion; therefore, Kepler's sun cannot be the same as Ptolemy's sun; therefore what Kepler observes (Kepler's sun) cannot be the same as what Ptolemy observes (Ptolemy's sun). But this argument is flawed because it fails to distinguish between the object that is observed (the sun) and the rival beliefs that the two observers have about it. Thus the argument commits the fallacy of equivocation. The phrase *Kepler's sun* in the first premise means *the sun as Kepler believes it to be* (or, more strictly, *the thing that would answer to Kepler's beliefs if they were true*), but in the conclusion, *Kepler's sun* means *the object (the sun) that Kepler observes (regardless of Kepler's or anyone else's beliefs about it)*.

Another way of appreciating the implausibility of the thesis that proponents of rival theories observe different objects is to imagine that Ptolemy and Kepler are discussing the relative merits of their respective theories as dawn arrives. Suppose, as our literal construal of Kuhn's account requires, that the thing that Ptolemy sees is distinct from the thing that Kepler sees. (Never mind that this seems to require two heavenly bodies to be in exactly the same place.) Were Ptolemy to convert to Copernicanism (thanks to Kepler's arguments), we would be forced to accept the conclusion that there is suddenly one less star in the heavens. This would confer on Ptolemy the magical power of being able to change the world simply by changing his beliefs.

Sometimes the theory-ladenness thesis is expressed by saying that there is no (or cannot be any) theory-neutral observation language. Thus, even though it might be conceded that Kepler and Ptolemy are looking at (and hence observing) the same object, the words they use to describe their respective observations must have different meanings, depending on the different theories that they hold. Insofar as this new form of the argument involves considerations about language and meaning, it will be discussed in the next section, on the meaning-variance argument. For now, we will focus on the claim that no proponent of a scientific theory can ever observe anything that would undermine or disconfirm that theory. Clearly, this is untenable. Even if it were true that every observation a scientist makes in the course of testing a theory involves terms from the theory that is being tested (and that the scientist accepts as true), it simply does not follow that the scientist will never find any observational evidence that disconfirms the theory. For we must distinguish between the concepts used to make a judgment and the assertions made using those concepts.[5] For example, Kepler believed that all planets move in ellipses. Suppose that there is no theory-neutral observation language and that Kepler had to use the terms *planet* and *ellipse* in any statement reporting his observations for the statement to have an evidential bearing on his theory. Even given these constraints Kepler could easily have refuted his own theory by reporting that the planet he observed did *not* move in an ellipse. (Just this happened with regard to some of Kepler's earlier hypotheses about the shape of planetary orbits before he hit on the idea of an ellipse). The necessity of using certain concepts in reporting one's observations in no way determines the truth or falsity of the judgments one makes using those concepts. Thus, while it is true that our theoretical beliefs and expectations sometimes influence our perceptions and so, to that extent, at least some of our perceptions are theory laden, we have been given no reason to think that what we perceive must always agree with what we believe.[6]

MEANING VARIANCE

The crucial premise in the meaning-variance argument is the thesis of the theory-dependence of meaning, according to which the theories scientists accept (and the paradigms associated with them) significantly affect what scientists mean by the theoretical terms they use. In order for this thesis to support Kuhn's denial of rationality and progress, it would have to entail that scientists committed to rival paradigms speak different languages. They may utter the same words, but the words have different meanings, so any logical comparison of their utterances is precluded, leaving adherents of rival theories simply talking past one another. Kuhn sometimes expresses this view by saying that rival paradigms and the terms they employ are *incommensurable*.

The term *incommensurable* literally means "having no common measure." It comes from geometry, where it was first used to express the fact that in an isosceles right-angled triangle whose side is one unit in length (a rational number), the hypotenuse (the square root of two—an irrational number) cannot be expressed as any ratio of rational numbers. The side and the hypotenuse lack a common measure. In a similar way, Kuhn claims that the proponents of rival paradigms lack a common language. In his later writings Kuhn has insisted that his view in no way precludes translation from one theoretical language to another. But even then, he claims, the translation of one theory into the idiom of another will not exactly preserve the meaning of the original. Thus, for example, Tycho Brahe could understand (and reject) the Copernican claim that the Earth is a planet moving around the sun. But, according to Kuhn, when Brahe uses the term *planet* in his rejection of Copernican theory, the word *planet* means something different for Brahe than it does for Copernicans such as Kepler. So, Kuhn concludes, there is no theory-neutral, paradigm-independent way of comparing the two theories.

Claims about meaning variance are notoriously hard to evaluate, mainly because there is no consensus about what meanings are in the first place. But this much is clear: whatever meanings are, if comparisons between theories are to involve logically valid arguments, then the terms such arguments employ must have the same, unequivocal meaning throughout. (Otherwise, the arguments in question would commit the fallacy of equivocation.) Thus, if meaning variance inevitably occurs whenever two theories differ, then valid logical comparisons between them would be impossible. In this way, some proponents of the meaning-variance thesis (such as Paul Feyerabend) have attacked models of reduction that require the old theory to be logically derivable from the theory that replaces it. (For more on Feyerabend, theory reduction, and meaning variance, see chapter 8.)

Many of Kuhn's critics have pointed out an apparent contradiction between Kuhn saying, on the one hand, that rival paradigms are incommensurable with respect to meanings and, on the other hand, that it is impossible to believe both paradigms at the same time. For, without commensurability of meaning, there cannot be logical incompatibility. Consider, for example, the word *planet*. Suppose that Brahe meant by *planet*, *satellite of the earth* and that Kepler meant by *planet*, *satellite of the sun*. Would they have been disagreeing when Brahe said, "All planets go around the earth," and Kepler said, "All planets go around the sun"? Since the word *planet* has different meanings in the two sentences, neither sentence contradicts the other. In short, there is no single proposition that one affirms and the other denies. Indeed, if the word *planet* were completely defined by Brahe and Kepler in the two ways indicated, then both Brahe and Kepler would have been saying something true but utterly trivial (Brahe: all planets [read: *satellites of the earth*] go around the earth; Kepler: all planets [read: *satellites of the sun*] go around the sun).

Clearly something has gone wrong in our telling of the story about Brahe and Kepler. It is absurd to suggest that they would have been asserting tautologies were they to have uttered the sentences attributed to them. The flaw in the story seems to lie in the assumption that there is nothing more to the meaning of the word *planet* than the role that the term plays in a particular theory. The word was used long before Brahe and Kepler to refer to astronomical bodies such as Venus, Mercury, and Jupiter. And it is this referential dimension of the term that enables Brahe and Kepler to make contrary claims about those bodies.

Expressed more formally, there are (at least) two aspects to the meaning of theoretical terms such as *planet, electron, gene,* and *oxygen*: their sense and their reference. The sense of a term in a particular theory is given partly by its use prior to its adoption by the theory (its antecedent meaning) and partly by its role in that theory. Only when a term is a neologism, a brand new term, will its sense be defined solely by the theory that employs it. The reference of a term is the set of things, processes, or events that the term picks out. In general, rival theories can make opposing claims about the world, despite some differences in the sense of some of the terms they use, *if* they make opposing claims about some of the things, processes, or events that the terms pick out. This is only the barest beginnings of an account of the meaning of theoretical terms, but at least it enables us to see how theories can be genuine rivals despite the contribution that theoretical context makes to meaning. In short, radical incommensurability fails because there is more to the meaning of most terms than simply the sense defined implicitly by a particular theory (the "more" being what we have called antecedent meaning) and there is more to meaning than sense (the "more" in this case being reference).

The upshot of our discussion thus far is that neither of Kuhn's two arguments—neither the one based on the theory-ladenness of observation nor the one based on the paradigm-dependence of meaning—is strong enough to support his radical conclusions about the lack of rationality and progress in scientific revolutions. We now turn to four arguments that have proven more durable. These arguments, endorsed by philosophers of science favorable to Kuhn and criticized by McMullin and Laudan, are sketched, more or less explicitly, in the second reading from Kuhn, "Objectivity, Value Judgment, and Theory Choice." For convenience, we have adopted the labels that Laudan gives to these arguments (although we have changed the order in which they occur in Laudan's article.) What is distinctive about these arguments is that none of them depends on controversial theses about observation and meaning. Rather, their focus is on the criteria used to choose between competing theories and paradigms. In short, all these arguments concern epistemic values.

Problem Weighting

Part of the innovation wrought by Kuhn lies in his insistence that we should assess theories, not by looking at their empirical or observational consequences, but rather by seeing how good they are at solving problems. Designing a theory to agree with observation is, Kuhn thinks, a relatively simple task, especially if one is not fussy about what the resulting "theory" looks like. But, he insists, the point is not first and foremost to devise theories that agree with observation but to devise theories that answer questions and solve problems; the unit of scientific achievement is not the true observational consequence but the successful puzzle solution. Good theories belong to paradigms that lead to a progressive accumulation of solved problems. Often the key issue in the ferment preceding a scientific revolution is whether a rival paradigm is a better problem solver or has the promise of being so once the full energies of science are devoted to its development. But then, Kuhn points out, we run into a difficulty. For a characteristic feature of many scientific revolutions is that the new paradigm cannot solve all the problems of its predecessor. Some problems that were solved by the old paradigm are left unsolved or simply ignored by its successor. So, Kuhn argues, no algorithm or calculation can determine a rational choice between paradigms. The scientist's choice depends on the relative weight assigned to particular problems, and this weighting will vary from one paradigm to another.

Shifting Standards

According to Kuhn, paradigms include standards for assessing theories, and these standards vary from paradigm to paradigm. For example, one group of scientists might insist that no theory should be accepted unless it makes novel predictions, whereas another group might extol a theory's ability to give a unified explanation of diverse phenomena, even though all of those phenomena were previously known. The reception of Darwin's theory in the nineteenth century depended, in part, on just such a disagreement about methodology. Judgments about the relative merits of theories vary depending on the set of standards used. No higher court can adjudicate these disagreements. So, here again, Kuhn concludes that there is no single, rational way of resolving paradigm conflicts.

The Ambiguity of Shared Standards

As Kuhn has emphasized in his later writings (see "Kuhn's Second Thoughts . . ." below), some epistemic values, some views about theoretical virtue, are widely shared by scientists. But, he thinks, appeal to these shared standards cannot suffice to determine the outcome of paradigm debates. One important reason for this is the differing interpretations given to these standards. For example, proponents of rival paradigms might agree that explanatory power is desirable but disagree about what constitutes an explanation. Just such a case arose in the early eighteenth century in connection with Newton's gravitational theory. Newton's theory gave a unified lawlike treatment of the phenomena with plenty of predictive power. But Newton's European critics (especially Leibniz and the followers of Descartes) did not regard Newton's theory as having explained anything. In their view, explanation required the provision of a plausible (or at least possible) mechanism or process that obeyed the laws of mechanics and could causally bring about the phenomena to be explained. Newton and his followers agreed that, by relying on instantaneous action at a distance,[7] Newton's theory probably could not be reconciled with any such mechanism. So the parties agreed that good theories should explain, but they differed markedly in their understanding of what explanations are.

Another example of the ambiguity of standards is provided by the Copernican revolution. (Kuhn's first book was on the Copernican revolution, and Kuhn frequently appeals to this episode as a source of illustration and support for his views. The accuracy and legitimacy of these appeals is examined in the McMullin article, discussed below.) Both sides in the debate regarded simplicity as an important and desirable property of a scientific theory, but they disagreed about how simplicity should be measured. Ptolemaic astronomers complained that the Copernican theory was no simpler than the Ptolemaic theory, because it used just as many (and, in some cases, more) circles to model the planetary orbits. Copernicans (such as Kepler and Galileo) saw a deeper kind of simplicity in the Copernican theory that had nothing to do with circle counting but concerned instead the ability of the Copernican theory to reflect mathematical patterns in the motions of the planets much more simply than had the Ptolemaic theory. As far as Kuhn is concerned, there is no objectively right answer as to which conception of simplicity is correct. Ultimately, he thinks, these kinds of preferences are matters of taste. Since differing interpretations of epistemic values are, Kuhn thinks, inevitable and irresolvable by rational debate, he concludes that shared standards cannot provide a common ground for settling paradigm conflicts.

THE COLLECTIVE INCONSISTENCY OF RULES

Typically, a paradigm will include several different methodological rules and standards for judging theories. These might include a preference for theories that are quantitatively accurate in their predictions, of wide explanatory scope, and parsimonious in their postulation of unobservable entities and processes. There is no guarantee that even the best available theory will satisfy all these requirements: typically, some requirements are satisfied to a greater degree than others, and some may not be satisfied at all. Now suppose that rival theories T_1 and T_2 are being judged, either by a single paradigm or by rival paradigms sharing the same rules and standards. Assume that the first theory, T_1, gives more accurate predictions than the second theory, T_2, but is inferior with regard to the scope of its explanations and has roughly the same degree of simplicity. The first rule—prefer theories that make quantitatively accurate predictions—favors T_1, the second rule—prefer theories of wide explanatory scope—favors its rival, T_2. The entire set of rules thus gives conflicting advice, and Kuhn thinks that there is no uniquely rational way of settling the matter. Again, he concludes that scientists must inevitably have recourse to personal, subjective, and psychological factors to tip the balance one way or the other. If two scientists reach opposite verdicts, neither can be accused of being irrational. The choice between the two theories is, in Kuhn's view, nonrational; it cannot be resolved simply by appealing to rules. So even if scientists share the same rules and even if they agree about how those rules should be interpreted, paradigm choice (and even theoretical choices within a given paradigm) will not be determined by those rules.

2.4 | Kuhn's Second Thoughts about the Rationality of Scientific Revolutions

In "Objectivity, Value Judgment, and Theory Choice," Kuhn responds to those critics who charge that *The Structure of Scientific Revolutions* depicts scientific revolutions as irrational affairs governed largely by subjective preference, idiosyncrasy, and social pressure. Kuhn denies that he said any such thing, even in the penultimate chapter of that book (chapter 11, "The Resolution of Revolutions") in which he compares scientific revolutions to gestalt switches and conversion experiences. In his defense, Kuhn makes three main points.

■ SHARED EPISTEMIC VALUES

He does not deny but in fact affirms that there are some important methodological rules—shared values—that all paradigms share. Kuhn goes so far as to claim that these values are essential to the scientific enterprise, thus qualifying them as constitutive values on his view.[8] Insofar as paradigm debates are conducted solely within the confines of these shared values, about which there is total agreement (at least in broad outline), these debates are rational and based on reasoned argument.

■ INTERSUBJECTIVE AGREEMENT

Kuhn still denies that there can be objective progress in science. He rejects cumulativity (which he thinks is essential to progress) and dismisses the idea that we can show that our theories are getting closer to some objective, theory-independent truth.[9] Nonetheless, he adds, this does not mean that science lacks objectivity. For objectivity consists, not in the correspondence of our theories to the world, but in the intersubjective agreement about those theories among members of the scientific community, based on their shared values. Kuhn identifies objectivity and rationality with a special sort of social consensus, a consensus based on the values that make science what it is.

■ NONRATIONAL FACTORS

For the reasons sketched in the previous section ("Six Kuhnian Arguments for Relativism"), Kuhn continues to insist that paradigm debates, by their very nature, cannot be settled solely by an appeal to evidence, reason, and argument. Reason is not irrelevant to these debates, and considerations based on shared values often play a major role in marshalling support for a new paradigm. But shared values can only take us so far, and rival paradigms usually involve other values that are not shared. So when scientists decide to switch (or not to switch) paradigms, nonrational factors play a vital role. One of the major failings of traditional philosophies of science, in Kuhn's view, is that they have falsely assumed that scientific reasoning about which theories to accept is entirely govered by universal, context-independent rules. Philosophers of science then compound their error by thinking that science is objective only because it is completely governed by these rules.

SHARED EPISTEMIC VALUES

In this section, we begin to discuss Kuhn's first point, that all paradigms share certain fundamental values that are essential to the nature of science. (Further discussion and criticism of all three of Kuhn's points appears in the sections on McMullin and Laudan.) Three separate questions can be raised about shared epistemic values. First, what are the values that Kuhn deems essential to science, and is Kuhn's characterization of them adequate? Second, is it true that all paradigms share these values? And, third, is it true, as Kuhn claims, that rational debate during paradigm conflicts is limited solely to arguments based on shared values, or can there be a rational debate about which values to adopt?

We shall focus on the first of these questions. In "Objectivity, Value Judgment, and Theory Choice," Kuhn lists five shared epistemic values: accuracy, consistency, scope, simplicity, and fruitfulness.

■ ACCURACY

Accuracy is a matter of the theory agreeing with experiments and observations that have already been made. Presumably, accuracy can vary in degree depending on the number of such agreements and the relative precision with which the theory predicts them. Characteristically, in his discussion of accuracy, Kuhn lumps together both prediction and explanation, and he allows for both quantitative and qualitative agreement. Kuhn's discussion makes clear that he understands the accuracy of a theory to concern not merely agreement with experience (what Kuhn calls *matching*), but also explanation (what Kuhn calls *accounting for* a phenomenon or regularity, although as just noted, Kuhn often treats explanation and prediction as if they were the same thing). In Kuhn's terminology, the key to judging the accuracy of a theory is to identify the number and importance of the problems it solves. This view acknowledges the qualitative dimension of accuracy (since not all puzzle solutions are quantitative) and the ambiguity of accuracy as an epistemic value (since opinions may differ about which are the most important problems for a theory to solve).

■ CONSISTENCY

Consistency has two distinct aspects: internal and external. A theory is internally consistent when it is free from logical contradiction. A theory is externally consistent when it does not contradict any other currently accepted theory. For example, when first introduced in the sixteenth century, the Copernican theory was inconsistent with Aristotelian physics, the then-accepted theory of motion and its causes. Similarly, Bohr's theory of the atom, which required electrons to revolve in stable orbits around a positively charged nucleus, was inconsistent with Maxwell's electrodynamic theory, according to which all accelerating charges must radiate and lose energy.

■ SCOPE

Scope concerns the breadth of the theory's logical consequences. Under the rubric of scope, Kuhn emphasizes the special importance of a theory's having surplus content: the theory should have consequences that go far beyond the set of observations, laws, and subtheories that it was originally designed to explain. (For more on the question of whether so-called novel predictions have special significance for confirmation, see chapter 4.)

■ SIMPLICITY

Simplicity has been interpreted in several different ways. Kuhn's official definition equates simplicity with unifying power: a simple theory, he contends, brings order to phenomena that would otherwise be considered separate and isolated. But this seems to characterize simplicity as another aspect of scope and to ignore the essential core of simplicity, namely, parsimony. Two theories might both give unified treatments of a wide range of phenomena (and thus have equally wide scope), but one of them might use fewer assumptions than the other and thus be judged simpler. Something like this latter notion of simplicity was crucial for the Copernican revolution. Unfortunately, Kuhn's own comments about that revolution in this context are subtly misleading. After remarking that the Copernican theory required just as much labor to compute planetary positions as did the Ptolemaic theory (and thus, by that measure of simplicity, was no simpler), Kuhn observes that, in order to explain gross qualitative features of planetary motion such as retrograde motion and the restricted elongation of the inferior planets, the Copernican theory required only one circle per planet, whereas Ptolemy needed two. What Kuhn says (on page 97) is literally true, but it misses the point. Granted, Copernicus required only one circle per planet to explain the phenomena in question, but the explanation in each case crucially involved another circle, namely, the earth's orbit around the sun. And in the Copernican theory, the earth, too, is a planet. So if we count the total number of planetary circles needed to explain these phenomena, we get the same answer (two) for both theories. But still, Copernicus's theory was simpler than Ptolemy's. Why? Because from the single postulate that the earth and all the other planets move in roughly circular orbits around the sun, all the phenomena follow as immediate geometrical consequences, without the need for any further assumptions. In Ptolemy's theory, additional postulates were needed to tie the motion of each planet to the motion of the sun in a manner that, in the framework of Ptolemy's theory, was completely arbitrary. Thus, Copernicus's theory was simpler than Ptolemy's because it derived the same observable phenomena from a smaller set of assumptions. (For further criticisms of Kuhn's interpretation of the Copernican revolution, see the discussion of McMullin in the next section of this commentary.)

■ FRUITFULNESS

Fruitfulness, as Kuhn remarks, has often been underemphasized in traditional accounts of science. As an epistemic value used for making decisions between rival theories, the issue is not fruitfulness per se, but the promise of fruitfulness. What counts is not how successful a theory will be in leading to the discovery of new phenomena and laws (and, more generally, in solving problems) but how promising the theory appears to be right now, before that promise is realized (or not, as the case may be). Because fruitfulness (or fertility, to use McMullin's term) concerns our hopes for the future, Kuhn (in *The Structure of Scientific Revolutions*) contends that it involves an element of what he calls *faith*, and thus to a greater extent than the other epistemic values, it eludes rational evaluation.

2.5 | McMullin's Criticisms of Kuhn

In "Rationality and Paradigm Change in Science," Ernan McMullin documents the ways in which Kuhn has toned down his subjectivist rhetoric since *The Structure of Scientific Revolutions*. On the one hand, Kuhn has emphasized the extent to which paradigm debate can be rational insofar as these debates are based on shared values. Kuhn assumes that values such as accuracy, scope, and the like are fixed and permanent features of science. But, on the other hand, Kuhn (as we saw earlier) continues to insist that these values can never determine the outcome of scientific revolutions. Furthermore, Kuhn persists in denying that any objective notion of progress (such as getting closer to the truth) can be applied to science across revolutionary divides. As McMullin notes, Kuhn's denial of progress follows from his conviction that even the shared epistemic values are ultimately arbitrary and subjective. Specifically, Kuhn thinks that it is impossible to show that these values are connected in any necessary way with truth or the likelihood of truth.

McMullin has three main concerns about Kuhn's later views. First, how plausible is Kuhn's claim that the epistemic values he identifies are shared by all scientific paradigms? Second, is Kuhn correct in insisting that epistemic values cannot be given any rational justification? Third, can Kuhn ignore the question of scientific realism (crucial to his denial of objective progress) and still make sense of scientific revolutions such as the Copernican revolution? More generally, is it sufficient for Kuhn simply to accept, as a sociological fact, the epistemic values that scientists employ in judging theories without offering any deeper rationale for those values?

Shared Values

With regard to the first of these questions, McMullin notes that Kuhn counts as revolutions a remarkably wide variety of scientific changes, ranging from the shallow (e.g., the discovery of X rays) to the radical (e.g., the Copernican revolution), with many in between (e.g., the victory of Lavoisier's oxygen theory over the phlogiston theory[10]). If one looks only at episodes from the shallow end of this range, then claiming that these revolutions left epistemic values largely untouched is quite plausible. While some of them involved fundamental shifts in ontology and concepts, the standards by which theories are judged changed little or not at all (although, of course, advocates of rival paradigms might well disagree about the interpretation and application of those standards). But *deep revolutions* (as McMullin calls them) do involve debates about standards. The very notion of what constitutes a good theory and what qualifies as evidence was at issue in the Copernican revolution and, to a lesser extent, in the quantum revolution of the twentieth century. Indeed, McMullin thinks that disagreements about standards and the merits of alternative theories can occur even within what Kuhn would classify as normal science. (McMullin cites paleontology, high-energy physics, and planetary science as examples.) Thus, disagreements about theories and standards are not limited solely to confrontations between incommensurable paradigms. They can occur, to a greater or lesser extent, even within a prevailing paradigm. A close look at the history of science reveals that the distinction between normal science and a scientific revolution is more a difference of degree than a difference of kind.

Given that scientific revolutions can and do differ in depth, why should Kuhn think that some standards (epistemic values) are immune to change? Why should the features that Kuhn identifies as shared values be permanent features of science, shared by all scientific paradigms, while other criteria for choosing among theories can change with paradigms? McMullin discusses Kuhn's attempts to answer these questions and judges his answer unsatisfactory. Kuhn himself denies that science has a fixed essence, a set of necessary and sufficient conditions that distinguish it from other human activities. Rather, in the manner of Wittgenstein's *Philosophical Investigations*, Kuhn regards our idea of science as a cluster concept. For a theory or paradigm to be scientific does not entail that it must satisfy all the values in the cluster, only that it satisfy (to some degree) a fair number of them. And activities other than science might share one or more of these values. But, as McMullin points out, this picture (of a cluster of values associated with the activities that we now recognize as scientific) is consistent with the possibility that values could be added to or deleted from that cluster in the future in such way that the cluster as a whole would retain its special identification with science. The fact that a particular cluster of criteria are constitutive of science at present does not imply that the membership of this cluster must be permanent and unchanging.

THE JUSTIFICATION OF VALUES

Must all paradigms satisfy the epistemic values singled out by Kuhn? There is one sort of answer to this question that Kuhn thinks is impossible to give— namely, a philosophical argument connecting these values with the aim (or aims) of science. To simplify matters, let us focus our attention on truth as one of the main aims of science. Kuhn says that we cannot give any philosophical justification for the epistemic values of science because, if we could, we would have solved the problem of induction. (The problem of induction is discussed in some detail in chapter 4.) For our present purposes suffice it to say that by "solving the problem of induction" Kuhn means "showing that our criteria for theory choice lead to true theories or to theories that have a high probability of being true."

Like Kuhn, McMullin is not sanguine about the possibility of solving the problem of induction by *a priori* philosophical reasoning. But McMullin thinks that the problem of induction might be solved, at least to a limited extent, by an appeal to experience. As he explains in the section of his paper entitled "How Might Epistemic Values Be Validated?" (120–23), some epistemic values of science (such as simplicity and fertility) are subordinate to the primary goals of predictive accuracy and explanatory power. Simplicity and fertility are not valued for their own sake or as ends in themselves but as means to further ends. Even if we cannot prove that predictive accuracy and explanatory power are indicators of truth, it is still an important empirical discovery that theories of greater simplicity and fertility tend to be more reliable predictors and better explainers. Indeed, McMullin thinks that it was an empirical discovery of the Copernican revolution that it was possible for an astronomical theory to be both an accurate predictor *and* a good explainer. (Prior to Copernicus, astronomers in the Ptolemaic tradition had pursued predictive accuracy at the expense of explanation, arguing that this was an inevitable limitation of their discipline. Explanation was relegated to the realm of Aristotelian physics.) So even if we cannot demonstrate that theories that are good predictors are more likely to be true, and even if we cannot prove that theories that have been good predictors in the past are likely to remain good predictors in the future, we can still give a rational justification of some of our epistemic values (such as simplicity and fertility) by appealing to the connection, learned from experience, between simplicity and predictive reliability. Prior to the Copernican revolution, simplicity and fertility were seldom included among the criteria of a good scientific theory. One of the important things that Copernicus and his followers taught us was that, as a matter of empirical fact, simplicity and fertility are reliable indicators of predictive and explanatory success. So to a limited extent, we can justify some of our epistemic values by appealing to the lessons of history and experience.

Rationality and Realism

Kuhn is not a scientific realist. He scrupulously avoids making any inference from a scientific theory satisfying our criteria of a good theory to that theory being true or even close to the truth. Kuhn evinces a kind of metaphysical queasiness about talking in a literal and straightforward manner about theories being true or the entities that they postulate being real. As McMullin points out, Kuhn's work is written from the standpoint of an instrumentalist. Kuhn thinks that he can set aside the issues of objective truth and real, theory-independent existence. As long as a theory is good at solving puzzles, no further questions need arise, as far as Kuhn is concerned. (One of Kuhn's main arguments against scientific realism, based on the radical shifts in ontology that often occur when one theory replaces another, is presented in detail in Laudan's paper, "A Confutation of Convergent Realism," in chapter 9.)

McMullin, on the other hand, is an unabashed scientific realist. Although he is understandably reluctant to embark on a full-fledged defense of scientific realism in his paper, he does raise serious questions about Kuhn's ability to explicate the rationality of scientific revolutions without taking realism seriously. McMullin is especially critical of Kuhn's account of the Copernican revolution. In his book *The Copernican Revolution*, Kuhn compares and contrasts the Ptolemaic and Copernican theories. He concludes that both theories had about the same predictive accuracy and that both were able to explain the major phenomena of planetary motion. Kuhn admits that astronomers like Copernicus, Kepler, and Galileo were attracted to Copernican theory because its explanations were more "natural," but he regards this preference as merely a matter of taste, with no objective or epistemic significance.

McMullin shows that Kuhn is dead wrong when he claims that the Copernican and Ptolemaic theories had the same degree of explanatory power. True, the two theories had about the same predictive accuracy, but the Copernican theory could explain many things that the Ptolemaic theory could not. McMullin gives several examples of such phenomena, including restricted elongation—the fact that Mercury and Venus (the so-called inferior planets) are never seen at more than a certain angular distance from the sun—and various features of retrograde motion, such as the fact that a planet undergoes retrograde motion only when that planet, the earth, and the sun lie in a straight line. These phenomena are immediate consequences of the Copernican theory's basic postulates.[11] By an appropriate adjustment of parameters, the Ptolemaic theory could represent these facts, but it could not explain them. For Ptolemy's theory, they were unexplained coincidences; for Copernicus's theory, they were necessary consequences of the basic structure of our planetary system.

Copernicus, Kepler, and Galileo took the explanatory success of the Copernican theory to be a powerful argument for its truth. They reasoned that it could not be an accident that their theory was able to explain so much; only by uncovering the true causes of the observed motions of the planets could the Copernican theory account for those motions in such an elegant, unified fashion. As McMullin emphasizes, merely "saving the phenomena" was not the issue. Any number of theories could be devised, at least in principle, that could do as good a job as the theories of Ptolemy and Copernicus in predicting where the planets would be seen on any given night. Unlike predictive accuracy, explanatory power was an indicator of truth. Kuhn's instrumentalism makes him unwilling to recognize the important distinction between prediction and explanation; as a consequence, he misportrays the rational basis on which Kepler and Galileo preferred the Copernican theory. The Copernicans preferred their theory not primarily because it was a better predictor or a better problem solver than its rival, but because they believed it to be true. And they believed it to be true largely because of its tremendous explanatory power.

McMullin also argues for the epistemic significance of other theoretical virtues, such as fertility. Again, his point is that these virtues make sense only when seen as indicators of truth. Only when we talk about truth and embrace scientific realism, can we appreciate why scientists take these virtues so seriously and why they have played such an important role in the history of science. For example, one of the most persuasive arguments for the Copernican theory was the runaway success of the program of research to which it led. In part, this success stemmed from the fact that the Copernican theory (unlike its Ptolemaic rival) enabled astronomers to calculate the distance of each planet from the sun (as a function of the distance between earth and the sun). This information about distances was crucial to the advances made by Kepler and, eventually, Newton. Again, the Copernican theory's ability to generate new and important lines of research was valued not merely for its own sake or on pragmatic grounds (by promising employment and possible fame to eager young scientists) but primarily on epistemic grounds, as an indicator of truth. Only a theory (or, more generally, a paradigm) that is true or close to the truth could possibly be successful in producing such a cornucopia of wonderful new results. (Whether such arguments from the success of theories to their probable truth are plausible is discussed in detail in chapter 9, "Empiricism and Scientific Realism.")

2.6 | Laudan's Criticisms of Kuhn

Laudan's criticisms of Kuhn in "Kuhn's Critique of Methodology," differ markedly from those of McMullin in part because Laudan does not share McMullin's enthusiasm for scientific realism. Consequently, he evaluates Kuhn's claims at the level of problem solving without raising the issue of realism. But Laudan agrees with McMullin that Kuhn has drawn false conclusions about the nature of scientific rationality using flawed arguments. Before discussing Laudan's specific criticisms of Kuhn, it will be helpful to contrast two accounts of scientific rationality: the hierarchical model and the reticulational model. (The terms are Laudan's, and his account of them can be found in chapters 2 and 3 of his book *Science and Values*; "Kuhn's Critique of Methodology" is taken from chapter 4 of that book.) Part of Kuhn's error, Laudan argues, is that he has uncritically adopted the hierarchical model of scientific rationality, which is fundamentally flawed.

Competing Models of Scientific Rationality

Paradigms can be thought of as having three sorts of components, each at a different level. Laudan calls these levels the factual, the methodological, and the axiological. At the factual level, paradigms provide a conceptual framework, a general account of the entities and processes that supposedly populate the world. Also at this level are particular scientific theories, framed in terms of this ontology. The next level up is the methodological, which contains rules for choosing among theories. These methodological rules can include specific directives about what kinds of techniques and instruments to use in designing experiments, rules about what constitutes confirming evidence for a theory, and general statements of epistemic values such as testability and simplicity. Although it is convenient to talk simply of rules in this context, these must be understood to include statements of general preference such as, "Other things being equal, prefer theories that are simpler than their rivals." As Kuhn insists, and Laudan agrees, rules of the latter type function as what Kuhn calls values: rather than dictating a specific choice among theories, they merely specify the virtues of a good theory, leaving the individual scientist to judge how they should be balanced against other, similar, values. At the third and highest level, the axiological, are the aims and goals of science.

This model is hierarchical in the following sense: when there is disagreement at one level, the disagreement is rationally resolved (if at all) by going to the next highest level. Thus, to choose among competing theories at the factual level, appeal is made to rules at the methodological level. It is the methodological rules that provide the rationale for decisions about theories. But what if disagreement persists because of a difference of opinion about which rules to adopt? In that case, scientists must have recourse to the axiological level. For disputes about methodological rules can be rationally resolved only by specifying the goals that the rules are supposed to serve. If scientists all shared the same views about the aims of science, then, in principle, the dispute could be settled by seeing which rules would best promote those aims. But if scientists disagree about aims, then no rational resolution is possible. If, for example, one group of scientists insists that empirical adequacy is the most important goal of scientific theorizing, while another group maintains, with equal vigor, that science should aim at providing intelligible explanations, then the two groups have no objective, rational way of settling their dispute. According to the hierarchical model, the rational resolution of axiological disagreements is impossible because it would require going to a higher level, and there is no such level.

Admittedly, this is a highly idealized and simplified sketch of the hierarchical model, but it suffices to illustrate two of its central features. First, the direction of justification is strictly top-down and linear: always from a higher level to a lower one, never the reverse. Second, at the highest level, the axiological, rational justification ceases. On the hierarchical model, views about the goals and aims of science are necessarily arbitrary and subjective. Although they are viewed as the ultimate source of justification for the decisions made at the lower levels, they themselves cannot be justified.

According to Laudan, it is Kuhn's misleading picture of scientific rationality that pushes him to conclude that reason and argument play only a limited role in scientific revolutions. For, once one accepts the hierarchical model, disagreements about the criteria for assessing theories will be incapable of rational resolution if, as Kuhn supposes, rival paradigms have competing notions of the aims and goals of science. Another error contributing to Kuhn's embrace of relativism is his holism, his insistence that paradigms are seamless wholes. Kuhn assumes that paradigms are package deals that can be accepted or rejected only in their entirety. Specifically, he denies that scientists can evaluate or choose among the individual components of a paradigm, especially at the methodological and axiological levels.

Laudan advocates what he calls the reticulational model of scientific rationality. Its two main features—its antiholism and its nonlinear conception of justification—directly contradict the hierarchical model. According to the reticulational model, the components of a paradigm can be discussed, rejected, or accepted piecemeal. Like McMullin, Laudan criticizes Kuhn's assumption that paradigms have a kind of holistic unity that would make such a dissection impossible. He supports his case with historical examples in which scientists clearly did pick and choose elements from the paradigms they were debating. Contrary to Kuhn, changes at one level do not have to be accompanied by changes at a higher level: for example, method-ological rules can change in response to a discovery that the old rules are not the best way to realize the aims of science; it is precisely because scientists have not changed their view of the aims of science that they modify their rules.

Laudan's notion of justification is far richer than the strictly top-down, linear notion of the hierarchical model. On his model the flow of justifica-tion can be bottom-up as well as top-down, from the factual to the method-ological as well as the other way around. He offers historical examples of how the success (and, in some cases, failure) of scientific theories led scien-tists to adopt (or abandon) particular methodological rules. Of particular importance is Laudan's claim that there can be, and occasionally has been, rational debate concerning which aims and goals of science are appropriate. Laudan gives as an example the abandonment of the Aristotelian ideal that science should aim at certain, infallible knowledge. By the end of the nine-teenth century just about every scientist agreed that science could not pos-sibly achieve that goal but should settle instead for highly probable, but fallible, belief. As Laudan points out, this major change at the axiological level was associated not with any one particular paradigm, nor with any one particular area of science. Rather, the shift grew out of the awareness that many of the best scientific theories (whether in physics, biology, or chemis-try) involved hypothetical elements that were impossible to prove with cer-tainty. In short, the widespread success of a whole range of theories at the factual level led to the realization that the traditional Aristotelian ideal could not be attained. Since the goal of certainty could not be achieved, it was rational to abandon it.

The Flaws in Kuhn's Arguments

In general, a philosophical argument can fail for three reasons: either it is invalid (i.e., its conclusion does not follow from its premises); it has at least one false premise (in which case the argument, even if valid, is unsound); or it is circular (in which case it begs the question by taking for granted the very point that it purports to prove.) In his paper, Laudan undertakes to expose the flaws in Kuhn's arguments for his conclusion that methodological rules and standards can never determine the choice between paradigms. Laudan refers to Kuhn's conclusion as the thesis of *local underdetermina-tion*. The alleged underdetermination is local because it concerns the choices made by scientists between the few paradigms that have actually been put forward for serious consideration. *Global underdetermination* (discussed at length in chapter 3) is the far more sweeping thesis that our rules can never pick out a particular theory from among all the possible alternatives as the only one that is supported by the available evidence. Suffice it to say for now that the two theses are different and that even if (as Laudan argues) Kuhn is wrong about the local thesis, it might still be the case that the global thesis is true. (Hence the need to address the global thesis and related issues in the next chapter.)

Kuhn's six arguments—appealing to theory-ladenness of observation, meaning variance, problem weighting, shifting standards, the ambiguity of shared standards, and the collective inconsistency of rules—were summarized earlier (191–98). The first two of Kuhn's arguments were criticized above. Laudan's response to the last three of these arguments is to accuse them of being unsound. Each of them, he says, has at least one false premise: standards do not always change from one paradigm to another; some shared standards (such as consistency and novel prediction) are not ambiguous; and not all sets of methodological rules give conflicting advice.

Laudan's criticism of the problem-weighting argument is more complex. The nub of Kuhn's argument is that proponents of rival paradigms must disagree about which problems are the most important even when they agree about the rules and standards by which theories should be assessed. As Laudan explains, *importance* can mean two different things in this context. A problem might be judged important for social or economic reasons, or simply because a particular scientist is interested in solving it. That would be a nonepistemic sense of *importance*, since it would have no direct bearing on whether a theory that solves the problem is true, confirmed, or rationally credible. Contrast that with a problem that is considered important because its solution would confirm a theory. Such a problem would be epistemically important; it would have what Laudan calls *probative* significance. The probative significance of a problem is an objective matter, and claims about it will be defended or attacked by appealing to epistemic and methodological rules—rules that deal with questions of empirical support, evidence, and confirmation. So in this second sense of *importance*, disputes about which problems are the important ones for a theory to solve will not necessarily degenerate into mere differences of subjective opinion. Thus, Laudan condemns Kuhn's problem-weighting argument as invalid. Once we focus on the epistemic sense of *importance*, it simply does not follow that proponents of rival paradigms must disagree about which problems are important or, when such disagreements do occur, that such conflicts cannot be resolved by reasoned argument based on shared standards.

Laudan concludes by discussing a case from the history of chemistry that both Kuhn and Gerald Doppelt have appealed to for support of their position. Laudan admits that Dalton's atomic theory failed to address many of the questions that older chemical paradigms had tried to answer, for Dalton's interests lay elsewhere. But, Laudan counters, Dalton's assessment of which problems of chemistry are central reflects his pragmatic interest in certain problems rather than others. This subjective interest should not be confused with the judgment about which problems have the greatest epistemic value. Indeed, as Laudan points out, the problems that the advocates of a paradigm are most eager to solve often have little or no confirming value. For according to some philosophers, a problem solution does little or nothing to confirm a theory if the solution is one that the theory was deliberately designed to yield. (For an extended discussion about the relation between confirmation and novel predictions, see chapter 4.)

The evaluation of Kuhn's work by McMullin and Laudan is largely negative. Both authors agree that Kuhn has failed to establish his more sweeping claims about the irrationality of scientific revolutions. Neither the history of science nor Kuhn's philosophical arguments show that scientific revolutions cannot be resolved by rational argument based on evidence and shared rules. By treating paradigms as indivisible wholes and by failing to appreciate the ways in which rules and aims can be rationally debated, Kuhn has seriously underestimated the role of reason in paradigm debates. In passing this verdict on Kuhn, it is important to remember that McMullin and Laudan are not claiming that reason, rules, and evidence always suffice to determine the outcome of such debates. Rather, they are objecting to Kuhn's assertion that they can never do so. Thus, according to McMullin and Laudan, Kuhn's thesis that subjective, psychological, and rhetorical factors must play a leading role in all scientific revolutions is ill supported and false.

Many philosophers of science agree that the more extreme, subjectivist elements in Kuhn's picture of science (especially the picture presented in the first edition of *The Structure of Scientific Revolutions*) are exaggerations. Nonetheless, many of those same philosophers of science find much of value in Kuhn's work. This is especially true of one aspect of Kuhn's approach that we have not yet discussed, namely, his view that science is essentially a social activity. One philosopher who has tried to develop this theme in Kuhn's work and relate it to traditional philosophical concerns about objectivity and rationality is Helen Longino.

2.7 | Longino on Contextualism and Objectivity

In "Values and Objectivity," Longino begins by distinguishing two notions of objectivity. The first concerns the outcome of scientific inquiry, namely scientific knowledge; the second focuses on the process by which that outcome is generated, namely scientific method. According to the first notion, science is objective to the extent that it is true—to the extent that it describes the world as it really is. According to the second notion, science is objective to the extent that its methods, especially its criteria for assessing theories, are neither arbitrary nor subjective. As Longino points out, it is commonly believed that science is objective in the first sense because it is believed to be objective in the second.

THE LOGICAL POSITIVIST VERSUS THE CONTEXTUALIST ACCOUNT OF SCIENTIFIC METHOD

Longino contrasts two accounts of scientific method: the logical positivist account and the contextualist account of "historically minded wholist critics," such as Kuhn and Feyerabend. The logical positivist account divides scientific activity into two distinct phases: the context of discovery and the context of justification. The context of discovery is the period during which scientists first come up with their ideas and hypotheses. Here it is allowed that psychological and subjective factors often play a major role. But once the theory or hypothesis has been formulated and presented for assessment, then the context of justification begins. In this context, positivists claim that psychological and subjective factors play no role whatever (or, rather, that they *ought* to play no such role, since positivists are strongly wedded to the value-neutrality thesis). On one extreme version of the logical positivist view (and the only one that Longino considers), assigning degrees of confirmation to hypotheses and deciding which theories to accept (or reject) is solely a matter of applying the relevant rules to the available evidence. This rule-based assessment could be performed by a single scientist; if several scientists assess the same theories on the basis of the same evidence, they should all get the same answer. The social dimension of science is irrelevant to its production of reliable knowledge. In principle, science could be done by a single individual.

Although popular in the first half of the twentieth century, the logical positivist account of scientific method is now discredited, in part because of the criticisms of Kuhn and others. One of those criticisms is that the logical positivist picture of science does not match what has actually happened during the history of science. During scientific revolutions especially, we do not find anything like the degree of consensus about scientific theories that the logical positivist account leads us to expect. Another criticism concerns the shortcomings of the logical positivist treatment of confirmation. One of the things that Kuhn has taught us, Longino explains, is that confirmation cannot be a matter of mechanically applying rules to evidence because what an individual scientist counts as evidence for a theory in the first place will depend on the other beliefs that the scientist holds. Background beliefs and assumptions are crucial in determining which hypotheses we accept as being confirmed by which evidence. Longino calls this the *contextualist analysis of evidence.*

THE CONTEXTUALIST ANALYSIS OF EVIDENCE

In chapter 3 of her book *Science as Social Knowledge,* Longino gives the following illustration of the contextualist analysis of evidence. Suppose that an eight-year-old child has red spots on her stomach. Why should her parents take the rash as evidence for measles rather than, say, as evidence that the moon is blue? Ordinarily, the explanation would be that they think that the rash is caused by the measles virus, and there is no other likely cause of that kind of rash. Given this background belief, the rash strongly confirms their diagnosis of measles. But they could have held different background beliefs. They might, for example, have believed that a rash of this kind is caused by a gastric ailment. In that case, they would have inferred a different hypothesis from the same evidence. Furthermore, the things that they regard as evidence can be described in a variety of ways: red spots on the stomach, spots (without mentioning their color), a rash (without mentioning its location), urticaria, an itchy tummy. Which hypotheses they take to be confirmed by the evidence also depends on how that evidence is described.

Two important features of Longino's contextualist analysis of evidence need to be stressed. First, what Longino calls *evidence* is a physical object, a physical process, or, most generally, a state of affairs. This is the way that the term *evidence* is used in a court of law (where it also includes testimony). It is not the way that most philosophers of science use the term when discussing confirmation. As we shall see in chapters 4 and 5, when analyzing confirmation most philosophers of science assume that the evidence for a theory or hypothesis is a set of statements. That is, having assumed that the objects (events, processes) we observe have been described in some language, these philosophers then go on to analyze confirmation as a relation between observation statements and theoretical statements. While the distinction between the objects of observation (what Longino calls *evidence*) and observation statements (what most other philosophers of science call *evidence*) might seem pedantic, it is crucial for understanding the nature of Longino's project. For many philosophers of science would agree with Longino that the first step, the step from sensory stimulation to observation report, depends on the observer's background beliefs and training, and the language that person decides to use. For example, the words uttered when looking through a telescope trained on the moon will vary from individual to individual. To that extent, confirmation involves a subjective component. But acknowledging this component is consistent with the view that the confirmation relation between observation statements and theoretical statements is entirely objective. Longino stresses the subjective element in the first step because she is interested in the entire causal and social process by which scientists make judgments about theories. As she describes it, her project is to *explain* this kind of scientific activity by understanding the mechanisms at work. Like Kuhn, Longino is more concerned with explaining scientific behavior (and exploring its consequences for objectivity and rationality) than with giving formal analyses of isolated concepts.

Second, in presenting her contextualist analysis of evidence, Longino is scrupulous about respecting the important distinction between two questions: "Does this piece of evidence confirm this hypothesis?" and "Why do scientists take this piece of evidence to confirm this hypothesis?" Longino confines the contextualist analysis of evidence to answering the second question, deliberately leaving the first question to one side. One reason Longino avoids the first and more traditional question is that she thinks that philosophical attempts to give formal analyses of confirmation have been failures.[12] Another reason stems from her concern with explanation. For it is unlikely that a sophisticated, formal model of confirmation—even if it should prove philosophically adequate—could account for the individual psychological processes or the collective social mechanisms by which scientists reach judgments about confirmation. Since her primary goal is to explain scientific behavior, it is the second question, not the first, that is relevant to her project.

Nonetheless, it is still possible that a convincing philosophical answer to the first question might be found. For example, one of the attractions of the Bayesian approach to confirmation (explored in chapter 5) is that it provides a formal framework within which background beliefs can play an important role. Some philosophers of science (such as Wesley Salmon) have argued that this approach provides a perfect vehicle for incorporating the subjective factors Kuhn and Longino emphasize. So even though we might agree with Longino that background beliefs play a central role in confirmation, confirmation may still be an objective relation between statements that include those background beliefs as an element.

In summary, according to the contextualist analysis of evidence, confirmation (or, more precisely, what we take to be confirming evidence for a theory or hypothesis) depends on background beliefs and assumptions, what Longino calls *contextual values*, which can vary from individual to individual. How we describe the evidence and which hypotheses we judge to be confirmed by that evidence (and to what degree) all depend on contextual values.

CRITERIA FOR OBJECTIVITY

Despite her emphasis on contextual values in science, Longino is far from denying that science is objective. In fact, she adamantly rejects the charges of rampant subjectivism and irrationality that have been hurled against Kuhn. She thinks that critics of Kuhn have made two fundamental mistakes: they have failed to distinguish between the two senses of objectivity—process versus product—discussed at the beginning of her article, and they have failed to take seriously the social character of scientific activity. Longino regards the latter as the key to guaranteeing some degree of objectivity in science.

As Longino points out, the contextualist analysis of evidence raises a serious question: if all judgments about confirmation and evidence depend upon the subjective beliefs of the individual researcher, how can science be objective? The problem is serious because subjective background beliefs (contextual values) often escape the notice of the person who holds them. And unlike advocates of the value-neutrality thesis, Longino does not think it possible to eliminate contextual values from science.

Longino believes that the answer to the question, "How can science be objective, when contextual values play an essential role in the judgments made by individual scientists?" lies in the realization that science is a social activity organized to permit and encourage what she calls *transformative criticism*—that is, criticism with the power to change contextual values should they prove ill-founded. Thus, science can be objective only to the extent that transformative criticism is allowed to flourish. In fact, she regards this social feature of science as necessary for objectivity. Longino writes:

> From a logical point of view, if scientific knowledge were to be understood as the simple sum of finished products of individual activity, then not only would there be no way to block or mitigate the influence of subjective preference but scientific knowledge itself would be a potpourri of merrily inconsistent theories. Only if the products of inquiry are understood to be formed by the kind of critical discussion that is possible among a plurality of individuals about a commonly accessible phenomenon, can we see how they count as knowledge rather than opinion. (153)

> I have argued both that criticism from alternative points of view is required for objectivity and that the subjection of hypotheses and evidential reasoning to critical scrutiny is what limits the intrusion of individual subjective preference into scientific knowledge. (154)

In the remainder of her paper, Longino describes four criteria that, she argues, are necessary for transformative criticism to flourish (155):

1 "There must be recognized avenues [such as journals and conferences] for the criticism of evidence, of methods, and of assumptions and reasoning." Scientists should get appropriate credit for their critical activities, just as they do for their original research.

2 "There must exist shared standards that critics can invoke." Longino's list is similar to Kuhn's, although, like Kuhn, she denies that these standards can serve to determine choices among theories. Nonetheless, these constitutive values (or shared standards) play an important role in constraining and correcting the influence of contextual values.

3 "The community as a whole must be responsive to such criticism." Longino does not think that individual scientists should recant the moment they are criticized, for science is often best served when scientists defend their work from criticism. But criticism must be capable, in the long run, of changing the beliefs of the scientific community as a whole.

4 "Intellectual authority must be shared equally among qualified practitioners." Longino alludes here to the ideas of the German philosopher Jürgen Habermas, who has stressed the importance of public debate and rational consensus for preventing the domination of

society by one group of interests.[13] An egregious example of the violation of this criterion was the suppression of Mendelian genetics in the Soviet Union in the 1930s.[14] Longino thinks that the exclusion of women and racial minorities from science in the United States also violates this criterion.

Longino believes that these criteria are an ideal to which science should strive rather than a description of how actual science is conducted. Thus, Longino is much clearer than Kuhn in distinguishing between the descriptive and the normative in her account of science. Kuhn asserts that normal science is relatively dogmatic, reaching consensus by training scientists in one particular paradigm and discouraging dissent. Longino advocates that scientists should encourage criticism and reach consensus by rational negotiation; when science falls short of this ideal, its objectivity is compromised. Thus, Longino has a much clearer basis from which to criticize scientific behavior than Kuhn does. And Longino does precisely that in the later chapters of her book, where she criticizes contemporary biological research into sex differences and the explanation of human and animal behavior. (This research and criticisms of it are reviewed in Kathleen Okruhlik's paper, "Gender and the Biological Sciences," discussed below.)

Another feature of Longino's criteria is that she regards them as necessary conditions for scientific objectivity. But are they necessary conditions in the strict logical sense? Arguably, it is an empirical matter whether or not (or to what extent) scientific objectivity is compromised when, say, women or racial minorities are excluded from science.[15] Were all branches of science, from mathematics to chemistry, less than fully objective throughout the entirety of their history because, until recently, women and minorities were almost completely excluded from them? There is also the question of whether Longino's criteria for scientific objectivity are sufficient. In her paper, "Gender and the Biological Sciences" (discussed below), Kathleen Okruhlik argues that no set of criteria concerned solely with choosing among already formulated theories can guarantee objectivity.

A final point concerns the notion of objectivity itself. Longino's criteria are designed to promote objectivity by trying to eliminate the bias, arbitrariness, and subjectivism of individual scientists from the judgments made by the scientific community as a whole. But what is the connection, if any, between this notion of objectivity (roughly, freedom from bias) and the notion of objectivity mentioned at the outset of Longino's paper, namely, objectivity as truth? Habermas, whom Longino discusses at length in her book (see especially pp. 200–201), connects the two by *defining* truth as whatever all participants would agree about, provided they all had an equal chance to engage in free and uncoerced communication and all had equal power to make their views rationally persuasive. But, as Longino remarks, this attempt to define truth as rational consensus faces a number of powerful objections. What happens when consensus changes over time? Does this mean that truth also changes? And what about propositions concerning which there will never be consensus, not even in the long run? Are these propositions neither true nor false? Perhaps it is more plausible to regard scientific consensus not as a definition of truth, but as a reliable indicator of truth. In that case, however, we would seem to need some independent way of ascertaining the truth in order to justify our belief that scientific consensus leads to the truth. And what could that independent way possibly be when science is our best and only guide? These are difficult questions, but they must be confronted by anyone who believes that science is not merely a game played for its own sake but a reliable guide to the truth. (This issue is discussed at length in chapter 9, "Empiricism and Scientific Realism.")

2.8 | Okruhlik on the Feminist Critique of Science

In the first section of her article, "Gender and the Biological Sciences," Kathleen Okruhlik discusses several cases from modern biology in which contextual beliefs about gender (sexist and androcentric bias) have influenced scientific judgment. For example, the general belief that females are passive and males active may have prevented the recognition of evidence that the ovum plays an active role in fertilization. As Longino has stressed, the issue is not whether a particular observation was made—microvilli extending from the ovum were seen and photographed as early as 1895— but how the observation is described and, as a consequence, what evidential significance it is deemed to have. For example, it makes a difference whether what is seen under the microscope is described merely as "projections around the ovum *accompanying* the penetration of a sperm cell," or as the ovum "*clasping* the sperm and *drawing* it in." Okruhlik summarizes the influence of contextual values by alluding to the theory-ladenness of observation, but it is probably best regarded as an illustration of Longino's contextualist analysis of evidence, discussed in the previous section of this commentary.

Other examples indicate that androcentric bias played (and presumably continues to play) a major role in evolutionary biology and anthropology by shaping the questions asked and the theories devised to answer them. Okruhlik frames this part of her discussion by referring to the thesis of the underdetermination of theory by data. She claims that, even after the data have been described in some particular way, several (perhaps, indefinitely many) empirically adequate hypotheses might still be available. If all the hypotheses are empirically adequate, then contextual values, not cognitive ones, are determining which hypothesis gets selected. Although often appealed to in discussions such as these, the underdetermination thesis is highly controversial. (This thesis is examined at length in chapter 3, "The Duhem-Quine Thesis and Underdetermination.") But despite doubts about the truth of the underdetermination thesis as a generalization about scientific theories, in some cases, such as the ones described by Okruhlik, contextual values of a sexist kind have undeniably influenced theory choice.

Okruhlik points out that contextual values also shape scientific judgment when scientists steadfastly refuse to give up a hypothesis despite the accumulation of contrary evidence. An entrenched belief that women are biologically determined to be intellectually inferior to men has prevented some (male) scientists researching sex differences from recognizing when pet hypotheses (such as the hypothesis that spatial ability is X-linked) have been refuted by the evidence. Such cases illustrate the ambiguity of falsification, or what is often referred to as the Duhem-Quine thesis (see chapter 3 for a fuller discussion).

VARIETIES OF FEMINIST CRITIQUE

Assuming that Okruhlik and the authors she cites have made a plausible case for the pervasive influence of sexist beliefs on decision making in several branches of the life sciences, what are we to make of this fact? And what implications does it have for the larger question of scientific objectivity? For example, do the cases of androcentric bias show that science is, by its very nature, sexist? Or can these cases be dismissed as regrettable instances of "bad science," mere deviations from the objectivity that normally prevails in science?

At this point Okruhlik introduces a helpful trio of categories that Sandra Harding has proposed for classifying feminist criticisms of science: feminist empiricism, standpoint epistemologies, and feminist post-modernism.

Feminist empiricism regards science as essentially objective by virtue of its methods: rules of evidence, confirmation, and falsification. If these rules were always followed correctly, "bad science" would never occur. The whole point of scientific methodology is to make the gender, race, and personality of the individual scientist irrelevant. Thus, feminist empiricists espouse the value-neutrality thesis and deny that contextual values should play any role in science.

If the arguments of Longino, Okruhlik, and others are sound, then contextual values cannot be eliminated from science and the value-neutrality thesis embraced by feminist empiricists is not plausible. But this leaves open other versions of feminist criticism in which the objectivity of science is construed, along the lines proposed by Longino, not as a feature of methodological rules, but as the outcome of a complex social process designed to encourage criticism. Okruhlik advocates this alternative in the second half of her paper.

Standpoint epistemologists such as Sandra Harding insist that contextual values are essential to science and that some contextual values are better than others. If science were done by women, from the standpoint of women, the theories produced would give us a more objective picture of the world. (Here, of course, *more objective* means *truer*.) As far as getting to the truth about the world is concerned, women are supposed to have certain advantages over men, either because women as a group have fewer contextual beliefs to distort their ways of seeing and thinking about the world or because the contextual beliefs of women as a group are more in tune with reality.

There are a number of problems with standpoint epistemology. First, it is difficult to specify, even in broad outline, what contextual beliefs and values women might have that would give them a natural advantage over men in discovering important scientific truths about the world. Helen Longino makes a stab at this in her paper "In Search of Feminist Epistemology" in which she assembles a list of theoretical virtues that are often emphasized in feminist writings.[16] Longino contrasts this list with the set of shared cognitive values laid out in Kuhn's "Objectivity, Value Judgment, and Theory Choice." Kuhn's list is: accuracy, consistency, scope, simplicity, and fruitfulness. Longino's "feminist" list is: empirical adequacy, novelty, ontological heterogeneity, complexity of relationship, applicability to current human needs, and diffusion of power. The first item on both lists is essentially the same thing under different names. Although the accuracy (empirical adequacy) of theories is clearly relevant to their truth, it obviously cannot explain why feminist science would be epistemically superior to science as we now know it. The last two virtues on the feminist list might be relevant to the claim that feminist science would be better in some nonepistemic, moral sense (for example, by doing more to improve the lives of human beings), but it is hard to see what they have to do with truth. That leaves novelty, ontological heterogeneity, and complexity of relationship. Despite the relish with which Sandra Harding has advocated breaking with all the assumptions on which science is currently based, it is difficult to recommend mere novelty as a guide to the truth. Ontological heterogeneity and complexity of relationship have been cited as important themes in the work of Barbara McClintock on the genetics of maize.[17] But like simplicity, these values are both hard to define and difficult to connect with truth in any straightforward way.

A second difficulty with standpoint epistemology has been identified by Okruhlik and others: it is doubtful that any single standpoint is shared by all women, regardless of their age, race, sexual orientation, or economic status. Given this multiplicity of standpoints, there is a tendency for standpoint epistemology to degenerate into what Harding calls *feminist postmodernism*.

Feminist postmodernism is a variety of epistemic anarchism. All pretense to objectivity is abandoned in favor of the position that there are many different standpoints, each telling a different story about how the world is, and no one of these stories is better than any other. Science is just one possible story. For anyone who wants to criticize science on the grounds that it displays gender bias, feminist postmodernism is not an option, since it eschews all normative judgments.

OKRUHLIK'S PROPOSAL

In the second half of her paper, Okruhlik argues that the contextual values influencing the generation of theories in the context of discovery will inevitably affect the content of science in the context of justification. For the context of justification involves choosing among the available theories, and so if all those available theories are infected with gender bias, then that bias will continue to contaminate the theory we select as the best. Contrary to popular opinion, the scientific method (understood as rules, criteria, or values for theory selection) cannot, by itself, remove sexist bias from the content of science. Even if every scientific decision is perfectly rational (in the sense that it scrupulously follows all the right rules), the resulting product may still be defective because social factors have influenced the context of discovery. Thus, Okruhlik concludes, philosophers of science must acknowledge that the context of discovery has an epistemic significance and that philosophy of science should not focus exclusively on the context of justification.

In the light of her analysis, Okruhlik proposes that we should improve the objectivity of science by changing its social organization. Among other things, this means encouraging women to pursue scientific careers, eliminating sex discrimination in promotion and hiring, making sure that research by women gets the recognition it deserves, and ensuring that the voices of women scientists are heard and taken seriously when they criticize the work of their male colleagues. For, like Longino, Okruhlik locates the objectivity of science, not in its methodological rules or its cognitive values, nor in the vain hope that contextual values could be eliminated at the level of the individual scientist, but in the community of scientists striving to encourage criticism by including a diversity of viewpoints.

As Okruhlik notes, her proposal falls somewhere between the first two types of feminist critique on Harding's list. With feminist empiricism it shares the conviction that science should aim at improving its objectivity and that established scientific method (based on the cognitive values of science) plays an important role in achieving that goal. But, unlike feminist empiricism, Okruhlik's proposal insists that contextual values influence the content of science and that the social arrangements of science in controlling that influence are of vital importance. As Okruhlik puts it, both feminist empiricism and standpoint epistemology err when they assume that the rationality of the scientific community is just individual rationality writ large or that science can be reformed at the level of individual psychology. Unlike standpoint epistemology, Okruhlik does not assume that women have distinctively feminist values that are superior to those of men. Nor does Okruhlik claim that including more women in science in the ways proposed is either necessary or sufficient for an improvement in scientific objectivity. But she thinks that such an improvement is likely, given the way science operates. In short, science will quite probably become more objective if her proposal is adopted, not by being transformed into some new kind of feminist science approached from a distinctively feminist standpoint, but by including the feminist standpoint along with others.

2.9 | Summary

In one way or another, all the readings in this chapter focus on Kuhn's *The Structure of Scientific Revolutions*. For it was Kuhn's book that made such a dramatic case for the role contextual values play in shaping scientific judgments about theories. Kuhn's views about the limits of cognitive values in determining the outcome of scientific revolutions have been influential and remain controversial.

As we have seen, Kuhn has two groups of arguments for his conclusion about the lack of rationality and objective progress in scientific revolutions. The first group includes the arguments from the theory-ladenness of observation and from meaning variance. Both have been widely criticized, and in his later publications Kuhn toned down considerably his reliance on these arguments. The second group includes a number of arguments claiming that cognitive values alone cannot determine a scientist's decision to embrace or reject a new paradigm. Many critics have accused *The Structure of Scientific Revolutions* of portraying science as irrational. In later papers, such as "Objectivity, Value Judgment, and Theory Choice," Kuhn responded to these critics by emphasizing the importance of shared, cognitive values in providing a basis for reasoned argument during paradigm debates. Nevertheless, he still insists that shared values can take us only so far in deciding between rival paradigms. Ultimately, all such decisions must rest on personal, subjective, and cultural factors that are beyond the reach of rationality.

McMullin and Laudan both criticize Kuhn's second group of arguments. McMullin is especially critical of Kuhn's assumptions that all paradigms share a common set of epistemic values, that epistemic values cannot be given any rational justification, and that we can make sense of scientific revolutions (such as the Copernican revolution) without addressing the question of scientific realism.

Laudan condemns all of Kuhn's arguments as unsound, tracing their defects to Kuhn's hierarchical model of scientific rationality. Laudan advocates the reticulational model, which allows for the rational justification of the components of paradigms at a variety of levels, recognizing that scientists can and do discuss, accept, and reject the components of paradigms piecemeal. Laudan argues that, contrary to Kuhn, some rules for evaluating rival theories are precise and unambiguous; moreover there is no reason to accept Kuhn's assumption that sets of such rules must be "collectively inconsistent." Standards for judging theories do not have to shift when paradigms change, nor does the importance attributed to certain problem-solutions by the adherents of rival theories have to reflect discordant judgments about epistemic support. In short, Laudan accuses Kuhn of having seriously over-estimated the degree to which debates between proponents of competing paradigms must depend on judgments by individual scientists that fall outside the scope of rational consensus about scientific methodology.

Longino offers a new approach to understanding the nature of scientific objectivity by defending the contextualist analysis of evidence. Which hypotheses we take to be confirmed by which evidence crucially depends on our background beliefs and assumptions. Nonetheless, science can be objective if its social organization permits and encourages criticism of these beliefs. Thus, she concludes, scientific objectivity is a matter of degree, depending not on the methodological rules by which individual scientists judge and select hypotheses, but on the way that the scientific community is organized.

Okruhlik discusses several recent cases in the life sciences that have led feminists to accuse science of sexist bias. Okruhlik thinks that the feminists are right: in some areas of biology and anthropology, androcentric bias has compromised scientific objectivity. Of particular importance is the role that contextual values play in shaping the formation of hypotheses in the context of discovery. In this way, regardless of the rationality of our procedure for selecting among theories, contextual values can infect the content of science. Like Longino, Okruhlik thinks that the solution lies in the reform of the social organization of science, making sure that it includes feminist and other standpoints. Okruhlik compares and contrasts her proposal with three varieties of feminist critique of science—feminist empiricism, standpoint epistemologies, and feminist postmodernism—and argues that her position falls somewhere between the first two.

■ | Notes

1. Also important is the question of how scientific theories have affected our views about morality. Of particular interest here is the effect that biological theories (such as Darwin's) have had on our ideas about egoism, altruism, and cooperation, and whether there is any sort of evolutionary ethics worthy of the name. Again, fascinating though this is, it falls outside the scope of this chapter.

2. Kuhn retracted this claim in his *Postscript*, where he concedes that "the members of all scientific communities, including the schools of the "pre-paradigm" period, share the sorts of elements which I have collectively labelled 'a paradigm.' What changes with the transition to maturity is not the presence of a paradigm but rather its nature. Only after the change is normal puzzle-solving research possible" (*The Structure of Scientific Revolutions*, 179).

3. For criticisms of Kuhn's appeal to the history of science, see Janet A. Kourany, "The Nonhistorical Basis of Kuhn's Theory of Science," *Nature and System* 1 (1979): 46–59, and Martin Curd "Kuhn, Scientific Revolutions and the Copernican Revolution," *Nature and System* 9 (1984): 1–14.

4. The example is adapted from Carl Kordig, "The Theory-Ladenness of Observation," *Review of Metaphysics* 24 (1971): 448–84.

5. This point is made with devastating clarity in Israel Scheffler, *Science and Subjectivity* (Indianapolis, Ind.: Bobbs-Merrill, 1967), 36–40.

6. And, of course, Kuhn himself agrees with us, since, using his terminology, the accumulation of anomalies, of observations that contradict our theoretical expectations, are a prerequisite to a scientific revolution. Thus, it cannot be Kuhn's view that perception is so theory laden as to preclude such observations.

7. Newton's theory requires that the force of gravity acts instantaneously between every pair of particles in the universe, regardless of their distance apart and the emptiness of the space that separates them. Among other things, this implies that, if the sun were suddenly to disappear, the effect on the earth would be immediate and the earth would start moving in a straight line at the very instant that the sun vanishes. In modern field theories of gravity (such as the general theory of relativity) it would take about eight minutes for the effect to reach the earth (since gravitational influence travels at the speed of light). It was precisely because action at a distance seems so spooky, so "miraculous," lacking as it does any intelligible mechanism, that Leibniz and others declared it to be scientifically illegitimate.

8. "Throughout this paper I have implicitly assumed that, whatever their initial source, the criteria or values deployed in theory choice are fixed once and for all, unaffected by their participation in transitions from one theory to another. Roughly speaking, but only very roughly, I take that to be the case. If the list of relevant values is kept short (I have mentioned five, not all independent) and if their specification is left vague, then such values as accuracy, scope, and fruitfulness are permanent attributes of science" (Kuhn, "Objectivity, Value Judgment, and Theory Choice," *The Essential Tension*, 335 [p. 105]).

9. As Peter Lipton put it, Kuhn is "Kant on wheels." Like Kant, Kuhn distinguishes between the world as it is in itself and the world as structured by our concepts, insisting that we can know only the latter world, a world we are partly responsible for creating. But unlike Kant, Kuhn thinks that this world changes whenever our scientific theories change. Thus, according to Kuhn, the world that scientists study at a particular time has only a limited lifespan. There is no single, objective notion of truth that can be applied to all such worlds and thus no objective sense in which science can be said to progress by getting closer to the truth. See Peter Lipton, "Coordinating Science," *Nature* 364 (1993): 770. Scientific realists find views such as Kuhn's false and abhorrent. See chapter 9.

10. See editorial footnote [p. 12].

11. Since all the planets, including the earth, circle the sun, the planets that lie inside the earth's orbit, namely Mercury and Venus, will never be seen at more than a certain angular distance from the sun. The planets move more slowly, the further their distance from the sun. Retrograde motion occurs either when the earth is overtaken on the inside by Mercury and Venus or when the earth overtakes the planets (such as Mars, Jupiter, and Saturn—the so-called superior planets) that lie outside of the earth's orbit. Necessarily, then, retrograde motion will occur only when the planet, the earth, and the sun lie in a straight line. For the superior planets, this will be when the planet lies opposite the sun as seen from the earth (what astronomers

call opposition); for the inferior planets, retrograde motion will occur only when the planet passes between the sun and the earth (what astronomers call conjunction).

12. See, for example, her criticism of Hempel's satisfaction criterion in chapter 2 of *Science as Social Knowledge.*

13. For a helpful introduction to Habermas's views about science, placing them in the context of thinkers such as Max Weber and Karl Marx, see Robert Hollinger, "From Weber to Habermas," in *Introductory Readings in the Philosophy of Science,* rev. ed., ed. E. D. Klemke, Robert Hollinger, and A. David Kline (Buffalo, N.Y.: Prometheus Books, 1988), 416–26.

14. Trofim Lysenko, a convinced believer in the Larmarckian doctrine of the inheritance of acquired characteristics, obtained dictatorial power over Soviet biology under Stalin. Scientists and teachers who disagreed with Lysenko lost their jobs and some, their lives. This grim episode set Soviet biology back for decades.

15. Longino's own formulation of her fourth criterion is somewhat problematic, since it specifies that intellectual authority in science should be shared among "qualified practitioners." Obviously, if "qualified practitioners" means something like "having received a first-rate scientific training," no violation of the criterion will have occurred if women or racial minorities are denied such training or discouraged from seeking it.

16. Helen E. Longino, "In Search of Feminist Epistemology," *Monist* 77 (1994): 427–85.

17. Barbara McClintock (1902–92) received the Nobel Prize for Physiology or Medicine in 1983 for her pioneering work on "jumping genes" in maize. Using classical genetical and cytological techniques, McClintock concluded that genetic elements in the chromosomes of maize can move around in response to changes in the chemical environment of the cell. Although she was not herself a feminist, McClintock's research has been regarded as a good illustration of a distinctively feminist approach to science. See, for example, Evelyn Fox Keller, A *Feeling for the Organism: The Life and Work of Barbara McClintock* (San Francisco: W. H. Freeman, 1983). It should be noted that, although McClintock's theory of transposable elements was ignored for several decades, she was far from being a scientific outsider. In fact, she was elected to the United States Academy of Sciences in 1944 and received many honors and awards for her experimental work in cytogenetics, including the Kimber Genetics Award (1967) and the National Medal of Science (1970). In 1981, she was the first recipient of the MacArthur Laureate Award, granting her a lifetime fellowship of $60,000 a year, tax free. Keller comments that "if Barbara McClintock's story illustrates the fallibility of science, it also bears witness to the underlying health of the scientific enterprise. Her eventual vindication demonstrates the capacity of science to overcome its own characteristic kinds of myopia, reminding us that its limitations do not reinforce themselves indefinitely" (Keller, 197). Despite respect for her experimental work, McClintock's theoretical ideas were hard for many scientists to accept in the 1950s and 1960s because they ran counter to the prevailing paradigm that genes are fixed, stable units of heredity and that information always flows from genes to proteins, never in the reverse direction.

3 |

The Duhem-Quine Thesis and Under-determination

INTRODUCTION

Probably no set of doctrines has had a greater influence on modern philosophy of science than those included under the designation of the Duhem-Quine thesis. Thinkers as diverse as Sandra Harding, Bas van Fraassen, Mary Hesse, David Bloor, Arthur Fine, Helen Longino, Thomas Kuhn, and Richard Rorty have invoked, in one form or another, a version of the Duhem-Quine thesis to reach conclusions about the limitations of empirical evidence and the rules of scientific method as a constraint on our acceptance or rejection of scientific theories. Some of these philosophers have argued that no scientific theory can ever be conclusively refuted (thus repudiating one of the central features of Popper's philosophy of science); others of them have concluded that, given the presumed gap separating theory from evidence and the presumed insufficiency of methodological rules in bridging that gap, we should never accept any theory as objectively true no matter how well it agrees with the available evidence (thus abandoning scientific realism in favor of some version of skeptical relativism). As we shall see, an astonishing variety of doctrines fall under the Duhem-Quine umbrella. The aim of the readings and commentary in this chapter is to identify these doctrines, to understand how they are related, and to provide a framework for assessing their credibility.

The chapter begins, appropriately enough, with readings from Pierre Duhem (1861–1916) and W. V. Quine (1908–2000). The selection by Duhem comes from his book *The Aim and Structure of Physical Theory*, first published (in French) in 1906 (although the papers on which it was based were written in the 1890s). The piece from Quine, "Two Dogmas of Empiricism" (1951), is a modern classic that has had a wide-ranging influence on many areas of philosophy, especially epistemology and the philosophy of science. As Donald Gillies points out in his article, "The Duhem Thesis and the Quine Thesis," the differences in the scope and focus of the arguments of Duhem and Quine are so great that it makes little sense, historically, to link the names of the two men together as espousing a single, common thesis. Nonetheless, Gillies thinks that, historical considerations aside, we can assemble from the different but related views of Duhem and Quine a plausible version of holism that can aptly be named the Duhem-Quine thesis.

In "Demystifying Underdetermination" Larry Laudan launches a spirited attack on all versions of the underdetermination thesis that have been espoused by Quine and his followers. Laudan is especially critical of those who—like Kuhn, Hesse, and Bloor—have used Duhem-Quine–style arguments to bolster their view that science is governed to a large degree by sociological forces (not logic and scientific method) and can be understood only by taking these historical and social factors into account. Laudan argues that once one distinguishes different versions of the underdetermination thesis, underdetermination shows itself to be either true but innocuous or dramatic and false. All too often, he warns, philosophers take for granted a radical version of the underdetermination thesis without giving anything like a plausible argument to support it.

It is one thing to argue that the currently available arguments for the underdetermination thesis fail; it is quite another to explain how and why it can be rational to retain a core theory in the face of anomalous evidence. This is what Colin Howson and Peter Urbach undertake to do in "The Duhem Problem," by defending a Bayesian solution to the Duhem problem first offered by Jon Dorling.

PIERRE DUHEM

Physical Theory
and Experiment

1 | **The Experimental Testing of a Theory Does Not
Have the Same Logical Simplicity in Physics as in
Physiology**

The sole purpose of physical theory is to provide a representation and classification of experimental laws; the only test permitting us to judge a physical
theory and pronounce it good or bad is the comparison between the consequences of this theory and the experimental laws it has to represent and classify. Now that we have minutely analyzed the characteristics of a physical
experiment and of a physical law, we can establish the principles that should
govern the comparison between experiment and theory; we can tell how we
shall recognize whether a theory is confirmed or weakened by facts.

When many philosophers talk about experimental sciences, they think
only of sciences still close to their origins, e.g., physiology or certain branches
of chemistry where the experimenter reasons directly on the facts by a
method which is only common sense brought to greater attentiveness but
where mathematical theory has not yet introduced its symbolic representations. In such sciences the comparison between the deductions of a theory
and the facts of experiment is subject to very simple rules. These rules were
formulated in a particularly forceful manner by Claude Bernard, who would
condense them into a single principle, as follows:

"The experimenter should suspect and stay away from fixed ideas, and
always preserve his freedom of mind.

"The first condition that has to be fulfilled by a scientist who is devoted
to the investigation of natural phenomena is to preserve a complete freedom of mind based on philosophical doubt."[1]

FROM Pierre Duhem, *The Aim and Structure of Physical Theory*, trans. Philip P.
Wiener (Princeton, N.J.: Princeton University Press, 1954), 180–95, 208–18.

If a theory suggests experiments to be done, so much the better: ". . . we can follow our judgment and our thought, give free rein to our imagination provided that all our ideas are only pretexts for instituting new experiments that may furnish us probative facts or unexpected and fruitful ones."[2] Once the experiment is done and the results clearly established, if a theory takes them over in order to generalize them, coordinate them, and draw from them new subjects for experiment, still so much the better: ". . . if one is imbued with the principles of experimental method, there is nothing to fear; for so long as the idea is a right one, it will go on being developed; when it is an erroneous idea, experiment is there to correct it."[3] But so long as the experiment lasts, the theory should remain waiting, under strict orders to stay outside the door of the laboratory; it should keep silent and leave the scientist without disturbing him while he faces the facts directly; the facts must be observed without a preconceived idea and gathered with the same scrupulous impartiality, whether they confirm or contradict the predictions of the theory. The report that the observer will give us of his experiment should be a faithful and scrupulously exact reproduction of the phenomena, and should not let us even guess what system the scientist places his confidence in or distrusts.

"Men who have an excessive faith in their theories or in their ideas are not only poorly disposed to make discoveries but they also make very poor observations. They necessarily observe with a preconceived idea and, when they have begun an experiment, they want to see in its results only a confirmation of their theory. Thus they distort observation and often neglect very important facts because they go counter to their goal. That is what made us say elsewhere that we must never do experiments in order to confirm our ideas but merely to check them. . . . But it quite naturally happens that those who believe too much in their own theories do not sufficiently believe in the theories of others. Then the dominant idea of these condemners of others is to find fault with the theories of the latter and to seek to contradict them. The setback for science remains the same. They are doing experiments only in order to destroy a theory instead of doing them in order to look for the truth. They also make poor observations because they take into the results of their experiments only what fits their purpose, by neglecting what is unrelated to it, and by very carefully avoiding whatever might go in the direction of the idea they wish to combat. Thus one is led by two parallel paths to the same result, that is to say, to falsifying science and the facts.

"The conclusion of all this is that it is necessary to obliterate one's opinion as well as that of others when faced with the decisions of the experiment; . . . we must accept the results of experiment just as they present themselves with all that is unforeseen and accidental in them."[4]

Here, for example, is a physiologist who admits that the anterior roots of the spinal nerve contain the motor nerve-fibers and the posterior roots the sensory fibers. The theory he accepts leads him to imagine an experiment: if he cuts a certain anterior root, he ought to be suppressing the mobility of a certain part of the body without destroying its sensibility; after making the section of this root, when he observes the consequences of his operation and when he makes a report of it, he must put aside all his ideas concerning the physiology of the spinal nerve; his report must be a raw description of the facts; he is not permitted to overlook or fail to mention any movement or quiver contrary to his predictions or to attribute it to some secondary cause unless some special experiment has given evidence of this cause; he must, if he does not wish to be accused of scientific bad faith, establish an absolute separation or watertight compartment between the consequences of his theoretical deductions and the establishing of the facts shown by his experiments.

Such a rule is not by any means easily followed; it requires of the scientist an absolute detachment from his own thought and a complete absence of animosity when confronted with the opinion of another person; neither vanity nor envy ought to be countenanced by him. As Bacon put it, he should never show eyes lustrous with human passions. Freedom of mind, which constitutes the sole principle of experimental method, according to Claude Bernard, does not depend merely on intellectual conditions, but also on moral conditions, making its practice rarer and more meritorious.

But if experimental method as just described is difficult to practice, the logical analysis of it is very simple. This is no longer the case when the theory to be subjected to test by the facts is not a theory of physiology but a theory of physics. In the latter case, in fact, it is impossible to leave outside the laboratory door the theory that we wish to test, for without theory it is impossible to regulate a single instrument or to interpret a single reading. We have seen that in the mind of the physicist there are constantly present two sorts of apparatus. one is the concrete apparatus in glass and metal, manipulated by him, the other is the schematic and abstract apparatus which theory substitutes for the concrete apparatus and on which the physicist does his reasoning. For these two ideas are indissolubly connected in his intelligence, and each necessarily calls on the other; the physicist can no sooner conceive the concrete apparatus without associating with it the idea of the schematic apparatus than a Frenchman can conceive an idea without associating it with the French word expressing it. This radical impossibility, preventing one from dissociating physical theories from the experimental procedures appropriate for testing these theories, complicates this test in a singular way, and obliges us to examine the logical meaning of it carefully.

Of course, the physicist is not the only one who appeals to theories at the very time he is experimenting or reporting the results of his experiments. The chemist and the physiologist when they make use of physical instruments, e.g., the thermometer, the manometer, the calorimeter, the galvanometer, and the saccharimeter, implicitly admit the accuracy of the theories justifying the use of these pieces of apparatus as well as of the theories giving meaning to the abstract ideas of temperature, pressure, quantity of heat, intensity of current, and polarized light, by means of which the concrete indications of these instruments are translated. But the theories used, as well as the instruments employed, belong to the domain of physics; by accepting with these instruments the theories without which their readings would be devoid of meaning, the chemist and the physiologist show their confidence in the physicist, whom they suppose to be infallible. The physicist, on the other hand, is obliged to trust his own theoretical ideas or those of his fellow-physicists. From the standpoint of logic, the difference is of little importance; for the physiologist and chemist as well as for the physicist, the statement of the result of an experiment implies, in general, an act of faith in a whole group of theories.

2 | An Experiment in Physics Can Never Condemn an Isolated Hypothesis but Only a Whole Theoretical Group

The physicist who carries out an experiment, or gives a report of one, implicitly recognizes the accuracy of a whole group of theories. Let us accept this principle and see what consequences we may deduce from it when we seek to estimate the role and logical import of a physical experiment.

In order to avoid any confusion we shall distinguish two sorts of experiments: experiments of *application*, which we shall first just mention, and experiments of *testing*, which will be our chief concern.

You are confronted with a problem in physics to be solved practically; in order to produce a certain effect you wish to make use of knowledge acquired by physicists; you wish to light an incandescent bulb; accepted theories indicate to you the means for solving the problem; but to make use of these means you have to secure certain information; you ought, I suppose, to determine the electromotive force of the battery of generators at your disposal; you measure this electromotive force: that is what I call an experiment of application. This experiment does not aim at discovering whether accepted theories are accurate or not; it merely intends to draw on these theories. In order to carry it out, you make use of instruments that these same theories legitimize; there is nothing to shock logic in this procedure.

But experiments of application are not the only ones the physicist has to perform; only with their aid can science aid practice, but it is not through them that science creates and develops itself; besides experiments of application, we have experiments of testing.

A physicist disputes a certain law; he calls into doubt a certain theoretical point. How will he justify these doubts? How will he demonstrate the inaccuracy of the law? From the proposition under indictment he will derive the prediction of an experimental fact; he will bring into existence the conditions under which this fact should be produced; if the predicted fact is not produced, the proposition which served as the basis of the prediction will be irremediably condemned.

F. E. Neumann assumed that in a ray of polarized light the vibration is parallel to the plane of polarization, and many physicists have doubted this proposition. How did O. Wiener undertake to transform this doubt into a certainty in order to condemn Neumann's proposition? He deduced from this proposition the following consequence: If we cause a light beam reflected at 45° from a plate of glass to interfere with the incident beam polarized perpendicularly to the plane of incidence, there ought to appear alternately dark and light interference bands parallel to the reflecting surface; he brought about the conditions under which these bands should have been produced and showed that the predicted phenomenon did not appear, from which he concluded that Neumann's proposition is false, viz., that in a polarized ray of light the vibration is not parallel to the plane of polarization.

Such a mode of demonstration seems as convincing and as irrefutable as the proof by reduction to absurdity customary among mathematicians; moreover, this demonstration is copied from the reduction to absurdity, experimental contradiction playing the same role in one as logical contradiction plays in the other.

Indeed, the demonstrative value of experimental method is far from being so rigorous or absolute: the conditions under which it functions are much more complicated than is supposed in what we have just said; the evaluation of results is much more delicate and subject to caution.

A physicist decides to demonstrate the inaccuracy of a proposition; in order to deduce from this proposition the prediction of a phenomenon and institute the experiment which is to show whether this phenomenon is or is not produced, in order to interpret the results of this experiment and establish that the predicted phenomenon is not produced, he does not confine himself to making use of the proposition in question; he makes use also of a whole group of theories accepted by him as beyond dispute. The prediction of the phenomenon, whose nonproduction is to cut off debate, does not derive from the proposition challenged if taken by itself, but from the proposition at issue joined to that whole group of theories; if the predicted phenomenon is not produced, not only is the proposition questioned at fault, but so is the whole theoretical scaffolding used by the physicist. The only thing the experiment teaches us is that among the propositions used to predict the phenomenon and to establish whether it would be produced, there is at least one error; but where this error lies is just what it does not tell us. The physicist may declare that this error is contained in exactly the proposition he wishes to refute, but is he sure it is not in another proposition? If he is, he accepts implicitly the accuracy of all the other propositions he has used, and the validity of his conclusion is as great as the validity of his confidence.

Let us take as an example the experiment imagined by Zenker and carried out by O. Wiener. In order to predict the formation of bands in certain circumstances and to show that these did not appear, Wiener did not make use merely of the famous proposition of F. E. Neumann, the proposition which he wished to refute; he did not merely admit that in a polarized ray vibrations are parallel to the plane of polarization; but he used, besides this, propositions, laws, and hypotheses constituting the optics commonly accepted: he admitted that light consists in simple periodic vibrations, that these vibrations are normal to the light ray, that at each point the mean kinetic energy of the vibratory motion is a measure of the intensity of light, that the more or less complete attack of the gelatine coating on a photographic plate indicates the various degrees of this intensity. By joining these propositions, and many others that would take too long to enumerate, to Neumann's proposition, Wiener was able to formulate a forecast and establish that the experiment belied it. If he attributed this solely to Neumann's proposition, if it alone bears the responsibility for the error this negative result has put in evidence, then Wiener was taking all the other propositions he invoked as beyond doubt. But this assurance is not imposed as a matter of logical necessity; nothing stops us from taking Neumann's proposition as accurate and shifting the weight of the experimental contradiction to some other proposition of the commonly accepted optics; as H. Poincaré has shown, we can very easily rescue Neumann's hypothesis from the grip of Wiener's experiment on the condition that we abandon in exchange the hypothesis which takes the mean kinetic energy as the measure of the light intensity; we may, without being contradicted by the experiment, let the vibration be parallel to the plane of polarization, provided that we measure the light intensity by the mean potential energy of the medium deforming the vibratory motion.

These principles are so important that it will be useful to apply them to another example; again we choose an experiment regarded as one of the most decisive ones in optics.

We know that Newton conceived the emission theory for optical phenomena. The emission theory supposes light to be formed of extremely thin projectiles, thrown out with very great speed by the sun and other sources of light; these projectiles penetrate all transparent bodies; on account of the various parts of the media through which they move, they undergo attractions and repulsions; when the distance separating the acting particles is very small these actions are very powerful, and they vanish when the masses between which they act are appreciably far from each other. These essential hypotheses joined to several others, which we pass over without mention, lead to the formulation of a complete theory of reflection and refraction of light; in particular, they imply the following proposition: The index of refraction of light passing from one medium into another is equal to the velocity of the light projectile within the medium it penetrates, divided by the velocity of the same projectile in the medium it leaves behind.

This is the proposition that Arago chose in order to show that the theory of emission is in contradiction with the facts. From this proposition a second follows: Light travels faster in water than in air. Now Arago had indicated an appropriate procedure for comparing the velocity of light in air with the velocity of light in water; the procedure, it is true, was inapplicable, but Foucault modified the experiment in such a way that it could be carried out; he found that the light was propagated less rapidly in water than in air. We may conclude from this, with Foucault, that the system of emission is incompatible with the facts.

I say the *system* of emission and not the *hypothesis* of emission; in fact, what the experiment declares stained with error is the whole group of propositions accepted by Newton, and after him by Laplace and Biot, that is, the whole theory from which we deduce the relation between the index of refraction and the velocity of light in various media. But in condemning this system as a whole by declaring it stained with error, the experiment does not tell us where the error lies. Is it in the fundamental hypothesis that light consists in projectiles thrown out with great speed by luminous bodies? Is it in some other assumption concerning the actions experienced by light corpuscles due to the media through which they move? We know nothing about that. It would be rash to believe, as Arago seems to have thought, that Foucault's experiment condemns once and for all the very hypothesis of emission, i.e., the assimilation of a ray of light to a swarm of projectiles. If physicists had attached some value to this task, they would undoubtedly have succeeded in founding on this assumption a system of optics that would agree with Foucault's experiment.

In sum, the physicist can never subject an isolated hypothesis to experimental test, but only a whole group of hypotheses; when the experiment is in disagreement with his predictions, what he learns is that at least one of the hypotheses constituting this group is unacceptable and ought to be modified; but the experiment does not designate which one should be changed.

We have gone a long way from the conception of the experimental method arbitrarily held by persons unfamiliar with its actual functioning. People generally think that each one of the hypotheses employed in physics can be taken in isolation, checked by experiment, and then, when many varied tests have established its validity, given a definitive place in the system of physics. In reality, this is not the case. Physics is not a machine which lets itself be taken apart; we cannot try each piece in isolation and, in order to adjust it, wait until its solidity has been carefully checked. Physical science is a system that must be taken as a whole; it is an organism in which one part cannot be made to function except when the parts that are most remote from it are called into play, some more so than others, but all to some degree. If something goes wrong, if some discomfort is felt in the functioning of the organism, the physicist will have to ferret out through its effect on the entire system which organ needs to be remedied or modified without the possibility of isolating this organ and examining it apart. The watchmaker to whom you give a watch that has stopped separates all the wheelworks and examines them one by one until he finds the part that is defective or broken. The doctor to whom a patient appears cannot dissect him in order to establish his diagnosis; he has to guess the seat and cause of the ailment solely by inspecting disorders affecting the whole body. Now, the physicist concerned with remedying a limping theory resembles the doctor and not the watchmaker.

3 | A "Crucial Experiment" Is Impossible in Physics

Let us press this point further, for we are touching on one of the essential features of experimental method, as it is employed in physics.

Reduction to absurdity seems to be merely a means of refutation, but it may become a method of demonstration: in order to demonstrate the truth of a proposition it suffices to corner anyone who would admit the contradictory of the given proposition into admitting an absurd consequence. We know to what extent the Greek geometers drew heavily on this mode of demonstration.

Those who assimilate experimental contradiction to reduction to absurdity imagine that in physics we may use a line of argument similar to the one Euclid employed so frequently in geometry. Do you wish to obtain from a group of phenomena a theoretically certain and indisputable explanation? Enumerate all the hypotheses that can be made to account for this group of phenomena; then, by experimental contradiction eliminate all except one; the latter will no longer be a hypothesis, but will become a certainty.

Suppose, for instance, we are confronted with only two hypotheses. Seek experimental conditions such that one of the hypotheses forecasts the production of one phenomenon and the other the production of quite a different effect; bring these conditions into existence and observe what happens; depending on whether you observe the first or the second of the predicted phenomena, you will condemn the second or the first hypothesis; the hypothesis not condemned will be henceforth indisputable; debate will be cut off, and a new truth will be acquired by science. Such is the experimental test that the author of the *Novum Organum* [Francis Bacon] called the *"fact of the cross,"* borrowing this expression from the crosses which at an intersection indicate the various roads.

We are confronted with two hypotheses concerning the nature of light; for Newton, Laplace, or Biot light consisted of projectiles hurled with extreme speed, but for Huygens, Young, or Fresnel light consisted of vibrations whose waves are propagated within an ether. These are the only two possible hypotheses as far as one can see: either the motion is carried away by the body it excites and remains attached to it, or else it passes from one body to another. Let us pursue the first hypothesis; it declares that light travels more quickly in water than in air; but if we follow the second, it declares that light travels more quickly in air than in water. Let us set up Foucault's apparatus; we set into motion the turning mirror; we see two luminous spots formed before us, one colorless, the other greenish. If the greenish band is to the left of the colorless one, it means that light travels faster in water than in air, and that the hypothesis of vibrating waves is false. If, on the contrary, the greenish band is to the right of the colorless one, that means that light travels faster in air than in water, and that the hypothesis of emissions is condemned. We look through the magnifying glass used to examine the two luminous spots, and we notice that the greenish spot is to the right of the colorless one; the debate is over; light is not a body, but a vibratory wave motion propagated by the ether; the emission hypothesis has had its day; the wave hypothesis has been put beyond doubt, and the crucial experiment has made it a new article of the scientific credo.

What we have said in the foregoing paragraph shows how mistaken we should be to attribute to Foucault's experiment so simple a meaning and so decisive an importance; for it is not between two hypotheses, the emission and wave hypotheses, that Foucault's experiment judges trenchantly; it decides rather between two sets of theories each of which has to be taken as a whole, i.e., between two entire systems, Newton's optics and Huygens' optics.

But let us admit for a moment that in each of these systems everything is compelled to be necessary by strict logic, except a single hypothesis; consequently, let us admit that the facts, in condemning one of the two systems, condemn once and for all the single doubtful assumption it contains. Does it follow that we can find in the "crucial experiment" an irrefutable procedure for transforming one of the two hypotheses before us into a demonstrated truth? Between two contradictory theorems of geometry there is no room for a third judgment; if one is false, the other is necessarily true. Do two hypotheses in physics ever constitute such a strict dilemma? Shall we ever dare to assert that no other hypothesis is imaginable? Light may be a swarm of projectiles, or it may be a vibratory motion whose waves are propagated in a medium; is it forbidden to be anything else at all? Arago undoubtedly thought so when he formulated this incisive alternative: Does light move more quickly in water than in air? "Light is a body. If the contrary is the case, then light is a wave." But it would be difficult for us to take such a decisive stand; Maxwell, in fact, showed that we might just as well attribute light to a periodical electrical disturbance that is propagated within a dielectric medium.

Unlike the reduction to absurdity employed by geometers, experimental contradiction does not have the power to transform a physical hypothesis into an indisputable truth; in order to confer this power on it, it would be necessary to enumerate completely the various hypotheses which may cover a determinate group of phenomena; but the physicist is never sure he has exhausted all the imaginable assumptions. The truth of a physical theory is not decided by heads or tails.

4 | Criticism of the Newtonian Method. First Example: Celestial Mechanics

It is illusory to seek to construct by means of experimental contradiction a line of argument in imitation of the reduction to absurdity; but the geometer is acquainted with other methods for attaining certainty than the method of reducing to an absurdity; the direct demonstration in which the truth of a proposition is established by itself and not by the refutation of the contradictory proposition seems to him the most perfect of arguments. Perhaps physical theory would be more fortunate in its attempts if it sought to imitate direct demonstration. The hypotheses from which it starts and develops its conclusions would then be tested one by one; none would have to be accepted until it presented all the certainty that experimental method can confer on an abstract and general proposition; that is to say, each would necessarily be either a law drawn from observation by the sole use of those two intellectual operations called induction and generalization, or else a corollary mathematically deduced from such laws. A theory based on such hypotheses would then not present anything arbitrary or doubtful; it would deserve all the confidence merited by the faculties which serve us in formulating natural laws.

It was this sort of physical theory that Newton had in mind when, in the "General Scholium" which crowns his *Principia*, he rejected so vigorously as outside of natural philosophy any hypothesis that induction did not extract from experiment; when he asserted that in a sound physics every proposition should be drawn from phenomena and generalized by induction.

The ideal method we have just described therefore deserves to be named the Newtonian method. Besides, did not Newton follow this method when he established the system of universal attraction, thus adding to his precepts the most magnificent of examples? Is not his theory of gravitation derived entirely from the laws which were revealed to Kepler by observation, laws which problematic reasoning transforms and whose consequences induction generalizes?

This first law of Kepler's, "The radial vector from the sun to a planet sweeps out an area proportional to the time during which the planet's motion is observed," did, in fact, teach Newton that each planet is constantly subjected to a force directed toward the sun.

The second law of Kepler's, "The orbit of each planet is an ellipse having the sun at one focus," taught him that the force attracting a given planet varies with the distance of this planet from the sun, and that it is in an inverse ratio to the square of this distance.

The third law of Kepler's, "The squares of the periods of revolution of the various planets are proportional to the cubes of the major axes of their orbits," showed him that different planets would, if they were brought to the same distance from the sun, undergo in relation to it attractions proportional to their respective masses.

The experimental laws established by Kepler and transformed by geometric reasoning yield all the characteristics present in the action exerted by the sun on a planet; by induction Newton generalized the result obtained; he allowed this result to express the law according to which any portion of matter acts on any other portion whatsoever, and he formulated this great principle: "Any two bodies whatsoever attract each other with a force which is proportional to the product of their masses and in inverse ratio to the square of the distance between them." The principle of universal gravitation was found, and it was obtained, without any use having been made of any fictive hypothesis, by the inductive method the plan of which Newton outlined.

Let us again examine this application of the Newtonian method, this time more closely; let us see if a somewhat strict logical analysis will leave intact the appearance of rigor and simplicity that this very summary exposition attributes to it.

In order to assure this discussion of all the clarity it needs, let us begin by recalling the following principle, familiar to all those who deal with mechanics: We cannot speak of the force which attracts a body in given circumstances before we have designated the supposedly fixed term of reference to which we relate the motion of all bodies; when we change this point of reference or term of comparison, the force representing the effect produced on the observed body by the other bodies surrounding it changes in direction and magnitude according to the rules stated by mechanics with precision.

That posited, let us follow Newton's reasoning.

Newton first took the sun as the fixed point of reference, he considered the motions affecting the different planets by reference to the sun; he admitted Kepler's laws as governing these motions, and derived the following proposition: If the sun is the point of reference in relation to which all forces are compared, each planet is subjected to a force directed toward the sun, a force proportional to the mass of the planet and to the inverse square of its distance from the sun. Since the latter is taken as the reference point, it is not subject to any force.

In an analogous manner Newton studied the motion of the satellites and for each of these he chose as a fixed reference point the planet which the satellite accompanies, the earth in the case of the moon, Jupiter in the case of the masses moving around Jupiter. Laws just like Kepler's were taken as governing these motions, from which it follows that we can formulate the following proposition: If we take as a fixed reference point the planet accompanied by a satellite, this satellite is subject to a force directed toward the planet varying inversely with the square of the distance. If, as happens with Jupiter, the same planet possesses several satellites, these satellites, were they at the same distance from the planet, would be acted on by the latter with forces proportional to their respective masses. The planet is itself not acted on by the satellite.

Such, in very precise form, are the propositions which Kepler's laws of planetary motion and the extension of these laws to the motions of satellites authorize us to formulate. For these propositions Newton substituted another which may be stated as follows: Any two celestial bodies whatsoever exert on each other a force of attraction in the direction of the straight line joining them, a force proportional to the product of their masses and to the inverse square of the distance between them. This statement presupposes all motions and forces to be related to the same reference point; the latter is an ideal standard of reference which may well be conceived by the geometer but which does not characterize in an exact and concrete manner the position in the sky of any body.

Is this principle of universal gravitation merely a generalization of the two statements provided by Kepler's laws and their extension to the motion of satellites? Can induction derive it from these two statements? Not at all. In fact, not only is it more general than these two statements and unlike them, but it contradicts them. The student of mechanics who accepts the principle of universal attraction can calculate the magnitude and direction of the forces between the various planets and the sun when the latter is taken as the reference point, and if he does he finds that these forces are not what our first statement would require. He can determine the magnitude and direction of each of the forces between Jupiter and its satellites when we refer all the motions to the planet, assumed to be fixed, and if he does he notices that these forces are not what our second statement would require.

The principle of universal gravity, very far from being derivable by generalization and induction from the observational laws of Kepler, formally contradicts these laws. If Newton's theory is correct, Kepler's laws are necessarily false.

Kepler's laws based on the observation of celestial motions do not transfer their immediate experimental certainty to the principle of universal weight, since if, on the contrary, we admit the absolute exactness of Kepler's laws, we are compelled to reject the proposition on which Newton based his celestial mechanics. Far from adhering to Kepler's laws, the physicist who claims to justify the theory of universal gravitation finds that he has, first of all, to resolve a difficulty in these laws: he has to prove that his theory, incompatible with the exactness of Kepler's laws, subjects the motions of the planets and satellites to other laws scarcely different enough from the first laws for Tycho Brahé, Kepler, and their contemporaries to have been able to discern the deviations between the Keplerian and Newtonian orbits. This proof derives from the circumstances that the sun's mass is very large in relation to the masses of the various planets and the mass of a planet is very large in relation to the masses of its satellites.

Therefore, if the certainty of Newton's theory does not emanate from the certainty of Kepler's laws, how will this theory prove its validity? It will calculate, with all the high degree of approximation that the constantly perfected methods of algebra involve, the perturbations which at each instant remove every heavenly body from the orbit assigned to it by Kepler's laws; then it will compare the calculated perturbations with the perturbations observed by means of the most precise instruments and the most scrupulous methods. Such a comparison will not only bear on this or that part of the Newtonian principle, but will involve all its parts at the same time; with those it will also involve all the principles of dynamics; besides, it will call in the aid of all the propositions of optics, the statics of gases, and the theory of heat, which are necessary to justify the properties of telescopes in their construction, regulation, and correction, and in the elimination of the errors caused by diurnal or annual aberration and by atmospheric refraction. It is no longer a matter of taking, one by one, laws justified by observation, and raising each of them by induction and generalization to the rank of a principle; it is a matter of comparing the corollaries of a whole group of hypotheses to a whole group of facts.

Now, if we seek out the causes which have made the Newtonian method fail in this case for which it was imagined and which seemed to be the most perfect application for it, we shall find them in that double character of any law made use of by theoretical physics: This law is symbolic and approximate.

Undoubtedly, Kepler's laws bear quite directly on the very objects of astronomical observation; they are as little symbolic as possible. But in this purely experimental form they remain inappropriate for suggesting the principle of universal gravitation; in order to acquire this fecundity they must be transformed and must yield the characters of the forces by which the sun attracts the various planets.

Now this new form of Kepler's laws is a symbolic form; only dynamics gives meanings to the words "force" and "mass," which serve to state it, and only dynamics permits us to substitute the new symbolic formulas for the old realistic formulas, to substitute statements relative to "forces" and "masses" for laws relative to orbits. The legitimacy of such a substitution implies full confidence in the laws of dynamics.

And in order to justify this confidence let us not proceed to claim that the laws of dynamics were beyond doubt at the time Newton made use of them in symbolically translating Kepler's laws; that they had received enough empirical confirmation to warrant the support of reason. In fact, the laws of dynamics had been subjected up to that time to only very limited and very crude tests. Even their enunciations had remained very vague and involved; only in Newton's *Principia* had they been for the first time formulated in a precise manner. It was in the agreement of the facts with the celestial mechanics which Newton's labors gave birth to that they received their first convincing verification.

Thus the translation of Kepler's laws into symbolic laws, the only kind useful for a theory, presupposed the prior adherence of the physicist to a whole group of hypotheses. But, in addition, Kepler's laws being only approximate laws, dynamics permitted giving them an infinity of different symbolic translations. Among these various forms, infinite in number, there is one and only one which agrees with Newton's principle. The observations of Tycho Brahé, so felicitously reduced to laws by Kepler, permit the theorist to choose this form, but they do not constrain him to do so, for there is an infinity of others they permit him to choose.

The theorist cannot, therefore, be content to invoke Kepler's laws in order to justify his choice. If he wishes to prove that the principle he has adopted is truly a principle of natural classification for celestial motions, he must show that the observed perturbations are in agreement with those which had been calculated in advance; he has to show how from the course of Uranus he can deduce the existence and position of a new planet, and find Neptune in an assigned direction at the end of his telescope. . . .

8 | Are Certain Postulates of Physical Theory Incapable of Being Refuted by Experiment?

We recognize a correct principle by the facility with which it straightens out the complicated difficulties into which the use of erroneous principles brought us.

If, therefore, the idea we have put forth is correct, namely, that comparison is established necessarily between the *whole* of theory and the *whole* of experimental facts, we ought in the light of this principle to see the disappearance of the obscurities in which we should be lost by thinking that we are subjecting each isolated theoretical hypothesis to the test of facts.

Foremost among the assertions in which we shall aim at eliminating the appearance of paradox, we shall place one that has recently been often formulated and discussed. Stated first by G. Milhaud in connection with the *"pure bodies"* of chemistry,[5] it has been developed at length and forcefully by H. Poincaré with regard to principles of mechanics;[6] Edouard Le Roy has also formulated it with great clarity.[7]

That assertion is as follows: Certain fundamental hypotheses of physical theory cannot be contradicted by any experiment, because they constitute in reality *definitions*, and because certain expressions in the physicist's usage take their meaning only through them.

Let us take one of the examples cited by Le Roy:

When a heavy body falls freely, the acceleration of its fall is constant. Can such a law be contradicted by experiment? No, for it constitutes the very definition of what is meant by "falling freely." If while studying the fall of a heavy body we found that this body does not fall with uniform acceleration, we should conclude not that the stated law is false, but that the body does not fall freely, that some cause obstructs its motion, and that the deviations of the observed facts from the law as stated would serve to discover this cause and to analyze its effects.

Thus, M. Le Roy concludes, "laws are verifiable, taking things strictly . . . , because they constitute the very criterion by which we judge appearances as well as the methods that it would be necessary to utilize in order to submit them to an inquiry whose precision is capable of exceeding any assignable limit."

Let us study again in greater detail, in the light of the principles previously set down, what this comparison is between the law of falling bodies and experiment.

Our daily observations have made us acquainted with a whole category of motions which we have brought together under the name of motions of heavy bodies; among these motions is the falling of a heavy body when it is not hindered by any obstacle. The result of this is that the words "free fall of a heavy body" have a meaning for the man who appeals only to the knowledge of common sense and who has no notion of physical theories.

On the other hand, in order to classify the laws of motion in question the physicist has created a theory, the theory of weight, an important application of rational mechanics. In that theory, intended to furnish a symbolic representation of reality, there is also the question of "free fall of a heavy body," and as a consequence of the hypotheses supporting this whole scheme free fall must necessarily be a uniformly accelerated motion.

The words "free fall of a heavy body" now have two distinct meanings. For the man ignorant of physical theories, they have their *real* meaning, and they mean what common sense means in pronouncing them; for the physicist they have a *symbolic* meaning, and mean "uniformly accelerated motion." Theory would not have realized its aim if the second meaning were not the sign of the first, if a fall regarded as free by common sense were not also regarded as uniformly accelerated, or *nearly* uniformly accelerated, since common-sense observations are essentially devoid of precision, according to what we have already said.

This agreement, without which the theory would have been rejected without further examination, is finally arrived at: a fall declared by common sense to be nearly free is also a fall whose acceleration is nearly constant. But noticing this crudely approximate agreement does not satisfy us; we wish to push on and surpass the degree of precision which common sense can claim. With the aid of the theory that we have imagined, we put together apparatus enabling us to recognize with sensitive accuracy whether the fall of a body is or is not uniformly accelerated; this apparatus shows us that a certain fall regarded by common sense as a free fall has a slightly variable acceleration. The proposition which in our theory gives its symbolic meaning to the words "free fall" does not represent with sufficient accuracy the properties of the real and concrete fall that we have observed.

Two alternatives are then open to us.

In the first place, we can declare that we were right in regarding the fall studied as a free fall and in requiring that the theoretical definition of these words agree with our observations. In this case, since our theoretical definition does not satisfy this requirement, it must be rejected; we must construct another mechanics on new hypotheses, a mechanics in which the words "free fall" no longer signify "uniformly accelerated motion," but "fall whose acceleration varies according to a certain law."

In the second alternative, we may declare that we were wrong in establishing a connection between the concrete fall we have observed and the symbolic free fall defined by our theory, that the latter was too simplified a scheme of the former, that in order to represent suitably the fall as our experiments have reported it the theorist should give up imagining a weight falling freely and think in terms of a weight hindered by certain obstacles like the resistance of the air, that in picturing the action of these obstacles by means of appropriate hypotheses he will compose a more complicated scheme than a free weight but one more apt to reproduce the details of the experiment; in short, . . . we may seek to eliminate by means of suitable "corrections" the "causes of error," such as air resistance, which influenced our experiment.

M. Le Roy asserts that we shall prefer the second to the first alternative, and he is surely right in this. The reasons dictating this choice are easy to perceive. By taking the first alternative we should be obliged to destroy from top to bottom a very vast theoretical system which represents in a most satisfactory manner a very extensive and complex set of experimental laws. The second alternative, on the other hand, does not make us lose anything of the terrain already conquered by physical theory; in addition, it has succeeded in so large a number of cases that we can bank with interest on a new success. But in this confidence accorded the law of fall of weights, we see nothing analogous to the certainty that a mathematical definition draws from its very essence, that is, to the kind of certainty we have when it would be foolish to doubt that the various points on a circumference are all equidistant from the center.

We have here nothing more than a particular application of the principle set down in Section 2 of this chapter. A disagreement between the concrete facts constituting an experiment and the symbolic representation which theory substitutes for this experiment proves that some part of this symbol is to be rejected. But which part? This the experiment does not tell us; it leaves to our sagacity the burden of guessing. Now among the theoretical elements entering into the composition of this symbol there is always a certain number which the physicists of a certain epoch agree in accepting without test and which they regard as beyond dispute. Hence, the physicist who wishes to modify this symbol will surely bring his modification to bear on elements other than those just mentioned.

But what impels the physicist to act thus is *not* logical necessity. It would be awkward and ill inspired for him to do otherwise, but it would not be doing something logically absurd; he would not for all that be walking in the footsteps of the mathematician mad enough to contradict his own definitions. More than this, perhaps some day by acting differently, by refusing to invoke causes of error and take recourse to corrections in order to reestablish agreement between the theoretical scheme and the fact, and by resolutely carrying out a reform among the propositions declared untouchable by common consent, he will accomplish the work of a genius who opens a new career for a theory.

Indeed, we must really guard ourselves against believing forever warranted those hypotheses which have become universally adopted conventions, and whose certainty seems to break through experimental contradiction by throwing the latter back on more doubtful assumptions. The history of physics shows us that very often the human mind has been led to overthrow such principles completely, though they have been regarded by common consent for centuries as inviolable axioms, and to rebuild its physical theories on new hypotheses.

Was there, for instance, a clearer or more certain principle for thousands of years than this one: In a homogeneous medium, light is propagated in a straight line? Not only did this hypothesis carry all former optics, catoptrics, and dioptrics, whose elegant geometric deductions represented at will an enormous number of facts, but it had become, so to speak, the physical definition of a straight line. It is to this hypothesis that any man wishing to make a straight line appeals, the carpenter who verifies the straightness of a piece of wood, the surveyor who lines up his sights, the geodetic surveyor who obtains a direction with the help of the pinholes of his alidade, the astronomer who defines the position of stars by the optical axis of his telescope. However, the day came when physicists tired of attributing to some cause of error the diffraction effects observed by Grimaldi, when they resolved to reject the law of the rectilinear propagation of light and to give optics entirely new foundations; and this bold resolution was the signal of remarkable progress for physical theory.

9 | On Hypotheses Whose Statement Has No Experimental Meaning

This example, as well as others we could add from the history of science, should show that it would be very imprudent for us to say concerning a hypothesis commonly accepted today: "We are certain that we shall never be led to abandon it because of a new experiment, no matter how precise it is." Yet M. Poincaré does not hesitate to enunciate it concerning the principles of mechanics.[8]

To the reasons already given to prove that these principles cannot be reached by experimental refutation, M. Poincaré adds one which seems even more convincing: Not only can these principles not be refuted by experiment because they are the universally accepted rules serving to discover in our theories the weak spots indicated by these refutations, but also, they cannot be refuted by experiment because *the operation which would claim to compare them with the facts would have no meaning.*

Let us explain that by an illustration.

The principle of inertia teaches us that a material point removed from the action of any other body moves in a straight line with uniform motion. Now, we can observe only relative motions; we cannot, therefore, give an experimental meaning to this principle unless we assume a certain point chosen or a certain geometric solid taken as a fixed reference point to which the motion of the material point is related. The fixation of this reference frame constitutes an integral part of the statement of the law, for if we omitted it, this statement would be devoid of meaning. There are as many different laws as there are distinct frames of reference. We shall be stating one law of inertia when we say that the motion of an isolated point assumed to be seen from the earth is rectilinear and uniform, and another when we repeat the same sentence in referring the motion to the sun, and still another if the frame of reference chosen is the totality of fixed stars. But then, one thing is indeed certain, namely, that whatever the motion of a material point is, when seen from a first frame of reference, we can always and in infinite ways choose a second frame of reference such that seen from the latter our material point appears to move in a straight line with uniform motion. We cannot, therefore, attempt an experimental verification of the principle of inertia; false when we refer the motions to one frame of reference, it will become true when selection is made of another term of comparison, and we shall always be free to choose the latter. If the law of inertia stated by taking the earth as a frame of reference is contradicted by an observation, we shall substitute for it the law of inertia whose statement refers the motion to the sun; if the latter in its turn is contraverted, we shall replace the sun in the statement of the law by the system of fixed stars, and so forth. It is impossible to stop this loophole.

The principle of the equality of action and reaction, analyzed at length by M. Poincaré,[9] provides room for analogous remarks. This principle may be stated thus: "The center of gravity of an isolated system can have only a uniform rectilinear motion."

This is the principle that we propose to verify by experiment. "Can we make this verification? For that it would be necessary for isolated systems to exist. Now, these systems do not exist; the only isolated system is the whole universe.

"But we can observe only relative motions; the absolute motion of the center of the universe will therefore be forever unknown. We shall never be able to know if it is rectilinear and uniform or, better still, the question has no meaning. Whatever facts we may observe, we shall hence always be free to assume our principle is true."

Thus many a principle of mechanics has a form such that it is absurd to ask one's self: "Is this principle in agreement with experiment or not?" This strange character is not peculiar to the principles of mechanics; it also marks certain fundamental hypotheses of our physical or chemical theories.[10]

For example, chemical theory rests entirely on the "law of multiple proportions"; here is the exact statement of this law:

Simple bodies A, B, and C may by uniting in various proportions form various compounds M, M', . . . The masses of the bodies A, B, and C combining to form the compound M are to one another as the three numbers a, b, and c. Then the masses of the elements A, B, and C combining to form the compound M' will be to one another as the numbers xa, yb, and zc (x, y, and z being three whole numbers).

Is this law perhaps subject to experimental test? Chemical analysis will make us acquainted with the chemical composition of the body M' not exactly but with a certain approximation. The uncertainty of the results obtained can be extremely small; it will never be strictly zero. Now, in whatever relations the elements A, B, and C are combined within the compound M', we can always represent these relations, with as close an approximation as you please, by the mutual relations of three products xa, yb, and zc, where x, y, and z are whole numbers; in other words, whatever the results given by the chemical analysis of the compound M', we are always sure to find three integers x, y, and z thanks to which the law of multiple proportions will be verified with a precision greater than that of the experiment. Therefore, no chemical analysis, no matter how refined, will ever be able to show the law of multiple proportions to be wrong.

In like manner, all crystallography rests entirely on the "law of rational indices" which is formulated in the following way:

A trihedral being formed by three faces of a crystal, a fourth face cuts the three edges of this trihedral at distances from the summit which are proportional to one another as three given numbers, the parameters of the crystal. Any other face whatsoever should cut these same edges at distances from the summit which are to one another as xa, yb, and zc, where x, y, and z are three integers, the indices of the new face of the crystal.

The most perfect protractor determines the direction of a crystal's face only with a certain degree of approximation; the relations among the three segments that such a face makes on the edges of the fundamental trihedral are always able to get by with a certain error; now, however small this error is, we can always choose three numbers x, y, and z such that the mutual relations of these segments are represented with the least amount of error by the mutual relations of the three numbers xa, yb, and zc; the crystallographer who would claim that the law of rational indices is made justifiable by his protractor would surely not have understood the very meaning of the words he is employing.

The law of multiple proportions and the law of rational indices are mathematical statements deprived of all physical meaning. A mathematical statement has physical meaning only if it retains a meaning when we introduce the word "nearly" or "approximately." This is not the case with the statements we have just alluded to. Their object really is to assert that certain relations are *commensurable* numbers. They would degenerate into mere truisms if they were made to declare that these relations are approximately commensurable, for any incommensurable relation whatever is always approximately commensurable; it is even as near as you please to being commensurable.

Therefore, it would be absurd to wish to subject certain principles of mechanics to *direct* experimental test; it would be absurd to subject the law of multiple proportions or the law of rational indices to this *direct* test.

Does it follow that these hypotheses placed beyond the reach of direct experimental refutation have nothing more to fear from experiment? That they are guaranteed to remain immutable no matter what discoveries observation has in store for us? To pretend so would be a serious error.

Taken in isolation these different hypotheses have no experimental meaning; there can be no question of either confirming or contradicting them by experiment. But these hypotheses enter as essential foundations into the construction of certain theories of rational mechanics, of chemical theory, of crystallography. The object of these theories is to represent experimental laws; they are schematisms intended essentially to be compared with facts.

Now this comparison might some day very well show us that one of our representations is ill adjusted to the realities it should picture, that the corrections which come and complicate our schematism do not produce sufficient concordance between this schematism and the facts, that the theory accepted for a long time without dispute should be rejected, and that an entirely different theory should be constructed on entirely different or new hypotheses. On that day some one of our hypotheses, which taken in isolation defied direct experimental refutation, will crumble with the system it supported under the weight of the contradictions inflicted by reality on the consequences of this system taken as a whole.[11]

In truth, hypotheses which by themselves have no physical meaning undergo experimental testing in exactly the same manner as other hypotheses. Whatever the nature of the hypothesis is, we have seen at the beginning of this chapter that it is never in isolation contradicted by experiment; experimental contradiction always bears as a whole on the entire group constituting a theory without any possibility of designating which proposition in this group should be rejected.

There thus disappears what might have seemed paradoxical in the following assertion: Certain physical theories rest on hypotheses which do not by themselves have any physical meaning.

10 | Good Sense Is the Judge of Hypotheses Which Ought to Be Abandoned

When certain consequences of a theory are struck by experimental contradiction, we learn that this theory should be modified but we are not told by the experiment what must be changed. It leaves to the physicist the task of finding out the weak spot that impairs the whole system. No absolute principle directs this inquiry, which different physicists may conduct in very different ways without having the right to accuse one another of illogicality. For instance, one may be obliged to safeguard certain fundamental hypotheses while he tries to reestablish harmony between the consequences of the theory and the facts by complicating the schematism in which these hypotheses are applied, by invoking various causes of error, and by multiplying corrections. The next physicist, disdainful of these complicated artificial procedures, may decide to change some one of the essential assumptions supporting the entire system. The first physicist does not have the right to condemn in advance the boldness of the second one, nor does the latter have the right to treat the timidity of the first physicist as absurd. The methods they follow are justifiable only by experiment, and if they both succeed in satisfying the requirements of experiment each is logically permitted to declare himself content with the work that he has accomplished.

That does not mean that we cannot very properly prefer the work of one of the two to that of the other. Pure logic is not the only rule for our judgments; certain opinions which do not fall under the hammer of the principle of contradiction are in any case perfectly unreasonable. These motives which do not proceed from logic and yet direct our choices, these "reasons which reason does not know" and which speak to the ample "mind of finesse" but not to the "geometric mind," constitute what is appropriately called good sense.

Now, it may be good sense that permits us to decide between two physicists. It may be that we do not approve of the haste with which the second one upsets the principles of a vast and harmoniously constructed theory whereas a modification of detail, a slight correction, would have sufficed to put these theories in accord with the facts. On the other hand, it may be that we may find it childish and unreasonable for the first physicist to maintain obstinately at any cost, at the price of continual repairs and many tangled-up stays, the worm-eaten columns of a building tottering in every part, when by razing these columns it would be possible to construct a simple, elegant, and solid system.

But these reasons of good sense do not impose themselves with the same implacable rigor that the prescriptions of logic do. There is something vague and uncertain about them; they do not reveal themselves at the same time with the same degree of clarity to all minds. Hence, the possibility of lengthy quarrels between the adherents of an old system and the partisans of a new doctrine, each camp claiming to have good sense on its side, each party finding the reasons of the adversary inadequate. The history of physics would furnish us with innumerable illustrations of these quarrels at all times and in all domains. Let us confine ourselves to the tenacity and ingenuity with which Biot by a continual bestowal of corrections and accessory hypotheses maintained the emissionist doctrine in optics, while Fresnel opposed this doctrine constantly with new experiments favoring the wave theory.

In any event this state of indecision does not last forever. The day arrives when good sense comes out so clearly in favor of one of the two sides that the other side gives up the struggle even though pure logic would not forbid its continuation. After Foucault's experiment had shown that light traveled faster in air than in water, Biot gave up supporting the emission hypothesis; strictly, pure logic would not have compelled him to give it up, for Foucault's experiment was *not* the crucial experiment that Arago thought he saw in it, but by resisting wave optics for a longer time Biot would have been lacking in good sense.

Since logic does not determine with strict precision the time when an inadequate hypothesis should give way to a more fruitful assumption, and since recognizing this moment belongs to good sense, physicists may hasten this judgment and increase the rapidity of scientific progress by trying consciously to make good sense within themselves more lucid and more vigilant. Now nothing contributes more to entangle good sense and to disturb its insight than passions and interests. Therefore, nothing will delay the decision which should determine a fortunate reform in a physical theory more than the vanity which makes a physicist too indulgent towards his own system and too severe towards the system of another. We are thus led to the conclusion so clearly expressed by Claude Bernard: The sound experimental criticism of a hypothesis is subordinated to certain moral conditions; in order to estimate correctly the agreement of a physical theory with the facts, it is not enough to be a good mathematician and skillful experimenter; one must also be an impartial and faithful judge.

■ | Notes

1. Claude Bernard, *Introduction à [l'étude de] la Médecine expérimentale* (Paris, 1865), p. 63. (Translator's note: Translated into English by H. C. Greene, *An Introduction to the Study of Experimental Medicine* [New York: Henry Schuman, 1949].)

2. Claude Bernard, *Introduction à [l'étude de] la Médecine expérimentale* (Paris, 1865), p. 64.

3. *ibid.*, p. 70.

4. *ibid.*, p. 67.

5. G. Milhaud, "La Science rationnelle," *Revue de Métaphysique et de Morale*, IV (1896), 280. Reprinted in *Le Rationnel* (Paris, 1898), p. 45.

6. H. Poincaré, "Sur les Principes de la Mécanique," *Bibliothèque du Congrès international de Philosophie*, III: *Logique et Histoire des Sciences* (Paris, 1901), p. 457; "Sur la valeur objective des théories physiques," *Revue de Métaphysique et de Morale*, X (1902), 263; *La Science et l'Hypothèse*, p. 110.

7. E. Le Roy, "Un positivisme nouveau," *Revue de Métaphysique et de Morale*, IX (1901), 143–144.

8. H. Poincaré, "Sur les Principes de la Mécanique," *Bibliothèque du Congrès international de Philosophie*, Sec. III: "Logique et Histoire des Sciences" (Paris, (1901), pp. 475, 491.

9. *ibid.*, pp. 472ff.

10. P. Duhem, *Le Mixte et la Combinaison chimique: Essai sur l'évolution d'une idée* (Paris, 1902), pp. 159–161.

11. At the International Congress of Philosophy held in Paris in 1900, M. Poincaré developed this conclusion: "Thus is explained how experiment may have been able to edify (or suggest) the principles of mechanics, but will never be able to overthrow them." Against this conclusion, M. Hadamard offered various remarks, among them the following: "Moreover, in conformity with a remark of M. Duhem, it is not *an* isolated hypothesis but the whole group of the hypotheses of mechanics that we can try to verify experimentally." *Revue de Métaphysique et de Morale*, VIII (1900), 559.

W. V. QUINE

Two Dogmas of Empiricism

Modern empiricism has been conditioned in large part by two dogmas. One is a belief in some fundamental cleavage between truths which are *analytic*, or grounded in meanings independently of matters of fact, and truths which are *synthetic*, or grounded in fact. The other dogma is *reductionism*: the belief that each meaningful statement is equivalent to some logical construct upon terms which refer to immediate experience. Both dogmas, I shall argue, are ill-founded. One effect of abandoning them is, as we shall see, a blurring of the supposed boundary between speculative metaphysics and natural science. Another effect is a shift toward pragmatism.

1 | Background for Analyticity

Kant's cleavage between analytic and synthetic truths was foreshadowed in Hume's distinction between relations of ideas and matters of fact, and in Leibniz's distinction between truths of reason and truths of fact. Leibniz spoke of the truths of reason as true in all possible worlds. Picturesqueness aside, this is to say that the truths of reason are those which could not possibly be false. In the same vein we hear analytic statements defined as statements whose denials are self-contradictory. But this definition has small explanatory value; for the notion of self-contradictoriness, in the quite broad sense needed for this definition of analyticity, stands in exactly the same need of clarification as does the notion of analyticity itself. The two notions are the two sides of a single dubious coin.

Kant conceived of an analytic statement as one that attributes to its subject no more than is already conceptually contained in the subject. This

From W. V. Quine, *From a Logical Point of View* (Cambridge, Mass.: Harvard University Press, 1953), 20–46. Originally published in *Philosophical Review* 60 (1951): 20–43.

formulation has two shortcomings: it limits itself to statements of subject-predicate form, and it appeals to a notion of containment which is left at a metaphorical level. But Kant's intent, evident more from the use he makes of the notion of analyticity than from his definition of it, can be restated thus: a statement is analytic when it is true by virtue of meanings and independently of fact. Pursuing this line, let us examine the concept of *meaning* which is presupposed.

Meaning, let us remember, is not to be identified with naming.[1] Frege's example of 'Evening Star' and 'Morning Star', and Russell's of 'Scott' and 'the author of *Waverly*', illustrate that terms can name the same thing but differ in meaning. The distinction between meaning and naming is no less important at the level of abstract terms. The terms '9' and 'the number of the planets' name one and the same abstract entity but presumably must be regarded as unlike in meaning; for astronomical observation was needed, and not mere reflection on meanings, to determine the sameness of the entity in question.

The above examples consist of singular terms, concrete and abstract. With general terms, or predicates, the situation is somewhat different but parallel. Whereas a singular term purports to name an entity, abstract or concrete, a general term does not; but a general term is *true* of an entity, or of each of many, or of none.[2] The *class* of all entities of which a general term is true is called the *extension* of the term. Now paralleling the contrast between the meaning of a singular term and the entity named, we must distinguish equally between the meaning of a general term and its extension. The general terms 'creature with a heart' and 'creature with kidneys', for example, are perhaps alike in extension but unlike in meaning.

Confusion of meaning with extension, in the case of general terms, is less common than confusion of meaning with naming in the case of singular terms. It is indeed a commonplace in philosophy to oppose intension (or meaning) to extension, or, in a variant vocabulary, connotation to denotation.

The Aristotelian notion of essence was the forerunner, no doubt, of the modern notion of intension or meaning. For Aristotle it was essential in men to be rational, accidental to be two-legged. But there is an important difference between this attitude and the doctrine of meaning. From the latter point of view it may indeed be conceded (if only for the sake of argument) that rationality is involved in the meaning of the word 'man' while two-leggedness is not; but two-leggedness may at the same time be viewed as involved in the meaning of 'biped' while rationality is not. Thus from the point of view of the doctrine of meaning it makes no sense to say of the actual individual, who is at once a man and a biped, that his rationality is essential and his two-leggedness accidental or vice versa. Things had essences, for Aristotle, but only linguistic forms have meanings. Meaning is what essence becomes when it is divorced from the object of reference and wedded to the word.

For the theory of meaning a conspicuous question is the nature of its objects: what sort of things are meanings? A felt need for meant entities may derive from an earlier failure to appreciate that meaning and reference are distinct. Once the theory of meaning is sharply separated from the theory of reference, it is a short step to recognizing as the primary business of the theory of meaning simply the synonymy of linguistic forms and the analyticity of statements; meanings themselves, as obscure intermediary entities, may well be abandoned.[3]

The problem of analyticity then confronts us anew. Statements which are analytic by general philosophical acclaim are not, indeed, far to seek. They fall into two classes. Those of the first class, which may be called *logically true*, are typified by:

(1) No unmarried man is married.

The relevant feature of this example is that it not merely is true as it stands, but remains true under any and all reinterpretations of 'man' and 'married'. If we suppose a prior inventory of *logical* particles, comprising 'no', 'un-', 'not', 'if', 'then', 'and', etc., then in general a logical truth is a statement which is true and remains true under all reinterpretations of its components other than the logical particles.

But there is also a second class of analytic statements, typified by:

(2) No bachelor is married.

The characteristic of such a statement is that it can be turned into a logical truth by putting synonyms for synonyms; thus (2) can be turned into (1) by putting 'unmarried man' for its synonym 'bachelor'. We still lack a proper characterization of this second class of analytic statements, and therewith of analyticity generally, inasmuch as we have had in the above description to lean on a notion of "synonymy" which is no less in need of clarification than analyticity itself.

In recent years Carnap has tended to explain analyticity by appeal to what he calls state-descriptions.[4] A state-description is any exhaustive assignment of truth values to the atomic, or noncompound, statements of the language. All other statements of the language are, Carnap assumes, built up of their component clauses by means of the familiar logical devices, in such a way that the truth value of any complex statement is fixed for each state-description by specifiable logical laws. A statement is then explained as analytic when it comes out true under every state description. This account is an adaptation of Leibniz's "true in all possible worlds." But note that this version of analyticity serves its purpose only if the atomic statements of the language are, unlike 'John is a bachelor' and 'John is married', mutually independent. Otherwise there would be a state-description which assigned truth to 'John is a bachelor' and to 'John is married', and consequently 'No bachelors are married' would turn out synthetic rather than analytic under the proposed criterion. Thus the criterion of analyticity in terms of state-descriptions serves only for languages devoid of extralogical synonym-pairs, such as 'bachelor' and 'unmarried man'—synonym-pairs of the type which give rise to the "second class" of analytic statements. The criterion in terms of state-descriptions is a reconstruction at best of logical truth, not of analyticity.

I do not mean to suggest that Carnap is under any illusions on this point. His simplified model language with its state-descriptions is aimed primarily not at the general problem of analyticity but at another purpose, the clarification of probability and induction. Our problem, however, is analyticity; and here the major difficulty lies not in the first class of analytic statements, the logical truths, but rather in the second class, which depends on the notion of synonymy.

2 | Definition

There are those who find it soothing to say that the analytic statements of the second class reduce to those of the first class, the logical truths, by *definition*; 'bachelor', for example, is *defined* as 'unmarried man'. But how do we find that 'bachelor' is defined as 'unmarried man'? Who defined it thus, and when? Are we to appeal to the nearest dictionary, and accept the lexicographer's formulation as law? Clearly this would be to put the cart before the horse. The lexicographer is an empirical scientist, whose business is the recording of antecedent facts; and if he glosses 'bachelor' as 'unmarried man' it is because of his belief that there is a relation of synonymy between those forms, implicit in general or preferred usage prior to his own work. The notion of synonymy presupposed here has still to be clarified, presumably in terms relating to linguistic behavior. Certainly the "definition" which is the lexicographer's report of an observed synonymy cannot be taken as the ground of the synonymy.

Definition is not, indeed, an activity exclusively of philologists. Philosophers and scientists frequently have occasion to "define" a recondite term by paraphrasing it into terms of a more familiar vocabulary. But ordinarily such a definition, like the philologist's, is pure lexicography, affirming a relation of synonymy antecedent to the exposition in hand.

Just what it means to affirm synonymy, just what the interconnections may be which are necessary and sufficient in order that two linguistic forms be properly describable as synonymous, is far from clear; but, whatever these interconnections may be, ordinarily they are grounded in usage. Definitions reporting selected instances of synonymy come then as reports upon usage.

There is also, however, a variant type of definitional activity which does not limit itself to the reporting of preëxisting synonymies. I have in mind what Carnap calls *explication*—an activity to which philosophers are given, and scientists also in their more philosophical moments. In explication the purpose is not merely to paraphrase the definiendum into an outright synonym, but actually to improve upon the definiendum by refining or supplementing its meaning.* But even explication, though not merely reporting a preëxisting synonymy between definiendum and definiens, does rest nevertheless on *other* preëxisting synonymies. The matter may be viewed as follows. Any word worth explicating has some contexts which, as wholes, are clear and precise enough to be useful; and the purpose of explication is to preserve the usage of these favored contexts while sharpening the usage of other contexts. In order that a given definition be suitable for purposes of explication, therefore, what is required is not that the definiendum in its antecedent usage be synonymous with the definiens, but just that each of these favored contexts of the definiendum, taken as a whole in its antecedent usage, be synonymous with the corresponding context of the definiens.

Two alternative definientia may be equally appropriate for the purposes of a given task of explication and yet not be synonymous with each other; for they may serve interchangeably within the favored contexts but diverge elsewhere. By cleaving to one of these definientia rather than the other, a definition of explicative kind generates, by fiat, a relation of synonymy between definiendum and definiens which did not hold before. But such a definition still owes its explicative function, as seen, to preëxisting synonymies.

There does, however, remain still an extreme sort of definition which does not hark back to prior synonymies at all: namely, the explicitly conventional introduction of novel notations for purposes of sheer abbreviation. Here the definiendum becomes synonymous with the definiens simply because it has been created expressly for the purpose of being synonymous with the definiens. Here we have a really transparent case of synonymy created by definition; would that all species of synonymy were as intelligible. For the rest, definition rests on synonymy rather than explaining it.

The word 'definition' has come to have a dangerously reassuring sound, owing no doubt to its frequent occurrence in logical and mathematical writings. We shall do well to digress now into a brief appraisal of the role of definition in formal work.

In logical and mathematical systems either of two mutually antagonistic types of economy may be striven for, and each has its peculiar practical utility. On the one hand we may seek economy of practical expression— ease and brevity in the statement of multifarious relations. This sort of economy calls usually for distinctive concise notations for a wealth of concepts. Second, however, and oppositely, we may seek economy in grammar and

---

* The *definiendum* is the word or phrase to be defined; the *definiens* is the definition (that which does the defining). As Quine notes, philosophers and scientists often seek to explicate—to clarify and analyze—the notions they define. Accordingly, the definiens of an explication will not be exactly equivalent in meaning to the definiendum but will offer an improved, more precise version of it. Sometimes this involves analyzing the original definiendum into several distinct concepts. See, for example, the attempts to explicate the notions of confirmation, explanation, and reduction in chapters 5, 6, and 8 of this volume.

vocabulary; we may try to find a minimum of basic concepts such that, once a distinctive notation has been appropriated to each of them, it becomes possible to express any desired further concept by mere combination and iteration of our basic notations. This second sort of economy is impractical in one way, since a poverty in basic idioms tends to a necessary lengthening of discourse. But it is practical in another way: it greatly simplifies theoretical discourse *about* the language, through minimizing the terms and the forms of construction wherein the language consists.

Both sorts of economy, though prima facie incompatible, are valuable in their separate ways. The custom has consequently arisen of combining both sorts of economy by forging in effect two languages, the one a part of the other. The inclusive language, though redundant in grammar and vocabulary, is economical in message lengths, while the part, called primitive notation, is economical in grammar and vocabulary. Whole and part are correlated by rules of translation whereby each idiom not in primitive notation is equated to some complex built up of primitive notation. These rules of translation are the so-called *definitions* which appear in formalized systems. They are best viewed not as adjuncts to one language but as correlations between two languages, the one a part of the other.

But these correlations are not arbitrary. They are supposed to show how the primitive notations can accomplish all purposes, save brevity and convenience, of the redundant language. Hence the definiendum and its definiens may be expected, in each case, to be related in one or another of the three ways lately noted. The definiens may be a faithful paraphrase of the definiendum into the narrower notation, preserving a direct synonymy[5] as of antecedent usage; or the definiens may, in the spirit of explication, improve upon the antecedent usage of the definiendum; or finally, the definiendum may be a newly created notation, newly endowed with meaning here and now.

In formal and informal work alike, thus, we find that definition—except in the extreme case of the explicitly conventional introduction of new notations—hinges on prior relations of synonymy. Recognizing then that the notion of definition does not hold the key to synonymy and analyticity, let us look further into synonymy and say no more of definition.

3 | Interchangeability

A natural suggestion, deserving close examination, is that the synonymy of two linguistic forms consists simply in their interchangeability in all contexts without change of truth value—interchangeability, in Leibniz's phrase, *salva veritate*.[6] Note that synonyms so conceived need not even be free from vagueness, as long as the vaguenesses match.

But it is not quite true that the synonyms 'bachelor' and 'unmarried man' are everywhere interchangeable *salva veritate*. Truths which become false under substitution of 'unmarried man' for 'bachelor' are easily constructed with the help of 'bachelor of arts' or 'bachelor's buttons'; also with the help of quotation, thus:

'Bachelor' has less than ten letters.

Such counterinstances can, however, perhaps be set aside by treating the phrases 'bachelor of arts' and 'bachelor's buttons' and the quotation ''bachelor'' each as a single indivisible word and then stipulating that the interchangeability *salva veritate* which is to be the touchstone of synonymy is not supposed to apply to fragmentary occurrences inside of a word. This account of synonymy, supposing it acceptable on other counts, has indeed the drawback of appealing to a prior conception of "word" which can be counted on to present difficulties of formulation in its turn. Nevertheless some progress might be claimed in having reduced the problem of synonymy to a problem of wordhood. Let us pursue this line a bit, taking "word" for granted.

The question remains whether interchangeability *salva veritate* (apart from occurrences within words) is a strong enough condition for synonymy, or whether, on the contrary, some heteronymous expressions might be thus interchangeable. Now let us be clear that we are not concerned here with synonymy in the sense of complete identity in psychological associations or poetic quality; indeed no two expressions are synonymous in such a sense. We are concerned only with what may be called *cognitive* synonymy. Just what this is cannot be said without successfully finishing the present study; but we know something about it from the need which arose for it in connection with analyticity in §1. The sort of synonymy needed there was merely such that any analytic statement could be turned into a logical truth by putting synonyms for synonyms. Turning the tables and assuming analyticity, indeed, we could explain cognitive synonymy of terms as follows (keeping to the familiar example): to say that 'bachelor' and 'unmarried man' are cognitively synonymous is to say no more nor less than that the statement:

(3) All and only bachelors are unmarried men

is analytic.[7]

What we need is an account of cognitive synonymy not presupposing analyticity—if we are to explain analyticity conversely with help of cognitive synonymy as undertaken in §1. And indeed such an independent account of cognitive synonymy is at present up for consideration, namely, interchangeability *salva veritate* everywhere except within words. The question before us, to resume the thread at last, is whether such interchangeability is a sufficient condition for cognitive synonymy. We can quickly assure ourselves that it is, by examples of the following sort. The statement:

(4) Necessarily all and only bachelors are bachelors

is evidently true, even supposing 'necessarily' so narrowly construed as to be truly applicable only to analytic statements. Then, if 'bachelor' and 'unmarried man' are interchangeable *salva veritate*, the result:

(5) Necessarily all and only bachelors are unmarried men

of putting 'unmarried man' for an occurrence of 'bachelor' in (4) must, like (4), be true. But to say that (5) is true is to say that (3) is analytic, and hence that 'bachelor' and 'unmarried man' are cognitively synonymous.

Let us see what there is about the above argument that gives it its air of hocus-pocus. The condition of interchangeability *salva veritate* varies in its force with variations in the richness of the language at hand. The above argument supposes we are working with a language rich enough to contain the adverb 'necessarily', this adverb being so construed as to yield truth when and only when applied to an analytic statement. But can we condone a language which contains such an adverb? Does the adverb really make sense? To suppose that it does is to suppose that we have already made satisfactory sense of 'analytic'. Then what are we so hard at work on right now?

Our argument is not flatly circular, but something like it. It has the form, figuratively speaking, of a closed curve in space.

Interchangeability *salva veritate* is meaningless until relativized to a language whose extent is specified in relevant respects. Suppose now we consider a language containing just the following materials. There is an indefinitely large stock of one-place predicates (for example, 'F' where 'Fx' means that x is a man) and many-place predicates (for example, 'G' where 'Gxy' means that x loves y), mostly having to do with extralogical subject matter. The rest of the language is logical. The atomic sentences consist each of a predicate followed by one or more variables 'x', 'y', etc.; and the complex sentences are built up of the atomic ones by truth functions ('not', 'and', 'or', etc.) and quantification.[8] In effect such a language enjoys the benefits also of descriptions and indeed singular terms generally, these being contextually definable in known ways.[9] Even abstract singular terms naming classes, classes of classes, etc., are contextually definable in case the assumed stock of predicates includes the two-place predicate of class membership.[10] Such a language can be adequate to classical mathematics and indeed to scientific discourse generally, except in so far as the latter involves debatable devices such as contrary-to-fact conditionals or modal adverbs like 'necessarily'.[11] Now a language of this type is extensional, in this sense: any two predicates which agree extensionally (that is, are true of the same objects) are interchangeable *salva veritate*.[12]

In an extensional language, therefore, interchangeability *salva veritate* is no assurance of cognitive synonymy of the desired type. That 'bachelor' and 'unmarried man' are interchangeable *salva veritate* in an extensional language assures us of no more than that (3) is true. There is no assurance here that the extensional agreement of 'bachelor' and 'unmarried man' rests on meaning rather than merely on accidental matters of fact, as does the extensional agreement of 'creature with a heart' and 'creature with kidneys'.

For most purposes extensional agreement is the nearest approximation to synonymy we need care about. But the fact remains that extensional agreement falls far short of cognitive synonymy of the type required for explaining analyticity in the manner of §1. The type of cognitive synonymy required there is such as to equate the synonymy of 'bachelor' and 'unmarried man' with the analyticity of (3), not merely with the truth of (3).

So we must recognize that interchangeability *salva veritate*, if construed in relation to an extensional language, is not a sufficient condition of cognitive synonymy in the sense needed for deriving analyticity in the manner of §1. If a language contains an intensional adverb 'necessarily' in the sense lately noted, or other particles to the same effect, then interchangeability *salva veritate* in such a language does afford a sufficient condition of cognitive synonymy; but such a language is intelligible only in so far as the notion of analyticity is already understood in advance.

The effort to explain cognitive synonymy first, for the sake of deriving analyticity from it afterward as in §1, is perhaps the wrong approach. Instead we might try explaining analyticity somehow without appeal to cognitive synonymy. Afterward we could doubtless derive cognitive synonymy from analyticity satisfactorily enough if desired. We have seen that cognitive synonymy of 'bachelor' and 'unmarried man' can be explained as analyticity of (3). The same explanation works for any pair of one-place predicates, of course, and it can be extended in obvious fashion to many-place predicates. Other syntactical categories can also be accommodated in fairly parallel fashion. Singular terms may be said to be cognitively synonymous when the statement of identity formed by putting '=' between them is analytic. Statements may be said simply to be cognitively synonymous when their biconditional (the result of joining them by 'if and only if') is analytic.[13] If we care to lump all categories into a single formulation, at the expense of assuming again the notion of "word" which was appealed to early in this section, we can describe any two linguistic forms as cognitively synonymous when the two forms are interchangeable (apart from occurrences within "words") *salva* (no longer *veritate* but) *analyticitate*. Certain technical questions arise, indeed, over cases of ambiguity or homonymy; let us not pause for them, however, for we are already digressing. Let us rather turn our backs on the problem of synonymy and address ourselves anew to that of analyticity.

4 | Semantical Rules

Analyticity at first seemed most naturally definable by appeal to a realm of meanings. On refinement, the appeal to meanings gave way to an appeal to synonymy or definition. But definition turned out to be a will-o'-the-wisp, and synonymy turned out to be best understood only by dint of a prior appeal to analyticity itself. So we are back at the problem of analyticity.

I do not know whether the statement 'Everything green is extended' is analytic. Now does my indecision over this example really betray an incomplete understanding, an incomplete grasp of the "meanings", of 'green' and 'extended'? I think not. The trouble is not with 'green' or 'extended', but with 'analytic'.

It is often hinted that the difficulty in separating analytic statements from synthetic ones in ordinary language is due to the vagueness of ordinary language and that the distinction is clear when we have a precise artificial language with explicit "semantical rules." This, however, as I shall now attempt to show, is a confusion.

The notion of analyticity about which we are worrying is a purported relation between statements and languages: a statement S is said to be *analytic for* a language L, and the problem is to make sense of this relation generally, that is, for variable 'S' and 'L'. The gravity of this problem is not perceptibly less for artificial languages than for natural ones. The problem of making sense of the idiom 'S is analytic for L', with variable 'S' and 'L', retains its stubbornness even if we limit the range of the variable 'L' to artificial languages. Let me now try to make this point evident.

For artificial languages and semantical rules we look naturally to the writings of Carnap. His semantical rules take various forms, and to make my point I shall have to distinguish certain of the forms. Let us suppose, to begin with, an artificial language L_0 whose semantical rules have the form explicitly of a specification, by recursion or otherwise, of all the analytic statements of L_0. The rules tell us that such and such statements, and only those, are the analytic statements of L_0. Now here the difficulty is simply that the rules contain the word 'analytic', which we do not understand! We understand what expressions the rules attribute analyticity to, but we do not understand what the rules attribute to those expressions. In short, before we can understand a rule which begins 'A statement S is analytic for language L_0 if and only if . . .', we must understand the general relative term 'analytic for'; we must understand 'S is analytic for L' where 'S' and 'L' are variables.

Alternatively we may, indeed, view the so-called rule as a conventional definition of a new simple symbol 'analytic-for-L_0', which might better be written untendentiously as 'K' so as not to seem to throw light on the interesting word 'analytic'. Obviously any number of classes K, M, N, etc. of statements of L_0 can be specified for various purposes or for no purpose; what does it mean to say that K, as against M, N, etc., is the class of the "analytic" statements of L_0?

By saying what statements are analytic for L_0 we explain 'analytic-for-L_0' but not 'analytic', not 'analytic for'. We do not begin to explain the idiom 'S is analytic for L' with variable 'S' and 'L', even if we are content to limit the range of 'L' to the realm of artificial languages.

Actually we do know enough about the intended significance of 'analytic' to know that analytic statements are supposed to be true. Let us then turn to a second form of semantical rule, which says not that such and such statements are analytic but simply that such and such statements are included among the truths. Such a rule is not subject to the criticism of containing the un-understood word 'analytic'; and we may grant for the sake of argument that there is no difficulty over the broader term 'true'. A semantical rule of this second type, a rule of truth, is not supposed to specify all the truths of the language; it merely stipulates, recursively or otherwise, a certain multitude of statements which, along with others unspecified, are to count as true. Such a rule may be conceded to be quite clear. Derivatively, afterward, analyticity can be demarcated thus: a statement is analytic if it is (not merely true but) true according to the semantical rule.

Still there is really no progress. Instead of appealing to an unexplained word 'analytic', we are now appealing to an unexplained phrase 'semantical rule'. Not every true statement which says that the statements of some class are true can count as a semantical rule—otherwise *all* truths would be "analytic" in the sense of being true according to semantical rules. Semantical rules are distinguishable, apparently, only by the fact of appearing on a page under the heading 'Semantical Rules'; and this heading is itself then meaningless.

We can say indeed that a statement is *analytic-for-L_0* if and only if it is true according to such and such specifically appended "semantical rules," but then we find ourselves back at essentially the same case which was originally discussed: 'S is analytic-for-L_0 if and only if. . . .' Once we seek to explain 'S is analytic for L' generally for variable 'L' (even allowing limitation of 'L' to artificial languages), the explanation 'true according to the semantical rules of L' is unavailing; for the relative term 'semantical rule of' is as much in need of clarification, at least, as 'analytic for'.

It may be instructive to compare the notion of semantical rule with that of postulate. Relative to a given set of postulates, it is easy to say what a postulate is: it is a member of the set. Relative to a given set of semantical rules, it is equally easy to say what a semantical rule is. But given simply a notation, mathematical or otherwise, and indeed as thoroughly understood a notation as you please in point of the translations or truth conditions of its statements, who can say which of its true statements rank as postulates? Obviously the question is meaningless—as meaningless as asking which points in Ohio are starting points. Any finite (or effectively specifiable infinite) selection of statements (preferably true ones, perhaps) is as much *a* set of postulates as any other. The word 'postulate' is significant only relative to an act of inquiry; we apply the word to a set of statements just in so far as we happen, for the year or the moment, to be thinking of those statements in relation to the statements which can be reached from them by some set of transformations to which we have seen fit to direct our attention. Now the notion of semantical rule is as sensible and meaningful as that of postulate, if conceived in a similarly relative spirit—relative, this time, to one or another particular enterprise of schooling unconversant persons in sufficient conditions for truth of statements of some natural or artificial language L. But from this point of view no one signalization of a subclass of the truths of L

is intrinsically more a semantical rule than another; and, if 'analytic' means 'true by semantical rules', no one truth of L is analytic to the exclusion of another.[14]

It might conceivably be protested that an artificial language L (unlike a natural one) is a language in the ordinary sense *plus* a set of explicit semantical rules—the whole constituting, let us say, an ordered pair; and that the semantical rules of L then are specifiable simply as the second component of the pair L. But, by the same token and more simply, we might construe an artificial language L outright as an ordered pair whose second component is the class of its analytic statements; and then the analytic statements of L become specifiable simply as the statements in the second component of L. Or better still, we might just stop tugging at our bootstraps altogether.

Not all the explanations of analyticity known to Carnap and his readers have been covered explicitly in the above considerations, but the extension to other forms is not hard to see. Just one additional factor should be mentioned which sometimes enters: sometimes the semantical rules are in effect rules of translation into ordinary language, in which case the analytic statements of the artificial language are in effect recognized as such from the analyticity of their specified translations in ordinary language. Here certainly there can be no thought of an illumination of the problem of analyticity from the side of the artificial language.

From the point of view of the problem of analyticity the notion of an artificial language with semantical rules is a *feu follet* [will-o'-the-wisp] *par excellence*. Semantical rules determining the analytic statements of an artificial language are of interest only in so far as we already understand the notion of analyticity; they are of no help in gaining this understanding.

Appeal to hypothetical languages of an artificially simple kind could conceivably be useful in clarifying analyticity, if the mental or behavioral or cultural factors relevant to analyticity—whatever they may be—were somehow sketched into the simplified model. But a model which takes analyticity merely as an irreducible character is unlikely to throw light on the problem of explicating analyticity.

It is obvious that truth in general depends on both language and extralinguistic fact. The statement 'Brutus killed Caesar' would be false if the world had been different in certain ways, but it would also be false if the word 'killed' happened rather to have the sense of 'begat'. Thus one is tempted to suppose in general that the truth of a statement is somehow analyzable into a linguistic component and a factual component. Given this supposition, it next seems reasonable that in some statements the factual component should be null; and these are the analytic statements. But, for all its a priori reasonableness, a boundary between analytic and synthetic statements simply has not been drawn. That there is such a distinction to be drawn at all is an unempirical dogma of empiricists, a metaphysical article of faith.

5 | The Verification Theory and Reductionism

In the course of these somber reflections we have taken a dim view first of the notion of meaning, then of the notion of cognitive synonymy, and finally of the notion of analyticity. But what, it may be asked, of the verification theory of meaning? This phrase has established itself so firmly as a catchword of empiricism that we should be very unscientific indeed not to look beneath it for a possible key to the problem of meaning and the associated problems.

The verification theory of meaning, which has been conspicuous in the literature from Peirce onward, is that the meaning of a statement is the method of empirically confirming or infirming [disconfirming] it. An analytic statement is that limiting case which is confirmed no matter what.

As urged in §1, we can as well pass over the question of meanings as entities and move straight to sameness of meaning, or synonymy. Then what the verification theory says is that statements are synonymous if and only if they are alike in point of method of empirical confirmation or information.

This is an account of cognitive synonymy not of linguistic forms generally, but of statements.[15] However, from the concept of synonymy of statements we could derive the concept of synonymy for other linguistic forms, by considerations somewhat similar to those at the end of §3. Assuming the notion of "word," indeed, we could explain any two forms as synonymous when the putting of the one form for an occurrence of the other in any statement (apart from occurrences within "words") yields a synonymous statement. Finally, given the concept of synonymy thus for linguistic forms generally, we could define analyticity in terms of synonymy and logical truth as in §1. For that matter, we could define analyticity more simply in terms of just synonymy of statements together with logical truth; it is not necessary to appeal to synonymy of linguistic forms other than statements. For a statement may be described as analytic simply when it is synonymous with a logically true statement.

So, if the verification theory can be accepted as an adequate account of statement synonymy, the notion of analyticity is saved after all. However, let us reflect. Statement synonymy is said to be likeness of method of empirical confirmation or infirmation. Just what are these methods which are to be compared for likeness? What, in other words, is the nature of the relation between a statement and the experiences which contribute to or detract from its confirmation?

The most naïve view of the relation is that it is one of direct report. This is *radical reductionism*. Every meaningful statement is held to be translatable into a statement (true or false) about immediate experience. Radical reductionism, in one form or another, well antedates the verification theory of meaning explicitly so called. Thus Locke and Hume held that every idea must either originate directly in sense experience or else be compounded of ideas thus originating; . . . we might rephrase this doctrine in semantical jargon by saying that a term, to be significant at all, must be either a name of a sense datum or a compound of such names or an abbreviation of such a compound. So stated, the doctrine remains ambiguous as between sense data as sensory events and sense data as sensory qualities; and it remains vague as to the admissible ways of compounding. Moreover, the doctrine is unnecessarily and intolerably restrictive in the term-by-term critique which it imposes. More reasonably, and without yet exceeding the limits of what I have called radical reductionism, we may take full statements as our significant units—thus demanding that our statements as wholes be translatable into sense-datum language, but not that they be translatable term by term.

This emendation would unquestionably have been welcome to Locke and Hume . . . , but historically it had to await an important reorientation in semantics—the reorientation whereby the primary vehicle of meaning came to be seen no longer in the term but in the statement. This reorientation, seen in Bentham and Frege, underlies Russell's concept of incomplete symbols defined in use;[16] also it is implicit in the verification theory of meaning, since the objects of verification are statements.

Radical reductionism, conceived now with statements as units, set itself the task of specifying a sense-datum language and showing how to translate the rest of significant discourse, statement by statement, into it. Carnap embarked on this project in the *Aufbau*.*

The language which Carnap adopted as his starting point was not a sense-datum language in the narrowest conceivable sense, for it included also the notations of logic, up through higher set theory. In effect it included the whole language of pure mathematics. The ontology implicit in it (that is, the range of values of its variables) embraced not only sensory events but classes, classes of classes, and so on. Empiricists there are who would boggle at such prodigality. Carnap's starting point is very parsimonious, however, in its extralogical or sensory part. In a series of constructions in which he exploits the resources of modern logic with much ingenuity, Carnap succeeds in defining a wide array of important additional sensory concepts which, but for his constructions, one would not have dreamed were definable on so slender a basis. He was the first empiricist who, not content with

---

* The reference is to Carnap's first major work, *Der Logische Aufbau der Welt* (1928), published two years after Carnap joined the Vienna Circle. It has been translated into English by Rolf A. George as *The Logical Structure of the World* (Berkeley, Calif.: University of California Press, 1969). Carnap's work was inspired by Wittgenstein's *Tractatus Logico-Philosophicus* (1921) and Russell's doctrine of logical atomism, especially Russell's injunction to philosophers that, wherever possible, they should replace inferences to unknown entities with logical constructions out of known entities. In the *Aufbau*, Carnap tried to reduce all the concepts used in science and everyday life to the sensory qualities given in immediate experience. As Quine later notes, Carnap's project was never completed and, by 1936, Carnap had given up the reductionist thesis that statements about the physical world can be translated into equivalent statements about immediate experience.

asserting the reducibility of science to terms of immediate experience, took serious steps toward carrying out the reduction.

If Carnap's starting point is satisfactory, still his constructions were, as he himself stressed, only a fragment of the full program. The construction of even the simplest statements about the physical world was left in a sketchy state. Carnap's suggestions on this subject were, despite their sketchiness, very suggestive. He explained spatio-temporal point-instants as quadruples of real numbers and envisaged assignment of sense qualities to point-instants according to certain canons. Roughly summarized, the plan was that qualities should be assigned to point-instants in such a way as to achieve the laziest world compatible with our experience. The principle of least action was to be our guide in constructing a world from experience.

Carnap did not seem to recognize, however, that his treatment of physical objects fell short of reduction not merely through sketchiness, but in principle. Statements of the form 'Quality q is at point-instant $x;y;z;t$' were, according to his canons, to be apportioned truth values in such a way as to maximize and minimize certain over-all features, and with growth of experience the truth values were to be progressively revised in the same spirit. I think this is a good schematization (deliberately oversimplified, to be sure) of what science really does; but it provides no indication, not even the sketchiest, of how a statement of the form 'Quality q is at $x;y;z;t$' could ever be translated into Carnap's initial language of sense data and logic. The connective 'is at' remains an added undefined connective; the canons counsel us in its use but not in its elimination.

Carnap seems to have appreciated this point afterward; for in his later writings he abandoned all notion of the translatability of statements about the physical world into statements about immediate experience. Reductionism in its radical form has long since ceased to figure in Carnap's philosophy.

But the dogma of reductionism has, in a subtler and more tenuous form, continued to influence the thought of empiricists. The notion lingers that to each statement, or each synthetic statement, there is associated a unique range of possible sensory events such that the occurrence of any of them would add to the likelihood of truth of the statement, and that there is associated also another unique range of possible sensory events whose occurrence would detract from that likelihood. This notion is of course implicit in the verification theory of meaning.

The dogma of reductionism survives in the supposition that each statement, taken in isolation from its fellows, can admit of confirmation or infirmation at all. My countersuggestion, issuing essentially from Carnap's doctrine of the physical world in the *Aufbau*, is that our statements about the external world face the tribunal of sense experience not individually but only as a corporate body.[17]

The dogma of reductionism, even in its attenuated form, is intimately connected with the other dogma—that there is a cleavage between the analytic and the synthetic. We have found ourselves led, indeed, from the latter problem to the former through the verification theory of meaning. More directly, the one dogma clearly supports the other in this way: as long as it is taken to be significant in general to speak of the confirmation and infirmation of a statement, it seems significant to speak also of a limiting kind of statement which is vacuously confirmed, *ipso facto*, come what may; and such a statement is analytic.

The two dogmas are, indeed, at root identical. We lately reflected that in general the truth of statements does obviously depend both upon language and upon extralinguistic fact; and we noted that this obvious circumstance carries in its train, not logically but all too naturally, a feeling that the truth of a statement is somehow analyzable into a linguistic component and a factual component. The factual component must, if we are empiricists, boil down to a range of confirmatory experiences. In the extreme case where the linguistic component is all that matters, a true statement is analytic. But I hope we are now impressed with how stubbornly the distinction between analytic and synthetic has resisted any straightforward drawing. I am impressed also, apart from prefabricated examples of black and white balls in an urn, with how baffling the problem has always been of arriving at any explicit theory of the empirical confirmation of a synthetic statement. My present suggestion is that it is nonsense, and the root of much nonsense, to speak of a linguistic component and a factual component in the truth of any individual statement. Taken collectively, science has its double dependence upon language and experience; but this duality is not significantly traceable into the statements of science taken one by one.

The idea of defining a symbol in use was, as remarked, an advance over the impossible term-by-term empiricism of Locke and Hume. The statement, rather than the term, came with Bentham to be recognized as the unit accountable to an empiricist critique. But what I am now urging is that even in taking the statement as unit we have drawn our grid too finely. The unit of empirical significance is the whole of science.

6 | Empiricism without the Dogmas

The totality of our so-called knowledge or beliefs, from the most casual matters of geography and history to the profoundest laws of atomic physics or even of pure mathematics and logic, is a man-made fabric which impinges on experience only along the edges. Or, to change the figure, total science is like a field of force whose boundary conditions are experience. A conflict with experience at the periphery occasions readjustments in the interior of the field. Truth values have to be redistributed over some of our statements. Reëvaluation of some statements entails reëvaluation of others, because of their logical interconnections—the logical laws being in turn simply certain further statements of the system, certain further elements of the field.

Having reëvaluated one statement we must reëvaluate some others, which may be statements logically connected with the first or may be the statements of logical connections themselves. But the total field is so underdetermined by its boundary conditions, experience, that there is much latitude of choice as to what statements to reëvaluate in the light of any single contrary experience. No particular experiences are linked with any particular statements in the interior of the field, except indirectly through considerations of equilibrium affecting the field as a whole.

If this view is right, it is misleading to speak of the empirical content of an individual statement—especially if it is a statement at all remote from the experiential periphery of the field. Furthermore it becomes folly to seek a boundary between synthetic statements, which hold contingently on experience, and analytic statements, which hold come what may. Any statement can be held true come what may, if we make drastic enough adjustments elsewhere in the system. Even a statement very close to the periphery can be held true in the face of recalcitrant experience by pleading hallucination or by amending certain statements of the kind called logical laws. Conversely, by the same token, no statement is immune to revision. Revision even of the logical law of the excluded middle has been proposed as a means of simplifying quantum mechanics;* and what difference is there in principle between such a shift and the shift whereby Kepler superseded Ptolemy, or Einstein Newton, or Darwin Aristotle?

For vividness I have been speaking in terms of varying distances from a sensory periphery. Let me try now to clarify this notion without metaphor. Certain statements, though *about* physical objects and not sense experience, seem peculiarly germane to sense experience—and in a selective way: some statements to some experiences, others to others. Such statements, especially germane to particular experiences, I picture as near the periphery. But in this relation of "germaneness" I envisage nothing more than a loose association reflecting the relative likelihood, in practice, of our choosing one statement rather than another for revision in the event of recalcitrant experience. For example, we can imagine recalcitrant experiences to which we would surely be inclined to accommodate our system by reëvaluating just the statement that there are brick houses on Elm Street, together with related statements on the same topic. We can imagine other recalcitrant experiences to which we would be inclined to accommodate our system by reëvaluating just the statement that there are no centaurs, along with kindred statements. A recalcitrant experience can, I have urged, be accommodated by any of various alternative reëvaluations in various alternative quarters of the total system; but, in the cases which we are now imag-

---

* The law of excluded middle asserts that, for any statement p, the statement p or not-p is a necessary truth. For a discussion of arguments against the law and the alleged relevance of quantum mechanics to its revisability, see the subsection, "Gillies' Criticisms of Duhem" in the commentary on this chapter.

ining, our natural tendency to disturb the total system as little as possible would lead us to focus our revisions upon these specific statements concerning brick houses or centaurs. These statements are felt, therefore, to have a sharper empirical reference than highly theoretical statements of physics or logic or ontology. The latter statements may be thought of as relatively centrally located within the total network, meaning merely that little preferential connection with any particular sense data obtrudes itself.

As an empiricist I continue to think of the conceptual scheme of science as a tool, ultimately, for predicting future experience in the light of past experience. Physical objects are conceptually imported into the situation as convenient intermediaries—not by definition in terms of experience, but simply as irreducible posits[18] comparable, epistemologically, to the gods of Homer. For my part I do, qua lay physicist, believe in physical objects and not in Homer's gods; and I consider it a scientific error to believe otherwise. But in point of epistemological footing the physical objects and the gods differ only in degree and not in kind. Both sorts of entities enter our conception only as cultural posits. The myth of physical objects is epistemologically superior to most in that it has proved more efficacious than other myths as a device for working a manageable structure into the flux of experience.

Positing does not stop with macroscopic physical objects. Objects at the atomic level are posited to make the laws of macroscopic objects, and ultimately the laws of experience, simpler and more manageable; and we need not expect or demand full definition of atomic and subatomic entities in terms of macroscopic ones, any more than definition of macroscopic things in terms of sense data. Science is a continuation of common sense, and it continues the common-sense expedient of swelling ontology to simplify theory.

Physical objects, small and large, are not the only posits. Forces are another example; and indeed we are told nowadays that the boundary between energy and matter is obsolete. Moreover, the abstract entities which are the substance of mathematics—ultimately classes and classes of classes and so on up—are another posit in the same spirit. Epistemologically these are myths on the same footing with physical objects and gods, neither better nor worse except for differences in the degree to which they expedite our dealings with sense experiences.

The over-all algebra of rational and irrational numbers is underdetermined by the algebra of rational numbers, but is smoother and more convenient; and it includes the algebra of rational numbers as a jagged or gerrymandered part.[19] Total science, mathematical and natural and human, is similarly but more extremely underdetermined by experience. The edge of the system must be kept squared with experience; the rest, with all its elaborate myths or fictions, has as its objective the simplicity of laws.

Ontological questions, under this view, are on a par with questions of natural science.[20] Consider the question whether to countenance classes as entities. This, as I have argued elsewhere,[21] is the question whether to quantify with respect to variables which take classes as values. Now Carnap [3] has maintained that this is a question not of matters of fact but of choosing a convenient language form, a convenient conceptual scheme or framework for science. With this I agree, but only on the proviso that the same be conceded regarding scientific hypotheses generally. Carnap ([3], p. 32n) has recognized that he is able to preserve a double standard for ontological questions and scientific hypotheses only by assuming an absolute distinction between the analytic and the synthetic; and I need not say again that this is a distinction which I reject.[22]

The issue over there being classes seems more a question of convenient conceptual scheme; the issue over there being centaurs, or brick houses on Elm Street, seems more a question of fact. But I have been urging that this difference is only one of degree, and that it turns upon our vaguely pragmatic inclination to adjust one strand of the fabric of science rather than another in accommodating some particular recalcitrant experience. Conservatism figures in such choices, and so does the quest for simplicity.

Carnap, Lewis, and others take a pragmatic stand on the question of choosing between language forms, scientific frameworks; but their pragmatism leaves off at the imagined boundary between the analytic and the synthetic. In repudiating such a boundary I espouse a more thorough pragmatism. Each man is given a scientific heritage plus a continuing barrage of sensory stimulation; and the considerations which guide him in warping his scientific heritage to fit his continuing sensory promptings are, where rational, pragmatic.

■ | **Notes**

1. See p. 9 of "On What There Is," in Quine [2].

2. See p. 10 of "On What There Is," and pp. 107–15 of "Logic and the Reification of Universals," both in Quine [2].

3. See pp. 11ff. of "On What There Is" and pp. 48 of "Meaning in Linguistics," both in Quine [2].

4. Carnap [1], pp. 9ff.; Carnap [2], pp. 70ff.

5. According to an important variant sense of 'definition', the relation preserved may be the weaker relation of mere agreement in reference; see p. 132 of "Notes on the Theory of Reference," in Quine [2]. But definition in this sense is better ignored in the present connection, being irrelevant to the question of synonymy.

6. Cf. Lewis [1], p. 373.

7. This is cognitive synonymy in a primary, broad sense. Carnap ([1], pp. 56ff.) and Lewis ([2], pp. 83ff.) have suggested how, once this notion is at hand, a narrower

sense of cognitive synonymy which is preferable for some purposes can in turn be derived. But this special ramification of concept-building lies aside from the present purposes and must not be confused with the broad sort of cognitive synonymy here concerned.

8. Pp. 81ff. of "New Foundations for Mathematical Logic," in Quine [2], contain a description of just such a language, except that there happens there to be just one predicate, the two-place predicate '∈'.

9. See pp. 5–8 of "On What There Is," in Quine [2]; also pp. 85f. of "New Foundations for Mathematical Logic" and 166f. of "Meaning and Existential Inference," both in Quine [2].

10. See p. 87 of "New Foundations For Mathematical Logic," in Quine [2].

11. On such devices see also "Reference and Modality," in Quine [2].

12. This is the substance of Quine [1], p. 121.

13. The 'if and only if' itself is intended in the truth functional sense. See Carnap [1], p. 14.

14. The foregoing paragraph was not part of the present essay as originally published. It was prompted by Martin (see References).

15. The doctrine can indeed be formulated with terms rather than statements as the units. Thus Lewis describes the meaning of a term as "*a criterion in mind*, by reference to which one is able to apply or refuse to apply the expression in question in the case of presented, or imagined, things or situations" (Lewis [2], p. 133).—For an instructive account of the vicissitudes of the verification theory of meaning, centered however on the question of meaning*fulness* rather than synonymy or analyticity, see Hempel (see References).

16. See p. 6 of "On What There Is," in Quine [2].

17. This doctrine was well argued by Duhem, pp. 302–28 (see References). Or see Lowinger, pp. 132–40 (see References).

18. Cf. pp. 17f. of "On What There Is," in Quine [2].

19. Cf. p. 18 of "On What There Is," in Quine [2].

20. "L'ontologie fait corps avec la science elle-même et ne peut en être separée." [Ontology is an integral part of science and cannot be separated from it.] Meyerson, p. 439 (see References).

21. See pp. 12f. of "On What There Is," and pp. 102ff. of "Logic and the Reification of Universals," both in Quine [2].

22. For an effective expression of further misgivings over this distinction, see White (see References).

■ | References

Carnap, Rudolf [1], *Meaning and Necessity* (Chicago: University of Chicago Press, 1947).

Carnap, Rudolf [2], *Logical Foundations of Probability* (Chicago: University of Chicago Press, 1950).

Carnap, Rudolf [3], "Empiricism, Semantics, and Ontology," *Revue internationale de philosophie* 4 (1950): 20–40. Reprinted in Linsky.

Duhem, Pierre, *La Théorie physique: son objet et sa structure* (Paris, 1906).

Hempel, C. G., "Problems and Changes in the Empiricist Criterion of Meaning," *Revue internationale de philosophie* 4 (1950): 41–63. Reprinted in Linsky.

Lewis, C. I. [1], *A Survey of Symbolic Logic* (Berkeley, 1918).

Lewis, C. I. [2], *An Analysis of Knowledge and Valuation* (LaSalle, Ill.: Open Court, 1946).

Linsky, Leonard (ed.), *Semantics and the Philosophy of Language* (Urbana: University of Illinois Press, 1952).

Lowinger, Armand, *The Methodology of Pierre Duhem* (New York: Columbia University Press, 1941).

Martin, R. M., "On 'Analytic'," *Philosophical Studies* 3 (1952): 42–47.

Meyerson, Émile, *Identité et realité* (Paris, 1908; 4th ed., 1932).

Quine, W. V. [1], *Mathematical Logic* (New York: Norton, 1940; Cambridge, Mass.: Harvard University Press, 1947; rev. ed., Cambridge, Mass.: Harvard University Press, 1951).

Quine, W. V. [2], *From a Logical Point of View* (Cambridge, Mass.: Harvard University Press, 1953).

White, Morton, "The Analytic and the Synthetic: An Untenable Dualism," in Sidney Hook (ed.), *John Dewey: Philosopher of Science and Freedom* (New York: Dial Press, 1950), 316–30. Reprinted in Linsky.

DONALD GILLIES

The Duhem Thesis and the Quine Thesis

In current writings on the philosophy of science, reference is often made to what is called 'the Duhem-Quine thesis'. Really, however, this is something of a misnomer; for, as we shall see, the Duhem thesis differs in many important respects from the Quine thesis. In this chapter I will expound the two theses in turn and explain how they differ. I will conclude the chapter by suggesting that the phrase 'the Duhem-Quine thesis' could be used to refer to a thesis which combines elements from both the Duhem thesis and the Quine thesis. . . .

1 | Preliminary Exposition of the Thesis. The Impossibility of a Crucial Experiment

Of Duhem's many significant contributions to the philosophy of science, perhaps the most important was his formulation of what I will call the *Duhem thesis*. With his usual clarity and incisiveness, Duhem states this thesis as a section heading thus:

> An Experiment in Physics Can Never Condemn an Isolated Hypothesis but Only a Whole Theoretical Group (1904–5, p. 183 [230]).*

Later in this section he expounds the thesis as follows:

> In sum, the physicist can never subject an isolated hypothesis to experimental test, but only a whole group of hypotheses; when the experiment is in

---

FROM Donald Gillies, *Philosophy of Science in the Twentieth Century* (Oxford: Blackwell Publishers, 1993), 98–116.

* All page references in square brackets are to the excerpts from Duhem and Quine in this volume.

disagreement with his predictions, what he learns is that at least one of the hypotheses constituting this group is unacceptable and ought to be modified; but the experiment does not designate which one should be changed, (p. 187 [233]).

In order to discuss the Duhem thesis, it will be useful to introduce the notion of an *observation statement*. . . . Let us take an observation statement to be a statement which can provisionally be agreed to be either true or false on the basis of observation and experiment.

According to the Duhem thesis, an isolated hypothesis in physics (h, say) can never be falsified by an observation statement, O. As a generalisation covering all the hypotheses of physics, this is somewhat doubtful. Physics does appear to contain some falsifiable hypotheses. Consider, for example, Kepler's first law that planets move in ellipses with the Sun at one focus. Suppose that we observe a large number of positions of a given planet and that these do not lie on an ellipse of the requisite kind. We have then surely falsified Kepler's first law. The schema of falsification can be written, where 'not-h' is short for 'It is not the case that h':

If h, then O, but not-O, therefore not-h.   (1)

This uses a logical law called *modus tollens*.

However, the Duhem thesis does apply to some hypotheses. . . . Consider, for example, Newton's first law of motion (T_1, say). . . . We cannot find an O such that schema (1) above holds when we substitute T_1 for h.*

Newton's full theory (T, say) consisted of three laws of motion (T_1, T_2, and T_3) and the law of gravity, T_4. So T was a conjunction of these four laws ($T = T_1$ & T_2 & T_3 & T_4). Even from T by itself, however, we cannot derive any observable consequences regarding the solar system. To do so, we need to add to T a number of auxiliary hypotheses: for example, that no other forces but gravitational ones act on the planets, that the interplanetary attractions are small compared with those between the Sun and the planets, that the mass of the Sun is very much greater than that of the planets, and so on. Let us call the conjunction of such auxiliary hypotheses which are appropriate in a given case A. We now have the schema:

---

* In the book from which this reading is excerpted, Gillies refers to an argument of Poincaré to support the claim that Newton's first law of motion is not falsifiable. Newton's first law has the form of a conditional statement: if there is no external force acting on a body, then the velocity of the body will not change. Thus, to falsify the law requires a body that accelerates even though it is free from any net external force. Poincaré argues that if we find such an apparent counterexample, we can always deny that the body is genuinely free from a net external force by attributing the acceleration to forces exerted by as-yet-undetected invisible molecules. In this way, the law will be protected from refutation. See Henri Poincaré, *Science and Hypothesis*, trans. W. J. Greenstreet (New York: Dover, 1952), 95–96; originally published in French (1902).

If T_1 & T_2 & T_3 & T_4 & A, then O, but not-O,
therefore not-(T_1 & T_2 & T_3 & T_4 & A). $\qquad$ (2)

Moreover, from not-(T_1 & T_2 & T_3 & T_4 & A) it follows that at least one of the set (T_1, T_2, T_3, T_4, or A) is false, but we cannot say which one.

As the history of science shows, it is often a very real problem in scientific research to decide which one of a group of hypotheses should be changed. Consider, for example, Adams and Leverrier's discovery of Neptune in 1846. From Newton's theory T together with auxiliary hypotheses, astronomers were able to calculate the theoretical orbit of Uranus (the most distant planet then known). This theoretical orbit did not agree with the observed orbit. This meant that either T or one of the auxiliary hypotheses was false. Adams and Leverrier conjectured that the auxiliary hypothesis concerning the number of planets was in error. They postulated a new planet Neptune beyond Uranus, and calculated the mass and position it would have to have to cause the observed perturbations in Uranus's orbit. Neptune was duly observed on 23 September 1846 only 52' away from the predicted position.[1]

This part of the story is quite well known, but there were some subsequent events which are also relevant to the Duhem thesis. Another difficulty which occupied astronomers at the time concerned the anomalous motion of the perihelion of Mercury, which was found to advance slightly faster than it should do according to standard theory. Leverrier tried the same approach that had proved successful in the case of the Uranus anomaly. He postulated a planet Vulcan nearer to the Sun than Mercury, with a mass, orbit, and so forth which would explain the advance in Mercury's perihelion. However, no such planet could be found.

The discrepancy here is very small. Newcomb in 1898 gave its value as $41.24'' \pm 2.09''$ per century; that is, less than an eightieth part of a degree per century. However, this tiny anomaly was explained with great success by the general theory of relativity (T'), which Einstein proposed in 1915 as a replacement for Newton's theory, T. The value of the anomalous advance of the perihelion of Mercury which followed from the general theory of relativity was $42.89''$ per century—a figure well within the bounds set by Newcomb. We see that, although the Uranus anomaly and the Mercury anomaly were *prima facie* very similar, success was obtained in one case by altering an auxiliary hypothesis, in the other by altering the main theory.

In the next section, Duhem goes on to draw an important consequence from his thesis. This section is in fact headed 'A "Crucial Experiment" Is Impossible in Physics' (1904–5, p. 188 [234]). Duhem uses the term *crucial experiment* in something like the sense given by Bacon in the *Novum Organum* to his 'fact of the cross'. He formulates this notion of crucial experiment as follows: 'Enumerate all the hypotheses that can be made to account for this group of phenomena; then, by experimental contradiction eliminate all except one; the latter will no longer be a hypothesis, but will become a certainty' (ibid.). However, there is an obvious objection to crucial experiments in this strong sense: namely, that we can never be sure that we have listed all the hypotheses capable of explaining a group of phenomena. Duhem makes this point as follows:

> [E]xperimental contradiction does not have the power to transform a physical hypothesis into an indisputable truth; in order to confer this power on it, it would be necessary to enumerate completely the various hypotheses which may cover a determinate group of phenomena; but the physicist is never sure he has exhausted all the imaginable assumptions. (p. 190 [235])

In view of this difficulty, it seems desirable to adopt a rather weaker sense of crucial experiment, which may be defined as follows. Suppose we have two competing theories T_1 and T_2. An experiment (E, say) is crucial between T_1 and T_2, if T_1 predicts that E will give the result O and T_2 predicts that E will give the result not-O. If we perform E, and O occurs, then T_2 is eliminated. If we perform E, and not-O occurs, then T_1 is eliminated. In any event, one of the two theories will be eliminated by E, which is thus crucial for deciding between them. It does not of course follow that the successful theory is necessarily true, because there may be some, as yet unthought of, theory, T_3, which differs from T_1 and T_2 but explains the whole matter much more satisfactorily.

Duhem's point is that if T_1 and T_2 are such that his thesis applies to them, then we cannot derive O from T_1 but only from T_1 and A, where A is a conjunction of auxiliary assumptions. So, if not-O is the result of the experiment, this does not demonstrate beyond doubt that T_1 should be eliminated in favour of T_2. It could be that one of the auxiliary hypotheses in A is at fault.

Duhem illustrates this by what is perhaps the most famous example of an alleged crucial experiment in the history of science: Foucault's experiment, which was designed to decide between the wave theory and the particle theory of light. The wave theory of light predicted that the velocity of light in water should be less than its velocity in air, whereas the particle theory predicted that the velocity of light in water should be greater than its velocity in air. Foucault devised a method for measuring the velocity of light in water, and found that it was actually less than the velocity of light in air. Here, then, we seem to have a crucial experiment which decides definitely in favour of the wave theory of light. Indeed, some of Foucault's contemporaries, notably Arago, did maintain that Foucault's experiment was a crucial experiment in just this sense.

Duhem pointed out, however, that to derive from the particle theory that the velocity of light in water is greater than its velocity in air, we need, not just the assumption that light consists of particles (the fundamental hypothesis of the particle theory), but many auxiliary assumptions as well. The particle theory could always be saved by altering some of these auxiliary assumptions. As Duhem puts it: '[F]or it is not between two hypotheses, the emission and wave hypotheses, that Foucault's experiment judges trenchantly; it decides rather between two sets of theories each of which has to be taken as a whole, i.e. between two entire systems, Newton's optics and Huygens' optics' (p. 189 [235]). So, according to Duhem, Foucault's experiment is not a crucial experiment in a strictly logical sense. Yet, as we shall see in the next section, there is another, weaker sense in which the experiment is crucial, even for Duhem.

2 | Duhem's Criticisms of Conventionalism.
His Theory of Good Sense (*le bon sens*)

Duhem is sometimes classified as a conventionalist as regards his philosophy of science, but he is certainly not a conventionalist in the sense of Le Roy and Poincaré. Indeed, he devotes two sections of his *Aim and Structure of Physical Theory* to criticising these thinkers very clearly and explicitly. He formulates their conventionalist position as follows: 'Certain fundamental hypotheses of physical theory cannot be contradicted by any experiment, because they constitute in reality *definitions*, and because certain expressions in the physicist's usage take their meaning only through them' (p. 209 [240]).

Duhem objects strongly to Poincaré's claim that the principles of Newtonian mechanics will never be given up, because they are the simplest conventions available and cannot be contradicted by experiment. According to Duhem, the study of the history of science makes any such claim highly dubious:

> [T]he history of science should show that it would be very imprudent for us to say concerning a hypothesis commonly accepted today: 'We are certain that we shall never be led to abandon it because of a new experiment, no matter how precise it is.' Yet M. Poincaré does not hesitate to make this assertion concerning the principles of mechanics. (p. 212 [243]; I have here slightly altered the standard English translation in the interests of clarity.)

Poincaré's mistake, according to Duhem, was to take each principle of mechanics singly and in isolation. It is indeed true that when a principle of mechanics—for example, Newton's first law of motion—is taken in this fashion, it cannot be either confirmed or refuted by experience. However, by adding other hypotheses to any such principle, we get a group of hypotheses which can be compared with experience. Moreover, if the group in question is contradicted by the results of experiment and observation, it is possible to change any of the hypotheses of the group. We cannot say with Poincaré that certain fundamental hypotheses, because they are appropriately simple conventions, are above question and can never be altered. This is how Duhem puts the matter:

[I]t would be absurd to wish to subject certain principles of mechanics to *direct* experimental test; . . .

Does it follow that these hypotheses placed beyond the reach of direct experimental refutation have nothing more to fear from experiment? That they are guaranteed to remain immutable no matter what discoveries observation has in store for us? To pretend so would be a serious error.

Taken in isolation these different hypotheses have no experimental meaning; there can be no question of either confirming or contradicting them by experiment. But these hypotheses enter as essential foundations into the construction of certain theories of rational mechanics . . . these theories . . . are schematisms intended essentially to be compared with facts.

Now this comparison might some day very well show us that one of our representations is ill-adjusted to the realities it should picture, that the corrections which come and complicate our schematism do not produce sufficient concordance between this schematism and the facts, that the theory accepted for a long time without dispute should be rejected, and that an entirely different theory should be constructed on entirely different or new hypotheses. On that day some one of our hypotheses, which taken in isolation defied direct experimental refutation, will crumble with the system it supported under the weight of the contradictions inflicted by reality on the consequences of this system taken as a whole. (pp. 215–16 [246])

Thus Duhem's position seems to me more accurately described as *modified falsification,* rather than *conventionalism.* Duhem claims that some hypotheses of physics, when taken in isolation, can defy direct experimental refutation. He is thus not a strict falsificationist. On the other hand, he denies that such a hypothesis is immune from revision in the light of experimental evidence. A hypothesis of this kind may be tested indirectly if it forms part of a system of hypotheses which can be compared with experiment and observation. Further, such a hypothesis may on some occasion 'crumble with the system it supported under the weight of contradictions inflicted by reality'. Duhem does not deny that 'among the theoretical elements . . . there is always a certain number which the physicists of a certain epoch agree in accepting without test and which they regard as beyond dispute' (p. 211 [242]). However, he is very concerned to warn scientists against adopting too dogmatic an attitude towards any of their assumptions. His point is that, in the face of recalcitrant experience, the best way forward may be to alter one of the most entrenched assumptions. As he says:

Indeed, we must really guard ourselves against believing forever warranted those hypotheses which have become universally adopted conventions, and whose certainty seems to break through experimental contradiction by throwing the latter back on more doubtful assumptions. The history of physics shows us that very often the human mind has been led to overthrow such principles completely, though they have been regarded by common consent for centuries as inviolable axioms, and to rebuild its physical theories on new hypotheses. (p. 212 [243])

Duhem gives as an example the principle that light travels in a straight line. This was accepted as correct for hundreds—indeed, thousands—of years, but was eventually modified to explain certain diffraction effects.

Duhem even cites Newton's law of gravity as a law which is only provisional and may be changed in future. Unfortunately this passage has been accidentally omitted from the English edition of the *Aim and Structure of Physical Theory*. It is here translated from the French edition:

Of all the laws of physics, the one best verified by its innumerable consequences is surely the law of universal gravity; the most precise observations on the movements of the stars have not been able up to now to show it to be faulty. Is it, for all that, a definitive law? It is not, but a provisional law which has to be modified and completed unceasingly to make it accord with experience. (p. 267)

The episode of the anomalous motion of the perihelion of Mercury fits Duhem's analysis perfectly. It would surely have seemed reasonable to explain such a small discrepancy between Newton's theory and observation by altering some auxiliary assumption. In fact, however, the anomaly was only explained satisfactorily when Newton's whole theory of gravity was replaced by Einstein's general theory of relativity. Indeed, from a logical point of view, Duhem's philosophy of science can be seen as offering support to the Einsteinian revolution in physics. It therefore comes as a surprise to discover that Duhem rejected Einstein's theory of relativity in the most violent terms. In his 1915 booklet *La Science allemande* ('German Science'), Duhem argues that Einstein's theory of relativity must be considered as an aberration due to the lack of sound judgement of the German mind and its disrespect for reality. Admittedly, this booklet was written at a time when bitter nationalistic feelings were being generated by the First World War. Indeed, it belongs to a genre known as 'war literature', and is actually a relatively mild example of this unfortunate species of writing. All the same, it is clear that Duhem did reject Einstein's theory of relativity in no uncertain terms.

So, as already observed, we find in both Duhem and Poincaré a contradiction between their philosophical views and their scientific practice. Duhem was led by philosophical considerations to the conclusion that Newtonian mechanics is provisional and may be altered in future; yet he repudiated the new Einsteinian mechanics.[2] Conversely, Poincaré suggested in his philosophical writings of 1902 that the principles of Newtonian mechanics were conventions so simple that they would never be given up; yet, only two years later, in 1904, he decided that Newtonian mechanics needed to be changed, and started work on the development of a new mechanics. Some light is thrown on these strange contradictions by one further element in the Duhem thesis which we have still to discuss. This is Duhem's theory of good sense (le bon sens).

Let us take the typical situation envisioned by the Duhem thesis. From a group of hypotheses, $\{h_1 \ldots h_n\}$, say, a scientist has deduced O. Experiment or observation then shows that O is false. It follows that at least one of $\{h_1 \ldots h_n\}$ is false. But which one or ones are false? Which hypothesis or hypotheses should the scientist try to change in order to reestablish the agreement between theory and experience? Duhem states quite categorically that logic by itself cannot help the scientist. As far as pure logic is concerned, the choice between the various hypotheses is entirely open. The scientist in reaching his decision must be guided by what Duhem calls 'good sense' (le bon sens):

> Pure logic is not the only rule for our judgements; certain opinions which do not fall under the hammer of the principle of contradiction are in any case perfectly unreasonable. These motives which do not proceed from logic and yet direct our choices, these 'reasons which reason does not know' and which speak to the ample 'mind of finesse' but not to the 'geometric mind,' constitute what is appropriately called good sense. (1904–5, p. 217 [247])

Duhem imagines two scientists who, when faced with the experimental contradiction of a group of hypotheses, adopt different strategies. Scientist A alters a fundamental theory in the group, whereas scientist B alters some of the auxiliary assumptions. Both strategies are logically possible, and only good sense can enable us to decide between the two scientists. Thus, in the dispute between the particle theory of light and the wave theory of light, Biot, by a continual alteration and addition of auxiliary assumptions, tenaciously and ingeniously defended the particle theory, whereas Fresnel constantly devised new experiments favouring the wave theory. In the end, however, the dispute was resolved.

> After Foucault's experiment had shown that light travelled faster in air than in water, Biot gave up supporting the emission hypothesis; strictly, pure logic would not have compelled him to give it up, for Foucault's experiment was *not* the crucial experiment that Arago thought he saw in it, but by resisting wave optics for a longer time Biot would have been lacking in good sense. (p. 218, [248])

This passage in effect qualifies some of Duhem's earlier remarks about crucial experiments. Let us take two theories, T_1 and T_2, which are both subject to the Duhem thesis; that is, which cannot be tested in isolation but only by adjoining further assumptions. In a strictly logical sense, there cannot be a crucial experiment which decides between T_1 and T_2. The good sense of the scientific community can, however, lead it to judge that a particular experiment, such as Foucault's experiment, is in practice crucial in deciding the scientific controversy in favour of one of the two contending theories.

In his 1991 book (particularly chapters 4–6), Martin argues that 'lifelong meditation on certain texts of Pascal shaped many of the most important and difficult features of Duhem's thought' (p. 101). In particular, Duhem's theory of good sense (*le bon sens*) was derived in part from Pascal. Indeed, in the passage introducing *le bon sens*, Duhem quotes part of Pascal's famous saying that the heart has its reasons which reason knows nothing of.[3]

Although Duhem was undoubtedly influenced by Pascal, it is possible to suggest factors of a more personal and psychological nature which may have led him to his theory of scientific good sense. As his writings on philosophy of science show, Duhem was a man of outstanding logical ability; yet, as a physicist, he was a failure. In almost every scientific controversy in which he was involved, he chose the wrong side, rejecting those theories such as atomism, Maxwell's electrodynamics, and Einstein's theory of relativity which were to prove successful and lead to scientific progress. Although Duhem stubbornly defended his erroneous scientific opinions, he must have known in his heart of hearts that he was not proving to be a successful scientist. Yet he must also have been aware of his own exceptional logical powers. This situation could only be explained by supposing that something in addition to pure logic was needed in order to become a successful scientist. Here, then, we have a possible psychological origin of Duhem's theory of scientific good sense: namely, that Duhem saw that good sense is necessary for a scientist precisely because he himself was lacking in good sense. Duhem's rejection of a new theory which agreed so well with his own philosophy of science (that is, Einstein's theory of relativity) is just another instance of that lack of good sense which unfortunately characterized Duhem's scientific career.

Poincaré, by contrast, was one of the great physicists of his generation, and was amply endowed with the scientific good sense which Duhem lacked. The contrast between the two men is particularly evident in their respective discussions of electrodynamics. As we have already remarked, Duhem attacked Maxwell's theory harshly, and advocated the ideas of Helmholtz. Poincaré devotes a chapter (the thirteenth) of his 1902 book to electrodynamics. He begins (pp. 225–38) by discussing the theories of Ampère and Helmholtz and by mentioning the difficulties which he finds in these theories. Then, on p. 239, he introduces Maxwell's theory with the words: 'Such were the difficulties raised by the current theories, when Maxwell with a stroke of the pen caused them to vanish.' Subsequent developments completely endorsed Poincaré's support for Maxwell, while Helmholtz's ideas on electrodynamics, so strenuously advocated by Duhem, are now remembered only by a few erudite historians of science. It was Poincaré's scientific good sense which led him, contrary to the principles of his own conventionalist philosophy of science of 1902, to a modification of Newtonian mechanics.

Duhem's theory of good sense seems to me correct, but, at the same time, more in the nature of a problem, or a starting-point for further analysis, than of a final solution to the difficulty with which it deals. What factors contribute to forming scientific good sense? Why are some highly intelligent individuals like Duhem lacking in good sense? These are important questions. . . . In the next section, however, I will turn to a consideration of the Quine thesis.

3 | The Quine Thesis

In his famous 1951 article, 'Two Dogmas of Empiricism', Quine puts forward, with a reference to Duhem, a thesis which is related to Duhem's. Nonetheless, it seems to me that Quine's thesis is sufficiently different from Duhem's to make the conflation of the two intellectually unsatisfactory.[4] I will next briefly describe the Quine thesis,[5] and explain how it differs from the Duhem thesis.

The first obvious difference between Quine and Duhem is that Quine develops his views in the context of a discussion about whether a distinction can be drawn between analytic and synthetic statements, whereas Duhem does not even mention (let alone discuss) the analytic/synthetic problem.

[There are] two ways of defining an analytic statement. The first [is] due to Kant, who actually introduced the analytic/synthetic distinction. According to Kant, a statement is analytic if its predicate is contained in its subject. This formulation presupposes an Aristotelian analysis of statements into subject and predicate. It is not surprising that Frege, who rejected Aristotelian logic and introduced modern logic, should have proposed a new way of defining an analytic statement. Frege defines an analytic statement as one which is reducible to a truth of logic by means of explicit definitions. These two ways of defining an analytic statement are both illustrated by the standard example of an analytic statement, namely 'All bachelors are unmarried'. But Quine defines analytic statement in yet a third way. He writes critically of 'a belief in some fundamental cleavage between truths which are *analytic*, or grounded in meanings independently of matters of fact, and truths which are *synthetic*, or grounded in fact' (1951, p. 20 [250]). In effect, Quine is here taking a sentence to be analytic if it is true in virtue of the meanings of the words it contains. This is the definition of 'analytic' which is adopted by most modern philosophers interested in the question. Once again it is admirably illustrated by the standard example: S = 'All bachelors are unmarried'. Someone who knows the meanings of 'all', 'bachelors', 'are', and 'unmarried' will at once recognize that S is true, without having to make any empirical investigations into matters of fact. Thus S is analytic.

All this seems very convincing; yet Quine denies that the distinction between analytic and synthetic is a valid one. He writes:

> It is obvious that truth in general depends on both language and extralinguistic fact. The statement 'Brutus killed Caesar' would be false if the world had been different in certain ways, but it would also be false if the word 'killed' happened rather to have the sense of 'begat'. Thus one is tempted to suppose in general that the truth of a statement is somehow analyzable into a linguistic component and a factual component. Given this supposition, it next seems reasonable that in some statements the factual component should be null; and these are the analytic statements. But, for all its a priori reasonableness, a boundary between analytic and synthetic statements, simply has not been drawn. That there is such a distinction to be drawn at all is an unempirical dogma of empiricists, a metaphysical article of faith. (1951, pp. 36–7 [261])

The empiricists to whom Quine refers are, of course, the empiricists of the Vienna Circle, especially Carnap. . . . [T]heir particular brand of empiricism (logical empiricism) did indeed involve drawing a distinction between analytic and synthetic statements. However, support for the distinction is not confined to some members of the empiricist camp. Kantians too support the distinction, which was indeed introduced by Kant himself.

But what has all this to do with the issues involving Duhem and conventionalism, which we have been discussing? We can begin to build a bridge by observing that the meanings given to sounds and inscriptions are determined purely by social convention. Indeed, the social conventions differ from one language to another. So if a sentence is true in virtue of the meanings of the words it contains (that is, is analytic), it is *a fortiori* true by convention. Thus if a law is analytic, it is true by convention. The converse may not hold, since it is conceivable that a law might be rendered true by a set of conventions which include not just linguistic conventions concerning the meanings of words but also perhaps conventions connected with measuring procedures.

Duhem used his thesis against the claim that a particular scientific law was true by convention. It is now obvious that exactly the same argument could be used against the claim that the law is analytic. Indeed, Quine does argue against the analytic/synthetic distinction along just these lines.[6]

But to carry his argument through, Quine makes a claim (the Quine thesis) which is much stronger than the Duhem thesis. The key difference between the two theses is clearly expressed by Vuillemin as follows: 'Duhem's thesis ("D-thesis") has a limited and special scope not covering the field of physiology, for Claude Bernard's experiments are explicitly acknowledged as crucial. Quine's thesis ("Q-thesis") embraces the whole body of our knowledge' (1979, p. 599).

Duhem does indeed place explicit limitations on the scope of his thesis. He writes: 'The Experimental Testing of a Theory Does Not Have the Same

Logical Simplicity in Physics as in Physiology' (1904–5, p. 180 [227]). He thinks that his thesis does not apply in physiology or in certain branches of chemistry, and defends it only for the hypotheses of physics. My own view is that Duhem is correct to limit the scope of his thesis, but wrong to identify its scope with that of a particular branch of science—namely, physics. There are in physics falsifiable laws—for example, Snell's law of refraction applied to glass—whereas physiology and chemistry no doubt contain hypotheses subject to the Duhem thesis. . . . For the moment, however, it is not of great importance where exact boundaries are drawn. The crucial point is that Duhem wanted to apply his thesis to some statements and not to others, whereas the Quine thesis is supposed to apply to any statement whatever.

This is closely connected with a second difference between the Duhem thesis and the Quine thesis. Duhem maintains that hypotheses in physics cannot be tested in isolation, but only as part of a group. However, his discussion makes clear that he places limits on the size of this 'group'. Quine, however, thinks that the group extends and ramifies until it includes the whole of human knowledge. Quine writes: 'The unit of empirical significance is the whole of science' (1951, p. 42 [265]); and again:

> The totality of our so-called knowledge or beliefs, from the most casual matters of geography and history to the profoundest laws of atomic physics or even of pure mathematics and logic, is a man-made fabric which impinges on experience only along the edges. Or, to change the figure, total science is like a field of force whose boundary conditions are experience. A conflict with experience at the periphery occasions readjustments in the interior of the field. . . . But the total field is so underdetermined by its boundary conditions, experience, that there is much latitude of choice as to what statements to reëvaluate in the light of any single contrary experience. No particular experiences are linked with any particular statements in the interior of the field, except indirectly through considerations of equilibrium affecting the field as a whole. (pp. 42–3 [265–66])

The Quine thesis is stronger than the Duhem thesis, and, in my view, less plausible. Let us take, as a concrete example, one of the cases analysed earlier. Newton's first law cannot, taken in isolation, be compared with experience. Adams and Leverrier, however, used this law as one of a group of hypotheses from which they deduced conclusions about the orbit of Uranus. These conclusions disagreed with observation. Now the group of hypotheses used by Adams and Leverrier was, no doubt, fairly extensive, but it did not include the whole of science. Adams and Leverrier did not, for example, mention the assumption that bees collect nectar from flowers in order to make honey, although such an assumption might well have appeared in a contemporary scientific treatise dealing with a question in biology. We agree, then, with Quine that a single statement may not always be (to use his terminology) a 'unit of empirical significance'. But this does not mean that

'The unit of empirical significance is the whole of science' (1951, p. 42 [265]). A group of statements which falls considerably short of the whole of science may sometimes be a perfectly valid unit of empirical significance.

Another difference between Duhem and Quine is that Quine does not have a theory of scientific good sense. Let us take, for example, Quine's statement: 'Any statement can be held true come what may, if we make drastic enough adjustments elsewhere in the system' (p. 43 [266]). It is easy to imagine how Duhem would have reacted to such an assertion when applied to a statement falling under his thesis. Duhem would have agreed that, from the point of view of *pure logic*, one can indeed hold a particular statement—for example, Newton's particle theory of light—to be true, come what may. However, someone who did so in certain evidential situations would be *lacking in good sense*, and indeed *perfectly unreasonable*.

Because Quine does not have a theory of good sense, he cannot give the Duhemian analysis which we have just sketched. Indeed, it is significant that his 1951 article, 'Two Dogmas of Empiricism', is reprinted in a collection entitled *From a Logical Point of View*. Where Quine does go beyond logic, it is towards pragmatism, though Quine's pragmatism is usually mentioned only in passing, rather than elaborated, as in the following passage: 'Each man is given a scientific heritage plus a continuing barrage of sensory stimulation, and the considerations which guide him in warping his scientific heritage to fit his continuing sensory promptings are, where rational, pragmatic' (p. 46 [268]).

Although the Duhem thesis is quite clearly distinct from the Quine thesis, it might still be possible—indeed, useful—to form a composite thesis containing some, but not all, elements from each of the two theses. The phrase *Duhem-Quine thesis* could then be validly used to denote this composite thesis. In the last section of this chapter, I will elaborate a suggestion along these lines.

4 | The Duhem-Quine Thesis

Let us say that the *holistic thesis* applies to a particular hypothesis if that hypothesis cannot be refuted by observation and experiment when taken in isolation, but only when it forms part of a theoretical group. The differences between the Duhem and Quine theses concern the range of hypotheses to which the holistic thesis is applied and the extent of the 'theoretical group' for a hypothesis to which the holistic thesis does apply. In discussing these differences, I have so far sided with Duhem against Quine. There is one point, however, on which I would like to defend Quine against Duhem Quine, as we have seen, extends the holistic thesis to mathematics and logic. Duhem, however, thought that mathematics and logic had a character quite different from that of physics. Crowe (1990) gives an excellent general account and critique of Duhem's views on the history and philosophy of mathematics. I will here confine myself to a brief account of some views concerning geometry and logic which Duhem expounded in his late work *La Science allemande* ('German Science').

Duhem begins his treatment of geometry with the following remarks:

> Among the sciences of reasoning, arithmetic and geometry are the most simple and, consequently, the most completely finished; . . .
> What is the source of their axioms? They are taken, it is usually said, from common sense knowledge *(connaissance commune)*: that is to say that any man sane of mind is sure of their truth before having studied the science of which they will be the foundations. (1915, pp. 4–5)

Duhem agrees with this point of view. In fact, he holds what in 1915 was a very old-fashioned opinion, that the axioms of Euclid are established as true by common-sense knowledge *(connaissance commune)* or common sense *(le sens commun)* or intuitive knowledge *(connaissance intuitive)*. A proposition from which Euclid's fifth postulate can be deduced is that, given a geometrical figure (say a triangle), there exists another geometrical figure similar to it but of a different size. Duhem argues that the intuitions of palaeolithic hunters of reindeer were sufficient to establish the truth of this proposition. As he says:

> One can represent a plane figure by drawing, or a solid figure by sculpture, and the image can resemble the model perfectly, even though they have different sizes. This is a truth which was in no way doubted, in palaeolithic times, by the hunters of reindeer on the banks of the Vézère. Now that figures can be similar without being equal, implies, as the geometric spirit demonstrates, the exact truth of Euclid's postulate. (pp. 115–16)

Naturally enough, this attitude to the foundations of geometry leads Duhem to criticize non-Euclidean geometry, and, in particular, Riemannian geometry. This is what he says:

> Riemann's doctrine is a *rigorous algebra*, for all the theorems which it formulates are very precisely deduced from its basic postulates; so it satisfies the geometric spirit. It is not a *true geometry*, for, in putting forward its postulates, it is not concerned that their corollaries should agree at every point with the judgements, drawn from experience, which constitute our intuitive knowledge of space; it is therefore repugnant to common sense. (p. 118)

It is perhaps no accident that the non-Euclidean geometer cited by Duhem (namely, Riemann) was a German; for, as already remarked, *La Science allemande*, written in 1915, was an example of the war literature of the time, designed to denigrate the enemy nationality. Duhem attacks German scientists by claiming that, while they possess the geometric spirit (*l'esprit géométrique*), their theories contradict common sense (*le sens commun*) or *l'esprit de finesse*, which is Duhem's new term for something like his old notion of good sense.

Given this general point of view, it is not surprising that we find Duhem condemning the theory of relativity. He speaks of 'the principle of relativity such as has been conceived by an Einstein, a Max Abraham, a Minkowski, a Laue' (p. 135). Forgetting the contributions of his own compatriot Poincaré, he denounces relativity as a typical aberration of the German mind. As he says:

> The fact that the principle of relativity confounds all the intuitions of common sense, does not arouse against it the mistrust of the German physicists— quite the contrary! To accept it is, by that very fact, to overturn all the doctrines where space, time, movement were treated, all the theories of mechanics and physics; such a devastation has nothing about it which can displease German thought; on the ground which it will have cleared of the ancient doctrines, the geometric spirit of the Germans will devote itself with a happy heart to rebuilding a whole new physics of which the principle of relativity will be the foundation. If this new physics, disdainful of common sense, runs counter to all that observation and experience have allowed to be constructed in the domain of celestial and terrestrial mechanics, the purely deductive method will only be more proud of the inflexible rigour with which it will have followed to the end the ruinous consequences of its postulate. (p. 136)

The development and acceptance of non-Euclidean geometry and relativity have rendered Duhem's attempt to found geometry on common sense untenable. It is surely now more reasonable to extend the holistic thesis from physics to geometry and to say that, in the face of recalcitrant observations, we have the option of altering postulates of geometry as well as postulates of physics. This is, after all, precisely what Einstein did when he devised his general theory of relativity.

The picture is the same when we turn from geometry to logic. . . . Duhem [claimed] that 'There is a general method of deduction; Aristotle has formulated its laws for all time (*pour toujours*)' (p. 58). Yet by 1915 the new logic of Frege, Peano, and Russell had clearly superseded Aristotelian logic. Moreover, Brouwer had criticized some of the standard logical laws, and suggested his alternative intuitionistic approach. Quine writes: 'Revision even of the logical law of the excluded middle has been proposed as a means of simplifying quantum mechanics' (1951, p. 43 [266]). Admittedly the new 'quantum logic' has not proved very successful in resolving the paradoxes of microphysics; but there is no reason in principle why a change of this kind should not prove efficacious in some scientific context. In artificial intelligence, non-standard logics (for example, non-monotonic logics) are being devised in order to model particular forms of intelligent reasoning, and this programme has met with some success. Thus it seems reasonable to extend the holistic thesis to include logic as well and to allow the possibility of altering logical laws as well as scientific laws to explain recalcitrant observations.

I am now in a position to formulate what I will call the *Duhem-Quine thesis*, which combines what seem to me the best aspects of the Duhem thesis and the Quine thesis. It will be convenient to divide the statement in two parts.

A The holistic thesis applies to any high-level (level 2) theoretical hypotheses, whether of physics or of other sciences, or even of mathematics and logic. (A incorporates ideas from the Quine thesis.)

B The group of hypotheses under test in any given situation is in practice limited, and does not extend to the whole of human knowledge. Quine's claim that 'Any statement can be held to be true come what may, if we make drastic enough adjustments elsewhere in the system' (1951, p. 43 [266]) is true from a purely logical point of view; but scientific good sense concludes in many situations that it would be perfectly unreasonable to hold on to particular statements. (B obviously follows the Duhem thesis rather than the Quine thesis.)

. . . The thesis seems to me to be both true and important. . . .

■ | Notes

1. 1 degree = 60', and 1' = 60". So 52' is slightly less than a degree.

2. Einstein may have been influenced by Duhem, however, as is suggested by Howard in his interesting 1990 article. Howard shows that Einstein was on very friendly terms with Friedrich Adler, who prepared the first German translation of the *Aim and Structure of Physical Theory*, which appeared in 1908. From the autumn of 1909, Einstein and his wife rented an apartment in Zurich just immediately upstairs from the Adlers, and Einstein and Adler would meet frequently to discuss philosophy and physics. So probably Einstein had read *Aim and Structure* by the end of 1909 at the latest.

3. Or rather, misquotes. Duhem writes: 'raisons que la raison ne connaît pas' (1904–5, French edn, p. 330), whereas Pascal's original *pensée* was 'Le coeur a ses raisons que la raison ne connaît point.' Giving quotations which are slightly wrong is often a sign of great familiarity with a particular author.

4. Vuillemin (1979) and Ariew (1984) give valuable discussions of the differences between the Duhem thesis and the Quine thesis. I found these articles very helpful when forming my own views on the subject.

5. Quine's views have altered over the years, but here we will discuss only the position found in his 1951 article.

6. It is possible, however, to use arguments not involving the Quine thesis against the analytic/synthetic distinction. I give two such arguments against the distinction, the argument from justification and the argument from truth, in my 1985 article.

■ | References

Ariew, R. 1984. "The Duhem Thesis," *British Journal for the Philosophy of Science* 35: 313–25.

Crowe, M. J. 1990. "Duhem and the History and Philosophy of Mathematics," *Synthese* 83: 431–47.

Duhem, P. 1904–5. *The Aim and Structure of Physical Theory.* English translation by Philip P. Wiener of the 2nd French edn. of 1914, Atheneum, 1962. French edn., Vrin, 1989.

Duhem, P. 1915. *La Science allemande.* A. Hermann et Fils.

Gillies, D. A. 1985. "The Analytic/Synthetic Problem," *Ratio* 27: 149–59.

Howard, D. 1990. "Einstein and Duhem," *Synthese* 83: 363–84.

Martin, R. N. D. 1991. *Pierre Duhem. Philosophy and History in the Work of a Believing Physicist.* Open Court.

Poincaré, H. 1902. *Science and Hypothesis.* English translation, Dover, 1952. French edn., Flammarion, 1968.

Poincaré, H. 1904. "L'État actuel et l'avenir de la physique mathématique." Lecture delivered 24 Sept. 1904 to the International Congress of Arts and Science, St. Louis, Missouri, and published in *Bulletin des sciences mathématiques* 28: 302–24. Reprinted in Poincaré, 1905, 91–111.

Poincaré, H. 1905. *The Value of Science.* English translation, Dover, 1958.

Quine, W. V. O. 1951. "Two Dogmas of Empiricism." Reprinted in *From a Logical Point of View,* 2nd rev. edn., Harper Torchbooks, 1961, 20–46.

Vuillemin, J. 1968. *Préface* to H. Poincaré, *La Science et l'hypothèse,* Flammarion, 1968, 7–19.

Vuillemin, J. 1979. "On Duhem's and Quine's Theses," *Grazer Philosophische Studien* 9: 69–96. Quotations are from reprint in *The Philosophy of W. V. Quine,* ed. L. E. Hahn and P. A. Schilpp, Library of Living Philosophers (La Salle, Ill.: Open Court, 1986), 595–618.

LARRY LAUDAN

Demystifying Underdetermination

Pure logic is not the only rule for our judgments; certain opinions which do not fall under the hammer of the principle of contradiction are in any case perfectly unreasonable.
—Pierre Duhem[1]

■ | Introduction

This essay begins with some good sense from Pierre Duhem. The piece can be described as a defense of this particular Duhemian thesis against a rather more familiar doctrine to which Duhem's name has often been attached. To put it in a nutshell, I shall be seeking to show that the doctrine of underdetermination, and the assaults on methodology that have been mounted in its name, founder precisely because they suppose that the logically possible and the reasonable are coextensive. Specifically, they rest on the assumption that, unless we can show that a scientific hypothesis cannot possibly be reconciled with the evidence, then we have no epistemic grounds for faulting those who espouse that hypothesis. Stated so baldly, this appears to be an absurd claim. That in itself is hardly decisive, since many philosophical (and scientific) theses smack initially of the absurd. But, as I shall show below in some detail, the surface implausibility of this doctrine gives way on further analysis to the conviction that it is even more untoward and ill argued than it initially appears. And what compounds the crime is that precisely this thesis is presupposed by many of the fashionable epistemologies of science of the last quarter century. Before this complex indictment can be made plausible, however, there is a larger story that has to be told.

There is abroad in the land a growing suspicion about the viability of scientific methodology. Polanyi, Wittgenstein, Feyerabend and a host of

FROM C. Wade Savage, ed., *Scientific Theories*, vol. 14, *Minnesota Studies in the Philosophy of Science* (Minneapolis: University of Minnesota Press, 1990), 267–97.

others have doubted, occasionally even denied, that science is or should be a rule-governed activity. Others, while granting that there are rules of the 'game' of science, doubt that those rules do much to delimit choice (e.g., Quine, Kuhn). Much of the present uneasiness about the viability of methodology and normative epistemology can be traced to a series of arguments arising out of what is usually called "the underdetermination of theories." Indeed, on the strength of one or another variant of the thesis of underdetermination, a motley coalition of philosophers and sociologists has drawn some dire morals for the epistemological enterprise.

Consider a few of the better-known examples: Quine has claimed that theories are so radically underdetermined by the data that a scientist can, if he wishes, hold on to *any* theory he likes, "come what may." Lakatos and Feyerabend have taken the underdetermination of theories to justify the claim that the only difference between empirically successful and empirically unsuccessful theories lay in the talents and resources of their respective advocates (i.e., with sufficient ingenuity, more or less *any* theory can be made to look methodologically respectable).[2] Boyd and Newton-Smith suggest that underdetermination poses several prima facie challenges to scientific realism.[3] Hesse and Bloor have claimed that underdetermination shows the *necessity* for bringing noncognitive, social factors into play in explaining the theory choices of scientists (on the grounds that methodological and evidential considerations alone are demonstrably insufficient to account for such choices).[4] H. M. Collins, and several of his fellow sociologists of knowledge, have asserted that underdetermination lends credence to the view that the world does little if anything to shape or constrain our beliefs about it.[5] Further afield, literary theorists like Derrida have utilized underdetermination as one part of the rationale for "deconstructionism" (in brief, the thesis that, since every text lends itself to a variety of interpretations and thus since texts underdetermine choice among those interpretations, texts have no determinate meaning).[6] This litany of invocations of underdeterminationist assumptions could be expanded almost indefinitely; but that is hardly called for, since it has become a familiar feature of contemporary intellectual discourse to endow underdetermination with a deep significance for our understanding of the limitations of methodology, and thus with broad ramifications for all our claims to knowledge—insofar as the latter are alleged to be grounded in trustworthy procedures of inquiry. In fact, underdetermination forms the central weapon in the relativistic assault on epistemology.

As my title suggests, I think that this issue has been overplayed. Sloppy formulations of the thesis of underdetermination have encouraged authors to use it—sometimes inadvertently, sometimes willfully—to support whatever relativist conclusions they fancy. Moreover, a failure to distinguish several distinct species of underdetermination—some probably viable, others decidedly not—has encouraged writers to lump together situations that ought to be sharply distinguished. Above all, inferences have been drawn from the fact of underdetermination that by no means follow from it. Because all that is so, we need to get as clear as we can about this slippery concept before we can decide whether underdetermination warrants the critiques of methodology that have been mounted in its name. That is the object of the next section of this paper. With those clarifications in hand, I will then turn in succeeding parts to assess some recent garden-variety claims about the methodological and epistemic significance of underdetermination.

Although this paper is one of a series whose larger target is epistemic relativism in general[7], my limited aim here is not to refute relativism in all its forms. It is rather to show that one important line of argument beloved of relativists, the argument from underdetermination, will not sustain the global conclusions that they claim to derive from it.

■ | Vintage Versions of Underdetermination

Humean Underdetermination

Although claims about underdetermination have been made for almost every aspect of science, those that interest philosophers most have to do specifically with claims about the underdetermination of *theories*. I shall use the term "theory" merely to refer to any set of *universal statements* that purport to describe the natural world.[8] Moreover, so as not to make the underdeterminationists' case any harder to make out than it already is, I shall—for purposes of this essay—suppose, with them, that single theories by themselves make no directly testable assertions. More or less everyone, relativist or non-relativist, agrees that "theories are underdetermined" in some sense or other; but the seeming agreement about that formula disguises a dangerously wide variety of different meanings.

Our first step in trying to make some sense of the huge literature on underdetermination comes with the realization that there are two quite distinct families of theses, both of which are passed off as "the" thesis of underdetermination. Within each of these "families," there are still further differentiating features. The generic and specific differences between these versions, as we shall see shortly, are not minor or esoteric. They assert different things; they presuppose different things; the arguments that lead to and from them are quite different. Nonetheless each has been characterized, and often, as "*the* doctrine of underdetermination."

The first of the two generic types of underdetermination is what I shall call, for obvious reasons, deductive or *Humean underdetermination* (HUD). It amounts to one variant or other of the following claim:

HUD   For any finite body of evidence, there are indefinitely many mutually contrary theories, each of which logically entails that evidence.

The arguments for HUD are sufficiently familiar and sufficiently trivial that they need no rehearsal here.* HUD shows that the fallacy of affirming the consequent is indeed a deductive fallacy (like so many other interesting patterns of inference in science); that the method of hypothesis is not logically probative; that successfully "saving the phenomena" is not a robust warrant for detachment or belief. I have no quarrels with either HUD or with the familiar arguments that can be marshaled for it. But when duly considered, HUD turns out to be an extraordinarily *weak* thesis about scientific inference, one that will scarcely sustain any of the grandiose claims that have been made on behalf of underdetermination.

Specifically, HUD is weak in two key respects: First, it addresses itself only to the role of *deductive logic* in scientific inference; it is wholly silent about whether the rules of a broader ampliative logic underdetermine theory choice. Secondly, HUD provides no motivation for the claim that *all* theories are reconcilable with any given body of evidence; it asserts rather that indefinitely many theories are so. Put differently, even if our doxastic policies were so lax that they permitted us to accept as rational any belief that logically entailed the evidence, HUD would not sanction the claim (which we might call the *"thesis of cognitive egalitarianism"*) that all rival theories are thereby equally belief-worthy or equally rational to accept.

Despite these crucial and sometimes overlooked limitations of its scope, HUD still has some important lessons for us. For instance, HUD makes clear that theories cannot be "deduced from the phenomena" (in the literal, non-Newtonian sense of that phrase). It thus establishes that the resources of deductive logic are insufficient, no matter how extensive the evidence, to enable one to determine for certain that any theory is true. But for anyone comfortable with the nowadays familiar mixture of (a) fallibilism about knowledge and (b) the belief that ampliative inference depends on modes of argument that go beyond deductive logic, none of that is either very surprising or very troubling.

As already noted, HUD manifestly does *not* establish that all theories are equally good or equally well supported, or that falsifications are inconclusive or that any theory can be held on to, come what may. Nor, finally, does it suggest, let alone entail, that the methodological enterprise is hopelessly flawed because methodological rules radically underdetermine theory selection. Indeed, consistently with HUD, one could hold (although I shall not) that the ampliative rules of scientific method fully determine theory choice. HUD says nothing whatever about whether ampliative rules of theory appraisal do or do not determine theory choice uniquely. What HUD teaches, and all that it licenses, is that if one is prepared to accept only those theories that can be proven to be true, then one is going to have a drastically limited doxastic repertoire.

---

* For a rehearsal of some trivial arguments for HUD, see the section "Humean Underdetermination" in the commentary on chapter 3.

Mindful of the some of the dire consequences (enumerated above) that several authors have drawn from the thesis of underdetermination, one is inclined to invoke minimal charity by saying that Humean underdetermination must not be quite what they have in mind. And I think we have independent evidence that they do not. I have dwelt on this weak form of underdetermination to start with because, as I shall try to show below, it is the only *general* form of underdetermination that has been incontrovertibly established. Typically, however, advocates of underdetermination have a much stronger thesis in mind. Interestingly, when attacked, they often fall back on the truism of HUD; a safe strategy since HUD is unexceptionable. They generally fail to point out that HUD will support none of the conclusions that they wish to draw from underdetermination. By failing to distinguish between HUD and stronger (and more controversial) forms of underdetermination, advocates of undifferentiated underdetermination thus piggyback their stronger claims on this weaker one. But more of that below.

The Quinean Reformulations of Underdetermination[9]

Like most philosophers, Quine of course accepts the soundness of HUD. But where HUD was silent on the key question of ampliative underdetermination, Quine (along with several other philosophers) was quick to take up the slack. In particular, Quine has propounded two distinct doctrines, both of which have direct bearing on the issues before us. The first, and weaker, of these doctrines I shall call *the nonuniqueness thesis*. It holds that: *for any theory, T, and any given body of evidence supporting T, there is at least one rival (i.e. contrary) to T that is as well supported as T.*[10] In his more ambitious (and more influential) moments, Quine is committed to a much stronger position, which I call *the egalitarian thesis*. It insists that: *every theory is as well supported by the evidence as any of its rivals.*[11] Quine nowhere explicitly expresses the egalitarian thesis in precisely this form. But it will be the burden of the following analysis to show that Quine's numerous pronouncements on the retainability of theories, in the face of virtually any evidence, presuppose the egalitarian thesis, and make no sense without it. What follows is not meant to be an exegesis of Quine's intentions; it is meant, rather, as an exploration of whether Quine's position on this issue will sustain the broad implications that many writers (sometimes including Quine himself) draw from it.

What distinguishes both the nonuniqueness thesis and the egalitarian thesis from HUD is that they concern ampliative rather than deductive underdetermination; that is, they centrally involve the notion of "empirical support," which is after all the central focus of ampliative inference. In this section and the first part of the next, I shall focus on Quine's discussion of these two forms of ampliative underdetermination (especially the egalitarian thesis), and explore some of their implications. The egalitarian thesis is sufficiently extreme—not to say epistemically pernicious—that I want to take some time showing that some versions of Quine's holism are indeed committed to it. I shall thus examine its status in considerable detail before turning in later sections to look at some other prominent accounts of ampliative underdetermination.

Everyone knows that Quine, in his "Two Dogmas of Empiricism," maintained that:

(0)   one may hold onto any theory whatever in the face of any evidence whatever.[12]

Crucial here is the sense of "may" involved in this extraordinary claim. If taken as asserting that human beings are psychologically capable of retaining beliefs in the face of overwhelming evidence against them, then it is a wholly uninteresting truism, borne out by every chapter in the saga of human folly. But if Quine's claim is to have any bite, or any philosophical interest, it must be glossed along roughly the following lines:

(1)   It is rational to hold onto any theory whatever in the face of any evidence whatever.

I suggest this gloss because I suppose that Quine means to be telling us something about scientific rationality; and it is clear that (0), construed descriptively, has *no* implications for normative epistemology. Combined with Quine's counterpart claim that one is also free to jettison any theory one is minded to, (1) appears to assert the *equirationality* of all rival theoretical systems. Now, what grounds does Quine have for asserting (1)? One might expect that he could establish the plausibility of (1) only in virtue of examining the relevant rules of rational theory choice and showing, if it could be shown, that those rules were always so ambiguous that, confronted with any pair of theories and any body of evidence, they could never yield a decision procedure for making a choice. Such a proof, if forthcoming, would immediately undercut virtually every theory of empirical or scientific rationality. But Quine *nowhere*, neither in "Two Dogmas . . ." nor elsewhere, engages in a general examination of ampliative rules of theory choice.

His specific aim in propounding (0) or (1) is often said to be to exhibit the ambiguity of falsification or of *modus tollens*. The usual reading of Quine here is that he has shown the impotence of negative instances to disprove a theory, just as Hume had earlier showed the impotence of positive instances to prove a theory. Indeed, it is this gloss that establishes the parallel between Quine's form of the thesis of underdetermination and HUD. Between them, they seem to lay to rest any prospect for a purely deductive logic of scientific inference.

But what is the status of (1)? I have already said that Quine nowhere engages in an exhaustive examination of various rules of rational theory choice with a view to showing them impotent to make a choice between all pairs of theories. Instead, he is content to examine a *single* rule of theory choice, what we might call the Popperian gambit. That rule says, in effect, "reject theories that have (known) falsifying instances." Quine's strategy is to show that this particular rule radically underdetermines theory choice. I intend to spend the bulk of this section examining Quine's case for the claim that this particular rule underdetermines theory choice. But the reader should bear in mind that even if Quine were successful in his dissection of this particular rule (which he is not), that would still leave unsettled the question whether other ampliative rules of detachment suffer a similar fate.*

How does he go about exhibiting the underdeterminative character of falsification? Well, Quine's explicit arguments for (1) in "Two Dogmas . . ." are decidedly curious. Confronted, for instance, with an apparent refutation of a claim that "there are brick houses on Elm Street," we can—he says— change the meaning of the terms so that (say) "Elm Street" now refers to Oak Street, which adventitiously happens to have brick houses on it, thereby avoiding the force of the apparent refutation. Now this is surely a Pickwick-ian sense of "holding onto a theory come what may," since what we are holding onto here is not what the theory asserted, but the (redefined) string of words constituting the theory.[13] Alternatively, says Quine, we can always change the laws of logic if need be. We might, one supposes, abandon *modus tollens*, thus enabling us to maintain a theory in the face of evidence that, under a former logical regime, was falsifying of it; or we could jettison *modus ponens* and thereby preclude the possibility that the theory we are concerned to save is "implicated" in any schema of inference leading to the awkward prediction. If one is loath to abandon such useful logical devices (and Quine is), other resources are open to us. We could, says Quine, dismiss the threatening evidence "by pleading hallucination."[14]

But are there no constraints on when it is reasonable to abandon selected rules of logic or when to label evidence specious (because the result of hallucination) or when to redefine the terms of our theories? Of course, it is (for all I know) humanly possible to resort to any of these stratagems, as a descriptivist reading of (0) might suggest. But nothing Quine has said thus far gives us any grounds to believe, as (1) asserts, that it will ever, let alone *always*, be rational to do so. Yet his version of the thesis of underdetermination, if he means it to have any implications for normative epistemology, requires him to hold that it is rational to use some such devices.[15] Hence he would appear to be committed to the view that epistemic rationality gives us no grounds for avoiding such maneuvers. (On Quine's view, the only considerations that we could possibly invoke to block such stratagems have to do with pragmatic, not epistemic, rationality.[16]) Thus far, the argument for ampliative underdetermination seems made of pretty trifling stuff.

---

* In inductive logic, rules of detachment are often called *acceptance rules*. They specify when it is permissible to accept a hypothesis as true, thus "detaching" the hypothesis *h* from the assertion $P(h/e) = r$, that *h* has probability *r* on evidence *e*.

But there is a fourth, and decidedly nontrivial, stratagem that Quine envisages for showing how our Popperian principle underdetermines theory choice. This is the one that has received virtually all the exegetical attention; quite rightly too, since Quine's arguments on the other three are transparently question begging because they fail to establish the rationality of holding onto any theory in the face of any evidence. Specifically, Quine proposes that a threatened statement or theory can always be immunized from the threat of the recalcitrant evidence by making suitable adjustments in our auxiliary theories. It is here that the familiar "Duhem-Quine thesis" comes to the fore. What confronts experience in any test, according to both Quine and Duhem, is an entire theoretical structure (later dubbed by Quine "a web of belief") consisting inter alia [among other things] of a variety of theories. Predictions, they claim, can never be derived from single theories but only from collectives consisting of multiple theories, statements of initial and boundary conditions, assumptions about instrumentation, and the like. Since (they claim) it is whole systems and whole systems alone that make predictions, when those predictions go awry it is theory complexes, not individual theories, that are indicted via *modus tollens*. But, so the argument continues, we cannot via *modus tollens* deduce the falsity of any component of a complex from the falsity of the complex as a whole. Quine put it this way:

> But the failure [of a prediction] falsifies only a block of theory as a whole, a conjunction of many statements. The failure shows that one or more of those statements is false, but it does not show which.[17]

Systems, complexes or "webs" apparently turn out to be unambiguously falsifiable on Quine's view; but the choice between individual theories or statements making up these systems is, in his view, radically underdetermined.

Obviously, this approach is rather more interesting than Quine's other techniques for saving threatened theories, for here we need not abandon logic, redefine the terms in our theories in patently ad hoc fashion, nor plead hallucinations. The thesis of underdetermination in this particular guise, which I shall call Quinean underdetermination (QUD), can be formulated as follows:

QUD   Any theory can be reconciled with any recalcitrant evidence by making suitable adjustments in our other assumptions about nature.

Before we comment on the credentials of QUD, we need to further disambiguate it. We especially need to focus on the troublesome phrase "can be reconciled with." On a weak interpretation, this would be glossed as "can be made logically compatible with the formerly recalcitrant evidence." I shall call this the *"compatibilist version of QUD."* On a stronger interpretation, it might be glossed as "can be made to function significantly in a complex that entails" the previously threatening evidence. Let us call this the *"entailment version of QUD."* To repeat, the compatibilist version says that any theory can be made *logically compatible* with any formerly threatening evidential report; the entailment interpretation insists further that any theory can be made to function essentially in a *logical derivation* of the erstwhile refuting instance.

The compatibilist version of QUD can be trivially proven. All we need do, given any web of belief and a suspect theory that is part of it, is to remove *(without replacement)* any of those ancillary statements within the web needed to derive the recalcitrant prediction from the theory. Of course, we may well lose enormous explanatory power thereby, and the web may lose much of its pragmatic utility thereby, but there is nothing in deductive logic that would preclude any of that.

The entailment version of QUD, by contrast, insists that there is always a set of auxiliary assumptions that can replace others formerly present, and that will allow the *derivation*, not of the wrongly predicted result, but of precisely what we have observed. As Grünbaum, Quinn, Laudan and others have shown,[18] neither Quine nor anyone else has ever produced a general existence proof concerning the availability either in principle or in practice of suitable (i.e., nontrivial) theory-saving auxiliaries. Hence the entailment version of QUD is without apparent warrant. For a time (circa 1962), Quine himself conceded as much.[19] That is by now a familiar result. But what I think needs much greater emphasis than it has received is the fact that, *even if nontrivial auxiliaries existed that would satisfy the demands of the entailment version of QUD, no one has ever shown that it would be rational to prefer a web that included them and the threatened theory to a rival web that dispensed with the theory in question.* Indeed, as I shall show in detail, what undermines *both* versions of QUD is that neither logical *compatibility* with the evidence nor logical *derivability* of the evidence is sufficient to establish that a theory exhibiting such empirical compatibility and derivability is rationally acceptable.

It will prove helpful to distinguish four different positive relations in which a theory (or the system in which a theory is embedded) can stand to the evidence. Specifically, a theory (or larger system of which it is a part) may:

- be logically compatible with the evidence;
- logically entail the evidence;
- explain the evidence;
- be empirically supported by the evidence.

Arguably, none of these relations reduces to any of the others; despite that, Quine's analysis runs all four together. But what is especially important for our purposes is the realization that *satisfaction of either the compatibility relation or the entailment relation fails to establish either an explanatory relation or a relation of empirical support.* For instance, theories may entail statements that they nonetheless do not explain; self-entailment being the most obvious example. Equally, theories may entail evidence statements, yet not be empirically supported by them (e.g., if the theory was generated by the algorithmic manipulation of the "evidence" in question).

So, when QUD tells us that any theory can be "reconciled" with any bit of recalcitrant evidence, we are going to have to attend with some care to what that reconciliation consists in. Is Quine claiming, for instance, that any theory can—by suitable modifications elsewhere—continue to function as part of an *explanation* of a formerly recalcitrant fact? Or is he claiming, even more ambitiously, that any formerly recalcitrant instance for a theory can be transformed into a *confirming instance* for it?

As we have seen, the only form of QUD that has been firmly established is compatibilist Quinean underdetermination (an interpretation that says a theory can always be rendered logically compatible with any evidence, provided we are prepared to give up enough of our other beliefs); so I shall begin my discussion there. Saving a prized, but threatened, theory by abandoning the auxiliary assumptions once needed to link it with recalcitrant evidence clearly comes at a price. Assuming that we give up those beliefs without replacement (and recall that this is the only case that has been made plausible), we not only abandon an ability to say anything whatever about the phenomena that produced the recalcitrant experience; we also now give up the ability to explain all the other things which those now-rejected auxiliaries enabled us to give an account of—with no guarantee whatever that we can find alternatives to them that will match their explanatory scope.

But further and deeper troubles lurk for Quine just around the corner. For it is not just explanatory scope that is lost; it is also *evidential support.* Many of those phenomena that our web of belief could once give an account of (and which presumably provided part of the good reasons for accepting the web with its constituent theories) are now beyond the resources of the web to explain and predict. That is another way of saying that the revised web, stripped of those statements formerly linking the theory in question with the mistaken prediction, now has substantially less empirical support than it once did; assuming, of course, that the jettisoned statements formerly functioned to do more work for us than just producing the discredited prediction.[20] Which clearly takes things from bad to worse. For now Quine's claim about the salvageability of a threatened theory turns out to make sense just in case the only criterion of theory appraisal is logical compatibility with observation. If we are concerned with issues like explanatory scope or empirical support, Quine's QUD in its compatibilist version cuts no ice whatsoever.

Clearly, what is wrong with QUD, and why it fails to capture the spirit of (1), is that it has dropped out any reference to the *rationality* of theory choices, and specifically theory rejections. It doubtless is possible for us to jettison a whole load of auxiliaries in order to save a threatened theory (where "save" now means specifically "to make it logically compatible with the evidence"), but Quine nowhere establishes the reasonableness or the rationality of doing so. And if it is plausible, as I believe it is, to hold that scientists are aiming (among other things) at producing theories with broad explanatory scope and impressive empirical credentials, then it has to be said that Quine has given us no arguments to suppose that any theory we like can be doctored up so as to win high marks on those scores.

This point underscores the fact that too many of the discussions of underdetermination in the last quarter century have proceeded in an evaluative vacuum. They imagine that if a course of action is logically possible, then one need not attend to the question of its rationality. But if QUD is to carry any epistemic force, it needs to be formulated in terms of the rationality of preserving threatened theories. One might therefore suggest the following substitute for QUD (which was itself a clarification of (1)):

(2) any theory can be rationally retained in the face of any recalcitrant evidence.

Absent strong arguments for (2) or its functional equivalents, Quinean holism, the Duhem-Quine thesis and the (non-Humean) forms of underdetermination appear to pose no threat in principle for an account of scientific methodology or rationality. The key question is whether Quine, or any of the other influential advocates of the methodological significance of underdetermination, have such arguments to make.

Before we attempt to answer that question, a bit more clarification is called for, since the notion of retainment, let alone rational retainment, is still less than transparent. I propose that we understand that phrase to mean something along these lines: to say that a theory can be rationally retained is to say that reasons can be given for holding that theory, or the system of which it is a part, as true (or empirically adequate) that are (preferably stronger than but) as least as strong as the reasons that can be given for holding as true (or empirically adequate) any of its *known* rivals. Some would wish to give this phrase a more demanding gloss; they would want to insist that a theory can be rationally held only if we can show that the reasons in its behalf are stronger than those for all its *possible* rivals, both extant and those yet-to-be-conceived. That stronger gloss, which I shall resist subscribing to, would have the effect of making it even harder for Quine to establish (2) than my weaker interpretation does. Because I believe that theory choice is generally a matter of comparative choice among extant alternatives, I see no reason why we should saddle Quine and his followers with having to defend (2) on its logically stronger construal. More to the point, if I can show that the arguments on behalf of the weaker construal fail, that indeed the weaker construal is false, it follows that its stronger counterpart fails as well, since the stronger entails the weaker. I therefore propose emending (2) as follows:

(2*) any theory can be shown to be as well supported by any evidence as any of its known rivals.

Quine never formulates this thesis as such, but I have tried to show that defending a thesis of this sort is incumbent on anyone who holds, as Quine does, that any theory can be held true, come what may. Duly considered, (2*) is quite a remarkable thesis, entailing as it does that all the known contraries to every known theory are equally well supported. Moreover, (2*) is our old friend, the egalitarian thesis. If correct, (2*) entails (for instance) that the flat-earth hypothesis is as sound as the oblate-spheroid hypothesis,[21] that it is as reasonable to believe in fairies at the bottom of my garden as not. But, for all its counter-intuitiveness, this is precisely the doctrine to which authors like Quine, Kuhn, and Hesse are committed.[22] (In saying that Quine is committed to this position, I do not mean that he would avow it if put to him directly; I doubt that very much. My claim rather is (a), that Quine's argument in "Two Dogmas . . ." commits him to such a thesis, and (b), that those strong relativists who look to Quine as having espoused and established the egalitarian thesis are exactly half right. I prefer to leave it to Quine exegetes to decide whether the positions of the *later* Quine allow him to be exonerated of the charge that his more recent writing run afoul of the same problem.)

One looks in vain in "Two Dogmas . . ." for even the whiff of an argument that would make the egalitarian thesis plausible. As we have seen, Quine's only marginally relevant points there are his suppositions (1) that any theory can be made logically compatible with any evidence (statement) and (2) that any theory can function in a network of statements that will entail any particular evidence statement.[23] But what serious epistemologist has ever held either (a) that bare logical compatibility with the evidence constituted adequate reason to accept a scientific theory,[24] or (b) that logical entailment of the evidence by a theory constituted adequate grounds for accepting a theory? One might guess otherwise. One might imagine that some brash hypothetico-deductivist would say that any theory that logically entailed the known evidence was acceptable. If one conjoins this doctrine with Quine's claim (albeit one that Quine has never made out) that every theory can be made to logically entail any evidence, then one has the makings of the egalitarian thesis. But such musings cut little ice, since no serious twentieth-century methodologist has ever espoused, without crucial qualifications, logical compatibility with the evidence or logical derivability of the evidence as a sufficient condition for detachment of a theory.[25]

Consider some familiar theories of evidence to see that this is so. Within Popper's epistemology, two theories, T_1 and T_2, that thus far have the same positive instances, e, may nonetheless be differentially supported by e. For instance, if T_1 predicted e before e was determined to be true, whereas T_2 is produced after e is known, then e (according to Popper) constitutes a good test of T_1 but no test of T_2. Bayesians too insist that rival (but nonequivalent) theories sharing the same known positive instances are not necessarily equally well confirmed by those instances. Indeed, if two theories begin with different prior probabilities, then their posterior probabilities must be different, *given the same positive instances.*[26] But that is just to say that even if two theories enjoy precisely the same set of known confirming instances, *it does not follow that they should be regarded as equally well confirmed by those instances.* All of which is to say that showing that rival theories enjoy the same "empirical support"—in any sense of that term countenanced by (2^*)—requires more than that those rivals are compatible with, or capable of entailing, the same "supporting" evidence. (2^*) turns out centrally to be a claim in the theory of evidence and, since Quine does not address the evidence relation in "Two Dogmas . . . ," one will not find further clarification of this issue there.[27]

Of course, "Two Dogmas . . ." was not Quine's last effort to grapple with these issues. Some of these themes recur prominently in *Word and Object,* and it is worth examining some of Quine's arguments about underdetermination to be found there. In that work, Quine explicitly if briefly addresses the question, already implicit in "Two Dogmas . . . ," whether ampliative rules of theory choice underdetermine theory choice.[28] Quine begins his discussion there by making the relatively mild claim that scientific methodology, along with any imaginable body of evidence, *might possibly* underdetermine theory choice. As he wrote:

> *conceivably* the truths about molecules are only partially determined by any ideal organon of scientific method plus all the truths that can be said in common sense terms about ordinary things:[29]

Literally, the remark in this passage in unexceptionable. Since we do not yet know what the final "organon of scientific method" will look like, it surely is "conceivable" that the truth status of claims about molecular structure might be underdetermined by such an organon. Three sentences later, however, this claim about the conceivability of ampliative underdetermination becomes a more ambitious assertion about the *likelihood* of such underdetermination:

> The incompleteness of determination of molecular behavior by the behavior of ordinary things . . . remains true even if we include all past, present and future irritations of all the far-flung surfaces of mankind, and probably *even if we throw in [i.e., take for granted] an in fact achieved organon of scientific method besides.*[30]

As it stands, and as it remains in Quine's text, this is no argument at all, but a bare assertion. But it is one to which Quine returns still later:

we have no reason to suppose that man's surface irritations even unto eternity admit of any systematization that is scientifically better or simpler than all possible others. It seems *likelier*, if only on account of symmetries or dualities, that countless alternative theories would be tied for first place.[31]

Quite how Quine thinks he can justify this claim of "likelihood" for ampliative underdetermination is left opaque. Neither here nor elsewhere does he show that *any* specific ampliative rules of scientific method[32] actually underdetermine theory choice—let alone that the rules of a "final methodology" will similarly do so. Instead, on the strength of the notorious ambiguities of simplicity (and by some hand-waving assertions that other principles of method may "plausibly be subsumed under the demand for simplicity"[33]—a claim that is anything but plausible), Quine asserts "in principle," that there is "probably" no theory that can uniquely satisfy the "canons of any ideal organon of scientific method."[34] In sum, Quine fails to show that theory choice is ampliatively underdetermined even by *existing* codifications of scientific methodology (all of which go considerably beyond the principle of simplicity), let alone by all possible such codifications.[35]

More important for our purposes, even if Quine were right that no ideal organon of methodology could ever pick out any theory as uniquely satisfying its demands, we should note—in the version of underdetermination contained in the last passage from Quine—how drastically he has apparently weakened his claims from those of "Two Dogmas. . . ." That essay, you recall, had espoused the egalitarian thesis that *any* theory can be reconciled with any evidence. We noted how much stronger that thesis was than the nonuniqueness thesis to the effect that there will always be some rival theories reconcilable with any finite body of evidence. But in *Word and Object*, as the passages I have cited vividly illustrate, *Quine is no longer arguing that any theory can be reconciled with any evidence;*[36] he is maintaining rather that, no matter what our evidence and no matter what our rules of appraisal, there will always remain the possibility (or the likelihood) that the choice will not be uniquely determined. But that is simply to say that there will (probably) always be at least one contrary to any given theory that fits the data equally well—a far cry from the claim, associated with QUD and (2*), that *all* the contraries to a given theory will fit the data equally well. In a sense, therefore, Quine appears in *Word and Object* to have abandoned the egalitarian thesis for the nonuniqueness thesis, since the latter asserts not the epistemic equality of all theories but only the epistemic equality of certain theories.[37] That surmise aside, it is fair to say that *Word and Object* does nothing to further the case for Quine's egalitarian view that "any theory can be held true come what may."

Some terminological codification might be useful before we proceed, since we have reached a natural breaking point in the argument. As we have seen, one can distinguish between (a) *descriptive* (0) and (b) *normative* (1, 2, 2*) forms of underdetermination, depending upon whether one is making a claim about what people are capable of doing or what the rules of scientific rationality allow.[38] One can also distinguish between (c) *deductive* and (d) *ampliative* underdetermination, depending upon whether it is the rules of deductive logic (HUD) or of a broadly inductive logic or theory of rationality that are alleged to underdetermine choice (QUD). Further, we can distinguish between the claims that theories can be reconciled with recalcitrant evidence via establishing (e) *compatibility* between the two or (f) a one-way *entailment* between the theory and the recalcitrant evidence or (g) equivalence of support between rival theories. Finally, one can distinguish between (h) the doctrine that choice is underdetermined between at least one of the contraries of a theory and that theory (*nonuniqueness*) and (i) the doctrine that theory choice is underdetermined between every contrary of a theory and that theory ("cognitive *egalitarianism*").

Using this terminology, we can summarize such conclusions as we have reached to this point: In "Two Dogmas . . . ," Quine propounded a thesis of normative, ampliative, egalitarian underdetermination. Whether we construe that thesis in its compatibilist or entailment versions, it is clear that Quine has said nothing that makes plausible the idea that every prima facie refuted theory can be embedded in a rationally acceptable (i.e., empirically well-supported) network of beliefs. Moreover, "Two Dogmas . . ." developed an argument for underdetermination for only one rationality principle among many, what I have been calling the Popperian gambit. This left completely untouched the question whether other rules of theory choice suffered from the same defects that Quine thought Popper's did. Perhaps with a view to remedying that deficiency, Quine argued—or, rather, alleged without argument—in *Word and Object* that *any* codification of scientific method would underdetermine theory choice. Unfortunately, *Word and Object* nowhere delivers on its claim about underdetermination.

But suppose, just for a moment, that Quine had been able to show what he claimed in *Word and Object*, to wit, the nonuniqueness thesis. At best, that result would establish that for any well-confirmed theory, there is in principle at least one other theory that will be equally well-confirmed by the same evidence. That is an interesting thesis to be sure, and possibly a true one, although Quine has given us no reason to think so. (Shortly, we shall examine arguments of other authors that seem to provide some ammunition for this doctrine.) But even if true, the nonuniqueness thesis will not sustain the critiques of methodology that have been mounted in the name of underdetermination. Those critiques are all based, implicitly or explicitly, on the strong, egalitarian reading of underdetermination. They amount to saying that the project of developing a methodology of science is a waste of time since, no matter what rules of evidence we eventually produce, those rules will do nothing to delimit choice between rival theories. The charge that methodology is toothless pivots essentially on the viability of QUD in its ampliative, egalitarian version. Nonuniqueness versions of the thesis of ampliative underdetermination at best establish that methodology will not allow us to pick out a theory as uniquely true, no matter how strong its evidential support. (*Word and Object*'s weak ampliative thesis of underdetermination, even if sound, would provide no grounds for espousing the strong underdeterminationist thesis implied by the "any theory can be held come what may" dogma.[39])

Theory choice may or may not be ampliatively underdetermined in the sense of the nonuniqueness thesis; that is an open question. But however that issue is resolved, that form of underdetermination poses no challenge to the methodological enterprise. What would be threatening to, indeed debilitating for, the methodological enterprise is if QUD in its egalitarian version were once established. Even though Quine offers no persuasive arguments in favor of normative, egalitarian, ampliative underdetermination, there are several other philosophers who appear to have taken up the cudgels on behalf of precisely such a doctrine. It is time I turned to their arguments.

■ | Ampliative Underdetermination

With this preliminary spade work behind us, we are now in a position to see that the central question about underdetermination, at least so far as the philosophy of science is concerned, is the issue of ampliative underdetermination. Moreover, as we have seen, the threat to the epistemological project comes, not from the nonuniqueness version of underdetermination, but from the egalitarian version. (That version states that any theory can be embedded in a system that will be as strongly supported by the evidence as any rival is supported by the same evidence.) The question is whether anyone has stronger arguments than Quine's for the methodological underdetermination of theory choice. Two plausible contenders for that title are Nelson Goodman and Thomas Kuhn. I shall deal briefly with them in turn.

Goodman's *Fact, Fiction and Forecast* is notorious for posing a particularly vivid form of ampliative underdetermination, in the form of the grue/green, and related, paradoxes of induction.* Goodman is concerned there to deliver what Quine had elsewhere merely promised, namely, a proof that the inductive rules of scientific method underdetermine theory choice in the face of any conceivable evidence. The general structure of Goodman's argument is too familiar to need any summary here. But it is important to characterize carefully what Goodman's result shows. I shall do so utilizing terminology we have already been working with. Goodman shows that one specific rule of ampliative inference (actually a whole family of rules bearing structural similarities to the straight rule of induction) suffers from this defect: Given any pair (or n-tuple) of properties that have previously always

---

* See Nelson Goodman, "The New Riddle of Induction," and the associated commentary in chapter 4.

occurred together in our experience, it is possible to construct an indefinitely large variety of contrary theories, all of which are compatible with the inductive rule: "If, for a large body of instances, the ratio of the successful instances of a hypothesis is very high compared to its failures, then assume that the hypothesis will continue to enjoy high success in the future." All these contraries will (along with suitable initial conditions) entail all the relevant past observations of the pairings of the properties in question. Thus, in one of Goodman's best-known examples, the straight rule will not yield an algorithm for choosing between "All emeralds are green" and "All emeralds are grue"; it awards them equally good marks.

There is some monumental question begging going on in Goodman's setting up of his examples. He supposes without argument that—since the contrary inductive extrapolations all have the same positive instances (to date)—the inductive logician must assume that the extrapolations from each of these hypotheses are all rendered equally likely by those instances. Yet we have already had occasion to remark that "possessing the same positive instances" and "being equally well confirmed" boil down to the same thing only in the logician's never-never land. (It was Whewell, Peirce and Popper who taught us all that theories sharing the same positive instances need not be regarded as equally well tested or equally belief-worthy.) But Goodman does have a point when he directs our attention to the fact that the straight rule of induction, as often stated, offers no grounds for distinguishing between the kind of empirical support enjoyed by the green hypothesis and that garnered by the grue hypothesis.

Goodman himself believes, of course, that this paradox of induction can be overcome by an account of the entrenchment of predicates. Regardless whether one accepts Goodman's approach to that issue, it should be said that strictly he does not hold that theory choice is underdetermined; on his view, such ampliative underdetermination obtains only if we limit our organon of scientific methodology to some version of the straight rule of induction.

But, for purposes of this paper, we can ignore the finer nuances of Goodman's argument since, even if a theory of entrenchment offered no way out of the paradox, and even if the slide from "possessing the same positive instances" to "being equally well confirmed" was greased by some plausible arguments, Goodman's arguments can provide scant comfort to the relativist's general repudiation of methodology. Recall that the relativist is committed, as we have seen, to arguing an egalitarian version of the thesis of ampliative underdetermination, i.e., he must show that all rival theories are equally well supported by any conceivable evidence. But there is nothing whatever in Goodman's analysis—even if we grant *all* its controversial premises—that could possibly sustain such an egalitarian conclusion. Goodman's argument, after all, does not even claim to show apropos of the straight rule that it will provide support for any and every hypothesis; his concern, rather, is to show that there will always be a family of contrary hypotheses between which it will provide no grounds for rational choice. The difference is crucial. If I propound the hypothesis that "All emeralds are red" and if my evidence base happens to be that all previously examined emeralds are green, then the straight rule is unambiguous in its insistence that my hypothesis be rejected. The alleged inability of the straight rule to distinguish between green- and grue-style hypotheses provides no ammunition for the claim that such a rule can make no epistemic distinctions whatever between rival hypotheses. If we are confronted with a choice between (say) the hypotheses that all emeralds are red and that all are green, then the straight rule gives us entirely unambiguous advice concerning which is better supported by the relevant evidence. Goodmanian underdetermination is thus of the nonuniqueness sort. When one combines that with a recognition that Goodman has examined but one among a wide variety of ampliative principles that arguably play a role in scientific decision making, it becomes clear that no global conclusions whatever can be drawn from Goodman's analysis concerning the general inability of the rules of scientific methodology for strongly delimiting theory choice.

But we do not have to look very far afield to find someone who does propound a strong (viz., egalitarian) thesis of ampliative underdetermination, one which, if sound, would imply that the rules of methodology were never adequate to enable one to choose between any rival theories, regardless of the relevant evidence. I refer, of course, to Thomas Kuhn's assertion in *The Essential Tension* to the effect that the shared rules and standards of the scientific community *always* underdetermine theory choice.[40] Kuhn there argues that science is guided by the use of several methods (or, as he prefers to call them, "standards"). These include the demand for empirical adequacy, consistency, simplicity, and the like. What Kuhn says about these standards is quite remarkable. He is not making the point that the later Quine and Goodman made about the methods of science; namely, that for any theory picked out by those methods, there will be indefinitely many contraries to it that are equally compatible with the standards. On the contrary, Kuhn is explicitly pushing the same line that the early Quine was implicitly committed to, viz., that the methods of science are inadequate ever to indicate that any theory is better than any rival, regardless of the available evidence. In the language of this essay, it is the egalitarian form of underdetermination that Kuhn is here proposing.

Kuhn, of course, does not use that language, but a brief rehearsal of Kuhn's general scheme will show that egalitarian underdetermination is one of its central underpinnings. Kuhn believes that there are divergent paradigms within the scientific community. Each paradigm comes to be associated with a particular set of practices and beliefs. Once a theory has been accepted within an ongoing scientific practice, Kuhn tells us, there is nothing that the shared standards of science can do to dislodge it. If paradigms do change, and Kuhn certainly believes that they do, this must be the result of "individual" and "subjective" decisions by individual researchers, not because there is anything about the methods or standards scientists share that ever requires the abandonment of those paradigms and their associated theories. In a different vein, Kuhn tells us that a paradigm always looks good by its own standards and weak by the standards of its rivals and that there never comes a point at which adherence to an old paradigm or resistance to a new one ever becomes "unscientific."[41] In effect, then, Kuhn is offering a paraphrase of the early Quine, but giving it a Wittgensteinean twist: "once a theory/paradigm has been established within a practice, it can be held on to, come what may." The shared standards of the scientific community are allegedly impotent ever to force the abandonment of a paradigm, and the specific standards associated with any paradigm will always give it the nod.

If this seems extreme, I should let Kuhn speak for himself. "[E]very individual choice between competing theories," he tells us, "depends on a mixture of objective and subjective factors, or of shared and individual criteria."[42] It is, in Kuhn's view, no accident that individual or subjective criteria are used alongside the objective or shared criteria, for the latter "are not by themselves sufficient to determine the decisions of individual scientists."[43] Each individual scientist "must complete the objective criteria [with 'subjective considerations'] before any computations can be done."[44] Kuhn is saying here that the shared methods or standards of scientific research are *always* insufficient to justify the choice of one theory over another.[45] That could only be so if (2*) or one of its functional equivalents were true of those shared methods.

What arguments does Kuhn muster for this egalitarian claim? Well, he asserts that all the standards that scientists use are ambiguous and that "individuals may legitimately differ about their application to concrete cases."[46] "Simplicity, scope, fruitfulness, and even accuracy can be judged quite differently . . . by different people."[47] He is surely right about some of this. Notoriously, one man's simplicity is another's complexity; one may think a new approach fruitful, while a second may see it as sterile. But such fuzziness of conception is precisely why most methodologists have avoided falling back on these hazy notions for talking about the empirical warrant for theories. Consider a different set of standards, one arguably more familiar to philosophers of science:

- prefer theories that are internally consistent;
- prefer theories that correctly make some predictions that are surprising given our background assumptions;
- prefer theories that have been tested against a diverse range of kinds of phenomena to those that have been tested only against very similar sorts of phenomena.

Even standards such as these have some fuzziness around the edges, but can anyone believe that, confronted with *any* pair of theories and *any* body

of evidence, these standards are so rough-hewn that they could be used indifferently to justify choosing either element of the pair? Do we really believe that Aristotle's physics correctly made the sorts of surprising predictions that Newton's physics did? Is there any doubt that Cartesian optics, with its dual insistence on the instantaneous propagation of light and that light traveled faster in denser media than in rarer ones, violated the canon of internal consistency?

Like the early Quine, Kuhn's wholesale holism commits him to the view that, consistently with the shared canons of rational acceptance, any theory or paradigm can be preserved in the face of any evidence. As it turns out, however, Kuhn no more has plausible arguments for this position than Quine had. In each case, the idea that the choice between changing or retaining a theory/paradigm is ultimately and always a matter of personal preference turns out to be an unargued dogma. In each case, if one takes away that dogma, much of the surrounding edifice collapses.

Of course, none of what I have said should be taken to deny that all forms of underdetermination are bogus. They manifestly are not. Indeed, there are several types of situations in which theory choice is indeed underdetermined by the relevant evidence and rules. Consider a few:

a) We can show that for some rules, and for certain theory pairs, theory choice is underdetermined for certain sorts of evidence. Consider the well-known case of the choice between the astronomical systems of Ptolemy and Copernicus. If the only sort of evidence available to us involves reports of line-of-sight positions of planetary position, and if our methodological rule is something like "Save the phenomena," then it is easy to prove that any line-of-sight observation that supports Copernican astronomy also supports Ptolemy's.[48] (It is crucial to add, of course, that if we consider other forms of evidence besides line-of-sight planetary position, this choice is not strongly underdetermined.)

b) We can show that for some rules and for some local situations, theory choice is underdetermined, regardless of the sorts of evidence available. Suppose our only rule of appraisal says, "Accept that theory with the largest set of confirming instances," and that we are confronted with two rival theories that have the same known confirming instances. Under these special circumstances, the choice is indeterminate.[49]

What is the significance of such limited forms of ampliative underdetermination as these? They represent interesting cases to be sure, but none of them—taken either singly or in combination—establishes the soundness of strong ampliative underdetermination as a general doctrine. Absent sound arguments for global egalitarian underdetermination (i.e., afflicting every theory on every body of evidence), the recent dismissals of scientific methodology turn out to be nothing more than hollow, anti-intellectual sloganeering.

I have thus far been concerned to show that the case for strong ampliative underdetermination has not been convincingly made out. But we can more directly challenge it by showing its falsity in specific concrete cases.

To show that it is ill conceived (as opposed to merely unproved), we need to exhibit a methodological rule, or set of rules, a body of evidence, and a local theory choice context in which the rules and the evidence would *unambiguously* determine the theory preference. At the formal level it is of course child's play to produce a trivial rule that will unambiguously choose between a pair of theories. (Consider the rule: "Always prefer the later theory.") But, unlike the underdeterminationists,[50] I would prefer real examples, so as not to take refuge behind contrived cases.

The history of science presents us with a plethora of such cases. But I shall refer to only one example in detail, since that is all that is required to make the case. It involves the testing of the Newtonian celestial mechanics by measurements of the "bulging" of the earth.[51] The Newtonian theory predicted that the rotation of the earth on its axis would cause a radical protrusion along the equator and a constriction at the poles—such that the earth's actual shape would be that of an oblate spheroid, rather than (as natural philosophers from Aristotle through Descartes had maintained) that of a uniform sphere or a sphere elongated along the polar axis. By the early eighteenth century, there were well-established geodesic techniques for ascertaining the shape and size of the earth (to which all parties agreed). These techniques involved the collection of precise measurements of distance from selected portions of the earth's surface. (To put it oversimply, these techniques generally involved comparing measurements of chordal segments of the earth's polar and equatorial circumferences.[52]) Advocates of the two major cosmogonies of the day, the Cartesians and the Newtonians, looked to such measurements as providing decisive evidence for choosing between the systems of Descartes and Newton.[53] At great expense, the Paris Académie des Sciences organized a series of elaborate expeditions to Peru and Lapland to collect the appropriate data. The evidence was assembled by scientists generally sympathetic to the Cartesian/Cassini hypothesis. Nonetheless, it was *their* interpretation, as well as everyone else's, that the evidence indicated that the diameter of the earth at its equator was significantly larger than along its polar axis. This result, in turn, was regarded as decisive evidence showing the superiority of Newtonian over Cartesian celestial mechanics. The operative methodological rule in the situation seems to have been something like this:

> when two rival theories, T_1 and T_2, make conflicting predictions that can be tested in a manner that presupposes neither T_1 nor T_2, then one should accept whichever theory makes the correct prediction and reject its rival.

(I shall call this rule R_1.) We need not concern ourselves here with whether R_1 is methodologically sound. The only issue is whether it underdetermines a choice between these rival cosmogonies. It clearly does not. Everyone in the case in hand agreed that the measuring techniques were uncontroversial; everyone agreed that Descartes's cosmogony required an earth that did not bulge at the equator and that Newtonian cosmogony required an oblately spheroidal earth.

Had scientists been prepared to make Quine-like maneuvers, abandoning (say) *modus ponens*, they obviously could have held on to Cartesian physics "come what may." But that is beside the point, for if one suspends the rules of inference, then there are obviously no inferences to be made. What those who hold that underdetermination undermines methodology must show is that methodological rules, even when scrupulously adhered to, fail to sustain the drawing of any clear preferences. As this historical case makes clear, the rule cited and the relevant evidence required a choice in favor of Newtonian mechanics.

Let me not be misunderstood. I am not claiming that Newtonian mechanics was "proved" by the experiments of the Académie des Sciences, still less that Cartesian mechanics was "refuted" by those experiments. Nor would I suggest for a moment that the rule in question (R_1) excluded all possible rivals to Newtonian mechanics. What is being claimed, rather, is that this case involves a certain plausible rule of theory preference that, when applied to a specific body of evidence and a specific theory choice situation, yielded (in conjunction with familiar rules of deductive logic and of evidential assessment) *unambiguous* advice to the effect that one theory of the pair under consideration should be *rejected*. That complex of rules and evidence *determined* the choice between the two systems of mechanics, for anyone who accepted the rule(s) in question.

■ | Underdetermination and the "Sociologizing of Epistemology"

If (as we saw in the first section) some scholars have been too quick in drawing ampliative morals from QUD, others have seen in such Duhem-Quine-style underdetermination a rationale for the claim that science is, at least in large measure, the result of social processes of "negotiation" and the pursuit of personal interest and prestige. Specifically, writers like Hesse and Bloor have argued that, because theories are deductively underdetermined (HUD), it is reasonable to expect that the adoption by scientists of various ampliative criteria of theory evaluation is the result of various social, "extra-scientific" forces acting on them. Such arguments are as misleading as they are commonplace.[54]

The most serious mistake they make is that of supposing that *any* of the normative forms of underdetermination (whether deductive or ampliative, weak or strong) entails anything whatever about what *causes* scientists to adopt the theories or the ampliative rules that they do. Consider, for instance, Hesse's treatment of underdetermination in her recent *Revolutions and Reconstructions in the Philosophy of Science*. She there argues that, since Quine has shown that theories are deductively underdetermined by the data, it follows that theory choice must be based, at least in part, on certain "non-logical," "extra-empirical" criteria for what counts as a good theory.[55] Quine himself would probably agree with that much. But Hesse then goes on to say that:

> it is only a short step from this philosophy of science to the suggestion that adoption of such [non-logical, extra-empirical] criteria, that can be seen to be different for different groups and at different periods, should be explicable by social rather than logical factors.

The thesis being propounded by these writers is that since the rules of deductive logic by themselves underdetermine theory choice, it is only natural to believe that the choice of ampliative criteria of theory evaluation (with which a scientist supplements the rules of deductive logic) are to be explained by "social rather than logical factors." It is not very clear from Hesse's discussion precisely what counts as a "social factor"; but she evidently seems to think—for her argument presupposes—that everything is either deductive logic or sociology. To the extent that a scientist's beliefs go beyond what is deductively justified, Hesse seems to insist, to that degree is it an artifact of the scientist's social environment. (Once again, we find ourselves running up against the belief—against which Duhem inveighs in the opening quotation—that formal logic exhausts the realm of the "rational.")

Hesse's contrast, of course, is doubly bogus. On the one side, it presupposes that there is nothing social about the laws of logic. But since those laws are formulated in a language made by humans and are themselves human artifacts fashioned to enable us to find our way around the world, one could hold that the laws of logic are at least in part the result of social factors. But if one holds, with Hesse, that the laws of formal logic are not the result of social factors, then what possible grounds can one have for holding that the practices that constitute ampliative logic or methodology are apt to be primarily sociological in character?

What Hesse wants to do, of course, is to use the fact of logical underdetermination (HUD) as an argument for taking a sociological approach to explaining the growth of scientific knowledge. There may or may not be good arguments for such an approach. But, as I have been at some pains to show in this essay, the underdetermination of theory choice by deductive logic is not among them.

There is another striking feature of her treatment of these issues. I refer to the fact that Hesse thinks that a semantic thesis about the relations between sets of propositions (and such is the character of the thesis of deductive underdetermination) might sustain *any* causal claim whatever about the factors that lead scientists to adopt the theoretical beliefs they do. Surely, whatever the causes of a scientist's acceptance of a particular (ampliative) criterion of theory evaluation may be (whether sociological or otherwise), the thesis of deductive underdetermination entails nothing whatever about the character of those causes. The Duhem-Quine thesis is, in all of its many versions, a thesis about the logical relations between certain statements; it is not about, nor does it directly entail anything about, the causal interconnections going on in the heads of scientists who believe those statements. Short of a proof that the causal linkages between propositional attitudes mirror the formal logical relations between propositions, theses about logical underdetermination and about causal underdetermination would appear to be wholly distinct from one another. Whether theories are deductively determined by the data, or radically underdetermined by that data; in neither case does *anything* follow concerning the contingent processes whereby scientists are caused to utilize extralogical criteria for theory evaluation.

The point is that normative matters of logic and methodology need to be sharply distinguished from empirical questions about the causes of scientific belief. None of the various forms of normative underdetermination that we have discussed in this essay entails anything whatever about the causal factors responsible for scientists adopting the beliefs that they do. Confusion of the idiom of good reasons and the idiom of causal production of beliefs can only make our task of understanding either of them more difficult.[56] And there is certainly no good reason to think (with Hesse and Bloor) that, because theories are deductively underdetermined, the adoption by scientists of ampliative criteria "should be explicable by social rather than logical factors." It may be true, of course, that a sociological account can be given for why scientists believe what they do; but the viability of that program has nothing to do with normative underdetermination. The slide from normative to causal underdetermination is every bit as egregious as the slide (discussed earlier) from deductive to ampliative underdetermination. The wonder is that some authors (e.g., Hesse) make the one mistake as readily as the other.

David Bloor, a follower of Hesse in these matters, produces an interesting variant on the argument from underdetermination. He correctly notes two facts about the history of science: sometimes a group of scientists changes its "system of belief," even though there is "no change whatsoever in their evidential basis."[57] "Conversely," says Bloor, "systems of belief can be and have been held stable in the face of rapidly changing and highly problematic inputs from experience."[58] Both claims are surely right; scientists do not necessarily require new evidence to change their theoretical commitments, nor does new evidence—even prima facie refuting evidence—always cause them to change their theories. But the conclusion that Bloor draws from these two commonplaces about belief change and belief maintenance in science comes as quite a surprise. For he thinks these facts show that *reasonable scientists are free to believe what they like, independently of the evidence.* Just as Quine had earlier asserted that scientists can hold any doctrine immune from refutation or, alternatively, they can abandon any deeply entrenched belief, so does Bloor hold that there is virtually no connection between beliefs and evidence. He writes: "So [sic] the stability of a system of belief [including science] is the prerogative of its users."[59] Here would seem to be underdetermination with a vengeance! But once the confident rhetoric is stripped away, this emerges—like the parallel Quinean holism on which it is modeled—as a clumsy non sequitur. The fact that scientists sometimes give up a theory in the absence of anomalies to it, or sometimes hold on to a theory in the face of prima facie anomalies for it, provides no license whatever for the claim that scientists can rationally hold on to any system of belief they like, just so long as they choose to do so.

Why do I say that Bloor's examples about scientific belief fail to sustain the general morals he draws from them? Quite simply because his argument confuses necessary with sufficient conditions. Let us accept without challenge the desiderata Bloor invokes: scientists sometimes change their mind in the absence of evidence that would seem to force them to, and scientists sometimes hang on to theories even when those theories are confronted by (what might appear to be) disquieting new evidence. What the first case shows, and all that it shows, is that the theoretical preferences of scientists are influenced by factors other than purely empirical ones. But that can scarcely come as a surprise to anyone. For instance, even the most ardent empiricists grant that considerations of simplicity, economy and coherence play a role in theory appraisal. Hence, a scientist who changes his mind in the absence of new evidence *may* simply be guided in his preferences by those of his standards that concern the nonempirical features of theory. Bloor's second case shows that new evidence is not necessarily sufficient to cause scientists to change their minds even when that evidence is prima facie damaging to their beliefs. Well, to a generation of philosophers of science raised to believe that theories proceed in a sea of anomalies, this is not exactly news either.

What is novel is Bloor's suggestion that one can derive from the conjunction of these home truths the thesis that scientists—quite independent of the evidence—can reasonably decide when to change their beliefs and when not to, irrespective of what they are coming to learn about the world. But note where the argument goes astray: it claims that because certain types of evidence are neither necessary nor sufficient to occasion changes of belief, it follows that no evidence can ever compel a rational scientist to change his beliefs. This is exactly akin to saying that, because surgery is not always necessary to cure gall stones, nor always sufficient to cure them, it follows that surgery is never the appropriate treatment of choice for gall stones. In the same way, Bloor argues that because beliefs sometimes change reasonably in the absence of new evidence and sometimes do not change in the face of new evidence, it follows that we are always rationally free to let our social interests shape our beliefs.

▪ | Conclusion

We can draw together the strands of this essay by stating a range of conclusions that seem to flow from the analysis:

- The fact that a theory is deductively underdetermined (relative to certain evidence) does *not* warrant the claim that it is ampliatively underdetermined (relative to the same evidence).

- Even if we can show in principle the nonuniqueness of a certain theory with respect to certain rules and evidence (i.e., even if theory choice is weakly underdetermined by those rules), it does not follow that that theory cannot be rationally judged to be better than its extant rivals (viz., that the choice is strongly underdetermined).

- The *normative* underdetermination of a theory (given certain rules and evidence) does not entail that a scientist's belief in that theory is causally underdetermined by the same rules and evidence, and vice versa.

- The fact that *certain* ampliative rules or standards (e.g., simplicity) may strongly underdetermine theory choice does not warrant the blanket (Quinean/Kuhnian) claim that all rules similarly underdetermine theory choice.

None of this involves a denial (a) that theory choice is always deductively underdetermined (HUD) or (b) that the nonuniqueness thesis may be correct. But one may grant all that and still conclude from the foregoing that no one has yet shown that established forms of underdetermination do anything to undermine scientific methodology as a venture, in either its normative or its descriptive aspect. The relativist critique of epistemology and methodology, insofar as it is based on arguments from underdetermination, has produced much heat but no light whatever.

▪ | Appendix

In the main body of the paper, I have (for ease of exposition) ignored the more *holistic* features of Quine's treatment of underdetermination. Thus, I have spoken about single theories (a) having confirming instances, (b) entailing observation statements, and (c) enjoying given degrees of evidential support. Most of Quine's self-styled advocates engage in similar simplifications. Quine himself, however, at least in most of his moods, denies that single theories exhibit (a), (b), or (c). It is, on his view, only *whole systems* of theories that link up to experience. So if this critique of Quine's treatment of underdetermination is to have the force required, I need to recast it so that a thoroughgoing holist can see its force.

The reformulation of my argument in holistic terms could proceed along the following lines. The nested or systemic version of the non-uniqueness thesis would insist that: *For any theory, T, embedded in a system, S, and any body of evidence, e, there will be at least one other system, S′* (containing a rival to T), *such that S′ is as well supported by e as S is.* The stronger, nested egalitarian thesis would read: *For any theory, T, embedded in a system, S, and any body of evidence, e, there will be systems, $S_1, S_2, \ldots, S_n$, each containing a different rival to T, such that each is as well supported by e as S.*

Both these doctrines suffer from the defects already noted afflicting their nonholistic counterparts. Specifically, Quine has not shown that, for any arbitrarily selected rival theories, T_1 and T_2, there are respective nestings for them, S_1 and S_2, that will enjoy equivalent degrees of empirical support. Quine can, with some degree of plausibility, claim that it will be possible to find systemic embeddings for T_1 and T_2 such that S_1 and S_2 will be logically compatible with all the relevant evidence. And it is even remotely possible, I suppose, that he could show that there were nestings for T_1 and T_2 such that S_1 and S_2 respectively entailed all the relevant evidence. But as we have seen, such a claim is a far cry from establishing that S_1 and S_2 exhibit equal degrees of empirical support. Thus, Quine's epistemic egalitarianism is as suspect in its holistic versions as in its atomistic counterpart.

■ | Notes

1. Pierre Duhem, *Aim and Structure of Physical Theory*, 217.

2. Lakatos once put the point this way:

   > A brilliant school of scholars (backed by a rich society to finance a few well-planned tests) might succeed in pushing any fantastic programme [however "absurd"] ahead, or, alternatively, if so inclined, in overthrowing any arbitrarily chosen pillar of "established knowledge." (In I. Lakatos and A. Musgrave, eds., *Criticism and the Growth of Knowledge* (Cambridge: Cambridge University Press, 1970), 187–88.)

3. See especially R. Boyd, "Realism, Underdetermination, and a Causal Theory of Evidence," *Noûs*, 7 (1973): 1–12. W. Newton-Smith goes so far as to entertain (if later to reject) the hypothesis that "given that there can be cases of the underdetermination of theory by data, realism . . . has to be rejected." ("The Underdetermination of Theories by Data," in R. Hilpinen, ed., *Rationality in Science* [Dordrecht: Reidel, 1980], 105.) (Compare John Worrall, "Scientific Realism and Scientific Change," *The Philosophical Quarterly*, 32 [1982]: 210–31.)

4. See chap. 2 of Hesse's *Revolutions and Reconstructions in the Philosophy of Science* (Notre Dame: Notre Dame Press, 1980) and D. Bloor's *Knowledge and Social Imagery* (London: Routledge, 1976) and "The Strengths [sic] of the Strong Programme," *Philosophy of the Social Sciences*, 11 (1981): 199–214.

5. See H. Collins's essays in the special number of *Social Studies of Science*, 11 (1981). Among Collins's many fatuous *obiter dicta*, my favorites are these:

▪ "the natural world in no way constrains what is believed to be" (*ibid.*, 54); and

▪ "the natural world has a small or nonexistent role in the construction of scientific knowledge" (3).

Collins's capacity for hyperbole is equaled only by his tolerance for inconsistency, since (as I have shown in "Collins's Blend of Relativism and Empiricism," *Social Studies of Science*, 12 [1982], 131–33) he attempts to argue for these conclusions by the use of empirical evidence! Lest it be supposed that Collins's position is idiosyncratic, bear in mind that the self-styled "arch-rationalist," Imre Lakatos, could also write in a similar vein, apropos of underdetermination, that:

> The direction of science is determined primarily by human creative imagination and not by the universe of facts which surrounds us. Creative imagination is likely to find corroborating novel evidence even for the most "absurd" programme, if the search has sufficient drive (Lakatos, *Philosophical Papers* [Cambridge: Cambridge University Press, 1978], vol. 1, 99.)

6. For a discussion of many of the relevant literary texts, see A. Nehamas, "The Postulated Author," *Critical Inquiry*, 8 (1981): 133–49.

7. See, for instance, my "The Pseudo-Science of Science?," *Philosophy of the Social Sciences*, 11 (1981): 173–98; "More on Bloor," *Philosophy of the Social Sciences*, 12 (1982): 71–74; "Kuhn's Critique of Methodology," in J. Pitt, ed., *Change and Progress in Modern Science* (Dordrecht: Reidel, 1985), 283–300; "Explaining the Success of Science: Beyond Epistemic Realism and Relativism," G. Gutting *et al.*, eds., *Science and Reality: Recent Work in the Philosophy of Science* (Notre Dame: Notre Dame Press, 1984), 83–105; "Are All Theories Equally Good?" in R. Nola ed., *Relativism and Realism in Science* (Dordrecht: Reidel, 1988), 117–39; "Cognitive Relativism," in R. Egidi, ed., *La Svolte Relativistica* (Rome: Franco Angeli, 1988) 203–24; "Relativism, Naturalism and Reticulation," *Synthese*, 71 (1987), 114–39; "Methodology's Prospects," in A. Fine and P. Machamer, eds., *PSA 1986*, vol. 2, (East Lansing, Mich.: Philosophy of Science Association) [347–54]; and "For Method: Or Against Feyerabend," in [J. R. Brown and J. Mittelstrass, eds., *An Intimate Relation: Studies in the History and Philosophy of Science: Presented to Robert E. Butts on his Sixtieth Birthday* (Dordrecht, Netherlands: Kluwer, 1989), 299–317].

8. There are, of course, more interesting conceptions of "theory" than this minimal one; but I do not want to beg any questions by imposing a foreign conception of theory on those authors whose work I shall be discussing.

9. Quine has voiced a preference that the view I am attributing to him should be called "the holist thesis," rather than a "thesis of underdetermination." (See especially his "On Empirically Equivalent Systems of the World," *Erkenntnis*, 9 (1975): 313–28.) I am reluctant to accept his terminological recommendation here, both because Quine's holism is often (and rightly) seen as belonging to the family of underdetermination arguments, and because it has become customary to use the term "underdetermination" to refer to Quine's holist position. I shall be preserving the spirit of Quine's recommendation, however, by insisting that we distinguish between what I call "nonuniqueness" (which is very close to what Quine himself calls "underdetermination") and egalitarianism (which represents one version of Quinean holism). (For a definition of these terms, see below [in text].)

10. It is important to be clear that Quine's nonuniqueness thesis is *not* simply a restatement of HUD, despite certain surface similarities. HUD is entirely a *logico-semantic* thesis about deductive relationships; it says nothing whatever about issues of empirical support. The nonuniqueness thesis, by contrast, is an *epistemic* thesis.

11. Obviously, the egalitarian thesis entails the nonuniqueness thesis, but not conversely.

12. Quine specifically put it this way: "Any statement can be held true come what may, if we make drastic enough adjustments elsewhere in the system [of belief]." ("Two Dogmas of Empiricism," in S. Harding, ed., *Can Theories Be Refuted?* (Dordrecht: Reidel, 1976), 60 [p. 266]. I am quoting from the version of Quine's paper in the Harding volume since I will be citing a number of other works included there.)

13. Grünbaum, in his *Philosophical Problems of Space and Time*, 2d ed., (Dordrecht: Reidel 1974), 590–610, has pointed to a number of much more sophisticated, but equally trivial, ways of reconciling an apparently refuted theory with recalcitrant evidence.

14. Quine in Harding (see note 12 above), 60 [p. 266]. In a much later, backtracking essay ("On Empirically Equivalent Systems of the World," *Erkenntnis*, 9 (1975): 313–28), Quine seeks to distance himself from the proposal, implied in "Two Dogmas . . . ," that it is always (rationally) possible to reject 'observation reports'. Specifically, he says that QUD "would be wrong if understood as imposing an equal status on all the statements in a scientific theory and thus denying the strong presumption in favor of the observation statements. It is this [latter] bias which makes science empirical" (*ibid.*, p. 314).

15. In fact, of course, Quine thinks that we generally do (should?) not use such stratagems. But his only argument for avoiding such tricks, at least in "Two Dogmas of Empiricism," is that they make our theories more complex and our belief systems less efficient. On Quine's view, neither of those considerations carries any epistemic freight.

16. Quine, in Harding (see note 12 above), 63 [p. 268]. I am not alone in finding Quine's notion of pragmatic rationality to be epistemically sterile. Lakatos, for instance, remarks of Quine's "pragmatic rationality": "I find it irrational to call this 'rational'" (Lakatos, *Philosophical Papers* [Cambridge: Cambridge University Press, 1978], vol. 1, 97n).

17. W. Quine, *Ontological Relativity and Other Essays* (New York: Columbia University Press, 1969), 79.

18. See especially A. Grünbaum, *Philosophical Problems of Space and Time* (Dordrecht: Reidel 1974), 585–92 and Larry Laudan, "Grünbaum on 'the Duhemian Argument'," in S. Harding, *Can Theories Be Refuted?* (Dordrecht: Reidel, 1975).

19. In a letter to Grünbaum, published in Harding (see note 12 above), p. 132, Quine granted that "the Duhem-Quine thesis" (a key part of Quine's holism and thus of QUD) "is untenable if taken nontrivially." Quine even goes so far as to say that the thesis is not "an interesting thesis as such." He claims that all he used it for was to motivate his claim that meaning comes in large units, rather than sentence-by-sentence. But just to the extent that Quine's QUD is untenable on any nontrivial reading, then so is his epistemic claim that any theory can rationally be held true come what may. Interestingly, as late as 1975, and despite his concession that the

D-Q thesis is untenable in its nontrivial version, Quine was still defending his holistic account of theory testing (see below in text).

20. And if they did not, the web would itself be highly suspect on other epistemic grounds.

21. Or, more strictly, that there is a network of statements that includes the flat-earth hypothesis and that is as well confirmed as any network of statements including the oblate-spheroid hypothesis.

22. Since I have already discussed Quine's views on these matters, and will treat Kuhn's in the next section, I will limit my illustration here to a brief treatment of Hesse's extrapolations from the underdetermination thesis. The example comes from Mary Hesse's recent discussion of underdetermination in her *Revolutions and Reconstructions in the Philosophy of Science*. She writes:

> Quine points out that scientific theories are never logically determined by data, and that there are consequently [sic] always in principle alternative theories that fit the data more or less adequately. (See note 4 above, 32–33)

Hesse appears to be arguing that, because theories are deductively underdetermined, it follows that numerous theories will always fit the data "more or less adequately." But this conclusion follows not at all from Quine's arguments, since the notion of "adequacy of fit" between a theory and the data is an epistemic and methodological notion, not a logical or syntactic one. I take it that the claim that a theory fits a given body of data "more or less *adequately*" is meant to be, among other things, an indication that the data lend a certain degree of support to the theory that they "fit." As we have already seen, there may be numerous rival theories that fit the data (say in the sense of entailing them); yet that implies nothing about equivalent degrees of support enjoyed by those rival theories. It would do so only if we subscribed to some theory of evidential support that held that "fitting the data" was merely a matter of entailing it, or approximately entailing it (assuming counterfactually that this latter expression is coherent). Indeed, it is generally true that *no* available theories exactly entail the available data; so sophisticated inductive-statistical theories must be brought to bear to determine which fits the data best. We have seen that Quine's discussion of underdetermination leaves altogether open the question whether there are always multiple theories that "fit the data" equally well, when that phrase is acknowledged as having extra-syntactic import. If one is to establish that numerous alternative theories "fit the data more or less adequately," then one must give arguments for such ampliative underdetermination that goes well beyond HUD and any plausible version of QUD.

23. I remind the reader again that neither Quine nor anyone else has successfully established the cogency of the entailment version of QUD, let alone the explanatory or empirical support versions thereof.

24. If it did, then we should have to say that patently nonempirical hypotheses like "The Absolute is pure becoming" had substantial evidence in their favor.

25. In his initial formulation of the qualitative theory of confirmation, Hempel toyed with the idea of running together the entailment relation and the evidential relation; but he went on firmly to reject it, not least for the numerous paradoxes it exhibits.

26. Consider, for sake of simplicity, the case where two theories each entail a true evidence statement, e. The posterior probability of each theory is a function of the

ratio of the prior probability of the theory to the prior probability of e. Hence if the two theories began with different priors, they must end up with different posterior probabilities, *even though supported by precisely the same evidence.*

27. It is generally curious that Quine, who has had such a decisive impact on contemporary epistemology, scarcely ever—in "Two Dogmas . . ." or elsewhere—discussed the rules of ampliative inference. So far as I can see, Quine generally believed that ampliative inference consisted wholly of hypothetico–deduction and a simplicity postulate!

28. As we shall eventually see, the kind of underdetermination advocated in *Word and Object* has no bearing whatever on (2*) or QUD.

29. W. V. Quine, *Word and Object* (Cambridge, Mass.: M.I.T. Press, 1960), 22, my italics.

30. *Ibid.*, p. 22, my italics. There is, of course, this difference between these two passages: The first says that commonsense talk of objects may conceivably underdetermine theory preferences, whereas the second passage is arguing for the probability that sensations underdetermine theory choice. In neither case does Quine give us an argument.

31. *Ibid.*, 23, my italics.

32. Except a vague version of the principle of simplicity.

33. *Ibid.*, 21.

34. *Ibid.*, 22–23.

35. In some of Quine's more recent writings (see especially his "On Empirically Equivalent Systems of the World," *Erkenntnis*, 9 (1975): 313–28), he has tended to soften the force of underdetermination in a variety of ways. As he now puts it, "The more closely we examine the thesis [of underdetermination], the less we seem to be able to claim for it as a 'theoretical thesis'" (*ibid.*, 326).

He does, however, still want to insist that "it retains significance in terms of what is practically feasible" (*ibid.*). Roughly speaking, Quine's distinction between theoretical and practical underdetermination corresponds to the situations we would be in if we had all the available evidence (theoretical underdetermination) and if we had only the sort of evidence we now possess (practical underdetermination). If the considerations that I have offered earlier are right, the thesis of practical Quinean underdetermination is as precarious as the thesis of theoretical underdetermination.

36. Quine does not repudiate the egalitarian thesis in *Word and Object*; it simply does not figure here.

37. In some of Quine's later gyrations (esp. his "On Empirically Equivalent Systems of the World") he appears to waver about the soundness of the nonuniqueness thesis, saying that he does not know whether it is true. However, he still holds on there to the egalitarian thesis, maintaining that it is "plausible" and "less beset with obscurities" than HUD (*ibid.*, 313). He even seems to think that nonuniqueness depends argumentatively on the egalitarian thesis, or at least, as he puts it, that the "holism thesis [egalitarianism] lends credence to the underdetermination theses [nonuniqueness]." (*ibid.*) This is rather like saying that the hypothesis that there are fairies at the bottom of my garden lends credence to the hypothesis that something is eating my carrots.

38. E.g., the difference between Quine's (0) and (1).

39. Quine's repeated failures to turn any of his assertions about normative under-determination into plausible arguments may explain why, since the mid-1970s, he has been distancing himself from virtually all the strong readings of his early writings on this topic. Thus, in his 1975 paper on the topic, he offers what he calls "my latest tempered version" of the thesis of underdetermination. It amounts to a variant of nonuniqueness thesis. ("The thesis of underdetermination . . . asserts that our system of the world is bound to have empirically equivalent alternatives . . ." *ibid.*, 327.) Significantly, Quine is now not even sure whether he believes this thesis: "This, for me, is [now] an open question" (*ibid.*).

40. What follows is a condensation of a much longer argument, which can be found, with appropriate documentation, in my "Kuhn's Critique of Methodology" (see note 7 above).

41. Apropos the resistance to the introduction of a new paradigm, Kuhn claims that the historian "will not find a point at which resistance becomes illogical or unscientific" (*The Structure of Scientific Revolutions*, Chicago: University of Chicago Press, 1962, 159).

42. Kuhn, *The Essential Tension* (Chicago: University of Chicago Press, 1970), 325 [p. 98]. My italics.

43. *Ibid.* [p. 97].

44. *Ibid.*, 329 [p. 101]. My italics.

45. In *Structure of Scientific Revolutions*, Kuhn had maintained that the refusal to accept a theory or paradigm "is not a violation of scientific standards" (159).

46. Kuhn, *The Essential Tension*, 322 [p. 95].

47. [Kuhn, "Reflections on My Critics," in I. Lakatos and A. Musgrave, eds., *Criticism and the Growth of Knowledge* (Cambridge: Cambridge University Press, 1970), 262.]

48. See, for instance, Derek Price, "Contra-Copernicus," in M. Clagett, ed., *Critical Problems in the History of Science* (Madison, 1959), 197–218.

49. A similar remark can be made about several of Popper's rules about theory choice. Thus, Miller and Tichý have shown that Popper's rule "accept the theory with greater verisimilitude" underdetermines choice between incomplete theories; and Grünbaum has shown that Popper's rule "prefer the theory with a higher degree of falsifiability" underdetermines choice between mutually incompatible theories. [David Miller, "Popper's Qualitative Theory of Verisimilitude," *British Journal for the Philosophy of Science* 25 (1974): 166–77; Pavel Tichý, "On Popper's Definitions of Verisimilitude," *British Journal for the Philosophy of Science* 25 (1974): 155–60; Adolf Grünbaum, "Is the Method of Bold Conjectures and Attempted Refutations Justifiably the Method of Science?" *British Journal for the Philosophy of Science* 27 (1976): 105–36.]

50. Recall Quine's claim that we can hang on to any statement we like by changing the meaning of its terms.

51. See, for instance, I. Todhunter, [A] *History of the [Mathematical] Theories of Attraction and the Figure of the Earth[: From the Time of Newton to that of Laplace*, 2 vols. (1873)] (New York: Dover, 1962).

52. Typically, astronomical measurements of angles subtended at meridian by stipulated stars were used to determine geodetic distances.

53. In fact, the actual choice during the 1730s, when these measurements were carried out, was between a Cassini-emended version of Cartesian cosmogony (which predicted an *oblong* form for the earth) and Newtonian cosmology (which required an *oblate* shape).

54. Indeed, most of so-called radical sociology of knowledge rests on just such confusions about what does and does not follow from underdetermination.

55. M. Hesse (see note 4 above), 33.

56. This is not to say, of course, that there are no contexts in which it is reasonable to speak of reasons as causes of beliefs and actions. But it is to stress that logical relations among statements cannot unproblematically be read off as causal linkages between propositional attitudes.

57. Bloor, "Reply to Buchdahl," *Studies in History and Philosophy of Science*, 13 (1982): 306.

58. *Ibid.*

59. *Ibid.* In his milder moments, Bloor attempts to play down the radicalness of his position by suggesting (in my language) that it is the nonuniqueness version of underdetermination rather than the egalitarian version that he is committed to. Thus, he says at one point that "I am not saying that any alleged law would work in any circumstances" ("Durkheim and Mauss Revisited," *Studies in History and Philosophy of Science*, 13 [1982]: 273). But if indeed Bloor believes that the stability of a system of belief *is* the prerogative of its users, then it seems he *must* hold that any "alleged law" could be made to work in any conceivable circumstances; otherwise, there would be some systems of belief that it was not at the prerogative of the holder to decide whether to hang on to.

COLIN HOWSON AND PETER URBACH

The Duhem Problem

The Duhem (sometimes called the Duhem-Quine) problem arises with philosophies of science of the type associated with Popper, which emphasize the power of certain evidence to refute a theory. According to Popper, falsifiability is the feature of a theory which makes it scientific. "Statements or systems of statements," he said, "in order to be ranked as scientific, must be capable of conflicting with possible, or conceivable, observations" (1963, p. 39 [p. 9]). And claiming to apply this criterion, he judged Einstein's gravitational theory scientific and Freud's psychology not. The term 'scientific' carries a strong flavour of commendation, which is, however, misleading in this context. For Popper could never demonstrate a link between his concept of scientificness and epistemic or inductive merit: a theory that is scientific in Popper's sense is not necessarily true, or probably true, nor can it be said either definitely or even probably to lead to the truth. There is little alternative then, in our judgment, to regarding Popper's demarcation between scientific and unscientific statements as without normative significance, but as a claim about the content and character of what is ordinarily termed science.

Yet as an attempt to understand the practice of science, Popper's ideas bear little fruit. First of all, the claim that scientific theories are falsifiable by "possible, or conceivable, observations" raises a difficulty, because an observation can only falsify a theory (in other words conclusively demonstrate its falsity) if it is itself conclusively certain. Yet as Popper himself appreciated, no observations fall into this category; they are all fallible. But unwilling to concede degrees of fallibility or anything of the kind, Popper took the view that observation reports that are admitted as evidence "are accepted as the result of a decision or agreement; and to that extent they are *conventions*" (1959, p. 106; our italics). It is unclear to what psychological attitude such

FROM Colin Howson and Peter Urbach, *Scientific Reasoning: The Bayesian Approach*, 3rd ed. (Chicago and La Salle, Ill.: Open Court, 2006), 103–14.

acceptance corresponds, but whatever it is, Popper's view pulls the rug from under his own philosophy, since it implies that no theory can really be falsified by evidence. Every 'falsification' is merely a convention or decision: "From a logical point of view, the testing of a theory depends upon basic statements whose acceptance or rejection, in its turn, depends upon our *decisions*. Thus it is *decisions* which settle the fate of theories" (1959, p. 108).

Watkins was one of those who saw that the Popperian position could not rest on this arbitrary basis, and he attempted to shore it up by arguing that some infallibly true observation statements do in fact exist. He agreed that a statement like 'the hand on this dial is pointing to the numeral 6' is fallible, since it is possible, however unlikely, that the person reporting the observation mistook the position of the hand. But he claimed that introspective perceptual reports, such as 'in my visual field there is now a silvery crescent against a dark blue background', "may rightly be regarded by their authors when they make them as infallibly true" (1984, pp. 79 and 248). But in our opinion Watkins was wrong, and the statements he regarded as infallible are open to the same skeptical doubts as any other observational report. We can illustrate this through the above example: clearly it is possible, though admittedly not very probable, that the introspector has misremembered and mistaken the shape he usually describes as a crescent, or the sensation he usually receives on reporting blue and silvery images. These and other similar sources of error ensure that introspective reports are not exempt from the rule that non-analytic statements are fallible.

Of course, the kinds of observation statement we have mentioned, if asserted under appropriate circumstances, would never be seriously doubted, for although they could be false, they have a force and immediacy that carries conviction: in the traditional phrase, they are 'morally certain'. But if they are merely indubitable, then whether or not a theory is regarded as refuted by observational data rests ultimately on a subjective feeling of certainty, a fact that punctures the objectivist pretensions of Popperian philosophy.

A second objection to Popper's falsifiability criterion, and the one upon which we shall focus for its more general interest, is that it deems unscientific most of those theories that are usually judged science's greatest achievements. This is the chief aspect of the well-known criticisms advanced by Polanyi (1962), Kuhn (1970), and Lakatos (1970), amongst others, but based on the arguments of Duhem (1905). They pointed out that notable theories of science are typically unfalsifiable by observation statements, because they only make empirical predictions in association with certain auxiliary theories. Should any such prediction turn out to be false, logic does not compel us to regard the principal theory as untrue, since the error may lie in one or more of the auxiliaries. Indeed, there are many occasions in the history of science when an important theory led to a false prediction but was not itself significantly impugned thereby. The problem that Duhem posed was this: *when several distinct theories are involved in deriving a false prediction, which of them should be regarded as false?*

▪ | Lakatos and Kuhn on the Duhem Problem

Lakatos and Kuhn both investigated scientific responses to anomalies and were impressed by the tendency they observed for the benefit of the doubt persistently to be given to particular, especially fundamental theories, and for one or more of the auxiliary theories regularly to be blamed for any false prediction. Lakatos drew from this observation the lesson that science of the most significant kind usually proceeds in what he called scientific research programmes, each comprising a central, or 'hard core', theory, and a so-called 'protective belt' of auxiliary theories. During the lifetime of a research programme, these elements are combined to yield empirical predictions, which are then experimentally checked; and if they turn out to be false, the auxiliary hypotheses act as a protective shield, as it were, for the hard core, and take the brunt of the refutation. A research programme is also characterised by a set of heuristic rules by which it develops new auxiliary hypotheses and extends into new areas. Lakatos regarded Newtonian physics as an example of a research programme, the three laws of mechanics and the law of gravitation comprising the hard core, and various optical theories, propositions about the natures and dispositions of the planets, and so forth, being the protective belt.

Kuhn's theory is similar to the methodology we have just outlined and probably inspired it in part. Broadly speaking, Kuhn's 'paradigm' is the equivalent of a scientific research programme, though his idea is developed in less detail.

Lakatos, following Popper, also added a normative element, something that Kuhn deliberately avoided. He held that it was perfectly all right to treat the hard core systematically as the innocent party in a refutation, provided the research programme occasionally leads to successful "novel" predictions or to successful, "non-ad hoc" explanations of existing data. Lakatos called such programmes "progressive".

> The sophisticated falsificationist [which Lakatos counted himself] . . . sees nothing wrong with a group of brilliant scientists conspiring to pack everything they can into their favourite research programme . . . with a sacred hard core. As long as their genius—and luck—enables them to expand their programme *'progressively'*, while sticking to its hard core, they are allowed to do it. (Lakatos 1970, p. 187)

If, on the other hand, the research programme persistently produces false predictions, or if its explanations are habitually ad hoc, Lakatos called it "degenerating". The notion of an ad hoc explanation—briefly, one that does not produce new and verified predictions—is central to attempts by the Popperian school to deal with the Duhem problem* . . . In appraising research

---

* Howson and Urbach's criticisms of the attempts by Popper and Lakatos to analyze ad hocness in terms of independent testability are discussed by Alan Chalmers in

programmes, Lakatos employed the tendentious terms 'progressive' and 'degenerating', but he never succeeded in substantiating their normative intimations, and in the end he seems to have abandoned the attempt and settled on the more modest claim that, as a matter of historical fact, progressive programmes were well regarded by scientists, while degenerating ones were distrusted and eventually dropped.

This last claim, it seems to us, contains a measure of truth, as evidenced by case studies in the history of science, such as those in Howson 1976. But although Lakatos and Kuhn identified and described an important aspect of scientific work, they could not explain it or rationalize it. So, for example, Lakatos did not say why a research programme's occasional predictive success could compensate for numerous failures, nor did he specify how many such successes are needed to convert a degenerating programme into a progressive one, beyond remarking that they should occur "now and then".

Lakatos was also unable to explain why certain theories are raised to the privileged status of hard core in a research programme while others are left to their own devices. His writings give the impression that the scientist is free to decide the question at will, by "methodological fiat", as he says. Which suggests that it is perfectly canonical scientific practice to set up any theory whatever as the hard core of a research programme, or as the central pattern of a paradigm, and to attribute all empirical difficulties to auxiliary hypoth-

---

"The Bayesian Approach" (in chapter 5 of this volume). Howson and Urbach also deploy a general argument against the view that ad hoc hypotheses are unacceptable. Suppose that a scientist realizes that a theory consisting of the conjunction of hypotheses H and A implies E and performs an experiment to check this prediction but finds instead E', which is incompatible with E. She then proposes a new theory $(H \& A')$ that implies E' but that makes no new predictions (or none that has been confirmed). Popper and Lakatos would judge the new theory to be ad hoc. Meanwhile, another scientist, working completely independently of the first, designs a different experiment to test $(H \& A)$, an experiment with only two possible outcomes: either E or ~E. The second scientist obtains the outcome ~E, rejects $(H \& A)$ as refuted, and then proposes $(H \& A')$ as a replacement. Upon realizing that $(H \& A')$ implies E', the second scientist performs an experiment and verifies the prediction. Popper and Lakatos would judge that the second scientist's new theory is not ad hoc. But the theory $(H \& A')$ is the same in both cases as is the observational result E'. Howson and Urbach regard the intrusion of subjective considerations (such as whether a particular scientist became aware of evidence before or only after proposing a theory) into the evaluation of theories on the basis of that evidence as "irrelevant and incongruous in a methodology with pretensions to objectivity." Howson and Urbach argue that theories dismissed as "ad hoc" are usually rejected because of the implausibility of the claims they make, not because of their inability to generate new predictions. See *Scientific Reasoning: The Bayesian Approach*, 118–23. For further discussions of ad hocness, see Jarrett Leplin, "The Assessment of Auxiliary Hypotheses," *British Journal for the Philosophy of Science* 33 (1982): 235–49 and Greg Bamford, "What Is the Problem of *Ad Hoc* Hypotheses?" *Science & Education* 8 (1999): 375–86.

eses. This is far from being the case. For these reasons and also because of difficulties with the notion of an ad hoc hypothesis, . . . neither Kuhn's theory of paradigms nor Lakatos's so-called 'sophisticated falsificationism' are in any position to solve the Duhem problem.

▪ | The Bayesian Resolution

The questions left unanswered in the Kuhn and Lakatos methodologies are addressed and resolved, as Dorling (1979) brilliantly showed, by referring to Bayes's theorem and considering how the individual probabilities of theories are severally altered when, as a group, they have been falsified.

We shall illustrate the argument through a historical example that Lakatos (1970, pp. 138–140; 1968, pp. 174–75) drew heavily upon. In the early nineteenth century, William Prout (1815, 1816), a medical practitioner and chemist, advanced the idea that the atomic weight of every element is a whole-number multiple of the atomic weight of hydrogen, the underlying assumption being that all matter is built up from different combinations of some basic element. Prout believed hydrogen to be that fundamental building block. Now many of the atomic weights recorded at the time were in fact more or less integral multiples of the atomic weight of hydrogen, but some deviated markedly from Prout's expectations. Yet this did not shake the strong belief he had in his hypothesis, for in such cases he blamed the methods that had been used to measure those atomic weights. Indeed, he went so far as to adjust the atomic weight of the element chlorine, relative to that of hydrogen, from the value 35.83, obtained by experiment, to 36, the nearest whole number. Thomas Thomson (1818, p. 340) responded in a similar manner when confronted with 0.829 as the measured atomic weight (relative to the atomic weight of oxygen) of the element boron, changing it to 0.875, "because it is a multiple of 0.125, which all the atoms seem to be". (Thomson erroneously took the relative atomic weights of hydrogen and oxygen as 0.125.)*

---

* Thomson assigned a value of unity to oxygen, not to hydrogen, on the grounds that oxygen combines with many more elements than does hydrogen, and often in different proportions. (The modern standard sets the atomic weight of carbon-12 as exactly 12, which makes the atomic weight of hydrogen come out to be nearly exactly 1, and the atomic weight of oxygen very close to 16.) As Howson and Urbach point out, Thomson believed, falsely, that relative to oxygen (set as unity), the atomic weight of hydrogen is 0.125, i.e., that the atomic weights of oxygen and hydrogen stand in the ratio of 8 to 1. That is because, Thomson, like Dalton, believed that the formula for water is HO (on the grounds that water is the simplest compound of the two elements) and the measured proportion by weight of hydrogen in water was known to be 1/9. This episode illustrates how calculations of atomic weights depend on assumptions about the composition of compounds that, in the early nineteenth century, were very hard to verify.

Prout's reasoning relative to chlorine, and Thomson's relative to boron, can be understood in Bayesian terms as follows: Prout's hypothesis t, together with an appropriate assumption a, asserting the accuracy (within specified limits) of the measuring techniques, the purity of the chemicals employed, and so forth, implies that the ratio of the measured atomic weights of chlorine and hydrogen will approximate (to a specified degree) a whole number. In 1815 that ratio was reported as 35.83—call this the evidence e—a value judged to be incompatible with the conjunction of t and a.

The posterior and prior probabilities of t and of a are related by Bayes's theorem, as follows:

$$P(t \mid e) = \frac{P(e \mid t)P(t)}{P(e)} \quad \text{and} \quad P(a \mid e) = \frac{P(e \mid a)P(a)}{P(e)}.$$

To evaluate the two posterior probabilities, it is necessary to quantify the various terms on the right-hand sides of these equations.

Consider first the prior probabilities of t and of a. J. S. Stas, a distinguished Belgian chemist whose careful atomic weight measurements were highly influential, gives us reason to think that chemists of the period were firmly disposed to believe in t, recalling that "In England the hypothesis of Dr Prout was almost universally accepted as absolute truth" and that when he started investigating the subject, he himself had "had an almost absolute confidence in the exactness of Prout's principle" (1860, pp. 42 and 44).

It is less easy to ascertain how confident Prout and his contemporaries were in the methods used to measure atomic weights, but their confidence was probably not great, in view of the many clear sources of error. For instance, errors were recognised to be inherent in the careful weighings and manipulations that were required; the particular chemicals involved in the experiments to measure the atomic weights were of questionable purity; and, in those pioneer days, the structures of chemicals were rarely known with certainty.[1] These various uncertainties were reinforced by the fact that independent measurements of atomic weights, based on the transformations of different chemicals, rarely delivered identical results.[2] On the other hand, the chemists of the time must have felt that that their atomic weight measurements were more likely to be accurate than not, otherwise they would hardly have reported them.[3]

For these reasons, we conjecture that $P(a)$ was in the neighbourhood of 0.6 and that $P(t)$ was around 0.9, and these are the figures we shall work with. We stress that these figures and those we shall assign to other probabilities are intended chiefly to show that hypotheses that are jointly refuted by an observation, may sometimes be disconfirmed to very different degrees, so illustrating the Bayesian resolution of Duhem's problem. Nevertheless, we believe that the figures we have suggested are reasonably accurate and sufficiently so to throw light on the historical progress of Prout's hypothesis. As

will become apparent, the results we obtain are not very sensitive to variations in the assumed prior probabilities.

The posterior probabilities of t and of a depend also on $P(e)$, $P(e \mid t)$, and
$P(e \mid a)$. Using the theorem of total probability, the first two of these terms
can be expressed as follows:

$$P(e) \quad = P(e \mid t)P(t) + P(e \mid \sim t)P(\sim t)$$

$$P(e \mid t) = P(e \mid t \ \& \ a)P(a \mid t) + P(e \mid t \ \& \ \sim a)P(\sim a \mid t).$$

We will follow Dorling in taking t and a to be independent, viz,
$P(a \mid t) = P(a)$ and hence, $P(\sim a \mid t) = P(\sim a)$. As Dorling points out (1996), this
independence assumption makes the calculations simpler but is not crucial
to the argument. Nevertheless, that assumption accords with many historical cases and seems clearly right here. For we put ourselves in the place of
chemists of Prout's day and consider how our confidence in his hypothesis
would have been affected by a knowledge that particular chemical samples
were pure, that particular substances had particular molecular structures,
specific gravities, and so on. It seems to us that it would not be affected at all.
Bovens and Hartmann (2003, p. 111) take a different view and have objected
to the assumption of independence in this context. Speaking in general
terms, they allege that "experimental results are determined by a hypothesis
and auxiliary theories that are often hopelessly interconnected with each
other."

> And these interconnections raise havoc in assessing the value of experimental
> results in testing hypotheses. There is always the fear that the hypothesis and
> the auxiliary theory really come out of the same deceitful family and that the
> lies of one reinforce the lies of the other.

We do not assert that theories are never entangled in the way that Bovens
and Hartmann describe, but for the reasons we have just cited, it seems to
us that the present situation is very far from being a case in point.

Returning to the last equation, if we incorporate the independence
assumption and take account of the fact that since the conjunction $t \ \& \ a$ is
refuted by e, $P(e \mid t \ \& \ a)$ must be zero, we obtain:

$$P(e \mid t) \quad = P(e \mid t \ \& \ \sim a)P(\sim a).$$

By parallel reasoning, we may derive the results:

$$P(e \mid a) \quad = P(e \mid \sim t \ \& \ a)P(\sim t)$$

$$P(e \mid \sim t) = P(e \mid \sim t \ \& \ a)P(a) + P(e \mid \sim t \ \& \ \sim a)P(\sim a).$$

So, provided the following terms are fixed, which we have done in a tentative way, to be justified presently, the posterior probabilities of *t* and of *a* can be calculated:

$P(e \mid \sim t \ \& \ a)$ = 0.01

$P(e \mid \sim t \ \& \sim a)$ = 0.01

$P(e \mid t \ \& \sim a)$ = 0.02.

The first of these gives the probability of the evidence if Prout's hypothesis is not true, but if the assumptions made in calculating the atomic weight of chlorine are accurate. Certain nineteenth-century chemists thought carefully about such probabilities, and typically took a theory of random distribution of atomic weights as the alternative to Prout's hypothesis (for instance, Mallet 1880); we shall follow this. Suppose it had been established for certain that the atomic weight of chlorine lies between 35 and 36. (The final results we obtain respecting the posterior probabilities of *t* and of *a* are, incidentally, unaffected by the width of this interval.) The random-distribution theory assigns equal probabilities to the atomic weight of an element lying in any 0.01-wide band. Hence, on the assumption that *a* is true, but *t* false, the probability that the atomic weight of chlorine lies in the interval 35.825 to 35.835 is 0.01. We have attributed the same value to $P(e \mid \sim t \ \& \sim a)$, on the grounds that if *a* were false, because, say, some of the chemicals were impure, or had been inaccurately weighed, then, still assuming *t* to be false, one would not expect atomic weights to be biased towards any particular part of the interval between adjacent integers.

We have set the probability $P(e \mid t \ \& \sim a)$ rather higher, at 0.02. The reason for this is that although some impurities in the chemicals and some degree of inaccuracy in the measurements were moderately likely at the time, chemists would not have considered their techniques entirely haphazard. Thus if Prout's hypothesis were true and the measurement technique imperfect, the measured atomic weights would be likely to deviate somewhat from integral values; but the greater the deviation, the less the likelihood, so the probability distribution of atomic weight measurements falling within the 35–36 interval would not be uniform, but would be more concentrated around the whole numbers.

Let us proceed with the figures we have proposed for the crucial probabilities. We note however that the absolute values of the probabilities are unimportant, for, in fact, only their relative values count in the calculation. Thus we would arrive at the same results with the weaker assumptions that $P(e \mid \sim t \ \& \ a) = P(e \mid \sim t \ \& \sim a) = \frac{1}{2} P(e \mid t \ \& \sim a)$. We now obtain:

$P(e \mid \sim t) = 0.01 \times 0.6 + 0.01 \times 0.4$ = 0.01

$P(e \mid t)$ $= 0.02 \times 0.4$ = 0.008

$$P(e \mid a) = 0.01 \times 0.1 \qquad\qquad = 0.001$$
$$P(e) \quad = 0.008 \times 0.9 + 0.01 \times 0.1 = 0.0082.$$

Finally, Bayes's theorem allows us to derive the posterior probabilities in which we are interested:

$$P(t \mid e) = 0.878 \text{ (Recall that } P(t) = 0.9)$$
$$P(a \mid e) = 0.073 \text{ (Recall that } P(a) = 0.6).$$

We see then that the evidence provided by the measured atomic weight of chlorine affects Prout's hypothesis and the set of auxiliary hypotheses very differently; for while the probability of the first is scarcely changed, that of the second is reduced to a point where it has lost all credibility.

It is true that these results depend upon certain—we have argued plausible—premises concerning initial probabilities, but this does not seriously limit their general significance, because quite substantial variations in the assumed probabilities lead to quite similar conclusions, as the reader can verify. So for example, if the prior probability of Prout's hypothesis were 0.7 rather than 0.9, the other assignments remaining unchanged, $P(t \mid e)$ would equal 0.65, and $P(a \mid e)$ would be 0.21. Thus, as before, Prout's hypothesis is still more likely to be true than false in the light of the adverse evidence, and the auxiliary assumptions are still much more likely to be false than true.

Successive pieces of adverse evidence may, however, erode the probability of a hypothesis so that eventually it becomes more likely to be false than true and loses its high scientific status. Such a process would correspond to a Lakatosian degenerating research programme or be the prelude to a Kuhnian paradigm shift. In the present case, the atomic weight of chlorine having been repeated in various, improved ways by Stas, whose laboratory skill was universally recognized, Mallet (1893, p. 45) concluded that "It may be reasonably said that probability is against the idea of any future discovery . . . ever making the value of this element agree with an integer multiple of the atomic weight of hydrogen". And in the light of this and other atomic weight measurements he regarded Prout's original idea as having been "shown by the calculus of probability to be a very improbable one". And Stas himself, who started out so very sure of its truth, reported in 1860 that he had now "reached the complete conviction, the entire certainty, as far as certainty can be attained on such a subject, that Prout's law . . . is nothing but an illusion" (1860, p. 45).

We conclude that Bayes's theorem provides a framework that resolves the Duhem problem, unlike the various non-probabilistic methodologies which philosophers have sought to apply to it. And the example of Prout's hypothesis, as well as others that Dorling (1979 and 1996) has analysed, show in our view, that the Bayesian model is essentially correct.

■ | Notes

1. The several sources of error were rehearsed by Mallet (1893).

2. For example, Thomson (1818, p. 340) reported two independent measurements—2.998 and 2.66—for the weight, relative to the atomic weight of oxygen, of a molecule of boracic (boric) acid. He required this value in order to calculate the atomic weight of boron from the weight of the boric acid produced after the element was combusted.

3. "I am far from flattering myself that the numbers which I shall give are all accurate; on the contrary, I have not the least doubt that many of them are still erroneous. But they constitute at least a nearer approximation to the truth than the numbers contained in the first table [which Thomson had published some years before]" (Thomson 1818, p. 339).

■ | References

Bovens, L., and S. Hartmann. 2003. *Bayesian Epistemology*. Oxford: Oxford University Press.

Dorling, J. 1979. Bayesian Personalism, the Methodology of Research Programmes, and Duhem's Problem. *Studies in History and Philosophy of Science*, Volume 10, 177–187.

———. 1996. Further Illustrations of the Bayesian Solution of Duhem's Problem. http://www.princeton.edu/~bayesway/Dorling/dorling.html

Duhem, P. 1905. *The Aim and Structure of Physical Theory*. Translated by P. P. Wiener, 1954. Princeton, N.J.: Princeton University Press.

Howson, C., ed. 1976. *Method and Appraisal in the Physical Sciences*. Cambridge: Cambridge University Press.

Kuhn, T. S. 1970 [1962]. *The Structure of Scientific Revolutions*. Second edition. Chicago: University of Chicago Press.

Lakatos, I. 1968. Criticism and the Methodology of Scientific Research Programmes. *Proceedings of the Aristotelian Society*, Volume 69, 149–86.

———. 1970. Falsification and the Methodology of Scientific Research Programmes. In *Criticism and the Growth of Knowledge*, edited by I. Lakatos and A. Musgrave. Cambridge: Cambridge University Press.

Mallett, J. W. 1880. Revision of the Atomic Weight of Aluminium. *Philosophical Transactions*, Volume 171, 1003–35.

———. 1893. The Stas Memorial Lecture. In *Memorial Lectures delivered before the Chemical Society 1893–1900*. Published 1901. London: Gurney and Jackson.

Polanyi, M. 1962. *Personal Knowledge*. Second edition. London: Routledge.

Popper, K. R. 1959. *The Logic of Scientific Discovery*. London: Hutchinson.

———. 1963. *Conjectures and Refutations*. London: Routledge.

Prout, W. 1815. On the Relation Between the Specific Gravities of Bodies in Their Gaseous State and the Weights of Their Atoms. *Annals of Philosophy*, Volume 6, 321–330. Reprinted in *Alembic Club Reprints*, No. 20, 1932 (Edinburgh: Oliver and Boyd), 25–37.

——. 1816. Correction of a Mistake in the Essay on the Relations Between the Specific Gravities of Bodies in Their Gaseous State and the Weights of Their Atoms. *Annals of Philosophy*, Volume 7, 111–13.

Stas, J. S. 1860. Researches on the Mutual Relations of Atomic Weights. *Bulletin de l'Académie Royale de Belgique*, 208–336. Reprinted in part in *Alembic Club Reprints*, No. 20, 1932 (Edinburgh: Oliver and Boyd), 41–47.

Thomson, T. 1818. Some Additional Observations on the Weights of the Atoms of Chemical Bodies. *Annals of Philosophy*, Volume 12, 338–50.

Watkins, J. W. N. 1984. *Science and Scepticism.* London: Hutchinson and Princeton, N.J.: Princeton University Press.

3 | COMMENTARY

3 | COMMENTARY

3.1 | Duhem's Holism and the Ambiguity of Falsification

Duhem begins "Physical Theory and Experiment" with some lengthy quotations from his fellow countryman, Claude Bernard (1813–1878). Bernard was hailed, then as now, as the father of experimental physiology. In his famous work on scientific method, *An Introduction to the Study of Experimental Medicine* (1865), he rejected the narrow inductivist conception of science that he associated (probably falsely) with writers such as Francis Bacon. The narrow inductivist sees science as a matter of simply generalizing from observation and experiment, untainted by hypotheses and theories.[1] According to the narrow inductivist, scientists should perform experiments and make observations without having any prior theory to guide them.

Bernard, quite rightly, regarded this sort of inductivism as absurd. Theories play an indispensable role in science, in part by suggesting new experiments and observations that can be used to confirm or refute those theories and thus advance our knowledge. But Bernard stressed the need for scientists to maintain an open mind and to adopt a skeptical attitude toward all theories when performing experiments (or making observations) to test a theory. When testing a theory, experimenters should leave to one side all theoretical preconceptions and beliefs in order to preserve their impartiality. Theory proposes (by suggesting experimental tests), but experiment disposes. Should experiment disagree with theoretical predictions, then the theory making those predictions ought to be rejected; should experiment agree with theoretical predictions, then the theory is confirmed. What is important is to preserve the objectivity and impartiality of observation and experiment by striving to prevent the intrusion of theoretical beliefs into the assessment of experimental results. Thus, Bernard advocates that scientists wear two hats: a theoretician's hat when outside the laboratory and an experimentalist's strictly nontheoretical hat when inside the laboratory.

THE ROLE OF AUXILIARY HYPOTHESES IN THE EXPERIMENTAL TESTING OF THEORIES

Duhem finds much to commend in Bernard's experimental method for physiology, but he thinks it is too simplistic for theory testing in physics. The crucial difference between applied sciences such as physiology and fundamental sciences such as physics is that the physiologist accepts as established fact (given by the supposedly infallible physicist) the theories that specify how the experimental apparatus and instruments work, whereas the physicist is keenly aware that the theories underlying the use of such apparatus and instruments are no less vulnerable than the specific theory that is being

333

tested. When the physicist performs an experiment to test a theory, it is not just that particular theory that is involved in the logic of testing, but a whole range of theories and assumptions about the working of the experimental apparatus and instruments.[2]

Thus Duhem is led to espouse his famous thesis of holism: when a physical theory is tested by experiment, it is not that theory alone, but a large collection of theories, auxiliary hypotheses, and assumptions that are being put to the test. Let the theory being tested be T, the auxiliary hypotheses and assumptions be $A_1, A_2, \ldots, A_n$, and the testable prediction be O_1. Duhem is claiming that T alone will not yield the prediction O_1; to deduce O_1 requires not only T but also $A_1, A_2, \ldots, A_n$. Using "$\vDash$" to stand for "entails" (or "is logically derivable from") and "$\nvDash$" to stand for "does not entail," Duhem's holist thesis is:

D1 $T \nvDash O_1$, and

D2 $(T \& A_1 \& A_2 \& \ldots \& A_n) \vDash O_1$.

From (D1) and (D2) it is but a short step to Duhem's central claim about the ambiguity of falsification. Suppose that we perform an experiment and find that O_1 is false. In other words, we discover that $\sim O_1$. Because T alone does not entail O_1, we cannot conclude that T is false. All that follows, logically, is that at least one of $T, A_1, A_2, \ldots, A_n$ is false, and logic alone will not tell us where to pin the blame. Schematically, Duhem's minimal thesis about the ambiguity of falsification is:

D3 $\sim O_1 \nvDash \sim T$, and

D4 $\sim O_1 \vDash \sim (T \& A_1 \& A_2 \& \ldots \& A_n)$.

Although Duhem's minimal thesis about the ambiguity of falsification (D3 and D4) is important, it is not, by itself, very exciting. It is important because it contradicts many oversimplified accounts of theory testing in science. And it is certainly not trivial. But it does not directly imply anything about the impossibility of falsifying a theory, since it may be possible for scientists to establish that all the members of the set $\{A_1, A_2, \ldots, A_n\}$ are true (or highly confirmed), thus pointing the arrow of falsification squarely at T. Even less does Duhem's minimal thesis imply anything about the possibility, in general, of always being able to retain belief in a theory no matter what the contrary evidence, simply by changing the auxiliary hypotheses.[3] To draw wider implications such as these from Duhem's minimal thesis requires further assumptions, many of which are quite controversial. (Which of these wider implications, if any, were endorsed by Duhem is also unclear.) Further claims about falsification (some by Duhem, some merely influenced by Duhem) are addressed in the later sections of this commentary. For the moment let us focus on Duhem's holism and his minimal claim about falsification.

The real issue is Duhem's holism, summarized in (D1) and (D2). For, if (D1) and (D2) are true, then (D3) and (D4)—Duhem's minimal thesis— follow as an immediate logical consequence. So, why should we agree with Duhem that theories in physics entail testable consequences only with the help of numerous auxiliary hypotheses, theories, and assumptions? Consider a particular theory; say Newton's theory of universal gravitation. According to Newton's gravitational theory, every piece of matter attracts every other piece of matter with a force that varies directly with the product of the masses involved and inversely with the square of their distance apart. The law can be summarized in the familiar form:

$$F_{12} = G \frac{m_1 m_2}{d^2}.$$

Clearly, Newton's gravitational theory does not logically imply any testable prediction about the position or velocity of a particular body, such as a planet, at a particular time. For that we need more information and some additional theories and hypotheses. For example, we need Newton's three laws of motion; we need to know the initial conditions, such as the mass of the sun, the position and velocity of the planet, and its distance from the sun at a particular time; and we need to assume that no other significant masses will affect the planet gravitationally or by any other means, such as by colli- sion. Moreover, the velocity of the planet is not something we can determine simply by looking. For that we need to use instruments, such as telescopes, cameras, and clocks. These instruments, in turn, require us to use theories to move from what we directly observe—things such as pointer readings, telescopic images, and streaks on a photographic plate—to judgments about position and velocity at particular times. Even then, what we report as *the* value of a reading on an instrument will often be the mean of several values spread over a small range. Thus, for Newton's gravitational theory, Duhem's holist thesis (D1 and D2) is quite plausible. But showing that holism is true of one theory does not show that holism is true of all theories. For that, a general argument is needed. The argument to which Duhem and others most frequently appeal is based on the theory-ladenness of observation.

DUHEM AND THE THEORY-LADENNESS OF OBSERVATION

Duhem defends the thesis of the theory-ladenness of observation in part 2, chapter 4 of *Aim and Structure*. His version of the thesis is restricted to experiments performed to test theories in physics. In Duhem's own words:

> An experiment in physics is not simply the observation of a phenomenon; it is, besides, the theoretical interpretation of this phenomenon. (144)

> An experiment in physics is the precise observation of the phenomena accom- panied by an *interpretation* of these phenomena; this interpretation substitutes

for the concrete data really gathered by observation abstract and symbolic representations which correspond to them by virtue of the theories admitted by the observer. (147)

We can best appreciate Duhem's claim by comparing the language of theories in physics with the language of observation. Physical theories typically employ terms such as *current, voltage, force, pressure, entropy*, and *temperature* in order to formulate functional laws in the form of mathematical equations. When an untrained person enters the physicist's laboratory, he does not "see" that a current of 0.75 amps is flowing through a resistor, or that the temperature of molten titanium indicated by an optical pyrometer is 1800 degrees Celsius. Rather, he would report seeing various pieces of equipment, some connected to others with wires, some with pointers moving around dials, others with eyepieces through which various images can be seen. What the physicist can deduce from the physical theory, in conjunction with other theories, background assumptions, and initial conditions, are predictions using the theoretical language of amps, volts, degrees Celsius, and so on. In order to connect these predictions with direct observation, the physicist needs to translate from the everyday language of the untrained observer to the theoretical language of the physicist. This translation is effected by using theories about how the measuring instruments work. In Duhem's terminology we convert practical facts into theoretical facts, where practical facts, reported in the observation language of the scientifically untrained observer, are what are "really observed" as Duhem puts it (151). In this way Duhem argues that theory plays an indispensable role in physical experiments and thus, contrary to Bernard, that physicists cannot leave behind all theory when they enter the laboratory.

Thus, the theory-ladenness thesis gives us a general argument for Duhem's holist thesis: for any theory that is not expressed entirely in the language of observation, that theory will logically imply observation statements (expressed entirely in the language of observation) only with the aid of further theories about how measuring instruments and other pieces of apparatus function.

WHY CRUCIAL EXPERIMENTS ARE IMPOSSIBLE IN PHYSICS

One of the most famous aspects of Duhem's philosophy of science is his denial of the possibility of crucial experiments in physics. The term *crucial experiment* (*experimentum crucis*—literally, "experiment of the cross") was coined by Francis Bacon in the seventeenth century. By the nineteenth century, it had come to signify an experiment that conclusively falsifies one of two rival theories or hypotheses, thus establishing its rival as well confirmed or true. It is important to bear in mind both of these characteristics: the power to refute one theory conclusively and, in virtue of this, the presumed ability to establish its rival as well confirmed or true.

Duhem gives two examples of experiments regarded as crucial by most scientists in the nineteenth century, both from optics. The first is Otto Wiener's experiment (published in 1890) on the direction of vibration of polarized light. By the middle of the nineteenth century, the wave theory of light had gained widespread acceptance, displacing the earlier particle theory of Newton, Laplace, and Biot. According to the wave theory, light is a transverse vibration in a medium, the ether, that occupies all of space and permeates material bodies. Just like waves traveling down a rope, the vibrations that constitute light are at right angles to the direction in which the wave is traveling. One of the directions perpendicular to the direction of travel is the plane of polarization. The plane of polarization is defined purely conventionally in terms of operations that can be carried out in the laboratory. The interesting theoretical question was, Do light waves vibrate parallel to the plane of polarization, or perpendicular (normal) to it? According to the theory of Augustin Fresnel, the vibration is perpendicular to the plane of polarization. According to the theory of F. E. Neumann and James MacCullagh, the vibration is parallel to it. Wiener's ingenious experiment distinguished between these two opposed predictions and verified the prediction made by Fresnel. This experimental result was widely interpreted at the time as decisively refuting the Neumann-MacCullagh theory and confirming Fresnel's theory. Indeed, some physicists regarded Wiener's experiment as showing that the vibration of light was not merely a mathematical concept or a convenient theoretical postulate but a physical reality with real, dynamical effects.

The other experiment cited by Duhem as one that nearly every nineteenth-century scientist regarded as crucial was Jean Foucault's accurate measurement of the velocity of light in air and water, which was seen as a crucial test between the wave theory of light and its rival, the particle theory. As Descartes and Newton had realized, if light consisted of tiny particles, then refraction would occur when light enters a medium denser than air because of short-range forces that briefly accelerate the particles. Consequently, the particle theory predicts that the velocity of light in water (or glass, or any medium denser than air) must be greater than the velocity of light in air. The wave theory makes precisely the opposite prediction. Both theories entail that the ratio of velocities equals the ratio of the refractive indices, but the wave theory predicts that light travels more slowly in water than it does in air. Foucault's measurements, showing that light travels more slowly in water than in air, were heralded as the definitive refutation of the particle theory and a spectacular confirmation of the wave theory.[4]

Duhem regards neither Wiener's nor Foucault's experiment as crucial. For in both cases, what is in question is not a single proposition but a whole system of hypotheses, theories, and assumptions that are needed to derive testable predictions. Thus, the negative result obtained, for example, in Wiener's experiment shows only that at least one of these elements is false, but it does not tell us which one. Duhem's holism entails that no experiment can conclusively falsify a physical theory, so, *a fortiori*, there can be no crucial experiments in physics. This is Duhem's minimal thesis about crucial experiments: logic alone does not dictate that we have to abandon a physical theory as false when its predictions disagree with experiment. In other words, Duhem's minimal thesis about crucial experments is simply a restatement of (D3).

Duhem contrasts the use of experiments to refute theories in physics with reductio ad absurdum reasoning (or the method of indirect proof) in mathematics and logic. In mathematics, many important theorems can be proven using this method. First, one assumes that the proposition one wants to prove is false. Then, one deduces from the falsity of the proposition a contradiction, using other mathematical and logical truths as additional premises. Unlike the refutation of physical theories by experiment, reductio ad absurdum reasoning is logically conclusive. The essential difference between the two is that when mathematicians use reductio ad absurdum reasoning, all the additional assumptions they adopt as premises are necessary truths. Since all these other premises are necessary truths (and thus, let us assume, known to be true with absolute certainty), when mathematicians validly deduce a contradiction from them in conjunction with the proposition that they assumed, for the sake of argument, to be false, it follows that the proposition in question must be true. But physicists deal not with necessary truths, but with empirical assumptions and hypotheses any number of which might be false. Thus, experiments in physics cannot unequivocally show that the theory being tested is false when it yields false predictions.

At the end of the third section of his article, dealing with the impossibility of crucial experiments, Duhem extends the analogy with reductio reasoning a little further and makes a point of considerable logical and methodological importance. Suppose, Duhem says, that experimental reasoning in physics *were* just like reductio reasoning in mathematics and that, when faced with the situation represented by (D4),

$$D4 \quad \sim O_1 \models \sim(T \& A_1 \& A_2 \& \ldots \& A_n),$$

all the auxiliary assumptions, A_1 through A_n, were truths "compelled to be necessary by strict logic" (235) just as they are in mathematics. In that case, we would have a logically conclusive refutation of theory T by the observation statement $\sim O_1$. And in mathematics, that would suffice to establish that $\sim T$ is a necessary truth. But in physics, the alternative to theory T is not its logical contrary $\sim T$ but some rival theory, T^*. The truth of T^* does not follow logically from the falsity of T. Symbolically:

$$D5 \quad \sim T \not\models T^*.$$

In optics, for example, the wave theory and the particle theory are not the only possible theories of the nature of light. Thus even if Foucault's experiment could falsify the wave theory conclusively, it still would not follow that his experiment was crucial because a crucial experiment is traditionally defined as one that not only falsifies one theory but also thereby establishes its rival as true (or well confirmed). As Duhem says at the end of section 3, in order for the experimental method in physics to have the same conclusive power as reductio reasoning in mathematics, physicists would have to be able "to enumerate completely the various hypotheses which may cover a determinate group of phenomena" (235), something that lies beyond their abilities.[5] Without such a complete enumeration, the inference to the truth of a physical theory would be no better than a toss of a coin; and, as Duhem reminds us, "The truth of a physical theory is not decided by heads or tails."

We have seen that Duhem has two reasons for denying the possibility of crucial experiments in physics. One is the familiar argument from the ambiguity of falsification, summarized in the holist thesis (D4). The other rests on the fact, summarized in (D5), that rival theories in physics are not logically exhaustive: while two rivals cannot both be true, they might both be false; thus, we cannot infer the truth of one from the falsity of the other.

Before leaving this section on crucial experiments, a few words about the significance of Duhem's position will set the scene for later discussion. Duhem took care to express his views in a qualified and guarded manner. For example, he restricted his discussion of holism to theories in physics where he thought special conditions obtained that are not found in other, less abstract, branches of science. Moreover, the target of Duhem's attack was the extreme position that experiments have the power to falsify theories conclusively, solely as a matter of deductive logic, and that, in virtue of this falsifying power, crucial experiments can establish a rival theory as true with certainty. In chastising this position as unwarranted, Duhem left open the possibility that experiments (in conjunction with other considerations) could lead rationally to the rejection of theories as false and that experiments (and more generally, successful predictions) could confirm theories to a significant degree. Only someone who (like Popper) thinks that the logic of science is limited solely to deductive logic and who dismisses inductive inference as a myth would be seriously nonplussed by Duhem's arguments.[6]

Duhem himself was a practicing physicist who regarded many scientific theories as having been refuted partly as a result of experiments. To be sure, Duhem did not think that any theory could be shown to be false in the same conclusive, deductive manner distinctive of mathematics and logic. But, nonetheless, theories do get refuted in science, and Duhem never denied the fact. (For example, Duhem agreed that Fresnel's wave theory of light triumphed over its opposition.) It is true that in *Aim and Structure*, Duhem speculated about how the proponents of a theory might try to protect their theory from refutation by modifying one or more of the auxiliary hypotheses with which it is associated. Consider a theory T that, in

conjunction with auxiliary hypotheses A_1 through A_n, entails the observation statement O_1. Experiment reveals that O_1 is false ($\sim O_1$). So we can no
longer believe that the system of T conjoined with A_1 through A_n is true.
But instead of giving up T, we replace the auxiliary hypothesis A_1, say, with
a new auxiliary, B_1, so that $(T\ \&\ B_1\ \&\ A_2\ \&\ \ldots\ \&\ A_n)$ is logically consistent
and no longer entails $\sim O_1$. In this way, theory T could be saved from refutation. But nowhere does Duhem claim that this strategy can always be *reasonably* employed to save any theory whatever from refutation.

For example, when discussing the possibility of reconciling the particle
theory of light with Foucault's experiments, Duhem writes: "If physicists
had attached some value to this task, they would undoubtedly have succeeded in founding on this assumption [that light is a swarm of particles] a
system of optics that would agree with Foucault's experiment" (233). The
key phrase here is "if physicists had attached some value to this task." For
Duhem believed that some ways of trying to save a theory in the face of
experiment would be unreasonable and lack scientific value. In order for
the replacement of the system $(T\ \&\ A_1\ \&\ A_2\ \&\ \ldots\ \&\ A_n)$ by the system $(T\ \&$
$B_1\ \&\ A_2\ \&\ \ldots\ \&\ A_n)$ to be reasonable, some constraints have to placed on
the new system $(T\ \&\ B_1\ \&\ A_2\ \&\ \ldots\ \&\ A_n)$. We have already mentioned one
obvious constraint: the new system must be logically consistent. Furthermore, the new auxiliary hypothesis B_1 cannot be a proposition that is known
or reasonably believed to be false. Presumably, too, the new hypothesis
must not be perniciously ad hoc. The mere logical possibility of saving T
from refutation by exchanging B_1 for A_1 does not imply that this will always
be a reasonable move or one that has any scientific value. Indeed, as we shall
see below, Duhem attacked the views of conventionalists (such as Henri
Poincaré) who argued that many fundamental hypotheses of physical theory
are immune to refutation.

Duhem's Critique of Inductivism: The Attack on Newtonian Method

One of the novelties of *Aim and Structure*, compared with the earlier papers
on which it was based, is Duhem's attack on the naive form of inductivism
espoused by Newton and others. As we have seen, Duhem rejected the
claim that the experimental method can provide physicists with anything
approaching absolute certainty: experiments cannot conclusively falsify theories, nor can crucial experiments conclusively prove that any theory is
true. Nonetheless, some famous scientists (such as Newton and Ampère)
claimed absolute certainty for their theories on the grounds that those theories were deduced directly from experimental laws. Specifically, Newton
claimed that his theory of universal gravitation could be broken down into
its separate postulates and each postulate could then be established as true
by "deduction from the phenomena," principally Kepler's laws of planetary
motion. Ampère, emulating Newton, made a similar claim for his theory of

electrodynamics. (Duhem criticizes Ampère in section 5, which has been omitted from our selection.)

Now, as Duhem realized, the term *deduction* was used rather loosely prior to the twentieth century to include not just valid deductive arguments but also inductive inferences and arguments to the best explanation. For example, Newton's four Rules of Reasoning in Philosophy, which play an important role in the derivation of the law of universal gravitational attraction in the *Principia*, are rules for inductive generalization, and both Newton and Duhem were clearly aware of this fact. So Duhem does not criticize Newton and others because their arguments are not deductive. Rather, Duhem does what any responsible critic should do: he offers his criticism in light of what his opponent means by the terms he uses. Duhem aims to show that even when we understand "deduction" to include inductive generalization and the like, it still is false that each postulate of a theory such as Newton's can be inferred ("deduced") from experimental laws.

Given Duhem's holist thesis (D1 and D2), no theory and hence no postulate of a theory can, by itself, yield any testable prediction. Thus, one might expect that Duhem would criticize Newton on the grounds that the separate parts of Newton's theory cannot be linked with observation (or with experimental laws derived from observation) by any form of inference. But Duhem gives a completely different argument against Newton, an argument that has proven to be a powerful weapon against the cruder forms of inductivism in the hands of philosophers such as Karl Popper, Imre Lakatos, and Paul Feyerabend. Duhem points out that if Newton's theory of universal gravitation is true, then Kepler's laws of planetary motion are false. Because of perturbation from other planets, no planet or planetary satellite moves in a perfect ellipse, as Kepler's laws demand. Similarly, the constant in Kepler's third law (vital for deriving the inverse square variation of gravitational attraction with distance) in fact is not constant but differs slightly for each planet, since it depends not only on the mass of the sun but also on the mass of the particular planet. Since Newton's theory and Kepler's laws are incompatible, Newton's theory cannot be derived from Kepler's laws either by deductive argument or inductive generalization.

Duhem's refutation of Newton's official account of his method is as simple as it is compelling. Duhem's own view of the relation between Newton's theory and the observed motions of the planets is holistic: only when Newton's theory is taken in its entirety, along with the laws of motion and the theories of the measuring instruments, can one logically deduce predictions about planetary motions that agree with observation.[7]

DUHEM'S REJECTION OF POINCARÉ'S CONVENTIONALISM

Although Duhem is often cited by those who insist that any theory can, in principle, be preserved in the face of recalcitrant evidence, Duhem himself denied that any theory or any theoretical postulate in physics is immune to refutation. Duhem was particularly concerned to deny the doctrine of conventionalism advocated by Henri Poincaré and others. Poincaré maintained that many of the fundamental postulates of physics such as Newton's laws of motion and the conservation principles are not empirical hypotheses at all but definitions adopted by convention. As such, no experiment or observation could ever refute them.

Both Duhem and Poincaré were writing before the revolutions in relativity theory and quantum mechanics that overthrew Newtonian mechanics and Euclidean geometry as the last word about the motions of bodies and the geometry of space. Thus, neither participant in the debate over conventionalism could appeal to history as we now can. Since Newtonian mechanics and Euclidean geometry (as a theory of physical space) have been refuted, it seems clear to us now that Duhem was right and Poincaré wrong. But at the time, Duhem could attack Poincaré only on general methodological grounds.

Duhem's position in sections 8 and 9 of his article is, as one might expect, holistic. He agrees with Poincaré that individual postulates such as Newton's first law of motion (the principle of rectilinear inertia) cannot be tested directly by any experiment. But, contrary to Poincaré, Duhem does not take this to imply that these and other laws are irrefutable or that they are really definitions in disguise. Newton's first law can be tested indirectly, as can any hypothesis in science, by being part of a system of hypotheses that makes testable predictions. In a discussion resembling that of Lakatos regarding the methodology of scientific research programmes (see "Science and Pseudoscience" in chapter 1), Duhem explains that an individual postulate can be refuted if the system to which it belongs fails to represent reality as well as some rival system of hypotheses that contains a different postulate. When the new system replaces the old, the postulate is, in effect, refuted.

In the short final section (section 10) of his article, Duhem concludes with some brief remarks about the importance of what he calls "good sense" in science. Logic alone can never force scientists to abandon one theory or adopt another. For with sufficient ingenuity, an ailing theory can usually be repaired or modified so as to avoid falsification by experimental evidence. But, Duhem writes, after a while

> . . . it may be that we find it childish and unreasonable . . . to maintain obstinately at any cost, at the price of continual repairs and many tangled-up stays, the worm-eaten columns of a building tottering in every part, when by razing these columns it would be possible to construct a simple, elegant, and solid system. (247)

Duhem offers the work of the French physicist Jean Biot (1774–1862) as an illustration of scientific good sense in operation. For many decades in the nineteenth century, Biot (following Newton and Laplace) defended the particle theory of light. But this became increasingly difficult as the wave theory of light scored a number of successes in the hands of Thomas Young (1773–1829) and Augustin Fresnel (1788–1827). By midcentury, the vast majority of physicists had switched to the wave theory, yet Biot persisted with the particle theory, making continual modifications to the theory, often of a complex and arbitrary nature, in order to keep pace with the wave theory. But then came Foucault's experiment showing that light traveled more slowly in water than in air. Logic could not force Biot to abandon the particle theory because, in principle, its associated auxiliary hypotheses could still be modified to agree with Foucault's result. Foucault's experiment was not, according to Duhem, a crucial experiment in the logical sense. But Biot's good sense told him that enough was enough. It was time to give up the particle theory as a lost cause in favor of the much more elegant and fruitful wave theory. Thus, the particle theory was refuted by being part of a system of hypotheses that was eventually replaced by another, better system.

3.2 | Quine's Attack on the Two Dogmas of Empiricism

Although only thirty-five years separate Duhem's death (in 1916) from the publication of Quine's "Two Dogmas of Empiricism" (in 1951), there is world of difference between the work of the two men. For the first half of his professional life, Duhem was a working physicist, specializing in thermodynamics. Later, he achieved renown as a historian and philosopher of science. Understandably, then, Duhem is interested primarily in science. His theses about holism and the ambiguity of falsification concern scientific theories, usually in physics, and they are supported by examples drawn from the sciences. Quine, by contrast, was a philosopher interested primarily in logic and the philosophy of language. Much of "Two Dogmas" is a reaction to the views about meaning and analyticity held by Rudolf Carnap, with whom Quine studied and interacted in the 1930s and 1940s. Quine's work is not specifically about science at all, based as it is on general considerations about language, meaning, and empiricism. But it has implications for the philosophy of science that Quine alludes to in some brief, oft-cited passages towards the end of his article. Significantly, Quine mentions Duhem only once, in a footnote that was added to his paper when it was reprinted in 1953.[8] In "Two Dogmas" Quine was not consciously developing Duhem's line of thought or reacting to what Duhem had written. We should not, then, expect to find obvious similarities between Quine and Duhem given the differences in their interests and approaches and Quine's ignorance of Duhem's work.[9] A detailed comparison of the views of Duhem and Quine can be found in the discussion of Donald Gillies' essay later in this commentary; Quine's doctrine of empirical underdetermination is discussed and criticized in the section on Larry Laudan's paper. In this section, our goal is to provide a helpful guide to Quine's somewhat technical and abstract paper, supplementing where necessary the discussion in Gillies' article. Since our primary concern is philosophy of science, not philosophy of language, we shall skip some of the more technical parts of Quine's paper. Our aim, here, is to sketch the shape of Quine's reasoning so as to bring out its relevance to the issues of holism, falsification, and underdetermination.

The Two Dogmas: Analyticity and Reductionism

As Quine explains in his opening paragraph, the two dogmas of empiricism that he wishes to criticize are analyticity and reductionism. Analyticity usually goes by the longer title of the analytic-synthetic distinction. It is the doctrine that some statements (analytic ones) are true (or false) solely in virtue of the meanings of their constituent terms and completely independently of empirical matters of fact.[10] Statements that are not analytic are said to be synthetic. Unlike their analytic brethren, synthetic statements are true (or false) depending on the way the world is. (For the sake of convenience and brevity, we shall assume that the analytic and synthetic statements under consideration are true, unless otherwise specified.)

Reductionism is the doctrine, popular among the logical positivists (and fellow travelers, such as operationalists) that every meaningful, synthetic statement is logically equivalent to some sentence containing only experiential (or observational) terms joined together with logical connectives (such as *and, or,* and *if . . . then . . .*). The classic phenomenalist version of reductionism (favored by empiricist philosophers from George Berkeley in the eighteenth century to A. J. Ayer and Rudolf Carnap in the twentieth) insists that everyday statements about physical objects (e.g., "there is a yellow table in the room") are logically equivalent to statements referring solely to the experiences of one or more people. In its positivist and operationalist versions, theoretical statements in scientific theories are supposed to be equivalent to statements referring only to observable objects and operations that can be performed on them. (For example, behavioral psychologists such as John Watson and B. F. Skinner claim that all theoretical statements involving mental terms such as *belief* and *anxiety* can be reduced to statements referring only to observable behavior.) In short, reductionism claims that all synthetic statements are logically equivalent to statements in some special empiricist language—either the subjective language of immediate experience or the objective language of physical things.

Quine claims that, at bottom, the two dogmas are identical. He gives his reasons for this judgment towards the end of section 5 of his paper. First, he notes that subscribing to the verification theory of meaning entails believing that each synthetic (empirical) statement is equivalent in meaning to some set of observation statements (or, for the phenomenalist, statements about sensory events). Even those empiricists who have given up the verification theory of meaning often adopt a view of confirmation according to which each synthetic statement is uniquely associated with a set of observation statements that can either confirm or disconfirm it. (Quine uses the word *infirm* to mean *disconfirm.*) In either case, whether one is a strict, old-fashioned reductionist about meaning or what Quine regards as a closet reductionist about confirmation, one is naturally led to posit a limiting class of true statements—the analytic ones—that have no empirical meaning and that are confirmed no matter how the world is. Since these statements lack any factual component, it is tempting to conclude that their truth depends entirely on their linguistic component, on the meanings of their constituent words. In short, Quine thinks that reductionism leads to the analytic-synthetic distinction because it tempts one into positing the existence of a special class of sentences, the analytic ones, whose factual component is nil.[11]

THE STATUS OF MATHEMATICAL TRUTHS

One can appreciate the seductive quality of the chain of reasoning that Quine describes as leading from reductionism to analyticity by considering how most of the members of the Vienna Circle viewed logic and mathematics. These members of the Vienna Circle were logical positivists and thus committed empiricists, verificationists, and reductionists. In the 1920s and 1930s logical positivists such as Hans Hahn, Rudolf Carnap, A. J. Ayer, and Carl Hempel were struck by the fact that we have certain knowledge of simple mathematical and logical truths that is completely independent of experience. How is this possible? How can a priori mathematical knowledge be reconciled with empiricism (according to which all empirical knowledge depends on experience)?

One answer had been given in the eighteenth century by Immanuel Kant, who coined the terms *analytic* and *synthetic*. Kant defined a true statement as analytic if and only if the concept of the predicate is contained in the concept of the subject. (In categorical statements such as "All dogs are brown," "Some dogs are brown," and "No dogs are brown," the subject term is *dogs* and the predicate term is *brown*.) Any statement that is not analytic is synthetic. To Kant it seemed obvious that many mathematical statements, especially in geometry, were not analytic but synthetic. A typical example of a mathematical truth Kant judged to be synthetic is the proposition that a straight line is the shortest distance between two points. As far as Kant was concerned, the concept of the shortest distance between two points is not included in the concept of a line being perfectly straight. (Kant reasoned that the former is a quantitative concept, involving numerical comparisons, but the latter, he thought, is a purely qualitative idea. Hence, the former is not contained in the latter.) Contrast this with the proposition that all material bodies are extended in space, which Kant judged to be analytic because the concept of spatial extension is contained in our concept of a material body. Thus Kant concluded that strict empiricists (such as David Hume) were wrong to assume that all synthetic propositions must be a posteriori (that is, derived from experience). Mathematics (and, Kant thought, some important general principles of science, metaphysics, and ethics) are synthetic, a priori truths.

Kant's way of drawing the analytic-synthetic distinction has been widely criticized, especially by philosophers such as Gottlob Frege (1848–1925). According to Frege, the main flaws in Kant's characterization of analyticity are its restriction to categorical statements and its vague and psychological character. Many statements have a complex internal structure that resists analysis in subject-predicate form. An example is Euclid's fifth postulate: through any point outside of a given straight line there is one and only one straight line that is parallel to that given line. Another example is the following complex but necessarily true proposition: if no person has more than N hairs on his or her head and if the population of a city is greater than N, then at least two people in that city must have exactly the same number of hairs on their heads. Since Kant's definition of analyticity in terms of the predicate being contained in the subject is applicable only to categorical statements, it cannot give a complete account of analyticity. But, Frege argues, Kant's account is defective even for categorical statements because it is vague and psychological in character. When we consider whether the subject of a categorical statement already contains its predicate, we are engaging in a form of psychological introspection. This procedure is too subjective and variable to serve as an adequate account of what makes a statement analytic. For, being subjective, it is possible that two people might use the same procedure and yet arrive at conflicting judgments about the status of a particular proposition.

Frege's alternative to Kant's proposal is that a proposition is analytic if and only if it can be proved from definitions using only the laws of logic. In other words, a statement is analytic if it is a tautology or can be reduced to a tautology by using definitions to substitute some terms for others meaning the same thing. Thus, for example, "all bachelors are bachelors" is a tautology and hence analytic; "all bachelors are unmarried" is analytic, since it reduces to a tautology when "unmarried male" is substituted for "bachelor." In contrast with Kant, Frege espoused the doctrine of logicism, the thesis that a significant part of mathematics, namely arithmetic, is ultimately reducible to logic. We have a priori knowledge of arithmetic because arithmetical truths are analytic; arithmetical truths are analytic because they reduce to logical truths. They are known to be true independently of experience because, by means of definitions of concepts such as *number* and *is the successor of,* they reduce to the laws of logic and we see immediately, by the use of our reason, that these laws of logic must be true. (Concerning geometry, Frege accepted Kant's view that geometrical truths are synthetic a priori.) But why do we have a priori knowledge of logic? At this point Frege appears to have been a Platonist. We have a priori knowledge of logic because the human mind has the ability to grasp necessary truths about abstract reality by reason alone.

The logical positivists were influenced significantly by Frege's criticisms of Kant and by Frege's logicist thesis about arithmetic. But there were certain aspects of Frege's position that they could not accept. For example, they could not tolerate the notion that any human knowledge could be synthetic a priori; thus they rejected the Kant-Frege line on geometry. Furthermore, the logical positivists found unacceptable any account of human knowledge—even our knowledge of logic—that involved some mystical form of faculty of reason or rational intuition. Taking their cue from Wittgenstein's *Tractatus Logico-Philosophicus*, they regarded simple logical tautologies, not as expressing profound abstract truths into which our reason has some special insight, but as definitional truths resulting solely from the meanings of the terms and symbols used to express them. Logical and arithmetical statements are also true and known with certainty to be true precisely because they have no factual content: they are analytic statements; their truth is completely determined by the meaning of the words and symbols they contain. In this way, the positivists sought to explain the nature of arithmetical truth and the certainty of arithmetical knowledge without appealing to Platonic forms (such as numbers conceived of as abstract entities to which mathematical statements might be thought to correspond) or any form of mystical intuition (by which Platonists believe we acquire mathematical and logical knowledge).

It is unclear whether the logical positivists accepted the analyticity of arithmetic only because they also accepted Frege's logicist thesis or whether they would have continued to assert the analyticity of arithmetic even if the logicist thesis were to have been abandoned. As Hempel explains in "The Nature of Mathematical Truth," the logicist thesis ran into a number of severe problems.[12] One of these problems was the recognition that to reduce arithmetic to logic requires a liberal dose of set theory, and some of the axioms of set theory (such as the axiom of infinity and the axiom of choice) are neither tautologies nor reducible to tautologies by definitional substitution. Without the logicist thesis, the later logical positivists and empiricists were left with the definition of analyticity criticized by Quine: an analytic statement is one whose truth (or falsity) is completely determined by the meaning of the words and symbols it contains. This definition permits the truths of arithmetic to be analytic, and hence to be known a priori, even if arithmetic is not reducible to logic.

Concerning geometry, the logical positivists were struck by two things. First, the success of Einstein's general theory of relativity seemed to show that whether Euclidean geometry is true of space is an empirical matter, not something we can know a priori as Kant and Frege had thought. Second, despite the success of Einstein's theory, it still seemed that Euclidean geometry, along with its non-Euclidean rivals, could be studied as a legitimate branch of mathematics in which theorems are proved by deducing from them from axioms. For example, it is a nontrivial geometrical truth that we can demonstrate, independently of experience, that in Euclidean geometry—but not in alternative geometries—the sum of the interior angles of a triangle is exactly 180 degrees. The solution, they thought, was to distinguish *pure* geometry from *empirical* (or *interpreted*) geometry. Pure geometry, which explores the logical consequences of various axiom sets and the logical relations between axiom sets, is analytic and its results, like those of arithmetic, are known independently of experience. Empirical geometry, which involves hypotheses about the nature of physical space that can only be verified or falsified through observation and experiment, is synthetic. For example, once we have specified a physical interpretation of the term *straight line*—as, say, the path of a light ray in a vacuum—it is an empirical hypothesis that straight lines obey Euclid's fifth postulate concerning parallel lines.[13]

One of the principal motivations for adopting the analytic-synthetic distinction among philosophers of science in the first half of the twentieth century was to forge an account of mathematical truth and knowledge that is consistent with empiricism. All the necessary truths of arithmetic and pure geometry were judged to be analytic because their truth rests solely on linguistic conventions, on the meanings of words and symbols. The factual component of their meaning is nil. Thus, mathematical truths such as "$2 + 5 = 7$," and logical truths such as "everything that is both F and G is F," are in the same category as "all vixens are foxes" and "no sexagenarians are forty-five years old." All alike are analytic truths, true solely in virtue of the meanings of the words and symbols used to express them. And, because they are analytic, we can know them a priori, independently of experience.

Outside of pure mathematics and logic, synthetic statements abound. Unlike analytic statements, their truth depends on both a linguistic and a factual component. Obviously, the truth of a statement such as "cyanide is poisonous" depends on the meaning, in the English language, of the words *cyanide* and *poisonous*, and the rules governing the copula *is*. But no amount of linguistic analysis can tell us whether the statement is true. For that, we need experience and observation. The factual component is a function of the way the world is, and thus all synthetic statements are a posteriori.

According to the logical positivists, the meaning of each synthetic statement is given by its implications for experience and observation. This is usually referred to as the verifiability principle (or criterion) of meaning. The central idea is that an individual statement, such as "iron rusts in air" or "there are brick houses on Elm Street," has a cognitive or empirical meaning only if it logically implies a group of statements that are about our immediate experience or the outcome of observation. This is the doctrine that Quine calls *reductionism*: each individual synthetic statement has a meaning that is given by a unique set of experiential or observational consequences that follow from it and that, if true, would verify it conclusively. Synthetic statements that lack experiential or observational consequences are judged to be meaningless, devoid of cognitive and empirical significance. Into this disreputable category, the positivists threw many of the statements from theology and metaphysics. In later, slightly more liberal versions of reductionism, the meaning of a synthetic statement was identified with the observational or experiential statements that could either confirm or disconfirm it (even if they could not verify or refute it conclusively).

So the broad picture that emerges is that our knowledge of the truths of logic and pure mathematics is a priori because all such statements are analytic; they are true solely in virtue of the meanings of the words and symbols they contain. Ultimately, these meanings depend on the conventions of language. Statements that are not analytic are synthetic. All knowledge of synthetic statements is a posteriori: there is no synthetic a priori knowledge, as Kant and Frege had thought. Moreover, the meaning of each synthetic statement is given by the set of experiential sentences that it implies or the set of such sentences that could either confirm or disconfirm it.

QUINE'S REJECTION OF THE ANALYTIC-SYNTHETIC DISTINCTION

In the first four sections of "Two Dogmas" Quine criticizes the major attempts to define analyticity. After some brief remarks about Carnap's definition of analyticity in terms of state-descriptions, he concentrates on the Fregean definition, according to which a statement is analytic if it is a tautology or can be reduced to a tautology by means of definitions.[14] The idea is simple enough: "No bachelor is married" is analytic because it can be reduced to the tautology "No unmarried man is married" by replacing *bachelor* with its definition, "unmarried man." In section 2 of his paper, Quine explains that, except when introducing brand new terms whose meaning we stipulate, definitions are acceptable only when they preserve the existing meanings of the terms in question: "unmarried man" defines *bachelor* (and thus can be substituted for it) only because the two expressions have the same meaning. So the search for an acceptable account of analyticity depends on our finding an acceptable account of synonymy (or sameness of meaning).

One approach to characterizing synonymy would regard two expressions as synonymous when they can be exchanged *salva veritate*, that is, without changing the truth or falsity of the sentences in which they occur. But as Quine argues in section 3 of his paper, such interchangeability, by itself, does not guarantee sameness of meaning. For "all and only bachelors are unmarried men" might be true for the same reason that "all creatures that have a heart are creatures that have kidneys" is true: the two expressions in each sentence have the same extension; they refer to exactly the same set of objects, but (at least in the case of hearts and kidneys) they do not mean the same thing. Something more is needed for the synonymy of X and Y, beyond the truth of the biconditional statement "X if and only if Y." We require not merely that the biconditional "X if and only if Y" is true but that it is necessarily true. But then, Quine argues, to assert "Necessarily, X if and only if Y" is to do nothing more or less than to assert that the sentence "X if and only if Y" is analytic, and thus that the terms X and Y are synonymous. So we have come full circle. In order to understand analyticity, we appeal to synonymy. To understand synonymy, we appeal to interchangeability *salva veritate*. But such interchangeability is not sufficient for sameness of meaning. So we must add the further condition that synonymous expressions be not only interchangeable *salva veritate* but necessarily so, and we are led back to analyticity. This attempt at characterizing synonymy has not led to an independent account of analyticity.[15]

In section 4 of his paper, Quine addresses another attempt to make clear the notion of analyticity. Perhaps the failure to give an independent characterization of synonymy (and the difficulty of deciding in many cases whether a given sentence is analytic) can be traced to the vagueness of ordinary language. The remedy for vagueness is precision, and this is what artificial languages aim to provide. Quine is particularly concerned with the artificial languages invented by Carnap and others in which semantical rules for a given language generate all the analytic sentences in that language. But here again, Quine thinks that we encounter circularity, for he argues that when we ask what distinguishes these semantical rules from other semantical rules (such as those specifying all the truths of the language), the only answer that seems to be forthcoming is that *these* semantical rules are the ones that pick out all and only the analytic sentences. Once again, in our attempt to define analyticity we end up presupposing the very notion we wish to define. Thus (after some further objections to the semantical rules strategy that we shall ignore), Quine concludes that trying to understand analyticity by means of artificial languages is a "*feu follet par excellence*" (261); it is like chasing a will-o'-the-wisp that vanishes as soon as we approach it, only to reappear somewhere else.

Given the failure of several important attempts to give an adequate characterization of analyticity, Quine suspects that the whole analytic-synthetic distinction is a dogma, an unsupported and perhaps unsupportable article of faith. He reinforces this suspicion in section 5 of "Two Dogmas," where he criticizes the verifiability principle (and its cousin, the confirmability principle) as a criterion of meaning for synthetic sentences. Just as it proved difficult to justify the claim that some sentences (the analytic ones) are true solely in virtue of their meaning, so, too, does it prove difficult to characterize the meaning of each synthetic statement in terms of some special set of experiential or observational statements. In Quine's view, this kind of reductionism is based on the same kind of mistaken view about meaning as the doctrine of analyticity. Both dogmas incorrectly regard meanings as things that attach to words and to statements independently of the other statements that we accept. This leads us to assume, wrongly, that we can decide whether a given statement is analytic (in virtue of the meanings of the words and symbols it contains), or whether two synthetic statements mean the same thing (in virtue of their experiential or observational consequences), without considering the role that these words and sentences play in our entire web of belief.

Quine's Holism and His Rejection
of A Priori Knowledge

Given the complexity and abstractness of Quine's paper, it is easy to lose sight of the fact that Quine's primary interest is epistemological. Quine wants to deny that there is any such thing as a priori knowledge. He spends so much time attacking the doctrine of analyticity because his opponents (such as the logical positivists) had argued that we know some statements (such as logical and mathematical truths) to be true independently of experience owing to their special semantic property of being analytic. If Quine can show that analyticity is a myth (an unsupported "dogma"), then one of the main supports of the epistemological thesis that we have a priori knowledge will have been destroyed.[16]

In the final section (6) of his paper, Quine sketches his own view of meaning and belief. Because Quine denies that any statement, even those in mathematics and logic, are a priori, he concludes that any of our beliefs could, in principle, be revised or abandoned in light of experience. Similarly, he contends here that any belief, even those traditionally regarded as synthetic, could be retained regardless of the outcome of observations and experiment. It is this aspect of Quine's position that calls to mind Duhem's holism and which has been the subject of considerable scrutiny by philosophers of science. Because Laudan discusses each of Quine's arguments for holism in "Demystifying Underdetermination," we shall defer consideration of those arguments until our discussion of Laudan's article below.

3.3 | Gillies on Duhem and Quine

In "The Duhem Thesis and the Quine Thesis," Donald Gillies explains the reasoning behind the versions of holism espoused by Duhem and Quine respectively. Gillies criticizes aspects of each and then concludes by advocating a modified version of a holist thesis that incorporates elements from both. We begin our discussion of Gillies' article with a brief summary of three important differences he identifies between the theses of Duhem and Quine.

Three Contrasts between the Duhem Thesis
and the Quine Thesis

One major difference between Duhem and Quine is the scope of their holist claims. Duhem restricted his thesis regarding the ambiguity of falsification to theories in physics, excluding less abstract sciences such as physiology and botany. Quine, notoriously, extends holism to any sentence whatever—"the totality of our so-called knowledge or beliefs" (265)—thus including not only all the empirical sciences, from the most to the least abstract, but also logic and mathematics. Specifically, in "Two Dogmas" Quine claims that,

(i) despite all the evidence we might accumulate, any statement whatever could be retained as true in the face of that evidence, and (ii) any sentence, even one from logic or mathematics, might be rejected as false.[17]

Closely related to the scope of their respective claims are their differing views concerning what Quine calls "the unit of empirical significance" (265), the smallest linguistic grouping that logically implies observational statements. Just like the logical positivists he criticizes, Quine ties meaning (empirical significance, empirical content) to observational consequences and regards observational statements as unproblematic as far as their meaning is concerned. But unlike the logical positivists, who took the empirically significant unit to be the individual sentence of a scientific theory, for Quine the unit is nothing less than the whole of science. Anything smaller, in his view, lacks empirical significance because it fails to imply anything about observation. Duhem's view is far less sweeping. As we have seen, Duhem regards some individual sentences from low-level sciences as empirically significant in their own right. But in abstract, high-level sciences such as physics, he considers the smallest empirically significant unit to be a group of theories, hypotheses, and assumptions from within that science. The important contrast is that, even when espousing holism with respect to theories in physics, Duhem's brand of holism is far less global than Quine's: for Duhem, even the largest unit of empirical significance (in physics) is far smaller than the whole of science or even the whole of physics.

Another contrast between Duhem and Quine concerns ways of saving a theory (or a single theoretical statement) from falsification by modifying auxiliary hypotheses and assumptions. While both Duhem and Quine agree that deductive logic alone cannot compel a scientist to reject a theory as false, Duhem emphasizes that beyond a certain point, trying to retain an ailing theory in the face of predictive failures would be unreasonable and contrary to scientific good sense. For his part, Quine is content merely to stress the impotence of deductive logic to force such a decision. Although Quine mentions that pragmatic factors will play a role, those factors are not described in any detail.

Gillies' Criticisms of Quine

Gillies' main criticism of Quine concerns Quine's sweeping claim that "the whole of science" (265) is the unit of empirical significance. As Gillies remarks, this claim is altogether implausible. When one examines cases in which a theory in physics (such as Newton's theory of gravitation) is combined with other physical theories and assumptions to generate observational predictions, theories from other sciences (such as biology, geology, and botany) play no role whatever in that derivation. On this issue at least, Duhem is closer to the truth when he restricts the scope of his holist thesis to a relatively small group of auxiliary hypotheses.

Gillies is concerned solely with the claims that Quine advanced in "Two Dogmas," but it is interesting to note that, in his later publications, Quine significantly tones down his thesis that the unit of empirical significance is the whole of science. Quine still insists that, in virtue of their use of logic and mathematics, the sciences are far more unified and integrated than is commonly realized, but he admits that "little is gained by saying that the unit is in principle the whole of science, however defensible this claim may be in a legalistic way."[18]

Another shortcoming of Quine's "Two Dogmas," according to Gillies, is Quine's almost total lack of interest in exploring ways in which the retention of a theory in the face of adverse evidence might be unreasonable even though it is not forced by deductive logic. (The arrow of falsification points at a whole group of theories, hypotheses, and assumptions without singling out the culprit.) In this respect, Duhem's insistence on the importance of good sense in science is a step in the right direction. (Although, as Gillies points out, Duhem himself seems to have been singularly lacking in such good sense, judging by the theories he rejected—relativity theory, the kinetic theory of gases, and the atomic theory.)

Finally, Gillies rejects Quine's contention that no scientific statement, not even experimental laws, can be tested (and thus falsified) without the aid of auxiliary hypotheses. Gillies thinks that within science, and even within physics, what he calls level 1 hypotheses are both confirmable and falsifiable. This claim will be assessed in the later section on Gillies' version of the Duhem-Quine thesis.

GILLIES' CRITICISMS OF DUHEM

Despite his criticisms of Quine, on one important matter Gillies thinks that Quine is right and Duhem wrong, namely, Quine's insistence that all of science, not just physics, falls within the scope of the holist thesis. There are really three aspects to this issue, which we will discuss in turn: the extension of the holist thesis to other sciences, mathematics, and logic.

OTHER SCIENCES

Gillies faults Duhem in two respects: for denying that there are any falsifiable laws in physics and, more importantly, for limiting his thesis exclusively to physics. Postponing the question of falsifiable laws to the following section on Gillies' version of the Duhem-Quine thesis, let us focus on the charge that Duhem incorrectly limits the scope of his holist thesis to physics.

It is true that the title of section 1 of Duhem's chapter proclaims that "the experimental testing of a theory does not have the same logical simplicity in physics as in physiology" (227), but Duhem's discussion reveals that the underlying logic is the same for both, at least when measuring instruments are involved. Ambiguity of falsification is avoided in physiological and chemical experiments involving measuring instruments only because physiologists and chemists (allegedly) accept many auxiliary hypotheses as established truths on the presumed infallible authority of physicists. In short, physiologists take for granted what physicists know to be based on theory. Thus, the difference lies in the psychology of testing, not in its logic. This remains consistent with Duhem's view (articulated in a section not excerpted here) that in some (perhaps many) experiments outside of physics, no measuring instruments are involved, and we can rely on what Duhem calls "a recital of concrete and obvious facts," without the need for any theoretical interpretation (*Aim and Structure*, 147).

In short, the logic of Duhem's argument leads to the inclusion of other sciences within the scope of his thesis, since the testing of laws and theories in those sciences requires the use of instruments and special apparatus. Perhaps Duhem's failure to appreciate that the ambiguity of falsification is not limited to physics was a consequence of a tendency among physicists (of whom Duhem was one) to underestimate the degree of theoretical abstraction involved in other sciences. In addition, Duhem might have thought that all the auxiliary hypotheses involved in the use of instruments and special apparatus belong to physics rather than to the sciences in whose service they are employed.

MATHEMATICS

Until recently, philosophers of science have focused on two main branches of mathematics: geometry and arithmetic. As Gillies points out, Duhem was convinced that Euclidean geometry had been conclusively established as true by our commonsense knowledge of the world. For example, Euclidean geometry (in which space is perfectly flat, having a constant, zero curvature) is the only geometry in which there exist similar figures of different sizes. In spaces of nonzero curvature, geometrical figures are similar if and only if they are equivalent (i.e., if and only if all their angles and sides have exactly the same magnitude). To Duhem, it seemed incontestable that simple, everyday observation reveals space to be Euclidean because of the "obvious" fact that there are similar figures of different sizes. Thus, Duhem summarily rejected the theory of general relativity—according to which the curvature of space varies—as a purely formal theory that could have no application to the real world. (Gillies conjectures that Duhem's rejection of relativity theory was colored by Duhem's fervent anti-German nationalism in the midst of the First World War.) A careful application of Duhem's own analysis of the logic of testing should have persuaded him that our everyday observations are consistent with the hypothesis that, while space has a small but nonzero curvature in the vicinity of the earth, the curvature might be greater elsewhere. Thus, Duhem was wrong to regard geometry (*applied* geometry) as immune from revision. Gillies says nothing about Duhem's views concerning arithmetic, but presumably Duhem was equally conservative about that as well.

LOGIC

Duhem was unaware of the modern system of logic developed by Frege, Bertrand Russell, and Alfred North Whitehead, believing as he did that logic began and ended with Aristotle. This ignorance, by itself, is not terribly significant because the new logic is not inconsistent with Aristotle's theory of the categorical syllogism (at least when the modern view that universal statements lack existential import is adopted). The new logic simply includes Aristotle's theory as a small set of special cases in a much wider system. What Duhem might be faulted for is his uncritical assumption that the laws of logic are immune from revision or rejection. As Gillies notes, two main reasons have been given for rejecting one or more of the laws of standard logic: intuitionism and quantum mechanics.

Intuitionism began as a movement in the philosophy of mathematics founded by the Dutch mathematician L. E. J. Brouwer (1881–1966). Brouwer was suspicious of mathematical arguments that profess to prove claims about infinite sets (such as the set of all numbers) without actually giving us a method for constructing an instance of that claim for any given member of the set (such as an arbitrarily chosen number). The arguments that Brouwer rejected are indirect proofs relying crucially on the law of double negation: p if and only if not-not-p. In a system of intuitionistic logic, this biconditional is replaced by the conditional: if p, then not-not-p. Brouwer accepted that reductio ad absurdum reasoning could establish that not-not-p by assuming not-p and deriving a contradiction; but he refused to acknowledge the validity of the further inference from not-not-p to p. For similar reasons, systems of intuitionistic logic do not contain the law of excluded middle (p or not-p). Brouwer motivated these restrictions based on his philosophical view that the proof of a mathematical existence claim must exhibit the entity that makes the claim true. Systems of intuitionist logic have been developed and defended by Brouwer's student, Arend Heyting (1898–1980), and more recently by the philosopher Michael Dummett (who appeals to verificationist-style arguments about how we acquire language and use it to communicate).[19]

It should be clear from our brief sketch that the rationale for intuitionism has nothing to do with observation, experiment, or the ambiguity of falsification, the considerations that normally apply to purported instances of the Duhem-Quine thesis. Rather, the arguments for intuitionism rest on controversial views about the status of mathematical objects, the relation between truth and proof, and—in the case of Dummett—on views about meaning and language. While intuitionistic logic has great formal interest for logicians, mathematicians have not adopted it, largely because they regard it as too restrictive.

Quantum logic was first proposed by the mathematicians Garrett Birkhoff and John von Neumann in 1936 as a solution to the paradoxes of quantum mechanics. Quine alludes to it in "Two Dogmas" as a possible instance of the revisability of logic, and, more recently, Hilary Putnam has championed it as showing that logic is empirical.[20] The essential feature of quantum logic is that it gives up the classical distribution laws for conjunction and disjunction. In standard classical logic, (1) logically implies (2).

1 $X \& (Y_1 \vee Y_2)$.

2 $(X \& Y_1) \vee (X \& Y_2)$.

In quantum logic, the inference from (1) to (2) is invalid. Take, for example, the classic two-slit experiment in which electrons are fired at a plate with two slits in it, thus forming an interference pattern on a screen. Consider a single electron passing through the apparatus, and let $X =$ "The electron is in region R on the screen," $Y_1 =$ "The electron went through slit number 1," and $Y_2 =$ "The electron went through slit number 2." Part of the oddity or "paradoxical" character of quantum systems and of quantum mechanics that correctly describes them is that while (1) is true, (2) is false. The pattern we obtain on the screen is typical of the interference pattern we would obtain if, instead of electrons, we had used light waves (or even water waves). If the electron had gone either through slit 1 (but not slit 2) or through slit 2 (but not slit 1), no interference pattern would have been formed.[21] To get the interference pattern, each single electron must act like a coherent wave front that passes through both slits and interferes with itself! To its supporters, the attraction of quantum logic is that it does not permit inferences to any conclusion that quantum mechanics and experiment reveal to be either false or unverifiable; in this way the paradoxes of quantum behavior are avoided. To its detractors, quantum logic does not solve any paradoxes—it simply shifts the mystery from physics to logic. Very few physicists or philosophers have argued that we should change the way that we reason about the world—a world that includes not just electrons and neutrinos but tables, chairs, and kangaroos—by giving up the distribution laws for disjunction and conjunction.

In summary we can say that a good case can be made for extending Duhem's thesis to sciences other than physics and for including geometry within its scope. But it is far less clear that we should follow Quine's "Two Dogmas" and Gillies by also including arithmetic and logic. Certainly no one, not even Putnam, has suggested that every law of classical logic could, in principle, be rejected. In fact, Putnam has argued that at least one logical law is a genuine a priori necessity, namely the law of noncontradiction.[22]

Gillies' Version of the Duhem-Quine Thesis

Gillies concludes his article by offering a revised version of the Duhem-Quine thesis that he considers both true and important. In order to clarify the relationship between Gillies' thesis and the views of Duhem and Quine, we have followed Gillies in dividing the thesis into several distinct parts.

(The only difference between our division and Gillies' is that we have split Gillies' second clause into the two distinct claims, B and C.)

A "The holist thesis applies to any high-level (level 2) theoretical hypotheses, whether of physics or of other sciences, or even of mathematics and logic." (286) (Contrary to Duhem, in light of Quine.)

B "The group of hypotheses under test in any given situation is in practice limited, and does not extend to the whole of human knowledge." (286) (Contrary to Quine, in light of Duhem.)

C "Quine's claim that 'Any statement can be held to be true come what may' . . . is true from a purely logical point of view; but scientific good sense concludes in many situations that it would be perfectly unreasonable to hold on to particular statements." (286) (Addition to Quine, in light of Duhem.)

Since we have already discussed B and C, we shall conclude this section on Gillies by taking a brief look at A.

An important feature of A is Gillies' restriction of the holist thesis to what he calls level 2 hypotheses. Earlier in his article, Gillies distinguishes between high-level (level 2) theoretical hypotheses and low-level (level 1) generalizations such as experimental laws. Gillies describes level 1 generalizations as those that can be falsified and confirmed directly by observation and experiment. Thus, Gillies thinks that Duhem's minimal thesis about falsification holds only when restricted to level 2 hypotheses (and higher). But clearly, Gillies runs the risk of trivializing the holist thesis if level 2 hypotheses are simply defined as those that satisfy the thesis and level 1 hypotheses are defined as those that do not. Hence, we need some independent characterization of the two types of hypothesis.

In his book *Philosophy of Science in the Twentieth Century*, from which Gillies' article is taken, he gives as examples of level 1 hypotheses Snell's law of refraction applied to glass, Kepler's first law, Planck's radiation law for black body radiation, and Einstein's equation for the photoelectric effect. Gillies' examples of level 2 hypotheses include Newton's first law of motion and Einstein's quantum theory of electromagnetic radiation. What Gillies seems to have in mind as level 1 hypotheses are what are often called *descriptive laws,* that is, generalizations using only variables such as position, velocity, temperature, and wavelength, whose values can be measured with instruments. As the name *descriptive laws* suggests, level 1 hypotheses are confined to describing the phenomena. Unlike level 2 hypotheses, level 1 hypotheses do not attempt to explain the phenomena they describe, nor do they deal with causes. But even so, this hardly establishes that level 1 hypotheses can be confirmed or falsified without using any theoretical assumptions whatever. For as long as measuring instruments are involved, or there is a need to verify that disturbing influences are absent, theories from other branches of physics will have to be used. As we indicated earlier, this is Duhem's central argument for his thesis of the ambiguity of falsification. Gillies appears to have simply assumed that because a law is descriptive (as opposed to being causal or explanatory) it can be tested without using any theories whatever. On this issue, at least, Duhem seems to have been closer to the truth.

3.4 | Laudan's Repudiation of Underdetermination

In "Demystifying Underdetermination" Larry Laudan takes aim at a variety of underdetermination theses that have become fashionable in the philosophy of science and epistemology since the publication of Quine's "Two Dogmas." As Laudan notes in his introduction, it is often unclear what specific claim about underdetermination a particular author is espousing and, partly as a consequence of this unclarity, relativistic consequences have been drawn from the presumed fact of underdetermination, vaguely construed, that do not follow from some of its more specific versions. Of particular concern to Laudan is the tactic of bait-and-switch: one is lured into accepting a fairly plausible version of underdetermination (the consequences of which are innocuous) only to discover that conclusions are then drawn that follow only from another, far less plausible version of underdetermination (the consequences of which are devastating).

Laudan's goal, then, is threefold: to clarify the different sorts of claims about underdetermination, to assess their plausibility, and to see what follows from them. To this end Laudan divides claims about underdetermination into two classes: the deductive and the ampliative. Deductive claims limit themselves to what can be established about the status of theories, given some evidence, using only deductive logic. Ampliative claims permit the use of nondeductive inferences as well. Nondeductive inferences are involved in judging how well the available evidence confirms a theory, whether one theory is more strongly confirmed by the evidence than a rival theory, and whether one theory explains the evidence better than its competitors. The classic example of a deductive thesis concerning underdetermination is the one associated with David Hume. All of the more recent versions of underdetermination, beginning with Quine and extending through Goodman, Kuhn, Hesse, and Bloor, are ampliative.[23]

One important preliminary matter needs to be addressed before proceeding to the heart of Laudan's article, namely, the concessions that Laudan is prepared to make to the philosophers whom he criticizes. To avoid needless debate, Laudan is willing to adopt an extremely broad notion of what constitutes a theory. He says that he will regard as a theory any set of universal statements that purport to describe the world. Presumably, he means that a theory should be empirical (it should make a definite claim about the world that could, logically, be false) and that it should contain at least one universal generalization (rather than, as his wording suggests, requiring that a theory contain nothing but universal statements). In addition, Laudan is willing to grant that single theories by themselves do not make any directly testable assertions. This, we will assume, means that Laudan accepts Duhem's holist theses (D1) and (D2): single theories yield testable predictions only in conjunction with auxiliary hypotheses and assumptions.

Humean Underdetermination (HUD)

Laudan defines deductive or Humean underdetermination as follows:

> HUD For any finite body of evidence, there are indefinitely many mutually contrary theories, each of which logically entails that evidence. (290)

Laudan comments that the arguments for HUD are familiar and trivial, and that HUD, while true, merely shows that the fallacy of affirming the consequent is indeed a fallacy: the mere fact that a particular theory (T_1) entails some evidence (E) does not logically imply that T_1 is true, since indefinitely many rival theories, T_2, T_3, and so on, have this same property. The weakness of HUD, in Laudan's view, is that it restricts itself solely to deductive logic, completely ignoring the possibility that some kind of ampliative inference (involving, say, inductive confirmation) might single out T_1 as more probable than its rivals. Moreover, even if one were to ignore this possibility and appeal to HUD to license some kind of relativist conclusion, at best what would follow is that all theories that entail E are equally worthy of belief, not that it would be rational to accept any theory whatever, regardless of its relation to E.

What are the trivial and familiar arguments for HUD to which Laudan refers? One common argument appeals to the logical possibility of generating indefinitely many theories, each of which logically entails E, simply by conjoining E with each member of a set of mutually contrary statements. Thus, we could have:

T_1 $(A \& E)$

T_2 $(B \& E)$

T_3 $(C \& E)$, and so on,

where, for example, A says that all electrons have a mass of exactly one gram, B says that all electrons have a mass of exactly two grams, and so on. Another trivial strategy involves using the truth-functional conditional connective $\supset$ in the following way:

H_1 $(A \supset E) \& A$

H_2 $(B \supset E) \& B$

H_3 $(C \supset E) \& C$, and so on.

Of course, no scientist would give these so-called theories the time of day, either because they are already known to be false or because they are utterly lacking in explanatory power and predictive capacity. But that is not the point. HUD is concerned solely with what is logically possible. Given a sufficiently liberal notion of what qualifies as a theory, these examples show that HUD is true—true but profoundly uninteresting.

There is one small problem with the wording of HUD, given the context of Laudan's article. Laudan is prepared to concede, for the sake of argument, that Duhem was right and that no single theory, by itself, makes any directly testable assertion. But HUD says that for any evidence, E, indefinitely many theories logically entail E, and in many cases, E will be a directly testable assertion. So HUD is inconsistent with Laudan's concession. The problem is easily remedied if we follow the suggestion in the appendix to Laudan's article (313–14) and rephrase HUD slightly as HUD*.

> HUD*   For any finite body of evidence, there are indefinitely many mutually contrary theories such that each theory, in conjunction with appropriate auxiliary hypotheses and assumptions, logically entails that evidence.

QUINEAN UNDERDETERMINATION (QUD)

Since Quine was one of the most influential proponents of underdetermination in the twentieth century, he is Laudan's main target and receives the lion's share of attention. Laudan begins by attributing to Quine two doctrines: the nonuniqueness thesis and the egalitarian thesis. (For convenience, we shall refer to these using the acronyms NUT and EGAL.)

> NUT   "For any theory, T, and any given body of evidence supporting T, there is at least one rival (i.e., contrary) to T that is as well supported (by that evidence) as T." (292)

> EGAL   "Every theory is as well supported by the evidence as any of its rivals." (292)

As Laudan notes, both of these doctrines are epistemic: both involve the notion of empirical support, which is a type of evidential warrant or justification—the sort of thing that, when there is enough of it, makes belief rational. In Laudan's terminology, both NUT and EGAL are varieties of ampliative underdetermination. Clearly, EGAL is the stronger claim: EGAL entails NUT, but not vice versa. Laudan claims that Quine is committed to both doctrines—explicitly to NUT and implicitly to EGAL, the sweeping doctrine that strong relativists take Quine to have made plausible in "Two Dogmas." Laudan thinks that the relativists are half right: right in attributing EGAL to Quine, wrong in thinking that Quine has done anything to render EGAL plausible.

Laudan's account of Quine's relation to EGAL is rather complicated. Laudan argues that the doctrines that Quine holds explicitly, in particular Quine's claim that

(0)   one may hold onto any theory whatever in the face of any evidence whatever (293),

implicitly commit Quine to EGAL. In Laudan's words, Quine's explicit endorsements of (0) "presuppose the egalitarian thesis, and make no sense without it" (292). Thus, Laudan is claiming not only that (0), when correctly interpreted, implies EGAL, but also, and more importantly, that unless Quine can show that EGAL is plausible, his case for (0) crumbles. It is important to realize that Laudan's second claim does not follow logically from his first claim but must be established on independent grounds. In general, one can be justified in holding a belief without first having to be justified in believing everything that is entailed by the belief. In the typical case of belief formation, one first believes some proposition on what one takes to be good grounds, and then one acquires a justified belief in the further propositions that follow from it. To insist that, in order to be justified in believing one proposition, one must first be justified in believing all the logical consequences of that proposition, is to place an unmeetable and hence unreasonable demand on what it takes for a belief to be justified. So Laudan must show that Quine's case for (0) vitally depends on accepting EGAL; Laudan cannot simply appeal to the fact that (0) has EGAL as one of its consequences.[24]

Laudan's account of why (0) depends on EGAL runs as follows: First, he notes that, in order for (0) to be at all philosophically interesting, it has to be read normatively, not descriptively. The issue is not whether people can, by whatever means, retain their belief in any theory come what may, but whether they can do so rationally. So, Laudan interprets (0) as (1):

(1)   It is rational to hold onto any theory whatever in the face of any evidence whatever. (293)

Has Quine established (1)? Laudan thinks that in order to do so, Quine would have to examine every single ampliative rule governing rational theory choice and show that none of them has the power to favor one theory over another. In fact, the only rule that Quine considers is the Popperian admonition to "reject theories that have (known) falsifying instances" (294).

Before discussing Laudan's criticism of Quine's conclusion that the Popperian falsification rule radically underdetermines theory choice, we pause to consider whether Quine is as negligent as Laudan charges in ignoring all the other rules governing the choice among rival theories. Someone interested in defending Quine might point out that Quine's concern is not whether we should accept a given theory, but whether evidence could make it irrational for us to continue to accept that theory, however irrationally we might have arrived at that theory in the first place. Thus, one might argue that ampliative rules concerning when it is rational to accept a theory as well confirmed by evidence are not relevant to Quine's inquiry and that he is justified in restricting his attention to the Popperian rule for theory rejection. But this attempt to defend Quine fails for at least two reasons. First, rules for theory acceptance clearly are relevant to theory rejection, for if one is rationally justified in accepting theory T_2, then one is justified in rejecting as false all the contraries of T_2. The fact that T_1 was there first, so to speak, should count for little if the rule says that, in light of evidence E, T_2 is the theory one should accept and T_1 is inconsistent with T_2. Second, there might be rules concerning theory rejection other than the Popperian one. For example, a rule might say that if several rival theories entail exactly the same evidence, then rationality requires one to reject as false the theory having the lowest predictive content, the least explanatory power, and the lowest degree of simplicity (if there is any theory that scores lowest in all three areas). The point is not to advocate such a rule as correct but merely to remind ourselves that the Popperian falsification rule—essentially, *modus tollens*—is not the only plausible rejection rule once we include ampliative inferences that go beyond deductive logic. Thus Laudan seems to be correct: there is a significant gap in Quine's argument when, in defense of (1), he confines himself solely to arguing for the inconclusive nature of Popper's rejection rule.

Even though Quine confines his attention solely to Popper's rejection rule, some of his attempts to defend (1) in the face of that rule are, as Laudan says, "pretty trifling stuff" (294). The possibility of changing the meanings of key words, of abandoning the laws of logic, and of pleading hallucination do not come close to showing that any theory can be retained rationally in the face of any evidence. Insofar as Quine has a case for (1), it rests entirely on his doctrine of underdetermination (QUD).

QUD   Any theory can be reconciled with any recalcitrant evidence by making suitable adjustments in our other assumptions about nature. (295)

Now, as Laudan points out, in order for QUD to be in any way relevant to (1), it must concern ways in which we can rationally reconcile any theory with any evidence. If QUD is talking merely about what is logically possible, however irrational, then Quinean underdetermination is an utterly trivial consequence of holism.[25] So we must add the crucial adverb *rationally* to modify the verb *reconcile*. Moreover, the notion of reconciliation is multiply ambiguous. Laudan distinguishes four senses of what it might mean to reconcile a theory with recalcitrant evidence. The theory (or, more precisely, the group consisting of the theory together with auxiliary hypotheses and assumptions) may:

- be logically compatible with the evidence;
- logically entail the evidence;
- explain the evidence;
- be empirically supported by the evidence. (296)

Which one of these does Quine have in mind? We can gain further insight into Laudan's criticism of QUD by considering the formal scheme we introduced earlier in this commentary, in the section on Duhem. Let the theory whose rational retention is at issue be T. We grant, along with Duhem, Quine, and Laudan, that T by itself does not yield testable predictions. For this we need auxiliary hypotheses and assumptions. Let these auxiliaries be $A_1, A_2, \ldots, A_n$, and let the testable prediction made by the entire group (T & A_1 & A_2 & $\ldots$ & A_n) be O_1. When we perform the experiment or make the observation, what we actually find is not O_1 but E, where E entails that O_1 is false. Schematically we have:

$$(T \ \& \ A_1 \ \& \ A_2 \ \& \ \ldots \ \& \ A_n) \vDash O_1,$$

$$E, \text{ and } (E \vDash {\sim}O_1).$$

We can now express relatively precise versions of Quine's underdetermination thesis, corresponding to the first two meanings of *reconcile* on Laudan's list:

QUD1   Any theory, T, can be rationally reconciled with any recalcitrant evidence, E, by deleting some of the original auxiliaries, and perhaps adding a new auxiliary, B, such that the new group (T & B & A_2 & $\ldots$ & A_n) does not entail anything that is inconsistent with E.

QUD2   Any theory, T, can be rationally reconciled with any recalcitrant evidence, E, by deleting some of the original auxiliaries, and adding a new auxiliary, B, such that the new group (T & B & A_2 & $\ldots$ & A_n) entails E.

Laudan makes four important points about QUD1 and QUD2.

First, neither consistency with E nor entailment of E guarantees that the new group will explain E or that the new group will be empirically supported by E. (Whether theories are automatically confirmed by their deductive consequences is addressed in chapter 4; explanation is discussed in chapter 6. We will here simply assume that Laudan is right about this: deductive entailment is sufficient neither for explanation nor for confirmation.) Second, neither Quine nor anyone else has proven that QUD2 is true.

In other words, no one has shown that there will always be rationally acceptable auxiliaries such that, when added to any theory T, the entire group will
entail any evidential statement. Third, deleting some of the original auxiliaries and perhaps adding the new auxiliary, B, does not guarantee that the
new group will have the same degree of explanatory power or the same
degree of empirical support as the original group. Indeed, deleting some of
the old auxiliaries will not only prevent the derivation of O_1, but is also
likely to diminish the predictive and explanatory power of the new group as
compared with the old. Fourth, the shift from the old group to the new
group is rational only if the new group has a significant degree of empirical
support. (Laudan also considers the role of other theoretical virtues, such as
simplicity and explanatory power, in making theoretical changes rational.)

So Laudan returns the Scottish verdict, not proven, in response to
Quine's underdetermination thesis (with more than a hint of skepticism
regarding its truth). Quine could render his thesis plausible only by giving a
convincing argument for the egalitarian thesis, EGAL. For only if every
theory (or, more relevantly, every group consisting of a theory together with
auxiliary hypotheses and assumptions) enjoyed the same degree of empirical support, could any theory whatever be rationally retained in the face of
any evidence whatever. But Quine does not argue for EGAL. He simply
assumes that EGAL is true without giving any argument that renders it
plausible.

In the remainder of his section on Quine, Laudan considers whether
Quine made a better case for underdetermination in his later book, *Word
and Object*. Laudan's judgment, again, is negative. He concludes that *Word
and Object* merely provides poor arguments for the weaker thesis of nonuniqueness (NUT). Since NUT does not entail EGAL, and it is EGAL (not
NUT) that is required for any strong and interesting version of relativism,
devotees of underdetermination need to look elsewhere for support. This
search for more convincing support for EGAL leads Laudan to consider the
views of Goodman, Kuhn, Hesse, and Bloor in the remainder of his
article.

Goodman's New Riddle of Induction

Laudan neatly sidesteps the debate over Goodman's new riddle of induction and its proposed solutions. (See chapter 4 for an extended discussion of
Goodman's new riddle and the problem of distinguishing, for the purposes
of confirmation, straight hypotheses from "grue-like" bent hypotheses.)
Laudan is concerned solely with the egalitarian thesis that, given some evidence, every theory entailing that evidence is as well supported as its rivals.
He makes two simple points. First, not even Goodman himself accepts that
bent hypotheses receive the same inductive support as straight hypotheses
from the observation of their instances. The whole point of the new riddle
is to find out what principles of ampliative inference should replace the

simple, but flawed, principle of enumerative induction (that is, from all observed As are B, infer that all As are B). Second, even if the bent and straight hypotheses were equally well supported by the evidence, the egalitarian thesis would not be vindicated, since bent hypotheses are only a subset of all the possible alternatives to any given hypothesis. As Laudan reminds us, nothing in Goodman's new riddle of induction suggests that, when we have observed a number of green emeralds, the hypothesis that all emeralds are red is just as well confirmed as the hypothesis that all emeralds are green.

KUHN'S THESIS OF LOCAL UNDERDETERMINATION

Laudan's criticism of Kuhn's version of the underdetermination thesis is closely related to Laudan's attack on Kuhn in his "Kuhn's Critique of Methodology" in chapter 2, and should be read in conjunction with that piece. Here, Laudan argues that Kuhn is implicitly committed to a restricted version of the egalitarian thesis, namely, that any theory can be shown to be as well supported by any evidence as any of its known rivals. This is the thesis designated (2*) earlier in Laudan's article. As Laudan notes, (2*) is weaker than the unrestricted egalitarian thesis we have called EGAL, since, unlike EGAL, (2*) refers only to the known rivals of a given theory, not to all possible rivals. In Laudan's terminology, Kuhn espouses a thesis of *local*, not *global*, underdetermination. Even when we restrict our attention to the relatively few rival theories competing for the allegiance of scientists at a given time, Kuhn's view is that shared methods and rules (what, in chapter 2, we called *cognitive* or *constitutive values*) are always insufficient to determine the choice. Only because he holds this view—the restricted egalitarian thesis (2*)—can Kuhn conclude that in all cases of theory choice the final decision must depend on subjective factors (what, in chapter 2, were called *contextual values*.)

Kuhn's argument for local underdetermination rests on his claim that the shared standards (or *values*, as Kuhn calls them) of accuracy, consistency, scope, simplicity, and fruitfulness are too vague to lead proponents of rival theories (or paradigms) to the same verdict. (See the section "Kuhn's Second Thoughts about the Rationality of Scientific Revolutions" in the commentary on chapter 2.) In response, Laudan concedes that the standards Kuhn has mentioned are vague and that, as a consequence, they are unlikely to point to a unique resolution that all scientists will accept. But, Laudan objects, Kuhn's list of shared standards does not include many of the rules that scientists commonly employ in choosing among rival theories. These rules include the admonition to prefer theories that are internally consistent[26] and the directive to prefer theories that make predictions that are surprising given our background assumptions. Unlike the items on Kuhn's list, these rules are quite precise. Certainly there is no reason to think that they can never suffice to pick out one theory as better than its rivals. Hence, Kuhn's thesis of local underdetermination is unproven.

Laudan admits that some theoretical choices are underdetermined relative to some restricted body of evidence and some subset of shared methodological rules. What he strenuously denies is Kuhn's thesis that all theoretical choices are necessarily underdetermined by any body of evidence and the set of all such rules. To show the falsity of Kuhn's thesis of local underdetermination, Laudan considers in some detail one important historical case: the disagreement in the early eighteenth century between the Newtonians and the Cartesians about the shape of the earth.

In the seventeenth century, Descartes argued that the effects of gravity on a body near the earth's surface are due entirely to a swirling vortex of transparent matter that presses down on the earth. The axis of rotation of this vortex coincides with the earth's north-south axis. The force exerted by this vortex will be greatest at the equator and least at the poles. Hence, according to Descartes and his followers, the earth is compressed around the equator. So the Cartesians predicted that the earth's radius at the equator should be slightly less than the earth's radius at either pole. The Newtonians made the opposite prediction. On Newton's theory, the effects of gravity are due to mutual attraction between a body and the earth, and this force varies inversely as the square of their distance apart. As the earth rotates once every twenty-four hours about its north-south axis, centrifugal force will counteract this gravitational attraction. The opposing centrifugal force will be greatest at the equator and least at the poles. Hence, according to Newton and his followers, the earth should bulge at the equator; the earth's radius at the equator should be slightly greater than the earth's radius at either pole. The Académie des Sciences in Paris sent surveying expeditions north and south, to Lapland and Peru, to measure the length of a terrestrial degree at both locations. The results were unequivocally in favor of Newton and against Descartes: a degree at the earth's surface was longer in Peru than in Lapland. As Newton had predicted, the earth is fatter at the equator.[27] Both the French and the English agreed on this verdict. Because of the findings of the two expeditions, Newton's theory was accepted and Descartes's theory rejected.

Laudan does not claim that the French expeditions performed a crucial experiment in the sense that they established conclusively that Newton's theory is true and Descartes's false. His point is more modest but no less telling against Kuhn: here is an actual historical case, involving proponents of rival paradigms, the Newtonian paradigm and the Cartesian paradigm, in which a single, shared, methodological rule led to a decision that both sides accepted as correct. The rule in question is, roughly, that when two rival theories make conflicting predictions that can be tested in a manner that does not presuppose the truth of either theory, then one should accept the theory that makes the correct prediction and reject its rival. One example cannot prove a universal generalization, but one counterexample can refute it. The case of the rival theories about the shape of the earth is a counterexample to Kuhn's thesis of local underdetermination. Hence, Kuhn's thesis is false.

THE STRONG PROGRAMME IN THE SOCIOLOGY OF SCIENCE

In his paper's last section, "Underdetermination and the 'Sociologizing of Epistemology,'" Laudan takes aim at Mary Hesse and David Bloor, who have argued that underdetermination implies that scientists' decisions about theories are caused by social factors and processes, rather than by reasoning and logic. David Bloor is a sociologist of science and one of the founding members of the Science Studies Unit at Edinburgh University. Along with his colleagues, Barry Barnes and David Edge, Bloor is a leading proponent of the so-called strong programme in the sociology of knowledge. The core tenets of the strong programme are described and defended in his book *Knowledge and Social Imagery*.[28] Mary Hesse is a distinguished philosopher of science who taught for many years in the History and Philosophy of Science Programme at Cambridge University. Although not herself an advocate of Bloor's version of the sociology of science, Hesse is sympathetic to certain aspects of the strong programme and her writings are often cited favorably by Bloor. Laudan is particularly concerned with Hesse's paper "The Strong Thesis of Sociology of Science," which appeared as the second chapter of her book, *Revolutions and Reconstructions in the Philosophy of Science*.[29]

Elsewhere, Laudan has vigorously attacked the strong programme, accusing it of error and confusion.[30] Since the issue before us now is not the truth or plausibility of the strong programme but merely the merits of two arguments (Hesse's and Bloor's) given in its defense, we need not discuss the strong programme itself in detail. A few brief comments will suffice before we address the specific arguments that Laudan attacks.

According to David Bloor, the defining characteristics of the strong programme are causality, impartiality, symmetry, and reflexivity.[31] Bloor claims that each of these conditions reflect procedural assumptions common to the established sciences. Only by adopting them, Bloor argues, can the study of scientific beliefs itself become "scientific."

The causal condition simply means that sociologists should identify the conditions that bring about scientific belief or states of knowledge, paying particular attention to those cultural and social features often presumed (falsely, in Bloor's view) to lie outside of science proper and the mechanisms within science by which belief is produced.[32]

Impartiality is the requirement that explanations of scientific beliefs should be "impartial with respect to truth and falsity, rationality and irrationality, success or failure." Bloor insists that a scientific belief's being true or rational does not, by itself, explain why a particular scientist holds that belief, and Bloor often repeats that rational beliefs stand just as much in need of explanation as do irrational ones. Thus, Bloor rejects the views held by philosophers such as R. G. Collingwood and Imre Lakatos that: (1) an action or belief can be explained only if it can be shown to be rational (Collingwood); (2) when we have shown that a person's action or belief is rational, no further explanation is needed or possible—the explanation of a human belief or action stops with the exhibition of its rationality (Collingwood and Lakatos); and (3) explanations in terms of sociological and psychological causes are appropriate only when the action or belief is irrational (Lakatos).[33] Since much of science is presumed to be rational, if (3) were true, very little would be left for the sociologist of science to do. Indeed, until the advent of the strong programme, most sociologists of science accepted that nearly all scientific beliefs are off-limits to sociology because they are rationally well founded: sociology may have much to tell us about the structure and organization of science, but it can do little or nothing to explain the content of scientific belief. Bloor denies this, claiming all of science, rational and irrational alike, as an appropriate subject of sociological explanation. Arguably, Bloor's second condition (impartiality) follows from his first condition (causality). If all explanation is causal and all beliefs (whether true or false, rational or irrational) have causes, then all beliefs (whether true or false, rational or irrational) can be explained in terms of those causes. This, of course, leaves it completely open as to what kinds of cause (perception, experience, reasoning, psychological conditioning, social pressure) are causally involved in the production of any particular belief. Hence the importance of Bloor's third condition, symmetry.

According to Bloor, symmetry requires that all scientific beliefs should receive the same style of explanation and that "the same type of cause would explain, say, true and false beliefs." Obviously, much depends on what Bloor means here by "the same type of cause." As Laudan has argued elsewhere, to claim that all beliefs, rational and irrational alike, are produced by the *same* causes—whether they be what the scientist takes to be good evidence, a neurophysiological mechanism, a psychological compulsion, education, peer pressure, political ideology, religion, class, or wealth—seems quite implausible. Contrary to Bloor's claim, symmetry is not a condition common to the well-established sciences. For example, nowhere in physics, chemistry, or geology do we find it stipulated in advance of empirical investigation that some broad class of phenomena must be the result of the same kind of cause in order for theories that attempt to explain those phenomena to qualify as scientific.[34]

The fourth and final condition, reflexivity, states that the patterns of explanation used by sociologists should be applicable to their own discipline. Bloor thinks that this condition must be imposed "because otherwise sociology would be a standing refutation of its own theories."

With some idea in hand of what the strong programme amounts to, let us consider Hesse and Bloor's arguments in favor of the strong programme's approach to explaining scientific belief and Laudan's criticism of those arguments. Hesse's argument begins from the fact of Humean underdetermination (HUD) and runs (with some embellishment) as follows:

1 HUD—scientific theories are deductively underdetermined by the data.

2 So, scientists must adopt extraempirical criteria for what counts as a good theory when deciding to accept one theory in preference to its empirically adequate rivals.

3 These extraempirical criteria differ over time and between groups.

4 Hence, the adoption of these criteria should be explained by social rather than logical factors.

5 Thus, the decision to accept particular scientific theories on the basis of these criteria must also be explained by social rather than logical factors.

Laudan has no quibble with premises (1) and (3) or the intermediate conclusion (2). His complaint is that (4)—the conclusion that the adoption of extraempirical criteria for evaluating theories must be socially caused— does not follow from them. He charges that Hesse is assuming, without argument, that anything that is not determined by deductive logic must be the product of social factors. But, he protests, why should we think that the decision to adopt a particular set of ampliative rules cannot be the result of reasoning? Only historical investigation can reveal what caused a particular scientist to adopt a particular set of ampliative rules. From the mere fact of deductive underdetermination of theories by data nothing at all follows about the nature of those causes: they may be sociological or they may not.

Although Laudan does not discuss it explicitly, he obviously thinks that premise (3)—the fact that different scientists in different periods have adopted different extraempirical criteria for choosing among theories—is irrelevant to Hesse's conclusion. The different values adopted, or the different weights attached to values such as simplicity, explanatory scope, and fertility, imply nothing about what caused individual scientists to adopt them. Indeed, in every case the adoption may have been the result of deductive reasoning, the difference in outcome being due solely to the different premises from which the scientists started. Another, more likely, possibility is that different scientists not only used different starting assumptions but also used both deductive and nondeductive reasoning to reach their conclusions about how rival scientific theories should be assessed.

Having dismissed Hesse's argument as a non sequitur, Laudan turns his attention to Bloor. The argument of Bloor's that Laudan criticizes is taken from a special issue of the journal *Studies in History and Philosophy of Science*.[35] The lead article in that issue, a piece by Bloor, was followed by responses by a number of critics to whom Bloor, in turn, replied. One of those critics was Gerd Buchdahl (an editor of the journal), and Bloor's argument comes from his reply to Buchdahl's criticisms.[36] Buchdahl was especially critical of the following passage from Bloor's paper:

What is it that can then account for the known stability of our explicit theoretical knowledge? For the sociologist the answer is simple.

Such stability as there is in a system of knowledge comes entirely from the collective decisions of its creators and users. That is to say: from the requirement that certain laws and classifications be kept intact, and all adjustments and alterations carried out elsewhere. . . . We need not assume that a protected law or classification is singled out because of any intrinsic properties like truth, self-evidence or plausibility. Of course, such properties will be imputed to them, but this will be a justification for the special treatment rather than the cause of it.[37]

Buchdahl took exception to Bloor's pronouncement that the "stability" of some laws in the sciences (that is, the fact these laws have been accepted by scientists over a long period of time) comes *entirely* from the decision of scientists (the law's "creators and users") to protect the law from refutation. To Buchdahl it seemed obvious that the truth of a law and the evidence for it can also have much to do with the law's stability. If the law were true, then the evidence would run in its favor; scientists would accept the law on the basis of that evidence; and thus the fact that the law is true would, contrary to Bloor, play a role in explaining why scientists have retained it. In his reply to Buchdahl on this point, Bloor writes:

If my use of the word "completely" gave the impression that sensory input had nothing whatever to do with the resulting system of knowledge, then it was indeed the wrong word. The fact is, however, that with no change whatsoever in their evidential basis, systems of belief can be and have been destabilized. Conversely they can be and have been held stable in the face of rapidly changing and highly problematic inputs from experience. So the stability of a system of a belief is the prerogative of its users.[38]

It is this argument of Bloor's that Laudan criticizes in his paper. Laudan charges that Bloor's argument, like Hesse's argument discussed earlier, is a glaring non sequitur: its conclusion does not follow from its premises. Laudan accepts that the two premises are true: scientists have sometimes changed their theoretical beliefs without that change being prompted by new evidence; scientists have also sometimes retained their theoretical beliefs despite new evidence that seemed to refute them. But from this it scarcely follows that "the stability of a belief is the prerogative of its users." The fact that in some cases evidence has not been decisive in causing scientists to change or retain their beliefs does not imply that evidence can never play such a role.

In judging Bloor's argument invalid, Laudan interprets Bloor's conclusion as asserting, in effect, that scientists can rationally hold onto any belief, regardless of the evidence against it. But in the final paragraph of his section on Bloor, Laudan interprets Bloor's conclusion as saying, instead, that no evidence can ever compel a rational scientist to change his beliefs. These two interpretations of Bloor's conclusion are not quite the same. For example, it might still be irrational for a scientist to retain a particular belief in the face of refuting evidence, even though that evidence does not compel rejection of the belief, either logically or causally. (Logical "compulsion" would presumably arise only if the evidence entailed that the belief is false, and, as Laudan keeps reminding us, there is more to rationality than deductive logic.) But the important point is Laudan's insistence that the belief change be a rational one. For, as he argued in evaluating Quine's version of the underdetermination thesis, without the restriction to the rational retention of belief, the contention that any belief can be retained "come what may" is trivial. Thus, Laudan concludes that Bloor's argument from underdetermination does nothing to make plausible the strong programme's claim that scientific beliefs are always caused by social forces and interests.

3.5 | The Duhem Problem: A Bayesian Solution

The Duhem problem is the Achilles' heel of Popper's naïve falsificationism. Very few scientific theories of any importance imply testable consequences all by themselves: auxiliary hypotheses (including other theories, assumptions, and initial conditions) are also needed to derive predictions. When those predictions fail—when Kuhnian "anomalies" arise—where should we pin the blame? As early as 1937, Hempel pointed out that the ambiguity of falsification is the inevitable consequence of the "holist" nature of theory testing.[39] Even Popper and his followers recognized that rejecting a core theory as soon as one of its predictions failed would be foolhardy and inconsistent with well-established scientific practice. Often the most reasonable response to predictive failure is to amend or replace one or more of the auxiliaries rather than to discard prematurely a theory that has worked well in the past and in which one has invested a considerable amount of time and effort (not to mention one's scientific reputation). As Howson and Urbach explain, both Kuhn and Lakatos attempt to make a virtue out of necessity by condoning a degree of conservatism when it comes to retaining theories that are central to a paradigm (Kuhn) or a scientific research programme (Lakatos). But neither author, in their view, has succeeded in articulating in a defensible way which theory-saving moves are permissible. Kuhn says virtually nothing about the matter; Lakatos (the "sophisticated falsificationist") allows significant changes to the "protective belt" of auxiliaries surrounding the programme's "hard core" just as long as the programme occasionally generates novel predictions as a result. How many such predictions are required to prevent a formerly progressive programme from becoming degenerating is left unspecified. While Howson and Urbach concede that Lakatos gives a useful framework for describing scientific change (as manifested in a series of case studies), they fault his methodology of research programmes for failing to "rationalize" or explain those changes. What is lacking, they insist, is a general normative account that shows why retaining a core theory in the face of anomalies is, under certain conditions, justified or rational. This is what they aim to provide using a Bayesianism analysis in which probabilities represent the degrees of belief in various propositions held by actual scientists. This analysis was first given in 1979 by Jon Dorling in his paper "Bayesian Personalism, The Methodology of Scientific Research Programmes, and Duhem's Problem."[40]

Jon Dorling's Analysis

The following account of Dorling's analysis presupposes a familiarity with probability theory and Bayes's theorem, topics that are explored at length in chapter 5. Also in chapter 5 is an article on the Bayesian approach to philosophy of science by Alan Chalmers in which he criticizes Dorling's solution to the Duhem problem. Chalmers discusses the same historical case study as that used by Howson and Urbach, namely, the theory of British physician and chemist William Prout (1785–1850) that all elements are built up out of hydrogen and, as a consequence, their atomic weights are whole-number multiples of the atomic weight of hydrogen. Prout's theory was popular among British chemists in the first half of nineteenth century. The underlying structural theory made sense (since hydrogen is the lightest element), and Prout's prediction of whole-number ratios fitted most of the then-known elements pretty well; but the measured values of the atomic weights of a few elements (especially chlorine) were inconsistent with Prout's theory.[41] Prout and his followers (such as the chemist Thomas Thomson) assumed that the fault must lie with the procedures used to measure atomic weights (and perhaps also with the purity of their samples). Thus they protected their theory, T, from refutation by anomaly, E, by blaming an auxiliary hypothesis, A. Howson and Urbach claim that far from this being irrational, it was actually the rational thing for them to do given their degrees of belief in T and A respectively, prior to the discovery of the anomalous atomic weights.

In his 1979 paper Dorling applied his analysis to a case study from nineteenth-century physics in which the observed secular acceleration of the Earth's moon disagreed with the value calculated from Newtonian theory by the British mathematician and astronomer John Couch Adams (1819–92). Newtonian theory at that time was a well-entrenched part of science with many predictive successes to its credit. As in the Prout case, initial confidence in theory T was high. Though it is hard to be precise about these things, in both cases we can (following Dorling, and Howson and Urbach) put $P(T) = 0.9$. One of the auxiliary hypotheses that Adams had assumed in his calculation was that tidal friction has a negligible effect on lunar acceleration. This auxiliary hypothesis, A, was believed to be more probable than not (otherwise Adams would not have used it) but not as probable as T. In both the Adams case and the Prout case, we can (again following Dorling, and Howson and Urbach) put $P(A) = 0.6$. Dorling then argues that given these initial degrees of belief and given the crucial fact that the anomalous result, E, could not be explained by any plausible rival theory to T available at the time, it follows from a Bayesian analysis that $P(T/E)$ is only slightly less than $P(T)$; but $P(A/E)$ plummets to a very low value. Thus, in the face of anomaly E, it was rational to reject A as false and retain T, with one's initial high confidence in that theory barely diminished.

Here is an outline of Dorling's formal argument, with some of the intermediate steps filled in.

$(T \,\&\, A) \vDash E_1$, but we observe E, and $E \vDash {\sim}E_1$.
Let $P(T) = 0.9$, and $P(A) = 0.6$

In order to calculate how observing E should change our degrees of belief in T and A, we need to evaluate $P(T/E)$ and $P(A/E)$ using Bayes's theorem.

Bayes's theorem says that $P(H/E) = [P(E/H)P(H)/P(E)]$.

So we need to calculate $P(E)$, $P(E/T)$, and $P(E/A)$.

1. $P(E) = P(E/T)P(T) + P(E/{\sim}T)P({\sim}T)$
2. $P(E/T) = P(E \,\&\, A/T) + P(E \,\&\, {\sim}A/T)$

Note: (2) follows from the theorem of total probability. If $P\,(a_1 \lor a_2 \ldots \lor a_n) = 1$, and $a_i \vDash {\sim}a_j$ for $i \neq j$, then $P(b) = \Sigma\, P\,(b \,\&\, a_i)$ for any sentence b.

3. $P(E/T) = P(E/\, T \,\&\, A)P(A/T) + P(E/\, T \,\&\, {\sim}A)P({\sim}A/T)$.   [from 2]

Assume that A and T are mutually independent. This assumption seems reasonable in the Prout case and the Adams case. So,

$P(A/T) = P(A)$ and $P({\sim}A/T) = P({\sim}A)$. Thus,

4. $P(E/T) = P(E/\, T \,\&\, A)P(A) + P(E/\, T \,\&\, {\sim}A)P({\sim}A)$.

But since $(T \,\&\, A) \vDash {\sim}E$, $P(E/\, T \,\&\, A) = 0$. So,

5. $P(E/T) = P(E/\, T \,\&\, {\sim}A)P({\sim}A)$.

Similarly, we can derive:

6. $P(E/{\sim}T) = P(E/\, {\sim}T \,\&\, A)P(A) + P(E/\, {\sim}T \,\&\, {\sim}A)P({\sim}A)$.
                    [from 4 by substituting ${\sim}T$ for T]
7. $P(E/A) = P(E/\, {\sim}T \,\&\, A)P({\sim}T)$
                    [from 5 by substituting A for T, and T for A]

If we let $P(T)=p$, and $P(A)=q$, $P(E/\sim T \& A)=x$, $P(E/T \& \sim A)=y$, and $P(E/\sim T \& \sim A)=z$, we can express $P(T/E)$ and $P(A/E)$ as functions of p, q, x, y, and z:

8. $P(T/E)=p(1-q)/X$, and
9. $P(A/E)=(x/y)q(1-p)/X$, where
10. $X=[p(1-q)+(1-p)(qx/y+(1-q)z/y)]$.

The expressions for $P(T/E)$ and in $P(A/E)$ in (8) and (9) are quite revealing. They show how the disconfirmatory power of the anomaly E can be so much greater for A than for T. Since

$$P(T/E)/P(T)=(1-q)/X, \text{ and}$$
$$P(A/E)/P(A)=(x/y)(1-p)/X,$$

the *ratio* of the effect of E on A to the effect of E on T is given by $(x/y)(1-p)/(1-q)$. When, as in Dorling's Adams case study, $y=50x$, $p=0.9$, and $q=0.6$, this means that the proportional reduction in the probability of A—as measured by $P(A/E)/P(A)$—is 200 times greater than the proportional reduction in the probability of T: the probability of T is scarcely affected by the anomaly E, but the probability of A declines from 0.6 to 0.003. In Howson and Urbach's Prout example, with $y=2x$, the probability of A is reduced from 0.6 to 0.073 (which is 8 times as large as the corresponding reduction in the probability of T.)[42]

As a formal analysis, Dorling's Bayesian argument is impeccable, and it certainly establishes the possibility that it can be rational to retain a core theory in the face of an anomaly. The important question is whether one can make plausible assignments to the probabilities involved. In particular how do we decide what numbers to assign to $P(E/\sim T \& A)=x$, $P(E/T \& \sim A)=y$, and $P(E/\sim T \& \sim A)=z$? As Dorling and Howson and Urbach point out, given assignments for $P(T)$ and $P(A)$, $P(T/E)$ and $P(A/E)$ depend only on the ratios of x to y, and of z to y, not on their absolute values. This gives us considerable leeway in assigning numbers to $P(E/\sim T \& A)$, $P(E/T \& \sim A)$, and $P(E/\sim T \& \sim A)$ since what matters are their relative values, not the specific numbers. Both Dorling, and Howson and Urbach, give very low values to $P(E/\sim T \& A)$ on the grounds that there were no rivals to T, available at the time in question, that could explain the anomalous result E if A were true. Thus, while logically the negation of T (the so-called "catchall hypothesis") is equivalent to a long disjunction of hypotheses that are contraries of T, in practice the range of hypotheses that were considered as alternatives to T by working scientists in the historical context was limited to just a few, none of which offered a plausible account of why E is true if A were correct. Howson and Urbach note that in the nineteenth century, chemists who rejected the Prout hypothesis typically assumed that atomic weights were randomly distributed. Since the reported value for chlorine was 35.83, this indicates an assumed accuracy to the second decimal place. Thus it seems reasonable to divide the interval between 35 and 36 into a hundred equal increments and thus estimate that the probability that the atomic weight falls within any particular one of those (on the assumption that T is false and A is true) to be 0.01. $P(E/\,T\,\&\,\sim\!A)$ is reckoned to be higher than $P(E/\,\sim\!T\,\&\,A)$ because there was a genuine likelihood, in both the Prout case and in the Adams case, that the T might be able to explain the anomalous result if the auxiliary assumption A were false. In order to assign a value to $P(E/\,\sim\!T\,\&\,\sim\!A)$ that is historically plausible, we have to determine if there were rivals to T at the time in question that could explain E if A were false. In the Prout case, Howson and Urbach judge this to be quite unlikely (hence they set $z=x$); in the Adams case, Dorling judges that this was much more likely than x (and thus sets $z = y$). Ultimately these assignments rest on historical judgments that reflect the degrees of belief of scientists whose decisions about theories we are trying to explain and justify. For the Bayesian solution to the Duhem problem, it is important to remember that only the *relative* magnitudes of these degrees of belief matter—the Bayesian solution is robustly insensitive to their actual values. Given the initial degrees of belief in T and A, the ratio that explains why the anomaly E disconfirms A much more than it disconfirms H is that between $P(E/\,\sim\!T\,\&\,A)$ and $P(E/\,T\,\&\,\sim\!A)$. The value of that ratio depends on the ability of the available alternatives to T to explain E on the supposition that A is true, and scientists' confidence in T to account for the anomaly on the supposition that A is false.

3.6 | Summary

When Pierre Duhem published his views on holism and the ambiguity of falsification at the turn of the twentieth century, he could not have guessed how they would be used and misused by later philosophers of science. Unlike the more recent doctrines that have flown under the banner of the Duhem-Quine thesis, Duhem's own version of holism is moderate, qualified, and plausible. Duhem argues that individual physical theories and postulates cannot be tested in isolation. Testable predictions (in the form of observation statements) can be made only when a single theory or postulate is combined with other theories, hypotheses, and assumptions. Observation or experiment may reveal that a prediction is false, but deductive logic (in the form of *modus tollens*) cannot tell us which theory, hypothesis, or assumption is at fault. All we can deduce is that at least one of these is false. To that extent, falsification is ambiguous.

Duhem was especially concerned to deny that there can be crucial experiments in physics. Since a crucial experiment would not only falsify one theory but also establish its rival theory as true or highly probable, clearly no experiment can be crucial if falsification is ambiguous. Moreover, even if an experiment could refute a particular theory, the rival theory need not be true or probable. In this respect the logic of testing in physics differs from the reductio ad absurdum reasoning used in logic and mathematics.

Duhem was equally critical of inductivism, arguing that no theory could be "deduced from the phenomena." Newton and Ampère had both claimed that each individual postulate of their theories could be inferred directly from observational facts or from experimental laws that, in turn, had been derived from such facts. Holism implies that this is impossible, since no single postulate by itself has any observational consequences. Duhem also pointed out that if Newton's gravitational theory is true, then the laws from which its postulates are allegedly deduced (Kepler's laws of planetary motion) must be false. Hence, again, inductivism fails.

Duhem was well aware that, because of the ambiguity of falsification, protecting a theory from refutation by making changes to the auxiliary hypotheses and assumptions used in testing that theory was possible, at least in principle. But he did not conclude from this that no theory can ever be falsified or that any scientific principle, however well entrenched, is immune to refutation. For example, Duhem criticized Poincaré's conventionalism, which held that many high-level principles in physics are not empirical hypotheses but definitions. Duhem agreed with Poincaré that individual principles (such as Newton's first law of motion) could not be tested in isolation. But he denied that it follows that these principles are nonempirical definitions. For, as with any other postulate in science, these principles can be tested indirectly, once they are combined with auxiliary hypotheses and assumptions. Thus, Duhem used his holist thesis to argue for the contention that all physical theories and postulates are testable and, in principle, falsifiable.

Duhem's account of how falsification can occur is rather sketchy and relies on what he calls scientific "good sense." Suppose there are two rival theories. One theory is fruitful in suggesting new experiments and its predictions prove successful. The other is less successful; with growing frequency, it and its associated auxiliaries have to be repaired and modified in ways that are often arbitrary, in order to protect the theory from refutation. Under these circumstances "good sense" tells the scientist that it is time to give up the ailing theory in favor of its more successful rival. Logic does not force the decision, but as time passes not adopting the more successful theory becomes increasingly unreasonable and irrational.

Quine was ignorant of Duhem's work when he wrote "Two Dogmas of Empiricism." Nonetheless, he used premises about holism and the ambiguity of falsification resembling Duhem's to reach controversial conclusions about meaning, analyticity, and a priori knowledge. In this paper, he attacked two empiricist dogmas held by the logical positivists of the Vienna Circle— reductionism and analyticity. Reductionism, in its purest form, is the thesis that each meaningful synthetic statement is logically equivalent to (can be reduced to) the set of experiential statements (observation statements) that it implies. The experiential statements can be verified directly. A synthetic statement is conclusively verified when all of its experiential consequences are verified. According to the verifiability principle of meaning embraced by the logical positivists, the meaning of a synthetic statement is identified with the set of its verifiable consequences. So, for example, if two statements have exactly the same consequences, then they are synonymous. Synthetic statements that have no empirical consequences are judged to be meaningless (or, at least, to lack any cognitive meaning). Analyticity is the doctrine that certain statements having no empirical consequences (such as those of logic and mathematics) are true solely in virtue of the meanings of the words and terms used to express them. The logical positivists argued that it is because the propositions of logic and mathematics are analytic that we have certain a priori knowledge of their truth.

Quine rejects reductionism on the grounds that no single statement by itself logically implies anything about experience or observation. Statements have observationally verifiable consequences only when combined with other statements. With regard to science, Quine proclaims that the whole of science is needed to derive testable consequences. (In Quine's terminology, the smallest unit of empirical significance is science as a whole.) Thus, in "Two Dogmas," Quine espouses a global version of holism. He criticizes the doctrine of analyticity on the grounds that no satisfactory account has been given of the notion that a statement is true solely in virtue of its meaning. In particular, he rejects both the characterization of analytic statements as those that reduce to tautologies by synonym substitution and Carnap's appeal to semantical rules in artificial languages. In light of these failures to explain analyticity, Quine suggests that there is no such thing. No statements are analytic, and none are a priori. In principle, any statement, even one from logic or mathematics, could be revised or abandoned in light of experience. Conversely, Quine also insists that any statement can be retained as true, regardless of experience, if we make drastic enough changes to the rest of our system of belief. Thus, he concludes by espousing a thesis of radical underdetermination: observation and evidence do not determine (or even constrain) which individual theories we decide to accept and which we choose to reject.

Gillies explores the similarities and differences between Duhem and Quine, criticizing aspects of each before offering a version of the Duhem-Quine thesis that he judges to be plausible. He rejects the global character of Quine's holism, faults Quine for paying little attention to rational constraints on theory choice besides deductive logic, and thinks that Quine is wrong in claiming that no single scientific statement can be tested without the aid of auxiliary hypotheses and assumptions. He criticizes Duhem for not extending his holist thesis to sciences other than physics. In particular, Gillies sides with Quine by arguing (against Duhem) that mathematical theories and even logical principles (such as the law of excluded middle) could be given up as false in response to observation and experiment.

In "Demystifying Underdetermination" Laudan attacks a variety of arguments that have been used by philosophers and sociologists of science who are skeptical about the rationality and objectivity of science. Typically, these arguments appeal to some version of the underdetermination thesis in an attempt to show that there must be an important sociological, non-rational dimension to the decisions that scientists make about theories. In assessing these arguments, Laudan emphasizes the importance of distinguishing between deductive (Humean) and ampliative versions of the underdetermination thesis. According to the deductive underdetermination thesis, indefinitely many theories are logically consistent with any given body of evidence and (in conjunction with other statements) deductively entail that evidence. Ampliative underdetermination comes in many varieties. The most far-reaching is what Laudan calls the egalitarian thesis, namely, that every theory is as well supported by the evidence as any of its rivals. The thesis is ampliative because it talks about support, which includes such things as inductive confirmation, explanatory power, and simplicity. These go beyond the relations of logical consistency and deductive entailment to which the deductive thesis of underdetermination confines itself.

Laudan insists that, if it is not to be utterly trivial, Quine's assertion that any theory can be held as true regardless of the evidence must be understood normatively—as claiming that any theory can be held *rationally* whatever the evidence. He argues that deductive underdetermination is insufficient to support Quine's assertion and that Quine's assertion (with the crucial stipulation about rationality) presupposes the truth of the egalitarian thesis. For if the egalitarian thesis were false, then there would be times when the evidence would support one theory much better than its rivals, and, under those circumstances, it would be irrational to retain the inferior theory. Moreover, he contends that the further arguments Quine gives in "Two Dogmas" do nothing to show that any theory whatever can be rationally retained in the face of recalcitrant evidence. Laudan concludes that neither in "Two Dogmas" nor in his other writings has Quine made the egalitarian thesis plausible.

In the remainder of his article, Laudan looks at the work of Goodman, Kuhn, Hesse, and Bloor to see whether any of them has made a plausible case for ampliative underdetermination (in the form of the egalitarian thesis). Without venturing a solution to Goodman's new riddle of induction, Laudan observes that even if it were true that some gruelike hypotheses are as well supported by our evidence as some nongruelike hypotheses, this hardly shows that any theory whatever enjoys the same support from that evidence. Kuhn defends a local version of the underdetermination thesis by restricting its scope to the known rivals of a given theory and arguing that shared methodological rules (involving values such as accuracy, scope, and simplicity) are too vague to pick out one theory as the best. Laudan counters by describing other rules, also widely shared by scientists, that are much less vague and which could, at least in principle, be used to resolve a contest between competing theories. Laudan illustrates his point with a historical example, showing how the Newtonians and the Cartesians settled their dispute about the shape of the earth by applying a simple methodological rule to the evidence they had gathered. Finally, Laudan examines the arguments of Hesse and Bloor in favor of the strong programme in the sociology of science. Laudan agrees with Hesse that deductive underdetermination implies that scientists must rely on extraempirical criteria in deciding among theories, but he rejects as gratuitous her implicit assumption that the adoption of such criteria must be caused by social factors. Similarly, Laudan finds no merit in Bloor's argument which, from premises about historical cases in which evidence has not determined the theoretical choices made by scientists, concludes that scientists can retain any theory or law regardless of any evidence. As in his criticism of Quine, so, too, in his response to Bloor, Laudan emphasizes the need to stipulate that it must be rational to retain the theory in the face of recalcitrant evidence if underdetermination is to be anything more than a boring triviality. Laudan concludes that none of the arguments he has examined do anything to make plausible an interesting version of the underdetermination thesis.

If one regards consistency with the probability calculus as the sole rational underpinning of inductive inference, then Howson and Urbach provide a demonstration that, under certain conditions, it can be rational for scientists to retain a preferred theory in the face of anomalous evidence by rejecting as false an auxiliary hypothesis implicated in the failed prediction. Following the Bayesian analysis of Jon Dorling, Howson and Urbach argue that, in at least some important historical cases, the conditions on degrees of belief required by Dorling's Bayesian analysis are satisfied and that this shows that something other than external social factors or Duhemian appeals to "good sense" can explain why scientists were justified in sticking with a theory with a proven track record when some of its predictions fail. The larger question of whether or not we should adopt the Bayesian framework for understanding inductive inference in the sciences is examined in chapter 5.

■ | Notes

1. In its most extreme form, inductivism claims that theories can be deduced (in some sense of *deduce*) from the observational phenomena. Duhem attacks this claim later in his article.

2. For a discussion of whether Duhem was correct in limiting his holist thesis to physics and excluding disciplines such as physiology and chemistry, see the later section, "Gillies' Criticisms of Duhem" (below).

3. Adolf Grünbaum was among the first to point out that Quine's thesis that any theory whatever can be retained in the face of contrary evidence (in a non-trivial way) cannot be established solely on the basis of Duhem's minimal thesis about the ambiguity of falsification. See Adolf Grünbaum, "The Duhemian Argument," *Philosophy of Science* 27 (1960): 75–87, and "The Falsifiability of Theories: Total or Partial? A Contemporary Evaluation of the Duhem-Quine Thesis," *Synthese* 14 (1962): 17–34. In fact, Grünbaum offers an example from physical geometry in

which he argues that Duhem's minimal thesis is true but Quine's thesis is false. That Duhem's minimal thesis is not trivial is illustrated by the fact that Gillies, for example, thinks that it is false as a general claim about all theoretical hypotheses in science. See the section "Gillies' Version of the Duhem-Quine Thesis" (below) for an evaluation of this claim.

4. At the time of Foucault's experiment, the particle theory had already been largely abandoned in favor of the wave theory of Young and Fresnel. Whether Foucault performed a crucial experiment is an epistemological judgment about the evidential significance of his work. It does not imply that Foucault's work was historically crucial in changing people's minds about the status of the rival theories (although it did have this effect on one important scientist, namely, Biot).

5. Duhem focuses exclusively on the conception of a crucial experiment as one having "the power to transform a physical hypothesis into an indisputable truth" [p. 235], effectively ignoring the weaker alternative that a crucial experiment might merely establish a hypothesis as well confirmed. In our example, the rival theory T^* predicts the experimental observation $\sim O_1$ but, presumably, Duhem would regard this single successful prediction as insufficient to establish T^* as well-confirmed. Many philosophers of science would disagree with Duhem on this point, arguing that, in the right circumstances, a single successful prediction, especially a prediction of something considered highly improbable before the experiment was performed, could significantly raise the probability of T^* and thus make it well confirmed. See chapters 4 and 5 for a fuller discussion of this and related issues concerning confirmation.

6. For Popper's dismissal of induction and inductive confirmation, see the papers by Popper and Salmon in chapter 4 and the sections "Popper's Rejection of Induction" and "Salmon's Criticism of Popper's Anti-Inductivism" in the accompanying commentary.

7. Duhem does not consider trying to offer a more refined interpretation of Newton's method in terms of idealizations that give successively better approximations of the data. Newton's use of idealized evidence is explored in Ronald Laymon, "Newton's Demonstration of Universal Gravitation and Philosophical Theories of Confirmation," in *Testing Scientific Theories: Minnesota Studies in the Philosophy of Science*, vol. 10, ed. J. Earman (Minneapolis: University of Minnesota Press, 1983), 179–99.

8. In "Two Dogmas in Retrospect," *Canadian Journal of Philosophy* 21 (1991): 265–74, Quine confesses that, at the time he wrote his paper, he did not know about Duhem. It was Carl Hempel and Philipp Frank who brought Duhem to his attention.

9. In his later publications, Quine has drawn attention to the differences between his views and those of Duhem, distinguishing between Duhem's holism and his own far more sweeping thesis of empirical underdetermination. See, for example, W. V. Quine, "On Empirically Equivalent Systems of the World," *Erkenntnis* 9 (1975): 313–28.

10. This definition of analyticity is controversial. The characterization in terms of meanings is the one that Quine adopts, following Carnap. For some alternative definitions, see the next section, on the status of mathematical truths.

11. Thus, the relation between the two dogmas is, strictly speaking, neither one of identity nor even of entailment. In saying that the two dogmas are "identical," Quine is claiming merely that anyone who accepts one of the dogmas (reductionism) will have a strong psychological propensity to accept the other (analyticity) as well.

12. Carl G. Hempel, "On the Nature of Mathematical Truth," *American Mathematical Monthly* 52 (1945): 543–56.

13. Getting clear about the precise nature of the distinction between pure geometry and empirical geometry is no easy matter. For a defense of the distinction, see Carl G. Hempel, "Geometry and Empirical Science," *American Mathematical Monthly* 52 (1945): 7–17, and Ernest Nagel, *The Structure of Science* (London: Routledge & Kegan Paul, 1961), 215–76. For some of the difficulties with the distinction, see Alan Hausman, "Non-Euclidean Geometry and Relative Consistency Proofs," in *Motion and Time, Space and Matter*, ed. P. K. Machamer and R. G. Turnbull (Columbus: Ohio State University Press, 1976), 418–35.

14. Quine's remarks about the limitations of Carnap's characterization of analytic statements as those that come out true in all state-descriptions can be appreciated by thinking of truth tables. Consider Quine's own example of the compound statement, "John is a bachelor, and John is married." Obviously, this is a conjunction of the two statements A = "John is a bachelor," and B = "John is married." If A and B are regarded as atomic statements (and thus as ingredients in a state-description of a possible world), then one possible assignment of truth values to these atomic statements would have them both true. Thus, there would be at least one state-description (at least one possible world) in which "John is a bachelor, and John is married" is true. Similarly, there would be at least one other state-description (at least one other possible world) in which "John is a bachelor, and John is married" is false. Thus, the compound statement would be misclassified as synthetic. The problem, as Carnap was clearly aware, is that before employing his definition of analyticity, we first must ensure that all the statements that we take as atomic are logically independent of one another. In particular, we must rule out as candidates for atomicity any pair of sentences that contain synonyms or antonyms such as "bachelor" and "married man." Thus, the problem with this approach is the same as the problem with the other approaches Quine criticizes, namely, it fails to give a satisfactory account of synonymy.

15. As many of Quine's critics—notably H. P. Grice and P. F. Strawson—have pointed out, it is quite common for there to be a circle of expressions such that if any one member of the circle is understood or explained, then any other member of the circle can be understood or explained in terms of it. The terms *morally wrong, blameworthy,* and *violation of moral rules* belong to one such family circle, and the terms *analytic, synonymous, necessary,* and *semantical rule* belong to another. In arguing, as Quine does, that we cannot make "satisfactory sense" of analyticity, he seems to be assuming that a satisfactory account of an expression must give necessary and sufficient conditions for that expression without using any other member of the family circle to which it belongs. As a necessary condition for an expression's making sense, Quine's assumptions appear to be unreasonable, for there are many expressions that cannot be formally defined except in terms of members from the same family group. See H. P. Grice and P. F. Strawson, "In Defense of a Dogma," *Philosophical Review* 65 (1956): 141–58.

16. This is not the same as attributing to Quine the following, invalid, argument: if mathematical truths are analytic, then they are a priori; they are not analytic; therefore, they are not a priori. Rather, Quine's remarks about the revisability of mathematics and logic in the final section of his paper are intended to show that, contrary to popular belief, mathematics and logic are not a priori and, hence, they are not analytic.

17. Understandably, it is Quine's claim about the possibility of retaining a sentence "come what may" that has excited the interest of philosophers of science. But given Quine's attack on a priori knowledge (via his repudiation of the analytic-synthetic distinction), his claim about falsifying logical and mathematical principles is at least as radical and closer to Quine's real concern in "Two Dogmas." From Quine's perspective, (i) and (ii) are opposite sides of the same coin, since each follows from (a) Quine's doctrine that there is no difference in kind, either semantic or epistemological, between the two types of sentence traditionally classified as analytic and synthetic, respectively; (b) Quine's global version of holism; and (c) Quine's assumptions about the limitations of scientific method in deciding matters of confirmation and disconfirmation. From a more traditional perspective, (i) and (ii) seem curiously at odds with one another, for (i) seems to assert that any sentence could, in principle, be "made" a priori by the decision never to give it up, while (ii) asserts that no sentence is a priori in the sense of being immune to revision.

18. W. V. Quine, "On Empirically Equivalent Systems of the World," 315. Quine thinks that we often fail to recognize the unity that logic and mathematics confer on the sciences because, in the grip of the analytic-synthetic distinction, we place logic and mathematics in a different category from the rest of science.

19. Classic articles by Brouwer and Dummett defending intuitionism can be found in Paul Benacerraf and Hilary Putnam, eds., *Philosophy of Mathematics*, 2d ed. (Cambridge: Cambridge University Press, 1983); see also Arend Heyting, *Intuitionism: An Introduction* (Amsterdam: North Holland, 1956), and John P. Burgess, *Philosophical Logic* (Princeton, N.J.: Princeton University Press, 2009), ch. 6.

20. Hilary Putnam, "Is Logic Empirical?" *Boston Studies in the Philosophy of Science*, vol. 5, ed. R. S. Cohen and M. W. Wartofsky (Dordecht, Netherlands: D. Reidel, 1968), 216–41; revised version reprinted as "The Logic of Quantum Mechanics," *Mathematics, Matter, and Method, Philosophical Papers*, vol. 1 (Cambridge: Cambridge University Press, 1975), 174–97. For Putnam, all the laws of the *true* logic are analytic, but we discover which they are empirically, by doing physics. Thus Putnam denies that all analytic truths are a priori. He thinks that the logical connectives in quantum logic are the same as those in classical logic because they satisfy many of the same basic principles. What quantum mechanics teaches us, according to Putnam, is that the classical distribution laws governing *and, not,* and *or* are not analytic.

21. One can appreciate even more vividly the failure of distribution in quantum logic by considering the following illustration. Let the variables X and P be the position and momentum of a simple system consisting of a single particle. Let X_1, $X_2, \ldots X_n$ be a complete list of all the possible values of position that the particle can have, and let $P_1, P_2, \ldots P_m$ be a complete list of all the possible values of its momentum. In quantum logic, the conjunction $(X_1 \vee X_2 \vee \ldots \vee X_n) \& (P_1 \vee P_2 \vee \ldots \vee P_m)$ is a necessary truth, a tautology. But the disjunction $(X_1 \& P_1) \vee \ldots \vee$

$(X_i \ \& \ P_j) \ldots \vee (X_n \ \& \ P_m)$ is a quantum logical contradiction. See Peter Gibbins, *Particles and Paradoxes: The Limits of Quantum Logic* (Cambridge: Cambridge University Press, 1987), 152.

22. Hilary Putnam, "There Is At Least One A Priori Truth," *Erkenntnis* 13 (1978): 153–70; reprinted in H. Putnam, *Realism and Reason* (New York: Cambridge University Press, 1983), 98–114. Putnam actually defends the a priori status, not of the classical law of noncontradiction (that no statement is both true and false) but of the weaker assertion that not every statement is both true and false (which, unlike the law of noncontradiction, allows for the possibility that some statements might have both truth values).

23. One feature of the organization of Laudan's article might confuse the reader. The first two sections after the introduction are labeled "Vintage Versions of Underdetermination" and "Ampliative Underdetermination," respectively. But the "Vintage" section includes discussions of both Hume and Quine, even though Quine is concerned largely, but not exclusively, with ampliative underdetermination. But there is a rationale for Laudan's inclusion of Quine in the same section as Hume: although Quine's versions of the underdetermination thesis are all ampliative, he never seriously discusses ampliative inference but confines himself almost entirely to the deductive relations of entailment and consistency. Indeed, Laudan argues that Quine comes perilously close to affirming doctrines of ampliative underdetermination solely on the basis of deductive underdetermination.

24. Although, of course, if Quine's (0) really does entail EGAL and we can show that EGAL is false, then we would have refuted (0). Laudan's aim in this paper is more modest, namely, to show that Quine has not given us any good reason to believe that (0) is true because he has failed to give us any good reason for accepting EGAL.

25. And indeed, as Laudan points out in a footnote, Quine himself conceded this in a letter to Adolf Grünbaum (dated 1 June 1962) published in Sandra Harding, ed., *Can Theories Be Refuted?* (Dordecht, Netherlands: D. Reidel, 1976), 132. Quine admits that his thesis in "Two Dogmas"—the thesis Laudan has labelled QUD—is "probably trivial" and that his only purpose in advancing it in that paper was to illustrate his doctrine of semantic holism, that individual sentences do not have empirical content except as members of much larger groups of beliefs.

26. Under the rubric of consistency, Kuhn includes both what Laudan calls internal consistency (freedom from logical contradiction) and what Kuhn calls external consistency (compatibility with other currently accepted theories). So another way of making Laudan's point is to say that one important component of what Kuhn calls consistency (namely, internal consistency) is not at all vague.

27. For Maupertuis's dispute with the Cassini brothers, see Mary Terrall, *The Man Who Flattened the Earth: Maupertuis and the Sciences in the Enlightenment* (Chicago, Ill.: University of Chicago Press, 2002). In the *Principia*, Book 3, Proposition 19, Newton calculated that the Earth's equatorial radius should exceed its polar radius by 17 miles. Newton's calculation assumed that the Earth's density is uniform—something that Newton recognized would be inaccurate if the density increases from the surface to the core. Michael Hoskin relates that Huygens (working in the Cartesian tradition) came to the same conclusion as Newton about the Earth's bulge but disagreed about the amount. The expeditions to Lapland and Peru verified the

bulge (and thus showed that the Cassinis' measurements were wrong); but they were unable to discriminate between the predictions of Newton and Huygens. Hoskin reproduces a splendid painting of P. L. M. Maupertuis (the leader of the northern expedition) showing him dressed in Lapp costume, pushing down on the Earth with his hand to make it bulge. See *The Cambridge Illustrated History of Astronomy*, ed. Michael A. Hoskin (Cambridge: Cambridge University Press), 169. In the *Principia*, Book 3, Proposition 20, Newton also calculates how the net acceleration due to gravity (and hence the weight of a body) must decrease as one goes from pole to equator, due to the Earth's rotation. This centrifugal effect, much larger than that due to the Earth's bulge, had been amply confirmed as early as 1672 on the island of Cayenne by Jean Richer using a pendulum that had been calibrated in Paris. Richer discovered that his pendulum would beat seconds in Cayenne only if shortened by a tenth of an inch.

28. David Bloor, *Knowledge and Social Imagery*, 2d ed. (Chicago, Ill.: University of Chicago, 1991). In the second edition of his book, Bloor weakened some of the claims attributed to the strong programme in the first edition. For a discussion of these changes and their significance, see Peter Slezak, "The Social Construction of Social Constructionism," *Inquiry* 37 (1994): 139–57. (The phrases *social constructionism*, *sociology of scientific knowledge*, and *social construction of knowledge* are often used to refer to the general movement to which the strong programme belongs.) For appraisals of the strong programme from standpoints different from those of Laudan and Slezak, see Thomas McCarthy, "Scientific Rationality and the 'Strong Program' in the Sociology of Knowledge," in *Construction and Constraint*, ed. E. McMullin (Notre Dame, Ind.: University of Notre Dame Press, 1988), 73–96, and Arthur Fine, "Science Made Up: Constructivist Sociology of Scientific Knowledge," in *The Disunity of Science*, ed. P. Galison and D. J. Stump (Stanford, Calif.: Stanford University Press, 1996), 231–54.

29. Hesse's conclusion about the strong thesis (as she calls it) is guarded. According to Hesse, the strong thesis does not imply that there is no difference between rational rules and mere social conventions, nor does it entail that there are laws of scientific and social development. "All that is implied is the possibility of finding some correlations, amounting to historical explanations in particular cases, between types of scientific theory and particular social provenance. It may be felt that the 'strong' thesis has now become so weak as to be indistinguishable from something any rationalist or realist could accept in regard to the development of science." Mary B. Hesse, "The Strong Thesis of Sociology of Science," in *Revolutions and Reconstructions in the Philosophy of Science*, Bloomington: Indiana University Press, 1980), 56–57.

30. For Laudan's attack on the strong programme, see Larry Laudan, "The Pseudo-Science of Science?" *Philosophy of the Social Sciences* 11 (1981): 173–98; David Bloor responded to Laudan in "The Strengths of the Strong Programme," *Philosophy of the Social Sciences* 11 (1981): 199–213; finally, Larry Laudan published "More on Bloor," *Philosophy of the Social Sciences* 12 (1982): 71–74.

31. This list of characteristics is given in David Bloor, *Knowledge and Social Imagery*, p. 7. All the quotations attributed to Bloor are from this page of his book.

32. Among the latter, Bloor is a champion of what he calls *negotiation*. Negotiation, as Bloor describes it, is the informal, interpretative process by which logic and

formal principles are applied to particular cases. Bloor thinks that the presentation of our reasoning in the form of deductive arguments is often an after-the-fact attempt to justify conclusions already reached on other grounds. Negotiation can also lead us to reject formal principles or even propositions in logic and mathematics that we formerly believed to be self-evident (such as the principle that the whole is always greater than the part, the denial of which can now serve as a definition of an infinite set). As far as Bloor is concerned, in logic and mathematics "there are no foundations other than social ones" (153). Here, as elsewhere in Bloor's manifesto for the strong programme, what professes to be simply the result of applying "the scientific method" to science itself actually rests on controversial claims about human psychology and a philosophical theory about a priori knowledge. For Bloor's account of negotiation, see David Bloor, *Knowledge and Social Imagery*, 131–56.

33. See R. G. Collingwood, *The Idea of History* (Oxford: Clarendon Press, 1946), and Imre Lakatos, "History of Science and Its Rational Reconstructions," in *Method and Appraisal in the Physical Sciences*, ed. C. Howson (Cambridge: Cambridge University Press, 1976), 1–39.

34. For these and other criticisms of the symmetry condition, see Larry Laudan, "The Pseudo-Science of Science?"

35. David Bloor, "Durkheim and Mauss Revisited: Classification and the Sociology of Knowledge," *Studies in History and Philosophy of Science* 13 (1982): 267–97.

36. Gerd Buchdahl, "Editorial Response to David Bloor," *Studies in History and Philosophy of Science* 13 (1982): 299–304, and David Bloor, "A Reply to Gerd Buchdahl," *Studies in History and Philosophy of Science* 13 (1982): 305–11.

37. David Bloor, "Durkheim and Mauss Revisited," 279–80.

38. David Bloor, "A Reply to Gerd Buchdahl," 306.

39. Carl G. Hempel, "The Problem of Truth," in Carl G. Hempel, *Selected Philosophical Essays*, ed. Richard Jeffrey (Cambridge: Cambridge University Press, 2000); originally published as "Le problème de la vérité," *Theoria* 3 (1937): 206–46.

40. Jon Dorling, "Bayesian Personalism, the Methodology of Scientific Research Programmes, and Duhem's Problem," *Studies in History and Philosophy of Science* 10 (1979): 177–87. See also Jon Dorling, "Further Illustrations of the Bayesian Solution of Duhem's Problem," unpublished, accessible at http:\\www.princeton.edu\~bayesway; and Richard Jeffrey, "Take Back the Day! Jon Dorling's Bayesian Solution of the Duhem Problem," *Philosophical Issues* 3 (1993): 197–207. For criticisms of the Bayesian solution of the Duhem problem, see Deborah C. Mayo, "Duhem's Problem, the Bayesian Way, and Error Statistics, or 'What's Belief Got to Do with It?'," *Philosophy of Science* 64 (1997): 222–44; Deborah C. Mayo, *Error and the Growth of Experimental Knowledge* (Chicago, Ill.: University of Chicago Press, 1996); and Alan F. Chalmers, *What Is This Thing Called Science?* 3rd ed. (Indianapolis, Ill.: Hackett, 1999), ch. 12, "The Bayesian Approach," reprinted in chapter 5.

41. We now know that Prout's theory about hydrogen being the building block of matter was close to the truth since the nucleus of hydrogen consists of a single proton and the nuclei of all heavier elements contain neutrons and protons. Since the mass of the neutron and the mass of the proton are almost exactly equal, all atomic

weights would be in whole-numbers ratios of the atomic weight of hydrogen were it not for the prevalence of naturally occurring isotopes in elements such as chlorine. Ernest Rutherford honored Prout in 1920 by naming the proton after him.

42. As Howson and Urbach point out, some choices of x and y can even result in T being *confirmed* by E.

4 |
Induction, Prediction, and Evidence

INTRODUCTION

Scientific laws and theories are universal generalizations that far outstrip the finite number of observations and experiments on which they are based. So it has seemed inevitable to philosophers of science from Aristotle to Carnap that some form of nondeductive inference must connect theories and laws with our evidence for them in some rational, justified way. Nearly always, induction and inductive inference are picked for this role. Scientific theories are confirmed, inductively, by evidence, and this inductive connection with evidence is, for the most part, what makes scientific theories reliable and our belief in them justified. Without inductive confirmation, science would be no better epistemically than blind guesswork, soothsaying, or wishful thinking.

Traditionally, the two main roles claimed for induction have been creative inference, which leads from evidence to the formulation of new theories (a logic of discovery) and confirmation, which connects evidence to theories after they have been formulated (a logic of justification).[1] These two roles have sometimes been connected by arguing that a theory generated in the right way—by inductive generalization from data—will automatically be justified without needing any new evidence to support it. This view has fallen into disfavor,[2] mainly because many theories in modern science appeal to entities and processes that are not, in any straightforward sense, observable.[3] Theories postulate genes, quarks, and electrons, for example, but it would be impossible to infer these theories inductively from observation reports in which such terms are entirely absent. Moreover, the mathematical complexity of modern scientific theories makes it hard to imagine how they could possibly be generated by inferences as simple as inductive generalization. So induction as a logic of discovery (and a fortiori as a discovery mechanism that brings justification with it) has been largely abandoned.

Most modern philosophers of science view induction exclusively as a logic of justification, to be applied to theories only after they have been generated. The most popular position is called the *hypothetico-deductive model*. On this model, a theory's origin—the way it was generated—is entirely irrelevant to its epistemic justification. All that matters is how the theory stands with respect to the things it explains and predicts. Many hypothetico-deductivists place a special emphasis on prediction, arguing that a theory is most powerfully confirmed when it successfully forecasts the outcome of new experiments and observations. Some of these philosophers go further, denying that the explanation of results already known has any power to confirm a theory. These philosophers, called *predictionists*, argue that only novel predictions count, inductively, as evidence. This debate within the ranks of the hypothetico-deductivists, between predictionism and more accommodating views of what can count as evidence, is one of several issues explored in this chapter.

Despite its apparent ubiquity in science, inductive reasoning is philosophically controversial. In "Induction," Peter Lipton explains why. He distinguishes between the problem of describing the general principles we follow when we reason inductively and the problem of showing that those inferences are justified. The second problem, often simply referred to as *the* problem of induction, originates with the skeptical argument of David Hume (1711–76), who notoriously concluded that it is impossible to show that induction is either reliable or reasonable. Karl Popper thinks that Hume was right about this: there is no justification of induction. In "The Problem of Induction," Popper rejects all inductive reasoning and, with it, the whole notion of inductive confirmation. In its place he offers a theory of scientific method—falsificationism—that relies solely on deduction. Wesley Salmon disputes the adequacy of Popper's falsificationism as an account of science. In "Rational Prediction," Salmon argues that, by dispensing with induction and inductive confirmation, Popper cannot justify our preference for using our best-tested theories when we need to make practical decisions about the future.

With Carl Hempel's "Criteria of Confirmation and Acceptability," the focus shifts away from the problem of justifying induction and towards articulating and assessing the various criteria that scientists use in weighing the merits of theories. These criteria include the quantity, diversity, and precision of evidence, as well as a theory's simplicity, its support from other (well-established) theories, and its ability to make novel predictions. Hempel pays particular attention to whether the value we place on simplicity is merely a subjective or pragmatic preference or whether it can be objectively justified in terms of inductive probability or greater falsifiability.

Many predictionists and explanationists argue that the power of a result to confirm a theory depends on when it was discovered. Predictionists insist that a result can confirm a theory only if it is discovered after the theory has been proposed. Explanationists assert exactly the opposite, arguing that only the explanation of previously known facts has the power to confirm a theory. In "Explanation v. Prediction: Which Carries More Weight?" Peter Achinstein rejects both views, arguing that the distinction between explanation and prediction is not, by itself, relevant to confirmation. Nonetheless, he reasons, historical information does sometimes determine whether a result is evidence for a theory and how strongly it confirms the theory. If, for example, an experiment is performed, then regardless of whether the outcome verifies a novel prediction or describes something previously known, how the experiment was conducted can affect confirmation.

The chapter concludes with an excerpt from Nelson Goodman's *Fact, Fiction, and Forecast*, in which Goodman introduces the unusual predicate *grue*. Goodman points out that past observations of green emeralds are instances not only of the "straight" hypothesis that all emeralds are green but also of the "bent" hypothesis that all emeralds are grue. Since he agrees that "all emeralds are grue" is *not* confirmed by the observation of green emeralds in the past, he concludes that no purely formal or syntactical theory of confirmation can capture the distinction between those hypotheses that are confirmable by their instances and those that are not. Goodman's view is that confirmable hypotheses must contain only predicates that are "projectible" and that projectibility is determined by our having used such predicates to make successful inductive inferences in the past.

■ | Notes

1. See the section "Why There Is No Logic of Discovery" in the commentary following the readings in this chapter.

2. For two contrasting accounts of where the failure of inductivism as an account of theory generation leaves the whole question of the relevance of generation to justification, see Thomas Nickles, "Justification and Experiment," in *The Uses of Experiment*, ed. David Gooding, Trevor Pinch, and Simon Schaffer (Cambridge: Cambridge University Press, 1989), 299–333, and Larry Laudan, "Why Was the Logic of Discovery Abandoned?" in *Scientific Discovery, Logic, and Rationality*, ed. Thomas Nickles (Dordrecht, Netherlands: D. Reidel, 1980), 173–83.

3. The concept of observability is explored at length in chapter 9. See especially the first three papers in that chapter by Maxwell, Van Fraassen, and Musgrave and the discussion of them in the accompanying commentary.

PETER LIPTON

Induction

■ | Underdetermination

We infer some claims on the basis of other claims: we move from premises
to a conclusion. Some inferences are deductive: it is impossible for the premises to be true but the conclusion false. All other inferences I call 'inductive',
using that term in the broad sense of non-demonstrative reasons. Inductive
inference is thus a matter of weighing evidence and judging probability, not
of proof. How do we go about making these judgments, and why should we
believe they are reliable? Both the question of description and the question
of justification arise from underdetermination. To say that an outcome is
underdetermined is to say that some information about initial conditions
and rules or principles does not guarantee a unique solution. The information that Tom spent five dollars on apples and oranges and that apples are
fifty cents a pound and oranges a dollar a pound underdetermines how
much fruit Tom bought, given only the rules of deduction. Similarly, those
rules and a finite number of points on a curve underdetermine the curve,
since there are many curves that would pass through those points.

Underdetermination may also arise in our description of the way a person learns or makes inferences. A description of the evidence, along with a
certain set of rules, not necessarily just those of deduction, may underdetermine what is learned or inferred. Insofar as we have described all the evidence and the person is not behaving erratically, this shows that there are
hidden rules. We can then study the patterns of learning or inference to try
to discover them. Noam Chomsky's argument from 'the poverty of the stimulus' is a good example of how underdetermination can be used to disclose
the existence of additional rules (1965: ch. 1, sec. 8, esp. 58–9). Children
learn the language of their elders, an ability that enables them to understand
an indefinite number of sentences on first acquaintance. The talk young

FROM Peter Lipton, *Inference to the Best Explanation*, 2d ed. (New York: Routledge,
2004), 5–19.

children hear, however, along with rules of deduction and any plausible general rules of induction, grossly underdetermine the language they learn. What they hear is limited and includes many ungrammatical sentences, and the little they hear that is well formed is compatible with many possible languages other than the one they learn. Therefore, Chomsky argues, in addition to any general principles of deduction and induction, children must be born with strong linguistic rules or principles that further restrict the class of languages they will learn, so that the actual words they hear are now sufficient to determine a unique language. Moreover, since a child will learn whatever language he is brought up in, these principles cannot be peculiar to a particular human language; instead, they must specify something that is common to all of them. For Chomsky, determining the structure of these universal principles and the way they work is the central task of modern linguistics.

Thomas Kuhn provides another well-known example of using underdetermination as a tool to investigate cognitive principles. He begins from an argument about scientific research strikingly similar to Chomsky's argument about language acquisition (1970; 1977, esp. ch. 12). In most periods in the history of a developed scientific specialty, scientists are in broad agreement about which problems to work on, how to attack them and what counts as solving them. But the explicit beliefs and rules scientists share, especially their theories, data, general rules of deduction and induction, and any explicit methodological rules, underdetermine these shared judgments. Many possible judgments are compatible with these beliefs and rules other than the ones the scientists make. So Kuhn argues that there must be additional field-specific principles that determine the actual judgments. Unlike Chomsky, Kuhn does not argue for principles that are either innate or in the form of rules, narrowly construed. Instead, scientists acquire through their education a stock of exemplars—concrete problem solutions in their specialty—and use them to guide their research. They pick new problems that look similar to an exemplar problem, they try techniques that are similar to those that worked in that exemplar, and they assess their success by reference to the standards of solution that the exemplars illustrate. Thus the exemplars set up a web of 'perceived similarity relations' that guide future research, and the shared judgments are explained by the shared exemplars. These similarities are not created or governed by rules, but they result in a pattern of research that mimics one that is rule governed. Just how exemplars do this work, and what happens when they stop working, provide the focus of Kuhn's account of science.

As I see it, Chomsky and Kuhn are both arguing for unacknowledged principles of induction, even though the inferences in the one case concern grammaticality rather than the world around us and even though the principles governing the inference in the other case are determined by exemplars rather than by rules (cf. Curd and Cover 1998: 497). In both cases, inferences are drawn that are not entailed by the available evidence. However, here underdetermination is taken to be a symptom of the existence of highly specialized principles, whether of language acquisition or of scientific research in a particular field at a particular time, since the underdetermination is claimed to remain even if we include general principles of induction among our rules. But it is natural to suppose that there are some general principles, and the same pattern of argument applies there. If an inference is inductive, then by definition it is underdetermined by the evidence and the rules of deduction. Insofar as our inductive practices are systematic, we must use additional principles of inference, and we may study the patterns of our inferences in an attempt to discover what those principles are and to determine what they are worth.

▪ | Justification

The two central questions about our general principles of induction concern description and justification. What principles do we actually use? Are these good principles to use? The question of description seems at first to take priority. How can we even attempt to justify our principles until we know what they are? Historically, however, the justification question came first. One reason for this is that the question of justification gets its grip from skeptical arguments that seem to apply to any principles that could account for the way we fill the gap between the evidence we have and the inference we make. It is the need for such principles rather than the particular form they take that creates the skeptical trouble.

The problem of justification is to show that our inferential methods are good methods, fit for purpose. The natural way to understand this is in terms of truth. We want our methods of inference to be 'truth-tropic', to take us towards the truth. For deduction, a good argument is one that is valid, a perfect truth conduit, where if the premises are true, the conclusion must be true as well. The problem of justification here would be to show that arguments we judge valid are in fact so. For induction, such perfect reliability is out of the question. By definition, even a good inductive argument is one where it is possible for there to be true premises but a false conclusion. Moreover, it is clear that the reasonable inductive inferences we make are not entirely reliable even in this world, since they sometimes sadly take us from truth to falsehood. Nevertheless, it remains natural to construe the task of justification as that of showing truth-tropism. We would like to show that those inductive inferences we judge worth making are ones that tend to take us from true premises to true conclusions.

A skeptical argument that makes the problem of justification pressing has two components, underdetermination and circularity. The first is an argument that the inferences in question are underdetermined, given only our premises and the rules of deduction; that the premises and those rules are compatible not just with the inferences we make, but also with other, incompatible inferences. This shows that the inferences in question really are inductive and, by showing that there are possible worlds where the principles we use take us from true premises to false conclusions, it also shows that there are worlds where our principles would fail us. Revealing this underdetermination, however, does not yet generate a skeptical argument, since we might have good reason to believe that the actual world is one where our principles are at least moderately reliable. So the skeptical argument requires a second component, an argument for circularity, which attempts to show that we cannot rule out the possibility of massive unreliability that underdetermination raises without employing the very principles that are under investigation, and so begging the question.

Although it is not traditionally seen as raising the problem of induction, Descartes's 'First Meditation' (1641) is a classic illustration of this technique. Descartes's goal is to cast doubt on the 'testimony of the senses', which leads us to infer that there is, say, a mountain in the distance, because that is what it looks like. He begins by arguing that we ought not to trust the senses completely, since we know that they do sometimes mislead us, 'when it is a question of very small and distant things'. This argument relies on underdetermination, on the fact that the way things appear does not entail the way they are; but it does not yet have the circularity component, since we can corroborate our inferences about small and distant things without circularity by taking a closer look (Williams 1978: 51–2). But Descartes immediately moves on from the small and the distant to the large and near. No matter how clearly we seem to see something, it may only be a dream, or a misleading experience induced by an evil demon. These arguments describe possible situations where even the most compelling sensory testimony is misleading. Moreover, unlike the worry about small and distant things, these arguments also have a circle component. There is apparently no way to test whether a demon is misleading us with a particular experience, since any test would itself rely on experiences that the demon might have induced. The senses may be liars, giving us false testimony, and we should not find any comfort if they also report that they are telling us the truth.

The demon argument begins with the underdetermination of observational belief by observational experience, construes the missing principle of inference on the model of inference from testimony, and then suggests that the reliability of this principle could only be shown by assuming it. Perhaps one of the reasons Descartes's arguments are not traditionally seen as raising the problem of justifying induction is that his response to his own skepticism is to reject the underdetermination upon which it rests. Descartes argues that, since inferences from the senses must be inductive and so raise a skeptical problem, our knowledge must instead have a different sort of foundation for which the problem of underdetermination does not arise. The *cogito* and the principles of clarity and distinctness that it exemplifies are supposed to provide the non-inductive alternative. Circularity is also avoided since the senses do not have to justify themselves, even if the threat of circularity notoriously reappears elsewhere, in the attempt to justify the principles of clarity and distinctness by appeal to an argument for the existence of God that is to be accepted because it itself satisfies those principles. Another reason why Descartes is not credited with the problem of induction may be that he does not focus directly on the principles governing inferences from experience, but rather on the fallibility of the conclusions they yield.

The moral Descartes draws from underdetermination and circularity is not that our principles of induction require some different sort of defence or must be accepted without justification, but that we must use different premises and principles, for which the skeptical problem does not arise. Thus he attempts to wean us from the senses. For a skeptical argument about induction that does not lead to the rejection of induction, we must turn to its traditional home, in the arguments of David Hume.

Hume also begins with underdetermination, in this case that our observations do not entail our predictions (1748: sec. IV). He then suggests that the governing principle of all our inductive inferences is that nature is uniform, that the unobserved (but observable) world is much like what we have observed. The question of justification is then the question of showing that nature is indeed uniform. This cannot be deduced from what we have observed, since the claim of uniformity itself incorporates a massive prediction. But the only other way to argue for uniformity is to use an inductive argument, which would rely on the principle of uniformity, leaving the question begged. According to Hume, we are addicted to the practice of induction, but it is a practice that cannot be justified.

To illustrate the problem, suppose our fundamental principle of inductive inference is 'More of the Same'. We believe that strong inductive arguments are those whose conclusions predict the continuation of a pattern described in the premises. Applying this principle of conservative induction, we would infer that the sun will rise tomorrow, since it has always risen in the past; and we would judge worthless the argument that the sun will not rise tomorrow since it has always risen in the past. One can, however, come up with a factitious principle to underwrite the latter argument. According to the principle of revolutionary induction, 'It's Time for a Change', and this sanctions the dark inference. Hume's argument is that we have no way to show that conservative induction, the principle he claims we actually use for our inferences, will do any better than intuitively wild principles like the principle of revolutionary induction. Of course conservative induction has had the more impressive track record. Most of the inferences from true premises that it has sanctioned have also had true conclusions. Revolutionary induction, by contrast, has been conspicuous in failure, or would have been, had anyone relied on it. The question of justification, however, does not ask which method of inference has been successful; it asks which one will be successful.

Still, the track record of conservative induction appears to be a reason to trust it. That record is imperfect (we are not aspiring to deduction), but very impressive, particularly as compared with revolutionary induction and its ilk. In short, induction will work because it has worked. This seems the only justification our inductive ways could ever have or require. Hume's disturbing observation was that this justification appears circular, no better than trying to convince someone that you are honest by saying that you are. Much as Descartes argued that we should not be moved if the senses give testimony on their own behalf, so Hume argued that we cannot appeal to the history of induction to certify induction. The trouble is that the argument that conservative inductions will work because they have worked is itself an induction. The past success is not supposed to prove future success, only make it very likely. But then we must decide which standards to use to evaluate this argument. It has the form 'More of the Same', so conservatives will give it high marks, but since its conclusion is just to underwrite conservatism, this begs the question. If we apply the revolutionary principle, it counts as a very weak argument. Worse still, by revolutionary standards, conservative induction is likely to fail precisely because it has succeeded in the past, and the past failures of revolutionary induction augur well for its future success (Skyrms 1986: ch. II). The justification of revolutionary induction seems no worse than the justification of conservative induction, which is to say that the justification of conservative induction looks very bad indeed.

The problem of justifying induction does not show that there are other inductive principles better than our own. Instead it argues for a deep symmetry: many sets of principles, most of them wildly different from our own and incompatible with each other, are yet completely on a par from a justificatory point of view. This is why the problem of justification can be posed before we have solved the problem of description. Whatever inductive principles we use, the fact that they are inductive seems enough for the skeptic to show that they defy justification. We fill the gap of underdetermination between observation and prediction in one way, but it could be filled in many other ways that would have led to entirely different predictions. We have no way of showing that our way is any better than any of the other ways that would certify their own reliability. Each is on a par in the sense that it can only argue for its principles by appeal to those very principles. And it is not just that the revolutionaries will not be convinced by the justificatory arguments of the conservatives: the conservatives should not accept their own defense either, since among their standards is one which says that a circular argument is a bad argument, even if it is in one's own aid. Even if I am honest, I ought to admit that the fact that I say so ought not to carry any weight. We have a psychological compulsion to favor our own inductive principles but, if Hume is right, we should see that we cannot even provide a cogent rationalization of our behavior.

It seems to me that we do not yet have a satisfying solution to Hume's challenge and that the prospects for one are bleak. (Though some seem unable to give up trying. See e.g. Lipton 2000 and forthcoming.) There are, however, other problems of justification that are more tractable. The peculiar difficulty of meeting Hume's skeptical argument against induction arises because he casts doubt on our inductive principles as a whole, and so any recourse to induction to justify induction appears hopeless. But one can also ask for the justification of particular inductive principles and, as Descartes's example of small and distant things suggests, this leaves open the possibility of appeal to other principles without begging the question. For example, among our principles of inference is one that makes us more likely to infer a theory if it is supported by a variety of evidence than if it is supported by a similar amount of homogenous data. This is the sort of principle that might be justified in terms of a more basic inductive principle, say that we have better reason to infer a theory when all the reasonable competitors have been refuted, or that a theory is only worth inferring when each of its major components has been separately tested. Another, more controversial, example of a special principle that might be justified without circularity is that, all else being equal, a theory deserves more credit from its successful predictions than it does from data that the theory was constructed to fit. This appears to be an inductive preference most of us have, but the case is controversial because it is not at all obvious that it is rational. On the one hand, many people feel that only a prediction can be a real test, since a theory cannot possibly be refuted by data it is built to accommodate; on the other, that logical relations between theory and data upon which inductive support exclusively depends cannot be affected by the merely historical fact that the data were available before or only after the theory was proposed. In any event, this is an issue of inductive principle that is susceptible to noncircular evaluation. . . . Finally, though a really satisfying solution to Hume's problem would have to be an argument for the reliability of our principles that had force against the inductive skeptic, there may be arguments for reliability that do not meet this condition yet still have probative value for those of us who already accept some forms of induction. . . .

■ | Description

We can now see why the problem of justification, the problem of showing that our inductive principles are reliable, did not have to wait for a detailed description of those principles. The problem of justifying our principles gets its bite from skeptical arguments, and these appear to depend only on the fact that these principles are principles of induction, not on the particular form they take. The crucial argument is that the only way to justify our principles would be to use an argument that relies on the very same principles, which is illegitimate; an argument that seems alas to work whatever the details of our inferences. The irrelevance of the details comes out in the symmetry of Hume's argument: just as the future success of conservative induction gains no plausibility from its past success, so the future success of revolutionary induction gains nothing from its past failures. As the practice varies, so does the justificatory argument, preserving the pernicious circularity. Thus the question of justification has had a life of its own: it has not waited for a detailed description of the practice whose warrant it calls into doubt.

By the same token, the question of description has fortunately not waited for an answer to the skeptical arguments. Even if our inferences were unjustifiable, one still might be interested in saying how they work. The problem of description is not to show that our inferential practices are reliable; it is just to describe them as they stand. One might have thought that this would be a relatively simple problem. First of all, there are no powerful reasons for thinking that the problem of description is insoluble, as there are for the problem of justification. There is no great skeptical argument against the possibility of description. It is true that any account of our principles will itself require inductive support, since we must see whether it jibes with our observed inductive practice. This, however, raises no general problem of circularity now that a general justification of induction is not the issue. Using induction to investigate the actual structure of our inductive practices is no more suspect than using observation to study the structure and function of the eye. Secondly, it is not just that a solution to the problem of describing our inductive principles should be possible, but that it should be fairly easy. After all, they are our principles, and we use them constantly. It thus comes as something of a shock to discover how extraordinarily difficult the problem of description has turned out to be. It is not merely that ordinary reasoners are unable to describe what they are doing: years of focused effort by epistemologists and philosophers of science have yielded little better. Again, it is not merely that we have yet to capture all the details, but that the most popular accounts of the gross structure of induction are wildly at variance with our actual practice.

Why is description so hard? One reason is a quite general gap between what we can do and what we can describe. You may know how to do something without knowing how you do it; indeed, this is the usual situation. It is one thing to know how to tie one's shoes or to ride a bike; it is quite another thing to be able to give a principled description of what it is that one knows. Chomsky's work on principles of language acquisition and Kuhn's work on scientific method are good cognitive examples. Their investigations would not be so important and controversial if ordinary speakers knew how they distinguished grammatical from ungrammatical sentences or normal scientists knew how they made their methodological judgments. Speakers and scientists employ diverse principles, but they are not conscious of them. The situation is similar in the case of inductive inference generally. Although we may partially articulate some of our inferences if, for example, we are called upon to defend them, we are not conscious of the diverse principles of inductive inference we constantly use.

Since our principles of induction are neither available to introspection, nor otherwise observable, the evidence for their structure must be indirect. The project of description is one of black box inference, where we try to reconstruct the underlying mechanism on the basis of the superficial patterns of evidence and inference we observe in ourselves. This is no trivial problem. Part of the difficulty is simply the fact of underdetermination. As the examples of Chomsky and Kuhn show, underdetermination can be a symptom of missing principles and a clue to their nature, but it is one that does not itself determine a unique answer. In other words, where the evidence and the rules of deduction underdetermine inference, that information also underdetermines the missing principles. There will always be many different possible mechanisms that would produce the same patterns, so how can one decide which one is actually operating? In practice, however, as epistemologists we usually have the opposite problem: we can not even come up with a single description that would yield the patterns we observe. The situation is the same in scientific theorizing generally. There is always more than one account of the unobserved and often unobservable world that would account for what we observe, but scientists' actual difficulty is often to come up with even one theory that fits the observed facts. On reflection, then, it should not surprise us that the problem of description has turned out to be so difficult. Why should we suppose that the project of describing our inductive principles is going to be easier than it would be, say, to give a detailed account of the working of a computer on the basis of the correlations between keys pressed and images on the screen?

Now that we are prepared for the worst, we may turn to some of the popular attempts at description. In my discussion of the problem of justification, I suggested that, following Hume's idea of induction as habit formation, we describe our pattern of inference as 'More of the Same'. This is pleasingly simple, but the conservative principle is at best a caricature of our actual practice. We sometimes do not infer that things will remain the same and we sometimes infer that things are going to change. When my mechanic tells me that my brakes are about to fail, I do not suppose that he is therefore a revolutionary inductivist. Again, we often make inductive inferences from something we observe to something invisible, such as from people's behavior to their beliefs or from the scientific evidence to unobservable entities and processes, and this does not fit into the conservative mold. 'More of the Same' might enable me to predict what you will do on the basis of what you have done (if you are a creature of habit), but it will not tell me what you are or will be thinking.

Faced with the difficulty of providing a general description, a reasonable strategy is to begin by trying to describe one part of our inductive practice. This is a risky procedure, since the part one picks may not really be describable in isolation, but there are sometimes reasons to believe that a particular part is independent enough to permit a useful separation. Chomsky must believe this about our principles of linguistic inference. Similarly, one might plausibly hold that, while simple habit formation cannot be the whole of our inductive practice, it is a core mechanism that can be treated in isolation. Thus one might try to salvage the intuition behind the conservative principle by giving a more precise account of the cases where we are willing to project a pattern into the future, leaving to one side the apparently more difficult problems of accounting for predictions of change and inferences to the unobservable. What we may call the instantial model of inductive confirmation may be seen in this spirit. According to it, a hypothesis of the form 'All As are B' is supported by its positive instances, by observed As that are also B (Hempel 1965: ch. 1). This is not, strictly speaking, an account of inductive *inference*, since it does not say either how we come up with the hypothesis in the first place or how many supporting instances are required before we actually infer it, but this switching of the problem from inference to support may also be taken as a strategic simplification. In any event, the underlying idea is that if enough positive instances and no refuting instances (As that are not B) are observed, we will infer the hypothesis, from which we may then deduce the prediction that the next A we observe will be B.

This model could only be a very partial description of our inductive principles but, within its restricted range, it strikes many people initially as a truism, and one that captures Hume's point about our propensity to extend observed patterns. Observed positive instances are not necessary for inductive support, as inferences to the unobserved and to change show, but they might seem at least sufficient. But the instantial model has been shown to be wildly over-permissive. Some hypotheses are supported by their positive instances, but many are not. Observing only black ravens may lead one to believe that all ravens are black, but observing only bearded philosophers would probably not lead one to infer that all philosophers are bearded. Nelson Goodman has generalized this problem, by showing how the instantial model sanctions any prediction at all if there is no restriction on the hypotheses to which it can be applied (Goodman 1983: ch. III). His technique is to construct hypotheses with factitious predicates. Black ravens provide no reason to believe that the next swan we see will be white, but they do provide positive instances of the artificial hypothesis that 'All raveswans are blight', where something is a raveswan just in case it is either observed before today and a raven, or not so observed and a swan, and where something is blight just in case it is either observed before today and black, or not so observed and white. But the hypothesis that all raveswans are blight entails that the next observed raveswan will be blight which, given the definitions, is just to say that the next swan will be white.

The other famous difficulty facing the instantial model arises for hypotheses that do seem to be supported by their instances. Black ravens support the hypothesis that all ravens are black. This hypothesis is logically equivalent to the contrapositive hypothesis that all non-black things are non-ravens: there is no possible situation where one hypothesis would be true but the other false. According to the instantial model, the contrapositive hypothesis is supported by non-black, non-ravens, such as green leaves. The rub comes with the observation that whatever supports a hypothesis also supports anything logically equivalent to it. This is very plausible, since support provides a reason to believe true, and we know that if a hypothesis is true, then so must be anything logically equivalent to it. But then the instantial model once again makes inductive support far too easy, counting green leaves as evidence that all ravens are black (Hempel 1965: ch. 1). . . . Something like conservative induction must play a role in both everyday and scientific inferences (cf. Achinstein 1992), but it can at best be only part of the story, and it has turned out to be tantalizingly difficult to articulate in a way that could even give it a limited role.

Another famous account of inductive support is the hypothetico-deductive model (Hempel 1966: chs 2, 3). On this view, a hypothesis or theory is supported when it, along with various other statements, deductively entails a datum. Thus a theory is supported by its successful predictions. This account has a number of attractions. First, although it leaves to one side the important question of the source of hypotheses, it has much wider scope than the instantial model, since it allows for the support of hypotheses that appeal to unobservable entities and processes. The big bang theory of the origin of the universe obviously cannot be directly supported; but along with other statements it entails that we ought to find ourselves today traveling through a uniform background radiation, like the ripples left by a rock falling into a pond. The fact that we do now observe this radiation (or effects of it) provides some reason to believe the big bang theory. Thus, even if a hypothesis cannot be supported by its instances, because its instances are not observable, it can be supported by its observable logical consequences. Secondly, the model enables us to co-opt our accounts of deduction for an account of induction, an attractive possibility since our understanding of deductive principles is so much better than our understanding of inductive principles. Lastly, the hypothetico-deductive model seems genuinely to reflect scientific practice, which is perhaps why it has become the scientists' philosophy of science.

In spite of all its attractions, our criticism of the hypothetico-deductive model here can be brief, since it inherits all the over-permissiveness of the instantial model. Any case of support by positive instances will also be a case of support by consequences. The hypothesis that all As are B, along with the premise that an individual is A, entails that it will also be B, so the thing observed to be B supports the hypothesis, according to the hypothetico-deductive model. That is, any case of instantial support is also a case of hypothetico-deductive support, so the model has to face the problem of insupportable hypotheses and the raven paradox. Moreover, the hypothetico-deductive model is similarly over-permissive in the case of vertical inferences to hypotheses about unobservables, a problem that the instantial model avoided by ignoring such inferences altogether. The difficulty is structurally similar to Goodman's problem of factitious predicates. Consider the conjunction of the hypotheses that all ravens are black and that all swans are white. This conjunction, along with premises concerning the identity of various ravens, entails that they will be black. According to the model, the conjunction is supported by black ravens, and it entails its own conjunct about swans. The model thus appears to sanction the inference from black ravens to white swans (cf. Goodman 1983: 67–8). Similarly, the hypothesis that all swans are white taken alone entails the inclusive disjunction that either all swans are white or there is a black raven, a disjunction we could establish by seeing a black raven, again giving illicit hypothetico-deductive support to the swan hypothesis. These maneuvers are obviously artificial, but nobody has managed to show how the model can be modified to avoid them without also eliminating most genuine cases of inductive support (cf. Glymour 1980: ch. II). Finally, in addition to being too permissive, finding support where none exists, the model is also too strict, since data may support a hypothesis which does not, along with reasonable auxiliary premises, entail them. . . .

We believe some things more strongly than others, and our next approach to the descriptive problem represents degrees of belief in terms of probabilities and so is able to exploit the probability calculus for an account of when evidence inductively confirms a hypothesis. The probabilities one assigns to some statements—which will all fall between zero for the impossible and one for the certain—will constrain the probabilities one assigns to others, in order to preserve something analogous to deductive consistency (Howson 2000). Thus, if one statement entails another, the probability given to the conclusion must be at least as great as the probability of the premise, since if the premise is true the conclusion must be true as well. One consequence of the probability calculus is Bayes's theorem, which forms the basis for this approach to confirmation. In its near-simplest form, the theorem gives the probability of hypothesis H given evidence E in terms of the probability of E given H and the prior probabilities of H and of E:

$$P(H/E) = P(E/H) \times P(H)/P(E)$$

The Bayesian approach to confirmation exploits this formula by imagining the scientist's situation just before E is observed. The probabilities of H and of E are equated with her degree of belief in those two statements—their 'prior' probability—and the probability of H given E is then taken to be the probability she should assign to H after observing E—the 'posterior' probability of H. The basic Bayesian view is that E confirms H just in case the posterior probability of H is higher than the prior probability of H. This is a natural thing to say, since this is the condition under which observing E raises the probability of H.

To use the Bayesian equation to calculate the posterior of H requires not just the priors of H and E; it also requires the probability of E given H. But in a situation where H entails E, then E must be true if H is, so the probability of E given H is one and posterior probability becomes a simple ratio:

$$P(H/E) = P(H)/P(E)$$

If H entails E, and the priors of H and E are neither zero nor one, the posterior of H must be greater than the prior of H, since we are dividing the prior by something less than one. Thus E will count as confirming H whenever E is a consequence of H. Moreover, this simple form of the equation makes clear that the degree of confirmation—the degree to which observing E raises the probability of H—will be greater as the prior probability of E is lower. In English, this is to say that there is high confirmation when your hypothesis entails an unlikely prediction that turns out to be correct, a very plausible claim.

The Bayesian account inevitably faces its own share of objections. These may be that the account is too permissive, too strict, or that it fails to describe a mechanism that could represent the way scientists actually determine the bearing of data on theory (Earman 1992; Howson and Urbach 1989). For example, since the Bayesian account has it that a hypothesis is confirmed by any of its logical consequences (so long as all the probabilities lie between zero and one), it seems to inherit the over-permissiveness of the hypothetico-deductive model. The account also threatens to be too strict, because of the problem of 'old evidence'. There is considerable dispute over whether evidence available before a hypothesis is formulated provides as strong confirmation as evidence only gathered afterwards to test a prediction. . . . [B]ut it is agreed by almost all that old evidence can provide *some* confirmation. On its face, however, the Bayesian account does not allow for this, since old evidence will have a prior probability of one, and so have no effect on the posterior probability of the hypothesis. Finally, there are a series of objections to the basic ingredients of the Bayesian scheme, that beliefs do not in fact come in degrees well represented by probabilities, that there is no proper source for the values of the priors, and that in the realistic cases where the hypothesis in question does not on its own entail the evidence, there is no plausible way that the scientist has to determine the probability of the evidence given the hypothesis. . . .

We have now briefly canvassed four attempts to tackle the descriptive problem: 'More of the Same', the instantial model, the hypothetico-deductive model and the Bayesian approach. At least on this first pass, all four appear both too permissive and too strict, finding inductive support where there is none and overlooking cases of genuine support. They do not give enough structure to the black box of our inductive principles to determine the inferences and judgments we actually make. This is not to say that these accounts describe mechanisms that would yield too many inferences: they would probably yield too few. A 'hypothetico-deductive box', for example, would probably have little or no inferential output, given the plausible additional principle that we will not make inferences we know to be contradictory. For every hypothesis that we would be inclined to infer on the basis of the deductive support it enjoys, there will be an incompatible hypothesis that is similarly supported, and the result is no inference at all, so long as both hypotheses are considered and their incompatibility recognized.

A fifth account of induction, the last I will consider in this section, focuses on causal inference. It is a striking fact about our inductive practice, both lay and scientific, that so many of our inferences depend on inferring from effects to their probable causes. This is something that Hume himself emphasized (Hume 1748: sec. IV). Causal inferences are legion, such as the doctor's inference from symptom to disease, the detective's inference from evidence to crook, the mechanic's inference from the engine noises to what is broken, and many scientific inferences from data to theoretical explanation. Moreover, it is striking that we often make a causal inference even when our main interest is in prediction. Indeed, the detour through causal theory on the route from data to prediction seems to be at the centre of many of the dramatic successes of scientific prediction. All this suggests that we might do well to consider an account of the way causal inference works as a central component of a description of our inductive practice.

The best known account of causal inference is John Stuart Mill's discussion of the 'methods of experimental inquiry' (Mill 1904: bk III, ch. VIII; cf. Hume 1739: bk I, pt 3, sec. 15). The two central methods are the Method of Agreement and especially the Method of Difference. According to the Method of Agreement, in idealized form, when we find that there is only one antecedent that is shared by all the observed instances of an effect, we infer that it is a cause (bk III, ch. VIII, sec. 1; hereafter as 'III.VIII. 1'). This is how we come to believe that hangovers are caused by heavy drinking. According to the Method of Difference, when we find that there is only one prior difference between a situation where the effect occurs and an otherwise similar situation where it does not, we infer that the antecedent that is only present in the case of the effect is a cause (III. VIII.2). If we add sodium to a blue flame, and the flame turns yellow, we infer that the presence of sodium is a cause of the new color, since that is the only difference between the flame before and after the sodium was added. If we once successfully follow a recipe for baking bread, but fail another time when we have left out the yeast and the bread does not rise, we would infer that the yeast is a cause of the rising in the first case. Both methods work by a combination of retention and variation. When we apply the Method of Agreement, we hold the effect constant, vary the background, and see what stays the same; when we apply the Method of Difference, we vary the effect, hold the background constant, and see what changes.

Mill's methods have a number of attractive features. Many of our inferences are causal inferences, and Mill's methods give a natural account of these. In science, for example, the controlled experiment is a particularly common and self-conscious application of the Method of Difference. The Millian structure of causal inference is also particularly clear in cases of inferential dispute. When you dispute my claim that C is the cause of E, you will often make your case by pointing out that the conditions for Mill's methods are not met; that is, by pointing out C is not the only antecedent common to all cases of E, or that the presence of C is not the only salient difference between a case where E occurs and a similar case where it does not. Mill's methods may also avoid some of the over-permissiveness of other accounts, because of the strong constraints that the requirements of varied or shared backgrounds place on their application. These requirements suggest how our background beliefs influence our inferences, something a good account of inference must do. The methods also help to bring out the roles in inference of competing hypotheses and negative evidence, . . . and the role of background knowledge . . . . Of course, Mill's methods have their share of liabilities, of which I will mention just two. First, they do not themselves apply to unobservable causes or to any causal inferences where the cause's existence, and not just its causal status, is inferred. Secondly, if the methods are to apply at all, the requirement that there be only a single agreement or difference in antecedents must be seen as an idealization, since this condition is never met in real life. We need principles for selecting from among multiple agreements or similarities those that are likely to be causes, but these are principles Mill does not himself supply. . . . [H]owever, Mill's methods can be modified and expanded in a way that may avoid these and other liabilities it faces in its simple form.

This chapter has set part of the stage for an investigation of our inductive practices. I have suggested that many of the problems those practices raise can be set out in a natural way in terms of the underdetermination that is characteristic of inductive inference. The underdetermination of our inferences by our evidence provides the skeptics with their lever, and so poses the problem of justification. It also elucidates the structure of the descriptive problem, and the black box inferences it will take to solve it. I have canvassed several solutions to the problem of description, partly to give a sense of some of our options and partly to suggest just how difficult the problem is. . . .

References

Achinstein, P. (1992), "Inference to the Best Explanation: Or, Who Won the Mill—Whewell Debate?" *Studies in the History and Philosophy of Science*, 23, 349–64.

Chomsky, N. (1965), *Aspects of the Theory of Syntax*, Cambridge, Mass.: MIT Press.

Curd, M. and Cover, J. A. (1998), "Lipton on the Problem of Induction," in M. Curd and J. A. Cover (eds.) *Philosophy of Science: The Central Issues*, New York: Norton, 495–505.

Descartes, R. (1641), *Meditations on First Philosophy*, Donald Cress (trans.), Indianapolis, Ind.: Hackett (1979).

Earman, J. (1992) *Bayes or Bust? A Critical Examination of Bayesian Confirmation Theory*, Cambridge, Mass.: MIT Press.

Glymour, C. (1980), *Theory and Evidence*, Princeton, N.J.: Princeton University Press.

Goodman, N. (1983), *Fact, Fiction, and Forecast*, 4th ed., Indianapolis, Ind.: Bobbs-Merrill.

Hempel, C. (1965), *Aspects of Scientific Explanation*, New York: Free Press.

Hempel, C. (1966), *The Philosophy of Natural Science*, Englewoods Cliffs, N.J.: Prentice-Hall.

Howson, C. (2000), *Hume's Problem: Induction and the Justification of Belief*, Oxford: Oxford University Press.

Howson, C. and Urbach, P. (1989), *Scientific Reasoning: The Bayesian Approach*, La Salle, Ill.: Open Court.

Hume, D. (1739), *A Treatise of Human Nature*, D. F. and M. J. Norton (eds.), Oxford: Oxford University Press (2000).

Hume, D. (1748), *An Enquiry Concerning Human Understanding*, T. Beauchamp (ed.), Oxford: Oxford University Press (1999).

Kuhn, T. (1970), *The Structure of Scientific Revolutions*, 2d ed., Chicago: University of Chicago Press.

Kuhn, T. (1977), *The Essential Tension*, Chicago: University of Chicago Press.

Lipton, P. (2000), "Tracking Track Records," *The Aristotelian Society*, Supplementary Volume LXXIV, 179–206.

Lipton, P. (forthcoming), *The Humean Predicament*, Cambridge: Cambridge University Press.

Mill, J. S. (1904), *A System of Logic*, 8th ed., London: Longmans, Green & Co.

Skyrms, B. (1986), *Choice and Chance*, 3d ed., Belmont, Calif.: Wadsworth.

Williams, B. (1978), *Descartes*, Harmondsworth: Penguin.

KARL POPPER

The Problem
of Induction

I | The Problem of Induction

According to a widely accepted view . . . the empirical sciences can be
characterized by the fact that they use 'inductive methods', as they are
called. According to this view, the logic of scientific discovery would be
identical with inductive logic, i.e. with the logical analysis of these inductive methods.

It is usual to call an inference 'inductive' if it passes from *singular statements* (sometimes also called 'particular' statements), such as accounts of the
results of observations or experiments, to *universal statements*, such as hypotheses or theories.

Now it is far from obvious, from a logical point of view, that we are
justified in inferring universal statements from singular ones, no matter how
numerous; for any conclusion drawn in this way may always turn out to be
false: no matter how many instances of white swans we may have observed,
this does not justify the conclusion that *all* swans are white.

The question whether inductive inferences are justified, or under what
conditions, is known as *the problem of induction*.

The problem of induction may also be formulated as the question of
how to establish the truth of universal statements which are based on experience, such as the hypotheses and theoretical systems of the empirical sciences. For many people believe that the truth of these universal statements
is '*known by experience*'; yet it is clear that an account of an experience—of
an observation or the result of an experiment—can in the first place be only
a singular statement and not a universal one. Accordingly, people who say
of a universal statement that we know its truth from experience usually
mean that the truth of this universal statement can somehow be reduced to
the truth of singular ones, and that these singular ones are known by expe-

FROM Karl Popper, *The Logic of Scientific Discovery* (New York: Basic Books,
1959), 27–34.

rience to be true; which amounts to saying that the universal statement is based on inductive inference. Thus to ask whether there are natural laws known to be true appears to be only another way of asking whether inductive inferences are logically justified.

Yet if we want to find a way of justifying inductive inferences, we must first of all try to establish a *principle of induction*. A principle of induction would be a statement with the help of which we could put inductive inferences into a logically acceptable form. In the eyes of the upholders of inductive logic, a principle of induction is of supreme importance for scientific method: '. . . this principle', says Reichenbach,* 'determines the truth of scientific theories. To eliminate it from science would mean nothing less than to deprive science of the power to decide the truth or falsity of its theories. Without it, clearly, science would no longer have the right to distinguish its theories from the fanciful and arbitrary creations of the poet's mind.'[1]

Now this principle of induction cannot be a purely logical truth like a tautology or an analytic statement. Indeed, if there were such a thing as a purely logical principle of induction, there would be no problem of induction; for in this case, all inductive inferences would have to be regarded as purely logical or tautological transformations, just like inferences in deductive logic. Thus the principle of induction must be a synthetic statement; that is, a statement whose negation is not self-contradictory but logically possible. So the question arises why such a principle should be accepted at all, and how we can justify its acceptance on rational grounds.

Some who believe in inductive logic are anxious to point out, with Reichenbach, that 'the principle of induction is unreservedly accepted by the whole of science and that no man can seriously doubt this principle in everyday life either'.[2] Yet even supposing this were the case—for after all, 'the whole of science' might err—I should still contend that a principle of induction is superfluous, and that it must lead to logical inconsistencies.

That inconsistencies may easily arise in connection with the principle of induction should have been clear from the work of Hume; also, that they can be avoided, if at all, only with difficulty. For the principle of induction must be a universal statement in its turn. Thus if we try to regard its truth as known from experience, then the very same problems which occasioned its introduction will arise all over again. To justify it, we should have to employ inductive inferences; and to justify these we should have to assume an inductive principle of a higher order; and so on. Thus the attempt to base

* Hans Reichenbach (1891–1953) was a leading figure (along with Carl Hempel) in the Berlin school of philosophy of science in the 1920s and early 1930s. Although closely associated with the Vienna Circle, Reichenbach rejected the verifiability principle of meaning and adopted the phrase *scientific philosophy* to distinguish his views from those of the logical positivists. His works on space and time, quantum mechanics, probability, and induction have been an important influence on modern philosophy of science.

the principle of induction on experience breaks down, since it must lead to an infinite regress.

Kant tried to force his way out of this difficulty by taking the principle of induction (which he formulated as the 'principle of universal causation') to be 'a priori valid'. But I do not think that his ingenious attempt to provide an a priori justification for synthetic statements was successful.

My own view is that the various difficulties of inductive logic here sketched are insurmountable. So also, I fear, are those inherent in the doctrine, so widely current today, that inductive inference, although not 'strictly valid', can attain some degree of 'reliability' or of 'probability'. According to this doctrine, inductive inferences are 'probable inferences'.[3] 'We have described', says Reichenbach, 'the principle of induction as the means whereby science decides upon truth. To be more exact, we should say that it serves to decide upon probability. For it is not given to science to reach either truth or falsity . . . but scientific statements can only attain continuous degrees of probability whose unattainable upper and lower limits are truth and falsity'.[4]

At this stage I can disregard the fact that the believers in inductive logic entertain an idea of probability. . . . I can do so because the difficulties mentioned are not even touched by an appeal to probability. For if a certain degree of probability is to be assigned to statements based on inductive inference, then this will have to be justified by invoking a new principle of induction, appropriately modified. And this new principle in its turn will have to be justified, and so on. Nothing is gained, moreover, if the principle of induction, in its turn, is taken not as 'true' but only as 'probable'. In short, like every other form of inductive logic, the logic of probable inference, or 'probability logic', leads either to an infinite regress, or to the doctrine of apriorism.

The theory to be developed in the following pages stands directly opposed to all attempts to operate with the ideas of inductive logic. It might be described as the theory of the deductive method of testing, or as the view that a hypothesis can only be empirically tested—and only after it has been advanced.

Before I can elaborate this view (which might be called 'deductivism', in contrast to 'inductivism'[5]) I must first make clear the distinction between the psychology of knowledge which deals with empirical facts, and the logic of knowledge which is concerned only with logical relations. For the belief in inductive logic is largely due to a confusion of psychological problems with epistemological ones. It may be worth noticing, by the way, that this confusion spells trouble not only for the logic of knowledge but for its psychology as well.

2 | Elimination of Psychologism

I said above that the work of the scientist consists in putting forward and testing theories.

The initial stage, the act of conceiving or inventing a theory, seems to me neither to call for logical analysis nor to be susceptible of it. The question how it happens that a new idea occurs to a man—whether it is a musical theme, a dramatic conflict, or a scientific theory—may be of great interest to empirical psychology; but it is irrelevant to the logical analysis of scientific knowledge. This latter is concerned not with *questions of fact* (Kant's *quid facti?*), but only with questions of *justification or validity* (Kant's *quid juris?*). Its questions are of the following kind. Can a statement be justified? And if so, how? Is it testable? Is it logically dependent on certain other statements? Or does it perhaps contradict them? In order that a statement may be logically examined in this way, it must already have been presented to us. Someone must have formulated it, and submitted it to logical examination.

Accordingly I shall distinguish sharply between the process of conceiving a new idea, and the methods and results of examining it logically. As to the task of the logic of knowledge—in contradistinction to the psychology of knowledge—I shall proceed on the assumption that it consists solely in investigating the methods employed in those systematic tests to which every new idea must be subjected if it is to be seriously entertained.

Some might object that it would be more to the purpose to regard it as the business of epistemology to produce what has been called a 'rational reconstruction' of the steps that have led the scientist to a discovery—to the finding of some new truth. But the question is: what, precisely, do we want to reconstruct? If it is the processes involved in the stimulation and release of an inspiration which are to be reconstructed, then I should refuse to take it as the task of the logic of knowledge. Such processes are the concern of empirical psychology but hardly of logic. It is another matter if we want to reconstruct rationally the *subsequent tests* whereby the inspiration may be discovered to be a discovery, or become known to be knowledge. In so far as the scientist critically judges, alters, or rejects his own inspiration we may, if we like, regard the methodological analysis undertaken here as a kind of 'rational reconstruction' of the corresponding thought-processes. But this reconstruction would not describe these processes as they actually happen: it can give only a logical skeleton of the procedure of testing. Still, this is perhaps all that is meant by those who speak of a 'rational reconstruction' of the ways in which we gain knowledge.

. . . My view of the matter, for what it is worth, is that there is no such thing as a logical method of having new ideas, or a logical reconstruction of this process. My view may be expressed by saying that every discovery contains 'an irrational element', or 'a creative intuition', in Bergson's sense.* In

---

* The French philosopher Henri Bergson (1859–1941) attacked materialism and mechanism, rejected science as a complete account of reality, and advocated vitalism, a worldview based on creative forces and intuition. His writings on memory and our subjective experience of time were quite influential, especially on authors such as Marcel Proust, but his postulation of the *élan vital*, a spiritual force that

a similar way Einstein speaks of the 'search for those highly universal laws . . . from which a picture of the world can be obtained by pure deduction. There is no logical path', he says, 'leading to these . . . laws. They can only be reached by intuition, based upon something like an intellectual love (*'Einfühlung'*) of the objects of experience'.[6]

3 | Deductive Testing of Theories

According to the view that will be put forward here, the method of critically testing theories, and selecting them according to the results of tests, always proceeds on the following lines. From a new idea, put up tentatively, and not yet justified in any way—an anticipation, a hypothesis, a theoretical system, or what you will—conclusions are drawn by means of logical deduction. These conclusions are then compared with one another and with other relevant statements, so as to find what logical relations (such as equivalence, derivability, compatibility, or incompatibility) exist between them.

We may if we like distinguish four different lines along which the testing of a theory could be carried out. First there is the logical comparison of the conclusions among themselves, by which the internal consistency of the system is tested. Secondly, there is the investigation of the logical form of the theory, with the object of determining whether it has the character of an empirical or scientific theory, or whether it is, for example, tautological. Thirdly, there is the comparison with other theories, chiefly with the aim of determining whether the theory would constitute a scientific advance should it survive our various tests. And finally, there is the testing of the theory by way of empirical applications of the conclusions which can be derived from it.

The purpose of this last kind of test is to find out how far the new consequences of the theory—whatever may be new in what it asserts—stand up to the demands of practice, whether raised by purely scientific experiments, or by practical technological applications. Here too the procedure of testing turns out to be deductive. With the help of other statements, previously accepted, certain singular statements—which we may call 'predictions'— are deduced from the theory; especially predictions that are easily testable or applicable. From among these statements, those are selected which are not derivable from the current theory, and more especially those which the current theory contradicts. Next we seek a decision as regards these (and other) derived statements by comparing them with the results of practical applications and experiments. If this decision is positive, that is, if the singular conclusions turn out to be acceptable, or *verified*, then the theory has,

---

drives biological evolution, discredited Bergson's vitalism in the eyes of most scientists and philosophers.

for the time being, passed its test: we have found no reason to discard it. But if the decision is negative, or in other words, if the conclusions have been *falsified*, then their falsification also falsifies the theory from which they were logically deduced.

It should be noticed that a positive decision can only temporarily support the theory, for subsequent negative decisions may always overthrow it. So long as a theory withstands detailed and severe tests and is not superseded by another theory in the course of scientific progress, we may say that it has 'proved its mettle' or that it is 'corroborated'.

Nothing resembling inductive logic appears in the procedure here outlined. I never assume that we can argue from the truth of singular statements to the truth of theories. I never assume that by force of 'verified' conclusions, theories can be established as 'true', or even as merely 'probable'. . . .

■ | Notes

1. H. Reichenbach, *Erkenntnis* 1, 1930, p. 186 (cf. also p. 64 f.).

2. Reichenbach *ibid.*, p. 67.

3. *Cf* J. M. Keynes, *A Treatise on Probability* (1921); O. Külpe, *Vorlesungen über Logic* (ed. by Selz, 1923); Reichenbach (who uses the term 'probability implications'), *Axiomatik der Wahrscheinlichkeitrechnung, Mathem. Zeitschr.* 34 (1932); and in many other places.

4. Reichenbach, *Erkenntnis* 1, 1930, p. 186.

5. Liebig (in *Induktion und Deduktion*, 1865) was probably the first to reject the inductive method from the standpoint of natural science; his attack is directed against Bacon. Duhem (in *La Théorie physique, son objet et sa structure*, 1906; English translation by P. P. Wiener: *The Aim and Structure of Physical Theory*, Princeton, 1954) held pronounced deductivist views. (But there are also inductivist views to be found in Duhem's book, for example in the third chapter, Part One, where we are told that only experiment, induction, and generalization have produced Descartes's law of refraction; *cf.* the English translation, p. 34.) See also V. Kraft, *Die Grundformen der Wissenschaftlichen Methoden*, 1925; and Carnap, *Erkenntnis* 2, 1932, p. 440.

6. Address on Max Planck's 60th birthday. The passage quoted begins with the words, 'The supreme task of the physicist is to search for those highly universal laws . . . ,' etc. (quoted from A. Einstein, *Mein Weltbild*, 1934, p. 168; English translation by A. Harris: *The World As I See It*, 1935, p. 125). Similar ideas are found earlier in Liebig, *op. cit.*; *cf.* also Mach, *Principien der Wärmelehre* (1896), p. 443 *ff.* The German word '*Einfühlung*' is difficult to translate. Harris translates: 'sympathetic understanding of experience'.

WESLEY C. SALMON

Rational Prediction

A colleague, to whom I shall refer (quite accurately) as "the friendly physicist," recently recounted the following incident. While awaiting takeoff on an airplane, he noticed a young boy sitting across the aisle holding onto a string to which was attached a helium-filled balloon. He endeavored to pique the child's curiosity. "If you keep holding the string just as you are now," he asked, "what do you think the balloon will do when the airplane accelerates before takeoff?" The question obviously had not crossed the youngster's mind before that moment, but after giving it a little thought, he expressed the opinion that the balloon would move toward the back of the cabin. "I don't think so," said the friendly physicist, "I think it will move forward." The child was now eager to see what would happen when the plane began to move. Several adults in the vicinity were, however, skeptical about the physicist's prediction; in fact, a stewardess offered to wager a miniature bottle of Scotch that he was mistaken. The friendly physicist was not unwilling and the bet was made. In due course, the airplane began to accelerate, and the balloon moved toward the front of the cabin. The child's curiosity was satisfied[1], the theory—that all objects which are free to move will move toward the back of the cabin when the plane accelerates—was falsified; and the friendly physicist enjoyed a free drink.

I have related this anecdote to point out that there are at least three— probably more—legitimate reasons for making predictions. First, we are sometimes curious about future happenings, and we want to satisfy that curiosity without waiting for the events in question to transpire. To do so, we may make wild guesses, we may employ superstitious methods of prediction, we may appeal to common sense, or we may use more sophisticated scien-

FROM A. Grünbaum and W. C. Salmon, eds., *The Limitations of Deductivism* (Berkeley, Calif.: University of California Press, 1988), 47–60. This article was originally published in the *British Journal for the Philosophy of Science* 32 (1981): 115–25 and incorporates some minor revisions made by the author when the paper was reprinted.

tific theories. Second, we sometimes make predictions for the sake of testing a theory. In the example at hand, the prediction regarding the motion of the balloon was a rather good test of the hypothesis that all objects free to move in the cabin will tend to move toward the rear when the airplane accelerates. The fact that objects heavier than air tend to fall toward the earth when they are unsupported, while objects lighter than air (such as helium-filled balloons) tend to move in the opposite direction, suggests that the behavior of a helium-filled balloon has a reasonable chance of falsifying the hypothesis about the behavior of all material objects in the air-filled cabin of the accelerating airplane, if it is indeed false. Third, we sometimes find ourselves in situations in which some practical action is required, and the choice of an optimal decision depends upon predicting future occurrences. Although wagering is by no means the only such type of practical decision making, it is a clear and comprehensible example. We all agree, I take it, that scientific theories often provide sound bases for practical prediction.

A central feature of Sir Karl Popper's philosophy is his thesis concerning the status of induction. Indeed, he begins his book *Objective Knowledge* with the statement: "I think that I have solved a major philosophical problem: the problem of induction. . . . This solution has been extremely fruitful, and it has enabled me to solve a good number of other philosophical problems" (1972, p. 1). His solution, as is well known, involves a complete rejection of induction. This claim has been advanced in many of his writings spanning several decades, and it is reiterated in his autobiography (1974*a*) and in his "Replies to My Critics" (1974*b*).

For some time it has seemed to me that the crucial test of an anti-inductivist philosophy of science would be its capacity to deal with the predictive aspects of scientific knowledge. In a paper (Salmon 1968*a*) presented at the 1965 International Colloquium on Philosophy of Science at Bedford College, London, I attempted to offer a severe challenge to Popper's views concerning induction by posing what I took to be a serious dilemma: On Popper's account, either science embodies essential inductive aspects or else science is lacking in predictive content.[2] In the published proceedings of the Bedford College Colloquium (Lakatos 1968), J. W. N. Watkins contributed an answer to my critique. He denied that scientific reasoning is inductively infected, and he argued that it can, nevertheless, provide a basis for rational prediction. In Popper's replies to his critics (1974*b*, pp. 1028–1030), he acknowledges that I have understood his views "fairly well," and he endorses Watkins's response. I take this as evidence that we have located a genuine disagreement—one which is reasonably free from purely verbal disputes or out-and-out misrepresentations—regarding Popper's anti-inductivist stand. The question involves what Popper calls "the pragmatic problem of induction." It is this issue that I want to pursue in the present paper; it concerns the problem of rational prediction. Although the issue may appear to be rather narrow, it seems to me to have pivotal importance with regard to the assessment of Popper's deductivism.

Let me attempt to formulate the basic difficulty as I see it. In its very simplest terms, Popper's account of scientific knowledge involves generalizations and their observational tests. If we find a *bona fide* counterexample to a generalization, we can say that it has been deductively refuted. To be sure, as Popper explicitly acknowledges, there may be difficulties in some cases in determining whether certain observations constitute genuine counterexamples to a generalization, but that does not undermine the claim that a genuine counterexample yields a deductive refutation. According to Popper, negative instances provide rational grounds for rejecting generalizations. If, however, we make observations and perform tests, but no negative instance is found, all we can say deductively is that the generalization in question has not been refuted. In particular, positive instances do not provide confirmation or inductive support for any such unrefuted generalization. At this stage, I claim, we have no basis for rational prediction. Taken in themselves, our observation reports refer to past events, and consequently they have no predictive content. They say nothing about future events. If, however, we take a general statement as a premise, and conjoin to it some appropriate observation statements about past or present events, we may be able to deduce a conclusion which says something about future occurrences and that, thereby, has predictive content. Popper himself gives this account of *the logic of prediction* (1947*b*, p. 1030).

The problem of rational prediction concerns the status of the general premise in such an argument. One may claim, as Popper does, that we ought not to use a generalization that has actually been refuted as a premise in a predictive argument of this sort, for we are justified in regarding it as false. We ought not to employ premises which are known to be false if we hope to deduce true predictions. The exclusion of refuted generalizations does not, however, tell us what general premise should be employed. Typically there will be an infinite array of generalizations which are compatible with the available observational evidence, and that are therefore, as yet, unrefuted. If we were free to choose arbitrarily from among all the unrefuted alternatives, we could predict anything whatever. If there were no rational basis for choosing from among all of the unrefuted alternatives, then, as I think Popper would agree, there would be no such thing as rational prediction. We are not in this unfortunate situation, Popper contends, for we do have grounds for preferring one unrefuted generalization to another: "My *solution* of the logical problem of induction was that we may have *preferences* for certain of the competing conjectures; that is, for those which are highly informative and which so far have stood up to eliminative criticism" (1974*b*, p. 1024). Popper's concept of corroboration is designed to measure the manner in which conjectures have stood up to severe criticism, including severe testing. This, I take it, is the crucial thesis—that *there is a rational basis for preferring one unrefuted generalization to another for use in a predictive argument.* If that is correct, then Popper can legitimately claim to have solved the problem of rational prediction.

If we are going to talk about preference among generalizations, then we have to be quite explicit about the purpose for which the generalization is to be used. In this context, we are discussing prediction, so the preference must be in relation to predictive capability. As Popper rightly insists, any generalization we choose will have predictive import in the sense that it will make statements about future events—more precisely, in a predictive argument as characterized above, it yields conclusions about future occurrences. But since all of the various unrefuted generalizations have predictive content in that sense, we must still ask on what basis the predictive content of one conjecture is rationally preferable to that of another conjecture.

At this stage of the discussion, it is important to recall the point of the opening story, namely, that predictions are made for various purposes. Thus, even if we agree that we want to select a generalization for predictive purposes, we must still specify what type of prediction is involved. Popper explicitly acknowledges (1974b, pp. 1024–1025) that there are two types of preference, "the theoretician's preference" and that of "the man of practical action." As I understand Popper's view, the theoretician is interested in formulating bold conjectures which have high content and in subjecting them to severe tests. Insofar as the theoretician is mainly interested in explanations of known phenomena, he may not be much involved in making any sorts of predictions. I suppose we might distinguish the theoretician's explanatory preference from the theoretician's predictive preference, recognizing that there is bound to be a close connection between preferences of these two kinds. When the theoretician is actually involved *qua theoretician* in making predictions, the purpose is to devise (and, perhaps, to instruct the experimentalist on how to conduct) a severe test. The purpose of predictions made in this theoretical context is to gain information that is useful in the evaluation of scientific theories. If the chief value of the scientific theories is explanatory, then it is not at all clear that a primary desideratum of the predictive argument is to arrive at a true prediction. As Popper has emphasized, and as all of us know, a false prediction can be valuable, since the realization (on the basis of observation) that it is false can be highly informative.

Having briefly characterized theoretical preference, let us now focus attention upon the kind of preference which is pertinent to the practical context, with special attention to the kinds of predictions which play a role in practical decision making. As I have remarked above, Popper claims that for theoretical purposes we prefer theories which are highly corroborated to those that are less well corroborated. I do not think this claim is unproblematic, but I do not propose arguing the matter here. My aim is to emphasize that, even if we are entirely justified in letting such considerations determine our theoretical preferences, it is by no means obvious that we are justified in using them as the basis for our preferences among generalizations which are to be used for prediction in the practical decision-making context. Popper and Watkins have maintained, however, that corroboration should play a crucial role in determining both theoretical preference and practical preference.

Since scientific theories are used for both theoretical and practical purposes—including prediction—and since, according to Popper, theory preference is based upon corroboration, I had mistakenly inferred (prior to 1968) that the appraisal of a theory in terms of corroboration must imply some attempt at an appraisal of the theory with respect to its future performance. If that were Popper's thesis, I had argued, then corroboration must involve some element of induction (or nondemonstrative inference of some sort), for past performance of the theory is taken to constitute a basis for some sort of claim about future performance. However, I have since been informed by Watkins (1968) and Popper (1974a) that I had misconstrued Popper's view. Statements about the corroboration of theories are no more than appraisals of their past performances; corroboration statements hold no predictions with respect to future performance. If they did, they would be inductive (as I had claimed); but they are not inductive, so they cannot be predictive.

This view of corroboration holds serious difficulties. Watkins and Popper agree, I take it, that statements that report observations of past and present events do not, in and of themselves, have any predictive content. Moreover, they maintain, statements about the corroboration of conjectures do not, in and of themselves, have any predictive content. Conjectures, hypotheses, theories, generalizations—call them what you will—do have predictive content. The problem is that there are many such statements, rich in predictive content, which make incompatible predictive claims when conjoined with true statements about past and present occurrences. The fact that a general statement has predictive content does not mean that what it says is true. In order to make a prediction, one must choose a conjecture that has predictive content to serve as a premise in a predictive argument. In order to make a *rational* prediction, it seems to me, one must make a *rational* choice of a premise for such an argument. But from our observational evidence and from the statements about the corroboration of a given conjecture, no predictive appraisal follows. Given two conjectures which, in a particular situation, will lead to incompatible predictions, and given the corroboration ratings of these two hypotheses, *nothing follows* about their comparative predictive capacities. Thus, it seems to me, corroboration—the ground for theoretical preference— furnishes no rational basis for preference of one conjecture to another *for purposes of practical prediction*. I am not complaining that we are not told *for sure* that one will make a correct prediction and that the other will not. I am complaining that no rational basis whatever has been furnished for a preference of *this* type.

In his reply to my Bedford College paper, Watkins acknowledges that there is an important distinction between theoretical and practical preferences, and he further acknowledges that the two kinds of appraisal may have quite different bases:

Now our methods of hypothesis-selection in practical life should be well suited to our practical aims, just as our methods of hypothesis-selection in theoretical science should be well suited to our theoretical aims; and the two kinds of method may very well yield different answers in a particular case (1968, p. 65).

He goes on to explain quite correctly how utility considerations may bear upon the practical situation. Then he considers the case in which utility does not play a decisive role:

Now suppose that, for a particular agent, the mutually incompatible hypotheses h_1 and h_2 are on a par utility-wise, and that in the situation in which he finds himself, he has *got* to act since 'inaction' would itself be one mode of action. Then if h_1 is the only alternative to h_2 before him, he *has* to choose one of them. Then it would be rational for him to choose the better corroborated one, the one which has withstood the more severe criticism, since he has nothing else to go on (pp. 65–66).

Watkins offers no further argument for supposing that corroboration provides a rational basis for *practical* preference. Moreover, the hint of an argument which he does supply appeals to a false premise. The agent does have other things "to go on." He could decide between the two hypotheses by the flip of a coin. He could count the numbers of characters in each of the two hypotheses in the particular formulation given, and choose the one that has fewer. He could choose the hypothesis which comes first lexicographically in the given formulation. What Watkins is suggesting, it seems to me, is not that the agent has "nothing else to go on" but rather that he has no other *rational* basis for preference. But such an argument would be patently question begging. Even if all other bases for choice were irrational, it would not follow that the one cited by Watkins is *ipso facto* rational. Indeed, if we take seriously Popper's statement, "I regarded (and I still regard) the degree of corroboration of a theory merely as a critical report on the quality of past performance: *it could not be used to predict future performance*" (1974a, p. 82), it is hard to see how corroboration can supply a rational basis for preference of a theory *for purposes of practical prediction*.

Whether my criticism of Popper's position is correct or incorrect, the issue I am raising has fundamental importance. For if it should turn out that Popper could not provide a tenable account of rational prediction, then— given his persistent emphasis upon objectivity and rationality—we could hardly credit his claim to have solved the problem of induction. Moreover, in his replies to his critics, Popper acknowledges the issue. With the comment, "Our corroboration statements have no predictive import, although they motivate and justify our *preference* for some theory over another" (1974b, pp. 1029–1030), he endorses the answer Watkins had furnished. Since I am not attempting to deal with the psychological problem of induction, I shall not dispute the claim that corroboration may *motivate* the preference of one theory to another. What I want to see is how corroboration could *justify* such a preference. Unless we can find a satisfactory answer to that question, it appears to me that we have no viable theory of *rational* prediction, and no adequate solution to the problem of induction.

In *Objective Knowledge*, Popper offers an answer to the basic question which seems closely related to that of Watkins:

> [A] *pragmatic belief in the results of science* is not irrational, because there is nothing more 'rational' than the method of critical discussion, which is the method of science. And although it would be irrational to accept any of its results as certain, there is nothing 'better' when it comes to practical action: there is no alternative method which might be said to be more rational (1972, p. 27).

This response appears to miss the point. The question is not whether other methods—for example, astrology or numerology—provide more rational approaches to prediction than does the scientific method. The question is whether the scientific approach provides a more rational basis for prediction, for purposes of practical action, than do these other methods. The position of the Humean skeptic would be, I should think, that none of these methods can be shown either more or less rational than any of the others. But if every method is equally lacking in rational justification, then there is no method which can be said to furnish a rational basis for prediction, for any prediction will be just as unfounded rationally as any other. If the Humean skeptic were right, we could offer the following parallel claim. A pragmatic belief in the predictions found in Chinese fortune cookies is not irrational, for there is nothing more rational. . . .

In his replies to his critics, Popper again addressed the problem, and he came more firmly to grips with it:

> But every action presupposes a set of expectations, that is, of theories about the world. Which theory shall the man of action choose? Is there such a thing as a *rational choice?*
>
> This leads us to the *pragmatic problems of induction*, which to start with, we might formulate thus:
>
> (a) Upon which theory should we rely for practical action, from a rational point of view?
>
> (b) Which theory should we prefer for practical action, from a rational point of view?
>
> My answer to (a) is: from a rational point of view, we should not 'rely' on any theory, for no theory has been shown to be true, or can be shown to be true (or 'reliable').

My answer to (b) is: we should *prefer* the best tested theory as a basis for action.

In other words, there is no 'absolute reliance'; but since we *have* to choose, it will be 'rational' to choose the best tested theory. This will be 'rational' in the most obvious sense of the word known to me: the best tested theory is the one which, in the light of our *critical discussion*, appears to be the best so far; and I do not know of anything more 'rational' than a well-conducted critical discussion (1974b, p. 1025)

Let us not be seduced by honeyed words. If we wish to claim that a theory "appears to be the best so far," we must ask, "Best for what purpose—theoretical explanation or practical prediction?" Since it is "the best tested theory" and it has been subjected to "critical discussion," then, in the light of the many statements by Popper and others about the lack of predictive import of corroboration, we must conclude, I believe, that the answer is, "Best for theoretical explanation." Perhaps I am being unduly obtuse, but I cannot see that any reason has been provided for supposing that such a theory is best *for practical prediction.*

I must confess to the feeling that we have been given the runaround. We begin by asking how science can possibly do without induction. We are told that the aim of science is to arrive at the best explanatory theories we can find. When we ask how to tell whether one theory is better than another, we are told that it depends upon their comparative ability to stand up to severe testing and critical discussion. When we ask whether this mode of evaluation does not contain some inductive aspect, we are assured that the evaluation is made wholly in terms of their comparative success up to now; but since this evaluation is made entirely in terms of past performance, it escapes inductive contamination because it lacks predictive import. When we then ask how to select theories for purposes of rational prediction, we are told that we should prefer the theory which is "best tested" and that "in the light of our *critical discussion*, appears to be the best so far," even though we have been explicitly assured that testing and critical discussion have no predictive import. Popper tells us, "I do not know of anything more 'rational' than a well-conducted critical discussion." I fail to see how it could be rational to judge theories *for purposes of prediction* in terms of a criterion which is emphatically claimed to be lacking in predictive import.[3]

Fearing that the point of his initial argument may have been missed, Popper attempts another formulation:

Let us forget momentarily about what theories we 'use' or 'choose' or 'base our practical actions on', and consider only the resulting *proposal* or *decision* (to do X; not to do X; to do nothing; or so on). Such a proposal can, we hope, be rationally criticized; and if we are rational agents we will want it to survive, if possible, the most testing criticism we can muster. *But such criticism will freely make use of the best tested scientific theories in our possession.* Consequently any proposal that ignores these theories (where they are relevant, I need hardly add) will collapse under criticism. Should any proposal remain, it will be rational to adopt it.

This seems to me all far from tautological. Indeed, it might well be challenged by challenging the italicized sentence in the last paragraph. Why, it might be asked, does rational criticism make use of the best tested although highly unreliable theories? The answer, however, is exactly the same as before. Deciding to criticize a practical proposal from the standpoint of modern medicine (rather than, say, in phrenological terms) is itself a kind of 'practical' decision (anyway it may have practical consequences). Thus the rational decision is always: adopt critical methods which have themselves withstood severe criticism (1974b, pp. 1025–1026).

I have quoted Popper *in extenso* to try to be quite sure not to misunderstand his answer. The italicised sentence in the first paragraph raises precisely the question which seems to me crucial. In the second paragraph, Popper admits the legitimacy of the question, and he offers an answer. When he says, "The answer . . . is exactly the same as before. . . . [T]he rational decision is always: adopt critical methods which have themselves withstood severe criticism," he seems to be saying that we should adopt his methodological recommendations, because they have "withstood severe criticism." But his answer is inappropriate in this context because our aim is precisely to subject his philosophical views, in the best Popperian spirit, to severe criticism.

In my reply to Watkins, I said, "Watkins acknowledges . . . that corroboration does have predictive import in practical decision making" (1968b, p. 97). Popper has objected to this way of putting the matter: "[O]ur *theories do have predictive import*. Our corroboration statements have no predictive import, although they motivate and justify our *preference* for some theory or other" (1974b, pp. 1029–1030). Let us grant that corroboration statements have no predictive *content*—indeed, that they are analytic, as Watkins remarks (1968, p. 63)—and that theories are the kinds of statements that do have predictive *content*. It does not follow, as Popper has claimed, that corroboration has no predictive *import*. The distinction between predictive content and predictive import is no mere verbal quibble; a fundamental substantive point is at issue. Statements whose consequences refer to future occurrences may be said to have predictive content; rules, imperatives, and directives are totally lacking in predictive content because they do not entail any statements at all. Nevertheless, an imperative—such as "No smoking, please"—may have considerable predictive import, for it may effectively achieve the goal of preventing the occurrence of smoking in a particular room in the immediate future.

Since corroboration, in some cases at least, provides the basis for deciding which theory (with its predictive content) is to be used for the purpose of making practical predictions, it seems to me that corroboration, even if it is lacking in predictive content, does have enormous predictive import. Perhaps this point can be put more clearly in the following way. *Statements* assessing the corroboration of theories have no predictive *content*, as Popper, Watkins, and others maintain. The *directive*—to choose more highly corroborated theories in preference to theories that are less well corroborated for purposes of practical prediction—has considerable predictive *import*. The problem, which it seems to me the anti-inductivists have failed to solve, is how to vindicate this directive for making predictions.[4] Without some sort of vindication for this directive, the problem of rational prediction remains unresolved.

I have wondered why it would seem evident to Popper that corroboration, as he construes it, should provide a guide to rational prediction. In his autobiography, he gives what appear to be indications of an answer.

> I regarded (and I still regard) the degree of corroboration of a theory merely as a critical report on the quality of past performance: *it could not be used to predict future performance.* . . . When faced with the *need to act*, on one theory or another, the rational choice was to act on that theory—if there was one— which so far had stood up to criticism better than its competitors had: there is no better idea of rationality than that of a readiness to accept criticism. Accordingly, the degree of corroboration of a theory was a rational guide to practice (1974a, p. 82).

A further elaboration of the theme informs us that

> when we think we have found an approximation to the truth in the form of a scientific theory which has stood up to criticism and to tests better than its competitors, we shall, as realists, accept it as a basis for practical action, simply because we have nothing better (or nearer to the truth) (ibid., pp. 120–121).

Realism is a position to which Popper has adhered since the time of his earliest philosophical activity; near the beginning of his autobiography he tells us that "a realist who believes in an 'external world' necessarily believes in the existence of a cosmos rather than a chaos; that is, in regularities" (ibid., p. 14). Thus, I am led to conjecture, it may be that Popper's adherence to the thesis that corroboration can provide a basis for rational prediction rests ultimately upon his realism, which embodies a version of a principle of uniformity of nature. If this suggestion is correct, we can still legitimately wonder whether Popper's epistemology is as far from traditional inductivism as he would have us believe.

To conclude this discussion, I should like to recall the point of my opening anecdote. It seems to me incorrect to suppose that the only concern of *theoretical* science is to make bold explanatory conjectures that can be tested and criticized. It is a mistake, I believe, to suppose that all prediction, aside from that involved in the testing of theories, is confined to contexts in which practical action is at stake. Theoretical science furnishes both explanations and predictions. Some of these predictions have practical consequences and others do not. When, for example, scientists assembled the first man-made atomic pile under the West Stands at the University of Chicago, they had to make a prediction as to whether the nuclear chain reaction they initiated could be controlled, or whether it would spread to surrounding materials and engulf the entire city—and perhaps the whole earth—in a nuclear holocaust. Their predictions had both theoretical and practical interest. Contemporary cosmologists, for another example, would like to *explain* certain features of our universe in terms of its origin in a "big bang"; many of them are trying to *predict* whether it will end in a "big crunch." In this case, the predictive question seems motivated by pure intellectual curiosity, quite unattached to concerns regarding practical decision making. Whether a helium-filled balloon will move forward in the cabin of an airplane when the airplane accelerates, whether a nuclear chain reaction—once initiated—will run out of control, and whether the universe will eventually return to a state of high density are all matters of legitimate scientific concern.

In this paper, I have attempted to argue that pure deductivism could not do justice to the problem of rational prediction in contexts of practical decision making. If we ask whether Popperian deductivism can adequately account for scientific predictions of the more theoretical varieties, then I suspect that we would have to go through all of the preceding arguments once more. The net result would be, I think, that science is inevitably inductive in matters of intellectual curiosity as well as practical prediction. It *may* be possible to excise all inductive ingredients from science, but if the operation were successful, the patient (science), deprived of all predictive import, would die.[5]

■ | Notes

1. His curiosity regarding *what* would happen was satisfied, though not his curiosity as to *why*.

2. Similar themes were developed in Salmon 1967, chap. 2, sec. 3.

3. The argument advanced in this paragraph bears a strong resemblance, I think, to one developed in Grünbaum 1976; see esp. p. 246.

4. This felicitous reformulation was suggested by Abner Shimony (if I did not misunderstand him) in the discussion following my presentation at the Popper Symposium.

5. A version of this paper was presented orally at the Symposium on the Philosophy of Sir Karl Popper, London School of Economics, July 14–16, 1980. This material is based upon work supported by the National Science Foundation (U.S.A.) under Grant No. SES-7809146.

▪ | References

Grünbaum, Adolf. 1976: "Is Falsifiability the Touchstone of Scientific Rationality? Karl Popper versus Inductivism." In R. S. Cohen, P. K. Feyerabend, and M. W. Wartofsky, eds., *Essays in Memory of Imre Lakatos*, pp. 213–252. Dordrecht: Reidel.

Lakatos, Imre, ed. 1968: *The Problem of Inductive Logic.* Amsterdam: North-Holland.

Popper, Karl R. 1972. *Objective Knowledge.* Oxford: Clarendon Press.

——. 1974a. "Autobiography." In P. A. Schilpp, ed., *The Philosophy of Karl Popper*, pp. 3–181. LaSalle, Ill.: Open Court.

——. 1974b. "Replies to My Critics." In P. A. Schilpp, ed., *The Philosophy of Karl Popper*, pp. 961–1197. LaSalle, Ill.: Open Court.

Salmon, Wesley C. 1967. *The Foundations of Scientific Inference.* Pittsburgh, Pa.: University of Pittsburgh Press. Originally published in R. C. Colodny, ed., *Mind and Cosmos*, pp. 135–275. Pittsburgh, Pa.: University of Pittsburgh Press, 1966.

——. 1968a. "The Justification of Inductive Rules of Inference." In Lakatos 1968, pp. 24–43.

——. 1968b. "Reply." In Lakatos 1968, pp. 74–97.

Watkins, J. W. N. 1968. "Non-Inductive Corroboration." In Lakatos 1968, pp. 61–66.

CARL G. HEMPEL

Criteria of Confirmation
and Acceptability

. . . A favorable outcome of even very extensive and exacting tests cannot provide conclusive proof for a hypothesis, but only more or less strong evidential support, or confirmation. How strongly a hypothesis is supported by a given body of evidence depends on various characteristics of the evidence, which we will consider presently. In appraising what might be called the scientific acceptability or credibility of a hypothesis, one of the most important factors to consider is, of course, the extent and the character of the relevant evidence available and the resulting strength of the support it gives to the hypothesis. But several other factors have to be taken into account as well; these, too, will be surveyed in this chapter. We shall at first speak in a somewhat intuitive manner of more or less strong support, of small or large increments in confirmation, of factors that increase or decrease the credibility of a hypothesis, and the like. At the end of the chapter, we will briefly consider whether the concepts here referred to admit of a precise quantitative construal.

1 | Quantity, Variety, and Precision
of Supporting Evidence

In the absence of unfavorable evidence, the confirmation of a hypothesis will normally be regarded as increasing with the number of favorable test findings. For example, each new Cepheid variable whose period and luminosity are found to conform to the Leavitt-Shapley law will be considered as adding to the evidential support of the law.* But broadly speaking, the

FROM Carl G. Hempel, *Philosophy of Natural Science* (Englewood Cliffs, N.J.: Prentice-Hall, 1966), 33–46.

* Cepheid variables are stars that fluctuate in brightness with a constant period of one to fifty days. (They are called Cepheids after the prototype Delta Cephei, a variable star discovered in 1784.) The Leavitt-Shapley law is named in honor of the

increase in confirmation effected by one new favorable instance will generally become smaller as the number of previously established favorable instances grows. If thousands of confirmatory cases are already available, the addition of one more favorable finding will raise the confirmation but little.

This remark must be qualified, however. If the earlier cases have all been obtained by tests of the same kind, but the new finding is the result of a different kind of test, the confirmation of the hypothesis may be significantly enhanced. For the confirmation of a hypothesis depends not only on the quantity of the favorable evidence available, but also on its variety: the greater the variety, the stronger the resulting support.

Suppose, for example, that the hypothesis under consideration is Snell's law, which states that a ray of light traveling obliquely from one optical medium into another is refracted at the separating surface in such a way that the ratio, $\sin \alpha/\sin \beta$, of the sines of the angles of incidence and of refraction is a constant for any pair of media. Compare now three sets of 100 tests each. In the first set, the media and the angle of incidence are kept constant: in each experiment, the ray passes from air into water at an angle of incidence of 30°; the angle of refraction is measured. Suppose that in all cases, $\sin \alpha/\sin \beta$ does have the same value. In the second set, the media are kept constant, but the angle α is varied: light passes from air into water at varying angles; β is measured. Again, suppose that $\sin \alpha/\sin \beta$ has the same value in all cases. In the third set, both the media and the angle α are varied: 25 different pairs of media are examined: for each pair, four different angles α are used. Suppose that for each pair of media, the four associated values of the ratio $\sin \alpha/\sin \beta$ are equal, while the ratios associated with different pairs have different values.

Each test set then presents a class of favorable outcomes, since the ratios associated with any particular pair of media are found to be equal, as implied by Snell's law. But the third set, which offers the greatest variety of positive instances, will surely be regarded as supporting the law much more strongly than the second, which provides supporting instances of much more limited variety; and the first set, it will be agreed, lends even less strong support to the general law. In fact, it might seem that in the first set, the same experiment is performed over and over again, and that the positive outcome in all 100 cases can support the hypothesis no more strongly than do the first two tests in the set, which bear out the constancy of the ratio. But this idea is mistaken. What

Americans astronomers Henrietta Swan Leavitt and Harlow Shapley, who discovered that the period of Cepheids is proportional to their intrinsic luminosity. Because the apparent brightness of any object falls off as the inverse square of its distance from the observer, the law can be used to calculate the distance to remote galaxies by observing the period and apparent brightness of the Cepheids they contain. To confirm the Leavitt-Shapley law by observing Cepheids requires an independent way of measuring distance to those stars. For stars that are not too far from the earth, this can be done by parallax measurements.

is repeated here 100 times is not literally the same experiment, for the successive performances differ in many respects, such as the distance of the apparatus from the moon, perhaps the temperature of the light source, the atmospheric pressure, and so on. What is "kept the same" is simply a certain set of conditions, including a fixed angle of incidence and one particular pair of media. And even if the first two or more measurements under these circumstances yield the same value for sin α/sin β, it is logically quite possible that subsequent tests under the specified circumstances should yield different values for the ratio. Thus even here, repeated tests with favorable outcome add to the confirmation of the hypothesis—though much less so than do tests that cover a wider variety of instances.

. . . Scientific theories are often supported by empirical findings of amazing variety. Newton's theory of gravitation and of motion implies, for example, the laws for free fall, for the simple pendulum, for the motion of the moon about the earth and of the planets about the sun, for the orbits of comets and of man-made satellites, for the motion of double stars about each other, for tidal phenomena, and many more. And all the diverse experimental and observational findings that bear out those laws lend support to Newton's theory.

The reason why diversity of evidence is so important a factor in the confirmation of a hypothesis might be suggested by the following consideration, which refers to our example of various tests for Snell's law. The hypothesis under test—let us call it S for short—refers to *all* pairs of optical media and asserts that for any pair, the ratio sin α/sin β has the same value for *all* associated angles of incidence and of refraction. Now, the more widely a set of experiments ranges over the diverse possibilities here covered, the greater will be the chances of finding an unfavorable instance if S should be false. Thus, the first set of experiments may be said to test more specifically a hypothesis S_1 that expresses only a small part of Snell's law—namely, that sin α/sin β has the same value whenever the optical media are air and water and α is 30°. Hence, if S_1 should be true, but S false, the first kind of test will never disclose this. Similarly, the second set of experiments tests a hypothesis S_2, which asserts distinctly more than S_1 but still not nearly as much as S—namely, that sin α/sin β has the same value for all angles α and the associated angles β if the media involved are air and water. Hence, if S_2 should be true, but S false, a test set of the second kind would never disclose this. Thus, the third set of experiments might be said to test Snell's law more thoroughly than the other two; an entirely favorable outcome accordingly lends stronger support to it.

As an additional illustration of the power of diversified evidence, we might note that if the diversity of the evidence is still further increased by varying the temperature of the optical media or by using monochromatic light of different wavelengths, then Snell's law in the classical form cited above is in fact found to be false.

But have we not overstated the case for diversified evidence? After all, some ways of increasing variety would be regarded as pointless, as incapable of raising the confirmation of a hypothesis. This verdict would apply, for example, if in our first test set for Snell's law the variety were increased by having the experiment performed at different places, during different phases of the moon, or by experimenters with different eye color. But to try such variations would not be unreasonable if as yet we had no knowledge, or only extremely limited knowledge, of what factors are likely to affect optical phenomena. At the time of the Puy-de-Dôme experiment, for example, the experimenters had no very definite ideas of what factors other than altitude might affect the length of the mercury column in the barometer; and when Pascal's brother-in-law and his associates performed the Torricelli experiment on the mountaintop and found the mercury column over three inches shorter than it had been at the foot of the mountain, they decided to repeat the experiment then and there, changing the circumstances in various ways.* As Périer says in his report: "I therefore tried the same thing five times more, with great accuracy, at different places on the top of the mountain, once under cover in the little chapel which is there, once exposed, once in a shelter, once in the wind, once in good weather, and once during the rain and the mists which came over us sometimes, having taken care to get rid of the air in the tube every time; and in all these trials there was found the same height of the quicksilver . . . ; this result fully satisfied us."[1]

Thus, the qualification of certain ways of varying the evidence as important and of other ways as pointless is based on the background assumptions we entertain—perhaps as a result of previous research—

---

* The Puy-de-Dôme experiment was performed at the instigation of Blaise Pascal (1623–62) in 1648. The first mercury barometer had been constructed by Galileo's pupil, Torricelli, in 1644. Torricelli, like Galileo, believed that the barometer acts as a simple mechanical balance, the twenty-nine-inch column of mercury and a vacuum on one side being supported by the pressure of the atmosphere on the other. Aristotelians insisted that the entire effect is nonmechanical and due to nature's abhorrence of a vacuum (*horror vacui*). Prior to the invention of the air pump (in the 1650s), it was hard to rule out the Aristotelian theory, since a conclusive refutation would require placing the entire apparatus in a sealed chamber, reducing the pressure (by pumping out the air), and then observing a steady decrease in the height of the mercury column. But Pascal realized that, even without an air pump, a crucial test of the rival theories was possible because atmospheric pressure decreases as one ascends a mountain. Thus, he persuaded his brother-in-law Périer, who lived near the Puy-de-Dôme mountain in central France, to perform the famous experiment alluded to by Hempel. For an eyewitness report of the experiment, see W. F. Magie, *A Source Book in Physics* (Cambridge, Mass.: Harvard University Press, 1963), 70–75. There is a good account of this and Pascal's other brilliant experiment—the "vacuum in a vacuum"—in Richard S. Westfall, *The Construction of Modern Science: Mechanisms and Mechanics* (New York: John Wiley and Sons, 1971), 43–49.

concerning the probable influence of the factors to be varied upon the phenomenon with which the hypothesis is concerned.

And sometimes when such background assumptions are questioned and experimental variations are accordingly introduced which, on the generally accepted view, are pointless, a revolutionary discovery may be the outcome. This is illustrated by the recent overthrow of one of the basic background assumptions of physics, the principle of parity. According to this principle, the laws of nature are impartial between right and left; if a certain kind of physical process is possible (i.e., if its occurrence is not precluded by the laws of nature), then so is its mirror image (the process as seen in a reflecting mirror), where right and left are interchanged. In 1956, Yang and Lee, who were trying to account for some puzzling experimental findings concerning elementary particles, suggested that the principle of parity is violated in certain cases; and their bold hypothesis soon received clear experimental confirmation.*

Sometimes, a test can be made more stringent, and its result the more weighty, by increasing the precision of the procedures of observation and measurement it involves. Thus, the hypothesis of the identity of inertial and gravitational mass—supported, for example, by the equality of the accelerations shown in free fall by bodies of different chemical constitution—has recently been re-examined with extremely precise methods; and the results, which have so far borne out the hypothesis, have greatly strengthened its confirmation.†

2 | Confirmation by "New" Test Implications

When a hypothesis is designed to explain certain observed phenomena, it will of course be so constructed that it implies their occurrence; hence, the fact to be explained will then constitute confirmatory evidence for it. But it is highly desirable for a scientific hypothesis to be confirmed also by "new" evidence—by facts that were not known or not taken into account when the hypothesis was formulated. Many hypotheses and theories in natural sci-

---

* The first experiment to confirm the Lee-Yang conjecture that parity is not conserved in weak interactions was performed in 1956–57 by a team led by Chien-Shiung Wu, professor of physics at Columbia University. For a clear account of the essential details of her experiment, see Martin Gardner, *The New Ambidextrous Universe*, 3d rev. ed. (New York: W. H. Freeman and Company, 1990), ch. 22.

† Hempel is referring to the experiments begun in 1959 by Robert H. Dicke and his colleagues at Princeton University, who verified to an accuracy of about 1 part in 10^{11} that gravity produces the same acceleration in all bodies regardless of their chemical composition. See Robert H. Dicke, "The Eötvös Experiment," *Scientific American* 205 (1961): 84–94. The hypothesis that inertial mass and gravitational mass are identical is usually referred to as the *principle of equivalence* and is one of the central axioms of Einstein's general theory of relativity.

ence have indeed received support from such "new" phenomena, with the result that their confirmation was considerably strengthened.

The point is well illustrated by an example that dates back to the last quarter of the nineteenth century, when physicists were searching for inherent regularities in the profusion of lines that had been found in the emission and absorption spectra of gases. In 1885, a Swiss school teacher, J. J. Balmer, proposed a formula that he thought expressed such a regularity for the wavelengths of a series of lines in the emission spectrum of hydrogen. On the basis of measurements that Ångström had made of four lines in that spectrum, Balmer constructed the following general formula:

$$\lambda = b\frac{n^2}{n^2 - 2^2}$$

Here, b is a constant, whose value Balmer determined empirically as 3645.6 Å, and n is an integer greater than 2. For $n = 3$, 4, 5, and 6, this formula yields values that agree very closely with those measured by Ångström; but Balmer was confident that the other values, too, would represent wavelengths of lines yet to be measured—or even yet to be found—in the hydrogen spectrum. He was unaware that some further lines had already been noted and measured. By now, 35 consecutive lines in the so-called Balmer series for hydrogen have been ascertained, and all of these have wavelengths that agree well with the values predicted by Balmer's formula.[2]

It is hardly surprising that such striking confirmation by correctly predicted "new" facts greatly enhances the credence we will be prepared to give to a hypothesis. A puzzling question arises in this context. Suppose for a moment that Balmer's formula had been constructed only after all the 35 lines now recorded in the series had been carefully measured. In this fictitious case, then, exactly the same experimental findings would be available that have in fact been obtained by measurements made in part before, and in much larger part after, the construction of the formula. Should that formula be considered as less well confirmed in the fictitious case than in the actual one? It might seem reasonable to answer in the affirmative, on these grounds: for *any* given set of quantitative data, it is possible to construct a hypothesis that covers them, just as for any finite set of points, it is possible to draw a smooth curve that contains them all. Thus, there would be nothing very surprising about the construction of Balmer's formula in our fictitious case. What *is* remarkable, and does lend weight to a hypothesis, is its fitting "new" cases: and Balmer's hypothesis has this accomplishment to its credit in the actual case, but not in the fictitious one. But this argument could be met with the reply that even in the fictitious case, Balmer's formula is not just some otherwise arbitrary hypothesis that is rigged to fit the 35 measured wavelengths: it is, rather, a hypothesis of striking formal simplicity; and the very fact that it subsumes those 35 wavelengths under a mathematically simple formula should lend it much higher credibility than could be accorded to a very complex formula fitting the same data. To state the idea in geometrical terms: if a set of points representing the results of measurements can be connected by a simple curve, we have much greater confidence in having discovered an underlying general law than if the curve is complicated and shows no perceptible regularity. (This notion of simplicity will be further considered, later on in this chapter.) Besides, from a logical point of view, the strength of the support that a hypothesis receives from a given body of data should depend only on what the hypothesis asserts and what the data are: the question of whether the hypothesis or the data were presented first, being a purely historical matter, should not count as affecting the confirmation of the hypothesis. This latter conception is certainly implicit in recently developed statistical theories of testing and also in some contemporary logical analyses of confirmation and induction, to which brief reference will be made at the end of this chapter.

3 | Theoretical Support

The support that may be claimed for a hypothesis need not all be of the inductive-evidential kind that we have considered so far: it need not consist entirely—or even partly—of data that bear out test implications derived from it. Support may also come "from above"; that is, from more inclusive hypotheses or theories that imply the given one and have independent evidential support. To illustrate: [consider the] hypothetical law for free fall on the moon, $s = 2.7 \ t^2$. Although none of its test implications have ever been checked by experiments on the moon, it has strong *theoretical support*, for it follows deductively from Newton's theory of gravitation and of motion (strongly supported by a highly diversified body of evidence) in conjunction with the information that the radius and the mass of the moon are .272 and .0123 of those of the earth and that the gravitational acceleration near the surface of the earth is 32.2 feet per second per second.*

Similarly, the confirmation of a hypothesis that does have inductive-evidential support will be further strengthened if, in addition, it acquires deductive support from above. This happened, for example, to Balmer's formula. Balmer had anticipated the possibility that the hydrogen spectrum might contain further series of lines, and that the wavelengths of all the lines might conform to a generalization of his formula; namely,

---

* Hempel wrote this in 1966, three years before the first moon landing by Apollo 11 astronauts Neil Armstrong and Edwin ("Buzz") Aldrin. Newton's theory entails that a, the acceleration due to gravity at the surface of any planet, is directly proportional to the planet's mass and inversely proportional to the square of its radius. For any planet of reasonable size, a will be nearly constant close to the planet's surface, and simple algebra yields the distance formula $s = \frac{1}{2}at^2$ for a falling body dropped from rest.

$$\lambda = b \frac{n^2}{n^2 - m^2}$$

Here, m is a positive integer, and n is any integer greater than m. For $m = 2$, this generalization yields Balmer's formula; whereas $m = 1, 3, 4, \ldots$ determine new series of lines. And indeed, the existence of the series corresponding to $m = 1, 3, 4,$ and 5 was later established by experimental exploration of the invisible infrared and ultraviolet parts of the hydrogen spectrum. Thus, there was strong evidential support for a more general hypothesis that implied Balmer's original formula as a special case, thus providing deductive support for it. And deductive support by a theory came in 1913, when the generalized formula—hence Balmer's original one, also—were shown by Bohr to be derivable from his theory of the hydrogen atom. This derivation greatly strengthened the support of Balmer's formula by fitting it into the context of quantum-theoretical conceptions developed by Planck, Einstein, and Bohr, which were supported by diverse evidence other than the spectroscopic measurements that lent inductive support to Balmer's formula.[3]

Correlatively, the credibility of a hypothesis will be adversely affected if it conflicts with hypotheses or theories that are accepted at the time as well-confirmed. In the *New York Medical Record* for 1877, a Dr. Caldwell of Iowa, reporting on an exhumation he claims to have witnessed, asserts that the hair and the beard of a man who had been buried clean-shaven, had burst the coffin and grown through the cracks.[4] Although presented by a presumptive eyewitness, this statement will be rejected without much hesitation because it conflicts with well-established findings about the extent to which human hair continues to grow after death. . . .

The principle here referred to [that conflict with a broadly supported theory militates against a hypothesis] must be applied with discretion and restraint, however. Otherwise, it could be used to protect any accepted theory against overthrow: adverse findings could always be dismissed as conflicting with a well-established theory. Science does not, of course, follow this procedure; it is not interested in defending certain pet conceptions against all possible adverse evidence. It aims, rather, at a comprehensive body of sound empirical knowledge, represented by a well-confirmed system of empirical statements, and it is accordingly prepared to give up or to modify whatever hypotheses it may have previously accepted. But findings that are to dislodge a well-established theory have to be weighty; and adverse experimental results, in particular, have to be repeatable. Even when a strong and useful theory has been found to conflict with an experimentally reproducible "effect", it may still continue to be used in contexts where it is not expected to lead into difficulties. For example, when Einstein propounded the theory of light quanta to account for such phenomena as the photoelectric effect, he noted that in dealing with the reflection, refraction, and polarization of light, the electromagnetic wave theory would probably never be replaced; and it is indeed still used in this context. A large-scale theory that has been successful in many areas will normally be abandoned only when a more satisfactory alternative theory is available—and good theories are difficult to come by.[5]

4 | Simplicity

Another aspect that affects the acceptability of a hypothesis is its simplicity, compared with that of alternative hypotheses that would account for the same phenomena.

Consider a schematic illustration. Suppose that investigation of physical systems of a certain type (Cepheids, elastic metal springs, viscous liquids, or whatever) suggests to us that a certain quantitative characteristic, v, of such systems, might be a function of, and thus uniquely determined by, another such characteristic, u (in the way in which the period of a pendulum is a function of its length). We therefore try to construct a hypothesis stating the exact mathematical form of the function. We have been able to check many instances in which u had one of the values 0, 1, 2, or 3; the associated values of v were regularly found to be 2, 3, 4, and 5, respectively. Suppose further that concerning these systems, we have no background knowledge that might bear on the likely form of the functional connection, and that the following three hypotheses have been proposed on the basis of our data:

$$H_1: v = u^4 - 6u^3 + 11u^2 - 5u + 2$$
$$H_2: v = u^5 - 4u^4 - u^3 + 16u^2 - 11u + 2$$
$$H_3: v = u + 2$$

Each of these fits the data: to each of the four u-values examined, it assigns exactly the v-value that has been found associated with it. In geometrical terms: if the three hypotheses are graphed in a plane coordinate system, then each of the resulting curves contains the four data-points (0,2), (1,3), (2,4), and (3,5).

Yet if, as has been assumed, we have no relevant background information that might indicate a different choice, we would no doubt favor H_3 over H_1 and H_2 on the ground that it is a simpler hypothesis than its rivals. This consideration suggests that if two hypotheses accord with the same data and do not differ in other respects relevant to their confirmation, the simpler one will count as more acceptable.

The relevance of the same basic idea to entire theories is often illustrated by reference to the Copernican heliocentric conception of the solar system, which was considerably simpler than the geocentric one it came to supersede, namely, Ptolemy's ingenious and accurate, but "gorgeously complicated system of main circles and sub-circles, with different radii, speeds, tilts, and different amounts and directions of eccentricity."[6]

Though, undeniably, simplicity is highly prized in science, it is not easy to state clear criteria of simplicity in the relevant sense and to justify the preference given to simpler hypotheses and theories.

Any criteria of simplicity would have to be objective, of course; they could not just refer to intuitive appeal or to the ease with which a hypothesis or theory can be understood or remembered, etc., for these factors vary from person to person. In the case of quantitative hypotheses like H_1, H_2, H_3, one might think of judging simplicity by reference to the corresponding graphs. In rectangular coordinates, the graph of H_3 is a straight line, whereas graphs of H_1 and H_2 are much more complicated curves through the four data-points. But this criterion seems arbitrary. For if the hypotheses are represented in polar coordinates, with u as the direction angle and v as the radius vector, then H_3 determines a spiral, whereas a function determining a "simple" straight line would be quite complicated.

When, as in our example, all the functions are expressed by polynomials, the order of the polynomial might serve as an index of complexity; thus H_2 would be more complex than H_1, which in turn would be more complex than H_3. But further criteria are needed when trigonometric and other functions are to be considered as well.

In the case of theories, the number of independent *basic assumptions* is sometimes suggested as an indicator of complexity. But assumptions can be combined and split up in many ways: there is no unambiguous way of counting them. For example, the statement that for any two points there is exactly one straight line containing them might be counted as expressing two assumptions rather than one: that there is at least one such line, and that there is at most one. And even if we could agree on the count, different basic assumptions might in turn differ in complexity and would then have to be weighed rather than counted. Similar remarks apply to the suggestion that the number of *basic concepts* used in a theory might serve as an index of its complexity. The question of criteria of simplicity has in recent years received a good deal of attention from logicians and philosophers, and some interesting results have been obtained, but no satisfactory general characterization of simplicity is available. As our examples suggest, however, there certainly are cases in which, even in the absence of explicit criteria, investigators would be in substantial agreement about which of two competing hypotheses or theories is the simpler.

Another intriguing problem concerning simplicity is that of justification: what reasons are there for following the *principle of simplicity*, as we might call it; that is, the maxim that the simpler of two otherwise equally confirmed rival hypotheses or theories is to be preferred, is to count as more acceptable?

Many great scientists have expressed the conviction that the basic laws of nature are simple. If this were known, there would indeed be a presumption that the simpler of two rival hypotheses is more likely to be true. But the assumption that the basic laws of nature are simple is of course at least as problematic as the soundness of the principle of simplicity and thus cannot provide a justification for it.

Some scientists and philosophers—among them Mach, Avenarius, Ostwald, and Pearson—have held that science seeks to give an economic or parsimonious description of the world, and that general hypotheses purporting to express laws of nature are economic expedients for thought, serving to compress an indefinite number of particular cases (e.g., many cases of free fall) into one simple formula (e.g., Galileo's law); and from this point of view, it seems entirely reasonable to adopt the simplest among several competing hypotheses. This argument would be convincing if we had to choose between different *descriptions of one and the same set of facts*; but in adopting one among several competing hypotheses, such as H_1, H_2, H_3 above, we also adopt the *predictions* it implies concerning as yet untested cases; and in this respect, the hypotheses differ widely. Thus, for $u = 4$, H_1, H_2, and H_3 predict the v-values 150, 30, and 6, respectively. Now, H_3 may be mathematically simpler than its rivals; but what grounds are there for considering it more likely to be true, for basing our expectations concerning the as yet unexamined case $u = 4$ on H_3 rather than on one of the competing hypotheses, which fit the given data with the same precision?

One interesting answer has been suggested by Reichenbach.[7] Briefly, he argues as follows: suppose that in our example v is indeed a function of u, $v = f(u)$. Let g be its graph in some system of coordinates; the choice is inessential. The true function f and its graph are, of course, unknown to the scientist who measures associated values of the two variables. Assuming, for the sake of the argument, that his measurements are exact, he will thus find a number of data-points that lie on the "true" curve g. Suppose now that in accordance with the principle of simplicity, the scientist draws the simplest, i.e., the intuitively smoothest, curve through those points. Then his graph, say g_1, may deviate considerably from the true curve, though it does share at least the measured data-points with the latter. But as the scientist determines more and more data-points and plots further simplest graphs, g_2, g_3, g_4, . . . , these will coincide more and more nearly with the true curve g, and the associated functions of f_2, f_3, f_4, . . . will approximate more and more closely the true functional connection f. Thus, observance of the principle of simplicity cannot be *guaranteed* to yield the function f in one step or even in many; but if there is a functional connection between u and v, the procedure will gradually lead to a function that approximates the true one to any desired degree.

Reichenbach's argument, which has here been stated in a somewhat simplified form, is ingenious; but its force is limited. For no matter how far the construction of successive graphs and functions may have gone, the procedure affords no indication at all of how close an approximation to the true function has been attained—if indeed there is a true function at all. (As we noted earlier, for example, the volume of a body of gas may seem to be, but is not in fact, a function of its temperature alone.) Moreover, the argument on grounds of convergence towards the true curve could be used also to justify certain other, intuitively complex and unreasonable methods of plotting graphs. For example, it is readily seen that if we were always to connect any two adjacent data-points by a semicircle whose diameter is the distance between the points, the resulting curves would eventually converge toward the true curve if there is one. Yet despite this "justification", this procedure would not be regarded as a sound way of forming quantitative hypotheses. Certain other nonsimple procedures, however—such as connecting adjacent data-points by hairpin loops whose length always exceeds a specified minimum value—are not justifiable in this fashion and can indeed be shown by Reichenbach's argument to be self-defeating. His idea is thus of distinct interest.

A very different view has been advanced by Popper. He construes the simpler of two hypotheses as the one that has greater empirical content, and he argues that the simpler hypothesis can therefore more readily be falsified (found out to be false), if indeed it should be false; and that this is of great importance to science, which seeks to expose its conjectures to the most thorough test and possible falsification. He summarizes his argument as follows: "Simple statements, if knowledge is our object, are to be prized more highly than less simple ones *because they tell us more; because their empirical content is greater, and because they are better testable.*"[8] Popper makes his notion of degree of simplicity as degree of falsifiability more explicit by means of two different criteria. According to one of them, the hypothesis that the orbit of a given planet is a circle is simpler than the hypothesis that it is an ellipse, because the former could be falsified by the determination of four positions that are found not to lie on a circle (three positions can always be connected by a circle), whereas the falsification of the second hypothesis would require the determination of at least six positions of the planet. In this sense, the simpler hypothesis is here the more readily falsifiable one, and it is also stronger because it logically implies the less simple hypothesis. This criterion surely contributes to clarifying the kind of simplicity that is of concern to science.

But Popper alternatively calls one hypothesis more falsifiable, and hence simpler, than another if the first implies the second and thus has greater content in a strictly deductive sense. However, greater content is surely not always linked to greater simplicity. To be sure, sometimes a strong theory, such as Newton's theory of gravitation and motion, will be regarded as simpler than a vast array of unrelated laws of more limited scope that are implied by it. But the desirable kind of simplification thus achieved by a theory is not just a matter of increased content; for if two unrelated hypotheses (e.g., Hooke's and Snell's laws) are conjoined, the resulting conjunction tells us more, yet is not simpler, than either component. Also, of the three hypotheses H_1, H_2, H_3 considered above, none tells us more than any of the others; yet they do not count as equally simple. Nor do those three hypotheses differ in point of falsifiability. If false, any one of them can be shown to be false with the same ease—namely, by means of one counter-instance; for example, the data-pair (4, 10) would falsify them all.

Thus, while all the different ideas here briefly surveyed shed some light on the rationale of the principle of simplicity, the problems of finding a precise formulation and a unified justification for it are not as yet satisfactorily solved.[9]

5 | The Probability of Hypotheses

Our survey of factors determining the credibility of scientific hypotheses shows that the credibility of a hypothesis H at a given time depends, strictly speaking, on the relevant parts of the total scientific knowledge at that time, including all the evidence relevant to the hypothesis and all the hypotheses and theories then accepted that have any bearing upon it; for as we have seen, it is by reference to these that the credibility of H has to be assessed. Strictly, therefore, we should speak of the *credibility of a hypothesis relative to a given body of knowledge;* the latter might be represented by a large set K of statements—all the statements accepted by science at the time.

The question naturally suggests itself whether it is possible to express this credibility in precise quantitative terms, by formulating a definition which, for any hypothesis H and any set K of statements, determines a number $c(H, K)$ expressing the degree of credibility that H possesses relative to K. And since we often speak of hypotheses as more or less probable, we might wonder further whether this quantitative concept could not be so defined as to satisfy all the basic principles of probability theory. In this case, the credibility of a hypothesis relative to any set K would be a real number no less than 0 and no greater than 1; a hypothesis that is true on purely logical grounds (such as 'Tomorrow it will rain in Central Park or it won't') would always have the credibility 1; and finally, for any two logically incompatible statements H_1 and H_2, the credibility of the hypothesis that one or the other of them is true would equal the sum of their credibilities: $c(H_1$ or $H_2, K) = c(H_1, K) + c(H_2, K)$.

Various theories for such probabilities have indeed been proposed.[10] They proceed from certain axioms like those just mentioned to a variety of more or less complex theorems that make it possible to determine certain probabilities *provided that others are already known;* but they offer no general definition of the probability of a hypothesis relative to given information.

And if the definition of the concept $c(H, K)$ is to take account of all the different factors we have surveyed, then the task is very difficult, to say the least; for as we saw, it is not even clear how such factors as the simplicity of a hypothesis, or the variety of its supporting evidence, are to be precisely characterized, let alone expressed in numerical terms.

However, certain illuminating and quite far-reaching results have recently been obtained by Carnap, who has studied the problem by reference to rigorously formalized model languages whose logical structure is considerably simpler than that required for the purpose of science. Carnap has developed a general method of defining what he calls the degree of confirmation for any hypothesis expressed in such a language with respect to any body of information expressed in the same language. The concept thus defined does satisfy all the principles of probability theory, and Carnap accordingly refers to it as the *logical* or *inductive probability* of the hypothesis relative to the given information.[11]

▪ | Notes

1. W. F. Magie, ed., A *Source Book in Physics*, p. 74.

2. A full and lucid account, on which this brief survey is based, will be found in Chap. 33 of G. Holton and D. H. D. Roller, *Foundations of Modern Physical Science* (Reading, Mass.: Addison-Wesley Publishing Co., 1958).

3. For details, see Holton and Roller, *Foundations of Modern Physical Science* Chap. 34 (especially section 7).

4. B. Evans, *The Natural History of Nonsense* (New York: Alfred A. Knopf, 1946), p. 133.

5. This point is suggestively presented and illustrated by reference to the phlogiston theory of combustion in Chap. 7 of J. B. Conant, *Science and Common Sense* [(New Haven: Yale University Press, 1951)]. A provocative general conception of the rise and fall of scientific theories is developed in T. S. Kuhn's book *The Structure of Scientific Revolutions* (Chicago: The University of Chicago Press, 1962).

6. E. Rogers, *Physics for the Inquiring Mind* (Princeton: Princeton University Press, 1960), p. 240. Chapters 14 and 16 of this work offer a splendid description and appraisal of the two systems; they give more substance to the claim of greater simplicity for Copernicus' scheme, but show also that it was able to account for various facts, known at Copernicus' time, that the Ptolemaic system could not explain.

7. H. Reichenbach, *Experience and Prediction* (Chicago: The University of Chicago Press, 1938), section 42.

8. K. R. Popper, *The Logic of Scientific Discovery* (London: Hutchinson, 1959), p. 142 (italics are quoted). Chapters VI and VII of this book, which offer many illuminating observations on the role of simplicity in science, contain the presentation of the ideas here referred to.

9. The reader who wishes to pursue these issues further will find the following discussions helpful: S. Barker, *Induction and Hypothesis* (Ithaca: Cornell University Press, 1957); "A Panel Discussion of Simplicity of Scientific Theories," *Philosophy of Science*, Vol. 28 (1961), 109–71; W.V.O. Quine, "On Simple Theories of a Complex World," *Synthese*, Vol. 15 (1963), 103–6.

10. One of them by the economist John Maynard Keynes, in his book, *A Treatise on Probability* (London: Macmillan & Company, Ltd., 1921).

11. Carnap has given a brief and elementary account of the basic ideas in his article "Statistical and Inductive Probability," reprinted in E. H. Madden, ed., *The Structure of Scientific Thought* (Boston: Houghton Mifflin Company, 1960), pp. 269–79. A more recent, very illuminating statement is given in Carnap's article, "The Aim of Inductive Logic" in E. Nagel, P. Suppes, and A. Tarski, eds., *Logic, Methodology and Philosophy of Science.* Proceedings of the 1960 International Congress (Stanford: Stanford University Press, 1962), pp. 303–18.

PETER ACHINSTEIN

Explanation v. Prediction: Which Carries More Weight?

1 | The Historical Thesis of Evidence

According to a standard view, predictions of new phenomena provide stronger evidence for a theory than explanations of old ones. More guardedly, a theory that predicts phenomena that did not prompt the initial formulation of that theory is better supported by those phenomena than is a theory by known phenomena that generated the theory in the first place. So say various philosophers of science, including William Whewell (1847) in the 19th century and Karl Popper (1959) in the 20th, to mention just two.

Stephen Brush takes issue with this on historical grounds. In a series of fascinating papers he argues that generally speaking scientists do not regard the fact that a theory predicts new phenomena, even ones of a kind totally different from those that prompted the theory in the first place, as providing better evidential support for that theory than is provided by already known facts explained by the theory. By contrast, Brush claims, there are cases, including general relativity and the periodic law of elements, in which scientists tend to consider known phenomena explained by a theory as constituting much stronger support than novel predictions.[1]

Both the predictionist and the explanationist are committed to an interesting historical thesis about evidence, viz.

> *Historical thesis:* Whether some claim e, if true, is evidence for an hypothesis h, or how strong that evidence is, depends on certain historical facts about e, h, or their relationship.

For example, whether, or the extent to which, e counts as evidence for h depends on whether e was known before or after h was formulated. Various historical positions are possible, as Alan Musgrave (1974) noted years ago in

FROM D. Hull, M. Forbes, and R. M. Burian, eds., *PSA 1994*, vol. 2 (East Lansing, Mich.: Philosophy of Science Association, 1994), 156–64.

a very interesting article. On a simple predictionist view (which Musgrave classifies as "purely temporal") e supports h only if e was not known when h was first proposed. On another view (which Musgrave attributes to Zahar (1973) and calls "heuristic"), e is evidence for h only if when h was first formulated it was not devised in order to explain e. On yet a third historical view (which Musgrave himself accepts), e is evidence for some theory T only if e cannot be explained by a "predecessor" theory, i.e., by a competing theory which was devised by scientists prior to the formulation of T. These views, and other variations, are all committed to the historical thesis.

Is the historical thesis true or false? I propose to argue that it is sometimes true, and sometimes false, depending on the type of evidence in question. Then I will consider what implications, if any, this has for the debate between Brush and the predictionists.

Before beginning, however, let me mention a curious but interesting fact about various well-known philosophical theories or definitions of evidence. As Laura Snyder (1994) points out in a perceptive paper entitled "Is Evidence Historical?", most such theories, including Carnap's (1962) a priori theory of confirmation, Hempel's (1945) satisfaction theory, Glymour's (1980) bootstrap account, and the usual hypothetico-deductive account, are incompatible with the historical thesis. They hold that whether, or the extent to which, e is evidence for, or confirms, h is an objective fact about e, h, and their relationship. It is in no way affected by the time at which h was first proposed, or e was first known, or by the intentions with which h was formulated. Defenders of these views must reject both the predictionist and the explanationist claims about evidence. They must say that whether, or the extent to which, e supports h has nothing to do with whether e was first formulated as a novel prediction from h or whether e was known before h and h was constructed to explain it.

Accordingly, we have two extreme or absolutist positions. There is the position, reflected in the historical thesis, that evidence is always historical (in the sense indicated). And there is a contrasting position, reflected in certain standard views, that evidence is never historical. Does the truth lie at either extreme? Or is it somewhere in the middle?

2 | Selection Procedures

Suppose that an investigator decides to test the efficacy of a certain drug D in relieving symptoms S. The hypothesis under consideration is

h: Drug D relieves symptoms S in approximately 95% of the cases.

The investigator may test drug D by giving it to persons suffering from S and by giving a placebo to other persons suffering from S (the "control group"). In deciding how to proceed, the investigator employs what I will call a "selection procedure," or rule, determining how to test, or obtain evidence for, an hypothesis, in this case determining which persons he will select for his studies and how he will study them.

For example, here is one of many possible selection procedures (SP) for testing h:

> SP 1: Choose a sample of 2000 persons of different ages, sexes, races, and geographical locations, all of whom have symptoms S in varying degrees; divide them arbitrarily into 2 groups; give one group drug D and the other a placebo; determine how many in each group have their symptoms relieved.

Now, suppose that a particular investigator uses this (or some other) selection procedure and obtains the following result:

> e: In a group of 1000 persons with symptoms S taking drug D, 950 persons had relief of S; in a control group of 1000 S-sufferers not taking D but a placebo none had symptoms S relieved.

The first thing to note about this example is that whether the report e supports hypothesis h, or the extent to which it does, depends crucially on what selection procedure was in fact used in obtaining e. Suppose that instead of SP1 the following selection procedure had been employed:

> SP 2: Choose a sample of 2000 females aged 5 all of whom have symptoms S in a very mild form; proceed as in SP1.

If result e had been obtained by following SP2, then e, although true, would not be particularly good evidence for h, certainly not as strong as that obtained by following SP1. The reason, of course, is that SP1, by contrast with SP2, gives a sample that is varied with respect to two factors that may well be relevant: age of patient and severity of symptoms. (Hypothesis h does not restrict itself to 5-year-old girls with mild symptoms, but asserts a cure-rate for the general population of sufferers with varying degrees of the symptoms in question.)

This means that if the result as described in e is obtained, then whether, or to what extent, that result confirms the hypothesis h depends crucially on what selection procedure was in fact used in obtaining e. That is, it depends on an historical fact about e: on how in fact e was obtained. If e resulted from following SP1, then e is pretty strong evidence for h; if e was obtained by following SP2, then e is pretty weak evidence for h, if it confirms it at all. Just by looking at e and h, and even by ascertaining that e is true, we are unable to determine to what extent, if any, e supports h. We need to invoke "history."

To nail down this point completely, consider a third selection procedure:

SP 3: Choose a sample of 2000 persons all of whom have S in varying degrees; divide them arbitrarily into 2 groups; give one group drugs D and D' (where D' relieves symptoms S in 95% of the cases and blocks possible curative effects of D when taken together); give the other group a placebo.

Consider once more result e (which, again, let us suppose, obtains). In this case e supports h not at all. And, again, whether this is so cannot be ascertained simply by examining the propositions e, h, or their "logical" relationship. We need to know an historical fact about e, viz. that the information it (truly) reports was obtained by following SP3.

So far then we seem to have support for the historical thesis about evidence. Can we generalize from examples like this to all cases? Can we say that for any true report e, and any hypothesis h, whether, or to what extent, e is evidence for h depends upon historical facts about how e was obtained? No, we cannot.

Consider another very simple case. Let e be the following report, which is true:

e = In last week's lottery, 1000 tickets were sold, of which John owned 999 at the time of the selection of the winner; this was a fair lottery in which one ticket was selected at random.

h = John won the lottery.

In an attempt to obtain information such as e to support h different rules or "selection procedures" might have been followed, e.g.,

SP 4: Determine who bought tickets, and how many, by asking lottery officials.

SP 5: Determine this by standing next to the person selling tickets.

SP 6: Determine this by consulting the local newspaper, which publishes this information as a service to its readers.

Let us suppose that following any of these selection procedures results in a true report e. (And, as in the symptoms case, we may suppose that following any of these procedures is a reasonable way to establish whether e is true.) But in this case, unlike the drug example, which selection procedure was in fact followed is completely irrelevant in determining whether, or to what extent, e is evidence for h. In this case, unlike the drug example, we do not need to know how information e was obtained to know that e (assuming it is true) is very strong evidence for h. Nor do we need to know any other historical facts about e, h, or their relationship. (In particular, contrary to both the predictionist and explanationist views, we do not need to know when e was first known relative to when h was first formulated; i.e., we do not need to know whether e was explained or predicted. But more of this later when these two historical views are examined more fully.) Accordingly, we have a case that violates the historical thesis of evidence.

Since examples similar to each of the two above can be readily constructed, we may conclude that there are many cases that satisfy the historical thesis of evidence, and many others that fail to satisfy it. Is there a general rule for deciding which do and which do not?

Perhaps our two examples will help generate such a rule. In the drug case the evidence report e is historical in an obvious sense: it reports the results of a particular study made at some particular time and place. But this is clearly not sufficient to distinguish the cases, since the evidence report e in the lottery case is also historical: it reports facts about a particular lottery, who bought tickets, and when. So, I submit, what distinguishes the cases is not the historical character of the evidence, but something else.

I shall say that a putative evidence statement e is *empirically complete* with respect to an hypothesis h if whether, or to what extent, e is evidence for, or confirms, h depends just on what e reports, what h says, and the relationship between them. It does not depend on any additional empirical facts—e.g., facts about when e or h were formulated, or with what intentions, or on any (other) facts about the world. In the drug example, e is not empirically complete with respect to h: whether, or to what extent, e supports h depends on how the sample reported in e was selected— empirical information not contained in e or h.[2] By contrast, in the lottery example, e is empirically complete with respect to h: whether, and to what extent, e supports h in this case does not depend on empirical facts in addition to e. To determine whether, and how much, e supports h in this case we do not need any further empirical investigation. To be sure, additional empirical inquiry may unearth new information e' which is such that both e and e' together do not support h to the same extent that e by itself does. But that is different. In the drug but not the lottery example information in addition to e is necessary to determine the extent to which e itself supports h. In the drug case we cannot legitimately say whether or to what extent the report e supports the efficacy of drug D unless we know how the patients described in e were selected. In the lottery case information about how purported evidence was obtained is irrelevant for the question of whether or how strongly that evidence, assuming its truth, supports h.

So we have one important difference between the two examples. Is this enough to draw a distinction between cases that satisfy the historical thesis of evidence and those that do not? Perhaps not. There may be cases in which e is empirically incomplete with respect to h, but in which empirical facts needed to complete it are not historical. Consider

e = Male crows are black.

h = Female crows are black.

One might claim that whether, or the extent to which, e supports h in this case depends on empirical facts in addition to e. If, e.g., other species of birds generally have different colors for different sexes, then e does not support h very much. If other species generally have the same color for both sexes, then e supports h considerably more. But these additional facts are not "historical," at least not in the clear ways of previous examples. (I construe "other species of birds generally have different colors for different sexes" to be making a general statement, and not to be referring to any particular historical period.) If this is granted, then we need to add a proviso to the completeness idea above.

There are cases (including our drug example) in which a putative evidence claim e is empirically incomplete with respect to an hypothesis h, where determining whether, or to what extent, e supports h requires determining the truth of some historical fact. I shall speak of these as historical evidence cases. They satisfy the historical thesis of evidence. By contrast, there are cases in which a putative evidence claim e is empirically complete with respect to hypothesis h (e.g., our lottery case); and there may be cases in which a putative evidence claim e, although empirically incomplete with respect to h, can be settled without appeal to historical facts (possibly the crow example). Cases of the latter two sorts violate the historical thesis of evidence. What implications, if any, does this hold for whether predictions or explanations provide better confirmation?

3 | Predictions v. Explanations

Let us return to the original question proposed by Brush. Do predictions of novel facts provide stronger evidence than explanations of old ones, as Whewell and Popper claim? Or is the reverse true? My answer is this: Sometimes a prediction provides better evidence for an hypothesis, sometimes an explanation does, and sometimes they are equally good. Which obtains has nothing to do with the fact that it is a prediction of novel facts or that it is an explanation of known ones.

To show this, let us begin with a case that violates the historical thesis of evidence. Here it should be easy to show that whether the putative evidence is known before or after the hypothesis is formulated is irrelevant for confirmation. Let the hypothesis [and evidence] be

> h = This coin is fair, i.e., if tossed in random ways under normal conditions it will land on heads approximately half the time in the long run.

> e = This coin is physically symmetrical, and in a series of 1000 random tosses under normal conditions it landed on heads approximately 500 times.

We might reasonably take e to be empirically complete with respect to h. Accordingly, whether e supports h, and the extent to which it does, does not depend on empirical facts other than e. In particular, it does not depend on when, how, or even whether e comes to be known, or on whether e was known first and h then formulated, or on whether h was conceived first and e then stated as a prediction from it. Putative evidence e supports hypothesis h and does so (equally well) whether or not e is known before or after h was initially formulated, indeed whether or not e is ever known to be true.

So let us focus instead on cases that satisfy the historical thesis of evidence. We might suppose that at least in such cases explanations (or predictions) are always better for confirmation. Return once again to our drug hypothesis:

h = Drug D relieves symptoms S in approximately 95% of the cases.

Consider now two evidence claims, the first a prediction about an unknown future event, the second a report about something already known:

e_1 = In the next clinical trial of 1000 patients who suffer from symptoms S and who take D approximately 950 will get some relief.

e_2 = In a trial that has already taken place involving 1000 patients with S who took D (we know that) approximately 950 got some relief.

On the prediction view, e_1 is stronger evidence for h than is e_2. On the explanation view it is the reverse. And to sharpen the cases let us suppose that e_2, by contrast to e_1, was not only known to be true prior to the formulation of h, but that h was formulated with the intention of explaining e_2. Which view is correct? Neither one.

Let us take the prediction case e_1 first. Whether, and to what extent, e_1 (if true) supports h depends on empirical facts in addition to e_1. In this case it depends on the selection procedure to be used in the next clinical trial. Suppose this selection procedure calls for choosing just 5-year-old girls with very mild symptoms who in addition to D are also taking drug D′ which ameliorates symptoms S in 95% of the cases and potentially blocks D from doing so. Then e_1 would be very weak evidence for h, if it supports it at all. This is so despite the fact that e_1 is a correct prediction from h, one not used in generating h in the first place. By contrast, suppose that the selection procedure used in the past trial mentioned in e_2 is much better with respect to h. For example, it calls for choosing humans of both sexes, of different ages, with symptoms of varying degrees, who are not also taking drug D′. Then e_2 would be quite strong evidence for h, much stronger than what is supplied by e_1. In such a case, a known fact explained by h would provide more support for h than a newly predicted fact would.

Obviously the situations here can be reversed. We might suppose that the selection procedure used to generate the prediction of e_1 is the one cited in the previous paragraph as being used to generate e_2 (and vice versa). In this situation a newly predicted fact would provide more support for h than an already explained one.

In these cases what makes putative evidence have the strength it does has nothing to do with whether it is being explained or predicted. It has to do with the selection procedure used to generate that evidence.[3] In one situation—whether it involves something that is explained or predicted—we have a putative evidence statement generated by a selection procedure that is a good one relative to h; in the other case we have a flawed selection procedure. This is what matters for confirmation—not whether the putative evidence is being explained or predicted.

4 | Brush Redux

Brush is clearly denying a general predictionist thesis. By contrast he cites cases in which scientists themselves regarded known evidence explained by a theory as stronger support for that theory than new evidence that was successfully predicted. And he seems to imply that this was reasonable. He offers an explanation for this claim, viz. that with explanations of the known phenomena, by contrast with successful predictions of the new ones, scientists had time to consider alternative theories that would generate these phenomena. Now, even if Brush does not do so, I want to extend this idea and consider a more general explanationist view that is committed to the following three theses that Brush invokes for some cases:

1 A selection procedure for testing a hypothesis h is flawed, or at least inferior to another, other things equal, if it fails to call for explicit consideration of competitors to h.

2 The longer time scientists have to consider whether there are plausible competitors to h the more likely they are to find some if they exist.

3 With putative evidence already known before the formulation of h scientists have (had) more time to consider whether there are plausible competitors to h than is the case with novel predictions.

I would challenge at least the first and third theses. In my first example, selection procedure 1 for the drug hypothesis does not call for explicitly considering competitors to that hypothesis. Yet it does not seem flawed on that account, or inferior to one that does. However, even supposing it were inferior, whether or not a selection procedure calls for a consideration of competitors is completely irrelevant to whether the putative evidence claim is a prediction or a known fact being explained. In the case of a prediction, no less than that of an explanation, the selection procedure may call for a consideration of competitors.

For example, in our drug case, where h is "Drug D relieves symptoms S in approximately 95% of the cases," and e is the prediction "In the next clinical trial of 1000 patients suffering from symptoms S who take D, approximately 950 will get some relief," the selection procedure to be used for the next clinical trial might include the rule

> In conducting this next trial, determine whether the patients are also taking some other drug which relieves S in approximately 95% of the cases and which blocks any effectiveness D might have.

Such a selection procedure calls for the explicit consideration of a competitor to explain e, viz. that it will be some other drug, not D, that will relieve symptoms S in the next trial. This is so even though e is a prediction. Moreover, to respond to the third thesis about time for considering competitors, an investigator planning a future trial can have as much time as she likes to develop a selection procedure calling for a consideration of a competing hypothesis. More generally, in designing a novel experiment to test some hypothesis h as much time may be spent in precluding competing hypotheses that will explain the test results as is spent in considering competing hypotheses for old data.

5 | Thomson v. Hertz

Finally, let me invoke an example more recognizably scientific. It involves a dispute between Heinrich Hertz and J. J. Thomson over the nature of cathode rays.[4] In experiments conducted in 1883 Hertz observed that the cathode rays in his experiments were not deflected by an electrical field. He took this to be strong evidence that cathode rays are not charged particles (as the English physicist William Crookes had concluded), but some type of ether waves. In 1897 J. J. Thomson repeated Hertz's experiments but with a much higher evacuation of gas in the cathode tube than Hertz had been able to obtain. Thomson believed that when cathode rays pass though a gas they make it a conductor, which screens off the electric force from the charged particles comprising the cathode rays.[5] This screening off effect will be reduced if the gas in the tube is more thoroughly evacuated. In Thomson's 1897 experiments electrical deflection of the cathode rays was detected, which Thomson took to be strong evidence that cathode rays are charged particles.

Here, however, I want to consider the evidential report of Hertz in 1883, not of Thomson in 1897. Let

e = In Hertz's cathode ray experiments of 1883 no electrical deflection of cathode rays was detected.

h = Cathode rays are not electrically charged.

Hertz took e to be strong evidence for h. In 1897 Thomson claimed, in effect, that Hertz's results as reported in e did not provide strong evidence for h, since Hertz's experimental set-up was flawed: He was employing insufficiently evacuated tubes. To use my previous terminology, Thomson was claiming that Hertz's selection procedure for testing h was inadequate.[6]

Here we can pick up on a point emphasized by Brush. Hertz, we might say, failed to use a selection procedure calling for considering a competitor to h to explain his results (viz. that cathode rays are charged particles, but that the tubes Hertz was using were not sufficiently evacuated to allow an electrical force to act on these particles). But—and this is the point I want to emphasize—in determining whether, or to what extent, Hertz's putative evidence e supports his hypothesis h, it seems to be irrelevant whether Hertz's e was a novel prediction from an already formulated hypothesis h or an already known fact to be explained by h. Hertz writes that in performing the relevant experiments he was trying to answer two questions:

> Firstly: Do the cathode rays give rise to electrostatic forces in their neighbourhood? Secondly: In their course are they affected by external electrostatic forces? (Hertz 1896, p. 249)

In his paper he did not predict what his experiments would show. Nor were the results of his experiments treated by him as facts known before he had formulated his hypothesis h. Once he obtained his experimental result he then claimed that they supported his theory:

> As far as the accuracy of the experiment allows, we can conclude with certainty that no electrostatic effect due to the cathode rays can be perceived. (p. 251)

To be sure, we might say that Hertz's *theory* itself predicted some such results, even if Hertz himself did not (i.e., even if Hertz did not himself draw this conclusion before getting his experimental results). But even if we speak this way, Hertz did not claim or imply that his experimental results provide better (or weaker) support for his theory because the theory predicted them before they were obtained. Nor did Thomson in his criticism of Hertz allude to one or the other possibility. Whichever it was—whether a prediction or an explanation or neither—Hertz (Thomson was claiming) should have used a better selection procedure. This is what is criticizable in Hertz, not whether he was predicting a novel fact or explaining a known one.

I end with a quote from John Maynard Keynes (1921, p. 305), whose book on probability contains lots of insights. Here is one:

> The peculiar virtue of prediction or predesignation is altogether imaginary. The number of instances examined and the analogy between them are the essential points, and the question as to whether a particular hypothesis happens to be propounded before or after their examination is quite irrelevant.

▪ | Notes

1. To what extent Brush wants to generalize this explanationist position is a question I leave for him to answer. There are passages in his writings that strongly suggest a more general position. For example: "There is even some reason to suspect that a successful explanation of a fact that other theories have already failed to explain satisfactorily (for example, the Mercury perihelion) is more convincing than the prediction of a new fact, at least until the competing theories have had their chance (and failed) to explain it" (1989, p. 1127). In what follows I consider a generalized explanationist thesis.

2. For Carnap (1962) and others, every e is empirically complete with respect to every h. For these writers, whether, and the extent to which, e confirms h is an a priori matter.

3. Cf. Mayo (1991).

4. See Achinstein (1991), Essays 10 and 11; also Buchwald (1994), ch. 10.

5. See Thomson (1897), p. 107.

6. Lord Rayleigh (1942, pp. 78–9), in a biography of Thomson, made the same claim: "He [Hertz] failed to observe this [electrical] effect, but the design of his experiment was open to certain objections which were removed in a later investigation by Perrin in 1895, directed to the same question. Perrin got definite evidence that the rays carried a negative charge. J. J. Thomson, in a modification of Perrin's experiment, showed that if the Faraday cylinder was put out of the line of fire of the cathode, it acquired a charge when, and only when, the cathode rays were so deflected by a magnet as to enter the cylinder." [Note Rayleigh's claim that Perrin (and Thomson) got "definite evidence" that cathode rays carry a negative charge, whereas, by implication, Hertz's experiments did not give "definite evidence" concerning the question of charge.]

Acknowledgement: I am indebted to Laura J. Snyder for very helpful discussions, and to Robert Rynasiewicz for trying to convince me of the error of my ways.

▪ | References

Achinstein, P. (1991), *Particles and Waves.* New York: Oxford University Press.

Brush, S. (1989), "Prediction and Theory Evaluation: The Case of Light Bending," *Science* 246:1124–29.

Buchwald, J. (1994), *The Creation of Scientific Effects.* Chicago: University of Chicago Press.

Carnap, R. (1962), *Logical Foundations of Probability.* Chicago: University of Chicago Press, 2nd ed.

Glymour, C. (1980), *Theory and Evidence.* Princeton: Princeton University Press.

Hempel, C. (1945), "Studies in the Logic of Confirmation," *Mind* 54: 1–26, 97–121.

Hertz, H. (1896), *Miscellaneous Papers.* London: Macmillan.

Keynes, J.M. (1921), A *Treatise on Probability.* London: Macmillan.

Mayo, D. (1991), "Novel Evidence and Severe Tests," *Philosophy of Science* 58: 523–52.

Musgrave, A. (1974), "Logical versus Historical Theories of Confirmation" *British Journal for the Philosophy of Science* 25: 1–23.

Popper, K. (1959), *The Logic of Scientific Discovery*. London: Hutchinson.

Rayleigh, R. (1942), *The Life of Sir J. J. Thomson*. Cambridge: Cambridge University Press.

Snyder, L. (1994), "Is Evidence Historical?" *Scientific Methods; Conceptual and Historical Problems*, P. Achinstein and L. Snyder, eds. Malabar, Florida: Krieger.

Thomson, J. J. (1897), "Cathode Rays," *The Electrician* 39: 104–108.

Whewell, W. (1847), *The Philosophy of the Inductive Sciences*. New York: Johnson Reprint, 1967.

Zahar, E. (1973), "Why Did Einstein's Programme Supercede Lorentz's?" *British Journal for the Philosophy of Science* 24: 95–123, 223–62.

Nelson Goodman

The New Riddle of Induction

Confirmation of a hypothesis by an instance depends rather heavily upon features of the hypothesis other than its syntactical form. That a given piece of copper conducts electricity increases the credibility of statements asserting that other pieces of copper conduct electricity, and thus confirms the hypothesis that all copper conducts electricity. But the fact that a given man now in this room is a third son does not increase the credibility of statements asserting that other men now in this room are third sons, and so does not confirm the hypothesis that all men now in this room are third sons. Yet in both cases our hypothesis is a generalization of the evidence statement. The difference is that in the former case the hypothesis is a *lawlike* statement; while in the latter case, the hypothesis is a merely contingent or accidental generality. Only a statement that is *lawlike*—regardless of its truth or falsity or its scientific importance—is capable of receiving confirmation from an instance of it; accidental statements are not. Plainly, then, we must look for a way of distinguishing lawlike from accidental statements.

So long as what seems to be needed is merely a way of excluding a few odd and unwanted cases that are inadvertently admitted by our definition of confirmation, the problem may not seem very hard or very pressing. We fully expect that minor defects will be found in our definition and that the necessary refinements will have to be worked out patiently one after another. But some further examples will show that our present difficulty is of a much graver kind.

Suppose that all emeralds examined before a certain time *t* are green.[1] At time *t*, then, our observations support the hypothesis that all emeralds are green; and this is in accord with our definition of confirmation. Our evidence statements assert that emerald *a* is green, that emerald *b* is green, and so on; and each confirms the general hypothesis that all emeralds are green. So far, so good.

From Nelson Goodman, *Fact, Fiction, and Forecast*, 4th ed. (Cambridge, Mass.: Harvard University Press, 1983), 72–81.

Now let me introduce another predicate less familiar than "green". It is the predicate "grue" and it applies to all things examined before t just in case they are green but to other things just in case they are blue. Then at time t we have, for each evidence statement asserting that a given emerald is green, a parallel evidence statement asserting that that emerald is grue. And the statements that emerald a is grue, that emerald b is grue, and so on, will each confirm the general hypothesis that all emeralds are grue. Thus according to our definition, the prediction that all emeralds subsequently examined will be green and the prediction that all will be grue are alike confirmed by evidence statements describing the same observations. But if an emerald subsequently examined is grue, it is blue and hence not green. Thus although we are well aware which of the two incompatible predictions is genuinely confirmed, they are equally well confirmed according to our present definition. Moreover, it is clear that if we simply choose an appropriate predicate, then on the basis of these same observations we shall have equal confirmation, by our definition, for any prediction whatever about other emeralds—or indeed about anything else.[2] As in our earlier example, only the predictions subsumed under lawlike hypotheses are genuinely confirmed; but we have no criterion as yet for determining lawlikeness. And now we see that without some such criterion, our definition not merely includes a few unwanted cases, but is so completely ineffectual that it virtually excludes nothing. We are left once again with the intolerable result that anything confirms anything. This difficulty cannot be set aside as an annoying detail to be taken care of in due course. It has to be met before our definition will work at all.

Nevertheless, the difficulty is often slighted because on the surface there seem to be easy ways of dealing with it. Sometimes, for example, the problem is thought to be much like the paradox of the ravens. We are here again, it is pointed out, making tacit and illegitimate use of information outside the stated evidence: the information, for example, that different samples of one material are usually alike in conductivity, and the information that different men in a lecture audience are usually not alike in the number of their older brothers. But while it is true that such information is being smuggled in, this does not by itself settle the matter as it settles the matter of the ravens. There the point was that when the smuggled information is forthrightly declared, its effect upon the confirmation of the hypothesis in question is immediately and properly registered by the definition we are using. On the other hand, if to our initial evidence we add statements concerning the conductivity of pieces of other materials or concerning the number of older brothers of members of other lecture audiences, this will not in the least affect the confirmation, according to our definition, of the hypothesis concerning copper or of that concerning this lecture audience. Since our definition is insensitive to the bearing upon hypotheses of evidence so related to them, even when the evidence is fully declared, the difficulty about accidental hypotheses cannot be explained away on the ground that such evidence is being surreptitiously taken into account.

A more promising suggestion is to explain the matter in terms of the effect of this other evidence not directly upon the hypothesis in question but *indirectly* through other hypotheses that *are* confirmed, according to our definition, by such evidence. Our information about other materials does by our definition confirm such hypotheses as that all pieces of iron conduct electricity, that no pieces of rubber do, and so on; and these hypotheses, the explanation runs, impart to the hypothesis that all pieces of copper conduct electricity (and also to the hypothesis that none do) the character of lawlikeness—that is, amenability to confirmation by direct positive instances when found. On the other hand, our information about other lecture audiences *dis*confirms many hypotheses to the effect that all the men in one audience are third sons, or that none are; and this strips any character of lawlikeness from the hypothesis that all (or the hypothesis that none) of the men in *this* audience are third sons. But clearly if this course is to be followed, the circumstances under which hypotheses are thus related to one another will have to be precisely articulated.

The problem, then, is to define the relevant way in which such hypotheses must be alike. Evidence for the hypothesis that all iron conducts electricity enhances the lawlikeness of the hypothesis that all zirconium conducts electricity, but does not similarly affect the hypothesis that all the objects on my desk conduct electricity. Wherein lies the difference? The first two hypotheses fall under the broader hypothesis—call it "*H*"—that every class of things of the same material is uniform in conductivity; the first and third fall only under some such hypothesis as—call it "*K*"—that every class of things that are either all of the same material or all on a desk is uniform in conductivity. Clearly the important difference here is that evidence for a statement affirming that one of the classes covered by *H* has the property in question increases the credibility of any statement affirming that another such class has this property; while nothing of the sort holds true with respect to *K*. But this is only to say that *H* is lawlike and *K* is not. We are faced anew with the very problem we are trying to solve: the problem of distinguishing between lawlike and accidental hypotheses.

The most popular way of attacking the problem takes its cue from the fact that accidental hypotheses seem typically to involve some spatial or temporal restriction, or reference to some particular individual. They seem to concern the people in some particular room, or the objects on some particular person's desk; while lawlike hypotheses characteristically concern all ravens or all pieces of copper whatsoever. Complete generality is thus very often supposed to be a sufficient condition of lawlikeness; but to define this complete generality is by no means easy. Merely to require that the hypothesis contain no term naming, describing, or indicating a particular thing or location will obviously not be enough. The troublesome hypothesis that all emeralds are grue contains no such term; and where such a term does occur, as in hypotheses about men in *this room*, it can be suppressed in favor of some predicate (short or long, new or old) that contains no such term but applies only to exactly the same things. One might think, then, of excluding not only hypotheses that actually contain terms for specific individuals but also all hypotheses that are equivalent to others that do contain such terms. But, as we have just seen, to exclude only hypotheses of which *all* equivalents contain such terms is to exclude nothing. On the other hand, to exclude all hypotheses that have *some* equivalent containing such a term is to exclude everything; for even the hypothesis

All grass is green

has as an equivalent

All grass in London or elsewhere is green.

The next step, therefore, has been to consider ruling out predicates of certain kinds. A syntactically universal hypothesis is lawlike, the proposal runs, if its predicates are 'purely qualitative' or 'non-positional'.[3] This will obviously accomplish nothing if a purely qualitative predicate is then conceived either as one that is equivalent to some expression free of terms for specific individuals, or as one that is equivalent to no expression that contains such a term; for this only raises again the difficulties just pointed out. The claim appears to be rather that at least in the case of a simple enough predicate we can readily determine by direct inspection of its meaning whether or not it is purely qualitative. But even aside from obscurities in the notion of 'the meaning' of a predicate, this claim seems to me wrong. I simply do not know how to tell whether a predicate is qualitative or positional, except perhaps by completely begging the question at issue and asking whether the predicate is 'well-behaved'—that is, whether simple syntactically universal hypotheses applying it are lawlike.

This statement will not go unprotested. "Consider", it will be argued, "the predicates 'blue' and 'green' and the predicate 'grue' introduced earlier, and also the predicate 'bleen' that applies to emeralds examined before time *t* just in case they are blue and to other emeralds just in case they are green. Surely it is clear", the argument runs, "that the first two are purely qualitative and the second two are not; for the meaning of each of the latter two plainly involves reference to a specific temporal position." To this I reply that indeed I do recognize the first two as well-behaved predicates admissible in lawlike hypotheses, and the second two as ill-behaved predicates. But the argument that the former but not the latter are purely qualitative seems to me quite unsound. True enough, if we start with "blue" and "green", then "grue" and "bleen" will be explained in terms of "blue" and "green" and a temporal term. But equally truly, if we start with "grue" and "bleen", then "blue" and "green" will be explained in terms of "grue" and "bleen" and a temporal term; "green", for example, applies to emeralds examined before time *t* just in case they are grue, and to other emeralds just in case they are bleen. Thus qualitativeness is an entirely relative matter and does not by itself establish any dichotomy of predicates. This relativity seems to be completely overlooked by those who contend that the qualitative character of a predicate is a criterion for its good behavior.

Of course, one may ask why we need worry about such unfamiliar predicates as "grue" or about accidental hypotheses in general, since we are unlikely to use them in making predictions. If our definition works for such hypotheses as are normally employed, isn't that all we need? In a sense, yes; but only in the sense that we need no definition, no theory of induction, and no philosophy of knowledge at all. We get along well enough without them in daily life and in scientific research. But if we seek a theory at all, we cannot excuse gross anomalies resulting from a proposed theory by pleading that we can avoid them in practice. The odd cases we have been considering are clinically pure cases that, though seldom encountered in practice, nevertheless display to best advantage the symptoms of a widespread and destructive malady.

We have so far neither any answer nor any promising clue to an answer to the question what distinguishes lawlike or confirmable hypotheses from accidental or non-confirmable ones; and what may at first have seemed a minor technical difficulty has taken on the stature of a major obstacle to the development of a satisfactory theory of confirmation. It is this problem that I call the new riddle of induction.

■ | Notes

1. Although the example used is different, the argument to follow is substantially the same as that set forth in my note, "A Query on Confirmation", *Journal of Philosophy* 43 (1946), 383–5.

2. For instance, we shall have equal confirmation, by our present definition, for the prediction that roses subsequently examined will be blue. Let "emerose" apply just to emeralds examined before *t*, and to roses examined later. Then all emeroses so far examined are grue, and this confirms the hypothesis that all emeroses are grue and hence the prediction that roses subsequently examined will be blue. The problem raised by such antecedents has been little noticed, but is no easier to meet than that raised by similarly perverse consequents. See further IV, 4 [of *Fact. Fiction, and Forecast*].

3. Carnap took this course in his paper "On the Application of Inductive Logic", *Philosophy and Phenomenological Research* 8 (1947), 133–47, which is in part a reply to my "A Query on Confirmation", cited in note 1. The discussion was continued in my note "On Infirmities of Confirmation Theory", *Philosophy and Phenomenological Research* 8 (1947), 149–51; and in Carnap's "Reply to Nelson Goodman", same journal, same volume, pp. 461–2.

4 | COMMENTARY

4 | COMMENTARY

4.1 | Lipton on the Problem of Induction

One important class of inductive arguments consists of inferences from a sample to a larger population from which the sample has been selected. For example, a biologist observes a number of reptiles (e.g., garter snakes, skinks, and tortoises) and observes (via dissection) that each of these creatures has a three-chambered heart. From this she infers (falsely) that all reptiles (including the many thousands that have not been and may never be observed) also have a three-chambered heart. Alternatively, the biologist might study several hundred lizards, find that 98 percent of them are carnivorous, and conclude that 98 percent of lizards are carnivorous. These arguments are typical examples of *inductive generalization*. When we verify some of the consequences of a scientific theory and infer that the theory is (probably) true, our inference from those verified consequences to the truth of the theory is like arguing from a sample—those verified consequences of the theory—to a population—the entire set of the theory's consequences. In concluding that all the theory's consequences are true we are, in effect, concluding that the theory is true.

Inductive arguments come in all shapes and sizes, inductive generalization being but one among them. Other types include *induction to a particular, statistical syllogism*, and *arguments from analogy*. Inductions to a particular conclude that the next observed A will be B from the premise that all observed As have been B. In a statistical syllogism, the same conclusion is drawn from the premise that a high percentage of As have been B. Arguments from analogy infer that an individual with properties R, S, and T will also have property U, from the fact that several other individuals with properties R, S, and T have been found to have property U.

Peter Lipton identifies two problems concerning induction: the problems of description and justification. The problem of description is the problem of identifying the general principles we follow when we make inductive inferences. The problem of justification is the problem of explaining why those general principles are reliable. Typically, general principles are judged reliable if arguments with true premises following those principles will lead to true conclusions most of the time. Lipton argues that both problems arise because in every inductive argument, the conclusion is *underdetermined* by the information in the premises. By their very nature, inductive arguments are ampliative: their conclusions go beyond anything that can be inferred from their premises using deductive logic.

Underdetermination

The justification problem—often just called the problem of induction in the philosophical literature—arises from the fact that inductive arguments are deductively underdetermined. Since the premises of inductive arguments do not entail their conclusions, no conclusion of an inductive argument is guaranteed to be true, even if all the premises of the argument are true and known to be true with absolute certainty. What, then, justifies our inferring the truth (or probable truth) of the conclusion of an inductive argument? As Lipton explains in the second section of his article, many philosophers (especially David Hume and Karl Popper) have judged that the correct answer to this question is, "Nothing." In other words, these philosophers are skeptics about the possibility of justifying induction.

Lipton's second problem, the problem of description, also arises from underdetermination. The problem of description, recall, is to say how inductive inferences in fact proceed—to describe the principles deployed in inductive reasoning, quite independently of whether these principles are reliable. For any group of reasoners and learners, a full description of the evidence and other known principles or rules available to them will often leave open what sort of inferences they in fact draw. Lipton gives two illustrations: Noam Chomsky's theory of how children acquire language and Thomas Kuhn's appeal to shared exemplars to explain how scientists reach agreement about scientific problems. In either case, the inferences made cannot be reconstructed adequately using only the principles of deductive logic and the simple, common forms of inductive argument (such as inductive generalization and arguments from analogy). Thus, in the case of language acquisition, Chomsky concludes that there must be additional principles and rules that children follow. Children are not consciously aware of these rules but they are implicit in their ability to master their native tongue rapidly and accurately. Children acquire the ability to construct indefinitely many new grammatical sentences on the basis of the sample of sentences they hear adults utter. Remarkably, not only are the sentences that form the "premises" of the child's inference relatively few but many of them are also ungrammatical. Nonetheless, young children acquire their first language amazingly quickly. Since any normal child can learn any human language, the implicit rules and principles must be common to all human languages (and thus Chomsky calls them a *universal grammar*) and the principles of this universal grammar must be innate.

Lipton says of both Chomsky and Kuhn that they are "arguing for unacknowledged principles of induction" (391). But the connection between Chomsky's theory of linguistic competence and the problem of description for induction is somewhat tenuous. Acquiring competence in a human language is unlike an inference in important ways. In an inference, one moves from the presumed truth of the premises to the presumed truth of the conclusion. What premises are presumed to be true in the case of language acquisition? The child is not arguing from the truth of the sentences she hears to the truth of the new sentences she constructs. Even if we imagine the child saying to herself (so to speak) something like "Many, but probably not all, the sentences I have heard are grammatically well formed," it is still unclear what conclusion about a particular language would follow from this premise, for Chomsky's point is that there are too many grammars consistent with the sentence inputs for inductive principles to be of any help. Indeed, the innate grammatical principles Chomsky posits are viewed as *imposing* generalizations on the data (sentence inputs), not as extracting grammatical generalizations from them (as talk of "inductive principles" of inference would suggest). On the face of it, it is unclear that Chomsky's fascinating and influential theory offers a cogent illustration of the problem of description for inductive inferences (even if it does illustrate how underdetermination gives rise to the problem of describing hidden principles generally).

A similar difficulty arises with respect to Lipton's appeal to Kuhn's views on scientific judgment as an illustration of the description problem. For as Kuhn is at pains to point out, he thinks that exemplars (shared examples) are necessary for producing consensus among scientists precisely because no set of rules could guide their reasoning. Consequently, it sounds odd for Lipton to characterize Kuhn as "arguing for unacknowledged principles of induction." (See chapter 2 for a detailed examination of Kuhn's views about scientific reasoning and judgment.)

THE PROBLEM OF JUSTIFICATION

In an essay on Francis Bacon, C. D. Broad described induction as "the glory of Science" but "the scandal of Philosophy."[1] Induction is scandalous because philosophers still cannot agree on the correct response to David Hume's skeptical conclusion that inductive reasoning lacks any kind of epistemic justification. Hume argued that any attempt to show that inductive inferences are (in Lipton's phrase) *truth-tropic* is doomed to fail. As far as Hume was concerned, our addiction to inductive reasoning is merely a blind, nonrational, animal faith: it can be explained psychologically, but if Hume is right, it cannot be justified philosophically.

Lipton describes the justification problem—the problem of answering Hume's skepticism about induction—as arising from underdetermination and circularity. We can illustrate this by considering one type of inductive inference, induction to a particular (IP):

All observed As have been B.
The next individual is an A.
The next individual will also be a B.

Now suppose that we attempt to show that IP is a justified form of inference. Hume's all-purpose argument that this is impossible runs as follows:

1 If IP can be shown to be justified, then there is an argument that
  shows it.

2 Arguments are either deductively valid or inductive.

3 No deductively valid argument can justify IP (because of
  underdetermination).

4 No inductive argument can justify IP (because of circularity).

5 IP cannot be shown to be justified.

This reconstruction of Hume's skeptical argument along the lines
mapped out by Lipton has several interesting features. The argument is undeniably valid: if the premises are true, then the conclusion also must be true.
So if all four premises are true, then the argument is sound. We shall focus on
each premise in turn. Because of the wider epistemological issues it raises, we
shall delay consideration of premise (1) until after we have looked at the other
premises.

Premise (2) would be necessarily true if inductive arguments were simply
defined as all those arguments that are not deductively valid. But this proposal would run counter to the way that we normally think and speak about
inductive arguments. As indicated earlier, inductive arguments are usually
regarded not simply as deductively invalid arguments, but as a proper subset
of deductively invalid arguments. The class of inductive arguments thus
includes universal and statistical generalization, statistical syllogism, induction to a particular, and so on, but it does not include the so-called deductive
fallacies, such as denying the antecedent and affirming the consequent. Typically, deductively fallacious arguments are those in which the premises give
no reason whatever for our believing that their conclusions are true. To keep
the deductive fallacies at arm's length, then, we should replace the term
argument(s) above in the first two premises with the more restrictive phrase
"argument(s) in which the premises give (or are widely believed to give) a
good reason for believing that their conclusions are true." Thus, premise (2)
would now say arguments that are (or, are widely believed to be) reliable or
truth-tropic are either deductively valid or inductive.

But even with this friendly amendment to premise (2), we are not quite
out of the woods. The problem is that some forms of argument that are
widely (but not universally) believed to be truth-tropic are neither deductively valid nor inductive. Arguments to the best explanation are a good
example. This type of argument plays an important role in the debate over
scientific realism, which is discussed in chapter 9. Some philosophers (such
as Gilbert Harman) have gone so far as to claim that inductive arguments
are reliable only when arguments to the best explanation underlie and justify them.[2] But while it may be initially plausible to claim (with Harman)

that "All observed As are B" is a good reason for believing that all As are B only because "All As are B" is the best explanation of why all observed As have been B, this sort of claim is not as plausible when it comes to other types of inductive argument, such as induction to a particular. For, surely, "The next A is a B" is not any sort of explanation of why all observed As have been B. One possible response to this objection would be to insist that IP is a reliable inference only because the conclusion that the next A is a B is derivable from the generalization that all As are B. The issue is a difficult one and cannot be resolved here. We shall simply say that the amended version of premise (2) would be true if all truth-tropic arguments that are not deductively valid are either explicitly inductive or ultimately reducible to inductive arguments. Alternatively, one could say that, insofar as premise (2) specifically concerns reliable arguments that can be used to justify IP (as opposed to inductive arguments in general), there is no nondeductively valid, noninductive argument (such as an argument to the best explanation) that can do the job.

Hume and his followers assert premises (3) and (4) because they have a certain view of what it would take to justify an inductive argument such as IP. Obviously, IP is not deductively valid. The principle of inference that connects its premises with its conclusion is: from "All observed As have been B" and "The next individual is A" infer "The next individual will also be B." So to justify IP, we would have to show that this principle of inference is reliable. Hume argued that this principle and all the other principles of inductive arguments rest on a presupposition about the world, namely that patterns we have observed to hold in the past will continue to hold in the future; or without referring specifically to time, that as-yet-unexamined individuals will resemble those we have examined; or more generally still, that populations tend to resemble the samples that are drawn from them. Of course, these versions of what is often called the *principle of the uniformity of nature* are terribly vague. Lipton calls it the *principle of conservative induction* or the principle of "More of the Same" (394). But the vagueness does not matter to Hume's argument. For Hume reasons that no matter how, exactly, we spell it out, the principle of the uniformity of nature is an empirical assumption. Logic alone does not guarantee its truth. Hence, if we are to justify our belief in the uniformity principle by means of a deductively valid argument, then that argument must have among its premises another empirical statement that is at least as general as the uniformity principle we are trying to justify. So no valid deductive argument can yield the justification we seek, since it will contain a premise that is at least as controversial as the principle we are trying to justify. If we try to justify the uniformity principle inductively, then that argument in turn will presuppose the uniformity principle. Thus, our attempt at justification would fail on grounds of circularity.

Hume's argument for premises (3) and (4) is ingenious and controversial.[3] Lipton gives a simple, intuitive gloss on premise (4)—the premise that denies that we can justify induction inductively—by comparing the principle of conservative induction, "More of the Same," with the principle of revolutionary induction, "It's Time for a Change." The revolutionary principle is counterinductive; it would license arguments of the following general form (CIP):

> All observed As have been B.
> The next individual is an A.
> The next individual will *not* be a B.

Lipton invites us to imagine that someone were to offer as a reason for trusting the next application of the principle of conservative induction the argument that, since it has always (or nearly always) worked in the past, it should be trusted in the future. Lipton remarks that this inductive argument (of the form IP) cannot be a sound justification for conservative induction because the counterinductivist could offer an exactly similar argument (of the form CIP) for the conclusion that induction will *fail* in the future. The counterinductivist would argue that since conservative induction has always (or nearly always) worked in the past, by CIP, it will fail on its next application. The counterinductivist can also appeal to the past failures of counterinduction as evidence that counterinduction will succeed in the future. Writes Lipton, "The justification of revolutionary induction seems no worse than the justification of conservative induction, which is to say that the justification of conservative induction looks very bad indeed" (395).

It is now time to consider the first premise of Hume's skeptical argument, which says that if IP can be shown to be justified, then there is an argument that shows it. This phrase about *showing* that IP is justified also appears in the conclusion of Hume's argument. Several epistemologists have emphasized the important differences among the following:[4] a belief *being* epistemically justified, an argument *conferring justification* on its conclusion, and using an argument to show someone that a belief is justified. We have deliberately phrased Hume's argument in terms of the last of these three notions. To get a sense of how these notion differ, forget about induction for a moment and consider the analogous problem of trying to justify deduction.[5] Let us focus specifically on the problem of justifying an elementary form of valid deductive argument, *modus ponens*:

> If P then Q
> P
> Q

Most people can just see that *modus ponens* is valid without needing any argument to support that judgment. It is a self-evident truth, recognizable a priori, that *modus ponens* is necessarily truth preserving. For beliefs that we recognize as being justified by a priori intuition, no argument is needed, but we could, if we chose, supply an argument, the conclusion of which asserts that *modus ponens* is a valid form of inference. Inevitably, such an argument would itself be deductively valid and would thus rely on *modus ponens* or some other simple form of valid argument. Since this new argument is valid and, let us assume, its premises are true and justified, it confers justification on its conclusion. Though we did not need such an argument to convince us that *modus ponens* is valid, nonetheless the new argument transmits justification from its premises to its conclusion. The argument justifies the conclusion that *modus ponens* is valid.

Now suppose we are confronted with a deductive skeptic who wants us to prove to him that *modus ponens* is valid by giving an argument. It is pretty obvious that no argument could possibly meet his demand. Inductive arguments would be too weak; valid deductive arguments would, as far as he is concerned, beg the question, since they would presuppose the very thing—the validity of deductive inference—that he is not yet prepared to accept. Nonetheless, since the deductive argument we provide is valid and its premises are justified, it does justify its conclusion that *modus ponens* is valid. Thus, the fact that an inference is valid, and, moreover, that an argument can justify the claim that the inference is valid, is consistent with the impossibility of using an argument to show that this is the case to someone who is a skeptic about the general class of arguments to which the inference belongs.

The same moral holds for inductive inference. Even if Hume's skeptical argument is flawless, its conclusion is strictly limited. Hume has not proven that inductive inference is unjustified. Nor has he proven that no argument can justify the belief that inductive inference is justified. At best, he has proven that no one can use an argument to show an inductive skeptic that inductive inference is justified. When put in this way, Hume's conclusion is far less troubling than it might otherwise appear. For valid deductive inference is in the same boat with regard to the deductive skeptic, yet no one regards deductive inference as a philosophical scandal or even as being epistemologically suspect.[6]

THE PROBLEM OF DESCRIPTION

As Lipton points out, Hume's all-purpose skeptical argument allowed philosophers to raise the justification problem without having to consider in any detail how inductive inferences are actually made. The latter problem, the problem of description, has proven to be remarkably intractable. Lipton canvasses five solutions: the more-of-the-same model, the instantial model, the hypothetico-deductive (H-D) model, the Bayesian approach to confirmation, and the causal model of Mill's methods.

The more-of-the-same model is not only extremely vague, but, Lipton argues, it also seems hard to reconcile with those inferences that predict change. He gives as an example the mechanic who predicts that the brakes on Lipton's car are about to fail. But the more-of-the-same model might still underlie the mechanic's inference about future brake failure if cars like Lipton's, with worn brake pads or leaking brake-fluid reservoirs, say, have invariably suffered brake failure in the past. Of greater difficulty for the more-of-the-same model are those inferences, common in the theoretical sciences, about hidden, perhaps unobservable, entities, drawn from observable data. Obviously these inferences cannot be understood merely as the extrapolation into the future of a pattern found to hold in the past.

Two responses suggest themselves. First, we could deny that inferences of this type are inductive and argue that they belong instead to another, noninductive category, such as arguments to the best explanation. Second, we might appeal to the distinction between the context of discovery and the context of justification, judging standard objections like those above as directed to the wrong context.

Giving a clear account of the distinction between the context of discovery and the context of justification is not easy, but the general idea is relatively straightforward.[7] On the one hand, there is the psychological question about how a scientific hypothesis first arises in the mind of the individual scientist. This often involves inferences both deductive and inductive. But it also often includes nonrational influences as well as a liberal amount of inspired guesswork. Anything that helps answer the psychological question about the origins of scientific hypotheses falls within the context of discovery. On the other hand, once a scientific hypothesis has been formulated, questions about what kind of evidence supports it, and to what extent, place us in the context of justification, where we are concerned with relations of inductive support and confirmation. Many philosophers of science (especially Karl Popper) have argued that philosophers should confine their attention to the context of justification, leaving the context of discovery to psychologists, historians, and sociologists. Thus, a second response to Lipton's objection to more-of-the-same is to deny that a description of inductive inference should provide a literal, blow-by-blow account of how scientific ideas and hypotheses are generated in the minds of individual scientists. Rather, it should concentrate on describing the relations that hold between hypotheses and the evidence that confirms them in the context of justification.

Both the instantial model and the H-D model are intended by their proponents as descriptions of the inductive principles that scientists follow in the context of justification. In its most basic form, the instantial model simply says that generalizations are inductively confirmed by their positive instances. For example, the hypothesis "All As are B" is confirmed by observed cases of As that are B. The usual objection to the instantial model is that it is both too weak and too strong. It is too weak because, as Goodman has argued, for any given piece of evidence there are just too many incompatible hypotheses that can claim that evidence as an instance.[8] Also, the so-called raven paradox discussed by Hempel seems to show that on the instantial model utterly irrelevant evidence, such as a white shoe, must confirm the hypothesis that all ravens are black.[9] The instantial model is judged too strong, since if it were a necessary condition for confirmation, far too many scientific theories, especially those involving unobservable entities and hidden causal mechanisms, could not be confirmed by observation.

The target of the last objection to the instantial model—its restriction to generalizations couched solely in the language of observation—naturally suggests the H-D model as a more realistic alternative. In its simplest formulation, the H-D model says that hypothesis H is confirmed by evidence E when H, in conjunction with some observation statements, entails E and E does not follow from those observation statements all by themselves. It places no limitation on the kinds of hypotheses that can be confirmed by evidence. But here again there are difficulties. Lipton mentions three. First, the H-D model is susceptible to the same grue and raven paradoxes as the instantial model. Second, the H-D model suffers from the problem of irrelevant conjunction (sometimes called the *tacking paradox*). For example, by conjoining "All crows are black" with "All swans are purple," it seems that the H-D model would license the inference from the observation of a black crow to the prediction of purple swans.[10] Third, by insisting that confirming evidence be entailed by any hypothesis that it supports, the model makes it impossible to confirm statistical hypotheses.

The descriptive adequacy of Lipton's fourth model, the Bayesian approach to confirmation, is controversial. As Lipton notes, the Bayesian account, like the H-D model, has been accused of being both too permissive and too strict. For a detailed discussion of the issues involved, see chapter 5.

Many of our inductive judgments depend on inferring likely causes from known effects. The fifth account of induction discussed by Lipton focuses on the causal model encapsulated in Mill's methods. In his *System of Logic* (1843), John Stuart Mill (1806–73) proposed five "methods of experimental inquiry" for establishing causes. Lipton focuses on the first two of these methods, the method of agreement and the method of difference.[11]

> The Method of Agreement: If several instances of P have only one antecedent condition, X, in common, then X is the cause of P.

> The Method of Difference: If two cases, one in which P occurs and the other in which P does not occur, have exactly the same antecedent conditions with one exception, namely, that X is present in the first but absent in the second, then X is the cause (or a necessary part of the cause) of P.

Both these and the other methods that Mill proposes are best regarded as applying to inferences in the context of justification rather than the context of discovery. In the discovery phase, we must first isolate the genuine cases of P and then identify all the factors (A, B, C, etc.) in each of these cases that might be causally relevant to P. Only after this hard work has been done can we use Mill's methods to eliminate rival hypotheses about the cause of P. This emphasis on refuting competing hypotheses is a distinctive feature of Mill's approach to induction (though it is also consistent with the Bayesian approach).

The method of agreement is designed to identify a necessary condition for P. In other words, it is intended to isolate a factor without which P could not occur. But in reality things can be more complicated than Mill's method suggests. For example, it may be true that, in four cases of P, the only single factor they all had in common was A. But that does not guarantee that A is necessary for P. It might be that two of the cases involved B, and the other two cases involved D so that the complex condition (B or D) is what is necessary for P and that A is in fact irrelevant. Moreover, the method of agreement says nothing about sufficient conditions for P. Again, B and D might each be sufficient to produce P, and A could be irrelevant.

The method of difference complements the method of agreement by describing a procedure aimed at identifying a sufficient condition for P. But as Mill himself realized, even when only one difference exists between a pair of cases, it does not follow that this factor is sufficient to produce P all by itself. For the distinguishing factor might, as Mill recognizes, merely be a necessary part of a sufficient condition for P. For example, suppose that the only difference between Adam (who gets ill) and Bob (who stays well) is that Adam was exposed to the disease organism, D. It does not follow that exposure to D will always produce illness in anyone exposed to it, for the illness might also depend on the person in question lacking immunity to D. All that we are entitled to assert is that D is probably a necessary part of a complex of factors that are jointly sufficient to produce P.

The limitations of Mill's two methods as a descriptive model of inductive inference are fairly obvious. At best they apply to cases in which all the factors of potential causal relevance are observable and can be enumerated prior to the application of Mill's methods. The vast majority of scientific cases are not like this, either because the theory to be confirmed is not causal or because the theory invokes entities and processes that are not observable. Furthermore, as Lipton points out, the antecedents of Mill's two methods require a single agreement or a single difference respectively. This condition is totally unrealizable in actual cases, where thousands of agreements and differences could be identified. Much of the real work in such cases is done by picking out from this vast array the relatively small number of factors that might be causally relevant. Since this selection is made before Mill's methods come into play, it means that Mill's methods do not capture important parts of the pattern of inductive inference.[12]

4.2 | Popper's Rejection of Induction

Sir Karl Popper (1902–94) was the most influential anti-inductivist philosopher of science of the twentieth century. Educated in Vienna during the heyday of logical positivism, Popper was one of the earliest and severest critics of the logical positivists' verifiability criterion of meaning and their attempt to use verifiability (and, later, confirmability) as a demarcation criterion between science and pseudoscience. Rejecting the positivists' criterion, Popper developed and proposed falsifiability (not verifiability or confirmability) as his own demarcation criterion. (See chapter 1 for a discussion of Popper's falsificationist demarcation criterion and its shortcomings.)

In "The Problem of Induction," Popper argues that there is no such thing as inductive confirmation. No scientific theory or law is made probable by evidence. Rather, science follows the method of conjectures and refutations. Scientists propose bold conjectures and then devise severe tests to falsify them if, indeed, they are false. A theory that withstands such tests is not confirmed or made probable by its successful predictions. Instead, it is what Popper calls *corroborated*. This, in very brief outline, is Popper's anti-inductivist falsificationist philosophy of science.

WHY INDUCTION CANNOT BE JUSTIFIED

Popper's case against induction is essentially Hume's, and Popper thinks it is unanswerable. There are just three small differences between Popper's skeptical argument and our reconstruction of Hume's argument in the preceding section of this commentary. First, instead of saying that any attempt to justify induction inductively must necessarily be circular, Popper says that it must lead to an infinite regress of inductive principles of ever higher order. Popper's idea is that when trying to justify a particular inductive principle or inference, we might avoid begging the question by employing an inductive principle or inference that is different from the one we are trying to justify. But this principle or inference, in turn, will be synthetic and hence will need to be justified by experience, thus requiring a further inductive argument, and so on. Popper assumes that such an infinite regress is vicious because the first inference will not be justified unless the second one is, and the second one will not be justified unless the third one is, and so on. Since the chain of justifying arguments can never be completed, the first inference never receives the justification it needs.

A second difference between Popper's argument and Hume's is that Popper considers, but nevertheless rejects, Kant's attempt to accord the principle of induction (in the guise of the "principle of universal causation") the status of an a priori synthetic truth. (Kant made his proposal after Hume's death.) If knowable a priori as Kant claimed, the causal principle would not need to be justified on the basis of experience. Indeed, Kant credited Hume, especially Hume's skeptical argument about induction, with having wakened him from his "dogmatic slumbers," from his uncritical assumption that the fundamental laws and principles of science could be justified solely on the basis of experience and perception.

Given the obscurity of Kant's arguments in the second Analogy of his *Critique of Pure Reason*, Popper wisely avoids giving them any detailed consideration in our selected reading. But elsewhere, Popper has criticized Kant's position in the following way.[13] Popper agrees with Hume and Kant that our belief in the uniformity of nature cannot be justified from the bottom up by making ever wider generalizations from the regularities we discover in nature because every one of these inferences already presupposes the point at issue, namely the uniformity of nature. Popper also agrees with Kant that the lawlikeness of nature must be imposed by us from the top down (and thus is prior to experience). But Popper sees no reason for Kant's rationalist confidence that our search for particular regularities must succeed or that our belief in those regularities can be justified a priori. According to Popper, Kant was misled by the tremendous success of Newtonian mechanics into believing that certain high-level generalizations about nature (such as those encapsulated in Newton's laws) must be true and that our search for such regularities must succeed. But as Popper puts it, all our knowledge is conjectural. Kant had confused the alleged psychological indispensability of certain of our beliefs about nature with an a priori guarantee that specific versions of those beliefs must be true.

A third wrinkle in Popper's argument that distinguishes it from our reconstruction of Hume's reasoning actually involves a point that Hume himself made. Hume anticipated that some inductivists (as we shall call them) might concede that we cannot prove with certainty that the conclusion of any inductive argument must be true. But, they might argue, we can nonetheless show that these conclusions are likely or highly probable relative to the premises from which they are inferred. Popper follows Hume in dismissing this strategy as unavailing, charging that this probabilistic version of inductivism leads either to "apriorism" or to a vicious infinite regress.

Why There Is No Logic of Discovery

Popper's main concern in section 2 of our reading, "Elimination of Psychologism," is to distinguish descriptive questions from normative ones. Empirical psychology, he says, may investigate how and why a new theory suggested itself to a scientist, but tracing this causal history is irrelevant to evaluating the theory. Only when the process of conceiving a new idea has been completed can the normative appraisal of what has been conceived begin. After the new theory has been formulated, its subsequent evaluation may in some cases mirror the actual thought processes of scientists and in that sense provide a rational reconstruction of them, but the epistemological evaluation is not intended as a psychological description and should not be judged by how well or ill it happens to reflect actual patterns of reasoning. Moreover, he concludes that there is no set of rules, no logical method for generating ideas. He agrees with Bergson and Einstein that scientific discoveries always involve an element of creative intuition that cannot be captured in any algorithm.

Despite the frequency with which this short section from *The Logic of Scientific Discovery* is cited, it is difficult to get straight what Popper is claiming and what his arguments are. Popper seems to be making two main points: that empirical psychology and history are irrelevant to epistemology and that there is no interesting version of a logic of discovery. These two points are independent, and neither depends on Popper's anti-inductivism. So, for example, inductivists such as Reichenbach and Hempel can endorse both positions while still rejecting Popper's inductive skepticism. Also, someone can accept Popper's first point (by insisting that epistemology cannot be naturalized) and yet insist that heuristic rules and maxims can guide the construction of new theories. All that anti-inductivist falsificationists (such as Popper) and inductivist hypothetico-deductivists (such as Hempel) rule out is the possibility of an algorithm for generating new theories that would guarantee the truth or high probability of those theories. But the rejection of such an infallibilist, mechanical conception of a logic of discovery leaves ample room for more modest versions, and not even Popper in his most outspoken moments would deny that reasoning (*deductive* reasoning) can play a role (however fallible and justificationally impotent) in the generation of new theories.[14]

FALSIFICATIONISM AND CORROBORATION

According to Popper the logic of science is purely deductive. The essence of rational scientific method is not to look for probable theories and try to confirm them with evidence, but to invent bold, improbable theories and try to refute them. Theories are tested by deducing predictions from them. If the predictions are false, then it follows deductively (by *modus tollens*) that the theory is false and should be discarded. If the predictions are verified, then the theory is not confirmed but corroborated. ("Being corroborated" and "having a high degree of corroboration" are terms of positive epistemic appraisal for Popper.) Withstanding a severe test proves a theory's mettle. Corroborated theories should not be accepted as true or even as probable, but nonetheless it is rational for us to rely on them at least until better theories come along. Eventually, by continually replacing falsified theories with ever bolder ones that have not yet been refuted, science progresses towards the truth. Or so Popper claims.

There are many problems with Popper's falsificationism as an account of scientific rationality. One problem is reconciling Popper's talk of "verifying" the predictions deduced from theories with his insistence that no empirical statement whatever can be established as true or highly probable. Popper's response is to say that predictions, in the form of "basic statements," are accepted by the conventional decision of the scientific community and that these decisions can be reversed later.[15] Popper adamantly denies that any basic statement is made probable by being grounded in or supported by the available evidence. Scientists simply make a free, unconstrained decision to accept the statement, at least until further notice. In a memorable and oft-quoted passage, Popper writes:

The empirical basis of objective science has thus nothing "absolute" about it. Science does not rest upon rock-bottom. The bold structure of its theories arises, as it were, above a swamp. It is like a building erected on piles. The piles are driven down from above into the swamp, but not down to any natural or "given" base; and when we cease our attempts to drive our piles into a deeper layer, it is not because we have reached firm ground. We simply stop when we are satisfied that they are firm enough to carry the structure, at least for the time being.[16]

But as many of Popper's critics have argued, this metaphor of the swamp and the doctrine it illustrates are deeply troubling, for despite Popper's claim that conventional acceptance occurs only at the level of basic statements, not at the level of theories, surely the refutation of a theory must be just as conventional as the basic statement that prompts it. Thus, it is hard to see how Popper can justifiably assert that science is rational and objectively progressive when it ultimately depends on purely conventional and arbitrary decisions.[17]

4.3 | Salmon's Criticism of Popper's Anti-Inductivism

As the preceding discussion has revealed, Popper's anti-inductivism creates difficulties for falsificationism by making the ungrounded acceptance of basic statements (and thus the refutation of theories) seem arbitrary and irrational. In his article, "Rational Prediction," Wesley Salmon also takes aim at Popper's rejection of induction. Salmon charges that, without induction, Popper cannot give a coherent justification for using corroborated theories to make what Salmon calls *rational predictions*.

The Problem of Rational Prediction

Although Popper uses the term *prediction* to refer exclusively to the consequences drawn from a theory in order to test it, Salmon reminds us that there are at least two other common reasons for making predictions. Sometimes, predictions are made solely to satisfy our intellectual curiosity about what will happen in the future. These predictions are of purely theoretical interest. More frequently, predictions are made not merely to satisfy our curiosity but because something of practical importance is at stake. For example, we might need to know how a high-rise apartment building will behave during a moderate earthquake to make sure that its design is safe. We could engage in blind guesswork or turn to clairvoyants for help, but the best guide is to rely on currently accepted scientific theories.

Salmon claims that, because Popper rejects any form of inductive confirmation, he is unable to do justice to rational predictions. (For the purposes of his article Salmon focuses on those rational predictions that are ingredients in the making of practical decisions.) If we are never justified in accepting any theory, and successful predictions and explanations can never count as confirming evidence for a theory, how can falsificationists such as Popper and his defender, John Watkins, explain why current scientific theories are a better guide to the future than blind guesswork, clairvoyance, or reading Chinese fortune cookies?

Popper and Watkins have responded to Salmon's challenge by appealing to the notion of corroboration. A theory is corroborated at a given time when it has survived our attempts to refute it up to that time; one theory has a higher degree of corroboration than another when it has survived more severe tests. Suppose we have to choose a theory as the basis for a practical prediction about the future from among a number of rivals none of which has yet been refuted. In this situation, it seems only rational to prefer the theory with the highest degree of corroboration. For in this kind of situation, Watkins says, we have nothing better to go on. Popper and Watkins argue that selecting in this way is not tantamount to making an inductive inference because judgments about corroboration are based solely on a theory's past performance: they say nothing and imply nothing about how well the theory will do in the future. Moreover, in choosing the best corroborated theory we are not "relying" on that theory in the sense of accepting it as true or even probable. We are simply making a choice.

THE INADEQUACY OF POPPERIAN CORROBORATION

Salmon finds the Popper-Watkins "solution" to the problem of rational prediction completely unsatisfactory. In a tone of patient but growing exasperation with the falsificationists, Salmon makes two main points. First, when trying to justify our selection of a theory for the purposes of practical prediction, it is not literally true to say, as Popper and Watkins do, that we have nothing else to go on. There are plenty of other ways to choose a theory. We could, for example, choose the theory which is shortest when written out in English or the one that gets the most votes on a TV call-in show. Presumably, what the falsificationists mean is not that there is nothing else to go on but that there is nothing else *rational* to go on. Yet even if all other methods are irrational, this does not entail that choosing the best corroborated scientific theory is rational. For it is perfectly possible that all methods are without exception equally irrational. Thus, in saying that there is nothing else to go on, Popper and Watkins must actually mean that there is nothing else to go on that is rational apart from corroboration. And that, Salmon complains, simply begs the question.

Second, Salmon sees an irreconcilable conflict between Popper's insistence that corroboration statements have no predictive import and his assertion that they justify our preference for one theory over another for purposes of rational prediction. If corroboration statements are solely about the past, how, without tacitly making an inductive inference, can they justify the choice of a theory on which to base predictions about the future?[18]

4.4 | Hempel on Criteria of Confirmation

In "Criteria of Confirmation and Acceptability," excerpted from his lucid introduction to the philosophy of science, *Philosophy of Natural Science*, Carl Hempel surveys the factors that scientists use in deciding whether a hypothesis is acceptable or probably true. These factors include inductive confirmation by the available evidence, simplicity, and support from other theories. Hempel attempts to clarify these factors, but unlike, for example, the Bayesian approach to scientific inference (discussed in chapter 5 of this book), he does not offer a systematic, unified theory to justify them. At the end of his article he briefly discusses the difficulties of providing such a theory within the framework of Carnap's system of inductive logic. Hempel presupposes that, because they involve universal generalizations, scientific hypotheses and laws cannot be conclusively verified, no matter how extensive the evidence in their favor. But like most other philosophers of science, Hempel rejects Popper's falsificationism and accepts that scientific hypotheses and laws can be made more or less probable by evidence.

Quantity, Variety, and Precision of Evidence

As a general rule, the more evidence there is for a hypothesis, the more strongly the hypothesis is confirmed. But as Hempel notes, there is a law of diminishing returns with regard to evidence. Repetitions of the same kind of experiment or observation, under the same conditions, rapidly lose their power to add substantially to the credibility of the hypothesis being tested, even when all the observations are exactly as the hypothesis predicts. What we desire in evidence is not mere quantity, but diversity. Different kinds of test, under widely varying conditions, have a greater power to confirm than do repetitions of the same kind of test under the same conditions. More generally, variety in evidence makes for strength in confirmation. Why is this?

Hempel argues that when the evidence is diverse, more of the hypothesis gets tested than when the evidence is narrow. He gives as an example, Snell's law in optics. For convenience, Hempel lets S stand for Snell's law, the hypothesis that, for any pair of transparent media, the ratio of the sine of the angle of incidence (*i*) to the sine of the angle of refraction (*r*) is a constant. This is a generalization about all associated angles of incidence and refraction, and all pairs of transparent media. Thus, confining our observations to air and water and to angles of incidence of 30 degrees would leave a large part of S untested. Let us call this restricted test, T_1. By varying our experiments to include a range of angles (test T_2) and many different media (T_3), more of S would be tested. But why should evidence collected by testing more of S lead to greater confirmation, assuming that the total number of data points—the quantity of the evidence for S—remains unchanged?

Hempel's answer appeals to falsifiability. Let S_1 be the hypothesis that the ratio of sin *i* to sin *r* is a constant for all samples of air and water when angle i is 30 degrees. S implies S_1, but S_1 could be true and S false, and the restricted test T_1 would not show it. In other words, if S_1 were true, T_1 could not falsify S. Similarly, if S_2 is the hypothesis that the ratio of sin i to sin r is a constant for all samples of air and water and all associated pairs of angles, then if S_2 were true, T_2 could not falsify S. Thus, Hempel concludes, the more thorough the test (that is, the greater the power of the test to falsify a hypothesis), the more support a favorable outcome gives to the hypothesis. More diverse evidence results from more powerful tests; the more powerful the test, the greater the confirming power of the evidence it generates.

Hempel's explanation of the power of diversified evidence gives an inductive twist to Popper's notion of the falsifying power of a test. But Hempel never spells out in detail why greater falsifying power implies greater confirming power. One way of connecting the two would involve the Bayesian approach to confirmation explored in chapter 5. For, as Hempel notes, we value diversity not for its own sake, but because we suspect it might affect the results of our experiments. To the Bayesian, diversity has evidential significance only when it is associated with competing hypotheses that have some initial degree of likelihood or probability. By disconfirming or falsifying some of these competitors, powerful tests thereby increase the probability of the surviving hypothesis (or hypotheses).

Hempel's discussion of the value of precision in evidence is brief. Increased precision of measurement has special significance when we are testing hypotheses that make identity claims (such as the identity of inertial and gravitational mass postulated by Einstein's general theory of relativity) or which predict a null result for a class of measurements (as does the special theory of relativity for all attempts to measure absolute velocities or to ascertain which inertial frames are at rest with respect to absolute space). For in these cases, the level of precision at which the results in question are supposed to hold is unlimited.

NOVEL PREDICTION

Hempel does not use the phrase *novel prediction*. Instead, he writes of "new" evidence and "new" phenomena. The word *new* is in quotes because the evidence or phenomena need not be new in the sense of not having been previously recorded or observed. Consider, for example, the case of the Swiss schoolteacher Johann Jakob Balmer (1825–98), discussed by Hempel. On the basis of four observations made by the Swedish physicist Anders Jonas Ångström (1814–74), Balmer published, in 1885, a general formula giving the wavelengths of a series of lines—now called the Balmer series—in the visible range of the emission spectrum of hydrogen. The original data set on which Balmer based his formula was already widely known by scientists at the time. Balmer then used his formula to predict further lines in the hydrogen spectrum, which were subsequently verified, thus confirming his formula. As Hempel notes, some of the additional lines had in fact already been observed in 1885, when Balmer put forward his theory, but Balmer and most other scientists were unaware of this fact. Thus, to call the later predictions new or novel does not mean that no one had previously observed them but rather that the results of these observations were not widely known. In particular, the results were not known to Balmer at the time he proposed his formula.

In analyzing the debate over novel prediction and confirmation, it is helpful to distinguish among the various senses in which a prediction might be considered novel. We shall introduce the phrase *epistemic novelty* to distinguish the restricted epistemic sense of predictive novelty—not widely known and not known by the person proposing the theory—from the wider sense of *temporal novelty*—not known by anyone at the time the theory was proposed. Obviously, both senses of *novelty* are epistemic and relative to time, but epistemic novelty is also relative to persons: even if two scientists were to propose the same theory at the same time, the theory could make predictions that were epistemically novel for one of the scientists, but not for the other, depending on what they knew and when they knew it.

Some philosophers of science have also deemed two further senses of *novelty* important. We shall call these *design-novelty* and *use-novelty*. A result is design-novel so long as the scientist did not deliberately construct the theory to yield the result in question. For example, when Newton devised his gravitational theory, he was perfectly well aware of the precession of the equinoxes, but this was not an explicit constraint on his theorizing. He did not, for example, go through several preliminary versions of his theory, rejecting those that did not give the right answer for the rate of precession. Contrast this with Einstein, who did reject earlier versions of the general theory of relativity because they failed to imply the correct value for the rate of precession of the perihelion of Mercury. Design-novelty is what Elie Zahar and others have called *heuristic novelty*: a result is novel in this sense if it "did not belong to the problem-situation which governed the construction of the hypothesis."[19] Use-novelty is a special case of design-novelty. A prediction lacks use-novelty when that prediction is used to fix the value of a free parameter in a theory or when the result in question is simply built into the theory. (The relevance of this sense of novelty for confirmation has been emphasized by John Worrall.[20])

There are simple logical relations among the four notions of novelty that we have distinguished: temporal novelty implies epistemic novelty, epistemic novelty implies design-novelty, and design-novelty implies use-novelty; but use-novelty does not imply design-novelty, design-novelty does not imply epistemic novelty, and epistemic novelty does not imply temporal novelty.

Several different questions can be raised about novel predictions and confirmation. Here are two of the most important:

> If a theory is deliberately designed to accommodate or explain phenomena that are already known, do those phenomena confirm the theory that entails them? (If *T* is designed to explain *E*, does *E* confirm *T*?)

Other things (such as quantity, diversity, and precision of evidence) being equal, does a theory derive greater confirmation from novel predictions than from the already-known phenomena that it explains or accommodates? (Is T confirmed more strongly by novel prediction D than it is by E?)

Since there are two questions and at least two possible answers to each (Yes, No, It all depends), at least four positions are possible. In practice, five positions have been held by significant numbers of scientists and philosophers of science. The most extreme position would answer "No," emphatically, to both questions, thus denying any form of inductive confirmation by evidence. This is Popperianism, which was discussed in the preceding sections of this commentary. Few philosophers of science regard such a complete rejection of inductive confirmation as plausible.

Two other extreme positions (but less extreme than Popperianism) are often referred to as *explanationism* and *predictionism*. Explanationism is the view that only the explanation of previously known and accepted results can confirm a theory and that novel predictions have no power to confirm. Thus the explanationist answers the first question with an emphatic "Yes" and the second question with an emphatic "No." Something close to this view has been defended by Stephen Brush and John Worrall. Predictionism is the opposite view that only novel predictions can confirm a theory and that the explanation of previously known and accepted results has no power to confirm. Thus predictionists respond to the first question and the second question with an emphatic "No" and an emphatic "Yes," respectively.

Between the two extremes of explanationism and predictionism are two more conciliatory positions. One popular choice is to answer "Yes" to both questions. This position, which we shall call *accommodationism*, allows that the explanation or accommodation of old evidence has some confirmatory value but insists that novel prediction must always have more. The other position, which we shall call *formalism*, agrees that accommodating old evidence can confirm but holds that, other things being equal, novel predictions are no better and no worse than old ones. Whether or not a particular novel prediction confirms better than an old one depends on its evidential content, not on its novelty. Formalism is driven by the insistence that confirmation depends solely on the formal, inductive, logical relation between propositions, regardless of when and by whom those propositions were known or to what purposes they were put. Our taxonomy of positions on the novel prediction debate is summarized in table 1 (on the next page).

Notice how explanationism, accommodationism, and formalism differ from each other. Explanationists deny that novel predictions are superior to explanations because they deny that novel predictions have any power to confirm. Accommodationists accept that novel predictions can confirm and think that the confirmation they provide is always superior to that of explanations. Formalists think that novelty per se is irrelevant: all that matters is the evidential content of the propositions involved, whether they be predictions or explanations. Sometimes novel predictions confirm better

	Question 1: If T is designed to explain E, does E confirm T?	Question 2: Is T confirmed more strongly by novel prediction D than it is by E?
Popperianism	No	No
Explanationism	Yes	No
Predictionism	No	Yes
Accomodationism	Yes	Yes
Formalism	Depends	Depends

Table 1

than explanations, sometimes they do not; it all depends on the propositions involved.

Hempel's answer to the first question is "Yes." Without elaborating, he simply asserts that it follows from the fact that theory T entails old evidence E that E must confirm T. Without placing further qualifications on the relationship between T and E, this assertion seems dubious. For example, if we were to construct a theory by conjoining quantum mechanics, Q, with the proposition E (which we already know to be true), most people would judge that the resulting theory (Q & E) is *not* confirmed by its true entailment, E. As the critics of the H-D model are fond of pointing out, entailment of true consequences is too weak to be a sufficient condition for confirmation.

With regard to the second question Hempel notes that most people think that the confirming power of novel predictions is greater than that of explanations or the accommodation of old evidence. He then raises "a puzzling question" (429). What if Balmer's equation, T, had been constructed only after the new evidence, D, was in, so that T was designed to fit all the evidence (E & D)? That T fits the evidence under these circumstances is hardly surprising, which seems to support the predictionist preference for new evidence. But, Hempel replies, something in this case is still surprising. The surprise is not that T fits (E & D), for with sufficient ingenuity, numerous theories can be invented that would fit any given evidence. But most of those theories would be horrendously complex and arbitrary. No, the surprise is that the theory, T, which fits (E & D) is such a simple mathematical formula. Thus, we would still have a reason for thinking that T is true, even though T was tailored to fit the evidence. What confirms T is not the fact that T fits (E & D) but that T is quite simple. Still, Hempel does not endorse this simplicity argument as a rejoinder to the predictionist because its plausibility depends on whether we can connect simplicity with truth, and that, in turn, depends on whether we can give an adequate objective characterization of simplicity—an issue Hempel takes up later in his article.

Hempel concludes this part of his discussion by noting that, quite apart from the arguments he has been exploring, from a logical point of view all that matters is the content of theory T and the evidence E, not when they were proposed or discovered. Again, he is not taking sides on the novel prediction issue but merely noting that if one adopts the inductive-logical approach to confirmation of Carnap and others, then the formalist view naturally follows. Deductive logic provides an appropriate analogy. Just as the deductive validity of an argument depends solely on the content of its premises and conclusion, so, too, does the inductive strength of an argument. The inductive probability of the conclusion given the premises is an objective, logical matter that is entirely unaffected by how the theory was constructed or when the truth of the premises was discovered.

THEORETICAL SUPPORT

Not all support for a theory or hypothesis comes from "below," from inductive confirmation by successful predictions; some comes from "above," from other well-established and well-accepted theories. The plausibility of new theories and hypotheses is often assessed by how well they stand up in the light of evidentially well-confirmed theories that we already accept. The strongest relation between one theory and another is deductive. Hempel gives as typical examples of deductive support from above the support that Balmer's formula received when a more general formula was confirmed of which Balmer's expression was a special case and the support that these equations received in turn from Bohr's well-confirmed theory of the hydrogen atom that explained them. Hempel uses the phrase *deductive support* in discussing these cases because of the deductive relation between the theories involved. But the support that one theory confers on another is still inductive. The basic idea is that if a theory, G, is logically derivable from some more general theory H and H is well-confirmed by a range of evidence, E, other than the evidence that confirms G, then G is also confirmed by E. Elsewhere in his writings on confirmation theory, Hempel has called this principle the *special consequence condition*:[21]

If E confirms H, and H entails G, then E confirms G.

The special consequence condition says, in effect, that whatever confirms a hypothesis H also confirms any statement that is deducible from H.[22] Imagining any confirmation theory that could dispense with this principle is difficult, since we presuppose it whenever we infer that the predictions of a well-confirmed theory are likely to be true. A similar looking principle that Hempel named the *converse consequence condition* also has an air of plausibility:

If E confirms G, and H entails G, then E confirms H.

Newton's gravitational theory is commonly regarded as being confirmed in virtue of its entailing (with certain additional hypotheses) Galileo's law of falling bodies and Kepler's laws of planetary motion. Galileo's law and Kepler's laws were well confirmed before Newton advanced his theory. So when Newton was able to derive these laws from his new theory, the theory immediately inherited a substantial measure of inductive confirmation. Indeed, on at least one model of scientific progress according to which scientific knowledge grows by a process of theoretical reduction, something like the converse consequence condition would seem to be an integral part of the logic of confirmation. (For an extended account and criticism of this model, see chapter 8.)

But as Hempel has shown, the special consequence condition and the converse consequence condition together have the unacceptable, in fact absurd, consequence that any observation report confirms any hypothesis whatsoever. To show that we cannot accept both conditions as principles of confirmation, consider the following example. Let E be the observation report, "This is a black crow." E confirms the hypothesis, G, that all crows are black. Let hypothesis H be the conjunction of two generalizations, G and J: all crows are black and all swans are purple. Trivially, H entails G. Therefore, by the converse consequence condition, E confirms H. But, equally trivially, H entails J. So by the special consequence condition, E confirms J. Thus, by employing both conditions, we reach the absurd consequence that "This is a black crow" confirms "All swans are purple."

The special consequence condition appears to be irreproachable. For, whatever the connection between probability and confirmation, any proposition must be at least as probable as the propositions from which it can be derived. Thus, it seems inevitable that inductive confirmation should be transmitted downwards to the logical consequences of any confirmed proposition.[23] Since we cannot embrace both conditions, we must give up the converse consequence condition. And this is the course that Hempel takes in his formal study of confirmation theory. Hempel thinks that the plausibility that the converse consequence condition seems to derive from classic cases of theory reduction is spurious. For in the classic cases, not merely are the well-established laws logically derivable from a more general theory, but, crucially in Hempel's view, the laws are *instances* of the more general theory. In his formal theory of confirmation, Hempel advocates the *satisfaction criterion*: a generalization, law, or theory is inductively confirmed by an observation report only if the report is an instance of the statement it confirms.[24] We cannot pursue the details of Hempel's theory here. Suffice it to say that it represents a significant departure from the crude form of the H-D model. For according to Hempel's satisfaction criterion, hypotheses, theories, and laws are no longer confirmed merely by the true predictions that can be deduced from them. For a theory to be confirmed, those predictions must be not merely logical consequences but instances of the hypotheses, theories, and laws in question. Thus, while "all crows are black" is confirmed by the report of a black crow, Hempel's theory denies that "this is a black crow" confirms "all crows are black and all swans are purple" because the report "this is a black crow" is not an instance of the conjunctive hypothesis. In simple terms, Hempel accepts that whatever confirms a theory also confirms any logically equivalent or weaker theory, but he denies that whatever confirms a theory must also confirm every stronger theory.

SIMPLICITY

The notion of simplicity has already been invoked in the section of Hempel's article on novel predictions. There, in response to the predictionist claim that a theory cannot be confirmed by evidence it was designed to explain, Hempel suggested a reply that assumed that simpler theories are more likely to be true than more complex ones, regardless of how they have been arrived at. Hempel now examines whether this and similar claims about simplicity can be justified.

Like most philosophers who discuss simplicity in scientific theories, Hempel takes curve fitting to be a paradigm illustration. In Hempel's example, the four data points are the four pairs of values of the variables u and v—$(0, 2)$, $(1, 3)$, $(2, 4)$, and $(3, 5)$—to which we wish to fit a curve. Many different hypotheses are consistent with these data. Intuitively, the simplest of them is the straight-line hypothesis $v = u + 2$ that Hempel calls H_3. But many other hypotheses expressing v as a function of u are possible—hypotheses involving additional terms and higher powers of u. In cases like this, the philosopher's task is twofold: first, to explicate the concept of simplicity so that hypotheses such as H_3 are judged correctly to be simpler than their rivals; second, to justify what Hempel calls the *principle of simplicity*: that, given two otherwise equally well-confirmed rival hypotheses, we should prefer the simpler; in other words, that simplicity enhances the credibility of a theory or hypothesis.

Hempel briefly surveys some of the attempts to analyze the notion of simplicity and the difficulties such attempts have encountered. A guiding assumption in this search has been that, whatever it ultimately amounts to, simplicity should be objective, not simply a matter of familiarity, convenience, or intuitive appeal that can vary from person to person or that is relative to context. Thus, estimating the complexity of algebraic equations by their appearance is an unreliable way to measure simplicity. So, too, is looking at the graphical representation of a curve in some coordinate system. In both cases, the judgment made will be relative to the variables or the coordinate system chosen. In Cartesian coordinates (x, y), a straight line will appear simple and a spiral complex; in polar coordinates (r, θ) the spiral will appear simple and the straight line complex. An alternative might have us counting axioms and judging that theory to be simplest whose axioms are fewest. But like the previous proposals, this fails due to arbitrariness: one can always reduce the number of axioms by conjoining statements.

Hempel discusses and criticizes four attempts to justify the simplicity principle. The first two are based on ideas that have been held by a number of scientists and philosophers; the last two derive specifically from Reichenbach and Popper, respectively.

■ THE SIMPLICITY-POSTULATE ARGUMENT

The simplicity postulate asserts that most of the fundamental laws of nature are simple. If we knew that to be true, then we would have an obvious rationale for embracing the principle of simplicity, which holds that simpler theories should be preferred because they are most likely to be true. The problem, as Hempel observes, is that we do not know that the simplicity postulate is true. In fact, the assumption that the laws of nature are simple is at least as controversial as the principle of simplicity itself and, hence, cannot be used to justify it without a great deal of further argument. Merely appealing to our past success in uncovering some relatively simple laws of nature will not suffice here, since the relevant question is precisely, Why should this past success be taken as good evidence that the laws of nature as a whole are simple? Obviously, it would be circular to appeal to the simplicity of the simplicity postulate as a reason for preferring it to more complex postulates.

One of these alternative postulates is that a principle of selection is driving us to find the simple laws and that the laws we have discovered thus far are not a representative sample of the laws of nature as a whole. Rather, the laws we have discovered are simple because those are the ones on which we have chosen to focus our attention. Both the simplicity postulate and the selection hypothesis explain the evidence, namely, the relative simplicity of the laws uncovered by science thus far. So what reason can we give for preferring the former to the latter that avoids begging the question?

■ THE ECONOMY-OF-SCIENCE ARGUMENT

In the generation of scientists and philosophers of science preceding the advent of logical positivism, positivistically inclined thinkers such as Ernst Mach, Karl Pearson, and Pierre Duhem insisted that the principal aim of science is not to explain but to find the most economical description of facts. On this view, theories are not hypotheses about the structure of reality; they are merely tools for summarizing information obtained from observation and experiment. Hence, the simplest theories should be preferred because they provide the most succinct, shorthand summary of the available data.

The flaw in this attempt to justify the simplicity postulate was exposed by Hans Reichenbach (among others) in his book, *Experience and Predic-*

tion, and Hempel's remarks in his article echo Reichenbach's criticism. Reichenbach introduced the phrases *descriptive simplicity* and *inductive simplicity* to stand for two very different notions of simplicity. Descriptive simplicity concerns empirically equivalent theories where the choice of one theory over another is, to quote Reichenbach, "nothing but a matter of taste or of economy."[25] Reichenbach gives several examples: the preference of scientists for the metric system over older systems for making measurements and performing calculations; the choice of one particular inertial system as the "rest system" in the special theory of relativity and the adoption of a particular definition of simultaneity in that theory; the decision to adopt Euclidean rather than non-Euclidean geometry as the geometry of space. But when, as in Hempel's illustration, we are trying to fit curves to a finite set of data points, the rival hypotheses are not empirically equivalent. They are hypotheses whose predictions about future measurements disagree. In curve-fitting problems, the issue is not, "Which hypothesis gives the most economical description of the facts already observed?" but rather, "Which hypothesis is the most reliable guide to observations that will be made in the future?" In short, inductive simplicity concerns what Wesley Salmon calls *rational prediction*, not economical description. Reichenbach (and Hempel) accuse positivists like Mach of confusing the two concepts of simplicity. Economy and convenience govern the choice between (equivalent) theories that differ merely in descriptive simplicity. But when (nonequivalent) theories differ in inductive simplicity, the choice depends on which theory makes the best predictions about the future.

■ REICHENBACH'S INDUCTIVE ARGUMENT

Reichenbach's defense of the principle of simplicity is an application of his well-known pragmatic vindication solution to the problem of induction. Whether one chooses the smoothest curve that fits a finite set of data points or uses the straight rule to estimate the properties of a population from an examination of a sample, Reichenbach admits that no method can be guaranteed to generate the correct answer any single time it is applied.[26] But, he argues, some methods have the important virtue of being self-corrective. It can be shown, solely by means of deductive arguments, that if they are applied repeatedly, they will get ever closer to the correct answer with the accumulation of evidence. The inductive straight rule and the principle of the simplest curve are examples of such self-corrective methods.

As Hempel notes, Reichenbach's argument is more complicated than Hempel's simplified presentation of it. The reason for this additional complexity sheds light on one of the philosophical difficulties with Reichenbach's approach. The problem is this: Many different, competing methods can be proven to be self-corrective in Reichenbach's sense. The property of asymptotic convergence is not unique to the straight rule and the principle of the simplest curve. What, then, justifies us in using these rules rather than others with the same virtue? In the case of curve fitting, the problem takes the following form. One self-corrective method—the method of linear interpolation—tells us to approximate the true function by drawing a chain of straight lines through the data points as they accumulate. But no self-respecting scientist would use that method. Instead, scientists insist on drawing smooth curves to link the data points. Why? Reichenbach suggests that it is because scientists assign a physical significance to the higher-order derivatives of the variables that define the curve. For example, if we plot distance against time, then the first-order derivative of distance with respect to time is velocity, and the second-order derivative is acceleration. Scientists want to discover laws relating those quantities to distance and time by means of continuous functions, too. So they insist on the method of the smoothest curve, rather than the method of linear interpolation, when choosing a curve to fit a set of paired values of distance and time. But even if this argument is convincing, a large number of smooth curves can still be drawn through any set of data points. Why should we prefer one curve to another that is, by whatever small degree, incompatible with it? At this point in his argument, Reichenbach concedes that there is no further inductive vindication for our preference. Inductive simplicity can take us only so far. Once we have narrowed our choice to the family of smooth curves, all of which converge asymptotically, any further preference among them is solely a matter of descriptive simplicity and convenience. In the long run, any member of the family will converge on the true function, so it does not matter which one we choose.

Hempel acknowledges that Reichenbach's argument does show something. For example, it shows that some ways of connecting data points are irrational because they lack the property of convergence. But, as Reichenbach concedes, the field of convergent methods is very wide. Reichenbach's appeal to the long run as the answer to this embarrassing variety carries little conviction, since he is unable to specify how quickly (or slowly) the convergence will take place. How satisfying is it as a philosophical solution to the problem of inductive inference to be told that a method is rational because it (along with thousands of other rival methods) will eventually lead us to the truth if only we persist for long enough (and if there is a truth to be discovered)? Most of the decisions we have to make using inductive reasoning concern the very short run. As John Maynard Keynes remarked, "in the long run we are all dead."

■ POPPER'S FALSIFIABILITY ANALYSIS

Popper's approach to the simplicity problem differs radically from that taken by the other philosophers discussed by Hempel. Since Popper denies the very possibility of inductive confirmation and insists that all theories, even the ones we currently accept, are almost certainly false, he can hardly claim that simplicity should be preferred because simple theories are more likely to be true. Instead, Popper links simplicity with falsifiability. Simple theories should be preferred, he tells us, because they are more falsifiable and thus easier to test. Popper also associates simplicity with empirical content, claiming that simple theories are more falsifiable because they have greater content. Inductivists and Bayesians defend the principle of simplicity by arguing that simpler theories are more likely to be true because they have higher prior probabilities.[27] Popper takes the opposite position, arguing that simpler theories have greater content and thus are more likely to be false. We can crudely summarize Popper's position regarding scientific theories as follows:

more falsifiable = simpler = greater empirical content = more improbable.

There are many fascinating and controversial aspects to Popper's network of claims concerning falsifiability, simplicity, and content. Some of these are touched on in Salmon's article, discussed earlier in this commentary. Here, we shall focus on two of Popper's claims: the link between simplicity and falsifiability, and the link between simplicity and content.

In order to link simplicity with falsifiability, we must first characterize each and then show that the one tracks the other. For curves specified by equations, Popper proposes that simplicity decreases with the number of freely adjustable parameters, the degree of the equation, and the order of the derivatives (if any) that occur within it. For example, the second-order equation of a circle is simpler than the second order equation of an ellipse because it has one less parameter (the radius of the circle replaces the major and minor axes of the ellipse). Falsifiability (testability) depends on the number of data points that would be needed to show that the theory is false. Four points that suffice to refute a circle hypothesis might still lie on an ellipse. Thus "all the planets move in circles" has a greater degree of falsifiability than "all the planets move in ellipses" because, if false, it can be falsified by a smaller number of data points. This example illustrates Popper's thesis that simpler theories have a greater degree of falsifiability.

Many objections have been raised against Popper's analysis of simplicity and falsifiability. One problem is its limitation to simple polynomial functions such as those specifying straight lines, circles, and similar curves. What about functions such as $y = \cos x$, and $y = \log (1 - x)$? Scientists would judge these to be relatively simple. Yet their power series expansions contain an infinite number of terms involving ever higher powers of x. Thus, Popper's theory would judge them all as being equally, infinitely, complex. Hempel's discussion raises two further problems. First, certain groups of hypotheses (for example, H_1, H_2, and H_3 from Hempel's curve-fitting example) can all be refuted by a single data point, even though some of the equations are of a higher order than others. So, it would seem, contrary to Popper, theories can differ greatly in simplicity and yet have the same degree of falsifiability. Second, when two hypotheses differ in their degree of falsifiability, the more falsifiable hypothesis is not always simpler. Consider, for example,

"All asteroids and comets move in circles" (K) and "All planets move in ellipses" (L). Because it takes fewer data points to refute a circle hypothesis than it does to refute an ellipse hypothesis, K is more falsifiable than L even though it is more complex. Of course, Popper might reply by insisting that L is more falsifiable than K because it has greater content.

The content of a scientific theory is notoriously hard to define. The natural idea is to regard a theory's content as a function of the number of the empirical consequences derivable from it: the larger the set of consequences, the greater the theory's content. But most, if not all, scientific theories have an infinite number of consequences. Thus we are faced with the daunting problem of trying to decide which infinite sets are "bigger" than others. Content comparison is only straightforward when one theory entails another. For, if H entails J, then J's consequence set is a subset of H's consequence set. And so long as H and J are not logically equivalent, the content of H must be greater than the content of J. But, as Hempel notes, logically stronger theories are not necessarily simpler. For example, the conjunction (H & J) entails J, but when H and J are unrelated theories, (H & J) is not simpler than J. Greater content does not imply greater simplicity. Moreover, in the curve-fitting problem, none of the rival hypotheses entails any of the others. Nonetheless, some are simpler than others, even though, intuitively, their content is the same, since each of them specifies an infinite set of pairs of values for the two variables u and v.

The Probability of Hypotheses

In the final section of his paper, Hempel considers whether it might be possible to construct a system of inductive logic in which the credibility of a hypothesis is a precise number between 0 and 1 satisfying the axioms of probability. Since the credibility of a hypothesis, H, depends not just on a narrow set of observational evidence, E, but on the other theories, statements, and laws that we accept, Hempel writes the credibility of H as $c(H, K)$, which is to be read as the credibility of H relative to the set K of all the statements that scientists accept at a given time (including E). Once we are given credibilities in the form of probabilities, it is easy to calculate further probabilities of disjunctions and conjunctions using the probability calculus. The philosophical challenge is to explain where the probabilities come from in the first place and what determines their values. Hempel mentions one influential attempt to do this, namely Rudolf Carnap's system (actually, infinitely many different systems) of logical probability (otherwise known as *inductive probability*, or *degree of confirmation*).

As Hempel recognizes, Carnap's work is highly formal, based as it is on an idealized model language. All the primitive predicates (of which there can be only finitely many) and individual constants (the terms that name individual objects, events, or, as Carnap prefers, spatiotemporal positions) must be completely specified at the outset and, for a significant class of these systems, the numerical values that the credibility function assigns to hypotheses depend on the number of primitive predicates in the language. Thus, if we wish to apply any of these Carnapian systems of inductive logic to science, either we must hold the language constant, refusing to allow any new predicates to be introduced, or we must recognize that the confirmation that, say, a hypothesis in physics receives from a particular body of evidence will change whenever new predicates are introduced, even if they are in completely different fields such as sociology or botany. Neither option is attractive, and in any case, the very idea that we could completely enumerate all the predicates and individual constants of a language adequate to express all the theories in present-day science is quite unrealistic.[28]

There is another reason for being skeptical about the usefulness of Carnap's formal systems of inductive logic for understanding the confirmation of scientific theories. As Hempel points out, the confirmation of theories depends on a number of factors, such as simplicity and diversity of evidence, that are extremely difficult to analyze precisely. Thus, it is very unlikely that they could ever be incorporated into an algorithm that would determine how the numerical credibility values would change in response to them. The Bayesian approach (explored in chapter 5) circumvents this problem by abandoning Carnap's attempt to assign inductive probabilities to hypotheses on logical, a priori grounds. Instead, Bayesians take the relevant probabilities as given, empirically, in the form of the degrees of belief that scientists have in their theories. Thus, factors such as simplicity and diversity of evidence exert their influence psychologically, and Bayesian confirmation theory takes it from there. This obviates the need for the kind of precise, logical analysis that has proven so elusive.

4.5 | Explanationism, Predictionism, and Evidence

The historical thesis of evidence asserts that if theory T entails a true statement E, then whether that statement confirms the theory depends on when the statement is known relative to when the theory that entails it is proposed.[29] Explanationists insist that E can confirm T only if E is known before T is proposed; predictionists insist that E can confirm T only if E is a novel prediction. Thus, predictionists hold that confirming evidence has to be "new" whereas explanationists hold that confirming evidence has to be "old."

THREE ARGUMENTS FOR PREDICTIONISM

There are three distinct arguments for the predictionist thesis that only new evidence can confirm a scientific theory.

■ THE NO-COINCIDENCE ARGUMENT

Predictionists and accommodationists (those who admit that old evidence can confirm but insist that novel predictions always have greater confirming power) often appeal to a form of inference to the best explanation in support of their doctrine. They point out that, when E is a novel prediction, there are two hypotheses that might explain why theory T correctly predicts E: the truth hypothesis and the coincidence hypothesis. Either T is true (and that is why E, along with all the other entailments of T, is true), or T is false and it is simply a coincidence that E happens to be true. The truth hypothesis is a much better explanation of the fact that T entails something true than is the coincidence hypothesis. Indeed, the coincidence hypothesis seems like no explanation at all. So when E is a novel prediction, E confirms T to a significant degree. But when the evidence that T was contrived to explain is old, a third and more likely explanation of why T entails that evidence enters the picture. That explanation, namely, the design hypothesis, asserts that T was deliberately constructed to explain E. Indeed, in this case, the coincidence hypothesis drops out as a candidate, leaving the contest between the design hypothesis and the truth hypothesis. The design hypothesis is a much better explanation than the truth hypothesis. So the predictionist concludes, when E is old evidence it lends no (or, as some accommodationists would concede, very little) support to T. The alternative hypotheses in the two cases are summarized in table 2.

Alternative explanations of the
fact that a theory, T, entails the
true statement E

Prediction of new evidence	T is true	Coincidence
Explanation of old evidence	T is true	Design

Table 2

The no-coincidence argument has several controversial aspects. One is the assumed connection between explanation and confirmation that underlies this sort of inference to the best explanation. The operative principle— that one proposition, P, confirms another, Q, only if Q is the best explanation for P—is not very plausible, in part, because confirmation comes in degrees but being the best explanation does not. Moreover, of the available explanations for P, Q, while the best, might be very poor. Under those circumstances it would seem unreasonable to regard P as confirming Q to any significant degree. These objections suggest the following revision: the degree to which P confirms Q depends on the strength of the explanation of P given by Q. The stronger the explanation of P by Q, the greater is the support that P lends to Q.

Another controversial aspect of the no-coincidence argument is the judgment about which hypothesis explains best in each of the two cases. Suppose that explanation requires either the logical derivability of the explanandum (what is to be explained) from the explanans (what does the explaining) or that the explanans confer on the explanandum a reasonably high probability. The explanandum is the conjunction of two statements: that T entails E, and that E is true. The first conjunct of the explanandum, that T entails E, follows deductively from all of the hypotheses under consideration. For if T entails E, then this is a necessary truth and is thus entailed by any hypothesis whatever. The second conjunct of the explanandum, the fact that E is true, is entailed by the truth hypothesis (together with the necessary truth that T entails E). Thus, whether E is a novel prediction or old evidence, the hypothesis that T is true would seem to be an excellent explanation in either case. So why should we agree with the predictionist that, when E is old evidence that T is introduced to explain, the design hypothesis is superior to the truth hypothesis? On the face of it, the design hypothesis provides little or no explanation of the second conjunct of the explanandum, namely, that E is true. Granted, the truth hypothesis beats out the coincidence hypothesis when E is a novel prediction. But then why does the truth hypothesis not similarly beat out the design hypothesis when E is old evidence?

■ THE FALSIFICATION ARGUMENT

Although Popper denies that there is any such thing as inductive confirmation, his principle of falsifiability is often invoked by inductivists. In the present debate between predictionism and explanationism, the relevant principle would be roughly as follows: if T entails E, then E can confirm (or "corroborate") T only if E could potentially falsify T. (Paul Horwich invokes a similar principle in his solution to the raven paradox discussed in chapter 5.) When E is already known to be true, it is assumed that E cannot falsify T. Therefore, novel predictions always confirm and explanations of old evidence can never do so.

The falsification argument hinges on the temporal novelty of the evidence: if E is already known, then it cannot serve to falsify T. But a number of philosophers of science, notably John Worrall, have argued that what really counts is not temporal novelty but use-novelty.[30] For example, if E was explicitly incorporated into the theory T, then it would not be use-novel and thus could not serve, even potentially, to falsify T. But E might have been known for centuries before T was proposed and yet still be able potentially to falsify T just as long as it was not used in the construction of T. For in that case, Worrall argues, T would be genuinely at risk from E. Worrall seems to have in mind *risk* in the sense of epistemic possibility, not logical possibility. From a logical point of view, either T entails E, or T entails something that is inconsistent with E (or T entails neither E nor anything that is inconsistent with E). Whichever entailment relation holds (or fails to hold), it does so of logical necessity, independently of what anyone believes or desires. But from an epistemic point of view, things are quite different. We want an explanation for some already known result E. We propose a theory T to explain it. Since we did not build E into T or use E to generate T, we do not know whether in fact E is derivable from T. For all we know, before we try to derive E from T, the theory might yield something that is inconsistent with E and thus fail the test. If we find that T does in fact entail E, then T has passed the test and E now supports T.

- THE OLD-EVIDENCE ARGUMENT

The old-evidence argument is usually given as an objection to Bayesianism rather than as support for predictionism. The argument is based on the positive-relevance view of evidence favored by Bayesians. (See the readings and commentary in chapter 5 for an extended discussion of this approach to confirmation and evidence.) According to the positive-relevance view, E is evidence for the theory T if and only if E raises the probability of T. When a scientist uses T to make a novel prediction E, we do not yet know whether E is true. Thus, the probability of E is less than 1. Assume that T entails E, so that the probability of E given T is 1: $P(E/T) = 1$. In that case, it follows immediately from Bayes's theorem that the probability of T given E is greater than the probability of T without E: $P(T/E) > P(T)$. (Again, see chapter 5 for much more on this and other applications of Bayes's theorem.) But when E is old evidence, E is already known to be true. So some philosophers have argued, in that case the probability of E is 1, and Bayes's theorem entails that E cannot change the probability of T. Hence, no matter how much old evidence T can explain, none of it can confirm T. This last claim has come to be known as the *problem of old evidence*. It should be noted that most Bayesians reject this claim and have offered various solutions to the problem of old evidence. (See chapter 5 for details.) The old-evidence argument for predictionism is highly controversial.

Two Arguments for Explanationism

Stephen Brush, in his article, "Prediction and Theory Evaluation: The Case of Light Bending," illustrates the explanationist thesis that novel predictions are powerless to confirm theories and that only the explanation of previously known and well-established results carries confirmational weight.[31] In fact, Brush (and other explanationists, such as Gingerich and Worrall) does not go quite this far. While Brush insists that explanations of old evidence are always confirmationally superior to the prediction of new results, he allows that novel predictions can confirm theories to some degree. Brush rests his case on two arguments: one appeals to the history of science, the other to a notion of reliable evidence.

■ THE HISTORY-OF-SCIENCE ARGUMENT

Many science textbooks and writings by philosophers of science assume that
scientists regard novel predictions as much stronger evidence for a theory
than the explanation of facts already known. Wesley Salmon, for example,
in his article "Rationality and Objectivity in Science *or* Tom Kuhn Meets
Tom Bayes" (reprinted in chapter 5 below), cites the Poisson bright spot as
a striking and unexpected prediction made by the wave theory of light. The
story as told by Salmon, Kuhn, Giere, and others runs as follows:[32] In 1819,
Augustin Fresnel, a supporter of the wave theory of light, won the prize
offered by the French Academy of Sciences for the best essay on diffraction.
At that time, most scientists in France (and elsewhere) accepted the corpuscular theory of light that had been developed by Isaac Newton. Very few
physicists thought that the wave theory of light could possibly be true. Of the
distinguished panel of scientists that judged Fresnel's entry, three (Laplace,
Biot, and Poisson) were committed corpuscularians; only one, Arago, favored
the wave theory. During his examination of Fresnel's paper, Poisson deduced
from Fresnel's wave theory of light the apparently absurd consequence that
if a small circular disk obstructs the light coming from a pinhole, there
should be a bright spot in the center of the disk's shadow and the spot should
be just as bright as if the disk were not there. Arago performed an experiment
to test this "absurd" consequence and, to everyone's amazement, found the
bright spot, just as Fresnel's theory predicted. Fresnel was immediately
awarded the prize and, almost overnight, the wave theory was accepted by
the scientific community, mainly on the strength of this stunning confirmation of Fresnel's theory.

Unfortunately for the predictionists, John Worrall has shown that Arago's experimental verification of the Poisson bright spot had nothing like the
dramatic effect claimed for it by most philosophers of science.[33] There was
no sudden conversion to the wave theory, and the commission that judged
Fresnel's paper was far more impressed by Fresnel's ability to explain, accurately and precisely, the known diffraction patterns cast by straight edges and
slits, than it was by the prediction of the bright spot. The usual story about
the dramatic confirming effect of the bright spot seems to be largely a myth.

Similarly mythical, according to Brush, is the story of Eddington's verification, in 1919, of Einstein's novel prediction that light from distant stars
will be deflected by the intense gravitational field of the sun. At the time,
physicists (as opposed to the general public) were just as impressed, if not
more impressed, by the ability of Einstein's general theory of relativity to
predict, accurately, the rate of precession of the perihelion of Mercury. The
problem of Mercury's orbit had been recognized for decades, whereas gravitational light bending was new and unexpected. But few, if any, scientists
gave light bending greater confirmational significance simply because it
was new. Most found Einstein's solution to the Mercury problem an achieve-

ment that was at least as, if not more, impressive than the prediction of light bending. In the view of most scientists, the power of evidence to confirm a theory has nothing to do with when the evidence was first acquired.

Thus, the history-of-science argument runs as follows: Contrary to the claims made by predictionists, scientists often accord greater weight to the explanation of known phenomena than to the prediction of new ones. This is borne out, even in those cases (such as the Poisson bright spot and gravitational light bending) to which predictionists appeal for support. Since philosophical claims about science have to square with the considered judgments of scientists in a reasonably wide variety of cases, explanationism is more plausible than predictionism.

■ THE RELIABLE-EVIDENCE ARGUMENT

Brush's appeal to the history of science as support for explanationism has a number of problems. History, that is *real* history not the simplified idealizations of textbook writers, is messy. As Brush readily admits, scientists were not unanimous in the light bending case: some did think that light bending was stronger evidence for Einstein's theory than the explanation of Mercury's orbit, and Brush, a scrupulous historian, quotes some of them in his article. Of course, Brush attempts to explain why they judged the matter incorrectly or why, in some cases, they did not literally mean what they wrote. But the fact remains, then as now, that some scientists endorse predictionism.

Another way in which history is messy concerns the "other things being equal" qualification that is implicit in the rival theses of predictionism and explanationism. Both camps recognize that, even if novel predictions are intrinsically superior to explanations as evidence, in an actual case a particular explanation may confirm more strongly than a particular novel prediction and a particular novel prediction, however startling and unexpected, may not confirm a theory to any significant extent. The reason for the latter is readily appreciated from a Bayesian perspective in which confirmation depends not only on the unexpectedness of the evidence predicted but also on the prior probability of the theory making the prediction. Thus, for example, in the case cited by Owen Gingerich,[34] Velikovsky made predictions about the high surface temperature of Venus and radio emissions from Jupiter that scientists at the time regarded as highly improbable, only to have these predictions verified a few years later. Much to Velikovsky's dismay these predictive successes did nothing to increase the scientific acceptability of his theory. To Bayesians this is not surprising: the prior probability of Velikovsky's theory was so low that no amount of evidence, not even the success of startling novel predictions, could raise that probability to any significant degree. Thus, the Velikovsky case is not a counterexample to predictionism because other things were not equal.

Given the problems with appealing to history for support of explanationism, what Brush needs is a philosophical argument showing why explanations should count more strongly than novel predictions, other things being equal. In his attempt to offer such an argument, he invokes the notion of reliable evidence. The phrase *reliable evidence* is not, perhaps, well chosen, since it suggests *probability of truth*, whereas Brush's brief remarks make clear that what he really means is something more like *confirming power*. In other words, as Brush sees it, the issue is not whether the evidence is known for sure to be genuine but whether the evidence, once known, has the power to confirm the theory.

Brush begins by noting that, prior to Einstein's general theory of relativity, not only was Mercury's orbit already well known, but, more importantly, no theory had been able to explain it. This failure did not lead to the rejection of these unsuccessful theories, since, at that time, scientists had nothing better to put in their place; Mercury's orbit was a classic example of a nonrefuting anomaly. But when Einstein solved the Mercury problem, the unsuccessful theories were rejected, and, as a consequence, Einstein's theory was substantially confirmed. Understood in this way, Brush is appealing to inductive confirmation by the elimination of alternatives. Contrast the Mercury problem with the prediction of light bending. Gravitational light bending was an entirely new phenomenon. Thus, at the time of Einstein's successful prediction whether the possible rival theories to Einstein's could also explain it was not known. (The only available rival theory that had been systematically explored was Newton's gravitational theory, and Newton's theory did indeed give the wrong answer.) So the novel prediction of light bending, though successful, had relatively little confirming power when compared with the solution of the Mercury problem. Only some years later, when rival theories had tried and failed to explain light bending, could light bending count as strong evidence for Einstein's theory.[35] On this interpretation of Brush's argument, it is not the explanation of previously known phenomena per se that counts strongly in favor of a theory but the theory's succeeding where its rivals have failed.[36]

4.6 | Achinstein's Rejection of Predictionism and Explanationism

The title of Peter Achinstein's article asks, "Explanation v. Prediction: Which Carries More Weight?" His answer is that it all depends on the evidence in question. Achinstein rejects the contention of the historical thesis that whether *E* confirms *H always* depends on when *E* is known relative to when *H* is proposed. Nonetheless, Achinstein thinks that whether *E* is evidence for *H* does *sometimes* depend on historical facts about how *E* was collected. Nonetheless, even when evidence is historical in this sense, Achinstein argues that it makes no difference whether *E* is an explanation of old evidence or a novel prediction. More generally, when extra information is relevant to evidence and confirmation in particular cases, sometimes the extra information is historical; sometimes it is not. What kind of information is relevant in particular cases is an empirical matter. Thus, Achinstein is sharply critical of confirmation theories such as Carnap's, which attempt to answer the question, "How strongly does E confirm H?" solely on a priori grounds.

SELECTION PROCEDURES

Achinstein's argument for his verdict that history sometimes is and sometimes is not relevant to evidence rests on the consideration of different cases. In the section of his article on selection procedures, he offers three such cases: the drug case, the lottery case, and the crow case. Their essential features are summarized in table 3.

	H	E
Drug case	Drug D relieves symptoms S in approximately 95 percent of cases.	In a group of 1,000 people with symptoms S taking drug D, 950 persons had relief of S; in a control group of 1,000 S-sufferers taking, not D, but a placebo, none had symptoms S relieved.
Lottery case	John won the lottery.	In last week's lottery, 1,000 tickets were sold of which John owned 999 at the time of the selection of the winner; this was a fair lottery in which one ticket was selected at random.
Crow case	Female crows are black.	Males crows are black.

Table 3

Achinstein argues that in the drug case, whether E is evidence for H (and certainly, how strong that evidence is) depends on the selection procedures used to choose the experimental and control groups. Without this historical information, we simply cannot tell how strongly E supports H or whether E supports H at all. Contrast this with the lottery case, in which E is strong evidence for H no matter what method was used to find out that E is true. Because E says that the lottery was fair, that the winning ticket was picked at random, and that John held all but one of the 1,000 tickets in the lottery, everything inductively relevant to H is already contained in E. Finally, in the crow case, like the drug case, whether E is evidence for H (and if so, how strong that evidence is) depends on extra, empirical information. But unlike the drug case, the extra information needed is ornithological, not historical. Using the terminology introduced by Achinstein, in the lottery case, E is empirically complete with respect to H. In the crow case, E is empirically incomplete with respect to H, but no historical facts are needed to settle claims about evidence and confirmation. So the lottery and crow cases are both counterexamples to the historical thesis of evidence.

PREDICTION VERSUS EXPLANATION

If the historical thesis is false, then both predictionism and explanationism (in their pure, strong forms defined earlier) must be false. But even if the predictionist is mistaken in asserting that explanations can never be evidence and the explanationist is mistaken in asserting that novel predictions can never be evidence, the interesting question remains whether, in cases in which history *is* relevant to confirmation, predictions or explanations provide the stronger evidence.

Obviously, in cases that violate the historical thesis (and hence in which temporal considerations are irrelevant to confirmation), E confirms (or fails to confirm) H regardless of whether evidence E is old or new. Achinstein reinforces this point with a coin-tossing example that, like the lottery case, is empirically complete. The more interesting cases are those in which the evidence E is historically incomplete. For an example of such a case, Achinstein returns to the drug case. The information contained in E can be construed either as E_1—a prediction about what will occur in the next trial—or as E_2—a report about a trial that has already taken place. And just to sharpen the contrast, Achinstein supposes that H (the drug hypothesis) was deliberately devised in order to explain E_2. So which is stronger evidence for H, E_1 or E_2?

Achinstein thinks that the correct answer to this question is "neither." For if (as Achinstein argued in the previous section) the confirming power of evidence E depends on the selection procedure used to generate it, then when our knowledge of E was first acquired makes no difference whatever. If E_1 is strong evidence for H because the trial will involve randomization over a diverse population, then E_2 must be equally strong evidence for H if the same selection procedure was used in a trial that already took place. The explanation-versus-prediction distinction is irrelevant.

Achinstein does not say what, in general, makes for a good selection procedure relative to a hypothesis H. But a theory implicit in his remarks is made explicit in an important article by Deborah Mayo that Achinstein cites in a footnote.[37] Mayo's proposal is that E is good evidence for H if and only if E is the outcome of a severe test of H. A test of H is severe if there is a very low probability that the test would have yielded a passing result such as E if H were false. A "passing result" in this context means a result that accords with H at least as well as E does. When H is a statistical hypothesis, as in Achinstein's drug case, a small range of outcomes (including E) would qualify as passing the test of H. When H is a deterministic hypothesis that entails E, then no other outcome could accord with H at least as well.

But in either case the crucial question in judging the severity of the test, and hence the power of E to confirm H, is "How likely is it that we would have obtained a result that is at least as good as E if H were false?" In the deterministic case, when H entails E, this boils down to requiring that $P(E/\sim H)$—the probability of E given that H is false—be very low. As Mayo points out, predictionists often assume, falsely, that this requirement will be violated whenever E has been used to generate H (and thus E is not use-novel or novel in any other of the senses we have distinguished).

Mayo gives a simple example to illustrate why the predictionists are mistaken in thinking that severe tests entail use-novelty. Suppose that the fender of a car has been dented while it was parked in a garage. The car's owner examines carefully the shape and size of the dent and uses this information, E, to infer H, the likely make of the car that caused the dent. If it is practically impossible for the dent to have its distinctive features unless it was created by a collision with a specific type of car tailfin, then E is good evidence for H, even though E was used to generate H.

Brush Redux

The word *redux* means "restored," but in the section "Brush Redux" Achinstein neither reinstates Brush's explanationist position nor defends it. (This is hardly surprising given Achinstein's preceding attack on explanationism and predictionism.) Instead, Achinstein criticizes two of the theses (the ones that Achinstein numbers 1 and 3) that Brush appeals to when trying to show why explanations of known phenomena confirm theories more strongly than do novel predictions. Thesis (1) says that a test of H is inferior if it fails to consider competitors to H, and thesis (3) says that scientists have more time to consider competitors when explaining known phenomena than when making novel predictions. Achinstein regards his drug case as a counterexample to thesis (1), since at least one of the selection procedures for generating E in that example (SP 1) results in a strong test of the drug hypothesis H, even though the procedure makes no explicit mention of competitors. And in any case, Achinstein continues, even if thesis (1) were true, there is no reason why it must favor explanation over prediction. If a strong test has to consider competitors to the hypothesis being tested, then, contrary to thesis (3), scientists will take the time needed for that consideration regardless of whether the evidence generated by the test is already known.

Achinstein's counterexample to thesis (1) is not entirely convincing. It is true that the selection procedure (SP 1) that he alludes to in criticizing thesis (1) does not mention competitors to H explicitly. (SP 1) says that the experimental and control groups should be chosen "arbitrarily" (441)—presumably, Achinstein means at random—from a diverse population of subjects. But the whole point of randomization and diversity is to diminish the chances that the results obtained could be due to any cause other than drug D that is being tested. The beauty of randomization is that we do not have to formulate explicitly the competing hypotheses to H in order to be pretty sure that we would not have obtained the result E unless H were true. So while the consideration of competitors does not figure in (SP 1) explicitly, it is an important part of the implicit rationale for adopting it. Moreover, when we turn our attention from simple statistical generalizations (as in the drug case) to hypotheses in physics and other sciences, it is hard to see how we could we begin to estimate $P(E/\sim H)$ for a test of H unless we had some idea of the likely ways in which H might be false. And that, surely, will often require the explicit consideration of the competitors to H.

THOMSON VERSUS HERTZ

Mindful of the highly artificial and simplified character of the cases on which so much of his argument depends, Achinstein concludes by giving a real-life example from the history of physics. When nineteenth-century physicists began applying a high electrical potential to gases at low pressures, it became evident that something was being emitted in straight lines from the negative electrode (the cathode). What was the nature of these cathode rays (as they were soon named)? There were two competing theories: either they were streams of negatively charged particles, or they were some kind of ether wave (electromagnetic radiation). In 1883, Heinrich Hertz (1857–94) reasoned as follows:[38] If they were charged particles, then they should be deflected both by a magnetic field and by an electrostatic field. Since the cathode rays did bend in response to a magnet but apparently did not bend in response to an electrostatic field, Hertz concluded that the rays must not be charged particles but some peculiar form of ether wave. (The waves were deemed peculiar because, unlike light rays, they were deflected by a magnetic field.[39]) In 1897, J. J. Thomson (1856–1940) challenged Hertz's conclusion on the grounds that Hertz's experiment was flawed. Because the pressure in Hertz's cathode tube had not been low enough, the gas molecules had become ionized by the radiation streaming from the electrodes. These ionized gas molecules effectively short-circuited the plates to which an electrostatic charge had been applied, thus destroying the electric field.[40] When the tube was evacuated more thoroughly, Thomson found the deflection that Hertz had been unable to detect.

Achinstein makes two points about this episode. First, Hertz was wrong to take his negative result E (the failure to detect any electrostatic deflection) as strong evidence for H (the hypothesis that cathode rays are not electrically charged). In fact, as Thomson later showed, Hertz's procedure for testing H was inadequate, and hence E is not evidence for H. Second, and more important in Achinstein's view, is that, as far as we can tell, Hertz did not propose hypothesis H until after he had performed his experiment. But no one (certainly not Thomson in 1897) suggested that, had Hertz first proposed H and then predicted E *before* his experiment, the confirmation (or, as it turns out, the lack of confirmation) that E confers on H would have been affected. The crucial flaw in Hertz's experiment, his use of a poor selection procedure, prevents E from supporting H, regardless of whether H was used to predict E in advance of the experiment or, as actually happened, H was introduced after the experiment to explain its outcome.

4.7 | Goodman's New Riddle of Induction

The predicate *grue* made its first appearance in the third chapter of Nelson Goodman's *Fact, Fiction, and Forecast* (1955).[41] Goodman defined this decidely weird predicate as follows: "it applies to all things examined before t just in case they are green but to other things just in case they are blue." (452) This is equivalent to the following (with the substitution of an upper case T for t): an object is grue if and only if either the object is green and examined before time T, or the object is blue and it is not the case that the object is examined before time T. Goodman's purpose in introducing this gerrymandered predicate was to make a telling point about induction and inductive confirmation. Conventional wisdom has it that a generalization such as "All emeralds are green" is confirmed by its instances. We inspect one emerald and find that it is green.[42] We observe another and find that it, too, is green; and so on. Soon, we infer from this uniform evidence that it is very probable that all emeralds are green, whether we have observed them or not; and we predict further that the emeralds we will observe in the future will also be green. But if all the green emeralds that we have inspected were examined before time T (where T is now or later), then it is equally true to say that all the emeralds we have observed have been grue. And if generalizations are confirmed by their instances, then our evidence confirms that all emeralds are grue. But if all emeralds are grue, then we predict that any emerald examined after time T will not be green but blue. (In everything that follows, "after T" is shorthand for "at or after T.") The green hypothesis (that all emeralds are green) and the grue hypothesis (that all emeralds are grue) are genuine rivals that make differing predictions about the future. Nonetheless, if generalizations are confirmed by their instances, our evidence confirms both hypotheses. Goodman calls this *the new riddle of induction*. The riddle is not to explain why we are rationally justified in accepting any generalization as true (or probable) when we have observed several instances of it and no counterinstances: that is the old riddle of induction, Hume's problem of justifying induction. Rather, the new riddle of induction is to explain why some generalizations (such as the green hypothesis) are confirmable by their instances and others (such as the grue hypothesis) are not.

DEFINITIONS OF GRUE

There are several different definitions of *grue* in the literature spawned by Goodman's book. Since there are competing definitions of grue in the literature, we shall use the term "grue" (without a subscript) either generically (as in the first clause in this sentence) or to refer specifically to Goodman's original definition.[43]

An important feature of Goodman's grue hypothesis that is sometimes misunderstood is that it does not imply that all grue emeralds must change color at T. The general form of the definition of grue is $[(Gx \& Ox) \lor (Bx \& \sim Ox)]$, where "$Ox$" stands for "$x$ is observed before time T."[44] The disjunction in the definition is inclusive so it is true that if emeralds changed color from green to blue at T (and were observed either before or after T), they would qualify as grue. But it is sufficient for an emerald to be grue that it is observed only before T and found to be green (and remains green); it also sufficient for an emerald to be grue that is observed only after T and found to be blue (and has always been blue). Thus, the grue hypothesis does not predict a change in observed color for once-examined emeralds. Of course, none of this is to deny that the grue hypothesis predicts that emeralds observed after T (*and not before*) will be found to be blue (just as Goodman says). For Goodman, the inductively interesting issue is the predictions we are warranted in making about the color of as-yet-unexamined emeralds after time T on the basis of (other) emeralds observed to be green before T.

Apart from its disjunctive character, a peculiarity of grue is that it is what we shall call a *lifetime* predicate (sometimes referred to as an "atemporal" or "tenseless" predicate). When lifetime predicates apply to an object, necessarily they characterize the object throughout the entire period of its existence as the thing that it is. Typical lifetime predicates are terms such as *human*, *fluorine*, and *soluble*. By contrast, many everyday predicates are *time-relative* (or so-called "time-slice" or "temporal" predicates): they apply to an object at a time (or during a period of time); and it is possible that they do not apply to the same object at another time. Typical examples are particular instances of color, spatial location, and weight. Obviously, a car can be yellow at one time but red at another (after receiving a spray job) or in my garage tonight and parked elsewhere tomorrow. Goodman defines grue as a lifetime predicate. Unlike color, grue is not the kind of predicate that can be true of an object at one time but not at another.

Of course, one can define a time-relative version of grue that is very closely related to the one above and matches what Goodman sometimes says in his text. Let us call it grue$_t$: x is grue$_t$ (that is, x is grue *at time t*) if and only if either x is observed at time t, t is before T, and x is green at t, or x is not observed prior to T and x is blue at t. In symbols: x is grue$_t = x$ is grue at $t =_{df} (Gxt \& Oxt \& t < T) \lor (Bxt \& \sim(Oxt \& t < T))$. On this definition of grue, an object x observed to be green prior to T (that is, x is grue$_{t_1}$, where $t_1 < T$) cannot remain grue if observed after T (that is, x will not be grue$_{t_2}$, where $t_2 \geq T$) unless it changes color. It is only with respect to such a time-relative conception of grue that it makes sense to ask whether an object that has been verified to be grue at a time earlier than T will remain grue when observed at a time later than T.

A popular variant of grue drops the reference to "being examined" or "being observed" that are features of Goodman's original definition. Thus, some authors (e.g., Barker and Achinstein, Elliott Sober) define *grue* in the following way: either x is green at time t, and t is before T, or x is blue at time t, and t is after T.[45] In symbols: x is grue at $t =_{df} (Gxt \ \& \ t < T) \lor (Bxt \ \& \ t \geq T)$. Like Goodman's grue$_t$, this notion of grue is time-relative; but, unlike grue$_t$, it depends on the color of an object at particular times regardless of its being observed or examined. We will return to this definition in the last section of this commentary. If we quantify the Barker-Achinstein time-relative definition of grue over all times t, then we generate a lifetime conjunctive predicate asserting that grue objects are green before T *and* blue thereafter. The four most common definitions of *grue* are illustrated in table 4.

Four Varieties of Grue

	Observed as green (or blue)	Colored green (or blue)
Grue as a tenseless lifetime predicate	*Goodman's core definition* Either observed before T and green, or otherwise and blue	*Conjunctive definition* Green before T and blue thereafter
Grue$_t$ as a time-relative predicate: grue at time t	*Goodman's implicit definition* Either observed at $t < T$ and green at t, or not observed prior to T and blue at t	*Barker and Achinstein* Either green at $t < T$, or blue at $t \geq T$

Table 4

Goodman's Solution

There have been many attempts to solve Goodman's new riddle of induction. These include Goodman's own theory of entrenchment, which he presented in *Fact, Fiction, and Forecast*. Very roughly, Goodman argues that hypotheses are confirmable by their instances only when they use predicates that are well entrenched, and predicates become entrenched when they have been used to make successful inductive inferences in the past.[46] Earlier in the third chapter (section 2) of *Fact, Fiction, and Forecast* Goodman explains how he thinks the old problem of induction should be dissolved. Just as in deductive logic, so in inductive logic, particular inferences are justified if they conform to accepted rules. If a rule yields an inference we are unwilling to accept, then we amend the rule; if an inference violates a rule we are unwilling to amend, then we reject the inference. Justification arises as a result of the back-and-forth process of mutual adjustment between general rules and particular inferences. In this way Goodman sees the problem of justification for induction as intimately related to the problem of description. To codify the rules governing our accepted inductive practice just is to give all the justification that induction needs or allows.

Goodman's solution to the new riddle has been criticized both with respect to his formal definition of entrenchment[47] and on more general philosophical grounds. Many philosophers remain unpersuaded that the historical and pragmatic character of Goodman's account can do justice to the normative dimension of the new riddle. Elliott Sober speaks for many when he writes:

> A predicate becomes entrenched when people use it to formulate predictions and generalizations. It has always been a mystery to me why the fact that people use a predicate should have any epistemic relevance. Why should our use of a predicate be evidence that this or that hypothesis is true? This naïve question is sometimes answered with the response that the "new" riddle of induction involves describing our inductive practices, not trying to justify them. I have my doubts about this descriptive claim as well. Is it really so obvious that human inference makers think a hypothesis with unentrenched predicates is less plausible than a hypothesis with entrenched predicates, all else being equal? But this reply to one side, I hope it is clear why Goodman's theory is the wrong *kind* of theory, at least if one is interested in normative questions of evidence and confirmation. To describe how well entrenched a predicate *now* is involves describing *past* events only. No such description can suffice to establish an epistemic asymmetry between a GREEN hypothesis and a GRUE hypothesis.[48]

There are also doubts about whether Goodman's entrenchment theory can give an adequate account of the confirmation of scientific hypotheses that introduce new terms into science such as *neutron, gene,* and *quasar.*[49] Goodman had anticipated this objection by insisting that what gets entrenched are not predicates but their extensions, that is, the things to which the predicates apply. He argued that novel terms such as *neutron* can inherit merited entrenchment from other, established predicates, by being related to them in appropriate ways.[50]

Positionality

Some authors, notably Barker and Achinstein,[51] have argued that there is a relevant epistemological asymmetry between the predicates *grue* and *green* because, unlike *green*, *grue* is positional: it makes reference to a particular time. Hence, while we can judge the colors of objects by simple inspection, we cannot decide whether an object is grue without knowing the date of our observation. And this, they claim, is what makes the grue hypothesis illegitimate as a candidate for induction, not the entrenchment of predicates.

A predicate is positional if, in order to find out whether it applies to an object, we have to know its relation to some other particular thing. When the relation is to a particular moment of time, such as the first instant of A.D. 2020, the positional predicate is a temporal one. A predicate is qualitative if it is not positional. Defined in this way, positionality is an epistemological concept because it makes essential reference to the way in which we can come to know something.

Barker and Achinstein argue that Goodman's theory of entrenchment is unnecessary since, they claim, the difference between projectible and unprojectible hypotheses can be explicated in terms of the difference between positional and qualitative predicates. Goodman had already dismissed this approach (which had been advocated by Carnap) in chapter 3 of *Fact, Fiction, and Forecast* (77–81, (453–55)). Goodman's rejection of the positionality theory of projectibility rests on the fact that the definitions of predicate pairs green/blue and grue/bleen are logically symmetric: the colors green and blue can be defined in terms of the "grulors" grue and bleen. An object is bleen if and only if either it is blue and examined before time *T*, or it is green and it is not the case that it is examined before time *T*. Thus, for example, an object is green if and only if either the object is grue and examined before time *T*, or the object is bleen and it is not the case that the object is examined before time *T*. From this, Goodman concludes that color and grulor are epistemologically symmetric. In other words, while color is qualitative and grulor positional for Mr. Green (who speaks regular English), color is positional and grulor qualitative for Mr. Grue (a grue-speaker). Thus, Goodman concludes that "qualitativeness [and positionality] is an entirely relative matter and does not by itself establish any dichotomy of predicates" (80, (454)). His point is that our recognition of predicates as being either positional or qualitative rests on our prior commitment to using predicates such as green/blue rather than grue/bleen. Thus, the positionality theory does not provide an independent alternative to the entrenchment theory, and, without it, it would not work at all.

Barker and Achinstein try to block Goodman's inference from logical symmetry to epistemological symmetry. Through a series of examples involving pictures, they undertake to show that grulor predicates are positional not only for Mr. Green, but also for Mr. Grue. Goodman's brief reply is that all of Barker and Achinstein's examples beg the question by requiring that the grulor grue be represented by two different pigments, thus assuming at the outset that grue is positional. Since all grue things are supposed to look alike grulor-wise to Mr. Grue, he would presumably represent the grulor grue by some device such as cross-hatching. But note that even if we follow this suggestion, we do not seem to get the symmetry Goodman needs because it now appears that grulor would be qualitative for both Mr. Grue and for Mr. Green, since both could judge the grulor of an object on the basis of its visual appearance without needing to know the date.[52]

POPPERIAN AND BAYESIAN SOLUTIONS

Some philosophers deny that generalizations are in fact always confirmed by their instances; Popperians, for example, argue that hypotheses are confirmed (or "corroborated") only when they survive severe tests, that is, tests that are designed to refute them.[53] Bayesians insist that the solution to Goodman's problem lies in paying attention to the degree to which hypotheses are confirmed by their instances. Degree of confirmation depends, in part, on the prior probability of the hypothesis relative to our background knowledge. Our background knowledge gives a very low prior probability to the grue hypothesis as compared with the green hypothesis. Hence, the grue hypothesis receives very little confirmation from the evidence, indeed none at all if its prior probability is zero.[54] (See chapter 5 for a discussion of the Bayesian approach to confirmation theory.)

DOES GOODMAN'S RIDDLE INVOLVE ANYTHING NEW?

In his article "Grue*" Frank Jackson makes two challenging claims that deserve careful consideration. First, he claims that no new riddle of induction arises if one adopts the time-relative definition of grue used by Barker and Achinstein and others, a definition that differs from Goodman's in making no mention of "being observed" or "examined." Second, Jackson claims that when one adopts Goodman's time-relative definition of grue the new riddle can be solved by making explicit a restriction on legitimate uses of the straight rule of inductive inference, a restriction that is usually taken for granted but that serves to block the inference to the grue hypothesis.

The Barker and Achinstein time-relative definition of grue says that x is grue at time t if and only if either x is green at time t, and t is before T, or x is blue at time t, and t is after T. Jackson cautions that we have to be very careful in applying the straight rule (SR) to this predicate. We have a sample of green emeralds at time t_1 (prior to T), and so those emeralds are also grue at time t_1. From this, SR permits us to infer that all emeralds are green at t_1 and also that all emeralds are grue at time t_1. No paradox there. The problem arises when we use SR, illegitimately, to draw the conclusion that all emeralds are green at t_2 and also that all emeralds are grue at time t_2 (where t_2 is later than T). The inference is illegitimate because being green (or grue) at t_2 is a different predicate from being green (or grue) at t_1. Of course, we often do infer inductively from the color of objects (observed now) that they (and relevantly similar but unobserved objects) will be the same color in the future. But in that case, we are treating the color predicate, not as time-relative, but as referring to a tenseless property of the temporal parts of the object considered as extended through time. If we regard grue in the same way, then all essential aspects of the grue paradox are the same as those that arise using Goodman's time-relative definition, since we are inferring from the sampled (or observed) time-slices of a 4-D object to a conclusion about the as-yet-unobserved time-slices of the same and similar objects.

The key to resolving the new riddle, in Jackson's view, is to realize that it is really all about methodological constraints on sampling rather than having anything specifically to do with time. Grue hypotheses are not licensed by the straight rule (SR) because of an implicit restriction that is satisfied in most normal situations but that is necessarily violated by the grue hypothesis. When we infer from a sample of As that are B to the conclusion that all As are B, we presuppose that the As in our sample would have been B even if they had not been sampled. This counterfactual conditional is satisfied in most normal cases since the sampling process does not affect our samples (either causally or logically). But in the case of grue, things are very different. We do not believe that the counterfactual conditional is satisfied because those things that are blue (and thus qualify as grue if observed after T) would not have been observed to be grue at the earlier time when our sample was drawn.

Jackson's resolution of the new riddle depends on accepting counterfactual conditionals as legitimate in a philosophical analysis of inductive confirmability, something that Goodman would reject. Jackson's original proposal has also undergone some refinements in response to counterexamples by Robert Pargetter and others.[55] But the core idea, that regular inductive reasoning from a sample to a population is governed by justifiable restrictions that are common in scientific practice, and that these restrictions are sufficient to rule out the grue hypothesis, remains an attractive solution to the new riddle of induction.

4.8 | Summary

The only thing concerning induction about which all philosophers agree is that inductive arguments are not deductively valid. After that, the controversy begins. Although we could simply define as inductive any argument in which the premises do not logically entail the conclusion, most philosophers follow the custom of identifying as inductive a select subclass of invalid arguments. This subclass can be picked out either by enumerating particular forms of argument (such as statistical syllogism and arguments from analogy) and designating them as inductive or by relying on the normative judgment that inductive arguments are those that confer on their conclusions a reasonably high degree of probability. As Peter Lipton explains, both strategies confront difficulties. The first strategy has to solve the problem of description, the second—the problem of justification.

The problem of description is the problem of articulating the general principles involved when we reason inductively. Although several simple forms of inductive argument have been identified, they are only a small part of the entire field of inductive reasoning, which includes both inferences to new hypotheses (in the context of discovery) and arguments used to support the claim that a hypothesis is probable, well confirmed, or acceptable (in the context of justification). Lipton examines five models of inductive inference and finds each wanting as a description of our inductive practice.

Even if the problem of description could be solved, the problem of justification would remain. For all inductive arguments, from the simplest to the most complex, face the problem posed by David Hume. Hume argued that it is impossible to show that inductive arguments lead from true premises to true conclusions with a reasonably high probability. A crucial step in Hume's reasoning is his insistence that to justify our future confidence in induction by appealing to its past success would be to beg the question: such an attempt at justification would be circular, since it, too, relies on inductive reasoning.

As explained in the commentary on Lipton, the challenge to induction raised by Hume may not be as threatening as it appears at first sight. Even if Hume's skeptical argument is sound, it does not establish that no inductive inference is justified or even that no inductive argument can justify our belief that induction is reliable. At best what Hume may have proved is that we cannot show an inductive skeptic, someone who is unwilling to admit that any nondeductive argument justifies its conclusion, that induction is reliable. When one reflects on the similar difficulty that would confront anyone trying to justify deductive reasoning to a deductive skeptic, it becomes less clear that Hume's problem is a serious challenge to the rationality of induction.

The many attempts to solve the problem of induction fall into four broad categories. Some philosophers claim that induction is rational and justified simply because of what *rational* and *justified* mean in the English language. Others reject this ordinary-language approach and give inductive arguments to support induction, arguing that Hume was wrong to condemn them as circular. John Stuart Mill (in the nineteenth century) and Max Black (in the twentieth) are advocates of this sort of inductive answer to Hume's skepticism. Still others (notably Reichenbach and Salmon) agree with Hume that inductive arguments in favor of induction are viciously circular and concede that inductive arguments cannot be shown to be justified. Nonetheless, these philosophers claim that our practice of using induction can be vindicated pragmatically by showing that it is our best tool for reaching the truth. The overall strategy of the vindication approach is to argue that, while there is no guarantee that induction will succeed, it can be shown (using only deductively valid arguments) that repeated applications of inductive reasoning will eventually take us closer to the truth, if indeed there is a truth of the matter to be reached.

Finally, there are Popper and his followers who think that induction can be neither justified nor vindicated. But they believe that science is rational despite this because it aims, not at the confirmation of hypotheses, but at their falsification. Science progresses by proposing ever bolder hypotheses (which Popper often calls *conjectures*) and then trying to refute them by comparing their deductive consequences with the results of observation and experiment (in the guise of what Popper calls *basic statements*). Theories that have survived severe tests and have not yet been refuted are *corroborated*. *Corroboration* is not the same as *confirmation*. Corroborated theories are neither probable nor confirmed, since, according to Popper, inductive probability and confirmation are myths. Rather, to say that a theory has been corroborated is to give a historical report on the severity of the trials that it has weathered.

Popper's falsificationism has been attacked from many angles. Some critics have charged that, because basic statements are simply accepted by convention in Popper's system, every decision either to reject or not to reject a hypothesis in light of those statements must ultimately be conventional and arbitrary. Others have challenged Popper's claim that following the method of bold conjectures and refutations will lead science in the direction of ever greater truth-content or verisimilitude. Wesley Salmon attacks Popper on the grounds that, without inductive confirmation, Popper (and his defender, John Watkins) cannot solve the problem of rational prediction, that is, the problem of explaining why we should use corroborated theories (as opposed to other as-yet-unfalsified theories) for making inferences about the future. Salmon focuses specifically on those inferences that are part of practical decision making. A key premise in Salmon's criticism of Popper is the assumption that, if corroboration is solely about a theory's past performance, then it cannot serve as a rational guide to the future unless one indulges in inductive reasoning. For, Salmon asks, how else can the past track record of a theory (which is all that judgments of corroboration amount to, according to Popper and Watkins) be relevant to our rational expectations about the future without our making an inference—an inductive inference—from the one to the other? Thus, Salmon concludes that we cannot make sense of science as a rational enterprise if we follow Popper. Scientists must make use of inferences that go beyond deduction, and Popper's falsificationism cannot be the whole story about scientific method.

Like Salmon and most other philosophers of science, Carl Hempel is convinced that Popper has failed to provide a complete and adequate model of scientific reasoning. Although Hempel acknowledges that falsification plays an important role in science, he contends that it is only part of the picture. The rest of the picture is confirmation, and Hempel undertakes a survey of the various criteria that scientists use in deciding when a theory is sufficiently well confirmed to be accepted as probable or true (at least until further notice). Hempel is especially interested in whether a convincing case can be made for these criteria as indicators of truth. He examines, in turn, quantity and variety of evidence, novelty of predictions, support from other theories, and simplicity. His discussion of these factors is inevitably rather sketchy, and he deliberately avoids taking a definite position in the debates concerning them.

Several points emerge from Hempel's survey of criteria of confirmability and acceptability. First, given the diversity of these criteria, it is not surprising that Hempel's discussion of them is rather piecemeal. He neither seeks nor offers a single, unified rationale for them all. In fact, at the end of his article, Hempel expresses serious doubts about whether these factors could be incorporated into a confirmation theory such as Carnap's in which there is a unique, a priori, degree of inductive probability that any piece of evidence confers on any theory. (In chapter 5, we examine the attempt by Bayesians to fit Hempel's criteria into a unified framework.)

Second, Hempel reveals that a genuine puzzle arises from the predictionist claim that novel predictions must always confirm a theory better than explanations of previously known facts, for predictionism entails that a result that strongly confirms a theory in one context (in which the result is new) might not confirm the theory to any appreciable degree in another context (in which the result is already known). The predictionist tries to dispel the air of paradox here by deploying the so-called no-coincidence or no-surprise argument: for a theory correctly to predict a novel result if the theory were not true would be very surprising, but for a theory designed to explain a previously known fact to get it right would be no surprise at all. Hempel briefly rehearses a possible rejoinder to the no-coincidence argument for predictionism that appeals to the notion of simplicity: that a theory yields the result it was designed to explain may not be surprising, but that the theory yielding the result in question is a simple one *is* surprising. This response raises the question, which Hempel addresses later in his article, of whether there is any connection between theoretical simplicity and probability of truth.

Third, the notion of theoretical support from "above" appears to be uncontroversial: whatever confirms a theory also confirms anything that can be derived from that theory, including other theories. That consequences inherit confirmation from any confirmed theory that entails them is presupposed whenever we gather evidence for a theory in order to have rational confidence in its predictions. But as Hempel has argued elsewhere, accepting top-down confirmation (in the form of the special consequence condition) means that we have to place some restrictions on bottom-up confirmation. One simple principle that might be thought to govern bottom-up confirmation is the converse consequence condition, according to which evidence that confirms a given theory also confirms any stronger theory. But the two conditions together entail that any evidence confirms any theory, which is unacceptable. Thus, in his other writings, Hempel has rejected the converse consequence condition and, with it, the naive version of the hypothetico-deductive (H-D) model. Contrary to the naive version of the H-D model, hypotheses are not automatically confirmed by their true deductive consequences. Rather, Hempel has argued, only the consequences of a theory that are instances of that theory have the power to confirm it.

Fourth, Hempel shows how difficult it is to characterize simplicity in precise terms and then to connect it with truth or probable truth. Rejecting the views of Mach and others who regarded simplicity as solely a matter of descriptive economy, Hempel explores in detail the contrasting theories of Reichenbach and Popper. Reichenbach, the inductivist, defends simplicity as a guide to truth by the same form of vindication argument that he offers in defense of induction. In either case, whether following the inductive straight rule (what Lipton calls "More of the Same") or choosing the simplest theory in light of the available evidence, Reichenbach claims that repeated applications of this procedure will allow us to converge on the correct answer in the long run. But, Reichenbach's critics object, convergence is a property shared by many different procedures, not just the inductive straight rule and the simplicity principle. And, besides, without some indication of how rapidly the convergence will occur (or, in other words, how long the "long run" is likely to be) Reichenbach's argument seems to be too weak to vindicate either the inductive straight rule or our preference for simplicity in theories. Popper, the falsificationist, denies that simpler theories are more probable but argues that they are to be preferred because they are more falsifiable and have greater empirical content. As Hempel explains, neither of these claims can withstand close examination.

Both predictionism and explanationism (the view that only the explanation of previously known results can confirm a theory) presuppose the historical thesis of evidence according to which confirmation depends on when evidence becomes known. Achinstein argues that the historical thesis is false and the explanation-versus-prediction distinction irrelevant to questions of evidence. Nonetheless, Achinstein acknowledges that historical information can sometimes affect the strength of evidence, especially when the results being used as evidence for a theory were obtained by performing an experiment. In cases involving drug trials, for example, the evidential value of the results crucially depends on how the experimental and control groups were selected. That one selection procedure rather than another was employed is a historical fact about the experiment. So regardless of whether the experiment was performed before or after the formulation of a theory— and hence regardless of whether the report of the outcome of the experiment is old or new evidence—the report's status as evidence depends on historical facts. Achinstein also gives an example from the history of physics to support his contention that selection procedures, not the explanation-versus-prediction distinction, are what determine whether an experimental outcome is or is not evidence for a theory.

Goodman's new riddle of induction challenges us to explain why we are justified in inductively inferring the green hypothesis rather than the grue hypothesis when all the emeralds that we have sampled to date have been found to be green. Goodman rejects the explanation (offered by Barker and Achinstein and others) that we have a legitimate epistemological preference for qualitative predicates over positional ones. He argues that positionality is not an intrinsic feature of predicates but depends on which other predicates we have already adopted. Green and blue can be defined in terms of grue and bleen, in exactly the same way that grue and bleen are defined in terms of green and blue. Thus, he insists, a native speaker of grue would regard grue as qualititative and green as positional. Goodman's own solution to the new riddle is his theory of entrenchment that regards some predicates as projectible (and others not) depending on their track record of successful use in past inductive inferences. Some philosophers find this naturalistic theory lacking both descriptively (as an account of our actual practice) and normatively (as an account of how we ought to reason). Recently some philosophers (following the lead of Frank Jackson) have argued that the methodological constraints on inductive reasoning from a sample to a population suffice to rule out the grue hypothesis without the need to invoke either the positionality of the grue predicate or its lack of entrenchment.

■ | **Notes**

1. C. D. Broad, "The Philosophy of Francis Bacon," in *Ethics and the History of Philosophy* (New York: Humanities Press, 1952), 117–43.

2. Gilbert Harman, "The Inference to the Best Explanation," *Philosophical Review* 74 (1965): 88–95. Richard Fumerton takes the opposite tack, arguing that arguments to the best explanation are good only insofar as they can be reduced to inductive (and deductive) reasoning. See Richard A. Fumerton, "Induction and Reasoning to the Best Explanation," *Philosophy of Science* 47 (1980): 589–600.

3. Hume's own version of the argument is different from ours, mainly because Hume divided arguments into the demonstrative and the experimental. (Hume himself never used the term *induction*.) Demonstrative arguments, for Hume, are deductively valid and all their premises are necessary truths or what Hume called *relations of ideas*. Experimental arguments have among their premises at least one empirical statement or what Hume called a *matter of fact or existence*. Thus, Hume would classify all inductive arguments as well as some deductively valid arguments as experimental. Since the principle of the uniformity of nature is empirical (a matter of fact or existence), it cannot be the conclusion of any demonstrative argument. (No empirical statement can be validly entailed by a set of premises unless at least one of those premises is also empirical.) So the uniformity principle can only be the conclusion of an experimental argument. Hume then argued that any such argument will beg the question if we attempt to use it to justify our belief in the uniformity principle.

4. See, for example, James Van Cleve, "Reliability, Justification, and the Problem of Induction," in P. French, T. Uehling, Jr., and H. Wettstein, eds., *Midwest Studies in Philosophy*, vol. 9 (Minneapolis: University of Minnesota Press, 1984), 555–67.

508 | Ch. 4 Induction, Prediction, and Evidence

5. The following analysis owes much to Susan Haack, "The Justification of Induction," *Mind* 85 (1976): 112–19; James Van Cleve, "Reliability, Justification, and the Problem of Induction"; and D. H. Mellor, "The Warrant of Induction," *Matters of Metaphysics* (Cambridge: University of Cambridge Press, 1991), 254–68.

6. For good introductory surveys of attempts to solve Hume's problem of justifying induction, see Wesley C. Salmon, *The Foundations of Scientific Inference* (Pittsburgh: University of Pittsburgh Press, 1967); Wesley C. Salmon, "Unfinished Business: The Problem of Induction," *Philosophical Studies* 33 (1978): 1–19; Brian Skyrms, *Choice and Chance: An Introduction to Inductive Logic*, 3d ed. (Belmont, Calif.: Wadsworth, 1986); and Richard G. Swinburne, ed., *The Justification of Induction* (Oxford: Oxford University Press, 1974).

7. Hans Reichenbach introduced the terms *context of discovery* and *context of justification* in the first chapter of his book, *Experience and Prediction* (Chicago: University of Chicago Press, 1938). For Reichenbach, the distinction was logical rather than temporal. Whether before or after a theory is formulated, anything describing psychological events belongs to the context of discovery, anything evaluating the logical relations among propositions belongs to the context of justification. Thus, for Reichenbach, there cannot be a "logic of discovery" in the sense of a "logic of the context of discovery," since the latter context is defined as being purely a psychological matter. Nonethless, Reichenbach thought it possible to give a rational reconstruction of the inferences involved in the discovery of scientific theories. See Martin Curd, "The Logic of Discovery: An Analysis of Three Approaches," in *Scientific Discovery, Logic, and Rationality*, ed. Thomas Nickles (Dordrecht, Netherlands: D. Reidel, 1980), 201–19.

8. See the discussion of Goodman's new riddle of induction later in this commentary.

9. For the raven paradox and an attempted solution, see "The Raven Paradox" in the commentary on chapter 5.

10. Let H be the hypothesis that all crows are black and all swans are purple, let E be the statement that this bird is black, and let O be the observation report that this bird is a crow. Although E does not follow from O all by itself, E is entailed by the conjunction of O with H. Therefore, according to the H-D model, given O, E confirms H. But if a hypothesis is confirmed, so too are the consequences derivable from it. H entails that all swans are purple. Thus, our observation of a black crow confirms the prediction of purple swans. For more on this problem, see "Theoretical Support" in the discussion of Hempel's article later in this commentary.

11. The other three methods are the joint method of agreement and difference, the method of residues, and the method of concomitant variations. (Despite its name, the joint method of agreement and difference has nothing to do with the method of difference.) There is a thorough treatment of Mill's methods and their limitations in Brian Skyrms, *Choice and Chance: An Introduction to Inductive Logic*. The account in this commentary derives from Lilly-Marlene Russow and Martin Curd, *Principles of Reasoning* (New York: St. Martin's Press, 1989), ch. 10.

12. Later in his book, *Inference to the Best Explanation* (from which the reading in this chapter is excerpted), Lipton changes and expands Mill's methods in an attempt to overcome some of these limitations.

13. See, for example, Karl Popper, *Conjectures and Refutations* (New York: Harper and Row, 1963), 47–48, 190–93.

14. For a good introduction to the debate over the possibility of a logic of discovery, see the lengthy introductory essay by Thomas Nickles in *Scientific Discovery, Logic, and Rationality*, 1–59, and Curd, "The Logic of Discovery: An Analysis of Three Approaches."

15. To forestall a possible misunderstanding, it should be recognized that Popper is not using the phrase *basic statement* to mean a statement that is completely justified or certain or one that can be conclusively verified by immediate observation. Rather, for Popper, basic statements are singular existential assertions about particular regions of space and time. Popper gives as an example: "There is a raven in space-time region *k*." Popper calls such statements *basic*, not because they have any special epistemological status, but because they are used to test and falsify theories. See Karl Popper, *The Logic of Scientific Discovery* (New York: Basic Books, 1959), ch. 5.

16. Popper, *The Logic of Scientific Discovery*, 111.

17. These and other searching criticisms of Popper are made with clarity and vigor in W. H. Newton-Smith, *The Rationality of Science* (Boston, Mass.: Routledge and Kegan Paul, 1981), ch. 3. See also Adolf Grünbaum, "Is the Method of Bold Conjectures and Attempted Refutations *Justifiably* the Method of Science?" *British Journal for the Philosophy of Science* 27 (1976): 105–36. Both Newton-Smith and Grünbaum dispute Popper's claim that his falsificationist methodology will lead to theories of greater truth-content or what Popper calls *verisimilitude*.

18. In a more recent publication, Watkins has conceded the force of Salmon's objections. See John Watkins, *Science and Scepticism* (Princeton, N.J.: Princeton University Press, 1984), ch. 9.

19. Elie Zahar, "Why Did Einstein's Programme Supersede Lorentz's? (I) and (II)," *British Journal for the Philosophy of Science* 24 (1973): 103. Alan Musgrave and others have criticized this notion of novelty as being too subjective, since it depends on the intentions of the person who designed the theory. See pp. 12–15 of Alan Musgrave, "Logical versus Historical Theories of Confirmation," *British Journal for the Philosophy of Science* 25 (1974): 1–23.

20. John Worrall, "Scientific Discovery and Theory-Confirmation," in *Change and Progress in Modern Science*, ed. Joseph C. Pitt (Dordrecht, Netherlands: D. Reidel, 1985), 301–31. The distinction between design-novelty and use-novelty is also discussed in Deborah G. Mayo, "Novel Evidence and Severe Tests," *Philosophy of Science* 58 (1991): 523–52.

21. See Carl G. Hempel, "Studies in the Logic of Confirmation," *Mind* 54 (1945): 1–26, 97–121; reprinted in C. G. Hempel, *Aspects of Scientific Explanation* (New York: Macmillan, 1965), 3–46.

22. It also entails the equivalence condition, an essential premise in the raven paradox. See "The Raven Paradox" in the commentary on chapter 5 for details.

23. This argument presupposes what, in the commentary on chapter 5, we call the absolute concept of confirmation. On this view, *E* confirms *H* if and only if the probability of *H* given *E* exceeds some suitably high threshold value, say, 0.9. On the relevance criterion of confirmation favored by Bayesians, the special consequence

condition fails. According to the relevance criterion, *E* confirms *H* if and only if *E* raises the probability of *H*. Consider the following example, given by Richard Jeffrey. Suppose that there is a simple eight-ticket lottery. *E*: the winner is ticket 2 or 3. *H*: the winner is ticket 3 or 4. *G*: the winner is neither ticket 1 nor 2. $P(H) = \frac{1}{4}$, and $P(H/E) = \frac{1}{2}$. Thus, *E* confirms *H*. Also, of course, *H* entails *G*. But $P(G) = \frac{3}{4}$, and $P(G/E) = \frac{1}{2}$. So *E* disconfirms *G*, thus violating the special consequence condition. See Richard C. Jeffrey, "Probability and the Art of Judgment," in *Observation, Experiment, and Hypothesis in Modern Physical Science*, ed. Peter Achinstein and Owen Hannaway: (Cambridge: MIT Press, 1985), 95–126. Elliott Sober gives an even simpler example to show that the relevance criterion is inconsistent with the special consequence condition. The information that the top card in a well-shuffled deck is red confirms the hypothesis that the top card is the jack of hearts (by raising its probability from 1/52 to 1/26); but that same information does not confirm that the top card is a jack (since its probability remains 1/13). See Elliott Sober, "Evolution without Naturalism," in *Oxford Studies in Philosophy of Religion*, vol. 3, ed. J. Kvanvig (Oxford: Oxford University Press, 2011).

24. More precisely, Hempel's satisfaction criterion asserts that a hypothesis *H* is confirmed by an observation report *B* if *H* is satisfied in the finite class of individuals mentioned (essentially) in the report. For further details, see Hempel, "Studies in the Logic of Confirmation."

25. Hans Reichenbach, *Experience and Prediction*, 374.

26. The straight rule is a version of the principle of more of the same. It says that when a certain fraction of As are *B* in a sample, infer that the same fraction of As are *B* in the population from which the sample was drawn.

27. See "Simplicity" in the commentary on chapter 5.

28. For this and other criticisms of Carnap's project, see Ernest Nagel, "Carnap's Theory of Induction," in *The Philosophy of Rudolf Carnap*, ed. P. A. Schilpp (LaSalle, Ill.: Open Court, 1963), 785–826.

29. For an insightful discussion of the historical thesis, to which this section is substantially indebted, see Laura J. Snyder, "Is Evidence Historical?" in Peter Achinstein and Laura J. Snyder, eds., *Scientific Methods: Conceptual and Historical Problems* (Malabar, Fla.: Krieger Publishing Company, 1994), 95–117.

30. See John Worrall, "Fresnel, Poisson and the White Spot: The Role of Successful Predictions in the Acceptance of Scientific Theories," in *The Uses of Experiment*, ed. David Gooding, Trevor Pinch, and Simon Schaffer (Cambridge: Cambridge University Press, 1989), 135–57, and the discussion of Zahar in Alan Musgrave, "Logical versus Historical Theories of Confirmation."

31. Stephen G. Brush, "Prediction and Theory Evaluation: The Case of Light Bending," *Science* 246 (1989): 1124–29. See also Stephen G. Brush, "Dynamics of Theory Change: The Role of Predictions," in *PSA 1994*, vol. 2, ed. D. Hull, M. Forbes, and R. M. Burian (East Lansing, Mich.: Philosophy of Science Association, 1994), 133–45. One case in which Brush thinks that novel predictions did play a crucial role in gaining acceptance for a scientific theory was Mendeleev's prediction of three new elements from the gaps in his periodic table. See Stephen G. Brush; "The Reception of Mendeleev's Periodic Law in America and Britain," *Isis* 87 (1996): 595-628. For a dissenting view, see Eric R. Scerri and John Worrall, "Pre-

diction and the Periodic Table," *Studies in History and Philosophy of Science* 32 (2001): 407-52.

32. Thomas S. Kuhn, *The Structure of Scientific Revolutions*, 2d ed. (Chicago, Ill.: University of Chicago Press, 1970), 155; Ronald N. Giere, "Testing Theoretical Hypotheses," in *Testing Scientific Theories, Minnesota Studies in the Philosophy of Science*, vol. 10, ed. J. Earman (Minneapolis: University of Minnesota Press, 1983), 279–84. Significantly, both accounts rely on the same source for their historical information: E. T. Whittaker, *A History of the Theories of Aether and Electricity*, vol. 1 (London: Thomas Nelson and Sons, 1910), 108.

33. See Worrall, "Fresnel, Poisson, and the White Spot."

34. Owen Gingerich, "Neptune, Velikovsky and the Name of the Game," *Scientific American* 275 (September 1996): 181–83.

35. Laura Snyder (in "Is Evidence Historical?") refers to this principle—that a successful prediction cannot be good ("reliable") evidence for a theory until other theories have also had a chance at explaining it—as Brush's *equal opportunity requirement*.

36. Alan Musgrave defends a similar condition as being the best version of the historical approach to confirmation. Musgrave attributes the core idea to Lakatos. See Musgrave, "Logical versus Historical Theories of Confirmation," 15–19.

37. Deborah G. Mayo, "Novel Evidence and Severe Tests."

38. Just three years later, in 1886, Hertz achieved international recognition for his experimental verification of the existence of radio waves predicted by Maxwell's theory of electromagnetism.

39. As Michael Faraday had discovered, a magnetic field will rotate the plane of polarization of polarized light, but it will not cause a light beam to bend. Moreover, as Achinstein points out in his book, *Particles and Waves*, Hertz also discovered in 1891 that cathode rays could penetrate thin layers of metal. These metal foils were far too thick to allow molecules or atoms to pass through them, but light was known to be able to penetrate thin films of gold. This was another result that Hertz took to be evidence against the particle hypothesis and in favor of the ether-wave hypothesis. See Peter Achinstein, *Particles and Waves* (New York: Oxford University Press, 1991), 283. For a different perspective on Hertz's experiment, see Jed Z. Buchwald, "Why Hertz Was Right about Cathode Rays," in *Scientific Practice: Theories and Stories of Doing Physics*, ed. Jed Z. Buchwald (Chicago: University of Chicago Press, 1995), 151–69.

40. Rom Harré, *Great Scientific Experiments* (Oxford: Phaidon, 1981), 176.

41. For a comprehensive annotated bibliography of writings about grue, see Douglas Stalker, ed., *Grue! The New Riddle of Induction* (La Salle, Ill.: Open Court, 1994). As Stalker notes, the word *gruebleen* was coined by James Joyce on page 23 of *Finnegan's Wake* (London: Faber and Faber, 1939), and Goodman had introduced a grue-like predicate as early as 1946, but the term *grue* itself was first used by Goodman in *Fact, Fiction, and Forecast* (Cambridge, Mass.: Harvard University Press, 1955), 74. Since then, to quote Stalker, "the new riddle of induction has become a well-known topic in contemporary analytic philosophy—so well-known that only a philosophical hermit wouldn't recognize the word grue" (p. 2).

42. We assume here, as is customary, that we can identify emeralds independently of their color and that it is not part of the definition of *emerald* that such gemstones have to be green. See J. J. Thomson's remarks about green beryls in "Grue," *Journal of Philosophy* 63 (1966): 289–309.

43. Even in Goodman's discussion there is minor potential ambiguity. According to the definition of *grue* given in the previous section, if an object is blue and not examined before *T*, then it is grue. This implies that blue objects that are never examined (either before or after *T*) are grue as well as those that are examined after *T* and found to be blue. But in a footnote (on page 74, [p. 452]) Goodman defines an *emerose* as applying "just to emeralds examined before time *t*, and to roses *examined later*" (italics added). This suggests that perhaps we should understand grue as applying to blue objects only if they are examined after *T* and found to be blue. The difference is minor since, on either definition, "all emeralds are grue" implies that all emeralds examined after *T* will be found to be blue.

44. Some authors treat grue as a conjunction, rather than as a disjunction. In symbols: x is grue $=_{df} (Gx \& t < T) \& (Bx \& t \geq T)$. Not only does this misinterpret Goodman's definition, it makes nonsense of his riddle since, in order to have sampled emeralds and observed them to be "grue" according to the conjunctive definition, we would have to have observed them as being green before *T* and blue after *T*. Under those circumstances we would be justified in inferring inductively that all emeralds will undergo a similar color change. It would be like predicting that unobserved bananas will change from green to yellow as they ripen. For details, see Frank Jackson, "Grue*" *Journal of Philosophy* 72 (1975): 113–31; reprinted in Stalker, *Grue!*

45. See, for example, the definition of grue in Stephen F. Barker and Peter Achinstein, "On the New Riddle of Induction," *Philosophical Review* 69 (1960): 511–22. Frank Jackson ("Grue*") attributes the tendency for some authors to slide between this definition of grue and the conjunctive definition to the fact that x will be grue in the conjunctive sense (and hence must change color at *T*) if x satisfies the definition of grue$_t$ at all times *t*.

46. The details of Goodman's theory of entrenchment are complex and subtle. See Nelson Goodman, *Fact, Fiction, and Forecast*, ch. 4.

47. See, for example, Andrzej Zabludowski, "Concerning a Fiction about How Facts Are Forecast," *Journal of Philosophy* 71 (1974): 97–112, reprinted in Stalker, *Grue!* Zabludowski's attack led to a lengthy exchange of papers that was terminated by the editors of the *Journal of Philosophy* in June 1982.

48. Elliott Sober, "A Bayesian Primer on the Grue Problem," in Stalker, *Grue!*, p. 237.

49. See Mary Hesse, "Ramifications of 'Grue'," *British Journal for the Philosophy of Science* 20 (1969): 13–25, and Paul Teller, "Goodman's Theory of Projection," *British Journal for the Philosophy of Science* 20 (1969): 219–38.

50. Bertolet criticizes Goodman's commitment to extensionalism in Rod J. Bertolet, "On the Merits of Entrenchment," *Analysis* 37 (1976): 29–31. Two predicates are extensionally equivalent if they apply to exactly the same set of physical objects, not only in the past and present, but also throughout the future. According to Goodman, a predicate *Q* can inherit merited entrenchment from past projections

of *P* if *Q* has the same extension as *P*. But how do we know that *grue* and *green* are not coextensive?

51. See Stephen F. Barker and Peter Achinstein, "On the New Riddle of Induction," *Philosophical Review* 69 (1960): 511–22, and Goodman's terse reply in "Positionality and Pictures," *Philosophical Review* 69 (1960): 523–25.

52. A different sort of reply to Barker and Achinstein can be found in Edward S. Shirley, "An Unnoticed Flaw in Barker and Achinstein's Solution to Goodman's New Riddle of Induction," *Philosophy of Science* 48 (1981): 611–17. Here the issue is the time at which an object depicted in a drawing or photograph is being observed. This involves some tricky questions about the precise definition of *grue*.

53. Popper denies that generalizations are ever confirmed by their instances because he rejects the whole notion of confirmation and inductive inference. See the section "Why All Theories Are Improbable" in the commentary on chapter 1, the papers by Popper and Salmon in this chapter, and the sections "Popper's Rejection of Induction" and "Salmon's Criticism of Popper's Anti-inductivism" in this commentary. Other authors accept the notion of inductive confirmation but argue that generalizations are not always confirmed by their instances. See, for example, L. Jonathan Cohen, *The Implications of Induction* (London: Methuen, 1970), and Roger D. Rosenkrantz, "Does the Philosophy of Induction Rest on a Mistake?" *Journal of Philosophy* 79 (1982): 78–97.

54. See, for example, Paul Horwich, *Probability and Evidence* (Cambridge: Cambridge University Press, 1982), and Colin Howson and Peter Urbach, *Scientific Reasoning: The Bayesian Approach*, 3d ed. (La Salle, Ill.: Open Court, 2006).

55. See Robert Pargetter, "Confirmation and the Nomological," *Canadian Journal of Philosophy* 10 (1980): 415–28, and Peter Godfrey-Smith, "Goodman's Problem and Scientific Methodology," *Journal of Philosophy* 100 (2003): 573–90.

5 |
Confirmation and Relevance: Bayesian Approaches

INTRODUCTION

Two years after his death in 1761, there appeared in the *Philosophical Trans-actions of the Royal Society* a paper entitled "An Essay Towards Solving a Problem in the Doctrine of Chances" by the Reverend Thomas Bayes: Presbyterian, fellow of the Royal Society, and mathematician.[1] From such inauspicious beginnings has arisen an astonishing variety of doctrines and theories that bear Bayes's name, ranging from work in the foundations of probability theory and statistical inference to models of scientific rationality and theories of confirmation. There are mathematicians, statisticians, and philosophers of every stripe and persuasion—objectivists, subjectivists, personalists, tempered personalists, falsificationists, eliminative inductivists, and hypothetico-deductivists—who call themselves "Bayesians." In this chapter we will examine some of the main Bayesian approaches to understanding the confirmation of scientific theories.

A considerable part of the attraction of the Bayesian approach is that, on the basis of a simple theorem of probability—Bayes's theorem—and a few assumptions about rationality and degrees of belief, Bayesianism promises a unified explanation of a wide range of accepted principles and truisms of scientific methodology. These include the special evidential value we attach to surprising predictions, our preference for simple hypotheses and our aversion to ad hoc ones when rival theories confront the same evidence, our conviction that diverse sets of evidence lend stronger support to theories than do narrow ones, and the fact that not every theory is confirmed equally well by the evidence it entails. Unlike the simple version of the hypothetico-deductive (H-D) method (explored in chapter 4), which regards these and other principles as separate and unconnected, Bayesianism sees them all as flowing from the central underlying logic of confirmation encapsulated in Bayes's equation. In addition, Bayesianism also offers to resolve various paradoxes of confirmation such as the raven paradox and Goodman's new riddle of induction. It is an offer that few modern philosophers of science have been able to refuse.

In "Rationality and Objectivity in Science *or* Tom Kuhn Meets Tom Bayes," Wesley C. Salmon attempts to reconcile Kuhn's historically based philosophy of science with the more traditional approach of the logical empiricists. The key to this reconciliation, Salmon argues, lies in paying proper attention to the role of prior probabilities in determining degrees of confirmation. Salmon candidly admits that some aspects of the application of Bayes's equation to confirmation, such as the calculation of the expectedness of evidence, present intractable problems. In response, Salmon advocates a modified version of Bayesianism that skirts these problems but still provides an algorithm for comparing rival theories when they are judged by the same evidence.

Deborah Mayo criticizes Salmon's "Bayesian Way" as part of her larger criticism of the entire Bayesian approach. As a bridge attempting to link Kuhn's historicism with logical empiricism, Mayo finds Bayesianism unsuited for the task because it would require either acceptance rules or decision-theoretic utilities. These would re-introduce the Kuhnian subjective factors that logical empiricists wish to keep at bay. Salmon's algorithm for choosing between rival theories is faulted because of its reliance on prior probabilities of hypotheses. Mayo argues that Salmon fails in his attempt to interpret these probabilities as the empirical frequency of success among similar hypotheses in the past. In its place Mayo offers a radically different frequency interpretation of the relevant probabilities, namely, one based on the ability of experiments and tests to detect when a hypothesis is false. This "error statistics" approach is defended as the right way to secure objectivity for the evidence on which scientific choices between theories ultimately depend. Alan Chalmers presents a survey of the achievements of Bayesianism in the philosophy of science but, like Mayo, eventually concludes that its subjectivism regarding prior probabilities makes it unacceptable as an account of scientific rationality.

Paul Horwich replies to Mayo, Chalmers, and other critics of Bayesianism in his article, "Therapeutic Bayesianism." He advocates the value of Bayesianism as a philosophical cure for the paradoxes of confirmation and for understanding the relevance of simplicity to inductive support. Horwich replies to critics who attack the foundational assumptions of the Bayesian approach, arguing that a certain sort of criticism is inappropriate—he calls it "misplaced scientism"—in response to a theory aiming to solve philosophical problems rather than to give a complete and accurate description of scientific method.

■ | **Note**

1. For a splendid account of Bayes's essay, see John Earman, *Bayes or Bust?* (Cambridge, Mass.: MIT Press, 1992). Earman describes himself as a "fallen Bayesian."

WESLEY C. SALMON

Rationality and Objectivity in Science or Tom Kuhn Meets Tom Bayes

Twenty-five years ago, as of this writing, Thomas S. Kuhn published *The Structure of Scientific Revolutions*.[1] It has been an extraordinarily influential book. Coming at the height of the hegemony of *logical empiricism*—as espoused by such figures as R. B. Braithwaite, Rudolf Carnap, Herbert Feigl, Carl G. Hempel, and Hans Reichenbach—it posed a severe challenge to the logistic approach that they practiced.[2] It also served as an unparalleled source of inspiration to philosophers with a historical bent. For a quarter of a century there has been a deep division between the logical empiricists and those who adopt the historical approach, and Kuhn's book was undoubtedly a key document in the production and preservation of this gulf.

At a 1983 meeting of the American Philosophical Association (Eastern Division), Kuhn and Hempel—the most distinguished living advocates for their respective viewpoints—shared the platform in a symposium devoted to Hempel's philosophy.[3] I had the honor to participate in this symposium. On that occasion Kuhn chose to address certain issues pertaining to the rationality of science that he and Hempel had been discussing for several years. It struck me that a bridge could be built between the differing views of Kuhn and Hempel if Bayes's theorem were invoked to explicate the concept of scientific confirmation.[4] At the time it seemed to me that this maneuver could remove a large part of the dispute between standard logical empiricism and the historical approach to philosophy of science on this fundamental issue.

I still believe that we have the basis for a new consensus regarding the choice among scientific theories. Although such a consensus, if achieved, would not amount to total agreement on every problem, it would represent a major rapprochement on an extremely fundamental issue. The purpose of

FROM C. Wade Savage, ed., *Scientific Theories*, vol. 14, *Minnesota Studies in the Philosophy of Science* (Minneapolis: University of Minnesota Press, 1990), 175–204.

the present essay is to develop this approach more fully. As it turns out, the project is much more complex than I thought in 1983.

1 | Kuhn on Scientific Rationality

A central part of Kuhn's challenge to the logical empiricist philosophy of science concerns the nature of theory choice in science. The choice between two fundamental theories (or paradigms), he maintains, raises issues that "cannot be resolved by proof." To see how they are resolved we must talk about "techniques of persuasion," or about "argument and counterargument in a situation in which there can be no proof." Such choices involve the exercise of the kind of judgment that cannot be rendered logically explicit and precise. Such statements, along with many others that are similar in spirit, led a number of critics to attribute to Kuhn the view that science is fundamentally irrational and lacking in objectivity.

Kuhn was astonished by this response, which he regarded as a serious misinterpretation. In his "Postscript—1969," in the second edition of *The Structure of Scientific Revolutions*, and in "Objectivity, Value Judgment, and Theory Choice"[5] he replies to these charges. What he had intended to convey was the claim that the decision by the community of trained scientists *constitutes* the best criterion of objectivity and rationality we can have. In order better to understand the nature of such objective and rational methods we need to look in more detail at the considerations that are actually brought to bear by scientists when they endeavor to make comparative evaluations of competing theories.

For purposes of illustration, Kuhn offers a (nonexhaustive) list of characteristics of good scientific theories that are, he claims:

> individually important and collectively sufficiently varied to indicate what is at stake. . . . These five characteristics—accuracy, consistency, scope, simplicity, and fruitfulness—are all standard criteria for evaluating the adequacy of a theory. . . . Together with others of much the same sort, they provide the shared basis for theory choice.[6]

Two sorts of problems arise when one attempts to use them.

> Individually the criteria are imprecise: individuals may legitimately differ about their applicability to concrete cases. In addition, when deployed together, they repeatedly prove to conflict with one another; accuracy may, for example, dictate the choice of one theory, scope the choice of its competitor.[7]

For reasons of these sorts—and others as well—individual scientists may, at a given moment, differ regarding a particular choice of theories. In the course of time, however, the interactions among individual members of the community of scientists produce a consensus for the group. Individual choices inevitably depend upon idiosyncratic and subjective factors; only the outcome of the group activity can be considered objective and fully rational.

One of Kuhn's major claims seems to be that observation and experiment, in conjunction with hypothetico-deductive reasoning, do not adequately account for the choice of scientific theories. This has led some philosophers to believe that theory choice is not rational. Kuhn, in contrast, has tried to locate the additional factors that are involved. These additional factors constitute a crucial aspect of scientific rationality.

2 | Bayes's Theorem

The first step in coming to grips with the problem of evaluating and choosing scientific hypotheses or theories[8] is the recognition of the inadequacy of the traditional hypothetico-deductive (H-D) schema as a characterization of the logic of science. According to this schema, we confirm a scientific hypothesis by deducing from it, in conjunction with suitable initial conditions and auxiliary hypotheses, an observational prediction that turns out to be true. The H-D method has a number of well-known shortcomings. (1) It does not take account of alternative hypotheses that might be invoked to explain the same prediction. (2) It makes no reference to the initial plausibility of the hypothesis being evaluated. (3) It cannot accommodate cases, such as the testing of statistical hypotheses, in which the observed outcome is not deducible from the hypothesis (in conjunction with the pertinent initial conditions and auxiliary hypotheses), but only rendered more or less probable.

In view of these and other considerations, many logical empiricists agreed with Kuhn regarding the inadequacy of hypothetico-deductive confirmation. A number—including Carnap and Reichenbach—appealed to Bayes's theorem, which may be written in the following form:*

$$P(T \mid E.B) = \frac{P(T \mid B)P(E \mid B.T)}{P(T \mid B)P(E \mid B.T) + P(\sim T \mid B)P(E \mid B. \sim T)}. \qquad 1$$

Let "T" stand for the theory or hypothesis being tested, "B" for our background information, and "E" for some new evidence we have just acquired. Then the expression on the left-hand side of the equation represents the probability of our hypothesis on the basis of the background information and the new evidence. This is known as the *posterior probability*. The right-

---

* Salmon uses a dot to stand for "and" and a vertical slash to indicate a conditional probability. Thus, *P(E|B.T)* should be read as "the probability of *E* given *B* and *T*." Other authors write this as *P(E/B&T)*.

hand side of the equation contains four probability expressions. Two of these, P(T | B) and P(~T | B), are called *prior probabilities*; they represent the probability, on the basis of background information alone, without taking account of the new evidence E, that our hypothesis is true or false respectively. Obviously the two prior probabilities must add up to one; if the value of one of them is known, the value of the other can be inferred immediately. The remaining two probabilities, P(E | T.B) and P(E |~T.B), are known as *likelihoods*; they are, respectively, the probability that the new evidence would occur if our hypothesis were true and the probability that it would occur if our hypothesis were false. The two likelihoods, in contrast to the two prior probabilities, must be established independently; the value of one does not automatically determine the value of the other. To calculate the posterior probability of our hypothesis, then, we need three separate probability values to plug into the right-hand side of Bayes's theorem—a prior probability and two likelihoods.

Before attempting to resolve any important issues concerning the nature of scientific reasoning, let us look at a simple and noncontroversial application of Bayes's theorem. Consider a factory that produces can openers at the rate of 6,000 per day. This factory has two machines, a new one that produces 5,000 can openers per day and an old one that produces 1,000 per day. Among the can openers produced by the new machine 1 percent are defective; among those produced by the old machine 3 percent are defective. We pick one can opener at random from today's production and find it defective. What is the probability that it was produced by the new machine?

We can get the answer to this question via Bayes's theorem. If we let "B" stand for the class of can openers produced in this factory today, "T" for the class of can openers produced by the new machine, and "E" for a can opener that is defective, then the probability we seek is the posterior probability P(T | B.E)—the probability that a defective can opener from today's production was produced by the new machine. The values of the prior probabilities and likelihoods have been given, namely:

$$P(T | B) = 5/6 \qquad\qquad P(\sim T | B) = 1/6$$
$$P(E | T.B) = 1/100 \qquad\qquad P(E | \sim T.B) = 3/100.$$

Plugging these values into equation (1) immediately yields P(T | B.E) = 5/8. Notice that the old machine has a greater probability of producing a defective can opener than does the new, but the probability that a defective can opener was produced by the new machine is greater than that it was produced by the old one. This results, obviously, from the fact that the new machine produces so many more can openers overall than does the old one.

One way to look at this example is to consider the hypothesis T that a given can opener was produced by the new machine. This is a causal hypothesis. Our background information B is simply that the can opener is part of today's production at this factory. On the basis of this prior information, we can evaluate the prior probability of T; it is 5/6. Now we add to our knowledge about this can opener the information E that it is defective. This knowledge is relevant to the hypothesis that it was produced by the new machine; the posterior probability is 5/8. Although one does not *need* to appeal to Bayes's theorem to establish this result,[9] the highly artificial example shows clearly just how Bayes's theorem can be used to ascertain the posterior probability of a simple causal hypothesis.

When we come to more realistic scientific cases, it is not so easy to see how to apply Bayes's theorem; the prior probabilities may seem particularly difficult. I believe that, in fact, they reflect the plausibility arguments scientists often bring to bear in their deliberations about scientific hypotheses. I shall discuss this issue in [section] 4; indeed, in subsequent sections we shall have to take a close look at all of the probabilities that enter into Bayes's theorem.

In this section I have been concerned to present Bayes's theorem and to make a few preliminary remarks about its application to the problem of evaluating scientific hypotheses. In the next section I shall try to spell out the connections between Bayes's theorem and Kuhn's views on the nature of theory choice. Before moving on to that discussion, however, I want to present two other useful forms in which Bayes's theorem can be given. In the first place, because of the theorem on total probability:

$$P(E \mid B) = P(T \mid B) \, P(E \mid T.B) + P(\sim T \mid B) \, P(E \mid \sim T.B), \qquad 2$$

equation (1) can obviously be rewritten as:

$$P(T \mid E.B) = \frac{P(T \mid B) P(E \mid B.T)}{P(E \mid B)} . \qquad 3$$

In the second place, equation (1) can be generalized to handle several alternative hypotheses, instead of just one hypothesis and its negation, as follows:

$$P(T_i \mid B.E) = \frac{P(T_i \mid B) \times P(E \mid T_i.B)}{\sum\limits_{j=1}^{k}[P(T_j \mid B) \times P(E \mid T_j.B)]}, \qquad 4$$

where $T_1 - T_k$ are mutually exclusive and exhaustive alternative hypotheses and $1 \leq i \leq k$.

Strictly speaking, (4) is the form that is needed for realistic historical examples—such as the corpuscular (T_1) and wave (T_2) theories of light in the nineteenth century. In that case, although we could construe T_1 and T_2 as mutually exclusive, we could not legitimately consider them exhaustive, for we cannot be sure that one or the other is true. Therefore, we would have to introduce T_3—what Abner Shimony has called the *catchall hypothesis*—which says that T_1 and T_2 are both false. $T_1 - T_3$ thus constitute a mutually exclusive and exhaustive set of hypotheses. This is the sort of situation that obtains when scientists are attempting to choose a correct hypothesis from among two or more serious candidates.

3 | Kuhn and Bayes

For purposes of discussion, Kuhn is willing to admit that "each scientist chooses between competing theories by deploying some Bayesian algorithm which permits him to compute a value for $P(T \mid E)$, i.e., for the probability of the theory T on the evidence E available both to him and the other members of his professional group at a particular period of time."[10] He then formulates the crucial issue in terms of the question of whether there is one unique algorithm used by all rational scientists, yielding a unique value for P, or whether different scientists, though fully rational, may use different algorithms yielding different values of P. I want to suggest a third possibility to account for the phenomena of theory choice—namely, that many different scientists might use the same algorithm, but nevertheless arrive at different values of P.

When one speaks of a Bayesian algorithm, the first thought that comes to mind is Bayes's theorem itself, as embodied in any of the equations (1), (3) or (4). We have, for instance:

$$P(T \mid E.B) = \frac{P(T \mid B) \times P(E \mid B.T)}{P(E \mid B)}, \qquad 3$$

which constitutes an algorithm in the most straightforward sense of the term. Let us call $P(E \mid B)$ the *expectedness* of the evidence. Given values for the prior probability, likelihood, and expectedness, the value of the posterior probability can be computed by trivial arithmetical operations.[11]

If we propose to use equation (3) as an algorithm, the obvious question is how to get values for the expressions on the right-hand side. Several answers are possible in principle, depending on what interpretation of the probability concept is espoused. If one adopts a Carnapian approach to inductive logic and confirmation theory, all of the probabilities that appear in Bayes's theorem can be derived a priori from the structure of the descriptive language and the definition of degree of confirmation.* Since it is extremely

---

* Salmon is referring to the formal approach to inductive logic pioneered by Rudolf Carnap. The Carnapian approach starts with a well-defined artificial language. Every well-formed sentence in that language gets assigned a numerical value for its prior probability via an a priori choice of a measure function over the state descriptions expressible in the language. (A state description is a statement that describes a

difficult to see how any genuine scientific case could be handled by means of the highly restricted apparatus available within that approach, not many philosophers are tempted to follow this line. Moreover, even if a rich descriptive language were available, it is not philosophically tempting to suppose that the probabilities associated with serious scientific theories are a priori semantic truths.

Two major alternatives remain. First, one might maintain that the probabilities on the right-hand side of (3)—especially the prior probability $P(T \mid B)$—are objective and empirical. I have attempted to defend the view that they refer, at bottom, to the frequencies with which various kinds of hypotheses or theories have been found successful.[12] Clearly, enormous difficulties are involved in working out that alternative; I shall return to the issue below. In the meantime, let us consider the other—far more popular—alternative.

The remaining alternative approach involves the use of personal probabilities. Personal probabilities are subjective in character; they represent subjective degrees of conviction on the part of the individual who has them, provided that they fulfill the condition of coherence.[13] Consider a somewhat idealized situation. Suppose that, in the presence of background knowledge B (which may include initial conditions, boundary conditions, and auxiliary hypotheses) theory T deductively entails evidence E. This is the situation to which the hypothetico-deductive method appears to be applicable. In this case, $P(E \mid T.B)$ must equal 1, and equation (3) reduces to:

$$P(T \mid E.B) = P(T \mid B)/P(E \mid B). \qquad\qquad 5$$

One might then ask a particular scientist for his or her plausibility rating of theory T on background knowledge B alone, quite irrespective of whether evidence E obtains or not. Likewise, the same individual may be queried regarding the degree to which evidence E is to be expected irrespective of the truth or falsity of T. According to the personalist, it should be possible—by direct questioning or by some less direct method—to elicit such *psychological* facts regarding a scientist involved in investigations concerning the theory in question. This information is sufficient to determine the degree of belief this individual should have in the theory T given the background knowledge B and the evidence E, namely, the posterior probability $P(T \mid E.B)$.

---

possible state of the world in as much detail as the language permits.) The axioms and definitions of probability theory can then be used to calculate the degree of inductive probability (confirmation or partial entailment) that any one sentence or conjunction of sentences confers on any other. As Salmon notes, it is very difficult to see how Carnap's formal treatment of inductive logic could be applied to real-life scientific theories. For a simple introduction to Carnap's approach, see Rudolf Carnap, "Statistical and Inductive Probability," in *The Structure of Scientific Thought*, ed. E. H. Madden (Boston: Houghton Mifflin, 1960), 269–79, and Henry E. Kyburg Jr., *Probability and Inductive Logic* (London: Collier-Macmillan, 1970), ch. 5.

In the more general case, when T and B do not deductively entail E, the procedure is the same, except that the value of P(E|T.B) must also be ascertained. In many contexts, where statistical significance tests can be applied, a value of the likelihood P(E|T.B) can be calculated, and the personal probability will coincide with the value thus derived. In any case, whether statistical tests apply or not, there is no *new* problem in principle involved in procuring the needed degree of confidence. This reflects the standard Bayesian approach in which all of the probabilities are taken to be personal probabilities.

In any case, whether one adopts an objective or a personalistic interpretation of probability, equation (3)—or some other version of Bayes's theorem—can be taken as an algorithm for evaluating scientific hypotheses or theories. Individual scientists, using the same algorithm, may arrive at different evaluations of the same hypothesis because they plug in different values for the probabilities. If the probabilities are construed as objective, different individuals may well have different estimates of these objective values. If the probabilities are construed as personal, different individuals may well have different subjective assessments of them. Bayes's theorem provides a mechanical algorithm, but the judgments of individual scientists are involved in procuring the values that are to be fed into it. This is a general feature of algorithms; they are not responsible for the data they are given.

4 | Prior Probabilities

In [section] 2 I remarked that the prior probabilities in Bayes's theorem can best be seen as embodying the kinds of plausibility judgments that scientists regularly make regarding the hypotheses with which they are concerned. Einstein, who was clearly aware of this consideration, contrasted two points of view from which a theory can be criticized or evaluated:

> The first point of view is obvious: the theory must not contradict empirical facts. . . . [it] is concerned with the confirmation of the theoretical foundation by the available empirical facts. The second point of view is not concerned with the relation of the material of observation but with the premises of the theory itself, with what may briefly but vaguely be characterized as the "naturalness" or "logical simplicity" of the premises. . . . The second point of view may briefly be characterized as concerning itself with the "inner perfection" of a theory, whereas the first point of view refers to the "external confirmation."[14]

Einstein's second point of view is the sort of thing I have in mind in referring to plausibility arguments or judgments concerning prior probabilities.

Plausibility considerations are pervasive in the sciences; they play a significant—indeed, *indispensable*—role. This fact provides the initial reason for appealing to Bayes's theorem as an aid to understanding the logic of evaluating scientific hypotheses. Plausibility arguments serve to enhance or diminish the probability of a given hypothesis prior to—i.e., without reference to—the outcome of a particular observation or experiment. They are designed to answer the question, "Is this the kind of hypothesis that is likely to succeed in the scientific situation in which the scientist finds himself or herself?" On the basis of their training and experience, scientists are qualified to make such judgments.

This point can best be explained, I believe, in terms of concrete examples. Since before the time of Newton, for instance, a well-known plausibility argument for the inverse square character of gravitational forces has been around. It is natural to think of the gravitational force emanating from a particle of matter as one that spreads spherically from it in a uniform manner. In the seventeenth and eighteenth centuries all competent physical scientists believed that physical space has a three-dimensional Euclidean structure. Since the surface of a Euclidean sphere increases as the square of the radius, it is reasonable to suppose that the force of gravity is diluted in just the same way, for the farther one goes from the particle, the greater the spherical surface over which the force must be spread.

A famous Canadian study of the effects of the consumption of large doses of saccharin provides another example.[15] A statistically significant association between heavy saccharin consumption and bladder cancer in a controlled experiment with rats lends considerable plausibility to the hypothesis that use of saccharin as an artificial sweetener in diet soft drinks increases the risk of bladder cancer in humans. This example, unlike the preceding one, is inherently statistical and does not have even the prima facie appearance of a hypothetico-deductive inference.

In order to come to a clearer understanding of the nature of prior probabilities, it will be necessary to look at them from the point of view of the personalist and that of the objectivist (frequency or propensity theorist).[16] The frightening thing about pure unadulterated personalism is that nothing prevents prior probabilities (and other probabilities as well) from being determined by all sorts of idiosyncratic and objectively irrelevant considerations. A given hypothesis might get an extremely low prior probability because the scientist considering it has a hangover, has had a recent fight with his or her lover, is in passionate disagreement with the politics of the scientist who first advanced the hypothesis, harbors deep prejudices against the ethnic group to which the originator of the hypothesis belongs, etc. What we want to demand is that the investigator make every effort to bring all of his or her *relevant* experience in evaluating hypotheses to bear on the question of whether the hypothesis under consideration is of a type likely to succeed, and to leave aside emotional irrelevancies.

It is rather easy to construct really perverse systems of belief that do not violate the coherence requirement. But we need to keep in mind the objectives of science. When we have a long series of events, such as tosses of fair

or biased coins, or radioactive decays of unstable nuclei, we want our subjective degrees of conviction to match what either a frequency theorist or a
propensity theorist would regard as the objective probability. Carnap was
profoundly correct in his notion that inductive or logical or epistemic probabilities should be reasonable estimates of relative frequencies.

A sensible personalist, I would suggest, is someone who wants his or
her personal probabilities to reflect objective fact. Betting on a sequence of
tosses of a coin, a personalist wants not only to avoid Dutch books,[17] but also
to stand a reasonable chance of winning (or of not losing too much too fast).
As I read it, the whole point of F. P. Ramsey's famous article on degrees of
belief is to consider what you get if your subjective degrees of belief match
the relevant frequencies.[18] One of the facts recognized by the sensible personalist is that whether the coin lands heads or tails is not affected by on
which side of the bed he or she got out that morning. If we grant that the
personalist's aim is to do as well as possible in betting on heads and tails, it
would be obviously counterproductive to allow the betting odds to be
affected by such irrelevancies.

The same general sort of consideration should be brought to bear on
the assignment of probabilities to hypotheses. Whether a particular scientist is dyspeptic on a given morning is irrelevant to the question of whether
a physical hypothesis that is under consideration is correct or not. Much
more troubling, of course, is the fact that any given scientist may be inadvertently influenced by ideological or metaphysical prejudices. It is obvious
that an unconscious commitment to capitalism or racism might seriously
affect theorizing in the behavioral sciences.

Similar situations may arise in the physical sciences as well; another
historical example will illustrate the point. In 1800 Alessandro Volta invented
the battery, thereby providing scientists with a way of producing steady electrical currents. It was not until 1820 that Hans Christian Oersted discovered
the effect of an electrical current on a magnetic needle. Why was there such
a delay? One reason was the previously established fact that a static electric
charge has no effect on a magnetic needle. Another reason that has been
mentioned is the fact that, contrary to the expectation if there were such an
effect, it aligns the needle perpendicular to the current carrying wire. As
Holton and Brush remark, "But even if one has currents and compass nee-
dles available, one does not observe the effect unless the compass is placed
in the right position so that the needle can respond to a force that seems to
act in a direction *around* the current rather than *toward* it."[19] I found it amusing when, on one occasion, a colleague set up the demonstration with the
magnetic needle oriented at right angles to the wire to show why the experiment fails if one begins with the needle in that position. When the current
was turned on, the needle rotated through 180 degrees; he had neglected to
take account of polarity. How many times, between 1800 and 1820, had the
experiment been performed without reversing the polarity? Not many. The

experiment had apparently not been tried by others because of Cartesian metaphysical commitments. It was undertaken by Oersted as a result of his proclivities toward *naturphilosophie.**

How should scientists go about evaluating the prior probabilities of hypotheses? In elaborating a view he calls *tempered personalism*—a view that goes beyond standard Bayesian personalism by placing further constraints on personal probabilities—Shimony[20] points out that experience shows that the hypotheses seriously advanced by serious scientists stand some chance of being successful. Science has, in fact, made considerable progress over the past four or five centuries, which constitutes strong empirical evidence that the probability of success among members of this class is nonvanishing. Likewise, hard experience has also taught us to reject claims of scientific infallibility. Thus, we have good reasons for avoiding the assignment of extreme values to the priors of the hypotheses with which we are seriously concerned. Moreover, Shimony reminds us, experience has taught that science is difficult and frustrating; consequently, we ought to assign fairly low prior probabilities to the hypotheses that have been explicitly advanced, allowing a fairly high prior for the catchall hypothesis—the hypothesis that we have not yet thought of the correct hypothesis. The history of science abounds with situations of choice among theories in which the successful candidate has not even been conceived at the time.

In *The Foundations of Scientific Inference*, I proposed that the problem of prior probabilities be approached in terms of an objective interpretation of probability, in particular, the frequency interpretation. I suggested three sorts of criteria that can be brought to bear in assessing the prior probabilities of hypotheses: formal, material, and pragmatic.

---

* By "Cartesian metaphysical commitments," Salmon is referring to the fact that, prior to the nineteenth century, the three inverse-square law forces that had been discovered—of gravity, electricity, and magnetism—were either attractions or repulsions that acted in a straight line between their sources. To most physicists it seemed inevitable that all forces must have this linear, push-and-pull character. Oersted's discovery, in 1820, that the interaction between a current-carrying wire and a magnet was rotatory and polar in nature, came as a profound shock. To observe the reversal of polarity noted by Salmon, put the wire above the compass needle and at right angles to it, with the current running from east to west. The needle will swivel through 180 degrees, clearly indicating the magnetic polarity of the force that circulates around the wire. Salmon also refers to the role of *Naturphilosophie* in Oersted's discovery. *Naturphilosophie* (philosophy of nature) was an antimechanistic, holistic, and idealist system of thought associated with the German philosopher Friedrich Schelling (1775–1854). Its main features were an insistence on the fundamental unity of nature (in the form of a few basic patterns or archetypes) and the appeal to opposed forces or "polarities" to explain diversity. It motivated Oersted to search for an underlying connection between electricity and magnetism and made him receptive to the possibility of polar forces that do not conform to the mechanistic, "Cartesian" push-pull pattern.

Pragmatic criteria have to do with the circumstances in which a new hypothesis originates. We have already seen an example of a pragmatic criterion in Shimony's observation that hypotheses advocated by serious scientists have nonvanishing chances of success. The opposite side of the same coin is provided by Martin Gardner, who offers an enlightening characterization of scientific cranks.[21] Since it is doubtful that a single useful scientific suggestion has ever been originated by anyone in that category, hypotheses advanced by people of that ilk have negligible chances of being correct. I recall when L. Ron Hubbard's *Dianetics* was first published. A psychologist friend, asked what he thought of it, said, "I can't condemn this book before reading it, but after I have read it, I will." When competent scientists offer hypotheses outside of their areas of specialization, we have a right to wonder whether appreciable plausibility accrues to such suggestions. Hubbard was, incidentally, an engineer with no training in psychology.

The *formal criteria* have to do not only with matters of internal consistency of a new hypothesis, but also with relations of entailment or incompatibility of the new hypothesis with accepted laws and theories. The fact that Immanuel Velikovsky's *Worlds in Collision*[22] contradicts many of the accepted basic laws of physics—e.g., the law of conservation of angular momentum—renders his 'explanations' of such biblically reported incidents as the parting of the waters of the Red Sea and the brief interruption of the rotation of the earth (the sun standing still) utterly implausible.

It should be recalled that among his five considerations for the evaluation of scientific theories—mentioned above—Kuhn includes consistency of the sort we are discussing. I take this as a powerful hint that one of the main issues Kuhn has raised about scientific theory choice involves the use of prior probabilities and plausibility judgments.

The *material criteria* have to do with the actual structure and content of the hypothesis or theory under consideration. The most obvious example is simplicity—another of Kuhn's five items. Simplicity strikes me as singularly important, for it has often been treated by scientists and philosophers as an a priori criterion. It has been suggested, for example, that the hypothesis that quarks are fundamental constitutents of matter loses plausibility as the number of different types of quarks increases, since it becomes less simple as a result.[23] It has also been advocated as a universal methodological maxim: *Search for the simplest possible hypothesis*. Only if the simpler hypotheses do not stand up under testing should one resort to more complex hypotheses.

Although simplicity has obviously been an important consideration in the physical sciences, its applicability in the social/behavioral sciences is problematic. In a recent article, "Slips of the Tongue," Michael T. Motley criticizes Freud's theory for being too simple—an oversimplification.

Further still, the categorical nature of Freud's claim that all slips have hidden meanings makes it rather unattractive. It is difficult to imagine, for example, that my six-year-old daughter's mealtime request to "help cut up my meef" was the result of repressed anxieties or anything of that kind. It seems more likely that she simply merged "meat" and "beef" into "meef." Similarly, about the only meaning one can easily read into someone's saying "roon mock" instead of "moon rock" is that the m and r got switched. Even so, how does it happen that words can merge or sounds can be switched in the course of speech production? And in the ease of my "pleased to beat you" error [to a competitor for a job], might Freud have been right?[24]

The most reasonable way to look at simplicity, I think, is to regard it as a highly relevant characteristic, but one whose applicability varies from one scientific context to another. Specialists in any given branch of science make judgments about the degree of simplicity or complexity that is appropriate to the context at hand, and they do so on the basis of extensive experience in that particular area of scientific investigation. Since there is no precise measure of simplicity as applied to scientific hypotheses and theories, scientists must use their judgment concerning the degree of simplicity a given hypothesis or theory possesses and concerning the degree of simplicity that is desirable in the given context. The kind of judgment to which I refer is not spooky; it is the kind of judgment that arises on the basis of training and experience. This experience is far too rich to be the sort of thing that can be spelled out explicitly. As Patrick Suppes[25] has pointed out, the assignment of prior probability by the Bayesian can be regarded as the best estimate of the chances of success of the hypothesis or theory on the basis of all relevant experience in that particular scientific domain. The personal probability represents, not an effort to contaminate science with subjective irrelevancies, but rather an attempt to facilitate the inclusion of all relevant evidence.

Simplicity is only one among many material criteria. Another closely related criterion—frequently employed in contemporary physics—is symmetry. Perhaps the most striking historical example is de Broglie's hypothesis regarding matter waves. Since light exhibits both particle and wave behavior, which are linked in terms of linear momentum, he suggested, why should not material particles, which obviously possess linear momentum, also have such wave characteristics as wavelength and frequency? Unbeknownst to de Broglie, experimental work by Davisson was, at that very time, providing positive evidence of wavelike behavior of electrons.

A third widely used material criterion is analogy, as illustrated by the saccharin study. The physiological analogy between rats and humans is sufficiently strong to lend considerable plausibility to the hypothesis that saccharin can cause bladder cancer in humans. I suspect that the use of arguments by analogy in science is almost always aimed at establishing prior probabilities. The formal criteria enable us to take account of the ways in which a given hypothesis fits *deductively* with what else we know. Analogy helps us to assess the degree to which a given hypothesis fits *inductively* with what else we know.

The moral I would draw concerning prior probabilities is that they can be understood as our best estimates of the frequencies with which certain kinds of hypotheses succeed. These estimates are rough and inexact; some philosophers might prefer to think of them in terms of intervals. If, however, one wants to construe them as personal probabilities, there is no harm in it, as long as we attribute to the subject who has them the aim of bringing to bear all his or her experience that is relevant to the success or failure of hypotheses similar to that being considered. The personalist and the frequentist need not be in any serious disagreement over the construal of prior probabilities.[26]

One point is apt to be immediately troublesome. If we are to use Bayes's theorem to compute values of posterior probabilities, it would appear that we must be prepared to furnish numerical values for the prior probabilities. Unfortunately, it seems preposterous to suppose that plausibility arguments of the kind we have considered could yield exact numerical values. The usual answer is that, because of a phenomenon known as "washing out of the priors" or "swamping of the priors," even very crude estimates of the prior probabilities will suffice for the kinds of scientific judgments we are concerned to make. Obviously, however, this sort of convergence depends upon agreement regarding the likelihoods.

5 | The Expectedness

The term "$P(E\,|\,B)$" occurring in the denominator of equation (3) is called the *expectedness* because it is the opposite of surprisingness. The smaller the value of $P(E\,|\,B)$, the more surprising E is; the larger the value of $P(E\,|\,B)$, the less surprising, and hence, the more expected E is. Since the expectedness occurs in the denominator, a smaller value tends to increase the value of the fraction. This conforms to a widely held intuition that the more surprising the predictions a theory can make, the greater is their evidential value when they come true.

A classic example of a surprising prediction that came true is the Poisson bright spot. If we ask someone who is completely naive about theories of light how probable it is that a bright spot appears in the center of the shadow of a brightly illuminated circular object (ball or disk), we would certainly anticipate the response that it is very improbable indeed. There is a good inductive basis for this answer. In our everyday lives we have all observed many shadows of opaque objects, and they do not contain bright spots at their centers. Once, when I demonstrated the Poisson bright spot to an introductory class, one student carefully scrutinized the ball bearing that cast the shadow because he strongly suspected that it had a hole through it.

Another striking example, to my mind, is the Cavendish torsion-balance experiment. If we ask someone who is totally ignorant of Newton's theory of universal gravitation how strongly they expect to find a force of attraction between a lead ball and a pith ball in a laboratory, I should think the answer, again, would be that it is very unlikely. There is, in this example as well, a sound inductive basis for the response. We are all familiar with the gravitational attraction of ordinary-size objects to the earth, but we do not have everyday experience of an attraction between two such relatively small (electrically neutral and unmagnetized) objects as those Cavendish used to perform his experiment. Newton's theory predicts, of course, that there will be a gravitational attraction between any two material objects. The trick was to figure out how to measure it.

As the foregoing two examples show, there is a possible basis for assigning a low value to the expectedness; it was made plausible by assuming that the subject was completely naive concerning the relevant physical theory. The trouble with this approach is that a person who wants to use Bayes's theorem—in the form of equation (3), say—cannot be totally innocent of the theory T that is to be evaluated, since the other terms in the equation refer explicitly to T. Consequently, we have to recognize the relationship between $P(E \mid B)$ and the prior probabilities and likelihoods that appear on the right-hand side in the theorem on total probability:

$$P(E \mid B) = P(T \mid B)\, P(E \mid T.B) + P(\sim T \mid B)\, P(E \mid \sim T.B). \qquad 2$$

Suppose that the prior probability of T is not negligible and that T, in conjunction with suitable initial conditions, entails E. Under these circumstances E cannot be totally surprising; the expectedness cannot be vanishingly small. Moreover, to evaluate the expectedness of E we must also consider its probability if T is false. By focusing on the expectedness, we cannot really avoid dealing with likelihoods.

There is a further difficulty. Suppose, for example, that the wave theory of light is true. It is surely *true enough* in the context of the Poisson bright spot experiment. If we want to evaluate $P(E \mid B)$ we must include in B the initial conditions of the experiment—the circular object illuminated by a bright light in such a way that the shadow falls upon a screen. Given the truth of the wave theory, the *objective probability* of the bright spot is one, for whenever those initial conditions are realized, the bright spot appears. It makes no difference whether we know that the wave theory is true, or believe it, or reject it, or have ever thought of it. Under the conditions specified in B the bright spot invariably occurs. Interpreted either as a frequency or a propensity, $P(E \mid B) = 1$. If we are to avoid trivialization in many important cases, the expectedness must be treated as a personal probability. To anyone who, like me, wants to base scientific theory preference or choice on objective considerations, this result poses a serious problem.

The net result is a twofold problem. First, by focusing on the expectedness, we do *not* escape the need to deal explicitly with the likelihoods. In [section] 6 I shall discuss the difficulties that arise when we focus on the likelihoods, especially the problem of the likelihood on the catchall hypothesis. Second, the expectedness defies interpretation as an objective probability. In [section] 7 I shall propose a strategy for avoiding involvement with either the expectedness or the likelihood on the catchall. That maneuver will, I hope, keep open the possibility of an objective basis for the evaluation of scientific hypotheses.

6 | Likelihoods

Equations (1), (3), and (4) are different forms of Bayes's theorem, and each of them contains a likelihood, $P(E \mid T.B)$, in the numerator. Two trivial cases can be noted at the outset. First, if the conjunction of theory T and background knowledge B are logically incompatible with evidence E, the likelihood equals zero, and the posterior probability, $P(T \mid E.B)$, automatically becomes zero.[27] Second, as we have already noticed, if T.B entails E, that likelihood equals one, and consequently drops out, as in equation (5).

Another easy case occurs when the hypothesis T involves various kinds of randomness assumptions, for example, the independence of a series of trials on a chance setup.[28] Consider, for example, the case of a coin that has been tossed 100 times, with the result that heads showed in 63 cases and tails in 37. We assume that the tosses are independent, but we are concerned whether the system consisting of the coin and tossing mechanism is biased. Calculation shows that the probability, given an unbiased coin and tossing mechanism, of the actual frequency of heads differing from ½ by 20 percent or more on 100 tosses (i.e., falling outside of the range 40 to 60) is about .05. Thus, the likelihood of the outcome on the hypothesis that the coin and mechanism are fair is less than .05. On the hypothesis that the coin has a 60 to 40 bias for heads, by contrast, the probability that the number of heads in 100 trials differs from 6/10 by less than 20 percent (i.e., lies within the 48 to 72 range) is well above .95. These are the kinds of likelihoods that would be used to compare the *null hypothesis* that the coin is fair with the hypothesis that it has a certain bias.[29] This example typifies a wide variety of cases, including the above-mentioned controlled experiment on rats and saccharin, in which statistical significance tests are applied. These yield a comparison between the probability of the observed result if the hypothesis is correct and the probability of the same result on a null hypothesis.

In still another kind of situation the likelihood $P(E \mid T.B)$ is straightforward. Consider, for example, the case in which a physician takes an X-ray for diagnostic purposes. Let T be the hypothesis that the patient has a particular disease and let E be a certain appearance on the film. From long medical experience it may be known that E occurs in 90 percent of all cases in which that disease is present. In many cases, as this example suggests, there may be accumulated frequency data from which the value of $P(E \mid T.B)$ can be derived.

Unfortunately, life with likelihoods is not always as simple as the fore-going cases suggest. Consider an important case, which I will present in a highly unhistorical way. In comparing the Copernican and Ptolemaic cosmologies, it is easy to see that the phases of Venus are critical. According to the Copernican system, Venus should exhibit a broad set of phases from a narrow crescent to an almost full disk. According to the Ptolemaic system, Venus should always present nearly the same crescent-shaped appearance. One of Galileo's celebrated telescopic observations was of the phases of Venus. The likelihood of such evidence on the Copernican system is unity; on the Ptolemaic it is zero. This is the decisive sort of case that we cherish.

The Copernican system did, however, face one serious obstacle. On the Ptolemaic system, because the earth does not move, the fixed stars should not appear to change their positions. On the Copernican system, because the earth makes an annual trip around the sun, the fixed stars should appear to change their positions in the course of the year. The very best astronomi-cal observations, including those of Tycho Brahe, failed to reveal any observ-able stellar parallax.[30] However, it was realized that, if the fixed stars are at a very great distance from the earth, stellar parallax, though real, would be too small to be observed. Consequently, the likelihood $P(E \mid T.B)$, where T is the Copernican system and E the absence of observable stellar parallax, is not zero. At the time of the scientific revolution, prior to the advent of New-tonian mechanics, there seemed no reasonable way to evaluate this likeli-hood. The assumption that the fixed stars are almost unimaginably distant from the earth was a highly ad hoc, and consequently implausible, auxiliary hypothesis to adopt just to save the Copernican system. Among other things, Christians did not like the idea that heaven was so very far away.

The most reasonable resolution of this anomaly was offered by Tycho Brahe, whose cosmology placed the earth at rest, with the sun and moon moving in orbits around the earth, but with all of the other planets moving in orbits around the sun. In this way both the observed phases of Venus and the absence of observable stellar parallax could be accommodated. Until Newton's dynamics came upon the scene, it seems to me, Tycho's system was clearly the best available theory.

In [section] 2 I suggested that the following form of Bayes's theorem is the most appropriate for use in actual scientific cases in which more than one hypothesis is available for serious consideration:

$$P(T_i \mid B.E) = \frac{P(T_i \mid B) \times P(E \mid T_i.B)}{\sum\limits_{j=1}^{k} [P(T_j \mid B) \times P(E \mid T_j.B)]}. \qquad 4$$

It certainly fits the foregoing example in which we compared the Ptolemaic, Copernican, and Tychonic systems. This equation involves a mutually exclusive and exhaustive set of hypotheses $T_1, \ldots, T_{k-1}, T_k$, where $T_1 - T_{k-1}$

are seriously entertained and T_k is the catchall. Thus, the scientist who wants to calculate the posterior probability of one particular hypothesis T_i on the basis of evidence E must ascertain likelihoods of three types: (1) the probability of evidence E given T_i, (2) the probability of that evidence on each of the other seriously considered alternatives T_j ($j \neq i$, $j \neq k$), and (3) the probability of that evidence on the catchall T_k.

In considering the foregoing example, I suggested that, although likelihoods in the first two categories are sometimes straightforward, there are cases in which they turn out to be quite problematic. We shall look at more examples in which they present difficulties as our discussion proceeds. But the point to be emphasized right now is the utter intractability of the likelihood on the catchall. The reason for this difficulty is easy to see. Whereas the seriously considered candidates are bona fide hypotheses, the catchall is a hypothesis only in a Pickwickian sense. It refers to all of the hypotheses we are *not* taking seriously, including all those that have not been thought of as yet; indeed, the catchall is logically equivalent to their disjunction. These will often include brilliant discoveries in the future history of science that will eventually solve our most perplexing problems.

Among the hypotheses hidden in the catchall are some that, in conjunction with present available background information, entail the present evidence E. On such as-yet-undiscovered hypotheses the likelihood is one. Obviously, however, the fact that its probability on one particular hypothesis is unity does not entail anything about its probability on some disjunction containing that hypothesis as one of its disjuncts. These considerations suggest to me that the likelihood on the catchall is totally intractable. To try to evaluate the likelihood on the catchall involves, it seems to me, an attempt to guess the future history of science. That is something we cannot do with any reliability.

In any situation in which a small number of theories are competing for ascendency it is tempting, though quite illegitimate, simply to ignore the likelihood on the catchall. In the nineteenth century, for instance, scientists asked what the probability of a given phenomenon is on the wave theory of light and what it is on the corpuscular theory. They did not seriously consider its probability if neither of these theories is correct. Yet we see, from the various forms in which Bayes's theorem is written, that either the expectedness or the likelihood on the catchall is an indispensable ingredient. In the next section I shall offer a *legitimate* way of eliminating those probabilities from our consideration.

7 | Choosing Between Theories

Kuhn has often maintained that in actual science the problem is never to evaluate one particular hypothesis or theory in isolation; it is always a matter of choosing from among two or more viable alternatives. He has emphasized that an old theory is never completely abandoned unless there is currently available a rival to take its place. Given that circumstance, it is a matter of choosing between the old and the new. On this point I think that Kuhn is quite right, especially as regards reasonably mature sciences. And this insight provides a useful clue on how to use Bayes's theorem to explicate the logic of scientific confirmation.

Suppose that we are trying to choose between T_1 and T_2, where there may or may not be other serious alternatives in addition to the catchall. By letting $i = 1$ and $i = 2$, we can proceed to write equation (4) for each of these candidates. Noting that the denominators of the two are identical, we can form their ratio as follows:

$$\frac{P(T_1 \mid E.B)}{P(T_2 \mid E.B)} = \frac{P(T_1 \mid B) \times P(E \mid T_1.B)}{P(T_2 \mid B) \times P(E \mid T_2.B)}. \qquad 6$$

No reference to the catchall hypothesis appears in this equation. Since the catchall is not a bona fide hypothesis, it is not a contender, and we need not try to calculate its posterior probability. The use of equation (6) frees us from the need to deal either with the expectedness of E or with its probability on the catchall.

Equation (6) yields a relation that can be regarded as a *Bayesian algorithm for theory preference*. Suppose that, prior to the emergence of evidence E, you prefer T_1 to T_2; that is, $P(T_1 \mid B) > P(T_2 \mid B)$. Then E becomes available. You should change your preference in the light of E if and only if $P(T_2 \mid E.B) > P(T_1 \mid E.B)$. From (6) it follows that

$$P(T_2 \mid E.B) > P(T_1 \mid E.B) \text{ iff } P(E \mid T_2.B)/P(E \mid T_1.B) > P(T_1 \mid B)/P(T_2 \mid B). \qquad 7$$

In other words, you should change your preference to T_2 if the ratio of the likelihoods is greater than the reciprocal of the ratio of the respective prior probabilities. A corollary is that, if both $T_1.B$ and $T_2.B$ entail E, so that:

$$P(E \mid T_1.B) = P(E \mid T_2.B) = 1,$$

the occurrence of E can never change the preference rating between the two competing theories.

At the end of [section] 4 I made reference to the well-known phenomenon of washing out of priors in connection with the use of Bayes's theorem. One might well ask what happens to this swamping when we switch from Bayes's theorem to the ratio embodied in equation (6).[31] The best answer, I believe, is this. If we are dealing with two hypotheses that are serious contenders in the sense that they do not differ too greatly in plausibility, the ratio of the priors will be of the order of unity. If, as the observational evidence accumulates, the likelihoods come to differ greatly, the ratio of the likelihoods will swamp the ratio of the priors. Recall the example of the tossed coin. Suppose we consider the prior probability of a fair device to be ten times as large as that of a biased device. If about the same proportion of heads occurs in 500 tosses as occurred in the aforementioned 100, the likelihood on the null hypothesis would be virtually zero and the likelihood on the hypothesis that the device has a bias approximating the observed frequency would be essentially indistinguishable from unity. The ratio of prior probabilities would obviously be completely dominated by the likelihood ratio.

8 | Plausible Scenarios

Although, by appealing to equation (6), we have eliminated the need to deal with the expectedness or the likelihood on the catchall, we cannot claim to have dealt adequately with the likelihoods on the hypotheses we are seriously considering, for their values are not always straightforwardly ascertainable. We have already mentioned one example, namely, the probability of absence of observable stellar parallax on the Copernican hypothesis. We noted that, by adding an auxiliary hypothesis to the effect that the fixed stars are located an enormous distance from the Earth, we could augment the Copernican hypothesis in such a way that the likelihood on this augmented hypothesis is one. But, for many reasons, this auxiliary assumption could hardly be considered plausible in that historical context. By now, of course, we have measured the parallax of relatively nearby stars, and from those values have calculated these distances. They are extremely far from us in comparison to the familiar objects in our solar system.

Consider another well-known example. During the seventeenth and eighteenth centuries the wave and corpuscular theories of light received considerable scientific attention. Each was able to explain certain important optical phenomena, and each faced fundamental difficulties. The corpuscular hypothesis easily explained how light could travel vast distances through empty space, and it readily explained sharp shadows. The theory of light as a longitudinal wave explained various kinds of diffraction phenomena, but failed to deal adequately with polarization. When, early in the nineteenth century, light was conceived as a transverse wave, the wave theory explained polarization as well as diffraction quite straightforwardly. And Huygens had long since shown how the wave theory could handle rectilinear propagation and sharp shadows. For most of the nineteenth century the wave theory dominated optics.

The proponent of the particle theory could still raise a serious objection. What is the likelihood of a wave propagating in empty space? Lacking a medium, the answer is zero. So wave theorists augmented their theory with the auxiliary assumption that all of space is filled with a peculiar substance known as the *luminiferous ether*. This substance was postulated to have precisely the properties required to transmit light waves.

The process I have been describing can appropriately be regarded as the discovery and introduction of *plausible scenarios*. A theory is confronted with an *anomaly*—a phenomenon that appears to have a small, possibly zero, likelihood given that theory. Proponents of the theory search for some auxiliary hypothesis that, if conjoined to the theory, renders the likelihood high, possibly unity. This move shifts the burden of the argument to the plausibility of the new auxiliary hypothesis. I mentioned two instances involved in the wave theory of light. The first was the auxiliary assumption that the wave is transverse. This modification of the theory was sufficiently plausible to be incorporated as an integral part of the theory. The second was the luminiferous ether. The plausibility of this auxiliary hypothesis was debated throughout the nineteenth, and into the twentieth, century. The ether had to be dense enough to transmit transverse waves (which require a denser medium than do longitudinal waves) and thin enough to allow astronomical bodies to move through it without noticeable diminution of speed. Attempts to detect the motion of the earth relative to the ether were unsuccessful. The Lorentz-Fitzgerald contraction hypothesis was an attempt to save the ether theory—that is, another attempt at a plausible scenario—but it was, of course, abandoned in favor of special relativity.

I am calling these auxiliaries *scenarios* because they are stories about how something could have happened, and *plausible* because they must have some degree of acceptability if they are to be of any help in handling problematic phenomena. The wave theory could handle the Poisson bright spot by deducing it from the theory. There seemed to be no plausible scenario available to the particle theory that could deal with this phenomenon. The same has been said with respect to Foucault's demonstration that the velocity of light is greater in air than it is in water.[32]

One nineteenth-century optician of considerable importance who did not adopt the wave theory, but remained committed to the Newtonian emission theory, was David Brewster.[33] In a "Report on the Present State of Physical Optics," presented to the British Association for the Advancement of Science in 1831, he maintained that the undulatory theory is "still burthened with difficulties and cannot claim our implicit assent."[34] Brewster freely admitted the unparalleled explanatory and predictive success of the wave theory; nevertheless, he considered it false.

Among the difficulties Brewster found with the wave theory, two might be mentioned. First, he considered the wave theory implausible, for the reason that it required "an *ether* invisible, intangible, imponderable, inseparable from all bodies, and extending from our own eye to the remotest verge of the starry heavens."[35] History has certainly vindicated him on that issue. Second, he found the wave theory incapable of explaining a phenomenon that he had discovered himself, namely, *selective absorption*—dark lines in the spectrum of sunlight that has passed through certain gases. Brewster points out that a gas may be opaque to light of one particular index of refraction in flint glass, while transmitting freely light whose refractive indices in the same glass are only the tiniest bit higher or lower. Brewster maintained that there was no plausible scenario the wave theorists could devise that would explain why the ether permeating the gas transmits two waves of very nearly the same wavelength, but does not transmit light of a very precise wavelength lying in between:

> There is no fact analogous to this in the phenomena of sound, and I can form no conception of a simple elastic medium so modified by the particles of the body which contains it, as to make such an extraordinary selection of the undulations which it stops or transmits. . . .[36]

Brewster never found a plausible scenario by means of which the Newtonian theory he favored could cope with absorption lines, nor could proponents of the wave theory find one to bolster their viewpoint. Dark absorption lines remained anomalous for both the wave and particle theories; neither could see a way to furnish them with high likelihood.

With hindsight we can say that the catchall hypothesis was looking very strong at this point. We recognize that the dark absorption lines in the spectrum of sunlight are closely related to the discrete lines in the emission spectra of gases, and that they, in turn, are intimately bound up with the problem of the stability of atoms. These phenomena played a major role in the overthrow of classical physics at the turn of the twentieth century.

I have introduced the notion of a plausible scenario to deal with problematic likelihoods. Likelihoods can cause trouble for a scientific theory for either of two reasons. First, if you have a pet theory that confers an extremely small—for all practical purposes zero—likelihood on some observed phenomenon, that is a problem for that favored theory. You try to come up with a plausible scenario according to which the likelihood will be larger—ideally, unity. Second, if there seems to be no way to evaluate the likelihood of a piece of evidence with respect to some hypothesis of interest, that is another sort of problem. In this case, we search for a plausible scenario that will make the likelihood manageable, whether this involves assigning it a high, medium, or low value.

What does this mean in terms of the Bayesian approach I am advocating? Let us return to:

$$\frac{P(T_1 \mid E.B)}{P(T_2 \mid E.B)} = \frac{P(T_1 \mid B) \times P(E \mid T_1.B)}{P(T_2 \mid B) \times P(E \mid T_2.B)}, \qquad 6$$

which contains two likelihoods. Suppose, as in nineteenth-century optics, that both likelihoods are problematic. As we have seen, we search for plausible scenarios A_1 and A_2 to augment T_1 and T_2 respectively. If the search has been successful, we can assess the likelihoods of E with respect to the augmented theories $A_1.T_1$ and $A_2.T_2$. Consequently, we can modify (6) so as to yield

$$\frac{P(A_1.T_1 \mid E.B)}{P(A_2.T_2 \mid E.B)} = \frac{P(A_1.T_1 \mid B) \times P(E \mid A_1.T_1.B)}{P(A_2.T_2 \mid B) \times P(E \mid A_2.T_2.B)}. \qquad 8$$

In order to use this equation to compare the posterior probabilities of the two augmented theories, we must assess the plausibilities of the scenarios, for the prior probabilities of both augmented theories—$A_1.T_1$ and $A_2.T_2$—appear in it. In [section] 4 I tried to explain how prior probabilities can be handled—that is, how we can obtain at least rough estimates of their values. If, as suggested, the plausible scenarios have made the likelihoods ascertainable, then we can use them in conjunction with our determinations of the prior probabilities to assess the ratio of the posterior probabilities. We have, thereby, handled the central issue raised by Kuhn, namely, what is the basis for preference between two theories.[37] Equation (8) is a Bayesian algorithm.

If either augmented theory, in conjunction with background knowledge B, entails E, then the corresponding likelihood is one and it drops out of (8). If both likelihoods drop out we have the special case in which:

$$\frac{P(A_1.T_1 \mid E.B)}{P(A_2.T_2 \mid E.B)} = \frac{P(A_1.T_1 \mid B)}{P(A_2.T_2 \mid B)}, \qquad 9$$

thereby placing the *whole* burden on the prior probabilities—the plausibility considerations. Equation (9) represents a simplified Bayesian algorithm that is applicable in this type of special case.

Another type of special case was mentioned above. If, as in our coin tossing example, the values of the prior probabilities do not differ drastically from one another, but the likelihoods become widely divergent as the observational evidence accumulates, there will be a washing out of the priors. In this case, the ratio of the posterior probabilities equals, for practical purposes, the ratio of the likelihoods.

The use of either (8) or (9) as an algorithm for theory choice does not imply that all scientists will agree on the numerical values or prefer the same theory. The evaluation of prior probabilities clearly demands the kind of scientific judgment whose importance Kuhn has rightly insisted upon. It should also be clearly remembered that these formulas provide no evaluations of individual theories; they furnish only comparative evaluations. Thus, instead of yielding a prediction regarding the chances of one particular theory being a component of "completed science," they compare existing theories with regard to their present merits.

9 | Kuhn's Criteria

Early in this paper I quoted five criteria that Kuhn mentioned in connection with his views on the rationality and objectivity of science. The time has come to relate them explicitly to the Bayesian approach I have been attempting to elaborate. In order to appreciate the significance of these criteria it is important to distinguish three aspects of scientific theories that may be called *informational virtues, confirmational virtues,* and *economic virtues.* Up to this point we have concerned ourselves almost exclusively with confirmation, for our use of Bayes's theorem is germane only to the confirmational virtues. But since Kuhn's criteria patently refer to the other virtues as well, we must also say a little about them.

Consider, for example, the matter of *scope.* Newton's three laws of motion and his law of universal gravitation obviously have greater scope than the conjunction of Galileo's law of falling bodies and Kepler's three laws of planetary motion. This means, simply, that Newtonian mechanics contains more information than the laws of Kepler and Galileo taken together. Given a situation of this sort, we prefer the more informative theory because it is a basic goal of science to increase our knowledge as much as possible. We might, of course, hesitate to choose a highly informative theory if the evidence for it were extremely limited or shaky, because the desire to be right might overrule the desire to have more information content. But in the case at hand that consideration does not arise.

In spite of its intuitive attraction, however, the appeal to scope is not altogether unproblematic. There are two ways in which we might construe the Galileo-Kepler-Newton example of the preceding paragraph. First, we might ignore the small corrections mandated by Newton's theory in the laws of Galileo and Kepler. In that case we can clearly claim greater scope for Newton's laws than for the conjunction of Galileo's and Kepler's laws, since the latter is entailed by the former but not conversely. Where an entailment relation holds we can make good sense of comparative scope.

Kuhn, however, along with most of the historically oriented philosophers, has been at pains to deny that science progresses by finding more general theories that include earlier theories as special cases. Theory choice or preference involves *competing* theories that are *mutually incompatible* or *mutually incommensurable.* To the best of my knowledge Kuhn has not offered any precise characterization of scope; Karl Popper, in contrast, has made serious attempts to do so. In response to Popper's efforts, Adolf Grünbaum has effectively argued that none of the Popperian measures can be usefully applied to make comparisons of scope among mutually incompatible competing theories.[38] Consequently, the concept of scope requires fundamental clarification if we are to use it to understand preferences among competing theories. However, since scope refers to information rather than confirmation, it plays no role in the Bayesian program I have been endeavoring to explicate. We can thus put aside the problem of explicating that difficult concept.

Another of Kuhn's criteria is *accuracy.* It can, I think, be construed in two different ways. The first has to do with informational virtues; the second with economic. On the one hand, two theories might both make true predictions regarding the same phenomena, but one of them might give us precise predictions where the other gives only predictions that are less exact. If, for example, one theory enables us to predict that there will be a solar eclipse on a given day, and that its path of totality will cross North America, it may well be furnishing correct information about the eclipse. If another theory gives not only the day, but also the time, and not only the continent, but also the precise boundaries, the second provides much more information, at least with respect to this particular occurrence. It is not that either is incorrect; rather, the second yields more knowledge than the first. However, it should be clearly noted—as it was in the case of scope—that these theories are not incompatible or incommensurable competitors (at least with respect to this eclipse), and hence do not illustrate the interesting type of theory preference with which Kuhn is primarily concerned.

On the other hand, one theory may yield predictions that are nearly, but not quite, correct, while another theory yields predictions that are entirely correct—or, at least, more nearly correct. Newtonian astrophysics does well in ascertaining the orbit of the earth, but general relativity introduces a correction of 3.8 seconds of arc per century in the precession of its perihelion.[39] Although the Newtonian theory is literally false, it is used in contexts of this sort because its inaccuracy is small, and the economic gain involved in using it instead of general relativity (the saving in computational effort) is enormous.

The remaining three criteria are *simplicity, consistency*, and *fruitfulness*; all of them have direct bearing upon the confirmational virtues. In the treatment of prior probabilities in [section] 4, I briefly mentioned simplicity as a factor having a significant bearing upon the plausibility of theories. More examples could be added, but I think the point is clear.

In the same section I also made passing reference to consistency, but more can profitably be said on that topic. Consistency has two aspects, internal consistency of a theory and its compatibility with other accepted theories. While scientists may be fully justified in *entertaining* collections of statements that contain contradictions, the goal of science is surely to accept only logically consistent theories.[40] The discovery of an internal inconsistency has a distinctly adverse effect on the prior probability of that theory, to wit, it must go straight to zero.

When we consider the relationships of a given theory to other accepted theories we again find two aspects. There are *deductive* relations of entailment and incompatibility, and there are *inductive* relations of fittingness and incongruity. The deductive relations are quite straightforward. Incompatibility with an accepted theory makes for implausibility; being a logical consequence of an accepted theory makes for a high prior probability. Although deductive subsumption of narrower theories under broader theories is probably something of an oversimplification of actual cases, nevertheless, the ability of an overarching theory to deductively unify diverse domains furnishes a strong plausibility argument.

When it comes to the inductive relations among theories, analogy is, I think, the chief consideration. I have already mentioned the use of analogy in inductively transferring results of experiments from rats to humans. In archaeology, the method of ethnographic analogy, which exploits similarities between extant primitive societies and prehistoric societies, is widely used. In physics, the analogy between the inverse square law of electrostatics and the inverse square law of gravitation provides an example of an important plausibility consideration.

Kuhn's criteria of consistency (broadly construed) and simplicity seem clearly to pertain to assessments of the prior probabilities of theories. They cry out for a Bayesian interpretation.

The final criterion in Kuhn's list is *fruitfulness*; it has many aspects. Some theories prove fruitful by unifying a great many apparently different phenomena in terms of a few simple principles. The Newtonian synthesis is, perhaps, the outstanding example; Maxwellian electrodynamics is also an excellent case. As I suggested above, this ability to accommodate a wide variety of facts tends to enhance the prior probability of a given theory. To attribute diverse success to happenstance, rather than basic correctness, is implausible.

Another sort of fertility involves the predictability of theretofore unknown phenomena. We might mention as familiar illustrations the prediction of the Poisson bright spot by the wave theory of light and the prediction of time dilation by special relativity. These are the kinds of instances in which, in an important sense, the expectedness is low. As we have noted, a small expectedness tends to increase the posterior probability of a hypothesis.

A further type of fertility relates directly to plausible scenarios; a theory is fruitful in this way if it successfully copes with difficulties with the aid of suitable auxiliary assumptions. Newtonian mechanics again provides an excellent example. The perturbations of Uranus were explained by postulating Neptune. The perturbations of Neptune were explained by postulating Pluto.[41] The motions of stars within galaxies and of galaxies within clusters are explained in terms of *dark matter*, concerning which there are many current theories. A theory that readily gives rise to plausible scenarios to deal with problematic likelihoods can boast this sort of fertility.

The discussion of Kuhn's criteria in this section is intended to show how adequately they can be understood within a Bayesian framework—insofar as they are germane to confirmation. If it is sound, we have constructed a fairly substantial bridge connecting Kuhn's views on theory choice with those of the logical empiricists—at least, those who find in Bayes's theorem a suitable schema for characterizing the confirmation of hypotheses and theories.

10 | Rationality vs. Objectivity

In the title of this essay I have used both the concept of *rationality* and that of *objectivity*. It is time to say something about their relationship. Perhaps the best way to approach the distinction between them is to enumerate various grades of rationality. In a certain sense one can be rational without paying any heed at all to objectivity. It is essentially a matter of good house-keeping as far as one's beliefs and degrees of confidence are concerned. As Bayesians have often emphasized, it is important to avoid logical contradictions in one's beliefs and to avoid probabilistic incoherence in one's degrees of conviction. If contradiction or incoherence are discovered, they must somehow be eliminated; the presence of either constitutes a form of irrationality. But the removal of such elements of irrationality can be accomplished without any appeal to facts outside of the subject's corpus of beliefs and degrees of confidence. To achieve this sort of rationality is to achieve a minimal standard that I have elsewhere called *static* rationality.[42]

One way in which additional facts may enter the picture is via Bayes's theorem. We have a theory T in which we have a particular degree of confidence. A new piece of evidence turns up—some objective fact E of which we were previously unaware—and we use Bayes's theorem to calculate a posterior probability of T. To accept this value of the posterior probability as one's degree of confidence in T is known as *Bayesian conditionalization*. Use of Bayes's theorem does not, however, guarantee objectivity. If the resulting posterior probability of T is one we are not willing to accept, we can make adjustments elsewhere to avoid incoherence. After all, the prior probabilities and likelihoods are simply personal probabilities, so they can be adjusted to achieve the desired result. If, however, the requirement of *Bayesian conditionalization* is added to those of static rationality we have a stronger type of rationality that I have called *kinematic*.[43]

The highest grade of rationality—what I have called *dynamic rationality*—requires much fuller reference to objective fact than is demanded by advocates of personalism. The most obvious way to inject a substantial degree of objectivity into our deliberations regarding choices of scientific theories is to provide an objective interpretation of the probabilities in Bayes's theorem. Throughout this discussion I have adopted that approach as thoroughly as possible. For instance, I have argued that prior probabilities can be given an objective interpretation in terms of frequencies of success. I have tried to show how likelihoods could be objective—by virtue of entailment relations, tests of statistical significance, or observed frequencies. When the likelihoods created major difficulties, I appealed to plausible scenarios. The result was that an intractable likelihood could be exchanged for a tractable prior probability—namely, the prior probability of a theory in conjunction with an auxiliary assumption.

We noted that the denominators of the right-hand sides of the various versions of Bayes's theorem—equations (1), (3), and (4)—contain either an expectedness or a likelihood on the catchall. It seems to me futile to try to construe either of these probabilities objectively. Consequently, in [section] 7 I introduced equation (6), which involves a ratio of two instances of Bayes's theorem, and from which the expectedness and the likelihood on the catchall drop out. Confining our attention, as Kuhn recommends, to comparing the merits of competing theories, rather than offering absolute evaluations of individual theories, we were able to eliminate the probabilities that most seriously defy objective interpretation.

11 | Conclusions

For many years I have been convinced that plausibility arguments in science have constituted a major stumbling block to an understanding of the logic of scientific inference. Kuhn was not alone, I believe, in recognizing that considerations of plausibility constitute an essential aspect of scientific reasoning, without seeing where they fit into the logic of science. If one sees confirmation solely in terms of the crude hypothetico-deductive method, there is no place for them. There is, consequently, an obvious incentive for relegating plausibility considerations to heuristics. If one accepts the traditional distinction between the *context of discovery* and the *context of justification*, it is tempting to place them in the former context. But Kuhn recognized, I think, that plausibility arguments enter into the justifications of choices of theories, with the result that he became skeptical of the value of that distinction. If, as I believe, plausibility considerations are simply evaluations of prior probabilities of hypotheses or theories, then it becomes apparent via Bayes's theorem that they play an indispensable role in the context of justification. We do not need to give up that important distinction.

At several places in this paper I have spoken of Bayesian algorithms, mainly because Kuhn introduced that notion into the discussion. I have claimed that such algorithms exist—and attempted to exhibit them—but I accord *very little* significance to that claim. The algorithms are trivial; what is important is the scientific judgment involved in assessing the probabilities that are fed into the equations. The algorithms give frameworks in terms of which to understand the role of the sort of judgment upon which Kuhn rightly placed great emphasis.

The history of science chronicles the successes and failures of attempts at scientific theorizing. If the Bayesian analysis I have been offering is at all sound, history of science—in addition to contemporary scientific experience, of course—provides a rich source of information relevant to the prior probabilities of the theories among which we are at present concerned to make objective and rational choices. This viewpoint captures, I believe, the point Kuhn made at the beginning of his first book:

> But an age as dominated by science as our own does need a perspective from which to examine the scientific beliefs which it takes so much for granted, and history provides one important source of such perspective. If we can discover the origins of some modern scientific concepts and the way in which they supplanted the concepts of an earlier age, we are more likely to evaluate intelligently their chances for survival.[44]

I suggested at the outset that an appeal to Bayesian principles could provide some aid in bridging the gap between Hempel's logical-empiricist approach and Kuhn's historical approach. I hope I have offered a convincing case. However that may be, there remain many unresolved issues. For instance, I have not even broached the problem of incommensurability of paradigms or theories. This is a major issue. For another example, I have assumed uncritically throughout the discussion that the various parties to disputes about theories share a common body B of background knowledge. It is by no means obvious that this is a tenable assumption. No doubt other points for controversy remain. I do not for a moment maintain that complete consensus would be in the offing even if both camps were to buy the Bayesian line I have been peddling. But I do hope that some areas of misunderstanding have been clarified.[45]

■ | Notes

1. Thomas S. Kuhn, *The Structure of Scientific Revolutions* (Chicago: University of Chicago Press, 1962; 2d ed., 1970). "Postscript—1969," added to the second edition, contains discussions of some of the major topics that are treated in the present essay.

2. Such philosophers are often characterized by their opponents as *logical positivists*, but this is an egregious historical inaccuracy. Although some of them had been members of or closely associated with the Vienna Circle in their earlier years, none of them retained the early positivistic commitment to phenomenalism and/or instrumentalism in their more mature writings. Reichenbach and Feigl, for example, were outspoken realists, and Carnap regarded physicalism as a tenable philosophical framework. Reichenbach never associated himself with positivism; indeed, he regarded his 1938 book, *Experience and Prediction* (Chicago: University of Chicago Press), as a refutation of logical positivism. I could go on and on. . . .

3. "Symposium: The Philosophy of Carl G. Hempel," *Journal of Philosophy* LXXX, 10 (Oct., 1983): 555–72.

4. I had offered a similar suggestion in "Bayes's Theorem and the History of Science," in *Minnesota Studies in the Philosophy of Science*, vol. 5, *Historical and Philosophical Perspectives of Science*, ed. Roger H. Stuewer (Minneapolis: University of Minnesota Press, 1970), 68–86.

5. Thomas S. Kuhn, *The Essential Tension* (Chicago: University of Chicago Press, 1977), 320–39. The response is given in greater detail in ["Objectivity, Value Judgment, and Theory Choice" (reprinted in chapter 2, this volume)] than it is in the Postscript.

6. "Objectivity, Value Judgment, and Theory Choice," 321–22 [pp. 95–96].

7. Ibid., p. 322 [p. 96].

8. Throughout this paper I shall use the terms "hypothesis" and "theory" more or less interchangeably. Kuhn tends to prefer "theory," while I tend to prefer "hypothesis," but nothing of importance hinges on this usage here.

9. As Adolf Grünbaum pointed out to me, if we assume that in a given day the actual frequency of defective can openers produced by the two machines matches precisely the respective probabilities, we can calculate the result as follows. The new machine produces 50 defective can openers and the old machine produces 30, so that 50 out of a total of 80 are produced by the new machine. However, it would be *incorrect* to assume that the frequencies match the probabilities each day; in fact, the probability of an exact match is quite small.

10. "Objectivity, Value Judgment, and Theory Choice," 328 [p. 101].

11. I remarked above that three probabilities are required to calculate the posterior probability—a prior probability and two likelihoods. Obviously, in view of (2), the theorem on total probability, if we have a prior probability, one of the likelihoods, and the expectedness, we can compute the other likelihood; likewise, if we have one prior probability and both likelihoods, we can compute the expectedness.

12. *The Foundations of Scientific Inference* (Pittsburgh: University of Pittsburgh Press, 1967), chap. 7.

13. A set of degrees of conviction is coherent provided that its members do not violate any of the conditions embodied in the mathematical calculus of probability.

14. From "Autobiographical Notes," in Paul A. Schilpp, ed., *Albert Einstein: Philosopher-Scientist* (Evanston, Ill: Library of Living Philosophers, 1949), 21–22.

15. This example is discussed in Ronald N. Giere, *Understanding Scientific Reasoning*, 2d ed. (New York: Holt, Rinehart, and Winston, 1984), 274–76.

16. I reject the so-called propensity interpretation of probability because, as Paul Humphreys pointed out, the probability calculus accommodates inverse probabilities of the type that occur in Bayes's theorem, but the corresponding inverse propensities do not exist. In the example of the can opener factory, each machine has a certain propensity to produce defective can openers, but it does not make sense to speak of the propensity of a given defective can opener to have been produced by the new machine.

17. A so-called Dutch book is a combination of bets such that, no matter what the outcome of the event upon which the wagers are made, the subject is bound to suffer a net loss.

18. "Truth and Probability," in Frank Plumpton Ramsey, *The Foundations of Mathematics*, ed. R. B. Braithwaite (New York: Humanities Press, 1950).

19. Gerald Holton and Stephen G. Brush, *Introduction to Concepts and Theories in Physical Science*, 2d ed. (Reading, Mass.: Addison-Wesley, 1973), 416, italics in original.

20. Abner Shimony, "Scientific Inference," in Robert G. Colodny, ed., *The Nature and Function of Scientific Theories* (Pittsburgh: University of Pittsburgh Press, 1970), 79–172.

21. Martin Gardner, *Fads and Fallacies in the Name of Science* (New York: Dover Publications, 1957), 7–15.

22. Doubleday & Company, 1950.

23. Haim Harari, "The Structure of Quarks and Leptons," *Scientific American* 248, (April 1983): 56–68.

24. *Scientific American*, 253 (Sept. 1985): 116. Adolf Grünbaum, *The Foundations of Psychoanalysis* (Berkeley/Los Angeles/London: University of California Press, 1984), 202–4, criticizes Motley's account of Freud's theory; he considers Motley's version a distortion, and points out that Freud's motivational explanations were explicitly confined to a very circumscribed set of slips. He defends Freud against Motley's criticism on the grounds that Freud's actual account has greater complexity than Motley gives it credit for.

25. "A Bayesian Approach to the Paradoxes of Confirmation," in Jaakko Hintikka and Patrick Suppes, eds., *Aspects of Inductive Logic* (Amsterdam: North Holland, 1966), 202–3.

26. I have discussed the relations between personal probabilities and objective probabilities in "Dynamic Rationality: Propensity, Probability, and Credence," in James H. Fetzer, ed., *Probability and Causality* (Dordrecht: Reidel 1988), 3–40.

27. As Duhem has made abundantly clear, in such cases we may be led to reexamine our background knowledge B, which normally involves auxiliary hypotheses, to see whether it remains acceptable in the light of the negative outcome E. Consequently, refutation of T is not usually as automatic as it appears in the simplified account just given. Nevertheless, the probability relation just stated is correct.

28. Exchangeability is the personalist's surrogate for randomness; it means that the subject would draw the same conclusion regardless of the order in which the members of an observed sample occurred.

29. Note that, in order to get the posterior probability—the probability that the observed results were produced by a biased device—the prior probabilities have to be taken into account.

30. Indeed, stellar parallax was not detected until the nineteenth century.

31. This question was, in fact, raised by Adolf Grünbaum in a private communication.

32. See, for example, Gerald Holton and Stephen G. Brush, *Introduction to Concepts and Theories in Physical Science*, 2d ed. (Reading, Mass.: Addison-Wesley, 1973), 392–93.

33. An excellent account of Brewster's position can be found in John Worrall, "Scientific Revolutions and Scientific Rationality: The Case of the Elderly Holdout," [in C. Wade Savage, ed., *Scientific Theories*, vol. 14, *Minnesota Studies in the Philosophy of Science* (Minneapolis: University of Minnesota Press, 1990)], pp. 319–36.

34. Quoted by Worrall, p. 321.

35. Quoted by Worrall, p. 322.

36. Quoted by Worrall, p. 323.

37. If more than two theories are serious candidates, the pairwise comparison can be repeated as many times as necessary.

38. Adolf Grünbaum, "Can a Theory Answer More Questions Than One of Its Rivals?" *British Journal for the Philosophy of Science* 27 (1976): 1–23.

39. Steven Weinberg, *Gravitation and Cosmology* (New York: John Wiley & Sons, 1972), 198. Note that this correction is smaller by an order of magnitude than the correction of 43 seconds of arc per century for Mercury.

40. See Joel Smith, *The Status of Inconsistent Statements in Scientific Inquiry* (doctoral dissertation, University of Pittsburgh, 1987).

41. Unfortunately, recent evidence regarding the mass of Pluto strongly suggests that Pluto is not sufficiently massive to explain the perturbations of Neptune. A different plausible scenario is needed, but I do not know of any serious candidates that have been offered.*

42. See "Dynamic Rationality" (note 26 above), 5–12, for a more detailed discussion of various grades of rationality. The term "static" was chosen to indicate the lack of any principled method for changing personal probabilities in the face of inconsistency or incoherence.

43. Ibid., esp. pp. 11–12.

44. Thomas S. Kuhn, *The Copernican Revolution* (Cambridge: Harvard University Press, 1957), 3–4.

45. I should like to express my deepest gratitude to Adolf Grünbaum and Philip Kitcher for important criticism and valuable suggestions with respect to an earlier version of this paper.

* Because Neptune moves so slowly—its period of revolution is almost 165 years— its complete orbit is difficult to plot with precision. The most plausible scenario now seems to be that the alleged perturbations in the orbit of Neptune are spurious. Certainly Pluto (now known to have a mass less than 1/500 of the mass of the earth) is incapable of producing the size of the presumed effects that motivated Percival Lowell to propose the existence of a trans-Neptunian "planet X," a false hypothesis that nonetheless led to Clyde Tombaugh's observational detection of Pluto in 1930.

Deborah G. Mayo

A Critique of Salmon's
Bayesian Way

. . . Salmon's discussion is an attempt to employ the Bayesian Way to solve a philosophical problem, this time to answer Kuhn's challenge as to the existence of an empirical logic for science. Reflecting on the deep division between the logical empiricists and those who adopt the "historical approach," a division owing much to Kuhn's *Structure of Scientific Revolutions*, Salmon (1990) proposes "that a bridge could be built between the differing views of Kuhn and Hempel if Bayes's theorem were invoked to explicate the concept of scientific confirmation" (p. 175 [518]). The idea came home to Salmon, he tells us, during an American Philosophical Association (Eastern Division 1983) symposium on Carl Hempel, in which Kuhn and Hempel shared the platform.[1] "At the time it seemed to me that this maneuver could remove a large part of the dispute between standard logical empiricism and the historical approach to philosophy of science" on the fundamental issue of confirmation (p. 175 [518]).

Granting that observation and experiment, together with hypothetico-deductive reasoning, fail adequately to account for theory choice, Salmon argues that the Bayesian Way can accommodate the additional factors Kuhn seems to think are required. In building his bridge, Salmon often refers to Kuhn's (1977) "Objectivity, Value Judgment, and Theory Choice." [See chapter 2, this volume.] It is a fitting reference: in that paper Kuhn himself is trying to build bridges with the more traditional philosophy of science, aiming to thwart charges that he has rendered theory choice irrational.

Deliberately employing traditional terminology, Kuhn attempts to assuage his critics. He assures us that he agrees entirely that the standard criteria—accuracy, consistency, scope, simplicity, and fruitfulness—play a vital role in choosing between an established theory and a rival (p. 322 [95]). . . . Kuhn charges that these criteria underdetermine theory choice:

From Deborah G. Mayo, *Error and the Growth of Experimental Knowledge* (Chicago: University of Chicago Press, 1996), 112–27.

they are imprecise, differently interpreted and differently weighed by different scientists. Taken together, they may contradict each other—one theory being most accurate, say, while another is most consistent with background knowledge. Hence theory appraisals may disagree even when agents ostensibly follow the same shared criteria. They function, Kuhn says, more like values than rules.

Here's where one leg of Salmon's bridge enters. The shared criteria of theory choice, Salmon proposes, can be cashed out, at least partly, in terms of prior probabilities. The conflicting appraisals that Kuhn might describe as resulting from different interpretations and weightings of the shared values, a Bayesian could describe as resulting from different assignments of prior probabilities. We have at least a partial bridge linking Bayes and Kuhn, but would a logical empiricist want to cross it?

Logical empiricists, it seems, would need to get around the Kuhnian position that the shared criteria are never sufficient to ground the choice between an accepted theory and a competitor, that consensus, if it occurs, always requires an appeal to idiosyncratic, personal factors beyond the shared ones. They would need to counter Kuhn's charge that in choosing between rival theories "scientists behave like philosophers,". . . .

Interestingly, Kuhn's single reference to a Bayesian approach is to combat criticism of his position. For the sake of argument, Kuhn says, suppose that scientists deploy some Bayesian algorithm to compute the posterior probabilities of rival theories on evidence and suppose that we could describe their choice between these theories as based on this Bayesian calculation (Kuhn 1977, 328). "Nevertheless," Kuhn holds that "the algorithms of individuals are all ultimately different by virtue of the subjective considerations with which each must complete the objective criteria before any computations can be done" (p. 329 [101]). So sharing Bayes's theorem does not count as a "shared algorithm" for Kuhn. Kuhn views his (logical empiricist) critic as arguing that since scientists often reach agreement in theory choice, the subjective elements are eventually eliminated from the decision process and the Bayesian posteriors converge to an objective choice. Such an argument, Kuhn says, is a non sequitur. In Kuhn's view, the variable priors lead different scientists to different theory choices, and agreement, if it does occur, results from sociopsychological factors, if not from unreasoned leaps of faith. Agreement, in other words, might just as well be taken as evidence of the further role of subjective and sociopsychological factors, rather than of their eventual elimination.

But perhaps building a logical empiricist bridge out of Bayesian bricks would not require solving this subjectivity problem. Perhaps Salmon's point is that by redescribing Kuhn's account in Bayesian terms, Kuhn's account need not be seen as denying science a logic based on empirical evidence. It can have a logic based on Bayes's theorem. It seems to me that much of the current appeal of the Bayesian Way reflects this kind of move: while allowing plenty of room for "extrascientific" factors, Bayes's theorem ensures at least some role for empirical evidence. It gives a formal model, we just saw, for reconstructing (after the fact) a given assignment of blame for an anomaly, and it may well allow for reconstructing Kuhnian theory choice. Putting aside for the moment whether a bridge from Bayes to Kuhn holds us above the water, let us see how far such a bridge would need to go.

Right away an important point of incongruity arises. While Kuhn talks of theory acceptance, the Bayesian talks only of probabilifying a theory—something Kuhn eschews. For the context of Kuhnian normal science, where problems are "solved" or not, this incongruity is too serious to remedy. But Salmon is talking about theory choice or theory preference, and here there seem to be ways of reconciling Bayes and Kuhn (provided radical incommensurabilities are put to one side), although Salmon does not say which he has in mind. One possibility would be to supplement the Bayesian posterior probability assessment with rules for acceptance or preference (e.g., accept or prefer a theory if its posterior probability is sufficiently higher than that of its rivals).

A second possibility would be to utilize the full-blown Bayesian decision theory. Here, averaging probabilities and utilities allows calculating the average or expected utility of a decision. The Bayesian rule is to choose the action that *maximizes expected utility.* Choosing a theory would then be represented in Bayesian terms as adopting the theory that the agent feels maximizes expected utility. If it is remembered that, according to Kuhn, choosing a theory means deciding to work within its paradigm, this second possibility seems more apt than the first. The utility calculation would provide a convenient place to locate the variety of values—those shared as well as those of "individual personality and biography"—that Kuhn sees as the basis for theory choice.

Even this way of embedding Kuhn in a Bayesian model would not quite reach the position Kuhn holds. In alluding to the Bayesian model, Kuhn (1977) concedes that he is tempering his position somewhat, putting to one side the problems of radically theory-laden evidence and incommensurability. Strictly speaking, comparing the expected utilities of choosing between theories describes a kind of comparison that Kuhn deems impossible for choosing between incommensurables. It is doubtful that a genuine Kuhnian conversion is captured as the result of a Bayesian conditionalization. Still, the reality of radical incommensurability has hardly been demonstrated. So let us grant that the subjective Bayesian Way, with the addition of some rule of acceptance such as that offered by Bayesian decision theory, affords a fairly good bridge between Bayes and a slightly-tempered Kuhn. Note also that the Kuhnian problems of subjectivity and relativism are rather well modeled—though not solved—by the corresponding Bayesian problems. The charge that Kuhn is unable to account for how scientists adjudicate disputes and often reach consensus seems analogous to the charge we put to the subjective Bayesian position. (For a good discussion linking Kuhn and Bayes, see Earman 1992, 192–93.)

But this is not Salmon's bridge. Our bridge pretty much reaches Kuhn, but the toll it exacts from the logical empiricist agenda seems too dear for philosophers of that school to want to cross it. Salmon's bridge is intended to be free of the kinds of personal interests that Kuhn allows, and as such it does not go as far as reaching Kuhn's philosophy of science. But that is not a mark against Salmon's approach, quite the opposite. A bridge that really winds up in Kuhnian territory is a bridge too far: a utility calculation opens theory choice to all manner of interests and practical values. It seems the last thing that would appeal to those wishing to retain the core of a logical empiricist philosophy. (It opens too wide a corridor for the enemy!) So let us look at Salmon's bridge as a possible link, not between a tempered Kuhn and Bayes, but between logical empiricism and a tempered Bayesianism. Before the last brick is in place, I shall question whether the bridge does not actually bypass Bayesianism altogether.

■ | Salmon's Comparative Approach and a Bayesian Bypass

Salmon endorses the Kuhnian position that theory choice, particularly among mature sciences, is always a matter of choosing between rivals. Kuhn's reason, however, is that he regards rejecting a theory or paradigm in which one had been working without accepting a replacement as tantamount to dropping out of science. Salmon's reason is that using Bayes's theorem comparatively helps cancel out what he takes to be the most troubling probability: the probability of the evidence e given not-T ("the catchall"). (Salmon, like me, prefers the term hypotheses to theories, but uses T in this discussion because Kuhn does. I shall follow Salmon in allowing either to be used.)

Because of some misinterpretations that will take center stage later, let us be clear here on the probability of evidence e on the catchall hypothesis.[2] Evidence e describes some outcome or information, and not-T, the catchall, refers to the disjunction of all possible hypotheses other than T, including those not even thought of, that might predict or be relevant to e. This probability is not generally meaningful for a frequentist, but is necessary for Bayes's theorem.[3] Let us call it the *Bayesian catchall factor* (with evidence e):[4]

Define the *Bayesian catchall factor* (in assessing T with evidence e) as

$P(e \mid \text{not-}T)$.

Salmon, a frequentist at heart, rejects the use of the Bayesian catchall factor.

> What is the likelihood of any given piece of evidence with respect to the catch-all? This question strikes me as utterly intractable; to answer it we would have to predict the future course of the history of science. (Salmon 1991, 329)

This recognition is a credit to Salmon, but since the Bayesian catchall factor is vital to the general Bayesian calculation of a posterior probability, his rejecting it seems almost a renunciation of the Bayesian Way. The central role of the Bayesian catchall factor is brought out in writing Bayes's theorem as follows:

$$P(T \mid e) = \frac{P(e \mid T)\, P(T)}{P(e \mid T)\, P(T) + P(e \mid \mathbf{not\text{-}}T)\, P(\mathbf{not\text{-}}T)}.$$

Clearly, the lower the value of the Bayesian catchall factor, the higher the posterior probability of T, because the lower its value, the less the denominator in Bayes's theorem exceeds the numerator. The subjectivist "solution" to Duhem turned on the agent assigning a very small value to the Bayesian catchall factor (where the evidence was the anomalous result e'), because that allowed the posterior of T to remain high despite the anomaly. Subjective Bayesians accept, as a justification for this probability assignment, that agents believe there to be no plausible rival to T that they feel would make them expect the anomaly e'. This is not good enough for Salmon.

In order to get around such a subjective assignment (and avoid needing to predict the future course of science), Salmon says we should restrict the Bayesian Way to looking at the ratio of the posteriors of two theories T_1 and T_2: In the ratio of the posteriors of the two theories, we get a canceling out of the Bayesian catchall factors (the probability of e on the catchall).[5] Let us see what the resulting comparative assessment looks like. Since the aim is no longer to bridge Kuhn, we can follow Salmon in talking freely about either theories or hypotheses. Salmon's Bayesian algorithm for theory preference is as follows (to keep things streamlined, I drop the explicit statement of the background variable B):

Salmon's Bayesian algorithm for theory preference (1990, 192 [536]):

Prefer T_1 to T_2 whenever $P(T_1 \mid e)/P(T_2 \mid e)$ exceeds 1, where:
$$\frac{P(T_1 \mid e)}{P(T_2 \mid e)} = \frac{P(T_1)\, P(e \mid T_1)}{P(T_2)\, P(e \mid T_2)}.$$

To start with the simplest case, suppose that both theories T_1 and T_2 entail e.[6] Then $P(e \mid T_1)$ and $P(e \mid T_2)$ are both 1. These two probabilities are the *likelihoods* of T_1 and T_2, respectively.[7] Salmon's rule for this special case becomes:

Special case: Salmon's rule for relative preference (where each of T_1 and T_2 entails e):
Prefer T_1 to T_2 whenever $P(T_1)$ exceeds $P(T_2)$.

Thus, in this special case, the relative preference is unchanged by evidence *e*. You prefer T_1 to T_2 just in case your prior probability of T_1 exceeds that of T_2 (or vice versa). Note that this is a general Bayesian result that we will want to come back to. In neutral terms, it says that if evidence is entailed by two hypotheses, then *that evidence* cannot speak any more for one hypothesis than another—according to the Bayesian algorithm.[8] If their appraisal differs, it must be due to some difference in prior probability assignments to the hypotheses. This will not be true on the error statistics model.

To return to Salmon's analysis, he proposes that where theories do not entail the evidence, the agent consider auxiliary hypotheses (A_1 and A_2) that, when coinjoined with each theory (T_1 and T_2, respectively), would entail the evidence. That is, the conjunction of T_1 and A_1 entails *e*, and the conjunction of T_2 and A_2 entails *e*. This allows, once again, the needed likelihoods to equal 1, and so to drop out. The relative appraisal of T_1 and T_2 then equals the ratio of the prior probabilities of the conjunctions of T_1 and A_1, and T_2 and A_2. We are to prefer that conjunction (of theory and auxiliary) that has the higher prior probability.[9] In short, in Salmon's comparative analysis the weight is taken from the likelihoods and placed on the priors, making the appraisal even more dependent upon the priors than the noncomparative Bayesian approach.

Problems with the Comparative Bayesian Approach

Bayesians will have their own problems with such a comparative Bayesian approach. How, asks Earman (1992, 172), can we plug in probabilities to perform the usual Bayesian decision theory? But Earman is reluctant to throw stones, confessing that "as a fallen Bayesian, I am in no position to chide others for acts of apostasy" (p. 171). Earman, with good reason, thinks that Salmon has brought himself to the brink of renouncing the Bayesian Way. Pursuing Salmon's view a bit further will show that he may be relieved of the yoke altogether.

For my part, the main problem with the comparative approach is that we cannot apply it until we have accumulated sufficient knowledge, by some non-Bayesian means, to arrive at the prior probability assignments (whether to theories or theories conjoined with auxiliaries). Why by some non-Bayesian means? Couldn't prior probability assessments of theories and auxiliaries themselves be the result of applying Bayes's theorem? They could, but only by requiring a *re*introduction of the corresponding assignments to the Bayesian catchall factors—the very thing Salmon is at pains to avoid. The problems of adjudicating conflicting assessments, predicting the future of science, and so on, remain.

Could not the prior probability assignments be attained by some more hard-nosed assessment? Here is where Salmon's view becomes most interesting. While he grants that assessments of prior probabilities, or, as he prefers, plausibilities, are going to be relative to agents, Salmon demands that the priors be constrained to reflect objective considerations.

> The frightening thing about pure unadulterated personalism is that nothing prevents prior probabilities (and other probabilities as well) from being determined by all sorts of idiosyncratic and objectively irrelevant considerations (Salmon 1990, 183 [526])

such as the agent's mood, political disagreements with or prejudices toward the scientists who first advanced the hypothesis, and so on.

> What we want to demand is that the investigator make every effort to bring all of his or her *relevant* experience in evaluating hypotheses to bear on the question of whether the hypothesis under consideration is of a type likely to succeed, and to leave aside emotional irrelevancies. (p. 183 [526])

Ever the frequentist, Salmon proposes that prior probabilities "can be understood as our best estimates of the frequencies with which certain kinds of hypotheses succeed" (p. 187 [531]).[10] They may be seen as personalistic so long as the agent is guided by "the aim of bringing to bear all his or her experience that is relevant to the success or failure of hypotheses similar to that being considered." According to Salmon, "On the basis of their training and experience, scientists are qualified to make such judgments" (p. 182 [526]).

But are they? How are we to understand the probability Salmon is after? The context may be seen as a single-universe one. The members of this universe are hypotheses similar to the hypothesis H being considered, presumably from the population of existing hypotheses. To assign the prior probability to hypothesis H, I imagine one asks oneself, What proportion of the hypotheses in this population are (or have been) successful? Assuming that H is a random sample from the universe of hypotheses similar to H, this proportion equals the probability of interest. Similar to hypothesis H in what respects? Successful in what ways? For how long? The reference class problem becomes acute.

I admit that this attempt at a frequentist prior (also found in Hans Reichenbach) has a strong appeal. My hunch is that its appeal stems from unconsciously equating this frequency with an entirely different one, and it is this different one that is really of interest and, at the same time, is really obtainable.

Let us imagine that one had an answer to Salmon's question: what is the relative frequency with which hypotheses relevantly similar to H are successful (in some sense)? Say the answer is that 60 percent of them are. If H can be seen as a fair sample from this population, we could assign H a probability of .6. Would it be of much help to know this? I do not see how. I want to know how often *this* hypothesis will succeed.

What might an error statistician mean by the probability that this hypothesis H will succeed? As always, for a frequentist, probability applies to outcomes of a type of experiment. (They are sometimes called "generic" outcomes.) The universe or population here consists of possible or hypothetical experiments, each involving an application of hypothesis H. Success is some characteristic of experimental outcomes. For example, if H is a hypothesized value of a parameter μ, a successful outcome might be an outcome that is within a specified margin of error of μ. The probability of success construed this way is just the ordinary probability of the occurrence of certain experimental outcomes. . . . Indeed, for the error theorist, the only kinds of things to which probabilities apply are things that can be modeled as experimental outcomes. Knowledge of H's probable success is knowledge of the probability distributions associated with applying H in specific types of experiments. Such knowledge captures the spirit of what C. S. Peirce would call the "experimental purport" of hypothesis H.

- ▪ TWO MEANINGS OF THE PROBABILITY THAT A HYPOTHESIS IS SUCCESSFUL.

Let us have a picture of our two probabilities. Both can be represented as one-urn models. In Salmon's urn are the members of the population of hypotheses similar to H. These hypotheses are to be characterized as successful or not, in some way that would need to be specified. The probability of interest concerns the relative frequency with which hypotheses in this urn are successful. This number is taken as the probability that H is successful.

In my urn are members of a population of outcomes (a sample space) of an experiment. Each outcome is defined as successful or not according to whether it is close to what H predicts relative to a certain experiment (for simplicity, omit degrees of closeness). The probability of interest concerns the relative frequency with which outcomes in this urn are successful. Hypothesis H can be construed as asserting about this population of outcomes that with high probability they will be successful (e.g., specifiably close to what H predicts). The logic of standard statistical inference can be pictured as selecting one outcome, randomly, from this "urn of outcomes" and using it to learn whether what H asserts is correct.

Take one kind of hypothesis already discussed, that a given effect is real or systematic and not artifactual. In particular, take Hacking's hypothesis H. . .

H: dense bodies are real structures in blood cells, not artifacts.

We have no idea what proportion of hypotheses similar to H are true, nor do we have a clue as to how to find out, nor what we would do if we did. In actuality, our interest is not in a probabilistic assignment to H, but in whether H is the case. We need not have infallible knowledge about H to learn about the correctness of H.

We ask: what does the correctness of H say about certain experimental results, ideally, those we can investigate? One thing Hacking's H says is that dense bodies will be detected even using radically different physical techniques, or at least that they will be detected with high reliability. Experimenting on dense bodies, in other words, will *not* be like experimenting on an artifact of the experimental apparatus. Or so H asserts. This leads to pinpointing a corresponding notion of success. We may regard an application of H as successful when it is specifiably far from what would be expected in experimenting on an artifact. More formally, a successful outcome can be identified as one that a specified experimental test would take as failing the artifact explanation. That H is frequently successful, then, asserts that the artifact explanation, that is, not-H, would very frequently be rejected by a certain statistical test.

The knowledge of the quality of the evidence in hand allows assessing whether there are good grounds for the correctness of H (in the particular respects indicated in defining success). In particular, knowledge that H passes extremely severe tests is a good indication that H is correct, that it will often be successful. Why? Because were the dense bodies an artifact, it would almost certainly not have produced the kinds of identical configurations that were seen, despite using several radically different processes.

What are needed, in my view, then, are arguments that H is correct, that experimental outcomes will very frequently be in accordance with what H predicts—that H will very frequently succeed, *in this sense.* These are the arguments for achieving experimental knowledge (knowledge of experimental distributions). We obtain such experimental knowledge by making use of probabilities—not of hypotheses but probabilistic characteristics of experimental testing methods (e.g., their reliability or severity). Where possible, these probabilities are arrived at by means of standard error probabilities (e.g., significance levels). In those cases, what I have in mind is well put in my favorite passage from Fisher:

> In relation to the test of significance, we may say that a phenomenon is experimentally demonstrable when we know how to conduct an experiment which will rarely fail to give us a statistically significant result. (Fisher 1947, 14)

A Natural Bridge between Salmon and Error Statistics

Whereas Salmon construes "the probability of H's success" as the relative frequency with which hypotheses similar to H are successful, the error statistician proposes that it be given a very different frequentist construal. For the error statistician, "the probability of H's success" or, more aptly, H's reliability, is viewed as an elliptical way of referring to the relative frequency with which H is expected to succeed in specifically defined experimental applications. There is no license to use the latter frequentist notion in applying Bayes's theorem. Nevertheless, it may be used in Salmon's comparative approach (where the likelihoods drop out), and doing so yields a very natural bridge connecting his approach to that of error statistics.

To see in a simple way what this natural bridge looks like, let the two hypotheses H_1 and H_2 entail evidence e (it would be adequate to have them merely fit e to some degree). Then, on Salmon's comparative Bayesian approach, H_1 is to be preferred to H_2 just in case the prior probability assessment of H_1 exceeds H_2. The assignment of the prior probabilities must not contain irrelevant subjective factors, says Salmon, but must be restricted to assessing whether the hypotheses are likely to be successful. Hypothesis H_1 is to be preferred to H_2 just in case H_1 is accorded a higher probability of success than H_2. Now let us substitute my error statistical construal of probable success (for some specified measure of "successful outcome"). Evaluating H's probable success (or H's reliability) means evaluating the relative frequency with which applications of H would yield results in accordance with (i.e., specifiably close to) what H asserts. As complex as this task sounds, it is just the kind of information afforded by experimental knowledge of H. The task one commonly sets for oneself is far less technically put. The task, informally, is to consider the extent to which specific obstacles to H's success have been ruled out. Here is where the kind of background knowledge I think Salmon has in mind enters. What training and experience give the experimenter is knowledge of the specific ways in which hypotheses can be in error, and knowledge of whether the evidence is so far sufficient to rule out those errors.

To put my point another way, Salmon's comparative approach requires only the two prior probabilities or plausibilities to be considered, effectively wiping out the rest of the Bayesian calculation. The focus is on ways of assessing the plausibilities of the hypotheses or theories themselves. However, Salmon's approach gives no specific directions for evaluating the plausibilities or probable success of the hypotheses. Interpreting probable success in the way I recommend allows one to work out these directions. Salmon's comparative appraisal of H_1 against a rival H_2 would become: prefer H_1 to H_2 just to the extent that the evidence gives a better indication of H_1's likely success than H_2's.

Further, the kinds of evidence and arguments relevant to judge H's success, in my sense, seem quite congenial to what Salmon suggests should go into a plausibility assessment. In one example Salmon refers explicitly to the way in which standard (non-Bayesian) significance tests may be used to give plausibility to hypotheses (Salmon 1990, 182 [526]). In particular, a statistically significant association between saccharin and bladder cancer in rats, he says, lends plausibility to the hypothesis H that saccharin in diet drinks increases the risk of bladder cancer in humans. Provided that errors of extrapolating from rats to humans and from high to low doses are satisfactorily ruled out, a statistically significant association may well provide evidence that H [is correct], *that H will be successful in our sense.* This success may be cashed out in different ways, because the truth of H has different implications. One implication of the correctness of H here is that were populations to consume such and such amount of saccharin the incidence of bladder cancer would be higher than if they did not. My point is that such experiments are evidence for the correctness of H in this sense. Such experiments do not provide the number Salmon claims to be after, the probability that hypotheses similar to the saccharin hypothesis are successful. So even if that probability were wanted (I claim it is not), that is not what the saccharin experiments provide.[11]

By allowing for this error statistical gloss of "H's probable success," the reader should not be misled into viewing our account as aiming to assign some quantitative measure to hypotheses—the reverse is true. My task here was to erect a bridge between an approach like Salmon's and the testing account I call error statistics. By demonstrating that the role Salmon gives to plausibility assessments is better accomplished by an assessment of the reliability of the tests hypotheses pass, I mean to show that the latter is all that is needed.

There are plenty of advantages to the testing account of scientific inference. First, by leading to accepting hypotheses as approximately correct, as well indicated, or as likely to be successful—rather than trying to assign some quantity of support or probability to hypotheses—it accords with the way scientists (and the rest of us) talk. Second, reporting the quality of the tests performed provides a way of communicating the evidence (summarizing the status of the problem to date) that is intersubjectively testable. A researcher might say, for example, that the saccharin rat study gives good grounds for holding that there is a causal connection with cancer in rats, but deny that the corresponding hypothesis about humans has been severely tested. This indicates what further errors would need to be ruled out (e.g., certain dose-response models are wrong).

Now it is open to a Bayesian to claim that the kinds of arguments and evidence that I might say give excellent grounds for the correctness of H, for accepting H, or for considering H to have passed a severe test can be taken as warranting a high prior probability assignment to H. For example, "there are excellent grounds for H" may be construed as "H has high prior probability" (say, around .9). (That Bayesians implicitly do this in their retrospective reconstructions of episodes is what gives their prior probability assessments their reasonableness.) Used in a purely comparative approach such as Salmon's, it might do no harm. However, there is nothing Bayesian left in this comparative approach! It is, instead, a quantitative sum-up of the quality of non-Bayesian tests passed by one hypothesis compared with those passed by another. (Whether such a non-Bayesian assessment of Bayesian priors could even be made to obey the probability calculus is not clear.)

To call such an approach Bayesian, even restricting it to comparisons, would be misleading. It is not just that the quantitative sum-up of H's warrant is not arrived at via Bayes's theorem. It is, as critics of error statistics are happy to demonstrate, that the principles of testing used in the non-Bayesian methods conflict with Bayesian principles.[12] . . . The Bayesian Way supposes, for any hypothesis one wishes to consider, that a Bayesian prior is available for an agent, and that an inference can be made. In general, however, there are not going to be sufficient (non-Bayesian) grounds to assign even a rough number to such hypotheses. We are back to the problem of making it too difficult to get started when, as is commonly the case, one needs a forward-looking method to begin learning something.

Bayesian Heretics, Fallen and Disgruntled Bayesians

The Bayesian landscape is littered with Bayesians who variably describe themselves or are described by others as fallen, heretical, tempered, nonstrict, or whatnot. Many Bayesians in this category came to the Bayesian Way in the movements led by Carnap and Reichenbach. Assigning probabilities to hypotheses was a natural way of avoiding the rigidities of a hypothetico-deductive approach. Inadequacies in the two main objective ways philosophers tried to define the prior probabilities—Carnapian logical or Reichenbachian frequentist—have left some in limbo: wanting to avoid the excesses of personalism and not sure how non-Bayesian statistics can help. Those Bayesians do not see themselves as falling under the subjectivist position that I criticized earlier. I invite them to try out the natural bridge proffered above, to see where it may lead.

What is really being linked by this bridge? Might it be said to link the cornerstone of logical empiricism on the one hand and the centerpiece of the New Experimentalism on the other? Such a bridge, as I see it, would link the (logical empiricist) view that the key to solving problems in philosophy of science is an inductive-statistical account of hypothesis appraisal with the view that the key to solving problems in philosophy of science is an understanding of the nature and role of experiment in scientific practice. It provides a way to *model* Kuhn's view of science—*where he is correct*—as well as a way to "*solve* Kuhn" where he challenges the objectivity and rationality of science.

In this chapter . . . I have brought out the main shortcomings of appeals to the Bayesian Way in modeling scientific inference and in solving problems about evidence and inference. Understanding these shortcomings also puts us in a better position to see what would be required of any theory of statistics that purports to take a leading role in an adequate philosophy of experiment. For one thing, we need an account that explicitly incorporates the intermediate theories of data, instruments, and experiment that are required to obtain experimental evidence in the first place. For another, the account must enable us to address the question of whether auxiliary hypotheses or experimental assumptions are responsible for observed anomalies from a hypothesis H, quite apart from how credible we regard hypothesis H. In other words, we need to be able to split off from some primary inquiry or test those questions about how well run the experiment was, or how well its assumptions were satisfied . . .

■ | Notes

1. See "Symposium: The Philosophy of Carl G. Hempel," *Journal of Philosophy* 80, no. 10 (October 1983): 555–72. Salmon's contribution is Salmon 1983.

2. To my knowledge, it was L. J. Savage who originated the term *catchall*.

3. See chapter 6 [of *Error and the Growth of Experimental Knowledge*].

4. I take this term from that of the Bayes factor, which is the ratio of the Bayesian catchall factor and $P(e \mid T)$.

5. This is because

$$P(T_i \mid e) = \frac{P(e \mid T_i) \, P(T_i)}{P(e \mid T_i) P(T_i) + P(e \mid \sim T_i) P(\sim T_i)}.$$

Note that the denominator equals $P(e)$. Since that is so for the posterior of T_1 as well as for T_2, the result of calculating the ratio is to cancel $P(e)$, and thereby cancel the probabilities of e on the catchalls.

6. While this case is very special, Salmon proposes that it be made the standard case by conjoining suitable auxiliaries to the hypotheses. I will come back to this in a moment.

7. Note that likelihoods of hypotheses are *not* probabilities. For example, the sum of the likelihoods of a set of mutually exclusive, collectively exhaustive hypotheses need not equal 1.

8. This follows from the likelihood principle . . . [As Mayo explains in chapter 10 of *Error and the Growth of Experimental Knowledge*, "Why You Cannot Be Just a Little Bit Bayesian," the likelihood principle (LP) defended by Bayesians asserts that evidential import of the outcome of an experiment, D, for a hypothesis, H, is contained entirely in the value of the likelihood $P(D/H)$. Thus, according to the LP, if D_1 and D_2 are the data obtained in two experiments to test a set of hypotheses, H_i, then D_1 and D_2 are evidentially equivalent if $P(D_1/H_i)$ and $P(D_2/H_i)$ are equal (or strictly proportional). LP is controversial because it implies that there is nothing wrong with "optional stopping," that is, deliberately continuing an experiment (by increasing the sample size) until one obtains a result that is deemed to be statistically significant. Mayo argues that LP is false because the strategy of "try and try again" is bound to reward experimenters with their desired result if they persist long enough, regardless of the truth or falsity of the hypothesis being tested. From the point of view of Mayo's non-Bayesian error-statistical approach, optional stopping is not a severe test of a hypothesis.]

9. For simplicity, we could just replace T_1, T_2, in the statement of the special case, with the corresponding conjunctions T_1 and A_1, and T_2 and A_2, respectively.

10. This was also Reichenbach's view. He did not consider that enough was known at present to calculate such a probability, but thought that it might be achievable in the future.

11. In the case of the saccharin hypothesis, it might look as if Salmon's frequentist probability is obtained. That, I think, is because of a tendency to slide from one

kind of probability statement to another. Consider hypotheses of the form x causes cancer in humans. They are all similar to H: saccharin causes cancer in humans. But what should be included in the reference set for getting Salmon's probability? Might x be anything at all? If so, then only a very tiny proportion would be successful hypotheses. That would not help in assessing the plausibility of H. I suggest that the only way this probability makes sense is if hypotheses "similar to H" refers to hypotheses similarly grounded or tested. In trying to specify the reference set in the case of the saccharin hypothesis we might restrict it to those causal hypotheses (of the required form) that have been shown to hold about as well as H. So it would consist of causal claims where a statistically significant correlation is found in various animal species, where certain dosage levels are used, where certain extrapolation models are applied (to go from animal doses to human doses as well as from rats to humans), where various other errors in identifying carcinogens are ruled out, and so on. Notice how these lead to a severity assessment.

A relative frequency question of interest that can be answered, at least qualitatively, is this: What is the relative frequency with which hypotheses of this sort (x causes cancer in humans) pass experimental tests $E_1, \ldots, E_n$ as well as H does, and yet do not succeed (turn out to be incorrect)? One minus this gives the severity of the test H passes. The question boils down to asking after the severity of the test H passes (where, as is common, several separate tests are taken together).

It does not matter that the hypotheses here differ. Error probabilities of procedures hold for applications to the same or *different* hypotheses. Neyman (e.g., 1977) often discussed the mistake in thinking they hold only for the former.

12. [T]he Bayesian Way follows the likelihood principle, which conflicts with error-probability principles. Quoting Savage: "Practically none of the 'nice properties' respect the likelihood principle" (1964, 184) The "nice properties" refer to error characteristics of standard statistical procedures, such as unbiasedness and significance levels . . .

■ | References

Earman, J. 1992. *Bayes or bust? A critical examination of Bayesian confirmation theory.* Cambridge, Mass.: MIT Press.

Fisher, R. A. 1947. *The design of experiments.* 4th ed. Edinburgh: Oliver and Boyd.

Kuhn, T. 1977. *The essential tension: Selected studies in scientific tradition and change.* Chicago: University of Chicago Press.

Neyman, J. 1977. Frequentist probability and frequentist statistics. *Synthese* 36: 97–131.

Salmon, W. 1983. Carl G. Hempel on the rationality of science. *Journal of Philosophy* 80 (no. 10): 555–62.

———. 1990. Rationality and objectivity in science, *or* Tom Kuhn meets Tom Bayes. In *Scientific theories,* edited by C. W. Savage, 175–204. Minnesota Studies in the Philosophy of Science, vol. 14. Minneapolis: University of Minnesota Press.

564 | Ch. 5 Confirmation and Relevance

———. 1991. The appraisal of theories: Kuhn meets Bayes. In *PSA 1990*, vol. 2, edited by A. Fine, M. Forbes, and L. Wessels, 325–32. East Lansing, Mich.: Philosophy of Science Association.

Savage, L. 1964. The foundations of statistics reconsidered. In *Studies in subjective probability*, edited by H. Kyburg and H. Smokler, 173–88. New York: John Wiley & Sons.

A L A N C H A L M E R S

The Bayesian Approach

■ | Introduction

Many of us had sufficient confidence in the prediction of the most recent
return of Halley's comet that we booked weekends in the country, far from
city lights and well in advance, in order to observe it. Our confidence proved
not to be misplaced. Scientists have enough confidence in the reliability of
their theories to send manned spacecraft into space. When things went
amiss in one of them, we were impressed, but perhaps not surprised, when
the scientists, aided by computers, were able to rapidly calculate how the
remaining rocket fuel could be utilised to fire the rocket motor in just the
right way to put the craft into an orbit that would return it to earth. These
stories suggest that perhaps the extent to which theories are fallible, stressed
by . . . philosophers . . . from Popper to Feyerabend, are misplaced or exaggerated. Can the Popperian claim that the probability of all scientific theories is zero be reconciled with them? It is worth stressing, in this connection,
that the theory used by the scientists in both of my stories was Newtonian
theory, a theory falsified in a number of ways at the beginning of this century according to the Popperian account (and most others). Surely something has gone seriously wrong.

One group of philosophers who do think that something has gone radically wrong, and whose attempts to put it right have become popular in the
last couple of decades, are the Bayesians, so called because they base their
views on a theorem in probability theory proved by the eighteenth-century
mathematician Thomas Bayes. The Bayesians regard it as inappropriate to
ascribe zero probability to a well-confirmed theory, and they seek some
kind of inductive inference that will yield non-zero probabilities for them in

FROM A. F. Chalmers, *What Is This Thing Called Science?*, 3d ed. (Indianapolis,
Ind.: Hackett Publishing Co., 1999), 174–92.

a way that avoids the difficulties of [inductivism].* For example, they would like to be able to show how and why a high probability can be attributed to Newtonian theory when used to calculate the orbit of Halley's comet or a spacecraft. An outline and critical appraisal of their viewpoint is given in this chapter.

■ | Bayes' Theorem

Bayes' theorem is about conditional probabilities, probabilities for propositions that depend on (and hence are conditional on) the evidence bearing on those propositions. For instance, the probabilities ascribed by a punter to each horse in a race will be conditional on the knowledge the punter has of the past form of each of the horses. What is more, those probabilities will be subject to change by the punter in the light of new evidence, when, for example, he finds on arrival at the racetrack that one of the horses is sweating badly and looking decidedly sick. Bayes' theorem is a theorem prescribing how probabilities are to be changed in the light of new evidence.

In the context of science the issue is how to ascribe probabilities to theories or hypotheses in the light of evidence. Let P(**h**/**e**) denote the probability of a hypothesis **h** in the light of evidence **e**, P(**e**/**h**) denote the probability to be ascribed to the evidence **e** on the assumption that the hypothesis **h** is correct, P(**h**) the probability ascribed to **h** in the absence of knowledge of **e**, and P(**e**) the probability ascribed to **e** in the absence of any assumption about the truth of **h**. Then Bayes' theorem can be written:

$$P(\mathbf{h}/\mathbf{e}) = P(\mathbf{h}) \cdot \frac{P(\mathbf{e}/\mathbf{h})}{P(\mathbf{e})}$$

P(**h**) is referred to as the *prior probability*, since it is the probability ascribed to the hypothesis prior to consideration of the evidence, **e**, and P(**h**/**e**) is referred to as the *posterior probability*, the probability after the evidence, **e**, is taken into account. So the formula tells us how to change the probability of a hypothesis to some new, revised probability in the light of some specified evidence.

The formula indicates that the prior probability, P(**h**), is to be changed by a scaling factor P(**e**/**h**)/P(**e**) in the light of evidence **e**. It can readily be

---

* In chapter 4 of *What Is This Thing Called Science?*, Chalmers defines *inductivism* as the view that "scientific knowledge is to be derived from the observable facts by some kind of inductive inference" (p. 49). Among the difficulties with inductivism discussed in that chapter are the problems of description and justification (see Peter Lipton, "Induction," chapter 4, this volume). Chalmers also argues that inductive generalizations from observable facts are incapable of yielding scientific knowledge about unobservable entities such as protons and genes.

seen how this is in keeping with common intuitions. The factor P(e/h) is a measure of how likely **e** is given **h**. It will take a maximum value of 1 if **e** follows from **h** and a minimum value of zero if the negation of **e** follows from **h**. (Probabilities always take values in between 1, representing certainty, and zero, representing impossibility.) The extent to which some evidence supports a hypothesis is proportional to the degree to which the hypothesis predicts the evidence, which seems reasonable enough. The term in the divisor of the scaling factor, P(e), is a measure of how likely the evidence is considered to be when the truth of the hypothesis, **h**, is not assumed. So, if some piece of evidence is considered extremely likely whether we assume a hypothesis or not, the hypothesis is not supported significantly when that evidence is confirmed, whereas if that evidence is considered very unlikely unless the hypothesis is assumed, then the hypothesis will be highly confirmed if the evidence is confirmed. For instance, if some new theory of gravitation were to predict that heavy objects fall to the ground, it would not be significantly confirmed by the observation of the fall of a stone, since the stone would be expected to fall anyway. On the other hand, if that new theory were to predict some small variation of gravity with temperature, then the theory would be highly confirmed by the discovery of that effect, since it would be considered most unlikely in the absence of the new theory.

An important aspect of the Bayesian theory of science is that the calculations of prior and posterior probabilities always take place against a background of assumptions that are taken for granted, that is, assuming what Popper called background knowledge. So, for example, when it was suggested in the previous paragraph that P(e/h) takes the value 1 when **e** follows from **h**, it was taken for granted that **h** was to be taken in conjunction with the available background knowledge. We have seen in earlier chapters that theories need to be augmented by suitable auxiliary assumptions before they yield testable predictions. The Bayesians take these considerations on board. Throughout this discussion it is assumed that probabilities are calculated against a background of assumed knowledge.

It is important to clarify in what sense Bayes' theorem is indeed a *theorem*. Although we will not consider the details here, we note that there are some minimal assumptions about the nature of probability which taken together constitute the so-called "probability calculus". These assumptions are accepted by Bayesians and non-Bayesians alike. It can be shown that denying them has a range of undesirable consequences. It can be shown, for example, that a gambling system that violates the probability calculus is "irrational" in the sense that it makes it possible for wagers to be placed on all possible outcomes of a game, race or whatever in such a way that the participants on one or other side of the betting transaction will win *whatever the outcome*. (Systems of betting odds that allow this possibility are called Dutch Books. They violate the probability calculus.) Bayes' theorem can be derived from the premises that constitute the probability calculus. In that sense, the theorem in itself is uncontentious.

So far, we have introduced Bayes' theorem, and have tried to indicate that the way in which it prescribes that the probability of a hypothesis be changed in the light of evidence captures some straightforward intuitions about the bearing of evidence on theories. Now we must press the question of the interpretation of the probabilities involved more strongly.

■ | Subjective Bayesianism

The Bayesians disagree among themselves on a fundamental question concerning the nature of the probabilities involved. On one side of the division we have the "objective" Bayesians. According to them, the probabilities represent probabilities that rational agents *ought* to subscribe to in the light of the objective situation. Let me try to indicate the gist of their position with an example from horse racing. Suppose we are confronted by a list of the runners in a horse race and we are given no information about the horses at all. Then it might be argued that on the basis of some "principle of indifference" the only rational way of ascribing probabilities to the likelihood of each horse winning is to distribute the probabilities equally among the runners. Once we have these "objective" prior probabilities to start with, then Bayes' theorem dictates how the probabilities are to be modified in the light of any evidence, and so the posterior probabilities that result are also those that a rational agent *ought* to accept. A major, and notorious, problem with this approach, at least in the domain of science, concerns how to ascribe objective prior probabilities to hypotheses. What seems to be necessary is that we list all the possible hypotheses in some domain and distribute probabilities among them, perhaps ascribing the same probability to each employing the principle of indifference. But where is such a list to come from? It might well be thought that the number of possible hypotheses in any domain is infinite, which would yield zero for the probability of each and the Bayesian game cannot get started. All theories have zero probability and Popper wins the day. How is some finite list of hypotheses enabling some objective distribution of nonzero prior probabilities to be arrived at? My own view is that this problem is insuperable, and I also get the impression from the current literature that most Bayesians are themselves coming around to this point of view. So let us turn to "subjective" Bayesianism.

For the subjective Bayesian the probabilities to be handled by Bayes' theorem represent subjective degrees of belief. They argue that a consistent interpretation of probability theory can be developed on this basis, and, moreover, that it is an interpretation that can do full justice to science. Part of their rationale can be grasped by reference to the examples I invoked in the opening paragraph of this chapter. Whatever the strength of the arguments for attributing zero probability to all hypotheses and theories, it is simply not the case, argue the subjective Bayesians, that people in general and scientists in particular ascribe zero probabilities to well-confirmed theories. The fact that I pre-booked my trip to the mountains to observe Halley's comet suggests that they are right in my case at least. In their work, scientists take many laws for granted. The unquestioning use of the law of refraction of light by astronomers and Newton's laws by those involved in the space program demonstrates that they ascribe to those laws a probability close, if not equal, to unity. The subjective Bayesians simply take the degrees of belief in hypotheses that scientists as a matter of fact happen to have as the basis for the prior probabilities in their Bayesian calculations. In this way they escape Popper's strictures to the effect that the probability of all universal hypotheses must be zero.

Bayesianism makes a great deal of sense in the context of gambling. We have noted that adherence to the probability calculus, within which Bayes' theorem can be proved, is a sufficient condition to avoid Dutch Books. Bayesian approaches to science capitalise on this by drawing a close analogy between science and gambling systems. The degree of belief held by a scientist in a hypothesis is analogous to the odds on a particular horse winning a race that he or she considers to be fair. Here there is a possible source of ambiguity that needs to be addressed. If we stick to our analogy with horse racing, then the odds considered to be fair by punters can be taken as referring either to their private subjective degrees of belief or to their beliefs as expressed in practice in their betting behaviour. These are not necessarily the same thing. Punters can depart from the dictates of the odds they believe in by becoming flustered at the race-track or by losing their nerve when the system of odds they believe in warrant a particularly large bet. Not all Bayesians make the same choice between these alternatives when applying the Bayesian calculus to science. For example, Jon Dorling (1979) takes the probabilities to measure what is reflected in scientific practice and Howson and Urbach (1989) take them to measure subjective degrees of belief. A difficulty with the former stand is knowing what it is within scientific practice that is meant to correspond to betting behaviour. Identifying the probabilities with subjective degrees of belief, as Howson and Urbach do, at least has the advantage of making it clear what the probabilities refer to.

Attempting to understand science and scientific reasoning in terms of the subjective beliefs of scientists would seem to be a disappointing departure for those who seek an objective account of science. Howson and Urbach have an answer to that charge. They insist that the Bayesian theory constitutes an *objective* theory of scientific inference. That is, given a set of prior probabilities and some new evidence, Bayes' theorem dictates in an objective way what the new, posterior, probabilities must be in the light of that evidence. There is no difference in this respect between Bayesianism and deductive logic, because logic has nothing to say about the source of the propositions that constitute the premises of a deduction either. It simply dictates what follows from those propositions once they are given. The Bayesian defence can be taken a stage further. It can be argued that the beliefs of individual scientists, however much they might differ at the outset, can be made to converge given the appropriate input of evidence. It is easy to see in an informal way how this can come about. Suppose two scientists start out by disagreeing greatly about the probable truth of hypothesis **h** which predicts otherwise unexpected experimental outcome **e**. The one who attributes a high probability to **h** will regard **e** as less unlikely than the one who attributes a low probability to **h**. So P(**e**) will be high for the former and low for the latter. Suppose now that **e** is experimentally confirmed. Each scientist will have to adjust the probabilities for **h** by the factor P(**e**/**h**)/P(**e**). However, since we are assuming that **e** follows from **h**, P(**e**/**h**) is 1 and the scaling factor is 1/P(**e**). Consequently, the scientist who started with a low probability for **h** will scale up that probability by a larger factor than the scientist who started with a high probability for **h**. As more positive evidence comes in, the original doubter is forced to scale up the probability in such a way that it eventually approaches that of the already convinced scientist. In this kind of way, argue the Bayesians, widely differing subjective opinions can be brought into conformity in response to evidence in an objective way.

■ | Applications of the Bayesian Formula

The preceding paragraph has given a strong foretaste of the kind of ways in which the Bayesians wish to capture and sanction typical modes of reasoning in science. In this section we will sample some more examples of Bayesianism in action.

. . . [T]here is a law of diminishing returns at work when testing a theory against experiment. Once a theory has been confirmed by an experiment once, repeating that same experiment under the same circumstances will not be taken by scientists as confirming the theory to as high a degree as the first experiment did. This is readily accounted for by the Bayesian. If the theory **T** predicts the experimental result **E** then the probability P(**E**/**T**) is 1, so that the factor by which the probability of **T** is to be increased in the light of a positive result **E** is 1/P(**E**). Each time the experiment is successfully performed, the more likely the scientist will be to expect it to be performed successfully again the subsequent time. That is, P(**E**) will increase. Consequently, the probability of the theory being correct will increase by a smaller amount on each repetition.

Other points in favour of the Bayesian approach can be made in the light of historical examples. Indeed, I suggest that it is the engagement by the Bayesians with historical cases in science that has been a key reason for the rising fortunes of their approach in recent years, a trend begun by Jon Dorling (1979). In [a previous] discussion of Lakatos's methodology we noted that according to that methodology it is the confirmations of a program that are important rather than the apparent falsifications, which can be blamed on the assumptions in the protective belt rather than on the hard core. The

Bayesians claim to be able to capture the rationale for this strategy. Let us see how they do it, by looking at a historical example utilised by Howson and Urbach (1989, pp. 97–102).

The example concerns a hypothesis put forward by William Prout in 1815. Prout, impressed by the fact that atomic weights of the chemical elements relative to the atomic weight of hydrogen are in general close to whole numbers, conjectured that atoms of the elements are made up of whole numbers of hydrogen atoms. That is, Prout saw hydrogen atoms as playing the role of elementary building blocks. The question at issue is what the rational response was for Prout and his followers to the finding that the atomic weight of chlorine relative to hydrogen (as measured in 1815) was 35.83, that is, not a whole number. The Bayesian strategy is to assign probabilities that reflect the prior probabilities that Prout and his followers might well have assigned to their theory together with relevant aspects of background knowledge, and then use Bayes' theorem to calculate how these probabilities change in light of the discovery of the problematic evidence, namely the non-integral value for the atomic weight of chlorine. Howson and Urbach attempt to show that when this is done the result is that the probability of Prout's hypothesis falls just a little, whereas the probability of the relevant measurements being accurate falls dramatically. In light of this it seems quite reasonable for Prout to have retained his hypothesis (the hard core) and to have put the blame on some aspect of the measuring process (the protective belt). It would seem that a clear rationale has been given for what in Lakatos's methodology appeared as "methodological decisions" that were not given any grounding. What is more, it would seem that Howson and Urbach, who are following the lead of Dorling here, have given a general solution to the so-called "Duhem-Quine problem". Confronted with the problem of which part of a web of assumptions to blame for an apparent falsification, the Bayesian answer is to feed in the appropriate prior probabilities and calculate the posterior probabilities. These will show which assumptions slump to a low probability, and consequently which assumptions should be dropped to maximise the chances of future success.

I will not go through the details of the calculations in the Prout case, or any of the other examples that Bayesians have given, but I will say enough to at least give the flavour of the way in which they proceed. Prout's hypothesis, h, and the effect of the evidence e, the non-integral atomic weight of chlorine, on the probability to be assigned to it is to be judged in the context of the available background knowledge, a. The most relevant aspect of the background knowledge is the confidence to be placed in the available techniques for measuring atomic weights and the degree of purity of the chemicals involved. Estimates need to be made about the prior probabilities of h, a and e. Howson and Urbach suggest a value of 0.9 for P(h), basing their estimate on historical evidence to the effect that the Proutians were very convinced of the truth of their hypothesis. They place P(a) somewhat lower at 0.6, on the grounds that chemists were aware of the problem of impurities, and that there were variations in the results of different measurements of the atomic weight of particular elements. The probability P(e) is assessed on the assumption that the alternative to **h** is a random distribution of atomic weights, so, for instance, P(e/not h & a) is ascribed a probability 0.01 on the grounds that, if the atomic weight of chlorine is randomly distributed over a unit interval it would have a one in a hundred chance of being 35.83. These probability estimates, and a few others like them, are fed into Bayes' theorem to yield posterior probabilities, P(h/e) and P(a/e), for **h** and **a**. The result is 0.878 for the former and 0.073 for the latter. Note that the probability for **h**, Prout's hypothesis, has fallen only a small amount from the original 0.9, whereas the probability of **a**, the assumption that the measurements are reliable, has fallen dramatically from 0.6 to 0.073. A reasonable response for the Proutians, conclude Howson and Urbach, was to retain their hypothesis and doubt the measurements. They point out that nothing much hinges on the absolute value of the numbers that are fed into the calculation so long as they are of the right kind of order to reflect the attitudes of the Proutians as reflected in the historical literature.

The Bayesian approach can be used to mount a criticism of some of the standard accounts of the undesirability of ad hoc hypotheses and related issues. Earlier . . . I proposed the idea, following Popper, that ad hoc hypotheses are undesirable because they are not testable independently of the evidence that led to their formulation. A related idea is that evidence that is used to construct a theory cannot be used again as evidence for that theory. From the Bayesian point of view, although these notions sometimes yield appropriate answers concerning how well theories are confirmed by evidence, they also go astray, and, what is more, the rationale underlying them is misconceived. The Bayesians attempt to do better in the following kinds of ways.

Bayesians agree with the widely held view that a theory is better confirmed by a variety of kinds of evidence than by evidence of a particular kind. There is a straightforward Bayesian rationale that explains why this should be so. The point is that there are diminishing returns from efforts to confirm a theory by a single kind of evidence. This follows from the fact that each time the theory is confirmed by that kind of evidence, then the probability expressing the degree of belief that it will do so in the future gradually increases. By contrast, the prior probability of a theory being confirmed by some new kind of evidence may be quite low. In such cases, feeding the results of such a confirmation, once it occurs, into the Bayesian formula leads to a significant increase in the probability ascribed to the theory. So the significance of independent evidence is not in dispute. Nevertheless, Howson and Urbach urge that, from the Bayesian point of view, if hypotheses are to be dismissed as ad hoc, the absence of independent testability is not the right reason for doing so. What is more, they deny that data used in the construction of a theory cannot be used to confirm it.

A major difficulty with the attempt to rule out ad hoc hypotheses by the demand for independent testability is that it is too weak, and admits hypotheses in a way that at least clashes with our intuitions. For instance, let us consider the attempt by Galileo's rival to retain his assumption that the moon is spherical in the face of Galileo's sightings of its mountains and craters by proposing the existence of a transparent, crystalline substance enclosing the observable moon. This adjustment cannot be ruled out by the independent testability criterion because it was independently testable, as evidenced by the fact that it has been refuted by the lack of interference from any such crystalline spheres experienced during the various moon landings. Greg Bamford (1993) has raised this, and a range of other difficulties with a wide range of attempts to define the notion of ad hocness by philosophers in the Popperian tradition, and suggests that they are attempting to define a technical notion for what is in effect nothing more than a common sense idea. Although Bamford's critique is not from a Bayesian point of view, the response of Howson and Urbach is similar, insofar as their view is that ad hoc hypotheses are rejected simply because they are considered implausible, and are credited with a low probability because of this. Suppose a theory **t** has run into trouble with some problematic evidence and is modified by adding assumption **a**, so that the new theory is (**t** & **a**). Then it is a straightforward result of probability theory that P(**t** & **a**) cannot be greater than P(**a**). From a Bayesian point of view, then, the modified theory will be given a low probability simply on the grounds that P(**a**) is unlikely. The theory of Galileo's rival could be rejected to the extent that his suggestion was implausible. There is nothing more to it, and nothing else needed.

Let us now turn to the case of the use of data to construct a theory, and the denial that that data can be considered to support it. Howson and Urbach (1989, pp. 275–80) give counter examples. Consider an urn containing counters, and imagine that we begin with the assumption that all of the counters are white and none of them coloured. Suppose we now draw counters 1,000 times, replacing the counter and shaking the urn after each draw, and that the result is that 495 of the counters are white. We now adjust our hypothesis to be that the urn contains white and coloured counters in equal numbers. Is this adjusted hypothesis supported by the evidence used to arrive at the revised, equal numbers, hypothesis? Howson and Urbach suggest, reasonably, that it is, and show why this is so on Bayesian grounds. The crucial factor that leads to the probability of the equal numbers hypothesis increasing as a result of the experiment that drew 495 white counters is the probability of drawing that number if the equal numbers hypothesis is false. Once it is agreed that that probability is small, the result that the experiment confirms the equal numbers hypothesis follows straightforwardly from the Bayesian calculus, even though the hypothesis was used in the construction of the data.

There is a standard criticism often levelled at the Bayesian approach that does strike at some versions of it, but I think the version defended by Howson and Urbach can counter it. To utilise Bayes' theorem it is necessary to be able to evaluate P(e), the prior probability of some evidence that is being considered. In a context where hypothesis h is being considered, it is convenient to write P(e) as P(e/h).P(h) + P(e/not h).P(not h), a straightforward identity in probability theory. The Bayesian needs to be able to estimate the probability of the evidence assuming the hypothesis is true, which may well be unity if the evidence follows from the hypothesis, but also the probability of the evidence should the hypothesis be false. It is this latter factor that is the problematic one. It would appear that it is necessary to estimate the likelihood of the evidence in the light of all hypotheses other than h. This is seen as a major obstacle, because no particular scientist can be in a position to know all possible alternatives to h, especially if, as some have suggested, this must include all hypotheses not yet invented. The response open to Howson and Urbach is to insist that the probabilities in their Bayesian calculus represent personal probabilities, that is, the probabilities that individuals, as a matter of fact, attribute to various propositions. The value of the probability of some evidence being true in the light of alternatives to h will be decided on by a scientist in the light of what that scientist happens to know (which will certainly exclude hypotheses not yet invented). So, for instance, when dealing with the Prout case, Howson and Urbach take the only alternative to Prout's hypothesis to be the hypothesis that atomic weights are randomly distributed on the basis of historical evidence to the effect that that is what the Proutians believed to be the alternative. It is the thoroughgoing nature of their move to subjective probabilities that makes it possible for Howson and Urbach to avoid the particular problem raised here.

In my portrayal of the elements of the Bayesian analysis of science, I have concentrated mainly on the position outlined by Howson and Urbach because it seems to me to be the one most free of inconsistencies. Because of the way in which probabilities are interpreted in terms of degrees of the beliefs actually held by scientists, their system enables non-zero probabilities to be attributed to theories and hypotheses, it gives a precise account of how the probabilities are to be modified in the light of evidence, and it is able to give a rationale for what many take to be key features of scientific method. Howson and Urbach embellish their system with historical case studies.

■ | Critique of Subjective Bayesianism

As we have seen, subjective Bayesianism, the view that consistently understands probabilities as the degrees of belief actually held by scientists, has the advantage that it is able to avoid many of the problems that beset alternative Bayesian accounts that seek for objective probabilities of some kind. For many, to embrace subjective probabilities is to pay too high a price for the luxury of being able to attribute probabilities to theories. Once we take probabilities as subjective degrees of belief to the extent that Howson and Urbach, for example, urge that we do, then a range of unfortunate consequences follow.

The Bayesian calculus is portrayed as an objective mode of inference that serves to transform prior probabilities into posterior probabilities in the light of given evidence. Once we see things in this way, it follows that any disagreements in science, between proponents of rival research programs, paradigms or whatever, reflected in the (posterior) beliefs of scientists, must have their source in the prior probabilities held by the scientists, since the evidence is taken as given and the inference considered to be objective. But the prior probabilities are themselves totally subjective and not subject to a critical analysis. They simply reflect the various degrees of belief each individual scientist happens to have. Consequently, those of us who raise questions about the relative merits of competing theories and about the sense in which science can be said to progress will not have our questions answered by the subjective Bayesian, unless we are satisfied with an answer that refers to the beliefs that individual scientists just happen to have started out with.

If subjective Bayesianism is the key to understanding science and its history, then one of the most important sources of information that we need to have access to in order to acquire that understanding are the degrees of belief that scientists actually do or did hold. (The other source of information is the evidence, which is discussed below.) So, for instance, an understanding of the superiority of the wave theory over the particle theory of light will require some knowledge of the degrees of belief that Fresnel and Poisson, for instance, brought to the debate in the early 1830s. There are two problems here. One is the problem of gaining access to a knowledge of these private degrees of belief. (Recall that Howson and Urbach distinguish between private beliefs and actions and insist that it is the former with which their theory deals, so we cannot infer beliefs of scientists from what they do, or even write.) The second problem is the implausibility of the idea that we need to gain access to these private beliefs in order to grasp the sense in which, say, the wave theory of light was an improvement on its predecessor. The problem is intensified when we focus on the degree of complexity of modern science, and the extent to which it involves collaborative work. . . . An extreme, and telling, example is provided by Peter Galison's (1997) account of the nature of the work in current fundamental particle physics, where very abstruse mathematical theories are brought to bear on the world via experimental work that involves elaborate computer techniques and instrumentation that requires state-of-the-art engineering for its operation. In situations like this there is no single person who grasps all aspects of this complex work. The theoretical physicist, the computer programmer, the mechanical engineer and the experimental physicist all have their separate skills which are brought to bear on a collaborative enterprise. If the progressiveness of this enterprise is to be understood as focusing on degrees of belief, then whose degree of belief do we choose and why?

The extent to which degrees of belief are dependent on prior probabilities in Howson and Urbach's analysis is the source of another problem. It would seem that, provided a scientist believes strongly enough in his or her theory to begin with (and there is nothing in subjective Bayesianism to prevent degrees of belief as strong as one might wish), then this belief cannot be shaken by any evidence to the contrary, however strong or extensive it might be. This point is in fact illustrated by the Prout study, the very study that Howson and Urbach use to support their position. Recall that in that study we assume that the Proutians began with a prior probability of 0.9 for their theory that atomic weights are equal multiples of the atomic weight of hydrogen and a prior probability of 0.6 for the assumption that atomic weight measurements are reasonably accurate reflections of actual atomic weights. The posterior probabilities, calculated in the light of the 35.83 value obtained for chlorine, were 0.878 for Prout's theory and 0.073 for the assumption that the experiments are reliable. So the Proutians were right to stick to their theory and reject the evidence. I point out here that the original incentive behind Prout's hypothesis was the near integral values of a range of atomic weights other than chlorine, measured by the very techniques which the Proutians have come to regard as so unreliable that they warrant a probability as low as 0.073! Does this not show that if scientists are dogmatic enough to begin with they can offset any adverse evidence? Insofar as it does, there is no way that the subjective Bayesian can identify such activity as bad scientific practice. The prior probabilities cannot be judged. They must be taken as simply given. As Howson and Urbach (1989, p. 273) themselves stress, they are "under no obligation to legislate concerning the methods people adopt for assigning prior probabilities".

Bayesians seem to have a counter to the Popperian claim that the probability of all theories must be zero, insofar as they identify probabilities with the degrees of belief that scientists happen, as a matter of fact, to possess. However, the Bayesian position is not that simple. For it is necessary for the Bayesians to ascribe probabilities that are *counterfactual*, and so cannot be simply identified with degrees of belief actually held. Let us take the problem of how past evidence is to count for a theory as an example. How can the observations of Mercury's orbit be taken as confirmation of Einstein's theory of general relativity, given that the observations preceded the theory by a number of decades? To calculate the probability of Einstein's theory in the light of this evidence, the subjective Bayesian is required, among other things, to provide a measure for the probability an Einstein supporter would have given to the probability of Mercury's orbit precessing in the way that it does *without a knowledge of Einstein's theory*. That probability is not a measure of the degree of belief that a scientist actually has but a measure of a degree of belief they would have had if they did not know what they in fact do know. The status of these degrees of belief, and the problem of how one is to evaluate them, pose serious problems, to put it mildly.

Let us now turn to the nature of "evidence" as it figures in subjective Bayesianism. We have treated the evidence as a given, something that is fed into Bayes' theorem to convert prior probabilities to posterior probabilities. However, . . . evidence in science is far from being straightforwardly given. The stand taken by Howson and Urbach (1989, p. 272) is explicit and totally in keeping with their overall approach.

> The Bayesian theory we are proposing is a theory of inference from data; we say nothing about whether it is correct to accept the data, or even whether your commitment to the data is absolute. It may not be, and you may be foolish to repose in it the confidence you actually do. The Bayesian theory of support is a theory of how the acceptance as true of some evidential statement affects your belief in some hypothesis. How you come to accept the truth of the evidence and whether you are correct in accepting it as true are matters which, from the point of view of the theory, are simply irrelevant.

Surely this is a totally unacceptable position for those who purport to be writing a book on *scientific reasoning*. For is it not the case that we seek an account of what counts as appropriate evidence in science? Certainly a scientist will respond to some evidential claim, not by asking the scientist making the claim how strongly he or she believes it, but by seeking information on the nature of the experiment that yielded the evidence, what precautions were taken, how errors were estimated and so on. A good theory of scientific method will surely be required to give an account of the circumstances under which evidence can be regarded as adequate, and be in a position to pinpoint standards that empirical work in science *should* live up to. Certainly experimental mental scientists have plenty of ways of rejecting shoddy work, and not by appealing to subjective degrees of belief.

Especially when they are responding to criticism, Howson and Urbach stress the extent to which both the prior probabilities and the evidence which need to be fed into Bayes' theorem are subjective degrees of belief about which the subjective Bayesian has nothing to say. But to what extent can what remains of their position be called a theory of scientific method? All that remains is a theorem of the probability calculus. Suppose we concede to Howson and Urbach that this theorem, as interpreted by them, is indeed a theorem with a status akin to deductive logic. Then this generous concession serves to bring out the limitation of their position. Their theory of scientific method tells us as much about science as the observation that science adheres to the dictates of deductive logic. The vast majority, at least, of philosophers of science would have no problem accepting that science takes deductive logic for granted, but would wish to be told much more.

■ | References

Bamford, G. (1993). "Popper's Explications of Ad Hocness: Circularity, Empirical Content and Scientific Practice", *British Journal for the Philosophy of Science*, 44, 335–55.

Dorling, J. (1979). "Bayesian Personalism and Duhem's Problem", *Studies in History and Philosophy of Science*, 10, 177–87.

Galison, P. (1997). *Image and Logic: A Material Culture of Physics*. Chicago, University of Chicago Press.

Howson, C. and Urbach, P. (1989) *Scientific Reasoning: The Bayesian Approach*. La Salle, Ill., Open Court.

Paul Horwich

Therapeutic Bayesianism

Belief is not an all-or-nothing matter. Rather, there are various *degrees* of conviction which may be represented by numbers between zero and one. Were we ideally rational, our full beliefs (of degree one) would comply with the laws of deductive logic; they would be consistent and closed under logical implication. And similarly, our *degrees* of belief should conform to the probability calculus.[1] This enrichment of epistemology—provided by the addition of degrees of belief and an appreciation of their probabilistic 'logic'—fosters progress with respect to many problems in the philosophy of science.

These statements form the core of a program, which I will call "therapeutic Bayesianism," whose primary goal is the solution of various puzzles and paradoxes that come from reflecting on scientific method. Its creed is that many of these problems are the product of oversimplification, and that the above-mentioned elementary probabilistic model of degrees of belief often contains just the right balance of accuracy and simplicity to enable us to command a clear view of the issues and see where we were going wrong.[2] This somewhat Wittgensteinian goal and creed distinguishes therapeutic Bayesianism from more systematic enterprises in which probabilistic degrees of belief play a prominent role: for example, Bayesian decision theory, Bayesian statistics, Bayesian psychology, Bayesian semantics, and Bayesian history of science. It is especially important to appreciate the difference between the problem-solving orientation of therapeutic Bayesianism—that of exploiting a simple, idealized model in order to help illuminate notorious philosophical perplexities—and the quite distinct project of providing a perfectly true and complete (descriptive or normative) *theory* of scientific practice. The latter task might well involve the postulation of belief-gradations, and might also

From P. A. French, T. E. Uehling, Jr., and H. K. Wettstein, eds., *Midwest Studies in Philosophy*, vol. 18 (Notre Dame, Ind.: University of Notre Dame Press, 1993), 62–77. This article originally appeared with the title "Wittgensteinian Bayesianism."

be done in the name of philosophy of science. However, its aims are quite different; and one must beware of judging one project by adequacy conditions appropriate to the other.[3]

Therapeutic Bayesianism is not self-evidently beneficial, but it does have some prima facie plausibility. Moreover, this plausibility is enhanced by substantial accomplishments, and, as we shall see, a great deal of the criticism it has received is misdirected—commonly for the reason just indicated. In this paper I would like to try to make a case for the program by discussing it from three, progressively abstract, points of view: substantial, foundational, and metaphilosophical. More specifically, there will follow sections on: (I) "The fruitfulness of therapeutic Bayesianism," in which I will sketch treatments of the 'raven' paradox and the question of diverse data and mention various other applications; (II) "Probabilistic foundations," in which the propriety of certain idealizations will be defended—particularly the representation of belief by numbers, the adoption of probabilistic canons of reason governing such beliefs, the definition of confirmation as increase in rational degree of belief, and the idea that induction may be codified in a confirmation function; and (III) "Misplaced scientism," in which I criticize a metaphilosophical perspective that does not properly distinguish science from the philosophy of science, and which overvalues the use of symbolic apparatus. Along the way, I shall respond to some criticisms of therapeutic Bayesianism that have recently been advanced.

I | The Fruitfulness of Therapeutic Bayesianism

A good illustration of therapeutic Bayesianism at work is its way of treating the notorious 'raven paradox'. It is plausible to suppose that any hypothesis of the form 'All Fs are G' would be supported by the observation of an F that is also G. But if this is generally true, then the discovery of a non-black non-raven (e.g., a white shoe) confirms that all non-black things are non-ravens; and thereby confirms the logically equivalent hypothesis, 'All ravens are black'—a seemingly bizarre conclusion. This is 'the paradox of confirmation'. The Bayesian approach to this problem is to argue that observing a known raven to be black will *substantially* confirm "All ravens are black," whereas observing that a known non-black thing is not a raven will confirm it only *negligibly*—the difference being explained, roughly speaking, by the fact that, given our background beliefs about the chances of coming across ravens and black things, the first of these observations is more surprising, more of a test of the hypothesis, and therefore more evidentially powerful, than the second. Thus, the paradoxical flavor of our conclusion comes from the not unnatural confusion of negligible support with no support at all—a confusion sustained by inattention to degrees of belief and their bearing on confirmation.

A formal version of this analysis proceeds from the following premises:

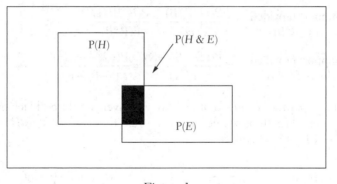

Figure 1

a That the amount of support for hypothesis *H* provided by evidence *E* is the factor by which the rational degree of belief in *H* is enhanced by the discovery of *E*—which is indicated by the ratio of subjective probabilities, P(*H/E*)/P(*H*), for a rational person.

b That a rational person's degrees of belief will ideally conform to the probability calculus; and, in particular, will obey Bayes's Theorem:

$$\frac{P(H/E)}{P(H)} = \frac{P(E/H)}{P(F)}$$

(To appreciate the intuitive plausibility of this theorem, note that it derives from the fact that the conditional probability of H given E is equal to the probability of the conjunction of H and E, divided by the probability of E: i.e., P(*H/E*) = P(*H* & *E*)/P(*E*). See figure 1. Therefore, since P(*H* & *E*) = P(*E* & *H*), we obtain P(*H/E*)P(*E*) = P(*E/H*)P(*H*), and hence Bayes's Theorem).

c That our degree of belief (prior to the investigation, and given the known scarcity of ravens) that a randomly selected non-black thing would turn out to be a non-raven is high.

d That our prior degree of belief (prior to the investigation, and given the known abundance of non-black things) that a randomly selected raven would turn out not to be black is substantial.

Now let us compare the support for the hypothesis, *H*, that all ravens are black, provided, first, by the discovery concerning a known raven that it is black (which is symbolized as *R* B*), and, second, by the discovery that a known non-black thing is not a raven (−*B* −R*). Applying premise (a) and then (b), we find:

$$\begin{array}{c} \text{Support provided} \\ \text{by } (R^*B) \end{array} = \frac{P(H/R^*B)}{P(H)} = \frac{P(R^*B/H)}{P(R^*B)}$$

$$\begin{array}{c} \text{Support provided} \\ \text{by } (-B^* - R) \end{array} = \frac{P(H/-B^*-R)}{P(H)} = \frac{P(-B^*-R/H)}{P(-B^*-R)}$$

But our hypothesis *entails* that any known raven would be black and any known non-black thing would not be a raven; therefore, $P(R^* B/H) = 1$ and $P(-B^* - R/H) = 1$. Therefore

$$\begin{array}{c} \text{Support from} \\ \text{raven found to} \\ \text{be black} \end{array} = \frac{P(H/R^*B)}{P(H)} = \frac{1}{P(R^*B)} = \begin{array}{c} 1/ \text{ prior degree of} \\ \text{belief that a known} \\ \text{raven would be black} \end{array}$$

$$\begin{array}{c} \text{Support from} \\ \text{non-black thing} \\ \text{found not to be} \\ \text{a raven} \end{array} = \frac{P(H/-B^*-R)}{P(H)} = \frac{1}{P(-B^*-R)} = \begin{array}{c} 1/ \text{ prior degree of belief} \\ \text{that a known non-black} \\ \text{thing would not be} \\ \text{a raven} \end{array}$$

Now one may assume (premise (c)) that a normal investigator of the hypothesis has prior background knowledge about the rough distribution of ravens and black things in his vicinity, and that this will lead him to expect that there is a very good chance that a randomly selected non-black thing will turn out not to be a raven. Thus $P(-B^* - R)$ is very nearly 1; and the amount of support for the hypothesis provided by observing that a non-black thing is not a raven is very little.

On the other hand, we would expect the background of investigation to dictate, in addition, (premise (d)) that the likelihood of a randomly selected raven being black is not especially high. After all, as far as we know at the outset of the research, there are many colors that the raven could perfectly well have. Thus $P(R^* B)$ is a good deal less than one. Therefore, the amount of support provided by observing that a known raven is black is substantial.

One might object to this reasoning that the final assumption is false, since the *objective* chances of finding that a raven is black are actually extremely high. However, this objection is based on a slip which is easy to identify. It confuses "probability" in the sense of *subjective degree of belief* and "probability" in the sense of *relative frequency*. All the probabilities mentioned in the argument are rational subjective probabilities, and it is under that construal that we may reasonably assume that $P(R^* B)$ is not near to 1. The feeling that this assumption is wrong derives from incorrectly reading $P(R^* B)$ as a relative frequency assertion. In that sense, since in fact almost all ravens are black, the probability that a randomly selected raven will be black is indeed very great. But this fact has no bearing on the argument.[4]

A similar objection is to deny that there could be any difference in evidential import between identifying a known raven as black and identifying a known black thing as a raven. Howson and Urbach,[5] for example, maintain that the only difference between these two data is the time order in which the elements of the observed fact are established. They think that in each case what is eventually known is the same, so there can be no variation in confirmation power between the two discoveries. Imagine, however, that an ornithologist instructs her assistant to go and find a black raven and bring it back to the lab for inspection. Surely, that inspection would count for nothing. And there is no paradox here, even though we might loosely speak of 'seeing a black raven' in all three cases. For a more precise characterization of the evidence shows that what is discovered in each case is not really the same. That a randomly selected raven turns out to be black, that a randomly selected black thing turns out to be a raven and that a randomly selected black raven turns out to be a black raven, are very different pieces of information, and it should not be surprising that they confirm our hypothesis to different degrees.

Therapeutic Bayesianism handles other issues in the philosophy of science similarly, putting a lot of weight on premises (a) and (b). By combining the idea of confirmation as enhancement of rational degree of belief, with the principle that rational degrees of belief should satisfy the probability calculus, we get a way of treating those problems that hinge upon considerations having to do with *degree of support*. Therefore the method has a wide scope. In particular, one can expect to shed light on why 'surprising' predictions have relatively great confirmation power, what is wrong with ad hoc hypotheses, whether prediction has more evidential value than mere accommodation of data, why a broad spectrum of facts can confirm a theory more than a narrow data set, why we base our judgments on as much data as possible, how statistical hypotheses can be testable despite their unfalsifiability, what is peculiar about 'grue-like' hypotheses, and various other problems.*

These issues are unified by their involvement with the notion of 'varying evidential quality'; and this is why traditional epistemology, with its fixation on all-or-nothing belief, is not able to resolve them. It is only to be expected that the introduction of degrees of belief, together with an understanding of the rational constraints to which they are subject, would open the way to progress. Of course, there is not the space here to fully substantiate this thesis by describing all these applications of therapeutic Bayesianism. Let me, however, give one further illustration of the approach.

How is it that a broad spectrum of different kinds of fact, when entailed by a hypothesis, will confirm it to a greater degree than a uniform, repetitive set of data? It is natural to answer as follows. To the extent that our observations cover a broad range of phenomena, they are capable of falsifying a large number of alternative hypotheses, which then bequeath substantial credibility to those hypotheses that survive. Now, this solution does not

---

* For a discussion of grue-like hypotheses, see Nelson Goodman, "The New Riddle of Induction" and the associated commentary in chapter 4.

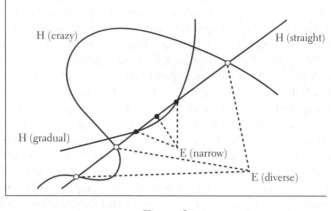

Figure 2

quite work. For a narrow data set can preclude just as *many* hypotheses as a diverse data set. Nevertheless, we can repair the solution by noting that there is a significant difference in the *kinds* of hypotheses that are excluded by the two sets of facts. We should notice that the diverse data tend to exclude more of the *simple* hypotheses than do the narrow data. Given a representation of simplicity in terms of high prior probability,[6] this suggests that diverse data tend to rule out more high-probability alternatives than narrow data. But if so, then a hypothesis that survives relatively diverse observations becomes more probable than one that is left in the running by a narrow set of data. (See figure 2).

In particular, the data points E(narrow) exclude (given experimental error) just as many alternatives to the line H(straight) as does E(diverse). Nonetheless E(diverse) confirms H(straight) more strongly than E(narrow) does, because E(diverse) is better than E(narrow) at excluding simple alternatives to H(straight)—for example, gradual curves—which have an initially high probability. Thus H(gradual) is ruled out by E(diverse) but not by E(narrow). On the other hand, the sort of hypothesis, like H(crazy), prohibited by E(narrow) yet not by E(diverse), is not very probable anyway; so excluding it does not greatly benefit those hypotheses that survive. Thus, with the help of a probabilistic representation of simplicity, we can begin to account for our methodological intuitions concerning diverse data.[7]

II | Probabilistic Foundations

In the last section I have tried to indicate something of the fruitfulness of therapeutic Bayesianism. Let me now consider various foundational questions that might be thought to cast doubt on the project:

1 Do people actually have numerical degrees of belief?

2 If so, can it be shown that *rational* degrees of belief conform to the probability calculus?

3 Is it correct to identify degree of confirmation with rational enhancement of subjective probability?

4 Are there objective facts of confirmation?

5 Does reason require *merely* that one's beliefs conform to the probability calculus? Or is it the case (as Carnap thought) that a rational system of beliefs is subject to several further constraints?

6 If further constraints are needed, then what are they?

On the first question, perhaps we should be agnostic. The successes of therapeutic Bayesianism will reinforce the evident fact that its basic principles are at least *roughly* correct. Thus we know that there are belief gradations of some sort, that there are rational constraints governing them (prohibiting, for example, a high degree of confidence in two contradictory propositions), and that confirmation is not wholly unrelated to increasing belief. Moreover, the Bayesian representation of these ideas has a great deal of plausibility. Consider a spectrum of situations in which we know that the propensities (as manifested by relative frequencies) of certain events are $x_1, x_2, \ldots, x_N$. For each such case there is a corresponding epistemic attitude—a degree of confidence—that the next trial will produce an event of the designated type. Presumably the appropriate attitude will vary with the relative frequency. Specifically, since the frequencies range over numbers between zero and one, so will the degrees of confidence.

However, despite the attractiveness of such considerations, one must of course acknowledge that the Bayesian framework might be wrong. The crude ideas that it represents should not be controversial. However, it is quite possible that the Bayesian articulation of those ideas is not absolutely right; and that, in particular, the assumption of precise-valued, numerical degrees of belief is incorrect.

Even so, such a model could be an excellent idealization, sufficing perfectly well for the primary purposes of therapeutic Bayesianism: namely, to dispel confusion, solve problems, and thereby improve our understanding of the scientific method. For the paradoxes are caused by forgetting the *crude* facts (that there are gradations of belief, etc.) or by failing to recognize their significance. And so the solutions will involve noticing that those rough ideas have been overlooked and coming to appreciate how they bear on the problems. This sort of treatment will not depend essentially on any particular theoretical refinements. The function of the Bayesian framework is merely to cast the crude, uncontroversial ideas into a form where their impact on our problems can have maximal clarity and force.

What then is the import of studies that cast doubt on the existence of numerical degrees of conviction and which develop more complex and allegedly more realistic conceptions of belief? Let me stress that this work falls well outside the focus of therapeutic Bayesianism, for there is no reason to believe that such improvements will help to solve the standard problems in the philosophy of science. Perhaps these developments are important in psychology, statistics, semantics, or decision theory; perhaps they will become important to philosophy when we have progressed enough in our understanding of science so that the *details* of an inductive logic become items of reasonable concern. But at this juncture, confusion is rampant, the traditional problems are still very much with us, and it seems rather unlikely that the slight gains in accuracy to be derived from a more realistic theory of belief would be worth the price—in terms of loss of simplicity—that we would have to pay for it.

Our treatment of the 'raven paradox' is a case in point. The problem was solved by exposing a certain misconception (that a non-black, non-raven would be irrelevant to our hypothesis), and by explaining why we are so tempted by that misconception: namely, that in forgetting about degrees of belief, we lose sight of the distinction between very slight confirmation and no confirmation at all. The simple Bayesian model of belief provides a sufficiently perspicuous representation of the situation to enable us to put this in a clear way. Further accuracy regarding the nature of belief would distract us from the main point, ruin the argument, and not help us to understand the basis of the paradox.

Let me give another example. The conflict between realism and instrumentalism with respect to the acceptance of scientific theories is fueled by a shared tendency to think in terms of all-or-nothing belief. The instrumentalist argues, in light of previous scientific revolutions, that it is foolishly optimistic to expect that our current theories are true and will not eventually be refuted. Whereas the realist complains that it is a distortion of science to distinguish rigidly between credible observation reports and incredible theoretical claims. However, once we see that the issue is not 'To believe or not to believe?' but rather, 'To what degree shall we believe?' then there is room for reconciliation. The crucial move is the elimination of the shared misconception. There is no reason to think that a fancy model of belief, even more accurate than the Bayesian idealization, would be any further help with the problem.

I do not mean to be suggesting that it is not worthwhile to investigate more sophisticated models of belief. On the contrary, I can readily imagine research programs—e.g., Bayesian psychology, or attempts to give a perfectly accurate description of scientific practice—in which this would be crucial. My point is that there is another enterprise—the one I am calling "therapeutic Bayesianism"—whose focus is on solving the traditional methodological puzzles and paradoxes, and for which the introduction of such complex models is likely to do more harm than good.

Suppose, then, that we do have numerical degrees of belief. Is there any way of justifying the Bayesian assumption that, to be rational, these degrees of belief must conform to the probability calculus? Although there are indeed various lines of reasoning which purport to establish this thesis, none is compelling. The best known of them is the 'Dutch book' argument[8] and it goes roughly as follows. Defining a person's degree of belief in a proposition as a function of the odds at which he is prepared to bet on its truth, it can be proved that if his degrees of belief do not satisfy the probability calculus then he will be prepared to accept a collection of bets which is guaranteed to lead to a loss. Therefore, since it would surely be irrational for him knowingly to put himself in such a no-win situation, it would be irrational to have a system of degrees of belief that violates the probability calculus. QED. However, the definition of 'degree of belief' that is employed in this argument presupposes that people maximize their expected utility. And there is a lot of room for skepticism about that assumption (and about the preference axioms to which it is equivalent). So the 'Dutch book' argument is far from airtight. Worse still, there is positive reason to think that its conclusion is false, for it requires logical omniscience. The probability of any logical truth is 1 and of any contradiction is 0. Yet it is surely quite rational to be less than perfectly confident in the truth of *some* logical truths—those that are especially hard to prove—and quite rational to give non-zero degrees of belief to contradictions that are hard to recognize as such.

The proper response to these difficulties is to repeat that the picture of rational degrees of belief obeying the probability calculus should be regarded as an *idealization* of the real normative facts. It is uncontroversial that one ought to be certain of elementary logical truths, and that one ought not be confident of the truth of obviously incompatible hypotheses. The probabilistic model of belief provides a sharp, perspicuous way of capturing these trivialities, and to the extent that it goes beyond them it need not be construed realistically.

A similar answer may be given to the third question concerning the definition of confirmation. In our discussion of the raven paradox we defined "the degree by which E confirms H" as "the ratio, P(H/E)/P(H), for a rational person." Evidently, this explication has at least *some* prima facie plausibility, and it certainly helps us to give a neat, compelling solution to the problem. Nonetheless, it is often argued that this particular explication is 'wrong'—yielding counterintuitive consequences—and that there are better definitions of confirmation which should be used instead.[9]

However, these criticisms have little relevance to the project of therapeutic Bayesianism. No doubt our explication leads to some strange-sounding consequences. No doubt it is strictly speaking false that the ordinary meaning of "confirms" is given by our explication. No doubt there are definitions (perhaps involving non-probabilistic notions) that come closer to what we ordinarily mean. But the object of therapeutic Bayesianism is not to give a theory of science. We are not trying to find the most accurate analyses of our concepts, but rather to use explications that are at least roughly right, and which are conducive to simple, convincing dissolutions of philosophical problems.

Since we assume that these problems are the product of confusion, it is desirable to look for ways of clarifying the issues, which have the proper blend of accuracy and simplicity. Of course it is possible to *over*simplify. But one can conclude that this has happened only after finding that the admittedly idealized models do not in fact help to solve our problems.[10]

On the fourth question—Are there objective facts of confirmation?—it seems evident that judgments of credibility and confirmation do purport to capture objective normative facts. They do not state what any individual's degrees of belief *actually* are, but rather they say something about what one's degrees of belief *ought* to be, or how they *ought* to change given the circumstances. Thus we should acknowledge non-subjective facts regarding confirmation.

A natural way of capturing this idea, due to Carnap, is to suppose that an attribution of probability to a hypothesis reflects the belief in an objective, logical fact about the degree to which one statement—a summary of the available evidence—probabilifies [makes probable] another statement—the hypothesis in question. Such logical facts might be codified in a confirmation function, $c(p/q) = x$, which would specify explicitly the degree, x, to which q confirms p, and would specify implicitly the degree to which one should believe p if the total evidence is q. Carnap says, for example:

> Probability-1 is the degree of confirmation of a hypothesis h with respect to an evidence statement e, e.g., an observation report. This is a logical, semantical concept. A sentence about this concept is based, not on observation of facts, but on logical analysis; if it is true, it is L-true (analytic). . . . Probability-2 [relative frequency] is obviously an objective concept. It is important to recognize that probability-1 is likewise objective.
>
> Let h be the sentence 'there will be rain tomorrow' and j the sentence 'there will be rain and wind tomorrow'. Suppose someone makes the statement in deductive logic: 'h follows logically from j'. . . . The statement 'the probability-1 of h on the evidence e is ⅕' has the same general character as the former statement. . . . Both statements express a purely logical relation between two statements. The difference between the two statements is merely this: while the first states a complete logical implication, the second states, so to speak, a partial logical implication; hence, while the first belongs to deductive logic, the second belongs to inductive logic.[11]

Thus, Carnap held that certain facts about confirmation are analytic and *objective*, and thought of inductive probability as a *partial* version of the logical relation of entailment.[12]

On the fifth question—Does reason impose constraints on belief *over and above* the requirement of conformity with the probability calculus?—there are grounds for sympathy with Carnap's view that it does. For it is hard to see how the probabilistic constraint alone can account for our intuitions about the relative plausibility of competing hypotheses that equally well fit the current data. In particular, it is hard to see how it can solve the 'grue' problem.[13]

Suppose that such further constraints are indeed required. Still, to take up the sixth question, it is no trivial matter to say what they are. Carnap tried out various constraints and employed them to derive confirmation functions for certain extremely simple formal languages. Unfortunately, these functions have the counterintuitive property that laws of nature are never able to acquire more than a negligible probability. And this shows a deficiency in Carnap's constructions: either the languages are too simple, or the constraints are wrong. However, one can certainly not conclude that anyone who endorses a Carnapian conception of logical probability *must* hold that general laws never attain a non-negligible probability. This is a non sequitur, arising from a failure to distinguish between the general conception of logical probability and the admittedly inadequate prototypes with which Carnap experimented.[14]

For the treatment of various problems it is helpful to suppose that inductive reasoning is represented by a specific (but unspecified) real-valued Carnapian confirmation function, c, allowing general laws to achieve a non-negligible credibility. In light of our responses to questions (1), (2), and (3), we see that it can be no objection to this procedure that our inductive practice is not in fact precisely described by a single c-function. For, once again, the intention is not to get at the exact truth, but merely to employ a useful idealization. Nor—as we have just said—is it fair to complain that *some* c-functions—those Carnap toyed with—always give zero probability to general laws. For we can suppose that 'the right c-function' is one of those that do *not* have that counterintuitive feature.

III | Misplaced Scientism

Much criticism of therapeutic Bayesianism arises from a conflation of philosophy and science. More exactly, it derives from a failure to recognize the legitimacy (even the existence) of non-scientific philosophical projects— those prompted, not by a desire to expose the whole truth regarding some domain, but by an interest in the resolution of paradoxes. Let me elaborate.

What one might call '*theory*-oriented philosophy of science' aims for a systematic account of the scientific method. The criteria of success are just those that pertain to theory construction *within* particular sciences: namely, empirical adequacy, scope, depth, simplicity, internal consistency, and coherence with the rest of our knowledge. More specifically, a perfect theory of the scientific method would be expected to conform with specific intuitions about the way that good science is done, to cover all aspects of methodology in detail, to expose fundamental principles enabling the complex, superficial aspects of scientific practice to be unified and explained, and to respect results in psychology and sociology. Thus it seems appropriate to regard theory-oriented philosophy of science as itself a department of science—a branch of naturalized epistemology. This characterization neither ignores nor denies that scientific methods are normative. A description of science will contain a codification of the basic norms which are implicit in the evaluation of theories. Moreover, it is quite possible that an identification of basic normative principles will result in the exposure of cases in which science is being done badly. One might thereby effect an improvement in the conduct of some science. Indeed, this may well be a motive for engaging in theory-oriented philosophy of science.

In contrast, the approach portrayed in this paper, 'problem-oriented philosophy of science', has very different goals, methods, and adequacy conditions. It aims at the resolution of deep puzzles and paradoxes that arise from reflection upon science. It includes in its domain, for example, the problem of induction, the paradox[es] of confirmation, the question of total evidence, and the issue of prediction versus accommodation. The problems here are not simply to fill various undesirable gaps in our knowledge about science. Characteristically, they are conceptual tensions, contradictions, absurd conclusions—that is to say, symptoms of confusion. We have somehow gone astray, and the task is to understand how this has happened and to get a clear view of the issue so that our misguided ways of thinking will be exposed and no longer seem so attractive.

These two approaches to the philosophy of science do not compete with one another. They are distinct projects with distinct objectives—not *wholly* unrelated to one another, but by no means simply parts of the same enterprise. Thus, it is not the case that the sort of full understanding provided by a successful theory-oriented philosophy of science would automatically solve the puzzles that form the domain of problem-oriented philosophy of science. For the resolution of a paradox requires a great deal more than just locating the wrong move in a fallacious argument. It is crucial to a proper resolution that one comes to see why that fallacy was natural. And it is important that one obtains a new perspective on the issue—a point of view from which the old and troublesome habits of thought no longer seem plausible. These elements of the solution do not simply fall out of a complete theory of science. (Similarly, the explanation of conjuring tricks does not follow from physics.) Thus, theory-oriented philosophy of science is not simply a more thorough, systematic, and ambitious project than problem-oriented philosophy of science.

Neither is it necessary, in order to solve problems, that one be in possession of an adequate theory of science. For confusions can be identified, understood, and removed without a theory of any particular depth or generality. Granted, assumptions about methodology will often be involved in the diagnosis and treatment of a problem, and if these were *wildly* false then it is unlikely that the discussion would be helpful. However, there is no reason why such assumptions should be *true* as long as their replacement with the truth would not undermine the solution that is based on them. Indeed it is quite possible that the perfect theory of science would be a very bad tool for solving problems. For the truth may be so complicated that it cannot provide the sort of simple and relevant perspective that is needed.

If the practice of conceptual troubleshooting is confused, as it often is, with the scientific search for a theory of science, then therapeutic Bayesianism will be wrongly subjected to all of the methodological requirements that are properly applied only in science. Let me describe some of the bad effects of this confusion.

One consequence, discussed above, of not seeing the distinctive aim of therapeutic Bayesianism is a tendency to misjudge the function of various helpful idealizations. Thus one commonly finds objections to the use of precise-valued degrees of belief, to the assumption that these should conform to the probability calculus, to the adoption of a particularly simple explication of confirmation, and to the idea that our inductive practice may be represented by a single Carnapian confirmation function. Doubtless some of these assumptions are, strictly speaking, false. (Just as it is false that a gas is made of point masses.) And in a different kind of study—one aimed at truth—it would be very important to discuss more realistic models. However, for the purposes of therapeutic Bayesianism it is important to use the simplest roughly accurate models of degrees of belief and of confirmation that will help to clarify the issues, and it is sufficient to proceed on the basis of their intuitive plausibility and to justify these models in retrospect in terms of their utility.

Secondly, a scientific understanding of confirmation aims for the truth, the *whole* truth, and nothing but the truth. Consequently, those wedded to this conception of the philosophy of science will find fault with studies that do not discuss every significant aspect of the phenomenon of confirmation. Consider, for example, *prior probability assignment*, the procedures for deciding, before data have been gathered, the various 'intrinsic' plausibilities of hypotheses; *belief-kinematics*, the way that systems of belief change over time in the light of new discoveries; or *direct inference*, the impact on our degrees of belief of a knowledge of empirical probabilities. These are fascinating topics, and a good theory of science must deal with them. But there is no reason why a paradox-oriented Bayesian program should incorporate a complete, systematic account of all such elements of methodology.[15]

In the third place, scientism in philosophy engenders a 'hyperformalist' fixation on symbolic technique—an overvaluation of logicomathematical machinery. Among the symptoms of this hyperformalist state are: (a) a blindness to the possibility of philosophical problems distinct from the scientific and mathematical issues that arise in statistics, decision theory, sociology of science, etc., further questions being dismissed as 'merely verbal';[16] (b) a dissatisfaction with informal discussions and conclusions; (c) an exaggerated concern with formal rigor for its own sake; and (d) an obsession with the elimination of any potential ambiguity or vagueness, leading to the feeling that the English language is too confusing and vague a medium for intellectual progress, and that it should, wherever possible, be replaced with mathematics or logic.

Thus, even if an approach employs formal techniques, as therapeutic Bayesianism clearly does, it may still be subjected to hyperformalist criticism. I think this is an unhealthy point of view—in philosophy generally, and particularly in the philosophy of science, where it is especially common. No doubt there are occasions when clarity is gained and confusion allayed with the help of formal apparatus. This, I believe, is one of the morals of Bayesianism's success. However, one can withdraw too quickly into the secure, regulated territory of a formal system. It is certainly a tempting relief from the frustrating vagaries of philosophy to be able to obtain definite, proven results and get clear answers to clear questions. But, unless we are very careful, these answers and results might have little to do with the problems that have traditionally motivated philosophy of science. Our methodological puzzles arise when we reflect informally about scientific practice; and they can be solved only with an appreciation of the misconceptions and confusions to which we are prone and an understanding of the ways in which they are fostered by the rich conceptual resources put at our disposal by natural language. It seems to me that only when that sort of understanding is eventually attained will we know what we are looking for in a fully fledged inductive logic; and then, perhaps, be in a better position to devise one. But this level of understanding will not be achieved by trying to express as many questions as possible within a formal system, proving some theorems, and dismissing the residue as intractable and uninteresting. At its worst, such scientistic hyperformalism betrays a lack of concern for truly philosophical problems. If "merely verbal" issues are any that do not make a scientific difference, and if only scientific problems are worth worrying about, then philosophy is truly an endangered enterprise.

I hope to have clarified what I believe is a valuable approach to the philosophy of science, and to have shown that many of the complaints about it derive from scientistic hyperformalism and are therefore misconceived. The goal is not a theory of science but the unravelling of puzzles surrounding our ideas about surprising data, prediction versus accommodation, ad hoc postulates, statistical hypotheses, our thirst for new data, the tenability of realism, and other aspects of methodology. And given some of the successes of therapeutic Bayesianism there is reason to have a fair amount of confidence in its basic principles.

Thus, the notion of rational degrees of belief conforming to the probability calculus has an important role in the philosophy of science. It would no doubt be easier to think in terms of all-or-nothing belief, but that oversimplification is part of what engendered our methodological puzzles in the first place. On the other hand, there are more complex and realistic conceptions of belief, but the cause of clarity is not served by using them. Therapeutic Bayesianism appears to offer the ideal compromise between accuracy and simplicity, enabling us to represent the issues starkly without neglecting the essential ingredients or clouding them with unnecessary details.[17]

■ | Notes

1. The axioms of elementary probability theory are as follows: (1) probabilities are less than or equal to one; (2) the probability of a necessary truth is equal to one; (3) if two statements are jointly impossible, then the probability that at least one of them is true is equal to the sum of their individual probabilities; and (4) the conditional probability of p given q equals the probability of the conjunction of p and q divided by the probability of q.

2. This project is attempted in my *Probability and Evidence* (Cambridge, 1982), henceforth abbreviated as *P & E*. The metaphilosophical outlook is inspired by Wittgenstein's *Philosophical Investigations*, paragraphs 88–133.

3. Bayesian programs of various kinds have been developed in the work of Rudolf Carnap, David Christensen, R. T. Cox, Bruno de Finetti, Ron Giere, I. J. Good, John Earman, Ellery Eells, Hartry Field, Allan Franklin, Ian Hacking, Mary Hesse, Jaakko Hintikka, Colin Howson, E. T. Jaynes, Richard Jeffrey, Harold Jeffreys, Mark Kaplan, J. M. Keynes, Henry Kyburg, Isaac Levi, Patrick Maher, Roger Rosenkrantz, Wesley Salmon, L. J. Savage, Teddy Seidenfeld, Abner Shimony, Brian Skyrms, Patrick Suppes, Peter Urbach, Bas van Fraassen and others. Much of this work (especially the studies by Good, Hesse, Howson & Urbach, and Earman) contains contributions to therapeutic Bayesianism. However, I cannot attribute to these philosophers the project that I have in mind by that label, since their work is oriented towards the discovery of a 'theory of science', and thus reflects a metaphilosophical point of view that is quite distinct from that of the program which I am calling "therapeutic Bayesianism."

4. Stephen Spielman's objection is based on the mistake described here: the identification of the probabilities with objective proportions. (See his review of *P & E*, *Journal of Philosophy* 81 [March 1984]: 168–73. Page references for Spielman are to this work).

To keep things relatively simple I have assumed that there are just two observations in question: namely, the discovery regarding a randomly selected raven that it is black and the discovery regarding a randomly selected non-black thing that it is not a raven. If we consider instead the discovery that a known black thing is a raven, or various other ways of seeing black ravens and non-black non-ravens, then the existence of confirmation depends on the presence of special additional background assumptions (e.g., that ravens are quite likely all to have the same color). Nonetheless a similar contrast between the degrees of confirmation provided by black ravens and non-black non-ravens may be established. In *P & E* I suggest that these other ways of seeing black ravens would provide *no* confirmation of the hypothesis. This is misleading. Sometimes our background theories include a belief in the projectibility of the generalization in question, and in that case all the ways of observing an instance of it will normally provide confirmation.

My treatment of the paradox is, in a couple of respects, different from Patrick Suppes's analysis ("A Bayesian Approach to the Paradoxes of Confirmation," in *Aspects of Inductive Logic*, edited by J. Hintikka and P. Suppes, [Amsterdam 1966]). In the first place, he does not distinguish between the discovery that a randomly selected object is a black raven and the discovery that a randomly selected raven is black; whereas it is a significant feature of my account that in certain circumstances only the latter datum would confirm the hypothesis. And secondly, he does not obtain his results from the basic principles of Bayesianism—the thesis that degrees of belief should conform to the probability calculus; rather, he starts with the assumption that surprising observations have greater confirmation power; and this, though correct, is much better derived than simply presupposed.

5. C. Howson and P. Urbach, *Scientific Reasoning: The Bayesian Approach* (La Salle, Ill., 1989).

6. An argument for associating simplicity with high prior probability is given in *P & E*, 70–71.

7. Teddy Seidenfeld maintains that this account goes in the "wrong direction." But he gives no grounds for that claim other than to note the above-mentioned deficiencies in our understanding of simplicity—our inability to solve either the descriptive or the normative problems surrounding it. And it seems to me that his observation is irrelevant in the absence of any reason to believe either (a) that we can get a satisfactory explication of simplicity in terms of evidential diversity, or (b) that the Bayesian account would not withstand a better grasp of simplicity. (See his review of *P & E*, *Philosophical Review* 93 [July 1984]: 474–83.)

8. For a good assessment of this argument and various others see John Earman's *Bayes or Bust?* (Cambridge, Mass., 1992). Bruno de Finetti perhaps deserves the credit for first having argued that degrees of belief *ought* to be 'coherent', i.e., conform to the probability calculus—though they *need* not be coherent if the believer is irrational ("Foresight: Its Logical Laws, Its Subjective Sources," translated in *Studies in Subjective Probability*, edited by H. E. Kyburg, Jr., and H. E. Smokler [New York, 1964]). I hesitate to credit Frank Ramsey's earlier paper (in *Foundations: Essays in Philosophy, Logic, Mathematics and Economics*, edited by D. H. Mellor [Atlantic Highlands, N.J., 1977]) with this result, since he defines "degrees of belief" in such a way that they *must* conform to the probability calculus. On Ramsey's account there is no room for the existence of someone who has degrees of belief that are not coherent.

9. For example, I. J. Good (in the *British Journal for the Philosophy of Science* 19 [1968]: 123–43) advocates:

$$\text{Weight of evidence} \atop \text{concerning } H \text{ provided by } E = \log \frac{P(E/H)}{P(E/-H)}$$

And Seidenfeld (op. cit), noting that on our account E might confirm both H_1 and H_2 yet disconfirm the conjunction (H_1 & H_2), suggests that confirmation cannot be defined in terms of probability alone.

10. A further complaint is that our definition of confirmation seems to go badly wrong when we apply it to measure the evidential value of *already known* data. For in that case $P(E) = 1$, therefore $P(H/E) = P(H)$. This problem for Bayesians was first posed by Clark Glymour (see his *Theory and Evidence* [Princeton, 1980]). It has been forcefully reiterated by James Woodward (in his review of *P & E*, *Erkenntnis* 23 (1985): 213–19) and treated thoroughly by John Earman (in *Bayes or Bust?*). In order to deal with it we should remember that the idea of the definition is to compare the credibility of a hypothesis, H, given the knowledge that E is true, with its credibility in the absence of such knowledge. Thus we should take the prior probability to be that which H would have had if the truth of E had not been discovered. Then, in order to assess E's confirmation power, we should consider what the abso-

lute subjective probability of E would have been in that counterfactual situation, and also what the conditional probability of E given H would have been. Then we can employ Bayes's Theorem to calculate the factor by which the prior probability of H would have been increased. Doubtless, there is substantial indeterminacy in the assessment of these counterfactual probabilities. But this is no objection, since we generally have no reason to expect the magnitude of E's confirmation power to be an especially determinate matter.

11. Rudolf Carnap, *Logical Foundations of Probability* (Chicago, 1962), 19, 31.

12. According to Spielman, this construal of Carnap is a "distorted caricature" (170), for "any careful reading of LFP [Logical Foundations of Probability] would show that Carnap never talks about 'objective relations of probabilification' or 'objective' relations of partial entailment" (170). Here I am at a loss to explain how Spielman could have arrived at his interpretation, and I can only refer the reader back to Carnap's work.

13. For further discussion of this point see *P & E*, 32–36 and 74–81 and Earman's *Bayes or Bust?*, chapter 6.

14. Spielman (171) falls into this error, complaining that one cannot endorse logical probability and yet still assume that laws can have a non-negligible credibility.

15. Thus Woodward writes: "The principal defect of *Probability and Evidence* is its unsystematic character. Horwich does not give us a fully worked out general theory of confirmation but rather a series of essays which offer solutions to various particular puzzles, where the interconnections among these solutions are by no means always clear" (214).

16. This is starkly revealed in Seidenfeld's dismissal of therapeutic Bayesianism on the grounds that it is no substitute for a combination of excellent, but highly technical, foundational studies in decision theory and statistical inference by Jeffrey, Fishburn, and Lindley—works that hardly touch upon the traditional philosophical puzzles that form the domain of therapeutic Bayesianism. In a similar vein, Spielman is bothered by the "fail(ure) to see that the only difference between an 'objectivist' account [of the 'grue' problem] and a personalist account would be verbal: an objectivist would say that we *ought* to assign a much higher probability to H_1 than to H_2, and a subjectivist says that this is what intelligent informed people in fact do" (170). Spielman thinks the issue between them is 'merely verbal'.

17. I have greatly benefited from James Woodward's thorough and perceptive criticism. I would also like to thank Ned Block, Susan Brison, Josh Cohen, Marcus Giaquinto, Mark Kaplan, and Judith Thomson for helping me to improve earlier drafts of this paper.

5 | Commentary

5 | COMMENTARY

5.1 | Bayes for Beginners

The Bayesian approach to confirmation differs from the hypothetico-deductive (H-D) approach (discussed in chapter 4), in several fundamental respects.[1] The aim of this first section is to explain and illustrate those differences as a prelude to the analysis of the readings in chapter 5 later in this commentary.

THE RELEVANCE CRITERION OF CONFIRMATION

A distinctive feature of the Bayesian approach is its reliance on the mathematical theory of probability. Unlike the H-D approach, which treats confirmation as a qualitative notion that might be made quantitative later on, Bayesians assume that confirmation is quantitative from the outset. Even such qualitative notions as evidence confirming a hypothesis and evidence confirming one hypothesis more strongly than it does another are analyzed in terms of probabilities with numerical values that lie between 0 and 1. Bayesians contend that this essentially quantitative approach to confirmation in terms of probability theory can solve the puzzles and paradoxes afflicting purely qualitative theories of confirmation, such as Hempel's satisfaction criterion.

The most common way that Bayesians connect confirmation with probability is by adopting the relevance criterion of confirmation, according to which a piece of evidence, E, confirms a hypothesis, H, if and only if E raises the probability of H:

Relevance Criterion of Confirmation	E confirms H if and only if $P(H/E) > P(H)$; E disconfirms H if and only if $P(H/E) < P(H)$.

For convenience, we shall often refer to $P(H)$ as the *prior probability of H*, and $P(H/E)$ as the *posterior probability of H*. $P(H/E)$ is a conditional probability and should be read as the *probability of H given E*.

It should be noted that, although it is defined in terms of quantitative probabilities, the notion of confirmation (often called *incremental confirmation*) defined by the relevance criterion is qualitative. The relevance criterion does not specify how degrees of confirmation should be measured; it merely gives a necessary and sufficient condition for that confirmation. Indeed, there is an ongoing dispute in the literature about whether numerical degrees of confirmation should be a function of the ratio of $P(H/E)$ to $P(H)$ or a function of the difference between them.[2] Regardless of where they stand on this issue, all Bayesians agree that the more E raises the probability of H, the more E confirms H.

The relevance criterion of confirmation differs significantly from the absolute criterion of confirmation, according to which E confirms H if and only if $P(H/E)$ exceeds some suitably high threshold value, say, 0.9. From the point of view of those who endorse the relevance criterion, the absolute criterion confuses confirmation with acceptance. High probability may be an appropriate condition for accepting a hypothesis, but it is not necessary for confirmation. Thus, adherents to the relevance criterion would consider H confirmed by E even though E raised the probability of H only a little, from, say, 0.2 to 0.4, and the posterior probability of H, $P(H/E)$, remained less than 0.5.

BAYES'S THEOREM AND THE AXIOMS OF PROBABILITY THEORY

Bayes's theorem (also called Bayes's rule, law, or equation) lies at the heart of the Bayesian approach to confirmation and gives that approach its name. In this section we shall be concerned solely with Bayes's theorem as a formal result in probability theory. As such, Bayes's theorem, like any mathematical theorem, is entirely uncontroversial. What is distinctively Bayesian about the Bayesian approach to confirmation is not merely its use of Bayes's theorem but its interpretation of the probabilities occurring in the theorem. The Bayesian interpretation of probabilities as subjective degrees of belief will be discussed later, in the section "Probabilities and Degrees of Belief."

Bayes's theorem is a deductive consequence of the three basic axioms of probability theory. Everything else in probability theory can also be deduced from these axioms, supplemented with definitions of notions such as conditional probability. Here are the axioms in their unconditional form.

Axiom 1   Every probability is a real number between 0 and 1: $0 \leq P(A) \leq 1$.

Axiom 2   If A is a necessary truth, then $P(A) = 1$.

Axiom 3   If A and B are mutually exclusive (that is, if it is impossible for both A and B to be true), then $P(A \vee B) = P(A) + P(B)$. This theorem is often referred to as the *special addition rule*.

Strictly speaking, the A, B, C, and so on that probability ranges over are propositions, but we shall, when convenient, talk about the probability of events, theories, classes of theories, and evidence.

Even though the set of axioms is small, several important rules that we shall use later on can be deduced from them.

Negation Rule $P(\sim A) = 1 - P(A)$.

Implication
Rule   If A logically entails B, then $P(B) \geq P(A)$.

Equivalence
Rule   If A and B are logically equivalent, then $P(A) = P(B)$.

General $P(A \vee B) = P(A) + P(B) - P(A\&B)$.
Addition Rule

The general addition rule is especially useful, since it applies regardless of whether A and B are exclusive. Obviously, when A and B are mutually exclusive, $P(A\&B)$ is 0 and the general addition rule reduces to the special addition rule (axiom 3).

One way to make the general addition rule intuitively obvious is to represent propositions by circles and to let the probability of each proposition equal the area of its circle.[3] When A and B are mutually exclusive, the A-circle and the B-circle do not overlap, and the probability of $(A \vee B)$ is simply the sum of $P(A)$—the area of the A-circle—and $P(B)$—the area of the B-circle (figure 1). When A and B are not exclusive, the circles overlap (figure 2).

Thus, to calculate $P(A \vee B)$ for figure 2 we add the areas of the two circles as before, but then we have to subtract $P(A\&B)$, which is the area of the overlap, to get the correct answer.[4]

To derive Bayes's theorem from the probability axioms, we need a definition of $P(A/B)$, the conditional probability of A given B.[5] It is:

Definition of
Conditional $P(A/B) = \dfrac{P(A\&B)}{P(B)}$, where $P(B) > 0$.
Probability

The rationale for adopting this definition of conditional probability can be appreciated by considering figure 2. Suppose that you are told that a dart has been thrown, randomly, at the figure and has landed somewhere inside the B-circle. Given that the dart is inside the B-circle, what is $P(A/B)$, the probability that the dart is also inside the A-circle? The answer is simple: it

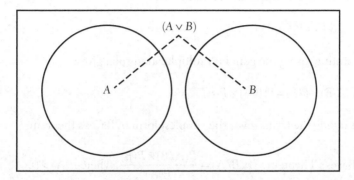

Figure 1

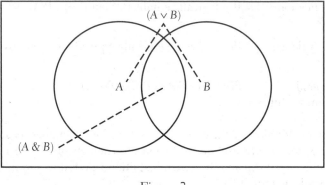

Figure 2

is the area common to both circles divided by the area of the B-circle. In other words, given that probabilities are proportional to areas, $P(A/B)$ is equal to $P(A\&B)$ divided by $P(B)$.

It is an immediate consequence of our definition of conditional probability that a general multiplication rule holds for the probability of any conjunction (or for the probability of the joint occurrence of any two events).

General Multiplication Rule $\quad P(A\,\&B) = P(A/B) \times P(B)$.

When $P(A/B) = P(A)$, A and B are said to be statistically independent of one another and the general multiplication rule simplifies to the special multiplication rule.

Special Multiplication Rule $\quad$ When A and B are independent, $$P(A\&B) = P(A) \times P(B).$$

It is only a short step from the general multiplication rule to Bayes's theorem. First, we note that, since (A&B) is logically equivalent to (B&A), it follows from the equivalence rule that:

$$P(A\&B) = P(B\&A).$$

Substituting, using the general multiplication rule, gives:

$$P(A/B) \times P(B) = P(B/A) \times P(A).$$

Rearranging the terms gives the simplest form of Bayes's theorem:

Bayes's Theorem $\quad P(B/A) = \dfrac{P(A/B) \times P(B)}{P(A)}$, where $P(A) > 0$.

Before trying to apply Bayes's theorem to scientific reasoning and the confirmation of theories by evidence, let us consider a simple example that illustrates its essential features. Imagine that a make of wheelchair, the Samson, is manufactured in just two plants in the United States. One factory is in Boston, the other in Chicago. The Boston plant makes four-fifths of all Samsons; the Chicago plant makes the rest. Of the Samsons manufactured in Boston, one-sixth have a special lightweight aluminum frame, whereas, three-quarters of the Samsons made in Chicago are of this type. You purchase a Samson wheelchair at an auction and discover that it has a lightweight aluminum frame. What is the probability that it was made in Boston?

Let A stand for having a lightweight aluminum frame, let B stand for being made in Boston, and let C stand for being made in Chicago. We want to calculate $P(B/A)$ using Bayes's theorem. The information we are given is that $P(B) = 4/5$, $P(C) = 1/5$, $P(A/B) = 1/6$, and $P(A/C) = 3/4$. Bayes's theorem tells us that

$$P(B/A) = \frac{P(A/B) \times P(B)}{P(A)}.$$

Thus, the numerator equals $1/6 \times 4/5 = 2/15$. But what about the denominator, $P(A)$? Samsons are made in only two places—Boston and Chicago—so the alternatives are $(A\&B)$ and $(A\&C)$, which are mutually exclusive and exhaustive. Thus, by axiom 3,

$$P(A) = P[(A\&B) \vee (A\&C)] = P(A\&B) + P(A\&C),$$

and then, using the general multiplication rule, we get:

$$P(A) = P(A/B) \times P(B) + P(A/C) \times P(C)$$
$$= (1/6) \times (4/5) + (3/4) \times (1/5) = 17/60.$$

Thus:

$$P(B/A) = (2/15)/(17/60) = (2/15) \times (60/17) = 8/17.$$

When using Bayes's theorem, we need not calculate the denominator, $P(A)$, from first principles every time. Instead, we can write down the answer immediately using the total probability rule:

Total Probability Rule $P(A) = P(A/B) \times P(B) + P(A/{\sim}B) \times P({\sim}B)$.

In the wheelchair example, there were just two exclusive alternatives: either the chair came from Boston (B) or from Chicago (C). So, in this example, we can calculate $P(A)$ by substituting C for ${\sim}B$ in the total probability rule:

$$P(A) = P(A/B) \times P(B) + P(A/C) \times P(C).$$

But it is easy to imagine a more complicated example in which aluminum chairs were also made in Detroit (D) and Evanston (E). In this case, the right-hand side of the equation for $P(A)$ will include two extra factors:

$$P(A/B) \times P(B) + P(A/C) \times P(C) + P(A/D) \times P(D) + P(A/E) \times P(E).$$

More generally, when $B_1, B_2, \ldots, B_n$ are n mutually exclusive and exhaustive hypotheses, we can express $P(A)$ as the sum of products as follows.

$$P(A) = P(A/B_1) \times P(B_1) + P(A/B_2) \times P(B_2) + \ldots + P(A/B_n) \times P(B_n).$$

This can be written more succinctly as:

$$P(A) = \sum_{i=1}^{i=n} P(A/B_i) \times P(B_i).$$

Bayes's Theorem and Scientific Reasoning

The wheelchair example used above to illustrate Bayes's theorem can be regarded as a simple analogue of scientific reasoning. Let the theory, T, be that our Samson wheelchair was made in Boston. Initially, before we discovered that the chair is made of aluminum, the probability that it came from Boston was 4/5, since 80 percent of all Samson wheelchairs are made there. In other words, in our example the prior probability of theory T, $P(T)$, is 4/5. Once we acquire evidence, E, that the chair is made of aluminum, we calculate that the posterior probability of T given evidence E, $P(T/E)$, is 8/17. So the probability of T has dropped, and E disconfirms T. (In fact, in this example, it is now slightly more probable than not that the chair came from Chicago rather than Boston.)

In light of this example, we can summarize the application of Bayes's theorem to scientific theories as follows:

Bayes's Equation $$P(T/E) = \frac{P(E/T) \times P(T)}{P(E)}.$$
(Version 1)

$P(T)$ is the prior probability of T, $P(E)$ is the probability of the evidence E (what Salmon calls the *expectedness of the evidence*), and $P(T/E)$, the probability of T conditional on E, is the posterior probability of T. The meaning of these terms is pretty straightforward and obvious, but the terminology for $P(E/T)$ can be misleading. $P(E/T)$, the probability of E conditional on T, is usually referred to as the *likelihood* of T (or, as some authors prefer, the likelihood of T on E). To repeat, the likelihood of T is $P(E/T)$, not $P(T)$, and similarly, the likelihood of T on E is $P(E/T)$, not $P(T/E)$. To avoid possible confusion, we will avoid using the term *likelihood* as a synonym for *probability* and use it solely to refer to $P(E/T)$ and similar expressions.[6]

When given some new evidence E for theory T, we revise our assessment of the theory's probability by using Bayes's equation to calculate $P(T/E)$ as a function of $P(T)$, $P(E/T)$, and $P(E)$. We then discard our old prior probability, $P(T)$, and replace it with $P(T/E)$. In this way, the posterior probability of T becomes our new prior probability:

$$P_{new}(T) = P(T/E).$$

Some authors, such as Salmon, explicitly include background knowledge, B, in Bayes's equation, thus making all the probabilities involved conditional. In the wheelchair example, B would include the information that the chairs are made either in Boston or Chicago and nowhere else. In real-life scientific reasoning, B would include the information about the world and other theories that scientists accept as true or highly probable. Including background knowledge, the equation becomes:

Bayes's Equation (Version 2)
$$P(T/E \& B) = \frac{P(E/T \& B) \times P(T/B)}{P(E/B)}.$$

We can also use the total probability rule to expand the denominator as follows.

$$P(T/E \& B) = \frac{P(E/T \& B) \times P(T/B)}{P(E/T \& B) \times P(T/B) + P(E/\sim T \& B) \times P(\sim T/B)}.$$

Since it is cumbersome to write out these equations when every probability is explicitly conditionalized upon the background information, we shall omit the reference to B whenever it is convenient. Thus, Bayes's equation can be expressed more simply as:

Bayes's Equation (Version 3)
$$P(T/E) = \frac{P(E/T) \times P(T)}{P(E/T) \times P(T) + P(E/\sim T) \times P(\sim T)}.$$

And in its most general form, when T_1, T_2, . . . , T_n are mutually exclusive and exhaustive, we have:

Bayes's Equation (Version 4)
$$P(T_k/E) = \frac{P(E/T_k) \times P(T_k)}{\sum_{i=1}^{i=n} P(E/T_i) \times P(T_i)}.$$

Finally, it is sometimes convenient to rearrange version 3 to make explicit the dependence of $P(T/E)$ on the so-called likelihood ratio $P(E/T)/P(E/\sim T)$.

Bayes's Equation
(Version 5)

$$P(T/E) = \cfrac{1}{1 + \cfrac{P(E/\sim T) \times P(\sim T)}{P(E/T) \times P(T)}}.$$

Intuitively, the less likely it is that evidence E would have been obtained if T were false, the greater is the effect of E in raising the probability of T (and thus confirming T).

Probabilities and Degrees of Belief

As a deduction from the axioms of probability theory, Bayes's theorem is beyond reproach. Controversy arises when Bayesian philosophers of science try to use it to understand and evaluate scientific decisions about theories. That requires interpreting what the probabilities in Bayes's equations mean in terms of the beliefs, judgments, and attitudes of scientists. The pure Bayesian line is that, subject to some important conditions, all probabilities are subjective degrees of belief and that rationality requires us to revise our beliefs using Bayes's theorem. The subjective (or so-called personalist) interpretation of probability and the alleged connection between rationality and Bayes's equation are controversial issues that are debated in the readings in this chapter. The purpose of this section is to explain and clarify the main issues involved in that debate.

The right-hand side of Bayes's equation (version 1) contains three sorts of probabilities: $P(T)$, $P(E)$, and $P(E/T)$—the prior probability of theory T, the expectedness of evidence E, and the likelihood of T. In many respects, likelihoods are the least controversial, since, even if $P(E/T)$ is a subjective degree of belief, it seems rational to base that probability on the objective relation between T and E. For example, if T is a deterministic theory that deductively entails E, then everyone agrees that the correct value to be assigned to $P(E/T)$ is I. Similarly, if T is a statistical theory, as in our wheelchair example, then T (together with background information B) will specify the probability that E is true on the condition that T is true.

Assigning values to $P(E)$ and $P(T)$ is more difficult. Later in this commentary we discuss the so-called problem of old evidence, that is, the problem of assigning to $P(E)$ some value other than 1 when E is already known to be true. Of course, one might try to use the second or third versions of Bayes's equation to calculate $P(E)$, but as Salmon points out in his article, to do so we would need to calculate $P(E/\sim T)$, and since $\sim T$ is simply the negation of T, not a specific theory, we cannot infer from $\sim T$ the value of $P(E/\sim T)$. The fourth version of Bayes's equation avoids the indeterminate character of $\sim T$ but only at the price of requiring a complete list of all the possible theories that predict E (or that assign to E some definite probability). In practice, only a small handful of rivals to T (possibly none) will be candidates for serious consideration. Scientists simply do not know what all the logically possible rival theories are. For further discussion of this problem, see the articles by Salmon and Chalmers and the sections on their articles later in this commentary.

Even if values for $P(E/T_i)$ were available for every single theory T_i that predicts E (either with certainty or some definite probability), the problem of determining the prior probabilities, $P(T_i)$, of each of these theories would remain. At this point subjectivist Bayesians say that the prior probability of a theory, T, is simply the actual degree of belief that a person has in T. Strictly speaking, on the subjectivist interpretation of probability, there is no such thing as *the* prior probability of T, since degrees of belief are relative to persons and it is perfectly possible that different people believe the same theory to different degrees, even given the same background information. According to subjectivist Bayesians, it is an illusion to think that there is one objectively correct answer to the question, "What is the prior probability of T?"

It seems incredible, on the face of it, that anything useful could be said about scientific inference and confirmation, starting from a basis as subjective as each person's degree of belief. Salmon, Mayo, and Chalmers discuss objections to the subjective interpretation of probability and other problems with the Bayesian approach. Salmon is mildly optimistic that these problems can be overcome; Mayo and Chalmers are skeptical.

5.2 | Salmon on Kuhn and Bayes

In a companion article, written at the same time as the piece in our book, Salmon relates that, when he first began reading Kuhn's *The Structure of Scientific Revolutions* (1962), he was so shocked by Kuhn's repudiation of the distinction between the context of discovery and the context of justification that he set the book aside without finishing it.[7] Later, in 1969, while preparing for a conference on the relation between the history of science and the philosophy of science, Salmon returned to Kuhn's book with renewed interest. Salmon conjectured that the reason for Kuhn's rejection of the traditional discovery-justification distinction lay in Kuhn's commitment to an inadequate conception of scientific justification, namely the H-D account. According to the H-D account, everything connected with the genesis of a scientific theory and its evaluation prior to being tested belongs to the context of discovery and, as such, is irrelevant the theory's epistemic justification; a theory is justified and its acceptance rationally warranted only when the theory has been confirmed, and a theory is confirmed if and only if the predictions deduced from it are observed to be true.

THE INADEQUACIES OF HYPOTHETICO-DEDUCTIVISM

As Salmon notes, a significant limitation of the H-D account of confirmation is that it ignores statistical theories. Statistical theories confine their predictions to assignments of probability to classes of events but do not logically imply that any particular event will occur. By regarding inductive confirmation, in effect, as the inverse of logical deduction, the H-D account excludes from its scope all those theories in which the relation between theory and evidence is not deductive but probabilistic. The Bayesian approach has no such limitation, since it permits the likelihood $P(E/T)$ to assume values less than 1.

Another limitation of the H-D account is that it provides no guidance for assigning degrees of confirmation to hypotheses, or even for making comparative judgments between them when they entail the same evidence. Bayes's equation is attractive because it can do justice to the differential confirmation of rival theories by the same evidence, by appealing to differences in the initial plausibility of those theories (and hence differences in their prior probabilities). A nice illustration of this is the Bayesian solution to the so-called tacking paradox or the problem of irrelevant conjunction. Suppose that theory T, in conjunction with background information B, entails the true observational prediction E. Now let I stand for some contingent statement that is logically independent of T and irrelevant to E. It follows trivially that if $(T\&B)$ entails E, then it must also be the case that $(T\&I\&B)$ entails E. Thus, according to the H-D account, E confirms both the original theory T and the augmented theory $(T\&I)$ and, moreover, confirms them by the same amount. The Bayesian analysis agrees with the H-D account that E confirms both theories but disagrees about the degree of confirmation. On the most popular version of the Bayesian analysis, the degree of confirmation of a theory by evidence is a function of the difference between the posterior probability of the theory given that evidence and the prior probability of the theory.[8] On this version, the Bayesian analysis entails that the degree of confirmation conferred by E on $(T\&I)$ must be less than the degree of confirmation that E confers on T alone. Here are the two expressions for the degrees of confirmation of T and $(T\&I)$ on the difference analysis.

$$P(T/E\&B) - P(T/B) = P(T/B) \times \left[\frac{1 - P(E/B)}{P(E/B)} \right].$$

$$P(T\&I/E\&B) - P(T\&I/B) = P(T\&I/B) \times \left[\frac{1 - P(E/B)}{P(E/B)} \right].$$

These expressions are derived from the second version of Bayes's equation, setting $P(E/T\&B)$ and $P(E/T\&I\&B)$ both equal to 1. The factor in the square brackets is the same for both theories, and their respective degrees of confirmation are proportional to their prior probabilities. Since $(T\&I)$ entails T, and I is a contingent statement that is independent of T, it follows from the special multiplication rule that $P(T\&I/B)$ must be less than $P(T/B)$. Thus, E confirms $(T\&I)$ by a smaller amount than it confirms T: adding the irrelevant conjunct I to T lowers the confirmation provided by E. In this respect, then, the Bayesian approach to confirmation is a decided improvement over the H-D account.

Although Salmon does not discuss it in his article, the Bayesian approach also promises a resolution of the Duhem problem (that is, the problem of assigning the blame for a failed prediction to a particular member of a group of hypotheses), for not only are some theories confirmed better by the same evidence, but Bayes's equation can also be used to explain how some components of a group of hypotheses and assumptions receive a much larger *dis*confirmation than other components when observations disagree with theoretical predictions.[9] Thus, the Bayesian approach can show us where the blame should be placed when a group of hypotheses and assumptions lead to a false prediction.

Salmon is not alone in thinking that the H-D account is inadequate. Popper, an early critic of hypothetico-deductivism, rejected it because he denied the whole notion of inductive confirmation. (See Popper, "The Problem of Induction," in chapter 4 for details.) Other critics, such as Carnap and Reichenbach, accepted that confirmation is an essential part of scientific rationality but insisted that it should be understood in terms of Bayes's theorem. Carnap interpreted the probabilities in Bayes's theorem as a priori logical probabilities; Reichenbach construed them as empirical frequencies. Both were objectivists about probability. More recently, an entire school of statisticians and philosophers of science has arisen—the personalists or Bayesians—that interprets the probabilities in Bayes's theorem subjectively, as degrees of belief.

Salmon thinks that we can use Bayes's theorem to reconcile Kuhn's historical approach to understanding science with the logical empiricism of philosophers such as Carnap, Reichenbach, and Hempel. The key is to incorporate Kuhn's values—criteria for theory assessment such as consistency, simplicity, and fruitfulness—into the Bayesian equation that defines confirmation. Variation in the interpretation of these values and the emphasis placed on them can give rise to differing judgments about the prior probability of a theory. Thus, scientists can reach different but equally rational judgments about how well a theory is confirmed by a particular piece of evidence using the same algorithm (Bayes's equation) because they insert different inputs into that algorithm in the form of different judgments about prior probability. Salmon contrasts this with Kuhn's own suggestion in "Objectivity, Value Judgment, and Theory Choice" (reprinted in chapter 2, above) that scientists reach different conclusions because they use different algorithms.

PRIOR PROBABILITIES

As Salmon emphasizes in his article, most of what is philosophically controversial about the Bayesian approach to confirmation depends on the interpretation of prior probabilities in Bayes's equation. Salmon distinguishes three such interpretations: the objective-logical, the objective-empirical, and the subjective. Salmon argues that the objective-logical interpretation of Carnap and others, according to which probabilities are assigned a priori to all statements, is hopelessly inadequate to the task of analyzing the probability of real-life scientific theories. That leaves the objective-empirical and the subjective interpretations.

In an earlier book, *The Foundations of Scientific Inference*, Salmon adopted Reichenbach's objective-empirical interpretation according to which probabilities are relative frequencies.[10] The basic idea is this: Every hypothesis is either true or false, but when a new hypothesis, *H*, is first proposed, we do not know which attribute (truth or falsity) it has. In order to make sense of the prior probability (or plausibility) of the single hypothesis, *H*, we have to place it in a reference class of similar hypotheses. Then, on the basis of past experience, we can see how often hypotheses in this class have turned out to be true. The ratio of the number of true hypotheses to the total number of hypotheses in the reference class is then taken to be the prior probability of *H*.

In rough outline, this procedure is supposed to be similar to the way frequency theorists handle the problem of the single case. A typical example is the problem of assigning a probability to whether a particular (asymmetrical) coin will show a head on its next toss. In the case of the coin, we estimate this probability by dividing the number of times the coin has come up heads by the number of times the coin has been tossed. Thus, very roughly, the frequency theorist would say that the probability of getting a head on the next toss of the coin is 0.55 if the frequency of heads converges to 0.55 as the number of tosses becomes ever larger. Applying this same approach to the prior probability of hypotheses is extremely difficult. Not least among these difficulties is the problem of specifying the appropriate reference class. What exactly does it mean to talk about hypotheses that are similar to *H*? Is it a matter of mathematical form, such as the use of inverse-square laws? And if so, why should the success of such laws in one domain of science (such as the study of gravity) make it more likely that they will succeed in another domain (such as the investigation of the strong force binding together particles in the atomic nucleus).[11]

Throughout his career, Salmon was highly critical of the unfettered subjectivism of the pure Bayesian or personalist interpretation of probabilities as degrees of belief. To Salmon, scientific judgments about confirmation should not depend in any way on the prejudices, emotions, or mood swings of individual researchers. The answer, he thinks, lies in what he (following Abner Shimony) calls *tempered personalism*. Tempered personalism places constraints on prior probabilities that go beyond mere coherence. Since experience has taught us that scientists have been moderately successful in the past, no hypothesis advanced by a serious scientist should be given a prior probability that is either 0 or vanishingly small. But, again, since experience tells us even the most promising hypotheses in the past have sometimes turned out to be false, the prior probability of any new hypothesis should be fairly low. The notion of *success* invoked in this discussion is crucial for understanding Salmon's proposal. For Salmon believes that when we assign a prior probability to a new hypothesis, we are trying to estimate the correct, objective probability that the hypothesis will turn out to be successful, and by *successful*, Salmon means *true*. Thus, while it may seem as if Salmon is making significant concessions to Kuhn when he admits consistency, analogy, and professional scientific standing as factors that play a legitimate role in determining the plausibility of new hypotheses, in fact his conception of probability is still, at bottom, objective and frequentist. It is, for example, only because past experience has taught us that hypotheses advanced by cranks very seldom turn out to be true that we should assign them a negligibly small prior probability. As Salmon himself puts it, "prior probabilities . . . can be understood as our best estimates of the frequencies with which certain kinds of hypotheses succeed. . . . The personalist and the frequentist need not be in any serious disagreement over the construal of prior probabilities" (531).

One final but important point: Salmon readily admits that it "seems preposterous" (531) that plausibility judgments based on values such as simplicity and symmetry could result in exact numbers for prior probabilities. Like many advocates of Bayes's equation, Salmon appeals to the washing out or swamping of priors to argue that their exact value really does not matter. For as soon as evidence begins to accumulate, the values for the posterior probability of a hypothesis converge. In the long run, the particular values adopted for the prior probability become irrelevant (so long as we avoid the extreme values of 0 and 1). But this convergence argument assumes that different scientists agree on the likelihoods, an assumption that Salmon defends later in his article.

THE EXPECTEDNESS OF EVIDENCE

Methodologists of science commonly hold that a theory receives greater confirmation from the successful prediction of something surprising than from the prediction of something expected. This issue is addressed, in part, in chapter 4 under the guise of the debate over novel predictions. Because the right-hand side of Bayes's equation has $P(E)$, the probability of the evidence E, as its denominator, it follows that, other things being equal, the lower the value of $P(E)$, the greater the value of $P(H/E)$. Thus, the more unexpected the prediction, the greater its confirming power if it should turn out to be true. But what is $P(E)$, the expectedness of the evidence, and how can it be measured?

Salmon uses the total probability rule to express $P(E)$ in terms of prior probabilities and likelihoods, writing all the probabilities involved as conditional on background knowledge B, where B includes initial conditions, boundary conditions, auxiliary hypotheses, and other relevant information.

$$P(E/B) = P(E/T\&B) \times P(T/B) + P(E/{\sim}T\&B) \times P({\sim}T/B).$$

If T is a deterministic theory, then T (in conjunction with auxiliary hypotheses and assumptions) entails E. In such a case, the likelihood $P(E/T\&B)$ equals 1 and $P(E/B)$, the expectedness of the evidence, must be at least as great as $P(T/B)$, the prior probability of T. But assigning an exact number to $P(E)$ is not easy, since it involves knowing the value of the likelihood $P(E/{\sim}T\&B)$, a problem that Salmon addresses in his section on likelihoods.

A second difficulty with $P(E/B)$ that Salmon acknowledges is a version of the problem of old evidence. This is discussed at some length later in this commentary in the section "The Problem of Old Evidence." For the moment we merely note that Salmon thinks that, given his characterization of background information B, the objective value of $P(E/B)$ must always be 1. Since B includes all the details about the experimental setup and the instruments used to observe E, the objective probability that E will occur under those conditions (assuming that the system in question is deterministic) is 1. Thus, Salmon concludes that the expectedness $P(E)$ can only be a subjective probability, reflecting the degree to which a particular scientist finds E psychologically surprising. Given Salmon's hostility towards subjectivism, the conclusion that "expectedness defies interpretation as an objective probability" (533) is highly unwelcome. At the end of his article, Salmon suggests a way to avoid this and a similar problem with the likelihood $P(E/{\sim}T\&B)$, while still permitting objective comparisons among rival theories.

LIKELIHOODS AND THE CATCH-ALL HYPOTHESIS

The main problem with likelihoods concerns the value of $P(E/{\sim}T\&B)$, which appears in the expression for $P(E)$. $P(E/{\sim}T\&B)$ is the probability that E is true given that theory T is false, but since $\sim T$ is not a specific theory, the corresponding likelihood is not well defined. Even when we have two competing theories, T_1 and T_2 (such as specific versions of the wave and particle theories of light), $P(E/{\sim}T_1)$ is not equal to $P(E/T_2)$. Although theories T_1 and T_2 are contraries, and thus T_1 entails $\sim T_2$ and T_2 entails $\sim T_1$, they are not contradictories; thus, $\sim T_1$ does not entail T_2, nor is T_1 logically equivalent to $\sim T_2$. It is possible that both T_1 and T_2 are false. Thus, if we write out the set of logically exclusive and exhaustive hypotheses, it will include not only T_1 and T_2, but also T_k, the so-called catch-all hypothesis. What is the catch-all hypothesis? Strictly speaking, it is not a single hypothesis at all but a lengthy disjunction of all the possible alternatives to T_1 and T_2, most of which we have never thought about. As Salmon says, trying to guess the ingredients of the catch-all would be like trying to predict the future of science. Even though some of these ingredient hypotheses entail E, this scarcely helps us to answer the question, What is $P(E/T_k)$, the likelihood of T_k? because T_k is the disjunction of *all* the possible alternatives, including those that do *not* entail E. And if we cannot answer this question, then we cannot calculate $P(E)$. Because Salmon regards the problem of calculating the likelihood of the catch-all as completely intractable, he proposes a method for choosing between theories that does not require the calculation of $P(E)$.

SALMON'S BAYESIAN ALGORITHM FOR THEORY PREFERENCE

Salmon agrees with Kuhn that theory choice in science is usually a comparative affair. Typically, the issue is not how well a particular piece of evidence confirms an individual theory, but how well that evidence favors one theory over its rivals. In any given scientific domain only a few theories—usually just two or three—will be competing for acceptance at any given time. Certainly the catch-all hypothesis is seldom a serious option. Thus, despite the intractability of calculating the likelihood of the catch-all and the expectedness of the evidence, the Bayesian approach can still reflect the realities of scientific practice if it can provide a comparative ranking of those hypotheses that are serious rivals. Salmon's proposal is that, in choosing between two theories, T_1 and T_2, on the basis of evidence E, we should compare the posterior probabilities $P(T_1/E\&B)$ and $P(T_2/E\&B)$. An attractive feature of this proposal is that, in forming the ratio of the posterior probabilities, the problematic term $P(E/B)$ cancels out.

$$\frac{P(T_1/E\&B)}{P(T_2/E\&B)} = \frac{P(E/T_1\&B) \times P(T_1/B)}{P(E/T_2\&B) \times P(T_2/B)}.$$

Assuming that T_1 and T_2 are the only candidates for serious consideration, Salmon's proposal is that, before the discovery of evidence E, scientists should prefer T_1 to T_2 if and only if the prior probability of T_1 is greater than the prior probability of T_2. After the discovery of E, scientists should change their preference from T_1 to T_2 if and only if the posterior probability of T_2 is greater than the posterior probability of T_1. It follows from the Bayesian expression for the ratio of the posterior probabilities that, after the discovery of E, scientists should prefer T_2 to T_1, if and only if

$$\frac{P(E/T_2\&B)}{P(E/T_1\&B)} > \frac{P(T_1/B)}{P(T_2/B)}$$

or, in other words, if and only if the ratio of the likelihoods is greater than the reciprocal of the ratios of the prior probabilities. Salmon refers to this as the *Bayesian algorithm for theory preference*.

Salmon's algorithm is both ingenious and attractive, but it also has its limitations and counterintuitive features. First, it should be clear that in "choosing" T_2 over T_1, we are not deciding to accept T_2 as true or well-confirmed. We are merely saying that, relative to evidence E, T_2 is better confirmed than T_1. For all we know, T_2 might be extremely improbable and unworthy of acceptance. It is important to remember that, in comparing the posterior probabilities of T_1 and T_2, we are not calculating—nor, if Salmon's pessimism is correct, can we ever calculate—the degree of confirmation of either hypothesis.[12] Thus, the judgment resulting from Salmon's algorithm is relatively weak, since it merely asserts that evidence E supports one theory better than its rival. The degree of that support is left entirely undetermined.[13]

Second (as noted by Wade Savage, the editor of the volume in which Salmon's article first appeared) Salmon's algorithm cannot, in its present form, give us any rational guidance when, as often happens, the body of evidence for T_1 is different from the body of evidence for T_2. For, obviously, $P(E)$ cancels out only when the evidence, E, is the same for both theories. Similarly, Salmon's algorithm does not permit us to judge whether one piece of evidence confirms a theory better than another piece of evidence.

Third, as Salmon himself notes, when both theories are deterministic and, in conjunction with B, entail the evidence E, the ratio of their posterior probabilities given E reduces to the ratio of their prior probabilities. Thus, according to Salmon's algorithm, no amount of evidence can change our initial preference ranking for such theories. For deterministic theories, the likelihoods become irrelevant and the prior probabilities (influenced by Kuhn's criteria for theory choice) dominate completely. Anyone who is critical of the vagueness of Kuhn's criteria and the difficulty of weighing and comparing them is unlikely to be impressed by this as a demonstration of a rational algorithm underlying scientific decisions about theories.

Salmon's response to this third point is contained in sections 8 and 9 of his paper. In section 9, "Kuhn's Criteria," Salmon distinguishes three types of theoretical virtue: informational, economic, and confirmational. Salmon argues that two of Kuhn's criteria—scope and accuracy—fall outside the confirmational category and are thus irrelevant to the prior plausibility of theories. This reduces the task of making the basis of plausibility judgments more precise by narrowing the focus to Kuhn's remaining three criteria, namely, simplicity, consistency, and fruitfulness.

In section 8, "Plausible Scenarios," Salmon explains that, when they are first formulated, important scientific theories often have great difficulty in explaining some puzzling phenomenon. He gives as examples the difficulty the absence of detectable stellar parallax posed for the Copernican theory and the problems of giving a coherent account of the optical ether and the phenomenon of selective absorption faced by the wave theory of light. Salmon's point is that the original versions of these theories did not logically entail the phenomena they had difficulty explaining. Indeed, the probability of E (a puzzling phenomenon) given T (the theory in question) was rather low. The challenge for supporters of T was to come up with what Salmon calls a *plausible scenario*, that is, a set of further assumptions, A, that when added to T, would significantly raise the probability of E. Ideally, when the augmented theory ($T\&A$) entails E, the likelihood $P(E/T\&A\&B)$ attains its maximum value of 1. In that case, applying Salmon's algorithm to the augmented version of T and its rivals makes crucial the prior probabilities of $(T\&A)$ and its competitors. This is where Kuhn's criteria of simplicity, consistency, and fruitfulness for assessing plausibility become important. Thus, when T is first proposed, likelihoods are not irrelevant because their low values motivate the search for plausible scenarios. If that search is successful, then likelihoods become irrelevant because judgments of prior probability are made by applying plausibility criteria, not to the original theory T, but to the new augmented version $(T\&A)$. So the allegation that, according to Salmon's algorithm, theory preference is permanently determined by the initial prior probabilities ignores the dynamics of scientific research.

5.3 | Mayo's Critique of Salmon's Bayesian Way

Deborah Mayo advocates the error-statistical approach to philosophy of science (sometimes called the testing account), an approach that is radically at odds with Bayesianism. In her article, Mayo criticizes Bayesianism for its failure to deliver an objective account of evidence that can be squared with scientific practice. She also faults Salmon's attempt to interpret the prior probabilities that figure in Bayes's equation as empirical frequencies.

SALMON'S BRIDGE: KUHN AND BAYES

Mayo doubts that Bayes's equation can be used to bridge the divide between Kuhn and the logical empiricists. Salmon's idea was to pack scientists' disagreements about how to interpret and weigh shared methodological values into their conflicting assessments of the prior probabilities of rival theories. As we have seen (in the discussion of Salmon), Kuhn had briefly considered this way of reducing theory choice to a "common algorithm" and rejected it. (See Kuhn, "Objectivity, Value Judgment, and Theory Choice," 94–110.) Kuhn makes two points. First, he insists that shared values are insufficient to determine the outcome when each individual scientist deliberates. Inevitably, he thinks, such verdicts must depend on personal values that are subjective and idiosyncratic. Thus, he concludes that each scientist has his own personal algorithm. Second, Kuhn is unimpressed by the argument that objective factors must outweigh subjective factors when scientific opinion concerning the merits of rival theories converges over time. The fact that individual judgments of probability converge and may even become unanimous does not imply that subjective factors have become irrelevant. Kuhn's preferred interpretation is that individual values always play a role even when scientists reach consensus about which theory is the best. Presumably Kuhn must think that the different subjective factors in each personal algorithm are tweaked and adjusted in ways that elude rational analysis so as to result in the same net judgment of a theory's overall worth. Perhaps wisely,

Mayo declines to resolve this particular dispute between Kuhn and Salmon. The disagreement turns on what one understands by the term *algorithm*. There is no incompatibility between acknowledging the role of subjective factors (as well as shared values) in the psychological processes that influence the judgments of individual scientists about prior probabilities while also insisting that those probabilities are revised in the light of new evidence in accordance with Bayes's equation. Indeed, the use of a Bayesian framework for belief revision promises to explain how rational consensus can emerge as evidence accumulates, something that Kuhn's account seems powerless to address. Moreover, as Mayo notes, Bayesianism also offers a solution to the Duhem problem at least when the required probabilities can be assigned "after the fact" (552).

A different kind of problem for Salmon's bridge is theory acceptance. In Kuhn's picture, scientists commit to a paradigmatic theory (or not), accept some theories and reject others, and act on their preferences. How can these notions be incorporated into the Bayesian framework? Bayes's theorem tells us how to calculate the posterior probabilities of theories on evidence. To go beyond these probabilities to reach judgments about acceptance (and rejection) would require either explicit probability-based rules (for example, as a supplement to Salmon's Bayesian algorithm for theory preference) or an application of decision theory. The latter, in turn, would require assigning numerical values to utilities. Those utilities could be purely epistemic (as in Hempel's analysis[14]) or, if they were to accommodate Kuhn's viewpoint, they would have to reflect subjective non-epistemic values that can vary in strength and kind from one scientist to another. As Mayo comments, this would amount to modeling Kuhnian subjectivism, but it would not address the normative questions characteristic of logical empiricism.

Problems with Salmon's Algorithm for Theory Preference

Salmon's algorithm turns on two probabilities: the likelihoods and the priors. For a pair of deterministic theories $(T_1$ and $A_1)$ and $(T_2$ and $A_2)$, both of which entail the same evidence, E, the likelihoods drop out leaving the priors as crucial. Where do our estimates of these priors come from? Mayo argues that either they are calculated by previous applications of Bayes's equation, or they are derived by non-Bayesian means. Previous applications of Bayes's equation would require judging the probability of the notorious catch-all hypotheses, "the very thing Salmon is at pains to avoid" (555). So Salmon is forced to entertain non-Bayesian means for assessing the priors. Given Salmon's abhorrence of subjectivism and the impossibility of a priori appeals, this means having recourse to the empirical frequency approach. The problem with this, as Mayo sees it, is that it assigns the hypothesis in question to a reference class of past hypotheses that are judged to be "relevantly similar." Why, she asks, is the frequency of success in this class of past hypotheses—even if it is known, and even if H is a random sample from that class—relevant to the probability that *this* particular hypothesis, H, will succeed?

There are really two questions here. One concerns the legitimacy of basing the estimate of a single-case prior probability of a specific hypothesis on the track record of the success of past hypotheses judged to be relevantly similar—a question that cuts to the heart of Reichenbach's and Salmon's version of the empirical frequency approach. The other question is, What does one mean by *success*? For Mayo, in the tradition of error statistics, these questions are linked. The *relevant* frequency is the frequency with which *this* hypothesis, H, would yield the outcome E (or sufficiently close to E to count as "success"), in repetitions of a particular experimental setup. Without the details of the kind of setup in which H is being tested, she insists that the *relevant* probability for assessing H is not well defined. This leads Mayo to suggest a "natural bridge" between Salmon's comparative Bayesian approach and her own error-statistical approach.

In comparing rival hypotheses H_1 and H_2 from the point of view of error statistics, what matters is our judgment of the probability that each hypothesis will yield a result close enough to the observed outcome, E, to count as "success." This judgment is something that is presumed to be entirely objective on the basis of our "background knowledge" that is on the basis of our presumed knowledge (i.e., our objectively justified beliefs that depend in no way on our commitments either to H_1 or to H_2). In real-life scientific cases this is relatively uncontroversial since it will normally involve "lower-level" theories about how measuring instruments function in light of theories that are well established and accepted as such by all the parties to the dispute.[15]

MAYO'S ERROR-STATISTICS APPROACH

The error-statistics approach in philosophy of science traces its origin to the theories of Ronald Fisher and, in modified form, to the theories of Jerzy Neyman and Egon Pearson, in the formal theory of statistical inference. The basic idea of Fisher's theory is very appealing, especially to philosophers of science with Popperian inclinations. Suppose that someone (Fisher's "lady tasting tea") claims that she can tell whether or not a cup of tea was poured with the milk already added as opposed to the milk being added afterward. If she is merely guessing—which is the so-called null hypothesis—we can readily calculate the probability that she will achieve a particular result when presented with, say, a hundred pairs of cups of tea, in which one cup in each pair is "milk before" and the other "milk after," presented in random order. If the subject is merely guessing, then in repeated trials we expect to observe outcomes normally distributed about a mean of fifty successes with a standard deviation of five. Any result on a single trial that was outside the range 40–60 would be judged so unlikely that we would reject the null hypothesis and accept that the subject really can tell the difference.[16]

To illustrate the logic of error statistics, Mayo cites an experiment discussed by Ian Hacking in his book *Representing and Intervening*. When red blood platelets are viewed under a low-powered electron microscope, tiny red dots called *dense bodies* are seen. Are the dense bodies really there in the cells, or are they merely an artifact of the microscope being used? The clincher is to see whether the same structures show up when viewed by different kinds of microscope—phase contrast, polarizing, and fluorescence, for example. When using equipment based on different physical principles, it is highly improbable that we would see the same thing if the dense bodies were not real (since each apparatus would have to produce the same illusion while operating in a completely different way).

In neither the tea-tasting example nor the dense body case is it relevant to know how frequently hypotheses that are "relevantly similar" have been "successful" in the past. The real issue is whether we can design an experiment that will make a severe test of the hypothesis in question. A severe test is one that the hypothesis is unlikely to pass if it is false (and likely to pass if it is true). Those probabilities are properties of the experimental setup, they can be calculated, and they define what we mean by good evidence. When a hypothesis passes a severe test, we have good grounds for believing that it will frequently be successful in the future in predicting experimental outcomes that agree with the hypothesis within a specified margin of error.

Mayo concludes her discussion by comparing her approach with Salmon's Bayesian algorithm. Both Mayo and Salmon are frequentists about probabilities and objectivists about evidence, but unlike Mayo, Salmon requires assignments of prior probabilities to hypotheses. Mayo considers how Salmon deals with the special circumstance in which rival theories entail the same evidence. In that case, the likelihoods drop out of Salmon's formula, leaving the choice between the hypotheses to be decided by their prior probabilities. The prior probability of a hypothesis is supposed to reflect its objective plausibility or likely success. Mayo suggests that in this special sort of case, it would make sense to base those plausibility judgments on the prior evidence—specifically the extent to which our background knowledge indicates that the hypotheses have already passed tests capable of ruling out error. In this way, we would not be assigning a probability to a hypothesis (as Salmon does), but we would provide a notion of the "probable success of a hypothesis" that could be used to decide between rival hypotheses in a way that is compatible with Salmon's approach. In truth, as Mayo acknowledges, if we were to do this, there would be nothing specifically "Bayesian" left in Salmon's account because our judgments of plausibility would rest on the strength of the *non-Bayesian tests* passed by the hypotheses being compared. Looking at things in this way has another advantage over Salmon's (misguided) attempt to assign prior probabilities to hypotheses: it focuses our attention on the strength of evidence for a hypothesis rather than the frequency of success in the class of past hypotheses that are presumed to be similar. In the case of the saccharin rat hypothesis, for example, the experimental data strongly support the claim that saccharin causes bladder cancer in rats. But those same experiments are not a severe test of the hypothesis that saccharin causes bladder cancer in humans, regardless of how often similar hypotheses about carcinogens in humans have been successful in the past. In fact, saccharin is now known *not* to cause bladder cancer in humans.[17]

MAYO AND HER CRITICS

Some Bayesians have accused Mayo of committing the base-rate fallacy when she judges that a test of a hypothesis is severe if it is one that the hypothesis is unlikely to pass if it is false (and likely to pass if it is true).[18] This supposed fallacy has been the focus of intense interest ever since Tversky and Kahnemann's empirical studies of human reasoning; and most philosophers and statisticians, Bayesians and non-Bayesians alike, agree that correct inductive reasoning should take base rates into account. We can illustrate the base-rate fallacy using a simple example. H is the hypothesis that an individual patient, Jones, has a particular disease. Jones's doctor orders a diagnostic test that always comes out positive (X) when the disease is present. When the disease is absent (H') a positive test outcome occurs only 5 percent of the time. What is the probability that Jones has the disease given his positive test result X? Bayesians insist that this question cannot be answered until we know the "base rate," which in this case is the frequency of the disease in the population to which Jones belongs. If the disease occurs in only 1 person in 1,000, then $P(H/X) = 0.02$. Yes, the hypothesis has been confirmed by the positive outcome; but, no, we should not believe that the patient has the disease since even after the test it is still 98 percent probable that the patient is disease free. Judged by Mayo's criterion of severity, the test with outcome X is a severe test of H since $P(X/H)/ P(X/H') \approx 20$, but this measure of evidence is an unreliable guide to belief because the base rate is so low. If we were to accept that H is true for every patient who obtained a positive test result, we would be wrong 98 percent of the time.

Mayo's response to this line of criticism is to accuse the Bayesians in turn of committing what she calls the "fallacy of instantiating probabilities."[19] The fallacy is committed when one takes a probability that *is* well-defined by a statistical hypothesis or model—in this case, simple random sampling from a population in which Z percent of the individuals in that population have R, the property of interest—and then (illicitly) attaches the probability $Z/100$ to the hypothesis that a particular individual from that population has R. When a patient has symptoms that lead him to see a doctor, he has not been randomly selected from a population. Indeed, in the absence of a formal sampling model, it is not at all obvious to what population Bayesians should regard the patient as belonging for the purposes of assigning a prior probability: the class of American males, the class of American males over forty, the class of males who smoke cigarettes, . . . ?

Mayo illustrates her response to critics by giving the example of Isaac, a high school student who takes a college-readiness test and achieves a high score, X, a score rarely achieved by people who are not ready for college, and a score that is invariably equaled or bettered by people who are. The example is deliberately constructed so that it is numerically similar to the case of the patient who gets a positive result on a diagnostic test for a disease. By Mayo's lights, Isaac's score of X is good evidence for the hypothesis, H, that Isaac is ready for college. This is because Isaac's score of X "fits" H and the test of college readiness that Isaac has passed is a severe one: it is very improbable (a probability of 0.05) that he would have gotten a score as high as this if he were not ready for college. So, we have:

H: Isaac is college-ready.

$H' = {\sim}H$: Isaac is not ready for college.

$P(X/H) \approx 1$, and $P(X/H') = 0.05$.[20]

Mayo then adds the information that Isaac comes from an underprivileged population (Mayo's "Fewready Town") in which the frequency of college readiness is very rare, say 1 in a 1,000. From this, Bayesians infer that $P(H) = 0.001$. Then, by applying Bayes's theorem they derive: $P(H/X) = 0.02$. As we saw in the diagnostic test example, Bayesians conclude that this posterior probability is so low that it is a counterexample to Mayo's account of evidence. Although the probability of Isaac's being college ready has increased by a factor of 20 (from 0.001 to 0.02) the rational thing to believe on the basis of X and the new information about the population to which Isaac belongs is that he is *not* ready for college.[21]

Mayo defends her conclusion that X is good evidence for H for anyone who takes the test, and condemns the Bayesian reasoning to the contrary as unsound. As in the diagnostic test example, she points out that there are many groups of students to which Isaac belongs—the set of young men, the set of students in a certain age range, the set of students with black hair, the set of students from California, and so on—in each of which the frequency of college readiness is different. It would appear arbitrary to pick just one of these groups as grounds for assigning a value to "the" prior probability that Isaac is ready for college. Once again, the Bayesian has committed the fallacy of instantiating probabilities.[22]

Peter Achinstein (a non-Bayesian) has responded to Mayo by saying that because he is concerned with epistemic probabilities—probabilities that reflect what it is rational to believe given one's available information—it is legitimate to take the value 0.001 as the probability that Isaac is ready for college before learning of his test score, if the only information one has is that 1 in 1,000 high school students in Fewready Town is college ready. Achinstein's response does not seem to meet Mayo's concern about what constitutes genuine evidence. To generate a posterior probability $P(H/X)$

that is greater than 0.5, Isaac would have to achieve a score that is attained by fewer than 1 in 1,000 non-college-ready students; to generate a posterior probability of 0.95, Isaac's score would have to be attained by fewer than 5 in 100,000 unprepared students. Why would Isaac have to do so much better than a student from "Fewdeficient Town" (a town in which most high students are prepared for college) in order to qualify as ready for college? Surely it is the severity of the test, and that alone, that provides the objective evidence of college readiness, regardless of "the" population to which one happens to belong. As Mayo points out, Achinstein would conclude that Isaac (and anyone else similarly placed) is unprepared for college on the basis of the same test performance that would lead to the opposite conclusion for any student from the more affluent neighborhoods of Fewdeficient Town. Not only does this fly in the face of a proper concern for the objectivity of evidence, it smacks of discrimination. It certainly seems odd to regard Isaac's high test score as "evidence" that he is *not* ready for college. Mayo has a good explanation for this: achieving a score of X is not a severe test of the hypothesis that someone is not ready for college, and hence it is not evidence for non-college-readiness.[23]

5.4 | Chalmers on the Bayesian Approach

Bayesianism in philosophy of science is an attempt to do justice to inductivism, the view that theories are made probable or confirmed by evidence. As Chalmers notes, even theories that have been falsified—such as Newtonian mechanics and gravitational theory—are often confidently relied on in a way strongly suggesting that scientists do not assign them a probability of 0. In its simplest form, Bayes's theorem says:

$$P(T/E) = \frac{P(E/T) \times P(T)}{P(E)}.$$

When T entails E, the likelihood $P(E/T) = 1$, and the posterior probability $P(T/E)$ is determined by the prior probability of T divided by the expectedness of the evidence. Assuming that the prior probability is not 0, this simple formula captures in a compelling way the widely held view that theories receive greater confirmation from evidence that was initially surprising and unexpected than from results that are already well known and considered unremarkable. Later in this commentary we discuss the problem of old evidence (that is, the problem of assigning a probability value other than 1 to evidence than is already known). The more immediate problem addressed by Chalmers is how to interpret the notion of probability and hence to assign values to prior probabilities.

Degrees of Belief

Bayesians regard probabilities as degrees of belief. Either these degrees of belief are objective, representing "probabilities that rational agents *ought* to subscribe to in the light of the objective situation" (568); or they are purely subjective and represent the degrees of belief that particular people happen to have regardless of their grounds or reasons for them. Chalmers quickly dismisses the objective interpretation as a nonstarter. To assign an objective prior probability to a hypothesis would seem to require that we list all the possible rival hypotheses to it in a given domain and then distribute probabilities over the entire list (according to some general principle such as giving the same probability to each entry). If the list is infinite, Chalmers can see no way to avoid Popper's conclusion that all hypotheses will end up with a prior probability of 0. (See the discussion of Popper's argument in the section "Why All Theories Are Improbable," in the commentary to chapter 1.) If the list is finite, what determines what ought to be included? Chalmers agrees with most Bayesians that the problems with the objective interpretation are "insuperable."

Subjective Bayesians (sometimes called personalists) begin with people's actual degrees of belief. That at least has the merit of yielding non-zero priors for the hypotheses in which scientists take an interest. But what are degrees of belief, and how can we measure them? There are basically two approaches: either we regard degrees of belief as internal psychological states that may be open to introspection or revealed in one's writings; or we identify a person's degrees of belief with his disposition to behave in certain ways in appropriate circumstances.

The pioneers of Bayesian theory (Frank P. Ramsey, Bruno De Finetti, Leonard J. Savage) took the second approach and interpreted degrees of belief in terms of people's behavior. (This has certain formal advantages in identifying degrees of belief with probabilities that are discussed below.) Ramsey proposed, for example, that a person's degree of belief in any proposition be measured by the least odds at which he would be willing to gamble on the proposition being true. In this way, we connect degrees of belief with something we can observe and measure, namely, betting behavior. Of course, this approach is feasible only with people who are currently alive to have their betting behavior investigated; and even then it is easy to imagine how there could be a mismatch between a person's actual subjective confidence in a proposition and the odds at which they are willing to bet on it when the stakes are high. In attributing degrees of belief to scientists in the past, the best one can do is to follow Dorling's recommendation by basing one's estimates on "scientific practice." Presumably that means drawing a conclusion about a scientist's degree of confidence in a particular theory from his tenacity in defending it, his willingness to invest time, energy, and prestige in developing the theory, what he says about it in print, and so on. From the perspective of the Bayesian *historian* of science this is tantamount to adopting the first approach, in which the real focus is on an individual's actual subjective degrees of belief rather than any real or inferred disposition to bet at various odds.

There is a second issue that has played a prominent role in the development of Bayesianism, and that is the identification of subjective degrees of belief with probabilities. In order to qualify as probabilities, the numbers assigned to degrees of belief have to satisfy the axioms of probability theory. Why should we think that the degrees of conviction that a person happens to have in various propositions will obey the probability axioms? Surely, people violate the axioms in many cases. For example, there might be a proposition Q, such that a person's degree of conviction in Q and his degree of conviction in $\sim Q$ do not add up exactly to 1.

The Bayesian response to this objection is the Dutch book argument. Professional gamblers say that a Dutch book has been made against someone if that person accepts a series of bets such that, no matter what the outcome, the person is guaranteed to lose money. No rational person would knowingly gamble in this way. The Dutch book theorem proves that a necessary and sufficient condition for avoiding a book being made against you is that your degrees of belief satisfy the axioms of probability theory. When this condition is satisfied, your degrees of belief are said to be coherent. Thus, when subjective Bayesians interpret probabilities as degrees of belief, a certain amount of idealization is involved. The degrees of belief in question are not necessarily the actual degrees of conviction that a particular person has, but rather the degrees of belief that she would have if she were ideally rational and her degrees of belief were coherent.

Bayesians view the coherence requirement as having the same status and rationale as the requirement of logical consistency. Rationality requires not only consistency with the laws of logic, but also consistency with the probability axioms. Both are necessary conditions for rational belief. What makes the Bayesian position thoroughly subjective is its insistence that coherence (which entails logical consistency) is not only necessary but also sufficient for rationality: no matter how crazy one's degrees of belief may seem to someone else, if they satisfy the probability axioms, then, according to the Bayesians, they cannot be condemned as irrational.

Suppose that Adam and Eve each have a different degree of conviction in the proposition that it will snow in Phoenix next July. Adam gives it a probability of 0.9, while Eve gives it only a 0.05 chance of being true. Coherence requires that Adam also assign a probability of 0.1 to the proposition that it will *not* snow in Phoenix next July. Similarly, if Eve's degrees of belief are coherent, then she thinks it is 0.95 likely that there will be no July snowfall in Phoenix next year.

As far as subjective Bayesians are concerned, both Adam and Eve are rational with respect to the proposition that it will snow in Phoenix next July and its negation. But there is more to coherence than merely satisfying the negation rule. One also has to satisfy the special and general addition rules for disjunctions (598–99), the special and general multiplication rules for conjunctions (600), the implication and equivalence rules (599), and axiom 2 concerning necessary truths (598). Axiom 2 is especially troubling, since it requires that every necessary truth, no matter how complex, be assigned a probability of 1. Thus, while it might seem as though coherence is a very weak condition (because it places no restrictions on the degree of belief that a rational person can assign to any particular contingent proposition), in fact coherence makes very strong demands on the degrees of conviction that can be assigned to the members of any reasonably sized set of propositions (where that set includes many contingent statements, many necessary statements, and all their truth-functional compounds). Moreover, every proposition in the set must be assigned a precise number in the interval from 0 to 1. We will return to the question of how well Bayesianism is able to capture the notion of scientific rationality, even with the imposition of the coherence condition, in the section "Subjectivism and Rationality" below.

Bayesianism and Scientific Method

Chalmers acknowledges that Bayesianism, especially as presented by Howson and Urbach, has the virtue of capturing some important features of scientific method. Not the least of these is Dorling's solution to the Duhem problem. Unlike other approaches (such as Lakatos's methodology of scientific research programmes), Bayesians can give a plausible account of how and why it can be rational for a scientist to respond to an anomaly by pinning the blame on an auxiliary hypothesis. (See the section on the Duhem problem in chapter 3.) Bayesianism are not forced to regard theoretical systems as seamless Quinean webs; they can explain how evidence can be strongly relevant to some parts of a system and yet have virtually no effect on other parts.

Bayesians can also explain why, even though scientists may start with radically different prior probabilities assigned to a hypothesis, their posterior probabilities (calculated using Bayes's theorem) are likely to converge as evidence accumulates. Chalmers gives a simple sketch of how this works. One scientist (whom we shall call "Mr. High") is an ardent believer in the truth of H; "Mr. Low" is not. So, their prior probabilities for H are far apart. But because of his confidence in the truth of H, Mr. High will also assign a relatively high probability to a prediction, E, that is deduced from H; whereas Mr. Low will reckon $P(E)$ to be much lower. Bayes's equation entails that the more unlikely the evidence, the greater is its confirming power. So, when the prediction E is verified, Mr. Low's value of $P(H/E)$ is markedly higher than his value of $P(H)$, whereas Mr. High's value of $P(H/E)$ is only slightly higher than his value of $P(H)$. In this way, they will tend to agree about the posterior probability of hypotheses in the light of confirming evidence.[24] Our preference for diversity of evidence is likewise amenable to a simple Bayesian treatment. (For a more thorough analysis, revealing some of the complexity of this topic, see the article by Horwich, discussed below.) Repetitions of the same experiment or very similar experiments tend to follow a "law of diminishing returns" as far as their confirming power is concerned: as our confidence in H rises, so does the probability we assign to predictions from H that are very similar to those that have already been verified. Greater confirming power accrues to new kinds of evidence because, prior to its verification, its probability is lower.

One area in which Bayesians disagree with many philosophers of science is with regard to whether a piece of evidence E can confirm T even though E has been deliberately used in the construction of T in the first place. Howson and Urbach give an urn example to show that this prejudice is unjustified. It seems eminently reasonable to infer that an urn contains one half red and one half white counters if we observe almost exactly those proportions in a sample of 10,000 counters drawn singly and at random from the urn and then replaced each time.[25]

Bayesians also bring an interesting perspective to bear on the traditional Popperian animosity toward hypotheses that are deemed to be "ad hoc." Despite numerous efforts, both within and without the Popperian fold, it has proven difficult to say with any degree of precision what makes a hypothesis ad hoc; moreover, given a characterization of ad hocness, to explain why it is necessarily a bad thing.[26] Chalmers cites the work of Greg Bamford who, after surveying many attempts to clarify the notion of ad hocness, has concluded that it is an intuitive notion that resists formal analysis.[27] Howson and Urbach have an attractively simple diagnosis of why, in the majority of cases, hypotheses characterized as ad hoc are not deemed worthy of scientific attention. It has nothing to do with independent testability or the tailoring of hypotheses to fit a particular case. Rather, it is because such hypotheses contain assumptions that, in the light of our background knowledge, are judged to be highly improbable or implausible. When one conjoins an existing theory, T, with an "ad hoc" modification A, the conjunction $(T \& A)$ cannot be more probable than A. So if the probability of A is low, the probability of $(T \& A)$ must be even lower.

THE PROBLEM OF OLD EVIDENCE

The problem of old evidence arises as follows. On the one hand, scientists often appeal to facts that are already known in support of new theories. Typical examples are Newton's use of Kepler's laws and Einstein's ability to derive the anomalous advance of the perihelion of Mercury from his general relativity theory. Indeed, there is a sense in which any piece of evidence, even evidence that has been newly discovered, must count as "old" if it is being used to confirm a theory, since, unless it is accepted as a well-established fact, it would be useless for the purposes of confirmation. On the other hand, if evidence is "old" in the sense of "known" or, at least "accepted as true," then it would seem that its probability must be 1. Thus, for old evidence, $P(E) = 1$, and regardless of the theory involved, the likelihood $P(E/T)$ must also be 1. But when these values are plugged into Bayes's equation, it follows immediately that E cannot confirm T, because the posterior probability of T given E will be equal to the prior probability of T. Thus, although scientific tradition says that old evidence can, in principle, confirm a theory, Bayesians seem forced to deny this.[28]

Bayesians can make a number of responses to this problem. One reply would be to insist that the scientific tradition is mistaken in thinking that old evidence has any power to confirm a theory, especially when the theory in question was deliberately designed to fit that evidence or explain it (as was the case with the examples of Newton and Einstein). This response is discussed and criticized in the commentary on chapter 4, in the section "Novel Prediction," and in the section on Achinstein.

If the scientific tradition is not mistaken, the Bayesian has two options: either deny that $P(E) = 1$ or adopt a new criterion for confirmation to replace the standard relevance criterion.

The core idea behind the first strategy is that the subjective probability of E should be reckoned, not with respect to one's actual background knowledge, B (which includes E), but with respect to what one's background knowledge would be if E were not yet known. The problem is to give an adequate account of how this counterfactual background knowledge should be defined and how $P(E)$ should be calculated in terms of it. There are two ways to go: either one tries to imagine what one's present background knowledge would have been if, contrary to fact, E were not known, or one considers what one's background knowledge was in the past, before one learned that E is true. Let us call these the *present proposal* and the *historical proposal*.

With regard to the present proposal, it is not sufficient merely to say that we should imagine "deleting" E and whatever entails it from B. We need to know, in some detail, how this process is to be carried out, especially when E has been thoroughly integrated into B. Part of the problem is that the content of $(B - E)$ varies depending on how the content of B is represented or axiomatized. Howson and Urbach give the following simple illustration: the set $\{a, b\}$ has the same logical content as the set $\{a, a \supset b\}$; but if a and b are contingent propositions that are consistent with each other, removing a from the first set will have a different result than removing a from the second set. Nonetheless Howson and Urbach cite recent work that has produced a formal solution to the problem. Moreover, they claim that it is sufficient from their subjectivist standpoint to take B to be represented in whatever way a particular individual represents her knowledge at a particular time. They also point out that whenever evidence is acquired it becomes "old," and so inevitably Bayesians must "go counterfactual" if any confirmation of hypotheses by that evidence is to be possible.[29]

Even if the present proposal can be made to yield an acceptable characterization of $(B - E)$, how to calculate the probability of E in terms of $(B - E)$ is still a problem. The usual suggestion is that we should use the total probability rule to express $P(E)$ as a sum of factors of the form $P(T_i) \times P(E/T_i)$. The problem, as Salmon explains in his discussion of the expectedness of evidence, lies in the intractability of the last term in this sum, which involves the catch-all hypothesis, T_k. And as Clark Glymour demonstrates, if one simply deletes the term involving the catch-all hypothesis from the calculation, one's prior and conditional probabilities will inevitably violate the probability axioms and generate absurdities.[30] Thus, calculating the probability of E given the catch-all hypothesis would seem to be impossible, and ignoring the catch-all leads to incoherence.

The historical proposal also has its difficulties. Glymour notes that even if we knew which historical period to choose as the right one, we cannot simply adopt as the value of $P(E)$ whatever degree of belief in E we would have had at that time and retain all our other actual present-day degrees of belief, for that would violate the coherence requirement. But this technical difficulty aside, there are serious philosophical objections to the historical proposal. How should we decide which historical period is the right one? How far back should we go? And when we have made that decision, does it really make sense to suppose that there is a degree of belief that each person would have had in E at that time? The difficulty of answering these questions is reason to be skeptical of the historical proposal. Given the equally severe problems with the present proposal, this would seem to recommend rejecting the first strategy—the strategy of denying that $P(E) = 1$—as a solution to the problem of old evidence.

The second strategy concedes that $P(E) = 1$, but offers a new criterion of confirmation to replace the relevance criterion. The second strategy takes its cue from the standard Bayesian assumption (via the coherence condition) that rational agents are logically omniscient. Bayesians assume, quite unrealistically, that rational agents know what all the logical consequences of their beliefs are and assign degrees of belief to them in accordance with the probability axioms. But none of us is even close to being that knowledgeable. In many of the cases involving old evidence, existing theories have been unable to explain some phenomenon (such as the anomalous behavior of Mercury's orbit). A new theory T (such as Einstein's) comes along, and we then discover that T entails E and is thus able to explain this phenomenon. As a result the new theory is confirmed. The important point is that, while the evidence E is "old," having been known for many years, we have just learned something new, namely, that T entails E, and it is this new piece of information, not E by itself, that confirms T.[31] Thus, we can summarize the second strategy by saying that the relevance criterion for confirmation, namely,

E confirms T if and only if $P(T/E \ \& \ B) > P(T/B)$,

should be replaced, at least for the case of old evidence, by the new criterion:

E confirms T if and only if $P(T/E \; \& \; B \; \& \; (T \vDash E)) > P(T/E \; \& \; B)$.

Expressed more simply, the new criterion says that:

E confirms T if and only if $P(T/T \vDash E) > P(T)$.

It would be misleading to say that, according to the new criterion, E confirms T, since the real confirmational work is being done, not by E, but by the discovery that T entails E. Some authors reject the new criterion because it evades the original problem of whether E literally confirms T. John Earman, for example, objects that the question of whether the astronomical data on Mercury's orbit (the perihelion advance) confirms Einstein's theory has not been answered; it has been replaced by the different question of whether Einstein's confidence in his theory was increased upon his learning that T entails E.[32] Moreover, as Earman points out, when most of us learn about Einstein's theory, among the first things we are told, even before we study the theory in detail, is that it entails the perihelion advance. So when we later derive this phenomenon from Einstein's theory, we already know that the entailment holds. Does that mean that, for us, the perihelion advance has ceased to be confirming evidence for the general theory of relativity? Other critics object that the new criterion cannot account for the fairly numerous cases in which a theory T was deliberately constructed so as to yield E as one of its consequences. For, in such cases, it can hardly be a discovery that T entails E when that is precisely what the theory was designed to do.

Subjectivism and Rationality

Chalmers concedes that Bayesians are able to accommodate many accepted views about confirmation, such as scientists' preference for variety in evidence and their general disdain for ad hoc hypotheses, and that they have offered an interesting solution to the Duhem problem. But these results are achieved by relying on prior probabilities that are purely subjective. How, Chalmers asks, can Bayesians claim to reconstruct the rationality of scientific inference from evidence to hypotheses if their entire framework depends on such arbitrary and nonrational degrees of belief? Surely, evidence must involve more than just what someone happens to believe, however strongly those beliefs are held or accepted as true. Chalmers takes particular exception to a passage from the first edition of Howson and Urbach's book, *Scientific Reasoning: The Bayesian Approach* (1989), in which the authors proclaim that their theory of inference is concerned solely with how what one accepts as evidence affects one's belief in some hypothesis. The provenance of an evidential statement—the reasons one may or may not have for it—are, they say, "simply irrelevant." Presumably what Howson and Urbach meant to say in their quote is that their sole concern is not with how *as a matter of fact* people change their beliefs—there is plenty of evidence that most people are not Bayesians[33]—but with how people *ought* to adjust their beliefs. Bayesianism is intended as a normative account of how people ought to reason, not a description of how they in fact reason.

In defense of their position, Urbach and Howson appeal to an analogy with deductive logic. In deductive logic our sole concern is whether or not arguments are valid. The actual truth (or falsity) of the premises and whether or not those premises are epistemically justified does not matter in the slightest. So, too, in (Bayesian) inductive logic, all that matters is whether one infers correctly from the premises that one starts with: the truth and epistemic status of those premises is irrelevant. Howson and Urbach insist that correct inferences must conform to Bayes's theorem using the Bayesian conditionalization rule $P_{new}(T) = P(T/E)$. How can they justify imposing this restriction on how one should change one's beliefs? The problem is that insisting on coherence among one's beliefs *at any given time* does not imply how those beliefs should be adjusted *over time*. Why is it irrational for someone to violate the Bayesian rule when revising her degrees of belief, as long as her entire set of beliefs is coherent after the revision? If a Bayesian were later to look at that coherent belief set, in ignorance of how the person's degrees of belief had been arrived at, the Bayesian would judge the set rational. The synchronic Dutch book argument (for the coherence of beliefs at a given time) seems to have no relevance to the diachronic conditionalization rule (for how probabilities should change over time). One popular Bayesian response to this challenge is to construct a further Dutch book argument that is explicitly diachronic.[34] The gist of the argument runs as follows. If one makes a series of bets, some of which depend on what one's degrees of belief will be in the future, and if one follows a rule other than the Bayesian conditionalization rule and this alternative rule is known to the person with whom one is betting, then the person with whom one is betting can always construct a Dutch book against one. A necessary and sufficient condition for avoiding a Dutch book under the conditions stated is that one uses nothing but the Bayesian conditionalization rule for changing one's degrees of belief. Just as the synchronic Dutch book argument is used to derive the coherence condition from the presumption of rationality, so, too, the diachronic Dutch book argument is supposed to show how rationality mandates the Bayesian rule for revising one's degrees of belief over time.

There is also a constraint that Bayesians have to impose on prior probabilities in order for learning from experience to be possible. Suppose one were to assign a prior probability of 0 to the theory, T; in that case, no amount of evidence could ever confirm it, since its posterior probability, $P(T/E)$, would always remain 0. It is important to realize that the coherence requirement does not solve this problem. Coherence requires that we conform to the probability axioms. The relevant axiom (axiom 2) dictates that all necessary truths have a probability of 1. In conjunction with the other axioms, this entails that all necessary falsehoods must have a probability of 0. But the axioms do not forbid us from also assigning a probability of 0 to any contingent proposition that we might judge to be impossible. The usual Bayesian response is to impose the further demand of *strict coherence*: a set of beliefs is strictly coherent if and only if it is coherent and no contingent proposition is assigned a probability (degree of conviction) of 0 or 1. (This is one response to the Popperian argument discussed in the section "Why All Theories Are Improbable," in the commentary on chapter 1.)

The value of $P(T/E)$ depends in part on $P(T)$; in other words, the amount by which evidence E confirms (or disconfirms) theory T will depend on the prior probabilities assigned to T by different scientists. Chalmers questions how this subjective influence on the degree of confirmation can be reconciled with scientific objectivity. Bayesians reply by appealing to a theorem about the *washing out* (or *swamping*) *of priors*. The theorem (alluded to earlier in Chalmers's article) shows that as evidence accumulates, the values of $P(T/E)$ calculated by individual scientists with different prior probabilities will tend to converge. In the long run, the initial divergence in the (subjective) prior probabilities diminishes and (objective) evidence dominates the calculation of confirmation. Anti-Bayesians are unimpressed by these "swamping" arguments, pointing out that they do not guarantee convergence or place any constraints on how quickly (or slowly) the convergence can occur.[35] Also, of course, the swamping argument assumes that the scientists involved agree about what they take to be evidence.

There is a final difficulty that has been stressed by critics such as Clark Glymour.[36] Bayesians define confirmation in terms of probability: a piece of evidence is positively relevant to a theory if and only if it raises the theory's probability. Suppose that we can separate the logical consequences of any theory into two classes: those that contain at least one theoretical term and the so-called observational statements that contain no theoretical terms. The implication rule entails that no theory can be more probable than its consequences. Why, then, would the Bayesian bother with theories at all, if observational statements always have a higher probability than any theory from which they can be derived? One standard answer is that, in addition to empirical warrant, we also value the explanatory power that only theories can provide. But even so, the Bayesian must divorce explanatory power from degree of confirmation. As far as the Bayesian is concerned, observational consequences alone can raise a theory's probability. Explanatory success can do nothing to increase the degree of warranted belief that we should have in a theory, and the probability of a theory can never exceed that of its observational consequences. What Glymour is saying, in effect, is that it is difficult for a Bayesian to be a scientific realist. Realists typically take the explanatory power of a theory as an important reason for thinking that the theory is true or highly probable. In this way, theories can become more credible than their observational consequences and thus achieve a level of warrant that Bayesians cannot account for.

5.5 | Horwich's Defense of Therapeutic Bayesianism

Paul Horwich thinks that many of the criticisms of Bayesianism by Gly-mour, Mayo, Chalmers, and others are unfair and misdirected, especially those hinging on the near impossibility of assigning numerically precise degrees of belief to scientists and the difficulty of giving rigorous Bayesian explications of every facet of scientific methodology. Bayesianism, Horwich insists, is not intended to be a complete, true, and detailed theory of scien-tific method. Rather, it should be viewed as a philosophical theory, a contri-bution to epistemology in general. Horwich follows Wittgenstein (that is, the later Wittgenstein of the *Philosophical Investigations*) in regarding phi-losophy as therapeutic: its goal is to dissolve problems and resolve paradoxes by untying the knots in our thinking. A good philosophical theory is one that unravels and corrects the misleading assumptions that have led us into confusion and error. The philosophical theories that can achieve this do not have to be especially complex and sophisticated; usually, a simple, ide-alized model will suffice. And so it is with therapeutic Bayesianism, accord-ing to Horwich. By replacing the traditional assumption that belief is an all-or-nothing affair with the idea that there are degrees of belief that con-form to the axioms of the probability calculus, we can solve a number of important problems in the philosophy of science. As illustrations of the power of therapeutic Bayesianism, Horwich offers a resolution of the raven para-dox and a rationale for our preference for diversity in evidence.

THE RAVEN PARADOX

The raven paradox was made a focus of philosophical inquiry by Carl Hem-pel in his seminal papers on confirmation theory, where he introduced "All ravens are black" as the paradigm of a scientific hypothesis, thus giving the raven paradox its name.[37] Although Hempel drew attention to several confir-mational paradoxes, we shall follow Horwich in concentrating solely on one of them, namely that the observation of a nonblack nonraven confirms the hypothesis "All ravens are black." Somewhat ironically, Hempel, like Hor-wich, does not regard this result as genuinely paradoxical, since he argues that, when correctly interpreted, observations of white shoes and green chairs (nonblack nonravens) do indeed confirm that all ravens are black. Let us first lay out the three assumptions that generate the alleged paradox and then examine Horwich's Bayesian resolution of it. For convenience, we shall use the letter H to stand for the hypothesis that all ravens are black.

The first assumption is that generalizations of the form "All Rs are B" are confirmed by their positive instances, that is, by anything that satisfies both the antecedent, R, and the consequent, B. The second assumption is that "All Rs are B" is logically equivalent to "All non-Bs are non-Rs." The third assumption is that if an observation report confirms hypothesis J, and J is logically equivalent to K, then the observation report also confirms K.

Putting these three assumptions together yields the result that the observation of a nonblack nonraven (such as a white shoe) confirms H, that all ravens are black. As Nelson Goodman has remarked, this would seem to open up wonderful prospects for indoor ornithology.[38]

One way to avoid the paradox is by rejecting one or more of the assumptions generating it. In fact Horwich, like many other Bayesians, rejects the first assumption, the positive-instance principle. The positive-instance principle entails that any observation report of the form "This is a black raven" necessarily confirms the hypothesis, H, that all ravens are black. But, Horwich objects, whether or not the observation of a black raven confirms H depends, at least in part, on the manner in which the evidence is collected. If, for example, we first pick a black object and then examine it to see whether it is a raven, then the discovery that the object is both black and a raven would do nothing to confirm H because the method employed to gather the evidence is incapable of generating a report that could falsify the hypothesis.[39] Thus Horwich replaces the first assumption with a slightly different principle: "All Rs are B" is confirmed by the positive instance $(Ra \ \& \ Ba)$ only if the method for obtaining the report involves first selecting an object that is an R and then examining it to see whether it is a B. Horwich symbolizes this as $(R^* B)$, using an asterisk to indicate the order in which the information was obtained.

Having revised the first assumption, Horwich accepts the other two assumptions and the conclusion of the raven paradox argument. Horwich argues that the appearance of paradox is an illusion that evaporates once we embrace the Bayesian doctrine of degrees of belief and the relevance criterion of confirmation. He notes that the conclusion of the paradox does not say how strongly the observation of a white shoe confirms H; it merely asserts that H is confirmed. The virtue of Bayesianism (as opposed to the H-D model or Glymour's bootstrap model of confirmation) is that it not only permits degrees of confirmation but also allows us to estimate their magnitude. Thus the way is open for the Bayesian to argue that the observation of a white shoe does confirm H but only to a very small degree, a degree that is much smaller than the confirmation conferred on H by the observation of a black raven. In this way, we have a psychological explanation of why many people regard the raven paradox as a genuine paradox when they first encounter it, for most of us are unable to distinguish a very low degree of confirmation from no confirmation at all.

As a measure of the degree to which E confirms H, Horwich adopts the ratio of the posterior probability of H to its prior probability. By Bayes's theorem, this is equal to $P(E/H)$ divided by $P(E)$—the likelihood of H divided by the expectedness of the evidence.

$$\text{Confirming power of } E = \frac{P(H/E)}{P(H)} = \frac{P(E/H)}{P(E)}.$$

As we have noted, Horwich insists that the manner in which evidence is collected is relevant to its confirming power, and he adopts a special notation to reflect this: $(R^* B)$ is the discovery that a randomly selected object that is already known to be a raven is black; $(\sim B^* \sim R)$ is the discovery that a randomly selected object that is already known to be non-black is a non-raven. The order in which the components of each observation are made—a feature that Horwich deems crucial to his analysis—is indicated by the asterisk. To repeat, the observation reports are not simply of the form $(Ra \ \& \ Ba)$ but include information about the method used to generate the report. That said, the likelihoods must both be equal to 1, for if all ravens are black, then the probability that a raven will turn out to be black is 1; similarly, given H, the probability that a nonblack thing will turn out to a nonraven is also 1. So the comparison of the confirming power of $(R^* B)$ with the confirming power of $(\sim B^* \sim R)$ reduces to the inverse ratio of their probabilities: it is the expectedness of evidence $(\sim B^* \sim R)$ divided by the expectedness of evidence $(R^* B)$.

$$\frac{\text{Confirming power of } (R^*B)}{\text{Confirming power of } (\sim B^* \sim R)} = \frac{P(\sim B^* \sim R)}{P(R^*B)}.$$

Horwich argues that, given some plausible assumptions about the relative numbers of ravens and black things in the world, our degrees of belief in $(R^* B)$ and $(\sim B^* \sim R)$ prior to our investigation of H can easily be estimated. We assume that, in the world as a whole, the number of ravens is pretty small whereas the number of nonblack things is very large: ravens are rare, but nonblack things (and black things, too, for that matter) are abundant. So, Horwich argues, if we randomly select something that is nonblack (which is a very large class of things) the chances that the thing will also be a nonraven must be high (since ravens are scarce and nonravens abundant), and hence our prior degree of belief in $(\sim B^* \sim R)$ must be high. Similarly, he argues, if we randomly select something that is a raven, then the chances that the thing will also be black must be low (given the abundance of nonblack things) and hence our prior degree of belief in $(R^* B)$ must be low. Thus, given the comparison of confirming powers noted above (as the inverse ratio of these probabilities), Horwich concludes that the ratio of $P(\sim B^* \sim R)$ to $P(R^* B)$ is high and so $(R^* B)$ confirms H to a much higher degree than does $(\sim B^* \sim R)$.

Horwich's argument for his Bayesian solution to the raven paradox is not entirely satisfactory. The problem with it can best be grasped by looking at figure 3, which depicts the background assumption that ravens are scarce and nonblack things abundant.

The R-rectangle represents the set of ravens; the area outside of the B-square represents the set of nonblack things. Since the non-B area is so much greater than the R area, Horwich is correct in saying that $P(\sim B^* \sim R)$ must be high. But that same background assumption, by itself, does not

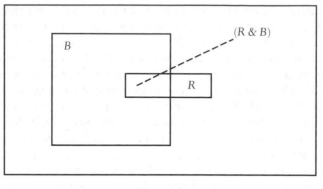

Figure 3

warrant the assertion that $P(R^* B)$ must be low. For $P(R^* B)$ depends, not on the size of the R and B areas, but on the fraction of the R-rectangle that overlaps with the B-square. In other words, we need to make some additional assumption about the fraction of ravens that are black. Remarkably, when we add that assumption and analyze the problem more carefully, it turns out that we can still vindicate Horwich's conclusion that black ravens confirm H more strongly than do white shoes. Here is the analysis.

Let x be the fraction of things in the universe that are ravens, let y be the fraction of things that are black, and let α be the fraction of things that, initially, are believed to be black ravens. In figure 3, the area of the R-rectangle is proportional to x, the area of the B-square is proportional to y, and the area of the intersection between R and B is proportional to α. (If assuming that we have precise beliefs about these numbers seems unrealistic, not to worry—in the final analysis nothing depends on their exact values.) Thus,

$$P(R^*B) = \frac{\alpha}{x}, \text{ and}$$
$$P(\sim B^* \sim R) = \frac{(1-y)-(x-\alpha)}{(1-y)}.$$

It follows from these expressions that a necessary and sufficient condition for either $(R^* B)$ or $(\sim B^* \sim R)$ to confirm H is that $x > \alpha$, that there are more ravens than black ravens. Thus, we have proven that $(\sim B^* \sim R)$ confirms H without having to rely on any assumption about the relative abundance of ravens and nonblack things. The only condition for $(\sim B^* \sim R)$ to confirm H is that, prior to the observation, we are not already convinced that all ravens are black. Moreover, we can compare the confirming power of the two reports.

$$\frac{\text{Confirming power of } (R^*B)}{\text{Confirming power of } (\sim B^* \sim R)} = \frac{P(\sim B^* \sim R)}{P(R^*B)} = \left(\frac{x}{\alpha}\right)\frac{(1-y)-(x-\alpha)}{(1-y)}.$$

Simple algebra shows that two conditions are necessary and sufficient for the expression on the right-hand side to be greater than 1: the first, as before, is that x is greater than α; the second is that $(1-y)$ is greater than x. So Horwich's judgment is vindicated: if we do not already believe that all ravens are black, $(R^* \, B)$ must support H more strongly than does $(\sim B^* \sim R)$ as long as we also believe that nonblack things are more abundant than ravens. Both sorts of observation must increase our confidence in H, but finding black ravens provides more powerful support.[40]

DIVERSITY OF EVIDENCE

Horwich's other illustration of the power of therapeutic Bayesianism is its explanation of the relation between diversity of evidence and strength of support. Why does diverse evidence, E_D, confirm a hypothesis more strongly than narrow evidence, E_N? In other words, why is it that, for any hypothesis H, $P(H/E_D) > P(H/E_N)$? One might attempt to vindicate the diversity principle by appealing to eliminative induction: as the set of data widens, more rival hypotheses are eliminated, thus enhancing the probability of the survivors. Although he believes that this answer is on the right track, Horwich thinks that it will not work as it stands because narrow data sets can eliminate just as many hypotheses as can broad ones. The relevant issue, in his view, is not the number but the kind of hypotheses that are excluded. Horwich reasons as follows:

1 Diverse data sets tend to eliminate more of the simpler hypotheses than do narrow data sets.

2 Simpler hypotheses have higher prior probabilities.

3 Thus, diverse data sets produce a greater increase in the probability of the surviving hypotheses than do narrow data sets.

In his book *Probability and Evidence*, Horwich gives a semiformal Bayesian treatment of this argument.[41] Let H_S be the straight-line hypothesis in the curve-fitting diagram in Horwich's article (584), and assume that H_S entails both E_D and E_N. Thus

$$P(E_D/H_S) = P(E_N/H_S) = 1.$$

Thus we can write:

$$\frac{P(H_S/E_D)}{P(H_S/E_N)} = \frac{P(E_N)}{P(E_D)} = \frac{P(H_S) + \Sigma \, P(E_N/H_i) \times P(H_i)}{P(H_S) + \Sigma \, P(E_D/H_i) \times P(H_i)}.$$

The numerator and the denominator on the right-hand side of this equation both contain a sum over all the rival hypotheses to H_S. These rival curves can be divided, roughly, into two classes: the crazy and the gradual (as Horwich labels them in his diagram). But the crazy curves have such low prior probabilities that their contribution to each sum can be ignored. That leaves H_S and the gradual curves, each of which is assumed to have a nonnegligible prior probability because of its relative simplicity. Any gradual curve that goes through E_D is likely also to go through E_N, but there are many gradual curves that go through E_N that go nowhere near E_D. So for most of the gradual curves, $P(E_D/H_i) < P(E_N/H_i)$. Thus, $P(E_N) > P(E_D)$, and so E_D lends greater support to H_S than does E_N.

Despite its ingenuity, Horwich's argument is not unassailable. Proving that simpler hypotheses must be more probable is difficult.[42] But Horwich might claim to have shown at least this much: given the widely accepted (but not uncontroversial) assumption that simpler hypotheses are more probable, there is a plausible Bayesian rationale for the diversity principle.[43]

PROBABILISTIC FOUNDATIONS

In section 2 of his paper, Horwich responds to critics who cast doubt on the foundational assumptions of Bayesianism: the existence of numerical degrees of belief, the coherence condition, the definition of degrees of confirmation in terms of probabilities, and the subjective nature of those probabilities. He acknowledges the difficulties. For example, he concedes that the Dutch book argument for coherence is "far from airtight" (587). Horwich regards the coherence requirement for degrees of belief as a normative idealization, a model of how rational agents ought to constrain their beliefs rather than as a description of how they actually do. Thus, in response to the subjectivity objection, Horwich cites Carnap in support of the contention that, of course, there are objective facts about confirmation even though the probabilities involved are degrees of belief. These facts are objective precisely because confirmation theory, even Bayesian confirmation theory, is normative: it specifies the degree of confidence that a rational agent should place in one proposition given the evidence of another; the right degree of confidence is the one that reflects the true, objective, relation of inductive confirmation between the two propositions.

Horwich also admits that there is room for reasonable disagreement concerning his definition of the degree of confirmation of H as the ratio of $P(H/E)$ to $P(H)$. Other authors, also Bayesians, have advocated different definitions. But Horwich insists that exactly how the degree of confirmation gets defined as a function of probabilities is irrelevant to therapeutic Bayesianism. For to resolve paradoxes and solve most philosophical problems of confirmation, all the therapeutic Bayesian requires is the relevance criterion, shared by all Bayesians, that hypotheses are confirmed if and only if their probability is raised by evidence. In the majority of the cases that interest philosophers, using a specific function to pin an exact number on the degree of confirmation is unnecessary.

MISPLACED SCIENTISM

In the last section of his paper, Horwich broadens his defense of Bayesianism in the philosophy of science by attacking what he calls "misplaced scientism" (589)—the mistake of applying to philosophical theories the standards we use to evaluate scientific theories. Horwich argues that science and philosophy are fundamentally different intellectual activities. Science aims, primarily, at achieving an accurate, true, comprehensive description of the natural world. Philosophy (according to Horwich's Wittgensteinian conception of it) aims at solving deep puzzles, dissolving paradoxes, and removing the contradictions and confusions that surround concepts such as confirmation, evidence, and simplicity. Given the very different aims of the two endeavors, there is no reason why their theories should be evaluated in the same way. A degree of vagueness, idealization, informality, and incompleteness that would not be tolerated in the sciences should not disqualify a theory (such as Bayesianism) in the philosophy of science if that theory does a good job of clearing up confusions and providing philosophical insight.

Horwich recognizes that the distinction between science and philosophy is not always sharp, especially in the philosophy of science, which runs the gamut from technical studies in logic and statistical inference to general epistemological theories that are quite informal. Nonetheless, Horwich thinks that we can distinguish two sorts of philosophy of science, theory oriented and problem oriented. Problem-oriented philosophy of science is philosophy properly so-called. It is, to use Horwich's term, "conceptual troubleshooting" (591). To criticize this sort of philosophy of science for being less than fully rigorous or for using highly idealized models is to miss the point. By contrast, theory-oriented philosophy of science is the search for a detailed, accurate, comprehensive account of scientific method. As such, it is really a continuation of science under the guise of naturalized epistemology. Thus, it has the same concern for rigor and completeness that we find in other branches of science. Here, the standards of science are appropriate.

Is the difference between science and philosophy, and between theory-oriented and problem-oriented philosophy of science, great enough to bear the weight that Horwich puts on it? One problem, as he recognizes, is that both sorts of activity have, as one of their goals, the provision of explanations via unifying theories, and often these theories involve simplified models. Idealization for the purpose of explanation is common in both fields. Similarly, in both areas we find the notion that there is some domain of phenomena and problems—whether the field be genetics, biochemistry, or the study of confirmation—of which an adequate theory must give an account. Horwich himself commends Bayesianism for its explanatory successes, for its ability to make sense of a broad range of methodological principles within one unifying framework. But at the same time, he wants us to discount the objections of those who criticize the Bayesian program for the things that it cannot explain by arguing that this is not an appropriate standard to apply to a philosophical theory. This has the appearance of special pleading. If explanatory successes count in favor of a theory, explanatory failures should count against it.

Horwich seems to be on firmer ground when he complains about the unreasonable demand for rigor in areas, such as problem-oriented philosophy of science, where it is unattainable, inappropriate, or unnecessary. His earlier example of the dispute over the definition of degrees of confirmation is a good example of such misplaced rigor. For most philosophical applications of Bayesianism to puzzles and paradoxes in confirmation theory, no precise confirmation function is needed. Similarly, as far as its philosophical applications are concerned, it is irrelevant that Bayesians cannot measure precise degrees of belief. In science, it would be unacceptable to build a theory around quantities that cannot be measured or calculated; in problem-oriented philosophy of science it scarcely matters.

5.6 | Summary

In the first section of this commentary, we identified the main elements of the Bayesian approach to confirmation theory. These are, in increasing order of philosophical contentiousness: Bayes's equation, the relevance criterion of confirmation, and the interpretation of the probabilities in that equation as degrees of belief.

Viewed simply as a theorem of the probability calculus, Bayes's equation is entirely uncontroversial. It is a deductive consequence of the axioms of probability, together with a definition of conditional probability. But in order to apply that equation to scientific reasoning, some connection has to be forged between probability and confirmation. The relevance criterion stipulates that a theory, T, is confirmed by evidence, E, whenever E raises the probability of T. Some philosophers of science balk at this. Peter Achinstein, for example, has protested that the relevance criterion entails the absurdity that whenever a world-class swimmer dives into a pool, we have confirmed the hypothesis that he will drown, and that this makes a travesty of the notion of confirming evidence. In reply, Bayesians insist on the distinction between a hypothesis being confirmed (which occurs whenever its probability is raised, however slightly) and a hypothesis being acceptable as highly confirmed or true (which requires a posterior probability close to 1). Obviously, no one would accept the hypothesis that Michael Phelps will drown simply because drowning is slightly more likely to occur when he goes swimming; but nonetheless, when Phelps goes swimming, this increases the chances that he will drown and thus, to a small degree, confirms the hypothesis.

Even when one accepts the relevance criterion, there is still the question of what $P(T)$, $P(E/T)$, $P(E)$, and $P(T/E)$ stand for in Bayes's equation. These four quantities are usually referred to as the prior probability of theory

T, the likelihood of *T*, the expectedness of evidence *E*, and the posterior probability of *T*. The official Bayesian view is that all of these quantities are subjective degrees of belief, that is, they are the actual degrees of belief of real people as measured, for example, by their willingness to bet at well-defined odds on various propositions. The appearance of subjectivism is mitigated somewhat by the stipulation that the people in question have to be rational and that rationality requires that their degrees of belief satisfy the probability axioms. The requirement of conformity to the probability axioms is called the *condition of coherence*, and Bayesians usually offer a Dutch book argument to justify it. The other constraint imposed by Bayesians is conditionalization: whatever a person's prior degree of belief in *T*, when new evidence, *E*, is acquired, that person should replace the old prior probability of *T* with the posterior probability of *T* given *E*, calculated using Bayes's theorem: posterior probabilities replace old prior probabilities. Again, Bayesians have appealed to Dutch book arguments in defense of this requirement.

Wesley Salmon is attracted to the Bayesian approach because it seems to represent a definite advance over the H-D model. Unlike the H-D model, the Bayesian approach can be applied to statistical theories, and it holds out the promise of being able to solve the Duhem problem (the problem of allocating blame within a group of hypotheses for a failed prediction) and the tacking paradox (otherwise known as the problem of irrelevant conjunction). Salmon also thinks that Bayesianism can be used to reconcile Thomas Kuhn's historically oriented approach to understanding science with logical empiricism. The key to this reconciliation, in Salmon's view, lies in the role of prior probabilities in determining degrees of confirmation. The factors influencing the judgment of scientists about theories—Kuhnian values such as consistency, simplicity, and fruitfulness—can lead to reasonable disagreements about how well a theory is confirmed by a piece of evidence by giving rise to different estimates of the theory's prior probability. Even though they may be using the same Bayesian algorithm, scientists generate different outputs (confirmation judgments) because their inputs (prior probabilities) differ.

Despite his enthusiasm for Bayes's theorem as a tool for understanding confirmation, Salmon rejects the unfettered subjectivism of many Bayesians. At bottom, he thinks, prior probabilities should be guided by experience. They should not reflect mere prejudice and subjective whim. But Salmon takes comfort in the fact that, in the long run, the initial choice of prior probabilities becomes irrelevant. For as evidence accumulates, the values of the posterior probability of a hypothesis converge, thus "swamping out" or "washing out" individual differences among the priors. One important condition for this sort of convergence is that we never assign to any hypothesis the extreme values of 0 or 1. Another condition is that we agree about the values of likelihoods.

In Salmon's view, the biggest difficulty in applying Bayes's theorem to scientific theories lies in the calculation of $P(E)$, the expectedness of the evidence. In fact, Salmon judges this problem to be insoluble. He therefore advocates a modified version of the Bayesian approach, employing an algorithm for theory preference that does not require us to calculate $P(E)$. This places a significant limitation on the scope of Salmon's analysis, for his algorithm permits the comparison of rival theories only when the evidence for each is the same.

Deborah Mayo deems Salmon's use of his Bayesian algorithm to reconcile logical empiricism with Kuhn a failure. The problem lies in the prior probabilities. For rival deterministic theories that make the same prediction, the priors dictate the values of Salmon's comparative measure of confirmation. Salmon attempts to avoid Kuhnian subjectivism by interpreting the prior probability of a hypothesis as the frequency with which similar hypotheses have been successful in the past. Not only is the notion of similarity hard to pin down, but it is also hard to see why that frequency should guide our expectations about how *this* hypothesis will fare in the future. Mayo advocates the non-Bayesian error-statistics approach to confirmation and what counts as evidence. The strength of evidence for a hypothesis is determined by the severity of the tests that it has passed. A severe test of a hypothesis is one that the hypothesis has a high probability of failing if the hypothesis should be false. Mayo argues that this well-defined, objective, empirical frequency-based notion of probability should be the basis for our decisions about which theories to accept and which to reject. The logic of error-statistics underlies common practice among non-Bayesian statisticians; it also guides how scientific experiments are designed and refined to test scientific theories, and explains why some parts of theories are better tested than other parts by the same evidence.

Alan Chalmers acknowledges that Bayesianism can give a rationale for several widely accepted methodological principles, but he doubts whether the Bayesian identification of probabilities with subjective degrees of belief is consistent with regarding science as an objective, rational enterprise. In particular, Chalmers criticizes Howson and Urbach for restricting the scope of inductive reasoning to what can be deduced using the probability calculus from the beliefs that scientists happen to have, regardless of how they were acquired. This, for Chalmers, is to ignore the normative dimension of philosophy of science that is concerned with what standards scientists *should* adopt and what *ought* to count as evidence.

A particularly difficult problem for the Bayesian account is the problem of old evidence. Scientific theories are often regarded as confirmed when they can explain some phenomenon or law that is already well known and accepted; but if the evidence is old in this sense, then it would seem that its probability should be 1. Unfortunately, making $P(E) = 1$ in Bayes's equation would prevent E from affecting the probability of any theory, T, that explains it. Thus, old evidence should not be able to confirm T.

Not everyone agrees that assigning old evidence a probability of 1 and thus denying it any power to confirm theories is a mistake. (See, for example, the discussion of predictionism in the sections of the commentary on chapter 4 devoted to Hempel and Achinstein.) But most Bayesians have tried to solve the problem by arguing either that old evidence should not be given a probability of 1 simply because it is already known or that we should adopt a new criterion of confirmation. There are two versions of the first strategy, which we have labeled the present proposal and the historical proposal. The present proposal advocates that we calculate $P(E)$ by trying to imagine what probability we would assign to E were E (and whatever entails it) to be deleted from our background knowledge. This operation of "deletion" is not easy to perform. It can also lead to violations of the probability axioms if we try to calculate $P(E)$ using the total probability rule to sum over all the hypotheses that can explain E. The historical proposal avoids this problem by taking as the value of $P(E)$ whatever the degree of belief in E was (or would have been) before it was discovered. This proposal, too, runs into difficulties, such as the problem of choosing the "right" historical period.

The second strategy takes a different tack. It suggests that the old evidence itself is not what confirms the theory. Rather, it is the discovery that the theory can explain the evidence (by entailing it) that provides the confirmation. Since scientists are not logically omniscient, they can genuinely discover that some theory T entails evidence E. Despite its ingenuity, like much else in the Bayesian system, the second strategy can be criticized for not getting to the epistemological heart of the matter; for there is more to the confirmational relevance of E to T than the mere logical derivability of E from T.

Paul Horwich undertakes to defend Bayesianism against its critics by arguing for its value as a philosophical theory that can solve conceptual problems. Horwich gives two illustrations of the power of what he calls "therapeutic Bayesianism": the resolution of the raven paradox and the explanation of why it is desirable to seek diversity in evidence.

The raven paradox follows from seemingly plausible principles of confirmation. Apparently, the observation of a nonblack thing that is not a raven should confirm the hypothesis that all ravens are black. Carl Hempel, who first drew attention to this result, regarded it as correct: the observation of a white shoe does confirm that all ravens are black, but only to a very small degree. Because the degree of confirmation is so very small, we have a psychological tendency to mistake it for no confirmation at all. Horwich attempts to vindicate Hempel's position by giving a simple Bayesian analysis incorporating assumptions about the relative size of the class of ravens as compared with the (very much larger) class of things that are not black.

Horwich uses a curve-fitting diagram to illustrate his Bayesian explanation of why diverse evidence confirms a hypothesis more strongly than does narrow evidence. The gist of his argument is that diverse evidence confirms more effectively because it eliminates a greater number of the simple hypotheses. A crucial premise in this argument is that simple hypotheses have higher prior probabilities. As we learned in chapters 4 and 5, it is very difficult to connect simplicity with truth or the probability of truth. But if Horwich's Bayesian argument is to have any power to convince non-Bayesians, the prior probabilities to which it appeals must be more than mere expressions of subjective preference.

In the remaining two sections of his paper, Horwich responds to some criticisms of the foundations of Bayesianism and concludes by attacking what he calls "misplaced scientism" (589)—the error of imposing on a philosophical theory the same standards of quantitative rigor that we demand of theories in science. Horwich recognizes these standards as reasonable in science and even within some branches of the philosophy of science, where the aim is to give an accurate, comprehensive description of some domain of phenomena. But when we are dealing with a philosophical theory, the primary function of which is to solve conceptual problems, Horwich regards these standards as inappropriate. Thus, he rejects as unwarranted some of the complaints made against Bayesianism, such as the difficulty of measuring precise degrees of belief or specifying a well-defined confirmation function. For many of the applications of the Bayesian theory in philosophy, Horwich judges this lack of completeness and rigor to be irrelevant.

■ | Notes

1. In the first edition we said that the Bayesian approach is "not inconsistent" with the H-D approach; but, as Hanspeter Fetz pointed out to us, this depends on how the H-D approach is characterized. If the H-D approach is understood, as it is in Hempel's article in chapter 4, so that it says merely that "G entails E" is *sufficient* for "E confirms G," then this permits other ways of confirming G such as Hempel's "top-down" special consequence condition [p. 477], according to which E confirms G, if E confirms H, and H entails G, whether or not G also entails E. It is an important similarity between the H-D approach and the Bayesian approach that (unlike Hempel's satisfaction criterion and Glymour's bootstrap theory of confirmation) they both regard "G entails E" as *sufficient* for "E confirms G." (Richard Jeffrey calls this the "converse entailment condition.") If proponents of the H-D approach insist that "G entails E" is not only sufficient but also necessary for "E confirms G," then it is inconsistent with the Bayesian approach, which allows E to confirm G (by raising its probability) even though G does not entail E.

2. An argument for preferring the difference analysis of degree of confirmation is given in footnote 8, below. The entire range of possible Bayesian measures of confirmation is analyzed in Vincenzo Crupi and Katya Tentori, "Irrelevant Conjunction: Statement and Solution of a New Paradox," *Philosophy of Science* 77 (2010): 1–13. Crupi and Tentori are interested in the ability of these measures to block the dilution of both the confirmation *and the disconfirmation* of a hypothesis through irrelevant conjunction. Their favored solution is a variant of a difference measure in which $P(h/e) - P(h)$ is divided by $(1 - P(h))$ when the difference is positive (confirmation) and divided by $P(h)$ when the difference is negative (disconfirmation). Intuitively, this measure, which they call $Z(h, e)$, reflects the relative increase in

certainty of *h* when *h* is confirmed (in terms of how far *e* has pushed $P(h/e)$ up from $P(h)$ as a fraction of the distance between $P(h)$ and 1), and the relative decrease in certainty when *h* is disconfirmed (in terms of how far *e* has pushed $P(h/e)$ down from $P(h)$ as a fraction of the distance between $P(h)$ and 0).

3. A small complication: since probabilities cannot exceed one, the area of each circle must be divided by the area of the rectangle that encloses all the circles. When there are just two circles, A and B, as in figure 2, the enclosing rectangle represents the proposition $(A \lor B \lor {\sim}A \lor {\sim}B)$. Since this is a necessary truth, its probability is one and so, too, is its area, when expressed in suitable units.

4. Here is a simple illustration. There are one hundred passengers on a cruise ship of whom forty speak French (and no English) and ten speak Swedish (and no English); all the rest speak only English. Exactly two passengers, Ingmar and Chantal, speak both French and Swedish. What is the probability that a randomly selected passenger speaks either French or Swedish? If we simply add $P(F)$ and $P(S)$ to get 0.5, then we will have counted Ingmar and Chantal twice, since they belong to both groups. Therefore, we must subtract 0.02 to get the correct answer of 0.48. Alternatively, we could calculate from the information given that the number of passengers who speak English is fifty-two (not fifty!) and hence that forty-eight out of a hundred is the fraction of passengers who speak either French or Swedish.

5. In some treatments of probability, the definition of $P(A/B)$ is regarded as a fourth axiom. See, for example, footnote 1 of Horwich's article.

6. Thus when someone refers to $P(E/H)$ as the likelihood of evidence *E* on hypothesis *H*, he is using the term *likelihood* informally, as a synonym for *probability*. Strictly speaking, $P(E/H)$ should be referred to as the likelihood of *H*, or the likelihood of *H* on *E*.

7. Wesley C. Salmon, "The Appraisal of Theories: Kuhn Meets Bayes," in *PSA 1990*, vol. 2 (East Lansing, Mich.: Philosophy of Science Association), 325. Evidently, Salmon did not read very far, since Kuhn first casts doubt on the traditional distinction between the two contexts on pages 8–9 of his book.

8. According to the other version, the degree of confirmation depends on the ratio of the two probabilities. But for deterministic theories, this version entails that theories such as *T* and *(T&I)* are confirmed to the same degree by *E*, since in both cases, $P(T/E\&B)$ divided by $P(T/B)$, and $P(T\&I/E\&B)$ divided by $P(T\&I/B)$, is simply the reciprocal of $P(E/B)$. This is a compelling reason for Bayesians to prefer the difference analysis of degree of confirmation.

9. See Jon Dorling, "Bayesian Personalism, the Methodology of Scientific Research Programmes, and Duhem's Problem," *Studies in History and Philosophy of Science* 10 (1979): 177–87. Dorling's Bayesian approach to the Duhem problem is explained in the article by Howson and Urbach in chapter 4 and by Chalmers in this chapter. See also the associated commentary in chapter 4 for further discussion.

10. Wesley C. Salmon, *The Foundations of Scientific Inference* (Pittsburgh, Pa.: University of Pittsburgh Press, 1967).

11. For these and other criticisms of the attempt by Reichenbach and Salmon to interpret prior probabilities as frequencies, see Mary B. Hesse, *The Structure of Scientific Inference* (Berkeley, Calif.: University of California Press, 1974), 106; Alvin Plantinga, "The Probabilistic Argument from Evil," *Philosophical Studies* 35

(1979): 1–53; and Deborah C. Mayo, "A Critique of Salmon's Bayesian Way," in this chapter.

12. Indeed, if the degree of confirmation is proportional to the difference between prior and posterior probabilities for each theory given evidence E, then in cases where T_1 and T_2 are statistical theories neither of which logically entails E, $P(E/B)$ *will not* cancel out when we divide the difference between $P(T_1/E\&B)$ and $P(T_1/B)$ by the difference between $P(T_2/E\&B)$ and $P(T_2/B)$. For statistical theories, the ratio of the degrees of confirmation is equal to Salmon's ratio of the posterior probabilities only when their likelihoods are the same; that is, only when $P(E/T_1\&B)$ equals $P(E/T_2\&B)$.

13. Just to emphasize that the comparative support measured by Salmon's algorithm is not a comparison of degrees of inductive confirmation, we point out that E might disconfirm both T_1 and T_2, and yet, by Salmon's algorithm, T_2 could be judged preferable to T_1, because the posterior probability of T_2 given E is greater than the posterior probability of T_1 given E.

14. See Carl G. Hempel, "Inductive Inconsistencies," *Synthese* 12 (1960): 439–69; reprinted in Carl G. Hempel, *Aspects of Scientific Explanation* (New York: The Free Press, 1965), 53–79.

15. This important point about the contrastive testing of scientific theories is emphasized in Elliott Sober, "Testability," *Proceedings and Addresses of the American Philosophical Association* 73 (1999): 542–64. Sober also applies it to the hypothesis of intelligent design and the hypothesis of evolution by natural selection.

16. Fisher was more cautious. For him, we do not accept hypotheses; we either reject or fail to reject the null hypothesis. The Neyman-Pearson approach includes rules for acceptance as well as rules for rejection.

17. Rat urine is relevantly different from human urine. In rats, saccharin (benzoic sulfilimine) combines with calcium phosphate and some proteins that rats have in high concentrations in their urine to produce tiny sharp crystals that abrade the lining of the bladder. The continual production of new cells to repair the damage causes tumors to form over time. These crystals are not formed in human consumers of saccharin. For more on the contrast between Mayo's error-statistical approach and Bayesianism, see Deborah G. Mayo, "Duhem's Problem, the Bayesian Way, and Error Statistics," *Philosophy of Science* 64 (1997): 222–44, and her "Response to Howson and Laudan," in the same volume, 323–33.

18. See, for example, Colin Howson, *Hume's Problem: Induction and the Justification of Belief* (Oxford: Clarendon Press, 2000), 106–8. For an especially clear presentation of the objection, see Harold Kincaid, "Scientific Realism and the Empirical Nature of Methodology: Bayesians, Error Statisticians and Statistical Inference," in Steve Clarke and Timothy D. Lyons, eds., *Recent Themes in the Philosophy of Science* (Dordrecht, Netherlands: Kluwer, 2002), 39–62.

19. Deborah G. Mayo, "Evidence as Passing Severe Tests: Highly Probable versus Highly Probed Hypotheses," in Peter Achinstein, ed., *Scientific Evidence: Philosophical Theories and Applications* (Baltimore: The Johns Hopkins University Press, 2005), 95–127.

20. Even this way of presenting Mayo's example is potentially misleading. Strictly speaking, Mayo does not accept this notation as properly representing her position

since, for her, $P(X/H)$ cannot be a conditional probability. This is because $P(X/H)$ is defined as $P(X \& H)/P(H)$ and Mayo rejects prior probabilities such as $P(H)$ as lacking a proper definition in her frequentist severity account. For this reason, in some of her articles, she prefers to write $P(X; H)$ instead of $P(X/H)$ to represent the probability of achieving the score X if H were true.

21. As Peter Achinstein points out, in the real-life case involving high school students from a poor Hispanic area of Los Angeles that inspired the 1988 movie *Stand and Deliver*, the Educational Testing Service required the students to retake their advancement placement calculus test under stringently monitored conditions. To their credit (and to the credit of their exceptional teacher), the students did just as well the second time as they did the first. That case, involving an entire class at the same school, differs significantly from the Isaac case since with Isaac there are no grounds to entertain any suspicion of collusion or cheating as an alternative explanation of his achieving a high score. The alternative explanation in Isaac's case is that he beat the odds by falling within the 5 percent of test takers who do very well despite their lack of preparation. See Peter Achinstein, "Mill's Sins or Mayo's Errors?" in Deborah G. Mayo and Aris Spanos, eds., *Error and Inference: Recent Exchanges on Experimental Reasoning, Reliability, and the Objectivity and Rationality of Science* (Cambridge: University of Cambridge Press, 2010), 170–88.

22. Mayo does not claim that there cannot be any legitimate frequentist prior for H: she speculates that perhaps one might be based on the "generic [genetic] and environmental factors that determine the chance of this deficiency." See Deborah G. Mayo, "Sins of the Epistemic Probabilist: Exchanges with Peter Achinstein," in Mayo and Spanos, eds., *Error and Inference*, p. 196.

23. As noted above, Achinstein is not a Bayesian. He insists that the mere raising of the posterior probability of a hypothesis above its prior probability by some information (which defines confirmation for Bayesians) is not sufficient to qualify that information as evidence. In his view, not only does $P(H/E)$ have to exceed some suitably high threshold (0.9, say), but there also has to be an appropriate "explanatory connection" between H and E. But, as Mayo points out, it is hard to see what connection exists between the hypothesis that Isaac is *not* ready for college and his high test score that would explain why he did so well. If both H and E are true, the "explanation" in this case could only be "luck of the draw," which seems no explanation at all. For a Bayesian approach that rejects both Mayo's definition of severity and the assumption that high posterior probabilities are sufficient to characterize severity, see Prasanta S. Bandyopadhyay and Gordon Brittan, Jr., "Acceptability, Evidence, and Severity," *Synthese* 148 (2006): 259–63.

24. Bayesian convergence arguments (otherwise known as "washout theorems" because they aim to show that prior probabilities are "washed out" or swamped as evidence accumulates) are not watertight. See Clark Glymour, *Theory and Evidence*, (Princeton, N.J.: Princeton University Press, 1980) ch. 2; John Earman, *Bayes or Bust? A Critical Examination of Bayesian Confirmation Theory* (Cambridge, Mass.: MIT Press, 1992); and Henry E. Kyburg Jr., "The Scope of Bayesian Reasoning," *PSA 1992*, vol. 2, D. Hull, M. Forbes, and K. Okruhlik, eds., 139–52 (East Lansing, Mich.: Philosophy of Science Association). Kyburg argues that convergence cannot be guaranteed because for any difference, D, between posteriors there is a choice of priors such that, given evidence E, the posteriors will differ by at least that amount. Why? Because $P(H/E)$ depends on the likelihood ratio $P(E/H)/P(E/\sim H)$ and that

ratio can vary between 0 and infinity. See version 5 of Bayes's equation in the section "Bayes for Beginners" in this commentary.

25. Though not a Bayesian, Deborah Mayo gives similar examples supporting the same conclusion in her book *Error and the Growth of Knowledge*. For a criticism of the Howson and Urbach theory of ad hocness, see Donald Gillies, "Bayesianism Versus Falsificationism," *Ratio* (New Series) 3 (1990): 82–98.

26. See Adolf Grünbaum, "Ad Hoc Auxiliary Hypotheses and Falsificationism," *British Journal for the Philosophy of Science* 29 (1976): 329–62, Imre Lakatos, "Falsification and the Methodology of Scientific Research Programmes," in *Criticism and the Growth of Knowledge*, I. Lakatos and A. Musgrave, eds. (Cambridge: Cambridge University Press, 1970), and Jarrett Leplin, "The Assessment of Auxiliary Hypotheses," *British Journal for the Philosophy of Science* 33 (1982): 235–49.

27. Greg Bamford, "Popper's Explications of Ad Hocness: Circularity, Empirical Content, and Scientific Practice," *British Journal for the Philosophy of Science* 44 (1993): 335–55. See also Greg Bamford, "What Is the Problem of Ad Hoc Hypotheses?" *Science & Education* 8 (1999): 375–86. Bamford points out that philosophical accounts of ad hocness often mix up epistemological factors with psychological ones. When they are first introduced, auxiliary hypotheses are often judged by their critics to be ad hoc because their proponents do not intend them to have consequences beyond saving a favored theory from refutation. That is a psychological fact about them. But, objectively, many such hypotheses are independently testable and are eventually tested, sometimes with favorable results.

28. This section follows the discussion in Clark Glymour, *Theory and Evidence* (1980). It was Glymour's analysis in his second chapter "Why I Am Not a Bayesian" that made the problem of old evidence a topic of interest among philosophers of science. In his chapter, Glymour systematically examines all the major problems with the Bayesian approach—including the weakness of standard arguments for the foundational assumptions of Bayesianism, and the difficulty in relating the simplicity of theories to their prior probability—and criticizes attempts that have been made to solve them. Apart from his specific criticisms, Glymour has a deeper complaint against Bayesianism, namely, its alleged inability to shed any interesting light on what makes evidence relevant to the theory it confirms. As a consequence, even when Bayesianism can be made compatible with methodological principles and truisms, Glymour charges that its account of them is superficial and unenlightening. In a similar vein, John Norton notes the "troubling elasticity" that enables Bayesianism to vindicate inductive norms in John Norton, "Challenges to Bayesian Confirmation Theory," in Prasanta S. Bandopadhyay and Malcolm Forster, eds., *Philosophy of Statistics, Handbook of the Philosophy of Science*, vol. 7 (Amsterdam: Elsevier, 2011).

29. Colin Howson and Peter Urbach, *Scientific Reasoning: The Bayesian Approach*, 3d ed. (La Salle, Ill.: Open Court, 2006), 297–301.

30. Glymour, *Theory and Evidence*, ch. 2.

31. Since the publication of Glymour's book, a number of authors have defended versions of the second strategy. See Daniel Garber, "Old Evidence and Logical Omniscience in Bayesian Confirmation Theory," and Richard Jeffrey, "Bayesianism with a Human Face," in *Testing Scientific Theories*, ed. John Earman, vol. 10, *Minnesota Studies in the Philosophy of Science* (Minneapolis: University of Min-

nesota Press, 1983), 99–131, 133–56; Ilkka Niiniluoto, "Novel Facts and Bayesianism," *British Journal for the Philosophy of Science* 34 (1983): 375–79. See also chapter 5 of John Earman, *Bayes or Bust?*

32. Earman, *Bayes or Bust?* 130–31.

33. See, for example, the papers in Daniel Kahneman, Paul Slovic, and Amos Tversky, eds., *Judgment under Uncertainty: Heuristics and Biases* (New York: Cambridge University Press, 1983).

34. Paul Teller, "Conditionalization and Observation," *Synthese* 26 (1973): 218–58. Teller attributes the diachronic Dutch book argument to David Lewis.

35. See the discussion of "washout theorems" in footnote 24.

36. Glymour, *Theory and Evidence*, ch. 2.

37. See especially, Carl G. Hempel, "Studies in the Logic of Confirmation," *Mind* 54 (1945): 1–26, 97–121; reprinted in *Aspects of Scientific Explanation* (New York: Macmillan, 1965), 3–46. For a thorough analysis of Hempel's attempt to resolve the raven paradox, see Branden Fitelson, "The Paradox of Confirmation," *Philosophy Compass* 1 (2006): 95–113.

38. Nelson Goodman, *Fact, Fiction, and Forecast*, 4th ed. (Cambridge, Mass.: Harvard University Press, 1983), 70–71.

39. For other Bayesian criticisms of the positive instance principle, see I. J. Good, "The White Shoe Is a Red Herring," *British Journal for the Philosophy of Science* 17 (1967): 322; Richard G. Swinburne, "The Paradoxes of Confirmation—A Survey," *American Philosophical Quarterly* 8 (1971): 318–30; Roger D. Rosenkrantz, *Inference, Method and Decision* (Dordrecht, Netherlands: D. Reidel, 1977), ch. 2; Roger D. Rosenkrantz, "Does the Philosophy of Induction Rest on a Mistake?" *Journal of Philosophy* 79 (1982): 78–97. One of Good's counterexamples to the positive instance principle relies on background information. Suppose we know that we are in one of two worlds: in one there are a hundred black crows, no crows that are not black, and a million other birds; in the other there are a thousand black crows, a single white crow, and a million other birds. A bird selected at random turns out to be a black crow. This is strong evidence that we are in the second world and thus the observation of a black crow undermines the hypothesis that all crows are black. (The evidence, E, for W_2 is strong as measured by Salmon's algorithm. For, if W_1 and W_2 have the same prior probability, $P(W_2/E)$ is about ten times greater than $P(W_1/E)$.)

40. For further illustrations of using background assumptions to resolve paradoxes in confirmation theory, see Richard G. Swinburne, "The Paradoxes of Confirmation—A Survey," and his book, *An Introduction to Confirmation Theory* (London: Methuen and Co., 1973).

41. Paul Horwich, *Probability and Evidence* (Cambridge: Cambridge University Press, 1982).

42. As Horwich notes in his article, he defends the simplicity postulate in his book, *Probability and Evidence*.

43. For a criticism of Horwich's argument, see Andrew Wayne, "Bayesianism and Diverse Evidence," *Philosophy of Science* 62 (1995): 111–21. Daniel Steel has replied to this criticism in "Bayesianism and the Value of Diverse Evidence," *Philosophy of*

Science 63 (1996): 666–74. Elliott Sober defends a "local" approach to simplicity in which the justification for preferring simpler hypotheses varies from problem to problem. See Elliott Sober, "What is the Problem of Simplicity?" in *Simplicity, Inference and Modelling: Keeping it Sophisticatedly Simple*, ed. Arnold Zellner, Hugo A. Keuzenkamp, and Michael McAleer (Cambridge: Cambridge University Press, 2002), 13–31. On this piecemeal account, simpler hypotheses do not always have higher prior probabilities but, depending on the case involved, they are nonetheless preferred either because they have higher likelihoods (that is, they confer a higher probability on the evidence) or because they have greater predictive accuracy (since simpler hypotheses reduce the risk of over-fitting the available data).

6 |
Models of
Explanation

INTRODUCTION

It is taken for granted in modern science that one of the main functions of scientific theories is to explain. What gets explained can vary widely from particular events and facts (e.g., why did the space shuttle *Challenger* explode? what caused the extinction of the dinosaurs? why do the equinoxes precess?) to empirical laws and theories (e.g., why do planets obey Kepler's second law? why do liquids maintain a constant temperature as they boil? why do elements from column 1 of the periodic table combine so vigorously with elements from column 7?). Laws and theories can be used to explain events and facts; higher-level laws and theories can be used to explain lower-level laws; and some theories can be used to explain others. In this chapter the main focus is on the explanation of particular events and facts. The explanation of laws and theories is addressed primarily in chapter 8, on reduction.

Customarily, the thing that gets explained (or a sentence describing it) is called the *explanandum*. (In Latin *explanandum* means simply "the thing to be explained.") The thing that does the explaining is called the *explanans* (again, a Latin word, meaning "that which explains"). Typically, in offering a model of explanation, a philosopher of science will tell us what conditions a group of sentences must satisfy in order to be an explanans and how they must be related to the explanandum in order to constitute a genuine explanation.[1]

Without doubt, the most influential claim about explanation in the twentieth century was the covering law thesis of Carl Hempel, according to which genuine explanations are arguments that must contain empirical laws. Hempel's two principal covering law models of explanation, defended and attacked in this chapter's readings, are the deductive-nomological (D-N) model and the inductive-statistical (I-S) model.[2]

In "The Value of Laws: Explanation and Prediction," Rudolf Carnap sheds light on the motivation for Hempel's covering law thesis. He relates conversations that took place in the early 1930s among members of the Vienna Circle who were dissatisfied with the entelechy theory advocated by the German biologist, Hans Driesch. Carnap and his colleagues diagnosed the main flaw in Driesch's theory to be its failure to provide testable laws, without which Driesch's theory could not really explain anything. Subsequently, Hempel included this requirement as a central condition of his two models of explanation, described in "Two Basic Types of Scientific Explanation."

According to Hempel's D-N model, the best scientific explanations are deductively valid arguments, with a statement of at least one law among the premises and a description of the event to be explained in the conclusion. If, instead of explaining an event, we had predicted it, the argument involved would have been very similar. This formal similarity between D-N explanations and predictions leads Hempel to espouse the doctrine discussed in "The Thesis of Structural Identity," that all adequate explanations are potentially predictions and all adequate predictions are potentially explanations. Critics of the D-N model have attacked both parts of this thesis, and Hempel responds to these criticisms—referred to, collectively, as the *symmetry objection*—in his article.

Critics have also charged that both Hempel's D-N and I-S models are flawed because they can be satisfied by arguments with premises that are explanatorily irrelevant to their conclusions. Usually, this irrelevance arises because the premises fail to identify the cause of the explanandum. These criticisms are discussed in the commentary.

Although Hempel proposed the D-N model in 1948, it was not until 1962 that he considered in detail those scientific explanations that use statistical laws to explain particular events. In "Inductive-Statistical Explanation," Hempel sets out his I-S model of these explanations. Hempel construes them as inductive arguments in which the explanans does not entail the explanandum but establishes it with reasonably high probability. From the outset, Hempel realized that I-S explanations face a problem that does not arise for D-N explanations, namely the problem of ambiguity: several statistical laws might be used to explain some particular event with each of these laws assigning a different probability to the event in question. Particularly disturbing, in Hempel's view, is the possibility that we could have explained the outcome regardless of whether or not the event actually occurred. Hempel's proposed solution to the problem of ambiguity for statistical explanations is his requirement of maximal specificity. But this requirement leads Hempel to argue that, unlike D-N explanations, I-S explanations are, by their very nature, relative to the state of scientific knowledge at a particular time.

Peter Railton repudiates several of the main features of Hempel's covering law models. In "A Deductive-Nomological Model of Probabilistic

Explanation," Railton rejects Hempel's contention that probabilistic explanations are arguments, along with Hempel's high-probability and maximal specificity requirements. Railton's own model of explanation—the deductive-nomological model of his title—has as its central feature the specification of the mechanism that brings about the event to be explained. The virtues of this account, according to Railton, are that it permits the explanation of improbable events and makes statistical explanations entirely objective.

One problem left unsolved by Hempel's model is the explanation of laws by other more general laws, a central feature of science. Railton addresses this problem by requiring the specification of a mechanism, but this would rule out as unexplanatory much of what occurs in branches of science (the so-called special sciences) outside of physics. Philip Kitcher's approach is to revive a notion that played a prominent role in the account of science of Kant in the eighteenth century (and, in the following century, in the writings of William Whewell) by seeing explanation as primarily the result of unification. Newton himself often replied to critics of his universal law of gravitation that, though he was reluctant to commit himself to an underlying mechanism for the action of gravity, he had nonetheless made significant scientific progress by unifying a diverse group of empirical laws under one overarching framework. Darwin made a similar response to the critics of his evolutionary theory in the nineteenth century.

In "Explanatory Unification" Kitcher articulates the idea that arguments explain when they are instances of patterns belonging to the "explanatory store," the set that maximizes unifying power over the sentences accepted by scientists at a given time. Unifying power is judged by three criteria: how few argument patterns are needed, the stringency of those patterns, and how many conclusions are generated. Kitcher responds to the asymmetry and irrelevance objections by arguing that these derivational patterns do not qualify for inclusion in the explanatory store. Kitcher also addresses the problem of "spurious unification" by defending a way to rule out patterns of very low stringency (even though they score very highly when judged by the other two criteria).

James Woodward's manipulationist model adopts a very different approach to explanation by focusing on the ways in which causal claims are tested in the areas such as biomedicine, psychology, sociology, and economics. In these "special sciences" (outside of physics), laws seldom hold with unlimited precision or with unrestricted scope. Nonetheless, these generalizations have a degree of invariance that enables them to explain phenomena within the range where they hold to a high degree of approximation. Eschewing any deep metaphysical account of the concept of causation, and drawing on recent research on causal networks, Woodward constructs his theory around the notion of an intervention—an event that isolates the causal dependence of one variable upon another. Though inspired by cases in which humans can literally "manipulate" causes, Woodward insists that the notion of causation implicit in his theory is entirely objective and applies just as well to cases in which no human intervention is possible. This involves an appeal to invariant generalizations as the grounds for counterfactual conditionals.

■ | Notes

1. To avoid circumlocution, we will follow Hempel in using the term *explanandum* to refer to either the event-to-be-explained or a sentence describing that event.

2. Hempel claims that his models explicate the concept of explanation not only in the physical sciences and biology, but also in the social sciences and history. The extension of Hempel's account to the explanation of human action and behavior is controversial and falls outside of the scope of the present volume. For a good introduction to the issues involved, see Carl G. Hempel, "The Function of General Laws in History," *Journal of Philosophy* 39 (1942): 35–42; reprinted in Carl G. Hempel, *Aspects of Scientific Explanation* (New York: Free Press, 1965), 231–43; Carl G. Hempel, "Aspects of Scientific Explanation," in *Aspects of Scientific Explanation*, sec. 6–10; William H. Dray, *Laws and Explanation in History* (Oxford: Oxford University Press, 1957); Patrick Gardiner, ed., *Theories of History* (New York: Free Press, 1959); Alan Donagan, "The Popper-Hempel Theory Reconsidered," in *Philosophical Analysis and History*, ed. W. H. Dray (New York: Harper and Row, 1966), 127–59; David Papineau, *For Science in the Social Sciences* (New York: St. Martin's Press, 1978); Merrilee H. Salmon, "Explanation in the Social Sciences," in *Scientific Explanation*, ed. Philip Kitcher and Wesley C. Salmon, vol. 13, *Minnesota Studies in the Philosophy of Science* (Minneapolis: University of Minnesota Press, 1989), 384–409; William H. Dray, "Explanation in History," in *Science, Explanation, and Rationality: Aspects of the Philosophy of Carl G. Hempel*, ed. James H. Fetzer (Oxford: Oxford University Press, 2000), 217–42.

Rudolf Carnap

The Value of Laws:
Explanation and Prediction

In the nineteenth century, certain Germanic physicists, such as Gustav Kirchhoff and Ernst Mach, said that science should not ask "Why?" but "How?" They meant that science should not look for unknown metaphysical agents that are responsible for certain events, but should only describe such events in terms of laws. This prohibition against asking "Why?" must be understood in its historical setting. The background was the German philosophical atmosphere of the time, which was dominated by idealism in the tradition of Fichte, Schelling, and Hegel. These men felt that a description of how the world behaved was not enough. They wanted a fuller understanding, which they believed could be obtained only by finding metaphysical causes that were behind phenomena and not accessible to scientific method. Physicists reacted to this point of view by saying: "Leave us alone with your why-questions. There is no answer beyond that given by the empirical laws." They objected to why-questions because they were usually metaphysical questions.

Today the philosophical atmosphere has changed. In Germany there are a few philosophers still working in the idealist tradition, but in England and the United States it has practically disappeared. As a result, we are no longer worried by why-questions. We do not have to say, "Don't ask why", because now, when someone asks why, we assume that he means it in a scientific, nonmetaphysical sense. He is simply asking us to explain something by placing it in a framework of empirical laws.

When I was young and part of the Vienna Circle, some of my early publications were written as a reaction to the philosophical climate of German idealism. As a consequence, these publications and those by others in the Vienna Circle were filled with prohibitory statements similar to the one

From Rudolf Carnap, *Philosophical Foundations of Physics*, ed. Martin Gardner (New York: Basic Books, 1966), 12–16. This reading incorporates corrections made by the editor, Martin Gardner, when Carnap's book was reprinted in 1995 by Dover Publications, Inc., under the title, *An Introduction to the Philosophy of Science*.

I have just discussed. These prohibitions must be understood in reference to the historical situation in which we found ourselves. Today, especially in the United States, we seldom make such prohibitions. The kind of opponents we have here are of a different nature, and the nature of one's opponents often determines the way in which one's views are expressed.

When we say that, for the explanation of a given fact, the use of a scientific law is indispensable, what we wish to exclude especially is the view that metaphysical agents must be found before a fact can be adequately explained. In prescientific ages, this was, of course, the kind of explanation usually given. At one time, the world was thought to be inhabited by spirits or demons who are not directly observable but who *act* to cause the rain to fall, the river to flow, the lightning to flash. In whatever one saw happening, there was something—or, rather, *somebody*—responsible for the event. This is psychologically understandable. If a man does something to me that I do not like, it is natural for me to make him responsible for it and to get angry and hit back at him. If a cloud pours water over me, I cannot hit back at the cloud, but I can find an outlet for my anger if I make the cloud, or some invisible demon behind the cloud, responsible for the rainfall. I can shout curses at this demon, shake my fist at him. My anger is relieved. I feel better. It is easy to understand how members of prescientific societies found psychological satisfaction in imagining agents behind the phenomena of nature.

In time, as we know, societies abandoned their mythologies, but sometimes scientists replace the spirits with agents that are really not much different. The German philosopher Hans Driesch, who died in 1941, wrote many books on the philosophy of science. He was originally a prominent biologist, famed for his work on certain organismic responses, including regeneration in sea urchins. He cut off parts of their bodies and observed in which stages of their growth and under what conditions they were able to grow new parts. His scientific work was important and excellent. But Driesch was also interested in philosophical questions, especially those dealing with the foundations of biology, so eventually he became a professor of philosophy. In philosophy also he did some excellent work, but there was one aspect of his philosophy that I and my friends in the Vienna Circle did not regard so highly. It was his way of *explaining* such biological processes as regeneration and reproduction.

At the time Driesch did his biological work, it was thought that many characteristics of living things could not be found elsewhere. (Today it is seen more clearly that there is a continuum connecting the organic and inorganic worlds.) He wanted to explain these unique organismic features, so he postulated what he called an "entelechy". This term had been introduced by Aristotle, who had his own meaning for it, but we need not discuss that meaning here. Driesch said, in effect: "The entelechy is a certain specific force that causes living things to behave in the way they do. But you must not think of it as a *physical* force such as gravity or magnetism. Oh, no, nothing like that."

The entelechies of organisms, Driesch maintained, are of various kinds, depending on the organism's stage of evolution. In primitive, single-celled organisms, the entelechy is rather simple. As we go up the evolutionary scale, through plants, lower animals, higher animals, and finally to man, the entelechy becomes more and more complex. This is revealed by the greater degree to which phenomena are integrated in the higher forms of life. What we call the "mind" of a human body is actually nothing more than a portion of the person's entelechy. The entelechy is much more than the mind, or, at least, more than the conscious mind, because it is responsible for everything that every cell in the body does. If I cut my finger, the cells of the finger form new tissue and bring substances to the cut to kill incoming bacteria. These events are not consciously directed by the mind. They occur in the finger of a one-month-old baby, who has never heard of the laws of physiology. All this, Driesch insisted, is due to the organism's entelechy, of which mind is *one* manifestation. In addition, then, to scientific explanation, Driesch had an elaborate theory of entelechy, which he offered as a *philosophical* explanation of such scientifically unexplained phenomena as the regeneration of parts of sea urchins.

Is this an explanation? I and my friends had some discussions with Driesch about it. I remember one at the International Congress for Philosophy, at Prague, in 1934. Hans Reichenbach and I criticized Driesch's theory, while he and others defended it. In our publications we did not give much space to this criticism because we admired the work Driesch had done in both biology and philosophy. He was quite different from most philosophers in Germany in that he really wanted to develop a scientific philosophy. His entelechy theory, however, seemed to us to lack something.

What it lacked was this: the insight that you cannot give an explanation without also giving a law.

We said to him: "Your entelechy—we do not know what you mean by it. You say it is not a physical force. What is it then?"

"Well", he would reply (I am paraphrasing his words, of course), "you should not be so narrow-minded. When you ask a physicist for an explanation of why this nail suddenly moves toward that bar of iron, he will tell you that the bar of iron is a magnet and that the nail is drawn to it by the force of magnetism. No one has ever seen magnetism. You see only the movement of a little nail toward a bar of iron."

We agreed. "Yes, you are right. Nobody has seen magnetism."

"You see", he continued, "the physicist introduces forces that no one can observe—forces like magnetism and electricity—in order to explain certain phenomena. I wish to do the same. Physical forces are not adequate to explain certain organic phenomena, so I introduce something that is force-like but is not a physical force because it does not act the way physical forces act. For instance, it is not spatially located. True, it acts on a physical organism, but it acts in respect to the entire organism, not just to certain parts of it. Therefore, you cannot say where it is located. There is no location. It is not a physical force, but it is just as legitimate for me to introduce it as it is for a physicist to introduce the invisible force of magnetism."

Our answer was that a physicist does not explain the movement of the nail toward the bar simply by introducing the word "magnetism". Of course, if you ask him why the nail moves, he may answer first by saying that it is due to magnetism; but if you press him for a fuller explanation, he will give you laws. The laws may not be expressed in quantitative terms, like the Maxwell equations that describe magnetic fields; they may be simple, qualitative laws with no numbers occurring in them. The physicist may say: "All nails containing iron are attracted to the ends of bars that have been magnetized." He may go on to explain the state of being magnetized by giving other nonquantitative laws. He may tell you that iron ore from the town of Magnesia (you may recall that the word "magnetic" derives from the Greek town of Magnesia, where iron ore of this type was first found) possesses this property. He may explain that iron bars become magnetized if they are stroked a certain way by naturally magnetic ores. He may give you other laws about conditions under which certain substances can become magnetized and laws about phenomena associated with magnetism. He may tell you that if you magnetize a needle and suspend it by the middle so that it swings freely, one end will point north. If you have another magnetic needle, you can bring the two north-pointing ends together and observe that they do not attract but repel each other. He may explain that if you heat a magnetized bar of iron, or hammer it, it will lose magnetic strength. All these are qualitative laws that can be expressed in the logical form, "if . . . then . . ." The point I wish to emphasize here is this: it is not sufficient, for purposes of explanation, simply to introduce a new agent by giving it a new name. You must also give laws.

Driesch did not give laws. He did not specify how the entelechy of an oak tree differs from the entelechy of a goat or giraffe. He did not classify his entelechies. He merely classified organisms and said that each organism had its own entelechy. He did not formulate laws that state under what conditions an entelechy is strengthened or weakened. Of course he described all sorts of organic phenomena and gave general rules for such phenomena. He said that if you cut a limb from a sea urchin in a certain way, the organism will not survive; if you cut it another way, the organism will survive, but only a fragmentary limb will grow back. Cut in still another way and at a certain stage in the sea urchin's growth, it will regenerate a new and complete limb. These statements are all perfectly respectable zoological laws.

"What do you add to these empirical laws", we asked Driesch, "if after giving them you proceed to tell us that all the phenomena covered by those laws are due to the sea urchin's entelechy?"

We believed that nothing was added. Since the notion of an entelechy does not give us new laws, it does not explain more than the general laws already available. It does not help us in the least in making new predictions. For these reasons we cannot say that our scientific knowledge has increased.

The concept of entelechy may at first seem to add something to our explanations; but when we examine it more deeply, we see its emptiness. It is a pseudoexplanation.

It can be argued that the concept of entelechy is not useless if it provides biologists with a new orientation, a new method of ordering biological laws. Our answer is that it would indeed be useful if by means of it we could formulate more general laws than could be formulated before. In physics, for example, the concept of energy played such a role. Nineteenth-century physicists theorized that perhaps certain phenomena, such as kinetic and potential energy in mechanics, heat (this was before the discovery that heat is simply the kinetic energy of molecules), the energy of magnetic fields, and so on, might be manifestations of one basic kind of energy. This led to experiments showing that mechanical energy can be transformed into heat and heat into mechanical energy but that the amount of energy remains constant. Thus, energy was a fruitful concept because it led to more general laws, such as the law of the conservation of energy. But Driesch's entelechy was not a fruitful concept in this sense. It did not lead to the discovery of more general biological laws.

In addition to providing *explanations* for observed facts, the laws of science also provide a means for *predicting* new facts not yet observed. The logical schema involved here is exactly the same as the schema underlying explanation. . . . Expressed symbolically:

1 $(x) (Px \supset Qx)$
2 Pa
3 Qa

First we have a universal law: for any object x, if it has the property P, then it also has the property Q. Second, we have a statement saying that object a has the property P. Third, we deduce by elementary logic that object a has the property Q. This schema underlies both explanation and prediction; only the knowledge situation is different. In explanation, the fact Qa is already known. We explain Qa by showing how it can be deduced from statements 1 and 2. In prediction, Qa is a fact *not yet known*. We have a law, and we have the fact Pa. We conclude that Qa must also be a fact, even though it has not yet been observed. For example, I know the law of thermal expansion. I also know that I have heated a certain rod. By applying logic in the way shown in the schema, I infer that if I now measure the rod, I will find that it is longer than it was before.

In most cases, the unknown fact is actually a future event (for example, an astronomer predicts the time of the next eclipse of the sun); that is why I use the term "prediction" for this second use of laws. It need not, however, be prediction in the literal sense. In many cases the unknown fact is simultaneous with the known fact, as is the case in the example of the heated rod. The expansion of the rod occurs simultaneously with the heating. It is only our observation of the expansion that takes place after our observation of the heating.

In other cases, the unknown fact may even be in the past. On the basis of psychological laws, together with certain facts derived from historical documents, a historian infers certain unknown facts of history. An astronomer may infer that an eclipse of the moon must have taken place at a certain date in the past. A geologist may infer from striations on boulders that at one time in the past a region must have been covered by a glacier. I use the term "prediction" for all these examples because in every case we have the same logical schema and the same knowledge situation—a known fact and a known law from which an unknown fact is derived.

In many cases, the law involved may be statistical rather than universal. The prediction will then be only probable. A meteorologist, for instance, deals with a mixture of exact physical laws and various statistical laws. He cannot say that it will rain tomorrow; he can only say that rain is very likely.

This uncertainty is also characteristic of prediction about human behavior. On the basis of knowing certain psychological laws of a statistical nature and certain facts about a person, we can predict with varying degrees of probability how he will behave. Perhaps we ask a psychologist to tell us what effect a certain event will have on our child. He replies: "As I see the situation, your child will probably react in this way. Of course, the laws of psychology are not very exact. It is a young science, and as yet we know very little about its laws. But on the basis of what is known, I think it advisable that you plan to . . . ". And so he gives us advice based on the best prediction he can make, with his probabilistic laws, about the future behavior of our child.

When the law is universal, then elementary deductive logic is involved in inferring unknown facts. If the law is statistical, we must use a different logic—the logic of probability. To give a simple example: a law states that 90 per cent of the residents of a certain region have black hair. I know that an individual is a resident of that region, but I do not know the color of his hair. I can infer, however, on the basis of the statistical law, that the probability his hair is black is $9/10$.

Prediction is, of course, as essential to everyday life as it is to science. Even the most trivial acts we perform during the day are based on predictions. You turn a doorknob. You do so because past observations of facts, together with universal laws, lead you to believe that turning the knob will open the door. You may not be conscious of the logical schema involved—no doubt you are thinking about other things—but all such deliberate actions presuppose the schema. There is a knowledge of specific facts, a knowledge of certain observed regularities that can be expressed as universal or statistical laws and provide a basis for the prediction of unknown facts. Prediction is involved in every act of human behavior that involves deliberate choice. Without it, both science and everyday life would be impossible.

Carl G. Hempel

Two Basic Types of Scientific Explanation

1 | Deductive-Nomological Explanation

In his book, *How We Think*,[1] John Dewey describes an observation he made one day when, washing dishes, he took some glass tumblers out of the hot soap suds and put them upside down on a plate: he noticed that soap bubbles emerged from under the tumblers' rims, grew for a while, came to a standstill, and finally receded inside the tumblers. Why did this happen? The explanation Dewey outlines comes to this: In transferring a tumbler to the plate, cool air is caught in it; this air is gradually warmed by the glass, which initially has the temperature of the hot suds. The warming of the air is accompanied by an increase in its pressure, which in turn produces an expansion of the soap film between the plate and the rim. Gradually, the glass cools off, and so does the air inside, with the result that the soap bubbles recede.

This explanatory account may be regarded as an argument to the effect that the event to be explained (let me call it the explanandum-event) was to be expected by reason of certain explanatory facts. These may be divided into two groups: (i) particular facts and (ii) uniformities expressed by general laws. The first group includes facts such as these: the tumblers had been immersed, for some time, in soap suds of a temperature considerably higher than that of the surrounding air; they were put, upside down, on a plate on which a puddle of soapy water had formed, providing a connecting soap film, etc. The second group of items presupposed in the argument includes the gas laws and various other laws that have not been explicitly suggested concerning the exchange of heat between bodies of different temperature, the elastic behavior of soap bubbles, etc. If we imagine these various presuppositions explicitly spelled out, the idea suggests itself of construing the explanation as a deductive argument of this form:

From "Explanation in Science and History," in *Frontiers of Science and Philosophy*, ed. R. G. Colodny (London and Pittsburgh: Allen and Unwin and University of Pittsburgh Press, 1962), 9–19, 32.

(D) $C_1, C_2, \ldots, C_k$
$$\frac{L_1, L_2, \ldots, L_r}{E}$$

Here, $C_1, C_2, \ldots, C_k$ are statements describing the particular facts invoked; $L_1, L_2, \ldots, L_r$ are general laws: jointly, these statements will be said to form the explanans. The conclusion E is a statement describing the explanandum-event; let me call it the explanandum-statement, and let me use the word "explanandum" to refer to either E or to the event described by it.

The kind of explanation thus characterized I will call *deductive-nomological explanation*; for it amounts to a deductive subsumption of the explanandum under principles which have the character of general laws: it answers the question "*Why* did the explanandum event occur?" by showing that the event resulted from the particular circumstances specified in C_1, $C_2, \ldots C_k$ in accordance with the laws $L_1, L_2, \ldots, L_r$. This conception of explanation, as exhibited in schema (D), has therefore been referred to as the covering law model, or as the deductive model, of explanation.[2]*

A good many scientific explanations can be regarded as deductive-nomological in character. Consider, for example, the explanation of mirror-images, of rainbows, or of the appearance that a spoon handle is bent at the point where it emerges from a glass of water: in all these cases, the explanandum is deductively subsumed under the laws of reflection and refraction. Similarly, certain aspects of free fall and of planetary motion can be accounted for by deductive subsumption under Galileo's or Kepler's laws.

In the illustrations given so far the explanatory laws had, by and large, the character of empirical generalizations connecting different observable aspects of the phenomena under scrutiny: angle of incidence with angle of reflection or refraction, distance covered with falling time, etc. But science raises the question "why?" also with respect to the uniformities expressed by such laws, and often answers it in basically the same manner, namely, by subsuming the uniformities under more inclusive laws, and eventually under comprehensive theories. For example, the question, "Why do Galileo's and Kepler's laws hold?" is answered by showing that these laws are but special consequences of the Newtonian laws of motion and of gravitation; and these, in turn, may be explained by subsumption under the more comprehensive general theory of relativity. Such subsumption under broader laws or theories usually increases both the breadth and the depth of our scientific understanding. There is an increase in breadth, or scope, because the new explanatory principles cover a broader range of phenomena; for example, Newton's principles govern free fall on the earth and on other celestial

---

* The phrase *covering law model* is appropriate because general laws must "cover" or subsume the explanandum. The adjective *nomological* in the phrase *deductive-nomological* is derived from the Greek word *nomos*, meaning "law."

bodies, as well as the motions of planets, comets, and artificial satellites, the movements of pendulums, tidal changes, and various other phenomena. And the increase thus effected in the depth of our understanding is strikingly reflected in the fact that, in the light of more advanced explanatory principles, the original empirical laws are usually seen to hold only approximately, or within certain limits. For example, Newton's theory implies that the factor g in Galileo's law, $s = \frac{1}{2} gt^2$, is not strictly a constant for free fall near the surface of the earth; and that, since every planet undergoes gravitational attraction not only from the sun, but also from the other planets, the planetary orbits are not strictly ellipses, as stated in Kepler's laws.

One further point deserves brief mention here. An explanation of a particular event is often conceived as specifying its *cause*, or causes. Thus, the account outlined in our first illustration might be held to explain the growth and the recession of the soap bubbles by showing that the phenomenon was *caused* by a rise and a subsequent drop of the temperature of the air trapped in the tumblers. Clearly, however, these temperature changes provide the requisite explanation only in conjunction with certain other conditions, such as the presence of a soap film, practically constant pressure of the air surrounding the glasses, etc. Accordingly, in the context of explanation, a cause must be allowed to consist in a more or less complex set of particular circumstances; these might be described by a set of sentences: $C_1, C_2, \ldots, C_k$. And, as suggested by the principle "Same cause, same effect," the assertion that those circumstances jointly caused a given event—described, let us say, by a sentence E—implies that whenever and wherever circumstances of the kind in question occur, an event of the kind to be explained comes about. Hence, the given causal explanation implicitly claims that there are general laws—such as $L_1, L_2, \ldots, L_r$ in schema (D)—by virtue of which the occurrence of the causal antecedents mentioned in $C_1, C_2, \ldots, C_k$ is a sufficient condition for the occurrence of the event to be explained. Thus, the relation between causal factors and effect is reflected in schema (D): causal explanation is deductive-nomological in character. (However, the customary formulations of causal and other explanations often do not explicitly specify all the relevant laws and particular facts: to this point, we will return later.)

The converse does not hold: there are deductive-nomological explanations which would not normally be counted as causal. For one thing, the subsumption of laws, such as Galileo's or Kepler's laws, under more comprehensive principles is clearly not causal in character: we speak of causes only in reference to *particular* facts or events, and not in reference to *universal facts* as expressed by general laws. But not even all deductive-nomological explanations of particular facts or events will qualify as causal; for in a causal explanation some of the explanatory circumstances will temporally precede the effect to be explained: and there are explanations of type (D) which lack this characteristic. For example, the pressure which a gas of specified mass possesses at a given time might be explained by reference to its temperature and its volume at the same time, in conjunction with the gas law which connects simultaneous values of the three parameters.[3]

In conclusion, let me stress once more the important role of laws in deductive-nomological explanation: the laws connect the explanandum event with the particular conditions cited in the explanans, and this is what confers upon the latter the status of explanatory (and, in some cases, causal) factors in regard to the phenomenon to be explained.

2 | Probabilistic Explanation

In deductive-nomological explanation as schematized in (D), the laws and theoretical principles involved are of *strictly universal form*: they assert that in *all* cases in which certain specified conditions are realized an occurrence of such and such a kind will result; the law that any metal, when heated under constant pressure, will increase in volume, is a typical example; Galileo's, Kepler's, Newton's, Boyle's, and Snell's laws, and many others, are of the same character.

Now let me turn next to a second basic type of scientific explanation. This kind of explanation, too, is nomological, i.e., it accounts for a given phenomenon by reference to general laws or theoretical principles; but some or all of these are of *probabilistic-statistical form*, i.e., they are, generally speaking, assertions to the effect that if certain specified conditions are realized, then an occurrence of such and such a kind will come about with such and such a statistical probability.

For example, the subsiding of a violent attack of hay fever in a given case might well be attributed to, and thus explained by reference to, the administration of 8 milligrams of chlor-trimeton. But if we wish to connect this antecedent event with the explanandum, and thus to establish its explanatory significance for the latter, we cannot invoke a universal law to the effect that the administration of 8 milligrams of that antihistamine will invariably terminate a hay fever attack: this simply is not so. What can be asserted is only a generalization to the effect that administration of the drug will be followed by relief with high statistical probability, i.e., roughly speaking, with a high relative frequency in the long run. The resulting explanans will thus be of the following type:

John Doe had a hay fever attack and took 8 milligrams of chlor-trimeton.

The probability for subsidence of a hay fever attack upon administration of 8 milligrams of chlor-trimeton is high.

Clearly, this explanans does not deductively imply the explanandum, "John Doe's hay fever attack subsided"; the truth of the explanans makes the truth of the explanandum not certain (as it does in a deductive-nomological explanation) but only more or less likely or, perhaps "practically" certain.

Reduced to its simplest essentials, a probabilistic explanation thus takes the following form:

$$(P) \quad \left. \begin{array}{c} Fi \\ \underline{p(O,F) \text{ is very high}} \\ Oi \end{array} \right\} \text{makes very likely}$$

The explanandum, expressed by the statement "Oi," consists in the fact that in the particular instance under consideration, here called i (e.g., John Doe's allergic attack), an outcome of kind O (subsidence) occurred. This is explained by means of two explanans-statements. The first of these, "Fi," corresponds to $C_1, C_2, \ldots, C_k$ in (D); it states that in case i, the factors F (which may be more or less complex) were realized. The second expresses a law of probabilistic form, to the effect that the statistical probability for outcome O to occur in cases where F is realized is very high (close to 1). The double line separating explanandum from explanans is to indicate that, in contrast to the case of deductive-nomological explanation, the explanans does not logically imply the explanandum, but only confers a high likelihood upon it. The concept of likelihood here referred to must be clearly distinguished from that of statistical probability, symbolized by "p" in our schema. A statistical probability is, roughly speaking, the long-run relative frequency with which an occurrence of a given kind (say, F) is accompanied by an "outcome" of a specified kind (say, O). Our likelihood, on the other hand, is a relation (capable of gradations) not between kinds of occurrences, but between statements. The likelihood referred to in (P) may be characterized as the strength of the inductive support, or the degree of rational credibility, which the explanans confers upon the explanandum; or, in Carnap's terminology, as the *logical*, or *inductive*, (in contrast to statistical) *probability* which the explanandum possesses relative to the explanans.

Thus, probabilistic explanation, just like explanation in the manner of schema (D), is nomological in that it presupposes general laws; but because these laws are of statistical rather than of strictly universal form, the resulting explanatory arguments are inductive rather than deductive in character. An inductive argument of this kind *explains* a given phenomenon by showing that, in view of certain particular events and certain statistical laws, its occurrence was to be expected with high logical, or inductive, probability.

By reason of its inductive character, probabilistic explanation differs from its deductive-nomological counterpart in several other important respects; for example, its explanans may confer upon the explanandum a more or less high degree of inductive support; in this sense, probabilistic explanation admits of degrees, whereas deductive-nomological explanation appears as an either-or affair: a given set of universal laws and particular statements either does or does not imply a given explanandum statement. A fuller examination of these differences, however, would lead us far afield and is not required for the purposes of this paper.[4]

One final point: the distinction here suggested between deductive-nomological and probabilistic explanation might be questioned on the ground that, after all, the universal laws invoked in a deductive explanation can have been established only on the basis of a finite body of evidence, which surely affords no exhaustive verification, but only more or less strong probability for it; and that, therefore, all scientific laws have to be regarded as probabilistic. This argument, however, confounds a logical issue with an epistemological one: it fails to distinguish properly between the *claim* made by a given law-statement and the *degree of confirmation*, or *probability*, which it possesses on the available evidence. It is quite true that statements expressing laws of either kind can be only incompletely confirmed by any given finite set—however large—of data about particular facts; but law-statements of the two different types make claims of different kind, which are reflected in their logical forms: roughly, a universal law-statement of the simplest kind asserts that *all* elements of an indefinitely large reference class (e.g., copper objects) have a certain characteristic (e.g., that of being good conductors of electricity); while statistical law-statements assert that in the long run, a specified proportion of the members of the reference class have some specified property. And our distinction of two types of law and, concomitantly, of two types of scientific explanation, is based on this difference in claim as reflected in the difference of form.

The great scientific importance of probabilistic explanation is eloquently attested to by the extensive and highly successful explanatory use that has been made of fundamental laws of statistical form in genetics, statistical mechanics, and quantum theory.

3 | Elliptic and Partial Explanations: Explanation Sketches

As I mentioned earlier, the conception of deductive-nomological explanation reflected in our schema (D) is often referred to as the covering law model, or the deductive model, of explanation: similarly, the conception underlying schema (P) might be called the probabilistic or the inductive-statistical, model of explanation. The term "model" can serve as a useful reminder that the two types of explanation as characterized above constitute ideal types or theoretical idealizations and are not intended to reflect the manner in which working scientists actually formulate their explanatory accounts. Rather, they are meant to provide explications, or rational reconstructions, or theoretical models, of certain modes of scientific explanation.

In this respect our models might be compared to the concept of mathematical proof (within a given theory) as construed in meta-mathematics. This concept, too, may be regarded as a theoretical model: it is not intended to provide a descriptive account of how proofs are formulated in the writings of mathematicians: most of these actual formulations fall short of rigorous and, as it were, ideal, meta-mathematical standards. But the theoretical model has certain other functions: it exhibits the rationale of mathematical proofs by revealing the logical connections underlying the successive steps; it provides standards for a critical appraisal of any proposed proof constructed within the mathematical system to which the model refers; and it affords a basis for a precise and far-reaching theory of proof, provability, decidability, and related concepts. I think the two models of explanation can fulfill the same functions, if only on a much more modest scale. For example, the arguments presented in constructing the models give an indication of the sense in which the models exhibit the rationale and the logical structure of the explanations they are intended to represent.

I now want to add a few words concerning the second of the functions just mentioned; but I will have to forgo a discussion of the third.

When a mathematician proves a theorem, he will often omit mention of certain propositions which he presupposes in his argument and which he is in fact entitled to presuppose because, for example, they follow readily from the postulates of his system or from previously established theorems or perhaps from the hypothesis of his theorem, if the latter is in hypothetical form; he then simply assumes that his readers or listeners will be able to supply the missing items if they so desire. If judged by ideal standards, the given formulation of the proof is elliptic or incomplete; but the departure from the ideal is harmless: the gaps can readily be filled in. Similarly, explanations put forward in everyday discourse and also in scientific contexts are often *elliptically formulated.* When we explain, for example, that a lump of butter melted because it was put into a hot frying pan, or that a small rainbow appeared in the spray of the lawn sprinkler because the sunlight was reflected and refracted by the water droplets, we may be said to offer elliptic formulations of deductive-nomological explanations; an account of this kind omits mention of certain laws or particular facts which it tacitly takes for granted, and whose explicit citation would yield a complete deductive-nomological argument.

In addition to elliptic formulation, there is another, quite important, respect in which many explanatory arguments deviate from the theoretical model. It often happens that the statement actually included in the explanans, together with those which may reasonably be assumed to have been taken for granted in the context at hand, explain the given explanandum only *partially*, in a sense which I will try to indicate by an example. In his *Psychopathology of Everyday Life*, Freud offers the following explanation of a slip of the pen that occurred to him: "On a sheet of paper containing principally short daily notes of business interest, I found, to my surprise, the incorrect date, 'Thursday, October 20th,' bracketed under the correct date of the month of September. It was not difficult to explain this anticipation as the expression of a wish. A few days before I had returned fresh from my vacation and felt ready for any amount of professional work, but as yet there were few patients.

On my arrival I had found a letter from a patient announcing her arrival on the 20th of October. As I wrote the same date in September I may certainly have thought 'X ought to be here already; what a pity about that whole month!,' and with this thought I pushed the current date a month ahead."[5]

Clearly, the formulation of the intended explanation is *at least incomplete* in the sense considered a moment ago. In particular, it fails to mention any laws or theoretical principles in virtue of which the subconscious wish, and the other antecedent circumstances referred to, could be held to explain Freud's slip of the pen. However, the general theoretical considerations Freud presents here and elsewhere in his writings suggests strongly that his explanatory account relies on a hypothesis to the effect that when a person has a strong, though perhaps unconscious, desire, then if he commits a slip of pen, tongue, memory, or the like, the slip will take a form in which it expresses, and perhaps symbolically fulfills, the given desire.

Even this rather vague hypothesis is probably more definite than what Freud would have been willing to assert. But for the sake of the argument let us accept it and include it in the explanans, together with the particular statements that Freud did have the subconscious wish he mentions, and that he was going to commit a slip of the pen. Even then, the resulting explanans permits us to deduce only that the slip made by Freud would, *in some way or other*, express and perhaps symbolically fulfill Freud's subconscious wish. But clearly, such expression and fulfillment might have been achieved by many other kinds of slip of the pen than the one actually committed.

In other words, the explanans does not imply, and thus fully explain, that the particular slip, say s, which Freud committed on this occasion, would fall within the narrow class, say W, of acts which consist in writing the words "Thursday, October 20th"; rather, the explanans implies only that s would fall into a wider class, say F, which includes W as a proper subclass, and which consists of all acts which would express and symbolically fulfill Freud's subconscious wish *in some way or other*.

The argument under consideration might be called a *partial explanation:* it provides complete, or conclusive, grounds for expecting s to be a member of F, and since W is a subclass of F, it thus shows that the explanandum, i.e., s falling within W, accords with, or bears out, what is to be expected in consideration of the explanans. By contrast, a deductive-nomological explanation of the form (D) might then be called *complete* since the explanans here does imply the explanandum.

Clearly, the question whether a given explanatory argument is complete or partial can be significantly raised only if the explanandum sentence is fully specified; only then can we ask whether the explanandum does or does not follow from the explanans. Completeness of explanation, in this sense, is relative to our explanandum sentence. Now, it might seem much more important and interesting to consider instead the notion of a complete explanation of some *concrete event*, such as the destruction of Pompeii, or the death of Adolf Hitler, or the launching of the first artificial satellite: we might want to regard a particular event as completely explained only if an explanatory account of deductive or of inductive form had been provided for all of its aspects. This notion, however, is self-defeating; for any particular event may be regarded as having infinitely many different aspects or characteristics, which cannot all be accounted for by a finite set, however large, of explanatory statements.

In some cases, what is intended as an explanatory account will depart even further from the standards reflected in the model schemata (D) and (P) above. An explanatory account, for example, which is not explicit and specific enough to be reasonably qualified as an elliptically formulated explanation or as a partial one, can often be viewed as an *explanation sketch*: it may suggest, perhaps quite vividly and persuasively, the general outlines of what, it is hoped, can eventually be supplemented so as to yield a more closely reasoned argument based on explanatory hypotheses which are indicated more fully, and which more readily permit of critical appraisal by reference to empirical evidence.

The decision whether a proposed explanatory account is to be qualified as an elliptically formulated deductive or probabilistic explanation, as a partial explanation, as an explanation sketch, or perhaps as none of these is a matter of judicious interpretation; it calls for an appraisal of the intent of the given argument and of the background assumptions that may be assumed to have been tacitly taken for granted, or at least to be available, in the given context. Unequivocal decision rules cannot be set down for this purpose any more than for determining whether a given informally stated inference which is not deductively valid by reasonably strict standards is to count nevertheless as valid but enthymematically formulated, or as fallacious, or as an instance of sound inductive reasoning, or perhaps, for lack of clarity, as none of these. . . .

■ | Notes

1. See Dewey, John. *How We Think*. Boston, New York, Chicago, 1910; Chapter VI.

2. For a fuller presentation of the model and for further references, see, for example, Hempel, C. G. and P. Oppenheim, "Studies in the Logic of Explanation," *Philosophy of Science* 15: 135–175 (1948). (Secs. 1-7 of this article, which contain all the fundamentals of the presentation, are reprinted in Feigl, H. and M. Brodbeck (eds.), *Readings in the Philosophy of Science*. New York, 1953.)—The suggestive term "covering law model" is W. Dray's; cf. his *Laws and Explanation in History*. Oxford, 1957; Chapter I. Dray characterizes this type of explanation as "subsuming what is to be explained under a general law" (*loc. cit.*, p. 1), and then rightly urges, in the name of methodological realism, that "the requirement of a *single* law be dropped" (*loc. cit.*, p. 24; italics, the author's): it should be noted, however, that, like the schema (D) above, several earlier publications on the subject (among them the article mentioned at the beginning of this note) make explicit provision for the inclusion of more laws than one in the explanans.

3. The relevance of the covering-law model to causal explanation is examined more fully in sec. 4 of Hempel, C. G., "Deductive-Nomological *vs.* Statistical Explanation." In Feigl, H., et al. (eds.), *Minnesota Studies in the Philosophy of Science,* vol. III. Minneapolis, 1962.

4. The concept of probabilistic explanation, and some of the peculiar logical and methodological problems engendered by it, are examined in some detail in Part II of the essay cited in note 3.

5. Freud, S. *Psychopathology of Everyday Life.* Translated by A. A. Brill. New York (Mentor Books) 1951; p. 64.

CARL G. HEMPEL

The Thesis of
Structural Identity

Since in a fully stated D-N explanation of a particular event the explanans logically implies the explanandum, we may say that the explanatory argument might have been used for a deductive prediction of the explanandum-event *if* the laws and the particular facts adduced in its explanans had been known and taken into account at a suitable earlier time. In this sense, a D-N explanation is a potential D-N prediction.

This point was made already in an earlier article by Oppenheim and myself,[1] where we added that scientific explanation (of the deductive-nomological kind) differs from scientific prediction not in logical structure, but in certain pragmatic respects. In one case, the event described in the conclusion is known to have occurred, and suitable statements of general law and particular fact are sought to account for it; in the other, the latter statements are given and the statement about the event in question is derived from them before the time of its presumptive occurrence. This conception, which has sometimes been referred to as the *thesis of the structural identity* (or of the symmetry) *of explanation and prediction*, has recently been questioned by several writers. A consideration of some of their arguments may help to shed further light on the issues involved.

To begin with, some writers[2] have noted that what is usually called a prediction is not an argument but a sentence. More precisely, as Scheffler has pointed out, it is a sentence-token, i.e., a concrete utterance or inscription of a sentence purporting to describe some event that is to occur after the production of the token.[3] This is certainly so. But in empirical science predictive sentences are normally established on the basis of available information by means of arguments that may be deductive or inductive in character; and the thesis under discussion should be understood, of course, to refer to explanatory and predictive *arguments*.

FROM *Aspects of Scientific Explanation* (New York: Free Press, 1965), 366–76.

Thus construed, *the thesis of structural identity* amounts to the conjunction of *two sub-theses*, namely (i) that *every adequate explanation is potentially a prediction* in the sense indicated above; (ii) that conversely *every adequate prediction is potentially an explanation.* I will now examine a number of objections that have been raised against the thesis, dealing first with those which, in effect, concern the first sub-thesis, and then with those concerning the second sub-thesis. I will argue that the first sub-thesis is sound, whereas the second one is indeed open to question. Though the following considerations are concerned principally with D-N explanation, some of them are applicable to other types of explanation as well. . . .

The first sub-thesis, as has already been noted, is an almost trivial truth in the case of D-N explanation, since here the explanans logically implies the explanandum. But it is supported also by a more general principle, which applies to other types of explanation as well, and which expresses, I would submit, a general *condition of adequacy for any rationally acceptable explanation of a particular event.* That condition is the following: Any rationally acceptable answer to the question 'Why did event X occur?' must offer information which shows that X was to be expected—if not definitely, as in the case of D-N explanation, then at least with reasonable probability. Thus, the explanatory information must provide good grounds for believing that X did in fact occur; otherwise, that information would give us no adequate reason for saying: "That explains it—that does show why X occurred." And an explanatory account that satisfies this condition constitutes, of course, a potential prediction in the sense that it could have served to predict the occurrence of X (deductively or with more or less high probability) if the information contained in the explanans had been available at a suitable earlier time.

The condition of adequacy just stated can be extended, in an obvious manner, to explanations concerned, not with individual events, but with empirical uniformities expressed by putative laws. But such explanations cannot well be spoken of as potential *predictions* since law-statements purport to express timeless uniformities and thus make no reference to any particular time, whether past, present, or future.[4]

It will hardly be necessary to emphasize that it is not, of course, the *purpose* of an explanation to provide grounds in support of the explanandum-statement; for . . . a request for an explanation normally *presupposes* that the explanandum-statement is true. The point of the preceding remarks is rather that an adequate explanation cannot help providing information which, if properly established, also provides grounds in support of the explanandum-statement. . . . We may say that an adequate answer to an explanation-seeking why-question is always also a potential answer to the corresponding epistemic why-question.

The converse, however, does not hold; the condition of adequacy is necessary but not sufficient for an acceptable explanation. For example, certain empirical findings may give excellent grounds for the belief that the orientation of the earth's magnetic field shows diurnal and secular variations, without in the least explaining why. Similarly, a set of experimental data may strongly *support* the assumption that the electric resistance of metals increases with their temperature or that a certain chemical inhibits the growth of cancer cells, without providing any *explanation* for these presumptive empirical regularities. The predictive inferences here involved are inductive rather than deductive; but what bars them from the status of potential explanations is not their inductive character . . . , but the fact that they invoke no laws or theoretical principles, no explanatory statements that make a general claim. Reliance on general principles, while perhaps not indispensable for prediction, is required in any explanation: such principles alone can give to whatever particular circumstances may be adduced the status of explanatory factors for the event to be explained.

Some of the objections recently raised against the thesis of the structural identity of explanation and prediction concern in effect the first of its two sub-theses, which has now been presented in some detail: the claim that any adequate explanatory argument is also potentially predictive. I will consider three objections to the effect that there are certain perfectly satisfactory explanations that do not constitute potential predictions.

Scriven has argued that the occurrence of an event X is sometimes quite adequately explained by means of a "proposition of the form 'The only cause of X is A' . . . for example, 'The only cause of paresis is syphilis';" this proposition enables us to explain why a certain patient has paresis by pointing out that he previously suffered from syphilis. And this explanation holds good, according to Scriven, even though only quite a small percentage of syphilitic patients develop paresis, so that "we must, on the evidence [that a given person has syphilis], still predict that [paresis] will *not* occur."[5] But if it does occur, then the principle that the only cause of paresis is syphilis can "provide and guarantee our explanation" in terms of antecedent syphilitic infection.[6] Thus we have here a presumptive explanation which indeed is not adequate as a potential prediction. But precisely because paresis is such a rare sequel of syphilis, prior syphilitic infection surely cannot by itself provide an adequate explanation for it. A condition that is nomically necessary for the occurrence of an event does not, in general, explain it; or else we would be able to explain a man's winning the first prize in the Irish sweepstakes by pointing out that he had previously bought a ticket, and that only a person who owns a ticket can win the first prize.

A second argument which, like Scriven's, has considerable initial plausibility has been advanced by Toulmin[7] by reference to "Darwin's theory, explaining the origin of species by variation and natural selection. No scientist has ever used this theory to foretell the coming-into-existence of creatures of a novel species, still less verified his forecast. Yet many competent scientists have accepted Darwin's theory as having great explanatory power." In examining this argument, let me distinguish what might be called the *story* of evolution from the *theory* of the underlying mechanisms of mutation and natural selection. The story of evolution, as a hypothesis about the gradual development of various types of organisms, and about the subsequent extinction of many of these, has the character of a hypothetical historical narrative *describing* the putative stages of the evolutionary process; it is the associated theory which provides what *explanatory insight* we have into this process. The story of evolution might tell us, for example, that at a certain stage in the process dinosaurs made their appearance and that, so much later, they died out. Such a narrative account does not, of course, explain why the various kinds of dinosaurs with their distinctive characteristics came into existence, nor does it explain why they became extinct. Indeed even the associated theory of mutation and natural selection does not answer the first of these questions, though it might be held to shed some light on the latter. Yet, even to account for the extinction of the dinosaurs, we need a vast array of additional hypotheses about their physical and biological environment and about the species with which they had to compete for survival. But if we have hypotheses of this kind that are specific enough to provide, in combination with the theory of natural selection, at least a probabilistic explanation for the extinction of the dinosaurs, then clearly the explanans adduced is also qualified as a basis for a potential probabilistic prediction. The undeniably great persuasiveness of Toulmin's argument would seem to derive from two sources, a widespread tendency to regard the basically descriptive story of evolution as explaining the various states of the process, and a similarly widespread tendency to overestimate the extent to which even the theory of mutation and natural selection can account for the details of the evolutionary sequence.

I now turn to a third objection to the claim that an adequate explanation is also a potential prediction. It is based on the observation that sometimes the only ground we have for asserting some essential statement in the explanans lies in the knowledge that the explanandum event did in fact occur. In such cases, the explanatory argument clearly could not have been used to predict that event. Consider one of Scriven's examples.[8] Suppose that a man has killed his wife whom he knew to have been unfaithful to him, and that his action is explained as the result of intense jealousy. The fact that the man was jealous might well have been ascertainable before the deed, but to explain the latter, we need to know that his jealousy was intense enough to drive him to murder; and this we can know only after the deed has actually been committed. Here then, the occurrence of the explanandum event provides the only grounds we have for asserting one important part of the explanans; the explanandum event could not therefore have been predicted by means of the explanatory argument. In another example,[9] Scriven considers an explanation to the effect that the collapse of a bridge was caused by metal fatigue. This account, he argues, might be supported by pointing out that the failure could have been caused only by an excessive load, by external damage, or by metal fatigue, and that the first two factors were not present in the case at hand, whereas there is evidence of metal fatigue. *Given the information that the bridge did in fact collapse*, this would establish not only that metal fatigue was at fault but that it was strong enough to cause the failure. While Scriven's notion of "the only possible cause" of a given event surely requires further elucidation, his example does afford another illustration of an explanatory account one of whose constituent hypotheses is supported only by the occurrence of the event to be explained—so that the latter could not have been predicted by means of the explanatory argument.

However, the point thus illustrated does not affect at all the conditional thesis that an adequate explanatory argument must be such that it could have served to predict the explanandum event *if* the information included in the explanans had been known and taken into account before the occurrence of that event. What Scriven's cases show is that sometimes we do not know independently of the occurrence of the explanandum event that all the conditions listed in the explanans are realized. However, this means only that in such cases our conditional thesis is counterfactual, i.e., that its if-clause is not satisfied, but not that the thesis itself is false. Moreover, Scriven's argument does not even show that in the kind of case he mentions it is logically or nomologically impossible (impossible by reason of the laws of logic or the laws of nature) for us to know the critical explanatory factor before, or independently of, the occurrence of the explanandum-event; the impossibility appears to be rather a practical and perhaps temporary one, reflecting present limitations of knowledge or technology.

But while it thus leaves our thesis unaffected, Scriven's observation is of methodological interest in its own right: it shows that sometimes an event is explained by means of hypotheses for some of which the fact of its occurrence affords the only available evidential support. This may happen, as we saw, when one of the explanatory hypotheses states that a certain relevant factor was strong enough to bring about the event in question; but the observation applies also to other cases. Thus the explanation . . . of the appearance and initial growth of the soap bubbles, includes in its explanans the assumption that a soap film had formed between the plate and the rims of the tumblers,* and practically the only evidence available in support of this explanatory assumption is the fact that soap bubbles did emerge from under the tumblers. Or consider the explanation of the characteristic dark lines in the absorption spectrum of a particular star. The key assumption in the explanans is that the star's atmosphere contains certain elements, such as hydrogen, helium, and calcium, whose atoms absorb radiation of the wave lengths corresponding to the dark lines; the explanation relies, of course, on many other assumptions, including the optical theory that forms the basis for spectroscopy, and the assumption that the apparatus used is a properly constructed spectroscope. But while these latter explanans statements are capable of independent test and corroboration, it may well be that the only

---

* See the first section of Hempel's "Two Basic Types of Scientific Explanation," (the preceding reading in this chapter) for more about why bubbles appear when a tumbler is taken from warm soapy water and inverted on a plate.

evidence available in support of the key explanatory hypothesis is the occurrence of the very lines whose appearance in the spectrum the argument serves to explain. Strictly speaking, the explanandum event here provides support for the key explanatory hypothesis only by virtue of the background theory, which connects the presence of certain elements in the atmosphere of a star with the appearance of corresponding absorption lines in its spectrum. Thus, the information that the explanandum event has occurred does not by itself support the explanatory hypothesis in question, but it constitutes, as we might say, an essential part of the only evidence available in support of that hypothesis.

Explanations of the kind here considered may be schematically characterized as arguments of the form (D-N) in which the information or assumption that E is true provides an indispensable part of the only available evidential support for one of the explanans statements, say, C_1. Let us call such explanations *self-evidencing*. It might be held that the actual occurrence of the explanandum event always provides some slight additional support even for an explanans whose constituent sentences have been accepted on the basis of independent evidence, and that in this sense every D-N explanation with true explanandum is in some measure self-evidencing; but we will apply this appellation to an explanatory account only if, at the time of its presentation, the occurrence of the explanandum event provides the only evidence, or an indispensable part of the only evidence, available in support of some of the explanans-statements.

An explanatory argument of the form (D-N) which is self-evidencing is not for that reason circular or pointless. To be sure, if the same argument were adduced in support of the assertion that the explanandum-event did occur (or, that E is true), then it would be open to the charge of epistemic circularity. If the argument is to achieve its objective then all the grounds it adduces in support of E—i.e., $C_1, C_2 \ldots, C_k; L_1, L_2, \ldots, L_r$—would have to be established independently of E; and this condition is violated here since the only ground we have for believing or asserting C_1 includes the assumption that E is true. But when the same argument is used for explanatory purposes it does not claim to establish that E is true; that is *presupposed* by the question 'Why did the event described by E occur?'. Nor need a self-evidencing explanation involve an explanatory circle. The information that the explanandum event has occurred is not included in the explanans (so that the occurrence of the event is not "explained by itself"); rather it serves, quite outside the explanatory context, as evidence supporting one of the explanans statements. Thus, an acceptable self-evidencing explanation benefits, as it were, by the wisdom of hindsight derived from the information that the explanandum event has occurred, but it does not misuse that information so as to produce a circular explanation.

An explanation that is self-evidencing may for that reason rest on a poorly supported explanans and may therefore have no strong claim to empirical soundness. But even this is not inevitable. In the case of the absorption spectrum of a star, for example, the previously accepted background information, including the relevant theories, may indicate that the dark lines observed occur *only* if the specified elements are present in the star's atmosphere; and then the explanandum, in conjunction with the background information, lends very strong support to the crucial explanatory hypothesis.

The notion of a self-evidencing explanation can, I think, shed some further light on the puzzle illustrated by the explanation of paresis in terms of antecedent syphilitic infection. Consider another illustration. Some cases of skin cancer are attributed to intensive ultraviolet irradiation. But this factor very often does not lead to cancer, so that the information that a person has been exposed to such radiation does not permit the prediction of cancer. Is that information alone nevertheless sufficient to explain the development of skin cancer when it does follow intensive irradiation? No doubt, an explanation will often be formulated so as to mention only the antecedent irradiation; but the underlying rationale surely must be more complex. Leaving aside the important quantitative aspects of the problem, the crucial point in that rationale can, I suggest, be schematically stated as follows: Some, though by no means all, individuals have the disposition to develop skin cancer upon exposure to strong ultraviolet irradiation; let us call these radiation-sensitive. Now, in the case of explanation, we know that the given individual was exposed to strong radiation (C_1) and did develop cancer of the skin in the affected area (E). But jointly, these two pieces of information lend support to the assumption that the individual is radiation-sensitive (C_2)—an hypothesis that is not supported in the case of prediction, where C_1 is available, but not E. And the two statements C_1 and C_2 (in combination with the general statement that sensitive individuals will develop skin cancer when exposed to intensive radiation) do provide an adequate explanans for E. If the explanation is thus construed as invoking C_2 in addition to C_1, it is seen to be self-evidencing, but also to possess an explanans which would provide an adequate basis for prediction if C_2 could be known in advance. That is impossible, of course, as long as the only available test for radiation-sensitivity consists in checking whether an individual does develop skin cancer upon intensive irradiation. But, clearly, it is conceivable that other, independent, tests of radiation-sensitivity might be found and then C_2 might well be established independently of, and even prior to, the occurrence of the event described by E.

In discussing the structural identity of explanation and prediction, I have so far considered only the first of the two sub-theses distinguished earlier, namely, the claim that every adequate explanation is also a potential prediction. I have argued that the objections raised against this claim fall short of their mark, and that the first sub-thesis is sound and can indeed serve as a necessary condition of adequacy for any explicitly stated, rationally acceptable explanation.

I turn now to the second sub-thesis, namely, that every adequate predictive argument also affords a potential explanation. This claim is open to question even in the case of certain predictive arguments that are of deductive-nomological character, as the following example illustrates. One of the early symptoms of measles is the appearance of small whitish spots, known as Koplik spots, on the mucous linings of the cheeks. The statement, L, that the appearance of Koplik spots is always followed by the later manifestations of the measles might therefore be taken to be a law, and it might then be used as a premise in D-N arguments with a second premise of the form 'Patient i has Koplik spots at time t', and with a conclusion stating that i subsequently shows the later manifestations of the measles. An argument of this type is adequate for predictive purposes, but its explanatory adequacy might be questioned. We would not want to say, for example, that i had developed high fever and other symptoms of the measles because he had previously had Koplik spots. Yet this case—and others similar to it—does not constitute a decisive objection, against the second sub-thesis. For the reluctance to regard the appearance of Koplik spots as explanatory may well reflect doubts as to whether, as a matter of universal law, those spots are always followed by the later manifestations of measles. Perhaps a local inoculation with a small amount of measles virus would produce the spots without leading to a full-blown case of the measles. If this were so, the appearance of the spots would still afford a usually reliable basis for predicting the occurrence of further symptoms, since exceptional conditions of the kind just mentioned would be extremely rare; but the generalization that Koplik spots are always followed by later symptoms of the measles would not express a law and thus could not properly support a corresponding D-N explanation.

The objection just considered concerns the explanatory potential of predictive arguments of the form (D-N). But the second sub-thesis, in its general form, which is not limited to D-N predictions, has further been challenged, particularly by Scheffler and by Scriven,[10] on the ground that there are other kinds of predictive argument that are adequate for scientific prediction, yet not for explanation. Specifically, as Scheffler notes, a scientific prediction may be based on a finite set of data which includes no laws and which would have no explanatory force. For example, a finite set of data obtained in an extensive test of the hypothesis that the electric resistance of metals increases with their temperature may afford good support for that hypothesis and may thus provide an acceptable basis for the prediction that in an as yet unexamined instance, a rise in temperature in a metal conductor will be accompanied by an increase in resistance. But if this event then actually occurs, the test data clearly do not provide an explanation for it. Similarly, a list of the results obtained in a long series of tossings of a given coin may provide a good basis for predicting the percentage of Heads and Tails to be expected in the next 1000 tossings of the same coin; but again, that list of data provides no explanation for the subsequent results. Cases like these raise the question of whether there are not sound modes of scientific prediction that proceed from particulars to particulars without benefit of general laws such as seem to be required for any adequate explanation. Now, the predictive arguments just considered are not deductive but probabilistic in character. . . . In regard to the second sub-thesis of the structural identity claim, let us note this much here: the predictions in our illustrations proceed from an observed sample of a population to another, as yet unobserved one; and on some current theories of probabilistic inference such arguments do not depend upon the assumption of general empirical laws. According to Carnap's theory of inductive logic,[11] for example, such inferences are possible on purely logical grounds; the information about the given sample confers a definite logical probability upon any proposed prediction concerning an as yet unobserved sample. On the other hand, certain statistical theories of probabilistic inference eschew the notion of purely logical probabilities and qualify predictions of the kind here considered as sound only on the further assumption that the selection of individual cases from the total population has the character of a random experiment with certain general statistical characteristics. But that assumption, when explicitly spelled out, has the form of a general law of statistic-probabilistic form; hence, the predictions are effected by means of covering laws after all. And though these laws do not have the strictly universal character of those invoked in D-N explanations and predictions, they can serve in an explanatory capacity as well. Thus construed, even the predictions here under discussion turn out to be (incompletely formulated) potential explanations. . . .

■ | Notes

1. Hempel and Oppenheim (1948), section 3.

2. See Scheffler (1957), section 1 and (1963), Part 1, sections 3 and 4; Scriven (1962), p. 177.

3. *Cf.* Scheffler (1957), section 1. . . .

4. This point is made, for example, by Scriven (1962), pp. 179 ff.

5. Scriven (1959a), p. 480, italics the author's.

6. *Loc. cit.* Barker has argued analogously that "it can be correct to speak of explanation in many cases where specific prediction is not possible. Thus, for instance, if the patient shows all the symptoms of pneumonia, sickens and dies, I can then explain his death—I know what killed him—but I could not have definitely predicted in advance that he was going to die; for usually pneumonia fails to be fatal." (1961, p. 271). This argument seems to me open to questions similar to those just raised in reference to Scriven's illustration. First of all, it is not clear just what would be claimed by the assertion that pneumonia killed the patient. Surely the mere information that the patient had pneumonia does not suffice to explain his death, precisely because in most cases pneumonia is not fatal. And if the explanans is taken to state that the patient was suffering from very severe pneumonia (and perhaps that he was elderly or weak) then it may well provide a basis at least for a probabilistic explanation of the patient's death—but in this case it obviously also permits prediction of his death with the same probability. For some further observations on Barker's argument, see the comments by Feyerabend and by Rudner, and Barker's rejoinders, in Feigl and Maxwell (1961), pp. 278–85. A detailed critical discussion

that sheds further light on Scriven's paresis example will be found in Grünbaum (1963) and (1963a), chapter 9; see also Scriven's rejoinder (1963).

7. Toulmin (1961), pp. 24–25. Scriven (1959a) and Barker (1961) have offered arguments in the same vein. For a critical discussion of Scriven's version, see Grünbaum (1963) and (1963a), chapter 9.

8. Scriven (1959), pp. 468–69.

9. Scriven (1962), pp. 181–87.

10. See Scheffler (1957), p. 296 and (1963), p. 42; Scriven (1959a), p. 480.

11. Carnap (1950), section 110.

■ | Bibliography

Barker, S. F., "The Role of Simplicity in Explanation," in Feigl and Maxwell (1961), 265–74.

Baumrin, B. (ed.). *Philosophy of Science. The Delaware Seminar.* Volume I, 1961–62 (New York: John Wiley & Sons, 1963).

Carnap, R. *Logical Foundations of Probability* (Chicago: University of Chicago Press, 1950; 2nd revised ed. 1962).

Feigl, H. and G. Maxwell (eds.). *Current Issues in the Philosophy of Science* (New York: Holt, Rinehart & Winston, 1961).

Feigl, H. and G. Maxwell (eds.). *Minnesota Studies in the Philosophy of Science,* Volume III (Minneapolis: University of Minnesota Press, 1962).

Gardiner, P. (ed.). *Theories of History* (New York: The Free Press, 1959).

Grünbaum, A., "Temporally Asymmetric Principles, Parity between Explanations and Prediction, and Mechanism vs. Teleology," in Baumrin (1963), 57–96.

Grünbaum, A. *Philosophical Problems of Space and Time* (New York: Knopf, 1963a).

Hempel, C. G. and P. Oppenheim, "Studies in the Logic of Explanation," *Philosophy of Science* 15 (1948): 135–75.

Scheffler, I., "Explanation, Prediction, and Abstraction," *British Journal for the Philosophy of Science* 7 (1957): 293–309.

Scheffler, I. *The Anatomy of Inquiry: Philosophical Studies in the Theory of Science* (New York: Alfred A. Knopf, 1963).

Scriven, M., "Truisms as the Grounds for Historical Explanations," in Gardiner (1959), 443–75.

Scriven, M., "Explanation and Prediction in Evolutionary Theory," *Science* 130 (1959a): 477–82.

Scriven, M., "Explanations, Predictions, and Laws," in Feigl and Maxwell (1962), 170–230.

Scriven, M., "The Temporal Asymmetry between Explanations and Predictions," in Baumrin (1963), 97–105.

Toulmin, S. *Foresight and Understanding* (London: Hutchinson, 1961; New York: Harper & Row, 1963).

Carl G. Hempel

Inductive-Statistical
Explanation

1 | Inductive-Statistical Explanation

As an explanation of why patient John Jones recovered from a streptococcus
infection, we might be told that Jones had been given penicillin. But if we
try to amplify this explanatory claim by indicating a general connection
between penicillin treatment and the subsiding of a streptococcus infection
we cannot justifiably invoke a general law to the effect that in all cases of
such infection, administration of penicillin will lead to recovery. What can
be asserted, and what surely is taken for granted here, is only that penicillin
will effect a cure in a high percentage of cases, or with a high statistical
probability. This statement has the general character of a law of statistical
form, and while the probability value is not specified, the statement indicates that it is high. But in contrast to the cases of deductive-nomological
and deductive-statistical explanation, the explanans consisting of this statistical law together with the statement that the patient did receive penicillin
obviously does not imply the explanandum statement, 'the patient recovered',
with deductive certainty, but only, as we might say, with high likelihood, or
near certainty. Briefly, then, the explanation amounts to this argument:

> 1a   The particular case of illness of John Jones—let us call it j—was an
> instance of severe streptococcal infection (Sj) which was treated with large
> doses of penicillin (Pj); and the statistical probability $p(R, S \cdot P)$ of recovery
> in cases where S and P are present close to 1; hence, the case was practi-
> cally certain to end in recovery (Rj).*

From *Aspects of Scientific Explanation* (New York: Free Press, 1965), 381–83,
394–403.

* Throughout this paper, Hempel uses a dot to stand for conjunction, a bar over a
letter to stand for negation, and a comma within parentheses to represent conditional probabilities. Thus, for example, $p(R, S \cdot \bar{P})$ means the probability of R given
S and not-P.

This argument might invite the following schematization:

1b $p(R, S \cdot P)$ is close to 1
   $\dfrac{Sj \cdot Pj}{}$

(Therefore:) It is practically certain (very likely) that Rj

In the literature on inductive inference, arguments thus based on statistical hypotheses have often been construed as having this form or a similar one. On this construal, the conclusion characteristically contains a modal qualifier such as 'almost certainly,' 'with high probability', 'very likely', etc. But the conception of arguments having this character is untenable. For phrases of the form 'it is practically certain that p' or 'It is very likely that p', where the place of 'p' is taken by some statement, are not complete self-contained sentences that can be qualified as either true or false. The statement that takes the place of 'p'—for example, 'Rj'—is either true or false, quite independently of whatever relevant evidence may be available, but it can be qualified as more or less likely, probable, certain, or the like only *relative to some body of evidence.* One and the same statement, such as 'Rj', will be certain, very likely, not very likely, highly likely, and so forth, depending upon what evidence is considered. The phrase 'it is almost certain that Rj' taken by itself is therefore neither true nor false; and it cannot be inferred from the premises specified in (1b) nor from any other statements.

The confusion underlying the schematization (1b) might be further illuminated by considering its analogue for the case of deductive arguments. The force of a deductive inference, such as that from 'all F are G' and 'a is F' to 'a is G', is sometimes indicated by saying that if the premises are true, then the conclusion is necessarily true or is certain to be true—a phrasing that might suggest the schematization:

All F are G
$\dfrac{a \text{ is } F}{}$

(Therefore:) It is necessary (certain) that a is G

But clearly the given premises—which might be, for example, 'all men are mortal' and 'Socrates is a man'—do not establish the sentence 'a is G' ('Socrates is mortal') as a necessary or certain truth. The certainty referred to in the informal paraphrase of the argument is relational: the statement 'a is G' is certain, or necessary, *relative to the specified premises*; i.e., their truth will guarantee its truth—which means nothing more than that 'a is G' is a logical consequence of those premises.

Analogously, to present our statistical explanation in the manner of schema (1b) is to misconstrue the function of the words 'almost certain' or 'very likely' as they occur in the formal wording of the explanation. Those words clearly must be taken to indicate that on the evidence provided by the explanans, or relative to that evidence, the explanandum is practically certain or very likely, i.e., that

    1c 'Rj' is practically certain (very likely) relative to the explanans containing the sentences '$p(R, S \cdot P)$ is close to 1' and '$Sj \cdot Pj$'.[1]

The explanatory argument misrepresented by (1b) might therefore suitably be schematized as follows:

    1d $p(R, S \cdot P)$ is close to 1

        $Sj \cdot Pj$

       ——————————— [makes practically certain]

        Rj

In this schema, the double line separating the "premises" from the "conclusion" is to signify that the relation of the former to the latter is not that of deductive implication but that of inductive support, the strength of which is indicated in square brackets.[2] . . .

2 | The Problem of Explanatory Ambiguity

Consider once more the explanation (1d) of recovery in the particular case j of John Jones's illness. The statistical law there invoked claims recovery in response to penicillin only for a high percentage of streptococcal infections, but not for all of them; and in fact, certain streptococcus strains are resistant to penicillin. Let us say that an occurrence, e.g., particular case of illness, has the property S^* (or belongs to the class S^*) if it is an instance of infection with a penicillin-resistant streptococcus strain. Then the probability of recovery among randomly chosen instances of S^* which are treated with penicillin will be quite small, i.e., $p(R, S^* \cdot P)$ will be close to 0 and the probability of nonrecovery, $p(\bar{R}, S^* \cdot P)$ will be close to 1. But suppose now that Jones's illness is in fact a streptococcal infection of the penicillin-resistant variety, and consider the following argument:

    2a $p(\bar{R}, S^* \cdot P)$ is close to 1

        $S^*j \cdot Pj$

       ——————————— [makes practically certain]

        $\bar{R}j$

This "rival" argument has the same form as (1d), and on our assumptions, its premises are true, just like those of (1d). Yet its conclusion is the contradictory of the conclusion of (1d).

Or suppose that Jones is an octogenarian with a weak heart, and that in this group, S^{**}, the probability of recovery from a streptococcus infection in response to penicillin treatment, $p(R, S^{**} \cdot P)$, is quite small. Then, there is the following rival argument to (1d), which presents Jones's nonrecovery as practically certain in the light of premises which are true:

> **2b** $p(\bar{R}, S^{**} \cdot P)$ is close to 1
> $$\frac{S^{**} j \cdot Pj}{\bar{R}_j} \quad \text{[makes practically certain]}$$

The peculiar logical phenomenon here illustrated will be called the *ambiguity of inductive-statistical explanation* or, briefly, of *statistical explanation*. This ambiguity derives from the fact that a given individual event (e.g., Jones's illness) will often be obtainable by random selection from any one of several "reference classes" (such as $S \cdot P$, $S^* \cdot P$, $S^{**} \cdot P$), with respect to which the kind of occurrence (e.g., R) instantiated by the given event has very different statistical probabilities. Hence, for a proposed probabilistic explanation with true explanans which confers near certainty upon a particular event, there will often exist a rival argument of the same probabilistic form and with equally true premises which confers near certainty upon the nonoccurrence of the same event. And any statistical explanation for the occurrence of an event must seem suspect if there is the possibility of a logically and empirically equally sound probabilistic account for its nonoccurrence. *This predicament has no analogue in the case of deductive explanation*; for if the premises of a proposed deductive explanation are true then so is its conclusion; and its contradictory, being false, cannot be a logical consequence of a rival set of premises that are equally true.

Here is another example of the ambiguity of I-S explanation: Upon expressing surprise at finding the weather in Stanford warm and sunny on a date as autumnal as November 27, I might be told, by way of explanation, that this was rather to be expected because the probability of warm and sunny weather (W) on a November day in Stanford (N) is, say, .95. Schematically, this account would take the following form, where 'n' stands for 'November 27':

> **2c** $p(W, N) = .95$
> $$\frac{Nn}{Wn} \quad [.95]$$

But suppose it happens to be the case that the day before, November 26, was cold and rainy, and that the probability for the immediate successors (S) of cold and rainy days in Stanford to be warm and sunny is .2; then the account (2c) has a rival in the following argument which, by reference to equally true premises, presents it as fairly certain that November 27 is not warm and sunny:

> **2d** $p(\bar{W}, S) = .8$
> $$\frac{Sn}{\bar{W}_n} \quad [.8]$$

In this form, the problem of ambiguity concerns I-S arguments whose premises are in fact true, no matter whether we are aware of this or not. But, as will now be shown, the problem has a variant that concerns explanations whose explanans statements, no matter whether in fact true or not, are *asserted or accepted* by empirical science at the time when the explanation is proffered or contemplated. This variant will be called *the problem of the epistemic ambiguity of statistical explanation*, since it refers to what is presumed to be known in science rather than to what, perhaps unknown to anyone, is in fact the case.

Let K_t be the class of all statements asserted or accepted by empirical science at time t. This class then represents the total scientific information, or "scientific knowledge" at time t. The word 'knowledge' is here used in the sense in which we commonly speak of the scientific knowledge at a given time. It is not meant to convey the claim that the elements of K_t are true, and hence neither that they are definitely known to be true. No such claim can justifiably be made for any of the statements established by empirical science; and the basic standards of scientific inquiry demand that an empirical statement, however well supported, be accepted and thus admitted to membership in K_t only tentatively, i.e., with the understanding that the privilege may be withdrawn if unfavorable evidence should be discovered. The membership of K_t therefore changes in the course of time; for as a result of continuing research, new statements are admitted into that class; others may come to be discredited and dropped. Henceforth, the class of accepted statements will be referred to simply as K when specific reference to the time in question is not required. We will assume that K is logically consistent and that it is closed under logical implication, i.e., that it contains every statement that is logically implied by any of its subsets.

The *epistemic ambiguity of I-S explanation* can now be characterized as follows: The total set K of accepted scientific statements contains different subsets of statements which can be used as premises in arguments of the probabilistic form just considered, and which confer high probabilities on logically contradictory "conclusions." Our earlier examples (2a), (2b) and (2c), (2d) illustrate this point if we assume that the premises of those arguments all belong to K rather than that they are all true. If one of two such rival arguments with premises in K is proposed as an explanation of an event considered, or acknowledged, in science to have occurred, then the conclusion of the argument, i.e., the explanandum statement, will accordingly belong to K as well. And since K is consistent, the conclusion of the rival argument will not belong to K. Nonetheless it is disquieting that we should be able to say: No matter whether we are informed that the event in question (e.g., warm and sunny weather on November 27 in Stanford) did occur or that it did not occur, we can produce an explanation of the reported outcome in either case; and an explanation, moreover, whose premises are scientifically established statements that confer a high logical probability upon the reported outcome.

This epistemic ambiguity, again, has no analogue for deductive explanation; for since K is logically consistent, it cannot contain premise-sets that imply logically contradictory conclusions.

Epistemic ambiguity also bedevils the predictive use of statistical arguments. Here, it has the alarming aspect of presenting us with two rival arguments whose premises are scientifically well established, but one of which characterizes a contemplated future occurrence as practically certain, whereas the other characterizes it as practically impossible. Which of such conflicting arguments, if any, are rationally to be relied on for explanation or for prediction?

3 | The Requirement of Maximal Specificity and the Epistemic Relativity of Inductive-Statistical Explanation

Our illustrations of explanatory ambiguity suggest that a decision on the acceptability of a proposed probabilistic explanation or prediction will have to be made in the light of all the relevant information at our disposal. This is indicated also by a general principle whose importance for inductive reasoning has been acknowledged, if not always very explicitly, by many writers, and which has recently been strongly emphasized by Carnap, who calls it *the requirement of total evidence.* Carnap formulates it as follows: "in the application of inductive logic to a given knowledge situation, the total evidence available must be taken as basis for determining the degree of confirmation."[3] Using only a part of the total evidence is permissible if the balance of the evidence is irrelevant to the inductive "conclusion," i.e., if on the partial evidence alone, the conclusion has the same confirmation, or logical probability, as on the total evidence.[4]

The requirement of total evidence is not a postulate nor a theorem of inductive logic; it is not concerned with the formal validity of inductive arguments. Rather, as Carnap has stressed, it is a maxim for the *application* of inductive logic; we might say that it states a necessary condition of rationality of any such application in a given "knowledge situation," which we will think of as represented by the set K of all statements accepted in the situation.

But in what manner should the basic idea of this requirement be brought to bear upon probabilistic explanation? Surely we should not insist that the explanans must contain all and only the empirical information available at the time. Not *all* the available information, because otherwise all probabilistic explanations acceptable at time t would have to have the same explanans, K_t; and not *only* the available information, because a proffered explanation may meet the intent of the requirement in not overlooking any relevant information available, and may nevertheless invoke some explanans statements which have not as yet been sufficiently tested to be included in K_t.

The extent to which the requirement of total evidence should be imposed upon statistical explanations is suggested by considerations such as the following. A proffered explanation of Jones's recovery based on the information that Jones had a streptococcal infection and was treated with penicillin, and that the statistical probability for recovery in such cases is very high, is unacceptable if K includes the further information that Jones's streptococci were resistant to penicillin, or that Jones was an octogenarian with a weak heart, and that in these reference classes the probability of recovery is small. Indeed, one would want an acceptable explanation to be based on a statistical probability statement pertaining to the narrowest reference class of which, according to our total information, the particular occurrence under consideration is a member. Thus, if K tells us not only that Jones had a streptococcus infection and was treated with penicillin, but also that he was an octogenarian with a weak heart (and if K provides no information more specific than that) then we would require that an acceptable explanation of Jones's response to the treatment be based on a statistical law stating the probability of that response in the narrowest reference class to which our total information assigns Jones's illness, i.e., the class of streptococcal infections suffered by octogenarians with weak hearts.[5]

Let me amplify this suggestion by reference to an example concerning the use of the law that the half-life of radon is 3.82 days in accounting for the fact that the residual amount of radon to which a sample of 10 milligrams was reduced in 7.64 days was within the range from 2.4 to 2.6 milligrams. According to present scientific knowledge, the rate of decay of a radioactive element depends solely upon its atomic structure as characterized by its atomic number and its mass number, and it is thus unaffected by the age of the sample and by such factors as temperature, pressure, magnetic and electric forces, and chemical interactions. Thus, by specifying the half-life of radon as well as the initial mass of the sample and the time interval in question, the explanans takes into account all the available information that is relevant to appraising the probability of the given outcome by means of statistical laws. To state the point somewhat differently: Under the circumstances here assumed, our total information K assigns the case under study first of all to the reference class say F_1, of cases where a 10 milligram sample of radon is allowed to decay for 7.64 days; and the half-life law for radon assigns a very high probability, within F_1, to the "outcome," say G, consisting in the fact that the residual mass of radon lies between 2.4 and 2.6 milligrams. Suppose now that K also contains information about the temperature of the given sample, the pressure and relative humidity under which it is kept, the surrounding electric and magnetic conditions, and so forth, so that K assigns the given case to a reference class much narrower than F_1, let us say, $F_1 F_2 F_3 \ldots F_n$. Now the theory of radioactive decay, which is equally included in K, tells us that the statistical probability of G within this narrower class is the same as within F_1. For this reason, it suffices in our explanation to rely on the probability $p(G, F_1)$.

Let us note, however, that "knowledge situations" are conceivable in which the same argument would not be an acceptable explanation. Suppose, for example, that in the case of the radon sample under study, the amount remaining one hour before the end of the 7.64-day period happens to have been measured and found to be 2.7 milligrams, and thus markedly in excess of 2.6 milligrams—an occurrence which, considering the decay law for radon, is highly improbable, but not impossible. That finding, which then forms part of the total evidence K, assigns the particular case at hand to a reference class, say F^*, within which, according to the decay law for radon, the outcome G is highly improbable since it would require a quite unusual spurt in the decay of the given sample to reduce the 2.7 milligrams, within the one final hour of the test, to an amount falling between 2.4 and 2.6 milligrams. Hence, the additional information here considered may not be disregarded, and an explanation of the observed outcome will be acceptable only if it takes account of the probability of G in the narrower reference class, i.e., $p(G, F_1 F^*)$. (The theory of radioactive decay implies that this probability equals $p(G, F^*)$, so that as a consequence the membership of the given case in F_1 need not be explicitly taken into account.)

The requirement suggested by the preceding considerations can now be stated more explicitly; we will call it the *requirement of maximal specificity for inductive-statistical explanations.* Consider a proposed explanation of the basic statistical form

3a $p(G, F) = r$
$$\frac{Fb}{Gb} = [r]$$

Let s be the conjunction of the premises, and, if K is the set of all statements accepted at the given time, let k be a sentence that is logically equivalent to K (in the sense that k is implied by K and in turn implies every sentence in K). Then, to be rationally acceptable in the knowledge situation represented by K, the proposed explanation (3a) must meet the following condition (the requirement of maximal specificity): If $s \cdot k$ implies[6] that b belongs to a class F_1, and that F_1 is a subclass of F, then $s \cdot k$ must also imply a statement specifying the statistical probability of G in F_1, say

$p(G, F_1) = r_1$

Here, r_1 must equal r unless the probability statement just cited is simply a theorem of mathematical probability theory.

The qualifying unless-clause here appended is quite proper, and its omission would result in undesirable consequences. It is proper because theorems of pure mathematical probability theory cannot provide an explanation of empirical subject matter. They may therefore be discounted when we inquire whether $s \cdot k$ might not give us statistical laws specifying the probability of G in reference classes narrower than F. And the omission of the clause would prove troublesome, for if (3a) is proffered as an explanation, then it is presumably accepted as a fact that Gb; hence 'Gb' belongs to K. Thus K assigns b to the narrower class $F \cdot G$, and concerning the probability of G in that class, $s \cdot k$ trivially implies the statement that $p(G, F \cdot G) = 1$, which is simply a consequence of the measure-theoretical postulates for statistical probability. Since $s \cdot k$ thus implies a more specific probability statement for G than that invoked in (3a), the requirement of maximal specificity would be violated by (3a)— and analogously by any proffered statistical explanation of an event that we take to have occurred—were it not for the unless-clause, which, in effect, disqualifies the notion that the statement '$p(G, F \cdot G) = 1$' affords a more appropriate law to account for the presumed fact that Gb.

The requirement of maximal specificity, then, is here tentatively put forward as characterizing the extent to which the requirement of total evidence properly applies to inductive-statistical explanations. The general idea thus suggested comes to this: In formulating or appraising an I-S explanation, we should take into account all that information provided by K which is of potential *explanatory* relevance to the explanandum event; i.e., all pertinent statistical laws, and such particular facts as might be connected, by the statistical laws, with the explanandum event.[7]

The requirement of maximal specificity disposes of the problem of epistemic ambiguity; for it is readily seen that of two rival statistical arguments with high associated probabilities and with premises that all belong to K, at least one violates the requirement of maximum specificity. Indeed, let

$$p(G,F) = r_1 \qquad\qquad p(\overline{G},H) = r_2$$
$$\frac{Fb}{Gb}\!=\![r_1] \qquad \text{and} \qquad \frac{Hb}{\overline{G}b}\!=\![r_2]$$

be the arguments in question, with r_1 and r_2 close to 1. Then, since K contains the premises of both arguments, it assigns b to both F and H and hence to $F \cdot H$. Hence if both arguments satisfy the requirement of maximal specificity, K must imply that

$$p(G, F \cdot H) = p(G, F) = r_1$$
$$p(\overline{G}, F \cdot H) = p(\overline{G}, H) = r_2$$

But $\quad p(G, F \cdot H) + p(\overline{G}, F \cdot H) = 1$

Hence $\quad r_1 + r_2 = 1$

and this is an arithmetic falsehood, since r_1 and r_2 are both close to 1; hence it cannot be implied by the consistent class K.

Thus, for I-S explanations that meet the requirement of maximal specificity the problem of epistemic ambiguity no longer arises. We are *never* in a position to say: No matter whether this particular event did or did not occur, we can produce an acceptable explanation of either outcome; and an explanation, moreover, whose premises are scientifically accepted statements which confer a high logical probability upon the given outcome.

While the problem of epistemic ambiguity has thus been resolved, ambiguity in the first sense discussed [in section 2] remains unaffected by our requirement; i.e., it remains the case that for a given statistical argument with true premises and a high associated probability, there may exist a rival one with equally true premises and with a high associated probability, whose conclusion contradicts that of the first argument. And though the set K of statements accepted at any time never includes all statements that are in fact true (and no doubt many that are false), it is perfectly possible that K should contain the premises of two such conflicting arguments; but as we have seen, at least one of the latter will fail to be rationally acceptable because it violates the requirement of maximal specificity.

The preceding considerations show that *the concept of statistical explanation for particular events is essentially relative to a given knowledge situation as represented by a class K of accepted statements.* Indeed, the requirement of maximal specificity makes explicit and unavoidable reference to such a class, and it thus serves to characterize the concept of "I-S explanation relative to the knowledge situation represented by K." We will refer to this characteristic as the *epistemic relativity of statistical explanation.*

It might seem that the concept of deductive explanation possesses the same kind of relativity, since whether a proposed D-N or D-S [deductive-statistical] account is acceptable will depend not only on whether it is deductively valid and makes essential use of the proper type of general law, but also on whether its premises are well supported by the relevant evidence at hand. Quite so; and this condition of empirical confirmation applies equally to statistical explanations that are to be acceptable in a given knowledge situation. But the epistemic relativity that the requirement of maximal specificity implies for I-S explanations is of quite a different kind and has no analogue for D-N explanations. For the specificity requirement is not concerned with the evidential support that the total evidence K affords for the explanans statements: it does not demand that the latter be included in K, nor even that K supply supporting evidence for them. It rather concerns what may be called the concept of a *potential* statistical explanation. For it stipulates that no matter how much evidential support there may be for the explanans, a proposed I-S explanation is not acceptable if its potential explanatory force with respect to the specified explanandum is vitiated by statistical laws which are included in K but not in the explanans, and which might permit the production of rival statistical arguments. As we have seen, this danger never arises for deductive explanations. Hence, these are not subject to any such restrictive condition, and the notion of a potential deductive explanation (as contradistinguished from a deductive explanation with well-confirmed explanans) requires no relativization with respect to K.

As a consequence, we can significantly speak of true D-N and D-S explanations: they are those potential D-N and D-S explanations whose premises (and hence also conclusions) are true—no matter whether this happens to be known or believed, and thus no matter whether the premises are included in K. But this idea has no significant analogue for I-S explanation since, as we have seen, the concept of potential statistical explanation requires relativization with respect to K.

■ | Notes

1. Phrases such as 'It is almost certain (very likely) that j recovers', even when given the relational construal here suggested, are ostensibly concerned with relations between propositions, such as those expressed by the sentences forming the conclusion and the premises of an argument. For the purpose of the present discussion, however, involvement with propositions can be avoided by construing the phrases in question as expressing logical relations between corresponding *sentences*, e.g., the conclusion-sentence and the premise-sentence of an argument. This construal, which underlies the formulation of (1c), will be adopted in this essay, though for the sake of convenience we may occasionally use a paraphrase.

2. In the familiar schematization of deductive arguments, with a single line separating the premises from the conclusion, no explicit distinction is made between a weaker and a stronger claim, either of which might be intended; namely (i) that the premises logically imply the conclusion and (ii) that, in addition, the premises are true. In the case of our probabilistic argument, (1c) expresses a weaker claim, analogous to (i), whereas (1d) may be taken to express a "proffered explanation" (the term is borrowed from I. Scheffler, 'Explanation, Prediction, and Abstraction', *British Journal for the Philosophy of Science* 7 (1957), sect. 1) in which, in addition, the explanatory premises are—however tentatively—asserted as true.

The considerations here outlined concerning the use of terms like 'probably' and 'certainly' as modal qualifiers of individual statements seem to me to militate also against the notion of categorical probability statement that C. I. Lewis sets forth in the following passage (italics the author's):

> Just as 'If D then (certainly) P, and D is the fact', leads to the categorical consequence, 'Therefore (certainly) P'; so too, 'If D then probably P, and D is the fact', leads to a categorical consequence expressed by 'It is probable that P'. And this conclusion is not merely the statement over again of the probability relation between 'P' and 'D'; any more than 'Therefore (certainly) P' is the statement over again of 'If D then (certainly) P'. 'If the barometer is high, tomorrow will probably be fair; and the barometer *is* high', categorically assures something expressed by 'Tomorrow will probably be fair'. This probability is still relative to the grounds of judgment; but if these grounds are actual, and contain all the available evidence which is pertinent, then it is not only categorical but may fairly be called *the* probability of the event in question (1946: 319).

This position seems to me to be open to just those objections suggested in the main text. If 'P' is a statement, then the expressions 'certainly P' and 'probably P' as envisaged in the quoted passage are not statements. If we ask how one would go about trying to ascertain whether they were true, we realize that we are entirely at a loss unless and until a reference set of statements or assumptions has been specified relative to which P may then be found to be certain, or to be highly probable, or neither. The expressions in question, then, are essentially incomplete; they are elliptic formulations of relational statements; neither of them can be the conclusion of an inference. However plausible Lewis's suggestion may seem, there is no analogue in inductive logic to *modus ponens*, or the "rule of detachment," of deductive logic, which, given the information that 'D' and also 'if D then P', are true statements, authorizes us to detach the consequent 'P' in the conditional premise and to assert it as a self-contained statement which must then be true as well.

At the end of the quoted passage, Lewis suggests the important idea that 'probably P' might be taken to mean that the total relevant evidence available at the time confers high probability upon P. But even this statement is relational in that it tacitly refers to some unspecified time, and, besides, his general notion of a categorical probability statement as a conclusion of an argument is not made dependent on the assumption that the premises of the argument include all the relevant evidence available.

It must be stressed, however, that elsewhere in his discussion, Lewis emphasizes the relativity of (logical) probability, and, thus, the very characteristic that rules out the conception of categorical probability statements.

Similar objections apply, I think, to Toulmin's construal of probabilistic arguments; cf. Toulmin (1958) and the discussion in Hempel (1960), sects. 1–3.

3. R. Carnap, *Logical Foundations of Probability* (Chicago, 1950), 211. The requirement is suggested, e.g., in the passage from Lewis quoted in n. [2]. Similarly Williams speaks of "the most fundamental of all rules of probability logic, that 'the' probability of any proposition is its probability in relation to the known premises and them only" (*The Ground of Induction* (Cambridge, Mass., 1947), 72).

I am greatly indebted to Professor Carnap for having pointed out to me in 1945, when I first noticed the ambiguity of probabilistic arguments, that this was but one of several apparent paradoxes of inductive logic that result from disregard of the requirement of total evidence.

S. F. Barker, *Induction and Hypothesis* (Ithaca, N.Y., 1957), 70–78, has given a lucid independent presentation of the basic ambiguity of probabilistic arguments, and a skeptical appraisal of the requirement of total evidence as a means of dealing with the problem. However, I will presently suggest a way of remedying the ambiguity of probabilistic explanation with the help of a rather severely modified version of the requirement of total evidence. It will be called the requirement of maximal specificity, and is not open to the same criticism.

4. Cf. Carnap, *Logical Foundations*, 211 and 494.

5. This idea is closely related to one used by H. Reichenbach, (cf. *The Theory of Probability* (Berkeley, Calif., and Los Angeles, 1949), sect. 72) in an attempt to show that it is possible to assign probabilities to individual events within the framework of a strictly statistical conception of probability. Reichenbach proposed that the probability of a single event, such as the safe completion of a particular scheduled flight of a given commercial plane, be construed as the statistical probability which the *kind* of event considered (safe completion of a flight) possesses within the narrowest

reference class to which the given case (the specified flight of the given plane) belongs, and for which reliable statistical information is available (e.g., the class of scheduled flights undertaken so far by planes of the line to which the given plane belongs, and under weather conditions similar to those prevailing at the time of the flight in question).

6. Reference to $s \cdot k$ rather than to k is called for because, as was noted earlier, we do not construe the condition here under discussion as requiring that all the explanans statements invoked be scientifically accepted at the time in question, and thus be included in the corresponding class K.

7. By its reliance on this general idea, and specifically on the requirement of maximal specificity, the method here suggested for eliminating the epistemic ambiguity of statistical explanation differs substantially from the way in which I attempted in an earlier study (Hempel, 'Deductive-Nomological vs. Statistical Explanation', esp. sect. 10) to deal with the same problem. In that study, which did not distinguish explicitly between the two types of explanatory ambiguity characterized earlier in this section, I applied the requirement of total evidence to statistical explanations in a manner which presupposed that the explanans of any acceptable explanation belongs to the class K, and which then demanded that the probability which the explanans confers upon the explanandum be equal to that which the total evidence, K, imparts to the explanandum. The reasons why this approach seems unsatisfactory to me are suggested by the arguments set forth in the present section. Note in particular that, if strictly enforced, the requirement of total evidence would preclude the possibility of any significant statistical explanation for events whose occurrence is regarded as an established fact in science: For any sentence describing such an occurrence is logically implied by K and thus trivially has the logical probability 1 relative to K.

■ | References

Barker, S. F., *Induction and Hypothesis* (Ithaca, N.Y.: Cornell University Press, 1957).

Carnap, R., *Logical Foundations of Probability* (Chicago: University of Chicago Press 1950; second, rev., edn. 1962).

Feigl, H., and Maxwell, G. (eds.), *Minnesota Studies in the Philosophy of Science*, iii (Minneapolis: University of Minnesota Press, 1962).

Hempel, C. G., 'Inductive Inconsistencies', in *Synthese* 12 (1960): 439–69, repr. in *Aspects of Scientific Explanation* (New York and London: Free Press and Collier Macmillan, 1965), 53–79.

——, 'Deductive-Nomological vs. Statistical Explanation', in Feigl and Maxwell (eds.), 98–169.

Lewis, C. I., *An Analysis of Knowledge and Valuation* (La Salle, Ill.: Open Court, 1946).

Reichenbach, H., *The Theory of Probability* (Berkeley, Calif., and Los Angeles: The University of California Press, 1949).

Scheffler, I., 'Explanation, Prediction, and Abstraction', *The British Journal for the Philosophy of Science* 7 (1957): 293–309.

Toulmin, S., *The Uses of Argument* (Cambridge: Cambridge University Press, 1958).

Williams, D. C., *The Ground of Induction* (Cambridge, Mass.: Harvard University Press, 1947).

PETER RAILTON

A Deductive-Nomological Model of Probabilistic Explanation

What if some things happen by chance—can they nonetheless be explained? How?

Some things *do* happen by chance, according to the dominant interpretation of our present physical theory, the probabilistic interpretation of quantum mechanics. Nonetheless, they can be explained: by that theory, in virtually the same way as deterministic phenomena—deductive-nomologically. At least, that is what I hope to show in this essay.

Our universe may not be deterministic, but all is not chaos. It is governed by laws of two kinds: probabilistic (such as the laws concerning barrier penetration and certain other quantum phenomena) and non probabilistic (such as the laws of conservation of mass-energy, charge, momentum, etc.).[1] Were the probabilism of laws of the first sort remediable by suitable elaboration of laws of the second sort, the universe would be deterministic after all, and the problem of explaining chance phenomena would no longer be with us. However, indications are that physical indeterminism is irremediable, and that the universe exhibits not only chances, but lawful chances. I will argue that we come to understand chance phenomena, even when the chance involved is extremely remote, by subsuming them under these irremediably probabilistic laws.

1 | Introductory Remarks on Explanation

Do I offer a deductive-nomological (D-N) model of probabilistic explanation because I believe that nomic subsumption always explains?—No. There are familiar-enough kinds of non-explanatory D-N arguments, for example, those that deduce the explanandum from nomically-related symptoms or after-the-fact conditions alone, citing no causes.

FROM *Philosophy of Science* 45 (1978): 206–26.

Yet it will not do simply to add to the D-N model a requirement that the explanans contain causes whenever the explanandum is a particular fact. First, some particular facts may be explained non-causally, for example, by subsumption under structural laws such as the Pauli exclusion principle.* Second, even where causal explanation is called for, the existence of general, causal laws that cover the explanandum has not always been sufficient for explanation: the search for explanation has also taken the form of a search for mechanisms that underlie these laws. 'Mechanisms', however, is not meant to suggest a parochial attitude toward the nomic connections— deterministic or otherwise—that tie the world together and make explanation possible.

An example may help clarify the notion of mechanism appealed to here. The following D-N argument suffices to forecast *that* nasty weather lies ahead, but not to explain *why* this is so:

S   The glass is falling.
    Whenever the glass falls the weather turns bad.
    The weather will turn bad. ([5], p. 106)

Now nothing works like a barometer for predicting the weather, but nothing like a barometer works for changing it. So it is often maintained that (S) lacks explanatory efficacy because barometers lack the appropriate causal efficacy. The following inference, then, remedies the lack of the first because "it proves that the fact is a fact by citing causes and not mere symptoms" ([5], p. 107):

C   The glass is falling.
    Whenever the glass is falling the atmospheric pressure is falling.
    Whenever the atmospheric pressure is falling the weather turns bad.
    The weather will turn bad. ([5], p. 106)

---

* Wolfgang Pauli proposed the exclusion principle in 1925 to solve a problem in spectroscopy by requiring that no two electrons bound in the same atom can have the same set of quantum numbers. In 1926, Fermi and Dirac (independently) generalized the principle by requiring that the wave function of a multi-electron system must be antisymmetric. Originally formulated for electrons, the exclusion principle is now known to apply to all fermions (particles having half-integer spin). Among other things, it explains why the electrons surrounding the nuclei of different elements are arranged in shells rather than all occupying the state of lowest energy closest to the nucleus. As Railton points out, this kind of explanation is not causal: the principle is a purely formal or structural law. Other structural laws include the conservation principles of energy and momentum, Einstein's principle that the velocity of light is constant in all inertial frames, and the second law of thermodynamics.

Yet as explanations go, (C) is also lacking: we remain in the dark as to *why* the weather will turn bad. No connection between cause and effect, no mechanism by which falling atmospheric pressure produces a change for the worse in the weather, has been revealed. I do not doubt that some account of this mechanism exists; my point is that its existence is what makes (C) superior to (S) for explanatory purposes.

(C), if moderated by boundary conditions and put less qualitatively, would supply us the capability to predict *and* control the weather (whenever, as in a laboratory simulator, we can manipulate the atmospheric pressure). While prediction and control may exhaust our practical problems in the natural world, the unsatisfactoriness of (C) shows that explanation is an activity not wholly practical in purpose. The goal of understanding the world is a theoretical goal, and if the world is a machine—a vast arrangement of nomic connections—then our theory ought to give us some insight into the structure and workings of the mechanism, above and beyond the capability of predicting and controlling its outcomes. Until supplemented with an account of the nomic links connecting changes in atmospheric pressure to changes in the weather, (C) will explain but poorly. Knowing enough to subsume an event under the right kind of laws is not, therefore, tantamount to knowing the *how* or *why* of it. As the explanatory inadequacies of successful practical disciplines remind us: explanations must be more than potentially-predictive inferences or law-invoking recipes.

Is the deductive-nomological model of explanation therefore unacceptable?—No, just incomplete. Calling for an account of the mechanism leaves open the nature of that account, and as far as I can see, the model explanations offered in scientific texts are D-N when complete, D-N sketches when not. What is being urged is that D-N explanations making use of true, general, causal laws may legitimately be regarded as unsatisfactory unless we can back them up with an account of the mechanism(s) at work. "An account of the mechanism(s)" is a vague notion, and one obviously admitting of degrees of thoroughness, but I will not have much to say here by way of demystification. If one sees what is lacking in (C)—a characterization, whether sketchy or blow-by-blow, of how it is that declining atmospheric pressure effects the changes we describe as "a worsening of the weather," that is, a more or less complete filling-in of the links in the causal chains—one has the rough idea.

The D-N probabilistic explanations to be given below do not explain by giving a deductive argument terminating in the explanandum, for it will be a matter of chance, resisting all but *ex post facto* demonstration. Rather, these explanations subsume a fact in the sense of giving a D-N account of the chance mechanism responsible for it, and showing that our theory implies the existence of some physical possibility, however small, that this mechanism will produce the explanandum in the circumstances given. I hope the remarks just made about the importance of revealing mechanisms have eased the way for an account of probabilistic explanation that focuses on the indeterministic mechanisms at work, rather than the "nomic expect-ability" of the explanandum.

2 | Hempel's Inductive-Statistical Model

For Hempel, a statistical explanation (what is called elsewhere in this paper 'a probabilistic explanation') is one that "makes essential use of at least one law or theoretical principle of statistical form" ([3], p. 380). Since Hempel distinguishes between statistical laws and mere statistical generalizations, and asserts that the former apply only where "peculiar, namely probabilistic, modes of connection" exist among the phenomena ([3], p. 377), his charac-terization permits statistical explanation only of genuinely indeterministic processes.[2] Were some process to have the appearance of indeterminism owing to arcane workings or uncontrolled initial conditions, then no "pecu-liar . . . probabilistic" modes of connection would figure essentially in explaining this "pseudo-random" process's outcomes. Not only would statisti-cal explanation be unnecessary for such a process, it would be impossible: no probabilistic *laws* would govern it.

For example, it has been observed that 99% of all cases of infectious mononucleosis involve lymph-gland swelling. The exceptions might be due to a process that randomly misfires 1% of the time. Or, they might arise from the operation of an unknown deterministic mechanism that works to inhibit swelling whenever a patient begins in a particular initial condition, which as a mere matter of fact is typical of 1% of the population. If initial conditions could be partitioned into two mutually exclusive and jointly exhaustive classes S and –S, such that all Ss by law eventually develop swelling, and all –Ss do not, the generalization "99% of all cases of infectious mononucleosis develop lymph-gland swelling" would have been shown to be no law, but merely a descriptive report of observed rela-tive frequencies. No law, it cannot support a statistical explanation. But discovering it not to be a law is just discovering that statistical explanation is uncalled for, since each case of mononucleosis will have been of type S or type –S from the outset.

On the other hand, suppose that no such partition of initial conditions exists. Then the presence or absence of swelling is presumably due to a "peculiar . . . probabilistic" connection between disease and symptom, that is, a real causal indeterminism with probability .99 in each case to produce swelling. The generalization in question would thus be nomological, creat-ing both the possibility and the necessity of statistical explanation.

Given such genuine statistical laws, how does Hempel claim statistical explanation should proceed? He begins his account by distinguishing two sorts of statistical explanation. The first, *deductive-statistical* (D-S) explana-tion, involves "the deductive subsumption of a narrow statistical uniformity under more comprehensive ones" ([3], p. 380). The second, he argues, is of a qualitatively different sort:

Ultimately . . . statistical laws are meant to be applied to particular occurrences and to establish explanatory and predictive connections among them.
([3], p. 381)

To make such laws relevant to "particular occurrences," Hempel believes
we must go beyond the reach of deduction, and so he proposes an inductive
model of statistical explanation.

Inductive-statistical (I-S) explanation proceeds by adducing statistical
laws and associated initial conditions relative to which the explanandum is
highly probable. High relative probability is required because, on Hempel's
view, statistical laws become explanatorily relevant to an individual chance
event only by giving us a basis upon which to inductively infer its occurrence with "practical certainty." Yet although an I-S explanation shows the
explanandum to have been "nomically expectable" relative to the explanans,
it does not permit detachment of a conclusion; it is less an inference than the
expression of an inferential relationship: the explanandum receives a high
degree of epistemic support from the explanans. If, for example, we learn
that Jones has contracted infectious mononucleosis, we may infer with practical certainty that he will develop lymph-gland swelling. The same inference serves as an I-S explanation of the swelling, should it occur. Should it
not occur, we would have no explanation for *this*, on Hempel's model.

However, further investigation of Jones' medical history might reveal
that he suffered mononucleosis once before, and failed to develop any
swelling. Let us suppose that such individuals have a much higher than
normal probability of *not* showing swelling in any later bouts with mononucleosis, say .9 rather than .01. This new law and new information about
Jones together permit an inference with practical certainty to the conclusion that he will *not* develop swelling, and thus support a corresponding I-S
explanation. Relative to these new facts, however, no I-S explanation would
be available should Jones, improbably, develop swelling. What are we to say
now about the previous I-S explanation, which had just the opposite result?
Hempel would reject it as no longer *maximally specific* relative to what we
believe about Jones' case. The requirement of maximal specificity is a complicated affair,[3] but the basic idea is that we refer each case to the narrowest
class of cases to which our present beliefs assign it in which the explanandum has a characteristically different probability. In Jones' case, the narrower class is clearly the class of those contracting mononucleosis for a
second time who failed to develop lymph-gland swelling the first time.

If more information about Jones or new discoveries about mononucleosis turn up, we may be forced to move on to still another explanation. I-S
explanations must be relativized to our current "epistemic situation," and are
subject to change along with it. Hempel notes that this sets off I-S explanations from D-N and D-S explanations in a fundamental way:

> . . . *the concept of statistical explanation for particular events is essentially*
> *relative to a given knowledge situation as represented by a class K of accepted*

statements. . . . [W]e can significantly speak of true D-N and D-S explanations: they are those potential D-N and D-S explanations whose premises (and hence also conclusions) are true—no matter whether this happens to be known or believed, and thus no matter whether the premises are included in K. But this idea has no significant analogue for I-S explanation. . . . ([3], pp. 402–3)

On Hempel's view, neither of the two contradictory explanations concerning Jones contains false premises, and the explananda in each case do indeed receive the degree of support indicated. It is just that we no longer regard the evidential relationship expressed by the first as explanatorily relevant. Were Jones to develop swelling after all, it would now have to be regarded as inexplicable.

What I take to be the two most bothersome features of I-S arguments as models for statistical explanation—the requirement of high probability and the explicit relativization to our present epistemic situation (bringing with it an exclusion of questions about the truth of I-S explanations)—derive from the inductive character of such inferences, not from the nature of statistical explanation itself. If a non-inductive model for the statistical explanation of particular facts is given, there need be no temptation to require high probability or exclude truth.

3 | Jeffrey's Criticism of I-S Explanation

Richard C. Jeffrey has criticized Hempel's account on the grounds that statistical explanation is not a form of inference at all, except when the probability of the explanandum is "so high as to allow us to reason, in *any* decision problem, as if its probability were 1" ([5], p. 105). For such exceptional, "beautiful" cases, Jeffrey accepts I-S inferences as explanatory because they provide virtual "proof that the phenomenon *does* take place" ([5], p. 106).

For unbeautiful cases, there is no way of proving (in advance) that the explanandum phenomenon will occur. According to Jeffrey, the explanation *why* such unbeauties come to be is a curt "By chance." He has more to say on *how* they come about:

. . . in the statistical case I find it strained to speak of knowledge *why* the outcome is such-and-such. I would rather speak of *understanding the process*, for the explanation is the same no matter what the outcome: it consists of a statement that the process is a stochastic one, following such-and-such a law.[4] ([5], p. 24)

Jeffrey is surely right, as against Hempel, that probable and improbable outcomes of indeterministic processes are equally explicable, and explicable in the same way. After all, why should it be explicable that a genuinely random wheel of fortune with 99 red stops and 1 black stop came to a halt on red, but inexplicable that it halted on black? Worse, on Hempel's view, halting at any *particular* stop would be inexplicable, even though the wheel must halt at some particular stop in order to yield the explicable outcome *red*.

But I fail to see how Jeffrey can defend his exemption of beautiful cases against a similar line of argument. If the burden in statistical explanation really lies with *"understanding the process* . . . no matter what the outcome,"* then why should it matter whether the outcome is so highly probable "as to allow us to reason, in *any* decision problem, as if its probability were 1?" The neglect Jeffrey shows here toward minute chances is appropriate for the practical task of decision-making (and perhaps explained by his generally subjectivist approach to probability), but we must not overlook them in the not-entirely-practical task of explaining. Virtually impossible events may occur, and they deserve and can receive the same explanation as the merely improbable or the virtually certain.

4 | A D-N Model of Probabilistic Explanation

I will present my account of probabilistic explanation by developing an example of just such "practically negligible"—but physically real and lawful— chance: alpha-decay in long-lived radioactive elements. The mean-life of the more stable radionuclides is so long as to make the probability for any particular nucleus of such an element to decay during our lifetimes effectively zero. But our nuclear theory shows that it is *not* zero, and explains how such rarities can occur.

On the account offered here, probabilistic explanations will be either true or false independent of our epistemic situation. Moreover, to explain, they must be true. Here I am following Hempel's usage in calling an explanatory argument *true* just in case it is valid and its premises are true ([3], p. 338). Such an explanation will *not* be true if the probabilistic laws it invokes are not true; in particular, it will not be true unless the process responsible for the explanandum is genuinely indeterministic. If alpha-decay is to serve as our paradigm for probabilistic explanation, we must be correct in assuming that the probabilistic wave-mechanical account of particle transmission through the nuclear potential barrier tells us all there is to know about the cause of alpha-decay. At least, it must be true that there are no hidden variables characterizing unknown initial conditions that suffice to account for alpha-decay deterministically. However, I take it to be uncontroversial that alpha-decay is an indeterministic process, if any is.

Let us suppose that we are given an individual instance of alpha-decay to explain: a nucleus of radionuclide uranium^{238}, call it 'u', has emitted an alpha-particle during the time interval lasting from t_0 to $t_0 + \theta$, where θ is very small and expressed in standard units. Since the mean-life of U^{238} is

6.5×10^9 years, the probability of observing a decay by u during this interval is exceedingly small, but unquestionably exists (witness the decay). This probability can be given precisely by using the radioactive decay constant λ_{238} characteristic of all atoms of U^{238}. Significantly, we need not know when in the course of the history of u time t_0 occurs: the probability of decay is unaffected by the age of the atom. Therefore, as long as decay has not yet occurred, individual "trials"—consisting of observing a single isolated radioactive nucleus for successive intervals of the same length—are statistically independent. Using these two facts we can determine the probability of decay for individual nuclei during any time interval chosen: it will be 1 minus the probability that any such nucleus *survives* the interval intact; for u, $(1 - \exp(-\lambda_{238} \cdot \theta))$.*

To obtain experimental confirmation of this value, we infer *from* the probability to decay of individual nuclei *to* statistical features of sample populations of nuclei, for example, half-life and mean-life. These predicted statistical features are then checked against actual observed relative frequencies in large populations over long intervals. *Physical* probabilities of the sort being considered here are therefore to be contrasted with *statistical* probabilities; the former express the strength of a certain physical possibility for a given system, while the latter reduce to claims about the (limiting) relative frequencies of traits in sample populations. Much well-founded doubt has been expressed about the applicability of statistical probabilities to single cases, but physical probabilities are *located* in the features of the single case. Therefore, we can understand our nuclear theory as implying strictly universal (physical) probability-attributing laws of the form:

1 All nuclei of radioelement E have probability $(1 - \exp(-\lambda_E \cdot t))$ to emit an alpha-particle during any time interval of length t, unless subjected to environmental radiation.

Because schema (1) is universal in form, its instances are candidates for law premises in deductive-nomological inferences concerning individual nuclei. Thus, for u:

2 a All nuclei of U^{238} have probability $(1 - \exp(-\lambda_{238} \cdot \theta))$ to emit an alpha-particle during any interval of length θ, unless subjected to environmental radiation.

   b u was a nucleus of U^{238} at time t_0, and was subjected to no environmental radiation before or during the interval $t_0 - (t_0 + \theta)$.

   c u had probability $(1 - \exp(-\lambda_{238} \cdot \theta))$ to emit an alpha-particle during the interval $t_0 - (t_0 + \theta)$.

---

* In this exponential decay formula, *exp* stands for e, the base of natural logarithms. Thus, the formula should be read as: 1 minus e raised to the power of minus lambda times theta.

(2), it appears, gives a D-N explanation only of the fact that u had such-and-such a probability to decay during the interval in question, but we should look a bit closer. I submit that (2), when supplemented as follows, is the probabilistic explanation of u's decay:

3   A derivation of (2a) from our theoretical account of the mechanism at work in alpha-decay.
    The D-N inference (2).
    A parenthetic addendum to the effect that u did alpha-decay during the interval $t_0 - (t_0 + \theta)$.

Am I merely making a virtue of necessity, and saying that since (3) contains all we can say about u's decay, (3) must explain it? In fact, there is a great deal more we could say about u's decay. Deliberately left out of (3) are innumerable details about the experimental apparatus (temperature, pressure, location, etc.), about the beliefs and expectations of those monitoring the experiment, and about the epistemic position of the scientific community at the time. These facts are omitted as *explanatorily irrelevant* to u's decay because they are *causally irrelevant* to the physical possibility for decay that obtained during the interval in question, and to whether or not that possibility was realized.[5] A full account of these notions of explanatory and causal relevance is not possible here, so instead I will go on to argue that what (3) comprises *is* explanatorily relevant, and explanatory.

I must begin this task with a defense of the nomological status of (2a), and of the legitimacy of treating it as a covering law for u's decay. The following criterion of nomologicality will be used: a law is a universal truth derivable from our theory without appeal to particular facts. This criterion of course lacks generality (what counts as theory if not the laws themselves?), fails to segregate natural from logical laws, picks out only so-called "universal" (as opposed to "local") laws, and is entirely too vague (how to distinguish "particular facts" from the rest?). But I trust it will do for now. The motive for excluding "particular facts" is that some true, universal statements derivable from our theory *plus* particular facts would not normally be regarded as universal laws, but would at best be "local laws," for example, "All *Homo neanderthalensis* live during the late Pleistocene age."

The generalization in question here, (2a), is derived by solving the Schrödinger wave equation for an alpha-particle of energy $\approx 4.2 \, \text{MeV}$ for the potential regions in and around the nucleus of an element with atomic number 92 and atomic weight 238, none of which are "particular facts," plus some simplifying assumptions about the structure of the nucleus and the distinctness of the alpha-particle within it prior to decay. While it is forbidden by classical physics for a low-energy particle like the $\approx 4.2 \, \text{MeV}$ alpha-particle associated with U^{238} to pass through the 24.2 MeV potential barrier surrounding so massive a nucleus, the quantum theory predicts that the probability amplitude for finding such an alpha-particle outside the potential barrier is non-zero. Thus a transmission coefficient for U^{238} alpha-particles is determined, which, given certain simplifying assumptions about the goings-on inside the nucleus, yields the probability that such a particle will tunnel out of the potential well "per unit time for one nucleus," namely, λ_{238} ([1], p. 175). (2a) thus neither reports a summary of past observations nor expresses a mere statistical uniformity that scattered initial conditions would lead us to anticipate. Instead, it is a law of irreducibly probabilistic form, assigning definite, physically determined probabilities to individual systems.

It follows that the derivation of conclusions from (2a) by universal instantiation and *modus ponens* is unexceptionable.* Were (2a) but a statistical generalization, properly understood as meaning "$(1 - \exp(- \lambda_{238} \cdot \theta))N$ of U^{238} nuclei in samples of sufficiently large size N, on average, decay during the interval $t_0 - (t_0 + \theta)$," it could not undergo universal instantiation, and would not permit detachment of a conclusion about the probability obtaining in a single case.

Further, if the wave equation does indeed tell us all there is to know about the mechanism involved in nuclear barrier penetration, it follows that nothing more can be said to explain why the observed decay of *u* took place, once we have shown how (2a) is derived from our account of this mechanism, and established that (2) is valid and that (3)'s parenthetic addendum is true.

Still, does (3) explain why the decay took place? It does not explain why the decay *had* to take place, nor does it explain why the decay *could be expected to* take place. And a good thing, too: there is no *had to* or *could be expected to* about the decay to explain—it is not only a chance event, but a very improbable one. (3) does explain why the decay *improbably* took place, which is how it did. (3) accomplishes this by demonstrating that there existed at the time a small but definite physical possibility of decay, and noting that, by chance, this possibility was realized. The derivation of (2a) that begins (3) shows, by assimilating alpha-decay to the chance process of potential barrier tunneling, how this possibility comes to exist. If alpha-decays are chance phenomena of the sort described, then once our theory has achieved all that (3) involves, it has explained them to the hilt, however unsettling this may be to *a priori* intuitions. To insist upon stricter subsumption of the explanandum is not merely to demand what (alas) cannot be, but what decidedly should not be: sufficient reason that one probability rather than another be realized, that is, chances without chance.

Because of the peculiar nature of chance phenomena, it is explanatorily relevant whether the probability in question was realized, even though there is no before-the-fact explanatory *argument*, deductive or inductive, to this conclusion. Indeed, it is the absence of such an argument that makes a place in probabilistic explanation for a parenthetic addendum concerning

---

* See the discussion of this derivation on page 779.

whether the possibility became actual in the circumstances given. These addenda may offend those who believe that explanations must always be arguments, but at the most general level explanations are *accounts*, not arguments. It so happens that for indeterministic phenomena inferences of a particular kind—D-N arguments meeting the desiderata suggested in section 1—*are* explanatory accounts, and this for good reasons. However, indeterministic phenomena are a different matter, and explanatory accounts of them must be different as well. If the present model is accepted, then almost all of the explanatory burden in probabilistic explanation can be placed on deductive arguments—those characterizing the indeterministic mechanism and those attributing a certain probability to the explanandum. But these arguments leave out a crucial part of the story: did the chance fact obtain?

The parenthetic addendum fills this gap in the account, and communicates information that is relevant to the causal origin of the explanandum by telling us that it came about as the realization of a particular physical possibility. Further, it permits us to chain probabilistic explanations together to make more comprehensive explanations, in which each link is able to bear the full explanatory burden for the fact it covers, and is capable of leading us on to the next fact in the causal sequence being explained. From (2) alone we cannot move directly to an account of what the alpha-particle did to a nearby photographic plate, but only to a probability (and a miserably low one) that this account will be true. The parenthetic addendum to (3) furnishes a non-probabilistic premise from which to begin an account of the condition of the photographic plate: the occurrence of an alpha-decay in the vicinity. Dropping off the addendum leaves an explanation, but it is a D-N explanation of the occurrence of a particular probability, not a probabilistic explanation of the occurrence of a particular decay.

The scheme for probabilistic explanation of particular chance facts by nomic subsumption that is being offered here, the *deductive-nomological-probabilistic* (D-N-P) model, is this. First we display (or truthfully claim an ability to display) a derivation from our theory of a law of essentially probabilistic form, complete with an account of how the law applies to the indeterministic process in question. The derived law is of the form:*

4a $(t)(x)[F_{x,t} \rightarrow \text{Prob } (G)_{x,t} = p]$

"At any time, anything that is F has probability p to be G."

Next, we adduce the relevant fact(s) about the case at hand, e:

4b F_{e,t_0}

"e is F at time t_0."

---

* We have changed Railton's notation for the universal quantifier.

and draw the obvious conclusion:

4c $\text{Prob}(G)_{e,t_0} = p$

"*e* has probability *p* to be G at time t_0."

To which we add parenthetically, and according to how things turn out:

4d $(G_{e,t_0} / {-} G_{e,t_0})$

"(*e* did/did not become G at t_0)."

Whether a D-N-P explanation is true will depend solely upon the truth-values of its premises and addendum, and the validity of its logic. I leave open what becomes of a D-N-P explanation that contains true laws, initial facts, and addendum, but botches the theoretical account of the laws invoiced. Let us simply say that the more botched, the less satisfactory the explanation.

The law premise (4a) will be true if all things at all times satisfy the conditional '$F_{x,t} \to \text{Prob}(G)_{x,t} = p$', using whatever reading of '$\to$' we decide upon for the analysis of natural laws in general. It will be false if there exists a partition of the *F*s into those with *physical* probability *r* to be G and those with *physical* probability *s* to be G, where $s \neq r \neq p$. Such a partition might exist according to some *other* interpretation of probability, but this would not affect the truth of (4a). For example, suppose that a coin toss meeting certain specifications is an indeterministic event with probability ½ of yielding heads. We now perform the experiment of repeating such a toss a great many times. Curiously, all and only even numbered tosses yield heads. This result supplies certain frequentists with grounds for saying that Prob(heads, even-numbered toss) = 1, while Prob(heads, odd-numbered toss) = 0.[6] But because all tosses met the specification laid down, the probability of heads was the same, ½, on each toss, despite the curious behavior. Such behavior may make us suspicious of our original claims about the indeterminacy of the process or about the physical probability it has of producing heads, but is no proof against them. Indeed, the original probability attribution requires us to assign a definite physical probability to just such an untoward sequence of outcomes, the occurrence of which therefore hardly contradicts this attribution.

The particular fact premise (4b) will be true iff [if and only if] *e* is an *F* during the time in question, and not either an F^* (with probability $r \neq p$ to be G) or an F^{**} (with probability $q = p$ to be G, but unlike an *F* in other respects). Using the (let us say) true law that all F^{**}s have probability $q = p$ to be G, and the falsehood that *e* is an F^{**}, we could derive a true conclusion, indistinguishable from (4c). Hence the requirement that the *premises* be true if the argument is to explain; and if we reason logically from true premises, the conclusion will take care of itself.

5 | Epistemic Relativity and Maximal Specificity Disowned

Have I kept my promise to give an account of probabilistic explanation free from relativization to our present epistemic situation?

Let us return to explanation (3), and admit that it is not the whole story: 23% of the alpha-particles emitted by U^{238} have kinetic energy 4.13 MeV, while the remaining 77% have 4.18 MeV. Therefore there are two different decay constants, and both are distinct from λ_{238}, used in (3). Hence we must be quite careful in stating what exactly (3) explains. It does *not* explain the particular *event* observed, for this was either a 4.13 or a 4.18 MeV decay, neither of which has probability λ_{238} in unit time. Instead, (3) explains the particular *fact about* the event observed that we set out to explain, namely, that an alpha-decay with unspecified energy (or direction, or angular momentum, etc.) took place at nucleus u during the time interval in question. This fact *does* have probability λ_{238} of obtaining in unit time, representing the sum of the two energy-correlated probabilities with which such a decay might occur.

If we should learn that the decay of u was of a 4.18 MeV alpha-particle, an explanation of *this* fact would have to be referred to the more specific class of decays with probability $\lambda_{238}^{4.18}$ in unit time. Is the maximal specificity requirement thereby resurrected? There is no need for it. (3) is not an unspecific explanation of this more specific fact, but a fallacious one. It would be logically corrupt to conclude from law (2a) that an individual U^{238} nucleus has probability $(1 - \exp(-\lambda_{238} \cdot \theta))$ to decay *with energy 4.18 MeV* during any interval of length θ, since (2a) says nothing whatsoever about decay energies. The only relevant conclusion to draw from (2a) is (2c), which remains true in the face of our more detailed knowledge about the event in question. Nor is law (2a) falsified by the discovery of a 23:77 proportional distribution of decay energies, and the associated difference in decay rates. For according to our nuclear theory, there is no difference in initial condition between a nucleus about to emit a 4.13 MeV alpha-particle and one about to emit a 4.18 MeV alpha-particle. It remains true that *all* U^{238} nuclei have probability λ_{238} to decay in unit time, but it is further true that all have probability $\lambda_{238}^{4.13}$ to decay one way, and probability $\lambda_{238}^{4.18}$ to decay another.

It must next be determined whether the existence of a difference in probability *due to* a difference in initial condition can be handled by the D-N-P model without appeal to a maximal specificity requirement. To permit consideration of possible epistemological complications, it will be assumed that neither the difference in probability nor the partition of initial conditions is known at the start.

Imagine that, although we do not know it, in virtue of certain permanent structural features 23% of all naturally-occurring U^{238} nuclei fall into a class P, and the remaining 77% into a class $-P$, such that only those in P

have any probability of emitting a 4.13 MeV alpha-particle, and only those in class $-P$ have any probability of emitting a 4.18 MeV alpha-particle. Suppose further that these two laws have been derived:

5 a All U^{238} nuclei of type P have probability $(1 - \exp(-\lambda_{238}^{4.13} \cdot t))$ to emit a 4.13 MeV alpha-particle during any time interval of length t, unless subjected to environmental radiation.

   b All U^{238} nuclei of type $-P$ have probability $(1 - \exp(-\lambda_{238}^{4.18} \cdot t))$ to emit a 4.18 MeV alpha-particle during any time interval of length t, unless subjected to environmental radiation.

Note that, by our assumptions, the specification of the kinetic energy of the particle (possibly) emitted may be dropped from (5a) and (5b) without altering the truth of either.

Until the structural differences between types P and $-P$ are discovered and understood, (3) will stand as the accepted explanation of u's decay. However, once (5a) and (5b) have become known, it will be clear from the fact that u's alpha-emission had kinetic energy 4.18 MeV that u must have been of type $-P$ prior to decay. Thus a more specific account of u's decay will be available to scientists, who, already familiar with the theoretical derivation of law (5b), offer the following truncated D-N-P version of this account:

6 a All nuclei of U^{238} of type $-P$ have probability $(1 - \exp(-\lambda_{238}^{4.18} \cdot \theta))$ to emit an alpha-particle during any time interval of length θ, unless subjected to environmental radiation.

   b u was a nucleus of U^{238} of type $-P$ at t_0, and was subjected to no environmental radiation before or during the interval $t_0 - (t_0 + \theta)$.

   c u had probability $(1 - \exp(-\lambda_{238}^{4.18} \cdot \theta))$ to emit an alpha-particle during the interval $t_0 - (t_0 + \theta)$.

   d (And it did.)

On the Hempelian model (modified so as to permit I-S explanations of improbable phenomena), there is no problem in accounting for the previous acceptability of the I-S counterpart of (3), or for its present unacceptability. (3) had been maximally specific relative to our previous beliefs about alpha-decay in U^{238}, but no longer is, and so is superseded by the more specific (relatively speaking) I-S counterpart of (6).

On the D-N-P model, too, there is no problem in accounting for the acceptability of (3) prior to the discovery of class $-P$ and law (5b): (3)'s premises (and, of course, addendum) were taken to be true. The question is whether, in light of current beliefs, (3) can be ruled out—and (6) ruled in—without invocation of Hempelian constraints. Resolution of the problem (3) and (6) pose through epistemic relativization and maximal specificity requirements seems to me unacceptable. If we were to attribute to nucleus u two unequal probabilities to alpha-decay in a specified way during a single time interval, adding, "Let's pick the most specifically defined value for explanatory purposes," we'd be showing an unseemly tolerance for contradiction in our nuclear theory—and why stop there? Better face up to the confrontation over truth between (3) and (6), and replace complex and unappealingly relativistic maximal specificity requirements with the simple requirement of truth. The D-N-P model does this. The current unacceptability of (3) is located not in premises insufficiently specific, but in premises insufficiently true, that is, false. Contrary to (3)'s purported covering law (2a), not all nuclei of U^{238} have probability λ_{238} to decay in unit time if unperturbed by radiation—in fact, none do. In spite of giving accurate expectation values for decay rates in large samples of U^{238}, (2a) is false, and so explanation (3) is ruled out as unsound. Explanation (6), on the other hand, meets the simple requirement of truth, and rules itself in.[7]

Problems about incomplete, misleading, or false beliefs do not bear on whether D-N-P explanations have unrelativized truth-values, but concern rather difficulties in *establishing* the truth-values they unrelativistically have. Relativization to our current epistemic situation comes into play only when we begin to discuss whether a given D-N-P explanation *seems* true. Whether it *is* true is another matter.

6 | Objections to the D-N-P Model

I cannot pretend to have said enough about deductive-nomological-probabilistic explanation to have characterized this model adequately. Such reservations as were expressed in section 1 about taking nomic subsumption under a causal law as sufficient for explanation are still in force, and little has been done—except by way of example—to show how the account offered here might accommodate them.

That the probabilistic laws invoked in D-N-P explanations are even (in some relevant sense) *causal* cannot be defended until a plausible account of physical probability has been worked out, a task well beyond the scope of this paper. Under a *propensity interpretation*, probability has the characteristics sought: a probability is the expression of the strength of a physical tendency in an individual chance system to produce a particular outcome; it is therefore straightforwardly applicable to single cases; and it is (in a relevant sense) causally responsible for that outcome whenever it is realized. However, propensities are notoriously unclear. For now I can at best assume that clarification is possible, point to a promising start in the attempt to do so—R. N. Giere, "Objective, Single-Case Probabilities and the Foundations of Statistics" ([2])—, and admit that the D-N-P model is viable only if sense can be made of propensities, or of objective, physical, lawful, single-case probabilities by any other name.

As for the requirement that explanations elucidate mechanisms, I can only repeat that an essential role is played in D-N-P explanations by the theoretical deduction of the probabilistic law(s) covering the explanandum.

In lieu of further exposition, I offer the beginnings of a defense, hoping thereby to sketch out the account a bit more fully in those areas most likely to be controversial.

BECAUSE IT APPLIES ONLY TO GENUINELY INDETERMINISTIC PROCESSES, OF WHICH THERE ARE FEW (IF ANY), D-N-P EXPLANATION IS TOO RESTRICTED IN SCOPE.

It is widely believed that the probabilities associated with standard gambling devices, classical thermodynamics, actuarial tables, weather forecasting, etc., arise not from any underlying physical indeterminism, but from an unknown or uncontrolled scatter of initial conditions. If this is right, then D-N-P explanation would be inapplicable to these phenomena even though they are among the most familiar objects of probabilistic explanation. I do not, however, find this troublesome: if something does not happen by chance, it cannot be explained by chance. The use of epistemic or statistical probabilities in connection with such phenomena unquestionably has instrumental value, and should not be given up. What must be given up is the idea that *explanations* can be based on probabilities that have no role in bringing the world's explananda about, but serve only to describe deterministic phenomena.[8] Whether there *are* any probabilities that enter into the mechanisms of nature is still debated, but the successes of the quantum-mechanical formalism, and the existence of "no hidden variable" results for it, place the burden of proof on those who would insist that physical chance is an illusion.

It could be objected more justly that D-N-P explanation is too broad, not too narrow, in scope. Once restrictions have been lifted from the value a chance may have in probabilistic explanation, virtually all explanations of particular facts must become probabilistic. All but the most basic regularities of the universe stand forever in peril of being interrupted or upset by intrusion of the effects of random processes. It might seem a fine explanation for a light's going out that we opened the only circuit connecting it with an electrical power source, but an element of chance was involved: had enough atoms in the vicinity of the light undergone spontaneous beta-decay at the right moment, the electrons emitted could have kept it glowing. The success of a social revolution might appear to be explained by its overwhelming popular support, but this is to overlook the revolutionaries' luck: if all the naturally unstable nuclides on earth had commenced spontaneous nuclear fission in rapid succession, the triumph of the people would never have come to pass.

No doubt this proliferation of probabilistic explanations is counterintuitive, but contemporary science will not let us get away with any other sort of explanation in these cases—it simply cannot supply the requisite non-probabilistic laws. Because they figure in the way things *work*, tiny probabilities appropriately figure in explanations of the way things *are*, even though they scarcely ever show up in the way things turn out.

THE D-N-P MODEL BREAKS THE LINK BETWEEN PREDICTION AND EXPLANATION.

Hempel has justified a "qualified thesis of the structural identity of explanation and prediction" with this principle:

> Any rationally acceptable answer to the question "Why did X occur?" must offer information which shows that X was to be expected—if not definitely, then at least with reasonable certainty. ([3], pp. 367–368)

Abundantly many D-N-P explanations—all those covering less than highly probable facts—violate this condition.

However, to abide by this condition and renounce explanations with meager probabilities I take to be worse. Why forgo the explanations of improbable phenomena offered by our theories, when these explanations provide as much of an account of why (and how) their explananda occur as do the explanations of "reasonably certain" phenomena that Hempel's condition sanctions?

Too restrictive as it stands, Hempel's condition may be taken in a way not incompatible with D-N-P explanation. A D-N-P explanation does yield one prediction that is perfectly strict, to the effect that a certain physical probability exists in the circumstances given. If this probability fails to obtain, or to have the value attributed to it, the explanation must be false. It is a complaint against the world, not against the D-N-P model, that a direct, non-statistical test for the presence or value of this probability may prove impossible. Remarkably, the mechanisms of the world leave room for spontaneous nuclear disintegrations. Equally remarkably, our physical theory gives us insight into how they come about, and assigns determinate probabilities to them. These probabilities are connected to the rest of our theory by laws that permit both prediction and (where means exist) control: if undisturbed, nucleus a will have probability p to alpha-decay (so we should expect a's decay with *epistemic* probability p); and if we wish to alter p, our theory tells us how a must be disturbed.

It has been objected to the view of probability taken in this paper that unless probability attributions are interpreted as predictions about how relative frequencies will *actually* come out in the long run, probabilistic laws lack empirical content. Thus if the relative frequency of decayed atoms in a large sample of some radioelement were, over a great length of time, to diverge significantly from the probability theoretically attributed to decay, that attribution would not be "borne out," that is, would be falsified. Otherwise, it is argued, probabilistic laws are compatible with all frequencies, and empirically vacuous.

But it is impossible for a world to "bear out" all of its probability-attributing laws in this sense. For these laws imply, among other things, that it is extremely unlikely that *all* actual long-run sequences will show a relative frequency near to the single-case probability. Therefore, the demand that all long-run decay rates nearly match all corresponding decay constants comes to a demand that nothing improbable show up in the long run, which is itself an improbability showing up in the long run. Intended to clear things up on the epistemological front, this proposal cannot even get out of its own way.

BY SPLITTING APART PROBABILISTIC EXPLANATION AND INDUCTION, THE D-N-P MODEL HAS LOST THE POINT OF PROBABILISTIC EXPLANATION.

Behind this objection lies the view that probabilistic (or statistical) explanation is an activity fundamentally unlike D-N explanation. A probabilistic explanation is seen as a piece of detective work. Unable to give a causal demonstration of the explanandum from evidence thus far assembled, we develop hypotheses, which are judged by how probable they are on the evidence, and whether they make the explanandum sufficiently probable. In the end, we put forward the most convincing inductive argument yet found—the one making the explanandum most antecedently probable, given what else we know about events leading up to it.

This view of probabilistic explanation confuses epistemic with objective probability, and induction with explanation. Perhaps responsible for this confusion is the similarity of the tasks of explaining a phenomenon, gathering support for such an explanation, and gathering before- or after-the-fact evidence for a phenomenon's occurrence. This confusion is abetted by misleading ways of talking about "strong" or "good" explanations. We should distinguish the following, (i) A *strong (good) explanation* is one that has great theoretical power, regardless of how well-confirmed it is or how probable it holds the explanandum to be. (ii) A *strong (good) candidate for explanation* is a proffered explanation with well-confirmed *premises*, regardless of how probable it holds the explanandum to be and irrespective of how theoretically powerful it happens to be. (iii) A *strong (good) reason for believing that the explanandum fact will obtain* is furnished by before-the-fact evidence that leads, via one's theory, to an expectation of the explanandum with high epistemic probability. (iv) A *strong (good) reason for believing that the explanandum fact obtained* is given by any evidence that lends high epistemic probability to the proposition that the explanandum fact is a fact. Strong after-the-fact evidence, even for very improbable events, may be easy to come by. Reasons of types (iii) and (iv) need have nothing to do with explanation, and may be based on symptoms (Will it rain today?—Harry's rheumatism is acting up) or even less causally relevant information (Was Sue upset?—Her brother is certain she would have been).

Although the link between probabilistic explanation and induction is looser on the D-N-P model than on the I-S model, this is no fault: on Hempel's account it was entirely too close. Measuring the strength or "acceptability" of an explanation by the magnitude of the probability it confers on the explanandum blurs the distinctions just made. Keeping (i)–(iv) distinct, the D-N-P model enables us to state quite simply the object of induction in explanation: given a particular fact, to find, and gather evidence for, an explanans that subsumes it; given a generalization, to find, and gather evidence for, a higher-level explanans that subsumes it; in all cases, then, to discover and establish a true and relevant explanans. The issue of showing the explanandum to have high (relative or absolute) probability is a red herring, distracting attention from the real issue: the truth or falsity, and applicability, of the laws and facts adduced in explanatory accounts.[9]

■ | Notes

1. Let us say rather loosely that a system is deterministic if, for any one instant, its state is physically compatible with only one (not necessarily different) state at each other instant. A system is indeterministic otherwise, but lawfully so if a complete description of its state at some one instant plus all true laws together entail a distribution of probabilities over possible states at later times.

2. Although there is some difficulty in reconciling all that is said in [3] with this conclusion. Hempel now accepts it (personal communication).

3. See, for example, [4].

4. A typographical error has been corrected.

5. Causal relevance is established here via the wave equation. I do not mean to suggest that *causal* relevance is the only explanatory kind; cf. the mention of structural laws in section 1.

   Some such notion of causal relevance appears to lie behind Salmon's "statistical-relevance" model of probabilistic explanation. Yet what matters is whether a factor enters into the probabilities present, not the statistics they produce.

6. Cf. the discussion of place selections and homogeneity in [6], sections 4 and 7. Salmon's criterion, which requires formal randomness, would here fail to distinguish a *randomly-produced* regular sequence from a *deterministically-produced* one. Notwithstanding formal similarities, only the latter is appropriately explained non-probabilistically.

7. Explanation (6) is true, however, only under the contrary-to-fact assumption—made for the sake of the example—of the existence of a class −*P*.

8. Of course, we might speak of statistical or epistemic probabilities as causes of, for example, beliefs. But if belief formation is not *physically* probabilistic, then

probabilistic explanation of it would be impossible, in spite of this sort of causal involvement on the part of statistical or epistemic probabilities.

9. I would like to thank C. G. Hempel, Richard C. Jeffrey, and David Lewis for helpful criticisms of earlier drafts. I am especially indebted to David Lewis for the idea that a propensity interpretation of probability sits best with the account of probabilistic explanation given here. I have greatly benefited from discussions of related matters with Sam Scheffler and David Fair.

■ | References

[1] Evans, R. D., *The Atomic Nucleus*. New York: McGraw-Hill, 1965.

[2] Giere, R. N., "Objective Single-Case Probabilities and the Foundations of Statistics." In *Logic, Methodology and Philosophy of Science*, vol. IV. Edited by P. Suppes et al. Amsterdam: North-Holland, 1973.

[3] Hempel, C. G., "Aspects of Scientific Explanation." In *Aspects of Scientific Explanation and Other Essays*. New York: Free Press, 1965.

[4] Hempel, C. G., "Maximal Specificity and Lawlikeness in Probabilistic Explanation." *Philosophy of Science* 35 (1968), 116–33.

[5] Jeffrey, R. C., "Statistical Explanation vs. Statistical Inference." In *Essays in Honor of C. G. Hempel.* Edited by N. Rescher et al. Dordrecht: D. Reidel, 1970.

[6] Salmon, W. C., "Statistical Explanation." In *Statistical Explanation and Statistical Relevance.* Edited by W. Salmon. Pittsburgh: University of Pittsburgh Press, 1971.

PHILIP KITCHER

Explanatory Unification

1 | The Decline and Fall of the Covering Law Model

One of the great apparent triumphs of logical empiricism was its official theory of explanation. In a series of lucid studies (Hempel 1965, Chapters 9, 10, 12; Hempel 1962; Hempel 1966), C. G. Hempel showed how to articulate precisely an idea which had received a hazy formulation from traditional empiricists such as Hume and Mill. The picture of explanation which Hempel presented, the *covering law model*, begins with the idea that explanation is derivation. When a scientist explains a phenomenon, he derives (deductively or inductively) a sentence describing that phenomenon (the *explanandum* sentence) from a set of sentences (the *explanans*) which must contain at least one general law.

Today the model has fallen on hard times. Yet it was never the empiricists' whole story about explanation. Behind the official model stood an unofficial model, a view of explanation which was not treated precisely, but which sometimes emerged in discussions of theoretical explanation. In contrasting scientific explanation with the idea of reducing unfamiliar phenomena to familiar phenomena, Hempel suggests this unofficial view: "What scientific explanation, especially theoretical explanation, aims at is not [an] intuitive and highly subjective kind of understanding, but an objective kind of insight that is achieved by a systematic unification, by exhibiting the phenomena as manifestations of common, underlying structures and processes that conform to specific, testable, basic principles" (Hempel 1966, p. 83; see also Hempel 1965, pp. 345, 444). Herbert Feigl makes a similar point: "The aim of scientific explanation throughout the ages has been *unification*, i.e., the comprehending of a maximum of facts and regularities in terms of a minimum of theoretical concepts and assumptions" (Feigl 1970, p. 12).

This unofficial view, which regards explanation as unification, is, I think, more promising than the official view. My aim in this paper is to develop

FROM *Philosophy of Science* 48 (1981): 507–31.

the view and to present its virtues. Since the picture of explanation which results is rather complex, my exposition will be programmatic, but I shall try to show that the unofficial view can avoid some prominent shortcomings of the covering law model.

Why should we want an account of scientific explanation? Two reasons present themselves. Firstly, we would like to understand and to evaluate the popular claim that the natural sciences do not merely pile up unrelated items of knowledge of more or less practical significance, but that they increase our understanding of the world. A theory of explanation should show us *how* scientific explanation advances our understanding. (Michael Friedman cogently presents this demand in his (1974).) Secondly, an account of explanation ought to enable us to comprehend and to arbitrate disputes in past and present science. Embryonic theories are often defended by appeal to their explanatory power. A theory of explanation should enable us to judge the adequacy of the defense.

The covering law model satisfies neither of these *desiderata*. Its difficulties stem from the fact that, when it is viewed as providing a set of necessary *and sufficient* conditions for explanation, it is far too liberal. Many derivations which are intuitively nonexplanatory meet the conditions of the model. Unable to make relatively gross distinctions, the model is quite powerless to adjudicate the more subtle considerations about explanatory adequacy which are the focus of scientific debate. Moreover, our ability to derive a description of a phenomenon from a set of premises *containing a law* seems quite tangential to our understanding of the phenomenon. Why should it be that exactly those derivations which employ laws advance our understanding?

The unofficial theory appears to do better. As Friedman points out, we can easily connect the notion of unification with that of understanding. (However, as I have argued in my (1976), Friedman's analysis of unification is faulty; the account of unification offered below is indirectly defended by my diagnosis of the problems for his approach.) Furthermore, as we shall see below, the acceptance of some major programs of scientific research— such as, the Newtonian program of eighteenth-century physics and chemistry, and the Darwinian program of nineteenth-century biology—depended on recognizing promises for unifying, and thereby explaining, the phenomena. Reasonable skepticism may protest at this point that the attractions of the unofficial view stem from its unclarity. Let us see.

2 | Explanation: Some Pragmatic Issues

Our first task is to formulate the problem of scientific explanation clearly, filtering out a host of issues which need not concern us here. The most obvious way in which to categorize explanation is to view it as an activity. In this activity we answer the actual or anticipated questions of an actual or anticipated audience. We do so by presenting reasons. We draw on the beliefs we hold, frequently using or adapting arguments furnished to us by the sciences.

Recognizing the connection between explanations and arguments, proponents of the covering law model (and other writers on explanation) have identified explanations as special types of arguments. But although I shall follow the covering law model in employing the notion of argument to characterize that of explanation, I shall not adopt the ontological thesis that explanations are arguments. Following Peter Achinstein's thorough discussion of ontological issues concerning explanation in his (1977), I shall suppose that an explanation is an ordered pair consisting of a proposition and an act type.[1] The relevance of arguments to explanation resides in the fact that what makes an ordered pair $(p$, explaining $q)$ an explanation is that a sentence expressing p bears an appropriate relation to a particular argument. (Achinstein shows how the central idea of the covering law model can be viewed in this way.) So I am supposing that there are acts of explanation which draw on arguments supplied by science, reformulating the traditional problem of explanation as the question: What features should a scientific argument have if it is to serve as the basis for an act of explanation?[2]

The complex relation between scientific explanation and scientific argument may be illuminated by a simple example. Imagine a mythical Galileo confronted by a mythical fusilier who wants to know why his gun attains maximum range when it is mounted on a flat plain, if the barrel is elevated at 45° to the horizontal. Galileo reformulates this question as the question of why an ideal projectile, projected with fixed velocity from a perfectly smooth horizontal plane and subject only to gravitational acceleration, attains maximum range when the angle of elevation of the projection is 45°. He defends this reformulation by arguing that the effects of air resistance in the case of the actual projectile, the cannonball, are insignificant, and that the curvature of the earth and the unevenness of the ground can be neglected. He then selects a kinematical argument which shows that, for fixed velocity, an ideal projectile attains maximum range when the angle of elevation is 45°. He adapts this argument by explaining to the fusilier some unfamiliar terms ('uniform acceleration', let us say), motivating some problematic principles (such as the law of composition of velocities), and by omitting some obvious computational steps. Both Galileo and the fusilier depart satisfied.

The most general problem of scientific explanation is to determine the conditions which must be met if science is to be used in answering an explanation-seeking question Q. I shall restrict my attention to explanation-seeking why-questions, and I shall attempt to determine the conditions under which an argument whose conclusion is S can be used to answer the question "Why is it the case that S?" More colloquially, my project will be that of deciding when an argument explains why its conclusion is true.[3]

We leave on one side a number of interesting and difficult issues. So, for example, I shall not discuss the general relation between explanation-seeking questions and the arguments which can be used to answer them, nor the pragmatic conditions governing the idealization of questions and the adaptation of scientific arguments to the needs of the audience. (For illuminating discussions of some of these issues, see Bromberger 1962.) Given that so much is dismissed, does anything remain?

In a provocative article (van Fraassen 1977), Bas van Fraassen denies, in effect, that there are any issues about scientific explanation other than the pragmatic questions I have just banished. After a survey of attempts to provide a theory of explanation he appears to conclude that the idea that explanatory power is a special virtue of theories is a myth. We accept scientific theories on the basis of their empirical adequacy and simplicity, and, having done so, we use the arguments with which they supply us to give explanations. This activity of applying scientific arguments in explanation accords with extra-scientific, "pragmatic", conditions. Moreover, our views about these extra-scientific factors are revised in the light of our acceptance of new theories: ". . . science schools our imagination so as to revise just those prior judgments of what satisfies and eliminates wonder" (van Fraassen 1977, p. 150). Thus there are no context-independent conditions, beyond those of simplicity and empirical adequacy which distinguish arguments for use in explanation.

van Fraassen's approach does not fit well with some examples from the history of science—such as the acceptance of Newtonian theory of matter and Darwin's theory of evolution—examples in which the explanatory promise of a theory was appreciated in advance of the articulation of a theory with predictive power. [See below, pp. 715–17.] Moreover, the account I shall offer provides an answer to skepticism that no "global constraints" (van Fraassen 1977, p. 146) on explanation can avoid the familiar problems of asymmetry and irrelevance, problems which bedevil the covering law model. I shall try to respond to van Fraassen's challenge by showing that there are certain context-independent features of arguments which distinguish them for application in response to explanation-seeking why-questions, and that we can assess theories (including embryonic theories) by their ability to provide us with such arguments. Hence I think that it is possible to defend the thesis that historical appeals to the explanatory power of theories involve recognition of a virtue over and beyond considerations of simplicity and predictive power.

Resuming our main theme, we can use the example of Galileo and the fusilier to achieve a further refinement of our problem. Galileo selects and adapts an argument from his new kinematics—that is, he draws an argument from a set of arguments available for explanatory purposes, a set which I shall call the *explanatory store*. We may think of the sciences not as providing us with many unrelated individual arguments which can be used in individual acts of explanation, but as offering a reserve of explanatory arguments, which we may tap as need arises. Approaching the issue in this way, we shall be led to present our problem as that of specifying the conditions which must be met by the explanatory store.

The set of arguments which science supplies for adaptation in acts of explanation will change with our changing beliefs. Therefore the appropriate *analysandum* is the notion of the store of arguments relative to a set of accepted sentences. Suppose that, at the point in the history of inquiry which interests us, the set of accepted sentences is K. (I shall assume, for simplicity's sake, that K is consistent. Should our beliefs be inconsistent then it is more appropriate to regard K as some tidied version of our beliefs.) The general problem I have set is that of specifying $E(K)$, the *explanatory store over K*, which is the set of arguments acceptable as the basis for acts of explanation by those whose beliefs are exactly the members of K. (For the purposes of this paper I shall assume that, for each K there is exactly one $E(K)$.)

The unofficial view answers the problem: for each K, $E(K)$ is the set of arguments which best unifies K. My task is to articulate the answer. I begin by looking at two historical episodes in which the desire for unification played a crucial role. In both cases, we find three important features: (i) prior to the articulation of a theory with high predictive power, certain proposals for theory construction are favored on grounds of their explanatory promise; (ii) the explanatory power of embryonic theories is explicitly tied to the notion of unification; (iii) particular features of the theories are taken to support their claims to unification. Recognition of (i) and (ii) will illustrate points that have already been made, while (iii) will point towards an analysis of the concept of unification.

3 | A Newtonian Program

Newton's achievements in dynamics, astronomy and optics inspired some of his successors to undertake an ambitious program which I shall call "dynamic corpuscularianism".[4] *Principia* had shown how to obtain the motions of bodies from a knowledge of the forces acting on them, and had also demonstrated the possibility of dealing with gravitational systems in a unified way. The next step would be to isolate a few basic force laws, akin to the law of universal gravitation, so that, applying the basic laws to specifications of the dispositions of the ultimate parts of bodies, all of the phenomena of nature could be derived. Chemical reactions, for example, might be understood in terms of the rearrangement of ultimate parts under the action of cohesive and repulsive forces. The phenomena of reflection, refraction and diffraction of light might be viewed as resulting from a special force of attraction between light corpuscles and ordinary matter. These speculations encouraged eighteenth-century Newtonians to construct very general hypotheses about inter-atomic forces—even in the absence of any confirming evidence for the existence of such forces.

In the preface to *Principia*, Newton had already indicated that he took dynamic corpuscularianism to be a program deserving the attention of the scientific community:

> I wish we could derive the rest of the phenomena of Nature by the same kind of reasoning from mechanical principles, for I am induced by many reasons to suspect that they may all depend upon certain forces by which the particles of bodies, by some causes hitherto unknown, are either mutually impelled towards one another, and cohere in regular figures, or are repelled and recede from one another. (Newton 1962, p. xviii. See also Newton 1952, pp. 401–2)

This, and other influential passages, inspired Newton's successors to try to complete the unification of science by finding further force laws analogous to the law of universal gravitation. Dynamic corpuscularianism remained popular so long as there was promise of significant unification. Its appeal began to fade only when repeated attempts to specify force laws were found to invoke so many different (apparently incompatible) attractive and repulsive forces that the goal of unification appeared unlikely. Yet that goal could still motivate renewed efforts to implement the program. In the second half of the eighteenth century Boscovich revived dynamic corpuscularian hopes by claiming that the whole of natural philosophy can be reduced to "one law of forces existing in nature."[5]

The passage I have quoted from Newton suggests the nature of the unification that was being sought. *Principia* had exhibited how one style of argument, one "kind of reasoning from mechanical principles", could be used in the derivation of descriptions of many, diverse, phenomena. The unifying power of Newton's work consisted in its demonstration that one *pattern* of argument could be used again and again in the derivation of a wide range of accepted sentences. (I shall give a representation of the Newtonian pattern in Section 5.) In searching for force laws analogous to the law of universal gravitation, Newton's successors were trying to generalize the pattern of argument presented in *Principia*, so that one "kind of reasoning" would suffice to derive all phenomena of motion. If, furthermore, the facts studied by chemistry, optics, physiology and so forth, could be related to facts about particle motion, then one general pattern of argument would be used in the derivation of all phenomena. I suggest that this is the ideal of unification at which Newton's immediate successors aimed, which came to seem less likely to be attained as the eighteenth century wore on, and which Boscovich's work endeavored, with some success, to reinstate.

4 | The Reception of Darwin's Evolutionary Theory

The picture of unification which emerges from the last section may be summarized quite simply: a theory unifies our beliefs when it provides one (or more generally, a few) pattern(s) of argument which can be used in the derivation of a large number of sentences which we accept. I shall try to develop this idea more precisely in later sections. But first I want to show how a different example suggests the same view of unification.

In several places, Darwin claims that his conclusion that species evolve through natural selection should be accepted because of its explanatory power, that ". . . the doctrine must sink or swim according as it groups and explains phenomena" (F. Darwin 1887; Vol. 2, p. 155, quoted in Hull 1974, p. 292). Yet, as he often laments, he is unable to provide any complete derivation of any biological phenomenon—our ignorance of the appropriate facts and regularities is "profound". How, then, can he contend that the primary virtue of the new theory is its explanatory power?

The answer lies in the fact that Darwin's evolutionary theory promises to unify a host of biological phenomena (C. Darwin 1964, pp. 243–4). The eventual unification would consist in derivations of descriptions of these phenomena which would instantiate a common pattern. When Darwin expounds his doctrine what he offers us is the pattern. Instead of detailed explanations of the presence of some particular trait in some particular species, Darwin presents two "imaginary examples" (C. Darwin 1964, pp. 90–96) and a diagram, which shows, in a general way, the evolution of species *represented by schematic letters* (1964, pp. 116–126). In doing so, he exhibits a pattern of argument, which, he maintains, can be instantiated, *in principle*, by a complete and rigorous derivation of descriptions of the characteristics of any current species. The derivation would employ the principle of natural selection—as well as premises describing ancestral forms and the nature of their environment and the (unknown) laws of variation and inheritance. In place of detailed evolutionary stories, Darwin offers *explanation-sketches*. By showing how a particular characteristic would be advantageous to a particular species, he indicates an explanation of the emergence of that characteristic in the species, suggesting the outline of an argument instantiating the general pattern.

From this perspective, much of Darwin's argumentation in the *Origin* (and in other works) becomes readily comprehensible. Darwin attempts to show how his pattern can be applied to a host of biological phenomena. He claims that, by using arguments which instantiate the pattern, we can account for analogous variations in kindred species, for the greater variability of specific (as opposed to generic) characteristics, for the facts about geographical distribution, and so forth. But he is also required to resist challenges that the pattern cannot be applied in some cases, that premises for arguments instantiating the pattern will not be forthcoming. So, for example, Darwin must show how evolutionary stories, fashioned after his pattern, can be told to account for the emergence of complex organs. In both aspects of his argument, whether he is responding to those who would limit the application of his pattern or whether he is campaigning for its use within a realm of biological phenomena, Darwin has the same goal. He aims to show that his theory should be accepted because it unifies and explains.

5 | Argument Patterns

Our two historical examples[6] have led us to the conclusion that the notion of an argument pattern is central to that of explanatory unification. Quite different considerations could easily have pointed us in the same direction. If someone were to distinguish between the explanatory worth of two arguments instantiating a common pattern, then we would regard that person as an explanatory deviant. To grasp the concept of explanation is to see that if one accepts an argument as explanatory, one is thereby committed to accepting as explanatory other arguments which instantiate the same pattern.

To say that members of a set of arguments instantiate a common pattern is to recognize that the arguments in the set are similar in some interesting way. With different interests, people may fasten on different similarities, and may arrive at different notions of argument pattern. Our enterprise is to characterize the concept of argument pattern which plays a role in the explanatory activity of scientists.

Formal logic, ancient and modern, is concerned in one obvious sense with patterns of argument. The logician proceeds by isolating a small set of expressions (the logical vocabulary), considers the schemata formed from sentences by replacing with dummy letters all expressions which do not belong to this set, and tries to specify which sequences of these schemata are valid patterns of argument. The pattern of argument which is taught to students of Newtonian dynamics is not a pattern of the kind which interests logicians. It has instantiations with different logical structures. (A rigorous derivation of the equations of motion of different dynamical systems would have a logical structure depending on the number of bodies involved and the mathematical details of the integration.) Moreover, an argument can only instantiate the Newtonian pattern if particular *non*logical terms, 'force', 'mass' and 'acceleration', occur in it in particular ways. However, the logician's approach can help us to isolate the notion of argument pattern which we require.

Let us say that a *schematic sentence* is an expression obtained by replacing some, but not necessarily all, the nonlogical expressions occurring in a sentence with dummy letters. A set of *filling instructions* for a schematic sentence is a set of directions for replacing the dummy letters of the schematic sentence, such that, for each dummy letter, there is a direction which tells us how it should be replaced. A *schematic argument* is a sequence of schematic sentences. A *classification* for a schematic argument is a set of sentences which describe the inferential characteristics of the schematic argument: its function is to tell us which terms in the sequence are to be regarded as premises, which are to be inferred from which, what rules of inference are to be used, and so forth.

We can use these ideas to define the concept of a *general argument pattern*. A general argument pattern is a triple consisting of a schematic argument, a set of sets of filling instructions containing one set of filling instructions for each term of the schematic argument, and a classification for the schematic argument. A sequence of sentences instantiates the general argument pattern just in case it meets the following conditions:

(i) The sequence has the same number of terms as the schematic argument of the general argument pattern.

(ii) Each sentence in the sequence is obtained from the corresponding schematic sentence in accordance with the appropriate set of filling instructions.

(iii) It is possible to construct a chain of reasoning which assigns to each sentence the status accorded to the corresponding schematic sentence by the classification.

We can make these definitions more intuitive by considering the way in which they apply to the Newtonian example. Restricting ourselves to the basic pattern used in treating systems which contain one body (such as the pendulum and the projectile) we may represent the schematic argument as follows:

(1) The force on α is β.
(2) The acceleration of α is γ.
(3) Force = mass • acceleration.
(4) (Mass of α) • $(\gamma) = \beta$
(5) $\delta = \theta$

The filling instructions tell us that all occurrences of 'α' are to be replaced by an expression referring to the body under investigation; occurrences of 'β' are to be replaced by an algebraic expression referring to a function of the variable coordinates and of time; 'γ' is to be replaced by an expression which gives the acceleration of the body as a function of its coordinates and their time-derivatives (thus, in the case of a one-dimensional motion along the x-axis of a Cartesian coordinate system, 'γ' would be replaced by the expression 'd^2x/dt^2'); 'δ' is to be replaced by an expression referring to the variable coordinates of the body, and 'θ' is to be replaced by an explicit function of time (thus the sentences that instantiate (5) reveal the dependence of the variable coordinates on time, and so provide specifications of the positions of the body in question throughout the motion). The classification of the argument tells us that (1)-(3) have the status of premises, that (4) is obtained from them by substituting identicals, and that (5) follows from (4) using algebraic manipulation and the techniques of the calculus.

Although the argument patterns which interest logicians are general argument patterns in the sense just defined, our example exhibits clearly the features which distinguish the kinds of patterns which scientists are trained to use. Whereas logicians are concerned to display all the schematic premises which are employed and to specify exactly which rules of inference are used, our example allows for the use of premises (mathematical assumptions) which do not occur as terms of the schematic argument and it does not give a complete description of the way in which the route from (4) to (5) is to go. Moreover, our pattern does not replace all nonlogical expressions by dummy letters. Because some nonlogical expressions remain, the pattern imposes special demands on arguments which instantiate it. In a different way, restrictions are set by the instructions for replacing dummy letters. The patterns of logicians are very liberal in both these latter respects. The conditions for replacing dummy letters in Aristotelian syllogisms, or first-order schemata, require only that some letters be replaced with predicates, others with names.

Arguments may be similar either in terms of their logical structure or in terms of the nonlogical vocabulary they employ at corresponding places. I think that the notion of similarity (and the corresponding notion of pattern) which is central to the explanatory activity of scientists results from a compromise in demanding these two kinds of similarity. I propose that scientists are interested in *stringent* patterns of argument, patterns which contain some nonlogical expressions and which are fairly similar in terms of logical structure. The Newtonian pattern cited above furnishes a good example. Although arguments instantiating this pattern do not have exactly the same logical structure, the classification imposes conditions which ensure that there will be similarities in logical structure among such arguments. Moreover, the presence of the nonlogical terms sets strict requirements on the instantiations and so ensures a different type of kinship among them. Thus, without trying to provide an exact analysis of the notion of stringency, we may suppose that the stringency of a pattern is determined by two different constraints: (1) the conditions on the substitution of expressions for dummy letters, jointly imposed by the presence of nonlogical expressions in the pattern and by the filling instructions; and (2) the conditions on the logical structure, imposed by the classification. If both conditions are relaxed completely then the notion of pattern degenerates so as to admit *any* argument. If both conditions are simultaneously made as strict as possible, then we obtain another degenerate case, a "pattern" which is its own unique instantiation. If condition (2) is tightened at the total expense of (1), we produce the logician's notion of pattern. The use of condition (1) requires that arguments instantiating a common pattern draw on a common nonlogical vocabulary. We can glimpse here that ideal of unification through the use of a few theoretical concepts which the remarks of Hempel and Feigl suggest.

Ideally, we should develop a precise account of how these two kinds of similarity are weighted against one another. The best strategy for obtaining such an account is to see how claims about stringency occur in scientific discussions. But scientists do not make explicit assessments of the stringency of argument patterns. Instead they evaluate the ability of a theory to explain and to unify. The way to a refined account of stringency lies through the notions of explanation and unification.

6 | Explanation as Unification

As I have posed it, the problem of explanation is to specify which set of arguments we ought to accept for explanatory purposes given that we hold certain sentences to be true. Obviously this formulation can encourage confusion: we must not think of a scientific community as *first* deciding what sentences it will accept and *then* adopting the appropriate set of arguments. The Newtonian and Darwinian examples should convince us that the promise of explanatory power enters into the modification of our beliefs. So, in proposing that $E(K)$ is a function of K, I do not mean to suggest that the acceptance of K must be temporally prior to the adoption of $E(K)$.

$E(K)$ is to be that set of arguments which best unifies K. There are, of course, usually many ways of deriving some sentences in K from others. Let us call a set of arguments which derives some members of K from other members of K a *systematization* of K. We may then think of $E(K)$ as the best systematization of K.

Let us begin by making explicit an idealization which I have just made tacitly. A set of arguments will be said to be *acceptable relative to* K just in case every argument in the set consists of a sequence of steps which accord with elementary valid rules of inference (deductive or inductive) and if every premise of every argument in the set belongs to K. When we are considering ways of systematizing K we restrict our attention to those sets of arguments which are acceptable relative to K. This is an idealization because we sometimes use as the basis of acts of explanation arguments furnished by theories whose principles we no longer believe. I shall not investigate this practice nor the considerations which justify us in engaging in it. The most obvious way to extend my idealized picture to accommodate it is to regard the explanatory store over K, as I characterize it here, as being supplemented with an extra class of arguments meeting the following conditions: (a) from the perspective of K, the premises of these arguments are approximately true; (b) these arguments can be viewed as approximating the structure of (parts of) arguments in $E(K)$; (c) the arguments are simpler than the corresponding arguments in $E(K)$. Plainly, to spell out these conditions precisely would lead into issues which are tangential to my main goal in this paper.

The moral of the Newtonian and Darwinian examples is that unification is achieved by using similar arguments in the derivation of many accepted sentences. When we confront the set of possible systematizations of K we should therefore attend to the *patterns* of argument which are employed in each systematization. Let us introduce the notion of a *generating set*: if Σ is a set of arguments then a generating set for Σ is a set of argument patterns

Π such that each argument in Σ is an instantiation of some pattern in Π. A generating set for Σ will be said to be *complete with respect to K* if and only if every argument which is acceptable relative to K and which instantiates a pattern in Π belongs to Σ. In determining the explanatory store E(K) we first narrow our choice to those sets of arguments which are acceptable relative to K, the systematizations of K. Then we consider, for each such set of arguments, the various generating sets of argument patterns which are complete with respect to K. (The importance of the requirement of completeness is to debar explanatory deviants who use patterns selectively.) Among these latter sets we select that set with the greatest unifying power (according to criteria shortly to be indicated) and we call the selected set the *basis* of the set of arguments in question. The explanatory store over K is that systematization whose basis does best by the criteria of unifying power.

This complicated picture can be made clearer, perhaps, with the help of a diagram.

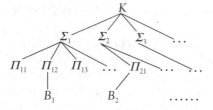

If B_k is the basis with the greatest unifying power, then $E(K) = \Sigma_k$.

Systematizations. Sets of arguments acceptable relative to K.

Complete generating sets. Π_{ij} is a generating set for Σ_i which is complete with respect to K.

Bases. B_i is the basis for Σ_i, and is selected as the best of the Π_{ij} on the basis of unifying power.

The task which confronts us is now formulated as that of specifying the factors which determine the unifying power of a set of argument patterns. Our Newtonian and Darwinian examples inspire an obvious suggestion: unifying power is achieved by generating a large number of accepted sentences as the conclusions of acceptable arguments which instantiate a few, stringent patterns. With this in mind, we define the *conclusion set* of a set of arguments Σ, $C(\Sigma)$, to be the set of sentences which occur as conclusions of some argument in Σ. So we might propose that the unifying power of a basis B_i with respect to K varies directly with the size of $C(\Sigma_i)$, varies directly with the stringency of the patterns which belong to B_i, and varies inversely with the number of members of B_i. This proposal is along the right lines, but it is, unfortunately, too simple.

The pattern of argument which derives a specification of the positions of bodies as explicit functions of time from a specification of the forces acting on those bodies is, indeed, central to Newtonian explanations. But not every argument used in Newtonian explanations instantiates this pattern.

Some Newtonian derivations consist of an argument instantiating the pattern followed by further derivations from the conclusion. Thus, for example, when we explain why a pendulum has the period it does we may draw on an argument which *first* derives the equation of motion of the pendulum and *then* continues by deriving the period. Similarly, in explaining why projectiles projected with fixed velocity obtain maximum range when projected at 45° to the horizontal, we first show how the values of the horizontal and vertical coordinates can be found as functions of time and the angle of elevation, use our results to compute the horizontal distance travelled by the time the projectile returns to the horizontal, and then show how this distance is a maximum when the angle of elevation of projection is 45°. In both cases we take further steps beyond the computation of the explicit equations of motion—and the further steps in each case are different.

If we consider the entire range of arguments which Newtonian dynamics supplies for explanatory purposes, we find that these arguments instantiate a number of different patterns. Yet these patterns are not entirely distinct, for all of them proceed by using the computation of explicit equations of motion as a prelude to further derivation. It is natural to suggest that the pattern of computing equations of motion is the *core* pattern provided by Newtonian theory, and that the theory also shows how conclusions generated by arguments instantiating the core pattern can be used to derive further conclusions. In some Newtonian explanations, the core pattern is supplemented by a *problem-reducing pattern*, a pattern of argument which shows how to obtain a further type of conclusion from explicit equations of motion.

This suggests that our conditions on unifying power should be modified, so that, instead of merely counting the number of different patterns in a basis, we pay attention to similarities among them. All the patterns in the basis may contain a common core pattern, that is, each of them may contain some pattern as a subpattern. The unifying power of a basis is obviously increased if some (or all) of the patterns it contains share a common core pattern.

As I mentioned at the beginning of this paper, the account of explanation as unification is complicated. The explanatory store is determined on the basis of criteria which pull in different directions, and I shall make no attempt here to specify precisely the ways in which these criteria are to be balanced against one another. Instead, I shall show that some traditional problems of scientific explanation can be solved without more detailed specification of the conditions on unifying power. For the account I have indicated has two important corollaries.

(A) Let Σ, Σ' be sets of arguments which are acceptable relative to K and which meet the following conditions:

   (i) the basis of Σ' is as good as the basis of Σ in terms of the criteria of stringency of patterns, paucity of patterns, presence of core patterns, and so forth.

    (ii) $C(\Sigma)$ is a proper subset of $C(\Sigma')$.
  Then $\Sigma \neq E(K)$.
(B) Let Σ, Σ' be sets of arguments which are acceptable relative to K and which meet the following conditions:
    (i) $C(\Sigma) = C(\Sigma')$
    (ii) the basis of Σ' is a proper subset of the basis of Σ.
  Then $\Sigma \neq E(K)$

(A) and (B) tell us that sets of arguments which do equally well in terms of some of our conditions are to be ranked according to their relative ability to satisfy the rest. I shall try to show that (A) and (B) have interesting consequences.

7 | Asymmetry, Irrelevance and Accidental Generalization

Some familiar difficulties beset the covering law model. The *asymmetry problem* arises because some scientific laws have the logical form of equivalences. Such laws can be used "in either direction". Thus a law asserting that the satisfaction of a condition C_1 is equivalent to the satisfaction of a condition C_2 can be used in two different kinds of argument. From a premise asserting that an object meets C_1, we can use the law to infer that it meets C_2; conversely, from a premise asserting that an object meets C_2, we can use the law to infer that it meets C_1. The asymmetry problem is generated by noting that in many such cases one of these derivations can be used in giving explanations while the other cannot.

    Consider a hoary example. (For further examples, see Bromberger 1966.) We can explain why a simple pendulum has the period it does by deriving a specification of the period from a specification of the length and the law which relates length and period. But we cannot explain the length of the pendulum by deriving a specification of the length from a specification of the period and the same law. What accounts for our different assessment of these two arguments? Why does it seem that one is explanatory while the other "gets things backwards"? The covering law model fails to distinguish the two, and thus fails to provide answers.

    The *irrelevance problem* is equally vexing. The problem arises because we can sometimes find a lawlike connection between an accidental and irrelevant occurrence and an event or state which would have come about independently of that occurrence. Imagine that Milo the magician waves his hands over a sample of table salt, thereby "hexing" it. It is true (and, I shall suppose, lawlike) that all hexed samples of table salt dissolve when placed in water. Hence we can construct a derivation of the dissolving of Milo's hexed sample of salt by citing the circumstances of the hexing. Although this derivation fits the covering law model, it is, by our ordinary lights, nonexplanatory.

(This example is given by Wesley Salmon in his (1970); Salmon attributes it to Henry Kyburg. For more examples, see Achinstein 1971.)

The covering law model explicitly debars a further type of derivation which any account of explanation ought to exclude. Arguments whose premises contain no laws, but which make essential use of accidental generalizations are intuitively nonexplanatory. Thus, if we derive the conclusion that Horace is bald from premises stating that Horace is a member of the Greenbury School Board and that all members of the Greenbury School Board are bald we do not thereby explain why Horace is bald. (See Hempel 1965, p. 339.) We shall have to show that our account does not admit as explanatory derivations of this kind.

I want to show that the account of explanation I have sketched contains sufficient resources to solve these problems.[7] In each case we shall pursue a common strategy. Faced with an argument we want to exclude from the explanatory store we endeavor to show that any set of arguments containing the unwanted argument could not provide the best unification of our beliefs. Specifically, we shall try to show either that any such set of arguments will be more limited than some other set with an equally satisfactory basis, or that the basis of the set must fare worse according to the criterion of using the smallest number of most stringent patterns. That is, we shall appeal to the corollaries (A) and (B) given above. In actual practice, this strategy for exclusion is less complicated than one might fear, and, as we shall see, its applications to the examples just discussed brings out what is intuitively wrong with the derivations we reject.

Consider first the irrelevance problem. Suppose that we were to accept as explanatory the argument which derives a description of the dissolving of the salt from a description of Milo's act of hexing. What will be our policy for explaining the dissolving of samples of salt which have not been hexed? If we offer the usual chemical arguments in these latter cases then we shall commit ourselves to an inflated basis for the set of arguments we accept as explanatory. For, unlike the person who explains *all* cases of dissolving of samples of salt by using the standard chemical pattern of argument, we shall be committed to the use of two different patterns of argument in covering such cases. Nor is the use of the extra pattern of argument offset by its applicability in explaining other phenomena. Our policy employs one extra pattern of argument without extending the range of things we can derive from our favored set of arguments. Conversely, if we eschew the standard chemical pattern of argument (just using the pattern which appeals to the hexing) we shall find ourselves unable to apply our favored pattern to cases in which the sample of salt dissolved has not been hexed. Moreover, the pattern we use will not fall under the more general patterns we employ to explain chemical phenomena such as solution, precipitation and so forth. Hence the unifying power of the basis for our preferred set of arguments will be less than that of the basis for the set of arguments we normally accept as explanatory.[8]

If we explain the dissolving of the sample of salt which Milo has hexed by appealing to the hexing then we are faced with the problems of explaining the dissolving of unhexed samples of salt. We have two options: (a) to adopt two patterns of argument corresponding to the two kinds of case; (b) to adopt one pattern of argument whose instantiations apply just to the cases of hexed salt. If we choose (a) then we shall be in conflict with [B], whereas choice of (b) will be ruled out by [A]. The general moral is that appeals to hexing fasten on a local and accidental feature of the cases of solution. By contrast our standard arguments instantiate a pattern which can be generally applied.[9]

A similar strategy succeeds with the asymmetry problem. We have general ways of explaining why bodies have the dimensions they do. Our practice is to describe the circumstances leading to the formation of the object in question and then to show how it has since been modified. Let us call explanations of this kind "origin and development derivations". (In some cases, the details of the original formation of the object are more important; with other objects, features of its subsequent modification are crucial.) Suppose now that we admit as explanatory a derivation of the length of a simple pendulum from a specification of the period. Then we shall either have to explain the lengths of *non*swinging bodies by employing quite a different style of explanation (an origin and development derivation) or we shall have to forego explaining the lengths of such bodies. The situation is exactly parallel to that of the irrelevance problem. Admitting the argument which is intuitively nonexplanatory saddles us with a set of arguments which is less good at unifying our beliefs than the set we normally choose for explanatory purposes.

Our approach also solves a more refined version of the pendulum problem (given by Paul Teller in his (1974)). Many bodies which are not currently executing pendulum motion *could* be making small oscillations, and, were they to do so, the period of their motion would be functionally related to their dimensions. For such bodies we can specify the *dispositional period* as the period which the body would have if it were to execute small oscillations. Someone may now suggest that we can construct derivations of the dimensions of bodies from specifications of their dispositional periods, thereby generating an argument pattern which can be applied as generally as that instantiated in origin and development explanations. This suggestion is mistaken. There are some objects—such as the Earth and the Crab Nebula— which *could not* be pendulums, and for which the notion of a dispositional period makes no sense. Hence, the argument pattern proposed cannot entirely supplant our origin and development derivations, and, in consequence, acceptance of it would fail to achieve the best unification of our beliefs.

The problem posed by accidental generalizations can be handled in parallel fashion. We have a general pattern of argument, using principles of physiology, which we apply to explain cases of baldness. This pattern is generally applicable, whereas that which derives ascriptions of baldness using the principle that all members of the Greenbury School Board are bald is not. Hence, as in the other cases, sets which contain the unwanted derivation will be ruled out by one of the conditions (A), (B).

Of course, this does not show that an account of explanation along the lines I have suggested would sanction only derivations which satisfy the conditions imposed by the covering law model. For I have not argued that an explanatory derivation need contain *any* sentence of universal form. What *does* seem to follow from the account of explanation as unification is that explanatory arguments must not use accidental generalization, and, in this respect, the new account appears to underscore and generalize an important insight of the covering law model. Moreover, our success with the problems of asymmetry and irrelevance indicates that, even in the absence of a detailed account of the notion of stringency and of the way in which generality of the consequence set is weighed against paucity and stringency of the patterns in the basis, the view of explanation as unification has the resources to solve some traditional difficulties for theories of explanation.

8 | Spurious Unification

Unfortunately there is a fly in the ointment. One of the most aggravating problems for the covering law model has been its failure to exclude certain types of self-explanation. (For a classic source of difficulties see Eberle, Kaplan and Montague 1961.) As it stands, the account of explanation as unification seems to be even more vulnerable on this score. The problem derives from a phenomenon which I shall call *spurious unification.*

Consider, first, a difficulty which Hempel and Oppenheim noted in a seminal article (Hempel 1965, Chapter 10). Suppose that we conjoin two laws. Then we can derive one of the laws from the conjunction, and the derivation conforms to the covering law model (unless, of course, the model is restricted to cover only the explanation of singular sentences; Hempel and Oppenheim do, in fact, make this restriction). To quote Hempel and Oppenheim:

> The core of the difficulty can be indicated briefly by reference to an example: Kepler's laws, K, may be conjoined with Boyle's law, B, to [yield] a stronger law $K\&B$; but derivation of K from the latter would not be considered as an explanation of the regularities stated in Kepler's laws; rather it would be viewed as representing, in effect, a pointless "explanation" of Kepler's laws by themselves. (Hempel 1965, p. 273 fn. 33)

This problem is magnified for our account. For, why may we not unify our beliefs *completely* by deriving all of them using arguments which instantiate the one pattern?

$$\frac{\alpha \mathbin{\&} B}{\alpha} \qquad [`\alpha' \text{ is to be replaced by any sentence we accept.}]$$

Or, to make matters even more simple, why should we not unify our beliefs by using the most trivial pattern of self-derivation?

$$\alpha \qquad [`\alpha' \text{ is to be replaced by any sentence we accept.}]$$

There is an obvious reply. The patterns just cited may succeed admirably in satisfying our criteria of using a few patterns of argument to generate many beliefs, but they fail dismally when judged by the criterion of stringency. Recall that the stringency of a pattern is assessed by adopting a compromise between two constraints: stringent patterns are not only to have instantiations with similar logical structures; their instantiations are also to contain similar nonlogical vocabulary at similar places. Now both of the above argument patterns are very lax in allowing any vocabulary whatever to appear in the place of 'α'. Hence we can argue that, according to our intuitive concept of stringency, they should be excluded as non-stringent.

Although this reply is promising, it does not entirely quash the objection. A defender of the unwanted argument patterns may *artificially* introduce restrictions on the pattern to make it more stringent. So, for example, if we suppose that one of *our* favorite patterns (such as the Newtonian pattern displayed above) is applied to generate conclusions meeting a particular condition C, the defender of the patterns just cited may propose that 'α' is to be replaced, not by any sentence, but by a sentence which meets C. He may then legitimately point out that his newly contrived pattern is as stringent as our favored pattern. Inspired by this partial success, he may adopt a general strategy. Wherever we use an argument pattern to generate a particular type of conclusion, he may use some argument pattern which involves self-derivation, placing an appropriate restriction on the sentences to be substituted for the dummy letters. In this way, he will mimic whatever unification we achieve. His "unification" is obviously spurious. How do we debar it?

The answer comes from recognizing the way in which the stringency of the unwanted patterns was produced. Any condition on the substitution of sentences for dummy letters would have done equally well, provided only that it imposed constraints comparable to those imposed by acceptable patterns. Thus the stringency of the restricted pattern seems accidental to it. This accidental quality is exposed when we notice that we can vary the filling instructions, while retaining the same syntactic structure, to obtain a host of other argument patterns with equally many instantiations. By contrast, the constraints imposed on the substitution of nonlogical vocabulary in the Newtonian pattern (for example) cannot be amended without destroying the stringency of the pattern or without depriving it of its ability to furnish us with many instantiations. Thus the constraints imposed in the

Newtonian pattern are essential to its functioning; those imposed in the unwanted pattern are not.

Let us formulate this idea as an explicit requirement. If the filling instructions associated with a pattern P could be replaced by different filling instructions, allowing for the substitution of a class of expressions of the same syntactic category, to yield a pattern P' and if P' would allow the derivation of *any* sentence, then the unification achieved by P is spurious. Consider, in this light, any of the patterns which we have been trying to debar. In each case, we can vary the filling instructions to produce an even more "successful" pattern. So, for example, given the pattern:

α ['α' is to be replaced by a sentence meeting condition C]

we can generalize the filling instructions to obtain

α ['α' is to be replaced by any sentence].

Thus, under our new requirement, the unification achieved by the original pattern is spurious.

In a moment I shall try to show how this requirement can be motivated, both by appealing to the intuition which underlies the view of explanation as unification and by recognizing the role that something like my requirement has played in the history of science. Before I do so, I want to examine a slightly different kind of example which initially appears to threaten my account. Imagine that a group of religious fanatics decides to argue for the explanatory power of some theological doctrines by claiming that these doctrines unify their beliefs about the world. They suggest that their beliefs can be systematized by using the following pattern:

God wants it to be the case that α. What God wants to be the case is the case.	['α' is to be replaced by any accepted sentence describing the physical world]
α	

The new requirement will also identify as spurious the pattern just presented, and will thus block the claim that the theological doctrines that God exists and has the power to actualize his wishes have explanatory power. For it is easy to see that we can modify the filling instructions to obtain a pattern that will yield any sentence whatsoever.

Why should patterns whose filling instructions can be modified to accommodate any sentence be suspect? The answer is that, in such patterns, the nonlogical vocabulary which remains is idling. The presence of that nonlogical vocabulary imposes no constraints on the expressions we can substitute for the dummy symbols, so that, beyond the specification that a place be filled by expressions of a particular syntactic category, the structure we impose by means of filling instructions is quite incidental. Thus the patterns in question do not genuinely reflect the contents of our beliefs. The explanatory store should present the order of natural phenomena which is exposed by what we think we know. To do so, it must exhibit connections among our beliefs beyond those which could be found among any beliefs. Patterns of self-derivation and the type of pattern exemplified in the example of the theological community merely provide trivial, omnipresent connections, and, in consequence, the unification they offer is spurious.

My requirement obviously has some kinship with the requirement that the principles put forward in giving explanations be testable. As previous writers have insisted that genuine explanatory theories should not be able to cater to all possible evidence, I am demanding that genuinely unifying patterns should not be able to accommodate all conclusions. The requirement that I have proposed accords well with some of the issues which scientists have addressed in discussing the explanatory merits of particular theories. Thus several of Darwin's opponents complain that the explanatory benefits claimed for the embryonic theory of evolution are illusory, on the grounds that the style of reasoning suggested could be adapted to any conclusion. (For a particularly acute statement of the complaint, see the review by Fleeming Jenkin, printed in Hull 1974, especially p. 342.) Similarly, Lavoisier denied that the explanatory power of the phlogiston theory was genuine, accusing that theory of using a type of reasoning which could adapt itself to any conclusion (Lavoisier 1862, Volume II p. 233). Hence I suggest that some problems of spurious unification can be solved in the way I have indicated, and that the solution conforms both to our intuitions about explanatory unification and to the considerations which are used in scientific debate.

However, I do not wish to claim that my requirement will debar all types of spurious unification. It may be possible to find other unwanted patterns which circumvent my requirement. A full characterization of the notion of a stringent argument pattern should provide a criterion for excluding the unwanted patterns. My claim in this section is that it will do so by counting as spurious the unification achieved by patterns which adapt themselves to any conclusion and by patterns which accidentally restrict such universally hospitable patterns. I have also tried to show how this claim can be developed to block the most obvious cases of spurious unification.

9 | Conclusions

I have sketched an account of explanation as unification, attempting to show that such an account has the resources to provide insight into episodes in the history of science and to overcome some traditional problems for the covering law model. In conclusion, let me indicate very briefly how my view of explanation as unification suggests how scientific explanation yields understanding. By using a few patterns of argument in the derivation of many beliefs we minimize the number of *types* of premises we must take as underived. That is, we reduce, in so far as possible, the number of types of facts we must accept as brute. Hence we can endorse something close to Friedman's view of the merits of explanatory unification (Friedman 1974, pp. 18–19).

Quite evidently, I have only *sketched* an account of explanation. To provide precise analyses of the notions I have introduced, the basic approach to explanation offered here must be refined against concrete examples of scientific practice. What needs to be done is to look closely at the argument patterns favored by scientists and attempt to understand what characteristics they share. If I am right, the scientific search for explanation is governed by a maxim, once formulated succinctly by E. M. Forster. Only connect.

Notes

A distant ancestor of this paper was read to the Dartmouth College Philosophy Colloquium in the Spring of 1977. I would like to thank those who participated, especially Merrie Bergmann and Jim Moor, for their helpful suggestions. I am also grateful to two anonymous referees for *Philosophy of Science* whose extremely constructive criticisms have led to substantial improvements. Finally, I want to acknowledge the amount I have learned from the writing and the teaching of Peter Hempel. The present essay is a token payment on an enormous debt.

1. Strictly speaking, this is one of two views which emerge from Achinstein's discussion and which he regards as equally satisfactory. As Achinstein goes on to point out, either of these ontological theses can be developed to capture the central idea of the covering law model.

2. To pose the problem in this way we may still invite the charge that *arguments* should not be viewed as the bases for acts of explanation. Many of the criticisms levelled against the covering law model by Wesley Salmon in his seminal paper on statistical explanation (Salmon 1970) can be reformulated to support this charge. My discussion in section 7 will show how some of the difficulties raised by Salmon for the covering law model do not bedevil my account. However, I shall not respond directly to the points about statistical explanation and statistical inference advanced by Salmon and by Richard Jeffrey in his (1970). I believe that Peter Railton has shown how these specific difficulties concerning statistical explanation can be accommodated by an approach which takes explanations to be (or be based on) arguments (see Railton 1978), and that the account offered in section 4 of his paper can be adapted, to complement my own.

3. Of course, in restricting my attention to why-questions I am following the tradition of philosophical discussion of scientific explanation: as Bromberger notes in section IV of his (1966) not all explanations are directed at why-questions, but attempts to characterize explanatory responses to why-questions have a special interest for the philosophy of science because of the connection to a range of

methodological issues. I believe that the account of explanation offered in the present paper could be extended to cover explanatory answers to some other kinds of questions (such as how-questions). But I do want to disavow the claim that unification is relevant to all types of explanation. If one believes that explanations are sometimes offered in response to what-questions (for example), so that it is correct to talk of someone explaining what a gene is, then one should allow that some types of explanation can be characterized independently of the notions of unification or of argument. I ignore these kinds of explanation in part because they lack the methodological significance of explanations directed at why-questions and in part because the problem of characterizing explanatory answers to what-questions seems so much less recalcitrant than that of characterizing explanatory answers to why-questions (for a similar assessment, see Belnap and Steel 1976, pp. 86–7). Thus I would regard a full account of explanation as a heterogeneous affair, because the conditions required of adequate answers to different types of questions are rather different, and I intend the present essay to make a proposal about how *part* of this account (the most interesting part) should be developed.

4. For illuminating accounts of Newton's influence on eighteenth-century research see Cohen (1956) and Schofield (1969). I have simplified the discussion by considering only *one* of the programs which eighteenth-century scientists derived from Newton's work. A more extended treatment would reveal the existence of several different approaches aimed at unifying science, and I believe that the theory of explanation proposed in this paper may help in the historical task of understanding the diverse aspirations of different Newtonians. (For the problems involved in this enterprise, see Heimann and McGuire 1971.)

5. See Boscovich (1966) Part III, especially p. 134. For an introduction to Boscovich's work, see the essays by L. L. Whyte and Z. Marković in Whyte (1961). For the influence of Boscovich on British science, see the essays of Pearce Williams and Schofield in the same volume, and Schofield (1969).

6. The examples could easily be multiplied. I think it is possible to understand the structure and explanatory power of such theories as modern evolutionary theory, transmission genetics, plate tectonics, and sociobiology in the terms I develop here.

7. More exactly, I shall try to show that my account can solve some of the principal versions of these difficulties which have been used to discredit the covering law model. I believe that it can also overcome more refined versions of the problems than I consider here, but to demonstrate that would require a more lengthy exposition.

8. There is an objection to this line of reasoning. Can't we view the arguments $<(x)$ $((Sx \& Hx) \rightarrow Dx), Sa \& Ha, Da>, <(x)((Sx \& \sim Hx) \rightarrow Dx), Sb \& \sim Hb, Db>$ as instantiating a common pattern? I reply that, insofar as we can view these arguments as instantiating a common pattern, the standard pair of comparable (low-level) derivations—$<(x)(Sx \rightarrow Dx), Sa, Da>, <(x)(Sx \rightarrow Dx), Sb, Db>$—share a more stringent common pattern. Hence, incorporating the deviant derivations in the explanatory store would give us an inferior basis. We can justify the claim that the pattern instantiated by the standard pair of derivations is more stringent than that shared by the deviant derivations, by noting that representation of the deviant pattern would compel us to broaden our conception of schematic sentence, and, even were we to do so, the deviant pattern would contain a "degree of freedom" which the standard pattern lacks. For a representation of the deviant "pattern" would take the form $<(x)((Sx \&$

$\alpha Hx) \to Dx$), Sa & αHa, $Da>$, where 'α' is to be replaced uniformly either with the null symbol or with '$\sim$'. Even if we waive my requirement that, in schematic sentences, we substitute for nonlogical vocabulary, it is evident that this "pattern" is more accommodating than the standard pattern.

9. However, the strategy I have recommended will not avail with a different type of case. Suppose that a deviant wants to explain the dissolving of the salt by appealing to some property which holds universally. That is, the "explanatory" arguments are to begin from some premise such as "$(x)((x$ is a sample of salt & x does not violate conservation of energy) $\to x$ dissolves in water)" or "$(x)((x$ is a sample of salt & $x=x)$ $\to x$ dissolves in water)." I would handle these cases somewhat differently. If the deviant's explanatory store were to be as unified as our own, then it would contain arguments corresponding to ours in which a redundant conjunct systematically occurred, and I think it would be plausible to invoke a criterion of simplicity to advocate dropping that conjunct.

References

Achinstein, P. (1971), *Law and Explanation.* Oxford: Oxford University Press.

Achinstein, P. (1977), "What is an Explanation?" *American Philosophical Quarterly* 14: pp. 1–15.

Belnap, N. and Steel, T. B. (1976), *The Logic of Questions and Answers.* New Haven: Yale University Press.

Boscovich, R. J. (1966), A *Theory of Natural Philosophy* (trans. J. M. Child). Cambridge: M.I.T. Press.

Bromberger, S. (1962), "An Approach to Explanation", in R. J. Butler (ed.), *Analytical Philosophy* (First Series). Oxford: Blackwell.

Bromberger, S. (1966), "Why-Questions", in R. Colodny (ed.), *Mind and Cosmos.* Pittsburgh: University of Pittsburgh Press.

Cohen, I. B. (1956), *Franklin and Newton.* Philadelphia: American Philosophical Society.

Darwin, C. (1964), *On the Origin of Species,* Facsimile of the First Edition, edited by E. Mayr. Cambridge: Harvard University Press.

Darwin, F. (1887), *The Life and Letters of Charles Darwin.* London: John Murray.

Eberle, R., Kaplan, D., and Montague, R. (1961), "Hempel and Oppenheim on Explanation", *Philosophy of Science* 28: pp. 418–28.

Feigl, H. (1970), "The 'Orthodox' View of Theories: Remarks in Defense as well as Critique", in M. Radner and S. Winokur (eds.), *Minnesota Studies in the Philosophy of Science,* Volume IV. Minneapolis: University of Minnesota Press.

Friedman, M. (1974), "Explanation and Scientific Understanding", *Journal of Philosophy LXXI*: pp. 5–19.

Heimann, P., and McGuire, J. E. (1971), "Newtonian Forces and Lockean Powers", *Historical Studies in the Physical Sciences* 3: pp. 233–306.

Hempel, C. G. (1962), "Deductive-Nomological vs. Statistical Explanation", in H. Feigl and G. Maxwell (eds.) *Minnesota Studies in the Philosophy of Science,* Volume III. Minneapolis: University of Minnesota Press.

Hempel, C. G. (1965), *Aspects of Scientific Explanation.* New York: The Free Press.

Hempel, C. G. (1966), *Philosophy of Natural Science.* Englewood Cliffs: Prentice-Hall.

Hull, D. (ed.) (1974), *Darwin and his Critics.* Cambridge: Harvard University Press.

Jeffrey, R. (1970), "Statistical Explanation vs. Statistical Inference", in N. Rescher (ed.), *Essays in Honor of Carl G. Hempel.* Dordrecht: D. Reidel.

Kitcher, P. S. (1976), "Explanation, Conjunction and Unification", *Journal of Philosophy,* LXXIII: pp. 207–12.

Lavoisier, A. (1862), *Oeuvres.* Paris.

Newton, I. (1952), *Opticks.* New York: Dover.

Newton, I. (1962), *The Mathematical Principles of Natural Philosophy* (trans. A. Motte and F. Cajori). Berkeley: University of California Press.

Railton, P. (1978), "A Deductive-Nomological Model of Probabilistic Explanation", *Philosophy of Science* 45: pp. 206–226.

Salmon, W. (1970), "Statistical Explanation", in R. Colodny (ed.), *The Nature and Function of Scientific Theories.* Pittsburgh: University of Pittsburgh Press.

Schofield, R. E. (1969), *Mechanism and Materialism.* Princeton: Princeton University Press.

Teller, P. (1974), "On Why-Questions", *Noûs* VIII: pp. 371–80.

van Fraassen, B. (1977), "The Pragmatics of Explanation", *American Philosophical Quarterly* 14: pp. 143–50.

Whyte, L. L. (ed.) (1961), *Roger Joseph Boscovich.* London: Allen and Unwin.

James Woodward

The Manipulability Conception of Causal Explanation

The manipulability conception plays an important role in the way that scientists themselves think about causal explanation but has received rather less attention from philosophers. The basic idea is nicely illustrated by a contrast drawn between descriptive and explanatory science in a paper by Robert Weinberg (1985) on recent developments in molecular biology. Weinberg tells us that "biology has traditionally been a descriptive science," but that because of recent advances, particularly in instrumentation and experimental technique, it is now appropriate to think of molecular biology as providing "explanations" and identifying "causal mechanisms." What does this contrast between description and explanation consist in? Weinberg explicitly links the ability of molecular biology to provide explanations with the fact that it provides information of a sort that can be used for purposes of manipulation and control. New experimental and instrumental techniques have played such a decisive role in the development of molecular biology into an explanatory science precisely because such techniques make it possible to intervene in and manipulate biological systems and to observe the results in ways that were not previously possible. Molecular biologists correctly think that "the invisible submicroscopic agents they study can explain, at one essential level, the complexity of life" because by manipulating those agents it is now "possible to change critical elements of the biological blue print at will" (p. 48).

From James Woodward, *Making Things Happen: A Theory of Causal Explanation* (New York: Oxford University Press, 2003), 9–24. Our excerpt is from the first chapter of *MTH*, "Introduction and Preview," in which Woodward illustrates, motivates, and sketches the manipulability conception in a "very rough (some may think reckless) way" (*MTH*, p. 9). Details, refinements, and qualifications are discussed at length in the later chapters of Woodward's *MTH*, references to which have been retained in this excerpt along with the author's numbering of the subheadings (1.4, 1.5, etc.) from the first chapter.

This passage suggests the underlying idea of my account of causal explanation: we are in a position to explain when we have information that is relevant to manipulating, controlling, or changing nature, in an "in principle" sense of manipulation characterized in [MTH] chapter 3. We have at least the beginnings of an explanation when we have identified factors or conditions such that manipulations or changes in those factors or conditions will produce changes in the outcome being explained. Descriptive knowledge, by contrast, is knowledge that, although it may provide a basis for prediction, classification, or more or less unified representation or systemization, does not provide information potentially relevant to manipulation. It is in this that the fundamental contrast between causal explanation and description consists. On this way of looking at matters, our interest in causal relationships and explanation initially grows out of a highly practical interest human beings have in manipulation and control; it is then extended to contexts in which manipulation is no longer a practical possibility. This interest is importantly different from a number of the other interests philosophers have associated with explanation, for example, from our interest in prediction or even in nomically grounded prediction, or from our interest in constructing theories that unify, systematize, and organize in various ways, or that trace spatiotemporally continuous processes. As we shall see, one can have information that is relevant to prediction (including prediction based on generalizations that many philosophers are prepared to regard as laws), or information about spatiotemporally continuous processes, or information that allows for the sort of unification and systemization that many philosophers have thought relevant to explanation, and yet lack the kind of information that is relevant to manipulation on which my account focuses. When this is the case, my view is that one doesn't have a (causal) explanation. Conversely, one can have information that is relevant to manipulation and hence to explanation, even though one lacks the other features described above. What one needs for manipulation is information about *invariant* relationships, and one can identify invariant relationships even in cases in which one doesn't know laws, cannot trace spatiotemporally continuous processes, or unify and systematize.

I said above that explanatory information is information that is potentially relevant to manipulation and control. It is uncontroversial, however, that causal relationships exist and that explanation is possible in circumstances in which actual manipulation is impossible, whether for practical or other sorts of reasons. For example, we construct causal explanations of past events and of large-scale cosmological events, and in neither case is manipulation of these phenomena possible. The notion of information that is relevant to manipulation thus needs to be understood modally or counterfactually: the information that is relevant to causally explaining an outcome involves the identification of factors and relationships such that *if* (perhaps contrary to fact) manipulation of these factors *were* possible, this would be a way of manipulating or altering the phenomenon in question. For example, it is currently believed that the explanation (1.3.1) of the mass extinctions at the end of the Cretaceous period has to do with the impact of a large asteroid and the killing effects of the dust it created. Clearly, we cannot now do anything to affect whether this impact occurred, and quite possibly humans could have done nothing to alter the impact even if they had existed with current levels of technology at the time of impact. My suggestion is that if this explanation is correct, it nonetheless will be true that if it had been possible to alter or prevent the impact, this would have altered the character of or prevented the extinction. Put differently, my idea is that one ought to be able to associate with any successful explanation a hypothetical or counterfactual experiment that shows us that and how manipulation of the factors mentioned in the explanation (the *explanans,* as philosophers call it) would be a way of manipulating or altering the phenomenon explained (the *explanandum).* Put in still another way, an explanation ought to be such that it can be used to answer what I call a *what-if-things-had-been-different question:* the explanation must enable us to see what sort of difference it would have made for the explanandum if the factors cited in the explanans had been different in various possible ways. We can also think of this as information about a pattern of counterfactual dependence between explanans and explanandum, provided the counterfactuals in question are understood appropriately. As we shall see, even when actual manipulation is impossible, it is heuristically useful to think of causal and explanatory claims in this way: it both clarifies their content and enables us to understand why they have many of their distinctive features.

On this view, our interest in causal explanation represents a sort of generalization or extension of our interest in manipulation and control from cases in which manipulation is possible to cases in which it is not, but in which we nonetheless retain a concern with what would or might happen to the outcome being explained if various possible changes were to occur in the factors cited in the explanans. If we had been unable to manipulate nature—if we had been, in Michael Dummett's (1964) example, intelligent trees capable only of passive observation—then it is a reasonable conjecture that we would never have developed the notions of causation and explanation and the practices associated with them that we presently possess. Once developed, these notions and practices were then extended to contexts in which actual manipulation was infeasible or impossible. This extension was very natural and perhaps inevitable because, as we shall see in [MTH] chapter 3, it is built into the notion of a relationship that is usable for purposes of manipulation and control that whether such a relationship holds does not depend on whether the manipulation in question can be actually carried out.

Although it will not be news to historians that the aim of manipulating or controlling nature has played a central role in the development of modern science, this aim has received relatively little attention from philosophers. Most philosophers have distinguished sharply between pure science and applied science or technology, and have regarded explanation as a characteristic aim of pure science and manipulation and control as aims of applied science. To the extent that philosophers have concerned themselves with applied science, they often have seen it as primarily focused on prediction and have failed to appreciate how different prediction is from control. To readers in the grip of this conventional picture of science, my association of our interest in explanation with our practical interest in control over nature will seem misguided and counterintuitive.

However, a variety of more recent developments in history and philosophy of science and in science studies challenges this sharp distinction between pure science and its application. Part of my intention in writing this book is to contribute to a conception of the role of causal explanation in science that fits with these new developments. I include among these developments recent work in the history of science that emphasizes how concerns with technological application have heavily influenced the content of the more theoretical parts of science (e.g., Smith and Wise 1989; Barkan 1999) and recent work by sociologists, philosophers, and historians on experimentation, which has emphasized in various ways how important our ability to intervene and manipulate nature is in the development of a scientific understanding of nature. Broadly similar ideas can be found in some of the recent philosophical literature on explanation, for example, in Paul Humphreys's (1989) work, with its criticisms of "passive empiricism." On the conception of science that I favor, two aims that are often regarded as quite separate— the "pure science" aim of representing nature in a way that is truthful and accurate and the "applied science" aim of representing nature in a way that permits manipulation and control—are deeply intertwined.

My association of our interest in explanation with our interest in manipulation and control will also seem less surprising when one reflects that a very central part of the commonsense notion of cause is precisely that causes are potential handles or devices for bringing about effects. We find this idea in the manipulability theories of causation defended by writers like Collingwood (1940), Gasking (1955), and von Wright (1971). As we shall see in [MTH] chapter 2, it is also widely endorsed by social scientists and statisticians, who have shown that this idea can play an important heuristic role in both elucidating the meaning of causal claims and clarifying how statistical evidence can be used to test them. Unfortunately, however, standard philosophical statements of the manipulability theory lead to accounts of causation that are unacceptably anthropocentric and subjectivist. I show in [MTH] chapter 2 how a manipulability account of causation/ explanation can be developed in a way that satisfies reasonable expectations about the objectivity of causal relationships.

1.4 | Causal Explanation, Invariance, and Intervention Illustrated

My discussion so far has been rather abstract. It will be useful to have a concrete example in front of us to illustrate some of the ideas to which I have

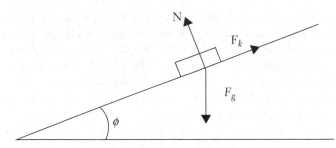

Figure 1.4.1

been referring. Consider (1.4.1) a block sliding down an inclined plane with acceleration a (fig. 1.4.1). What accounts for or explains the motion of the block? The standard textbook analysis proceeds as follows. The block is subject to three forces: a gravitational force due to the weight of the block; a normal force N, which is perpendicular to the plane; and a force due to friction, which opposes the motion of the block. The frictional force F_k obeys the relationship

(1.4.2) $F_k = \mu_k N$

where μ_k is the coefficient of kinetic friction. The gravitational force F_g is directed toward the center of mass of the earth and obeys the relationship

(1.4.3) $F_g = mg$

where g is a constant that represents the acceleration due to gravity near the surface of the earth and m is the mass of the block.

From the diagram in figure 1.4.1, we see that there is a force directed down the plane of magnitude $mg \sin \phi$. The normal force $N = mg \cos \phi$ and so the frictional force $F_g = \mu_k mg \cos \phi$. The net force on the block along the plane is the resultant of these two forces, which is

(1.4.4) $F_{net} = mg \sin \phi - \mu_k mg \cos \phi$

The acceleration of the block is thus given by

(1.4.5) $a = g \sin \phi - \mu_k g \cos \phi$

How does this explanation work? What is it about the information just conveyed that makes it explanatory? The account I defend takes this and other explanations to provide understanding by exhibiting a pattern of counterfactual dependence between explanans and explanandum—a pattern of counterfactual dependence of the special sort associated with relationships that are potentially exploitable for purposes of manipulation and control. In particular, the above explanation allows us to see how changing the values of various variables in (1.4.2)–(1.4.5) would result in changes in what it is that we are trying to explain: the acceleration of the block. For example, we see from (1.4.5) how changing the angle of elevation of the plane will change the acceleration of the block: as the angle is increased, the acceleration will be greater. Similarly, the above explanation shows us how, if we were to move the apparatus to a stronger or weaker gravitational field (i.e., if we were to change the value of the "constant" g), the acceleration of the block would change. We can also see, by way of contrast, that changing the value of m will make no difference to the motion of the block. The suggestion that I develop below is that this what-if-things-had-been-different information is intimately connected to our judgments of causal and explanatory relevance: we see whether and how some factor or event is causally or explanatorily relevant to another when we see whether (and if so, how) changes in the former are associated with changes in the latter.

It is a familiar point, however, that not every case of counterfactual dependence corresponds to a causal or explanatory relationship. There is a straightforward interpretation of the notion of counterfactual dependence according to which the joint effects of a common cause such as the reading of a barometer B and the occurrence/nonoccurrence S of a storm produced by atmospheric pressure A are counterfactually dependent on each other; that is, according to which, counterfactuals such as

(1.4.6) If the barometer reading were to fall, a storm would occur

are true. Nonetheless, the barometer reading does not cause or explain the occurrence/nonoccurrence of the storm. Some way must be found of distinguishing the kind of counterfactual dependence that is associated with causal and explanatory relevance from the counterfactual dependence of S on B. My solution to this problem appeals to two closely interrelated notions: *intervention* and *invariance*.

For reasons that will become clearer in [MTH] chapter 2, a manipulability theory is most naturally formulated in terms of variables—quantities or magnitudes that can take more than one value. Causal relationships, of course, have to do with patterns of dependence that hold in the world, rather than with relationships between numbers or other abstracta, but in the interest of avoiding cumbersome circumlocutions, I will often speak of causal relationships as obtaining between variables or their values, trusting that it is obvious enough how to sort out what is meant. Suppose, then, that X and Y are variables. The notion of an intervention attempts to capture, in non-anthropomorphic language that makes no reference to notions like human agency, the conditions that would need to be met in an ideal experimental manipulation of the value of X performed for the purpose of determining whether X causes Y. A more precise characterization is provided in [MTH] chapter 3, but the intuitive idea is that an intervention on X with respect to Y changes the value of X in such a way that if any change occurs in Y, it occurs only as a result of the change in the value of X and not from some other source. This requires, among other things, that the intervention not be correlated with other causes of Y except for those causes of Y (if any) that are causally between the intervention and X or between X and Y, and that the intervention not affect Y via a route that fails to go through X.

Suppose that, in the barometer/storm example, we make use of some process that sets the value of B at a high or low reading, in a way that is causally and probabilistically independent of the value of A. This would constitute an intervention on B with respect to S. What we expect is that the value of S would not change if such an intervention were to be performed on B; that is, that counterfactuals of the form

(1.4.7)   If the value of B were to be changed as a result of an intervention, then the value of S would change

are false. My view is that the sorts of counterfactuals that matter for purposes of causation and explanation are just such counterfactuals that describe how the value of one variable would change under interventions that change the value of another. Thus, as a rough approximation, a necessary and sufficient condition for X to cause Y or to figure in a causal explanation of Y is that the value of Y would change under some intervention on X in some background circumstances (which may include interventions on other variables besides X and Y; see [MTH] chapter 2). It is because there is no intervention on B that will change the value of S that B does not cause or figure in an explanation of S. The difference between interventionist counterfactuals such as (1.4.7) and noninterventionist counterfactuals such as (1.4.6) corresponds, roughly, to the difference between what Lewis ([1973] 1986) calls nonbacktracking and backtracking counterfactuals. But, as will become clear in [MTH] chapter 3, this correspondence is only rough; Lewis's view differs from my own both in the judgments it reaches about the truth values of particular counterfactual claims and, more fundamentally, in its basic motivation.

Although interventions on B (the barometer reading) will not change S (whether or not the storm occurs), there are at least some interventions or experimental manipulations of the mass of the block, the angle of elevation of the plane, the value of g, and so on, such that, if they were to occur, the acceleration of the block would change in just the way described by the generalizations (1.4.2)–(1.4.5). In other words, in contrast to the correlation between B and S, these equations do correctly tell us how the values of the acceleration would change under (some) interventions. In my view, it is this that accounts for the difference in explanatory status between (1.4.1) (the explanation of the motion of the block) and the "explanation" of the occurrence of the storm in terms of the barometer reading.

The notion of *invariance* is closely related to the notion of an intervention. A generalization G (relating, say, changes in the value of X to changes in the value of Y) is invariant if G would continue to hold under some intervention that changes the value of X in such a way that, according to G, the value of Y would change—"continue to hold" in the sense that G correctly describes how the value of Y would change under this intervention. A necessary and sufficient condition for a generalization to describe a causal relationship is that it be invariant under some appropriate set of interventions. The generalization describing the correlation between B and S in the above example will break down under all interventions on B: it is a noninvariant generalization. This corresponds to the fact that B does not cause S, and we cannot appeal to this correlation to explain the occurrence or nonoccurrence of the storm.

By contrast, the generalizations (1.4.2) and (1.4.3) in the inclined plane example are invariant under at least a certain range of interventions that change the value of N. For some such changes, μ_k will be (at least approximately) a constant that is independent of N, and hence (1.4.2) will correctly describe how F_k will change under this intervention. Similarly, the relationship $F = mg \sin \phi$ will continue to correctly describe how the component of the weight of the block directed along the plane will change under at least some interventions that change the value of ϕ and of m.

When a relationship is invariant under at least some interventions in this way, it is potentially usable for purposes of manipulation and control—potentially usable in the sense that, though it may not as a matter of fact be possible to carry out an intervention on X, it is nonetheless true that *if* an intervention on X were to occur, this would be a way of manipulating or controlling the value of Y. Thus, in the above examples, we may control the value of F_k by controlling the value of N; we may manipulate the component of the gravitational force on the block directed along the plane by manipulating m and ϕ and so on. By contrast, one cannot manipulate whether or not a storm occurs by altering the position of a barometer dial. It is in this sense that the account I favor associates causal and explanatory relationships with relationships that tell us about the distinctive patterns of counterfactual dependence that are associated with manipulation and control.

The notion of invariance under interventions is intended to do the work (the work of distinguishing between causal and merely accidental generalizations) that is done by the notion of a *law of nature* in other philosophical accounts. In the above example, although the generalizations (1.4.2) and (1.4.3) represent or describe causal relationships, they are highly problematic candidates for laws of nature. The generalization (1.4.3) $F_g = mg$ lacks many of the features standardly ascribed to laws; it holds only approximately, even near the surface of the earth, and fails to hold even approximately at sufficiently large distances from the earth's surface. It is obviously contingent on the earth's having a particular mass and radius. The relationship (1.4.2) $F_k = \mu_k N$ is, if anything, a less appealing candidate for a law.

The textbook from which I took this example goes out of its way not to describe (1.4.2) as a law, but instead describes it as a nonfundamental and only approximate "empirical relationship" (Frautshi, Olenick, Apostol, and Goodstein 1986, pp. 179ff). The particular value of the coefficient of kinetic friction between a pair of surfaces is the net result of a large number of extremely complicated contact forces and depends, in ways that are still not well understood from the perspective of fundamental theory, on the detailed characteristics of the two surfaces and will change if the characteristics of those surfaces are altered. Thus, even if (1.4.2) correctly describes the frictional force on the block for some particular experimental setup, we could easily disrupt that generalization by, for example, greasing the surface of the plane or abrading it with sandpaper. This is very different from the behavior of paradigmatic laws of nature.

As (1.4.2) and (1.4.3) illustrate, whether or not a generalization is invariant is surprisingly independent of whether it satisfies many of the traditional criteria for lawfulness, such as exceptionlessness, breadth of scope, and degree of theoretical integration. This is in large measure a consequence of the way in which invariance is defined: for a generalization to be invariant, all that is required is that it be stable under *some* (range of) changes and interventions.[1] It is not required that it be invariant under *all* possible changes and interventions. Thus, the generalization (1.4.2), describing the relationship between frictional and normal force, is invariant as long as it would continue to hold under some interventions that change the value of N. Its invariance is not undermined by the fact that there are other interventions on N (e.g., increasing N to a very large value) or other sorts of changes (greasing the contact surface) under which (1.4.2) would break down. I argue below that the notion of invariance is better suited than the notion of lawfulness for capturing the distinctive features of many of the generalizations that describe causal relationships and figure in explanations. The notion of law imports many features that generalizations like (1.4.2) do not seem to possess. Calling (1.4.2) a law immediately places us in the dialectically awkward position of needing to explain why that characterization is appropriate, even though (1.4.2) lacks many of the features usually ascribed to laws, whereas describing (1.4.2) as invariant generates no such difficulties. Rather than thinking of all causal generalizations as laws, I suggest that we should think of laws as just one kind of invariant generalization.

1.5 | Some Varieties of Causal Claims

I have been claiming that causal relationships share a common feature: they are invariant relationships that are potentially exploitable for purposes of manipulation and control. But within this broad genus there are additional distinctions that can be drawn among varieties of causal relationships and among dimensions along which causal relationships may vary. For example, one way in which the explanation (1.4.2) differs from the explanation (1.3.1) of the extinction of the dinosaurs is that the former is not (at least as I have presented it) an explanation of any specific episode involving a block sliding down a plane. Instead, what is explained is a generic pattern in the motion of blocks down planes—what James Bogen and I (Bogen and Woodward 1988) call a *phenomenon*. By contrast, what is explained in (1.3.1) is a specific episode of extinction. Causal claims or explanations of particular events like (1.3.1) are often called *token causal claims* or *singular causal explanations* and, as we shall see, they have a number of distinctive features. They share with other varieties of causal explanation the general feature that they are to be understood in terms of counterfactuals about what will happen under interventions, but the relevant counterfactuals differ in various ways from those associated with explanations like (1.4.2). Similarly, type-level causal claims like those in (1.4.2) may either have to do with what I call in [*MTH*] chapter 2 *total* or *net* causes or with *direct* or *contributing* causes, and these also will differ in the counterfactuals with which they are associated. In general, we may illuminate the content of different sorts of causal claims and causal notions by describing the different hypothetical experiments with which they are associated.

There is yet another important dimension along which causal claims may differ. Consider the claim

(1.5.1)   Depressing the gas pedal on my car causes it to accelerate.

Assuming that my car is functioning normally, the manipulationist view that I advocate judges that this is a true causal claim and one to which I might appeal to explain why my car has accelerated on some particular occasion. Nonetheless, as the econometrician Haavelmo (1944) observed decades ago, there is an obvious sense in which it is explanatorily shallow in comparison with the sort of theory of the internal mechanism of the car that might be provided by an automotive engineer; I don't have any very deep understanding of why my car moves as it does, if I know only (1.5.1). This example illustrates the point that causal explanations differ in the degree or depth of understanding they provide. One would like a theory of causal explanation to provide some insight into what makes one causal explanation deeper or better than another. The theory I develop attempts to do this by tracing differences in explanatory depth to differences in the degree of invariance of the generalizations to which the explanation appeals (i.e., differences in the range of changes and interventions over which those generalizations are stable) and to differences in the range of what-if-things-had-been-different questions that the generalization answers.

1.6 | Causal Explanation as a Practical Activity

I said above that explanations and causal inference pervade our lives. I mean by this that these are widespread, everyday activities in which most human beings, including people in other cultures quite different from our own, engage. They are, if you like, "natural" human activities, rather than activities that only scientists or philosophers with a taste for extravagant metaphysics engage in. Few cultures have developed the systematic procedures for investigating nature that we think of as science, but all cultures, including those in the distant past, have been curious about causal and explanatory relationships and have accumulated a great deal of causal knowledge of a mundane sort: for example, that exposure to fire or heat causes pain and tissue damage, that the impact of a moving rock can cause another object to break or move, and that crops require water to grow. I take this fact to suggest (though I readily acknowledge that it hardly conclusively establishes the correctness of) several additional ideas about how to understand our explanatory practices. First, it suggests that the construction of explanations and the acquisition of causal knowledge must have, at least sometimes, some practical point or payoff. There must be some benefit, other than the satisfaction of idle curiosity, that is sometimes provided by these activities. One way of interpreting different theories of causation and explanation is to think of them as providing different answers to the question of what this benefit might be. As already intimated, the theory that I defend takes the distinctive benefit associated with causal and explanatory knowledge to have to do with manipulation and control.

Second, we should expect there to be some sort of continuity between the everyday practices and kinds of causal/explanatory knowledge that are present in all human societies and the more systematic and sophisticated practices and kinds of causal/explanatory knowledge that characterize contemporary science. We should expect continuity on both substantive and methodological levels. On a substantive level, and in contrast to the tendency among some philosophers to regard all folk or commonsense beliefs as fundamentally mistaken, I see causal explanation in science as building on and requiring causal knowledge of a more mundane, everyday sort. If we did not have prior causal knowledge of the sort possessed by craftspeople and artisans about how to manipulate the natural world, or if we did not know various simple causal truths (e.g., massive objects fall when unsupported, solid macroscopic objects cannot pass through one another, gases and liquids diffuse unless surrounded by solid containers), we could not construct instruments or carry out or interpret even simple experiments. Similarly, on a methodological level, we should expect causal explanations in different areas of science to share at least some structural features with causal explanations in more ordinary contexts. We should see scientists who construct such explanations as attempting to satisfy some of the same explanatory goals and interests as people who construct explanations in ordinary life, but as achieving those goals by appealing to knowledge that is more rigorous, detailed, and systematic. Explanation in science does not give us something that is fundamentally different in kind from explanation in more ordinary contexts, but rather, as it were, a better version of the latter.

Given that we should expect some continuity between causal explanation in ordinary life and the sorts of explanation provided by sophisticated scientific theories, what form should we expect this continuity to take? There are at least two possibilities. One is to use structures and categories characteristic of some forms of scientific explanation to understand explanation in ordinary life. This is the approach taken by most philosophers, including defenders of the DN model and adherents of unificationist models of explanation such as Philip Kitcher (1989; see chapter 8). For example, as the account of the motion of the block down the plane (1.4.1) illustrates, explanations in science often (although by no means always) rely on explicit chains of deductive reasoning in a way that explanations in nonscientific contexts often do not. Thus, one might try to capture the continuity between ordinary and scientific explanations by understanding the former as having the structure of an "implicit" deductive argument or as explanatory in virtue of "tacitly" invoking or relying on such arguments—arguments that are explicit or overt in the case of scientific explanations. Similarly, some explanations in physics appeal to generalizations that are (or so I argue below) appropriately described as laws of nature. One might try to capture the continuity between such explanations and other sorts of explanation by arguing that all causal explanations "tacitly" rely on "backing" laws even when they do not explicitly appeal to generalizations at all or do not appeal to any generalization that might naturally be described as a law.

For a variety of reasons described in more detail below [in MTH, chapter 4], I find this approach unpersuasive. The "backing" relationship that allegedly holds between garden-variety causal claims and laws is difficult to make clear. Moreover, there are crucial features of the content of causal claims and causal explanations that are not captured by the "instantiation of laws" picture just described. I argue below that the DN view of the relationship between ordinary and scientific explanation gets matters exactly backwards. All human cultures have produced causal explanations, but the notion of a deductively valid argument and the notion of a law of nature are complex and sophisticated products of a very specific intellectual and scientific tradition. Rather than trying to understand all varieties of causal explanation in terms of these specialized categories, we should instead begin with a more general notion of causal explanation, understood in manipulationist terms, and then attempt to understand explanations that appeal to explicit chains of deductive reasoning and laws of nature as one specific variety within this genus.

1.7 | Accounts of Causation and Explanation Can Be Illuminating without Being Reductionist

There is another set of issues that deserves comment. There is a very widespread tendency in philosophical discussions of causation and explanation to assume that any interesting account of these notions must be "reductive." Just what this means is rarely made clear, but I take the general idea to be that concepts like "cause" and "explanation" belong to an interrelated family or circle of concepts that also includes notions like "law," "physical possibility," and other modally committed notions. (The circle may also include "counterfactual dependence," depending on how this notion is understood.) An account is reductive if it analyzes concepts in this family solely in terms of concepts that lie outside of it. It is also usually assumed that the acceptable concepts in terms of which a reductive analysis might be framed must satisfy certain epistemological and metaphysical constraints of an "empiricist" stripe; for example, it is assumed that the analysis should appeal only to notions that are "actualist" or nonmodal in the sense that they have to do with what actually happens or will happen, rather than what must or can happen. So-called regularity theories of law and causation are among the most familiar theories that are reductive in this sense, but there are other examples as well, such as Lewis's counterfactual theory. It is frequently assumed that any account of causation or explanation that fails to be reductive will be "circular" and hence unilluminating.

The account that I present is not reductive, and I am skeptical that any reductive account will turn out to be adequate. However, little if anything in my positive account turns on whether this skepticism is correct. Indeed, I would be delighted if someone were able to show how the nonreductive characterizations of cause and explanation that I provide might be replaced by reductive characterizations. By contrast, it is crucial to my argument that an account of causation and explanation can be worthwhile and illuminating without being reductive. Of course, the only really convincing way of showing this is to actually produce the account in question and to allow the reader to see that it *is* illuminating. Nonetheless, some general remarks at this point may be helpful in allaying misgivings about the nonreductive character of the theory that follows.

It is perfectly true that some nonreductionist theories of causation/explanation (e.g., *c* causes/explains *e* if *c* produces or generates *e*), with no further account of "production" or "generation," are completely unilluminating. But not all nonreductive theories are trivial in the way just illustrated. One way in which nonreductive theories can be interesting and controversial rather than trivial and empty is by virtue of conflicting with other reductive or nonreductive theories and suggesting different assessments of particular explanations. For example, according to the manipulationist account of explanation that I defend, explanations that involve action at a distance or otherwise fail to trace spatiotemporally continuous causal processes nonetheless can be genuinely explanatory. Theories such as Salmon's causal/mechanical model reach the opposite conclusion (see [*MTH*] chapters 3 and 8). Which of these approaches is more plausible seems quite independent of whether either achieves a successful reduction.

A second point that is worth keeping in mind is this: even if we opt for a nonreductive account according to which some notions in the circle of concepts that includes "cause," "explanation," and so on are explained in terms of other notions in that circle, we still face many nontrivial choices about exactly how the various notions in this circle should be connected with or used to elucidate one another—choices that can be made in more or less defensible ways. For example, although I offer a nonreductive treatment of the kinds of counterfactuals that are relevant to elucidating "cause," "law," and "explanation," I also argue that the counterfactuals on which the philosophical tradition has tended to focus in elucidating these notions are the wrong counterfactuals for this purpose. Again, the correctness of this claim seems completely independent of questions of reduction. As another illustration of the same point, I argue below that to elucidate certain kinds of causal claims, including claims about direct causal relationships and singular causal claims, one must appeal to counterfactuals with complex antecedents—counterfactuals that describe what will happen under combinations of manipulations or interventions, rather than under single manipulations. These sorts of counterfactuals have rarely been used by philosophers to elucidate causal claims, although one may think of them as implicit in some treatments of causation devised by nonphilosophers. Whether these are the right counterfactuals to look at in elucidating these causal claims or whether other counterfactuals would be more appropriate is again completely independent of the issue of reduction.

My general view is that in their enthusiasm for reductive accounts, philosophers have often misdescribed or oversimplified the content of the causal and explanatory claims they have hoped to reduce. We need more careful description of just what such claims say (and of the regularities and counterfactuals associated with them). Only after this has been done should we investigate what sorts of reductions are possible.

Yet another point is that even in the absence of a fully reductive account of causation and explanation, it may be possible to test or elucidate the content of particular causal and explanatory claims and show how they can be tested by appealing to other particular causal/explanatory claims and noncausal information such as correlational claims. In other words, we may be able to test or elucidate the claim that C_1 causes E_1 by appealing to our antecedent knowledge that some *other* causal claim (e.g., C_2 causes E_2) is true, along with other noncausal information, perhaps about correlations. The theory I propose has this sort of structure: it holds that we may test or elucidate the claim that C causes E by appealing to what will happen to E under an intervention on C. The notion of an intervention is itself a causal notion— among other things, it involves the idea of an intervention variable I that *causes* a change in *C*—but the causal relationships to which we need to appeal in characterizing what it is for *I* to be an intervention on *C* are different from the causal relationship between *C* and *E* that we are trying to elucidate. An account of this sort is not reductive, because it doesn't explain causation in terms of the concepts that lie outside of the circle of concepts to which "cause" belongs, but it is not viciously circular in the way that explaining "cause" in terms of a primitive notion of "production" would be.

1.8 | Epistemic Constraints on Explanation

A theory of causal explanation should also satisfy plausible epistemological constraints. It is true and important that we need to respect the distinction between issues about the content of causal or explanatory claims—issues about what such claims mean or what they say—and epistemological issues about how we test such claims or determine whether they are true. However, in our enthusiasm for this distinction, we should not overlook the equally important need to tell an integrated and plausible story about how these two sorts of issues fit together. In particular, our theory of the content of causal and explanatory claims should be accompanied by some epistemological story that makes it understandable how human beings can sometimes learn whether claims with that content are true or false from evidence that is actually available to them. This story should enable us to understand how (or to what extent) widely accepted procedures for testing causal and explanatory claims—for example, controlled experimentation and the causal modeling techniques discussed in [MTH] chapter 7—work. To put the point negatively, if our theory of what causal and explanatory claims say leaves it a complete mystery how we ever find out whether claims with that content are true or if procedures like controlled experiments that we ordinarily think of as testing such claims have no discernible bearing on whether claims with that content are true, this is an indication that something is fundamentally wrong. Either our theory about what such claims say or our views about how we should determine whether such claims are true or both need to be rethought.

Causal relationships are features of the world: they are "out there" in nature. By contrast, explanation is an activity carried out by humans and conceivably by some other animals, having to do with the discovery and provision of *information*, information about causal relationships. This leads, I argue, to an additional epistemic constraint on explanation that has no counterpart if our concern is just with causal claims. The constraint is, roughly, that explanatory information must be epistemically accessible information. It must be information that can be recognized, surveyed, or appreciated—in short, information that can contribute to understanding. The significance of this constraint is explored in [MTH] chapter 4.

1.9 | Desiderata

Drawing together the various strands of this discussion, I offer by way of summary the following nonexhaustive list of goals/constraints for a theory of causation and explanation, ordered from less to more controversial.[2]

The theory should be descriptively adequate in the sense that it captures relevant features of paradigmatic explanations in science and ordinary life. It should give us some insight into how such explanations work in the sense that it identifies the features or structures in virtue of which they convey understanding.

If the theory recognizes different varieties or sorts of causal explanation (as the theory that I propose does), it should show us what these have in common: why it is that they all count as species of the genus "causal explanation."

The theory should allow us to evaluate explanations. It should help us to distinguish between better and less good explanations, and it should enable us to understand the grounds on which such normative assessments are made. It should distinguish causal and explanatory clams from claims that are merely descriptive.

The theory should elucidate the successes and failures of previous philosophical theories of causal explanation; it should solve problems that are not adequately dealt with in previous theories.

The theory should have adequate epistemological underpinnings. If the theory tells us that an explanation works by conveying certain information or by possessing a certain structure, then there should be some plausible accompanying epistemological story that makes it clear how people who use the explanation can learn about this information or structure, how they can check whether the claims embedded in the explanation are correct, and so on. More generally, if an explanation provides understanding by conveying certain information, then this information should be epistemically accessible to those who use the explanation (cf. [MTH] chapter 4). Relatedly, the theory should enable us to make sense of widely accepted procedures for testing causal and explanatory claims.

Notes

1. The requirement that an invariant generalization must be stable under some interventions is a relatively weak, although far from vacuous, requirement. Various ways of strengthening this requirement without requiring invariance under all possible interventions are considered in chapter 6.

2. A similar list of desiderata is offered in Hughes (1993).

References

Barkan, D. 1999. *Walter Nernst and the Transition to Modern Physical Science.* Cambridge, UK: Cambridge University Press.

Bogen, J. and J. Woodward. 1998. "Saving the Phenomena." *Philosophical Review* 97: 303–52.

Collingwood, R. 1940. *An Essay on Metaphysics.* Oxford: Clarendon.

Dummett, M. 1964. "Bringing about the Past." *Philosophical Review* 73: 338–59.

Frautschi, S., R. Olenick, T. Apostol, and D. Goodstein. 1986. *The Mechanical Universe: Mechanics and Heat, Advanced Edition.* Cambridge: Cambridge University Press.

Gasking, D. 1955. "Causation and Recipes." *Mind* 64: 479–87.

Haavelmo, T. 1944. "The Probability Approach in Econometrics." *Econometrica* 12 (Supplement): 1–118.

Hughes, R. I. G. 1993. "Theoretical Explanation." In *Midwest Studies in Philosophy,* Vol. 18: *Philosophy of Science,* ed. P. French, T. Uehling, and H. Wettstein. Notre Dame, IN: University of Notre Dame Press.

Humphreys, P. 1989. *The Chances of Explanation.* Princeton, NJ: Princeton University Press.

Kitcher, P. 1989. "Explanatory Unification and the Causal Structure of the World." In *Scientific Explanation,* ed. P. Kitcher and W. Salmon, 410–505. Minneapolis: University of Minnesota Press.

Lewis, D. [1973] 1986. "Causation." Reprint with postcripts, *Philosophical Papers,* vol. 2, 159–213. Oxford: Oxford University Press.

Smith, C. and N. Wise. 1989. *Industry and Empire: A Biographical Study of Lord Kelvin.* Cambridge: Cambridge University Press.

von Wright, G. 1971. *Explanation and Understanding.* Ithaca, NY: Cornell University Press.

Weinberg, R. 1985. "The Molecules of Life." *Scientific American* 253.4: 48–57.

6 | COMMENTARY

6 | COMMENTARY

6.1 | Hempel's D-N Model

The phrase *deductive-nomological* is an apt description of the model that Hempel first proposed in his classic paper published in 1948.[1] For according to Hempel, many scientific explanations are deductively valid arguments having at least one statement of an empirical law in their premises. (The adjective *nomological* is derived from the Greek word *nomos*, meaning "law.") Only later, in 1962, did Hempel focus his attention on statistical explanation.[2] He then proposed his I-S model, according to which statistical explanations are inductive arguments with at least one statement of an empirical statistical law in their premises. Thus, both models are instances of the covering law thesis that all explanations are arguments (either deductively valid or inductively strong) that must involve laws. Before discussing Hempel's models in detail, it will be helpful to consider the underlying motivation for Hempel's proposals.

CARNAP ON THE MOTIVATION FOR HEMPEL'S D-N MODEL

As Rudolf Carnap points out in "The Value of Laws: Explanation and Prediction," scientists and philosophers have not always valued theories for their explanatory power. In the second half of the nineteenth century, for example, philosopher-scientists such as Ernst Mach and Pierre Duhem insisted that the proper function of scientific theories was not to explain phenomena but merely to classify and summarize experimental laws.[3] Duhem defines explanation as follows: "To explain (explicate, *explicare*) is to strip reality of the appearances covering it like a veil, in order to see the bare reality itself."[4] Duhem then argues that if scientific theories are intended to explain (in his own sense of stripping reality bare), then they will inevitably make science subordinate to metaphysics. The kinetic theory of gases and the wave theory of light, for example, posit the existence of atoms and the optical aether as an essential part of the explanations they provide. But neither atoms nor the aether can be observed directly. At best, claims about these "bare realities" can be tested only indirectly, by seeing what experimental laws they entail. Moreover, these experimental laws are compatible with a wide range of different assumptions about atoms and the aether. Thus, testing will be unable to confirm any particular theory (or the unobservable entities it posits) relative to its rivals. So, Duhem concludes, the assertion that atoms exist or that there really is an aether are metaphysical claims that empirical science can neither confirm nor refute.[5] Explanation is not, and cannot be, one of the aims of scientific theories.

Carnap rejects the claim of Duhem, Mach, and other positivists that science cannot explain; nevertheless he is sympathetic to their skepticism about theories involving metaphysical assumptions about unobservable entities and causes. Carnap insists that scientific theories can give genuine explanations but only if the explanations involve testable, empirical laws. Consider the example Carnap discusses of the German biologist and philosopher Hans Driesch (1867–1941). Driesch, who had done pioneering work on embryology and limb regeneration in sea urchins, discovered in 1891 that prior to the fifth division of a fertilized sea-urchin's egg any cell was capable of developing into a complete embryo. These and other findings led Driesch to espouse a "vitalistic" account, according to which an inner force or purpose in each living organism—what Driesch called an *entelechy*—is responsible for maintaining the integrity of the organism, directing development, and regenerating lost parts. In a number of works, Driesch championed his entelechy theory as an explanation for biological and psychological phenomena. He saw an important similarity between an embryo developing into an adult organism and a person voluntarily deciding to perform one action rather than another. In both cases, he argued, we cannot predict what will happen on the basis of the laws of physics and chemistry. We can explain what has happened after the event, but only because we then know how things have turned out. Driesch regarded the power of a human being to choose which action to perform as a refined manifestation of the teleological, directive power of the entelechy that is in every living creature. And what is an entelechy? It is a nonphysical, nonmaterial, nonspatial inner force that directs everything that an organism does, from cellular processes to voluntary actions. Every living thing has an entelechy, each species has its own distinctive kind, and it is possession of an entelechy that distinguishes living things from machines.

Despite their respect for Driesch's scientific work and his genuine desire that philosophy take science seriously, Carnap and Reichenbach could not accept that Driesch's entelechy theory really explained anything. In defending the entelechy theory against this accusation, Carnap recalls Driesch retorting that his introduction of the term *entelechy* to explain the behavior of organisms was no different from physicists introducing the term *magnetism* to explain the behavior of magnets and bits of iron. After all, we can neither see nor touch the force of magnetism. All we can actually observe is the motion of bodies, which magnetism was posited to explain. Similarly, Driesch maintained, it is legitimate for him to introduce entelechies to explain biological phenomena. Carnap responded by pointing out that the cases are relevantly different. For when physicists introduced the term *magnetism*, they did not simply posit the existence of an unobservable entity; they also specified laws that magnetized bodies must obey. These laws can be used to make predictions that can be tested by experiment and observation. Driesch's entelechy theory specifies no such laws and is thus completely lacking in predictive power. Therefore, Carnap concludes, Driesch's theory does not give genuine explanations.

Our discussion of Carnap's criticisms of Driesch's entelechy theory brings to light one of the central motivations for Hempel's D-N model of explanation, namely, the requirement that every genuine scientific explanation include at least one empirical law in its explanans. Another closely related theme is Carnap and Hempel's insistence that explanation and prediction go hand in hand: every genuine explanation must be capable of predicting its explanandum.[6] Later we will focus our attention on Hempel's defense of what he calls the *thesis of structural identity*, that, in all formal respects, explanations are the same as predictions, and predictions the same as explanations.

Hempel's Conditions of Adequacy for D-N Explanation

The requirement that the explanans include at least one empirical law is obviously not sufficient for an explanation for two reasons. First, just including a law, any law, is not enough; the law must be essential to the derivation of the explanandum. Clearly, we would not be much impressed by a purported explanation in which the laws mentioned were entirely irrelevant to the event or phenomenon needing explanation. Second, all by themselves laws do not entail that any specific thing will happen. Thus, when we seek to explain the occurrence of an event, we must also include in the explanans statements of various initial conditions that, in conjunction with the laws, logically imply the explanandum.

We thus arrive at the general scheme of Hempel's D-N model of explanation, which he summarizes in the selection "Two Basic Types of Scientific Explanation":

$$C_1, C_2, \ldots, C_k \quad \text{Statements of particular facts and initial conditions}$$
$$L_1, L_2, \ldots, L_r \quad \text{General laws}$$
$$\left. \right\} \quad \text{Explanans}$$

E	Description of the event, law, or fact to be explained	Explanandum

In his 1948 article, Hempel gave the criteria for a D-N explanation in the form of four conditions of adequacy, which he divided into two groups: logical and empirical.

- LOGICAL CONDITIONS OF ADEQUACY FOR D-N EXPLANATION

    R1 The explanandum must be a logical consequence of the explanans.

    R2 The explanans must contain general laws, and these must be essential for the derivation of the explanandum.

    R3 The explanans must have empirical content; that is, it must be capable, at least in principle, of test by experiment or observation.

- EMPIRICAL CONDITION OF ADEQUACY FOR D-N EXPLANATION

    R4 The sentences in the explanans must be true.

Before discussing Hempel's defense of the thesis of structural identity (that is, his thesis that explanation and prediction are formally identical) and his other model of explanation (the I-S model), a few brief remarks about Hempel's criteria of adequacy for a D-N explanation will be helpful in understanding criticisms of his views. We offer brief comments on each criterion in turn.

Why does Hempel require that the explanans logically entail the explanandum? The reason is simple. If we have genuinely explained why the explanandum event has occurred, then we must have given sufficient grounds for expecting that the event in question would occur. After all, the question "Why did E_1 occur?" usually arises when other outcomes seem possible: why did E_1 occur, rather than, say, E_2 or E_3? A completely satisfactory answer to this question would thus show that E_1 was the only event among the possible alternatives that could have occurred. Ideally, then, an explanation must be a logically valid argument, for in a valid argument, if the premises are true, then the conclusion also has to be true. The best kind of explanation is one in which, given the information in the explanans, the explanandum has to be true. If it were still possible, given the explanans, for the explanandum to be false, then we would not have explained, fully and completely, why the explanandum event occurred. As we shall see later in this commentary, Hempel is forced to relax this deductive standard of explanation in order to accommodate statistical explanations of particular events.

Our discussion of Carnap's criticisms of Driesch has already touched on the motivation for requiring that every explanation contain at least one law. Two additional features of this requirement are noteworthy. First, Hempel does not require that the laws be causal. As far as Hempel is concerned, there can be perfectly satisfactory scientific explanations solely in terms of Snell's law, Hooke's law, the equation of state for an ideal gas, or the like. These laws, expressed in the form of equations, are sometimes called *functional laws* because they specify the mathematical function relating the value of one variable (such as pressure) to the value of other variables (such as temperature and volume). (For more on functional laws, see the section "The 'Missing Values' Problem for Functional Laws" in the commentary on chapter 7.) Thus, Hempel deliberately does not limit scientific explanation to causal explanation. A second noteworthy feature of (R2) and the other logical conditions of adequacy is that Hempel does not require that the explanans contain any statements of initial conditions. Of course, such initial conditions must be included in any deductive explanation of the occurrence of a particular event, but Hempel wants his model to cover not only the explanation of events but also the explanation of laws. That is, he wants his model to apply to cases in which one law or a set of general laws (such as Newton's laws of motion and gravitational attraction) explain another law (such as Kepler's second law) by deductively implying it. (The explanation of one law by another, more general, law or theory is discussed in chapter 8. In the present chapter, the main focus is on the explanation of events.)

Condition (R2) mentions general laws, which would include laws of mathematics. While laws of mathematics often play an important role in scientific explanations, Hempel requires that at least one of the laws in the explanans be empirical. It is perfectly legitimate for the explanans to include mathematical laws (such as the Pythagorean theorem or the principles of algebra), but it must also include at least one empirical law that, unlike the laws of mathematics, can be tested by observation or experiment. Strictly speaking, (R3) is redundant, since it follows from (R1): anything that logically implies the explanandum, which has empirical content, must itself have empirical content. But such is the importance of (R3) to the spirit of Hempel's conception of explanation that it is listed as a separate requirement.

The three conditions (R1), (R2), and (R3) are grouped together under the heading of logical conditions of adequacy in Hempel's 1948 article because he thinks that we can tell whether something satisfies them without our needing any empirical information about the world.[7] But the last condition, (R4), is different. This condition requires the sentences in the explanans to be true. But for most if not all of these sentences their truth or falsity can only be determined empirically.[8] Elsewhere, Hempel calls a set of sentences that satisfies the logical conditions of adequacy a *potential explanans*. It is only when those sentences are actually true that we have a genuine explanation. As far as the D-N model is concerned, Hempel's notion of a correct or genuine explanation is strongly objective. A group of scientists may believe that they have given an explanation, and they may indeed have considerable justification for their belief, but unless the potential explanans they provide is actually true, no genuine explanation has been given. Thus, for example, Hempel would deny that the phlogiston theory of chemistry explained why metals burn in air (calcination). Why? Because the phlogiston theory is, as a matter of fact, false. Of course, we can say that phlogiston theory explained the calcination of metals, and this is often the simplest way of expressing ourselves. But if we accept Hempel's empirical truth condition (R4) for explanation, then what we really mean is that the phlogiston theory would have explained the burning of metals in air if it had been true.

Elliptically Formulated Explanations

Hempel recognizes that the explanations scientists actually give often fall short of the conditions of adequacy for D-N explanation. But Hempel does not see this failure of his model to describe actual scientific practice as in any way refuting the D-N model as an account of scientific explanation. Partly this is because Hempel's model is intended to be normative rather than merely descriptive: his aim is to articulate an ideal of what a good scientific explanation *should* be like, not simply to summarize or describe the explanations that scientists actually give. Hempel arrives at his model not by ignoring scientific practice, but by reflecting on paradigmatic examples of scientific explanation (such as those mentioned in his articles) and then isolating their essential features. With his D-N model clearly formulated, Hempel thinks he can account for why many actual scientific explanations fail to satisfy his model by appealing to pragmatic factors.

In many respects, Hempel's attempt to clarify the concept of explanation is similar to the analysis of the concept of proof given by logicians and philosophers of mathematics. The actual proofs that mathematicians write down often leave out steps (considered obvious or trivial), but this in no way undermines or refutes the strict, formal concept of proof that serves as our ideal.

The most common way in which the explanations given by scientists deviate from Hempel's D-N model is by being incomplete. For example, why does ice float on water? Most scientists would accept as a satisfactory explanation the assertion that, unlike most substances, water expands when it freezes. This is an instance of what Hempel calls an *elliptically formulated explanation*. Some laws or facts are left out of the explanans of such explanatory arguments. Why? Because the relevant laws and facts (such as Archimedes' principle of buoyancy and the inverse relation between density and volume) are so well known and accepted (at least by other scientists) that to write all of them down would be tedious and a waste of time. Consequently, the explanandum does not actually follow logically from the explanans, but the missing material can easily be filled in, and once it is filled in, the amended explanans does logically imply the explanandum. So this kind of deviation from the D-N model is innocuous and justified on pragmatic grounds. Far from refuting the D-N model, such examples actually confirm it, since the D-N account is able to explain in a plausible way why so many actual explanations are incomplete, and in making them more complete we come closer to the D-N ideal.

PARTIAL EXPLANATIONS AND EXPLANATION SKETCHES

There are other, more radical, kinds of explanatory incompleteness discussed by Hempel, namely *partial explanations* and *explanation sketches*. In a partial explanation, even when all the missing premises are added to the explanans, it still does not logically imply the explanandum. What does follow from the supplemented explanans is not the original explanandum but something more general. For example, we might seek to explain why some kinds of apple turn from green to red as they ripen. Suppose that we give an explanation but, even when our explanans is filled out, what it actually entails is that apples change color as they ripen. Since the change from green to red in some kinds of apple is a special case of this more general sort of change, to that extent, Hempel is prepared to call such explanations "partial." But strictly speaking, a partial explanation does not actually explain the explanandum it was invoked to explain. Hempel offers a similar example from the *Psychopathology of Everyday Life* in which Freud attempts to explain why he wrote the wrong date in his diary. What Freud actually explained (even when his explanans is filled out with the psychological facts and the presumed laws of Freudian psychology that are missing from his account) is not the particular slip he made about the date, but why he performed an action that, in some way or other, symbolically represented the fulfillment of one of his subconscious wishes.

The final category of incomplete explanations, *explanation sketches*, consists of those explanatory accounts that are so vague and incomplete—in short, so sketchy—that they fail to qualify as either elliptical or partial explanations. At best, an explanation sketch provides a general outline that might prove capable of being developed into a satisfactory explanation at some future time.

6.2 | Hempel's D-N Model and the Thesis of Structural Identity

In "The Thesis of Structural Identity," Hempel formulates his thesis as the conjunction of two claims or subtheses:

> every adequate explanation is potentially a prediction; and
>
> every adequate prediction is potentially an explanation.

Because Hempel's thesis asserts that there is a symmetry between explanation and prediction, his thesis of structural identity is often called the *symmetry thesis*, and criticisms of the thesis are instances of the *symmetry objection*. In his article, Hempel attempts to defend both parts of his thesis of structural identity against alleged counterexamples.

In order to understand Hempel's defense, several features of his thesis need to be underscored. First, although many instances of the symmetry objection focus on deductive explanations, Hempel believes that his thesis applies to both D-N and I-S explanations.

Second, Hempel uses the term *prediction* in a special sense. Ordinarily, we would say that Jones has predicted the outcome of the Super Bowl if he announces the name of the winning team before the game is played. In this sense, a prediction is simply a statement that need not be accompanied by any supporting reasons or argument. But since Hempel is primarily interested in the explanations and predictions given by scientists, he deliberately restricts the meaning of the term *prediction* to *predictive argument*. Without this restriction, his thesis would be obviously false.

Third, Hempel intends his models of explanation to cover both the explanation of laws and the explanation of events. But it makes little sense to talk about the prediction of laws, since laws are not the sort of things that happen at any particular time. So Hempel implicitly restricts his symmetry thesis to the explanation and prediction of events.

Finally, Hempel recognizes that we can sometimes use laws and theories to make inferences about what has happened in the past on the basis of initial conditions and particular facts that hold at a later time. These inferences are often called *postdictions* or *retrodictions*, and scientists ordinarily regard them as a species of prediction, for just like predictions about the outcome of an experiment, postdictions involve the deduction from a law or theory of something whose truth is not yet known. Hempel excludes postdictions from the scope of his symmetry thesis for an obvious reason. Even though we can use Newtonian mechanics to infer from the present positions of the sun, the moon, and the earth that a total solar eclipse occurred two thousand years ago, most philosophers cannot accept that the way things are now explains the way things were two thousand years ago; present events cannot explain past events. So the symmetry thesis is concerned solely with predictive arguments in Hempel's restricted sense, that is, with arguments in which all the initial conditions and particular facts hold at times prior to the event described in the conclusion.[9]

With these preliminaries in place, we may now consider objections to the symmetry thesis. Counterexamples to the first subthesis must be adequate explanations that are not potentially predictions; counterexamples to the second subthesis must be adequate predictions that are not potentially explanations. We begin with three objections to the first subthesis—alleged examples of adequate explanations that could not have been used to predict the events that they explain. For convenience, we shall assign them titles.

THE SYPHILITIC MAYOR

Michael Scriven asks us to consider a certain medical patient—the mayor of his town, let us suppose—whose name is Jones. Consider the question, Why did the mayor, Jones, contract paresis? Paresis is a form of general paralysis that affects only those who have had untreated syphilis for many years. This being so, the answer to our question is, Because Jones had untreated syphilis, and the only cause of paresis is syphilis. This, allegedly, is an explanation of why Jones now has paresis. But since only 10 percent of untreated syphilitics go on to develop paresis, Jones's syphilis could not have been used to predict his paresis. Indeed, since 90 percent of syphilitics do not contract paresis, we would have predicted exactly the opposite.

Hempel's reply to this objection is short and sweet: no adequate explanation has been given. Merely to cite one condition that is necessary (but not sufficient) for the occurrence of an event—even if that condition is based on a law—is not to explain that event. Presumably, it is a law that all people who die have needed to breathe oxygen while they were alive. But we do not suppose that we have explained why someone has died merely by pointing out that the person was an oxygen breather. Similarly, to use one of Hempel's own examples, we do not think that we have explained why a particular person won the Irish sweepstakes simply by adducing the fact that the person had bought a ticket.

Many people have judged Hempel's reply to be unsatisfactory, in part because, in the syphilitic mayor example, we know of no other factors that can influence the chance of someone's being afflicted with paresis. There is thus an understandable tendency to regard this example as an event of low probability being explained by a statistical law. So what may in fact be at issue here is whether we should insist, as Hempel does, that any statistical explanation confers high probability on its explanandum. Hempel's high-probability requirement will be discussed later when we examine the I-S model and the criticisms of it.

Evolutionary Theory

Hempel's critics have raised several versions of the evolutionary theory objection. In its simplest form, the objection runs as follows:

> Darwin explained the origin of species using this theory of natural selection working on random variations. Scientists accept that Darwin's theory offers genuine explanations, yet no scientist has been able to use Darwin's theory to predict the coming-into-existence of any new species. Thus, evolutionary theory explains but it does not predict.

In reply, Hempel stresses the importance of distinguishing between what he calls the *story* of evolution and the *theory* of evolution. The story of evolution is a narrative describing the sequence of species that have arisen and become extinct since life first appeared on earth. Even if this narrative is completely true, it has no explanatory import whatever. It is merely a description of what has happened in the past. The theory of evolution, by contrast, employs generalizations about heredity, mutation, and selection plus a host of detailed assumptions about environmental conditions and ecological relations. At best, this theory can offer only partial, probabilistic explanations of general facts about species survival and extinction. What it cannot do, at least in the present incomplete state of our biological knowledge, is explain why any particular species came into existence when it did. In short, by arguing that evolutionary theory explains considerably less than one might have supposed, Hempel denies that evolutionary theory explains what it is unable to predict. Biologists may have reasonably good explanations of why, following the extinction of the dinosaurs, other species were able to flourish and evolve, but they cannot yet explain why, say, a particular species of rat or aardvark evolved when it did, with the characteristics it did. The random nature of the variations on which natural selection works precludes explaining or predicting anything very detailed about the coming-to-be of new species. Like the syphilitic-mayor objection, then, this one fails because it is based on a false assumption (about the supposed extent of explanation offered by evolutionary theory).

A closely related objection to Hempel's theory, also involving evolutionary theory, was raised by Michael Scriven.[10] Even after a new trait has appeared in an organism, it is difficult to know whether that trait is adaptive. In particular, it is hard to judge the magnitude of the advantage, if any, that a new trait (such as a larger shell) confers on its possessor. Moreover, there is an ineliminable element of chance involved in determining which individuals will actually survive, for even the fittest giraffe might be killed by lightning before being able to mate and produce offspring. But once natural selection has operated for some time on many thousands or millions of individual organisms, it is much easier to identify which traits are adaptive. Clearly, we must confront the threat of a circularity of definition here—namely, that of simply identifying the fittest organisms with those that actually survive.[11] But that aside, we must grant that sometimes the information needed to explain an event can be obtained only by making inferences from the fact that the event in question has actually occurred. Hempel refers to explanations having this feature as *self-evidencing* explanations. This notion plays a major role in Hempel's treatment of the next objection, also proposed by Scriven, of a collapsing bridge.

SCRIVEN'S BRIDGE

Sometimes the only ground we have for asserting that some statement in the explanans is true lies in our knowledge that the explanandum event did, in fact, occur. This feature is at work in Scriven's example of the collapsed bridge. The collapse tells us not only that metal fatigue occurred, but also that it was serious enough to cause the failure of the entire structure. Similarly with the man who kills his wife out of jealousy or the patient who develops skin cancer after exposure to ultraviolet light: in all such cases, we could not have predicted the relevant events, but we can nevertheless explain them after they occur. Here again, contrary to the symmetry thesis, we seem to have explanations that are not predictions.

Hempel agrees that Scriven's bridge is a case in which we would not have had all the information necessary for predicting the collapse prior to its occurrence. But, Hempel insists, this does not mean that Scriven has given a counterexample to the first subthesis because, when interpreted correctly, that subthesis makes merely a *conditional* claim, namely:

> if all the information in the explanans had been known and taken into account before the occurrence of the explanandum event, then the event could have been predicted.

What Scriven has done is to show that, in some cases, the antecedent of this conditional is not satisfied. But this does not show that the conditional is false.

Hempel introduces the term *self-evidencing* to describe those explanations in which the information that the explanandum statement, E, is true provides a crucial evidential support for one of the particular statements in the explanans, C_1. Hempel insists that such explanations are not circular, since they are not being used to establish that E is true. As with any explanation, we already know (or presume that we know) that E is true, that the explanandum event happened as described. Thus, although we are using E as part (or even the whole) of the evidence for C_1, we are not then using C_1 as evidence for E. Hence, there is no epistemic circularity. Moreover, although we are using C_1 to explain E, we are not also using E to explain C_1. Hence, there is no explanatory circularity.

Hempel concedes that the second subthesis of the symmetry thesis is less secure than the first. That is, Hempel admits that there seem to be adequate predictions that are not potentially explanations. His own example, involving the association between Koplik spots and the measles, serves to illustrate the problem.

Koplik Spots

Koplik spots are tiny, whitish spots that appear on the inside of one's cheeks about a week before one succumbs to a full-blown case of measles. Supposing that the appearance of Koplik spots is *always* followed by the measles, the connection between them may be judged lawlike. Such a law can then be used to predict that a patient with Koplik spots will have measles a week later. Still, the Koplik spots do not explain why the patient will develop full-blown measles in a week's time.

Hempel suggests that our reluctance to regard the Koplik-spots argument as explanatory likely reflects our doubts about whether measles do in fact always follow the spots as a matter of universal law. Perhaps, he conjectures, we could produce Koplik spots by injecting a small quantity of the measles virus into someone's cheek without the spots then being followed by full-blown measles. But this response to the Koplik spots case is not entirely satisfactory. For even if the relation between Koplik spots and the measles is not one of universal law but merely one of high probability, it remains unclear why the resulting argument would not satisfy the conditions of Hempel's I-S model and hence qualify as a statistical explanation. (Hempel's I-S model is discussed in section 6.3 of this commentary.)

To some of his critics, Hempel's models of explanation seem vulnerable to the Koplik spots example because the models include no condition that mentions causation. Why do the Koplik spots fail to explain the later case of full-blown measles? Because the spots are not the cause of the measles. Rather, the spots and the full-blown measles are both joint effects of a common cause, namely, infection with the measles virus. In just the same way we can use the falling reading on a barometer to predict that a storm is approaching, but we do not take the barometer reading to be the cause of the storm's approach, nor do we take it to be an explanation of its approach. Again, we have prediction without explanation. The falling reading on the barometer and the approach of the storm are joint effects of a common cause, namely, a drop in atmospheric pressure.

The Flagpole and the Pendulum

Although Hempel does not consider these particular examples, critics of the symmetry thesis often cite them. A flagpole of height H casts a shadow of length S. Given the law that light travels in straight lines, and the elevation of the sun, θ, we can deduce the length of the shadow from the height of the flagpole. Thus, we have both an adequate prediction of S and, let us assume, an explanation of why the shadow has that particular length by using the equation $H = S \tan \theta$. But given that same equation, we could have just as easily deduced the height of the flagpole from the length of its shadow. Although perfectly fine as a prediction, this would not be accepted as an explanation: the length of the shadow does not explain the flagpole's height. The pendulum example is similar. From the period of a simple pendulum (the time it takes to perform one complete oscillation) we can deduce (i.e., predict) the length of the pendulum, but we do not think that the pendulum's period explains its length. Thus, on the face of it, the flagpole and the pendulum seem to provide plausible counterexamples to the second subthesis: not every adequate prediction is potentially an explanation.

6.3 | Hempel's I-S Model

As we have seen, Hempel's D-N model construes explanations as deductive arguments. But Hempel was aware all along that some scientific explanations could not be reconstructed in this fashion. This is especially true of theories in physics and genetics that use probabilistic laws to explain particular events. In 1962, Hempel turned his attention to these other, nondeductive arguments and formulated his inductive-statistical (I-S) model of explanation.

Hempel's Conditions of Adequacy for I-S Explanation

Although by 1962 Hempel no longer required that all explanations deductively entail the events they explain, he continued to defend the covering law thesis: just like D-N explanations, I-S explanations must have at least one (in this case statistical) law among their premises. Hempel also retained a substantial part of the concept of explanation that motivated the D-N model.

Like their D-N counterparts, I-S explanations are still arguments, and while the conclusions of I-S arguments no longer follow from their premises with logical necessity, Hempel insisted that the explanans must make the explanandum highly probable: the higher the probability, the stronger the argument and the better the explanation. The strength of an I-S explanation is measured by the inductive probability of its conclusion relative to its premises. Hempel assumed that this strength is equal to the numerical value of the probability given by the statistical law in the explanans. Thus, given the statistical law that Fs are very likely to be (or be followed by) Gs, and given some particular fact that a is F, it is very likely that a is G. An I-S explanation therefore has the following schematic form:

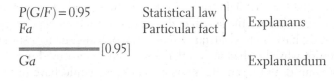

$$P(G/F) = 0.95 \qquad \text{Statistical law} \left.\right\}$$
$$Fa \qquad\qquad \text{Particular fact} \left.\right\} \quad \text{Explanans}$$
$$\overline{\overline{Ga}} = [0.95] \qquad\qquad\qquad \text{Explanandum}$$

Note that the conclusion of this I-S argument is not "a is almost certain to be G" but the unqualified statement "a is G." It is the fact that a is G that this inductive argument purports to explain. Hempel insists that expressions such as "a is almost certain to be G" are incomplete and thus neither true nor false. All meaningful, empirical statements of probability must be qualified as being relative to some body of evidence. It is either true or false that a is or will be G: what the full I-S explanation expresses is that the truth of Ga is very probable relative to the statistical law and the particular facts.

From everything we have said thus far, D-N explanations would seem to be a limiting case of I-S explanations: when the argument is deductively valid, the inductive probability of the conclusion relative to the premises is 1. Hempel denies this. He sees I-S explanations as essentially different from D-N explanations because, he insists, I-S explanations must be relativized to a particular "knowledge situation" (686). This epistemic relativity of I-S explanations arises from the requirement of maximal specificity (RMS), which Hempel imposes on all statistical explanations of particular events as a way of solving a certain problem of ambiguity that infects such explanations.[12] Hempel's RMS, and the problem of ambiguity it is meant to solve, are discussed in the next section. For the present, we will summarize the conditions of adequacy for an I-S explanation of a particular fact or event.

- ■ LOGICAL CONDITIONS OF ADEQUACY FOR I-S EXPLANATION

    S1 The explanandum must follow from the explanans with high inductive probability.

S2 The explanans must contain at least one statistical law, and this must be essential for the derivation of the explanandum.

S3 The explanans must have empirical content; that is, it must be capable, at least in principle, of test by experiment or observation.

- EMPIRICAL CONDITIONS OF ADEQUACY FOR I-S EXPLANATION

S4 The sentences in the explanans must be true.

S5 The statistical law in the explanans must satisfy the requirement of maximal specificity.

THE PROBLEM OF AMBIGUITY IN STATISTICAL EXPLANATION

As explained in his article, "Inductive-Statistical Explanation," Hempel proposed his RMS in response to the problem of ambiguity in statistical explanation. The problem of ambiguity can be regarded as an instance of a more general difficulty arising whenever we wish to use statistical information about classes of cases or events to decide the probability of a single case or event. This is the so-called problem of the single case: for any event that we wish to explain, there will be many different reference classes to which the event could be assigned; each choice of a reference class will present us with a different statistical law, and often these laws will have different probabilities associated with them. Let us consider one of Hempel's own examples.

We wish to explain why a particular day, n, November 27 in Stanford, has the property, W, of being warm and sunny. Thus the explanandum is Wn Among the many reference classes to which n belongs is (on the one hand) the class, N, of November days in Stanford, and the probability of warm weather on such a day, $P(W/N)$, is 0.95. So if we assign n to the reference class N, the high-probability condition is met and, apparently, we have an I-S explanation of Wn. That is, we have explained why November 27 in Stanford was warm and sunny by pointing out that n belongs to class N and citing the statistical law that says that the probability of warm and sunny days in that class is very high. But November 26 in Stanford was cold and rainy, and so n also belongs (on the other hand) to a different reference class, S, of immediate successors of cold and rainy days in Stanford. Assume, with Hempel, that $P(W/S) = 0.2$. Thus, $P(\sim W/S) = 0.8$, which we may agree qualifies as high. Now if, contrary to fact, n had not been warm and sunny, then we could have used the law $P(\sim W/S) = 0.8$ to explain why November 27 in Stanford was not warm and sunny. This, to Hempel, is intolerable. Of course, since n was warm and sunny, we could not use $P(\sim W/S) = 0.2$ to explain Wn. But, nonetheless, without some further condition of adequacy for I-S explanations, we are in the position of being able to "explain" the weather on November 27 in Stanford either way, whether the day was warm and sunny or not. For Hempel, this possibility of "explaining" an event whether or not it occurred means that no genuine explanation has been given at all.

We can state the problem of ambiguity as follows: given an I-S explanation with true premises of some explanandum, Ga, there will often be another I-S explanation with true premises and conclusion, $\sim Ga$. For convenience, let us call these two arguments (1) and (2):

Argument 1

$$P(G/F) = 0.95$$
$$\frac{Fa}{Ga} \quad [0.95]$$

Argument 2

$$P(\sim G/H) = 0.96$$
$$\frac{Ha}{\sim Ga} \quad [0.96]$$

Now, it might be thought that ambiguity is not really a problem at all. Since an explanation is sought only once the explanandum event has occurred, we would never actually accept both (1) and (2) as correct explanations. But Hempel judges this way of dismissing the problem to be unsatisfactory. Why? Because it is still the case that the premises of both (1) and (2) are true, and, let us assume, both are contained in our body of knowledge, K. (More will be said about Hempel's notion of a body of knowledge presently.) So if a had turned out not to be G, we could have given an equally strong explanation for $\sim Ga$. In what sense, then, have we explained the fact that Ga if by appeal to truths in K we could just as well have explained the fact that $\sim Ga$? To put the point in what may be a clearer form: if the essence of explanation is nomic expectability (that is, predictability based on laws), then we cannot accept that (2) would be just as good an explanation as (1) if a had failed to be G. Clearly, there is a connection between Hempel's concerns over explanatory ambiguity and his commitment to the thesis of structural identity (the symmetry thesis) between predictions and explanations. K contains the premises of (1) and the premises of (2). Thus, if we are justified in predicting Ga, we would also be justified in predicting $\sim Ga$. The fact that Ga turns out to be true and $\sim Ga$ false cannot be given as a reason for judging one of these predictions justified and the other not. We cannot without further restrictions allow that both predictions are justified, for we would then be justified in accepting a contradiction. Although K remains consistent under logical implication, it would not remain consistent under unrestricted inductive inference. Arguments like (1) and (2) would inevitably generate what Hempel has elsewhere called "inductive inconsistencies."[13]

We might try to avoid the problem of ambiguity by adopting Carnap's requirement of total evidence on all applications of inductive logic. This requirement demands that we use all the evidence available in determining degrees of confirmation (inductive probability), allowing us to use a part of the total evidence only if the evidence we ignore is irrelevant to the conclusion (i.e., only if the conclusion has the same probability given the relevant part of the evidence as it has given the total evidence). Hempel believes that this suggestion is on the right track. The task is to refine Carnap's requirement so that it addresses the specific problem of ambiguity in statistical explanation. As Hempel notes, we should not interpret Carnap's requirement of total evidence as demanding that we use all the information available to us. For in that case all probabilistic explanations offered at a given time would have the same (very large) explanans. Moreover, when we offer an explanation, we already know that the explanandum event has occurred. Including this fact in the premises would make the argument trivially deductively valid (not inductive) and nonexplanatory (since no law in the premises would be essential to the deduction of the conclusion). So we must limit Carnap's requirement of total evidence to just that evidence that is of potential explanatory relevance to the explanandum-event. In short, we must express the desired condition in such a way that I-S arguments satisfying it will use the right reference class for the purposes of explaining the explanandum-event. Here is Hempel's proposal.

HEMPEL'S REQUIREMENT OF MAXIMAL SPECIFICITY (RMS)

Consider our standard statistical explanation schema:

$$P(G \,/\, F) = r$$
$$\frac{Fb}{Gb} = [r]$$

Let S be the conjunction of the premises, and let K be the total set of statements accepted at the time the explanation is proposed. Hempel's RMS stipulates that if $(S \ \& \ K)$ implies that b belongs to a class F_1, and that F_1 is a subclass of F, then $(S \ \& \ K)$ must imply a statement specifying the probability of G in subclass F_1, say $P(G/F_1) = r_1$. Here, r_1 must equal r unless the probability statement $P(G/F_1) = r_1$ is simply a theorem of mathematical probability theory.

As its name indicates, the RMS insists that we assign the explanandum event, Gb, to the most specific reference class (the maximally specific reference class) to which it is known to belong. To understand better how the requirement works, we may consider again the explanation of why the weather in Stanford on November 27 was warm and sunny and the problem of epistemic ambiguity arising from the rival arguments (1) and (2). (Recall that argument (1) begins with the probability that, relative to the class N of days in November, day n has the property W of being warm; argument (2) begins with the probability that, relative to the class S of days succeeding cold and rainy days in Stanford, day n has the property $\sim W$ of failing to be warm and sunny.) In this example, we have stipulated that our body of knowledge, K, includes the premises of both argument (1) and argument (2). Consider argument (1), offered as an explanation of why that November day was warm and sunny. Is argument (1) an admissible I-S explanation of its conclusion, once we add the RMS? The conjunction of the premises of (1) and K implies that n, November 27 in Stanford, belongs to the class (N & S), which is a subclass of N. Now, either the conjunction of the premises of (1) and K implies the probability of W (a warm and sunny day) in (N & S), or it does not. If the value of $P(W/N$ & $S)$ is not given, then the RMS is not satisfied. In that case, neither (1) nor (2) will qualify as an adequate I-S explanation. Alternatively, if the conjunction of the premises of (1) and K does imply that $P(W/N$ & $S) = r_1$, then we have to consider what the value of r_1 is. On the one hand, if $r_1 = r$, where r is the numerical value of $P(W/N)$, then S is statistically (and explanatorily) irrelevant to W, and argument (1) satisfies the RMS. If there is no other reference class more specific than N to which n is known to belong, then argument (1) stands as an acceptable I-S explanation of its explanandum, Wn. On the other hand, if $r_1 \neq r$, and if the statement that $P(W/N$ & $S) = r_1$ is not simply a theorem of mathematical probability theory, then the RMS is not satisfied, and argument (1) is disqualified as an acceptable I-S explanation. In this case, some other argument referring n to the more specific class (N & S) may constitute an acceptable I-S explanation of Wn only if it, in turn, satisfies the RMS. The reason for the "unless" clause in the RMS is thus easy to appreciate: the class (N & W) is clearly a subset of N, but $P(W/N$ & $W) = 1$ by the probability calculus alone, and so the probability of W in the subclass (N & W) must differ from the probability of W in the class N. Obviously, without the "unless" clause, no inductive argument could ever satisfy the RMS.

As we have seen, Hempel explicitly relativizes the RMS to a particular knowledge situation, K. What is K? It is the class of all the sentences that are accepted as true by empirical science at a given time. Thus, K could (and quite likely does) contain some false sentences, and the contents of K will change over time.[14] This feature of Hempel's RMS has profound consequences for his concept of an I-S explanation. It means that, for Hempel, there is no such thing as an objective, "correct" inductive explanation independent of the scientific context. By relativizing the RMS to the beliefs of scientists at a given time, Hempel is admitting that inductive explanations (unlike their D-N counterparts) are fundamentally relative and subjective: they depend on the beliefs of scientists for their very existence. Hempel calls this feature of inductive explanations the *epistemic relativity of statistical explanation*.

6.4 | The Irrelevance Objection to Hempel's Models of Explanation

Many philosophers of science have judged Hempel's models of explanation unacceptable because they fail to rule out cases in which the information in the premises is explanatorily irrelevant to the conclusion they imply. There are two sets of such counterexamples. The first set concerns the explanation of laws; the second set (using examples from Peter Achinstein and Wesley Salmon) deals with the explanation of particular facts.[15] We shall focus on the second set.

THE ARSENIC EATER

Achinstein invites us to consider the ill-fated Jones, who eats at least a pound of arsenic and dies within twenty-four hours. Suppose that it is a law of nature that anyone who eats that much arsenic will be dead within a day. From this law and the initial condition that Jones ingests more than a pound of arsenic, we can deduce that Jones dies within twenty-four hours of his eating the arsenic. Thus, we have here an argument that satisfies all the conditions of Hempel's D-N model and that seems to be a good explanation of why Jones died. But then we learn that Jones did not die of arsenic poisoning but was run over by a bus shortly after his poisonous meal. Clearly, the D-N argument citing the lethal properties of arsenic now seems to fail as an explanation. Even though the premises of the D-N argument are true and make essential use of a (true) law to validly entail that Jones dies within twenty-four hours, those premises do not explain why Jones died. Why? Because it was the bus that killed Jones, not the arsenic. The premises of the D-N argument are explanatorily irrelevant to the explanandum.[16]

THE BIRTH-CONTROL PILLS

It is poor Jones yet again who figures in a counterexample to Hempel's model, this time from Wesley Salmon. John Jones takes birth-control pills regularly and fails to become pregnant. This is hardly surprising. Yet it is, presumably, a law that any man who takes birth-control pills regularly will fail to become pregnant, and John Jones is such a man. So we have a D-N argument with Jones's failure to become pregnant as its conclusion. But clearly this argument does not explain why Jones failed to become pregnant. Jones failed to become pregnant because he is a man, not because he is a man who took birth-control pills regularly. The explanans of the D-N argument does assert that Jones is a man, but it also includes the irrelevant information about the oral contraceptives. As Salmon argues, it is the presence of this additional, irrelevant information that robs the D-N argument of its explanatory power.

THE HEXED SALT

Salmon and Achinstein's examples can also be adapted to provide counterexamples to Hempel's I-S model. Suppose, for example, that ordinary table salt has a high probability (say, 0.95) of dissolving when stirred into cold water for five minutes. I take some salt and place a "dissolving spell" on it. It is now a sample of hexed salt.[17] It is a law that all hexed salt dissolves in water with a probability of 0.95. But although this law can be used to predict that my sample of hexed salt will dissolve in water, it does not explain why it does so. As with Salmon's birth-control pills example, the I-S argument based on the hexed-salt law contains irrelevant information.

Notice that in the hexed-salt example, Hempel's RMS is satisfied. There is no more specific reference class to which the hexed salt could be assigned that would make any difference to its chances of dissolving. Indeed, what has gone wrong in this example is that the salt has already been assigned to a reference class (the class of things that are hexed salt) that is *too* specific. The solution to the problem might seem to lie in a simple modification of Hempel's RMS. Instead of requiring that the explanandum be referred to the most specific class that makes a difference to the probability, we should instead assign the explanandum to the widest, least specific class that satisfies the RMS. Salmon has named this new requirement, *the requirement of the maximal class of maximal specificity.* Unfortunately, this proposal (which Hempel advocated in 1968) will not work.[18]

To see why it fails, suppose, for the sake of argument, that baking soda (sodium bicarbonate) has a probability of 0.95 of dissolving when stirred into cold water for five minutes. This is exactly the same as the probability of salt (sodium chloride) dissolving under the same circumstances. Some white powder (that we know to be salt) is stirred into cold water and after five minutes all of it has dissolved. What explains the fact that the powder dissolved? According to the requirement of the maximal class of maximal specificity, we must seek the widest class that satisfies the RMS. The class of things that are salt satisfies the RMS; so does the class of things that are baking soda; and so, too, does the disjunctive class of things that are either salt or baking soda. On the assumption that we know of no other chemicals that dissolve in water with a probability of 0.95, it follows that the class of things that are either salt or baking soda is the widest class that satisfies the RMS. Thus, according to the requirement of the maximal class of maximal specificity, what explains the dissolving of the powder is not the fact that it is salt but the fact that it is either salt or baking soda. On this proposal, to say that the powder is salt is irrelevant—just as irrelevant as saying that it is hexed or that it was mined in Utah. And that just seems wrong. Surely, what explains the dissolving of the powder—at least at some level of explanation—is the fact that it is salt. Thus, while the RMS fails because it permits the explanandum to be referred to classes that are too specific, the new proposal fails because, in some cases, it requires that the explanandum be referred to classes that are too wide.

A PROPOSED CURE FOR THE IRRELEVANCE PROBLEM: THE CAUSAL CONDITION

Reflection on the irrelevance problem (and on the symmetry problem, discussed earlier) leads naturally to the idea that an adequate explanation for a particular fact must include a description of the cause of that fact. As Salmon says, such a proposal would "put the *cause* back into *because*."[19] Thus, we need to consider whether Hempel's model can be repaired by adding an empirical causal condition—a condition requiring that the explanans contain a description of the cause of the explanandum and that this description play an essential role in the derivation of the conclusion of the explanatory argument.

As a remedy for the irrelevance and symmetry problems, the causal condition sounds promising. But as Timothy McCarthy has shown, the causal condition is no guarantee that the resulting argument will be genuinely explanatory.[20] Consider a version of the first of McCarthy's counterexamples, regarding an attempt to explain why a particular forest caught fire. Let us suppose that the actual cause of the forest fire was a lightning strike. Our D-N argument (substituting for the *A*s, *B*s, *C*s, and *D*s in McCarthy's formula) runs as follows:

1 All metals are conductors.

2 The forest was struck by lightning, and this screw is metallic.

3 Either this screw is not a conductor, or the forest was not struck by lightning, or the forest caught fire.

---

4 The forest caught fire.

Admittedly, this example is highly artificial. But the point is that, even though it satisfies Hempel's requirements and the causal condition, this argument surely fails to explain why the forest caught fire, because it is circular.

The circularity in the forest fire argument can be diagnosed in the following way. Premise (3) is a disjunctive statement containing three unrelated disjuncts. In order for premise (3) to be true, at least one of its disjuncts must be true. From premises (1) and (2) we can deduce that the first two disjuncts are false. So to know that the third premise is true, we must know that the third disjunct is true. But the third disjunct is simply a restatement of the conclusion. Thus, in a fairly obvious sense, the forest fire argument is viciously circular. It was this problem of vicious circularity that Jaegwon Kim attempted to avoid by imposing a further condition, namely, that the explanandum (the conclusion) not entail any of the conjuncts in the singular premises when those premises are written in conjunctive normal form.[21] Kim's condition rules out the forest fire argument because the third premise of that argument is already in conjunctive normal form and is entailed by the explanandum. McCarthy responded by devising a new argument that satisfies Hempel's conditions, the causal condition, and Kim's condition but that still fails to explain its conclusion.

The purpose of the causal condition, in meeting the irrelevance and symmetry objections, is to tighten the connection between explanans and explanandum so that their deductive relation exists by virtue of some actual causal connection. But notice that the proposed causal condition merely requires that the premises of an explanatory argument mention the event, c, which, as a matter of fact, is the cause of the explanandum event, e. The causal condition does not require that the premises contain a statement that says "c is the cause of e." McCarthy's counterexamples show that, in this form, the causal condition is too weak: a mere mention of the cause does not secure the explanatory relevance we seek. But a suitable strengthening of the causal condition is not easily accomplished. Suppose we strengthen the causal condition so that it now does require a statement in the premises that says "c is the cause of e." The consequences of this adjustment for the theory of explanation are radical. For it now emerges that any explanation of a particular event, e, would reduce to a very simple argument that has but a single (true) premise, namely: "c is the cause of e." Moreover, on such a construal of explanation, no law is explicitly mentioned. As David Hillel-Ruben has remarked, this argument is so trivial that the thesis of Hempel (Carnap, Mill, Aristotle, and many others) that explanations are arguments is thrown into doubt.[22] Later we will consider one way in which a causal model has been developed, by James Woodward, that gives a systematic account of how causes are identified and discusses the role of laws and generalizations weaker than traditional laws in scientific explanations.

6.5 | Railton on What Is Wrong with Hempel's Models of Explanation

In his article "A Deductive-Nomological Model of Probabilistic Explanation," Peter Railton begins by discussing what he takes to be the fundamental flaws in Hempel's two models of explanation. After diagnosing these flaws, Railton presents his own account, the D-N-P model of explanation.

Nomic Subsumption

Throughout his writings on explanation, Hempel's guiding assumption is that to explain some event or phenomenon is essentially a matter of subsuming the explanandum under a law (whether universal or statistical). For Hempel, explanation just is nomic subsumption. But as we have seen from the earlier discussion of the thesis of structural identity and the symmetry objection criticizing that thesis, not every case of nomic subsumption is an explanation. We can, for example, use the laws of optics and geometry to predict the height of the flagpole from the length of its shadow and the elevation of the sun, but this prediction does not explain why the flagpole has the height it does. Nomic subsumption is not sufficient for explanation.

According to Railton, what is lacking in D-N arguments that fail to explain is an account of the underlying mechanisms responsible for causing the fact to be explained. Without providing some such account, many D-N arguments are too superficial to explain their conclusions. To explain why this *A* is also a *B*, we need to do more than cite the law that all *A*s are *B*s; for even if the law in question is causal, the explanation would be incomplete without an account of the mechanism (or mechanisms) at work. This condition, that explanations provide a mechanism, plays a central role in Railton's account of probabilistic explanation. Railton insists that explanations require mechanisms even when the phenomena in question are not deterministic but irreducibly probabilistic.

INDUCTIVE ARGUMENTS

Railton rejects two prominent features of Hempel's I-S model of probabilistic explanation: the requirements of high probability and maximal specificity. Railton thinks that both of these objectionable features stem from Hempel's insistence that the statistical explanation of particular facts be a type of *inductive* argument. Railton's proposal, in light of this diagnosis, is simple and radical: probabilistic explanations are not inductive arguments. Indeed, on Railton's D-N-P model, probabilistic explanations are not arguments at all; they include arguments, but none of the arguments they include is inductive, and the explanations are not themselves arguments.

To appreciate Railton's diagnosis of the source of those two features he finds worrisome in Hempel's I-S model, consider, first, the high-probability requirement. An inductive argument's strength depends on the degree of probability that its premises confer on its conclusion—the higher the probability, the stronger the argument. If statistical explanations are inductive arguments, then the high-probability requirement immediately follows: a statistical explanation will be strong only if it establishes that the explanandum occurs with suitably high probability. Now consider the RMS. Inductive arguments are notoriously sensitive to the addition of information to their premises. (In this respect, inductive arguments differ from deductively valid arguments. If an argument is deductively valid, then it remains valid, no matter what is added to its premises.) If Smith is a twenty-year-old woman, and 90 percent of such women survive to celebrate their fortieth birthday, then, relative to those premises, the probability of the conclusion that Smith will live at least another twenty years is quite high. But once we add the information that Smith has amyotrophic lateral sclerosis (Lou Gehrig's disease, a motor neuron disease), the probability of her long-term survival plummets. Thus, we generate the problem of ambiguity that Hempel's RMS is intended to solve. Again, the root of the problem is the inductive nature of the arguments involved.

THE HIGH-PROBABILITY REQUIREMENT

Assuming that Railton has correctly traced the problems with Hempel's high-probability and maximal-specificity requirements to their inductive source, are these features as damaging to Hempel's account as Railton claims? The high-probability requirement has often been criticized (by Salmon, Jeffrey, and others) on the grounds that it rules out the possibility of explaining improbable events. Railton gives the example of a genuinely random wheel of fortune with 99 red stops and 1 black stop. The stipulation that the wheel is genuinely random means that no factor, even in principle, can affect the outcome once the wheel has been set spinning. Each stop thus has exactly the same probability of being chosen when the wheel comes to rest. In a setup like this, it seems absurd to insist that we can explain why the wheel halts at a red stop but not why the wheel halts at the black one. Surely, the explanation is equally good in either case, regardless of the outcome— there is an irreducibly indeterministic mechanism that generates red with probability 99/100 and black with probability 1/100. Hempel's high probability requirement seems to conflate *inductive strength* with *explanatory value*. It is the second that is of concern in seeking an adequate account of explanation. The predictive inference to the conclusion that the wheel will stop at black is much weaker, inductively, than the inference to the conclusion that the wheel will stop at red. But the explanation of why the wheel stops at black is just as good as the explanation of why the wheel stops at red, since, in either case, the explanatory statements provide us with as much of an understanding of the underlying mechanism as it is possible to have. Explanations should be judged by the completeness of the explanatory information they provide, not by the strength of the inferences they permit. To think otherwise is, in Railton's words, to confuse explanation with induction.

The Maximal Specificity Requirement

Railton rejects Hempel's requirement of maximal specificity because it relativizes the concept of probabilistic explanation. If Hempel were right, then there could be no correct or true probabilistic explanation of anything, for all such explanations would be relative to the state of scientific knowledge (and ignorance) at the time they were proposed. For Railton this is unacceptable because it would deprive probabilistic explanation of its objectivity.[23] It is important to note here that Railton is not denying that which explanations we believe to be correct depends on our beliefs. In just the same way, which propositions we believe to be true also depends on our beliefs. But, Railton insists, what it is for something to be a correct explanation cannot depend on our beliefs and neither can what it is for a proposition to be true. Indeed, it is only if being a correct explanation and being true are belief-independent that we can make sense of our beliefs that particular propositions are true or that particular explanations are correct.

Part of the reason for Hempel's error, Railton thinks, is that Hempel has failed to take seriously his own distinction between statistical descriptions (that just happen to be true) and genuine probabilistic laws. Suppose, on the one hand, that determinism were true. In this case, all laws would be universal; there would be no probabilistic laws and thus no correct or true probabilistic explanations. In such a purely deterministic world, statistical explanations would be merely a stopgap measure until we could discover the true, objective D-N explanations of the things that happen in that world. The statistical "laws" mentioned in such explanations would not be real laws at all but merely expressions of our ignorance at a given time. While it may seem as if, in a completely deterministic world, the conditions for Hempel's I-S model of explanation would be satisfied (capturing our best probabilistic explanations in varying states of ignorance), in fact they would not be satisfied, simply because in such a world there could be no genuine probabilistic laws at all.

On the other hand, suppose that the world is not deterministic but is governed by at least some true probabilistic laws. Once we realize that true probabilistic laws require genuine indeterminism, the epistemic relativity thesis collapses and with it goes the motivation for Hempel's RMS. Remember that the RMS says that, for the purposes of explaining why o has property G, we should assign o to the most specific reference class of which we have knowledge that would make a difference to the probability that o is G. (It is the reference in RMS to our state of knowledge, that is, to our beliefs at a given time, that makes probabilistic explanation epistemically relative on Hempel's account.) But if it is a genuine law that the probability that o is G, given that o is F, is 0.95, then there cannot be any more specific reference class than F to which o could be assigned that would make a lawlike difference to the probability that o is G. If it is a genuine probabilistic law that $P(G/F) = 0.95$, then, if o is F, the real, objective probability that o is G is 0.95, regardless of what anyone thinks or believes. Genuine probabilistic laws require indeterminism, and such indeterminism guarantees the objectivity of statistical explanation.

6.6 | Railton's Deductive-Nomological Model of Probabilistic Explanation

Railton's main concern is the explanation of what he calls *lawful* chance phenomena. Chance phenomena (such as radioactive decay) are lawful when they obey a statistical or probabilistic law. Like Railton, we shall focus on the explanation of particular events that are brought about by chance mechanisms in a lawful way.

Several key elements of Railton's D-N-P model of explanation have emerged already from the preceding discussion of Railton's critique of Hempel. Those elements can be summarized as follows:

- all explanations are objective; none of them is relative to a set of beliefs or to the state of scientific knowledge at a particular time;

- explanations are not arguments, nor should they be evaluated as if they were arguments; explanations are accounts that provide relevant information;

- explanations (whether probabilistic or not) require not only laws but also an account of the underlying mechanism(s);

- probabilistic explanations require genuine probabilistic laws; genuine probabilistic laws require indeterminism;

- there is no high-probability requirement for probabilistic explanation; improbable events can be explained just as well as highly probable events.

Railton's Conditions of Adequacy for D-N-P Explanation

Like Hempel's D-N model, Railton's D-N-P model can be set out in schematic form. (We have made minor changes to Railton's notation.)

Explanandum Ge, t_0 : e's having property G at time t_0.

Explanans

a A theoretical derivation of a probabilistic law of the form (b).   Theoretical derivation

b $(t)(x) [Fx,t \rightarrow P(Gx,t)=r]$ Probabilistic law

c Fe,t_0 Initial condition

──────────────────────────

d $P(Ge,t_0)=r$ Statement of a single-case propensity

e (Ge,t_0) Parenthetic addendum

Deductive argument (bracketing b, c, d)

The explanandum, the event to be explained, is e's having property G at time t_0, which Railton writes as Ge, t_0. For example, the explanandum might be that a particular wheel of fortune, e, stopped on black at time t_0. The statement that this event occurred appears again at the end of the explanans, but only as a parenthetic addendum. Putting Ge, t_0 in parentheses in line (e) is Railton's way of indicating that the explanandum is not the conclusion of the explanans, nor is it inferred from the explanans. Rather, the parenthetic addendum is put there simply to remind us that the explanandum event did, in fact, occur. Remember, although the explanans contains arguments—the deduction of (d) from (b) and (c), plus whatever arguments are involved in the theoretical derivation of the probabilistic law in line (a)—the explanans as a whole is not itself an argument.

The probabilistic law written schematically in line (b) should be read as follows: for all times t, and all things x, if x has property F at time t, then x has probability r of having property G at that time. For example, if F is the property of being a genuinely random wheel of fortune with 99 red stops and 1 black stop, and G is the property of stopping on black after the wheel has been set spinning, then the value of r is 1/100. (In this example, the probability does not depend on time and so the variable t can be dropped from the statement of the law.)

Line (d) is derived from lines (b) and (c) in two steps: by universal instantiation followed by modus ponens. Again, it helps to consider our simple example of the wheel of fortune. The probabilistic law for the wheel of fortune is a universal generalization: it says (ignoring the time variable) that "for any object whatever, if that object is a genuinely random wheel with 99 red stops and 1 black stop, then the probability that the wheel will stop on black is 1/100." Universal instantiation allows us deduce that this generalization holds for a particular object, such as e. So we get the conditional statement: "if e is a genuinely random wheel with 99 red stops and 1 black stop, then the probability that e will stop on black is 1/100." In our wheel of fortune example, line (c) tells us that the antecedent of this conditional statement is true: e is indeed a genuinely random wheel with 99 red stops and 1 black stop. Thus, by modus ponens we can deduce further that the probability that e will stop on black is 1/100. As Railton remarks about an exactly similar deduction (argument (2) on page 698), it is vital for the first step of this derivation that (b) be a genuine universal law, for only then can conclusions be deduced from it by universal instantiation. If (b) were merely a statistical generalization saying, in our example, that in a very large sample of N wheels, on average, N/100 of them stop on black, nothing could be validly deduced about the probability that this particular wheel e will stop on black.

Line (a) of the explanans expresses Railton's requirement that any adequate explanation must specify an underlying mechanism that causally brings about the event to be explained. While the derivation of the probabilistic law of the form (b) is indeed meant to issue from scientific theory, Railton has in mind a quite liberal notion of what counts as a mechanism. For example, in the quantum-mechanical derivation of the law of radioactive decay, which Railton gives as his central illustration, the alpha particle escapes from the nucleus by what is called *tunneling* or *barrier penetration*. But tunneling is not a visualizable mechanism or any kind of process that is continuous in space and time. Like other distinctively quantum phenomena, it involves the discontinuous changes in energy and momentum of a system obeying the fundamental laws of quantum mechanics. The alpha particle, for example, only has a distinct existence as a particle, with a well-defined trajectory and momentum, after it has emerged from the nucleus. To refer to tunneling as a mechanism requires using the word *mechanism* in a very broad sense to mean, roughly, *a law-governed process (not necessarily spatially continuous or extended in time) that causally brings about the event we are interested in.* Newton's force of gravity, acting instantaneously at large distances through a vacuum, would qualify as a mechanism. The important thing, for Railton, is the theoretical derivation, not the intuitive idea of a mechanism as an assembly of pulleys, wheels, and strings.

Propensities and Probabilities

The deductively valid argument appearing explicitly in a D-N-P explanation is the inference from (b) and (c) to (d). Thus, just as in Hempel's D-N model, this argument contains a statement of law and a statement of initial conditions in its premises. The crucial difference between the two models lies in the nature of the law and the resulting difference in the conclusion. In a D-N-P explanation, the law is probabilistic and expresses the propensity, $P(Gx,t)$, for an object or system, x, to have the property G at time t. What is a propensity? It is a property of the single system x, a lawlike tendency or physical probability for that system to behave in a certain way. The conclusion that $P(Ge,t_0)=r$, which is deduced from (b) and (c), is a single-case probability; it is the probability that this particular e is G at the time in question. This notion of a propensity as a single-case probability is crucial to Railton's model. As Railton says in his paper, "the D-N-P model is viable only if sense can be made of propensities, or of objective, physical, lawful, single-case probabilities by any other name" (705).

We can try to grasp what propensities (or single-case probabilities) are supposed to be by considering the simple example of a perfectly symmetrical coin. Because of the perfectly symmetrical distribution of its mass and the laws of mechanics, the coin has a physical probability of exactly ½ of yielding a head on any toss. Even if the coin is never tossed, it has that propensity. If the coin is tossed and lands tails, its propensity to yield heads on the next toss remains ½. Propensities are not the same as frequencies; rather, they are properties that explain why we observe certain kinds of outcomes with the frequency that we do. Thus, on Railton's model, if (as we have supposed) the coin has a lawlike, indeterministic tendency of ½ to yield a head on any toss, and if we toss the coin eight times and get, much to our surprise, eight heads in a row, then the full, complete explanation of that outcome is that the coin had a propensity of $(½)^8$ to behave in that way when tossed. The full explanation of the outcome has been given when the true chance of its occurrence, however low, has been given.

Like probabilities, propensities range in numerical value from 0 to 1. But propensities are not the same as probabilities. The difference between probabilities and propensities can be illustrated in the following example. Suppose that we have a 50-50 mixture of two radioactive elements, A and B, each with a different, known propensity to emit alpha particles. We know how likely it is that the A atoms will emit an alpha particle in a given time interval, and we know how likely it is that the B atoms will emit an alpha particle in the same time interval. Suppose that an alpha particle has been emitted from the mixture. From the information provided we can, using Bayes's theorem, calculate the probability that the alpha particle came from the A atoms, and we can also calculate the probability that the alpha particle came from the B atoms. But neither of these inverse probabilities can be a propensity, for an alpha particle existing at a given time cannot have a propensity to have been produced from a particular source at an earlier time.[24] Propensities are indeterministic causal tendencies. Like causes in general, propensities have a forward temporal direction, from the past and present into the future. Unlike probabilities, propensities can never run from the present to the past. Obviously much more work is needed before we can have a clear notion of what propensities are. Railton's own suggestions are the barest beginnings of this project.

RAILTON'S RESPONSE TO SOME OBJECTIONS

Toward the end of his article, Railton raises several objections to his D-N-P model of explanation and responds to them. One of these objections (the first) is particularly instructive. The criticism is that Railton's model of probabilistic explanation is at once too narrow and too broad. On the one hand, the D-N-P model may be judged too narrow because very few phenomena outside of quantum mechanics are genuinely indeterministic. Sciences such as evolutionary biology, genetics, epidemiology, economics, sociology, fluid mechanics, and meteorology all deal with phenomena that are deterministic but very complicated. Statistics and probability are used by these sciences, but largely because we are ignorant of the many thousands (often, millions) of initial conditions actually determining the behavior in which we are interested. Thus, on Railton's model, none of these uses of probability and statistics would be explanatory; the calculations, predictions, and arguments involved are not D-N-P explanations. On the other hand, Railton's model is accused of being too broad because, if any genuinely indeterministic propensities whatsoever are at work, however minuscule, then a proper D-N-P explanation must take account of them. For example, when an ice cube is placed in a glass of warm water, we expect it to melt. Normally, we would think that we have explained this melting by showing that it follows from the laws of classical thermodynamics (such as the second law stating that, in isolated systems, entropy always increases). But ice and water are composed of molecules, and the molecules of atoms; since atoms are subject to quantum-mechanical laws, there is a small probability that those molecules could move in such a way that the ice does not melt but instead becomes even colder while the water becomes hotter. The probability that this might happen is very, very small, but it is not 0. Thus, the propensity of the ice to melt when placed in water is not exactly 1. (So the second law of classical thermodynamics is not really a law at all because it is, strictly speaking, false that entropy must always increase.) Consequently, a proper explanation of the melting has to be a D-N-P explanation in which the exact numerical value of the propensity is calculated. (This, of course, is a very difficult task and not one we are normally able to perform.) Moreover, as Railton points out, just about everything that happens in the world, even at the macroscopic level is, to some very small degree, contingent on what happens at the atomic level. So, even an explanation of human behavior or the motions of the planets would have to include a fiendishly complicated quantum-mechanical calculation of the relevant propensity before it could qualify as a proper, probabilistic explanation.

Railton's response to both arms of the objection is to concede the point it makes, but to deny that this reflects badly on his model. Thus, Railton willingly embraces the conclusion that if a system is genuinely deterministic, then however complicated it might be, no real explanation of its behavior can be probabilistic. Whatever else we are doing when we appeal to probabilities or statistics in contexts like this—whether it is predicting, approximating, or estimating—it is not explaining, and we should not pretend that it is. Likewise, if the system is indeterministic, to whatever small degree, then we cannot shirk our explanatory responsibility by ignoring that fact. In the final analysis, the aim of explanation is to achieve a true understanding of the ways things really are. Being conscientious about that aim makes the task of explanation—genuine explanation—extremely difficult. The fault, if fault there be, lies in the complexity of the world, not in the demands of the D-N-P model.

6.7 | Kitcher's Unification Model

In a series of influential papers, Michael Friedman and Philip Kitcher have argued for the importance of unification in science and its intimate connection with explanation. Indeed, most philosophers of science now writing about explanation regard the unification model, along with various versions of the causal model as the two main successors to Hempel's covering law account.[25] As with much else in the philosophy of science, the difficulty lies in the details. Can one give a precise explication of the notion of unification that reveals its connection with explanation and does justice to scientific practice while, in addition, being able to defend the account against the objections (asymmetry, irrelevance, etc.) afflicting the covering law model?[26]

In terms of the helpful taxonomy introduced by Wesley Salmon, the unification approach to explanation shares, along with Hempel's covering law model, an "inferential" conception of explanation by seeing explanations as essentially arguments or inferences having the explanandum (the thing to be explained) as their conclusion. As Salmon notes, Hempel deliberately left room in his D-N model for the explanation of laws (in addition to the explanation of particular facts) but, in the "notorious footnote 33" of Hempel and Oppenheim's 1948 paper Hempel confessed his inability to solve the conjunction problem.[27] That problem is how to rule out as non-explanatory the trivial deduction of, say, Galileo's law (G) from its arbitrary conjunction with Boyle's law (B) when the valid argument from (G & B) to G satisfies all the conditions of Hempel's D-N model. The unification approach takes the explanation of laws as its primary focus and the solution of the conjunction problem as one of its important tasks.

Another distinctive feature of the unification approach noted by Salmon is its "top-down" focus on the most general unifying theories in science as providing the best, and perhaps the only, acceptable explanations. Kitcher, for example, willingly embraces the top-down "Kantian" aspect of his unification model by arguing that what we regard as laws of nature and the processes we identify as causal are the outcome of accepting unifying theories rather than notions to which we have independent and prior access. As Kitcher has declared, "the 'because' of causation is *always* derivative from the 'because' of explanation" (italics added).[28] This differs from the "bottom-up" approach of Salmon's causal-mechanical model with its emphasis on the particular, local, causal mechanisms that link events. It also contrasts with Hempel's willingness to regard explanations as correct and genuine, even when the laws in the premises are low-level empirical generalizations that may not be integrated into a more comprehensive theoretical framework.

FRIEDMAN'S MODEL

Michael Friedman gave the first formal analysis of a unification model of explanation,[29] which was criticized by Philip Kitcher.[30] Then, in his "Explanatory Unification," Kitcher gave a different unificationist account that continues to be the focus of active discussion.

The starting point for unificationists is that understanding is the essence of explanation, and that scientific understanding is achieved when a general theory is able to tie together a group of laws in one unified framework. The core of Friedman's approach lies in the notion that in paradigm cases—when, say, the kinetic theory of gases enables the derivation of the Boyle-Charles law, Gay-Lussac's law, and Graham's law—explanatory unification is achieved by reducing the number of laws we must accept as brute independent facts. A crucial feature of Friedman's formal analysis is his notion of a partition. A *partition* of a sentence S is any set of sentences that when conjoined are equivalent to S and such that every member of the set is acceptable independently of S.[31] For example: In the case of the arbitrary conjunction of Galileo's law and Boyle's law, S is equivalent to the conjunction (G & B), and clearly, G and B are each acceptable independently of S. (We have grounds for accepting G that are not also grounds for accepting (G & B), and similarly for B.) Thus <G, B> is a partition of S. If there is no partition for S, then Friedman says that S is K-*atomic*. (The "K" in "K-atomic" refers to the set of lawlike sentences, K, that are accepted by the scientific community at a given time.) Friedman then gives a set of necessary and sufficient conditions,

D1, for one sentence to explain another. In (relatively) plain English, D1 reads: S_1 explains S_2 if and only if (1) S_1 logically implies S_2; (2) S_2 is acceptable independently of S_1; (3) when we add S_1 to the set of S_1's independently acceptable consequences, the number of K-atomic sentences in the smallest equivalent set of such sentences is thereby reduced. D1 blocks cases of spurious unification because an arbitrary conjunction does not reduce the set of its independently acceptable consequences.[32] In cases of genuine unifying explanation, we add a theory to a collection of lawlike sentences (that, up until then, were each accepted on independent grounds) and end up with a new (larger) set of sentences in which fewer members have to be accepted on independent grounds.

Although it was not evident when it was first published, Philip Kitcher proved that Friedman's initial formal characterization of explanation by unification (see D1 above) entails that sentences can explain only if they are K-atomic. While this certainly would solve the conjunction problem, it would also rule out many paradigm cases of genuine scientific explanation. Kitcher gives the example of the usual derivation of the law for the adiabatic expansion of an ideal gas, which has both the Boyle-Charles law and the first law of thermodynamics in its explanans. Since those laws are acceptable independently of one another, their conjunction is not K-atomic; nonetheless they provide a genuine explanation of the adiabatic expansion law. Kitcher's other counterexamples involve the explanation of complex phenomena, such as why lightning flashes are followed by thunderclaps, which uses laws of electricity, thermodynamics, acoustics, etc., each of which is independently acceptable.

More generally, Salmon has given reasons for doubting whether there can be any K-atomic sentences of lawlike form at all.[33] Newton's law of universal gravitation is one of Salmon's examples. We can divide up pairs of bodies into three groups: large-large, large-small, and small-small (where Salmon defines "large" as "having a mass at least as great as the smallest planet"). In Newton's day scientists had evidence for inverse-square-law attraction only for the first two groups, not for the third. So the threefold division would not serve as a partition. Thus Newton's law would have been K-atomic. But after Henry Cavendish's torsion balance experiment in the eighteenth century, Newton's law would not be K-atomic because we then had, for the first time, independent evidence for the small-small class of objects.[34] More generally, any law of the form "All F are G" can be regarded as equivalent to the conjunction of "All (F & H) are G" and "All (F and ~H) are G" (or equivalent to conjunctions with even more complex antecedents) on purely logical grounds. Ruling out these candidates as legitimate partitions on the grounds that they are "gerrymandered" or fail to reflect "natural kinds" would, in Salmon's view, be a difficult task without invoking a prior notion of explanation.

ARGUMENT PATTERNS

Friedman tried to measure unification in terms of getting the most conclusions out of the fewest independently acceptable premises; but, as we have seen, it ran into technical problems. Moreover, there is more to the notion of unification than simply reducing the number of brute givens. By appealing to two historical case studies—the attempt (ultimately unsuccessful) to apply Newton's "dynamic corpuscularianism" to chemistry and optics in the eighteenth century, and the reception of Darwin's evolutionary theory in the nineteenth century—Kitcher argues that unification involves using the same patterns of derivation, over and over again. In the Newtonian case, this meant trying to find the forces acting between microscopic particles and then deducing equations of motion to explain chemical and optical phenomena in the same way that had proven so successful with the planets and other macroscopic objects in the *Principia*. In the Darwinian case, it meant using Darwin's "branching tree" model of descent with modification supplemented by natural selection as its principal causal mechanism to explain a host of otherwise unconnected biological generalizations. In both cases, the promise of explanatory power resides in the applicability of a core pattern of reasoning to a wide range of cases. Thus, Kitcher proposes that explanation depends on using derivations that strike the best balance between generating the most conclusions with the fewest patterns of argument. In his later writings, Kitcher argues that one virtue of viewing theories in terms of the patterns of derivation associated with them is that it allows us to understand the relation between theories that succeed one another in a given domain (as "explanation extension"), even though the formal conditions for reduction (derivability of the laws and equations of the predecessor theory from those of the successor theory, in the standard model) are not met.[35]

Kitcher's model of explanation rests on his notion of an argument pattern. A challenge for Kitcher's approach is to define the notion of a pattern such that it is neither too broad (so that just about anything would qualify as an instance) nor too narrow (in which case perhaps only one actual derivation might qualify as an instance). To do this, Kitcher defines an argument pattern as having three components ("an ordered triple"): a set of schematic sentences, filling instructions, and a classification. A schematic sentence is any sentence in which all its nonlogical terms have been replaced by dummy letters (variables). The filling instructions stipulate the kinds of terms one is allowed to substitute for the variables in the schematic sentences. The classification tells us the essential logical and inferential structure of any argument constructed out of the schematic sentences; in other words, the classification specifies the premises, conclusion, and rules of inference of any instance of a schematic argument pattern.

There are three main criteria for assessing the unifying power of a set of argument patterns:

- the number of argument patterns in the set (the fewer the better),
- their stringency (the more stringent the better), and
- the number of conclusions in K that the set generates (the more, the better).

For convenience, we shall refer to these criteria as *paucity, stringency,* and *fecundity,* respectively. Stringency depends on how restrictive the filling instructions and the classification are in specifying what is to count as an instance of the argument pattern. Obviously there is a tension between paucity and stringency on the one hand, and fecundity on the other, since more conclusions will be derivable if argument patterns are added or existing patterns are reduced in their stringency. But Kitcher refrains from saying how one should weigh these factors or resolve trade-offs among them.[36]

When confronted with a derivation of some explanandum, how do we decide whether or not it is a genuine explanation? In order for a derivation to be an explanation it must belong to E(K), the *explanatory store over K*.[37] The explanatory store, E(K), is the set of derivations that best unifies K. What is K? It is the set of beliefs that are accepted at a given time in the history of science (corrected and idealized if necessary so that K is consistent and deductively closed). The simplest case to consider is when we have a fixed body of accepted beliefs, K, and a fixed language, L, and we wish to compare two sets of derivations, S and S', with respect to their unifying power over K. The set with the greatest unifying power will be chosen as the explanatory store over K.[38] That set unifies the best which generates *the broadest class of consequences* using the *least number* ("paucity") of the *most stringent* patterns.

Deductive Chauvinism

Like Hempel, Kitcher advocates the inferential approach to explanation. In fact, Kitcher is even more of a deductive chauvinist than Hempel. The standard deductive chauvinist line on probabilistic arguments is that, insofar as they are acceptable, it is only because they point the way to a complete, proper, deductive explanation. As long as the premises of the argument are logically consistent with not-E, the argument does not explain why E occurred. Thus, deductive chauvinists like Kitcher deny that quantum mechanics explains why *this* particular electron tunneled through the potential barrier (rather than being reflected like its exactly similar cousin).

Kitcher concedes that there are many other things that quantum mechanics does explain, since it gives answers to other sorts of explanatory questions such as "How possibly—?" and "Why is the probability of tunneling = 0.9?" Wesley Salmon had given the example of a soldier ("Herman") who dies of cancer many years after watching "Old Smoky" and other atomic tests. Salmon wanted to allow that statistical evidence proves that the cause of Herman's death was exposure to radioactive fallout—the same kind of evidence that has clinched the case for smoking as a cause of lung cancer.[39] Kitcher's response is more guarded. If the causes of Herman's cancer are genuinely indeterministic, then Kitcher insists that there is no explanation of why Herman got the disease. Despite this, Kitcher allows that there is an explanation of why Herman had a greater *risk* of cancer than someone in the general population, and this is sufficient to make the government partially responsible for Herman's death (because the actions of the government were responsible for raising that risk). But there is no explanation of why Herman got cancer whereas his buddy Arnold, similarly exposed, did not. As Kitcher acknowledges (citing an objection from Stephen Stich), this account of responsibility needs to be able to handle cases of preemption. If Herman dies of a heart attack (from causes unrelated to his exposure to radioactive fallout) the government is not responsible for his death even though, according to Kitcher, the government still caused Herman's elevated probability of contracting cancer. This kind of problem is similar to the objections of Achinstein and others to Hempel's D-N model. Even though everyone who ingests a pound of arsenic dies within twenty-four hours, this law does not explain Jones's death if he is killed by being run over by a bus within twelve hours of eating the arsenic.

There is also the "percolation objection" to deductive chauvinism from quantum mechanics. How can we ignore, for the purposes of explanation, the pervasive influence of quantum-mechanical probabilities? Even though quantum behavior usually produces only small deviations from classical behavior for macroscopic objects, nonetheless if quantum mechanics is true (and classical mechanics false), surely the real explanation of everything we observe at the macroscopic level must ultimately be probabilistic. (This is the position defended by Peter Railton.) Kitcher responds to this objection by appealing to the idealization of macro-phenomena. We can legitimately ignore small deviations in just the same way that we ignore the effects of air resistance when calculating the trajectory of a projectile using Galileo's law of falling bodies and the principle of inertia. Kitcher's apparent acceptance of pragmatic considerations would appear to undermine his claim to have provided the ultimate, objective, correct account of what it is to explain a lawlike regularity.

THE IRRELEVANCE AND ASYMMETRY OBJECTIONS

Kitcher's strategy for dealing with problem cases is to argue that derivations lacking explanatory cogency do not qualify for inclusion in the explanatory store over our current beliefs.[40] Thus, in response to irrelevance objections such as the hexed salt, Kitcher points out that the standard pattern (based, in this case, on our well-accepted chemical theory of ionic dissociation in water) enables us to derive conclusions not just about hexed salt but unhexed salt as well. Either the deviant "hexed" pattern includes the standard pattern, or it does not. If it does, then it has inferior unifying power (because it includes one extra pattern); if it does not, then it has a smaller (more restricted) consequence set. Either way, the standard pattern wins out.

One problem with this response, which Kitcher addresses in his later paper, is that we have to exclude the gerrymandering of patterns. Kitcher invents the example of "hexable salt."[41] Not all salt is hexed (by having some incantation muttered over it) but all salt is hexable. It could be hexed. All *hexable* salt dissolves in water. So why doesn't this explain why a particular sample of salt dissolved in water? Kitcher's response is that $E(K)$ contains a deeper set of derivations of dissolving behavior than does the "hexable" derivation. The hexable derivation is an isolated pattern that brings with it no additional ability to generate testable claims. Of course, we could amend all of standard chemistry so that all soluble substances are hexable; but in that case "hexability" would be an "idle wheel."

Extreme cases of irrelevance arise from patterns of derivation that would fit every candidate explanation regardless of context. The most egregious would be self-derivation patterns of the form: α, therefore α (where we can replace "α" with any sentence that we accept). Let us call this excessively accommodating argument pattern *Alpha*. Kitcher refers to patterns such as Alpha as instances of "spurious unification" (727). On the face of it, we can rule out these patterns on the grounds that they fail the stringency requirement by being maximally unrestrictive with respect to what one is allowed to substitute for "α." In his paper, Kitcher explains how someone might attempt to evade this prohibition by restricting the instances substituted for "α" to precisely those that appear as the conclusions of arguments generated by a bona fide set of patterns. In that case, the number of conclusions credited to Alpha will be the same as those associated with the respectable set. Moreover, in virtue of its extreme paucity, Alpha will qualify as having greater unifying power. Kitcher blocks this move by imposing a further restriction: if the substitution instructions for some pattern P could be replaced by a different set that would allow the derivation of any sentence whatever, then P is a spurious unifier. Kitcher's rationale for this block is that it correctly prohibits argument forms in which the non-logical vocabulary places no restriction on what can be deduced. One problem with this position is that it would seem to prohibit any deductively valid argument form in which the conclusion is a dummy variable.[42]

The asymmetry objections to Hempel's model involve examples such as towers and their shadows, and pendulums and their period of oscillation. It is not open to Kitcher to rule out these cases by appealing to causation as a necessary condition for explanation. Instead, Kitcher argues that they can be dealt with by comparing the unifying power of two different kinds of accounts of why towers and pendulums have their respective dimensions. In the tower example, an "origin-and-development" account of why the tower has its current height (an instance of the OD pattern of argument) is superior to a "shadow-based" account (the SB pattern of argument) because OD has a much wider consequence set than SB.[43] SB can derive only (some) dimensions of objects that are casting shadows. OD can derive most if not all the dimensions of objects, whether or not they are illuminated. Hence, OD but not SB belongs in the explanatory store over K.

But, as Kitcher points out, the asymmetry problem cuts more deeply. The tower may be in the dark but it *could* have been illuminated by a specific kind of light source; and if it had been so illuminated, it would have cast such-and-such a shadow on a suitable surface such that we could calculate the height of the tower from the length of its shadow. The tower may be said to possess a dispositional property. The dispositional-shadow pattern has far more instantiations (just as many?) as the OD pattern. Thus, the explanatory store, $S(K)$, that includes the dispositional-shadow pattern seems to be at least as good as the explanatory store, $E(K)$, that has only OD. So, why is $E(K)$ superior to our new, deviant $S(K)$? Kitcher's response is that there is no single dispositional property that does the work in $S(K)$. Rather $S(K)$ contains an open-ended, heterogeneous, gerrymandered, disjunctive property in order to cover objects that are unilluminated, transparent objects, objects that are themselves a source of illumination, objects of astronomical dimensions, and so on.[44]

Eric Barnes has proposed several counterexamples to Kitcher's unificationist solution of the asymmetry problem.[45] Barnes invites us to consider a closed mechanical system governed by the time-symmetric laws of Newtonian mechanics. An (idealized) example would be perfectly elastic billiard balls colliding on a frictionless billiard table. There are as many retrodictive Newtonian arguments (allowing earlier states to be inferred from later ones) as predictive arguments (that proceed forward in time). Thus, the retrodictive derivations are instances of a pattern that has the same degree of unification as the predictive derivations. But usually we regard only predictive patterns as explanatory. Does Kitcher have to insist that the unification model should trump our usual assumption of causal asymmetry (that causes must precede their effects)?

Barnes's other examples involve retrodictive inferences for which competing predictive inferences seem to be absent, thus apparently entitling them to be added to $E(K)$. Typical cases would be inferences about a human being having recently strolled on the beach (inferred, inductively, from footprints in the wet sand) or inferences about an extinct dinosaur (inferred from fossil remains). None of these "backward" evidentiary inferences can be regarded as explanatory, and yet, arguably, they belong among the argument patterns that best unify K. Although in both cases, we must add an extra derivational pattern to our repertoire in order to generate the conclusions in question, this increase in the number of patterns is outweighed, for the purposes of assessing unification, by the large increase in the number of similar such conclusions it enables us to obtain. Moreover, at the level of precision involved, these conclusions are not obtainable using existing patterns.[46]

Woodward's "Winner-Take-All" Objection

A virtue that Kitcher claims for his model is its ability, at least in principle, to accommodate degrees of explanatory power. Kitcher thinks that it is important that these comparative judgments can be made prospectively, that is, before the theory has made many predictions and before those predictions have been tested. In his view, it is often on the basis of assessments of their explanatory potential that scientists decide which theories to pursue and develop. Thus, Kitcher argues, van Fraassen's attempt to reduce explanatory virtue to prediction and pragmatic factors is flawed.

Intuitively, it seems that we should regard some explanations as deeper than others, depending on the degree of unification wrought by their premises. Thus, we might regard Newton's gravitational theory as providing a better explanation of the trajectory of projectiles than that provided by Galileo's law of falling bodies. But the most unified theory is the only one that Kitcher recognizes as explanatory. Everything else is deemed to be completely non-explanatory. Kitcher seems forced to take this "winner-take-all" position (as James Woodward calls it) by his response to the irrelevance and asymmetry objections. Indeed given the likelihood that scientific progress will reveal theories that are more unifying in the future than any presently available, we seem obliged to concede that it is likely that none of our current best theories are genuinely explanatory, if Kitcher's model is correct.[47]

Correct Explanation

Kitcher makes the concept of explanation—what he later calls an *acceptable* explanation—relative to K, the sentences scientists accept at a given time, whether or not they are true. While this might capture the notion of what scientists justifiably believe or accept as a correct explanation at a given time, it cannot serve to define what a correct explanation is. (See the discussion, earlier in this commentary, of Railton's criticism of Hempel for holding that the concept of statistical explanation is essentially relative to our epistemic situation.) We might try to define the notion of a correct explanation by stipulating that K is the set of all true sentences. But this won't work because such a K would contain all true causal statements, all true statements about what explains what, and so on. That would trivialize the entire unification theory.

Kitcher approaches this problem by considering an objection he attributes to David Lewis.[48] What if the world is not unified, but instead consists of a vast number of different types of causal processes having no common structure? If the world were this messy, the unification approach would lead us in the wrong direction, away from a correct appreciation of how the world really works.

Kitcher responds by first defining an *ideal Hume corpus*: a set of beliefs using only projectable predicates in a language that includes all true statements in that language using no causal, explanatory, or counterfactual concepts. Kitcher then formalizes Lewis's objection as three propositions.

(1) *F* is explanatorily relevant to *P*, according to the unification approach, provided that there is a derivation of *P* belonging to the best unifying systematization of an ideal Hume corpus such that there is a premise of the derivation in which reference to *F* is made.

(2) If *F* is causally relevant to *P*, then *F* is explanatorily relevant to *P*.

(3) It is possible (and may, for all we know, be true) that there is a factor *F* that is causally relevant to some phenomenon *P* such that derivations of a description of *P* belonging to the best unifying systematization of an ideal Hume corpus would not contain any premise making reference to *F*.

In other words, according to (3) it is possible that the "obsessive unifier" would insist on producing derivations of *P* that make no mention of the real causal factors at work: "unity is imposed where none is really to be found."[49]

We could respond by saying that (1) and (2) are merely contingent (and hence logically compatible with (3), which talks of what is possible); but that would be contrary to the spirit of the unification approach as an analysis of explanation. What we must face is the apparent possibility that there is, at best, merely a contingent connection between unification and causal structure. (Wesley Salmon, and other proponents of the causal model, would be unfazed by this since they regard unification as a contingent consequence of causal explanation.) Kitcher bites the bullet and denies that (3) is true. He denies that there are "causal truths that are independent of our search for order in the phenomena."[50] Kitcher gives a Peircian twist to his Kantian theme by invoking *the limit of the rational development of scientific practice.* "Correct explanations are those derivations that appear in the explanatory store in the limit of the rational development of scientific practice."[51]

When we now modify (3) to (3′), Kitcher rejects (3′) as false.

(3′) It is possible (and may, for all we know, be true) that there is a factor *F* that is causally relevant to some phenomenon *P* such that no derivation occurring in the explanatory store in the limit of scientific practice derives a description of *P* from premises that make no reference to *F*.

The only concession that Kitcher is prepared to make to Lewis is to acknowledge that there is no guarantee that science, even in the ideal limit, will achieve any particular degree of unification; but he insists that there is "no sense to the notion of causal relevance independent of that of explanatory relevance."[52]

6.8 | Woodward's Manipulationist Model

In an oft-quoted passage from a famous essay, Bertrand Russell remarked that in philosophy the law of causality is rather like the British monarchy— tolerated because it is believed, falsely, to be harmless.[53] One of the serious points of Russell's essay (written in 1912) was to argue that "the reason why physics has ceased to look for causes is because, in fact, there are no such things." Undoubtedly the word "cause" seldom appears in the theories of modern physics. Most physical laws are expressed in the form of equations, and very few if any of these equations are identified explicitly as causal. But when we turn to the special sciences, especially the social and behavioral sciences (psychology, economics, etc.) but also biomedicine, ecology, and the less rarefied parts of physics and chemistry dealing with macroscopic objects and everyday phenomena, the picture is very different. Here an important part of scientific activity consists in testing and evaluating causal claims, discriminating between genuine causal relations and non-causal correlations, diagnosing cases of common causes and confounding causes, and investigating cases of causal prevention, preemption, and overdetermination. It is this activity, involving the special sciences and our common-sense notion of causation, that James Woodward seeks to clarify and understand through his manipulationist (or "interventionist") model. In a manner not uncommon in the philosophy of science, his model is intended to be both descriptive (reflecting what scientists do) and normative (recommending a philosophical account that makes sense of scientific activity and that can, if necessary, be used to criticize it).[54]

SALMON'S AND WOODWARD'S APPROACHES COMPARED

Philosophers of science have offered several different models for understanding causal explanation, notably the causal-mechanical (C-M) account of Wesley Salmon. There are important differences between Salmon's approach and Woodward's. Salmon sees causal explanation as fundamentally a matter of situating the explanandum event in the network of spatio-temporally continuous causal processes that precede the event and lead up to it. Like most philosophers who have addressed the concept of causation, Salmon's aim was to give a reductive analysis; in his case, the goal was to define what a genuine causal process is and how it differs from a mere pseudo-process. Because Salmon intended to provide an analysis, this meant avoiding (on pain of circularity) any appeal to prior causal notions. Salmon was also reluctant, as are many modern empiricists, to rely on counterfactuals in his analysis. In its final version (following the "conserved quantity" theory developed by Phil Dowe), the C-M model seems best suited to micro-reductive and mechanistic explanations in physics rather than to non-mechanical explanations at the macroscopic scale in the special sciences. Also, by focusing on continuous causal pathways leading up to an event,

Salmon's model has been faulted for losing sight of the element of relevance that distinguishes explanations from non-explanations. For example, we can presumably trace a continuous causal path (at the biochemical level) from John Jones's ingestion of birth-control pills to his failure to become pregnant without feeling the least inclination to regard this as a case of explanation.[55]

Woodward takes his lead from the way in which causal claims are tested and distinguished from mere correlations in the special sciences. Two simple paradigm cases are instructive. Consider a classic case (see (740), 1.4.6) in which two events are joint effects of a common cause, say the reading on a barometer (B) and an approaching storm (S), both being caused by a fall in atmospheric pressure (A).

How do we show that there is no direct causal link between B and S despite their correlation? If B were the cause of S, then manipulating the value of B would be matched by corresponding changes in S. Thus a simple method to show the absence of a causal link between them would be to vary the readings on the barometer (perhaps by altering the pressure in its immediate vicinity or by adjusting its mechanism directly) and noting that there is no corresponding change in the likelihood of a storm. The presumption here, of course, is that the method we use to change the value of B has no effect on the value of A. By intervening on B, we break the causal link between A and B (represented by an arrow in figure 1) and thus show the absence of a causal link between B and S. Because B and S are not causally related, we cannot use B to explain S. Similarly, if we were to test the claim that a certain drug, D, causes a lowering of blood pressure, P, we would perform a randomized clinical trial to see whether administering the drug to an experimental group of subjects is matched by a corresponding drop in blood pressure in that group as compared with a control group. The purpose of the randomization is to try to ensure, as far as possible, that there are no other causes of P at work that might confound the association between D and P. Once we have established that D is the cause of P we have thereby explained why P occurs when D occurs.

It is by reflecting on simple, uncontroversial examples such as these, that Woodward motivates his manipulationist account of causal explanation. We will explore some of the details later. For the moment we note some general respects in which Woodward's approach differs from Salmon's. First, Woodward does not require an account of the continuous underlying

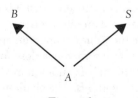

Figure 1

processes at work between two variables in order for someone to establish that they are causally related or to offer that relation as an explanation. In that respect, Woodward's approach is far closer to the realities of scientific practice than Salmon's, especially in the special sciences. Second, Woodward does not intend to offer a philosophical analysis of the concept of causation. In fact Woodward doubts that any analysis of causation can be given that would reduce it to concepts that are exclusively non-causal.[56] Thus, Woodward tolerates the use of causal concepts (such as the notion of an "intervention") in his account of when causal relations obtain. Third, Woodward is willing to use counterfactuals in his exposition, regarding their truth-conditions as well understood and uncontroversial in the scientific contexts he is seeking to understand. Nonetheless, Woodward agrees with Salmon (and most other philosophers of science) that the notions of cause and explanation that he is exploring are intended to be fully objective.[57] Despite the frequent use of the terms *manipulation* and *intervention* with their connotations of human agency in ordinary language, Woodward explicitly denies that his account is restricted to the alterations and changes that human beings can produce. As far as he is concerned, an intervention can occur in nature without the involvement of any human, and interventions can occur that no human is able to produce. In the first chapter of *MTH* (from which our excerpt is taken), he gives the example of the asteroid collision widely believed to have caused a mass extinction at the end of the Cretaceous period (736). Any account of causation needs to be able to make sense of such claims even though they fall outside any possibility of human control or manipulation. Although we often have a practical interest in discovering causal relations so that we can produce (or prevent) certain kinds of events by manipulating their causes, this should not obscure our general intellectual interest in explaining why things happen when such manipulation is impractical or impossible.

INTERVENTIONS

In his book *MTH*, Woodward refines and gives precise definitions of the concepts of *intervention, direct cause,* and *contributing cause* that figure prominently in his account. As we have already indicated, Woodward does not regard the interrelated nature of these definitions as a flaw since he is not claiming that any one of these notions is more fundamental than the others. His aim is not analysis or ontology but the clarification of methodology. The basic idea guiding his approach is that A is the cause of B if and only if the value of B would change if an intervention were to alter the value of A. In the simplest case, where the values in question are either "event occurs" or "event does not occur," we have something close to the idea underlying the counterfactual, third "definition" of causation in Hume's *Enquiry*: if A were not to happen, then neither would B. The challenge is to restrict the conditions under which the value of A is altered (or the occurrence of A is prevented) so that it eliminates confounding causes and delivers the right verdict about the causal relation between *A* and *B*. Leaving aside some refinements that Woodward defends in chapter 3 of *MTH*, his basic characterization of an intervention is as follows.

(IV) *I* is an intervention on *X* with respect to *Y* if and only if (1) *I* causes a change in *X*, (2) *I* does not cause a change in *Y* via some route that does not go through *X*, and (3) there is no cause of *I* that affects *Y* via a route that does not go through *X*.

Given this notion of an intervention on *X* (with respect to *Y*) as an event that causes a change in *X* but has no causal effect on *Y* except that (if any) resulting from the change in *X*, the corresponding definition of the (direct) causal relation (again, leaving out some qualifications and refinements) is:

(M) *X* causes *Y* if and only if were an intervention to occur that changes the value of *X*, then *Y* (or the probability distribution of *Y*) would change in some regular, stable way (at least in some range of relevant circumstances).

LAWS, INVARIANCE, AND STABILITY

The reference in definition (M) to stability in some range of relevant circumstances underscores a major difference between Woodward and other philosophers regarding the connection between causation and laws. One mainstream view (endorsed by but not limited to followers of Hume) is that causation depends on laws, and that laws are generalizations that are maximally invariant under change of circumstances. This view (which Woodward rejects) regards a causal relation between *X* and *Y* as depending on there being a law connecting *X*s and *Y*s, and requires that laws hold without exception under all nomically possible conditions (that is, under conditions consistent with no other law being violated). Woodward's view is that causal explanation requires the *X-Y* generalization to have an appropriate degree of invariance (in order for it to support the relevant counterfactual conditionals) but not that the generalization's degree of invariance be maximal or that it hold under all nomically possible conditions.[58]

Many of the reliable generalizations that scientists call *laws* cease to hold under circumstances that other laws permit. For example, even though Newton's law of universal gravitation holds to a high degree of approximation under most circumstances, it breaks down in strong gravitational fields. Similarly, Maxwell's equations cease to be reliable at the subatomic level. Woodward thinks that this is sufficient grounds for recognizing as laws generalizations that hold to a high degree of approximation under a wide range of conditions (including those produced by interventions), even though those generalizations are not precisely accurate under those conditions and even if they no longer hold to the same high degree of approximation under other conditions that are consistent with other laws. In short, laws do not have to be exactly true as long as they hold reliably under a wide range of conditions; moreover, they do not have to hold even with a high degree of approximation under all nomically possible conditions. What matters for causality is that the law have a high degree of invariance under testing interventions within a significant range of the values of the variables mentioned in the law.[59]

THE CIRCULARITY OBJECTION

Some of the criticisms of Woodward's manipulationist theory have understandably focused on whether it provides an adequate account of how causal knowledge is possible. This concern stems from the fact that interventions are themselves events that cause other events. Thus, it would seem, in order for us to know that intervention I_1 causes a change in the value of variable X, we must know that I_1 satisfies definition (M); that in turn requires counterfactual knowledge about a further intervention I_2 that is connected to I_1 such that changes in the former would produce changes in the latter.[60] Thus an epistemic regress looms: the identification of any event as an intervention requires already knowing that a prior event is an intervention, and so on *ad infinitum*. The regress is vicious because it would, apparently, prevent us from ever coming to know of any event that it is an intervention and thus deprive us of causal knowledge. Unlike some other objections that have been advanced against Woodward's account, this objection is epistemological, not ontological.[61]

In reply to this criticism, Woodward emphasizes that all causal reasoning takes place against a background that already includes causal beliefs. He gives the example of a randomized clinical trial of a new drug for a disease in which tossing a coin, C, is used to assign subjects to either the treatment group or the control group and thus establish that T, treatment with the drug, is the cause of R, recovery from the disease. Surely, in this kind of case, we know that C is an intervention variable without having to apply (IV) and (M) directly, and thus without running afoul of a regress. The question of course, is how do we know this? Knowledge requires justification. What justifies us in believing that C is an intervention variable as defined in (IV)? One possibility (explored by Michael Baumgartner) is that we rely on a different theory of causation, say a probabilistic account such as that first offered by Patrick Suppes.[62] The idea is that we would be using this alternative account as a heuristic criterion in roughly the same way that one might use macroscopic tests of a yellow metal to ascertain that it is probably gold (and thus satisfies the fundamental physical and chemical definition of gold in terms of its atomic number). Even if this strategy is accepted, Baumgartner insists that at some point we need to verify that events that satisfy the alternative theory also satisfy Woodward's theory; and it is hard to see how that can be accomplished without applying Woodward's theory directly, without relying on the alternative account. (In terms of the gold analogy: at some point we need to verify that yellow metals that pass our heuristic tests really are gold.) But if the regress objection is correct, a direct application of Woodward's theory is impossible; and so the strategy fails.

There is a second line that Woodward might take against the regress objection. In his review of *MTH*, Clark Glymour distinguishes between two kinds of philosophical theories: the *Socratic* and the *Euclidean*.[63] Socratic theories are common in traditional areas of ethics, epistemology, and metaphysics. They aspire to analyze ("explicate") what certain core concepts such as moral wrongness, knowledge, and causation mean by providing necessary and sufficient conditions; they consider and respond to alleged counterexamples, and they criticize the shortcomings of rival analyses. As their name indicates, Euclidean theories are usually presented more or less formally in the manner of Euclid's *Elements*, beginning with definitions and axioms. Glymour gives Frege's work in logic, Kolmogorov's axiomatization of probability, and Lewis's theory of counterfactuals as examples. The definitions and axioms of Euclidean theories are assumed to capture, in an adequate way, fundamental aspects of the subject matter at hand but, unlike the definitions of Socratic theories, the definitions of Euclidean theories are not offered as analyses of the meaning of concepts or as revealing deep metaphysical truths. The main order of business is then to deduce consequences (theorems) from the axioms and definitions in the hope of learning something interesting. Perhaps Woodward's manipulationist theory could be regarded in this way, with (M) and (IV) regarded as definitions, and then further axiomatic *causal* assumptions added (such as, that the use of a randomizing device such as C is an intervention variable in randomized clinical trials) to see what causal conclusions can be deduced. In this way, the theory could be seen as displaying the logic of causal reasoning by showing what causal outputs can be deduced from what causal inputs. Baumgartner gives the following example both to illustrate this approach to interpreting Woodward's theory and to show its limitations. He uses the notion of screening off: if $P(C/B) \neq P(C)$, then A screens off B from C if $P(C/B \,\&\, A) = P(C/A)$. Suppose that we are told only the following about four variables: X screens off Y from Z, I from Z, and I from Y. The most that can be deduced from this probabilistic information is that the system has one of the following four causal structures illustrated in figure 2.[64]

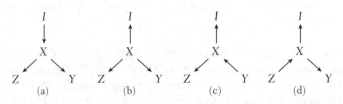

Figure 2 (from Baumgartner 2009, p. 190)

If we were now to assume that I is an intervention variable for X with respect to Y, then it follows from Woodward's theory that the causal structure must be (a) and cannot be any of the other three. While this may seem a significant achievement, it must be remembered that choosing I as an intervention variable is an assumption. We could just as well have selected Y (as an intervention variable for X with respect to I), in which case (c) would been singled out. Alternatively, we could have picked Z, in which case (d) would have been singled out. Using Woodward's manipulationist theory to derive a justified conclusion about what causes what in this case requires having justified grounds for choosing among I, Y, and Z as an intervention variable, something that the regress objection has called into doubt.

COUNTERFACTUALS AND LAWS

According to Woodward, laws and other generalizations (often expressed in the form of equations) explain when (and only when) they support subjunctive conditionals of the form "If the value of X were to be x, then the value of Y would be y." He insists that for explanatory equations we can decide the truth of such conditionals when they are counterfactual (when X was not in fact caused to change in value by an intervention) and even when the changes are such that it would have been humanly impossible to alter the value of X just as long as the generalizations (equations) in question are stable or invariant under a suitably wide range of interventions. Woodward eschews offering a general theory of the truth conditions for counterfactual conditionals (such as, for example, the theory of Lewis and Stalnaker), insisting that as a practical scientific matter, we know what judgments to make as long as we are able to test the relevant generalizations in principle, even if we do not do so in fact. Nonetheless, Woodward emphasizes that the truth-value of what he calls "active counterfactuals" (the testable ones) is an objective matter; their truth (or falsity) does not depend on which experiments we choose to perform or on our technological capabilities. What, then, are the truth-makers for active counterfactuals? Some passages in *MTH* strongly suggest that it is causal relations that play this role:

> According to the manipulationist account, given that C causes E, which counterfactual claims involving C and E are true will always depend on which other *causal* claims involving variables besides C and E are true in the situation under discussion. For example, it will depend on whether other causes of E besides C are present.[65]

For some of Woodward's critics, such as Stathis Psillos, this connection between causal connections and the truth-values for counterfactual conditionals is too tight to be philosophically enlightening.[66] While it seems perfectly reasonable to assert that which counterfactuals involving C and E are true depends on the causal structure of the network in which C and E

are embedded, there is a threat of epistemological circularity when an answer to the question "Does C cause E?" depends on a counterfactual conditional (via definition M) whose truth depends in turn on further causal claims about the network that contains C and E that again rest on counterfactuals. What we seem to need, and Woodward does not provide, is an independent account of lawlikeness that can ground the relevant counterfactuals without our first having to know the truth of causal claims.[67]

Psillos raises a further objection that Woodward had anticipated in *MTH* (see pages 127–33) and to which Woodward offered an answer. It involves causal claims for which interventions seem out of the question. For example, we accept as true the following causal claim: Changes in the earth-moon distance and the resulting changes in the gravitational attraction exerted by the moon cause tidal changes in the motions of the earth's oceans. Woodward concedes that the truth of this causal claim cannot rest on an interventionist counterfactual because realizing the antecedent of the relevant conditional is physically impossible. Nonetheless, Woodward says that the counterfactual is meaningful (and thus has an objective truth-value) despite the physical impossibility of realizing its antecedent by means of an intervention, just as long as there is some basis for assessing its truth ("conceptual or logical"). For Psillos, Woodward's reply is unsatisfactory because it allows that there is more to causation as an intrinsic relation (and hence more to causal explanation) than invariance under actual and counterfactual interventions; but Woodward has not spelled out what it is. In defense of Woodward's general position one might point out that changes in the earth-moon distance have occurred and continue to occur, with verifiable consequences for the rate of the earth's rotation (and the moon's rate of revolution) as a result of tidal friction. (See the discussion of the secular acceleration of the moon in the paper by Dorling cited in chapter 3.) Even in the case of the law that nothing can travel faster than the speed of light, there is ample experimental evidence, derived from attempts to attain that speed with electrons and other charged particles and failing. The limit for this strategy for tying laws to interventions is reached, as Woodward acknowledges, when we consider laws regarding the universe as a whole. In that case, for "conceptual or logical" reasons, the notion of an intervention fails to apply, even in principle.

6.9 | Summary

In the second half of the twentieth century, most of the philosophical debate concerning the nature of scientific explanation centered on Hempel's covering law thesis and his two models of explanation, the deductive-nomological (D-N) and the inductive-statistical (I-S). According to the covering law thesis, explanations are arguments (either deductively valid or inductively strong) having among their premises at least one statement of an empirical law. As Wesley Salmon has pointed out, Hempel's covering law thesis is characteristic of an epistemic conception of scientific explanation, since it takes explanation to be essentially an inference showing that the explanandum event was to be expected. Given the information in the premises of the explanatory argument, the explanandum event could have been predicted, either with certainty (D-N) or with high probability (I-S). This capability of being used as a prediction is seen by many philosophers of science as the hallmark of a good explanation. For Hempel, Carnap, Nagel, and others, being able, at least in principle, to predict the event to be explained guarantees that the explanation has testable, empirical content and gets to the heart of what distinguishes genuine explanation from mere pseudoexplanation.

Hempel's commitment to the epistemic conception of explanation is made explicit in his advocacy of the thesis of structural identity, that all adequate explanations are potentially predictions and that all adequate predictions are potentially explanations. In other words, Hempel insists that there is no formal, objective difference between explanation and prediction. Whether an argument is an explanation or a prediction depends on pragmatic factors, such as when the argument is put forward and the intentions of the scientist who presents the argument. Predictive arguments are advanced prior to the events mentioned in their conclusions either to test the theory from which the laws are taken or, if the theory and its laws are well accepted, to provide a reliable basis upon which to plan for the future. Explanations are advanced after the events mentioned in their conclusions in order to achieve a theoretical understanding of why those events occurred. We make predictions in order to anticipate the future; we give explanations in order to understand the past. But, according to Hempel, in either case, the formal structure of our reasoning is the same and is captured by the D-N and I-S models.

Attacks on the thesis of structural identity (by Michael Scriven and others) go under the general heading of the symmetry objection. They consist of two kinds of alleged counterexamples: explanations that are not predictions and predictions that are not explanations. Hempel has strenuously opposed the first kind of alleged counterexample, often arguing that the presumed explanation is no such thing. But Hempel concedes that the second subthesis, that all adequate predictions are potentially explanations, is not as secure as its converse.

The irrelevance objection is another important class of criticisms of Hempel's two models of explanation. Various philosophers (including Wesley Salmon and Peter Achinstein) have given arguments that, while they satisfy all of Hempel's conditions for an adequate explanation, do not appear to explain their conclusions. The birth-control pills and hexed-salt examples offer typical cases in which the presence of irrelevant information in the premises robs the arguments of their explanatory power. Because of his commitment to the epistemic conception of explanation, Hempel has denied that these arguments lack explanatory power, but this response seems implausible. What seems to go wrong in these examples is that the lawlike premise in these arguments, while true, fails to identify the cause of the explanandum event. Consequently, Baruch Brody has proposed that we amend Hempel's D-N model by stipulating that the premises of an explanatory argument must contain a description of the event that is the cause of the explanandum. But Brody's proposal has been roundly refuted by counterexamples devised by Timothy McCarthy. It is not enough that the premises contain a description of the event that is, as a matter of fact, the cause of the explanandum event; rather, the premises must explicitly identify that event as the cause. But in that case, the underlying structure of explanations is devastatingly simple—it consists simply of the inference from "c is the cause of e" to the conclusion "e"—and explanations need make no explicit mention of any laws. Thus, we are led to the view that explanations are not arguments but single causal statements and that laws, while vitally important for many types of explanation, are not an essential part of the sentence that explains.

Other important criticisms of Hempel concern his I-S model of probabilistic explanation. In order to avoid the problem of ambiguity for statistical explanation, Hempel introduced the requirement of maximal specificity (RMS). But the RMS is relativized to what Hempel calls "a given knowledge situation" (682), thus making statistical explanations depend for their very existence on the body of beliefs that scientists accept at a particular time. This epistemic relativity of statistical explanation in the I-S model stands in marked contrast to the objective, nonrelative character of explanations in the D-N model. Moreover, because Hempel construes all explanations as formally equivalent to predictions, I-S arguments are taken to explain their conclusions only when those conclusions follow with high inductive probability from their premises. Since the strength of the inductive relation between premises and conclusion depends directly on the value of the probability that appears in the statistical lawlike premise of such arguments, it follows that, on Hempel's model, it is impossible to explain improbable events. Both of these features—the epistemic relativity of I-S arguments and the high-probability requirement for explanation—have struck critics (such as Salmon and Railton) as serious flaws in Hempel's account.

Peter Railton has proposed his deductive-nomological model of probabilistic explanation (the D-N-P model) as an alternative to Hempel's account. Unlike Hempel's I-S model, Railton's D-N-P model regards probabilistic explanations as fully objective (not relative to any set of scientific beliefs) and permits the explanation of improbable events. Railton denies that explanations *are* arguments, even though he insists that all explanations must *contain* a deductive argument based on a law. Central to Railton's D-N-P model is the requirement that explanations specify the causal mechanism that brings about the event (or the kind of event) referred to in the explanandum. Genuine probabilistic explanations can be given only when the mechanism at work is indeterministic, and when an indeterministic mechanism is involved, any genuine explanation must be probabilistic. Railton interprets probabilistic laws, not as generalizations about the frequency of certain kinds of event, but as statements of single-case propensities, such as the chance of getting heads when a particular coin is tossed. The propensity of an object or system is its causal tendency to behave in a particular way. Like probabilities, the strength of a propensity can vary on a scale from 0 to 1, but, unlike the empirical frequency interpretation of probability, propensities are just as much physical properties of individual things and systems as their mass and electric charge; in fact Railton sometimes refers to propensities as physical probabilities.

Apart from the difficulty of understanding exactly what propensities are and how they are related to frequencies, Railton's D-N-P model has a number of consequences that may strike one as counterintuitive. Many of these counterintuitive features stem from Railton's demand that explanation properly so-called must be based on the deepest theoretical understanding of nature that we have. Thus, for example, if a process (such as the melting of an ice cube in warm water) involves the slightest chance that the ice cube not melt but become colder, then Railton requires that any explanation of the melting involve a theoretical calculation of the exact propensity (not quite equal to 1) of the cube to melt. This sets a very high standard for probabilistic explanation. Also, as previously noted, on the D-N-P model probabilistic explanations are legitimate only when there is genuine, rock-bottom, physical indeterminacy involved. Thus, any attempt to explain the behavior of complex but ultimately deterministic systems by means of statistical generalizations is ruled out (thereby disqualifying many of the arguments currently accepted as explanatory in the physical, biological, and social sciences).

Because of its emphasis on causation and causal mechanisms (whether deterministic or indeterministic), the model of explanation advocated by Railton is a good example of what Salmon has called the *ontic* conception of explanation.[68] Unlike the epistemic approach to explanation, the ontic approach does not regard explanations as arguments (even though they may include or involve arguments). What matters on the ontic account is not whether the explanandum can be predicted with high probability, but whether, regardless of the probability of the explanandum, we can give a correct description of the underlying causal mechanism that brought about the event we wish to explain. The ontic conception of explanation allows us to give complete explanations of events even when those events are highly improbable, as would be the case if nature is governed by indeterministic laws, as quantum mechanics supposes. Seen from the point of view of the ontic conception, the demand for high probability imposed by the epistemic conception ultimately stems from that conception's commitment to determinism as the final truth about the structure of the world. Seen from the point of view of the epistemic conception, Railton's insistence that we calculate objective, true propensities in all cases places an unreasonably high demand on what counts as an explanation. The epistemic approach (at least in Hempel's version of it) permits genuine statistical explanations even if the world is deterministic through and through, but it makes their status as explanations relative to human beliefs at a particular time. The ontic approach (in Railton's version) takes probabilistic explanations to be just as fully objective and nonrelative as those based on deterministic laws, but insists that such explanations can be given only when the mechanisms involved are fundamentally indeterministic.

Unlike Railton, Philip Kitcher endorses the epistemic conception of explanation according to which explanations are fundamentally arguments. Kitcher articulates a unification model that is primarily concerned with explaining general laws rather than particular facts, and offers a solution to the conjunction problem that Hempel had left unresolved. Earlier, Michael Friedman had proposed a unification model that avoided the conjunction problem; but Kitcher showed that it had unacceptable consequences, such as ruling out all arguments whose premises contain laws each of which could be accepted on independent grounds. Kitcher's model focuses instead on patterns of derivation. These patterns generate arguments when applied to the set of all sentences, K, accepted by scientists at a particular time. Kitcher argues that the set of generational patterns that enables the largest number of conclusions to be derived from the smallest set of the most stringent patterns is maximally unifying. This set he calls the explanatory store over K. A particular argument explains its conclusion if and only if it is an instance of a pattern from the explanatory store.

Kitcher solves the unification problem by pointing out that the valid argument form "From X and Y, infer Y" is maximally unrestrictive with regard to what can be substituted for X and Y. In most cases, where Y is a statement of a scientific law (Boyle's law, for example), there is a much more stringent pattern (in this case, one based on the kinetic theory of gases) that enables that law and several other laws to be derived. Thus, the conjunction pattern does not belong to the explanatory store, and so the trivial derivations using it do not qualify as explanations. In a similar way, Kitcher tackles the asymmetry and irrelevance problems by arguing that those patterns also fail to belong to the explanatory store.

Questions have been raised about whether Kitcher's solution to the asymmetry problem is adequate, especially in cases such as closed dynamical systems obeying time-symmetric laws. For it would seem in those cases that retrodictive patterns would have the same unifying power as predictive ones; yet few people accept that the present can explain the past. More generally, it has been questioned whether Kitcher's "top-down" account can capture adequately the causal relation that seems crucial to many explanations.

Causation and explanation are often thought to go hand in hand. Drawing on pioneering work on causal inference by Judea Pearl and by Peter

Spirtes, Clark Glymour, and Richard Scheines, James Woodward develops a manipulationist theory of causation and causal explanation. While the goal of Pearl and Spirtes et al. is to provide efficient and reliable methods for extracting causal conclusions from statistical data, Woodward's aim is to articulate what causation amounts to when such methods are employed. The resulting theory is intended to reflect how working scientists think about causes when they investigate causal claims (and distinguish genuine causal relations from mere correlations) not just in physics and chemistry, but also in medicine, sociology, and economics.

Woodward does not identify interventions with human agency since, unlike earlier manipulationist theorists (who *were* reductionists), he insists that causal relations are objective: they exist even when it is impossible for human beings to manipulate the relevant variables; and interventions can occur without human involvement. Although Woodward characterizes causes in terms of interventions, his theory is not reductionist since, as he is at pains to point out, interventions are themselves a kind of cause. Because of this, some critics question whether Woodward can give a satisfactory account of how we can come to know that certain events are interventions in the first place since this always seems to presuppose prior causal knowledge.

There is an interesting contrast between Salmon's causal-mechanical model and Woodward's manipulationist model. For Salmon, we explain events by describing how they fit into the network of causal processes that precede them. Salmon's model is reductionist—it defines causes without appealing to other causal concepts—and Salmon deliberately avoids having to rely on counterfactuals in articulating his theory. In Woodward's non-reductionist model, counterfactuals play a crucial role in explanation: to explain what happened involves specifying how things would have been different had the preceding conditions been altered. For that purpose we use generalizations (usually in the form of equations between variables) that have a degree of invariance over a range of possible interventions. Strictly invariant laws (as in Hempel's D-N model) are not necessary. Another issue raised by Woodward's theory is whether it can explain how we assign truth-values to counterfactuals and the invariant generalizations that support them since this, too, would seem to require prior causal knowledge.

■ | Notes

1. Carl G. Hempel and Paul Oppenheim, "Studies in the Logic of Explanation," *Philosophy of Science* 15 (1948): 567–79. Reprinted, with a postscript, in *Aspects of Scientific Explanation* (New York: Free Press, 1965), 245–95.

2. Carl G. Hempel, "Deductive-Nomological vs. Statistical Explanation," in *Minnesota Studies in the Philosophy of Science*, vol. 3, ed. H. Feigl and G. Maxwell (Minneapolis: University of Minnesota Press, 1962), 98–169.

3. "A physical theory is not an explanation. It is a system of mathematical propositions, deduced from a small number of principles, which aim to represent as simply,

as completely, and as exactly as possible a set of experimental laws" (Pierre Duhem, *The Aim and Structure of Physical Theory*, trans. Philip P. Wiener [1914; Princeton, N.J.: Princeton University Press, 1954], 19).

4. *Aim and Structure*, p. 7.

5. Before judging too harshly Mach and Duhem's skepticism about atoms and the aether, we should remember that the atoms we now believe in are utterly different from those described in the classical "billiard-ball" models of the nineteenth century and that the aether has been discarded entirely. What remains of the kinetic theory of gases and the wave theory of light of the nineteenth century are testable equations and laws, and these alone, according to the positivists, have permanent scientific value.

6. As Heather Douglas argues, there are two issues here. One is the importance (which Douglas defends) of assessing scientific explanations according to their ability to generate new testable predictions. The other (which Douglas rejects) is Hempel's requirement that every particular explanation be structurally equivalent to a predictive argument. See Heather Douglas, "Reintroducing Prediction to Explanation," *Philosophy of Science* 76 (2009): 444–63.

7. Strictly speaking, then, (R2) should be phrased in terms of lawlike sentences, rather than laws. It is commonly assumed that we can judge whether a sentence is lawlike simply by examining its logical form and the predicates it uses. But to judge that a lawlike sentence is true—and hence expresses a law—requires empirical information.

8. Since the explanans can include laws of mathematics, some of the sentences it contains might not be empirical; but (R3) guarantees that every explanans must contain at least one empirical sentence.

9. In their original 1948 paper on the D-N model, Hempel and Oppenheim did not require that the so-called antecedent conditions (labeled "statements of particular facts and initial conditions" on page 756) occur before the explanandum event, thus leaving open the possibility (which Hempel later defended, in *Aspects*, 353–54) that an event might be explained by referring to later events. See the interesting discussion of whether retrodictive arguments can be explanatory in Wesley C. Salmon, "On the Alleged Temporal Anisotropy of Explanation," in *Philosophical Problems of the Internal and External Worlds*, ed. John Earman et al. (Pittsburgh and Konstanz: University of Pittsburgh Press and Universitätsverlag Konstanz, 1993), 229–48; reprinted in Wesley C. Salmon, *Causality and Explanation* (New York: Oxford University Press, 1998), ch. 10.

10. Michael Scriven, "Explanation and Prediction in Evolutionary Theory," *Science* 130 (1959): 477–82.

11. See "Popper and the Theory of Evolution" in the commentary on chapter 1.

12. There is no problem of ambiguity for D-N explanations. If a set of true premises deductively entails that *a* is *G*, then no other set of true premises can deductively entail that *a* is not *G*.

13. Carl G. Hempel, "Inductive Inconsistencies," *Synthese* 12 (1960): 439–69. Reprinted in *Aspects of Scientific Explanation*, 53–79. Hempel is assuming that one would be justified in believing whatever one knows follows deductively from one's justified beliefs—in this case, that one would be justified in believing the conjunction of *A* and *B*, if one were justified (inductively) in believing *A* and one were

justified (inductively) in believing *B*. Doubts about this closure principle have been discussed in connection with the lottery paradox devised by Henry E. Kyburg, Jr. in his *Probability and the Logic of Rational Belief* (Middletown, CT: Wesleyan University Press, 1961). In a fair lottery of 1000 tickets, each ticket has only a 0.001 chance of being the winner. Assume that that probability is low enough to justify one in believing of any single ticket that it will lose. But if one were justified in believing that every ticket will lose, then that would contradict one's belief that the lottery is fair. For a comprehensive review of closure principles concerning justification and knowledge, see Steven Luper, "The Epistemic Closure Principle," *The Stanford Encyclopedia of Philosophy (Winter 2011 Edition)*, Edward N. Zalta (ed.), URL=<http://plato.stanford.edu/archives/win2011/entries/closure-epistemic>.

14. Hempel is using the term *knowledge* in a common, nontechnical sense to mean, roughly, *justified belief*. It is in this sense that we talk of the current state of scientific knowledge while recognizing that some of the beliefs in that body of knowledge are quite likely to be false. When epistemologists use the term in its technical sense, they insist that truth is a necessary condition for knowledge.

15. For the first set, see Ardon Lyon, "The Relevance of Wisdom's Work for the Philosophy of Science: A Study of the Concept of Scientific Explanation," in *Wisdom: Twelve Essays*, ed. Renford Bamborough (Oxford: Blackwell, 1974), 218–48; and Baruch Brody, "Towards an Aristotelian Theory of Explanation," *Philosophy of Science* 39 (1972): 20–31. For the second set, see Peter Achinstein, *The Nature of Explanation* (New York: Oxford University Press, 1983), 168, 170–71; and Wesley Salmon, "Statistical Explanation," in *The Nature and Function of Scientific Theories*, ed. Robert G. Colodny (Pittsburgh: University of Pittsburgh Press, 1970), 173–231, reprinted in Wesley C. Salmon, Richard C. Jeffrey, and James G. Greeno, *Statistical Explanation and Statistical Relevance* (Pittsburgh: University of Pittsburgh Press, 1970).

16. David Hillel-Ruben considers a number of ways in which the D-N model might, using resources already present in Hempel's theory, be defended against the arsenic-and-bus counterexample and argues that they are unsuccessful. See David Hillel-Ruben, *Explaining Explanation* (London: Routledge, 1990), ch. 6.

17. The example and the phrase *hexed salt* were first proposed by Henry Kyburg. See Henry E. Kyburg, Jr., "Comments," *Philosophy of Science* 32 (1965): 147–51.

18. Carl G. Hempel, "Maximal Specificity and Lawlikeness in Probabilistic Explanation," *Philosophy of Science* 35 (1968): 116–33. The following criticism of the proposal is adapted from John Meixner, "Homogeneity and Explanatory Depth," *Philosophy of Science* 46 (1979): 366–81.

19. Wesley C. Salmon, *Scientific Explanation and the Causal Structure of the World* (Princeton, N.J.: Princeton University Press, 1984), 96.

20. Timothy McCarthy, "On an Aristotelian Model of Scientific Explanation," *Philosophy of Science* 44 (1977): 159–66.

21. Jaegwon Kim, "On the Logical Conditions of Deductive Explanation," *Philosophy of Science* 30 (1963): 286–91.

22. David Hillel-Ruben, *Explaining Explanation* (London: Routledge, 1990), 197. Ruben defends the view that full explanations are typically not arguments contain-

ing laws as premises but, rather, singular sentences using terms such as *because*, *reason*, and *causes*. This view has been developed, albeit in different ways, by figures such as Michael Scriven, Wesley Salmon, and Peter Achinstein (among others).

23. This line of criticism is also advanced in J. Alberto Coffa, "Hempel's Ambiguity," *Synthese* 28 (1974): 141–63.

24. This criticism originated with Paul Humphreys. See his "Why Propensities Cannot Be Probabilities," *Philosophical Review* 94 (1985): 557–70.

25. With the "mechanisms" account as a possible third. See Peter Machamer, Lindley Darden, and Carl F. Craver, "Thinking about Mechanisms," *Philosophy of Science* 67 (2000): 1–25. Causal models have been advocated by Wesley Salmon, *Scientific Explanation and the Causal Structure of the World* (Princeton, N.J.: Princeton University Press, 1984); Paul Humphreys, *The Chances of Explanation: Causal Explanation in the Social, Medical, and Physical Sciences* (Princeton, N.J.: Princeton University Press, 1989); Phil Dowe, *Physical Causation* (Cambridge: Cambridge University Press, 2000); and James Woodward, *Making Things Happen: A Theory of Causal Explanation* (New York: Oxford University Press, 2003).

26. For general accounts of unification relevant to this project, see Tim Maudlin, "On the Unification of Physics," *Journal of Philosophy* 93 (1996): 129–44; and Margaret Morrison, *Unifying Scientific Theories* (Cambridge: Cambridge University Press, 2000). Maudlin agrees with Richard Feynman that there is more to genuine unification than simply combining different laws into one equation. Real unification, at least in physics, requires showing that forces and phenomena previously regarded as distinct are aspects of the same underlying physical reality. Morrison stresses the great diversity of conceptions of unification in science: some have explanatory significance; many do not. Earlier writers, from Newton to Whewell, usually tied unification to confirmation (rather than to explanation per se), and this trend continues in recent debates over inference to the best explanation in which the ability of a theory to give a unified account of diverse phenomena is thought to provide a reason for thinking that the theory is true or probably true. Sometimes there is a fine distinction between confirmation and explanation in discussions of unification. Those who (like Friedman and Kitcher) see understanding as paramount are usually focusing on explanation.

27. See Wesley C. Salmon, "Scientific Explanation: How We Got from There to Here," in Wesley C. Salmon, *Causality and Explanation* (New York: Oxford University Press, 1998), 308.

28. Philip Kitcher, "Explanatory Unification and the Causal Structure of the World," in *Scientific Explanation, Minnesota Studies in the Philosophy of Science*, ed. Philip Kitcher and Wesley C. Salmon, vol. 13 (Minneapolis: University of Minnesota Press, 1989), 477. (Hereafter cited as "Kitcher (1989).") Kitcher regards it as an advantage of his approach that, because it is non-causal, it also applies to mathematical explanation as well as to explanations that appeal to high-level general principles in the sciences.

29. Michael Friedman, "Explanation and Scientific Understanding," *Journal of Philosophy* 71 (1974): 5–19. See also Michael Friedman, "Theoretical Explanation," in *Reduction, Time and Reality*, ed. Richard Healey (Cambridge: Cambridge University Press, 1981), 1–16.

30. Philip Kitcher, "Explanation, Conjunction, and Unification," *Journal of Philosophy* 73 (1976): 207–12.

31. By "acceptable," Friedman means roughly "have good grounds for accepting as true." Suppose that A entails B; then grounds sufficient for accepting A will also be sufficient grounds for accepting B. In that case, A would not be acceptable independently of B. Similarly, if A entails B, and the grounds for accepting a sentence, C, are also grounds for accepting A, then the grounds for accepting C would also be grounds for accepting B. Taking the contrapositive, this is equivalent to: Given that A entails B, if C is acceptable independently of B (i.e., C is not grounds for accepting B), then C is acceptable independently of A (i.e., C is not grounds for accepting A).

32. As it turns out, Friedman was not happy with D1, judging it to be too strong. He reasons that if S_1 explains S_2 (in accordance with D1) and we were then to conjoin an irrelevant law S_3 with S_1, the resulting conjunction (S_1 and S_3) would not satisfy D1. He comments, "This seems undesirable—why should the conjunction of a completely irrelevant law to a good explanation destroy its explanatory power?" (p. 197). So, Friedman proposes D1′ which says, in effect, that S_1 explains S_2 if and only if S_1 has a partition containing a sentence that satisfies D1. Kitcher criticizes Friedman's D1′ on the grounds that because D1′ requires that S_1, in order to explain, have a partition, D1′ entails that no K-atomic sentence can explain. This, Kitcher argues, is too strong. Salmon also rejects D1′ on the straightforward grounds that it would allow the explanans to contain irrelevant information.

33. See Wesley C. Salmon, "Four Decades of Scientific Explanation," in *Scientific Explanation, Minnesota Studies in the Philosophy of Science*, ed. Philip Kitcher and Wesley C. Salmon, vol. 13 (Minneapolis: University of Minnesota Press, 1989), 96–99.

34. Here is another of Salmon's examples. Suppose the ancient Greeks were such male chauvinist pigs that they accepted the statement "All humans are mortal" by inductively inferring that generalization solely on the basis of "All men are mortal." So they could accept "All men are mortal" independently of "All humans are mortal." Having inferred that all humans are mortal, they then deduce that all women are mortal, but their grounds for accepting it rest entirely on that deduction. So, for them, "All men are mortal" is acceptable independently of "All humans are mortal" but "All women are mortal" is not. Thus, <All men are mortal, All women are mortal> is (for them) not a partition of "All humans are mortal." Thus "All humans are mortal" is (for them) K-atomic. But once ancient Greek men became more enlightened and began noticing that women die, they would have grounds for accepting "All women are mortal" that are independent of "All humans are mortal." In that case, "All humans are mortal" would not be K-atomic. This relativization of K-atomicity to the vagaries of epistemological history (the order in which evidence is obtained) seems objectionable as a basis for an objective account of explanation.

35. Kitcher (1989), 447–48. For more on reduction and objections to the standard model, see chapter 8.

36. A general problem for Kitcher's approach lies in deciding how to classify patterns of derivation and how to count the number of conclusions attributed to each pattern. Consider Kitcher's example of Galileo and his calculation of the angle for a cannon ball's maximum range. We could use a very similar type of derivation to predict the range for any angle whatever. Does this mean there is just one pattern of

derivation here? If there is just one pattern, how many conclusions does it generate? The number of ranges that can be computed is potentially infinite. Does that mean that the pattern of derivation generates an infinite number of conclusions? Or is this problem avoided by stipulating that there is just one conclusion of these similar arguments, namely, a general equation specifying range as a function of angle and initial velocity?

37. In his later work (1989), Kitcher distinguishes between an explanatory derivation being *acceptable* and its being *correct*. See the section "Correct Explanation" below.

38. In this later work, Kitcher addresses the more difficult case of scientific change in which our beliefs and language change. See Kitcher (1989), 488–94.

39. Kitcher (1989), 457–59.

40. Ibid., 482.

41. Ibid., 483–84.

42. For this and similar objections to Kitcher's restriction as a solution to the problem of spurious unification, see Victor Gijsbers, "Why Unification Is Neither Necessary nor Sufficient for Explanation," *Philosophy of Science* 74 (2007): 481–500.

43. Kitcher (1989), 485.

44. Ibid., 486–87.

45. Eric Barnes, "Explanatory Unification and the Problem of Asymmetry," *Philosophy of Science* 59 (1992): 558–71.

46. Todd Jones has replied to Barnes by arguing that existing patterns (such as the Darwinian evolutionary pattern in the dinosaur case, and the family of origin-and-development derivations in the footprint case) can yield either "speculative" accounts, based on "educated guesses" about prior conditions, or "partial derivations" that can generate some but not all aspects of the conclusions in question. In the Newtonian case, since the predictive derivations can do everything that the retrodictive derivations can, there is unificatory virtue in not adding the latter to our repertoire, especially since the predictive derivations belong to the wider origin-and-development family. Todd Jones, "How the Unification Theory of Explanation Escapes Asymmetry Problems," *Erkenntnis* 43 (1995): 229–40.

47. See James Woodward, *Making Things Happen* (New York: Oxford University Press, 2003), ch. 8; and James Woodward, "Scientific Explanation," *The Stanford Encyclopedia of Philosophy* (*Spring 2010 Edition*), ed. Edward N. Zalta, http://plato.stanford.edu/archives/spr2010/entries/scientific-explanation.

48. Kitcher (1989), 494.

49. Ibid., 496.

50. Ibid., 497.

51. Ibid., 498.

52. Ibid., 499.

53. "The law of causality, I believe, like much that passes muster among philosophers, is a relic of a bygone age, surviving, like the monarchy, only because it is erroneously supposed to do no harm." Bertrand Russell, "On the Notion of Cause,"

reprinted in *Mysticism and Logic* (London: George Allen & Unwin, 1918), 180. Russell gives a metaphysical argument for why the concept of causation is incoherent: either the cause, *C*, is temporally continuous with its effect, *E*, or there is a gap between them; if they are continuous, then there is always another event, closer to *E* than *C*, that can be regarded as causing *E*; if there is a gap between *C* and *E*, then it is not inevitable that *E* will follow *C* because another event, later than *C*, could intervene and prevent *E* from occurring.

54. Woodward draws an analogy with the jointly descriptive and normative character of philosophical work in ethics in James Woodward, "Causation with a Human Face," in *Causation, Physics, and the Constitution of Reality: Russell's Republic Revisited* ed. Huw Price and Richard Corry (Oxford: Oxford University Press, 2007), 66–105.

55. For this and other criticisms of Salmon's C-M model, see Woodward, *MTH*; and James Woodward, "Scientific Explanation," *The Stanford Encyclopedia of Philosophy (Spring 2010 Edition)*, ed. Edward N. Zalta, http://plato.stanford.edu /archives/spr2010/entries/scientific-explanation. As Woodward points out, Salmon concedes that the C-M model needs to be supplemented by statistical relevance information in Wesley Salmon, "Causality and Explanation: A Reply to Two Critiques," *Philosophy of Science* 64 (1997): 461–77. See also Wesley Salmon, "Causality without Counterfactuals," *Philosophy of Science* 61 (1994): 297–312.

56. Other non-reductive theories of causation, to which Woodward acknowledges a great debt, are Judea Pearl, *Causality, Models, Reasoning, and Inference* (Cambridge: Cambridge University Press, 2000); and Peter Spirtes, Clark Glymour, and Richard Scheines, *Prediction, Causation, and Search*, 2nd ed. (Cambridge, Mass.: MIT Press, 2000).

57. Exceptions to this are Peter Menzies and Huw Price, who argue for an anthropomorphic, subjective concept of causation as manipulation in their "Causation as a Secondary Quality," *British Journal for the Philosophy of Science* 44 (1993): 187–203. See also John Norton, "Causation as Folk Science," in *Causation, Physics, and the Constitution of Reality: Russell's Republic Revisited* ed. Huw Price and Richard Corry (Oxford: Oxford University Press, 2007), 11–44; and James Woodward, "Causation and Manipulability," *The Stanford Encyclopedia of Philosophy (Winter 2008 Edition)*, ed. Edward N. Zalta, http://plato.stanford.edu/archives/ win2008/entries/causation-mani.

58. Not requiring that laws be maximally invariant and thus hold under all nomically possible conditions is one important respect in which Woodward's account differs from that of Marc Lange. See Marc Lange, *Laws and Lawmakers: Science, Metaphysics, and the Laws of Nature* (New York: Oxford University Press, 2009).

59. In his writings since *MTH*, Woodward distinguishes between two ways that generalizations (laws) can break down outside of their usual range of application by giving them different labels. First, a law can cease to hold reliably for extreme values of the variables that are mentioned explicitly in the law. What matters for causality is that the law be *invariant* under testing interventions within a significant range of those values (because, Woodward argues, this is what underwrites counterfactual conditionals). Second, a law can cease to hold reliably under some changes of background conditions that are not mentioned explicitly in the law. This would be a failure of the law's *stability*. See James Woodward, "Causation with a Human

Face," in *Causation, Physics, and the Constitution of Reality: Russell's Republic Revisited* ed. Huw Price and Richard Corry (Oxford: Oxford University Press, 2007), 66–105. It is not clear that this distinction can be made out in a way that serves Woodward's purposes. Newton's law mentions explicitly distance and mass, but not electrical charge. Does this mean that it fails to be "invariant" when dealing with test bodies that are close to extremely massive stars but "unstable" when the gravitating bodies are both charged? Along the same lines: Does Boyle's law (which says that PV = constant, at constant temperature) fail to be "invariant" at very high pressures (at which most gases condense into liquids) but fail to be "stable" at very high temperatures (at which gases dissociate into plasma)? Or is the mention of constant temperature as a condition for the law to hold sufficient to make this also a case of failure of "invariance"?

60. See Michael Baumgartner, "Interdefining Causation and Intervention," *Dialectica* 63 (2009): 175–94. Baumgartner denies that this problem arises for the interventionist theories of Judea Pearl (2000) or Peter Spirtes et al. (2000) because they take causation to be an undefined primitive; unlike Woodward, they do not "interdefine" causation and intervention. This section of our commentary is substantially indebted to Baumgartner's article.

61. See, for example, Michael Strevens, "Review of Woodward, *Making Things Happen*," *Philosophy and Phenomenological Research* 74 (2007): 233–49, and "Comments on Woodward, *Making Things Happen*," *Philosophy and Phenomenological Research* 77 (2008): 171–92. In his "Response to Strevens," *Philosophy and Phenomenological Research* 77 (2008): 193–212, Woodward emphasizes that his account is not circular in the sense that one already needs to know that X causes Y before one can determine that I is an intervention on X with respect to Y; but it does require knowledge of the causal relations between other variables in the system. As discussed below, Woodward claims that the beauty of randomized experiments is that they are cases in which one can know one has performed an intervention without requiring further causal knowledge.

62. Michael Baumgartner, "Interdefining Causation and Intervention," *Dialectica* 63 (2009): 175–94. As Baumgartner notes, Suppes's theory has since been refined by Igal Kvart and others. The basic idea of this type of theory is that causes raise the probability of their effects. In slightly fuller detail, C is the cause of T if C is positively correlated with T, C occurs prior to T, and there is no other event, X, that screens off T from C by making $P(T/X) = P(T/C \& X)$.

63. Clark Glymour, "Review of *Making Things Happen*," *British Journal for the Philosophy of Science* 55 (2004): 779–90.

64. As several of Woodward's critics have pointed out, the causal Bayes nets theories of Pearl and of Spirtes et al. make stronger, contingent assumptions about causal patterns than does Woodward's manipulationist theory. It is these assumptions (e.g., the causal Markov condition and the so-called faithfulness condition) that enable them to derive causal conclusions from purely probabilistic information (about correlations and the like) in contexts where no intervention variable is obvious. See Michael Strevens, "Comments on Woodward, *Making Things Happen*," *Philosophy and Phenomenological Research* 77 (2008): 190. The causal Markov condition says that the direct causes of X (i.e., its immediate causal ancestors) screen it off from (i.e., make it probabilistically independent of) all other variables

except its effects. The faithfulness condition is violated when X is a contributing cause of Y even though X and Y are uncorrelated (because X just happens to inhibit Y by a separate causal pathway that exactly cancels out X's positive causal effect on Y).

65. James Woodward, *MTH*, 136.

66. Stathis Psillos, "Causal Explanation and Manipulation," in *Rethinking Explanation* ed. Johannes Persson and Petri Ylikoski (Dordrecht: Springer, 2007), 93–107.

67. For Psillos that account is provided by the Mill-Ramsey-Lewis "best systems" theory of laws. See Stathis Psillos, *Causation and Explanation* (Chesham: Acumen, 2002), ch. 5, and the commentary on ch. 7.

68. Wesley C. Salmon, *Scientific Explanation and the Causal Structure of the World* (Princeton, N.J.: Princeton University Press, 1984), chs. 1 and 4.

7 | Laws of Nature

INTRODUCTION

Laws play a central role in scientific reasoning. As we saw in chapters 1 and 4, some philosophers of science think that using laws to explain things is an essential part of what it means to be genuinely scientific, and support for the view that scientific explanation must involve laws is widespread (though not unanimous). Many also believe that we are justified in trusting scientific inferences because these predictions rest, in part, on well-confirmed laws. Our expectations about the behavior of systems, instruments, and materials are reasonable to the extent that they are based on a correct understanding of the laws that govern them. Undoubtedly, much scientific activity is devoted to discovering laws, and one of the most cherished forms of scientific immortality is to join the ranks of Boyle, Newton, and Maxwell by having a law (equation or functional relation) linked with one's name. But despite the crucial importance of laws in science, it is difficult to find a general account of what sort of things laws are that can do justice to everything we take to be true of them.[1]

In this chapter, two important and influential ways of understanding laws—the regularity approach and the necessitarian approach—will be discussed and criticized.[2] In terms roughly hewn, the regularity approach says that laws describe the way things actually behave, that they are nothing more than a special kind of descriptive summary of what has happened and what will happen. The necessitarian approach insists that laws are more than just summaries, that they tell us not merely how things actually behave, but also, more importantly, how they must behave. For the necessitarians, both the universality and the necessity of laws are objective, real features of the world (although necessitarians disagree among themselves about the nature of that necessity).[3]

Modern adherents of the regularity approach trace their origins back to David Hume and his constant-conjunction analysis of causation. In "What Is a Law of Nature?" A. J. Ayer gives a sympathetic account of the epistemological considerations that drove Hume to deny that causal necessity is objective and hence to espouse the simple version of the regularity theory of laws. According to this simple version, a law of nature is nothing more, objectively, than a true universal generalization. Ayer explains the severe problems that afflict the simple Humean theory, including the problem of laws that lack instances and the problem of distinguishing between those generalizations that are genuine laws and those that are true merely by accident. Ayer's tentative solution to these problems is to add epistemic conditions to the regularity analysis of lawlikeness. Thus, according to Ayer's epistemic regularity theory, a law is a true universal generalization about which we have certain beliefs and attitudes and that plays a characteristic role in science.

In "Laws of Nature," Fred Dretske deals what he considers a lethal blow to Ayer's epistemic regularity theory. In its place, Dretske advocates a theory according to which laws are (express or describe) relations of necessitation between universals. Thus, instead of regarding laws as generalizations about events, Dretske regards them as singular statements about the properties events share. Dretske shows how his universals theory of laws can solve several of the difficulties facing the regularity theory.

Like other recent advocates of the universals theory of laws, and unlike earlier proponents of the necessitarian approach, Dretske insists that laws of nature are contingent, not necessary. This creates difficulties for Dretske's theory, since it requires that the nomic necessitation relation between universals hold contingently, not necessarily, and it is hard to see how merely contingent relations could obtain among abstract entities such as universals. One possible response to this problem is suggested by Saul Kripke and Hilary Putnam, who use their new theory of reference to argue that many laws of nature are not contingent but metaphysically necessary. D. H. Mellor criticizes the Kripke-Putnam argument in "Necessities and Universals in Natural Laws." Mellor also attacks the universals theory on the grounds that it cannot accommodate laws that have no instances. In this way, Mellor attempts to cast the regularity theory in a more favorable light by revealing the deficiencies of its rivals.

Despite their disagreement about whether laws involve an element of necessity, the regularity and necessitarian approaches share the conviction that laws of nature describe important facts about reality. That realist assumption about laws is challenged by Nancy Cartwright in the final piece in this chapter. In "Do the Laws of Physics State the Facts?" Cartwright argues that most of the laws physicists use to explain things are not even approximately true. They are false and are known to be false. Nonetheless, they provide excellent explanations. Cartwright argues that there is an irreconcilable tension between the goal of accurate description and the goal of explanation.

When lawlike statements are altered to make them describe the way bodies actually behave, they lose their explanatory power.

■ | Notes

1. One important issue concerning laws that is not treated in this chapter (but is addressed by Woodward in chapter 6, and by Fodor in chapter 8) is whether there are genuine laws in the special sciences, wherein generalizations are typically invariant only within a limited range and are said to hold *ceteris paribus*. See Marc Lange, "Who's Afraid of *Ceteris-Paribus* Laws? (or: How I Learned to Stop Worrying and Love Them," *Erkenntnis* 57 (2002): 407–23; and James Woodward, "Invariance," in *Making Things Happen: A Theory of Causal Explanation* (New York: Oxford University Press, 2003), ch. 6.

2. An older approach to understanding laws of nature—instrumentalism—has largely fallen into disfavor (though Ronald Giere and Bas van Frassen have recently made attempts to revive it). Instrumentalists (such as Ernst Mach, Karl Pearson, Ludwig Wittgenstein, and Stephen Toulmin) hold that laws are neither true nor false; they are simply tools that scientists use to summarize data and to make inferences. Gilbert Ryle once described this view by characterizing laws as "inference tickets." According to instrumentalists, neither the necessity nor the universality of laws is an objective feature of the world; both are human inventions that we impose on the world for the purposes of representation and prediction. The main problem with instrumentalism is that, if laws are neither true nor false, then it is difficult to make sense of their being tested, confirmed, and refuted. See Ronald N. Giere, "The Skeptical Perspective: Science without Laws of Nature," in *Laws of Nature: Essays on the Philosophical, Scientific and Historical Dimensions*, ed. Friedel Weinert (New York: Walter de Gruyter, 1995), 120–38; Bas C. van Fraassen, *Laws and Symmetry* (Oxford: Clarendon Press, 1989); and Alan Musgrave, "Wittgensteinian Instrumentalism," *Theoria* 47 (1981): 65–105.

3. Regularity theorists (of different types) include A. J. Ayer, R. B. Braithwaite, Rudolf Carnap, Richard Feynman, Carl Hempel, Ernest Nagel, Hans Reichenbach, Norman Swartz, and Peter Urbach. Necessitarians (of different types) include D. M. Armstrong, John Bigelow, John Carroll, Fred Dretske, W. C. Kneale, Christopher Swoyer, and Michael Tooley. See the bibliography at the end of this volume for references.

A. J. AYER

What Is a Law of Nature?

There is a sense in which we know well enough what is ordinarily meant by a law of nature. We can give examples. Thus it is, or is believed to be, a law of nature that the orbit of a planet around the sun is an ellipse, or that arsenic is poisonous, or that the intensity of a sensation is proportionate to the logarithm of the stimulus, or that there are 303,000,000,000,000,000,000,000,000 molecules in one gram of hydrogen. It is not a law of nature, though it is necessarily true, that the sum of the angles of a Euclidean triangle is 180 degrees, or that all the presidents of the third French Republic were male, though this is a legal fact in its way, or that all the cigarettes which I now have in my cigarette case are made of Virginian tobacco, though this again is true and, given my tastes, not wholly accidental. But while there are many such cases in which we find no difficulty in telling whether some proposition, which we take to be true, is or is not a law of nature, there are cases where we may be in doubt. For instance, I suppose that most people take the laws of nature to include the first law of thermodynamics, the proposition that in any closed physical system the sum of energy is constant: but there are those who maintain that this principle is a convention, that it is interpreted in such a way that there is no logical possibility of its being falsified, and for this reason they may deny that it is a law of nature at all. There are two questions at issue in a case of this sort: first, whether the principle under discussion is in fact a convention, and secondly whether its being a convention, if it is one, would disqualify it from being a law of nature. In the same way, there may be a dispute whether statistical generalizations are to count as laws of nature, as distinct from the dispute whether certain generalizations, which have been taken to be laws of nature, are in fact statistical. And even if we were always able to tell, in the case of any given proposition, whether or not it had the form of a law of nature, there would still remain the problem of making clear what this implied.

FROM *The Concept of a Person* (New York: St. Martin's Press, 1963), 209–34. This article was first published in *Revue Internationale de Philosophie* 36 (1956): 144–65.

The use of the word 'law', as it occurs in the expression 'laws of nature', is now fairly sharply differentiated from its use in legal and moral contexts: we do not conceive of the laws of nature as imperatives. But this was not always so. For instance, Hobbes in his *Leviathan* lists fifteen 'laws of nature' of which two of the most important are that men 'seek peace, and follow it' and 'that men perform their covenants made': but he does not think that these laws are necessarily respected. On the contrary, he holds that the state of nature is a state of war, and that covenants will not in fact be kept unless there is some power to enforce them. His laws of nature are like civil laws except that they are not the commands of any civil authority. In one place he speaks of them as 'dictates of Reason' and adds that men improperly call them by the name of laws: 'for they are but conclusions or theorems concerning what conduceth to the conservation and defence of themselves: whereas Law, properly, is the word of him, that by right hath command over others'. 'But yet,' he continues, 'if you consider the same Theorems, as delivered in the word of God, that by right commandeth all things; then they are properly called Laws.'[1]

It might be thought that this usage of Hobbes was so far removed from our own that there was little point in mentioning it, except as a historical curiosity; but I believe that the difference is smaller than it appears to be. I think that our present use of the expression 'laws of nature' carries traces of the conception of Nature as subject to command. Whether these commands are conceived to be those of a personal deity or, as by the Greeks, of an impersonal fate, makes no difference here. The point, in either case, is that the sovereign is thought to be so powerful that its dictates are bound to be obeyed. It is not as in Hobbes's usage a question of moral duty or of prudence, where the subject has freedom to err. On the view which I am now considering, the commands which are issued to Nature are delivered with such authority that it is impossible that she should disobey them. I do not claim that this view is still prevalent; at least not that it is explicitly held. But it may well have contributed to the persistence of the feeling that there is some form of necessity attaching to the laws of nature, a necessity which, as we shall see, it is extremely difficult to pin down.

In case anyone is still inclined to think that the laws of nature can be identified with the commands of a superior being, it is worth pointing out that this analysis cannot be correct. It is already an objection to it that it burdens our science with all the uncertainty of our metaphysics, or our theology. If it should turn out that we had no good reason to believe in the existence of such a superior being, or no good reason to believe that he issued any commands, it would follow, on this analysis, that we should not be entitled to believe that there were any laws of nature. But the main argument against this view is independent of any doubt that one may have about the existence of a superior being. Even if we knew that such a one existed, and that he regulated nature, we still could not identify the laws of nature with his commands. For it is only by discovering what were the laws of nature that we could know what form these commands had taken. But this implies that we have some independent criteria for deciding what the laws of nature are. The assumption that they are imposed by a superior being is therefore idle, in the same way as the assumption of providence is idle. It is only if there are independent means of finding out what is going to happen that one is able to say what providence has in store. The same objection applies to the rather more fashionable view that moral laws are the commands of a superior being: but this does not concern us here.

There is, in any case, something strange about the notion of a command which it is impossible to disobey. We may be sure that some command will never in fact be disobeyed. But what is meant by saying that it cannot be? That the sanctions which sustain it are too strong? But might not one be so rash or so foolish as to defy them? I am inclined to say that it is in the nature of commands that it should be possible to disobey them. The necessity which is ascribed to these supposedly irresistible commands belongs in fact to something different: it belongs to the laws of logic. Not that the laws of logic cannot be disregarded; one can make mistakes in deductive reasoning, as in anything else. There is, however, a sense in which it is impossible for anything that happens to contravene the laws of logic. The restriction lies not upon the events themselves but on our method of describing them. If we break the rules according to which our method of description functions, we are not using it to describe anything. This might suggest that the events themselves really were disobeying the laws of logic, only we could not say so. But this would be an error. What is describable as an event obeys the laws of logic: and what is not describable as an event is not an event at all. The chains which logic puts upon nature are purely formal: being formal they weigh nothing, but for the same reason they are indissoluble.

From thinking of the laws of nature as the commands of a superior being, it is therefore only a short step to crediting them with the necessity that belongs to the laws of logic. And this is in fact a view which many philosophers have held. They have taken it for granted that a proposition could express a law of nature only if it stated that events, or properties, of certain kinds were necessarily connected; and they have interpreted this necessary connection as being identical with, or closely analogous to, the necessity with which the conclusion follows from the premisses of a deductive argument; as being, in short, a logical relation. And this has enabled them to reach the strange conclusion that the laws of nature can, at least in principle, be established independently of experience: for if they are purely logical truths, they must be discoverable by reason alone.

The refutation of this view is very simple. It was decisively set out by Hume. 'To convince us', he says, 'that all the laws of nature and all the operations of bodies, without exception, are known only by experience, the following reflections may, perhaps, suffice. Were any object presented to us, and were we required to pronounce concerning the effect, which will result from it, without consulting past observation: after what manner, I beseech you, must the mind proceed in this operation? It must invent or imagine some event, which it ascribes to the object as its effect: and it is plain that this invention must be entirely arbitrary. The mind can never find the effect in the supposed cause, by the most accurate scrutiny and examination. For the effect is totally different from the cause, and consequently can never be discovered in it.'[2]

Hume's argument is, indeed, so simple that its purport has often been misunderstood. He is represented as maintaining that the inherence of an effect in its cause is something which is not discoverable in nature; that as a matter of fact our observations fail to reveal the existence of any such relation: which would allow for the possibility that our observations might be at fault. But the point of Hume's argument is not that the relation of necessary connection which is supposed to conjoin distinct events is not in fact observable: it is that there could not be any such relation, not as a matter of fact but as a matter of logic. What Hume is pointing out is that if two events are distinct, they are distinct: from a statement which does no more than assert the existence of one of them it is impossible to deduce anything concerning the existence of the other. This is, indeed, a plain tautology. Its importance lies in the fact that Hume's opponents denied it. They wished to maintain both that the events which were coupled by the laws of nature were logically distinct from one another, and that they were united by a logical relation. But this is a manifest contradiction. Philosophers who hold this view are apt to express it in a form which leaves the contradiction latent: it was Hume's achievement to have brought it clearly to light.

In certain passages Hume makes his point by saying that the contradictory of any law of nature is at least conceivable; he intends thereby to show that the truth of the statement which expresses such a law is an empirical matter of fact and not an *a priori* certainty. But to this it has been objected that the fact that the contradictory of a proposition is conceivable is not a decisive proof that the proposition is not necessary. It may happen, in doing logic or pure mathematics, that one formulates a statement which one is unable either to prove or disprove. Surely in that case both the alternatives of its truth and falsehood are conceivable. Professor W. C. Kneale, who relies on this objection,[3] cites the example of Goldbach's conjecture that every even number greater than two is the sum of two primes. Though this conjecture has been confirmed so far as it has been tested, no one yet knows for certain whether it is true or false: no proof has been discovered either way. All the same, if it is true, it is necessarily true, and if it is false, it is necessarily false. Suppose that it should turn out to be false. We surely should not be prepared to say that what Goldbach had conjectured to be true was actually inconceivable. Yet we should have found it to be the contradictory of a necessary proposition. If we insist that this does prove it to be inconceivable, we find ourselves in the strange position of having to hold that one of two alternatives is inconceivable, without our knowing which.

I think that Professor Kneale makes his case: but I do not think that it is an answer to Hume. For Hume is not primarily concerned with showing that a given set of propositions, which have been taken to be necessary, are not so really. This is only a possible consequence of his fundamental point that 'there is no object which implies the existence of any other if we consider these objects in themselves, and never look beyond the idea which we form of them,'[4] in short, that to say that events are distinct is incompatible with saying that they are logically related. And against this Professor Kneale's objection has no force at all. The most that it could prove is that, in the case of the particular examples that he gives, Hume might be mistaken in supposing that the events in question really were distinct: in spite of the appearances to the contrary, an expression which he interpreted as referring to only one of them might really be used in such a way that it included a reference to the other.

But is it not possible that Hume was always so mistaken; that the events, or properties, which are coupled by the laws of nature never are distinct? This question is complicated by the fact that once a generalization is accepted as a law of nature it tends to change its status. The meanings which we attach to our expressions are not completely constant: if we are firmly convinced that every object of a kind which is designated by a certain term has some property which the term does not originally cover, we tend to include the property in the designation; we extend the definition of the object, with or without altering the words which refer to it. Thus, it was an empirical discovery that loadstones attract iron and steel: for someone who uses the word 'loadstone' only to refer to an object which has a certain physical appearance and constitution, the fact that it behaves in this way is not formally deducible. But, as the word is now generally used, the proposition that loadstones attract iron and steel is analytically true: an object which did not do this would not properly be called a loadstone. In the same way, it may have become a necessary truth that water has the chemical composition H_2O. But what then of heavy water which has the composition D_2O? Is it not really water? Clearly this question is quite trivial. If it suits us to regard heavy water as a species of water, then we must not make it necessary that water consists of H_2O. Otherwise, we may. We are free to settle the matter whichever way we please.

Not all questions of this sort are so trivial as this. What, for example, is the status in Newtonian physics of the principle that the acceleration of a body is equal to the force which is acting on it divided by its mass? If we go by the text-books in which 'force' is defined as the product of mass and acceleration, we shall conclude that the principle is evidently analytic. But are there not other ways of defining force which allow this principle to be empirical? In fact there are, but as Henri Poincaré has shown,[5] we may then find ourselves obliged to treat some other Newtonian principle as a convention.* It would appear that in a system of this kind there is likely to be

---

* See chapter 6 of *La science et l'hypothèse* (Paris: E. Flammarion, 1902); *Science and Hypothesis*, trans. W. J. Greenstreet (New York: Dover, 1952). Poincaré reasons that any attempt to verify the second law, $F = ma$, by experiment—even on a single

a conventional element, but that, within limits, we can situate it where we choose. What is put to the test of experience is the system as a whole.

This is to concede that some of the propositions which pass for laws of nature are logically necessary, while implying that it is not true of all of them. But one might go much further. It is at any rate conceivable that at a certain stage the science of physics should become so unified that it could be wholly axiomatized: it would attain the status of a geometry in which all the generalizations were regarded as necessarily true. It is harder to envisage any such development in the science of biology, let alone the social sciences, but it is not theoretically impossible that it should come about there too. It would be characteristic of such systems that no experience could falsify them, but their security might be sterile. What would take the place of their being falsified would be the discovery that they had no empirical application.

The important point to notice is that, whatever may be the practical or aesthetic advantages of turning scientific laws into logically necessary truths, it does not advance our knowledge, or in any way add to the security of our beliefs. For what we gain in one way, we lose in another. If we make it a matter of definition that there are just so many million molecules in every gram of hydrogen, then we can indeed be certain that every gram of hydrogen will contain that number of molecules: but we must become correspondingly more doubtful, in any given case, whether what we take to be a gram of hydrogen really is so. The more we put into our definitions, the more uncertain it becomes whether anything satisfies them: this is the price that we pay for diminishing the risk of our laws being falsified. And if it ever came to the point where all the 'laws' were made completely secure by being treated as logically necessary, the whole weight of doubt would fall upon the statement that our system had application. Having deprived ourselves of the power of expressing empirical generalizations, we should have to make our existential statements do the work instead.

If such a stage were reached, I am inclined to say that we should no longer have a use for the expression 'laws of nature', as it is now understood. In a sense, the tenure of such laws would still be asserted: they would be smuggled into the existential propositions. But there would be nothing in the system that would count as a law of nature: for I take it to be characteristic of a law of nature that the proposition which expresses it is not logically

---

body of constant mass—requires a way of measuring forces independently of the accelerations they cause and of ascertaining when two forces are equal in magnitude. This, he argues, must presuppose the truth of the third law (that action and reaction are equal and opposite). Thus, he concludes that if the second law is empirical, then the third law must be treated as a definition. Poincaré also argues that if the second law is treated not as an empirical law but as a definition of force, then it can be applied to more than one body only if the masses of different bodies can be compared. This, too, he argues, presupposes Newton's third law, since when two bodies act on each other, the ratio of their masses is defined as the inverse ratio of their accelerations (assuming that no other bodies are acting on them).

true. In this respect, however, our usage is not entirely clear-cut. In a case where a sentence has originally expressed an empirical generalization, which we reckon to be a law of nature, we are inclined to say that it still expresses a law of nature, even when its meaning has been so modified that it has come to express an analytic truth. And we are encouraged in this by the fact that it is often very difficult to tell whether this modification has taken place or not. Also, in the case where some of the propositions in a scientific system play the rôle of definitions, but we have some freedom in deciding which they are to be, we tend to apply the expression 'laws of nature' to any of the constituent propositions of the system, whether or not they are analytically true. But here it is essential that the system as a whole should be empirical. If we allow the analytic propositions to count as laws of nature, it is because they are carried by the rest.

Thus to object to Hume that he may be wrong in assuming that the events between which his causal relations hold are 'distinct existences' is merely to make the point that it is possible for a science to develop in such a way that axiomatic systems take the place of natural laws. But this was not true of the propositions with which Hume was concerned, nor is it true, in the main, of the sciences of to-day. And in any case Hume is right in saying that we cannot have the best of both worlds; if we want our generalizations to have empirical content, they cannot be logically secure; if we make them logically secure, we rob them of their empirical content. The relations which hold between things, or events, or properties, cannot be both factual and logical. Hume himself spoke only of causal relations, but his argument applies to any of the relations that science establishes, indeed to any relations whatsoever.

It should perhaps be remarked that those philosophers who still wish to hold that the laws of nature are 'principles of necessitation'[6] would not agree that this came down to saying that the propositions which expressed them were analytic. They would maintain that we are dealing here with relations of objective necessity, which are not to be identified with logical entailments, though the two are in certain respects akin. But what are these relations of objective necessity supposed to be? No explanation is given except that they are just the relations that hold between events, or properties, when they are connected by some natural law. But this is simply to restate the problem; not even to attempt to solve it. It is not as if this talk of objective necessity enabled us to detect any laws of nature. On the contrary it is only *ex post facto*, when the existence of some connection has been empirically tested, that philosophers claim to see that it has this mysterious property of being necessary. And very often what they do 'see' to be necessary is shown by further observation to be false. This does not itself prove that the events which are brought together by a law of nature do not stand in some unique relation. If all attempts at its analysis fail, we may be reduced to saying that it is *sui generis* [altogether unique]. But why then describe it in a way which leads to its confusion with the relation of logical necessity?

A further attempt to link natural with logical necessity is to be found in the suggestion that two events E and I are to be regarded as necessarily connected when there is some well-established universal statement U, from which, in conjunction with the proposition *i*, affirming the existence of I, a proposition *e*, affirming the existence of E, is formally deducible.[7] This suggestion has the merit of bringing out the fact that any necessity that there may be in the connection of two distinct events comes only through a law. The proposition which describes 'the initial conditions' does not by itself entail the proposition which describes the 'effect': it does so only when it is combined with a causal law. But this does not allow us to say that the law itself is necessary. We can give a similar meaning to saying that the law is necessary by stipulating that it follows, either directly or with the help of certain further premisses, from some more general principle. But then what is the status of these more general principles? The question what constitutes a law of nature remains, on this view, without an answer.

■ | II

Once we are rid of the confusion between logical and factual relations, what seems the obvious course is to hold that a proposition expresses a law of nature when it states what invariably happens. Thus, to say that unsupported bodies fall, assuming this to be a law of nature, is to say that there is not, never has been, and never will be a body that being unsupported does not fall. The 'necessity' of a law consists, on this view, simply in the fact that there are no exceptions to it.

It will be seen that this interpretation can also be extended to statistical laws. For they too may be represented as stating the existence of certain constancies in nature: only, in their case, what is held to be constant is the proportion of instances in which one property is conjoined with another or, to put it in a different way, the proportion of the members of one class that are also members of another. Thus it is a statistical law that when there are two genes determining a hereditary property, say the colour of a certain type of flower, the proportion of individuals in the second generation that display the dominant attribute, say the colour white as opposed to the colour red, is three quarters. There is, however, the difficulty that one does not expect the proportion to be maintained in every sample. As Professor R. B. Braithwaite has pointed out, 'when we say that the proportion (in a non-literal sense) of the male births among births is 51 per cent, we are not saying of any particular class of births that 51 per cent [of them] are births of males, for the actual proportion might differ very widely from 51 per cent in a particular class of births, or in a number of particular classes of births, without our wishing to reject the proposition that the proportion (in the non-literal sense) is 51 per cent.'[8] All the same the 'non-literal' use of the word 'proportion' is very close to the literal use. If the law holds, the proportion must remain in the neighborhood of 51 per cent, for any sufficiently large class of cases: and the deviations from it which are found in selected sub-classes must be such as the application of the calculus of probability would lead one to expect. Admittedly, the question what constitutes a sufficiently large class of cases is hard to answer. It would seem that the class must be finite, but the choice of any particular finite number for it would seem also to be arbitrary. I shall not, however, attempt to pursue this question here. The only point that I here wish to make is that a statistical law is no less 'lawlike' than a causal law. Indeed, if the propositions which express causal laws are simply statements of what invariably happens, they can themselves be taken as expressing statistical laws, with ratios of 100 per cent. Since a 100 per cent ratio, if it really holds, must hold in every sample, these 'limiting cases' of statistical laws escape the difficulty which we have just remarked on. If henceforth we confine our attention to them, it is because the analysis of 'normal' statistical laws brings in complications which are foreign to our purpose. They do not affect the question of what makes a proposition lawlike; and it is in this that we are mainly interested.

On the view which we have now to consider, all that is required for there to be laws in nature is the existence of *de facto* constancies. In the most straightforward case, the constancy consists in the fact that events, or properties, or processes of different types are invariably conjoined with one another. The attraction of this view lies in its simplicity: but it may be too simple. There are objections to it which are not easily met.

In the first place, we have to avoid saddling ourselves with vacuous laws. If we interpret statements of the form 'All S is P' as being equivalent, in Russell's notation, to general implications of the form '$(x)(\Phi x \supset \Psi x)$,' we face the difficulty that such implications are considered to be true in all cases in which their antecedent is false.* Thus we shall have to take it as a universal truth both that all winged horses are spirited and that all winged horses are tame; for assuming, as I think we may, that there never have been or will be any winged horses, it is true both that there never have been or will be any that are not spirited, and that there never have been or will be any that are not tame.† And the same will hold for any other property that we care to

---

* Throughout this reading, we have added parentheses to Ayer's formulas. The universal generalization "$(x)(\Phi x \supset \Psi x)$" should be read as "for all x, if x has property Φ, then x has property Ψ." Because of the way that the truth-functional connective "$\supset$" is defined, any conditional formula of the form "$(p \supset q)$" is true whenever its antecedent, p, is false, regardless of whether the consequent, q, is true or false. Hence, Ayer's remark about winged horses in the next sentence.

† In predicate logic, "$(x)(\Phi x \supset \Psi x)$" is logically equivalent to "$\sim(\exists x)(\Phi x \ \& \sim \Psi x)$." This negation of an existential generalization says "it is not the case that there exists anything, x, such that x has property Φ and lacks property Ψ." Consequently, when nothing has property Φ—as in Ayer's example of winged horses—both statements are true, regardless of the nature of property Ψ.

choose. But surely we do not wish to regard the ascription of any property whatsoever to winged horses as the expression of a law of nature.

The obvious way out of this difficulty is to stipulate that the class to which we are referring should not be empty. If statements of the form 'All S is P' are used to express laws of nature, they must be construed as entailing that there are S's. They are to be treated as the equivalent, in Russell's notation, of the conjunction of the propositions '$(x)(\Phi x \supset \Psi x)$ and $(\exists x)\Phi x$'. But this condition may be too strong. For there are certain cases in which we do wish to take general implications as expressing laws of nature, even though their antecedents are not satisfied. Consider, for example, the Newtonian law that a body on which no forces are acting continues at rest or in uniform motion along a straight line. It might be argued that this proposition was vacuously true, on the ground that there are in fact no bodies on which no forces are acting; but it is not for this reason that it is taken as expressing a law. It is not interpreted as being vacuous. But how then does it fit into the scheme? How can it be held to be descriptive of what actually happens?

What we want to say is that if there *were* any bodies on which no forces were acting then they *would* behave in the way that Newton's law prescribes. But we have not made any provision for such hypothetical cases: according to the view which we are now examining, statements of law cover only what is actual, not what is merely possible. There is, however, a way in which we can still fit in such 'non-instantial' laws. As Professor C. D. Broad has suggested,[9] we can treat them as referring not to hypothetical objects, or events, but only to the hypothetical consequences of instantial laws. Our Newtonian law can then be construed as implying that there are instantial laws, in this case laws about the behaviour of bodies on which forces are acting, which are such that when combined with the proposition that there are bodies on which no forces are acting, they entail the conclusion that these bodies continue at rest, or in uniform motion along a straight line. The proposition that there are such bodies is false, and so, if it is interpreted existentially, is the conclusion, but that does not matter. As Broad puts it, 'what we are concerned to assert is that this false conclusion is a necessary consequence of the conjunction of a certain false instantial supposition with certain true instantial laws of nature'.

This solution of the present difficulty is commendably ingenious, though I am not sure that it would always be possible to find the instantial laws which it requires. But even if we accept it, our troubles are not over. For, as Broad himself points out, there is one important class of cases in which it does not help us. These cases are those in which one measurable quantity is said to depend upon another, cases like that of the law connecting the volume and temperature of a gas under a given pressure, in which there is a mathematical function which enables one to calculate the numerical value of either quantity from the value of the other. Such laws have the form '$x = Fy$', where the range of the variable y covers all possible values of the quantity in question. But now it is not to be supposed that all these values are actually to be found in nature. Even if the number of different temperatures which specimens of gases have or will acquire is infinite, there still must be an infinite number missing. How then are we to interpret such a law? As being the compendious assertion of all its actual instances? But the formulation of the law in no way indicates which the actual instances are. It would be absurd to construe a general formula about the functional dependence of one quantity on another as committing us to the assertion that just these values of the quantity are actually realized. As asserting that for a value n of y, which is in fact not realized, the proposition that it is realized, in conjunction with the set of propositions describing all the actual cases, entails the proposition that there is a corresponding value m of x? But this is open to the same objection, with the further drawback that the entailment would not hold. As asserting with regard to any given value n of y that either n is not realized or that there is a corresponding value m of x? This is the most plausible alternative, but it makes the law trivial for all the values of y which happen not to be realized. It is hard to escape the conclusion that what we really mean to assert when we formulate such a law is that there is a corresponding value of x to every *possible* value of y.

Another reason for bringing in possibilities is that there seems to be no other way of accounting for the difference between generalizations of law and generalizations of fact. To revert to our earlier examples, it is a generalization of fact that all the Presidents of the Third French Republic are male, or that all the cigarettes that are now in my cigarette case are made of Virginian tobacco. It is a generalization of law that the planets of our solar system move in elliptical orbits, but a generalization of fact that, counting the earth as Terra, they all have Latin names. Some philosophers refer to these generalizations of fact as 'accidental generalizations', but this use of the word 'accidental' may be misleading. It is not suggested that these generalizations are true by accident, in the sense that there is no causal explanation of their truth, but only that they are not themselves the expression of natural laws.

But how is this distinction to be made? The formula '$(x)(\Phi x \supset \Psi x)$' holds equally in both cases. Whether the generalization be one of fact or of law, it will state at least that there is nothing which has the property Φ but lacks the property Ψ. In this sense, the generality is perfect in both cases, so long as the statements are true. Yet there seems to be a sense in which the generality of what we are calling generalizations of fact is less complete. They seem to be restricted in a way that generalizations of law are not. Either they involve some spatio-temporal restriction, as in the example of the cigarettes *now* in my cigarette case, or they refer to particular individuals, as in the example of the presidents of France. When I say that all the planets have Latin names, I am referring definitely to a certain set of individuals, Jupiter, Venus, Mercury, and so on, but when I say that the planets move in elliptical orbits I am referring indefinitely to anything that has the properties that constitute being a planet in this solar system. But it will not do to say that generalizations of fact are simply conjunctions of particular statements, which definitely refer to individuals; for in asserting that the planets have Latin names, I do not individually identify them: I may know that they have Latin names without being able to list them all. Neither can we mark off generalizations of law by insisting that their expression is not to include any reference to specific places or times. For with a little ingenuity, generalizations of fact can always be made to satisfy this condition. Instead of referring to the cigarettes that are now in my cigarette case, I can find out some general property which only these cigarettes happen to possess, say the property of being contained in a cigarette case with such and such markings which is owned at such and such a period of his life by a person of such and such a sort, where the descriptions are so chosen that the description of the person is in fact satisfied only by me and the description of the cigarette case, if I possess more than one of them, only by the one in question. In certain instances these descriptions might have to be rather complicated, but usually they would not: and anyhow the question of complexity is not here at issue. But this means that, with the help of these 'individuating' predicates, generalizations of fact can be expressed in just as universal a form as generalizations of law. And conversely, as Professor Nelson Goodman has pointed out, generalizations of law can themselves be expressed in such a way that they contain a reference to particular individuals, or to specific places and times. For, as he remarks, 'even the hypothesis "All grass is green" has as an equivalent "All grass in London or elsewhere is green"'.[10] Admittedly, this assimilation of the two types of statement looks like a dodge; but the fact that the dodge works shows that we cannot found the distinction on a difference in the ways in which the statement can be expressed. Again, what we want to say is that whereas generalizations of fact cover only actual instances, generalizations of law cover possible instances as well. But this notion of possible, as opposed to actual, instances has not yet been made clear.

If generalizations of law do cover possible as well as actual instances, their range must be infinite; for while the number of objects which do throughout the course of time possess a certain property may be finite, there can be no limit to the number of objects which might possibly possess it: for once we enter the realm of possibility we are not confined even to such objects as actually exist. And this shows how far removed these generalizations are from being conjunctions: not simply because their range is infinite, which might be true even if it were confined to actual instances, but because there is something absurd about trying to list all the possible instances. One can imagine an angel's undertaking the task of naming or describing all the men that there ever have been or will be, even if their number were infinite, but how would he set about naming, or describing, all the possible men? This point is developed by F. P. Ramsey who remarks that the variable hypothetical '$(x)\Phi x$' resembles a conjunction (a) in that it contains all lesser, *i.e.* here all finite conjunctions, and appears as a sort of infinite product. (*b*) When we ask what would make it true, we inevitably answer that it is true if and only if every *x* has Φ; *i.e.* when we regard it as a proposition capable of the two cases truth and falsity, we are forced to make it a conjunction which we cannot express for lack of symbolic power'.[11] But, he goes on, 'what we can't say we can't say, and we can't whistle it either', and he concludes that the variable hypothetical is not a conjunction and that 'if it is not a conjunction, it is not a proposition at all'. Similarly, Professor Ryle, without explicitly denying that generalizations of law are propositions, describes them as 'seasonal inference warrants',[12] on the analogy of season railway-tickets, which implies that they are not so much propositions as rules. Professor Schlick also held that they were rules, arguing that they could not be propositions because they were not conclusively verifiable; but this is a poor argument, since it is doubtful if any propositions are conclusively verifiable, except possibly those that describe the subject's immediate experiences.

Now to say that generalizations of law are not propositions does have the merit of bringing out their peculiarity. It is one way of emphasizing the difference between them and generalizations of fact. But I think that it emphasizes it too strongly. After all, as Ramsey himself acknowledges, we do want to say that generalizations of law are either true or false. And they are tested in the way that other propositions are, by the examination of actual instances. A contrary instance refutes a generalization of law in the same way as it refutes a generalization of fact. A positive instance confirms them both. Admittedly, there is the difference that if all the actual instances are favourable, their conjunction entails the generalization of fact, whereas it does not entail the generalization of law: but still there is no better way of confirming a generalization of law than by finding favourable instances. To say that lawlike statements function as seasonal inference warrants is indeed illuminating, but what it comes to is that the inferences in question are warranted by the facts. There would be no point in issuing season tickets if the trains did not actually run.

To say that generalizations of law cover possible as well as actual cases is to say that they entail subjunctive conditionals. If it is a law of nature that the planets move in elliptical orbits, then it must not only be true that the actual planets move in elliptical orbits; it must also be true that if anything were a planet it would move in an elliptical orbit: and here 'being a planet' must be construed as a matter of having certain properties, not just as being identical with one of the planets that there are. It is not indeed a peculiarity of statements which one takes as expressing laws of nature that they entail subjunctive conditionals: for the same will be true of any statement that contains a dispositional predicate. To say, for example, that this rubber band is elastic is to say not merely that it will resume its normal size when it has been stretched, but that it would do so if ever it were stretched: an object may be elastic without ever in fact being stretched at all. Even the statement that this is a white piece of paper may be taken as implying not only how the piece of paper does look but also how it would look under certain conditions, which may or may not be fulfilled. Thus one cannot say that generalizations of fact do not entail subjunctive conditionals, for they may very well contain dispositional predicates: indeed they are more likely to do so than not: but they will not entail the subjunctive conditionals which are entailed by the corresponding statements of law. To say that all the planets have Latin names may be to make a dispositional statement, in the sense that it implies not so much that people do always call them by such names but that they would so call them if they were speaking correctly. It does not, however, imply with regard to anything whatsoever that if it were a planet it would be called by a Latin name. And for this reason it is not a generalization of law, but only a generalization of fact.

There are many philosophers who are content to leave the matter there. They explain the 'necessity' of natural laws as consisting in the fact that they hold for all possible, as well as actual, instances: and they distinguish generalizations of law from generalizations of fact by bringing out the differences in their entailment of subjunctive conditionals. But while this is correct so far as it goes, I doubt if it goes far enough. Neither the notion of possible, as opposed to actual, instances nor that of the subjunctive conditional is so pellucid that these references to them can be regarded as bringing all our difficulties to an end. It will be well to try to take our analysis a little further if we can.

The theory which I am going to sketch will not avoid all talk of dispositions; but it will confine it to people's attitudes. My suggestion is that the difference between our two types of generalization lies not so much on the side of the facts which make them true or false, as in the attitude of those who put them forward. The factual information which is expressed by a statement of the form 'for all x, if x has Φ then x has Ψ', is the same whichever way it is interpreted. For if the two interpretations differ only with respect to the possible, as opposed to the actual values of x, they do not differ with respect to anything that actually happens. Now I do not wish to say that a difference in regard to mere possibilities is not a genuine difference, or that it is to be equated with a difference in the attitude of those who do the interpreting. But I do think that it can best be elucidated by referring to such differences of attitude. In short I propose to explain the distinction between generalizations of law and generalizations of fact, and thereby to give some account of what a law of nature is, by the indirect method of analysing the distinction between treating a generalization as a statement of law and treating it as a statement of fact.

If someone accepts a statement of the form '$(x)(\Phi x \supset \Psi x)$' as a true generalization of fact, he will not in fact believe that anything which has the property Φ has any other property that leads to its not having Ψ. For since he believes that everything that has Φ has Ψ, he must believe that whatever other properties a given value of x may have they are not such as to prevent its having Ψ. It may be even that he knows this to be so. But now let us suppose that he believes such a generalization to be true, without knowing it for certain. In that case there will be various properties X, X_1 . . . such that if he were to learn, with respect to any value of a of x, that a had one or more of these properties as well as Φ, it would destroy, or seriously weaken his belief that a had Ψ. Thus I believe that all the cigarettes in my case are made of Virginian tobacco, but this belief would be destroyed if I were informed that I had absent-mindedly just filled my case from a box in which I keep only Turkish cigarettes. On the other hand, if I took it to be a law of nature that all the cigarettes in this case were made of Virginian tobacco, say on the ground that the case had some curious physical property which had the effect of changing any other tobacco that was put into it into Virginian, then my belief would not be weakened in this way.

Now if our laws of nature were causally independent of each other, and if, as Mill thought, the propositions which expressed them were always put forward as being unconditionally true, the analysis could proceed quite simply. We could then say that a person A was treating a statement of the form 'for all x, if Φx then Ψx' as expressing a law of nature, if and only if there was no property X which was such that the information that a value a of x had X as well as Φ would weaken his belief that a had Ψ. And here we should have to admit the proviso that X did not logically entail not-Ψ, and also, I suppose, that its presence was not regarded as a manifestation of not-Ψ; for we do not wish to make it incompatible with treating a statement as the expression of a law that one should acknowledge a negative instance if it arises. But the actual position is not so simple. For one may believe that a statement of the form 'for all x, if Φx then Ψx' expresses a law of nature while also believing, because of one's belief in other laws, that if something were to have the property X as well as Φ it would not have Ψ. Thus one's belief in the proposition that an object which one took to be a loadstone attracted iron might be weakened or destroyed by the information that the physical composition of the supposed loadstone was very different from what one had thought it to be. I think, however, that in all such cases, the information which would impair one's belief that the object in question had the property Ψ would also be such that, independently of other considerations, it would seriously weaken one's belief that the object ever had the property Φ. And if this is so, we can meet the difficulty by stipulating that the range of properties which someone who treats 'for all x, if Φx then Ψx' as a law must be willing to conjoin with Φ, without his belief in the consequent being weakened, must not include those the knowledge of whose presence would in itself seriously weaken his belief in the presence of Φ.

There remains the further difficulty that we do not normally regard the propositions which we take to express laws of nature as being unconditionally true. In stating them we imply the presence of certain conditions which we do not actually specify. Perhaps we could specify them if we chose, though we might find it difficult to make the list exhaustive. In this sense a generalization of law may be weaker than a generalization of fact, since it may admit exceptions to the generalization as it is stated. This does not mean, however, that the law allows for exceptions: if the exception is acknowledged to be genuine, the law is held to be refuted. What happens in the other cases is that the exception is regarded as having been tacitly provided for. We lay down a law about the boiling point of water, without bothering to mention that it does not hold for high altitudes. When this is pointed out to us, we say that this qualification was meant to be understood. And so in other instances. The statement that if anything has Φ it has Ψ was a loose formulation of the law; what we really meant was that if anything has Φ but not X, it has Ψ. Even in the case where the existence of the exception was not previously known, we often regard it as qualifying rather than refuting the law. We say, not that the generalization has been falsified, but that it was inexactly stated. Thus, it must be allowed that someone whose belief in the presence of Ψ, in a given instance, is destroyed by the belief that Φ is accompanied by X may still be treating '$(x)(\Phi x \supset \Psi x)$' as expressing a law of nature if he is prepared to accept '$(x)((\Phi x \cdot {\sim}Xx) \supset \Psi x)$' as a more exact statement of the law.

Accordingly I suggest that for someone to treat a statement of the form 'if anything has Φ it has Ψ' as expressing a law of nature, it is sufficient (i) that subject to a willingness to explain away exceptions he believes that in a non-trivial sense everything which in fact has Φ has Ψ (ii) that his belief that something which has Φ has Ψ is not liable to be weakened by the discovery that the object in question also has some other property X, provided (a) that X does not logically entail not-Ψ (b) that X is not a manifestation of not-Ψ (c) that the discovery that something had X would not in itself seriously weaken his belief that it had Φ (d) that he does not regard the statement 'if anything has Φ and not-X it has Ψ' as a more exact statement of the generalization that he was intending to express.

I do not suggest that these conditions are necessary, both because I think it possible that they could be simplified and because they do not cover the whole field. For instance, no provision has been made for functional laws, where the reference to possible instances does not at present seem to me eliminable. Neither am I offering a definition of natural law. I do not claim that to say that some proposition expresses a law of nature entails saying that someone has a certain attitude towards it; for clearly it makes sense to say that there are laws of nature which remain unknown. But this is consistent with holding that the notion is to be explained in terms of people's attitudes. My explanation is indeed sketchy, but I think that the distinctions which I have tried to bring out are relevant and important: and I hope that I have done something towards making them clear.

■ | Notes

1. *Leviathan*, Part I, Chap. xv.

2. *An Enquiry Concerning Human Understanding*, iv, 1.25.

3. *Probability and Induction*, pp. 79 ff.

4. *A Treatise of Human Nature*, i, iii, vi.

5. Cf. *La Science et l'hypothèse* [Science and Hypothesis], pp. 119–29.

6. Cf. Kneale, *op. cit.*

7. Cf. K. Popper, "What Can Logic Do for Philosophy?" *Supplementary Proceedings of the Aristotelian Society*, Vol. XXII: and papers in the same volume by W. C. Kneale and myself.

8. *Scientific Explanation*, pp. 118–29.

9. "Mechanical and Teleological Causation," *Supplementary Proceedings of the Aristotelian Society*, XIV, 98 ff.

10. *Fact, Fiction and Forecast*, p. 78.

11. *Foundations of Mathematics*, p. 238.

12. "'If', 'So,' and 'Because',"" in *Philosophical Analysis* (Essays edited by Max Black), p. 332.

Fred I. Dretske

Laws of Nature

It is tempting to identify the laws of nature with a certain class of universal truths. Very few empiricists have succeeded in resisting this temptation. The popular way of succumbing is to equate the fundamental laws of nature with what is asserted by those universally true statements of non-limited scope that embody only qualitative predicates.[1] On this view of things a law-like statement is a statement of the form "$(x)(Fx \supset Gx)$" or "$(x)(Fx \equiv Gx)$" where "F" and "G" are purely qualitative (nonpositional). Those law-like statements that are true express laws. "All robins' eggs are greenish blue," "All metals conduct electricity," and "At constant pressure any gas expands with increasing temperature" (Hempel's examples) are law-like statements. If they are true, they express laws. The more familiar sorts of things that we are accustomed to calling laws, the formulae and equations appearing in our physics and chemistry books, can supposedly be understood in the same way by using functors in place of the propositional functions "Fx" and "Gx" in the symbolic expressions given above.[*]

From *Philosophy of Science* 44 (1977): 248–68.

[*] Although it does not affect any of the philosophical issues debated in this chapter, Dretske's remark about functors raises an interesting question, namely, whether predicate logic has the resources to represent quantitative laws adequately. It seems most unlikely that a so-called functional law written as an equation involving several variables, each of which takes real numbers as values, can be properly regarded as having the simple form $(x)(Fx \supset Gx)$, where F and G are qualitative predicates. The functors mentioned by Dretske attempt to solve this problem by converting the equation into a function that is then treated as a predicate. Consider the ideal gas law, $PV = nRT$. The functor in this case might be "is identical with the value of nRT divided by V" and the law would read (roughly), "For all x, if x is the value of the pressure of an ideal gas, then x is identical with the value of nRT divided by V." Not only is this clumsy but also, by focusing on pressure in the antecedent, it obscures the interdependence of the variables: the ideal gas law is not about pressure; it is about all the variables and their functional relation. Alternatively, we might define the predicate "obeys the equation $PV = nRT$" and then portray the ideal gas law as "For all x, if x is an ideal gas, then x obeys the equation $PV = nRT$." But this says

I say that it is tempting to proceed in this way since, to put it bluntly, conceiving of a law as having a content greater than that expressed by a statement of the form $(x)(Fx \supset Gx)$ seems to put it beyond our epistemological grasp.[2] We must work with what we are given, and what we are given (the observational and experimental data) are facts of the form: this F is G, that F is G, all examined F's have been G, and so on. If, as some philosophers have argued,[3] law-like statements express a kind of nomic necessity between events, something *more* than that F's are, as a matter of fact, always and everywhere, G, then it is hard to see what kind of evidence might be brought in support of them. The whole point in acquiring instantial evidence (evidence of the form "This F is G") in support of a law-like hypothesis would be lost if we supposed that what the hypothesis was actually asserting was some kind of nomic connection, some kind of modal relationship, between things that were F and things that were G. We would, it seems, be in the position of someone trying to confirm the *analyticity* of "All bachelors are unmarried" by collecting evidence about the marital status of various bachelors. This kind of evidence, though relevant to the *truth* of the claim that all bachelors are unmarried, is powerless to confirm the *modality* in question. Similarly, if a hypothesis, in order to qualify as a law, must express or assert some form of necessity between F's and G's, then it becomes a mystery how we ever manage to confirm such attributions with the sort of instantial evidence available from observation.

Despite this argument, the fact remains that laws are *not* simply what universally true statements express, not even universally true statements that embody purely qualitative predicates (and are, as a result, unlimited in scope). This is not particularly newsworthy. It is commonly acknowledged that law-like statements have some peculiarities that prevent their straightforward assimilation to universal truths. That the concept of a law and the concept of a universal truth are different concepts can best be seen, I think, by the following consideration: assume that $(x)(Fx \supset Gx)$ is true and that the predicate expressions satisfy all the restrictions that one might wish to impose in order to convert this universal statement into a statement of law.[4] Consider a predicate expression "K" (eternally) coextensive with "F"; i.e., (x) $(Fx \equiv Kx)$ for all time. We may then infer that if $(x)(Fx \supset Gx)$ is a universal truth, so is $(x)(Kx \supset Gx)$. The class of universal truths is closed under the operation of coextensive predicate substitution. Such is *not* the case with laws. If it is a law that all F's are G, and we substitute the term "K" for the term "F" in this law, the result is not necessarily a law. If diamonds have a refractive index of 2.419 (law) and "is a diamond" is coextensive with "is mined in kimberlite (a dark basic rock)" we cannot infer that *it is a law* that things mined in kimberlite have a refractive index of 2.419. Whether this is a law or not depends on whether the co-extensiveness of "is a diamond" and

---

merely that all ideal gases obey the ideal gas law, which can hardly be regarded as a perspicuous *representation* of the law.

"is mined in kimberlite" is *itself* law-like. The class of laws is not closed under the same operation as is the class of universal truths.

Using familiar terminology we may say that the predicate positions in a statement of law are *opaque* while the predicate positions in a universal truth of the form $(x)(Fx \supset Gx)$ are *transparent*.* I am using these terms in a slightly unorthodox way. It is not that when we have a law, "All F's are G," we can alter its truth value by substituting a coextensive predicate for "F" or "G". For if the statement is true, it will remain true after substitution. What happens, rather, is that the expression's status *as a law* is (or may be) affected by such an exchange. The matter can be put this way: the statement

(A) All F's are G (understood as $(x)(Fx \supset Gx)$)

has "F" and "G" occurring in transparent positions. Its truth value is unaffected by the replacement of "F" or "G" by a coextensive predicate. The same is true of

(B) It is universally true that F's are G.

If, however, we look at

(C) It is a law that F's are G,

we find that "F" and "G" occur in opaque positions. If we think of the two prefixes in (B) and (C), "it is universally true that . . ." and "it is a law that . . . ," as operators, we can say that the operator in (B) does not, while the operator in (C) does, confer opacity on the embedded predicate positions. To refer to something as a statement of law is to refer to it as an expression in which the descriptive terms occupy opaque positions. To refer to something as a universal truth is to refer to it as an expression in which

---

* *Transparent* and *opaque* are terms used in the theory of reference. Consider the true sentence, "Blue whales live in water." If we replace the expression "blue whales" with a phrase that designates the same class of animals—"the largest mammals on earth," for example—then the sentence must remain true. Philosophers of language say that "Blue whales live in water" is a transparent context because its truth value cannot be altered by the substitution of coreferring expressions. Contrast this with "John knows that blue whales live in water." This is an opaque context. John might be ignorant of the fact that blue whales are the largest mammals on earth. Hence, he could know that blue whales live in water without also knowing that the largest mammals on earth live in water. Thus, in this case, the substitution of a coreferring expression could change a true sentence into a false one. Modal contexts (that is, sentences involving possibility or necessity) can also create opacity. "Blue whales are necessarily whales" is true but "The largest mammals on earth are necessarily whales" is false because it is contingent, not necessary, that the world's largest mammals happen to be whales. Similarly, Dretske argues, sentences of the form "It is a law that all Fs are G" are referentially opaque.

the descriptive terms occupy transparent positions. Hence, our concept of a law differs from our concept of a universal truth.[5]

Confronted by a difference of this sort, many philosophers have argued that the distinction between a natural law and a universal truth was not, fundamentally, an *intrinsic* difference. Rather, the difference was a difference in the *role* some universal statements played within the larger theoretical enterprise. Some universal statements are more highly integrated into the constellation of accepted scientific principles, they play a more significant role in the explanation and prediction of experimental results, they are better confirmed, have survived more tests, and make a more substantial contribution to the regulation of experimental inquiry. But, divorced from this context, stripped of these *extrinsic* features, a law is nothing but a universal truth. It has the same empirical content. Laws are to universal truths what shims are to slivers of wood and metal; the latter *become* the former by being *used* in a certain way. There is a *functional* difference, nothing else.[6]

According to this reductionistic view, the peculiar opacity (described above) associated with laws is not a manifestation of some intrinsic difference between a law and a universal truth. It is merely a symptom of the special status or function that some universal statements have. The basic formula is: law = universal truth + X. The "X" is intended to indicate the special function, status or role that a universal truth must have to qualify as a law. Some popular candidates for this auxiliary idea, X, are:

1 High degree of confirmation,

2 Wide acceptance (well established in the relevant community),

3 Explanatory potential (can be used to explain its instances),

4 Deductive integration (within a larger system of statements),

5 Predictive use.

To illustrate the way these values of X are used to buttress the equation of laws with universal truths, it should be noted that each of the concepts appearing on this list generates an opacity similar to that witnessed in the case of genuine laws. For example, to say that it is a law that all F's are G may possibly be no more than to say that it is well established that $(x)(Fx \supset Gx)$. The peculiar opacity of laws is then explained by pointing out that the class of expressions that are well established (or highly confirmed) is not closed under substitution of coextensive predicates: one cannot infer that $(x)(Kx \supset Gx)$ is well established just because "Fx" and "Kx" are coextensive and $(x)(Fx \supset Gx)$ is well established (for no one may know that "Fx" and "Kx" *are* coextensive). It may be supposed, therefore, that the opacity of laws is merely a manifestation of the underlying fact that a universal statement, to qualify as a law, must be well established, and the opacity is a result of this epistemic condition. Or, if this will not do, we can suppose that one of the other notions mentioned above, or a combination of them, is the source of a law's opacity.

This response to the alleged uniqueness of natural laws is more or less standard fare among empiricists in the Humean tradition. Longstanding (=venerable) epistemological and ontological commitments motivate the equation: law=universal truth+X. There is disagreement among authors about the differentia X, but there is near unanimity about the fact that laws are a *species* of universal truth.

If we set aside our scruples for the moment, however, there is a plausible explanation for the opacity of laws that has not yet been mentioned. Taking our cue from Frege, it may be argued that since the operator "it is a law that . . ." converts the otherwise transparent positions of "All F's are G" into opaque positions, we may conclude that this occurs because within the context of this operator (either explicitly present or implicitly understood) the terms "F" and "G" do not have their usual referents. There is a shift in what we are talking about. To say that *it is a law* that F's are G is to say that "All F's are G" is to be understood (in so far as it expresses a law), not as a statement about the extensions of the predicates "F" and "G," but as a singular statement describing a relationship between the universal properties F-ness and G-ness. In other words, (C) is to be understood as having the form:

$$6 \quad F\text{-ness} \to G\text{-ness}.[7]$$

To conceive of (A) as a universal truth is to conceive of it as expressing a relationship between the extensions of its terms; to conceive of it as a law is to conceive of it as expressing a relationship between the properties (magnitudes, quantities, features) which these predicates express (and to which we may refer with the corresponding abstract singular term). The opacity of laws is merely a manifestation of this change in reference. If "F" and "K" are coextensive, we cannot substitute the one for the other in the *law* "All F's are G" and expect to preserve truth; for the law asserts a connection between F-ness and G-ness and there is no guarantee that a similar connection exists between the properties K-ness and G-ness just because all F's are K and *vice versa*.[8]

It is this view that I mean to defend in the remainder of this essay. Law-like statements are singular statements of fact describing a relationship between properties or magnitudes. Laws are the relationships that are asserted to exist by true law-like statements. According to this view, then, there is an *intrinsic* difference between laws and universal truths. Laws imply universal truths, but universal truths do not imply laws. Laws are (expressed by) *singular* statements describing the relationships that exist between universal qualities and quantities; they are not universal statements about the particular objects and situations that exemplify these qualities and quantities. Universal truths are not transformed into laws by acquiring some of the extrinsic properties of laws, by being used in explanation or prediction, by being made to support counterfactuals, or by becoming well established. For, as we shall see, universal truths *cannot* function in these ways. They *cannot* be made to perform a service they are wholly unequipped to provide.

In order to develop this thesis it will be necessary to overcome some metaphysical prejudices, and to overcome these prejudices it will prove useful to review the major deficiencies of the proposed alternative. The attractiveness of the formula: law = universal truth + X, lies, partly at least, in its ontological austerity, in its tidy portrayal of what there is, or what there must be, in order for there to be laws of nature. The antidote to this seductive doctrine is a clear realization of how utterly hopeless, epistemologically and functionally hopeless, this equation is.

If the auxiliary ideas mentioned above (explanation, prediction, confirmation, etc.) are deployed as values of X in the reductionistic equation of laws with universal truths, one can, as we have already seen, render a satisfactory account of the opacity of laws. In this particular respect the attempted equation proves adequate. In what way, then, does it fail?

(1) and (2) are what I will call "epistemic" notions; they assign to a statement a certain epistemological status or cognitive value. They are, for this reason alone, useless in understanding the nature of a law.[9] Laws do not begin to be laws only when we first become aware of them, when the relevant hypotheses become well established, when there is public endorsement by the relevant scientific community. The laws of nature are the same today as they were one thousand years ago (or so we believe); yet, some hypotheses are highly confirmed today that were not highly confirmed one thousand years ago. It is certainly true that we only begin to *call* something a law when it becomes well established, that we only recognize something as a statement of law when it is confirmed to a certain degree, but that something is a law, that some statement does in fact express a law, does not similarly await our appreciation of this fact. We discover laws, we do not invent them—although, of course, some invention may be involved in our manner of expressing or codifying these laws. Hence, the status of something as a statement of law does not depend on its epistemological status. What does depend on such epistemological factors is our ability to identify an otherwise qualified statement *as true* and, therefore, *as a statement of law*. It is for this reason that one cannot appeal to the epistemic operators to clarify the nature of laws; they merely confuse an epistemological with an ontological issue.

What sometimes helps to obscure this point is the tendency to conflate laws with the verbal or symbolic expression of these laws (what I have been calling "statements of law"). Clearly, though, these are different things and should not be confused. There are doubtless laws that have not yet [received] (or will never receive) symbolic expression, and the same law may be given different verbal codifications (think of the variety of ways of expressing the laws of thermodynamics). To use the language of "propositions" for a moment, a law is the proposition expressed, not the vehicle we use to express it. The *use* of a sentence *as an expression of law* depends on epistemological considerations, but the law itself does not.

There is, furthermore, the fact that whatever auxiliary idea we select for understanding laws (as candidates for X in the equation: law = universal truth + X), if it is going to achieve what we expect of it, should help to account for the variety of other features that laws are acknowledged to have. For example, it is said that laws "support" counterfactuals of a certain sort. If laws are universal truths, this fact is a complete mystery, a mystery that is usually suppressed by using the word "support." For, of course, universal statements do not *imply* counterfactuals in any sense of the word "imply" with which I am familiar. To be told that all F's are G is not to be told anything that implies that if this x were an F, it would be G. To be told that all dogs born at sea have been and will be cocker spaniels is *not* to be told that we would get cocker spaniel pups (or no pups at all) if we arranged to breed dachshunds at sea. The only reason we might *think* we were being told this is because we do not expect anyone to assert that all dogs born at sea *will be* cocker spaniels unless they know (or have good reasons for believing) that this is true; and we do not understand *how* anyone could *know* that this is true without being privy to information that insures this result—without, that is, knowing of some bizzare law or circumstance that *prevents* anything but cocker spaniels from being born at sea. Hence, *if* we accept the claim at all, we do so with a certain presumption about what our informant must know in order to be a serious claimant. We assume that our informant knows of certain laws or conditions that *insure* the continuance of a past regularity, and it is this presumed knowledge that we exploit in endorsing or accepting the counterfactual. But the simple fact remains that the statement "All dogs born at sea have been and will be cocker spaniels" does not *itself* support or imply this counterfactual; at best, we support the counterfactual (if we support it at all) on the basis of what the claimant is supposed to know in order to advance such a universal projection.

Given this incapacity on the part of universal truths to support counterfactuals, one would expect some assistance from the epistemic condition if laws are to be analyzed as well established universal truths. But the expectation is disappointed; we are *left* with a complete mystery. For if a statement of the form "All F's are G" does not support the counterfactual, "If this (non-G) were an F, it would be G," it is clear that it will not support it just because it is well established or highly confirmed. The fact that all the marbles in the bag are red does not support the contention that if this (blue) marble were in the bag, it would be red; but neither does the fact that we *know* (or it is highly confirmed) that all the marbles in the bag are red support the claim that if this marble were in the bag it would be red. And making the universal truth *more universal* is not going to repair the difficulty. The fact that all the marbles in the universe are (have been and will be) red does not imply that I *cannot* manufacture a blue marble; it implies that I *will not*, not that I cannot or that if I were to try, I would fail. To represent laws on the model of one of our epistemic operators, therefore, leaves wholly unexplained one of the most important features of laws that we are trying to understand. They are, in this respect, unsatisfactory candidates for the job.

Though laws are not merely well established general truths, there is a related point that deserves mention: laws are the *sort* of thing that can become well established prior to an exhaustive enumeration of the instances to which they apply. This, of course, is what gives laws their predictive utility. Our confidence in them increases at a much more rapid rate than does the ratio of favorable examined cases to total number of cases. Hence, we reach the point of confidently using them to project the outcome of unexamined situations while there is still a substantial number of unexamined situations to project.

This feature of laws raises new problems for the reductionistic equation. For, contrary to the argument in the second paragraph of this essay, it is hard to see how confirmation is possible for universal truths. To illustrate this difficulty, consider the (presumably easier) case of a general truth of *finite* scope. I have a coin that you have (by examination and test) convinced yourself is quite normal. I propose to flip it ten times. I conjecture (for whatever reason) that it will land heads all ten times. You express doubts. I proceed to "confirm" my hypothesis. I flip the coin once. It lands heads. Is this evidence that my hypothesis is correct? I continue flipping the coin and it turns up with nine straight heads. Given the opening assumption that we are dealing with a fair coin, the probability of getting all ten heads (the probability that my hypothesis is true) is now, after examination of 90% of the total population to which the hypothesis applies, exactly .5. If we are guided by probability considerations alone, the likelihood of all ten tosses being heads is now, after nine favorable trials, a toss-up. After nine favorable trials it is no more reasonable to believe the hypothesis than its denial. In what sense, then, can we be said to have been accumulating evidence (during the first nine trials) that all would be heads? In what sense have we been confirming the hypothesis? It would appear that the probability of my conjecture's being true never exceeds .5 until we have exhaustively examined the entire population of coin tosses and found them *all* favorable. The probability of my conjecture's being true is either: (i) too low ($\leq .5$) to invest any confidence in the hypothesis, or (ii) so high ($= 1$) that the hypothesis is useless for prediction. There does not seem to be any middle ground.

Our attempts to confirm universal generalizations of nonlimited scope is, I submit, in exactly the same impossible situation. It is true, of course, that after nine successful trials the probability that all ten tosses will be heads is greatly increased over the initial probability that all would be heads. The initial probability (assuming a fair coin) that all ten tosses would be heads was on the order of .002. After nine favorable trials it is .5. In this sense I have increased the probability that my hypothesis is true; I have raised its probability from .002 to .5. The important point to notice, however, is that this sequence of trials did not alter the probability that the *tenth* trial would be heads. The probability that the unexamined instance would be favorable remains exactly what it was before I began flipping the coin. It was originally .5 and it is now, after nine favorable trials, still .5. I am in no better position now, after extensive sampling, to predict the outcome of the tenth toss than I was before I started. To suppose otherwise is to commit the converse of the Gambler's Fallacy.

Notice, we could take the first nine trials as evidence that the tenth trial would be heads *if* we took the results of the first nine tosses as evidence that the coin was biased in some way. Then, on *this* hypothesis, the probability of getting heads on the last trial (and, hence, on all ten trials) would be greater than .5 (how much greater would depend on the conjectured degree of bias and this, in turn, would presumably depend on the extent of sampling). This new hypothesis, however, is something quite different than the original one. The original hypothesis was of the form: $(x)(Fx \supset Gx)$, all ten tosses will be heads. Our new conjecture is that there is a physical asymmetry in the coin, an asymmetry that tends to yield more heads than tails. We have succeeded in confirming the general hypothesis (all ten tosses will be heads), but we have done so via an intermediate hypothesis involving *genuine laws* relating the physical make-up of the coin to the frequency of heads in a population of tosses.

It is by such devices as this that we create for ourselves, or some philosophers create for themselves, the *illusion* that (apart from supplementary *law-like* assumptions) general truths can be confirmed by their instances and therefore qualify, in this respect, as laws of nature. The illusion is fostered in the following way. It is assumed that confirmation is a matter of *raising the probability of a hypothesis*.[10] On this assumption any general statement of finite scope can be confirmed by examining its instances and finding them favorable. The hypothesis about the results of flipping a coin ten times can be confirmed by tossing nine straight heads, and this confirmation takes place without *any* assumptions about the coin's bias. Similarly, I confirm (to some degree) the hypothesis that all the people in the hotel ballroom are over thirty years old when I enter the ballroom with my wife and realize that *we* are both over thirty. In both cases I raise the probability that the hypothesis is true over what it was originally (before flipping the coin and before entering the ballroom). But this, of course, isn't confirmation. Confirmation is not simply raising the probability that a hypothesis is true, it is raising the probability that the unexamined cases resemble (in the relevant respect) the examined cases. It is *this* probability that must be raised if genuine confirmation is to occur (and if a confirmed hypothesis to be useful in *prediction*), and it is precisely this probability that is left unaffected by the instantial "evidence" in the above examples.

In order to meet this difficulty, and to cope with hypotheses that are *not* of limited scope,[11] the reductionist usually smuggles into his confirmatory proceedings the very idea he professes to do without: *viz.*, a type of law that is not merely a universal truth. The general truth then gets confirmed but *only* through the mediation of these supplementary laws. These supplementary assumptions are usually introduced to *explain* the regularities manifested in the examined instances so as to provide a basis for projecting these regularities to the unexamined cases. The only way we can get a purchase on the unexamined cases is to introduce a hypothesis which, while *explaining* the data we already have, *implies* something about the data we do not have. To suppose that our coin is biased (first example) is to suppose something that contributes to the explanation of our extraordinary run of heads (nine straight) and simultaneously implies something about the (probable) outcome of the tenth toss. Similarly (second example) my wife and I may be attending a reunion of some kind, and I may suppose that the other people in the ballroom are old classmates. This hypothesis not only explains our presence, it implies that most, if not all, of the remaining people in the room are of comparable age (well over thirty). In both these cases the generalization can be confirmed, but only via the introduction of a law or circumstance (combined with a law or laws) that helps to explain the data already available.

One additional example should help to clarify these last remarks. In sampling from an urn with a population of colored marbles, I can confirm the hypothesis that all the marbles in the urn are red by extracting at random several dozen red marbles (and no marbles of any other color). This is a genuine example of confirmation, not because I have raised the probability of the hypothesis that all are red by reducing the number of ways it can be false (the same reduction would be achieved if you *showed* me 24 marbles from the urn, all of which were red), but because the hypothesis that all the marbles in the urn are red, together with the fact (law) that you cannot draw nonred marbles from an urn containing only red marbles, *explains* the result of my random sampling. Or, if this is too strong, the law that assures me that random sampling from an urn containing a substantial number of nonred marbles would reveal (in all likelihood) at least one nonred marble lends its support to my confirmation that the urn contains only (or mostly) red marbles. Without the assistance of such auxiliary laws a sample of 24 red marbles is powerless to confirm a hypothesis about the total population of marbles in the urn. To suppose otherwise is to suppose that the *same* degree of confirmation would be afforded the hypothesis if you, whatever your deceitful intentions, showed me a carefully selected set of 24 red marbles from the urn. This *also* raises the probability that they are all red, but the trouble is that it does not (due to your unknown motives and intentions) raise the probability that the unexamined marbles resemble the examined ones. And it does not raise this probability because we no longer have, as the best available explanation of the examined cases (all red), a hypothesis that implies that the remaining (or most of the remaining) marbles are also red. Your careful selection of 24 red marbles from an urn containing many different colored marbles is an equally good explanation of the data and it does *not* imply that the remainder are red. Hence, it is not just the fact that we have 24 red marbles in our sample class (24 positive instances and no negative instances) that confirms the general hypothesis that all the marbles in the urn are red. It is this data *together with a law* that confirms it, a law that (together with the hypothesis) explains the data in a way that the general hypothesis alone cannot do.

We have now reached a critical stage in our examination of the view that a properly qualified set of universal generalizations can serve as the fundamental laws of nature. For we have, in the past few paragraphs, introduced the notion of *explanation*, and it is this notion, perhaps more than any other, that has received the greatest attention from philosophers in their quest for the appropriate X in the formula: law = universal truth + X. R. B. Braithwaite's treatment ([3]) is typical. He begins by suggesting that it is merely deductive integration that transforms a universal truth into a law of nature. Laws are simply universally true statements of the form $(x)(Fx \supset Gx)$ that are derivable from certain higher level hypotheses. To say that $(x)(Fx \supset Gx)$ is a statement of law is to say, not only that it is true, but that it is *deducible from* a higher level hypothesis, H, in a well established scientific system. The fact that it must be deducible from some higher level hypothesis, H, confers on the statement the opacity we are seeking to understand. For we may have a hypothesis from which we can derive $(x)(Fx \supset Gx)$ but from which we cannot derive $(x)(Kx \supset Gx)$ despite the coextensionality of "F" and "K." Braithwaite also argues that such a view gives a satisfactory account of the counterfactual force of laws.

The difficulty with this approach (a difficulty that Braithwaite recognizes) is that it only postpones the problem. Something is not a statement of law simply because it is true and deducible from some well-established higher level hypothesis. For every generalization implies another of smaller scope (e.g. $(x)(Fx \supset Gx)$ implies $(x)(Fx \cdot Hx \supset Gx)$), but this fact has not the slightest tendency to transform the latter generalization into a law.* What is required is that the higher level hypothesis *itself* be law-like. You cannot give to others what you do not have yourself. But now, it seems, we are back where we started from. It is at this point that Braithwaite begins talking about the higher level hypotheses having *explanatory force* with respect to the hypotheses subsumed under them. He is forced into this maneuver to account for the fact that these higher level hypotheses—not themselves law-like on his characterization (since not themselves derivable from still higher level hypotheses)—are capable of conferring lawlikeness on their consequences. The higher level hypotheses are laws because they explain; the lower level hypotheses are laws because they are deducible from laws. This fancy twist smacks of circularity. Nevertheless, it represents a conversion to *explanation* (instead of *deducibility*) as the fundamental feature of laws,

---

* Dretske uses a dot (instead of an ampersand) to stand for *and*.

and Braithwaite concedes this: "A hypothesis to be regarded as a natural law must be a general proposition which can be thought to *explain* its instances" ([3], p. 302) and, a few lines later, "Generally speaking, however, a true scientific hypothesis will be regarded as a law of nature if it has an explanatory function with regard to lower-level hypotheses or its instances." Deducibility is set aside as an incidental (but, on a Hempelian model of explanation, an important) facet of the more ultimate idea of explanation.

There is an added attraction to this suggestion. As argued above, it is difficult to see how instantial evidence can serve to confirm a universal generalization of the form: $(x)(Fx \supset Gx)$. If the generalization has an infinite scope, the ratio "examined favorable cases/total number of cases" never increases. If the generalization has a finite scope, or we treat its probability as something other than the above ratio, we may succeed in raising its probability by finite samples, but it is never clear how we succeed in raising the probability that the unexamined cases resemble the examined cases without invoking laws as auxiliary assumptions. And this is the very notion we are trying to analyze. To this problem the notion of explanation seems to provide an elegant rescue. If laws are those universal generalizations that explain their instances, then following the lead of a number of current authors (notably Harman ([8], [9]); also see Brody ([4])) we may suppose that universal generalizations can be confirmed because confirmation is (roughly) the converse of explanation; E confirms H if H explains E. *Some* universal generalizations can be confirmed; they are those that explain their instances. Equating laws with universal generalizations having explanatory power therefore achieves a neat economy: we account for the confirmability of laws in terms of the explanatory power of those generalizations to which laws are reduced.

To say that a law is a universal truth having explanatory power is like saying that a chair is a breath of air used to seat people. You cannot make a silk purse out of a sow's ear, not even a very good sow's ear; and you cannot *make* a generalization, not even a purely universal generalization, explain its instances. The fact that *every* F is G fails to explain why *any* F is G, and it fails to explain it, not because its explanatory efforts are too feeble to have attracted our attention, but because the explanatory attempt is never even made. The fact that all men are mortal does not explain why you and I are mortal; it *says* (in the sense of *implies*) that we are mortal, but it does not even suggest *why* this might be so. The fact that all ten tosses will turn up heads is a fact that logically guarantees a head on the tenth toss, but it is not a fact that explains the outcome of this final toss. On one view of explanation, *nothing* explains it. Subsuming an instance under a universal generalization has exactly as much explanatory power as deriving Q from $P \cdot Q$. None.

If universal truths of the form $(x)(Fx \supset Gx)$ could be *made* to explain their instances, we might succeed in making them into natural laws. But, as far as I can tell, no one has yet revealed the secret for endowing them with this remarkable power.

This has been a hasty and, in some respects, superficial review of the doctrine that laws are universal truths. Despite its brevity, I think we have touched upon the major difficulties with sustaining the equation: law = universal truth + X (for a variety of different values of "X"). The problems center on the following features of laws:

a A statement of law has its descriptive terms occurring in opaque positions.

b The existence of laws does not await our identification of them *as* laws. In this sense they are objective and independent of epistemic considerations.

c Laws can be confirmed by their instances and the confirmation of a law raises the probability that the unexamined instances will resemble (in the respect described by the law) the examined instances. In this respect they are useful tools for prediction.

d Laws are not merely summaries of their instances; typically, they figure in the explanation of the phenomena falling within their scope.

e Laws (in some sense) "support" counterfactuals; to know a law is to know what would happen if certain conditions were realized.

f Laws tell us what (in some sense) must happen, not merely what has and will happen (given certain initial conditions).

The conception of laws suggested earlier in this essay, the view that laws are expressed by singular statements of fact describing the relationships between properties and magnitudes, proposes to account for these features of laws in a single, unified, way: (a)–(f) are all manifestations of what might be called "ontological ascent," the shift from talking about individual objects and events, or collections of them, to the quantities and qualities that these objects exemplify. Instead of talking about green and red things, we talk about the *colors* green and red. Instead of talking about gases that have a volume, we talk about the volume (temperature, pressure, entropy) that gases have. Laws eschew reference to the things that have length, charge, capacity, internal energy, momentum, spin, and velocity in order to talk about these quantities themselves and to describe *their* relationship to each other.

We have already seen how this conception of laws explains the peculiar opacity of law-like statements. Once we understand that a law-like statement is not a statement about the extensions of its constituent terms, but about the intensions (= the quantities and qualities to which we may refer with the abstract singular form of these terms), then the opacity of laws to *extensional* substitution is natural and expected. Once a law is understood to have the form:

6 $F\text{-ness} \rightarrow G\text{-ness}$

the relation in question (the relation expressed by "→") is seen to be an *extensional* relation between *properties* with the terms "*F*-ness" and "*G*-ness" occupying *transparent* positions in (6). Any term referring to the same quality or quantity as "*F*-ness" can be substituted for "*F*-ness" in (6) without affecting its truth or its law-likeness. Coextensive terms (terms referring to the same *quantities* and *qualities*) can be freely exchanged for "*F*-ness" and "*G*-ness" in (6) without jeopardizing its truth value. The tendency to treat laws as some kind of intensional relation between extensions, as something of the form (x) $(Fx \boxed{N} \to Gx)$ (where the connective is some kind of modal connective), is simply a mistaken rendition of the fact that laws are extensional relations between intensions.

Once we make the ontological ascent we can also understand the modal character of laws, the feature described in (e) and (f) above. Although true statements having the form of (6) are not themselves *necessary* truths, nor do they describe a modal relationship between the respective qualities, the contingent relationship between properties that is described imposes a modal quality on the particular events falling within its scope. This *F must* be *G*. Why? Because *F*-ness is linked to *G*-ness; the one property yields or generates the other in much the way a change in the thermal conductivity of a metal yields a change in its electrical conductivity. The pattern of inference is:

I   *F*-ness → *G*-ness
    This is *F*
    _____
    This must be *G*.

This, I suggest, is a valid pattern of inference. It is quite unlike the fallacy committed in (II):

II $(x)(Fx \supset Gx)$
    This is *F*
    _____
    This must be *G*.

The fallacy here consists in the absorption *into* the conclusion of a modality (entailment) that belongs to the relationship *between* the premises and the conclusion. There is no fallacy in (I), and this, I submit, is the source of the "physical" or "nomic" necessity generated by laws. It is this which explains the power of laws to tell us what *would* happen if we did such-and-such and what *could not* happen whatever we did.

I have no proof for the validity of (I). The best I can do is an analogy. Consider the complex set of legal relationships defining the authority, responsibilities, and powers of the three branches of government in the United States. The executive, the legislative, and the judicial branches of government have, according to these laws, different functions and powers.

There is nothing *necessary* about the laws themselves; they could be changed. There is no law that prohibits scrapping all the present laws (including the constitution) and starting over again. Yet, given these laws, it follows that the President *must* consult Congress on certain matters, members of the Supreme Court *cannot* enact laws nor declare war, and members of Congress *must* periodically stand for election. The legal code lays down a set of relationships between the various *offices* of government, and this set of relationships (between the abstract offices) impose legal constraints on the individuals who occupy these offices—constraints that we express with such modal terms as "cannot" and "must." There are certain things the individuals (and collections of individuals—e.g., the Senate) can and cannot do. *Their* activities are subjected to this modal qualification whereas the framework of laws from which this modality arises is itself modality-free. The President (e.g., Ford) *must* consult the Senate on matter M, but the relationship between the *office* of the President and that *legislative body* we call the Senate that makes Gerald Ford's action obligatory is not *itself* obligatory. There is no law that says that this relationship between the office of President and the upper house of Congress must (legally) endure forever and remain indissoluble.

In matters pertaining to the offices, branches and agencies of government the "can" and "cannot" generated by laws are, of course, legal in character. Nevertheless, I think the analogy revealing. Natural laws may be thought of as a set of relationships that exist between the various "offices" that objects sometimes occupy. Once an object occupies such an office, its activities are constrained by the set of relations connecting that office to other offices and agencies; it *must* do some things, and it *cannot* do other things. In both the legal and the natural context the modality at level n is generated by the set of relationships existing between the entities at level $n + 1$. Without this web of higher order relationships there is nothing to support the attribution of constraints to the entities at a lower level.

To think of statements of law as expressing relationships (such as class inclusion) between the extensions of their terms is like thinking of the legal code as a set of universal imperatives directed to a set of particular individuals. A law that tells us that the United States President must consult Congress on matters pertaining to M is not an imperative issued to Gerald Ford, Richard Nixon, Lyndon Johnson, *et al.* The law tells us something about the duties and obligations attending the *Presidency*; only indirectly does it tell us about the obligations of the Presidents (Gerald Ford, Richard Nixon, *et al.*). It tells us about their obligations in so far as they are occupants of this office. If a law was to be interpreted as of the form: "For all x, if x is (was or will be) President of the United States, then x must (legally) consult Congress on matter M," it would be incomprehensible why Sally Bickle, were she to be president, would have to consult Congress on matter M. For since Sally Bickle never was, and never will be, President, the law, understood as an imperative applying to *actual* Presidents (past, present and future) does not apply to her. Even if there is a possible world in which she becomes

President, this does not make her a member of that class of people to which the law applies; for the law, under this interpretation, is directed to that class of people who become President in *this* world, and Sally is not a member of this class. But we all know, of course, that the law does not apply to individuals, or sets of individuals, in this way; it concerns itself, in part, with the offices that people occupy and only indirectly with individuals in so far as they occupy these offices. And this is why, if Sally Bickle were to become President, if she occupied this office, she would have to consult Congress on matters pertaining to M.[12]

The last point is meant to illustrate the respect and manner in which natural laws "support" counterfactuals. Laws, being relationships between properties and magnitudes, go *beyond* the sets of things in *this* world that exemplify these properties and have these magnitudes. Laws tell us that quality F is linked to quality G in a certain way; hence, if object O (which has neither property) were to acquire property F, it would also acquire G in virtue of this connection between F-ness and G-ness. A statement of law asserts something that allows us to entertain the prospect of alterations in the extension of the predicate expressions contained in the statement. Since they make no reference to the extensions of their constituent terms (where the extensions are understood to be the things that are F and G in this world), we can hypothetically alter these extensions in the antecedent of our counterfactual ("if this were an F . . .") and use the connection asserted in the law to reach the consequent (". . . it would be G"). Statements of law, by talking about the relevant properties rather than the sets of things that have these properties, have a far wider scope than any true generalization about the actual world. Their scope extends to those possible worlds in which the extensions of our terms differ but the connections between properties remains invariant. This is a power that no universal generalization of the form $(x)(Fx \supset Gx)$ has; this statement says something about the actual F's and G's in *this* world. It says absolutely nothing about those possible worlds in which there are *additional* F's or *different* F's. For this reason it cannot imply a counterfactual. To do this we must ascend to a level of discourse in which what we talk about, and what we say about what we talk about, remains the *same* through alterations in extension. This can only be achieved through an ontological ascent of the type reflected in (6).

We come, finally, to the notion of explanation and confirmation. I shall have relatively little to say about these ideas, not because I think that the present conception of laws is particularly weak in this regard, but because its very real strengths have already been made evident. Laws figure in the explanation of their instances because they are not merely summaries of these instances. I can explain why this F is G by describing the relationship that exists between the properties in question. I can explain why the current increased upon an increase in the voltage by appealing to the relationship that exists between the flow of charge (current intensity) and the voltage (notice the definite articles). The period of a pendulum decreases when you shorten the length of the bob, not because all pendulums do that, but because the period and the length are related in the fashion $T = 2\pi\sqrt{L/g}$. The principles of thermodynamics tell us about the relationships that exist between such quantities as energy, entropy, temperature and pressure, and it is for this reason that we can use these principles to explain the increase in temperature of a rapidly compressed gas, explain why perpetual motion machines cannot be built, and why balloons do not spontaneously collapse without a puncture.

Furthermore, if we take seriously the connection between explanation and confirmation, take seriously the idea that to confirm a hypothesis is to bring forward data for which the hypothesis is the best (or one of the better) competing explanations, then we arrive at the mildly paradoxical result that laws can be confirmed *because* they are more than generalizations of that data. Recall, we began this essay by saying that if a statement of law asserted anything more than is asserted by a universally true statement of the form (x) $(Fx \supset Gx)$, then it asserted something that was beyond our epistemological grasp. The conclusion we have reached is that *unless* a statement of law goes beyond what is asserted by such universal truths, unless it asserts something that cannot be completely verified (even with a complete enumeration of its instances), it cannot be confirmed and used for predictive purposes. It cannot be confirmed because it cannot explain; and its inability to explain is a symptom of the fact that there is not enough "distance" between it and the facts it is called upon to explain. To get this distance we require an ontological ascent.

I expect to hear charges of Platonism. They would be premature. I have not argued that there are universal properties. I have been concerned to establish something weaker, something conditional in nature: *viz.*, universal properties exist, and there exists a definite relationship between these universal properties, *if* there are any laws of nature. If one prefers desert landscapes, prefers to keep one's ontology respectably nominalistic, I can and do sympathize. I would merely point out that in such barren terrain there are no laws, nor is there anything that can be dressed up to look like a law. These are inflationary times, and the cost of nominalism has just gone up.[13]*

■ | Notes

1. This is the position taken by Hempel and Oppenheim ([10]).

2. When the statement is of nonlimited scope it is already beyond our epistemological grasp in the sense that we cannot *conclusively* verify it with the (necessarily)

---

* This essay was written during the 1970s, a decade of high inflation. Nominalists deny that universals have any real existence, insisting that general terms such as *red*, *giraffe*, and *electrically charged* do not refer to universal properties, abstract objects, or Platonic forms. Typically, nominalists view the meaning of general terms as deriving from particular resemblances between particular things.

finite set of observations to which traditional theories of confirmation restrict themselves. When I say (in the text) that the statement is "beyond our epistemological grasp" I have something more serious in mind than this rather trivial limitation.

3. Most prominently, William Kneale in [12] and [13].

4. I eliminate quotes when their absence will cause no confusion. I will also, sometimes, speak of laws and statements of law indifferently. I think, however, that it is a serious mistake to conflate these two notions. Laws are what is expressed by true lawlike statements (see [1], p. 2, for a discussion of the possible senses of "law" in this regard). I will return to this point later.

5. Popper ([17]) vaguely perceives, but fails to appreciate the significance of, the same (or a similar) point. He distinguishes between the structure of terms in laws and universal generalizations, referring to their occurrence in laws as "intensional" and their occurrence in universal generalizations as "extensional." Popper fails to develop this insight, however, and continues to equate laws with a certain class of universal truths.

6. Nelson Goodman gives a succinct statement of the functionalist position: "As a first approximation then, we might say that a law is a true sentence used for making predictions. That laws are used predictively is of course a simple truism, and I am not proposing it as a novelty. I want only to emphasize the Humean idea that rather than a sentence being used for prediction because it is a law, it is called a law because it is used for prediction; and that rather than the law being used for prediction because it describes a causal connection, the meaning of the causal connection is to be interpreted in terms of predictively used laws" ([7], p. 20–21). Among functionalists of this sort I would include Ayer ([2]), Nagel ([16]), Popper ([17]), Mackie ([14]), Bromberger ([6]), Braithwaite ([3]), Hempel ([10], [11]), and many others. Achinstein is harder to classify. He says that laws express regularities that can be cited in providing analyses and explanations ([1], p. 9), but he has a rather broad idea of regularities: "regularities might also be attributed to properties" ([1], pages 19, 22).

7. I attach no special significance to the connective "→." I use it here merely as a dummy connective or relation. The kind of connection asserted to exist between the universals in question will depend on the particular law in question, and it will vary depending on whether the law involves quantitative or merely qualitative expressions. For example, Ohm's Law asserts for a certain class of situations a constant ratio (R) between the magnitudes E (potential difference) and I (current intensity), a fact that we use the "=" sign to represent: $E/I = R$. In the case of simple qualitative laws (though I doubt whether there are many genuine laws of this sort) the connective "→" merely expresses a link or connection between the respective qualities and may be read as "yields." If it is a law that all men are mortal, then humanity yields mortality (humanity → mortality). Incidentally, I am not denying that we can, and do, express laws as simply "All F's are G" (sometimes this is the only convenient way to express them). All I am suggesting is that when lawlike statements are presented in this form it may not be clear what is being asserted: a law or a universal generalization. When the context makes it clear that a relation of law is being described, we can (without ambiguity) express it as "All F's are G" for it is then understood in the manner of (6).

8. On the basis of an argument concerned with the restrictions on predicate expressions that may appear in laws, Hempel reaches a similar conclusion but he interprets

it differently. "Epitomizing these observations we might say that a lawlike sentence of universal nonprobabilistic character is not about these classes or extensions *under certain* descriptions" ([11], p. 128). I guess I do not know what being *about* something *under a description* means unless it amounts to being about the property or feature expressed by that description. I return to this point later.

9. Molnar ([15]) has an excellent brief critique of attempts to analyze a law by using epistemic conditions of the kind being discussed.

10. Brody argues that a qualitative confirmation function need not require that any *E* that raises the degree of confirmation of *H* thereby (qualitatively) confirms *H*. We need only require (perhaps this is also too much) that if *E* does qualitatively confirm *H*, then *E* raises the degree of confirmation of *H*. His arguments take their point of departure from Carnap's examples against the special consequence and converse consequence condition ([4], pages 414–418). However this may be, I think it fair to say that most writers on confirmation theory take a *confirmatory* piece of evidence to be a piece of evidence that *raises* the probability of the hypothesis for which it is confirmatory. How well it must be confirmed to be acceptable is another matter of course.

11. If the hypothesis is of nonlimited scope, then its scope is not known to be finite. Hence, we cannot know whether we are getting a numerical increase in the ratio: examined favorable cases/total number of cases. If an increase in the probability of a hypothesis is equated with a (known) increase in this ratio, then we cannot raise the probability of a hypothesis of nonlimited scope in the simple-minded way described for hypotheses of (known) finite scope.

12. If the law was interpreted as a universal imperative of the form described, the most that it would permit us to infer about Sally would be a counteridentical: If Sally were one of the Presidents (i.e. identical with either Ford, Nixon, Johnson, . . .), then she would (at the appropriate time) have to consult Congress on matters pertaining to *M*.

13. For their helpful comments my thanks to colleagues at Wisconsin and a number of other universities where I read earlier versions of this paper. I wish, especially, to thank Zane Parks, Robert Causey, Martin Perlmutter, Norman Gillespie, and Richard Aquilla for their critical suggestions, but they should not be blamed for the way I garbled them.

■ | References

[1] Achinstein, P. *Law and Explanation*. Oxford: Clarendon Press, 1971.

[2] Ayer, A. J. "What Is a Law of Nature?" In [5], pages 39–54.

[3] Braithwaite, R. B. *Scientific Explanation*. Cambridge: Cambridge University Press, 1957.

[4] Brody, B. A. "Confirmation and Explanation." *Journal of Philosophy* 65 (1968): 282–299. Reprinted in [5], pages 410–426.

[5] Brody, B. A. *Readings in the Philosophy of Science*. Englewood Cliffs, N.J.: Prentice Hall, 1970.

[6] Bromberger, S. "Why-Questions." In [5], pages 66–87.

[7] Goodman, N. *Fact, Fiction and Forecast*. London: The Athlone Press, 1954.

[8] Harman, G. "The Inference to the Best Explanation." *Philosophical Review* 74 (1965): 88–95.

[9] Harman, G. "Knowledge, Inference and Explanation." *Philosophical Quarterly* 18 (1968): 164–173.

[10] Hempel, C. G., and Oppenheim, P. "Studies in the Logic of Explanation." In [5], pages 8–27.

[11] Hempel, C. G. "Maximal Specificity and Lawlikeness in Probabilistic Explanations." *Philosophy of Science* 35 (1968): 116–133.

[12] Kneale, W. "Natural Laws and Contrary-to-Fact Conditionals." *Analysis* 10 (1950): 121–125.

[13] Kneale, W. *Probability and Induction*. Oxford: Oxford University Press, 1949.

[14] Mackie, J. L. "Counterfactuals and Causal Laws." In *Analytical Philosophy*. (First Series). Edited by R. J. Butler. Oxford: Basil Blackwell, 1966.

[15] Molnar, G. "Kneale's Argument Revisited." *Philosophical Review* 78 (1969): 79–89.

[16] Nagel, E. *The Structure of Science*. New York: Harcourt Brace, 1961.

[17] Popper, K. "A Note on Natural Laws and So-Called 'Contrary-to-Fact Conditionals.'" *Mind* 58 (1949): 62–66.

D. H. Mellor

Necessities and Universals in Natural Laws

1 | Prologue

How do laws of nature differ from cosmic coincidences? This is a question very familiar to philosophers of science, and answers of two sorts still vie for their allegiance. One sort locates the difference in what laws say, the other "in the different roles which they play in our thinking", as Braithwaite's *Scientific Explanation* put it (1953: 295). In Chapter 9 of that book, Braithwaite developed and defended a classic answer of the second sort: the difference, he says there, lies in why we believe laws, not in what they say. In the quarter century since then, other answers of the same sort have been devised: Hesse presents one in [Hesse (1980)]. But since then also, answers of the first sort have again come into fashion. The revived fashion has mostly been for reading laws as saying how things must be; but some, more recently, have read them instead as relating not things but properties of things to each other. Hesse notes these fashions and rejects them, to my mind rightly, but she does not elaborate her reasons. It seems to me therefore that I can best complement her article by inspecting these fashions' argumentative cut, to see if they do indeed fit better than her and Braithwaite's Humean gear. Only first I shall build the problem up in my own way, to provide a lay figure to hang the garments on.

2 | The Problem

Certified laws of nature are the primary products of scientific thought and observation. They embody the generalized knowledge which science yields; they supply explanations and predictions of events; and they underlie the design of most modern artefacts. To take just three obvious examples: our

FROM D. H. Mellor, ed., *Science, Belief and Behaviour: Essays in Honour of R. B. Braithwaite* (Cambridge: Cambridge University Press, 1980), 105–25.

human life has been much altered in this century by the discovery and applications of laws governing plant genetics, aerodynamics and electromagnetic radiation.

Laws differ widely in their subject matter, importance and complexity. What they have in common is generality. A law says that *all* things or events of some kind have a certain property or are related in a certain way to something else. If the law is statistical, the property is having a chance of having some other property or of being related to something else. It is, for example, a law that all light has the property of going at the same speed in a vacuum; and it is a statistical law that all atoms of the most common isotope of radium have the same chance (fifty-fifty) of turning into something else within their half life of 1622 years.

What needs certifying about a law is its truth. We cannot know that all light goes at the same speed in a vacuum unless it truly does so. Its constant speed will not serve to explain or predict anything if its speed is not in fact constant. And it is unsafe to base the design of artefacts on what is not the case. We know of course that even a certified law may turn out to be false. But without good reason to think it true, we lack good reason to employ it as we do. This is why we do not call something a law unless we think it true, so that a false generalization cannot be a law, although it may be "lawlike": *i.e.*, such that it would be a law if only it were true.

Certifying the truth of some laws presents no problem. These are the analytic laws, those whose truth follows from the meanings of the terms they are couched in. There are more reasons than one for laws being analytic. A law may be analytic because it is used to define one of its terms. Newton's laws of motion, for example, may well be analytic because between them they define the Newtonian concepts of force and mass. Or a law, not originally analytic, may become so successful and theoretically important that its terms change their meaning to make it analytic. For this reason it is now arguably analytic that light is electromagnetic radiation, although that could not have been the case when the electromagnetic theory was first conjectured to apply to light. Then, we could easily have envisaged observing light to go faster or slower, for example, than the theory can be shown (by measuring the ratio of electromagnetic to electrostatic units) to require electromagnetic radiation to go. Nowadays we should take such an observation to show some error in the theory rather than question the law that light is electromagnetic radiation.

But even if some laws are analytic, most laws are not, and these are the ones that concern me. It does not follow from the meanings of the terms involved that radium's half life is 1622 years, nor that benzene is as insoluble in water as it is. Nothing semantic prevents a little more benzene sometimes dissolving in water, or some piece of radium having a rather different half life. How then can we certify the truth of what the law says, namely that these things never happen? We cannot see that they never do, if only because at no time can we see that they never will do in the future. We cannot directly perceive the truth of nonanalytic laws. At most, our senses can show us some of a law's past instances, and then only instances of laws about relatively observable properties of things and events. We can observe the speed of this or that ray of light and, perhaps indirectly, the half life of this or that piece of radium; but not all the things and events, past, present and to come, to which the law applies.

The problem then arises why a supposed law should be expected to hold in instances as yet unobserved; in short, Hume's problem of induction. Unlike Popper and his followers, I believe that induction does present a genuine and serious problem, which needs solution and has not yet been solved; although I believe Braithwaite's (1953: Ch. 8) attempted solution is along the right lines.* But wherever its solution lies, Hume's problem does not arise only incidentally for laws of nature. On the contrary, it is an inevitable concomitant to their role in supplying predictions. To make a prediction is to anticipate, rightly or wrongly, the result of making an observation; to say or just to expect, for example, that a bomb will explode before we see it do so. Whatever purports, as a law does, to justify such an expectation necessarily arouses Hume's problem. Only a generalization certified by observing all its instances would be free of inductive pretensions, and such a generalization is not much use for predicting things. It might indeed have some use: one might accept it on someone else's authority and use it to predict some instances one had not observed oneself. But real laws are used amongst other things to predict the results of future observations, and these are not yet available to anyone to certify the law with (see Mellor 1979). Real laws therefore undeniably need inductive support.

The other philosophical problem which laws of nature present is the one that concerns us. It is less obvious than the problem of induction, but perhaps more tractable: what exactly do laws say? I have taken them to be generalizations, and there is not much doubt of that. The debatable question is whether laws are more than generalizations, and if so, what more. Now if giving laws one content rather than another made the problem of induction soluble for them, this would be a strong argument for giving them that content. But since I believe no such solution is presently available for any credible content, I must look to other arguments. Hume's problem does, however, provide a reason for preferring weak readings of natural laws. The less a law says, the less there is to be certified in claiming it to be true.

The weakest reading seems to be the obvious one I have already given:

1 All *F*s are *G*s,

where *F* and *G* are properties of things or events. They may be relational, comparative or quantitative properties; in statistical laws *G* will be some

---

* The problem of induction is discussed in the readings by Lipton and Popper in chapter 4 and in the accompanying commentary. For Mellor's later views on induction, see D. H. Mellor, "The Warrant of Induction," in *Matters of Metaphysics* (Cambridge: Cambridge University Press, 1991), 254–68.

determinate chance of having another property. (1) is of course a very simple form of law, but it will do; it has all the relevantly problematic features of more complex forms. But before discussing its supposed deficiencies, some preliminary points need to be made clear.

First, as my examples have already illustrated, the 'are' in (1) is to be taken tenselessly. The law applies to all F items in the universe, past and future as well as present. The laws of radioactivity do not just give radium's present half life; they say what it always was and always will be. Now some of what we take to be physical constants, such as the half life of radio-elements, might indeed turn out to depend on the age of the universe. But then the true laws of radioactivity would say what the dependence was. Those laws would, like all other true laws, apply at all times; the values of our supposed constants at particular epochs being merely special cases of the general laws.

Secondly, I take it that anything in the universe is definitely either F or not F, either G or not G. This is not an uncontentious claim. Some have been led to deny it of so-called "vague" properties like being bald, because of its seemingly absurd consequences (for example, that at some point adding just one hair to a bald man's head gets rid of his baldness). Others have been led to deny it of some things and events in the future, either because they want the future to be open, at least in some respects, to being made definite by human decision and consequent human action or because of problems raised by quantum mechanics. They think it cannot now be the case, for example, that I shall definitely either be dead or be alive next year, if it is still open to me and others to settle the matter by what we decide to do between now and then. I think that these are both inadequate grounds for denying that everything is definitely F or not F, but I shall not argue the point here. (On the first, see Cargile 1969; on the second see Mellor 1981. I also think my being wrong in either case would make little difference to the ensuing discussion, but I shall not argue that either.)

Thirdly, I exclude from the range of F and G factitious properties such as Goodman's (1965) notorious "grue" (= green if the item is inspected before a specified time, otherwise = blue). I hope and believe criteria can be given to rule out these phoney properties (see for example Hesse 1974: Ch. 3); but in any event all parties agree that they are phoney, and I shall take their exclusion for granted.*

I should however emphasize that I do not mean to restrict F and G to physical, as opposed to psychological or social, properties. Some philosophers (*e.g.* Davidson 1970; McGinn 1978) deny the existence of laws relating nonphysical properties; but largely because they mistake laws to involve necessities of the kind I shall be concerned to dispute and which they correctly perceive to be absent from mental and social generalizations. Anyway the point should be left open here; so if I stick to physical examples, it is

---

* For a discussion of Goodman's "grue" problem, see chapter 4 of this volume.

only to avoid irrelevant controversy, and not because I think there are no others.

With this preamble, we may now ask what, if anything, is wrong with (1) as a reading of laws of nature. To see what seems to be wrong, we must look at (1)'s consequences in special cases, particularly the case, on which Braithwaite concentrates, where nothing in the world is F.

One might imagine that it did not matter what follows from (1) when nothing is F, but it does. Let us call a law 'vacuous' in that case. Many important laws are vacuous in this sense. The most famous one is Newton's first law of motion, that bodies acted on by no forces are at rest or move at a constant speed in a straight line. The law is central to Newtonian mechanics, but Newton's own gravitational theory implies its vacuity, since the theory says that all bodies exert gravitational forces on each other. No doubt Newton's laws of motion are peculiar, since as already remarked they may well be analytic. But Newton's first law illustrates a vacuity which is shared by many laws that are in no way analytic. There is in particular a multitude of nonanalytic laws quantifying over determinate values of continuously variable determinables: for example, the laws relating the vapour pressure of substances to their temperature. Each determinate value of these determinables yields another law as a special case, such as the law giving the boiling point of water at atmospheric pressure. Now there are infinitely many different temperatures and pressures, and hence infinitely many of these derived laws, all with mutually incompatible antecedents (nothing can be wholly at two different temperatures or pressures at the same time). Although the temperature and pressure of any given mass of water will vary continuously with time, there are many temperatures which no mass of water ever reaches: temperatures, for example, so high that water would decompose before it reached them. At any rate, so far as these derived laws are concerned, it is entirely accidental whether any water ever is at the temperatures and pressures they apply to. Consequently they must certainly be so construed as to make equal sense whether they happen to be vacuous or not (cf. Ayer 1956: 224–5 [824–25]).

In particular, it seems obvious that mere vacuity should not settle the truth of a law regardless of its content. But a lack of Fs makes 'All Fs are As' true for any A, including both A=G and A=not-G. If there never is any water at some temperature T, statements crediting all water at that temperature with any pressure whatever all come out true. That seems absurd; so vacuous laws should be read as saying something other than 'All Fs are Gs'. The question is what.

The obvious answer is that a vacuous law says

2 If there were Fs, they would be Gs.

But there are objections to (2). One is that it appears to imply that there are no Fs, whereas laws, even if they happen to be vacuous, certainly do not claim to be. We could in reply say that (2) is not to be read as having this implication; and this stipulation can indeed be given some independent rationale. A case can be made for saying that the implication is not part of what (2) says, but follows rather from applying general rules of discourse: namely, not to mislead, and to be as informative as possible (see Mackie 1973: 75–7). These rules dictate that one should not say (1) when the law is known to be vacuous, since (1) is no more true then than is any other generalization starting 'All Fs are . . .' To pick out as a law the generalization which relates F especially to G in these circumstances, one must have some reason other than its truth. The reason may not be specified, but the fact that there is one is signalled by using (2) instead of (1). Consequently, even if the law says no more than (1), (2) would normally be used when, but only when the law is known to be vacuous. So (2) will indeed signal its user's knowledge of the law's vacuity, even though that is no part of what (2) is being used to say.

This is one of the arguments which can be used to defend Humean accounts of laws as saying no more than (1). It still leaves the problems of saying what reason there is to link F and G as a law does when there are no Fs, and why (2) should be the right way to signal this reason. These are among the problems that have exercised Braithwaite and his Humean successors. But since my concern here is with their rivals, I shall concentrate instead on recent attempts to solve the problem of vacuous laws by giving (2) some assertible content over and above (1).

Laws, I have remarked, do not claim to be vacuous, even if they are; and ideally, they should say the same thing whether they are vacuous or not. It will hardly do to make laws say (2) if they are vacuous and (1) if they are not. A law cannot say (1) in both cases, we are supposing; can it say (2)? We have dealt with the obvious objection by removing (2)'s counterfactual implication (that there are no Fs), which would have made all nonvacuous lawlike generalizations false regardless of their content. What can be said positively in favour of the suggestion?

Consider the universe of non-F things or events of which a vacuous law says that if they were Fs they would be Gs. It is surely immaterial to this supposed fact about these things or events that there happens to be nothing else which is F. So perhaps we should take the *non*vacuous law also to say of every non-F thing or event that if it were F it would be G. But again, the law itself does not assert that these things or events are not F. It should say the same of all things or events, whether they are F or not. Let us therefore take a law to say of every thing or event x that

3 If x were F it would be G.

(Those who believe in possible as well as actual things and events may take 'x' to range over them too.) I shall take the problem for our non-Humeans to be that of saying what (3) means in this case.

I shall not demand of them a general analysis of so-called 'subjunctive' or 'counterfactual' conditionals like (3). A general analysis would of course have to cover those that we are supposing to give the content of natural laws. But I am not convinced that other uses of these conditionals are homogeneous enough with this one to shed much light on it. In most other uses, for example, (3) might very well imply that x is not F, which we have seen it cannot do here. Or again, to make (3) true of an x, it may often suffice for that x to be F and also G. Lewis's influential analysis, for example, takes this more or less for granted, and the way he reluctantly accommodates possible exceptions (1973: 29) will certainly not cope with natural laws. Yet natural laws must be exceptions: it might be a coincidence that an x is both F and G, and not a matter of natural law at all. So in this case it must take more than that to make (3) true of any x. And as our consideration of vacuous laws has shown, the extra cannot be that all other Fs are Gs too, for there might just as well be no other Fs. So whether the law is vacuous or not, the truth of (1) will not suffice to make (3) true of everything. But what more than (1) can a law say?

3 | Possible Worlds

The traditional non-Humean answer is that natural laws are or express necessities of some kind: what makes (3) true of everything is that Fs not merely are Gs, they have to be; (1) is not merely true, it is necessarily so. Conceptions of law as what Kneale (1949) called 'principles of necessitation' are of course by no means new. The problem with them is to justify the idea of necessity they invoke and to show how it explains the universal truth of (3). Of late years, the development of so-called "possible world semantics" has made that problem look more tractable, and thus encouraged a revival of the idea that natural laws are necessary truths. It has done this by providing a systematic way of saying what makes statements of necessity (and of possibility) true. So in particular we might hope to find in it an acceptable way of saying what makes necessary natural laws true.

The basic concept of this semantics is that of a possible world. A possible world is a way the world might be, or might have been. There are many such ways, and therefore many possible worlds, of which the actual world is just one. Possible worlds are distinguished by what the facts are supposed to be in them: if the supposed facts differ at all, so do the worlds. I might, for instance, die in various ways, and, for each way, at various ages. So there are many possible worlds in which I expire of, say, cirrhosis (or my counterpart in that world does so; see Lewis 1973: 39), and these differ amongst other things according to my or my counterpart's age at the time. In general, a statement which might be true, but fails to specify every detail of the universe, will be true in many possible worlds, differing amongst themselves in the details left unspecified.

Having in some such manner as this grasped the idea of possible worlds, and reified them, one can turn round and give, as the truth conditions of a statement, the set of possible worlds in which it is true. That is how possible world semantics offers to give the meaning of various kinds of modal statements, and in particular of statements of necessity and possibility. How enlightening this conceptual round trip is, from what might be the case, to what is the case in a possible world, and back again, is a very moot point, but one that can be waived while we see how well the concept copes with the supposed necessity of natural laws.

It follows at once from the definition of a possible world that a statement which might be true is one that is true in some possible world. Hence statements which have to be true are those which are true in all possible worlds. In particular, for (1) to be necessarily true is for it to hold in all the worlds there might be or might have been. Is that really what a natural law claims?

Suppose it is: does that solve the problem of vacuous laws and explain (3)'s being true of everything in the actual world? Consider again the case where there are no actual Fs. The law does not say there are none, and it is tempting to suppose there always might have been. If that were so, then, on this view of laws, (1) would have to be true not only in this world, but also in worlds containing Fs where its truth would not be the trivial consequence of vacuity it is here. And that would certainly distinguish (1) as a law from other vacuously true generalizations.

But this account depends on the possibility of there being Fs; and, on this view of laws, there will often be no such possibility. I have cited the example of high temperature instances of the vapour pressure law for water that are vacuous because water decomposes before it reaches those temperatures. Now, that water decomposes below these temperatures is itself a natural law and so, on this view of them, necessary. Consequently these high temperature instances of the vapour pressure law not only are vacuous, they have to be. There could be no water at such temperatures. But that is to say there are no possible worlds in which these instances are not vacuous; and therefore none in which the truth of this instance of (1) is other than a trivial consequence of vacuity.

So the idea of laws being true in all possible worlds does not solve the problem of vacuous laws. Nor, for much the same reason, does it explain why (3) is true of everything in this world. Again, it would if anything, a, in this world might have been F even if it isn't. Then there would be possible worlds in which a (or some counterpart of a) is in fact F; and in all these worlds it, like every other F, is G. Where that is so, it seems to me undeniable that (3) is true of a. However, for any F there will be many as of which it is quite incredible that they might have been F. Take the law that in a vacuum all light goes at a constant speed—which is to say that all photons do. It is true then, of anything at all, that it would go at that speed in a vacuum if it were a photon. But this is not to say of everything that it might have been a photon. There is no possible world in which I am (or any counterpart of me is) a photon; and *a fortiori* none in which, as a photon, I (or any counterparts of me) travel at the speed of light. That is not, I believe, what makes this instance of (3) true of me. Yet I believe it is true of me, since I believe the law; and there is surely no inconsistency in my combining these beliefs.

For subjunctive conditionals like (3) to be true, their antecedents do not have to be possible. This is blatantly obvious in *reductio ad absurdum* proofs, where the truth of a subjunctive conditional is actually used to prove that its antecedent is *not* possible. One and one cannot make three precisely because, if they were to, something impossible would be the case. It should be almost as obvious that conditionals which give the content of natural laws likewise do not imply the possibility of their antecedents being true. The vapour pressure example shows at least that they cannot both do this and themselves be necessary truths. And I have given elsewhere (1974: 173) the example of safety precautions at a nuclear power station, which are supposed to make impossible the conditions under which, as a matter of natural law, the fuel would explode. It is ridiculous to maintain that the success of these precautions would disprove the very law that makes them necessary.

I am not sure why (3) should be so often thought to imply that *x* might be *F*. The reason may well be the same for taking (3) to imply that *x* is not *F*: namely, that it is customary to reserve subjunctive conditionals for use when their antecedents are believed to be false but possible. We see, however, that this custom is not invariable, and have in any case seen reason [858–59] not to make such a custom part of a conditional's meaning. So however natural the thought may be, it is mistaken, at least of the conditionals implied by natural laws. But the mistake is very widespread and of long standing, and it has had serious consequences. It has bedevilled the analysis of disposition statements, as I argued in §9 of my (1974). It has likewise afflicted discussions of free will, in which 'I could have done *X*' is frequently equated with 'I would (or could) have done *X* had I chosen to'. But it obviously does not follow from the latter that I could have done *X*, since it obviously does not follow that I could have chosen to.

The common confounding of conditional statements with statements of possibility has thus had ill effects in more than one area of philosophy. The ill effect here has been that possible world semantics have been mistakenly thought to give sense to the idea that natural laws are, or assert, some kind of necessity.

4 | Natural Necessity

Laws might however still be necessary even if possible world semantics fails to say what makes them so. What makes (3) true of everything might still be that nothing could be both *F* and not *G*, whether or not it could be *F*. But it is not at all obvious that this is so. Subjunctive conditionals are not in general made true by necessities. Suppose that if I were to go to London I would go by train. This does not mean that I could not go any other way, merely that I would not. Lewis's (1973) treatment of subjunctive conditionals recognizes this fact about them: the consequent does not have to be true in all the possible worlds the antecedent is true in, only in those most like the real world.

Still, I have insisted that conditionals like (3) which follow from laws are a special case. In particular, it does not suffice for their truth that their antecedents and consequents are true; whereas my going to London by train may well make it true that, were I to go, I would go that way. So perhaps (3) does need some necessity to make it true of everything, even if conditionals in general do not.

But most natural laws seem to be contingent. Apart from those that are definitions, and those whose success has made them analytic, any law might have been false. We could have come across a counterexample to it; and we still could, even if we never will or would. That seems at any rate to be why we need to test our supposed laws by observation: things could be other than the law says, so we need to look and see whether or not they are. I believe, for example, that light could have gone in a vacuum at other than its constant speed, even if no photon ever does and even if nothing, were it a photon, ever would. So on the face of it, conditionals like (3) no more exhibit necessity than does the conditional about my going to London by train.

Attempts have been made to explain away the apparent contingency of natural laws. One attempt, which need not detain us long, distinguishes logical necessity and possibility from their natural or physical counterparts. It is logically possible for Fs not to be G, but not naturally or physically possible. But all 'physically possible' means is 'consistent with natural law'. So to say that something is physically necessary is merely to say that some law entails it. Whether the law says it has to happen, and whether the law itself has to be true, remain entirely open questions.

A more serious attempt distinguishes between metaphysical and epistemic necessities (Kripke 1971: 150–1; Dummett 1973: 121); that is, between being necessary and being knowable *a priori*. Laws appear to be contingent because they cannot be known *a priori*. They cannot be proved in the way the truths of logic and mathematics can. We need to look and see what the world's laws are, and it may always turn out that what was thought to be a law really is not one. The Fs we have seen to be G may mislead us into believing they all are, even though some future ones are not. It is consistent with all we have seen that there should be Fs which are not G. That is the epistemic possibility of a supposed law being false; and something like it exists in mathematics. There too, special cases may mislead us into believing a mathematical generalization to which there are in fact counterexamples. Now, recognizing this possibility in mathematics does not diminish our belief in the necessity of mathematical truths: if the generalization *is* true, it could not have been otherwise. It is likewise conceivable that natural laws, if true, are necessarily so, even though we may be mistaken in what we suppose the true laws to be.

The apparent contingency of natural laws could undoubtedly be explained away like this if there were good reason to think them necessary: but is there? The analogy with mathematics certainly does not give one. If Goldbach's conjecture proves true, any attempt to suppose it false will eventually lead to contradiction (that of course being one way of proving it). In that case no consistent description could be given of a world in which the conjecture was true. That is, there is no such possible world. We might therefore explain the conjecture's necessity, if true, as truth in all possible, *i.e.* coherently conceivable worlds; since conceivability is a notion arguably more basic than necessity and intelligible independently of it. But no such case can be made for the corresponding conception of natural necessity. As Hume insisted, there is no difficulty in conceiving a natural law to be false: since it is not analytic, no contradiction ensues. A perfectly coherent description can be given of a world containing *F*s that are not *G*. The only ground for thinking such a world impossible would be that the law which would be false in it is not only true but necessary; and this is the very fact that needs to be established and explained.

5 | Essences

Arguments have recently appeared for the metaphysical necessity of some laws, namely those specifying essential properties of natural kinds. An essential property of a kind is one which nothing of that kind can lack. So if being *G* is of the essence of a kind *F*, the law that all *F*s are *G*s will be a necessary truth. The exemplars most widely touted by advocates of essences concern the microstructure of kinds: the atomic number of gold, the molecular constitution of water, the genetic makeup of plant and animal species, and the mean kinetic energy of gas particles at a given temperature. The question is: why suppose that these, or any other, properties of kinds are essential?

Two sorts of arguments have lately been adduced for essences, and hence for the necessity of the corresponding laws. Both employ possible world semantics; neither therefore proves more than that some generalizations hold in all possible worlds, and we have seen in [section] 3 that this is not enough to serve our turn. But the arguments repay scrutiny nonetheless, since there is more to them than the possible world jargon they are couched in.

One argument, due to Putnam (1975), infers essences from a mechanism for fixing what things or events a kind predicate ('*F*') applies to, *i.e.* its extension. This mechanism fixes what things are, or might be, *F*s in two stages. First, there are archetypal actual *F*s (*e.g.* paradigm specimens of gold or water): things that have to be *F* if anything is. Second, to be *F*, anything else has to have a suitable 'same-kind' relation to these archetypes. What this means is that it has to share some property with them—apart of course from the property F. What the same-kind relation is, for any category of kinds, it is for empirically testable scientific theories to say. The relations are not discoverable *a priori*, and in particular they do not follow from the meanings of the predicates involved: the laws giving the essences of kinds are not supposed to be analytic. But any shared property G which a same-kind relation picks out will be an essential property of the kind since, Putnam assumes (1975: 232), the relation is an equivalence relation holding across all possible worlds. Thus not only are actual Fs all Gs, all possible Fs are: so (1) in this instance is true in all possible worlds.

I have elaborated elsewhere (1977) my reasons for rejecting this argument. Briefly, the extensions of real natural kinds do not in the first place depend on archetypes in the way Putnam's mechanism requires. And, in the second place, even if they did, his mechanism would still not produce essences. To produce an essence, the same-kind relation must be transitive, in order to ensure that all possible Fs share the *same* property G with each other. But the mechanism does not need a transitive relation, since what makes things Fs in other possible worlds is their sharing some property other than F with the archetypal Fs in this one, and there is nothing to say this shared property must be the same in every possible world. For Putnam to claim the same-kind relation to be transitive, which he does in taking it to be an equivalence relation, is for him gratuitously to assume the essentialist conclusion he is out to prove. His mechanism in fact gives us no reason to think any instance of (1) true in all possible worlds. And since in any case only those giving essences are in question, Putnam's theory, even if it worked, would not solve the general problem of distinguishing laws from universal coincidences.

The same of course is true of Kripke's (1971, 1972) argument for essences; but that too we must look at, since solving our problem for some laws would at least be better than solving it for none. Kripke's argument is quite different from Putnam's. Kripke takes laws giving essences to be identity statements: the law 'Water is H_2O' he takes to say not merely that anything, were it water, would be H_2O, but that being water and being H_2O are one and the same property. But identity is a necessary relation, in the sense that nothing could fail to be identical to itself. So being water and being H_2O are the same property in all possible worlds, not only in this one. Nothing that could be water could fail, were it water, to be H_2O.

This of course is the merest sketch of Kripke's argument. He has, for example, also to show that 'water' and 'H_2O' are what he calls 'rigid designators', *i.e.* that they refer to the same stuff in any possible world it exists in. Otherwise the identity statement, since it is not analytic, might be true without being necessary, as for example 'Water is the most powerful solvent' is. ('The most powerful solvent' is not a rigid designator: it refers to whatever the most powerful solvent is, which in a world restricted largely to oil products would not be water.) As I have abbreviated Kripke's argument, so I shall abbreviate the objections raised in my (1977). The chief objection is that the argument, like Putnam's, blatantly begs the question: for being water and being H_2O to be the same property at all, never mind necessarily, the predicates '. . . is water' and '. . . is H_2O' must already be coextensive in all possible worlds. This is not a conclusion to be derived from the necessity of the identity: it is built into the identity as a premise. Granted, 'Water is H_2O' states a true law, and it has the form of an identity statement. But it is clearly only a variant of 'All water is H_2O', which does not have that form. At any rate, the identity of these properties only follows if 'water' and 'H_2O' are rigid designators, *i.e.* could not refer to different properties. But to believe this, one needs already to believe what the argument from this premiss is supposed to show: namely, that there could not have been some samples of water of a different molecular constitution.

Kripke, like Putnam, fails to establish the existence of essences. The microstructural exemplars which give their doctrine its spurious appeal indeed have a special status in science, but not the status of essences. They are special because they are central to our current scientific theories; but that, I have argued elsewhere (1977: §7), is quite a different matter from being necessary features of the world.

6 | Universals

The properties of being water and being H_2O do not stand in the necessary relation of identity. Perhaps, however, as Armstrong (1978a: ch. 24), Dretske (1977) and Tooley (1977) have suggested, these universals stand in some contingent relation which makes it a law that all water is H_2O. This relation, that is to hold between the properties F and G whenever 'All F are G' is a law, Armstrong and Tooley call 'nomic necessitation'; I shall call it 'N'. F and G have to be differently related if the law is that no F are G or that all F have a chance p of being G. But N will do for now: if it works, the other relations will; if not, nor will they.

This suggestion requires a realist view of at least those universals which are related by natural laws: for N to relate F and G, these properties must exist. That of course is debatable, but suppose for the moment it is true. Then FNG is by definition the fact that makes 'All Fs are G's a law. This is a contingent fact, and not only because F and G might not exist. F and G could quite well exist without 'All Fs are Gs' being a law: laws do not relate every property to every other property. Being water and being at 100°C, for example, are properties that enter into laws, yet no law relates them to each other. But though it is not, it might have been a law that all water is at 100°C; just as it might not have been a law that all water boils at that temperature at atmospheric pressure. Apart from analytic laws, therefore, it is quite contingent that N relates any particular F and G.

To do its job, N has not only to make Gs out of actual Fs, it has to make (3) true of everything, *i.e.* to be such that anything, were it F, would be G. Since this is all it has to do and be, one might think that postulating N is more a relabelling of the problem than a solution to it. But that would not be a fair response. There is a dearth of candidates for making (3) universally true, as we take it to be. If F, G and N would between them make it true, that may well, as Tooley (1977: 262) urges, be reason enough to believe in them.

After all, we already invoke properties to make conditionals true. The inertial mass, m kilogrammes, of a thing a at time t makes true all the conditionals of the form

4 If a were subjected at t to a force of f newtons, it would then accelerate in the direction of the force at f/m metres/second².

Any of these conditionals is in fact a generalization about events, namely that, if they were subjectings of a to a (specific) force f at t, they would be (or be shortly followed by) accelerations of a of magnitude f/m. These generalizations are just like the conditionals (3) entailed by laws, except that they are restricted to the individual a. We think them true, and a fact is needed to make them so: and the requisite fact is that a has mass m at time t. This is all the property of inertial mass amounts to: a truth-maker, as Tooley puts it, for conditionals like (4); and I have argued elsewhere (1974) that all properties of things are just truth-makers for such conditionals. But if we believe in properties F and G because they are needed (and suffice) to make conditionals like (4) true, why jib at accepting N when it is likewise needed and (with F and G) suffices to make conditionals like (3) true?

Here, however, the crucial difference between (3) and (4) emerges: (4) entails that a exists. Without a, 'a' would have no reference, and I do not see what (4) could then mean, nor how in particular it could be true. So if (4) is true, a exists; so the fact that a has mass m is always available to make (4) true. But it is by no means so clear that F and G exist whenever (3) is true of everything. The law that all F are G, it is agreed, may be vacuous: and if it is, there are no Fs. Now if, as many (including Armstrong) suppose, properties and other universals need instances, then without Fs there will be no F. But without F there will be no fact FNG to make (3) true of everything; and the problem of accounting for vacuous laws will remain unsolved.

Perhaps then universals need no instances. Concepts certainly do not: we can have the concept of a unicorn without there being unicorns. But universals are not concepts: concepts, if anything, are parts of our thought or our language; whereas universals, if anything, are parts of the world whether or not it contains any thought or language or concepts. No doubt concepts are closely related to universals, but it is not safe to assume that universals can dispense with instances just because concepts can. That remains an open question.

Tooley takes it for granted that universals need no instances, since he uses a particular example of a vacuous law to argue by elimination "that it must be facts about universals that serve as the truth-makers for basic laws without positive instances" (1977: 672), going on to ask rhetorically: "if facts about universals constitute the truth-makers for some laws, why shouldn't they constitute the truth-makers for all laws?" Armstrong, by contrast, holds a "Principle of Instantiation: For each N-adic universal, U, there exist at least N particulars such that they U" (1978: 113); but he offers nonetheless to cope with Tooley's example of a vacuous law. Now if Armstrong really can supply enough universals to make vacuous laws true, without violating his principle of instantiation, we may not have to decide whether universals do in fact need instances; but can he?

In Tooley's example, as Armstrong puts it, just two out of several types of particle happen never to meet; so the law governing their interaction is vacuous. Nevertheless particles of other types meet, so that the universal, meeting (M), exists; as do these two mutually evasive particle types (A and J). Armstrong can claim therefore that the law, despite being vacuous, "holds in virtue of the universals [A, J, M] being what they are" (1978a: 157). This solution, however, is only available for special cases of vacuous laws. For a start, it only works here because particles of other types do meet, thereby ensuring the existence of the universal M. Now if the law governing A and J particle interactions does not ensure their meeting, it can hardly ensure the meeting of other types of particles; and if A and J particles can fail to meet, so can others. If no particles ever met, the laws of all their interactions would still be true, but the universal M would not, for Armstrong, exist to make them so.

More seriously, not only might there be no meetings, there might be no A or no J type particles. Yet the law could still be true and important, even if there were nothing it applied to. We have seen that to be the case with Newton's first law of motion, and the vacuous instances of vapour pressure laws. For these, and indeed for the bulk of vacuous laws, Armstrong's principle of instantiation does deprive him of the universals he needs as truth-makers. The Tooley case he discusses happens to be of the only sort he can cope with, and it is worth drawing out what makes cases of this sort amenable to Armstrong's treatment: namely, that in them there exist things with properties which make certain generalizations true. These are in fact generalizations of conditionals like (4) above. Consider that for many determinate values of the determinable force f, (4) is vacuous: a can only be subjected to one (net) force at any one time, and there will be many forces a never experiences. Yet a's always having mass m suffices to make all these vacuous generalizations true. And so it is with Armstrong's A and J particles. While they exist, they are disposed to interact as the law says, whether they ever actually meet or not. Armstrong's universals A and J are just conjunctions of such dispositions (see Mellor 1974), and can thus be truth-makers for those laws whose vacuity results merely from the dispositions of actual things failing to display themselves. But not for the more important cases in which laws are made vacuous by the non-existence of the things themselves; and hereafter I will reserve the term 'vacuous' for such cases.

Since vacuous laws will in fact defeat the Armstrong-Dretske-Tooley account if universals need instances, we have after all to consider whether they do. I follow Ramsey in taking particulars and universals to be simply parts of facts picked out in order to generalize. For example, "It is not '*aRb*' but '(*x*)*xRb*' which makes *Rb* prominent. In writing '(*x*)*xRb*' we use the expression '*Rb*' to collect together the set of propositions *xRb* which we want to assert to be true" (Ramsey 1925: 28–9). To recover a proposition from this set, we need to know what an instance of *xRb* is, *i.e.* we need criteria for identifying the items, such as *a*, which have been quantified over. But we do not need separate criteria to identify *Rb*. Given *a*, *Rb* is just the remainder of the fact *aRb*. If it were an independently identifiable constituent, then, as Ramsey says (1925: 23–4), *a*(*Rb*) would differ from (*aR*)*b* and *a*(*R*)*b*, because these facts would have different constituents; and this is absurd.

Similarly, if we form the doubly general '(*x*)(*y*)*xRy*', we must regard the universal *R* as just the common part of all the facts thus collected: the fact *aRb* minus *a* and *b*. And since in forming the law that all *F*s are *G*s we at least collect whatever facts such as *Fa* there may be, the universal *F* must likewise just be the common part of all such facts. At least in laws, there-fore, a universal must be regarded as derived from the particulars which are its instances and the facts that they are so. To regard them, as extreme real-ists do, as a primitive kind of entity, distinct from particulars but able to combine with them to yield facts, is to put the universal cart before the factual horse. It does nothing but pose such ancient but manifestly dotty conundrums as: why there are these two different kinds of entity, particu-lars and universals, and what the difference between them is; why two enti-ties of the same kind (two particulars or two universals) cannot combine to form a fact; what the relation is between a particular and a universal that are so combined. The last question on its own is fatal to this view, since any answer to it immediately generates Bradley's (1897: Ch. 3) notoriously vicious regress; a regress not avoided just by Armstrong's ingenuous device of calling the relation in question a "union . . . closer than relation" (1978a: 3).

The fact is, as Ramsey showed, that we have no *a priori* reason to sup-pose that universals are fundamentally different in kind from particulars. What we think of as particulars are merely the kinds of entity we can most readily individuate, typically by appeal to their spatio-temporal location (*cf.* Braithwaite 1926); and a universal is just the common residue of a set of facts about such individuals. So there is really no mystery about what relates particular to universal in a fact, nor about why a fact has to contain at least one of each. Nor is there any general reason why residual universals cannot themselves be individuated and so admitted in their own right as entities to be quantified over—thus, for example, leaving the particular *a* as the com-mon residue of a set of facts about *a*'s properties. Nominalism therefore is not the only, nor the most sensible, alternative to an extreme realism about universals.

From this Ramseyan account of universals it does however follow that they need instances: Armstrong's principle of instantiation is quite right. In the law that all *F*s are *G*s, the property *F* is just the residue of such facts as *Fa*. If the law is vacuous, there are no such facts; and no facts leave no residue. If there are no *F*s, there is no *F*. So there will be no fact *FNG* to make such a vacuous law true, and the Armstrong-Dretske-Tooley theory fails. Whether there are "real connections of universals", as Ramsey put it, I do not know: like him, "I cannot deny it; for I can understand nothing by such a phrase; what we call causal laws I find to be nothing of the sort" (Ramsey 1929: 148).

7 | Epilogue

I have considered two attempts, seriously undertaken of late years, to make natural laws say more than generalizations; both fail. The law that all *F*s are *G*s is given the force it needs neither by taking it to say that 'All *F*s are *G*s' is true in all possible worlds or is in some other sense necessary, nor by taking it to assert a contingent relation between *F* and *G*. Neither construal can cover the crucial case of vacuous laws which Braithwaite rightly stresses. There are no doubt likewise aspects of the problems of laws which solutions of Braithwaite's Humean cut also have difficulty covering; only they, to my mind, are more readily patched up. Those patches, however, must be woven elsewhere.[1]

▪ | Note

1. This article was written during my tenure of a Radcliffe Fellowship, for which I am indebted to the Radcliffe Trust.

▪ | References

Armstrong, D. M. 1978. *Nominalism and Realism*. Cambridge.

Armstrong, D. M. 1978a. *A Theory of Universals*. Cambridge.

Ayer, A. J. 1956. What is a law of nature? In his 1963 *The Concept of a Person and other essays*, pp. 209–34. London.

Bradley, F. H. 1897. *Appearance and Reality*, 2nd edn. Oxford.

Braithwaite, R. B. 1926. Universals and the 'method of analysis'. *Aristotelian Society Supplementary Volume* 6, 27–38.

Braithwaite, R. B. 1953. *Scientific Explanation*. Cambridge.

Cargile, James. 1969. The Sorites paradox. *The British Journal for the Philosophy of Science* 19, 193–202.

Davidson, Donald. 1970. Mental events. In *Experience and Theory*, ed. L. Foster and J. W. Swanson, pp. 79–101. London.

Dretske, Fred I. 1977. Laws of nature. *Philosophy of Science* 44, 248–68.

Dummett, Michael. 1973. *Frege: Philosophy of Language*. London.

Goodman, Nelson. 1965. *Fact, Fiction, and Forecast,* 2nd edn. New York.

Hesse, Mary. 1974. *The Structure of Scientific Inference.* London.

Hesse, Mary. 1980. A revised regularity view of scientific laws. In *Science, Belief and Behaviour,* ed. D. H. Mellor, pp. 87–103. Cambridge.

Kneale, William. 1949. *Probability and Induction.* Oxford.

Kripke, Saul A. 1971. Identity and necessity. In *Identity and Individuation,* ed. M. K. Munitz, pp. 135–64. New York.

Kripke, Saul A. 1972. Naming and necessity. In *Semantics of Natural Language,* ed. Donald Davidson and Gilbert Harman, pp. 253–355; 763–9. Dordrecht.

Lewis, David. 1973. *Counterfactuals.* Oxford.

Mackie. J. L. 1973. *Truth, Probability and Paradox.* Oxford.

McGinn, Colin. 1978. Mental states, natural kinds and psychophysical laws I. *Aristotelian Society Supplementary Volume* 52, 195–220.

Mellor, D. H. 1974. In defence of dispositions. *Philosophical Review* 83, 157–81.

Mellor, D. H. 1977. Natural kinds. *The British Journal for the Philosophy of Science* 28, 299–312.

Mellor, D. H. 1979. The possibility of prediction. *Proceedings of the British Academy* 65.

Mellor, D. H. 1981. McTaggart, fixity and coming true. *Reduction, Time and Reality,* ed. R. Healey. Cambridge.

Putnam, Hilary. 1975. The meaning of 'meaning'. In his *Mind, Language and Reality,* pp. 215–71. Cambridge.

Ramsey, F. P. 1925. Universals. In his 1978 *Foundations: Essays in Philosophy, Logic, Mathematics, and Economics,* ed. D. H. Mellor, pp. 17–39. London.

Ramsey, F. P. 1929. General propositions and causality. In his 1978 *Foundations: Essays in Philosophy, Logic, Mathematics and Economics,* ed. D. H. Mellor, pp. 133–51. London.

Tooley, Michael. 1977. The nature of laws. *Canadian Journal of Philosophy* 7, 667–98.

Nancy Cartwright

Do the Laws of Physics State the Facts?

■ | Introduction

There is a view about laws of nature that is so deeply entrenched that it doesn't even have a name of its own. It is the view that laws of nature describe facts about reality. If we think that the facts described by a law obtain, or at least that the facts which obtain are sufficiently like those described in the law, we count the law true, or true-for-the-nonce, until further facts are discovered. I propose to call this doctrine the *facticity* view of laws. (The name is due to John Perry.)

It is customary to take the fundamental explanatory laws of physics as the ideal. Maxwell's equations, or Schrödinger's, or the equations of general relativity are paradigms, paradigms upon which all other laws—laws of chemistry, biology, thermodynamics, or particle physics—are to be modeled. But this assumption confutes the facticity view of laws. For the fundamental laws of physics do not describe true facts about reality. Rendered as descriptions of facts, they are false; amended to be true, they lose their fundamental, explanatory force.

To understand this claim, it will help to contrast biology with physics. J. J. C. Smart ([10]: chapter 3) has argued that biology . . . has no genuine laws of its own. It resembles engineering. Any general claim about a complex system, such as a radio or a living organism, will be likely to have exceptions. The generalizations of biology, or engineering's rules of thumb, are not true laws because they are not exceptionless.* If this is a

From *Pacific Philosophical Quarterly* 61 (1980): 75–84.

* Smart's argument for why there are no laws in biology "in the strict sense" presupposes that laws—genuine laws—have to be universal in scope. Thus, to use Smart's own example, the generalization that albinotic mice always breed true, while probably true, does not qualify as a law because the term *mouse* contains an implicit reference to a particular planet, namely, the earth. (Mice are those creatures that stand in appropriate kinship relations to animals here on earth. No creature on

good reason [for regarding biology as a second-rate science], then it must be physics which is the second-rate science. Not only do the laws of physics have exceptions; unlike biological laws, they are not even true for the most part, or approximately true.

The view of laws with which I begin—"Laws of nature describe facts about reality"—is a pedestrian view that, I imagine, any scientific realist will hold. It supposes that laws of nature tell how objects of various kinds behave: how they behave some of the time, or all of the time, or even (if we want to prefix a necessity operator) how they must behave. What is critical is that they talk about objects—real concrete things that exist here in our material world, things like quarks, or mice, or genes; and they tell us what these objects do.

Biological laws provide good examples. For instance, here is a generalization taken from a Stanford text on chordates: (Alexander [1]: 179)

> The gymnotoids [American knife fish] are slender fish with enormously long anal fins, which suggest the blade of a knife of which the head is a handle. They often swim slowly with the body straight by undulating this fin. They [presumably "always" or "for the most part"] are found in Central and South America. . . . Unlike the characins they ["usually"?] hide by day under river banks or among roots, or even bury themselves in sand, emerging only at night.

The fundamental laws of physics, by contrast, do not tell what the objects in their domain do. If we try to think of them in this way, they are simply false, not only false but deemed false by the very theory which maintains them. But if physics' basic, explanatory laws do not describe how things behave, what do they do? Once we have given up facticity, I don't know what to say. Richard Feynman, in *The Character of Physical Law*, offers an idea, a metaphor. Feynman tells us "There is . . . a rhythm and a pattern between the phenomena of nature which is not apparent to the eye, but only to the eye of analysis; and it is these rhythms and patterns which we call Physical Laws" ([3]: 13). Most philosophers will want to know a lot more about how these rhythms and patterns function. But at least Feynman does not claim that the laws he studies describe the facts.

I say that the laws of physics do not provide true descriptions of reality. This sounds like an anti-realist doctrine. Indeed it is, but to describe the

---

another planet, however mouselike, could possibly be a mouse, given the way that evolutionary biologists define species.) We could remove the implicit reference to the earth by defining *mouse* in terms of a set of properties that all mice have and that, on this planet, are possessed only by mice. But then it is quite likely that, somewhere in the vast reaches of the universe, there are animals with those properties that do not breed true. Smart concludes that "if the propositions of biology are universal in scope, then such laws are very likely not universally true. If they are not falsified by some queer species or phenomenon on earth they are very likely falsified elsewhere in the universe" (*Philosophy and Scientific Realism*, 54). Smart draws the same conclusion about psychology.

claim in this way may be misleading. For anti-realist views in the philosophy of science are traditionally of two kinds. Bas van Fraassen [4] is a modern advocate of one of these versions of anti-realism; Hilary Putnam ([8], [9]) of the other. Van Fraassen is a sophisticated instrumentalist. He worries about the existence of unobservable entities, or rather, about the soundness of our grounds for believing in them; and he worries about the evidence which is supposed to support our theoretical claims about how these entities behave.* But I have no quarrel with theoretical entities; and for the moment I am not concerned with how we know what they do. What is troubling me here is that our explanatory laws don't tell us what they do. It is in fact part of their explanatory role not to tell.

Hilary Putnam in his new version of transcendental realism also maintains that the laws of physics don't represent facts about reality. But this is because nothing—not even the most commonplace claim about the cookies which are burning in the oven—represents facts about reality. If anything did, Putnam would probably think that the basic equations of modern physics did best. This is the claim that I reject. I think we can allow that all sorts of statements represent facts of nature, including the generalizations one learns in biology or engineering. It is just the fundamental explanatory laws that don't truly represent. Putnam is worried about meaning and reference and how we are trapped in the circle of words. I am worried about truth and explanation, and how one excludes the other.

I | Explanation by Composition of Causes, and the Trade-Off of Truth and Explanatory Power

Let me begin with a law of physics everyone knows—the law of universal gravitation. This is the law which Feynman uses for illustration; he endorses the view that this law is "the greatest generalization achieved by the human mind" (Feynman [3]: 14).

Law of Gravitation: $F = Gmm'/r^2$.

In words, Feynman tells us ([3]: 14)

The Law of Gravitation is that two bodies exert a force between each other which varies inversely as the square of the distance between them, and varies directly as the product of their masses.

Does this law truly describe how bodies behave?

Assuredly not. Feynman himself gives one reason why. "Electricity also exerts forces inversely as the square of the distance, this time between charges . . ." ([3]: 30). It's not true that for *any* two bodies the force between

* See chapter 9 for a detailed examination of van Fraassen's brand of antirealism.

them is given by the law of gravitation. Some bodies are charged bodies, and the force between them is not Gmm'/r^2. Rather it is some resultant of this force with the electric force which Feynman refers to.

For bodies which are both massive and charged, the law of universal gravitation and Coulomb's law (the law which gives the force between two charges) interact to determine the final force. But neither law by itself truly describes how the bodies behave. No charged objects will behave just as the law of universal gravitation says; and any massive objects will constitute a counterexample to Coulomb's law. These two laws are not true; worse, they are not even approximately true. In the interaction between the electrons and the protons of an atom, for example, the Coulomb effect swamps the gravitational one, and the force which actually occurs is very different from that described by the law of gravity.

There is an obvious rejoinder: I have not given a complete statement of these two laws, only a shorthand version. The Feynman version has an implicit *ceteris paribus* modifier in front which I have suppressed. Speaking more carefully, the law of universal gravitation is something like this:

If there are no forces other than gravitational forces at work, *then* two bodies exert a force between each other which varies inversely as the square of the distance between them, and varies directly as the product of their masses.

I will allow that this law is a true law, or at least one which is held true within a given theory. But it is not a very useful law. One of the chief jobs of the law of gravity is to help explain the forces which objects experience in various complex circumstances. *This* law can explain in only very simple, or ideal circumstances. It can account for why the force is as it is when just gravity is at work; but it is of no help for cases in which both gravity and electricity matter. Once the *ceteris paribus* modifier has been attached, the law of gravity is irrelevant to the more complex and interesting situations.

This unhappy feature is characteristic of explanatory laws. I said that the fundamental laws of physics do not represent the facts, whereas biological laws and principles of engineering do. This statement is both too strong and too weak. Some laws of physics do represent facts, and some laws of biology—particularly the explanatory laws—do not. The failure of facticity does not have so much to do with the nature of physics, but rather with the nature of explanation. We think that nature is governed by a small number of simple, fundamental laws. The world is full of complex and varied phenomena, but these are not fundamental. They arise from the interplay of simpler processes obeying the basic laws of nature.

This picture of how nature operates to produce the subtle and complicated effects we see around us is reflected in the explanations that we give: we explain complex phenomena by reducing them to their simpler components. This is not the only kind of explanation we give, but it is an important and central kind. I shall use the language of John Stuart Mill, and call this *explanation by composition of causes* (Mill [7]: Book III, Ch. VI).

It is characteristic of explanations by composition of causes that the laws they employ fail to satisfy the requirement of facticity. The force of these explanations comes from the presumption that the explanatory laws "act" in combination just as they would "act" separately. It is critical, then, that the laws cited have the same form, in or out of combination. But this is impossible if the laws are to describe the actual behavior of objects. The actual behavior is the resultant of simple laws in combination. The effect which occurs is not an effect dictated by any one of the laws separately. In order to be true in the composite case, the law must describe one effect (the effect which actually happens); but to be explanatory, it must describe another. There is a trade-off here between truth and explanatory power.

II | How Vector Addition Introduces Causal Powers

Our example, where gravity and electricity mix, is an example of the composition of forces. We know that forces add vectorially. Doesn't vector addition provide a simple and obvious answer to my worries? When gravity and electricity are both at work, two forces are produced, one in accord with Coulomb's law, the other according to the law of universal gravitation. Each law is accurate. Both the gravitational and the electric force are produced as described; the two forces then add together, vectorially, to yield the total "resultant" force.

The vector addition story is, I admit, a nice one. But it is just a metaphor. We add forces (or the numbers that represent forces) when we do calculations. Nature does not "add" forces. For the "component" forces are not there, in any but a metaphorical sense, to be added; and the laws which say they are there must also be given a metaphorical reading. Let me explain in more detail.

The vector addition story supposes that Feynman has left something out in his version of the law of gravitation. The way he writes it, it sounds as if the law describes the *resultant* force exerted between two bodies, rather than a component force—the force which is *produced between the two bodies in virtue of their gravitational masses* (or, for short, the force *due to gravity*). A better way to state the law would be

> Two bodies produce a force between each other (the force due to gravity) which varies inversely as the square of the distance between them, and varies directly as the product of their masses.

Similarly, for Coulomb's law

> Two charged bodies produce a force between each other (the force due to electricity) which also varies inversely as the square of the distance between them, and varies directly as the product of their charges.

These laws, I claim, do not satisfy the facticity requirement. They appear, on the face of it, to describe what bodies do: in the one case, the two bodies produce a force of size Gmm'/r^2, in the other, they produce a force of size qq'/r^2. But this cannot literally be so. For the force of size Gmm'/r^2 and the force of size qq'/r^2 are not real, occurrent forces. In interaction, a single force occurs—the force we call the "resultant"—and this force is neither the force due to gravity nor the electric force. On the vector addition story, the gravitational and the electric force are both produced, yet neither exists.

Mill would deny this. He thinks that in cases of the composition of causes, each separate effect does exist—it exists as *part* of the resultant effect, just as the left half of a table exists as part of the whole table. Mill's paradigm for composition of causes is mechanics. He says:

> In this important class of cases of causation, one cause never, properly speaking, defeats or frustrates another; both have their full effect. If a body is propelled in two directions by two forces, one tending to drive it to the north, and the other to the east, it is caused to move in a given time exactly as far in *both* directions as the two forces would separately have carried it. (Mill [7]: Book III, Ch. VI)

Mill's claim is unlikely. Events may have temporal parts, but not parts of the kind Mill describes. When a body has moved along a path due northeast it has travelled neither due north nor due east. The first half of the motion can be a part of the total motion; but no pure north motion can be a part of a motion which always heads northeast. (We learn this from Judith Jarvis Thomson's *Acts and Other Events*.) The lesson is even clearer if the example is changed a little: a body is pulled equally in opposite directions. It doesn't budge an inch, but on Mill's picture it has been caused to move both several feet to the left and several feet to the right. I realize, however, that intuitions are strongly divided on these cases, so in the next section I will present an example for which there is no possibility for seeing the separate effects of the composed causes as part of the effect which actually occurs.

It is implausible to take the force due to gravity and the force due to electricity literally as parts of the actually occurring force. Is there no way, then, to make sense of the story about vector addition? I think there is, but it involves giving up the facticity view of laws. We can preserve the truth of Coulomb's law and the law of gravitation by making them about something other than the facts—the laws can describe the causal powers that bodies have.

Hume taught that "the distinction, which we often make betwixt *power* and the *exercise* of it, is . . . without foundation" (Hume [5]: Part III, Section XIV). It is just Hume's illicit distinction that we need here: the law of gravitation claims that two bodies have the *power* to produce a force of size Gmm'/r^2. But they don't always succeed in the *exercise* of it. What they actually produce depends on what other powers are at work, and on what compromise is finally achieved among them. This may be the way we do sometimes imagine the composition of causes. But if so, the laws we use talk not about what bodies do, but about what powers they possess.

The introduction of causal powers will not be seen as a very productive starting point in our current era of moderate empiricism. Without doubt, we do sometimes think in terms of causal powers, so it would be foolish to maintain that the facticity view must be correct and the use of causal powers a total mistake. Still, facticity cannot be given up easily. We need an account of what laws are that connects them, on the one hand, with standard scientific methods for confirming laws, and on the other, with the use they are put to for prediction, construction, and explanation. If laws of nature are presumed to describe the facts, then there are familiar, detailed philosophic stories to be told about why a sample of facts is relevant to their confirmation, and how they help provide knowledge and understanding of what happens in nature. Any alternative account of what laws of nature do and what they say must serve at least as well; and no story I know of causal powers makes a very good start.

III | A Real Example of the Composition of Causes

The ground state of the carbon atom has five distinct energy levels [see figure 7.1]. Physics texts commonly treat this phenomenon sequentially, in three stages. I shall follow the discussion of Albert Messiah in volume II of *Quantum Mechanics* [6]. In the first stage, the ground state energy is calculated by a central field approximation; and the single line (a) is derived. For some purposes, it is accurate to assume that only this level occurs. But some problems require a more accurate description. This can be provided by noticing that the central field approximation takes account only of the *average* value of the electrostatic repulsion of the inner shell electrons on the two outer electrons. This defect is remedied at the second stage by considering the effects of a term which is equal to the difference between the exact Coulomb interaction and the average potential used in stage one. This corrective potential "splits" the single line (a) into three lines depicted in (b).

Still, the treatment is inaccurate because it neglects spin effects. Each electron has a spin, or internal angular momentum, and the spin of the electron couples with its orbital angular momentum to create an additional potential. The additional potential arises because the spinning electron has an intrinsic magnetic moment, and "an electron moving in [an electrostatic] potential 'sees' a magnetic field" ([6]: 552). About the results of this potential Messiah tells us, "Only the ^3P state is affected by the spin-orbit energy term; it gets split into three levels: $^3P_0$, $^3P_1$ and $^3P_2$" ([6]: 706). Hence the five levels pictured in (c).

The philosophic perplexities stand out most at the last stage. The five levels are due to a combination of a Coulomb potential [which produces

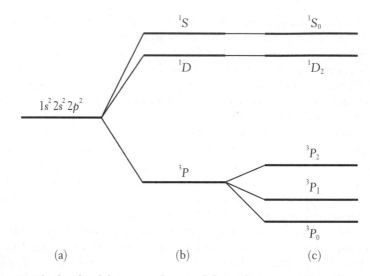

Fig. 7.1. The levels of the ground state of the carbon atom; (a) in the central field approximation ($V_1 = V_2 = 0$); (b) neglecting spin-orbit coupling ($V_2 = 0$); (c) including spin-orbit coupling. (Messiah [6])

three energy levels], and a potential created by spin-orbit coupling [that] "splits" the lowest of these again into three. *That* is the explanation of the five levels. But how can we state the laws that it uses?

For the Coulomb effect we might try

> Whenever a Coulomb potential is like that in the carbon atom, the three energy levels pictured in (b) occur.

(The real law will of course replace "like that in the carbon atom" by a mathematical description of the Coulomb potential in carbon; and similarly for "the three energy levels pictured in (b).") The carbon atom itself provides a counterexample to this law. It has a Coulomb potential of the right kind; yet the five levels of (c) occur, not the three levels of (b).

We might, in analogy with the vector addition treatment of composite forces, try instead

> The energy levels produced by a Coulomb potential like that in the carbon atom are the three levels pictured in (b).

But (as with the forces "produced by gravity" in our earlier example) the levels which are supposed to be produced by the Coulomb potential are levels that don't occur. In actuality, five levels occur, and they do not include the three levels of (b). In particular, as we can see from Messiah's diagram, the lowest of the three levels—the ^3P—is not identical with any of the five. In the case of the composition of motions, Mill tried to see the "component" effects as parts of the actual effect. But that certainly will not work here. The ^3P level in (b) may be "split," and hence "give rise to," the $^3P_0$, $^3P_1$, and $^3P_2$ levels in (c); but it is certainly not a part of any of these levels.

It is hard to state a true factual claim about the effects of the Coulomb potential in the carbon atom. But quantum theory does guarantee that a certain *counterfactual* is true: the Coulomb potential, if it *were* the only potential at work, would produce the three levels in (b). Clearly this counterfactual bears on our explanation. But we have no model of explanation that shows how. The covering law model shows how statements of fact are relevant to explaining a phenomenon.* But how is a truth about energy levels which would occur in quite different circumstances relevant to the levels which do occur in these? We think the counterfactual is important; but we have no account of how it works.

IV | Composition of Causes Versus Explanation by Covering Law

The composition of causes is not the only method of explanation which can be employed. There are other methods, and some of these are compatible with the facticity view of laws. Standard covering law explanations are a prime example.

Sometimes these other kinds of explanation are available even when we give an explanation which tells what the component causes of a phenomenon are. For example, in the case of Coulomb's law and the law of gravity, we know how to write down a more complex law (a law with a more complex antecedent) which says exactly what happens when a system has both mass and charge. Mill thinks that such "super" laws are always available for mechanical phenomena. In fact, he thinks, "This explains why mechanics is a deductive or demonstrative science, and chemistry is not" (Mill [7]: Book III, Ch. VI).

I want to make three remarks about these super laws and the covering explanations they provide: First, super laws aren't always available; second, even when they are available, they often don't explain much; third—and most importantly—even when other good explanations are to hand, if we fail to describe the component processes that go together to produce a phenomenon, we lose a central and important part of our understanding of what makes things happen.

1 There are a good number of complex scientific phenomena which we are quite proud to be able to explain. For many of these explanations,

* See Carl G. Hempel, "Two Basic Types of Scientific Explanation," in chapter 6.

however, super covering laws are not available to us. I argue this in "The Truth Doesn't Explain Much" (Cartwright [2]). Sometimes in these situations we have every reason to believe that a super law exists. (God has written it somewhere in the Book of Nature.) In other cases we have no good empirical reason to suppose even this much. (Nature may well be underdetermined; God failed to write laws for every complex situation.) Nevertheless, after we have seen what occurs in a specific case, we are often able to understand how various causes contributed to bring it about. We do explain, even without knowing the super laws. We need a philosophical account of laws and explanations which covers this very common scientific practice, and which shows why these explanations are good ones.

2 Sometimes super laws, even when they are available to cover a case, may not be very explanatory. This is an old complaint against the covering law model of explanation: "Why does the quail in the garden bob its head up and down in that funny way whenever it walks?" . . . "Because they all do." In the example of spin-orbit coupling it does not explain the five energy levels that appear in a particular experiment to say "All carbon atoms have five energy levels."

3 Often, of course, a covering law for the complex case will be explanatory. This is especially true when the antecedent of the law does not just piece together the particular circumstances that obtain on the occasion in question, but instead gives a more abstract description which fits with a general body of theory. In the case of spin-orbit coupling . . . quantum mechanics provides general theorems about symmetry groups, and Hamiltonians, and degeneracies, from which we could expect to derive, covering law style, the energy levels of carbon from the appropriate abstract characterization of its Hamiltonian, and the symmetries it exhibits.

Indeed, we can do this; and if we don't do it, we will fail to see that the pattern of levels in carbon is a particular case of a general phenomenon which reflects a deep fact about the effects of symmetries in nature. On the other hand, to do only this misses the detailed causal story of *how* the splitting of spectral lines by the removal of symmetry manages to get worked out in each particular case.

This two-faced character is a widespread feature of explanation. Even if . . . there is a single set of super laws which unifies all the complex phenomena one studies in physics . . . our current picture may yet provide the ground for these laws: what the unified laws dictate to happen, happens *because of* the combined action of laws from separate domains, like the law of gravity and Coulomb's law. Without these laws, we would miss an essential portion of the explanatory story. Explanation by subsumption under super, unified covering laws would be no replacement for the composition of causes. It would be a complement. To understand how the consequences of the unified laws are brought about would require separate operation of the law of gravity, Coulomb's law, and so forth; and the failure of facticity for these contributory laws would still have to be faced.

V | Conclusion

There is a simple, straightforward view of laws of nature which is suggested by scientific realism, the facticity view: laws of nature describe how physical systems behave. This is by far the commonest view, and a sensible one; but it doesn't work. It doesn't fit explanatory laws, like the fundamental laws of physics. Some other view is needed if we are to account for the use of laws in explanation; and I don't see any obvious candidate which is consistent with the realist's reasonable demand that laws describe reality and state facts which might well be true. There is, I have argued, a trade-off between factual content and explanatory power. We explain certain complex phenomena to be the result of the interplay of simple, fundamental laws. But what do these fundamental laws *say*? To play the role in explanation we demand of them, these laws must have the same form when they act together as when they act singly. In the simplest case, the consequences which the laws prescribe must be exactly the same in interaction, as the consequences which would obtain if the law were operating alone. But then, what the law states cannot literally be true, for the consequences which would occur if it acted alone are not the consequences which actually occur when it acts in combination.

If we state the fundamental laws as laws about what happens when only a single cause is at work, then we can suppose the law to provide a true description. The problem arises when we try to take that law and use it to explain the very different things which happen when several causes are at work. This is the point of "The Truth Doesn't Explain Much" (Cartwright [2]). There is no difficulty writing down laws which we suppose to be true: "*If* there are no charges, no nuclear forces . . . *then* the force between two masses of size m and m' separated by a distance r is Gmm'/r^2." We count this law true—what it says will happen, does happen—or at least happens to within a good approximation. But this law doesn't explain much. It is irrelevant to cases where there are electric or nuclear forces at work. The laws of physics, I concluded, to the extent that they are true, don't explain much. We could know all the true laws of nature, and still not know how to explain composite cases. Explanation must rely on something other than law.

But this view is absurd. There aren't two vehicles for explanation—laws for the rare occasions when causes occur separately; and another secret, nameless device for when they occur in combination. Explanations work in the same way whether one cause is at work, or many. "Truth Doesn't Explain" raises perplexities about explanation by composition of causes; and it concludes that explanation is a very peculiar scientific activity, which commonly does not make use of laws of nature. But scientific explanations do use laws. It is the laws themselves which are peculiar. The lesson to be learned is that the laws which explain by composition of causes fail to satisfy the facticity requirement. If the laws of physics are to explain how phenomena are brought about, they cannot state the facts.[1]

■ | Note

1. This paper was given as part of a symposium on "Explanation and Scientific Realism" at the School of Philosophy, University of Southern California, March 1980.

■ | References

[1] Alexander, R. McNeill. *The Chordates* (Cambridge University Press, 1975).

[2] Cartwright, Nancy. "The Truth Doesn't Explain Much." *American Philosophical Quarterly*, 17 (1980). [Reprinted in N. Cartwright, *How the Laws of Physics Lie* (Oxford: Clarendon Press, 1983).]

[3] Feynman, Richard. *The Character of Physical Law* (Cambridge, Mass.: MIT Press, 1967).

[4] van Fraassen, Bas. *The Scientific Image* (Oxford University Press, 1980).

[5] Hume, David. *A Treatise of Human Nature* (edited by L.A. Selby Bigge. Oxford: Clarendon Press, 1978).

[6] Messiah, Albert. *Quantum Mechanics* (Amsterdam: North Holland, 1961).

[7] Mill, John Stuart. *A System of Logic* (London: John W. Parker and Son, 1856).

[8] Putnam, Hilary. *Meaning and the Moral Sciences* (London: Routledge and Kegan Paul, 1978).

[9] Putnam, Hilary. "Models and Reality." [In H. Putnam, *Realism and Reason*, vol. 3, *Philosophical Papers* (Cambridge: Cambridge University Press, 1983), pp. 1–25.]

[10] Smart, J. J. C. *Philosophy and Scientific Realism* (London: Routledge and Kegan Paul, 1963).

7 | COMMENTARY

7 | COMMENTARY

7.1 | The Regularity Theory: Ayer and Hume

A. J. Ayer defends a version of the regularity theory of laws in his article, "What Is a Law of Nature?" As Ayer notes, the regularity theory has its origins in David Hume's analysis of causation.

HUME ON CAUSATION

In his *Treatise of Human Nature* (1739), David Hume (1711–76) advocated what is called the *constant-conjunction* (or *regularity*) *theory of causation.* According to Hume, the claim that one event, *a* (of type A), caused another event, *b* (of type B), means only that A-events are, as a matter of empirical fact, always followed by B-events. (From now on, we shall refer to events of type A simply as As and to events of type B as Bs.) Objectively speaking, the causal relation between *a* and *b* is nothing more than the constant conjunction of As and Bs. If As always have been and always will be followed by Bs, then As cause Bs and, in particular, *a* caused *b*. Hume denied any objective necessity "out there in the world" between As and Bs in virtue of which A-events produce B-events or make B-events occur. Our conviction that effects do not merely happen to follow causes but in some sense must necessarily occur given the appropriate cause results from a purely subjective feeling in our minds when we experience an A (or imagine an A) and expect a B to follow. Thus, on Hume's view, we regard As as the cause of Bs when past experience has induced in us the expectation that As will always be followed by Bs in the future.

Hume's constant-conjunction theory stands our commonsense view of causation on its head. For, typically, we think that we first discover the fact that As causally necessitate Bs, and then, on the basis of that discovery, understand why As have always been followed by Bs and predict that As always will be followed by Bs. As Ayer explains, part of the case for Hume's radical reversal of our usual thinking about causality rests on his skeptical, empiricist analysis of the idea of necessity. If there is a necessary connection between causes and their effects, then the necessity is either logical or nonlogical. Hume denies that the connection can be logical. If it were logical, then effects could be deduced from causes and we could know, prior to experience, that one kind of event (the cause) must invariably be associated with another kind of event (the effect). But no such deduction and no such a priori knowledge of effects is possible. For any cause, it is logically possible that its usual effect not follow it; moreover, our knowledge of causal relations is derived solely from experience. Thus, Hume and Ayer argue that the kind of necessity involved in causation cannot be logical.

Curiously, Ayer does not then consider Hume's case against the second alternative, namely that some kind of nonlogical necessity links causes with their effects. Hume's argument here is epistemological.[1] He considers the two sources most commonly claimed as the origin of our idea of causal necessity—namely, perception and the will—and argues that neither gives us any experience of necessity. When Hume examines his perceptual experiences, he cannot find in them any element (what Hume calls "an impression") of necessity. For example, when one billiard ball collides with another, what we literally see, according to Hume, is simply one motion followed by another, not the first ball making the second ball move. With regard to the willing of our actions, Hume denies that we know, independently of experience, which acts of will must be followed by which actions. Indeed, we are completely ignorant of the immediate effects of willing (presumably it is some change in the brain that then causes impulses to be transmitted by our nerves to our muscles), and only by experience do we learn which parts of our bodies we can control and which we cannot. As the possibility of sudden paralysis demonstrates, there is no logical or physical guarantee that a particular act of willing will be followed by a particular motion of one's body.

Having failed to locate any impression of necessity in either our perceptual experiences or in our ability to will actions, Hume concludes that the source of our idea of nonlogical necessity must be purely subjective. Hence, Hume sees nonlogical necessity as an imaginative fiction, something originating from our patterns of inference and expectation, which we then project onto nature and mistake for something objective.

So much by way of background in Hume's theory of causation: let us return to laws. The regularity theory of laws is often called the *Humean theory* because it, too, denies that any sort of objective nonlogical necessity connects the items appearing in a law. In its simplest form, the regularity theory says that laws of nature are nothing more than true universal generalizations. If it is a law that all copper conducts electricity, then what makes it a law is the fact that all pieces of copper, past, present, and future, conduct electricity. If it is law that all metals expand when heated, then it is a law because, as a matter of fact, that is how all metals always behave. According to the regularity theory, the objective content of laws is exhausted by what actually happens in the world.

THE PROBLEM OF VACUOUS LAWS

There are a number of serious problems with the simple version of the regularity theory of laws. These problems are discussed by Ayer, and briefly by Dretske and Mellor, in the readings in this chapter. First, there is the problem of avoiding what Ayer calls "vacuous laws." Modern logicians regard the generalization "All As are Bs" as logically equivalent to "It is false that there is an A that is not a B." Thus, "All copper conducts electricity" is true if and only if it is not the case that there is a piece of copper that is not a conductor.

If we translate the same generalization using predicate logic, then we get (x) $(Cx \supset Ex)$. Literally, for any x, if x is copper, then x conducts electricity. But just as before, this generalization is logically equivalent to the statement "it is not the case that there is a piece of copper that is not a conductor." In symbols we would write $\sim(\exists x)$ $(Cx \,\&\, \sim\!Ex)$: it is not the case that there exists any x, such that x is copper and x is not a conductor. Now the important point to notice is that if, as a matter of fact, there is no copper in the universe, then it automatically becomes true that it is not the case that there is a piece of copper that is not a conductor. But this last statement is logically equivalent to the generalization "All copper conducts electricity." According to the simple version of the regularity theory, laws are true universal generalizations. Thus, any generalization that is automatically true simply because there are no instances of its antecedent is a law. For example, according to the simple version of the regularity theory, it is a law that all perpetual motion machines weigh ten tons, that all mermaids contain chlorophyll, and that all particles traveling faster than light are red. Clearly, this is absurd. How can the regularity theorist avoid the problem of vacuous laws?

The most obvious response is to add a further condition, an existential condition, to the regularity analysis: a true universal generalization is a law provided there actually are objects satisfying the generalization. In symbols, "All Cs are Es" is a law if and only if $(x)(Cx \supset Ex) \,\&\, (\exists x)(Cx)$. This eliminates the counterexamples mentioned in the previous paragraph, since there are no mermaids, perpetual motion machines, or particles traveling faster than light. Hence, on the amended regularity analysis, no vacuously true generalizations will qualify as laws.

THE PROBLEM OF NONINSTANTIAL LAWS

Despite the advantage of modifying the regularity theory to avoid the problem of vacuous laws, Ayer and many other regularity theorists argue that the existential condition is too strong, because it rules out some of the most important laws in science. The classic example is Newton's first law of motion, which states the principle of rectilinear inertia:

> All bodies on which no net external force is acting either remain at rest or move at uniform velocity in a straight line.

Now it seems reasonable to assume that, since all bodies exert a gravitational pull on all the other bodies in the universe, no bodies are ever free from net external forces. Nonetheless, scientists accept Newton's first law not because it lacks instances but because it expresses an important truth about the world. Newton's first law is thus an example of a nonvacuous but noninstantial law. Here are some other examples of nonvacuous noninstantial laws: if two perfectly elastic bodies were to collide, the total kinetic energy of the system would be the same before and after the impact (discussed by C. D. Broad in the article referred to by Ayer); in a perfectly reversible process, the entropy remains constant; all lumps of plutonium weighing more than one million tons conduct electricity.

The natural inclination is to handle noninstantial laws in terms of subjunctive or counterfactual conditionals, that is, in terms of how objects of a certain kind would behave if there were such objects. But this approach does not seem to be open to the regularity theorist. For regularity theorists take statements of law to describe how actual objects behave, not how possible objects would behave if, contrary to fact, they were to exist.

Broad's proposal (reported by Ayer) for reconciling noninstantial laws with the regularity theory suggests that we distinguish ultimate laws of nature from derivative laws. Thus, all ultimate laws are taken to be instantial, but laws derived from one or more such ultimate laws may not be. Consider, for example, Newton's second law:

> If a net force, F, acts of a body of mass, m, then the body experiences an acceleration, $a = F/m$.

We can derive Newton's first law from his second law on the supposition that no net force is acting on a body. For according to Newton's second law, if the net force on a body were zero, then the acceleration of the body would also be zero. So Newton's first law is a noninstantial, derivative law because it can be derived from the instantial, ultimate second law. Similarly, there are ultimate, instantial laws of impact and motion which entail the nonvacuous, noninstantial law that, if two perfectly elastic bodies were to collide, kinetic energy would be conserved. But as Ayer notes, even if we could always find ultimate laws that would reconcile noninstantial laws with the regularity theory, that theory would still encounter severe problems with what are sometimes called *functional laws*.

The Missing-Values Problem for Functional Law

Functional laws assert a functional relation between two or more variables in the form of a mathematical equation. For example, Hooke's law, $F = kx$, says that the force, F, exerted by a spring is directly proportional to x, the amount the spring is stretched. Similarly, Hubble's law $V = HD$, says that the velocity, V, with which galaxies are moving away from each other is directly proportional to D, their distance apart. The ideal gas law, $PV = nRT$, asserts that the pressure times the volume of n moles of gas is proportional to the absolute temperature of the gas. In all these functional laws, the magnitude of the variables range over infinitely many values, only a small finite number of which will ever be realized. For example, no gas will actually be heated to all possible temperatures, nor will every spring be stretched to all possible lengths. Nonetheless, the ideal gas law tells us what the pressure of the gas would be at a temperature of 1 million degrees, and Hooke's law says what force a spring would exert if it were stretched to a hundred times its normal length. Thus, the missing values problem leads us inexorably to using subjunctive conditionals to express the content of laws in counterfactual situations. Once again, the problem for the regularity theorist is making sense of these counterfactual conditionals while still regarding laws as descriptions of what actually happens in the world.

The Problem of Accidental Generalizations

Closely related to the missing-values problem for functional laws is the more general difficulty of distinguishing between genuine laws and so-called accidental generalizations (what Ayer calls "generalizations of fact" as contrasted with "generalizations of law"). Consider one of Fred Dretske's examples. Suppose that, as a matter of brute fact, the only dogs that have been or ever will be born at sea are cocker spaniels. Thus, "all dogs born at sea are cocker spaniels" emerges as a true, universal generalization. But, clearly, we would not on this basis predict that, if the dog on board our ship were a dachshund, then she would give birth to cocker spaniel puppies. Our expectations about this and other counterfactual situations depend, not on accidental generalizations, but on genuinely lawful ones. We rely on the biological law that purebred dogs produce dogs of the same breed (at least, when mated with dogs of the same breed). But the simple version of the regularity theory cannot distinguish between those universal generalizations that are laws and those that are not. So the simple version of the regularity theory is inadequate.

7.2 | Ayer's Epistemic Regularity Theory

Because of the difficulty in distinguishing between laws and accidental generalizations, most proponents of the regularity theory advocate a more sophisticated version of the theory according to which laws are true, universal generalizations with some additional features. As Ayer puts it, "the difference between our two types of generalization lies not so much on the side of the facts which make them true or false, as in the attitude of those who put them forward" (829). In his paper "Laws of Nature," Dretske summarizes the sophisticated version of the regularity theory with the formula

law = universal truth + X,

where the usual candidates for X include:

- our willingness to use the generalization in question to make predictions, especially about counterfactual situations;
- our acceptance of the generalization as well confirmed even though we have examined only a relatively small, finite number of its instances;

- the role that the generalization plays in a deductively organized system of (scientific) statements; and
- our recognition that the generalization (unlike a mere generalization of fact) explains its instances.

Because all of these candidates for X involve our beliefs and epistemic attitudes, this sophisticated version of the regularity theory is often called the *epistemic regularity theory*. Ayer's own proposal falls into this category. In Ayer's own words:

> Accordingly I suggest that for someone to treat a statement of the form 'if anything has Φ it has Ψ' as expressing a law of nature, it is sufficient (i) that subject to a willingness to explain away exceptions he believes that in a non-trivial sense everything which in fact has Φ has Ψ (ii) that his belief that something which has Φ has Ψ is not liable to be weakened by the discovery that the object in question also has some other property X, provided (a) that X does not logically entail not-Ψ (b) that X is not a manifestation of not-Ψ (c) that the discovery that something had X would not in itself seriously weaken his belief that it had Φ (d) that he does not regard the statement 'if anything has Φ and not-X it has Ψ' as a more exact statement of the generalization that he was intending to express. (831)

DRETSKE'S CRITICISM OF AYER'S THEORY

As Ayer himself acknowledges (in his concluding paragraph), his proposal completely ignores the missing-values problem for functional laws. Even more striking is Ayer's candid admission that his proposal cannot be construed as an attempt to define what laws are. A definition of a concept would give both necessary and sufficient conditions, but Ayer offers only sufficient conditions for lawlikeness. In other words, Ayer recognizes that many things could turn out to be laws, even though they fail to satisfy his conditions. As he says, "I do not claim that to say that some proposition expresses a law of nature entails saying that someone has a certain attitude towards it; for clearly it makes sense to say that there are laws of nature which remain unknown" (831).

Ayer's candid admission about the limitations of his epistemic regularity analysis of laws—that is, its inability to countenance the existence of unknown laws—is a powerful objection to the whole approach. Dretske thinks the objection is decisive. As long as X includes factors that refer essentially to human beliefs, attitudes, and practices, the epistemic regularity approach entails that there are no unknown laws. As Dretske sees it, the epistemic regularity approach has confused an epistemological issue (why we believe something is a law of nature) with an ontological issue (what sort of thing a law of nature is). A more promising approach, in Dretske's view, is to address the ontological issue directly. First we should understand what laws of nature are and then (but only then) explain why we adopt towards them the attitudes that we do.

7.3 | The MRL Objective Regularity Theory

Before turning to Dretske's theory, there is an objective version of the regularity theory that deserves consideration because it answers some of Dretske's potent objections to Ayer's subjective version. This theory of laws is due principally to David Lewis, following earlier proposals by Frank Plumpton Ramsey and John Stuart Mill. Like Ayer's epistemic regularity analysis of laws, the Mill-Ramsey-Lewis theory (hereafter MRL) is thoroughly Humean in spirit: laws are true, contingent, universal generalizations that differ from non-lawlike ("merely accidental") truths by possessing an additional feature, X. Unlike Ayer's account, MRL avoids any subjective or epistemic elements (such as the beliefs, attitudes, and dispositions of scientists) in its characterization of X. In this way, proponents of MRL secure immunity from Dretske's "unknown laws" objection while remaining true to the "bottom-up" empiricist view that laws supervene on what Lewis calls the "Humean mosaic" of particular facts and events.

Another important motivation for MRL comes from an objection to the subjective theory (again voiced by Dretske) that is analogous to the "*Euthyphro* problem" in ethics. The subjective theory says that truths become laws because we use them in certain ways and have certain characteristic attitudes toward them. But, so the objection runs, a philosophically adequate account of lawhood should explain why it is appropriate and reasonable to use some generalizations (the ones we identify as laws) in this way. What warrants our confidence in using laws (but not other true generalizations) to predict, explain, and support counterfactual inferences? Obviously it is no answer to this question simply to say that without such an endorsement the generalizations in question would not qualify as laws.

So what is X, according to the MRL account of laws? Lewis proposed that a universal truth is a law if and only if it belongs to the best deductive system, that is, to a set of similar truths that give the simplest and strongest (most informative, most comprehensive) account of everything that happens and obtains in the world.[2] "Belonging to the best system" is a relational property; it is not something that a single generalization has in isolation. Thus, in order to judge at a given time whether a generalization is a law requires seeing how it fits into the entire network of propositions accepted by scientists at that time. If the generalization does fit at a given time, there is no guarantee that it is a genuine law since "the best system" is an ideal that scientists aspire to rather than something they have already achieved. Thus, MRL has no difficulty acknowledging "unknown laws" waiting to be discovered. Similarly, MRL can readily accommodate generalizations that were regarded as laws by scientists in the past, but that are no longer accorded that status (even if we continue to regard them as true). Finally, MRL has a plausible explanation of why we value laws and rely on them.

As Lewis notes, strength and simplicity often compete: a long catalog of unconnected truths may say a lot about the world but score very low as regards simplicity; a very simple, short list may be quite uninformative. Although some philosophers have argued that judgments of simplicity are inherently subjective and have questioned whether simplicity can be balanced with strength in an entirely objective way, the majority of those who reject MRL are prepared to concede the objectivity of these features of Lewis's account. This is because there is a more fundamental objection to MRL, first raised by Michael Tooley and John Carroll, that cuts to the heart of the best-system account of lawhood.[3] Here is a simplified version of that objection.

Imagine a very austere possible world, PW1, in which there are just three types of particle (A, B, and C), which interact only when they collide. In PW1, as a matter of contingent fact, there are no A-B collisions. Suppose further that A-A, A-C, B-B, B-C, and C-C interactions are governed by five separate and independent laws. According to MRL, there is no A-B law in this possible world (because adding any generalization about A-B interactions to the set of five laws will diminish simplicity with no compensating increase in strength). Now imagine a different possible world, PW2, in which As and Bs do interact according to their own distinct law L_{AB}. According to MRL, L_{AB} is a law in PW2 but not in PW1; that seems reasonable to a Humean because PW1 and PW2 are distinct possible worlds in which different things happen. But suppose that some of the laws of PW2 are indeterministic in such a way as to make it a matter of chance whether or not As and Bs ever collide. In that case, L_{AB} would remain a law even in another possible world—call it PW3—which is identical with PW1 with regard to all the events involving its particles. The problem for MRL is that, from a Humean perspective, PW3 and PW1 are the *same* possible world, since they contain all the same events and particular matters of fact. But, according to the (non-Humean) intuitions of many philosophers, PW3 and PW1 differ in one crucial respect: in PW3 L_{AB} is a law but in PW1 it is not. Like many metaphysical arguments, this objection to MRL is not conclusive; but it does focus our attention on a conviction about laws driving the universals theory advocated by Dretske and others, namely, that they are something more, something "over and above" particular things and events.

7.4 | Dretske's Universals Theory

Dretske proposes a necessitarian analysis of laws. (Similar proposals have been defended by D. M. Armstrong and Michael Tooley.) Dretske thinks that the law that we would usually express by saying "All Fs are G" really has the form

F-ness → G-ness,

where F-ness and G-ness are the properties of being F and G. The term *F-ness* refers to a universal, the property a thing must have in order to be F. Dretske suggests that we read the connective "→" as "yields" or "brings with it." (Tooley calls it "nomic necessitation," and Armstrong usually calls it "necessitation.") The Dretske-Armstrong-Tooley approach is called the *universals theory* because it regards laws of nature as being fundamentally about relations between universals (properties). Statements of laws of nature, on this view, are not universal generalizations about particulars but singular statements about universals.

Notice here one prima facie advantage the universals theory might, on a certain view of universals, be thought to have over the regularity theory. Adherents of the regularity view had difficulty explaining why laws, which are taken only to describe the way objects are, nevertheless support their counterfactuals. On the universals theory laws can support contrary-to-fact possibilities because universals are taken to be properties that can be variously possessed, or not, by objects that do, or could, exist. Given a law expressing (for example) the relation between electrical charge and magnetic field, we might reasonably go on to speak of the magnetic properties of vertebrates if they were electrically charged; likewise, if laws do indeed express relations among universals, we might reasonably state how a body would behave if, contrary to fact, no net force were acting on it.

EXTENSIONS AND INTENSIONS

One of the keys to understanding the universals theory of laws is to appreciate the difference between the *extension* and the *intension* of a predicate. A predicate is any term that, like an adjective, can be used to describe a thing. For example, the words *cat, elastic,* and *copper* are all predicates. The extension of a predicate is the set of objects (animals, regions of space) to which the term correctly applies. For example, the extension of *cat* is the set of all the objects that the term *cat* denotes, namely all the cats in the world.

It is more difficult to explain what the intension of a term is, and philosophers have differed in their accounts of it. The basic idea is that the intension of a predicate is its meaning (or what Mill called its *connotation*). The intension of *cat* is whatever the term *cat* means—the property of felinicity if you like or, perhaps, the concept of catness. It is evident that two terms can mean different things (i.e., have different intensions) but apply to exactly the same set of objects (i.e., have the same extension). For example, all mammals (even whales and porpoises) have hair somewhere on their bodies, and mammals are the only animals that have hair. So the terms *mammal* and *hairy* (in the sense of having at least some hair) are coextensive, since they pick out the same group of objects. But clearly the intensions of these terms differ. Even though all and only mammals have hair, the term *mammal* does not mean *hairy*.

Now, suppose that it is a law of nature, a biological law, that all mammals have mammary glands. (It is biologically necessary for mammals, which suckle their young, to have milk-secreting glands.) If it is a universal truth that all mammals have mammary glands, then it must also be a universal truth that all Ys have mammary glands, where Y is any term that is coextensive with *mammal*. So, in particular, it is a universal truth that all hairy animals have mammary glands. But, as Dretske points out, from the fact that "All mammals have mammary glands" is a law and *hairy* is coextensive with *mammal*, it does not follow that it is a law that all hairy animals have mammary glands. Laws, on Dretske's account, are *opaque* in a way that normal statements about universals may not be: statements that we can deduce from laws by the substitution of terms will themselves be laws only if the terms in question have the same intension. As Dretske puts it, "laws imply universal truths, but universal truths do not imply laws" (837).

DRETSKE'S NECESSITARIAN VIEW

One of the subtleties of Dretske's universals theory, which distinguishes it from earlier necessitarian analyses of laws, is the contrast he draws between treating laws as intensional relations between extensions and treating them as extensional relations between intensions. Many necessitarians in the past have espoused the first position. That is, they have supposed that laws are opaque because of the special, intensional, nature of the relation connecting the extensions of terms in statements of laws. Thus, on this older necessitarian view, the reason for the lawlike character of "all copper conducts electricity" is the special strong relation of (nonlogical) necessitation connecting things that are copper to things that conduct electricity. On Dretske's alternative view, there is no need for such a special strong relation of necessitation. The items linked by the law are not physical objects or events but properties (universals, intensions). According to Dretske, a law asserts that one property (an intension) is invariably associated with another property (another intension). As long as one were to substitute another term that picks out the same property in a lawlike statement, the new statement thus generated would also be a law. When we construe terms intensionally (as denoting properties), they will be coextensive whenever they denote the same property. Thus, on Dretske's view, the relation symbolized by "→" is extensional, while the things linked by the relation (namely properties, universals) are intensional.

We can summarize the difference between the two brands of necessitarianism by contrasting how they would represent the law that all Fs are G. Earlier necessitarians would write, "(x) (Fx $\boxed{N}$→ Gx)"—to be read as, "anything that is F must (in some nonlogical, physical sense of *must*) be G." Dretske would write, "F-ness → G-ness"—to be read as, "the property of being an F necessitates the property of being a G."

There are several problems with Dretske's universals theory of laws, to which critics such as Bas van Fraassen and D. H. Mellor have drawn attention. Van Fraassen calls two of these problems the *identification problem* and the *inference problem*. The identification problem is the challenge— posed to defenders of the universals theory—of giving an adequate account of the necessitation relation that allegedly holds between those universals comprising a law. The second (and related) inference problem concerns the inferential relation between laws and their instances. The universals theorist insists that not only does the *FG*-law logically imply that all *F*s are *G*, but also that the *FG*-law explains the "mustness" or necessity that we think holds between the particular things that are instances of the law. Presumably, then, the *FG*-law implies either that it is necessary that all *F*s are *G* or that if some particular thing is *F* then it must also be *G*. As we shall see, it is difficult to make sense of either inference if, as the universals theory supposes, laws of nature are themselves not necessary but contingent.

THE IDENTIFICATION PROBLEM

The identification problem is that of giving an account of the necessitation relation that, according to theories such as Dretske's, connects the universals that make up a law. Recall from our earlier discussion that empiricists such as Hume deny that we can make sense of any notion of objective necessity differing from logical necessity. As far as Hume was concerned, nomic necessity is simply subjective, a feeling; there is no objective, nonlogical necessity that connects objects, events, or universals. What if Hume were right about this? Does this rule out the possibility that laws involve a logically necessary connection between universals?

If the necessitation relation between the universals in a law were logical, then all laws of nature would be logically necessary. In that case, if "All *F*s are *G*" is a law, it would be logically impossible for there to be an *F* that was not a *G*. To many empiricist philosophers, this is a sufficient reason to reject reading nomic necessity as logical necessity, since we usually assume that laws of nature are contingent, not necessary, truths. But a degree of caution is advisable here. As Mellor points out in his "Necessities and Universals in Natural Laws," some philosophers (notably Kripke and Putnam) have argued that, although all the laws of empirical science are discovered through empirical research, nonetheless many of those laws are not contingent but necessary. This issue is explored later in this commentary, in the section "Mellor's Defense of the Regularity Theory."

THE INFERENCE PROBLEM

The inference problem is that of explaining the "mustness," or necessity, that necessitarians believe to hold between the particular things or events that are instances of a law. For example, if necessitarians believe that it is a law that gold has an atomic number of 79 and that a particular piece of metal is gold, then they would infer that the piece of metal in question *must* have an atomic number of 79. As Dretske acknowledges, one of the main challenges for his universals theory of laws is to explain the "mustness" that appears in the conclusion of this kind of inference. On Dretske's theory, we can write out the inference as follows:

F-ness → G-ness
a is F
_____

a must be G.

Where does the *must* in the conclusion come from? Presumably, it derives from some *must* implicit in the first premise, which is Dretske's way of representing the law that all Fs are G. In ordinary English, we could state the law using *must* in two different ways. We could say either "It must be the case that, if anything is F, then it is G" or "Anything that is F must be G."

Clearly, the first way of stating the law will not help. From the premises "It must be the case that, if x is F, then x is G" and "a is F," we cannot validly infer the conclusion that a must be G. All that follows is that, as matter of contingent fact, a is G.[4]

So we have to read the law as Dretske intends it to be read, as saying "Anything that is F must be G." But as far as Dretske is concerned, this is just another way of saying that the property of being F necessitates the property of being G. In other words, according to Dretske's theory, the law thus stated is not a generalization about objects or events (which are particulars); it is a singular statement about properties (which are universals). To repeat, on Dretske's view, the law does not say "Each individual thing that is F must also be G." Rather, it says that the property of F-ness necessitates, or brings with it, the property of G-ness. How, then, given Dretske's understanding of what the FG-law asserts, can we use it to deduce the conclusion that a particular thing must be G? As Dretske admits, the only inference that is uncontroversially valid is the inference from "F-ness → G-ness" to "All Fs are G." But "All Fs are G" does not validly imply that anything that is F must also be G. (Were this inference valid, the distinction between accidental generalizations and laws would collapse.)

In response to this difficulty, Dretske offers an analogy with the offices and branches of a government. There is a legal code in the United States that lays down what powers pertain to the office of the president, the two houses of Congress, and the Supreme Court, and how these branches of government are related to one another. The code itself is contingent; the Constitution of the United States could have been different. But given that Constitution, it is now true of anyone who holds the office of president that that person must consult Congress and receive its approval before declaring war. The law is not about the particular people who hold the various offices; the law is about the powers and duties of the offices themselves and the relations between them. But because the law is what it is, anyone who holds a particular office *must* behave in a certain way.

An analogy is not a proof, and Dretske does not claim to have proven anything. Some critics (such as van Fraassen) have stressed the disanalogies between a legal code and laws of nature. With a legal code, we understand the origin of the law's prescriptive force, namely the commitment of citizens to enforce it. But what is the origin of the analogously prescriptive force of laws of nature? Given the mystery that seems to surround the notion of contingent relations between universals, other advocates of the universals theory have candidly admitted that necessitation is an inexplicable basic concept that the theory is forced to postulate. The justification for accepting the concept lies in the superiority claimed for the universals theory as a whole in accounting for the essential features of laws.

MELLOR'S CRITICISM OF DRETSKE: CAN LAWS BE NONINSTANTIAL?

Recall the prima facie advantage discussed earlier that the universals theory might be thought to have over the regularity theory: that is, if universals can exist without being instantiated, then noninstantial laws can be accommodated easily by the universals theory but not so easily by the regularity theory. Mellor argues that this supposed advantage is illusory because the universals involved in laws must, after all, have instances. He reaches this conclusion by adopting the account of universals championed by the British mathematician and philosopher Frank Plumpton Ramsey (1903–30).

To appreciate the motivation for Ramsey's proposal, consider the traditional view that there are particulars—individual objects such as *this* apple and *that* apple—which exist in space and time—and universals—properties such as redness and greenness—which are neither spatial nor temporal. Somehow or other, on the traditional view, these two very different kinds of entity—particular objects and universal properties—combine to form facts, such as the fact that this apple is red. But how can this combination be possible given that particulars and universals are so utterly dissimilar? Taking his cue from Wittgenstein's *Tractatus*, Ramsey proposed that the world consists, not of particular things and universal properties, but of facts.[5] According to Ramsey, universals and particulars should be regarded, not as two fundamentally different sorts of thing each having an independent existence, but as mutually necessary parts of particular facts. On this view (which is a version of nominalism), particulars and universals alike are aspects of facts. For example, the particular denoted by *this apple* is what is common to all the particular facts about this apple, while the universal denoted by *red* is what is common to all the particular facts about red things.

Mellor agrees with Ramsey that if we make the mistake of thinking of universals and particulars as having some kind of independent real existence, then we will be led into such "dotty conundrums" (368) as worrying about how two such radically different kinds of thing can combine to form a fact. On Ramsey's view, universals are just the part that is common to all the facts of the relevant class. The relevant class of facts for the *FG*-law must include facts such as *Fa*. Therefore, *F*-ness, the universal in the *FG*-law, must have instances if the law is to be genuine. Genuine laws, on such an account of universals, cannot be noninstantial.

7.5 | Mellor's Defense of the Regularity Theory

In "Necessities and Universals in Natural Laws," Mellor defends the Humean regularity theory of laws. A key element of that theory is the insistence that laws of nature are contingent, not necessary, truths, and this insistence is one of the main differences between the regularity theory and traditional versions of the necessitarian theory. Until quite recently, it was thought that, whatever the defects of the regularity theory as an adequate account of laws, it was at least correct about the contingency of laws. And as we have seen, the conviction that laws of nature are contingent has proven to be a stumbling block for the newer, universals version of the necessitarian theory.[6]

The main arguments for the contingency of laws derive from Hume. Take any scientific law of the form "All *F*s are *G*." Even though the law is true, we can easily conceive that something could be *F* without that thing also being *G*. Since whatever is conceivable is possible, the *FG*-law could be false. Thus, it and all other similar laws are contingent. Moreover, laws of nature are discovered by empirical research, and this seems to be the only way we can find out which lawlike generalizations are true and which false. If laws of nature were necessary, then we should be able to discover them through a priori reasoning. But we cannot do this. Therefore, the laws of nature are contingent.

It is now widely acknowledged that Hume's arguments are unsound. Conceivability is not an infallible guide to possibility.[7] Moreover, as Saul Kripke has demonstrated, the fact that a proposition is a posteriori does not entail that it is contingent.[8] The classic example is the simple identity claim that Hesperus is (that is, is identical with) Phosphorus. The discovery that the two names *Hesperus* and *Phosphorus* refer to one and the same physical object (sometimes seen around sunset, at other times seen around sunrise) was empirical. No amount of a priori reasoning about the meanings of the names *Hesperus* and *Phosphorus* could have revealed this truth to us. Nonetheless, the fact that the object picked out by the name *Hesperus* (namely, the planet Venus) is identical with the object picked out by the name *Phosphorus* (namely, the planet Venus) is a necessary truth, since every object is necessarily identical with itself.

In understanding Kripke's view about the necessity of identity claims, it is important to appreciate the distinction between metaphysical necessity and logical necessity. As with logical necessity, it is impossible for a metaphysically necessary proposition to be false (that is, there is no possible world in which such a proposition is false). But unlike a logically necessary proposition, its necessary truth is not guaranteed solely by logic and definitions. The proposition that Hesperus is identical with Phosphorus is metaphysically necessary, but not logically necessary.

Kripke's case for the necessity of identity claims rests on his theory of reference, specifically his theory of rigid designation. Obviously, if it were possible for the names *Hesperus* and *Phosphorus* to refer to different planets, then the assertion that Hesperus is identical with Phosphorus would not be necessary. By saying that the names *Hesperus* and *Phosphorus* are rigid designators, Kripke is denying that this is possible. There is not space here to consider Kripke's ingenious (and, to many minds, convincing) arguments for his thesis that simple names such as *Hesperus* and *Phosphorus* are rigid designators.[9] Suffice it to say that part of the plausibility of his thesis rests on the fact that names such as *Hesperus* and *Phosphorus* are indeed simple names with no connotative meaning. So it is tempting to think that their reference, the things they refer to by virtue of some initial baptism, completely exhausts whatever meaning they have.

Of immediate concern to us is whether something like Kripke's view can be extended to laws of nature. The problem is that only a few scientific laws make identity claims, and when they do, those claims concern classes of objects and properties, rather than single objects designated by simple names. Saul Kripke and Hilary Putnam have argued that, as with the identity of Hesperus and Phosphorus, a significant class of laws of nature are metaphysically necessary, namely, those laws attributing essential properties to natural kinds. It is these arguments that Mellor tries to refute on behalf of the regularity theory.

KRIPKE AND PUTNAM ON NATURAL KINDS AND ESSENCES

Natural kinds have traditionally been thought to include such things as chemical elements and compounds like sodium and hydrochloric acid, and biological species like tigers and elm trees. The basic idea is that each member of such classes shares a common nature in virtue of which it belongs to that relevant kind. The intended contrast is with artificial kinds or groups, such as all animals weighing over fifty pounds or all compounds whose chemical name in English begins with the letter A, where there is nothing else that the items in these groups need have in common. Essential properties are those properties of a thing that it cannot exist without—or, perhaps, properties that a thing of some kind cannot lack while remaining of that kind. If the essential properties of tigers are P_1, P_2, and P_3, then nothing that lacks one or more of these properties can belong to the natural kind *tiger*. Thus, the doctrine of natural kinds and the notion of essential properties go hand in hand: natural kinds are classes of things sharing a common nature or core set of essential properties.

Kripke and Putnam usually take the essential properties of things to be their microstructural properties. They say, for example, that being H_2O is an essential property of water; having atomic number 79 is an essential property of gold; having a particular set of genes is essential for membership in a biological species. Consider the example of gold. Undoubtedly, modern scientists believe that having atomic number 79 is a fundamental property of gold. According to Kripke and Putnam, gold is, by its very nature, something that has 79 protons in its nucleus and 79 orbiting electrons. It is possible that some pieces of gold may not be hard, shiny, or yellow, but it is impossible that any piece of gold—any piece of that very element—could fail to have atomic number 79. So according to Kripke and Putnam, the law that gold has an atomic number of 79 is a necessary truth: given the essential nature of gold, it could not be otherwise.[10] Of course, the discovery of the law is empirical, but it does not follow from this that the law is contingent; statements can be necessary and yet a posteriori. Consider our earlier example that Hesperus is identical with Phosphorus. Given what the names *Hesperus* and *Phosphorus* designate (namely, the planet Venus), the identity could not fail to hold, since one thing could not be two distinct things; nonetheless, it was an empirical discovery that *Hesperus* and *Phosphorus* name one and the same planet.

Mellor rejects the Kripke-Putnam doctrine of natural kinds and essences, charging that their arguments are both unsound and question begging. Since Kripke and Putnam give different arguments and hold slightly different versions of the essentialist theory, we will focus exclusively on Putnam's account in what follows.

Putnam's task is to explain why the extension of natural-kind terms such as *water*, *gold*, and *tiger* must include all and only those things that have the same essential properties. Let us focus on the term *water*. The traditional theory of meaning, stemming from John Locke (1632–1704) and developed by Gottlob Frege (1848–1925), maintains that extensions are determined by intensions: the intension (or meaning) of *water* is a list of properties that define what we mean by *water*. Since people were talking meaningfully about water long before the advent of modern atomic theory, the intension of *water* does not include "being H_2O." Presumably, *water* means something like "the colorless, odorless liquid that fills lakes and rivers, falls from the sky, and is the most common solvent." According to this traditional theory of meaning, the term *water* need not refer to the same substance, with the same microstructural properties, in all possible worlds. The most common solvent that is odorless and colorless and fills the lakes and rivers of a possible world might be something other than water—say, XYZ. So Putnam has to give us a new theory of reference in which natural kind terms, such as *water*, *gold*, and *tiger*, do not have their extensions fixed by their intensions.

Putnam's New Theory of Reference

Putnam's new theory of reference comes in two parts. First, he contends that a natural kind term gets its reference fixed in the actual world by means of archetypal specimens and not by mere description or intension. In this picture, we fix the reference of *water* in our world by pointing to samples of that particular liquid and (for all practical purposes) saying "Water is *this* kind of stuff, the stuff in our lakes and streams, the stuff that falls from the sky when it rains." The key here is that we fix the reference by employing the relation of *same stuff* or *same kind*—a deep relation among all those things sharing, not merely superficial characteristics, but important microstructural features. (In the present case, these important properties are taken to be the molecular properties of the compound H_2O.) We thus collectively refer to all and only liquids standing in that same-kind relation to our particular sample.

Second, Putnam claims that we can then extend the reference of *water* to other possible worlds by saying (for all practical purposes) "Something is water, in any possible world, if and only if it is the same kind of stuff as this sample here." On such a view, then, nothing—in any world—can be water unless it stands in the same-kind relation to this liquid here in our glass or pitcher. That is to say, *water* refers to a single kind of stuff in every possible world: necessarily, whatever is water is H_2O. In this way, it emerges as a necessarily true law of nature that if x is water, then x contains hydrogen and oxygen.

MELLOR'S CRITICISMS OF PUTNAM'S THEORY OF REFERENCE

Mellor rejects both parts of Putnam's account. First, he denies that terms like *water* and *gold* get their reference fixed in this world by means of archetypes (rather than, say, by means of descriptions). One group of cases that Mellor thinks refutes Putnam's view are those in which there are no archetypes to point at when the term is first introduced. In an earlier paper (entitled "Natural Kinds"), he writes:

> Consider elements high in the periodic table, that do not occur in nature and have never been made. We have names for them, but there may never be archetypes to constrain our use of the names. Even if specimens eventually appear, the discovery, creation or synthesis of previously unknown fundamental particles, elements and compounds can surely be *predicted*. The term 'neutrino' applied to just the same particles when it was used to predict their existence as it has applied to since their discovery. Ostensive reference (say to a bubble chamber photograph) could not have fixed its extension then; why suppose exactly the same extension is fixed that way now?[11]

Second, Mellor thinks that Putnam's analysis of the same-kind relation across possible worlds begs the question. The issue, remember, is whether the term *water* must refer to exactly the same kind of stuff, with exactly the same essential nature, in all possible worlds. Putnam's procedure relies on "important" physical properties. But there is nothing in Putnam's procedure per se that guarantees that these important properties will be exactly the same in every possible world. Suppose, for example, that all samples of water share ten important properties but that water could lack any one of them.[12] Now imagine two possible worlds, PW1 and PW2. A liquid in PW1 shares nine of these important properties with water in the actual world. Let the property missing in PW1 be P_1. A liquid in PW2 also shares nine of these properties with water in the actual world, but in this case the missing property is P_2. When compared with water in the actual world, the liquid in PW1 and the liquid in PW2 both qualify as water. But the two liquids would not seem to be essentially the same, since they differ in two important properties. So Putnam's procedure for fixing the reference of *water* in other possible worlds does not entail essentialism; it does not guarantee that all the things referred to by the term *water* must have exactly the same essential properties in all possible worlds.

Putnam would reject Mellor's criticism because Putnam assumes that the same-kind relation across possible worlds is an equivalence relation and all equivalence relations are transitive. If the same-kind relation is transitive, then if the liquid in PW1 is the same kind of stuff as water in the actual world and if water in the actual world is the same kind of stuff as the liquid in PW2, then the liquid in PW1 has to be the same kind of stuff as the liquid in PW2. Thus, Putnam's assumption that the same-kind relation is an equivalence relation entails that all things that belong to the same kind have to share exactly the same set of important properties.

Mellor complains that this move of Putnam's begs the question because it illicitly takes for granted the very point at issue, namely, whether water has a fixed essence. In Mellor's words, "for Putnam to claim the same-kind relation to be transitive, which he does in taking it to be an equivalence relation, is for him gratuitously to assume the essentialist conclusion he is out to prove" (864).

Mellor's criticism seems to be correct. Putnam's new theory of reference does not, by itself, guarantee the truth of essentialism. Essentialism requires that the same-kind relation be an equivalence relation and hence transitive across possible worlds. Putnam's theory of reference does not, by itself, entail that the same-kind relation is transitive because, according to that theory, things in other possible worlds count as water by comparing them with archetypes in the actual world, not by comparing them with each other. Insisting that the same-kind relation be an equivalence relation entails that all possible things that are water must have the same fundamental properties as archetypes in the actual world. But that is simply to insist on the truth of essentialism, not to give an independent argument for it. Mellor's second criticism does not prove that Putnam's theory is false, but it does leave Putnam the task of responding to Mellor's first criticism (that archetypes do not seem necessary for fixing the reference of natural kind terms). And even if Putnam were to succeed in rebutting Mellor here, we would still be a long way from the conclusion that all laws of nature are necessary truths, for Putnam's theory applies only to natural kinds (such as molecules, animals, and plants), and it is not obvious that all laws of nature are about such kinds. (For example, consider the laws of thermodynamics and the laws of motion. None of these laws is about any specific class or category of objects having a common nature. They apply to all objects and systems *regardless* of their nature.)

7.6 | Cartwright's Antirealism about Fundamental Laws

It is noteworthy that at the end of his article, "Laws of Nature," Dretske limits himself to a conditional claim. He does not purport to have shown that there actually *are* universals and contingent relations of nomic necessitation between them. Rather, he asserts, *if* there are any laws of nature, then the universals account of them is correct. This raises the question that Cartwright addresses in "Do the Laws of Physics State the Facts?" Namely, are there laws of nature? Undoubtedly there are many generalizations in science, and many of these are called laws. But can any of these so-called laws play the role traditionally assigned to them, especially that of explanation, while at the same time being true? Do our explanatory laws truly describe how bodies actually behave? Cartwright thinks not.

Cartwright defends the following disjunction: either laws are false but can be used to explain things, or laws are true but are useless for explanation. There is, in her words, "a trade off between factual content and explanatory power" (881). Thus, Cartwright admits that some laws—those referred to as phenomenological—can be fairly accurate descriptions of how bodies actually behave. But these phenomenological laws achieve their descriptive accuracy at the price of being highly qualified and thus highly restricted in their scope. It is the unqualified, fundamental laws of wide explanatory scope that are the target of Cartwright's antirealism. (For an extended discussion of antirealism, see chapter 9 of this volume.)

Much of Cartwright's case against fundamental laws rests on her illustration involving the laws of gravitational attraction and electrostatic attraction (and repulsion). If the gravitational law were true, then it would describe how bodies behave. In particular the law would predict the real, actual forces that act on gravitating bodies. But nearly all gravitating bodies are also electrically charged, and the smaller the body, the greater the role that charge plays in determining its behavior. Thus, taken at face value, the gravitational law seems false; it does not state correctly the actual net force acting on all gravitating bodies because the actual net force experienced by most bodies depends jointly on their mass and electrical charge.

One way of trying to retain the gravitational law as a true description of how bodies actually behave would be to limit its scope to just those bodies on which only gravitation is acting—to bodies that are not charged (or affected in any way by nongravitational forces). But that would drastically limit the number of bodies that the law describes and would render the law virtually useless for most explanatory purposes. Thus, Cartwright concludes either the laws we use for explanation are false, or the laws are true but virtually useless for explanation: we cannot be realists about explanatory, fundamental laws.

A Response to Cartwright's Antirealist Argument

One way of responding to Cartwright's argument for antirealism about fundamental laws is to distinguish between the actual net force acting on a body and the component force due to gravity. Indeed, this is how most of us are taught physics and mechanics. First we use individual laws to calculate the component forces acting on a body due to gravitation, electricity, tension in a spring, and so on. Then we use vector addition (the parallelogram law) to sum these forces and derive the net force acting on the body. If the net force is not zero, then the body will accelerate in accordance with Newton's second law of motion.

Cartwright denies that this response is an adequate defense of realism about fundamental laws. Her main objection is that only the net force, the actual force that determines the acceleration of a body, is real. Nature does not add component forces. Indeed, component forces are fictitious; they are not real forces at all. Why not? Because if they were real, then they would act in addition to the net force and thus give the wrong prediction about the body's motion.

Cartwright's double-counting objection to the reality of component forces seems to depend on the assumption that real forces produce actual accelerations and are measured by those accelerations. Thus, Cartwright would say that if a body were pulled in opposite directions by equal component forces and thus remained at rest, then there is no real force acting at all. If this is indeed her view, then it seems mistaken. After all, a ball on which equal and opposite forces are acting is in a different state (a state of tension) from a ball that is free from external forces. Component forces can have real effects, even if they do not produce a net acceleration.[13]

One merit of Cartwright's paper is that it makes clear the price one must pay to be a realist about many laws in physics. Individual laws describe component forces, but component forces do not determine how bodies move (only net forces do that). So individual laws do not describe how bodies actually move. At best they describe, not actual behavior, but tendencies to behave. They specify capacities or dispositions of bodies by telling us how they would move if they were free from all other forces. And this, clearly, is a far cry from the traditional empiricist view of Hume and his followers, who would limit laws to describing what actually happens in the world. Taking laws seriously and realistically seems to require an ontology of powers and dispositions that is inconsistent with the regularity theory of laws.[14]

7.7 | Summary

Most philosophers attempting to explain what laws are adopt one of two approaches: the regularity approach or the necessitarian approach. Inspired by Hume's analysis of causation, regularity theorists (such as Ayer) insist that laws are simply descriptions or summaries of what actually happens in the world. In its simplest form, the regularity theory insists that the law that all Fs are G says nothing more about the world than does the generalization that, as a matter of fact, all Fs are G. This simple version of the regularity theory has a number of problems, such as the problem of accounting for those scientific laws that, like Newton's first law, have no instances. But the severest objection to the simple version of the regularity theory is that it cannot distinguish between genuine laws and accidental generalizations. In response to this difficulty, many regularity theorists impose further conditions on a true generalization before it can qualify as a law. Because these extra conditions refer to the beliefs and attitudes of scientists towards the generalization in question, this more sophisticated version of the regularity theory is often called the epistemic regularity theory. Ayer advocates one version of the epistemic regularity theory. But as Ayer acknowledges, the epistemic regularity theory makes laws inherently subjective by taking the lawlike status of a generalization to depend on whether scientists treat it as a law. Obviously, scientists cannot treat a generalization as a law if they have not yet discovered it. So the epistemic regularity theory entails that there are no unknown laws. This conclusion, that there are no unknown laws, is so counterintuitive that Dretske offers it as a sufficient reason for abandoning the entire regularity approach.

Dretske advocates a version of the necessitarian approach. Unlike older versions of this approach (which regarded laws as asserting a special relation of necessity between objects or events), Dretske's theory regards laws as relations between properties. Since properties are universals, Dretske's proposal is called the *universals theory*. One problem with the universals theory concerns the special relation that is supposed to hold between the universals that make up a law. Usually we think of the relations between universals as being logical relations. But if the necessitation relation were logical, then laws of nature would be necessary truths. Dretske himself rejects this, insisting that laws of nature are contingent. Against this view of the contingency of laws, some philosophers, notably Kripke and Putnam, have defended the claim that a wide class of natural laws, namely those that attribute essential properties to natural kinds, are metaphysically necessary. The case for the Kripke-Putnam doctrine rests largely on a new theory about how terms that refer to natural kinds get their meaning. Mellor, a regularity theorist, criticizes Putnam's new theory of reference on the grounds that, as a defense of essentialism, it is question begging. If Mellor is right, then Putnam has not given a compelling reason for regarding many laws of nature as metaphysically necessary. But this still leaves necessitarians such as Dretske with the task of explaining how the special nonlogical relation characteristic of laws can hold contingently between universals. This difficulty—the problem of identification—is closely related to a second difficulty—the problem of inference. The problem of inference is the problem of explaining how, according to the necessitarian, we can validly infer that a particular thing that is F must also be G (a conclusion about a particular object) from the law that F-ness necessitates G-ness (a premise about universals). Dretske offers an analogy between laws of nature and legal codes (such as the Constitution of the United States) that define the powers and relations between the branches of government. The laws are about offices and institutions, but they imply that the people who hold those offices or serve in those institutions must do certain things. Dretske admits that the analogy is imperfect and confesses that the necessitation relation is not easy to understand or explain. Nevertheless, he thinks that the regularity theory is so flawed that the universals theory must be on the right track, despite its difficulties.

One major difference between the regularity theory and older versions of the necessitarian theory is that the necessitarian thinks that the regularity theory is too weak. The regularity theorist regards laws merely as true descriptions of how objects actually behave. The necessitarian insists that laws do more than this: they not only describe how the world is, they also assert how the world must be. But both theorists agree that a generalization must be true in order for it to be a law. Nancy Cartwright challenges this assumption. Cartwright argues that many of the laws we use to explain things are, in fact, false because they do not describe what actually happens in the world. Laws about electrostatic, magnetic, and gravitational forces, for example, do not as a rule describe how bodies actually move. Rather, they specify how bodies would move were certain ideal conditions realized. But such conditions, as a matter of fact, hardly ever obtain. Although Cartwright's argument involves some controversial claims about the nonreality of component forces, one of her conclusions seems quite plausible, namely, that many laws specify the tendencies and dispositions of bodies rather than their actual behavior. Thus, neither the regularity approach nor the older necessitarian view can be deemed adequate, since both approaches entail that laws describe how bodies do or must behave. Universals theories such as Dretske's are not inconsistent with Cartwright's position but much more needs to be said by their supporters about the nature of the two ingredients in a law—the universals (properties) and the (contingent) relation of necessitation that holds between them—and about the relation between laws and the behavior of particular objects.

■ | Notes

1. See David Hume, "Of the Idea of Necessary Connection," Section 7 of *An Enquiry Concerning Human Understanding* (1748).

2. See David Lewis, *Counterfactuals* (Oxford: Blackwell, 1973), 72–77. Lewis quotes from Ramsey's 1928 paper "Universals of Law and Fact," in which Ramsey proposed the view (which he later rejected) that laws are "consequences of those propositions which we would take as axioms if we knew everything and organized it as simply as possible in a deductive system." Lewis refines this notion as follows: "a contingent generalization is a *law of nature* if and only if it appears as a theorem (or axiom) in each of the true deductive systems that achieves a best combination of simplicity and strength. A generalization is a law at a world *i*, likewise, if and only if it appears as a theorem in each of the best deductive systems true at *i*." Lewis's refinement emphasizes the contingency of laws: which generalizations are laws depends on the possible world under consideration. It also covers what happens if (with respect to a particular possible world) two or more deductive systems tie for first place. In that case a generalization is a law at that world only if it appears in every one of the "best" systems for that world. If, for a particular world, there are several best systems having no generalizations in common, then there are no laws in that world.

3. Michael Tooley, "The Nature of Laws," *Canadian Journal of Philosophy* 7 (1977): 667–98; John Carroll, *Laws of Nature* (Cambridge: Cambridge University Press, 1994).

4. It is important not to be misled by language here. Rather than saying (i) "It must be the case that, if Jones was elected, then he received the most votes," we might say, more idiomatically, (ii) "If Jones was elected, then he must have received the most votes." But if we add to (ii) that (iii) "Jones was indeed elected," we cannot conclude from this that "Jones *must have*—that is, *necessarily*—received the most votes." Since, in this example, (ii) is merely an equivalent way of saying (i), all that validly follows from (ii) and (iii) is that, as a matter of contingent fact, Jones received the most votes. So, too, in Dretske's example. All that follows from "It must be the case that, if *x* is *F*, then *x* is *G*" and "*a* is *F*" is that, as a matter of contingent fact, *a* is *G*, not that *a* must be *G*.

5. See the opening propositions of Ludwig Wittgenstein, *Tractatus Logico-Philosophicus*, trans. D. F. Pears and B. F. McGuinness (London: Routledge and Kegan Paul, 1961), p. 7.

6. For a lively attack by a fellow necessitarian on the contention (by Dretske, Armstrong, and others) that nomic necessitation between the universals of a law of nature is always contingent, see Martin Tweedale, "Universals and Laws of Nature," *Philosophical Topics* 13 (1982): 25–44.

7. Part of the difficulty with Hume's conceivability test is its vagueness. For it is one thing to conceive that something is water without also conceiving that it contains oxygen. Anyone who is ignorant of chemistry could do this. But can I conceive that something is water and that it does not contain oxygen? The problem is the unclarity surrounding the notion that I could properly be said to be conceiving of water—real, actual water—when I imagine a liquid that looks like water but does not contain oxygen.

8. See Saul A. Kripke, *Naming and Necessity* (Cambridge, Mass.: Harvard University Press, 1972).

9. For a good introduction see Kripke, *Naming and Necessity*; and Stephen P. Schwartz, ed., *Naming, Necessity, and Natural Kinds* (Ithaca, N.Y.: Cornell University Press, 1977).

10. A word of caution: Kripke and Putnam recognize that our best scientific theories might be mistaken. So, strictly speaking, their claim is conditional: if modern science is right about the nature of gold, then the law that gold has an atomic number of 79 is a necessary truth.

11. D. H. Mellor, "Natural Kinds," *British Journal for the Philosophy of Science* 28 (1977): 306.

12. The example and argument are from Mellor's "Natural Kinds." If Mellor's example seems implausible, consider biological species. Surely, not all tigers have exactly the same set of genes; but, presumably, tigers share a sufficiently large number of genes (or very similar genes) to be members of the same species. Similarly, we know that the stuff we usually call *water* is in fact a mixture of three compounds—hydrogen oxide, deuterium oxide, and tritium oxide—and although hydrogen oxide is by far the most common of the three in the actual world, it is logically possible that the ratio might vary. Moreover, it is at least imaginable that advances in elementary particle physics might reveal that not even all molecules of hydrogen oxide have exactly the same internal structure.

13. For more on Cartwright's double-counting objection to the reality of component forces, see Lewis Creary, "Causal Explanation and the Reality of Natural Component Forces," *Pacific Philosophical Quarterly* 62 (1981): 148–57; A. David Kline and Carl A. Matheson, "How the Laws of Physics Don't Even Fib," in *PSA 1986*, ed. A. Fine and P. Machamer (East Lansing, Mich.: Philosophy of Science Association, 1986), 1: 33–41; and Jessica Wilson, "The Causal Argument Against Component Forces," *Dialetica* 63 (2009): 525-54. Cartwright replies to Creary in Nancy Cartwright, *How the Laws of Physics Lie* (Oxford: Clarendon Press, 1983), 62–67.

14. Cartwright develops such a view, which she traces back to John Stuart Mill, in Nancy Cartwright, *Nature's Capacities and Their Measurement* (Oxford: Clarendon Press, 1989). There has been a debate about whether Cartwright's realism about capacities (which includes such things as dispositions, powers, and tendencies) is consistent with her antirealism about fundamental laws (which she continues to maintain). See Alan Chalmers, "So the Laws of Physics Needn't Lie," *Australasian Journal of Philosophy* 71 (1993): 196–205; Steve Clarke, "The Lies Remain the Same: A Reply to Chalmers," *Australasian Journal of Philosophy* 73 (1995): 152–55; and Alan Chalmers, "Cartwright on Fundamental Laws: A Response to Clarke," *Australasian Journal of Philosophy* 74 (1996): 150–52.

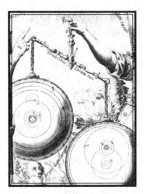

8 | Intertheoretic Reduction

INTRODUCTION

Reduction is a perennial theme in philosophy, where it crops up regularly in fields as diverse as epistemology, metaphysics, the philosophy of mind, and the philosophy of mathematics. But reduction has a special significance for the philosophy of science because several philosophers (notably Ernest Nagel) have claimed that reduction holds the key to understanding an important kind of progress in science.

According to Nagel and others, when scientists discover a new theory that is able to reduce or absorb one or more already well-established and accepted theories, then we have a paradigm instance of objective, cumulative progress. This account of scientific progress is sometimes called the *Russian doll model*; for just as each doll contains smaller dolls inside it, so, too, each scientific theory includes the verified content of the earlier theories it has absorbed. In this way, science is supposed to display a steady increase in empirical content and explanatory power. In the nineteenth century, for example, classical thermodynamics, with its distinctive concepts of heat, entropy, and temperature, became integrated into—was absorbed by—statistical mechanics, a new and more powerful branch of physics applying probability theory to molecular motion. In a similar way, twentieth-century molecular biology has managed to incorporate most of classical genetics, largely through the Watson-Crick discovery of the structure of DNA.

In "Issues in the Logic of Reductive Explanations," Nagel distinguishes between two sorts of intertheoretic reduction—homogeneous and inhomogeneous—and specifies the formal conditions that each must satisfy. Nagel views reduction as a type of explanation and accepts Hempel's deductive-nomological model of scientific explanation. (See chapter 6 for details of Hempel's model.) Thus, Nagel requires that in both homogeneous and inhomogeneous reduction the reduced theory be logically derivable from the reducing theory. In inhomogeneous reductions, there are terms in the reduced theory that are absent from the reducing theory. Consequently, these reductions also need *rules of correspondence* or *bridge laws* to make possible the derivation that Nagel requires. Another important constraint on reduction in Nagel's model is that the meanings of theoretical terms not change when reduction occurs.

Nagel's formal model of reduction has not gone unchallenged. Principal among its critics has been Paul Feyerabend, whose attack on Nagel is conveyed in his paper, "How to Be a Good Empiricist—a Plea for Tolerance in Matters Epistemological." Much of Feyerabend's case against Nagel rests on Feyerabend's controversial holist thesis that the meaning of any theoretical term always depends on the theory in which it occurs.

Obviously there is more to science than just physics. The so-called special sciences include biology, chemistry, earth and atmospheric science, and a wide range of behavioral and social sciences (such as psychology, anthropology, and economics). In "Special Sciences, or: the Disunity of Science as a Working Hypothesis," Jerry Fodor argues that these sciences are not reducible to physics, even in principle. They are autonomous: they have their own laws and provide legitimate explanations at a higher level than that of fundamental particles and processes in terms of legitimate macro-level kinds that are multiply realizable. Nonetheless, Fodor argues, this autonomy is consistent with token physicalism, the view that all events are physical.

Many philosophers, even those who reject Feyerabend's views about meaning, agree that Nagel's account of reduction is flawed because it does not reflect what actually goes on in science.

In "1953 and All That: A Tale of Two Sciences," Philip Kitcher casts a critical eye on the claim, oft-repeated, that classical genetics has been reduced to molecular biology. Kitcher argues that modern biology fails to match any of the available models of reduction. Instead, he advocates a new way of thinking about scientific theories in order to understand the relations between a theory and its successor. Echoing some of the views of Thomas Kuhn, Kitcher recommends that theories should be understood in terms of practices, problems, and patterns of reasoning. Only then, he thinks, can we do justice to the complexity of the relationship between theories in modern biology.

Ernest Nagel

Issues in the Logic of Reductive Explanations

A recurrent theme in the long history of philosophical reflection on science is the contrast—voiced in many ways by poets and scientists as well as philosophers—between the characteristics commonly attributed to things on the basis of everyday encounters with them, and the accounts of those things given by scientific theories that formulate some ostensibly pervasive executive order of nature. This was voiced as early as Democritus, when he declared that while things are customarily said to be sweet or bitter, warm or cold, of one color rather than another, in truth there are only the atoms and the void. The same contrast was implicit in Galileo's distinction, widely accepted by subsequent thinkers, between the primary and secondary qualities of bodies. It was dramatically stated by Sir Arthur Eddington in terms of currently held ideas in physics, when he asked which of the two tables at which he was seated was "really there"—the solid, substantial table of familiar experience, or the insubstantial scientific table which is composed of speeding electric charges and is therefore mostly "emptiness."*

From *Teleology Revisited* (New York: Columbia University Press, 1974), 95–113.

* See the introduction to Eddington's Gifford Lectures, published as *The Nature of the Physical World* (New York: Macmillan, 1928). Eddington (1882–1944) was Plumian Professor of Astronomy at Cambridge and made important contributions to relativity theory and astrophysics. In his popular writings, Eddington argued for the idealist conclusion that "the stuff of the world is mind-stuff" partly on the grounds that modern physics (relativity and quantum theory) has refuted materialism. For example, Eddington claimed that modern science had shown that the first of his two tables—the solid, substantial table of everyday life—is a fiction and that what really exists is the scientific table, which is mostly empty space occupied by electrons and other insubstantial particles. Beginning with Lizzie Susan Stebbing (1885–1943), philosophers have been critical of Eddington's attempt to derive idealist conclusions from the success of reductionist theories in modern physics. See L. S. Stebbing, *Philosophy and the Physicists* (London: Methuen, 1937). For some related criticisms, see "NOA's Ark—Fine for Realism," (in chapter 9), where Alan Musgrave diagnoses Eddington's mistake as the error of supposing that to explain

Formulations of the contrast vary and have different overtones. In some cases, as in the examples I have cited, the contrast is associated with a distinction between what is allegedly only "appearance" and what is "reality"; and there have been thinkers who have denied that so-called "common sense" deals with ultimate reality, just as there have been thinkers who have denied that the statements of theoretical science do so. However, a wholesale distinction between appearance and reality has never been clearly drawn, especially since these terms have been so frequently used to single out matters that happen to be regarded as important or valuable; nor have the historical controversies over what is to count as real and what as appearance thrown much light on how scientific theories are related to the familiar materials that are usually the points of departure for scientific inquiry. In any case, the contrast between the more familiar and manifest traits of things and those which scientific theory attributes to them need not be, and often is not, associated with the distinction between the real and the apparent; and in point of fact, most current philosophies of science, which in one way or another occupy themselves with this contrast, make little if any use of that distinction in their analyses.

But despite important differences in the ways in which the contrast has been formulated, I believe they share a common feature and can be construed as being addressed to a common problem. They express the recognition that certain relations of dependence between one set of distinctive traits of a given subject matter are allegedly explained by, and in some sense "reduced" to, assumptions concerning more inclusive relations of dependence between traits or processes not distinctive of (or unique to) that subject matter. They implicitly raise the question of what, in fact, is the logical structure of such reductive explanations—whether they differ from other sorts of scientific explanation, what is achieved by reductions, and under what conditions they are feasible. These questions are important for the understanding of modern science, for its development is marked by strong reductive tendencies, some of whose outstanding achievements are often counted as examples of reduction. For example, as a consequence of this reductive process, the theory of heat is commonly said to be but a branch of Newtonian mechanics, physical optics a branch of electromagnetic theory, and chemical laws a branch of quantum mechanics. Moreover, many biological processes have been given physicochemical explanations, and there is a continuing debate as to the possibility of giving such explanations for the entire domain of biological phenomena. There have been repeated though still unsuccessful attempts to exhibit various patterns of men's social behavior as examples of psychological laws.

It is with some of the issues that have emerged in proposed analyses of reductive explanations that this paper is concerned. I will first set out in

---

something (such as the solidity of table) is to explain it away (that is, to show that it does not really exist).

broad outlines what I believe is the general structure of such explanations; then examine some difficulties that have recently been raised against this account. . . .

■ | 1

Although the term "reduction" has come to be widely used in philosophical discussions of science, it has no standard definition. It is therefore not surprising that the term encompasses several sorts of things which need to be distinguished. But before I do this, a brief terminological excursion is desirable. Scientists and philosophers often talk of deducing or inferring one phenomenon from another (e.g., of deducing a planet's orbital motion), of explaining events or their concatenations (e.g., of explaining the occurrence of rainbows), and of reducing certain processes, things, or their properties to others (e.g., of reducing the process of heat conduction to molecular motions). However, these locutions are elliptical, and sometimes lead to misconceptions and confusions. For strictly speaking, it is not phenomena which are deduced from other phenomena, but rather *statements* about phenomena from other statements. This is obvious if we remind ourselves that a given phenomenon can be subsumed under a variety of distinct descriptions, and that phenomena make no assertions or claims. Consequently, until the traits or relations of a phenomenon which are to be discussed are indicated, and predications about them are formulated, it is literally impossible to make any deductions from them. The same holds true for the locutions of explaining or reducing phenomena. I will therefore avoid these elliptic modes of speech hereafter, and talk instead of deducing, explaining, or reducing statements about some subject matter.

Whatever else may be said about reductions in science, it is safe to say that they are commonly taken to be explanations, and I will so regard them. In consequence, I will assume that, like scientific explanations in general, every reduction can be construed as a series of statements, one of which is the conclusion (or the statement which is being reduced), while the others are the premises or reducing statements. Accordingly, reductions can be conveniently classified into two major types: homogeneous reductions, in which all of the "descriptive" or specific subject matter terms in the conclusion are either present in the premises also or can be explicitly defined using only terms that are present; and inhomogeneous reductions, in which at least one descriptive term in the conclusion neither occurs in the premises nor is definable by those that do occur in them. I will now characterize in a general way what I believe to be the main components and the logical structure of these two types of reduction, but will also state and comment upon some of the issues that have been raised by this account of reduction.

A frequently cited example of homogeneous reduction is the explanation by Newtonian mechanics and gravitational theory of various special laws concerning the motions of bodies, including Galileo's law for freely falling bodies near the earth's surface and the Keplerian laws of planetary motion. The explanation is homogeneous, because on the face of it at any rate, the terms occurring in these laws (e.g., distance, time, and acceleration) are also found in the Newtonian theory. Moreover, the explanation is commonly felt to be a reduction of those laws, in part because these laws deal with the motions of bodies in restricted regions of space which had traditionally been regarded as essentially dissimilar from motions elsewhere (e.g., terrestrial as contrasted with celestial motions), while Newtonian theory ignores this traditional classification of spatial regions and incorporates the laws into a unified system. In any event, the reduced statements in this and other standard examples of homogeneous reduction are commonly held to be deduced logically from the reducing premises. In consequence, if the examples can be taken as typical, the formal structure of homogenous reductions is in general that of deductive explanations. Accordingly, if reductions of this type are indeed deductions from theories whose range of application is far more comprehensive and diversified than that of the conclusions derived from them, homogeneous reductions appear to be entirely unproblematic, and to be simply dramatic illustrations of the well understood procedure of deriving theorems from assumed axioms.

However, the assumption that homogeneous reductions are deductive explanations has been recently challenged by a number of thinkers, on the ground that even in the stock illustrations of such reductions the reduced statements do not in general follow from the explanatory premises. For example, while Galileo's law asserts that the acceleration of a freely falling body near the earth's surface is constant, Newtonian theory entails that the acceleration is not constant, but varies with the distance of the falling body from the earth's center of mass. Accordingly, even though the Newtonian conclusion may be "experimentally indistinguishable" from Galileo's law, the latter is in fact "inconsistent" with Newtonian theory. Since it is this theory rather than Galileo's law that was accepted as sound, Galileo's law was therefore *replaced* by a different law for freely falling bodies, namely the law derived from the Newtonian assumptions. A similar outcome holds for Kepler's third planetary law. The general thesis has therefore been advanced that homogeneous reductions do not consist in the deduction or explanation of laws, but in the total *replacement* of incorrect assumptions by radically new ones which are believed to be more correct and precise than those they replace. This thesis raises far-reaching issues, and I will examine some of them presently. But for the moment I will confine my comments on it to questions bearing directly on homogeneous reductions.

a) It is undoubtedly the case that the laws derivable from Newtonian theory do not coincide exactly with some of the previously entertained hypotheses about the motions of bodies, though in other cases there may be such coincidence. This is to be expected. For it is a widely recognized function of comprehensive theories (such as the Newtonian one) to specify the conditions under which antecedently established regularities hold, and to indicate, in the light of those conditions, the modifications that may have to be made in the initial hypotheses, especially if the range of application of the hypotheses is enlarged. Nevertheless, the initial hypotheses may be reasonably close approximations to the consequences entailed by the comprehensive theory, as is indeed the case with Galileo's law as well as with Kepler's third law. (Incidentally, when Newtonian theory is applied to the motions of just two bodies, the first and second Keplerian laws agree fully with the Newtonian conclusions). But if this is so, it is correct to say that in homogeneous reductions the reduced laws are either derivable from the explanatory premises, or are good approximations to the laws derivable from the latter.

b) Moreover, it is pertinent to note that in actual scientific practice, the derivation of laws from theories usually involves simplifications and approximations of various kinds, so that even the laws which are allegedly entailed by a theory are in general only approximations to what is strictly entailed by it. For example, in deriving the law for the period of a simple pendulum, the following approximative assumptions are made; the weight of the pendulum is taken to be concentrated in the suspended bob; the gravitational force acting on the bob is assumed to be constant, despite variations in the distance of the bob from the earth's center during the pendulum's oscillation; and since the angle through which the pendulum may oscillate is stipulated to be small, the magnitude of the angle is equated to the sine of the angle. The familiar law that the period of a pendulum is proportional to the square root of its length divided by the constant of acceleration is therefore derivable from Newtonian theory only if these various approximations are taken for granted.

More generally, though no statistical data are available to support the claim, there are relatively few deductions from the mathematically formulated theories of modern physics in which analogous approximations are not made, so that many if not all the laws commonly said by scientists to be deducible from some theory are not strictly entailed by it. It would nevertheless be an exaggeration to assert that in consequence scientists are fundamentally mistaken in claiming to have made such deductions. It is obviously important to note the assumptions, including those concerning approximations, under which the deduction of a law is made. But it does not follow that given those assumptions a purported law cannot count as a consequence of some theory. Nor does it follow that if in a proposed homogeneous reduction the allegedly reduced law is only an approximation to what is actually entailed by the reducing theory when *no* approximative assumptions are added to the latter, the law has not been reduced but is being replaced by a radically different one.

c) Something must also be said about those cases of homogeneous reduction in which the law actually derivable from the reducing theory makes use of concepts not employed in the law to be reduced. Thus, while according to Kepler's third (or harmonic) law, the squares of the periods of the planets are to each other as the cubes of their mean distances from the sun, the Newtonian conclusion is that this ratio is not constant for all the planets but varies with their *masses*. But the notion of mass was introduced into mechanics by Newton, and does not appear in the Keplerian law; and although the masses of the planets are small in comparison with the mass of the sun, and the Keplerian harmonic law is therefore a close approximation to the Newtonian one, the two cannot be equated. Nevertheless, while the two are not equivalent, neither are they radically disparate in content or meaning. On the contrary, the Newtonian law identifies a causal factor in the motions of the planets which was unknown to Kepler.

■ | 2

I must now turn to the second major type of reductive explanations. Inhomogeneous reductions, perhaps more frequently than homogeneous ones, have occasioned vigorous controversy among scientists as well as philosophers concerning the cognitive status, interpretation, and function of scientific theories; the relations between the various theoretical entities postulated by these theories, and the familiar things of common experience; and the valid scope of different modes of scientific analysis. These issues are interconnected, and impinge in one way or another upon questions about the general structure of inhomogeneous reductions. Since none of the proposed answers to these issues has gained universal assent, the nature of such reductions is still under continuing debate.

Although there are many examples of inhomogeneous reductions in the history of science, they vary in the degree of completeness with which the reduction has been effected. In some instances, all the assumed laws in one branch of inquiry are apparently explained in terms of a theory initially developed for a different class of phenomena; in others, the reduction has been only partial, though the hope of completely reducing the totality of laws in a given area of inquiry to some allegedly "basic" theory may continue to inspire research. Among the most frequently cited illustrations of such relatively complete inhomogeneous reductions are the explanation of thermal laws by the kinetic theory of matter, the reduction of physical optics to electromagnetic theory, and the explanation (at least in principle) of chemical laws in terms of quantum theory. On the other hand, while some processes occurring in living organisms can now be understood in terms of physicochemical theory, the reducibility of all biological laws in a similar manner is still a much disputed question.

In any case, the logical structure of inhomogeneous reductive explanations is far less clear and is more difficult to analyze than is the case with homogeneous reductions. The difficulty stems largely from the circumstance that in the former there are (by definition) terms or concepts in the reduced laws (e.g., the notion of heat in thermodynamics, the term "light wave" in optics, or the concept of valence in chemistry) which are absent from the reducing theories. Accordingly, if the overall structure of the explanation of laws is taken to be that of a deductive argument, it seems impossible to construe inhomogeneous reductions as involving essentially little more than the logical derivation of the reduced laws (even when qualifications about the approximative character of the latter are made) from their explanatory premises. If inhomogeneous reductions are to be subsumed under the general pattern of scientific explanations, it is clear that additional assumptions must be introduced as to how the concepts characteristically employed in the reduced laws, but not present in the reducing theory, are connected with the concepts that do occur in the latter.

Three broad types of proposals for the structure of inhomogeneous reductions can be found in the recent literature of the philosophy of science. The first, which for convenience will be called the "instrumentalist" analysis, is usually advocated by thinkers who deny a cognitive status to scientific laws or theories, regarding them as neither true nor false but as rules (or "inference tickets") for inferring so-called "observation statements" (statements about particular events or occurrences capable of being "observed" in some not precisely defined sense) from other such statements. According to this view, for example, the kinetic theory of gases is not construed as an account of the composition of gases. It is taken to be a complex set of rules for predicting, among other things, what the pressure of a given volume of gas will be if its temperature is kept constant but its volume is diminished. However, the scope of application of a given law or theory may be markedly more limited than the scope of another. The claim that a theory T (e.g., the corpus of rules known as thermodynamics) is reduced to another theory T' (e.g., the kinetic theory of gases) would therefore be interpreted as saying that all the observation statements which can be derived from given data with the help of T can also be derived with the help of T', but not conversely. Accordingly, the question to which this account of inhomogeneous reduction is addressed is not the ostensibly asserted content of the theories involved in reduction, but the comparative ranges of observable phenomena to which two theories are applicable.

Although this proposed analysis calls attention to an important function of theories and provides a rationale for the reduction of theories, its adequacy depends on the plausibility of uniformly interpreting general statements in science as rules of inference. Many scientists certainly do not subscribe to such an interpretation, for they frequently talk of laws as true and as providing at least an approximately correct account of various relations of dependence among things. In particular, this interpretation precludes the explanation of macro-states of objects in terms of unobservable microprocesses postulated by a theory. Moreover, the proposal is incomplete in a number of ways: it has nothing to say about how theoretical terms in laws (e.g., "electron" or even "atom") may be used in connection with matters of observation, or just how theories employing such notions operate as rules of inference; and it ignores the question of how, if at all, the concepts of a reduced theory are related to those of the reducing one, or in what way statements about a variety of observable things may fall within the scope of both theories. In consequence, even if the proposed analysis is adequate for a limited class of reductive explanations, it does not do justice to important features characterizing many others.

The second proposed analysis of inhomogeneous reductions (hereafter to be referred to—perhaps misleadingly—as the "correspondence" proposal) is also based on several assumptions. One of them is that the terms occurring in the conclusion but not in the premises of a reduction have "meanings" (i.e., uses and applications) which are determined by the procedures and definitions of the discipline to which reduced laws initially belong, and can be understood without reference to the ideas involved in the theories to which the laws have been reduced. For example, the term "entropy" as used in thermodynamics is defined independently of the notions characterizing statistical mechanics. Furthermore, the assumption is made that many subject matter terms common to both the reduced and reducing theories— in particular, the so-called observation terms employed by both of them to record the outcome of observation and experiment—are defined by procedures which can be specified independently of these theories and, in consequence, have "meanings" that are neutral with respect to the differences between the theories. For example, the terms "pressure" and "volume change" which occur in both thermodynamics and the kinetic theory of gases are used in the two theories in essentially the same sense. It is important to note, however, that this assumption is compatible with the view that even observation terms are "theory impregnated," so that such terms are not simply labels for "bare sense-data," but predicate characteristics that are not immediately manifest and are defined on the basis of various theoretical commitments. For example, if the expression "having a diameter of five inches" is counted as an observation predicate, its application to a given object implicitly involves commitment to some theory of spatial measurement as well as to some laws concerning the instrument used in making the measurement. Accordingly, the point of the assumption is not that there are subject-matter terms whose meanings or uses are independent of *all* theories, but rather that every such term has a meaning which is fixed by *some* theory but independent of others. A third assumption underlying the correspondence analysis of inhomogeneous reductions is that, like homogeneous reduction, and with similar qualifications referring to approximations, they embody the pattern of deductive explanations.

In view of these assumptions, it is clear that if a law (or theory) T is to be reduced to a theory T′ not containing terms occurring in T, T′ must be supplemented by what have been called "rules of correspondence" or "bridge laws," which establish *connections* between the distinctive terms of T and certain terms (or combinations of terms) in T′. For example, since the second law of thermodynamics talks of the transfer of heat, this law cannot be deduced from classical mechanics, which does not contain the term "heat," unless the term is connected in some way with some complex of terms in mechanics. The statement of such a connection is a correspondence rule. However, because of the first of the above three assumptions, a correspondence rule cannot be construed as an explicit definition of a term distinctive of T, which would permit the elimination of the term on *purely logical grounds* in favor of the terms in T'. Thus, the notion of entropy as defined in thermodynamics can be understood and used without any reference to notions employed in theories about the microstructure of matter; and no amount of logical analysis of the concept of entropy can show the concept to be constituted out of the ideas employed in, say, statistical mechanics. If this is indeed the case (as I believe it is), then the theory T is not derivable from (and hence not reducible to) the theory T', although T may be derivable from T' when the latter is conjoined with an appropriate set of bridge laws.

What then is the status of the correspondence rules required for inhomogeneous reduction? Different articulations of the theories involved in a reduction, as well as different stages in the development of inquiry into the subject matter of the theories, may require different answers; but I will ignore these complications. In general, however, correspondence rules formulate *empirical hypotheses*—hypotheses which state certain relations of dependence between things mentioned in the reduced and reducing theories. The hypotheses are, for the most part, not testable by confronting them with observed instances of the relations they postulate. They are nevertheless not arbitrary stipulations, and as with many other scientific laws their factual validity must be assessed by comparing various consequences entailed by the system of hypotheses to which they belong with the outcome of controlled observations. However, bridge laws have various forms; and while no exhaustive classification of their structure is available, two sorts of bridge laws must be briefly described.

a) A term in a reduced law may be a predicate which refers to some distinctive *attribute* or characteristic of things (such as the property of having a certain temperature or of being red) that is not denoted by any of the predicates of the reducing theory. In this case the bridge law may specify the conditions, formulated in terms of the ideas and assumptions of the reducing theory, under which the attribute occurs. For example, the kinetic theory of gases formulates its laws in terms of such notions as molecule, mass, and velocity, but does not employ the thermodynamical notion of temperature. However, a familiar bridge law states that a gas has a certain temperature when the mean kinetic energy of its molecules has a certain magnitude. In some cases, bridge laws of the sort being considered may specify conditions for the occurrence of an attribute which are necessary as well as sufficient; in other cases the conditions specified may be sufficient without being necessary; and in still other cases, the conditions stated may only be necessary. In the latter case, however, laws involving the attribute will, in general, not be deducible from the proposed reducing theory. (Thus, though some of the necessary conditions for objects having colors can be stated in terms of ideas belonging to physical optics in its current form, the physiological equipment of organisms which must also be present for the occurrence of colors cannot be described in terms of those ideas. Accordingly, if there are any laws about color relations, they are not reducible to physical optics.)

In any case, such bridge laws are empirical hypotheses concerning the *extensions* of the predicates mentioned in these correspondence rules—that is, concerning the classes of individual things or processes designated by those predicates. An attribute of things connoted by a predicate in a reduced law may indeed be quite different from the attribute connoted by the predicates of the reducing theory; but the class of things possessing the former attribute may nevertheless coincide with (or be included in) the class of things which possess the property specified by a complex predicate in the reducing theory. For example, the statement that a liquid is viscous is not equivalent in meaning to the statement that there are certain frictional forces between the layers of molecules making up the liquid. But if the bridge laws connecting the macro-properties and the microstructure of liquids is correct, the extension of the predicate "viscous" coincides with (or is included in) the class of individual systems with that microstructure.

b) Let me now say something about a second sort of correspondence rule. Although much scientific inquiry is directed toward discovering the determining conditions under which various traits of things occur, some of its important achievements consist in showing that things and processes initially assumed to be distinct are in fact the same. A familiar example of such an achievement is the discovery that the Morning Star and the Evening Star are not different celestial objects but are identical. Similarly, although the term "molecule" designates one class of particles and the term "atom" designates another class, molecules are structures of atoms, and in particular a water molecule is an organization of hydrogen and oxygen atoms denoted by the formula "H_2O"; and accordingly, the extension of the predicate "water molecule" is the same as the class of things designated by the formula. Correspondence rules of the second sort establish analogous identifications between classes of individuals or "entities" (such as spatiotemporal objects, processes, and forces) designated by different predicates. An oft cited example of such rules is a bridge law involved in the reduction of physical optics to electromagnetic theory. Thus, prior to Maxwell, physicists postulated the existence of certain physical propagations designated as "light waves," while electromagnetic theory was developed on the assumption that there are electromagnetic waves. An essential step in the reduction of optics to electrodynamics was the introduction by Maxwell of the hypothesis (or bridge law) that these are not two *different* processes but a *single* one, even though electromagnetic waves are not always manifested as visible light. Analogous bridge laws are assumed when a flash of lightning is said to be a surge of electrically charged particles, or when the evaporation of a liquid is asserted to be the escape of molecules from its surface; and while the full details for formulating a similar bridge law are not yet available, the hope of discovering them underlies the claim that a biological cell is a complex organization of physicochemical particles.

Correspondence rules of the second kind thus differ from rules of the first, in that unlike the latter (which state conditions, often in terms of the ideas of a micro-theory, for the occurrence of traits characterizing various things, often macroscopic ones), they assert that certain logically nonequivalent expressions describe identical entities. Although both sorts of rules have a common function in reduction and both are in general empirical assumptions, failure to distinguish between them is perhaps one reason for the persistence of the mistaken belief that reductive explanations establish the "unreality" of those distinctive traits of things mentioned in reduced laws.

■ | 3

This account of inhomogeneous reduction has been challenged by a number of recent writers who have advanced an alternate theory which rejects the main assumptions of both the instrumentalist and the correspondence analyses, and which I will call the "replacement" view. Since I believe the correspondence account to be essentially correct, I shall examine the fundamental contention of the replacement thesis, as presented by Professor Paul Feyerabend, one of its most vigorous proponents.

Feyerabend's views on reduction rest upon the central (and, on the face of it, sound) assumption that "the meaning of every term we use depends upon the theoretical context in which it occurs."[1] This claim is made not only for "theoretical" terms like "neutrino" or "entropy" in explicitly formulated scientific theories, but also for expressions like "red" or "table" used to describe matters of common observation (i.e., for observation terms). Indeed, Feyerabend uses the word "theory" in a broad sense, to include such things as myths and political ideas.[2] He says explicitly that "even everyday languages, like languages of highly theoretical systems, have been introduced in order to give expression to some theory or point of view, and they therefore contain a well-developed and sometimes very abstract ontology."[3] "The description of every single fact," he declares, is "dependent on *some* theory."[4] He further maintains that "theories are meaningful independent of observations; observational statements are not meaningful unless they have been connected with theories."[5] There is, therefore, no "observation core," even in statements of perception, that is independent of theoretical interpretation,[6] so that strictly speaking each theory determines its own distinctive set of observation statements. And while he allows that two "low level" theories which fall within the conceptual framework of a comprehensive "background theory" may have a common interpretation for their observation statements, two "high level" theories concerning the nature of the basic elements of the universe "may not share a single observational statement."[7] It is therefore allegedly an error to suppose that the empirical adequacy of a theory can be tested by appeal to observation statements whose meanings are independent of the theory, and which are neutral as between that theory and some alternative competing theory. "The methodological unit to which we must refer when discussing questions of test and empirical context, is constituted by a *whole set of partly overlapping, factually adequate, but mutually inconsistent theories.*"[8]

Moreover, a change in a theory is accompanied by a change in the meanings of all its terms, so that theories constructed on "mutually inconsistent principles" are in fact "incommensurable."[9] Thus, if T is classical celestial mechanics, and T′ is the general theory of relativity, "the meanings of all descriptive terms of the two theories, primitive as well as defined terms, will be different," the theories are incommensurable, and "not a single descriptive term of T can be incorporated into T′."[10] In consequence, Feyerabend believes the correspondence account of inhomogeneous reduction is basically mistaken in supposing that allegedly reduced laws or theories can be derived from the reducing theory with the help of appropriate bridge laws:

> What happens . . . when transition is made from a theory T′ to a wider theory T (which, we shall assume, is capable of covering all the phenomena that have been covered by T′) is something much more radical than incorporation of the *unchanged* theory T′ (unchanged, that is, with respect to the meanings of its main descriptive terms as well as to the meanings of the terms of its observation language) into the context of T. What does happen is, rather, a *complete replacement* of the ontology (and perhaps even of the formalism) of T′ by the ontology (and the formalism) of T and a corresponding change of the meanings of the descriptive elements of the formalism of T′ (provided these elements and this formalism are still used). This replacement affects not only the theoretical terms of T′ but also at least some of the observational terms which occurred in its test statements. . . . In short: introducing a new theory involves changes of outlook both with respect to the observable and with respect to the unobservable features of the world, and corresponding changes in the meaning of even the most "fundamental" terms of the language employed.[11]*

Accordingly, if these various claims are warranted, there is not and cannot be any such thing as the reduction of laws or theories; and the examples often cited as instances of reduction are in fact instances of something else: the exclusion of previously accepted hypotheses from the corpus of alleged

---

* Unlike Nagel, Feyerabend uses *T′* to denote the reduced theory and *T* to denote the reducing theory.

scientific knowledge, and the substitution for them of incommensurably different ones.

But are these claims warranted? I do not believe they are. Feyerabend is patently sound in maintaining that no single statement or any of its constituent terms has a meaning in isolation, or independently of various rules or conventions governing its use. He is no less sound in noting that the meaning of a word may change when its range of application is altered. However, these familiar truisms do not support the major conclusion he draws from them. The presentation of his thesis suffers from a number of unclarities (such as what is to count as a change in a theory, or what are the criteria for changes in meaning), which cloud the precise import of some of his assertions. I shall, however, ignore these unclarities here[12] and will comment briefly only on two difficulties in Feyerabend's argument.

a) It is a major task of scientific inquiry to assess the adequacy of proposed laws to the "facts" of a subject matter as established by observation or experiment, and to ascertain whether the conclusions reached are consistent with one another. However, if two proposed theories for some given range of phenomena share no term with the same meaning in each of them, so that the theories have completely different meanings (as Feyerabend believes is commonly the case), it is not evident in what sense two such theories can be said to be either compatible or inconsistent with one another: for relations of logical opposition obtain only between statements whose terms have common meanings. Moreover, it is also difficult to understand how, if the content of observation statements is determined by the theory which is being tested (as Feyerabend maintains), those statements can serve as a basis for deciding between the theory and some alternative to it. For according to his analysis those observation statements will automatically corroborate the theory that happens to be used to interpret observational data, but will be simply irrelevant in assessing the empirical validity of an alternative theory. Theories thus appear to be self-certifying, and to be beyond the reach of criticism based on considerations that do not presuppose them. This outcome is reminiscent of Karl Mannheim's claim that truth in social matters is "historically relative": there are no universally valid analyses of social phenomena, since every such analysis is made within some distinctive social perspective which determines the meaning as well as the validity of what is said to be observed, so that those who do not share the same perspective can neither reach common conclusions about human affairs, nor significantly criticize each others' findings.*

* See Karl Mannheim, *Ideologie und Utopie* (Bonn: F. Cohen, 1929); *Ideology and Utopia*, trans. L. Wirth and E. Shils (New York: Harcourt Brace and Co., 1936). Mannheim (1893–1947) was the founding father of the sociology of knowledge, the discipline concerned with the social causes of beliefs as opposed to the reasons that people might have for them. As Nagel notes, Mannheim denied the possibility of objective "scientific" knowledge of social phenomena on the grounds that all judgments about society are relative to the historical period in which they are made. For criticism of a modern version of the sociology of scientific knowledge,

Feyerabend attempts to escape from such skeptical relativism by involving what he calls the "pragmatic theory of observation." In this theory, it is still the case that the meaning of an observation statement varies with the theory used to interpret observations. However, it is possible to describe the observational and predictive statements an investigator utters as *responses* to the situations which "prompt" the utterances, and to compare the order of these responses with the order of the physical situations that prompt them, so as to ascertain the agreements or disagreements between the two orders.[13] But if this account of the role of observation statements in testing theories is to outflank the relativism Feyerabend wants to avoid, the *secondary* statements (they are clearly observation statements) about the responses (or primary observation statements) of investigators cannot have meanings dependent on the theory being tested, and must be invariant to alternative theories. However, if secondary statements have this sort of neutrality, it is not evident why only such observation statements can have this privileged status.

b) Feyerabend has difficulties in providing a firm observational basis for objectively evaluating the empirical worth of proposed hypotheses. The difficulties stem from what I believe is his exaggerated view that the meaning of every term occurring in a theory or in its observation statements is wholly and uniquely determined by that theory, so that its meaning is radically changed when the theory is modified. For theories are not quite the monolithic structures he takes them to be—their component assumptions are, in general, logically independent of one another, and their terms have varying degrees of dependence on the theories into which they enter. Some terms may indeed be so deeply embedded in the totality of assumptions constituting a particular theory that they can be understood only within the framework of the theory: e.g., the meaning of "electron spin" appears to be inextricably intertwined with the characteristic ideas of quantum theory. On the other hand, there are also terms whose meanings seem to be invariant in a number of different theories: e.g., the term "electric charge" is used in currently accepted theories of atomic structure in the same sense as in the earlier theories of Rutherford and Bohr. Similar comments apply to observation terms, however these may be specified. Accordingly, although both "theoretical" and "observational" terms may be "theory laden," it does not follow that there can be no term in a theory which retains its meaning when it is transplanted into some other theory.

More generally, it is not clear how, on the replacement view of reduction, a theory T can be at the same time more inclusive than, and also have a meaning totally different from, the theory T′ it allegedly replaces—especially since according to Feyerabend the replacing theory will entail

---

see the discussion of the strong programme in Laudan's "Demystifying Underdetermination" in chapter 3 and in the accompanying commentary.

"that all the concepts of the preceding theory have extension zero, or . . . it introduces rules which cannot be interpreted as attributing specific properties to objects within already existing classes, but which change the system of classes itself."[14] Admittedly, some of the laws and concepts of the "wider theory" often differ from their opposite numbers in the earlier theory. But even in this case, the contrasted items may not be "incommensurable." Thus, the periodic table classifies chemical elements on the basis of certain patterns of similarity between the properties of the elements. The description (or theoretical explanation) of those properties has undergone important changes since the periodic table was first introduced by Mendeleev. Nevertheless, though the descriptions differ, the classification of the elements has remained fairly stable, so that fluorine, chlorine, bromine, and iodine, for example, continue to be included in the same class. The new theories used in formulating the classification certainly do not entail that the concepts of the preceding ones have zero extension. But it would be difficult to understand why this is so if, because of differences between the descriptions, the descriptions were totally disparate.

Consider, for example, the argument that thermodynamics is not reducible to statistical mechanics, on the ground that (among other reasons) entropy is a statistical notion in the latter theory but not in the former one: since the meaning of the word "entropy" differs in the two theories, entropy laws in statistical mechanics are not derivable from entropy laws in thermodynamics (and in fact are said to be incompatible). Admittedly, the connotation of the word "entropy" in each of the two theories is not identical; and if the correspondence account of reduction were to claim that they are the same, it would be patently mistaken. But the fact remains that the two theories deal with many phenomena common to both their ranges; and the question is how is this possible? In brief, the answer seems to be as follows. The word "entropy" in thermodynamics is so defined that its legitimate application is limited to physical systems satisfying certain specified conditions, e.g., to systems such as gases, whose internal motions are not too "tumultuous" (the word is Planck's), a condition which is not satisfied in the case of Brownian motions. These conditions are relaxed in the definition of "entropy" in statistical mechanics, so that the extension of the Boltzmann notion of entropy includes the extension of the Clausius notion. In consequence, despite differences in the connotations of the two definitions, the theories within which they are formulated have a domain of application in common, even though the class of systems for which thermodynamical laws are approximately valid is more restricted than is the class for the laws of statistical mechanics. But it is surely not the case that the latter theory implies that the Clausius definition of entropy has a zero extension or that the laws of thermodynamics are valid for no physical systems whatsoever.

This difficulty of the replacement view in explaining how the "wider" theory, which allegedly replaces a "narrower" one, may nevertheless have a domain of common application, does not arise in the correspondence account of reduction. For the bridge laws upon which the latter sets great store are empirical hypotheses, not logically true statements in virtue of the connotations of the terms contained in them. Bridge laws state what relations presumably obtain between the *extensions* of their terms, so that in favorable cases laws of the "narrower" theory (with suitable qualifications about their approximate character) can be deduced from the "wider" theory, and thereby make intelligible why the two theories may have a common field of application. Accordingly, although I will not pretend that the correspondence account of reduction is free from difficulties or that I have resolved all of them, on the whole it is a more adequate analysis than any available alternative to it. . . .

■ | Notes

1. Paul Feyerabend, "Problems of Empiricism," in R. G. Colodny, ed., *Beyond the Edge of Certainty* (Englewood Cliffs: Prentice-Hall, Inc., 1965), p. 180.

2. Paul Feyerabend, "Reply to Criticism," *Boston Studies in the Philosophy of Science* 2 (1962), p. 252.

3. Paul Feyerabend, "Explanation, Reduction, and Empiricism," *Minnesota Studies in the Philosophy of Science* 3 (1962), p. 76.

4. Feyerabend, "Problems of Empiricism," p. 175.

5. Ibid., p. 213.

6. Ibid., p. 216.

7. Ibid.

8. Ibid., p. 175.

9. Ibid., p. 227.

10. *Boston Studies in the Philosophy of Science* p. 231; cf. also Feyerabend, "On the 'Meaning' of Scientific Terms," *Journal of Philosophy* 62 (1965), p. 271.

11. Feyerabend, "Explanation, Reduction and Empiricism," pp. 28–9, 59.

12. Many of them are noted by Dudley Shapere in his "Meaning and Scientific Change," in R. G. Colodny, ed., *Mind and Cosmos* (Pittsburgh: University of Pittsburgh Press, 1966).

13. Feyerabend, "Problems of Empiricism," p. 21; and Feyerabend, "Explanation, Reduction, and Empiricism," p. 24.

14. Feyerabend, "On the 'Meaning' of Scientific Terms," *Journal of Philosophy* 62 (1965), p. 268.

Paul K. Feyerabend

How to Be a Good Empiricist—
A Plea for Tolerance in
Matters Epistemological

"Facts?" he repeated. "Take a drop more grog, Mr. Franklin, and you'll get over the weakness of believing in facts! Foul play, Sir!"
—Wilkie Collins, *The Moonstone* [Ch. 4]

1 | Contemporary Empiricism Liable to Lead to Establishment of a Dogmatic Metaphysics

Today empiricism is the professed philosophy of a good many intellectual enterprises. It is the core of the sciences, or so at least we are taught, for it is responsible both for the existence and for the growth of scientific knowledge. It has been adopted by influential schools in aesthetics, ethics, and theology. And within philosophy proper the empirical point of view has been elaborated in great detail and with even greater precision. This predilection for empiricism is due to the assumption that only a thoroughly observational procedure can exclude fanciful speculation and empty metaphysics as well as to the hope that an empiristic attitude is most liable to prevent stagnation and to further the progress of knowledge. It is the purpose of the present paper to show that empiricism in the form in which it is practiced today cannot fulfill this hope.

Putting it very briefly, it seems to me that the contemporary doctrine of empiricism has encountered difficulties, and has created contradictions which are very similar to the difficulties and contradictions inherent in some versions of the doctrine of democracy. The latter are a well-known phenomenon. That is, it is well known that essentially totalitarian measures

From Bernard Baumrin, ed., *Philosophy of Science, The Delaware Seminar*, vol. 2 (New York: Interscience Publishers, 1963), 3–39.

are often advertised as being a necessary consequence of democratic principles. Even worse—it not so rarely happens that the totalitarian character
of the defended measures is not explicitly stated but covered up by calling
them 'democratic,' the word 'democratic' now being used in a new, and
somewhat misleading, manner. This method of (conscious or unconscious)
verbal camouflage works so well that it has deceived some of the staunch-
est supporters of true democracy. What is not so well known is that modern empiricism is in precisely the same predicament. That is, some of the
methods of modern empiricism which are introduced in the spirit of anti-
dogmatism and progress are bound to lead to the establishment of a dogmatic metaphysics and to the construction of defense mechanisms which
make this metaphysics safe from refutation by experimental inquiry. It is
true that in the process of establishing such a metaphysics the words 'empirical' or 'experience' will frequently occur; but their sense will be as distorted
as was the sense of 'democratic' when used by some concealed defenders
of a new tyranny.[1] This, then, is my charge: Far from eliminating dogma
and metaphysics and thereby encouraging progress, modern empiricism
has found a new way of making dogma and metaphysics respectable, viz.,
the way of calling them 'well-confirmed theories,' and of developing a
method of confirmation in which experimental inquiry plays a large though
well controlled role. In this respect, modern empiricism is very different
indeed from the empiricism of Galileo, Faraday, and Einstein, though it
will of course try to represent these scientists as following its own paradigm
of research, thereby further confusing the issue.[2]

   From what has been said above it follows that the fight for tolerance in
scientific matters and the fight for scientific progress must still be carried on.
What has changed is the denomination of the enemies. They were priests,
or 'school-philosophers,' a few decades ago. Today they call themselves 'philosophers of science,' or 'logical empiricists.'[3] There are also a good many
scientists who work in the same direction. I maintain that all these groups
work against scientific progress. But whereas the former did so openly and
could be easily discerned, the latter proceed under the flag of progressivism
and empiricism and thereby deceive a good many of their followers. Hence,
although their presence is noticeable enough they may almost be compared
to a fifth column, the aim of which must be exposed in order that its detri-
mental effect be fully appreciated. It is the purpose of this paper to contribute to such an exposure.

   I shall also try to give a positive methodology for the empirical sciences
which no longer encourages dogmatic petrification in the name of experience. Put in a nutshell, the answer which this method gives to the question
in the title is: You can be a good empiricist only if you are prepared to work
with many alternative theories rather than with a single point of view and
'experience.' This plurality of theories must not be regarded as a preliminary stage of knowledge which will at some time in the future be replaced
by the One True Theory. Theoretical pluralism is assumed to be an *essential feature* of all knowledge that claims to be objective. Nor can one rest

content with a plurality which is merely abstract and which is created by denying now this and now that component of the dominant point of view. Alternatives must rather be developed in such detail that problems already 'solved' by the accepted theory can again be treated in a new and perhaps also more detailed manner. Such development will of course take time, and it will not be possible, for example, at once to construct alternatives to the present quantum theory which are comparable to its richness and sophistication. Still, it would be very unwise to bring the process to a standstill in the very beginning by the remark that some suggested new ideas are undeveloped, general, metaphysical. *It takes time to build a good theory* (a triviality that seems to have been forgotten by some defenders of the Copenhagen point of view of the quantum theory);* and it also takes time to develop an alternative to a good theory. The *function* of such concrete alternatives is, however, this: They provide means of criticizing the accepted theory in a manner which goes *beyond* the criticism provided by a comparison of that theory 'with the facts': however closely a theory seems to reflect the facts, however universal its use, and however necessary its existence seems to be to those speaking the corresponding idiom, its factual adequacy can be asserted only *after* it has been confronted with alternatives *whose invention and detailed development must therefore precede any final assertion of practical success and factual adequacy.* This, then, is the methodological justification of a plurality of *theories*: Such a plurality allows for a much sharper criticism of accepted ideas than does the comparison with a domain of 'facts' which are supposed to sit there independently of theoretical considerations. The function of unusual *metaphysical* ideas which are built up in a nondogmatic fashion and which are then developed in sufficient detail to give an (alternative) account even of the most common experimental and observational situations is defined accordingly: They play a decisive role in the criticism and in the development of what is generally believed and

---

* Feyerabend is referring to the instrumentalist interpretation of quantum mechanics advocated by Niels Bohr, Werner Heisenberg, Max Born, and other pioneers of quantum physics in the 1920s. Named after the city in which Bohr worked, the Copenhagen interpretation remains the orthodox textbook account of quantum mechanics to this day. Its elements include Bohr's doctrine of complementarity, the insistence that the wave function contains a complete description of reality, and a profound reluctance to say anything about the properties of quantum systems when they are not being observed or measured. Feyerabend was a lifelong critic of "the Copenhagen mafia" (as he sometimes referred to them). As Feyerabend explains later in this reading, he rejected as unwarranted dogma Heisenberg's insistence that, while limited by the uncertainty principle, classical physics must always be used to describe the result of any experiment. See Werner Heisenberg, *Physics and Philosophy* (New York: Harper and Row, 1958); David Bohm, *Causality and Chance in Modern Physics* (New York: Harper and Row, 1961); Dugald Murdoch, *Niels Bohr's Philosophy of Physics* (Cambridge: Cambridge University Press, 1987); and James T. Cushing, *Quantum Mechanics: Historical Contingency and the Copenhagen Hegemony* (Chicago: University of Chicago Press, 1994) for good accounts of the Copenhagen interpretation and alternatives to it.

'highly confirmed'; and they have therefore to be present at *any* stage of the development of our knowledge.[4] A science that is free from *metaphysics* is on the best way to become a *dogmatic* metaphysical system. So far the summary of the method I shall explain, and defend, in the present paper.

It is clear that this method still retains an essential element of *empiricism*: The decision between alternative theories is based upon *crucial experiments*. At the same time it must *restrict* the range of such experiments. Crucial experiments work well with theories of a low degree of generality whose principles do not touch the principles on which the ontology of the chosen observation language is based. They work well if such theories are compared with respect to a much more general background theory which provides a stable meaning for the observation sentences. However, this background theory, like any other theory, is itself in need of criticism. Criticism must use alternatives. Alternatives will be the more efficient the more radically they differ from the point of view to be investigated. It is bound to happen, then, that the alternatives do not share a single statement with the theories they criticize. Clearly, a crucial experiment is now impossible. It is impossible, not because the experimental device is too complex, or because the calculations leading to the experimental prediction are too difficult; it is impossible because there is no statement capable of expressing what emerges from the observation. This consequence, which severely restricts the domain of empirical discussion, cannot be circumvented by any of the methods which are currently in use and which all try to work with relatively stable observation languages. It indicates that the attempt to make empiricism a universal basis of all our factual knowledge cannot be carried out. The discussion of this situation is beyond the scope of the present paper.

On the whole, the paper is a concise summary of results which I have explained in a more detailed fashion in the following essays: "Explanation, Reduction, and Empiricism"; "Problems of Microphysics"; "Problems of Empiricism"; "Linguistic Philosophy and the Mind-Body Problem."[5] All the relevant acknowledgements can be found there. Let me only repeat here that my general outlook derives from the work of K. R. Popper (London) and David Bohm (London) and from my discussions with both. It was severely tested in discussion with my colleague, T. S. Kuhn (Berkeley). It was the latter's skillful defense of a scientific conservatism which triggered two papers, including the present one. Criticism by A. Naess (Oslo), D. Rynin (Berkeley), Roy Edgley (Bristol), and J. W. N. Watkins (London) have been responsible for certain changes I made in the final version.

2 | Two Conditions of Contemporary Empiricism

In this section I intend to give an outline of some assumptions of contemporary empiricism which have been widely accepted. It will be shown in the sections to follow that these apparently harmless assumptions which have been explicitly formulated by some logical empiricists, but which also seem to guide the work of a good many physicists, are bound to lead to exactly the results I have outlined above: dogmatic petrification and the establishment, on so-called 'empirical grounds,' of a rigid metaphysics.

One of the cornerstones of contemporary empiricism is its *theory of explanation*. This theory is an elaboration of some simple and very plausible ideas first proposed by Popper[6] and it may be introduced as follows: Let T and T' be two different scientific theories, T' the theory to be explained, or the explanandum, T the explaining theory, or the explanans. Explanation (of T') consists in the *derivation* of T' from T and initial conditions which specify the domain D' in which T' is applicable.* Prima facie, this demand of derivability seems to be a very natural one to make for "otherwise the explanans would not constitute adequate grounds for the explanation".[7] It implies two things: first, that the consequences of a satisfactory explanans, T, inside D' must be compatible with the explanandum, T'; and secondly, that the main descriptive terms of these consequences must either coincide, with respect to their meanings, with the main descriptive terms of T', or at least they must be related to them via an empirical hypothesis. The latter result can also be formulated by saying that the meaning of T' must be unaffected by the explanation. "It is of the utmost importance," writes Professor Nagel,[8] emphasizing this point, "that the expressions peculiar to a science will possess meanings that are fixed by its *own* procedures, and are therefore intelligible in terms of its own rules of usage, whether or not the science has been, or will be [explained in terms of] the other discipline."

Now if we take it for granted that more general theories are always introduced with the purpose of explaining the existent successful theories, then every new theory will have to satisfy the two conditions just mentioned. Or, to state it in a more explicit manner,

1 only such theories are then admissible in a given domain which either *contain* the theories already used in this domain, or which are at least *consistent* with them inside the domain[9]; and

2 meanings will have to be invariant with respect to scientific progress; that is, all future theories will have to be phrased in such a manner that their use in explanations does not affect what is said by the theories, or factual reports to be explained.

These two conditions I shall call the *consistency condition* and the *condition of meaning invariance*, respectively.

---

* Throughout this paper, Feyerabend uses T to stand for the reducing (explaining) theory, and T' for the theory that is reduced (explained). This is exactly the opposite of Nagel's notation: for Nagel it is T' that reduces (explains) T.

Both conditions are *restrictive* conditions and therefore bound profoundly to influence the growth of knowledge. I shall soon show that the development of actual science very often violates them and that it violates them in exactly those places where one would be inclined to perceive a tremendous progress of knowledge. I shall also show that neither condition can be justified from the point of view of a tolerant empiricism. However, before doing so I would like to mention that both conditions have occasionally entered the domain of the sciences and have been used here in attacks against new developments and even in the process of theory construction itself. Especially today, they play a very important role in the construction as well as in the defense of certain points of view in microphysics.

Taking first an earlier example, we find that in his *Wärmelehre*, Ernst Mach[10] makes the following remark:

> Considering that there is, in a purely mechanical system of absolutely elastic atoms no real analogue for the *increase of entropy*, one can hardly suppress the idea that a violation of the second law . . . should be possible if such a mechanical system were the *real* basis of thermodynamic processes.

And referring to the fact that the second law is a highly confirmed physical law, he insinuates (in his *Zwei Aufsaetze*[11]) that for this reason the mechanical hypothesis must not be taken too seriously. There were many similar objections against the kinetic theory of heat.[12] More recently, Max Born has based his arguments against the possibility of a return to determinism upon the consistency condition and the assumption which we shall here take for granted, that wave mechanics is incompatible with determinism.

> If any future theory should be deterministic it cannot be a modification of the present one, but must be entirely different. How this should be possible without sacrificing a whole treasure of well established results [i.e., without contradicting highly confirmed physical laws and thereby violating the consistency condition] I leave the determinist to worry about.[13]

Most members of the so-called Copenhagen school of quantum theory would argue in a similar manner. For them the idea of complementarity and the formalism of quantization expressing this idea do not contain any hypothetical element as they are "uniquely determined by the facts."[14] Any theory which contradicts this idea is factually inadequate and must be removed. Conversely, an explanation of the idea of complementarity is acceptable only if it either contains this idea, or is at least consistent with it. This is how the consistency condition is used in arguments against theories such as those of Bohm, de Broglie, and Vigier.[15]

The use of the consistency condition is not restricted to such general remarks, however. A decisive part of the existing quantum theory *itself*, viz., the projection postulate,[16] is the result of the attempt to give an account of the definiteness of macro objects and macro events that is in accordance with the consistency condition. The influence of the condition of meaning invariance goes even further.

> The Copenhagen interpretation of the quantum theory [writes Heisenberg[17]] starts from a paradox. Any experiment in physics, whether it refers to the phenomena of daily life or to atomic events, is to be described in the terms of classical physics. . . . *We cannot and should not replace these concepts by any others* [my italics]. Still the application of these concepts is limited by the relations of uncertainty. We must keep in mind this limited range of applicability of the classical concepts while using them, but we cannot and should not try to improve them.

This means that the meaning of the classical terms must remain invariant with respect to any future explanation of microphenomena. Microtheories have to be formulated in such a manner that this invariance is guaranteed. The principle of correspondence and the formalism of quantization connected with it were explicitly devised for satisfying this demand. Altogether, the quantum theory seems to be the first theory after the downfall of the Aristotelian physics that has been quite explicitly constructed with an eye both on the consistency condition and the condition of (empirical) meaning invariance. In this respect it is very different indeed from, say, relativity which violates both consistency and meaning invariance with respect to earlier theories. Most of the arguments used for the defense of its customary interpretation also depend on the validity of these two conditions and they will collapse with their removal. An examination of these conditions is therefore very topical and bound deeply to affect present controversies in microphysics. I shall start this investigation by showing that some of the most interesting developments of physical theory in the past have violated both conditions.

3 | These Conditions Not Invariably Accepted by Actual Science

The case of the consistency condition can be dealt with in a few words: it is well known (and has also been shown in great detail by Duhem[18]) that Newton's theory is inconsistent with Galileo's law of the free fall and with Kepler's laws;* that statistical thermodynamics is inconsistent with the second law of the phenomenological theory; that wave optics is inconsistent with geometrical optics; and so on. Note that what is being asserted here is *logical* inconsistency; it may well be that the differences of prediction are

---

* See section 4 of the reading "Physical Theory and Experiment" in chapter 3 and the section "Duhem's Critique of Inductivism: The Attack on Newtonian Method" in the accompanying commentary.

too small to be detectable by experiment. Note also that what is being asserted is not the inconsistency of, say, Newton's theory and Galileo's law, but rather the inconsistency of *some consequences* of Newton's theory in the domain of validity of Galileo's law, and Galileo's law. In this last case the situation is especially clear. Galileo's law asserts that the acceleration of the free fall is a constant, whereas application of Newton's theory to the surface of the earth gives an acceleration that is not a constant but *decreases* (although imperceptibly) with the distance from the center of the earth. Conclusion: If actual scientific procedure is to be the measure of method, then the consistency condition is inadequate.

The case of meaning invariance requires a little more argument, not because it is intrinsically more difficult, but because it seems to be much more closely connected with deep-rooted prejudices. Assume that an explanation is required, in terms of the special theory of relativity, of the classical conservation of mass in all reactions in a closed system S. If m', m'', m''', . . . , m^i, . . . are the masses of the parts P', P'', P''', . . . , P^i, . . . of S, then what we want is an explanation of

$$\Sigma m^i = \text{const.} \tag{1}$$

for all reactions inside S. We see at once that the consistency condition cannot be fulfilled: According to special relativity Σm^i will vary with the velocities of the parts relative to the coordinate system in which the observations are carried out, and the total mass of S will also depend on the relative potential energies of the parts. However, if the velocities and the mutual forces are not too large, then the variation of Σm^i predicted by relativity will be so small as to be undetectable by experiment. Now let us turn to the *meanings* of the terms in the relativistic law and in the corresponding classical law. The first indication of a possible change of meaning may be seen in the fact that in the classical case the mass of an aggregate of parts equals the sum of the masses of the parts:

$$M(\Sigma P^i) = \Sigma M(P^i).$$

This is not valid in the case of relativity where the relative velocities and the relative potential energies contribute to the mass balance. That the relativistic concept and the classical concept of mass are very different indeed becomes clear if we also consider that the former is a *relation*, involving relative velocities, between an object and a coordinate system, whereas the latter is a *property* of the object itself and independent of its behavior in coordinate systems.* True, there have been attempts to give a relational analysis even of the classical concept (Mach). None of these attempts, however,

---

* In claiming that relativity theory treats mass as a relation, Feyerabend is assuming a definition of "relativistic mass" according to which a particle's mass depends on

leads to the relativistic idea with its velocity dependence on the coordinate system, which idea must therefore be added even to a *relational* account of classical mass. The attempt to identify the classical mass with the relativistic rest mass is of no avail either. For although both may have the same numerical value, the one is still dependent on the coordinate system chosen (in which it is at rest and has that specific value), whereas the other is not so dependent. We have to conclude, then, that $(m)_c$ and $(m)_r$ mean very different things and that $(\Sigma m^i)_c =$ const. and $(\Sigma m^i)_r =$ const. are very different assertions. This being the case, the derivation from relativity of either equation (1) or of a law that makes slightly different quantitative predictions with Σm^i used in the classical manner, will be possible only if a further premise is added which establishes a relation between the $(m)_c$ and the $(m)_r$. Such a 'bridge law'—and this is a major point in Nagel's theory of reduction—is a hypothesis

> according to which the occurrence of the properties designated by some expression in the premises of the [explanans] is a sufficient, or a necessary and sufficient condition for the occurrence of the properties designated by the expressions of the [explanandum].[19]

Applied to the present case this would mean the following: Under certain conditions the occurrence of relativistic mass of a given magnitude is accompanied by the occurrence of classical mass of a corresponding magnitude; this assertion is inconsistent with another part of the explanans, viz., the theory of relativity. After all, this theory asserts that there are no invariants which are directly connected with mass measurements and it thereby asserts that '$(m)_c$' does not express real features of physical systems. Thus we inevitably arrive at the conclusion that mass conservation cannot be explained in terms of relativity (or 'reduced' to relativity) without a violation of meaning invariance. And if one retorts, as has been done by some critics of the ideas expressed in the present paper,[20] that meaning invariance is an essential part of both reduction and explanation, then the answer will simply be that equation (1) can neither be explained by, nor reduced to relativity. Whatever the *words* used for describing the situation, the *fact* remains that actual science does not observe the requirement of meaning invariance.

This argument is quite general and is independent of whether the terms whose meaning is under investigation are observable or not. It is therefore stronger than may seem at first sight. There are some empiricists who would admit that the meaning of theoretical terms may be changed in the course of scientific progress. However, not many people are prepared to extend meaning *variance* to observational terms also. The idea motivating this

---

its velocity relative to an inertial frame. For further discussion, see the commentary at the end of this chapter.

attitude is, roughly, that the meaning of observational terms is uniquely determined by the procedures of observation such as looking, listening, and the like. These procedures remain unaffected by theoretical advance.[21] Hence, observational meanings, too, remain unaffected by theoretical advance. What is overlooked, here, is that the 'logic' of the observational terms is not exhausted by the procedures which are connected with their application 'on the basis of observation.' As will turn out later, it also depends on the more general ideas that determine the 'ontology' (in Quine's sense) of our discourse. These general ideas may change without any change of observational procedures being implied. For example, we may change our ideas about the nature, or the ontological status (property, relation, object, process, etc.) of the color of a self-luminescent object without changing the methods of ascertaining that color (looking, for example). Clearly, such a change is bound profoundly to influence the meanings of our observational terms.

All this has a decisive bearing upon some contemporary ideas concerning the interpretation of scientific theories. According to these ideas, theoretical terms receive their meanings via correspondence rules which connect them with an observational language *that has been fixed in advance* and independently of the structure of the theory to be interpreted. Now, our above analysis would seem to show that *if we interpret scientific theories in the manner accepted by the scientific community*, then most of these correspondence rules will be either false, or nonsensical. They will be *false* if they *assert* the existence of entities denied by the theory; they will be *nonsensical* if they *presuppose* this existence. Turning the argument around, we can also say that the attempt to interpret the calculus of some theory that has been voided of the meaning assigned to it by the scientific community with the help of the double language system, will lead to a very different theory. Let us again take the theory of relativity as an example: It can be safely assumed that the physical thing language of Carnap, and any similar language that has been suggested as an observation language, is not Lorentz-invariant. The attempt to interpret the *calculus* of relativity on *its* basis therefore cannot lead to the *theory* of relativity as it was understood by Einstein. What we shall obtain will be at the very most *Lorentz's interpretation* with its inherent asymmetries. This undesirable result cannot be evaded by the *demand* to use a different and more adequate observation language. The double language system assumes that theories which are not connected with some observation language do not possess an interpretation. The demand assumes that they do, and asks to choose the observation language most suited to it. It reverses the relation between theory and experience that is characteristic for the double language method of interpretation, which means, it gives up this method. Contemporary empiricism, therefore, has not led to any satisfactory account of the meanings of scientific theories.[22]

What we have shown so far is that the two conditions of Section 2 are frequently violated in the course of scientific practice and especially at periods of scientific revolution. This is not yet a very strong argument. True: There are empirically inclined philosophers who have derived some satisfaction from the assumption that they only make explicit what is implicitly contained in scientific practice. It is therefore quite important to show that scientific practice is not what it is supposed to be by them. Also, strict adherence to meaning invariance and consistency would have made impossible some very decisive advances in physical theory such as the advance from the physics of Aristotle to the physics of Galileo and Newton. However, how do we know (independently of the fact that they do exist, have a certain structure, and are very influential—a circumstance that will have great weight with opportunists only[23]) that the sciences are a desirable phenomenon, that they contribute to the advancement of knowledge, and that their analysis will therefore lead to reasonable methodological demands? And did it not emerge in the last section that meaning invariance and the consistency condition *are* adopted by some scientists? Actual scientific practice, therefore, cannot be our last authority. We have to find out whether consistency and meaning invariance are *desirable* conditions and this quite independently of who accepts and praises them and how many Nobel prizes have been won with their help.[24] Such an investigation will be carried out in the next sections.

4 | Inherent Unreasonableness of Consistency Condition

Prima facie, the case of the consistency condition can be dealt with in very few words. Consider for that purpose a theory T' that successfully describes the situation in the domain D'. From this we can infer (a) that T' agrees with a *finite* number of observations (let their class be F); and (b) that it agrees with these observations inside a margin M of error only.[25] Any alternative that contradicts T' outside F and inside M is supported by exactly the same observations and therefore acceptable if T' was acceptable (we shall assume that F are the only observations available). The consistency condition is much less tolerant. It eliminates a theory not because it is in disagreement with the *facts*; it eliminates it because it is in disagreement with *another theory*, with a theory, moreover, whose confirming instances it shares. *It thereby makes the as yet untested part of that theory a measure of validity.* The only difference between such a measure and a more recent theory is age and familiarity. Had the younger theory been there first, then the consistency condition would have worked in its favor. In this respect the effect of the consistency condition is rather similar to the effect of the more traditional methods of transcendental deduction, analysis of essences, phenomenological analysis, linguistic analysis. It contributes to the preservation of the old and familiar not because of any inherent advantage in it—for example, not because it has a better foundation in observation than has the newly suggested alternative, or because it is more elegant—but just because it is old and familiar. This is not the only instance where on closer inspection a rather surprising similarity emerges between modern empiricism and some of the school philosophies it attacks.

Now it seems to me that these brief considerations, although leading to an interesting *tactical* criticism of the consistency condition, do not yet go to the heart of the matter. They show that an alternative of the accepted point of view which shares its confirming instances cannot be *eliminated* by factual reasoning. They do not show that such an alternative is *acceptable*; and even less do they show that it *should be used*. It is bad enough, so a defender of the consistency condition might point out, that the accepted point of view does not possess full empirical support. Adding new theories *of an equally unsatisfactory character* will not improve the situation; nor is there much sense in trying to *replace* the accepted theories by some of their possible alternatives. Such replacement will be no easy matter. A new formalism may have to be learned and familiar problems may have to be calculated in a new way. Textbooks must be rewritten, university curricula readjusted, experimental results reinterpreted. And what will be the result of all the effort? Another theory which, from an empirical point of view, has no advantage whatever over and above the theory it replaces. The only real improvement, so the defender of the consistency condition will continue, derives from the *addition of new facts*. Such new facts will either support the current theories, or they will force us to modify them by indicating precisely where they go wrong. In both cases they will precipitate real progress and not only arbitrary change. The proper procedure must therefore consist in the confrontation of the accepted point of view with as many relevant facts as possible. The exclusion of alternatives is then required for reasons of expediency: Their invention not only does not help, but it even hinders progress by absorbing time and manpower that could be devoted to better things. And the function of the consistency condition lies precisely in this. It eliminates such fruitless discussion and it forces the scientist to concentrate on the facts which, after all, are the only acceptable judges of a theory. This is how the practicing scientist will defend his concentration on a single theory to the exclusion of all empirically possible alternatives.[26]

It is worthwhile repeating the reasonable core of this argument: Theories should not be changed unless there are pressing reasons for doing so. The only pressing reason for changing a theory is disagreement with facts. Discussion of incompatible facts will therefore lead to progress. Discussion of incompatible alternatives will not. Hence, it is sound procedure to increase the number of relevant facts. It is not sound procedure to increase the number of factually adequate, but incompatible alternatives. One might wish to add that formal improvements such as increase of elegance, simplicity, generality, and coherence should not be excluded. But once these improvements have been carried out, the collection of facts for the purpose of test seems indeed to be the only thing left to the scientist.

5 | Relative Autonomy of Facts

And this it is—provided these facts *exist, and are available independently of whether or not one considers alternatives to the theory to be tested.* This assumption on which the validity of the argument in the last section depends in a most decisive manner I shall call the assumption of the relative autonomy of facts, or the autonomy principle. It is not asserted by this principle that the discovery and description of facts is independent of *all* theorizing. But it *is* asserted that the facts which belong to the empirical content of some theory are available whether or not one considers alternatives to *this* theory. I am not aware that this very important assumption has ever been explicitly formulated as a separate postulate of the empirical method. However, it is clearly implied in almost all investigations which deal with questions of confirmation and test. All these investigations use a model in which a *single* theory is compared with a class of facts (or observation statements) which are assumed to be 'given' somehow. I submit that this is much too simple a picture of the actual situation. Facts and theories are much more intimately connected than is admitted by the autonomy principle. Not only is the description of every single fact dependent on *some* theory (which may, of course, be very different from the theory to be tested). There exist also facts which cannot be unearthed except with the help of alternatives to the theory to be tested, and which become unavailable as soon as such alternatives are excluded. This suggests that the methodological unit to which we must refer when discussing questions of test and empirical content is constituted by a *whole set of partly overlapping, factually adequate, but mutually inconsistent theories.* In the present paper only the barest outlines will be given of such a test model. However, before doing this I want to discuss an example which shows very clearly the function of alternatives in the discovery of facts.

As is well known, the Brownian particle is a perpetual motion machine of the second kind and its existence refutes the phenomenological second law.* It therefore belongs to the domain of relevant facts for this law. Now,

---

* A perpetual motion machine of the second kind—or, as it is often called, perpetual motion of the second kind—is any machine, device, or system that can produce work merely by taking heat from a body (that is, by cooling the body, transforming the heat energy into work, and producing no other effect). The pioneers of classical thermodynamics (Clausius, Kelvin) took the impossibility of such a machine to be the empirical foundation of the second law of thermodynamics, just as the impossibility of perpetual motion of the first kind—the production of work without the transformation of energy—was regarded as the foundation of the first law (energy conservation). The second law asserts that entropy can never decrease in a isolated system. Brownian particles violate the second law because their random motion is acquired at the expense of very brief spontaneous decreases of entropy in the fluid in which they are suspended. (Even if the definition of perpetual motion of the second kind were to stipulate that the machine must operate continuously, in a cycle, it is still possible that Brownian particles might satisfy that definition,

could this relation between the law and the Brownian particle have been discovered in a *direct* manner, i.e., could it have been discovered by an investigation of the observational consequences of the phenomenological theory that did not make use of an alternative account of heat? This question is readily divided into two: (1) Could the *relevance* of the Brownian particle have been discovered in this manner? (2) Could it have been demonstrated that it actually *refutes* the second law? The answer to the first question is that we do not know. It is impossible to say what would have happened had the kinetic theory not been considered by some physicists. It is my guess, however, that in this case the Brownian particle would have been regarded as an oddity much in the same way in which some of the late Professor Ehrenhaft's astounding effects[27] are regarded as an oddity, and that it would not have been given the decisive position it assumes in contemporary theory. The answer to the second question is simply—No. Consider what the discovery of the inconsistency between the Brownian particle and the second law would have required! It would have required (*a*) measurement of the exact *motion* of the particle in order to ascertain the changes of its kinetic energy plus the energy spent on overcoming the resistance of the fluid; and (*b*) it would have required precise measurements of temperature and heat transfer in the surrounding medium in order to ascertain that any loss occurring here was indeed compensated by the increase of the energy of the moving particle and the work done against the fluid. Such measurements are beyond experimental possibilities.[28] Neither is it possible to make precise measurements of the heat transfer; nor can the path of the particle be investigated with the desired precision. Hence a 'direct' refutation of the second law that considers only the phenomenological theory and the 'facts' of Brownian motion is impossible. And, as is well known, the actual refutation was brought about in a very different manner. It was brought about via the kinetic theory and Einstein's utilization of it in the calculation of the statistical properties of the Brownian motion.[29] In the course of this procedure the phenomenological theory (*T'*) was incorporated into the wider context of statistical physics (*T*) *in such a manner that the consistency condition was violated*; and *then* a crucial experiment was staged (investigations of Svedberg and Perrin).

It seems to me that this example is typical for the relation between fairly general theories, or points of view, and 'the facts.' Both the relevance and the refuting character of many very decisive facts can be established only with the help of other theories which, although factually adequate, are yet not in agreement with the view to be tested. This being the case, the production of such refuting facts may have to be preceded by the invention and articulation of alternatives to that view. Empiricism demands that the

---

though the probability that their random motions acquire the necessary degree of coordination, purely by chance, is extraordinarily low. Because this probability is so very low, Brownian motion cannot be harnessed to produce usable work.)

empirical content of whatever knowledge we possess be increased as much as possible. Hence *the invention of alternatives in addition to the view that stands in the center of discussion constitutes an essential part of the empirical method.* Conversely, the fact that the consistency condition eliminates alternatives now shows it to be in disagreement with empiricism and not only with scientific practice. By excluding valuable tests it decreases the empirical content of the theories which are permitted to remain (and which, as we have indicated above, will usually be the theories which have been there first); and it especially decreases the number of those facts which could show their limitations. This last result of a determined application of the consistency condition is of very topical interest. It may well be that the refutation of the quantum-mechanical uncertainties presupposes just such an incorporation of the present theory into a wider context which is no longer in accordance with the idea of complementarily and which therefore suggests new and decisive experiments. And it may also be that the insistence, on the part of the majority of contemporary physicists, on the consistency condition will, if successful, forever protect these uncertainties from refutation. This is how modern empiricism may finally lead to a situation where a certain point of view petrifies into dogma by being, in the name of experience, completely removed from any conceivable criticism.

6 | The Self-Deception Involved in all Uniformity

It is worthwhile to examine this apparently empirical defense of a dogmatic point of view in somewhat greater detail. Assume that physicists have adopted, either consciously or unconsciously, the idea of the uniqueness of complementarily and that they therefore elaborate the orthodox point of view and refuse to consider alternatives. In the beginning such a procedure may be quite harmless. After all, a man can do only so many things at a time and it is better when he pursues a theory in which he is interested rather than a theory he finds boring. Now assume that the pursuit of the theory he chose has led to successes and that the theory has explained in a satisfactory manner circumstances that had been unintelligible for quite some time. This gives empirical support to an idea which to start with seemed to possess only this advantage: It was interesting and intriguing. The concentration upon the theory will now be reinforced, the attitude towards alternatives will become less tolerant. Now if it is true, as has been argued in the last section, that many facts become available only with the help of such alternatives, then the refusal to consider them *will result in the elimination of potentially refuting facts.* More especially, it will eliminate facts whose discovery would show the complete and irreparable inadequacy of the theory.[30] Such facts having been made inaccessible, the theory will appear to be free from blemish and it will seem that "all evidence points with merciless definiteness in the . . . direction . . . [that] all the processes involving . . .

unknown interactions conform to the fundamental quantum law."[31] This will further reinforce the belief in the uniqueness of the current theory and in the complete futility of any account that proceeds in a different manner. Being now very firmly convinced that there is only one good microphysics, the physicists will try to explain even adverse facts in its terms, and they will not mind when such explanations are sometimes a little clumsy. By now the success of the theory has become public news. Popular science books (and this includes a good many books on the philosophy of science) will spread the basic postulates of the theory; applications will be made in distant fields. More than ever the theory will appear to possess tremendous empirical support. The chances for the consideration of alternatives are now very slight indeed. The final success of the fundamental assumptions of the quantum theory and of the idea of complementarily will seem to be assured.

At the same time it is evident, on the basis of the considerations in the last section, that this appearance of success *cannot in the least be regarded as a sign of truth and correspondence with nature.* Quite the contrary, the suspicion arises that the absence of major difficulties is a result of the decrease of empirical content brought about by the elimination of alternatives, and of facts that can be discovered with the help of these alternatives only. In other words, *the suspicion arises that this alleged success is due to the fact that in the process of application to new domains the theory has been turned into a metaphysical system.* Such a system will of course be very 'successful' not, however, because it agrees so well with the facts, but because no facts have been specified that would constitute a test and because some such facts have even been removed. Its 'success' *is entirely manmade.* It was decided to stick to some ideas and the result was, quite naturally, the survival of these ideas. If now the initial decision is forgotten, or made only implicitly, then the survival will seem to constitute independent support, it will reinforce the decision, or turn it into an explicit one, and in this way close the circle. This is how empirical 'evidence' may be *created* by a procedure which quotes as its justification the very same evidence it has produced in the first place.

At this point an 'empirical' theory of the kind described (and let us always remember that the basic principles of the present quantum theory and especially the idea of complementarity are uncomfortably close to forming such a theory) becomes almost indistinguishable from a myth. In order to realize this, we need only consider that on account of its all-pervasive character a myth such as the myth of witchcraft and of demonic possession will possess a high degree of confirmation on the basis of observation. Such a myth has been taught for a long time; its content is enforced by fear, prejudice, and ignorance as well as by a jealous and cruel priesthood. It penetrates the most common idiom, infects all modes of thinking and many decisions which mean a great deal in human life. It provides models for the explanation of any conceivable event, conceivable, that is, for those who have accepted it.[32] This being the case, its key terms will be fixed in an unambiguous manner and the idea (which may have led to such a procedure in the first place) that they are copies of unchanging entities and that change of meaning, if it should happen, is due to human mistake—this idea will now be very plausible. Such plausibility reinforces all the maneuvres which are used for the preservation of the myth (elimination of opponents included). The conceptual apparatus of the theory and the emotions connected with its application having penetrated all means of communication, all actions, and indeed the whole life of the community, such methods as transcendental deduction, analysis of usage, phenomenological analysis which are means for further solidifying the myth will be extremely successful (which shows, by the way, that all these methods which have been the trademark of various philosophical schools old and new, have one thing in common: They tend to *preserve* the *status quo* of the intellectual life).[33] Observational results too, will speak in favor of the theory as they are formulated in its terms. It will seem that at last the truth has been arrived at. At the same time it is evident that all contact with the world has been lost and that the stability achieved, the semblance of absolute truth, *is nothing but the result of an absolute conformism.*[34] For how can we possibly test, or improve upon, the truth of a theory if it is built in such a manner that any conceivable event can be described, and explained, in terms of its principles? The *only* way of investigating such all-embracing principles is to compare them with a different set of *equally all-embracing* principles—but this way has been excluded from the very beginning. The myth is therefore of no objective relevance, it continues to exist solely as the result of the effort of the community of believers and of their leaders, be these now priests or Nobel prize winners. *Its 'success' is entirely manmade.* This, I think, is the most decisive argument against any method that encourages uniformity, be it now empirical or not. Any such method is in the last resort a method of deception. It enforces an unenlightened conformism, and speaks of truth; it leads to a deterioration of intellectual capabilities, of the power of imagination, and speaks of deep insight; it destroys the most precious gift of the young, their tremendous power of imagination, and speaks of education.

To sum up: *Unanimity of opinion may be fitting for a church, for the frightened victims of some (ancient, or modern) myth, or for the weak and willing followers of some tyrant; variety of opinion is a feature necessary for objective knowledge; and a method that encourages variety is also the only method that is compatible with a humanitarian outlook.* To the extent to which the consistency condition (and, as will emerge, the condition of meaning invariance) delimits variety, it contains a theological element (which lies, of course, in the worship of 'facts' so characteristic for nearly all empiricism).

7 | Inherent Unreasonableness of Meaning Invariance

What we have achieved so far has immediate application to the question whether the meaning of certain key terms should be kept unchanged in the course of the development and improvement of our knowledge. After all, the meaning of every term we use depends upon the theoretical context in which it occurs. Hence, if we consider two contexts with basic principles which either contradict each other, or which lead to inconsistent consequences in certain domains, it is to be expected that some terms of the first context will not occur in the second context with exactly the same meaning. Moreover, if our methodology demands the use of mutually inconsistent, partly overlapping, and empirically adequate theories, then it thereby also demands the use of conceptual systems which are mutually *irreducible* (their primitives cannot be connected by bridge laws which are meaningful *and* factually correct) and it demands that meanings of terms be left elastic and that no binding commitment be made to a certain set of concepts.

It is very important to realize that such a tolerant attitude towards meanings, or such a change of meaning in cases where one of the competing conceptual systems has to be abandoned need not be the result of directly accessible observational difficulties. The law of inertia of the so-called *impetus theory* of the later Middle Ages[35] and Newton's own law of inertia are in perfect quantitative agreement: Both assert that an object that is not under the influence of any outer force will proceed along a straight line with constant speed. Yet despite this fact, the adoption of Newton's theory entails a conceptual revision that forces us to abandon the inertial law of the impetus theory, not because it is quantitatively incorrect but *because it achieves the correct predictions with the help of inadequate concepts.* The law asserts that the *impetus* of an object that is beyond the reach of outer forces remains constant.[36] The impetus is interpreted as an inner *force* which pushes the object along. Within the impetus theory such a force is quite conceivable as it is assumed here that forces determine *velocities* rather than accelerations. The concept of impetus is therefore formed in accordance with a law (forces determine velocities) and this law is inconsistent with the laws of Newton's theory and must be abandoned as soon as the latter is adopted. This is how the progress of our knowledge may lead to conceptual revisions for which no direct observational reasons are available. The occurrence of such changes quite obviously refutes the contention of some philosophers that the invariance of *usage* in the trivial and uninteresting contexts of the private lives of not too intelligent and inquisitive people indicates invariance of *meaning* and the superficiality of all scientific changes. It is also a very decisive objection against any crudely operationalistic account of both observable terms and theoretical terms.

What we have said applies even to singular statements of observation. Statements which are empirically adequate, and which are the result of observation (such as 'here is a table') may have to be reinterpreted, not because it has been found that they do not adequately express what is seen, heard, felt, but because of some changes in sometimes very remote parts of the conceptual scheme to which they belong. Witchcraft is again a very good example. Numerous eyewitnesses claim that they have actually *seen* the devil, or *experienced* demonic influence. There is no reason to suspect that they were lying. Nor is there any reason to assume that they were sloppy observers, for the phenomena leading to the belief in demonic influence are so obvious that a mistake is hardly possible (possession; split personality; loss of personality; hearing voices; etc.). These phenomena are well known today.[37] In the conceptual scheme that was the one generally accepted in the 15th and 16th centuries, the only way of describing them, or at least the way that seemed to express them most adequately, was by reference to demonic influences. Large parts of this conceptual scheme were changed for philosophical reasons and also under the influence of the evidence accumulated by the sciences. Descartes' materialism played a very decisive role in discrediting the belief in spatially localizable spirits. The language of demonic influences was no part of the new conceptual scheme that was created in this manner. It was for this reason that a reformulation was needed, and a reinterpretation of even the most common 'observational' statements. Combining this example with the remarks at the beginning of the present section, we now realize that according to the method of classes of alternative theories a lenient attitude must be taken with respect to the meanings of all the terms we use. We must not attach too great an importance to 'what we mean' by a phrase, and we must be prepared to change whatever little we have said concerning this meaning as soon as the need arises. Too great concern with meanings can only lead to dogmatism and sterility. Flexibility, and even sloppiness in semantical matters is a prerequisite of scientific progress.[38]

8 | Some Consequences

Three consequences of the results so far obtained deserve a more detailed discussion. The first consequence is an evaluation of *metaphysics* which differs significantly from the standard empirical attitude. As is well known, there are empiricists who demand that science start from observable facts and proceed by generalization, and who refuse the admittance of metaphysical ideas at any point of this procedure. For them, only a system of thought that has been built up in a purely inductive fashion can claim to be genuine knowledge. Theories which are partly metaphysical, or 'hypothetical,' are suspect, and are best not used at all. This attitude has been formulated most clearly by Newton[39] in his reply to Pardies' second letter concerning the theory of colors:

> if the possibility of hypotheses is to be the test of truth and reality of things, I see not how certainty can be obtained in any science; since numerous hypotheses may be devised, which shall seem to overcome new difficulties.

This radical position, which clearly depends on the demand for a theoretical monism, is no longer as popular as it used to be. It is now granted that metaphysical considerations may be of importance when the task is to *invent* a new physical theory; such invention, so it is admitted, is a more or less irrational act containing the most diverse components. Some of these components are, and perhaps must be, metaphysical ideas. However, it is also pointed out that as soon as the theory has been developed in a formally satisfactory fashion and has received sufficient confirmation to be regarded as empirically successful, it is pointed out that in the very same moment it can *and must* forget its metaphysical past; metaphysical speculation must *now* be replaced by empirical argument.

> On the one side I would like to emphasize [writes Ernst Mach on this point[40]] that *every and any* idea is admissible as a means for research, provided it is helpful; still, it must be pointed out, on the other side, that it is very necessary from time to time to free the presentation of the *results* of research from all inessential additions.

This means that empirical considerations are still given the upper hand over metaphysical reasoning. Especially in the case of an inconsistency between metaphysics and some highly confirmed empirical theory it will be decided, *as a matter of course*, that the theory or the result of observation must stay, and that the metaphysical system must go. A very simple example is the way in which materialism is being judged by some of its opponents. For a materialist the world consists of material particles moving in space, of collections of such particles. Sensations, as introspected by human beings, do not look like collections of particles, and their observed existence is therefore assumed to refute and thereby to remove the metaphysical doctrine of materialism. Another example which I have analyzed in "Problems of Microphysics" is the attempt to eliminate certain very general ideas concerning the nature of microentities on the basis of the remark that they are inconsistent "with an immense body of experience" and that "to object to a lesson of experience by appealing to metaphysical preconceptions is unscientific."[41]

The methodology developed in the present paper leads to a very different evaluation of metaphysics. Metaphysical systems are scientific theories in their most primitive stage. If they *contradict* a well-confirmed point of view, then this indicates their usefulness as an alternative to this point of view. Alternatives are needed for the purpose of criticism. Hence, metaphysical systems which contradict observational results or well-confirmed theories *are most welcome* starting points of such criticism. Far from being misfired attempts at anticipating, or circumventing, empirical research which were deservedly exposed by a reference to experience, they are the only means at our disposal for examining those parts of our knowledge which have already become observational and which are therefore inaccessible to a criticism 'on the basis of observation.'

A second consequence is that a new attitude has to be adopted with respect to the *problem of induction*. This problem consists in the question of what justification there is for asserting the truth of a statement S given the truth of another statement, S′, whose content is smaller than the content of S. It may be taken for granted that those who want to justify the truth of S also assume that after the justification the truth of S will be *known*. Knowledge to the effect that S implies the *stability* of S (we must not change, remove, criticize, what we know to be true). The method we are discussing at the present moment cannot allow such stability. It follows that the problem of induction at least in some of its formulations, is a problem whose solution leads to undesirable results. It may therefore be properly termed a pseudo problem.

The third consequence, which is more specific, is that *arguments from synonymy* (or from coextensionality), far from being that measure of adequacy as which they are usually introduced, are liable severely to impede the progress of knowledge. Arguments from synonymy judge a theory or a point of view not by its capability to mimic the world but rather by its capability to mimic the descriptive terms of another point of view which for some reason is received favorably. Thus for example, the attempt to give a materialistic, or else a purely physiological, account of human beings is criticized on the grounds that materialism, or physiology, cannot provide synonyms for 'mind,' 'pain,' 'seeing red,' 'thinking of Vienna,' in the sense in which these terms are used either in ordinary English (provided there is a well-established usage concerning these terms, a matter which I doubt) or in some more esoteric mentalistic idiom. Clearly, such criticism silently assumes the principle of meaning invariance, that is, it assumes that the meanings of at least some fundamental terms must remain unchanged in the course of the progress of our knowledge. It cannot therefore be accepted as valid.[12]

However, we can, and must go, still further. The ideas which we have developed above are strong enough not only to *reject* the demand for synonymy, wherever it is raised, but also to *support* the demand for irreducibility (in the sense in which this notion was used at the beginning of Section 7). The reason is that irreducibility is a presupposition of high critical ability on the part of the point of view shown to be irreducible. An outer indication of such irreducibility which is quite striking in the case of an attack upon commonly accepted ideas is the feeling of *absurdity*: We deem absurd what goes counter to well-established linguistic habits. The absence, from a newly introduced set of ideas, of synonymy relations connecting it with parts of the accepted point of view; the feeling of absurdity therefore indicate that the new ideas are fit for the purpose of criticism, i.e., that they are fit for either leading to a strong *confirmation* of the earlier theories, or else to a very revolutionary *discovery*: absence of synonymy, clash of meanings, absurdity are desirable. Presence of synonymy, intuitive appeal, agreement with customary modes of speech, far from being *the* philosophical virtue, indicates that not much progress has been made and that the business of investigating what is commonly accepted *has not even started*.

9 | How to Be a Good Empiricist

The final reply to the question put in the title is therefore as follows. A good empiricist will not rest content with the theory that is in the center of attention and with those tests of the theory which can be carried out in a direct manner. Knowing that the most fundamental and the most general criticism is the criticism produced with the help of alternatives, he will try to invent such alternatives.[43] It is, of course, impossible at once to produce a theory that is formally comparable to the main point of view and that leads to equally many predictions. His first step will therefore be the formulation of fairly general assumptions which are not yet directly connected with observations; this means that his first step will be the invention of a new *metaphysics*. This metaphysics must then be elaborated in sufficient detail in order to be able to compete with the theory to be investigated as regards generality, details of prediction, precision of formulation.[44] We may sum up both activities by saying that a good empiricist must be a critical metaphysician. Elimination of all metaphysics, far from increasing the empirical content of the remaining theories, is liable to turn these theories into dogmas. The consideration of alternatives together with the attempt to criticize each of them in the light of experience also leads to an attitude where meanings do not play a very important role and where arguments are based upon assumptions of fact rather than analysis of (archaic, although perhaps very precise) meanings. The effect of such an attitude upon the development of human capabilities should not be underestimated either. Where speculation and invention of alternatives is encouraged, bright ideas are liable to occur in great number and such ideas may then lead to a change of even the most 'fundamental' parts of our knowledge, i.e., they may lead to a change of assumptions which either are so close to observation that their truth seems to be dictated by 'the facts,' or which are so close to common prejudice that they seem to be 'obvious,' and their negation 'absurd.' In such a situation it will be realized that neither 'facts' nor abstract ideas can ever be used for defending certain principles come what may. Wherever facts play a role in such a dogmatic defense, we shall have to suspect foul play (see the opening quotation)—the foul play of those who try to turn good science into bad, because unchangeable, metaphysics. In the last resort, therefore, being a good empiricist means being critical, and basing one's criticism not just on an abstract principle of skepticism but upon *concrete suggestions* which indicate in every single case how the accepted point of view might be further tested and further investigated and which thereby prepare the next step in the development of our knowledge.[45]

■ | Notes

1. K. R. Popper, *The Open Society and Its Enemies*, Princeton University Press, Princeton, New Jersey, 1953.

2. It is very interesting to see how many so-called empiricists, when turning to the past completely fail to pay attention to some very obvious facts which are incompatible with their empiristic epistemology. Thus Galileo has been represented as a thinker who turned away from the empty speculations of the Aristotelians and who based his own laws upon facts which he had carefully collected beforehand. Nothing could be further from the truth. *The Aristotelians could quote numerous observational results in their favor.* The Copernican idea of the motion of the earth, on the other hand, did not possess independent observational support, at least not in the first 150 years of its existence. Moreover, it was inconsistent with facts and highly confirmed physical theories. And *this* is how modern physics started: not as an observational enterprise *but as an unsupported speculation that was inconsistent with highly confirmed laws.* For details and further references see P. K. Feyerabend, "Realism and Instrumentalism—Comments on the Logic of Factual Support," in M. Bunge, Ed., *The Critical Approach to Science and Philosophy*, The Free Press, New York, 1964.

3. One might be inclined to add those who base their pronouncements upon an analysis of what they call 'ordinary language.' I do not think they deserve to be honored by a criticism. Paraphrasing Galileo, one might say that they "deserve not even that name, for they do not talk plainly and simply but are content to adore the shadows, philosophizing not with due circumspection but merely from having memorized a few ill-understood principles."

4. It is nowadays frequently assumed that "if one considers the history of a special branch of science, one gets the impression that non-scientific elements . . . relatively frequently occur in the earlier stages of development, but that they gradually retrogress in later stages and even tend to disappear in such advanced stages which become ripe for more or less thorough formalization" (H. J. Groenewold, *Synthese*, 1957, p. 305). Our considerations in the text would seem to show that such a development is very undesirable and can only result in a well-formalized, precisely expressed, and completely petrified metaphysics.

5. These essays were published in Volume III of the *Minnesota Studies in the Philosophy of Science*; in Volumes I and II of the *Pittsburgh Studies in the Philosophy of Science*; and in *Problems of Philosophy, Essays in Honor of Herbert Feigl*, respectively.

6. See K. R. Popper, *Logic of Scientific Discovery*, New York, 1959, Section 12. (This is a translation of his *Logik der Forschung* published in 1935.) The decisive feature of Popper's theory, a feature which was not at all made clear by earlier writers on the subject of explanation, is the emphasis he puts on the initial conditions and the implied possibility of two kinds of laws, viz., (1) laws concerning the temporal sequence of events, and (2) laws concerning the space of initial conditions. In the case of the quantum theory, the laws of the second kind provide very important information about the nature of the elementary particles and it is to *them* and *not* to the laws of motion that reference is made in the discussions concerning the interpretation of the uncertainty relations. In general relativity, the laws formulating the initial conditions concern the structure of the universe at large and only by overlooking them could it be believed that a purely relational account of space would be possible. For the last point, cf. E. L. Hill, "Quantum Physics and the Relativity Theory," in H. Feigl and G. Maxwell, Eds., *Current Issues in the Philosophy of Science*, Holt, Rinehart and Winston, New York, 1961.

7. C. G. Hempel, "Studies in the Logic of Explanation," reprinted in H. Feigl and M. Brodbeck, Eds., *Readings in the Philosophy of Science*, New York, 1953, p. 321.

8. E. Nagel, "The Meaning of Reduction in the Natural Sciences," reprinted in A. C. Danto and S. Morgenbesser, Eds., *Philosophy of Science*, New York, 1960, p. 301.

9. It has been objected to this formulation that theories which are consistent with a given explanandum may still contradict each other. This is quite correct, but it does not invalidate my argument. For as soon as a single theory is regarded as sufficient for explaining all that is known (and represented by the other theories in question), it will have to be consistent with all these other theories.

10. E. Mach, *Wärmelehre*, Leipzig, 1897, p. 364.

11. E. Mach, *Zwei Aufsaetze*, Leipzig, 1912.

12. For a discussion of these objections, cf. ter Haar's review article in *Reviews of Modern Physics*, 1957.

13. M. Born, *Natural Philosophy of Cause and Chance*, Oxford University Press, New York, 1948, p. 109.

14. L. Rosenfeld, "Misunderstandings about the Foundations of the Quantum Theory," in S. Körner, Eds., *Observation and Interpretation*, London, 1957, p. 42.

15. Cf. the discussions in *Observation and Interpretation* (see note 14).

16. For details and further literature, cf. Section 11 of my paper "Problems of Microphysics," in R. Colodny, Eds., *Frontiers of Science and Philosophy*, University of Pittsburgh Press, Pittsburgh, 1962, pp. 189–283.

17. W. Heisenberg, *Physics and Philosophy*, New York, 1958, p. 44.

18. P. Duhem, *La Théorie Physique: Son Objet, Sa Structure*, Paris, 1914, Chapters IX and X. See also K. R. Popper, "The Aim of Science," *Ratio*, Vol. I (1957) 24–35.

19. E. Nagel, "The Meaning of Reduction in the Natural Sciences," loc. cit., p. 302.

20. Cf. Section 4.7 of M. Scriven's paper "Explanations, Predictions, and Laws," in Vol. III of the *Minnesota Studies in the Philosophy of Science*. Similar objections have been raised by Viktor Kraft (Vienna) and David Rynin (Berkeley).

21. For an exposition and criticism of this idea cf. my 'Attempt at a Realistic Interpretation of Experience,' *Proceedings of the Aristotelian Society*, New Series, LVIII, 143–170 (1958).

22. It must be admitted, however, that Einstein's original interpretation of the special theory of relativity is hardly ever used by contemporary physicists. For them the theory of relativity consists of two elements: (1) the Lorentz transformations; and (2) mass-energy equivalence. The Lorentz transformations are interpreted purely formally and are used to make a selection among possible equations. This interpretation does not allow one to distinguish between Lorentz's original point of view and the entirely different point of view of Einstein. According to it Einstein achieved a very minor *formal* advance (this is the basis of Whittaker's attempt to 'debunk' Einstein). It is also very similar to what application of the double language model would yield. Still, an undesirable philosophical procedure is not improved by the support it gets from an undesirable procedure in physics. (The above comment on the contemporary attitude towards relativity was made by E. L. Hill in discussions at the Minnesota Center for the Philosophy of Science.)

23. In about 1925 philosophers of science were bold enough to stick to their theses even in those cases where they were inconsistent with actual science. They meant to be *reformers* of science, and not *imitators*. (This point was explicitly made by Mach in his controversy with Planck. Cf. again his *Zwei Aufsaetze*.) In the meantime they have become rather tame (or beat) and are much more prepared to change their ideas in accordance with the latest discoveries of the historians, or the latest fashion of the contemporary scientific enterprise. This is very regrettable, indeed, for it considerably decreases the number of the rational critics of the scientific enterprise. And it also seems to give unwanted support to the Hegelian thesis (which is now implicitly held by many historians and philosophers of science) that what exists has a 'logic' of its own and is for that very reason reasonable.

24. Even the most dogmatic enterprise allows for discoveries (cf. the 'discovery' of so-called 'white Jews' among German physicists during the Nazi period). Hence, before hailing a so-called discovery, we must make sure that the system of thought which forms its background is not of a dogmatic kind.

25. The indefinite character of all observations has been made very clear by Duhem, *La Théorie Physique: Son Objet, Sa Structure*, Chap. IX. For an alternative way of dealing with this indefiniteness, cf. S. Körner, *Conceptual Thinking*, New York, 1960.

26. More detailed evidence for the existence of this attitude and for the way in which it influences the development of the sciences may be found in T. S. Kuhn, *The Structure of Scientific Revolutions*, University of Chicago Press, Chicago, 1962. The attitude is extremely common in the contemporary quantum theory. 'Let us enjoy the successful theories we possess and let us not waste our time with contemplating what *would* happen if *other* theories were used'—this seems to be the motto of almost all contemporary physicists (cf. W. Heisenberg, *Physics and Philosophy*, pp. 56, 144) and philosophers (cf. N. R. Hanson, "Five Cautions for the Copenhagen Critics," *Philosophy of Science*, XXVI, 325–337 [1959]). It may be traced back to Newton's papers and letters (to Hooke, and Pardies) on the theory of color. See also [note] 23.

27. Having witnessed these effects under a great variety of conditions, I am much more reluctant to regard them as mere curiosities than is the scientific community of today. Cf. also my edition of Ehrenhaft's lectures, *Einzelne Magnetische Nord- und Südpole und deren Auswirkung in den Naturwissenschaften*, Vienna, 1947. [See Gerald Holton, "Subelectrons, Presuppositions, and the Millikan-Ehrenhaft Dispute," in *The Scientific Imagination* (Cambridge: Cambridge University Press, 1978), 25–83.]

28. R. Fürth, *Zeitschrift für Physik*, 81, 143–162 (1933).

29. For these investigations, cf. A. Einstein, *Investigations on the Theory of the Brownian Movement*, New York, 1956, which contains all the relevant papers by Einstein and an exhaustive bibliography by R. Fürth. For the experimental work, cf. J. Perrin, *Die Atome* Leipzig, 1920. For the relation between the phenomenological theory and the kinetic theory, cf. also M. v. Smoluchowski, "Experimentell nachwiesbare, der üblichen Thermodynamik widersprechende Molekularphänomene," *Physikalische Zeitschrift*, XIII, 1069 (1912); and K. R. Popper, "Irreversibility, or, Entropy since 1905," *British Journal for the Philosophy of Science*, VIII, 151 (1957). Despite Einstein's epoch-making discoveries and von Smoluchowski's

splendid presentation of their effect (for the latter cf. also *Oeuvres de Marian Smo-luchowski*, Cracouvie, 1927, Vol. II, pp. 226 ff., 316 ff., 462 ff., and 530 ff.), the present situation in thermodynamics is extremely unclear, especially in view of the continued presence of the ideas of reduction which we criticized in the text above. To be more specific, it is frequently attempted to determine the entropy balance of a complex *statistical* process by reference to the (refuted) *phenomenological* law after which procedure fluctuations are superimposed in a most artificial fashion. For details cf. Popper, *loc. cit.* [For further discussion, see Paul Feyerabend, "On the Possibility of a *Perpetuum Mobile* of the Second Kind," *Mind, Matter, and Method*, ed. P. K. Feyerabend and G. Maxwell (Minneapolis: University of Minnesota Press, 1966), 409–12.]

30. The quantum theory can be adapted to a great many difficulties. It is an open theory in the sense that apparent inadequacies can be accounted for in an *ad hoc* manner, by *adding* suitable operators, or elements in the Hamiltonian, rather than by recasting the whole structure. A refutation of its basic formalism (i.e., of the formalism of quantization, and of noncommuting operators in a Hilbert space or a reasonable extension of it) would therefore demand proof to the effect that *there is no conceivable adjustment of the Hamiltonian, or of the operators used* which makes the theory conform to a given fact. It is clear that such a general statement can only be provided by an *alternative theory* which of course must be detailed enough to allow for independent, and crucial tests.

31. L. Rosenfeld, "Misunderstandings about the Foundations of the Quantum Theory," loc. cit., p. 44.

32. For a very detailed description of a once very influential myth, cf. C. H. Lea, *Materials for a History of Witchcraft*, 3 Vols., New York, 1957, as well as *Malleus Malleficarum*, translated by Montague Summers (who, by the way, counts it "among the most important, wisest [sic!], and weightiest books of the world"), London, 1928.

33. Quite clearly, analysis of usage, to take only one example, presupposes certain regularities concerning this usage. The more people differ in their fundamental ideas, the more difficult will it be to uncover such regularities. Hence, analysis of usage will work best in a closed society that is firmly held together by a powerful myth such as was the philosophy in the Oxford of about 10 years ago.

34. Schizophrenics very often hold beliefs which are as rigid, all-pervasive, and unconnected with reality, as are the best dogmatic philosophies. Only such beliefs come to them naturally whereas a professor may sometimes spend his whole life in attempting to find arguments which create a similar state of mind.

35. For details and further references, cf. Section 6 of my "Explanation, Reduction, and Empiricism," *loc. cit.*

36. We assume here that a dynamical rather than a kinematic characterization of motion has been adopted. For a more detailed analysis cf. again the paper referred to in the previous note.

37. For very vivid examples, cf. K. Jaspers, *Allgemeine Psychopathologie*, Berlin, 1959, pp. 75–123.

38. Mae West is by far preferable to the precisionists: "I ain't afraid of pushin' grammar around so long as it sounds good" (*Goodness Had Nothing to Do with It*, New York, 1959, p. 19).

39. I. B. Cohen, Ed., *Isaac Newton's Papers and Letters on Natural Philosophy*, Harvard University Press, Cambridge, Massachusetts, 1958, p. 106.

40. 'Der Gegensatz zwischen der mechanischen und der phaenomenologischen Physik,' *Wärmelehre*, Leipzig, 1896, pp. 362 ff.

41. L. Rosenfeld, "Misunderstandings about the Foundations of the Quantum Theory," loc. cit., p. 42.

42. For details concerning the mind-body problem, cf. my "Materialism and the Mind-Body Problem," *Review of Metaphysics*, Sept. 1963.

43. In my paper 'Realism and Instrumentalism' I have tried to show that this is precisely the method which has brought about such spectacular advances of knowledge as the Copernican Revolution, the transition to relativity and to quantum theory.

44. Cf. Section 13 of my "Realism and Instrumentalism," [loc. cit.].

45. For support of research the author is indebted to the National Science Foundation and the Minnesota Center for the Philosophy of Science.

Jerry A. Fodor

Special Sciences
(or: The Disunity of Science
as a Working Hypothesis)

A typical thesis of positivistic philosophy of science is that all true theories in the special sciences should reduce to physical theories in the long run. This is intended to be an empirical thesis, and part of the evidence which supports it is provided by such scientific successes as the molecular theory of heat and the physical explanation of the chemical bond. But the philosophical popularity of the reductivist program cannot be explained by reference to these achievements alone. The development of science has witnessed the proliferation of specialized disciplines at least as often as it has witnessed their reduction to physics, so the widespread enthusiasm for reduction can hardly be a mere induction over its past successes.

I think that many philosophers who accept reductivism do so primarily because they wish to endorse the generality of physics *vis à vis* the special sciences: roughly, the view that all events which fall under the laws of any science are physical events and hence fall under the laws of physics.[1] For such philosophers, saying that physics is basic science and saying that theories in the special sciences must reduce to physical theories have seemed to be two ways of saying the same thing, so that the latter doctrine has come to be a standard construal of the former.

In what follows, I shall argue that this is a considerable confusion. What has traditionally been called 'the unity of science' is a much stronger, and much less plausible, thesis than the generality of physics. If this is true it is important. Though reductionism is an empirical doctrine, it is intended to play a regulative role in scientific practice. Reducibility to physics is taken to be a *constraint* upon the acceptability of theories in the special sciences, with the curious consequence that the more the special sciences succeed, the more they ought to disappear. Methodological problems about psychology, in particular, arise in just this way: the assumption that the subject-matter of psychology is part of the subject-matter of physics is taken to imply

From *Synthese* 28 (1974): 97–115.

that psychological theories must reduce to physical theories, and it is this latter principle that makes the trouble. I want to avoid the trouble by challenging the inference.

I

Reductivism is the view that all the special sciences reduce to physics. The sense of 'reduce to' is, however, proprietary. It can be characterized as follows.[2]

Let

(1) $S_1x \rightarrow S_2x$

be a law of the special science S. ((1) is intended to be read as something like 'all S_1 situations bring about S_2 situations'. I assume that a science is individuated largely by reference to its typical predicates, hence that if S is a special science 'S$_1$' and 'S$_2$' are not predicates of basic physics. I also assume that the 'all' which quantifies laws of the special sciences needs to be taken with a grain of salt; such laws are typically *not* exceptionless. This is a point to which I shall return at length.) A necessary and sufficient condition of the reduction of (1) to a law of physics is that the formulae (2) and (3) be laws, and a necessary and sufficient condition of the reduction of S to physics is that all its laws be so reducible.[3]

(2a) $S_1x \leftrightarrows P_1x$
(2b) $S_2x \leftrightarrows P_2x$
(3) $P_1x \rightarrow P_2x.$

'P$_1$' and 'P$_2$' are supposed to be predicates of physics, and (3) is supposed to be a physical law. Formulae like (2) are often called 'bridge' laws. Their characteristic feature is that they contain predicates of both the reduced and the reducing science. Bridge laws like (2) are thus contrasted with 'proper' laws like (1) and (3). The upshot of the remarks so far is that the reduction of a science requires that any formula which appears as the antecedent or consequent of one of its proper laws must appear as the reduced formula in some bridge law or other.[4]

Several points about the connective '$\rightarrow$' are in order. First, whatever other properties that connective may have, it is universally agreed that it must be transitive. This is important because it is usually assumed that the reduction of some of the special sciences proceeds via bridge laws which connect their predicates with those of intermediate reducing theories. Thus, psychology is presumed to reduce to physics via, say, neurology, biochemistry, and other local stops. The present point is that this makes no difference to the logic of the situation so long as the transitivity of '$\rightarrow$' is assumed.

Bridge laws which connect the predicates of S to those of S* will satisfy the constraints upon the reduction of S to physics so long as there are other bridge laws which, directly or indirectly, connect the predicates of S* to physical predicates.

There are, however, quite serious open questions about the interpretations of '→' in bridge laws. What turns on these questions is the respect in which reductivism is taken to be a physicalist thesis.

To begin with, if we read '→' as 'brings about' or 'causes' in proper laws, we will have to have some other connective for bridge laws, since bringing about and causing are presumably *asymmetric*, while bridge laws express symmetric relations. Moreover, if '→' in bridge laws is interpreted as any relation other than identity, the truth of reductivism will only guarantee the truth of a weak version of physicalism, and this would fail to express the underlying ontological bias of the reductivist program.

If bridge laws are not identity statements, then formulae like (2) claim at most that, by law, x's satisfaction of a P predicate and x's satisfaction of an S predicate are causally correlated. It follows from this that it is nomologically necessary that S and P predicates apply to the same things (i.e., that S predicates apply to a subset of the things that P predicates apply to). But, of course, this is compatible with a non-physicalist ontology since it is compatible with the possibility that x's satisfying S should not itself *be* a physical event. On this interpretation, the truth of reductivism does *not* guarantee the generality of physics *vis à vis* the special sciences since there are some events (satisfactions of S predicates) which fall in the domains of a special science (S) but not in the domain of physics. (One could imagine, for example, a doctrine according to which physical and psychological predicates are both held to apply to organisms, but where it is denied that the event which consists of an organism's satisfying a psychological predicate is, in any sense, a physical event. The up-shot would be a kind of psychophysical dualism of a non-Cartesian variety; a dualism of events and/or properties rather than substances.)

Given these sorts of considerations, many philosophers have held that bridge laws like (2) ought to be taken to express contingent event identities, so that one would read (2a) in some such fashion as 'every event which consists of x's satisfying S_1 is identical to some event which consists of x's satisfying P_1 and vice versa'. On this reading, the truth of reductivism would entail that every event that falls under any scientific law is a physical event, thereby simultaneously expressing the ontological bias of reductivism and guaranteeing the generality of physics *vis à vis* the special sciences.

If the bridge laws express event identities, and if every event that falls under the proper laws of a special science falls under a bridge law, we get the truth of a doctrine that I shall call 'token physicalism'. Token physicalism is simply the claim that all the events that the sciences talk about are physical events. There are three things to notice about token physicalism.

First, it is weaker than what is usually called 'materialism'. Materialism claims *both* that token physicalism is true *and* that every event falls under the laws of some science or other. One could therefore be a token physical-ist without being a materialist, though I don't see why anyone would bother.

Second, token physicalism is weaker than what might be called 'type physicalism', the doctrine, roughly, that every *property* mentioned in the laws of any science is a physical property. Token physicalism does not entail type physicalism because the contingent identity of a pair of events presum-ably does not guarantee the identity of the properties whose instantiation constitutes the events; not even where the event identity is nomologically necessary. On the other hand, if every event is the instantiation of a prop-erty, then type physicalism does entail token physicalism: two events will be identical when they consist of the instantiation of the same property by the same individual at the same time.

Third, token physicalism is weaker than reductivism. Since this point is, in a certain sense, the burden of the argument to follow, I shan't labour it here. But, as a first approximation, reductivism is the conjunction of token physicalism with the assumption that there are natural kind predi-cates in an ideally completed physics which correspond to each natural kind predicate in any ideally completed special science. It will be one of my morals that the truth of reductivism cannot be inferred from the assump-tion that token physicalism is true. Reductivism is a sufficient, but not a necessary, condition for token physicalism.

In what follows, I shall assume a reading of reductivism which entails token physicalism. Bridge laws thus state nomologically necessary contin-gent event identities, and a reduction of psychology to neurology would entail that any event which consists of the instantiation of a psychological property is identical with some event which consists of the instantiation of some neurological property.

Where we have got to is this: reductivism entails the generality of phys-ics in at least the sense that any event which falls within the universe of discourse of a special science will also fall within the universe of discourse of physics. Moreover, any prediction which follows from the laws of a spe-cial science and a statement of initial conditions will also follow from a theory which consists of physics and the bridge laws, together with the statement of initial conditions. Finally, since 'reduces to' is supposed to be an asymmetric relation, it will also turn out that physics is *the* basic science; that is, if reductivism is true, physics is the only science that is general in the sense just specified. I now want to argue that reductivism is too strong a constraint upon the unity of science, but that the relevantly weaker doc-trine will preserve the desired consequences of reductivism: token physical-ism, the generality of physics, and its basic position among the sciences.

II

Every science implies a taxonomy of the events in its universe of discourse. In particular, every science employs a descriptive vocabulary of theoretical and observation predicates such that events fall under the laws of the science by virtue of satisfying those predicates. Patently, not every true description of an event is a description in such a vocabulary. For example, there are a large number of events which consist of things having been transported to a distance of less than three miles from the Eiffel Tower. I take it, however, that there is no science which contains 'is transported to a distance of less than three miles from the Eiffel Tower' as part of its descriptive vocabulary. Equivalently, I take it that there is no natural law which applies to events in virtue of their being instantiations of the property *is transported to a distance of less than three miles from the Eiffel Tower* (though I suppose it is conceivable that there is some law that applies to events in virtue of their being instantiations of some distinct but co-extensive property). By way of abbreviating these facts, I shall say that the property *is transported . . .* does not determine a *natural kind*, and that predicates which express that property are not natural kind predicates.

If I knew what a law is, and if I believed that scientific theories consist just of bodies of laws, then I could say that P is a natural kind predicate relative to S iff [if and only if] S contains proper laws of the form $P_x \rightarrow \alpha_x$ or $\alpha_x \rightarrow P_x$; roughly, the natural kind predicates of a science are the ones whose terms are the bound variables in its proper laws. I am inclined to say this even in my present state of ignorance, accepting the consequence that it makes the murky notion of a natural kind viciously dependent on the equally murky notions *law* and *theory*. There is no firm footing here. If we disagree about what is a natural kind, we will probably also disagree about what is a law, and for the same reasons. I don't know how to break out of this circle, but I think that there are interesting things to say about which circle we are in.

For example, we can now characterize the respect in which reductivism is too strong a construal of the doctrine of the unity of science. If reductivism is true, then *every* natural kind is, or is co-extensive with, a physical natural kind. (Every natural kind *is* a physical natural kind if bridge laws express property identities, and every natural kind is co-extensive with a physical natural kind if bridge laws express event identities.) This follows immediately from the reductivist premise that every predicate which appears as the antecedent or consequent of a law of the special sciences must appear as one of the reduced predicates in some bridge law, together with the assumption that the natural kind predicates are the ones whose terms are the bound variables in proper laws. If, in short, some physical law is related to each law of a special science in the way that (3) is related to (1), then every natural kind predicate of a special science is related to a natural kind predicate of physics in the way that (2) relates 'S$_1$' and 'S$_2$' to 'P$_1$' and 'P$_2$'.

I now want to suggest some reasons for believing that this consequence of reductivism is intolerable. These are not supposed to be knock-down reasons; they couldn't be, given that the question whether reductivism is too strong is finally an *empirical* question. (The world could turn out to be such that every natural kind corresponds to a physical natural kind, just as it could turn out to be such that the property *is transported to a distance of less than three miles from the Eiffel Tower* determines a natural kind in, say, hydrodynamics. It's just that, as things stand, it seems very unlikely that the world *will* turn out to be either of these ways.)

The reason it is unlikely that every natural kind corresponds to a physical natural kind is just that (a) interesting generalizations (e.g., counterfactual supporting generalizations) can often be made about events whose physical descriptions have nothing in common, (b) it is often the case that *whether* the physical descriptions of the events subsumed by these generalizations have anything in common is, in an obvious sense, entirely irrelevant to the truth of the generalizations, or to their interestingness, or to their degree of confirmation or, indeed, to any of their epistemologically important properties, and (c) the special sciences are very much in the business of making generalizations of this kind.

I take it that these remarks are obvious to the point of self-certification; they leap to the eye as soon as one makes the (apparently radical) move of taking the special sciences at all seriously. Suppose, for example, that Gresham's 'law' really is true.* (If one doesn't like Gresham's law, then any true generalization of any conceivable future economics will probably do as well.) Gresham's law says something about what will happen in monetary exchanges under certain conditions. I am willing to believe that physics is general *in the sense that it implies that any event which consists of a monetary exchange* (hence any event which falls under Gresham's law) *has a true description in the vocabulary of physics and in virtue of which it falls under the laws of physics.* But banal considerations suggest that a description which covers all such events must be wildly disjunctive. Some monetary exchanges involve strings of wampum. Some involve dollar bills. And some involve signing one's name to a check. What are the chances that a disjunction of physical predicates which covers all these events (i.e., a disjunctive predicate which can form the right hand side of a bridge law of the form '*x* is a monetary exchange ⇆ ...') expresses a physical natural kind? In particular, what are the chances that such a predicate forms the antecedent or consequent of some proper law of physics? The point is that monetary exchanges

---

* Sir Thomas Gresham (1519–79) was the financial advisor to Queen Elizabeth who recommended that she mint new coins to replace those that her predecessors (Henry VIII and Edward VI) had debased by substituting inferior metals for gold and silver. Gresham's law states "bad money drives out good," where "drives out" means "drives out of circulation." Good money is money whose face (or nominal) value equals its market (or commodity) value. For gold coins, their market value is the price that they would fetch on the open market as bullion. The market value of bad money is less than its nominal value; consequently, people tend to hoard good money and circulate bad money. All modern money is "bad money" in Gresham's sense, and its universality is often taken to be strong confirmation of Gresham's law. Economists disagree about whether Gresham's law can be legitimately extended beyond currency to other financial instruments such as stocks, bonds, and mortgages.

have interesting things in common; Gresham's law, if true, says what one of these interesting things is. But what is interesting about monetary exchanges is surely not their commonalities under *physical* description. A natural kind like a monetary exchange *could* turn out to be co-extensive with a physical natural kind; but if it did, that would be an accident on a cosmic scale.

In fact, the situation for reductivism is still worse than the discussion thus far suggests. For, reductivism claims not only that all natural kinds are co-extensive with physical natural kinds, but that the co-extensions are nomologically necessary: bridge laws are *laws*. So, if Gresham's law is true, it follows that there is a (bridge) law of nature such that 'x is a monetary exchange $\leftrightarrows$ x is P', where P is a term for a physical natural kind. But, surely, there is no such law. If there were, then P would have to cover not only all the systems of monetary exchange that there *are*, but also all the systems of monetary exchange that there *could be*; a law must succeed with the counterfactuals. What physical predicate is a candidate for 'P' in 'x is a nomologically possible monetary exchange iff [if and only if] P_x'?

To summarize: an immortal econophysicist might, when the whole show is over, find a predicate in physics that was, in brute fact, co-extensive with 'is a monetary exchange'. If physics is general—if the ontological biases of reductivism are true—then there must *be* such a predicate. But (a) to paraphrase a remark Donald Davidson made in a slightly different context, nothing but brute enumeration could convince us of this brute co-extensivity, and (b) there would seem to be no chance at all that the physical predicate employed in stating the coextensivity is a natural kind term, and (c) there is still less chance that the co-extension would be lawful (i.e., that it would hold not only for the nomologically possible world that turned out to be real, but for any nomologically possible world at all).

I take it that the preceding discussion strongly suggests that economics is not reducible to physics in the proprietary sense of reduction involved in claims for the unity of science. There is, I suspect, nothing special about economics in this respect; the reasons why economics is unlikely to reduce to physics are paralleled by those which suggest that psychology is unlikely to reduce to neurology.

If psychology is reducible to neurology, then for every psychological natural kind predicate there is a co-extensive neurological natural kind predicate, and the generalization which states this co-extension is a law. Clearly, many psychologists believe something of the sort. There are departments of 'psycho-biology' or 'psychology and brain science' in universities throughout the world whose very existence is an institutionalized gamble that such lawful co-extensions can be found. Yet, as has been frequently remarked in recent discussions of materialism, there are good grounds for hedging these bets. There are no firm data for any but the grossest correspondence between types of psychological states and types of neurological states, and it is entirely possible that the nervous system of higher organisms characteristically achieves a given psychological end by a wide variety of neurological means. If so, then the attempt to pair neurological structures with psychological functions is foredoomed. Physiological psychologists of the stature of Karl Lashley have held precisely this view.

The present point is that the reductivist program in psychology is, in any event, *not* to be defended on ontological grounds. Even if (token) psychological events are (token) neurological events, it does not follow that the natural kind predicates of psychology are co-extensive with the natural kind predicates of any other discipline (including physics). That is, the assumption that every psychological event is a physical event does not guarantee that physics (or, *a fortiori*, any other discipline more general than psychology) can provide an appropriate vocabulary for psychological theories. I emphasize this point because I am convinced that the make-or-break commitment of many physiological psychologists to the reductivist program stems precisely from having confused that program with (token) physicalism.

What I have been doubting is that there are neurological natural kinds co-extensive with psychological natural kinds. What seems increasingly clear is that, even if there is such a co-extension, it cannot be lawlike. For, it seems increasingly likely that there are nomologically possible systems other than organisms (namely, automata) which satisfy natural kind predicates in psychology, and which satisfy no neurological predicates at all. Now, as Putnam has emphasized, if there are any such systems, then there are probably vast numbers, since equivalent automata can be made out of practically anything.* If this observation is correct, then there can be no serious hope that the class of automata whose psychology is effectively identical to that of some organism can be described by *physical* natural kind predicates (though, of course, if token physicalism is true, that class can be picked out by some physical predicate or other). The upshot is that the classical formulation of the unity of science is at the mercy of progress in the field of computer simulation. This is, of course, simply to say that that formulation was too strong. The unity of science was intended to be an empirical hypothesis, defeasible by possible scientific findings. But no one had it in mind that it should be defeated by Newell, Shaw and Simon.†

---

* See Hilary Putnam, "Minds and Machines," in Sidney Hook, ed., *Dimensions of Mind* (New York: New York University Press, 1960), 148–78. In the 1960s and 70s, Putnam and Fodor were influential proponents of functionalism in the philosophy of mind, according to which mental states (such as beliefs, desires, and sensations) are functional types individuated by their causal relations to other mental states, stimuli, and behavior. In his paper, Putnam draws an analogy between the functionalist conception of mental states and the logical states of a Turing machine, from which he concludes that mental states could, in principle, be realized by a wide variety of different physical (and perhaps also non-physical) systems.

† Allen Newell, James C. Shaw, and Herbert A. Simon were pioneers in artificial intelligence and computer science, developing programs for playing chess and proving theorems in propositional logic. For a survey of their work, see Allen Newell and Herbert A. Simon, "Computer Science as Empirical Inquiry: Symbols and

I have thus far argued that psychological reductivism (the doctrine that every psychological natural kind is, or is co-extensive with, a neurological natural kind) is not equivalent to, and cannot be inferred from, token physicalism (the doctrine that every psychological event is a neurological event). It may, however, be argued that one might as well take the doctrines to be equivalent since the only possible *evidence* one could have for token physicalism would also be evidence for reductivism: namely, the discovery of type-to-type psychophysical correlations.

A moment's consideration shows, however, that this argument is not well taken. If type-to-type psychophysical correlations would be evidence for token physicalism, so would correlations of other specifiable kinds.

We have type-to-type correlations where, for every n-tuple of events that are of the same psychological kind, there is a correlated n-tuple of events that are of the same neurological kind. Imagine a world in which such correlations are *not* forthcoming. What is found, instead, is that for every n-tuple of type identical psychological events, there is a spatiotemporally correlated n-tuple of type *distinct* neurological events. That is, every psychological event is paired with some neurological event or other, but psychological events of the same kind may be paired with neurological events of different kinds. My present point is that such pairings would provide as much support for token physicalism as type-to-type pairings do *so long as we are able to show that the type distinct neurological events paired with a given kind of psychological event are identical in respect of whatever properties are relevant to type-identification in psychology.* Suppose, for purposes of explication, that psychological events are type identified by reference to their behavioral consequences.[5] Then what is required of all the neurological events paired with a class of type homogeneous psychological events is only that they be identical in respect of their behavioral consequences. To put it briefly, type identical events do not, of course, have *all* their properties in common, and type distinct events must nevertheless be identical in *some* of their properties. The empirical confirmation of token physicalism does not depend on showing that the neurological counterparts of type identical psychological events are themselves type identical. What needs to be shown is only that they are identical in respect of those properties which determine which kind of *psychological* event a given event is.

Could we have evidence that an otherwise heterogeneous set of neurological events have these kinds of properties in common? Of course we could. The neurological theory might itself explain why an n-tuple of neurologically type distinct events are identical in their behavioral consequences, or, indeed, in respect of any of indefinitely many other such relational properties. And, if the neurological theory failed to do so, some science more basic than neurology might succeed.

Search," in John Haugeland, ed., *Mind Design II*. 2nd ed. (Cambridge, Mass.: MIT Press, 1997), 81–110.

My point in all this is, once again, not that correlations between type homogeneous psychological states and type heterogeneous neurological states would prove that token physicalism is true. It is only that such correlations might give us as much reason to be token physicalists as type-to-type correlations would. If this is correct, then the epistemological arguments from token physicalism to reductivism must be wrong.

It seems to me (to put the point quite generally) that the classical construal of the unity of science has really misconstrued the *goal* of scientific reduction. The point of reduction is *not* primarily to find some natural kind predicate of physics co-extensive with each natural kind predicate of a reduced science. It is, rather, to explicate the physical mechanisms whereby events conform to the laws of the special sciences. I have been arguing that there is no logical or epistemological reason why success in the second of these projects should require success in the first, and that the two are likely to come apart *in fact* wherever the physical mechanisms whereby events conform to a law of the special sciences are heterogeneous.

III

I take it that the discussion thus far shows that reductivism is probably too strong a construal of the unity of science; on the one hand, it is incompatible with probable results in the special sciences, and, on the other, it is more than we need to assume if what we primarily want is just to be good token physicalists. In what follows, I shall try to sketch a liberalization of reductivism which seems to me to be just strong enough in these respects. I shall then give a couple of independent reasons for supposing that the revised doctrine may be the right one.

The problem all along has been that there is an open empirical possibility that what corresponds to the natural kind predicates of a reduced science may be a heterogeneous and unsystematic disjunction of predicates in the reducing science, and we do not want the unity of science to be prejudiced by this possibility. Suppose, then, that we allow that bridge statements may be of the form

(4) $Sx \rightleftarrows P_1x \lor P_2x \lor \ldots \lor P_nx,$

where '$P_1 \lor P_2 \lor \ldots \lor P_n$' is *not* a natural kind predicate in the reducing science. I take it that this is tantamount to allowing that at least some 'bridge laws' may, in fact, not turn out to be laws, since I take it that a necessary condition on a universal generalization being lawlike is that the predicates which consitute its antecedent and consequent should pick out natural kinds. I am thus supposing that it is enough, for purposes of the unity of science, that every law of the special sciences should be reducible to physics by bridge statements which express true empirical generalizations.

Bearing in mind that bridge statements are to be construed as a species of identity statements, (4) will be read as something like 'every event which consists of x's satisfying S is identical with some event which consists of x's satisfying some or other predicate belonging to the disjunction '$P_1 \vee P_2 \vee \ldots \vee P_n$.''

Now, in cases of reduction where what corresponds to (2) is not a law, what corresponds to (3) will not be either, and for the same reason. Namely, the predicates appearing in the antecedent or consequent will, by hypothesis, not be natural kind predicates. Rather, what we will have is something that looks like (5).

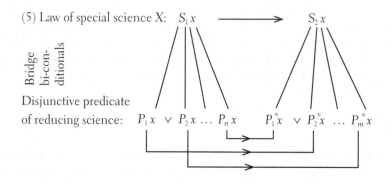

(5) Law of special science X: $S_1 x \longrightarrow S_2 x$

Bridge bi-conditionals

Disjunctive predicate of reducing science: $P_1 x \vee P_2 x \ldots P_n x \qquad P_1^* x \vee P_2^* x \ldots P_m^* x$

That is, the antecedent and consequent of the reduced law will each be connected with a disjunction of predicates in the reducing science, and, if the reduced law is exceptionless, there will be laws of the reducing science which connect the satisfaction of each member of the disjunction associated with the antecedent to the satisfaction of some member of the disjunction associated with the consequent. That is, if $S_1 x \to S_2 x$ is exceptionless, then there must be some proper law of the reducing science which either states or entails that $P_1 x \to P^*$ for some P^*, and similarly for $P_2 x$ through $P_n x$. Since there must be such laws, it follows that each disjunct of '$P_1 \vee P_2 \vee \ldots \vee P_n$' is a natural kind predicate, as is each disjunct of '$P_1^* \vee P_2^* \vee \ldots \vee P_m^*$.'

This, however, is where push comes to shove. For, it might be argued that if each disjunct of the P disjunction is lawfully connected to some disjunct of the P^* disjunction, it follows that (6) is itself a law.

(6) $P_1 x \vee P_2 x \vee \ldots \vee P_n x \to P_1^* x \vee P_2^* x \vee \ldots \vee P_m^* x$.

The point would be that (5) gives us $P_1 x \to P_2^* x$, $P_2 x \to P_m^* x$, etc., and the argument from a premise of the form $(P \supset R)$ and $(Q \supset S)$ to a conclusion of the form $(P \vee Q) \supset (R \vee S)$ is valid.

What I am inclined to say about this is that it just shows that 'it's a law that——' defines a non-truth functional context (or, equivalently for these purposes, that not all truth functions of natural kind predicates are themselves natural kind predicates). In particular, that one may not argue from 'it's a law that P brings about R' and 'it's a law that Q brings about S' to 'it's a law that P or Q brings about R or S'. (Though, of course, the argument from those premises to 'P or Q brings about R or S' *simpliciter* is fine.) I think, for example, that it is a law that the irradiation of green plants by sunlight causes carbohydrate synthesis, and I think that it is a law that friction causes heat, but I do not think that it is a law that (either the irradiation of green plants by sunlight or friction) causes (either carbohydrate synthesis or heat). Correspondingly, I doubt that 'is either carbohydrate synthesis or heat' is plausibly taken to be a natural kind predicate.

It is not strictly mandatory that one should agree with all this, but one denies it at a price. In particular, if one allows the full range of truth functional arguments inside the context 'it's a law that——', then one gives up the possibility of identifying the natural kind predicates of a science with those predicates which appear as the antecedents or the consequents of its proper laws. (Thus (6) would be a proper law of physics which fails to satisfy that condition.) One thus inherits the need for an alternative construal of the notion of a natural kind, and I don't know what that alternative might be like.

The upshot seems to be this. If we do not require that bridge statements must be laws, then either some of the generalizations to which the laws of special sciences reduce are not themselves lawlike, or some laws are not formulable in terms of natural kinds. Whichever way one takes (5), the important point is that it is weaker than standard reductivism; it does not require correspondences between the natural kinds of the reduced and the reducing science. Yet it is physicalistic on the same assumption that makes standard reductivism physicalistic (namely, that the bridge statements express true token identities). But these are precisely the properties that we wanted a revised account of the unity of science to exhibit.

I now want to give two reasons for thinking that this construal of the unity of science is right. First, it allows us to see how the laws of the special sciences could reasonably have exceptions, and, second, it allows us to see why there are special sciences at all. These points in turn.

Consider, again, the model of reduction implicit in (2) and (3). I assume that the laws of basic science are strictly exceptionless, and I assume that it is common knowledge that the laws of the special sciences are not. But now we have a painful dilemma. Since '$\rightarrow$' expresses a relation (or relations) which must be transitive, (1) can have exceptions only if the bridge laws do. But if the bridge laws have exceptions, reductivism loses its ontological bite, since we can no longer say that every event which consists of the instantiation of an S predicate is identical with some event which consists of the instantiation of a P predicate. In short, given the reductionist model, we cannot consistently assume that the bridge laws and the basic laws are exceptionless while assuming that the special laws are not. But we cannot accept the violation of the bridge laws unless we are willing to vitiate the ontological claim that is the main point of the reductivist program.

We can get out of this (*salve* the model) in one of two ways. We can give up the claim that the special laws have exceptions or we can give up the claim that the basic laws are exceptionless. I suggest that both alternatives are undesirable. The first because it flies in the face of fact. There is just no chance at all that the true, counter-factual supporting generalizations of, say, psychology, will turn out to hold in strictly each and every condition where their antecedents are satisfied. Even where the spirit is willing, the flesh is often weak. There are always going to be behavioral lapses which are physiologically explicable but which are uninteresting from the point of view of psychological theory. The second alternative is only slightly better. It may, after all, turn out that the laws of basic science have exceptions. But the question arises whether one wants the unity of science to depend upon the assumption that they do.

On the account summarized in (5), however, everything works out satisfactorily. A nomologically sufficient condition for an exception to $S_1 x \rightarrow S_2 x$ is that the bridge statements should identify some occurrence of the satisfaction of S_1 with an occurrence of the satisfaction of a P predicate which is not itself lawfully connected to the satisfaction of any P^* predicate. (I.e., suppose S_1 is connected to a P' such that there is no law which connects P' to any predicate which bridge statements associate with S_2. Then any instantiation of S_1 which is contingently identical to an instantiation of P' will be an event which constitutes an exception to $S_1 x \rightarrow S_2 x$.) Notice that, in this case, we need assume no exceptions to the laws of the *reducing* science since, by hypothesis, (6) *is not a law.*

In fact, strictly speaking, (6) has no status in the reduction at all. It is simply what one gets when one universally quantifies a formula whose antecedent is the physical disjunction corresponding to S_1 and whose consequent is the physical disjunction corresponding to S_2. As such, it will be true when $S_1 \rightarrow S_2$ is exceptionless and false otherwise. What does the work of expressing the physical mechanisms whereby n-tuples of events conform, or fail to conform, to $S_1 \rightarrow S_2$ is not (6) but the laws which severally relate elements of the disjunction $P_1 \vee P_2 \vee \dots \vee P_n$ to elements of the disjunction $P_1^* \vee P_2^* \vee \dots \vee P_m^*$. When there *is* a law which relates an event that satisfies one of the P disjuncts to an event which satisfies one of the P^* disjuncts, the pair of events so related conforms to $S_1 \rightarrow S_2$. When an event which satisfies a P predicate is *not* related by law to an event which satisfies a P^* predicate, that event will constitute an exception to $S_1 \rightarrow S_2$.† The point is that none

---

† When Fodor revised his paper for inclusion in *Representations: Philosophical Essays on the Foundations of Cognitive Science* (Cambridge, Mass.: MIT Press, 1997), he added the caption "Laws of reducing science" at the bottom left of diagram (5). He also added $P'x$ alongside $P_n x$ with a vertical sloping line connecting $P'x$ to $S_1 x$; but the horizontal line with an arrow running from $P'x$ (representing a law of the reducing science) does not connect with any of the P^* predicates that underlie $S_2 x$. This illustrates how exceptions to the special science law $S_1 x \rightarrow S_2 x$ can arise.

of the laws which effect these several connections need themselves have exceptions in order that $S_1 \rightarrow S_2$ should do so.

To put this discussion less technically: we could, if we liked, *require* the taxonomies of the special sciences to correspond to the taxonomy of physics by insisting upon distinctions between the natural kinds postulated by the former wherever they turn out to correspond to distinct natural kinds in the latter. This would *make* the laws of the special sciences exceptionless if the laws of basic science are. But it would also lose us precisely the generalizations which we want the special sciences to express. (If economics were to posit as many *kinds* of monetary systems as there are kinds of physical realizations of monetary systems, then the generalizations of economics *would* be exceptionless. But, presumably, only vacuously so, since there would be no generalizations left to state. Gresham's law, for example, would have to be formulated as a vast, open disjunction about what happens in monetary system$_1$ or monetary system$_n$ under conditions which would themselves defy uniform characterization. We would not be able to say what happens in monetary systems *tout court* since, by hypothesis, 'is a monetary system' corresponds to no natural kind predicate of physics.)

In fact, what we do is precisely the reverse. We allow the generalizations of the special sciences to *have* exceptions, thus preserving the natural kinds to which the generalizations apply. But since we know that the *physical* descriptions of the natural kinds may be quite heterogeneous, and since we know that the physical mechanisms which connect the satisfaction of the antecedents of such generalizations to the satisfaction of their consequents may be equally diverse, we expect both that there will be exceptions to the generalizations and that these exceptions will be 'explained away' at the level of the reducing science. This is one of the respects in which physics really is assumed to be bedrock science; exceptions to *its* generalizations (if there are any) had better be random, because there is nowhere 'further down' to go in explaining the mechanism whereby the exceptions occur.

This brings us to why there are special sciences at all. Reductivism as we remarked at the outset, flies in the face of the facts about the scientific institution: the existence of a vast and interleaved conglomerate of special scientific disciplines which often appear to proceed with only the most token acknowledgment of the constraint that their theories must turn out to be physics 'in the long run'. I mean that the acceptance of this constraint, *in practice*, often plays little or no role in the validation of theories. Why is this so? Presumably, the reductivist answer must be *entirely* epistemological. If only physical particles weren't so small (if only brains were on the *outside*, where one can get a look at them), *then* we would do physics instead of paleontology (neurology instead of psychology; psychology instead of economics; and so on down). There is an epistemological reply; namely, that even if brains were out where they can be looked *at*, as things now stand, we wouldn't know what to look *for*: we lack the appropriate theoretical apparatus for the psychological taxonomy of neurological events.

If it turns out that the functional decomposition of the nervous system corresponds to its neurological (anatomical, biochemical, physical) decomposition, then there are only epistemological reasons for studying the former instead of the latter. But suppose there is no such correspondence? Suppose the functional organization of the nervous system crosscuts its neurological organization (so that quite different neurological structures can subserve identical psychological functions across times or across organisms). Then the existence of psychology depends not on the fact that neurons are so sadly small, but rather on the fact that neurology does not posit the natural kinds that psychology requires.

I am suggesting, roughly, that there are special sciences not because of the nature of our epistemic relation to the world, but because of the way the world is put together: not all natural kinds (not all the classes of things and events about which there are important, counterfactual supporting generalizations to make) are, or correspond to, physical natural kinds. A way of stating the classical reductionist view is that things which belong to different physical kinds *ipso facto* can have no projectible descriptions in common; that if x and y differ in those descriptions by virtue of which they fall under the proper laws of physics, they must differ in those descriptions by virtue of which they fall under any laws at all. But why should we believe that this is so? Any pair of entities, however different their physical structure, must nevertheless converge in indefinitely many of their properties. Why should there not be, among those convergent properties, some whose lawful interrelations support the generalizations of the special sciences? Why, in short, should not the natural kind predicates of the special sciences *cross-classify* the physical natural kinds?[6]

Physics develops the taxonomy of its subject-matter which best suits its purposes: the formulation of exceptionless laws which are basic in the several senses discussed above. But this is not the only taxonomy which may be required if the purposes of science in general are to be served: e.g., if we are to state such true, counterfactual supporting generalizations as there are to state. So, there are special sciences, with their specialized taxonomies, in the business of stating some of these generalizations. If science is to be unified, then all such taxonomies must apply *to the same things*. If physics is to be basic science, then each of these things had better be a physical thing. But it is not further required that the taxonomies which the special sciences employ must themselves reduce to the taxonomy of physics. It is not required, and it is probably not true.

References

Block, N. and Fodor, J., 'What Psychological States Are Not', *Philosophical Review* 81 (1972), 159–181.

Chomsky, N., *Aspects of the Theory of Syntax*, MIT Press, Cambridge, 1965.

Notes

I wish to express my gratitude to Ned Block for having read a version of this paper and for the very useful comments he made.

1. I shall usually assume that sciences are about events, in at least the sense that it is the occurrence of events that makes the laws of a science true. But I shall be pretty free with the relation between events, states, things and properties. I shall even permit myself some latitude in construing the relation between properties and predicates. I realize that all these relations are problems, but they aren't my problem in this paper. Explanation has to *start* somewhere, too.

2. The version of reductionism I shall be concerned with is a stronger one than many philosophers of science hold; a point worth emphasizing since my argument will be precisely that it is too strong to get away with. Still, I think that what I shall be attacking is what many people have in mind when they refer to the unity of science, and I suspect (though I shan't try to prove it) that many of the liberalized versions suffer from the same basic defect as what I take to be the classical form of the doctrine.

3. There is an implicit assumption that a science simply *is* a formulation of a set of laws. I think this assumption is implausible, but it is usually made when the unity of science is discussed, and it is neutral so far as the main argument of this paper is concerned.

4. I shall sometimes refer to 'the predicate which constitutes the antecedent or consequent of a law'. This is shorthand for 'the predicate such that the antecedent or consequent of a law consists of that predicate, together with its bound variables and the quantifiers which bind them'. (Truth functions of elementary predicates are, of course, themselves predicates in this usage.)

5. I don't think there is any chance at all that this is true. What is more likely is that type-identification for psychological states can be carried out in terms of the 'total states' of an abstract automaton which models the organism. For discussion, see Block and Fodor (1972).

6. As, by the way, the predicates of natural languages quite certainly do. For discussion, see Chomsky (1965).

PHILIP KITCHER

1953 and All That:
A Tale of Two Sciences

Must we geneticists become bacteriologists, physiological chemists and physicists, simultaneously with being zoologists and botanists? Let us hope so.

—H. J. Muller, 1922[1]

1 | The Problem

Toward the end of their paper announcing the molecular structure of DNA, James Watson and Francis Crick remark, somewhat laconically, that their proposed structure might illuminate some central questions of genetics.[2] Thirty years have passed since Watson and Crick published their famous discovery. Molecular biology has indeed transformed our understanding of heredity. The recognition of the structure of DNA, the understanding of gene replication, transcription and translation, the cracking of the genetic code, the study of gene regulation, these and other breakthroughs have combined to answer many of the questions that baffled classical geneticists. Muller's hope—expressed in the early days of classical genetics—has been amply fulfilled.

Yet the success of molecular biology and the transformation of classical genetics into molecular genetics bequeath a philosophical problem. There are two recent theories which have addressed the phenomena of heredity. One, *classical genetics*, stemming from the studies of T. H. Morgan, his colleagues and students, is the successful outgrowth of the Mendelian theory of heredity rediscovered at the beginning of this century. The other, *molecular genetics*, descends from the work of Watson and Crick. What is the relationship between these two theories? How does the molecular theory illuminate the classical theory? How exactly has Muller's hope been fulfilled?

FROM *Philosophical Review* 93 (1984): 335–73.

There used to be a popular philosophical answer to the problem posed in these three connected questions: classical genetics has been reduced to molecular genetics. Philosophers of biology inherited the notion of reduction from general discussions in philosophy of science, discussions which usually center on examples from physics. Unfortunately attempts to apply this notion in the case of genetics have been vulnerable to cogent criticism. Even after considerable tinkering with the concept of reduction, one cannot claim that classical genetics has been (or is being) reduced to molecular genetics.[3] However, the antireductionist point is typically negative.[4] It denies the adequacy of a particular solution to the problem of characterizing the relation between classical genetics and molecular genetics. It does not offer an alternative solution.

My aim in this paper is to offer a different perspective on intertheoretic relations. The plan is to invert the usual strategy. Instead of trying to force the case of genetics into a mold, which is alleged to capture important features of examples in physics, or resting content with denying that the material can be forced, I shall try to arrive at a view of the theories involved and the relations between them that will account for the almost universal idea that molecular biology has done something important for classical genetics. In so doing, I hope to shed some light on the general questions of the structure of scientific theories and the relations which may hold between successive theories. Since my positive account presupposes that something is wrong with the reductionist treatment of the case of genetics, I shall begin with a diagnosis of the foibles of reductionism.

2 | What's Wrong with Reductionism?

Ernest Nagel's classic treatment of reduction[5] can be simplified for our purposes. Scientific theories are regarded as sets of statements.[6] To reduce a theory T_2 to a theory T_1, is to deduce the statements of T_2 from the statements of T_1. If there are nonlogical expressions which appear in the statements of T_2, but do not appear in the statements of T_1, then we are allowed to supplement the statements of T_1 with some extra premises connecting the vocabulary of T_1 with the distinctive vocabulary of T_2 (so-called *bridge principles*). Intertheoretic reduction is taken to be important because the statements which are deduced from the reducing theory are supposed to be explained by this deduction.

Yet, as everyone who has struggled with the paradigm cases from physics knows all too well, the reductions of Galileo's law to Newtonian mechanics and of the ideal gas laws to the kinetic theory do not exactly fit Nagel's model. Study of these examples suggests that, to reduce a theory T_2 to a theory T_1, it suffices to deduce the laws of T_2 from a suitably modified version of T_1, possibly augmented with appropriate extra premises.[7] Plainly, this sufficient condition is dangerously vague.[8] I shall tolerate its vagueness, proposing that we understand the issue of reduction in genetics by using the examples from physics as paradigms of what "suitable modifications" and "appropriate extra premises" are like. Reductionists claim that the relation between classical genetics and molecular biology is sufficiently similar to the intertheoretical relations exemplified in the examples from physics to count as the same type of thing: to wit, as intertheoretical reduction.

It may seem that the reductionist thesis has now become so amorphous that it will be immune to refutation. But this is incorrect. Even when we have amended the classical model of reduction so that it can accommodate the examples that originally motivated it, the reductionist claim about genetics requires us to accept three theses:

R1 Classical genetics contains general laws about the transmission of genes which can serve as the conclusions of reductive derivations.

R2 The distinctive vocabulary of classical genetics (predicates like '① is a gene', '① is dominant with respect to ②') can be linked to the vocabulary of molecular biology by bridge principles.

R3 A derivation of general principles about the transmission of genes from principles of molecular biology would explain why the laws of gene transmission hold (to the extent that they do).

I shall argue that each of the theses is false, offering this as my diagnosis of the ills of reductionism.

Before offering my criticisms, it may help to explain why reductionism presupposes (R1)–(R3). If the relation between classical genetics and molecular biology is to be like that between the theory of ideal gases and the kinetic theory (say), then we are going to need to find general principles, identifiable as the central laws of classical genetics, which can serve as the conclusions of reductive derivations. (We need counterparts for the Boyle-Charles law.) These will be general principles about genes, and, because classical genetics seems to be a theory about the inheritance of characteristics, the only likely candidates are laws describing the transmission of genes between generations. [So reductionism leads to (R1).] If we are to derive such laws from molecular biology, then there must be bridge principles connecting the distinctive vocabulary figuring in the laws of gene transmission (presumably expressions like '① is a gene', and perhaps '① is dominant with respect to ②') with the vocabulary of molecular biology. [Hence (R2).] Finally, if the derivations are to achieve the goal of intertheoretical reduction then they must explain the laws of gene transmission. [(R3).]

Philosophers often identify theories as small sets of general laws. However, in the case of classical genetics, the identification is difficult and those who debate the reducibility of classical genetics to molecular biology often proceed differently. David Hull uses a characterization drawn from Dobzhansky: classical genetics is "concerned with gene differences; the operation employed to discover a gene is hybridization: parents differing in some trait are crossed and the distribution of the trait in hybrid progeny is observed."[9] This is not unusual in discussions of reduction in genetics. It is much easier to identify classical genetics by referring to the subject matter and to the methods of investigation, than it is to provide a few sentences that encapsulate the content of the theory.

Why is this? Because when we read the major papers of the great classical geneticists or when we read the textbooks in which their work is summarized, we find it hard to pick out *any* laws about genes. These documents are full of informative statements. Together, they tell us an enormous amount about the chromosomal arrangement of particular genes in particular organisms, about the effect on the phenotype of various mutations, about frequencies of recombination, and so forth.[10] In some cases, we might explain the absence of formulations of general laws about genes (and even of reference to such laws) by suggesting that these things are common knowledge. Yet that hardly accounts for the nature of the textbooks or of the papers that forged the tools of classical genetics.

If we look back to the pre-Morgan era, we do find two general statements about genes, namely Mendel's Laws (or "Rules"). Mendel's second law states that, in a diploid organism which produces haploid gametes, genes at different loci will be transmitted independently; so, for example, if A, a and B, b are pairs of alleles at different loci, and if an organism is heterozygous at both loci, then the probabilities that a gamete will receive any of the four possible genetic combinations, AB, Ab, aB, ab, are all equal.[11] Once it was recognized that genes are (mostly) chromosomal segments, (as biologists discovered soon after the rediscovery of Mendel's laws), we understand that the law will not hold in general: alleles which are on the same chromosome (or, more exactly, close together on the same chromosome) will tend to be transmitted together because (ignoring recombination)[12] one member of each homologous pair is distributed to a gamete.[13]

Now it might seem that this is not very important. We could surely find a correct substitute for Mendel's second law by restricting the law so that it only talks about genes on nonhomologous chromosomes. Unfortunately, this will not quite do. There can be interference with normal cytological processes so that segregation of nonhomologous chromosomes need not be independent.[14] However, my complaint about Mendel's second law is not that it is incorrect: many sciences use laws that are clearly recognized as approximations. Mendel's second law, amended or unamended, simply becomes irrelevant to subsequent research in classical genetics.

We envisaged amending Mendel's second law by using elementary principles of cytology, together with the identification of genes as chromosomal segments, to correct what was faulty in the unamended law. It is the fact that the application is so easy and that it can be carried out far more generally that makes the "law" it generates irrelevant. We can understand the transmission of genes by analyzing the cases that interest us from a cytological perspective—by proceeding from "first principles," as it were. Moreover, we can adopt this approach whether the organism is haploid, diploid or polyploid, whether it reproduces sexually or asexually, whether the genes with which we are concerned are or are not on homologous chromosomes, whether or not there is distortion of independent chromosomal segregation at meiosis. Cytology not only teaches us that the second law is false; it also tells us how to tackle the problem at which the second law was directed (the problem of determining frequencies for pairs of genes in gametes). The amended second law is a restricted statement of results obtainable using a general technique. What figures largely in genetics after Morgan is the technique, and this is hardly surprising when we realize that one of the major research problems of classical genetics has been the problem of discovering the distribution of genes *on the same chromosome*, a problem which is beyond the scope of the amended law.

Let us now turn from (R1) to (R2), assuming, contrary to what has just been argued, that we can identify the content of classical genetics with general principles about gene transmission. (Let us even suppose, for the sake of concreteness, that the principles in question are Mendel's laws—amended in whatever way the reductionist prefers.) To derive these principles from molecular biology, we need a bridge principle. I shall consider first statements of the form

(*) (x) (x is a gene ↔ Mx)

where 'Mx' is an open sentence (possibly complex) in the language of molecular biology. Molecular biologists do not offer any appropriate statement. Nor do they seem interested in providing one. I claim that no appropriate bridge principle can be found.

Most genes are segments of DNA. (There are some organisms— viruses—whose genetic material is RNA; I shall henceforth ignore them.) Thanks to Watson and Crick, we know the molecular structure of DNA. Hence the problem of providing a statement of the above form becomes that of saying, in molecular terms, which segments of DNA count as genes.

Genes come in different sizes, and, for any given size, we can find segments of DNA of that size that are not genes. Therefore genes cannot be identified as segments of DNA containing a particular number of nucleotide pairs. Nor will it do to give a molecular characterization of those codons (triplets of nucleotides) that initiate and terminate transcription, and take a gene to be a segment of DNA between successive initiating and terminating codons. In the first place, mutation might produce a *single* allele containing within it codons for stopping and restarting transcription.[15] Secondly, and much more importantly, the criterion is not general since not every gene is transcribed on mRNA.

The latter point is worth developing. Molecular geneticists recognize regulatory genes as well as structural genes. To cite a classic example, the operator region in the *lac* operon of *E. coli* serves as a site for the attachment of protein molecules, thereby inhibiting transcription of mRNA and regulating enzyme production.[16] Moreover, it is becoming increasingly obvious that genes are not always transcribed, but play a variety of roles in the economy of the cell.[17]

At this point, the reductionist may try to produce a bridge principle by brute force. Trivially, there are only a finite number of terrestrial organisms (past, present and future) and only a finite number of genes. Each gene is a segment of DNA with a particular structure and it would be possible, in principle, to provide a detailed molecular description of that structure. We can now give a molecular specification of the gene by enumerating the genes and disjoining the molecular descriptions.[18] The point made above, that the segments which we count as genes do not share any structural property can now be put more precisely: any instantiation of (*) which replaces 'M' by a structural predicate from the language of molecular biology will insert a predicate that is essentially disjunctive.

Why does this matter? Let us imagine a reductionist using the enumerative strategy to deduce a general principle about gene transmission. After great labor, it is revealed that all actual genes satisfy the principle. I claim that more than this is needed to reduce a *law* about gene transmission. We envisage laws as sustaining counterfactuals, as applying to examples that might have been but which did not actually arise. To reduce the law it is necessary to show how possible but nonactual genes would have satisfied it. Nor can we achieve the reductionist's goal by adding further disjuncts to the envisaged bridge principle. For although there are only finitely many *actual* genes, there are indefinitely many genes which *might* have arisen.

At this point, the reductionist may protest that the deck has been stacked. There is no need to produce a bridge principle of the form (*). Recall that we are trying to derive a general law about the transmission of genes, whose paradigm is Mendel's second law. Now the gross logical form of Mendel's second law is:

(1) (x) (y) ((Gx & Gy) → Axy).

We might hope to obtain this from statements of the forms

(2) (x) (Gx → Mx)

(3) (x) (y) ((Mx & My) → Axy)

where 'Mx' is an open sentence in the language of molecular biology. Now there will certainly be true statements of the form (2): for example, we can take 'Mx' as 'x is composed of DNA [or] x is composed of RNA'. The question is whether we can combine some such statement with other appropriate premises—for example, some instance of (3)—so as to derive, and thereby explain (1). No geneticist or molecular biologist has advanced any suitable premises, and with good reason. We discover true statements of the form (2) by hunting for weak necessary conditions on genes, conditions which have to be met by genes but which are met by hordes of other biological entities as well. We can only hope to obtain *weak* necessary conditions because of the phenomenon that occupied us previously: from the molecular standpoint, genes are not distinguished by any common structure. Trouble will now arise when we try to show that the weak necessary condition is jointly sufficient for the satisfaction of the property (independent assortment at meiosis) that we ascribe to genes. The difficulty is illustrated by the example given above. If we take 'Mx' to be 'x is composed of DNA [or] x is composed of RNA' then the challenge will be to find a general law governing the distribution of all segments of DNA and RNA!

I conclude that (R2) is false. Reductionists cannot find the bridge principles they need, and the tactic of abandoning the form (*) for something weaker is of no avail. I shall now consider (R3). Let us concede both of the points that I have denied, allowing that there are general laws about the transmission of genes and that bridge principles are forthcoming. I claim that exhibiting derivations of the transmission laws from principles of molecular biology and bridge principles would not explain the laws, and, therefore, would not fulfill the major goal of reduction.

As an illustration, I shall use the envisaged amended version of Mendel's second law. Why do genes on nonhomologous chromosomes assort independently? Cytology provides the answer. At meiosis, chromosomes line up with their homologues. It is then possible for homologous chromosomes to exchange some genetic material, producing pairs of recombinant chromosomes. In the meiotic division, one member of each recombinant pair goes to each gamete, and the assignment of one member of one pair to a gamete is probabilistically independent of the assignment of a member of another pair to that gamete. Genes which occur close on the same chromosome are likely to be transmitted together (recombination is not likely to occur between them), but genes on nonhomologous chromosomes will assort independently.

This account is a perfectly satisfactory explanation of why our envisaged law is true to the extent that it is. (We recognize how the law could fail if there were some unusual mechanism linking particular nonhomologous chromosomes.) To emphasize the adequacy of the explanation is not to deny that it could be extended in certain ways. For example, we might want to know more about the mechanics of the process by which the chromosomes are passed on to the gametes. In fact, cytology provides such information. However, appeal to molecular biology would not deepen our understanding of the transmission law. Imagine a successful derivation of the law from principles of chemistry and a bridge principle of the form (*). In charting the details of the molecular rearrangements the derivation would only blur the outline of a simple cytological story, adding a welter of irrelevant detail. Genes on nonhomologous chromosomes assort independently because non-homologous chromosomes are transmitted independently at meiosis, and, so long as we recognize this, we do not need to know what the chromosomes are made of.

In explaining a scientific law, L, one often provides a deduction of L from other principles. Sometimes it is possible to explain some of the principles used in the deduction by deducing them, in turn, from further laws. Recognizing the possibility of a sequence of deductions tempts us to suppose that we could produce a better explanation of L by combining them, producing a more elaborate derivation in the language of our ultimate premises. But this is incorrect. What is relevant for the purposes of giving one explanation may be quite different from what is relevant for the purposes of explaining a law used in giving that original explanation. This general point is illustrated by the case at hand. We begin by asking why genes on nonhomologous chromosomes assort independently. The simple cytological story rehearsed above answers the question. That story generates *further* questions. For example, we might inquire why nonhomologous chromosomes are distributed independently at meiosis. To answer this question we would describe the formation of the spindle and the migration of chromosomes to the poles of the spindle just before meiotic division.[19] Once again, the narrative would generate yet further questions. Why do the chromosomes "condense" at prophase? How is the spindle formed? Perhaps in answering these questions we would begin to introduce the chemical details of the process. Yet simply plugging a molecular account into the narratives offered at the previous stages would *decrease* the explanatory power of those narratives. What is relevant to answering our original question is the fact that nonhomologous chromosomes assort independently. What is relevant to the issue of why nonhomologous chromosomes assort independently is the fact that the chromosomes are not selectively oriented toward the poles of the spindle. (We need to eliminate the doubt that, for example, the paternal and maternal chromosomes become separated and aligned toward opposite poles of the spindle.) In neither case are the molecular details relevant. Indeed, adding those details would only disguise the relevant factor.

There is a natural reductionist response. The considerations of the last paragraphs presuppose far too subjective a view of scientific explanation. After all, even if *we* become lost in the molecular details, beings who are cognitively more powerful than we could surely recognize the explanatory force of the envisaged molecular derivation. However, this response misses a crucial point. The molecular derivation forfeits something important.

Recall the original cytological explanation. It accounted for the transmission of genes by identifying meiosis as a process of a particular kind: a process in which paired entities (in this case, homologous chromosomes) are separated by a force so that one member of each pair is assigned to a descendant entity (in this case, a gamete). Let us call processes of this kind

PS-processes. I claim first that explaining the transmission law requires identifying PS-processes as forming a natural kind to which processes of meiosis belong, and second that PS-processes cannot be identified as a kind from the molecular point of view.

If we adopt the familiar covering law account of explanation, then we shall view the cytological narrative as invoking a law to the effect that processes of meiosis are PS-processes and as applying elementary principles of probability to compute the distribution of genes to gametes from the laws that govern PS-processes. If the illumination provided by the narrative is to be preserved in a molecular derivation, then we shall have to be able to express the relevant laws as laws in the language of molecular biology, and this will require that we be able to characterize PS-processes as a natural kind from the molecular point of view. The same conclusion, to wit that the explanatory power of the cytological account can be preserved only if we can identify PS-processes as a natural kind in molecular terms, can be reached in analogous ways if we adopt quite different approaches to scientific explanation—for example, if we conceive of explanation as specifying causally relevant properties or as fitting phenomena into a unified account of nature.

However, PS-processes are heterogeneous from the molecular point of view. There are no constraints on the molecular structures of the entities which are paired or on the ways in which the fundamental forces combine to pair them and to separate them. The bonds can be forged and broken in innumerable ways: all that matters is that there be bonds that initially pair the entities in question and that are subsequently (somehow) broken. In some cases, bonds may be formed directly between constituent molecules of the entities in question; in others, hordes of accessory molecules may be involved. In some cases, the separation may occur because of the action of electromagnetic forces or even of nuclear forces; but it is easy to think of examples in which the separation is effected by the action of gravity. I claim, therefore, that PS-processes are realized in a motley of molecular ways. (I should note explicitly that this conclusion is independent of the issue of whether the reductionist can find bridge principles for the concepts of classical genetics.)

We thus obtain a reply to the reductionist charge that we reject the explanatory power of the molecular derivation simply because we anticipate that our brains will prove too feeble to cope with its complexities.[20] The molecular account objectively fails to explain because it cannot bring out that feature of the situation which is highlighted in the cytological story. It cannot show us that genes are transmitted in the ways that we find them to be because meiosis is a PS-process and because any PS-process would give rise to analogous distributions. Thus (R3)—like (R1) and (R2)—is false.

3 | The Root of the Trouble

Where did we go wrong? Here is a natural suggestion. The most fundamental failure of reductionism is the falsity of (R1). Lacking an account of theories which could readily be applied to the cases of classical genetics and molecular genetics, the attempt to chart the relations between these theories was doomed from the start. If we are to do better, we must begin by asking a preliminary question: what is the structure of classical genetics?

I shall follow this natural suggestion, endeavoring to present a picture of the structure of classical genetics which can be used to understand the intertheoretic relations between classical and molecular genetics.[21] As we have seen, the main difficulty in trying to axiomatize classical genetics is to decide what body of statements one is attempting to axiomatize. The history of genetics makes it clear that Morgan, Muller, Sturtevant, Beadle, McClintock, and others have made important contributions to genetic theory. But the statements occurring in the writings of these workers seem to be far too specific to serve as parts of a general theory. They concern the genes of particular kinds of organisms—primarily paradigm organisms, like fruit flies, bread molds, and maize. The idea that classical genetics is simply a heterogeneous set of statements about dominance, recessiveness, position effect, nondisjunction, and so forth, in *Drosophila, Zea mays, E. coli, Neurospora*, etc. flies in the face of our intuitions. The statements advanced by the great classical geneticists seem more like *illustrations* of the theory than *components* of it. (To know classical genetics it is not necessary to know the genetics of any particular organism, not even *Drosophila melanogaster.*) But the only alternative seems to be to suppose that there are general laws in genetics, never enunciated by geneticists but reconstructible by philosophers. At the very least, this supposition should induce the worry that the founders of the field, and those who write the textbooks of today, do a singularly bad job.

Our predicament provokes two main questions. First, if we focus on a particular time in the history of classical genetics, it appears that there will be a set of statements about inheritance in particular organisms, which constitutes the corpus which geneticists of that time accept: what is the relationship between this corpus and the version of classical genetic theory in force at the time? (In posing this question, I assume, contrary to fact, that the community of geneticists was always distinguished by unusual harmony of opinion; it is not hard to relax this simplifying assumption.) Second, we think of genetic theory as something that persisted through various versions: what is the relation among the versions of classical genetic theory accepted at different times (the versions of 1910, 1930, and 1950, for example) which makes us want to count them as versions of the same theory?

We can answer these questions by amending a prevalent conception of the way in which we should characterize the state of a science at a time. The corpus of statements about the inheritance of characteristics accepted at a given time is only one component of a much more complicated entity that I shall call the *practice* of classical genetics at that time. There is a common language used to talk about hereditary phenomena, a set of accepted statements in that language (the corpus of beliefs about inheritance mentioned above), a set of questions taken to be the appropriate questions to ask about hereditary phenomena, and a set of patterns of reasoning which are instantiated in answering some of the accepted questions; (also: sets of experimental procedures and methodological rules, both designed for use in evaluating proposed answers; these may be ignored for present purposes). The practice of classical genetics at a time is completely specified by identifying each of the components just listed.[22]

A pattern of reasoning is a sequence of *schematic sentences*, that is sentences in which certain items of nonlogical vocabulary have been replaced by dummy letters, together with a set of *filling instructions* which specify how substitutions are to be made in the schemata to produce reasoning which instantiates the pattern.[23] This notion of pattern is intended to explicate the idea of the common structure that underlies a group of problem-solutions.*

The foregoing definitions enable us to answer the two main questions I posed above. Beliefs about the particular genetic features of particular organisms illustrate or exemplify the version of genetic theory in force at the time in the sense that these beliefs figure in particular problem-solutions generated by the current practice. Certain patterns of reasoning are applied to give the answers to accepted questions, and, in making the application, one puts forward claims about inheritance in particular organisms. Classical genetics persists as a single theory with different versions at different times in the sense that different practices are linked by a chain of practices along which there are relatively small modifications in language, in accepted questions, and in the patterns for answering questions. In addition to this condition of historical connection, versions of classical genetic theory are bound by a common structure: each version uses certain expressions to characterize hereditary phenomena, accepts as important questions of a particular form, and offers a general style of reasoning for answering those questions. Specifically, throughout the career of classical genetics, the theory is directed toward answering questions about the distribution of characteristics in successive generations of a genealogy, and it proposes to answer those questions by using the probabilities of chromosome distribution to compute the probabilities of descendant genotypes.

The approach to classical genetics embodied in these answers is supported by reflection on what beginning students learn. Neophytes are not taught (and never have been taught) a few fundamental theoretical laws from which genetic "theorems" are to be deduced. They are introduced to some technical terminology, which is used to advance a large amount of information about special organisms. Certain questions about heredity in these organisms are posed and answered. Those who understand the theory

---

* For more on patterns of reasoning and their role in Kitcher's view of theoretical explanation as unification, see Philip Kitcher, "Explanatory Unification and the Causal Structure of the World," in P. Kitcher and W. C. Salmon, eds., *Scientific Explanation, Minnesota Studies in the Philosophy of Science*, vol. 13 (Minneapolis: University of Minnesota Press, 1989), 410–505, and Philip Kitcher, *The Advancement of Science* (Oxford: Oxford University Press, 1993), ch. 2.

are those who know what questions are to be asked about hitherto unstudied examples, who know how to apply the technical language to the organisms involved in these examples, and who can apply the patterns of reasoning which are to be instantiated in constructing answers. More simply, successful students grasp general patterns of reasoning which they can use to resolve new cases.

I shall now add some detail to my sketch of the structure of classical genetics, and thereby prepare the way for an investigation of the relations between classical genetics and molecular genetics. The initial family of problems in classical genetics, the family from which the field began, is the family of *pedigree problems*. Such problems arise when we confront several generations of organisms, related by specified connections of descent, with a given distribution of one or more characteristics. The question that arises may be to understand the given distribution of phenotypes, or to predict the distribution of phenotypes in the next generation, or to specify the probability that a particular phenotype will result from a particular mating. In general, classical genetic theory answers such questions by making hypotheses about the relevant genes, their phenotypic effects and their distribution among the individuals in the pedigree. Each version of classical genetic theory contains one or more problem-solving patterns exemplifying this general idea, but the detailed character of the pattern is refined in later versions, so that previously recalcitrant cases of the problem can be accommodated.

Each case of a pedigree problem can be characterized by a set of *data*, a set of *constraints*, and a question. In any example, the data are statements describing the distribution of phenotypes among the organisms in a particular pedigree, or a diagram conveying the same information. The level of detail in the data may vary widely: at one extreme we may be given a full description of the interrelationships among all individuals and the sexes of all those involved; or the data may only provide the numbers of individuals with specific phenotypes in each generation; or, with minimal detail, we may simply be told that from crosses among individuals with specified phenotypes a certain range of phenotypes is found.

The constraints on the problem consist of general cytological information and descriptions of the chromosomal constitution of members of the species. The former will include the thesis that genes are (almost always)[24] chromosomal segments and the principles that govern meiosis. The latter may contain a variety of statements. It may be pertinent to know how the species under study reproduces, how sexual dimorphism is reflected at the chromosomal level, the chromosome number typical of the species, what loci are linked, what the recombination frequencies are, and so forth. As in the case of the data, the level of detail (and thus of stringency) in the constraints can vary widely.

Lastly, each problem contains a question that refers to the organisms described in the data. The question may take several forms: "What is the expected distribution of phenotypes from a cross between *a* and *b*?" (where

a, b are specified individuals belonging to the pedigree described by the data), "What is the probability that a cross between a and b will produce an individual having P?" (where a, b are specified individuals of the pedigree described by the data and P is a phenotypic property manifested in this pedigree), "Why do we find the distribution of phenotypes described in the data?" and others.

Pedigree problems are solved by advancing pieces of reasoning that instantiate a small number of related patterns. In all cases the reasoning begins from a *genetic hypothesis*. The function of a genetic hypothesis is to specify the alleles that are relevant, their phenotypic expression, and their transmission through the pedigree. From that part of the genetic hypothesis that specifies the genotypes of the parents in any mating that occurs in the pedigree, together with the constraints on the problem, one computes the expected distribution of genotypes among the offspring. Finally, for any mating occurring in the pedigree, one shows that the expected distribution of genotypes among the offspring is consistent with the assignment of genotypes given by the genetic hypothesis.

The form of the reasoning can easily be recognized in examples— examples that are familiar to anyone who has ever looked at a textbook or a research report in genetics.[25] What interests me is the style of reasoning itself. The reasoning begins with a genetic hypothesis that offers four kinds of information: (a) specification of the number of relevant loci and the number of alleles at each locus; (b) specification of the relationships between genotypes and phenotypes; (c) specification of the relations between genes and chromosomes, of facts about the transmission of chromosomes to gametes (for example, resolution of the question whether there is disruption of normal segregation) and about the details of zygote formation; (d) assignment of genotypes to individuals in the pedigree. After showing that the genetic hypothesis is consistent with the data and constraints of the problem, the principles of cytology and the laws of probability are used to compute expected distributions of genotypes from crosses. The expected distributions are then compared with those assigned in part (d) of the genetic hypothesis.[26]

Throughout the career of classical genetics, pedigree problems are addressed and solved by carrying out reasoning of the general type just indicated. Each version of classical genetic theory contains a pattern for solving pedigree problems with a method for computing expected genotypes which is adjusted to reflect the particular form of the genetic hypotheses that it sanctions. Thus one way to focus the differences among successive versions of classical genetic theory is to compare their conceptions of the possibilities for genetic hypotheses. As genetic theory develops, there is a changing set of conditions on admissible genetic hypotheses. Prior to the discovery of polygeny and pleiotropy (for example), part (a) of any adequate genetic hypothesis was viewed as governed by the requirement that there would be a one-one correspondence between loci and phenotypic traits.[27] After the discovery of incomplete dominance and epistasis, it was recognized that part (b) of an adequate hypothesis might take a form that had not previously been allowed: one is not compelled to assign to the heterozygote a phenotype assigned to one of the homozygotes, and one is also permitted to relativize the phenotypic effect of a gene to its genetic environment.[28] Similarly, the appreciation of phenomena of linkage, recombination, nondisjunction, segregation distortion, meiotic drive, unequal crossing over, and crossover suppression, modify conditions previously imposed on part (c) of any genetic hypothesis. In general, we can take each version of classical genetic theory to be associated with a set of conditions (usually not formulated explicitly) which govern admissible genetic hypotheses. While a general form of reasoning persists through the development of classical genetics, the patterns of reasoning used to resolve cases of the pedigree problem are constantly fine-tuned as geneticists modify their views about what forms of genetic hypothesis are allowable.

So far I have concentrated exclusively on classical genetic theory as a family of related patterns of reasoning for solving the pedigree problem. It is natural to ask if versions of the theory contain patterns of reasoning for addressing other questions. I believe that they do. The heart of the theory is the theory of *gene transmission*, the family of reasoning patterns directed at the pedigree problem. Out of this theory grow other subtheories. The theory of *gene mapping* offers a pattern of reasoning which addresses questions about the relative positions of loci on chromosomes. It is a direct result of Sturtevant's insight that one can systematically investigate the set of pedigree problems associated with a particular species. In turn, the theory of gene mapping raises the question of how to identify mutations, issues which are to be tackled by the *theory of mutation*. Thus we can think of classical genetics as having a central theory, the theory of gene transmission, which develops in the ways I have described above, surrounded by a number of satellite theories that are directed at questions arising from the pursuit of the central theory. Some of these satellite theories (for example, the theory of gene mapping) develop in the same continuous fashion. Others, like the theory of mutation, are subject to rather dramatic shifts in approach.

4 | Molecular Genetics and Classical Genetics

Armed with some understanding of the structure and evolution of classical genetics, we can finally return to the question with which we began. What is the relation between classical genetics and molecular genetics? When we look at textbook presentations and the pioneering research articles that they cite, it is not hard to discern major ways in which molecular biology has advanced our understanding of hereditary phenomena. We can readily identify particular molecular explanations which illuminate issues that were treated incompletely, if at all, from the classical perspective. What proves puzzling is the connection of these explanations to the theory of classical genetics. I hope that the account of the last section will enable us to make the connection.

I shall consider three of the most celebrated achievements of molecular genetics. Consider first the question of *replication*. Classical geneticists believed that genes can replicate themselves. Even before the experimental demonstration that all genes are transmitted to all the somatic cells of a developing embryo, geneticists agreed that normal processes of mitosis and meiosis must involve gene replication. Muller's suggestion that the central problem of genetics is to understand how mutant alleles, incapable of performing wild-type functions in producing the phenotype, are nonetheless able to replicate themselves, embodies this consensus. Yet classical genetics had no account of gene replication. A molecular account was an almost immediate dividend of the Watson-Crick model of DNA.

Watson and Crick suggested that the two strands of the double helix unwind and each strand serves as the template for the formation of a complementary strand. Because of the specificity of the pairing of nucleotides, reconstruction of DNA can be unambiguously directed by a single strand. This suggestion has been confirmed and articulated by subsequent research in molecular biology.[29] The details are more intricate than Watson and Crick may originally have believed, but the outline of their story stands.

A second major illumination produced by molecular genetics concerns the characterization of mutation. When we understand the gene as a segment of DNA we recognize the ways in which mutant alleles can be produced. "Copying errors" during replication can cause nucleotides to be added, deleted or substituted. These changes will often lead to alleles that code for different proteins, and which are readily recognizable as mutants through their production of deviant phenotypes. However, molecular biology makes it clear that there can be *hidden* mutations, mutations that arise through nucleotide substitutions that do not change the protein produced by a structural gene (the genetic code is redundant) or through substitutions that alter the form of the protein in trivial ways. The molecular perspective provides us with a general answer to the question, "What is a mutation?" namely that a mutation is the modification of a gene through insertion, deletion or substitution of nucleotides. This general answer yields a basic method for tackling (in principle) questions of form, "Is *a* a mutant allele?" namely a demonstration that *a* arose through nucleotide changes from alleles that persist in the present population. The method is frequently used in studies of the genetics of bacteria and bacteriophage, and can sometimes be employed even in inquiries about more complicated organisms. So, for example, there is good biochemical evidence for believing that some alleles which produce resistance to pesticides in various species of insects arose through nucleotide changes in the alleles naturally predominating in the population.[30]

I have indicated two general ways in which molecular biology answers questions that were not adequately resolved by classical genetics. Equally obvious are a large number of more specific achievements. Identification of the molecular structures of particular genes in particular organisms has enabled us to understand why those genes combine to produce the phenotypes they do. One of the most celebrated cases is that of the normal allele for the synthesis of human hemoglobin and the mutant allele that is responsible for sickle-cell anemia.[31] The hemoglobin molecule—whose structure is known in detail—is built up from four amino-acid chains (two "α-chains" and two "β-chains"). The mutant allele results from substitution of a single nucleotide with the result that one amino acid is different (the sixth amino acid in the β-chains). This slight modification causes a change in the interactions of hemoglobin molecules: deoxygenated mutant hemoglobin molecules combine to form long fibres. Cells containing the abnormal molecule become deformed after they have given up their oxygen, and because they become rigid, they can become stuck in narrow capillaries, if they give up their oxygen too soon. Individuals who are homozygous for the mutant gene are vulnerable to experience blockages of blood flow. However, in heterozygous individuals, there is enough normal hemoglobin in blood cells to delay the time of formation of the distorting fibres, so that the individual is physiologically normal.

This example is typical of a broad range of cases, among which are some of the most outstanding achievements of molecular genetics. In all of the cases, we replace a simple assertion about the existence of certain alleles which give rise to various phenotypes with a molecular characterization of those alleles from which we can derive descriptions of the phenotypes previously attributed.

I claim that the successes of molecular genetics which I have just briefly described—and which are among the accomplishments most emphasized in the biological literature—can be understood from the perspective on theories that I have developed above. The three examples reflect three different relations among successive theories, all of which are different from the classical notion of reduction (and the usual modifications of it). Let us consider them in turn.

The claim that genes can replicate does not have the status of a central law of classical genetic theory.[32] It is not something that figures prominently in the explanations provided by the theory (as, for example, the Boyle-Charles law is a prominent premise in some of the explanations yielded by phenomenological thermodynamics). Rather, it is a claim that classical geneticists took for granted, a claim presupposed by explanations, rather than an explicit part of them. Prior to the development of molecular genetics that claim had come to seem increasingly problematic. If genes can replicate, how do they manage to do it? Molecular genetics answered the worrying question. It provided a theoretical demonstration of the possibility of an antecedently problematic presupposition of classical genetics.

We can say that a theory presupposes a statement p if there is some problem-solving pattern of the theory, such that every instantiation of the pattern contains statements that jointly imply the truth of p. Suppose that, at a given stage in the development of a theory, scientists recognize an argument from otherwise acceptable premises which concludes that it is impossible that p. Then the presupposition p is problematic for those scientists. What they would like would be an argument showing that it is possible that p and explaining what is wrong with the line of reasoning which appears to threaten the possibility of p. If a new theory generates an argument of this sort, then we can say that the new theory gives a theoretical demonstration of the possibility of an antecedently problematic presupposition of the old theory.

A less abstract account will help us to see what is going on in the case of gene replication. Very frequently, scientists take for granted in their explanations some general property of entities that they invoke. Their assumption can come to seem problematic if the entities in question are supposed to belong to a kind, and there arises a legitimate doubt about whether members of the kind can have the property attributed. A milder version of the problem arises if, in all cases in which the question of whether things of the general kind have the property can be settled by appealing to background theory, it turns out that the answer is negative. Under these circumstances, the scientists are committed to regarding their favored entities as unlike those things of the kind which are amenable to theoretical study with respect to the property under discussion. The situation is worse if background theory provides an argument for thinking that *no* things of the kind can have the property.

Consider now the case of gene replication. For any problem-solution offered by any version of the theory of gene transmission (the central subtheory of classical genetic theory), that problem-solution will contain sentences implying that the alleles which it discusses are able to replicate. Classical genetics presupposes that a large number of identifiable genes can replicate. This presupposition was always weakly problematic because genes were taken to be complicated molecules and, in all cases in which appeal to biochemistry could be made to settle the issue of whether a molecular structure was capable of replication, the issue was decided in the negative. Muller exacerbated the problem by suggesting that mutant alleles are damaged molecules (after all, many of them were produced through x-ray bombardment, an extreme form of molecular torture). So there appeared to be a strong argument against the possibility that any mutant allele can replicate. After the work of Watson, Crick, Kornberg, and others, there was a theoretical demonstration of the allegedly problematic possibility. One can show that genes can replicate by showing that any segment of DNA (or RNA) can replicate. (DNA and RNA are the genetic materials. Establishing the power of the genetic material to replicate bypasses the problem of deciding which segments are genes. Thus the difficulties posed by the falsity of [R2] are avoided.) The Watson-Crick model provides a characterization of the (principal) genetic material, and when this description is inserted into standard patterns of chemical reasoning one can generate an argument whose conclusion asserts that, under specified conditions, DNA replicates. Moreover, given the molecular characterization of DNA and of mutation, it is possible to see that although mutant alleles are "damaged" molecules, the kind of damage (insertion, deletion, or substitution of nucleotides) does not affect the ability of the resultant molecule to replicate.

Because theoretical demonstrations of the possibility of antecedently problematic presuppositions involve derivation of conclusions of one theory from the premises supplied by a background theory, it is easy to assimilate them to the classical notion of reduction. However, on the account I have offered, there are two important differences. First, there is no commitment to the thesis that genetic theory can be formulated as (the deductive closure of) a conjunction of laws. Second, it is not assumed that all general statements about genes are equally in need of molecular derivation. Instead, one particular thesis, a thesis that underlies all the explanations provided by classical genetic theory, is seen as especially problematic, and the molecular derivation is viewed as addressing a specific problem that classical geneticists had already perceived. Where the reductionist identifies a general benefit in deriving all the axioms of the reduced theory, I focus on a particular derivation of a claim that has no title as an axiom of classical genetics, a derivation which responds to a particular explanatory difficulty of which classical geneticists were acutely aware. The reductionist's global relation between theories does not obtain between classical and molecular genetics, but something akin to it does hold between special fragments of these theories.[33]

The second principal achievement of molecular genetics, the account of mutation, involves a conceptual refinement of prior theory. Later theories can be said to provide conceptual refinements of earlier theories when the later theory yields a specification of entities that belong to the extensions of predicates in the language of the earlier theory, with the result that the ways in which the referents of these predicates are fixed are altered in accordance with the new specifications. Conceptual refinement may occur in a number of ways. A new theory may supply a descriptive characterization of the extension of a predicate for which no descriptive characterization was previously available; or it may offer a new description which makes it reasonable to amend characterizations that had previously been accepted.[34] In the case at hand, the referent of many tokens of 'mutant allele' was initially fixed through the description "chromosomal segment producing a heritable deviant phenotype." After Bridges's discovery of unequal crossing-over at the *Bar* locus in *Drosophila*, it was evident to classical geneticists that this descriptive specification covered cases in which the internal structure of a gene was altered and cases in which neighboring genes were transposed. Thus it was necessary to retreat to the less applicable description "chromosomal segment producing a heritable deviant phenotype as the result of an internal change within an allele." Molecular genetics offers a precise account of the internal changes, with the result that the description can be made more informative: mutant alleles are segments of DNA that result from prior alleles through deletion, insertion, or substitution of nucleotides. This re-fixing of the referent of 'mutant allele' makes it possible in principle to distinguish cases of mutation from cases of recombination, and thus to resolve those controversies that frequently arose from the use of 'mutant allele' in the later days of classical genetics.[35]

Finally, let us consider the use of molecular genetics to illuminate the action of particular genes. Here we again seem to find a relationship that initially appears close to the reductionist's ideal. Statements that are invoked as premises in particular problem-solutions—statements that ascribe particular phenotypes to particular genotypes—are derived from molecular characterizations of the alleles involved. On the account of classical genetics offered in Section 3, each version of classical genetic theory includes in its schema for genetic hypotheses a clause which relates genotypes to phenotypes. . . . Generalizing from the hemoglobin example, we might hope to discover a pattern of reasoning within molecular genetics that would generate as its conclusion the schema for assigning phenotypes to genotypes.

It is not hard to characterize the relation just envisioned. Let us say that a theory T′ provides an *explanatory extension* of a theory T just in case there is some problem-solving pattern of T one of whose schematic premises can be generated as the conclusion of a problem-solving pattern of T′. When a new theory provides an explanatory extension of an old theory, then particular premises occurring in explanatory derivations given by the old theory can themselves be explained by using arguments furnished by the new theory. However, it does not follow that the explanations provided by the old theory can be improved by replacing the premises in question with the pertinent derivations. What is relevant for the purposes of explaining some statement S may not be relevant for the purposes of explaining a statement S′ which figures in an explanatory derivation of S.

Even though reductionism fails, it may appear that we can capture part of the spirit of reductionism by deploying the notion of explanatory extension. The thesis that molecular genetics provides an explanatory extension of classical genetics embodies the idea of a global relationship between the two theories, while avoiding two of the three troubles that were found to beset reductionism. That thesis does not simply assert that some specific presupposition of classical genetics (for example, the claim that genes are able to replicate) can be derived as the conclusion of a molecular argument, but offers a general connection between premises of explanatory derivations in classical genetics and explanatory arguments from molecular genetics. It is formulated so as to accommodate the failure of (R1) and to honor the picture of classical genetics developed in Section 3. Moreover, the failure of (R2) does not affect it. If we take the hemoglobin example as a paradigm, we can justifiably contend that the explanatory extension does not require any general characterization of genes in molecular terms. All that is needed is the possibility of deriving phenotypic descriptions from molecular characterizations of the structures of *particular* genes. Thus, having surmounted two hurdles, our modified reductionist thesis is apparently within sight of success.

Nevertheless, even born-again reductionism is doomed to fall short of salvation. Although it is true that molecular genetics belongs to a cluster of theories which, taken together, provide an explanatory extension of classical genetics, molecular genetics, on its own, cannot deliver the goods. There are some cases in which the ancillary theories do not contribute to the explanation of a classical claim about gene action. In such cases, the classical claim can be derived and explained by instantiating a pattern drawn from molecular genetics. The example of human hemoglobin provides one such case. But this example is atypical.

Consider the way in which the hemoglobin example works. Specification of the molecular structures of the normal and mutant alleles, together with a description of the genetic code, enables us to derive the composition of normal and mutant hemoglobin. Application of chemistry then yields descriptions of the interactions of the proteins. With the aid of some facts about human blood cells, one can then deduce that the sickling effect will occur in abnormal cells, and, given some facts about human physiology, it is possible to derive the descriptions of the phenotypes. There is a clear analogy here with some cases from physics. The assumptions about blood cells and physiological needs seem to play the same role as the boundary conditions about shapes, relative positions and velocities of planets that occur in Newtonian derivations of Kepler's laws. In the Newtonian explanation we can see the application of a general pattern of reasoning—the derivation of explicit equations of motion from specifications of the forces acting—which yields the general result that a body under the influence of a centrally directed inverse square force will travel in a conic section; the general result is then applied to the motions of the planets by incorporating pieces of astronomical information. Similarly, the derivation of the classical claims about the action of the normal and mutant hemoglobin genes can be seen as a purely chemical derivation of the generation of certain molecular structures and of the interactions among them. The chemical conclusions are then applied to the biological system under consideration by introducing three "boundary conditions": first, the claim that the altered molecular structures only affect development to the extent of substituting a different molecule in the erythrocytes (the blood cells that transport hemoglobin); second, a description of the chemical conditions in the capillaries; and third, a description of the effects upon the organism of capillary blockage.

The example is able to lend comfort to reductionism precisely because of an atypical feature. In effect, one concentrates on the *differences* among the phenotypes, takes for granted the fact that in all cases development will proceed normally to the extent of manufacturing erythrocytes—which are, to all intents and purposes, simply sacks for containing hemoglobin molecules—and compares the difference in chemical effect of the cases in which the erythrocytes contain different molecules. *The details of the process of development can be ignored.* However, it is rare for the effect of a mutation to be so simple. Most structural genes code for molecules whose presence or absence make subtle differences. Thus, typically, a mutation will affect the distribution of chemicals in the cells of a developing embryo. A likely result is a change in the timing of intracellular reactions, a change that may, in turn, alter the shape of the cell. Because of the change of shape, the geometry of the embryonic cells may be modified. Cells that usually come into contact may fail to touch. Because of this, some cells may not receive the molecules necessary to switch on certain batteries of genes. Hence the chemical composition of these cells will be altered. And so it goes.[36]

Quite evidently, in examples like this, (which include most of the cases in which molecular considerations can be introduced into embryology) the reasoning that leads us to a description of the phenotype associated with a genotype will be much more complicated than that found in the hemoglobin case. It will not simply consist in a chemical derivation adapted with the help of a few boundary conditions furnished by biology. Instead, we shall encounter a sequence of subarguments: molecular descriptions lead to specifications of cellular properties, from these specifications we draw conclusions about cellular interactions, and from these conclusions we arrive at further molecular descriptions. There is clearly a pattern of reasoning here which involves molecular biology and which extends the explanations furnished by classical genetics by showing how phenotypes depend upon genotypes—but I think it would be folly to suggest that the extension is provided by molecular genetics alone.

In Section 2, we discovered that the traditional answer to the philosophical question of understanding the relation that holds between molecular genetics and classical genetics, the reductionist's answer, will not do. Section 3 attempted to build on the diagnosis of the ills of reductionism, offering an account of the structure and evolution of classical genetics that would improve on the picture offered by those who favor traditional approaches to the nature of scientific theories. In the present section, I have tried to use the framework of Section 3 to understand the relations between molecular genetics and classical genetics. Molecular genetics has done something important for classical genetics, and its achievements can be recognized by seeing them as instances of the intertheoretic relations that I have characterized. Thus I claim that the problem from which we began is solved.

So what? Do we have here simply a study of a particular case—a case which has, to be sure, proved puzzling for the usual accounts of scientific theories and scientific change? I hope not. Although the traditional approaches may have proved helpful in understanding some of the well-worn examples that have been the stock-in-trade of twentieth-century philosophy of science, I believe that the notion of scientific practice sketched in Section 3 and the intertheoretic relations briefly characterized here will both prove helpful in analyzing the structure of science and the growth of scientific knowledge *even in those areas of science where traditional views have seemed most successful.*[37] Hence the tale of two sciences which I have been telling is not merely intended as a piece of local history that fills a small but troublesome gap in the orthodox chronicles. I hope that it introduces concepts of general significance in the project of understanding the growth of science.

5 | Anti-Reductionism and the Organization of Nature

One loose thread remains. The history of biology is marked by continuing opposition between reductionists and anti-reductionists. Reductionism thrives on exploiting the charge that it provides the only alternative to the mushy incomprehensibility of vitalism. Anti-reductionists reply that their opponents have ignored the organismic complexity of nature. Given the picture painted above, where does this traditional dispute now stand?

I suggest that the account of genetics which I have offered will enable reductionists to provide a more exact account of what they claim, and will thereby enable anti-reductionists to be more specific about what they are denying. Reductionists and anti-reductionists agree in a certain minimal physicalism. To my knowledge, there are no major figures in contemporary biology who dispute the claim that each biological event, state or process is a complex physical event, state, or process. The most intricate part of ontogeny or phylogeny involves countless changes of physical state. What antireductionists emphasize is the organization of nature and the "interactions among phenomena at different levels." The appeal to organization takes two different forms. When the subject of controversy is the proper form of evolutionary theory, then anti-reductionists contend that it is impossible to regard all selection as operating at the level of the gene.[38] What concerns me here is not this area of conflict between reductionists and their adversaries, but the attempt to block claims for the hegemony of molecular studies in understanding the physiology, genetics, and development of organisms.[39]

A sophisticated reductionist ought to allow that, in the current practice of biology, nature is divided into levels which form the proper provinces of areas of biological study: molecular biology, cytology, histology, physiology, and so forth. Each of these sciences can be thought of as using certain language to formulate the questions it deems important and as supplying patterns of reasoning for resolving those questions. Reductionists can now set forth one of two main claims. The stronger thesis is that the explanations provided by any biological theories can be reformulated in the language of molecular biology and be recast so as to instantiate the patterns of reasoning supplied by molecular biology. The weaker thesis is that molecular biology provides an explanatory extension of the other biological sciences.

Strong reductionism falls victim to the considerations that were advanced against (R3). The distribution of genes to gametes is to be explained, not by rehearsing the gory details of the reshuffling of the molecules, but through the observation that chromosomes are aligned in pairs just prior to the meiotic division, and that one chromosome from each matched pair is transmitted to each gamete. We may formulate this point in the biologists' preferred idiom by saying that the assortment of alleles is to be understood at the cytological level. What is meant by this description is that there is a pattern of reasoning which is applied to derive the description of the assortment of alleles and which involves predicates that characterize cells and their large-scale internal structures. That pattern of reasoning is to be objectively preferred to the molecular pattern which would be instantiated by the derivation that charts that complicated rearrangements of individual molecules because it can be applied across a range of cases which would look heterogeneous from a molecular perspective. Intuitively, the cytological pattern makes connections which are lost at the molecular level, and it is thus to be preferred.

So far, anti-reductionism emerges as the thesis that there are *autonomous levels of biological explanation*. Anti-reductionism construes the current division of biology not simply as a temporary feature of our science stemming from our cognitive imperfections but as the reflection of levels of organization in nature. Explanatory patterns that deploy the concepts of cytology will endure in our science because we would foreswear significant unification (or fail to employ the relevant laws, or fail to identify the causally relevant properties) by attempting to derive the conclusions to which they are applied using the vocabulary and reasoning patterns of molecular biology. But the autonomy thesis is only the beginning of anti-reductionism. A stronger doctrine can be generated by opposing the weaker version of sophisticated reductionism.

In Section 4, I raised the possibility that molecular genetics may be viewed as providing an explanatory extension of classical genetics through deriving the schematic sentence that assigns phenotypes to genotypes from a molecular pattern of reasoning. This apparent possibility fails in an instructive way. Anti-reductionists are not only able to contend that there are autonomous levels of biological explanation. They can also resist the weaker reductionist view that explanation always flows from the molecular level up. Even if reductionists retreat to the modest claim that, while there are autonomous levels of explanation, descriptions of cells and their constituents are always explained in terms of descriptions about genes, descriptions of tissue geometry are always explained in terms of descriptions of cells, and so forth, anti-reductionists can resist the picture of a unidirectional flow of explanation. Understanding the phenotypic manifestation of a gene, they will maintain, requires constant shifting back and forth across levels. Because developmental processes are complex and because changes in the timing of embryological events may produce a cascade of effects at several different levels, one sometimes uses descriptions at higher levels to explain what goes on at a more fundamental level.

For example, to understand the phenotype associated with a mutant limb-bud allele, one may begin by tracing the tissue geometry to an underlying molecular structure. The molecular constitution of the mutant allele gives rise to a nonfunctional protein, causing some abnormality in the internal structures of cells. The abnormality is reflected in peculiarities of cell shape, which, in turn, affects the spatial relations among the cells of the embryo. So far we have the unidirectional flow of explanation which the reductionist envisages. However, the subsequent course of the explanation is different. Because of the abnormal tissue geometry, cells that are normally in contact fail to touch; because they do not touch, certain important molecules, which activate some batteries of genes, do not reach crucial cells; because the genes in question are not "switched on" a needed morphogen is not produced; the result is an abnormal morphology in the limb.

Reductionists may point out, quite correctly, that there is some very complex molecular description of the entire situation. The tissue geometry is, after all, a configuration of molecules. But this point is no more relevant than the comparable claim about the process of meiotic division in which alleles are distributed to gametes. Certain genes are not expressed because of the geometrical structure of the cells in the tissue: *the pertinent cells are too far apart.* However this is realized at the molecular level, our explanation must bring out the salient fact that it is the presence of a gap between cells that are normally adjacent that explains the nonexpression of the genes. As in the example of allele transmission at meiosis, we lose sight of the important connections by attempting to treat the situation from a molecular point of view. As before, the point can be sharpened by considering situations in which radically different molecular configurations realize the crucial feature of the tissue geometry: situations in which heterogeneous molecular structures realize the breakdown of communication between the cells.

Hence, embryology provides support for the stronger anti-reductionist claim. Not only is there a case for the thesis of autonomous levels of explanation, but we find examples in which claims at a more fundamental level (specifically, claims about gene expression) are to be explained in terms of claims at a less fundamental level (specifically, descriptions of the relative positions of pertinent cells). Two anti-reductionist biologists put the point succinctly:

> . . . a developmental program is not to be viewed as a linearly organized causal chain from genome to phenotype. Rather, morphology emerges as a consequence of an increasingly complex dialogue between cell populations, characterized by their geometric continuities, and the cells' genomes, characterized by their states of gene activity.[40]

A corollary is that the explanations provided by the "less fundamental" biological sciences are not extended by molecular biology alone.

994 | CH. 8 INTERTHEORETIC REDUCTION

It would be premature to claim that I have shown how to reformulate the anti-reductionist appeals to the organization of nature in a completely precise way. My conclusion is that, to the extent that we can make sense of the present explanatory structure within biology—that division of the field into subfields corresponding to levels of organization in nature—we can also understand the anti-reductionist doctrine. In its minimal form, it is the claim that the commitment to several explanatory levels does not simply reflect our cognitive limitations; in its stronger form, it is the thesis that some explanations oppose the direction of preferred reductionistic explanation. Reductionists should not dismiss these doctrines as incomprehensible mush unless they are prepared to reject as unintelligible the biological strategy of dividing the field (a strategy which seems to me well understood, even if unanalyzed).

The examples I have given seem to support both anti-reductionist doctrines. To clinch the case, further analysis is needed. The notion of explanatory levels obviously cries out for explication, and it would be illuminating to replace the informal argument that the unification of our beliefs is best achieved by preserving multiple explanatory levels with an argument based on a more exact criterion for unification. Nevertheless, I hope that I have said enough to make plausible the view that, despite the immense value of the molecular biology that Watson and Crick launched in 1953, molecular studies cannot cannibalize the rest of biology. Even if geneticists must become "physiological chemists" they should not give up being embryologists, physiologists, and cytologists.[41]

Notes

1. "Variation due to change in the individual gene," reprinted in J. A. Peters ed., *Classic Papers in Genetics* (Englewood Cliffs, N.J.: Prentice-Hall, 1959), pp. 104–116. Citation from p. 115.

2. "Molecular Structure of Nucleic Acids," *Nature* 171 (1953), pp. 737–738; reprinted in Peters, op. cit., pp. 241–243. Watson and Crick amplified their suggestion in "Genetic Implications of the Structure of Deoxyribonucleic Acid," *Nature* 171 (1953), pp. 934–937.

3. The most sophisticated attempts to work out a defensible version of reductionism occur in articles by Kenneth Schaffner. See, in particular, "Approaches to Reduction," *Philosophy of Science* 34 (1967), pp. 137–147; "The Watson-Crick Model and Reductionism," *British Journal for the Philosophy of Science* 20 (1969), pp. 325–348; "The Peripherality of Reductionism in the Development of Molecular Biology," *Journal of the History of Biology* 7 (1974), pp. 111–139; and "Reductionism in Biology: Prospects and Problems," R. S. Cohen *et al.* eds., *PSA 1974* (Boston: D. Reidel, 1976), pp. 613–632. See also Michael Ruse, "Reduction, Replacement, and Molecular Biology," *Dialectica* 25 (1971), pp. 38–72; and William K. Goosens, "Reduction by Molecular Genetics," *Philosophy of Science* 45 (1978), pp. 78–95. A

variety of anti-reductionist points are made in David Hull, "Reduction in Genetics—Biology or Philosophy?" *Philosophy of Science* 39 (1972), pp. 491–499; and Chapter 1 of *Philosophy of Biological Science* (Englewood Cliffs, N.J.: Prentice-Hall, 1974); in Steven Orla Kimbrough, "On the Reduction of Genetics to Molecular Biology," *Philosophy of Science* 46 (1979), pp. 389–406 and in Ernst Mayr, *The Growth of Biological Thought* (Cambridge, Mass.: Harvard University Press, 1982), pp. 59–63.

4. Typically, though not invariably. In a suggestive essay, "Reductive Explanation: A Functional Account," (R. S. Cohen *et al.* op. cit. pp. 671–710), William Wimsatt offers a number of interesting ideas about intertheoretic relations and the case of genetics. Also provocative are Nancy Maull's "Unifying Science Without Reduction," *Studies in the History and Philosophy of Science* 8 (1977), pp. 143–171; and Lindley Darden and Nancy Maull, "Interfield Theories," *Philosophy of Science* 44 (1977), pp. 43–64. My chief complaint about the works I have cited is that unexplained technical notions—"mechanisms," "levels," "domain," "field," "theory"—are invoked (sometimes in apparently inconsistent ways), so that no precise answer to the philosophical problem posed in the text is ever given. Nevertheless, I hope that the discussion of the later sections of this paper will help to articulate more fully some of the genuine insights of these authors, especially those contained in Wimsatt's rich essay.

5. E. Nagel, *The Structure of Science* (New York: Harcourt Brace, 1961), Chapter 11. A simplified presentation can be found in Chapter 8 of C. G. Hempel, *Philosophy of Natural Science* (Englewood Cliffs, N.J.: Prentice-Hall, 1966).

6. Quite evidently, this is a weak version of what was once the "received view" of scientific theories, articulated in the works of Nagel and Hempel cited in the previous note. A sustained presentation and critique of the view is given in the Introduction to F. Suppe, ed., *The Structure of Scientific Theories* (Urbana: University of Illinois Press, 1973). The fact that the standard model of reduction presupposes the thesis that theories are reasonably regarded as sets of statements has been noted by Clark Glymour, "On Some Patterns of Reduction," *Philosophy of Science* 36 (1969), pp. 340–353, 342; and by Jerry Fodor (*The Language of Thought*, New York: Crowell, 1975, p. 11, footnote 10). Glymour endorses the thesis; Fodor is skeptical about it.

7. Philosophers often suggest that, in reduction, one derives *corrected* laws of the reduced theory from an *unmodified* reducing theory. But this is not the way things go in the paradigm cases: one doesn't correct Galileo's law by using Newtonian mechanics; instead, one neglects "insignificant terms" in the Newtonian equation of motion for a body falling under the influence of gravity; similarly, in deriving the Boyle-Charles law from kinetic theory (or statistical mechanics), it is standard to make idealizing assumptions about molecules, and so obtain the exact version of the Boyle-Charles law; subsequently, corrected versions are generated by "subtracting" the idealizing procedures. Although he usually views reduction as deriving a corrected version of the reduced theory, Schaffner notes that reduction might sometimes proceed by modifying the reducing theory ("Approaches to Reduction," p. 138; "The Watson-Crick Model and Reductionism," p. 322). In fact, the point was already made by Nagel, op. cit.

8. In part, because modification might produce an inconsistent theory that would permit the derivation of anything. In part, because of the traditionally vexing problem of the proper form for bridge principles in heterogeneous reductions in physics.

The former problem is discussed in Glymour, op. cit., p. 352 and in Dudley Shapere, "Notes towards a Post-Positivistic Interpretation of Science" (P. Achinstein and S. Barker, eds. *The Legacy of Logical Positivism*, Baltimore: Johns Hopkins University Press, 1971). For discussion of the latter issue, see Larry Sklar, "Types of Inter-Theoretic Reduction," *British Journal for the Philosophy of Science* 18 (1967), pp. 109–124; Robert Causey, "Attribute Identities in Microreductions," *Journal of Philosophy* 69 (1972), pp. 407–422; and Berent Enç, "Identity Statements and Micro-Reductions," *Journal of Philosophy* 73 (1976), pp. 285–306. The concerns that I shall raise are orthogonal to these familiar areas of dispute.

9. Hull, *Philosophy of Biological Science*, p. 23, adapted from Theodosius Dobzhansky, *Genetics of the Evolutionary Process* (New York: Columbia University Press, 1970), p. 167. Similarly molecular genetics is said to have the task of "discovering how molecularly characterized genes produce proteins which in turn combine to form gross phenotypic traits" (Hull *ibid.*: see also James D. Watson, *Molecular Biology of the Gene*, Menlo Park, Ca., W. A. Benjamin, 1976, p. 54).

10. The phenotype/genotype distinction was introduced to differentiate the observable characteristics of an organism from the underlying genetic factors. In subsequent discussions the notion of phenotype has been extended to include properties which are not readily observable (for example, the capacity of an organism to metabolize a particular amino acid). The expansion of the concept of phenotype is discussed in my paper "Genes," *British Journal for the Philosophy of Science* 33 (1982), pp. 337–359.

11. A *locus* is the place on a chromosome occupied by a gene. Different genes which can occur at the same locus are said to be *alleles*. In diploid organisms, chromosomes line up in pairs just before the meiotic division that gives rise to gametes. The matched pairs are pairs of *homologous chromosomes*. If different alleles occur at corresponding loci on a pair of homologous chromosomes, the organism is said to be *heterozygous* at these loci.

12. *Recombination* is the process (which occurs before meiotic division) in which a chromosome exchanges material with the chromosome homologous with it. Alleles which occur on one chromosome may thus be transferred to the other chromosome, so that new genetic combinations can arise.

13. Other central Mendelian claims also turn out to be false. The Mendelian principle that if an organism is heterozygous at a locus then the probabilities of either allele being transmitted to a gamete are equal falls afoul of cases of meiotic drive. (A notorious example is the *t*-allele in the house mouse, which is transmitted to 95% of the sperm of males who are heterozygous for it and the wild-type allele; see R. C. Lewontin and L. C. Dunn, "The Evolutionary Dynamics of a Polymorphism in the House Mouse," *Genetics* 45 (1960), pp. 705–722.) Even the idea that genes are transmitted across the generations, unaffected by their presence in intermediate organisms, must be given up once we recognize that intra-allelic recombination can occur.

14. To the best of my knowledge, the mechanisms of this interference are not well understood. For a brief discussion, see J. Sybenga, *General Cytogenetics* (North-Holland, 1972), pp. 313–314. In this paper, I shall use "segregation distortion" to refer to cases in which there is a propensity for nonhomologous chromosomes to assort together. "Meiotic drive" will refer to examples in which one member of a pair of homologous chromosomes has a greater probability of being transmitted to

a gamete. The literature in genetics exhibits some variation in the use of these terms. Let me note explicitly that, on these construals, both segregation distortion and meiotic drive will be different from *nondisjunction*, the process in which a chromosome together with the whole (or a part) of the homologous chromosome is transmitted to a gamete.

15. This point raises some interesting issues. It is common practice in genetics to count a segment of DNA as a single gene if it was produced by mutation from a gene. Thus many mutant alleles are viewed as DNA segments in which modification of the sequence of bases has halted transcription too soon, with the result that the gene product is truncated and nonfunctional. My envisaged case simply assumes that a second mutation occurs further down the segment so that transcription starts and stops in two places, generating two useless gene products. The historical connection with the original allele serves to identify the segment as one gene.

Conversely, where there is no historical connection to any organism, one may have qualms about counting a DNA segment as a gene. Suppose that, in some region of space, a quirk of nature brings together the constituent atoms for the white eye mutant in *Drosophila melanogaster*, and that the atoms become arranged in the right way. Do we have here a *Drosophila* gene? If the right answer is "No" then it would seem that a molecular structure only counts as a gene given an appropriate history. I hasten to add that "appropriate histories" need not simply involve the usual biological ways in which organisms transmit, replicate and modify genes: one can reasonably hope to synthesize genes in the laboratory. The case seems analogous to questions that arise about personal identity. If a person's psychological features are replicated by a process that sets up the "right sort of causal connection" between person and product, then we are tempted to count the product as the surviving person. Similarly, if a molecular structure is generated in a way that sets up "the right sort of causal connection" between the structure and some prior gene then it counts as a gene. In both cases, causal connections of "the right sort" may be set up in everyday biological ways and by means of deliberate attempts to replicate a prior structure.

16. So-called *structural genes* direct the formation of proteins by coding for RNA molecules. They are "transcribed" to produce *messenger* RNA (mRNA) which serves as a more immediate "blueprint" for the construction of the protein. Transcription is started and stopped through the action of regulatory genes. In the simplest regulatory system (that of the *lac* operon) an area adjacent to the structural gene serves as a "dumping ground" for a molecule. When concentration of the protein product becomes too high, the molecule attaches to this site and transcription halts; when more protein is required, the cell produces a molecule that removes the inhibiting molecule from the neighborhood of the structural gene, and transcription begins again. (For much more detail, see Watson, op. cit., Chapter 14, and M. W. Strickberger, *Genetics* (New York: Macmillan, 1976), Chapter 29.)

17. The situation is complicated by the existence of "introns"—segments within genes whose products under transcription are later excised—and by the enormous amount of repetitive DNA that most organisms seem to contain. Moreover, the regulatory systems in eukaryotes appear to be much more complicated than the prokaryote systems (of which the *lac* operon is *one* paradigm). For a review of the situation, as of a few years ago, see Eric H. Davidson, *Gene Expression in Early Development* (New York: Academic Press, 1976).

18. The account will be even more complicated if we honor the suggestion of footnote 15, and suppose that, for a molecular structure to count as a gene, it must be produced in the right way.

19. Early in the process preceding meiotic division the chromosomes become more compact. As meiosis proceeds, the nucleus comes to contain a system of threads that resembles a spindle. Homologous chromosomes line up together near the center of the spindle, and they are oriented so that one member of each pair is slightly closer to one pole of the spindle, while the other is slightly closer to the opposite pole.

20. The point I have been making is related to an observation of Hilary Putnam's. Discussing a similar example, Putnam writes: "The same explanation will go in any world (whatever the microstructure) in which those *higher level structural features* are present"; he goes on to claim that "explanation is superior not just subjectively but methodologically, . . . if it brings out relevant laws." (Putnam, "Philosophy and our Mental Life," in *Mind, Language, and Reality* (Cambridge, Cambridge University Press, 1975), pp. 291–303, p. 296). The point is articulated by Alan Garfinkel (*Forms of Explanation*, New Haven: Yale University Press, 1981), and William Wimsatt has also raised analogous considerations about explanation in genetics.

It is tempting to think that the independence of the "higher level structural features" in Putnam's example and in my own can be easily established: one need only note that there are worlds in which the same feature is present without any molecular realization. So, in the case discussed in the text, PS-processes might go on in worlds where all objects were perfect continua. But although this shows that PS-processes form a kind which could be realized without molecular reshuffling, we know that all *actual* PS-processes do involve such reshufflings. The reductionist can plausibly argue that *if* the set of PS-processes with molecular realizations is itself a natural kind, then the explanatory power of the cytological account can be preserved by identifying meiosis as a process of this narrower kind. Thus the crucial issue is not whether PS-processes form a kind with nonmolecular realizations, but whether those PS-processes which have molecular realizations form a kind that can be characterized from the molecular point of view. Hence, the easy strategy of responding to the reductionist must give way to the approach adopted in the text. (I am grateful to the editors of *The Philosophical Review* for helping me to see this point.)

21. It would be impossible in the scope of this paper to do justice to the various conceptions of scientific theory that have emerged from the demise of the "received view." Detailed comparison of the perspective I favor with more traditional approaches (both those that remain faithful to core ideas of the "received view" and those that adopt the "semantic view" of theories) must await another occasion.

22. My notion of a practice owes much to some neglected ideas of Sylvain Bromberger and Thomas Kuhn. See, in particular, Bromberger, "A Theory about the Theory of Theory and about the Theory of Theories" (W. L. Reese ed., *Philosophy of Science, The Delaware Seminar*, New York, 1963); "Questions," (*Journal of Philosophy* 63 (1966), pp. 597–606); Kuhn, *The Structure of Scientific Revolutions* (Chicago: University of Chicago Press, 1962), Chapters II–V. The relation between the notion of a practice and Kuhn's conception of a paradigm is discussed in Chapter 7 of my book *The Nature of Mathematical Knowledge* (New York: Oxford University Press, 1983).

23. More exactly, a general argument pattern is a triple consisting of a sequence of schematic sentences (a *schematic argument*), a set of filling instructions (directions as to how dummy letters are to be replaced), and a set of sentences describing the inferential characteristics of the schematic argument (a *classification* for the schematic argument). A sequence of sentences instantiates the general argument pattern just in case it meets the following conditions: (i) the sequence has the same number of members as the schematic argument of the general argument pattern; (ii) each sentence in the sequence is obtained from the corresponding schematic sentence in accordance with the appropriate filling instructions; (iii) it is possible to construct a chain of reasoning which assigns to each sentence the status accorded to the corresponding schematic sentence by the classification. For some efforts at explanation and motivation, see my "Explanatory Unification," *Philosophy of Science* 48 (1981), pp. 507–531 (reprinted in chapter 6).

24. Sometimes particles in the cytoplasm account for hereditary traits. See Strickberger, op. cit., pp. 257–265.

25. For examples, see Strickberger op. cit. Chapters 6–12, 14–17, especially Chapter 11; Peters, op. cit.; and H. L. K. Whitehouse, *Towards an Understanding of the Mechanism of Heredity* (London: Arnold, 1965).

26. The comparison will make use of standard statistical techniques, such as the chi-squared test.

27. *Polygeny* occurs when many genes affect one characteristic; *pleiotropy* occurs when one gene affects more than one characteristic.

28. *Incomplete dominance* occurs when the phenotype of the heterozygote is intermediate between that of the homozygotes; *epistasis* occurs when the effect of a particular combination of alleles at one locus depends on what alleles are present at another locus.

29. See Watson, op. cit., Chapter 9; and Arthur Kornberg, *DNA Synthesis* (San Fransisco: W. H. Freeman, 1974).

30. See G. P. Georghiou, "The Evolution of Resistance to Pesticides," *Annual Review of Ecology and Systematics* 3 (1972), pp. 133–168.

31. See Watson, op. cit., pp. 189–193; and T. H. Maugh II, "A New Understanding of Sickle Cell Emerges," *Science* 211 (1981), pp. 265–267.

32. However, one might claim that "Genes can replicate" is a law of genetics, in that it is general, lawlike, and true. This does not vitiate my claim that the structure of classical genetics is not to be sought by looking for a set of general laws, for the law in question is so weak that there is little prospect of finding supplementary principles which can be conjoined with it to yield a representation of genetic theory. I suggest that "Genes can replicate" is analogous to the thermodynamic "law," "Gases can expand," or to the Newtonian "law," "Forces can be combined." If the only laws that we could find in thermodynamics and mechanics were weak statements of this kind we would hardly be tempted to conceive of these sciences as sets of laws. I think that the same point goes for genetics.

33. A similar point is made by Kenneth Schaffner in [*Discovery and Explanation in Biology and Medicine* (Chicago: University of Chicago Press, 1993)]. Schaffner's terminology is different from my own, and he continues to be interested in the prospects

of global reduction, but there is considerable convergence between the conclusions that he reaches and those that I argue for in the present section.

34. There are numerous examples of such modifications from the history of chemistry. I try to do justice to this type of case in "Theories, Theorists, and Theoretical Change," *The Philosophical Review* 87 (1978), pp. 519–547 and in "Genes."

35. Molecular biology also provided significant refinement of the terms 'gene' and 'allele'. See "Genes."

36. For examples, see N. K. Wessels, *Tissue Interactions and Development* (Menlo Park, Ca.: W. A. Benjamin, 1977), especially Chapters 6, 7, 13–15; and Donald Ede, *An Introduction to Developmental Biology* (London: Blackie, 1978), especially Chapter 13.

37. I attempt to show how the same perspective can be fruitfully applied to other examples in "Explanatory Unification," Sections 3 and 4; *Abusing Science* (Cambridge: MIT Press, 1982) Chapter 2; and "Darwin's Achievement," [in Nicholas Rescher, ed., *Reason and Rationality in Science* (Washington, D.C.: University Press of America, 1985) 123–85.]

38. The extreme version of reductionism is defended by Richard Dawkins in *The Selfish Gene* (New York: Oxford University Press, 1976) and *The Extended Phenotype* (San Francisco: W. H. Freeman, 1982). For an excellent critique, see Elliott Sober and Richard C. Lewontin, "Artifact, Cause, and Genic Selection," *Philosophy of Science* 49 (1982), pp. 157–180. More ambitious forms of anti-reductionism with respect to evolutionary theory are advanced in S. J. Gould, "Is a new and general theory of evolution emerging?" *Paleobiology*, 6 (1980), pp. 119–130; N. Eldredge and J. Cracraft, *Phylogenetic Patterns and the Evolutionary Process* (New York: Columbia University Press, 1980); and Steven M. Stanley, *Macroevolution* (San Francisco: W. H. Freeman, 1979). A classic early source of some (but not all) later anti-reductionist themes is Ernst Mayr's *Animal Species and Evolution* (Cambridge, Harvard University Press, 1963), especially Chapter 10.

39. Gould's *Ontogeny and Phylogeny* (Harvard, 1977) provides historical illumination of both areas of debate about reductionism. Contemporary anti-reductionist arguments about embryology are expressed by Wessels (op. cit.) and Ede (op. cit.). See also G. Oster and P. Alberch, "Evolution and Bifurcation of Developmental Programs," *Evolution* 36 (1982), pp. 444–459.

40. Oster and Alberch, op. cit., p. 454. The diagram on p. 452 provides an equally straightforward account of their anti-reductionist position.

41. Earlier versions of this paper were read at Johns Hopkins University and at the University of Minnesota, and I am very grateful to a number of people for comments and suggestions. In particular, I would like to thank Peter Achinstein, John Beatty, Barbara Horan, Patricia Kitcher, Richard Lewontin, Kenneth Schaffner, William Wimsatt, an anonymous reader and the editors of *The Philosophical Review*, all of whom have had an important influence on the final version. Needless to say, these people should not be held responsible for residual errors. I am also grateful to the American Council of Learned Societies and the Museum of Comparative Zoology at Harvard University for support and hospitality while I was engaged in research on the topics of this paper.

8.1 | Nagel on the Logic of Reduction

As Ernest Nagel points out in the opening paragraphs of "Issues in the Logic of Reductive Explanations," science has been and continues to be marked by strong reductive tendencies. Many of the classic cases of scientific reduction have brought about remarkable advances in our understanding of the world. For example, the reduction of classical thermodynamics to statistical mechanics brought with it new insights into the nature of heat and entropy; likewise, the reduction of optics to electromagnetic theory deepened our understanding of light and the behavior of light rays. The reductive explanation of chemical bonding by quantum mechanics revolutionized organic chemistry and biochemistry by revealing the nature of the covalent bond and the forces that determine the shape of molecules. Some people think that genetics and psychology are in the process of being reduced to biochemistry, and they expect great things as a consequence. Nagel wants to understand these and other classic cases of scientific reduction by articulating the structure of what he calls *reductive explanations*.

Before exploring Nagel's analysis of reduction, two preliminary points are worth highlighting. First, however much the contrast between commonsense and scientific descriptions of the natural world might recommend it, we should avoid approaching the issue of reduction in terms of a distinction between appearance and reality. Although used by philosophers in the past, the appearance-reality distinction is too imprecise to bear the weight of an analysis of reduction. And in any case, the distinction is not of the sort we need. For the relation of chemistry to quantum mechanics, or thermodynamics to kinetic theory and statistical mechanics, is not one of appearance to reality. Even if some of our prescientific views of the natural world turn out to be false, even if they prove to be less about reality and more about mere appearance, nevertheless many of our commonsense judgments are true. Thus "steam is hotter than ice" does not report the full facts about steam and ice by virtue of which that sentence is true, but for all that, it can scarcely be reckoned a judgment of mere appearance.

Second, whatever its details, reduction properly so-called is typically construed as a relation (or family of relations) between some reducing theory T' and a reduced theory T. While it would take us too far afield here to discuss their exact nature, theories are commonly regarded as a set of statements—axioms, laws, empirical hypotheses—about the world. However natural it may be, then, to speak of certain phenomena, properties, or objects as reducing to the various phenomena, properties, or objects described by a reducing theory, it is strictly speaking the theories themselves, or particular expressions or concepts figuring in them, that stand in the reduction relation. Thus, when we find ourselves saying, for example, that heat reduces to the random motion of molecules, this is best regarded as an elliptical, short-hand expression of the view that *claims* about heat reduce to *claims* about molecular motion.

The second point just noted leads us to the core elements of reduction. Nagel has these elements in mind when he refers to the logic (or elsewhere, the formal conditions) of reduction.[1] As Nagel's talk of *reductive explanation* suggests, reducibility is commonly viewed as a kind of explanatory relation, whereby the reduced, secondary science may be accounted for as a special case or branch of some more general or inclusive primary science. Thus, unless the fundamental axioms, laws, and hypotheses of a science have been explicitly formulated as statements and their constituent terms given determinate meanings, it will be impossible to say whether some scientific theory does in fact reduce to another. This is particularly so given the deductive-nomological model of explanation that Nagel, Hempel, and many others accept. For according to that model, explanations are arguments (see chapter 6 above). Accordingly, the reduced and reducing theories must be expressed in meaningful statements from which the premises and conclusion of an explanatory reduction may be drawn. On Nagel's account, a reduction is made when the fundamental claims of a secondary science or theory T are shown to be logical consequences of the fundamental claims of the primary, reducing theory T'. The conclusions of such arguments are said to be reduced statements, while the premises are called reducing statements. This, then, is the core idea in Nagel's treatment of intertheoretic reduction—that the reduced claims of the secondary theory T be logically derivable from premises made up wholly of claims from the reducing, primary theory T'.

The central feature of Nagel's account of reduction—logical derivability—has two important consequences. First, the claims of T and T' must be logically consistent: since nothing incompatible with the fundamental claims of a theory can be logically derived from them, the claims of a reduced theory T must be consistent with those of the reducing theory T'. Second, any term shared by the reduced and reducing theory must have a common meaning if it appears in both the premises and the conclusion of a reductive explanation. For example, suppose that two sets of laws, one set from thermodynamics and the other from mechanics, both contain the terms *volume* and *pressure*. We could reduce thermodynamics to molecular mechanics using those laws only if *volume* and *pressure* have the same sense in both theories. If they do not, the supposed derivation would be invalid.[2]

Homogeneous and Inhomogeneous Reduction

The requirement of logical derivability raises an interesting question. Typically, reduction is associated with scientific progress or change—with cases in which an existing theory established in one domain of inquiry can be explained by a theory developed for some other, broader domain. The subject matter of the secondary, reduced theory would, in such cases, appear to be qualitatively distinct from that of the primary, reducing theory. Indeed, many of the concepts figuring in classical thermodynamics (e.g., heat and temperature) make no appearance as such in statistical mechanics; much of the descriptive vocabulary in chemistry is absent from quantum mechanics; and so on. But given that a secondary, reduced theory T employs in its fundamental laws predicates unique to that science, predicates that appear nowhere in the basic terminology of the primary, reducing theory T', it is unclear how we can understand reduction as an explanatory derivation of the former theory from the latter. Irrelevant exceptions aside, no term can appear in the conclusion of a valid argument unless it is contained somewhere in the premises.[3]

Following Nagel, it has become common to distinguish "easy" cases of reduction from the "hard" cases. In the easy cases, the descriptive or subject-matter terms of the reduced theory are either present in or can be explicitly defined by terms in the reducing theory. In the hard cases, some term or terms in the reduced theory are neither present in nor definable by other terms in the reducing theory. The former cases are called *homogeneous reductions,* the latter—*inhomogeneous reductions* (or, as some authors prefer, *heterogeneous reductions*). A brief review of Nagel's account of homogeneous reduction will suffice, before we turn our attention to the more difficult and interesting case of inhomogeneous reduction.

A reduction is homogeneous if the descriptive vocabulary of the reduced theory T is a proper subset of the vocabulary of the reducing theory T'. Now all else being equal, were T simply a fragment of T', straightforwardly deducible from T' as a mere part of it, it would be implausible to regard the relationship between the two as reductive. Typical examples of homogeneous reduction are, rather, cases in which an earlier theory T, developed as an independent account of some domain of natural phenomena, comes to be explained as a consequence of a later, newer theory T' of some broader domain of phenomena. Thus, Galileo's law of falling bodies and Kepler's laws of planetary motion may be said to reduce to Newtonian mechanics and gravitational theory: not only do the fundamental concepts of those earlier-formulated laws (distance, velocity, acceleration, etc.) appear in Newton's own account, but the restricted and apparently dissimilar kinds of motion to which Galileo's and Kepler's laws apply (terrestrial, celestial) are also given a fundamentally unified treatment by Newton's general account of motion. Therefore, if the laws of terrestrial free fall and celestial motion can be derived from Newton's theory, then Nagel's formal conditions for a homogeneous reduction are met and those laws are said to be reducible to Newton's theory. (We shall take up some difficulties with this account presently.)

Inhomogeneous reduction occurs when the descriptive vocabulary of the reduced theory T is not a proper subset of the vocabulary of the reducing theory T'. We already encountered a standard example of inhomogeneous reduction—the explanatory reduction of thermodynamics, with its distinctive concepts of heat and temperature, to statistical mechanics. And we noted, too, the main respect in which an analysis of such a reduction is less straightforward than that of homogeneous cases: inhomogeneous reductions look unfit for the deductive explanation model, since it is unclear how the reduced laws in the conclusion of such an explanatory argument could contain terms or concepts absent from the reducing premises. If reduction is to be construed on the deductive model, and if inhomogeneous cases are a species of reduction, then it would seem that the premises of such reductive arguments must be supplemented with assumptions expressing how terms or concepts otherwise absent from the reducing theory T' are connected with those unique to the reduced theory T.

The requirement of supplying assumptions to connect the terms in T and T' presumes that scientific theories and their fundamental laws possess truth values. One broad approach to the nature of reduction, discussed by Nagel, denies this presumption. According to the *instrumentalist* account, theories and their laws are themselves neither true nor false; rather, they serve as rules or instruments through which one can infer certain experimental or observational claims from other such claims. On an instrumentalist reading, for example, the kinetic theory of gases is not a genuine description (of gases as systems of rapidly moving molecules), nor is it a true generalization of relations that hold among such objects; instead, it is a rule prescribing how one is to represent various observable states of affairs and how, given certain empirical data, one can infer other observational claims. As such, theories and their laws are not truths from which conclusions can be validly derived but functional devices or "inference tickets" in accordance with which claims may be drawn from other claims. Thus, the relevant issue for inhomogeneous reduction, on this account, is neither the cognitive content of the premises in a deductive explanation nor the logical or semantic relation of those premises to the conclusion. Rather, it is the relation between the sets of observational inferences licensed by the reduced and reducing theories. On the instrumentalist analysis of inhomogeneous reduction, T reduces to T' when all the observational claims that T licenses one to infer from some set of data are also licensed by T', but not conversely.

NAGEL'S CORRESPONDENCE ANALYSIS OF INHOMOGENEOUS REDUCTION

While Nagel acknowledges that the instrumentalist analysis captures an important relation between the observational inferences licensed by theories when one reduces another, he denies that it gives an acceptable account of inhomogeneous reduction. Quite aside from the fact that scientists do in fact speak as if theories and laws are true or approximately true, the instrumentalist account fails to provide any unified story of how the theoretical concepts of one theory are related to those of another in a way that explains why various observational statements do (or do not) fall within the scope of reduced and reducing theories. Instead, Nagel proposes what he calls the *correspondence analysis of inhomogeneous reduction*. Like the instrumentalist proposal, it rests on certain crucial assumptions.

First, the correspondence analysis presumes that expressions belonging to some theory or domain of science have determinate meanings fixed by its own methods of explication. Of course, the meaning of certain terms in a theory or science may sometimes be specified with the aid of terms in some other theory or science, but it does not follow from this that the meaning of every term in a theory can be given by terms in some distinct theory. In particular, then, if the terms that are unique to the fundamental laws of a reduced theory T appear in the conclusion of a reductive explanation but not in the reducing theory T' itself, then those terms from T must have determinate meanings that are independent of T'. For example, *heat, temperature*, and *entropy* must have determinate senses specifiable by their role in thermodynamics, and these meanings must be explicable independently of statistical mechanics.

Second, the correspondence analysis presumes that observational terms shared by the fundamental laws of a reduced theory T and reducing theory T' possess meanings capable of being specified apart from T and T'—that is, they have meanings that are neutral with respect to whatever it is that distinguishes T and T'.[4] Recall, for example, that thermodynamics and mechanics alike use the terms *volume* and *pressure*. The second assumption is that these terms possess a univocal sense that can be explicated independently of either T or T'.

Third, and finally, Nagel's correspondence analysis of inhomogeneous reduction presumes that reductions are a kind of deductive explanation.

BRIDGE LAWS

The logical-derivability condition of the correspondence analysis implies that, if a reduced theory T contains terms not appearing in the reducing theory T', then T (or one or another fundamental law of T) can be the conclusion of an explanatory deduction only if the reducing premises of T' are supplemented by some statement or statements expressing a connection between the vocabulary of T' and the terms unique to T. These statements, called *bridge laws*, are crucial to the success of inhomogeneous reduction in Nagel's explanatory-derivation model.[5] If the ideal gas law (relating the temperature of a gas to its pressure and volume) is to be derived from the kinetic theory of gases (in which the concept of temperature makes no explicit appearance), then clearly some bridge law connecting the temperature of a system to an appropriate property of its constituent molecules must be added to the premises of the derivation.

But while the need for bridge laws in inhomogeneous reductions seems relatively clear, their status remains obscure. What kind of statements are bridge laws, and how should they be understood? One proposal is that they express logical or semantic connections between the meanings of the relevant reduced and reducing taxonomies. On this account, the bridge laws of inhomogeneous reductions are explicit definitions of the distinctive terms of a reduced theory T. But Nagel's first assumption—that terms unique to a reduced theory T have determinate meanings independent of the reducing theory T'—rules against this construal. Presumably, terms such as *heat* and *temperature* were meaningful before the advent of the kinetic theory, being fixed by the rules and habits of usage in thermodynamics; and, presumably, such terms can remain explicable by these procedures and retain their meanings, even if the laws in which they figure come to be reductively explained by molecular mechanics. These plausible assumptions would seem to be jeopardized if bridge laws were understood to express analytic connections, whereby the meanings of terms in the reduced theory T are logically tied to the meanings of terms in the reducing theory T'.

Nagel's own view about the status of bridge laws is that they express not logical but material (i.e., factual) connections between the states of affairs signified by the relevant terms in question. On this account, bridge laws are empirical hypotheses about the world. Nagel describes two forms that such connecting hypotheses may take: either they express relations between predicates and their extensions, or they express identities between objects or processes.

To understand Nagel's first alternative, it is helpful to employ the distinction between the extension of a predicate—the set of objects or processes to which the term applies—and the intension of a predicate—the universal property that the term connotes. (See "Extensions and Intensions" in the commentary on chapter 7.) Suppose that the reduced theory T contains a predicate, the term *blue*, connoting a property (blueness) that is not connoted by any of the predicates of the reducing theory T'. Nonetheless, T' might give conditions under which instances of blueness will arise. For example, T' might predict that light of a certain range of wavelengths is selectively transmitted by fluctuations in the density of air in the earth's atmosphere, thus explaining why the sky is blue. In this example, the reducing theory T' does not contain the word *blue* but it provides a condition that is sufficient (though, in this case, not necessary) for the property of blueness to be manifested. The bridge law explicitly connects the predicate from T' (*selective transmission of light of a certain range of wavelengths*) with the predicate from T (*blue*) by saying that the former is sufficient for the latter. Nagel claims that bridge laws of this kind (involving either sufficient conditions, as in the present example, or necessary and sufficient conditions, as in other examples) are empirical hypotheses.[6] He interprets these bridge laws as asserting that the extension of the reduced predicate is included in (or, in the case of necessary and sufficient conditions, coincides with) the extension of the reducing predicate. Thus, in our simple example, Nagel regards the bridge law as making the empirical claim that things that transmit light of a certain range of wavelengths are a subset of blue things.

Nagel's second alternative is that some bridge laws show that apparently distinct objects or processes are in fact identical. This type of bridge law would also be empirical. Just as it is an empirical discovery that the morning star is (i.e., is identical with) the evening star, so it is an empirical hypothesis that, say, light waves are (i.e., are identical with) electromagnetic waves of a certain range of frequencies, thus enabling the reduction of optics to electrodynamics. Bridge laws of this second kind do not simply specify conditions for the manifestation of properties connoted by predicates from the reduced theory; rather, they express an identity between entities initially reckoned as distinct.

8.2 | Feyerabend's Criticisms of Nagel's Model of Reduction

In "How to Be a Good Empiricist—A Plea for Tolerance in Matters Epistemological," Paul Feyerabend criticizes Nagel's model of intertheoretic reduction by arguing that it is inconsistent with an enlightened empiricist view of scientific change.[7] Feyerabend's objections are part of a larger attack on the logical empiricist movement in twentieth-century philosophy of science.[8] Although he is sympathetic with much of the empiricist tradition in philosophy, especially the empiricists' use of observation and experience to combat dogmatic metaphysics, Feyerabend thinks that modern philosophy of science has taken a wrong turn and has itself become dogmatic and a barrier to progress. He advocates a different brand of empiricism, a "good" empiricism that, unlike the "bad" empiricism of the logical empiricists, will discourage dogmatism and promote scientific progress. According to Feyerabend, a nondogmatic empiricism must welcome a plurality of theories, not for the purpose of advancing toward a final, true theory that corresponds with "the facts," but because a tolerance of alternative theories offers the best means of criticizing and evaluating the theories we currently accept. Thus, Feyerabend has two goals: to undermine Nagel's reduction-based approach to scientific change and to defend a more liberal, theoretically pluralistic empiricist methodology.

We are already familiar with what Feyerabend calls a "cornerstone" of contemporary empiricism: a reductive explanation of T consists in the derivation of T from T'. We have also encountered two implications of this general account: first, that T' and any of its consequences within the domain of the reduced theory T must be consistent—i.e., logically compatible—with T, and, second, that either the descriptive terms of the reducing theory T' must be shared with those of the reduced theory T, or they must be explicitly connected to them via empirical bridge laws. These two implications may be stated as conditions that a new, more general theory must satisfy if it is to be accepted as explaining an existing theory in a given domain. According to the consistency condition, only those theories are admissible in a given domain that either contain the theories already used in that domain or are consistent with the theories in that domain. According to the condition of meaning invariance, meanings must not change with scientific progress; future theories must be expressed in such a way that they leave unaltered the claims made by the theories they explain.

In connection with his second goal—defending a more liberal methodology—Feyerabend is prepared to grant that many scientists and philosophers of science accept the consistency and meaning-invariance conditions as constraints on scientific progress. Postponing the question of their truth or falsity as descriptions of actual scientific change, Feyerabend wants to ask, Should they be accepted? Do they represent a reasonable empiricist methodology of science? Are they necessary conditions for scientific progress?

The Apparent Unreasonableness of Consistency and Meaning Invariance

According to Feyerabend, the consistency condition is unreasonably strong because it restricts the acceptance of new, alternative theories quite arbitrarily—on the basis of their relation to whatever theory happens to be accepted at the time. Suppose that a theory T is accepted as correctly accounting for the phenomena of some domain D, on the basis of a set F of confirming instances; and suppose a new theory T^* shares those confirming instances but contradicts T outside of F. On Feyerabend's reading, the consistency condition recommends against accepting T^* simply on the basis of age and familiarity, not on the basis of any disagreement with the facts: had T^* rather than T been there first, the consistency condition would have worked against T rather than in its favor.

Feyerabend believes that considerations of this sort show that the consistency condition is not only unreasonable but also counterproductive. For suppose that an alternative (T^*) that shares the confirming instances of an accepted theory (T) cannot be eliminated by appeal to those factual instances. Defenders of the consistency condition insist that neither do those considerations recommend accepting T^*—in light of the empirical evidence, the alternative T^* is no better or worse than T. A genuine advance beyond T will arise, not from the invention of alternatives to T, but only from the emergence of fresh empirical facts, which by either supporting or requiring a modification of T, will bring about scientific progress. Feyerabend concedes that theories should not be changed unless some disagreement with the facts demands it. But why, asks Feyerabend, should we think that such facts exist and are available in the absence of alternatives to T? According to Feyerabend, not only is the description of every fact dependent upon some theory or other, but many facts relevant to testing an existing theory are simply unavailable for discovery without considering alternatives to the existing theory.

To illustrate his claim that the discovery of new facts often depends on taking alternative theories seriously, Feyerabend gives the example of

Brownian motion. Brownian motion is named after the English botanist Robert Brown (1773–1858) who first observed it in 1827. Using a microscope, Brown noticed that pollen grains suspended in water move about chaotically, rotating and darting from place to place in a random manner. Some scientists attributed Brownian motion to external vibrations disturbing the fluid, while others thought that it was caused by the action of light. This phenomenon is now taken to be a convincing proof of the kinetic theory of heat and a demonstration that the second law of thermodynamics is false. But for seventy years or more after Brown's discovery, its significance as a refutation of the second law was not appreciated because no one was able to give a convincing explanation of the phenomenon using the kinetic theory.[9] When predictions based on the kinetic theory were tested, they failed to confirm the theory.[10] Finally, Einstein in 1905, and the Polish physicist, Marian Smoluchowski (1872–1917) in 1906, were able to give a mathematical treatment of the phenomenon that made a number of testable, quantitative predictions. These predictions were confirmed by the French physical chemist Jean Perrin (1870–1942) and others.

Feyerabend interprets the history of Brownian motion in the following way:[11] Taken simply as an observed phenomenon, Brownian motion could not undermine the universal truth of the second law of thermodynamics.[12] Only when viewed from the perspective of the kinetic theory and statistical mechanics, and given an adequate mathematical analysis by Einstein and Smoluchowski, could the relevant facts of Brownian motion show that the second law is not universal but statistical.[13] Since unearthing facts and recognizing their relevance to an accepted theory T is dependent upon the existence of alternative theories as well confirmed as T but in disagreement with it, Feyerabend believes that the invention of alternatives is crucial to a properly empiricist methodology of science. Because the consistency condition proscribes theories in disagreement with T, it can only contribute to intolerance and dogma.

Feyerabend argues that the consistency condition also gives rise to a kind of self-deception about the success of current theories. By rejecting competitors to T, potentially refuting facts are suppressed. Consequently, judgments about the success of T are heightened, reinforcing even further the belief that investigating alternatives to T would be time ill spent. The alleged "success" of T is, thus, altogether fabricated, not a sign of its correspondence with nature, but simply of its staying power in the absence of empirical counterevidence. The accepted theory comes thereby to take on the character of myth, which by its very nature serves to perpetuate itself and infect how we think, talk, and explain. The result is conformism—a uniformity and unanimity of scientific belief born of self-deceit. Only by tolerating and encouraging variety of opinion can science avoid both the dogma and the self-deception common in many religious and political institutions.

Feyerabend extends his criticism of the consistency condition to an attack on meaning invariance. Tolerance of alternative theories brings with it a tolerance of alternative meanings, for changes in conceptual and theoretical systems produce changes in the interpretation of scientific terms. If our methodology permits mutually inconsistent, empirically adequate theories, then it must also permit mutually irreducible theories whose terms are unconnected by anything like factually correct bridge laws. This elasticity of meaning need not be occasioned by observational difficulties with an existing theory. For example, Feyerabend points out that the impetus theory of late scholastic and early modern science did not make any false predictions. But adopting Newton's theory of inertia produced better explanations of a wider range of phenomena. Such progress does not require that theoretical terms (such as *force* in the impetus and inertial theories) remain invariant in meaning when theories change. Since the reinterpretation of terms often accompanies theoretical change, the condition of meaning invariance is an unreasonable constraint on scientific methodology that would stifle progress.

In light of his rejection of the consistency and meaning-invariance conditions, Feyerabend notes three consequences of the methodology he defends. (See section 8 of his paper.) The first consequence is that we may relax our judgments about the propriety of metaphysical claims in formulating scientific theories. In the past, many empiricists have supposed that science must begin only with observable facts and proceed by generalizing from them, refusing to admit any intrusion of metaphysics. Contemporary empiricism allows metaphysical claims a role in new theories but with the proviso that confirmation of the theory will eventually enable those claims to be purged. Feyerabend recommends an even more permissive attitude. On his view, metaphysical claims make up very primitive scientific theories. To this extent, even if they are inconsistent with accepted, well-confirmed, full-blown theories of nature, they serve precisely the critical and evaluative role that alternative theories should be encouraged to perform.

The second consequence of Feyerabend's methodology concerns the long-standing problem of how to justify inductive reasoning. We reason inductively when we make inferences from a sample to an entire population or from our past observations of some objects to a generalization about all objects of that type. Assume that a belief constitutes knowledge when it is true and justified. Since we must not change or give up what we know to be true, knowledge implies stability. Feyerabend claims that, since the methodology he endorses cannot allow such stability, the problem of justifying inductively formed beliefs is in fact a pseudoproblem: there is no such stability, hence no such knowledge, hence no such justification to be had.

Finally, the third consequence of Feyerabend's position is that so-called arguments from synonymy are likely to impede theoretical progress. Arguments from synonymy would evaluate a theory, not on the extent of its correspondence with the world, but on the extent to which its terms have the same meaning as (are synonymous with) the terms in some other theory that we already accept. Feyerabend gives the example of a neurophysiological theory that attempts to explain the mental life of persons solely in material terms. Proponents of synonymy will say that such a theory must fail, because it cannot provide synonyms for "pain," "thinking of Syracuse" or "feeling sad" in purely physical, neurochemical terms. Being in a certain kind of brain state, for example, does not mean the same thing as "thinking of Syracuse;" it is not what I mean when I say that I am thinking of Syracuse. Feyerabend asserts that "clearly, such criticism silently assumes the principle of meaning invariance" (947). Since Feyerabend rejects meaning invariance, he also rejects this kind of criticism as an unreasonable constraint on new theories: the terms they use should not have to mean exactly the same thing as terms already in use.[14]

In discussing this last consequence, Feyerabend endorses an even stronger result. He not only rejects synonymy; he also advocates irreducibility. By *irreducibility* Feyerabend means the absence of anything like Nagel's correspondence relation among theories. Now, it is one thing to claim that the conditions of consistency and meaning invariance are too restrictive; it is quite another to argue that in fact scientific change fails to meet the conditions of consistency and meaning invariance, that, in other words, Nagel's correspondence account of reduction is simply mistaken, and no theory is reducible to any other. We thus return to the first part of Feyerabend's paper, where he defends this latter, stronger claim of irreducibility.

THE APPARENT VIOLATION OF THE CONSISTENCY CONDITION

In our discussion of homogeneous reduction, we encountered the example of Galileo's law of falling bodies and Kepler's laws of planetary motion being said to reduce to Newtonian mechanics and gravitational theory. The reduction is homogeneous, since the fundamental concepts of those earlier-formulated laws (distance, velocity, acceleration, etc.) also appear in Newton's theory; moreover, the distinct and restricted domains to which Galileo's and Kepler's laws apply (terrestrial, celestial) are given a unified treatment by Newton's theory, from which Galileo's and Kepler's laws are said to be derivable. Feyerabend draws our attention to a point first made by Pierre Duhem (and repeated often by Karl Popper), namely, that the consistency condition required for such a derivation is apparently not met. (See section 4 of Pierre Duhem, "Physical Theory and Experiment" in chapter 3.) More precisely, the relevant consequences of Newton's theory are logically inconsistent with Galileo's law. Galileo's law asserts that the acceleration of a body in free fall near the earth's surface is a constant, yet an application of Newtonian theory to terrestrial free fall entails that the acceleration must vary with the distance of the body to the earth's center of mass. A similar result holds for Kepler's third law.[15] What these and other historical cases show is that actual examples of scientific progress frequently violate the consistency condition. Furthermore, since the consistency condition is necessary for the success of a derivation, reduction is not a matter of deductive explanation at all. If anything is to be learned from such cases, it is not that a new, more general theory T' is one from which the fundamental laws of the earlier theory T can be derived as a special instance, but rather that T' simply differs from T and replaces it.

Nagel's response to this charge (in part 1 of the reading "Issues in the Logic of Reductive Explanations") is to acknowledge that the consequences derivable from a reducing theory T' may not always exactly coincide with the reduced laws of T but that this does not refute his model of reduction. It is a commonplace that newer, more comprehensive theories sometimes indicate the special conditions under which older, more restricted laws hold, in light of which some adjustment to the older laws may be necessary. Still, the earlier laws of T are generally close approximations to the consequences of T'. This being so, we can retain the deductive model of explanatory reduction at its core, taking homogeneous reduction to consist in the derivation of either the reduced laws or their close approximations from the reducing theory.

This result should not surprise us, for scientists often use simplifications and approximations when they derive laws from theories. For example, Nagel points out that in deriving the law for the period of a pendulum, we assume that the weight of the pendulum is located at a single point in the suspended bob, that the gravitational force on the bob is constant, and so on. Although such approximative assumptions are common, Nagel thinks that it would be absurd to say that scientists are mistaken in claiming to have derived laws from their theories. Given such assumptions, we count the laws as consequences of the theories; likewise, even if a reduced law of T is only a close approximation of what we actually derive from the reducing theory T', it would be misleading to say that the law had not been reductively explained in the way Nagel's model describes.

The Apparent Violation of the Condition of Meaning Invariance

Feyerabend argues that, like the consistency condition, the condition of meaning invariance is violated by actual scientific practice. His primary example is the relation of classical to relativistic mechanics and the concept of mass that appears in each theory. If we try to use the special theory of relativity to explain why classical, Newtonian mass is conserved in a closed system, the consistency condition is again violated, for even if our two theories give experimentally indistinguishable results for the mass of *particles* at low velocities, Feyerabend points out that, according to special relativity, the mass of a *composite* object (even one at rest in a given frame) depends not only on the masses of its constituent particles but also on their velocities (their kinetic energy) and the potential energy resulting from the forces between them. In classical mechanics, mass is additive; in relativity theory it is not. Hence, Feyerabend concludes, the condition of meaning invariance must also be violated since the concepts of mass in the two theories are not the same.

Moreover, he argues, while classical mass m_c is a (one-place) *property* of an object, independent of coordinate system, relativistic mass m_r is a (two-place) *relation*, involving relative velocities, between an object and a coordinate system.[16] In light of this, Feyerabend claims that we must conclude that *mass* as it appears in classical mechanics and *mass* as it figures in relativity theory do not have the same meaning. As before, he contends that the successor theory T' has not absorbed or incorporated T, as Nagel's account requires, but straightforwardly differs from T and replaces it. As Feyerabend says elsewhere:

> What happens when transition is made from a restricted theory T' to a wider theory T (which is capable of covering all the phenomena which have been covered by T') is something much more radical than incorporation of the *unchanged* theory T' into the wider context of T. What happens is rather a *complete replacement* of the ontology of T' by the ontology of T, and a corresponding change in the meanings of all descriptive terms of T' (provided these terms are still employed).[17]

(Contrary to the notation used in this commentary and in Nagel, Feyerabend uses T' to stand for the reduced theory and T to stand for the reducing theory.)

It might be objected that Feyerabend's examples of meaning change fail to show anything so strong as this. At most they show that actual scientific practice sometimes violates the condition of meaning invariance for highly theoretical terms (such as *mass*), leaving the meaning of the observational terms shared by T and T' unchanged. But Feyerabend intends his arguments to apply to observational terms as well. The significance of denying meaning invariance for observational terms is best judged against the backdrop of a brief account of the distinction between observational and theoretical terms.

One way of construing the theoretical-observational distinction is to see it as marking off the descriptive predicates that appear in observational (or experimental) laws from the predicates that appear in theoretical laws. Observational laws are generalizations about phenomena accessible to the senses; accordingly, they are expressed in the language of observation and refer to perceived or perceivable objects, processes, properties, and so on. Theoretical laws, on the other hand, are not generalizations about observable phenomena but rather help to give unified explanations of observable phenomena and their laws. Since theoretical laws typically make reference to unobservable things and properties, they are expressed in a more abstract language. Statements of the freezing and boiling points of water as a function of pressure are observational laws, and *freezing point, boiling point, temperature*, and *pressure* are observational terms. A law expressing the degrees of rotational freedom for helium molecules is a theoretical law, and *degree of rotational freedom* and *helium molecule* are theoretical terms. (For an extended account of the theoretical-observational distinction, see chapter 9.)

To appreciate the force of Feyerabend's denial of meaning invariance for observational as well as theoretical terms, consider the empiricist view (held by Nagel and others) of how observational and theoretical terms get their meaning. The meaning of both sorts of terms is supposed to come from one source: experience. Observational terms get their meaning directly from observation and from test procedures that can be applied independently of any theoretical context. So even though theories change, the meaning of observational terms remains constant. Theoretical terms also get their meaning from experience, but indirectly: the meaning of a theoretical term is supposed to "seep up" from the experimental laws derivable from the theory in which the term occurs. So, in general, the meaning of theoretical terms changes as theories change, but if rival theories predict many of the same experimental laws, there will be a considerable overlap in the empirical meanings of the theoretical terms involved. Thus, observation terms can be used to state experimental laws that competing theories then seek to predict and explain. Moreover, because observation terms are independent of any theory, those experimental laws can serve as the independent basis for evaluating the explanatory and predictive success of those theories. By denying meaning invariance both to theoretical and observational terms, Feyerabend effectively cuts off *any* shared meanings between earlier and later theories and denies to experimental laws their role as neutral, theory-independent arbiters of theoretical success.

What leads Feyerabend to deny meaning invariance to observational terms? On his account, the content of such terms is not exhausted by observational procedures but also depends on our theories about the nature of the things we are observing. Thus, even if our observational procedures for ascertaining some perceivable aspect of an object do not change from the use of T to the adoption of T', nevertheless T and T' may diverge in how they represent the nature of that aspect of the object—as a property, as a relation, as a process, and so on. And this divergence will affect the meaning of our observational terms. If Feyerabend is correct about this, then the traditional distinction between the meaning of observational and theoretical terms is jeopardized.

Feyerabend's primary example of meaning variance is Einstein's special theory of relativity. (Kuhn also uses the same example in his "The Nature and Necessity of Scientific Revolutions," reprinted in chapter 2.) Feyerabend claims that the meaning of the theoretical term *length* is different in classical mechanics than it is in special relativity. If this is true, then classical mechanics is not reductively absorbed by Einstein's theory but is completely replaced by a wholly different theory. Specifically, Feyerabend denies that we can assign any common empirical meaning to the term *length* in the two theories by connecting it to a neutral observation language. For, he argues, any such observation language will, at best, yield a representation of length contraction that corresponds to Lorentz's electromagnetic theory, not to Einstein's relativity theory.[18]

Understanding Feyerabend's claim requires that we say something about the difference between Einstein's theory and Lorentz's. Both theories yield the same set of equations (the Lorentz transformations) and both predict the length contraction of moving objects. But the interpretation of length contraction in the two theories is radically different. Lorentz's theory is a classical theory in which any contraction of a physical object must be due to a physical force compressing the object and making it shorter. Lorentz (following Fitzgerald) hypothesized that all objects are composed of electrons and molecules and that all the forces that hold matter together are electrical. When an object moves through the electromagnetic aether (or, in other words, when an object moves with respect to absolute space), these forces become stronger, causing the object to shrink. Thus, on Lorentz's theory, the real length of an object is the length it has when it is at rest with respect to absolute space, and length contraction is a physical effect caused by the motion of electrons in absolute space.

Einstein's theory is totally different from Lorentz's. Einstein bases his theory on the empirical fact that the speed of light is the same in all frames of reference, together with an operational analysis of the concepts of length, time, and simultaneity. His derivation of the Lorentz transformations involves no hypotheses about the structure of matter and it dispenses with the notion of absolute space. On Einstein's theory, all objects moving with constant velocity with respect to an inertial frame of reference must appear shorter when measured in that frame of reference, regardless of their composition. Thus, in Einstein's theory there is no absolute space and no physical forces causing length contraction.

Feyerabend's point is that our ordinary, commonsense views about objects are embedded in our observational language. For example, Feyerabend presumes that it is part of what we ordinarily mean by observational terms such as *physical object* and *length* that physical objects will shrink only when a physical force compresses them. If this is true, then there is no way that we can interpret the theoretical terms in Einstein's special theory of relativity using such an observational language. We may assign empirical meanings to the terms in the Lorentz-Fitzgerald transformation equations, but the interpretation we give will be Lorentz's interpretation, not Einstein's. In brief, Einstein's theory involves a radical reinterpretation of our ordinary concepts of length and time. Thus, we should not expect that an observational language in which the classical, Newtonian concepts of length and time are embedded will be able to assign empirical meaning to the theoretical terms in Einstein's theory.

According to Feyerabend, even the most innocent-seeming observational language is theory dependent, and each theory determines its own distinctive set of observational statements. Thus, he concludes that the traditional empiricist distinction between observational and theoretical terms fails to give a satisfactory account of meaning in scientific theories and that the standard reductive picture of the relation between theories fails along with it.

Nagel's reply to the "replacement" account of intertheoretic reduction (part 3 of his "Issues in the Logic of Reductive Explanations") begins with a summary of Feyerabend's position. Since the meaning of every term— observational and theoretical alike—depends upon the particular theory in which it occurs, there is no theory-independent observational core to ground the interpretation of any sentences making up a theory; rather, the theory itself is meaningful independent of observation. Accordingly, a change from one theory to another brings with it a change in the meanings of all terms, and hence no theory T can be reductively derived as an explanatory consequence of T'.

The standard empiricist view is that scientists should formulate theories adequate to the facts of natural phenomena as given by theory-neutral observation. Suppose Feyerabend's view to the contrary is correct. That is, suppose a transition is made from some existing theory T about a domain D of phenomena to a distinct theory T', and suppose that the vocabularies of T and T' share no meanings in common. Nagel's first objection has two parts. First, if Feyerabend's account is correct, then no sense can be made of the claim that T and T' are either consistent or inconsistent (compatible or incompatible), since consistency (and compatibility) are logical relations obtaining between statements whose terms have shared meanings. (If you assert "x is F" and I assert "x is not F," we have not made inconsistent claims unless you mean by "F" what I mean by "F".) Second, if the meaning of observation terms in T is a function of T itself, then any laws in which those terms figure cannot be at odds with the observational claims of any alternative theory relative to which we may wish to assess the empirical adequacy of T. If there are no objective, theory-independent grounds for fixing the meaning of any terms in a theory, then every theory is equally immune from criticism by adherents to every other theory.

Nagel's second objection is a further attack on the intelligibility of Feyerabend's position. Feyerabend insists that when a transition is made from an existing theory to a new theory "capable of covering all the [same] phenomena" (922), the theories share no meanings in common. But how, Nagel asks, in the absence of common meanings, can we compare the relative scopes of the theories and their degree of overlap? If the descriptive means by which one is able to circumscribe the relevant phenomena are theory laden, and thus undergo meaning changes in the transition from T to T', it is unclear what neutral account is available for expressing the notion that T' is capable of covering all the phenomena that have been covered by T. (Expressing this notion is no difficulty for Nagel's correspondence account, since bridge laws are understood as empirical hypotheses about how the extensions of the terms of T are related to those of the terms of T'.)

There is a further issue concerning Feyerabend's thesis of meaning variance. In chapter 7, we encountered the distinction between a term's intension and its extension—or, as it is nowadays more commonly expressed, between the sense and the reference of a term. To which of these traditional aspects of meaning is Feyerabend's denial meant to apply? Reasons (however good or bad they may be) for claiming that, say, *mass* as it figures in classical mechanics and *mass* as it appears in relativity theory have different meanings (i.e., have different senses) are not yet reasons for believing that each of these homophones diverge in reference: classical mechanics and relativity theory may, for all that, refer to a single physical magnitude, diverging on the properties their hypotheses ascribe to it. Thus, even if one endorses a position according to which senses of terms are closely tied to the theories in which they figure, we may yet have grounds for believing, as Nagel's remarks imply, that there is an objective and common domain of natural phenomena to which both T and T' may be said to apply.

We can summarize the contrast between Nagel's and Feyerabend's views on meaning by returning to Feyerabend's criticism of synonymy as a constraint on new theories. (See pages 943–45) Feyerabend protests that synonymy—the requirement that the meaning of terms in established theories not change when new theories are introduced—would stifle innovation. He reaches this conclusion by adopting a holist theory of meaning according to which any new claim about, say, temperature or mass, automatically changes what the terms *temperature* and *mass* mean. As we have seen, Nagel rejects the holist theory of meaning because it would render impossible any comparison between theories (including the ones that Feyerabend makes). In its place, Nagel relies on the traditional distinction between a term's intension (sense) and its extension (reference). Even if the intension of a term varies with theoretical context, its extension—the things or processes that the term refers to—can remain the same. In this way, contra Feyerabend, rival theories can make competing claims about the same objects.

Nagel's account of reduction leaves open the possibility that two theories might use the same term not only with different intensions but also with different extensions. In such a case, any reductive relation between the theories must be inhomogeneous, secured by bridge laws. To illustrate this possibility, let us return to the disputed example of classical mechanics and relativity theory and allow, for the sake of argument, that the extension of the term *mass* is not the same in both theories. Nagel would insist that we introduce two new terms that are typographically distinct, m_c and m_r, say, and then try to express the relation between the extensions of those terms in a bridge law. Remember that the bridge law is not a definition or a statement of synonymy. Rather, it asserts an empirical hypothesis about how the extension of m_c is related to the extension of m_r. Whether or not such a bridge law can be found in this case is a matter of dispute. Nonetheless, the point remains that, contrary to Feyerabend, Nagel's condition of meaning invariance is not a barrier to theoretical innovation once it is understood as Nagel intended, namely, as a prohibition of referential ambiguity when theories are compared.

8.3 | Fodor on the Autonomy of the Special Sciences

In the heyday of logical positivism and logical empiricism, the doctrine of the "unity of science" was often associated with the reductivist thesis that all the special sciences—chemistry, biology, psychology, economics, geology, and so on—are reducible to physics.[19] Slightly more precisely, the reductivist (or "reductionist") thesis is that all the true theories of the special sciences must be reducible (in Nagel's deductive sense) to the true theories of fundamental physics (which is concerned with the most basic particles, forces, and fields, and the most general laws in nature). This version of reductivism is meant to apply not only to scientific theories outside of physics but also to theories within physics other than the most fundamental. For the sake of brevity in what follows, we shall use *physics* to mean fundamental physics unless specified otherwise.

As Fodor remarks, given the proliferation of sciences outside of physics, many with their own distinctive concepts and alleged laws, the reductivist thesis could hardly be an inductive inference from a general scientific trend. Admittedly there are cases in which reductionism *as a research strategy* appears to have been vindicated—Fodor mentions the quantum-mechanical explanation of the chemical bond as an example—but the main support for the reductivist thesis (construed as an empirical claim) derives from two sources: a commitment to physicalism and the "layer-cake" picture of how the world is structured.[20]

PHYSICALISM

Physicalism asserts that all events are physical events. (Like Fodor, we are making the simplifying assumption that the sciences are about events, about things that happen at particular places and times.) Later in his article, Fodor distinguishes between token physicalism and type physicalism. For the moment we are discussing token physicalism. Physical events are, by definition, subject to the laws of physics. Another way of putting this is to say that according to physicalism, no event violates any law of physics. This is an uncontroversial claim that reductionists and almost all anti-reductionists accept. Thus it is hard to see how physicalism could provide convincing grounds for accepting the much stronger thesis of reductivism. Physicalism is a necessary condition for reductivism, but it is hardly sufficient. If a special science describes phenomena using laws that violate the laws of physics, then those special laws cannot be logically derivable from physical laws; but token physicalism might be true—most scientists and philosophers of science think it is true—and yet reductivism false.

The "levels" picture of the way that natural systems are organized is often thought to support reductivism. On this picture there is a hierarchy with elementary particles at the most basic level followed (in the case of biology) by atoms, molecules, cells, organs, organisms, species, and ecosystems. At every higher level, it is assumed that the systems and entities at that level consist of "nothing but" systems and entities at each of the lower levels arranged and coordinated in distinctive ways, all the way down to the bottom, so to speak. This picture gives rise to an argument for what Kenneth Schaffner has called "layer-cake" reductionism: the laws at a higher level should be reducible to those at lower levels because the entities at the higher level are composed entirely out of lower-level entities that obey lower-level laws. The argument is a tempting one, but it makes a crucial assumption, namely, that the predicates used in making lawlike generalizations at higher levels can be defined in terms of the predicates used in theories and laws at lower levels all the way down to the level of physics. Fodor argues that this assumption amounts to what he calls *type physicalism*—the thesis that the kinds and types that figure in the special sciences are identical to the kinds and types found in physics. He then challenges this assumption by means of the multiple realizability argument.

TOKEN AND TYPE PHYSICALISM

Adopting Nagel's condition of logical derivability for reducibility, we get the schematic picture of reduction presented by Fodor (955):

(1) $S_1x \rightarrow S_2x$

(2a) $S_1x \leftrightarrow P_1x$

(2b) $S_2x \leftrightarrow P_2x$

(3) $P_1x \rightarrow P_2x$

where (1) is a law of a special science (such as psychology, economics, or biology) and (2a) and (2b) are the bridge laws that permit the logical derivability of (1) from the physical law expressed in (3). S_1 and S_2 are predicates of a special science, such as psychology, and P_1 and P_2 are physical predicates. The derivation is schematic since any actual derivation from (3) to (1) would presumably involve additional laws and predicates at several intermediate levels such as biochemistry and neurophysiology. The arrow connective "$\rightarrow$" used to represent laws must be transitive for the derivation to go through. The arrow might, for example, stand for causation, as in the psychological law that distractions cause an increase in the response time to a perceptual stimulus. But if it does stand for causation, then Fodor insists that, for the purposes of reduction, it is not strong enough also to represent the two-way arrow "↔" in the bridge laws. Something stronger, such as the relation of identity, is required to link the predicates. Why is this? Sticking with the case of the reducibility of psychology, Fodor argues that mere causal correlation between (instances of) a psychological predicate (e.g., being in pain) and (instances of) a neurophysiological predicate (e.g., stimulation of C-fibers) would be consistent with Cartesian dualism in which psychological predicates inhere in non-physical immaterial minds. In that case, physical laws would not apply to psychological events and the generality of physics would fail. Thus Fodor thinks the reductivist thesis requires bridge laws such as (2a) and (2b) to be construed as at least expressing event identities: every event that instantiates the psychological predicate S_1 is identical with (i.e., numerically the same as) an event that instantiates the physical predicate P_1; and every event that instantiates the physical predicate P_1 is identical with an event that instantiates the psychological predicate S_1. In other words, reductivism entails token physicalism. But is token physicalism sufficient for reductivism? Fodor argues that it is not, that something stronger is required, namely, type physicalism asserting (in the case of psychology) the identity of psychological properties with physical properties. Fodor's argument relies on the view that statements of laws such as (1) and (3) express a nomologically necessary connection between the properties (or natural kinds) referred to by their predicate terms. If the terms S_1 in (1) and P_1 in (3) stand for properties, they must stand for the same properties in bridge law (2a); otherwise the so-called bridge law would not be a law. Although Fodor admits some doubt about whether he (or anyone else) has the full story of what makes something a law of nature, he is confident of two things: first, that whatever lawlikeness amounts to it must be the same in the bridge laws linking two sciences as it is in the laws within a science; and second, that the Humean account of laws as true universal generalizations is too meager to distinguish laws from accidental generalizations or to do justice to the inferential role that laws play within the sciences. (For further discussion of laws of nature and problems with the Humean account, see chapter 7.)

The Multiple Realizability Argument

The nub of Fodor's argument against reductivism rests on the claim that, as a matter of empirical fact, the properties and kinds that typically figure in the laws of the special sciences are multiply realizable. As a paradigm example he takes Gresham's law from economics: bad money drives out good. An indefinite variety of physical things can count as money in an economic system, ranging from coins of different metals and banknotes to commodities such as salt, shells, and cigarettes. In every case, if Gresham's law is true, if some of the items that serve as money are adulterated or debased in value—Gresham had in mind the clipping of coins—people will want to pass on their "bad" money to others as quickly as possible while retaining their "good" money. In this way an increasing fraction of the money in circulation will be "bad" money.

While money is a natural kind within economics, there is no finite disjunction of chemical or physical kinds with which it can be identified.[21] Thus, Fodor concludes that there are no bridge laws linking the economic property with any physical property. Hence reductivism must fail for Gresham's law and similar laws from the special sciences. It is important to note that Fodor is not basing his conclusion on the presumption that no disjunction of natural kinds, even a finite disjunction, can be a natural kind in principle. Rather it is the open-ended, non-finite, gerrymandered nature of the disjunction, and the heterogeneity of the disjuncts, that leads him to deny that there is any physical kind that can be identified with the economic property of being money.[22]

One objection that Fodor considers at length concerns the supposition that the special science law $S_1x \rightarrow S_2x$ holds without exception and that, since physicalism is true, for each individual physical state P_i that underlies S_1 in a particular case there is another physical state P_j^* underlying S_2 that follows from P_i in a lawlike way. Here is Fodor's picture of the situation (964).

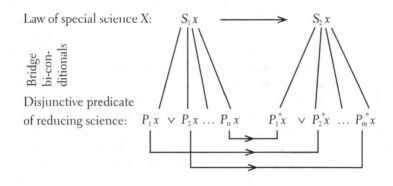

Law of special science X: $S_1 x$ $S_2 x$

Bridge bi-con-ditionals

Disjunctive predicate of reducing science: $P_1x \lor P_2x \ldots P_nx$ $P_1^*x \lor P_2^*x \ldots P_m^*x$

So we have a list of physical laws $P_1x \rightarrow P_2^*x$, $P_2x \rightarrow P_m^*x$, and so on. Suppose we disjoin the antecedents of these laws and disjoin their consequents so as to construct the rather untidy expression (numbered (6) on page 964):

$$P_1x \lor P_2x \lor \ldots \lor P_nx \rightarrow P_1^*x \lor \ldots \lor P_m^*x.$$

Why can't we regard (6) as a genuine law? After all it was derived from a list of physical laws by a logical operation that would be perfectly valid if the arrow connective stood for material implication. Fodor's reply is to insist that the "it is a law that . . ." operator creates an opaque context. Just as Fred Dretske argues (see chapter 7) that "it is a law that all Xs are Ys" does not imply that "it is a law that all Zs are Ys" even though Z and X are co-referring predicates, so Fodor denies that the conjunction of "it is a law that all As are Bs" and "it is a law that all Cs are Ds" implies that "it is a law that all things that are either As or Cs are Bs or Ds." As Fodor points out, if we were to reject this and accept (6) as a law then part of the price we would have to pay would be to give up our reliance on laws as our best (and perhaps our only reliable) guide to natural kinds. Fodor thinks that no one will be willing to pay that price. Thus, he concludes that the so-called bridge laws are not genuine laws: they express token identities between events, not type identities between properties or kinds.

LAWS WITH EXCEPTIONS

One apparent difference between the laws of the special sciences and the laws of fundamental physics is that the former usually have exceptions whereas the latter do not. Sometimes this difference is marked by saying that provisos (or *ceteris paribus* clauses) are attached to laws in the special sciences, indicating that they hold "for the most part" or "other things being equal." Since Fodor published his paper in 1974, much has been written on this topic defending several different views about the nature of provisos and whether they mark a genuine difference between physics and other sciences.[23] Fodor's view is that because the laws of the special sciences hold only *ceteris paribus* while the laws of physics hold without exception, this poses a destructive dilemma for reductivism. Either the deducibility condition for reduction is satisfied or it is not. If it is not satisfied, then reductivism is false. If it does hold, then at least one of the bridge laws involved in the deduction must have exceptions; otherwise it would be impossible to deduce special science laws (that have exceptions) from physical laws (that have no exceptions). But then, as Fodor puts it, "reductivism loses its ontological bite" (965), for there will be events instantiating S-predicates that are not identical with events instantiating P-predicates. Neither denying that S-laws have exceptions nor insisting (on the strength of reductivism's presumed truth) that P-laws must have exceptions strikes Fodor as a plausible way of avoiding the conclusion that reductivism is false.

Fodor argues that it is precisely their liability to exceptions that enables the laws of psychology, economics, and so on to offer kinds of explanation that would not be possible were reductivism true. Since the properties that figure in S-laws are open-ended disjunctions of a heterogeneous collection of P-properties, if there were no S-laws (with exceptions), then to achieve any explanation at all in the special sciences would require digging down to the "bedrock" of fundamental physics (in every case) and uncovering which particular P-event (underlying S_1) caused which particular P^*-event (underlying S_2). Not only would this task be daunting but also, even if it could be accomplished in a particular case, it would give us no insight into and provide no grounds for inferring what is likely to happen in other cases in which S_1 is instantiated by an entirely different microphysical arrangement. Luckily for us, the world is replete with nomologically stable and inferentially reliable macro-level properties despite the "unimaginably complicated to-ings and fro-ings of bits and pieces at the extreme *micro*level."[24]

8.4 | Kitcher on Reduction, Classical Genetics, and Molecular Biology

Nagel and others have proposed models of the reduction relation that is supposed to hold among theories when scientific progress occurs. As we have seen, Fodor gives a "multiple realizability" argument for doubting the reducibility of laws in the special sciences to the laws of fundamental physics. But Fodor's argument leaves open the question, which is ultimately an empirical matter, of whether reductions have been achieved at intermediate levels between the laws of one special science and those of another or between branches of the same special science. Some philosophers such as Feyerabend are skeptical about the existence (and methodological propriety) of any such reduction relation among scientific theories, regardless of their respective levels, but Feyerabend's arguments depend on a questionable holist theory of meaning that many philosophers reject. Thus, the philosopher of science who remains sympathetic to the idea of reduction is left to decide in what historical cases, and to what extent, some form of reduction has taken place. Typically, the examples used by philosophers of science are drawn from the physical sciences, usually physics and chemistry. But the life sciences should not be ignored. They, too, provide important cases of scientific progress in which the philosopher of science has to judge whether, and to what extent, one theory has been reduced to another.

In his paper "1953 and All That: A Tale of Two Sciences," Philip Kitcher presents some new ideas on the nature of theories and the relations among them, by asking whether or not two central theories in biology—classical genetics and molecular biology—stand in a reductive relation to one another. Classical genetics is the familiar textbook account of genetics developed in the first decades of the twentieth century by T. H. Morgan and his associates through their research on the fruit fly *Drosophila*. Morgan's work extended and refined Mendel's pioneering factor theory of heredity by identifying chromosomes as the carriers of Mendel's factors. These factors were soon renamed *genes*. Molecular genetics began with Watson and Crick's discovery of the structure of DNA in 1953.

In addition to expanding our discussion of intertheoretic reduction to include the life sciences, Kitcher's paper illustrates how philosophers of science strive to improve our understanding of intertheoretic relations by looking in some detail at particular scientific theories. Among philosophers of biology, a relatively common position has been that classical genetics reduces to molecular genetics. Kitcher rejects this position. But before taking up his arguments against reductionism in genetics, let us briefly review some of the central notions at work in classical and molecular biology.

Classical genetics regards genes as segments of chromosomes, each having its relative position or locus on the string of genes making up the chromosome. The chromosomes of diploid organisms are paired, members of these matched pairs being called *homologous chromosomes*. The genes that can occur at some locus on a chromosome are called *alleles*. If different alleles occur at the same locus on a pair of homologous chromosomes, the organism is said to be *heterozygous at that locus*.

There are two different kinds of cell division: mitosis and meiosis. When a cell divides by mitosis, it produces two cells that are genetically identical to the original cell; each pair of chromosomes is duplicated exactly. Mitosis is the cloning process by which multicellular organisms grow and many single-celled organisms reproduce. Unlike mitosis, meiosis involves a reduction in the number of chromosomes and is the process by which sex cells (gametes) are formed in the ovaries and testes. At meiosis, the pairs of chromosomes line up, and each member separates from its partner to form two gametes, each containing half the required number of chromosomes. Fertilization occurs when the male and female gametes unite to yield a cell with a full chromosomal complement.

Although the new individual that results from fertilization receives half its chromosomes from each parent, the paternal and maternal genes in homologous pairs of chromosomes do not remain separate for very long. Through several mechanisms they soon become mixed up together and arranged in a different order. For example, homologous chromosomes often wrap around each other, break, and then recombine to change the arrangement of genes. By such crossing over and recombination, chromosomes come to contain both maternal and paternal genes. The likelihood and frequency with which some gene comes to have new neighbors on the chromosome through recombination is a function of that gene's proximity to others and to the site of crossover. The organism's genetic makeup or genotype together with environmental effects give rise to the organism's outward, or phenotypic, characteristics. Classical genetics is thus a study of hereditary phenomena, explaining phenotypic characteristics in terms of gene-chromosome relations, gene loci and the alleles at these loci, gene transmission in gamete and zygote production, the dominance and recessiveness of genes, and so on.

The scope of molecular biology is somewhat different. It is, in effect, a study of biosynthesis at the molecular level, a theory about the molecular structure of fundamental genetic units, how they are formed and modified, and how, above all, these genetic units produce the proteins that ultimately determine phenotypic characteristics. The molecular analogue to the classical Mendelian gene is a segment of deoxyribonucleic acid (DNA)—a long spiral-shaped molecule composed of two strands of sugar and phosphate molecules connected by two pairs of complementary bases (adenine and thymine, guanine and cytosine).

The DNA in genes has two main functions: replication and transcription. In gene replication, the strands of DNA split apart, each strand serving as the template for the formation of two new DNA molecules. In transcription, the DNA serves in the production of a similar molecule called *ribonucleic acid* (RNA). RNA, whose single-strand sequence is uniquely determined by the DNA sequence, is responsible for the synthesis of proteins in ribosomes, through a process of translation. Proteins are synthesized by stringing together amino acids in long chains. By virtue of their particular sequence of amino acids, proteins take on complex three-dimensional secondary structures that enable them to perform their many functions: regulating the transcription process itself, producing other proteins via RNA, and serving as enzymatic substrates for biosynthetic pathways. Thus, the molecular structure of DNA—that is to say, the sequence of its base pairs—uniquely determines (in concert with the environment) all phenotypic characteristics of an organism. By deciphering the genetic code, molecular biologists discovered which triplets of DNA base pairs code for which amino acids and thus which sequences of base pairs (genes) produce which proteins.

THE APPARENT FAILURE OF REDUCTIONISM IN GENETICS

Kitcher approaches the issue of reduction in genetics[25] by allowing that some core, received account of intertheoretic reduction is able to capture many of the standard examples of this phenomenon in the physical sciences. Reductionists claim that the relation of classical to molecular genetics is sufficiently similar to these examples to count as yet another instance of this reductive phenomenon. Given the standard model of reduction as a derivational explanation by T' of reduced laws of T with the help of bridge laws connecting the distinctive vocabularies of T and T', the reductionist position requires us to accept three fundamental theses:

R1 Classical genetics contains general laws about the transmission of genes which can serve as the conclusions of reductive derivations.

R2 The distinctive vocabulary of classical genetics (predicates like "① is a gene," and "① is dominant with respect to ②") can be linked to the vocabulary of molecular genetics by bridge principles.

R3 A derivation of general principles about the transmission of genes from principles of molecular biology would explain why the laws of gene transmission hold (to the extent that they do). (972)

In the first part of his paper, Kitcher argues against reductionism, claiming that (R1), (R2), and (R3) are false. Let us consider these in turn.

(R1) As we noted earlier in this commentary, philosophers of science commonly identify theories with some manageable set of fundamental laws. Kitcher's first observation is that classical genetics is not presented by philosophers of science in this way at all (even by reductionists). Rather, it is presented by reference to the particular subject matter and to particular methods of investigation. The explanation for this fact is that nothing approaching genuine laws can be found in the research papers and textbooks themselves. It is true that, before Morgan's research in the early decades of this century, geneticists did make reference to "Mendel's laws." But these "laws" have since been proved false.

Kitcher's second point is taken from a particular example of the failure of these so-called laws. Mendel's second law claims that genes at different loci in diploid organisms are transmitted independently when gametes are formed by meiotic division. But since genes are segments of chromosomes, alleles on the same chromosome will tend to be transmitted together. In response to the reductionist claim that such mistakes in the laws can be corrected, Kitcher points out that it is not so much that the law is false as that the law itself plays no role in classical genetic research. Classical genetic researchers use cytological techniques of a quite general sort to see what can be learned from them. So any change these researchers might make to the second law would be simply a special-case result of the general cytological methods, with the core problem of discovering how genes are distributed on the same chromosome remaining beyond the scope of that amended law.

(R2) Suppose that the reductionist could, after all, specify the content of classical genetics in terms of some set of fundamental laws. Deriving these from molecular biology would require the use of bridge laws, of the following sort:

$$(^*) \ (x) \ (x \text{ is a gene} \leftrightarrow Mx),$$

where Mx is a (possibly complex) predicate of molecular genetics.[26] Kitcher claims that molecular biologists do not offer any such claims as $(^*)$ and that no such bridge laws can be found. We said earlier that the molecular analogues to classical (Mendelian) genes are segments of DNA. But which segments of DNA count as genes? Genes come in many lengths, and many DNA segments of those lengths are not genes; so "number of nucleotide pairs" will not suffice. Nor can we reckon a gene to be any segment of DNA between nucleotide triplets (codons) that initiate and terminate transcriptions; mutations can produce single alleles containing within them terminating and reinitiating triplets, and moreover not every gene is transcribed onto the RNA figuring in protein synthesis. In short, no general structural statement can designate which DNA segments count as genes.

Despite the impossibility of defining the term *gene* in molecular terms, the reductionist might insist that, since there are only finitely many organisms and hence finitely many genes, it should be possible in principle to supply a list containing molecular descriptions of all such DNA segments. We might then construe Mx as the large disjunction of all such molecular descriptions. Here, Kitcher argues that the reductionist will have failed to provide us with anything sufficient to reduce a classical genetic law about genes or gene transmission. Laws, presumably, apply not only to actual cases but to cases that might have but did not actually arise: laws, we say, support their counterfactuals. (For more on this feature of laws, see the commentary on chapter 7.) Thus, to reduce a law about genes is to show how possible but nonactual genes would satisfy it. But the disjunctive approach just described, expressed in terms of all actual gene-structures, fails to do this.

Now the reductionist might suggest that biconditional bridge laws such as (*) are stronger than we need. For if we are after the derivation of the law

(1) $(x)(y)$ $([Gx \ \& \ Gy] \rightarrow Axy)$

(an amended version of Mendel's second law, say, about the independent assortment of genes), then perhaps we can derive (1) from (2)—some statement expressing merely necessary molecular conditions for being a gene— together with another lawlike premise (3):

(2) (x) $(Gx \rightarrow Mx)$

(3) $(x)(y)$ $([Mx \ \& \ My] \rightarrow Axy)$.

Molecular genetics can presumably supply weaker claims of the form (2) to the effect that x is a gene only if it is composed of either DNA or RNA.[27] But Kitcher argues that there is no more hope of providing a strong molecular-biological law governing the independent assortment of all DNA and RNA segments, as in (3), than there is of providing a law of form (*) itself.

(R3) Suppose that the reductionist can provide fundamental laws of classical genetics and suitable bridge principles for deriving them. Our reduction model requires that the derivation of these laws explains them. Consider again an amended version of Mendel's second law. Why do genes of nonhomologous chromosomes assort and transmit independently? Cytology explains it perfectly in terms of the behavior of chromosomes during meiosis. Kitcher argues that, while we can provide molecular genetic details involved in the cytological account, even to the point of a successful derivation of the transmission law, doing so would not elaborate or further explain the transmission law itself. Indeed, the further details would only obscure a perfectly good cytological explanation with irrelevant information.

The reductionist might object here that, even if further molecular details only obscure the explanation, this obscurity is simply an artifact of our limited cognitive abilities; the explanatory force of the molecular details could be grasped by more powerful intellects. Kitcher's response is that this

misses the main point of his argument—it is not that the molecular derivation is too complicated but that it forfeits something present in the cytological explanation. The cytological account treats meiosis as belonging to a
natural kind, namely, the kind of process that separates paired items and
assigns them to distinct entities. In the meiotic instance of this process,
homologous chromosomes are separated and assigned to distinct gametes.
Kitcher's point is that explaining the transmission law requires identifying
this pair-separation (or PS) process as a natural kind to which meiosis
belongs and that, from a molecular point of view, the PS-process forms no
natural kind. From the viewpoint of molecular mechanics, PS-processes are
a mixed bag; there are no general constraints on the molecular structures of
items that can be paired or on how forces combine to pair or separate them.
Since PS-processes, from the molecular viewpoint, are realized in so many
disparate ways, a molecular account must fail to capture that feature of an
adequate explanation of gene transmission that marks it as a species of a
general kind. Hence, the cytological explanation, which uniformly accounts
for a wide range of cases (meiosis among them) that from the molecular
viewpoint are heterogeneous, is the objectively better explanation.[28]

Not all philosophers of science would agree with Kitcher's denial of the
explanatory thesis (R3). C. Kenneth Waters, for example, has explicitly
rejected what he calls the "gory details" objection to reductionism[29]—a reference to Kitcher's claim that "the distribution of genes to gametes is to be
explained, not by rehearsing the gory details of the reshuffling of the molecules, but through the observation that chromosomes are aligned in pairs
just prior to meiotic division, and that one chromosome from each matched
pair is transmitted to each gamete" (991–92). What force is there to Kitcher's point that the very broad natural kind he calls PS-processes, of which
meiotic segregation is an instance, is heterogeneous from the molecular
perspective? In claiming that pair separation may occur because of any
number of forces, including the action of electromagnetic, nuclear, and
even gravitational forces, Kitcher is surely conceiving of a natural kind that
includes processes quite unlike anything that occurs during meiosis.
Waters's first point is that, whatever analogies might hold between meiosis
and other pair-separation processes of quite different sorts, it does not yet
follow that cytological theory offers a unified explanation of a broad range
of phenomena that from the molecular viewpoint appear heterogeneous.[30]
Of course, phenomena do exist that classical genetics can explain uniformly, phenomena that involve a relatively wide range of molecular mechanisms. Under the category of dominance, for example, classical genetics
includes genes coding for structural proteins and enzymes, as well as regulatory genes. Waters's second point is that even if the molecular mechanisms
by which these different sorts of genes are expressed are quite different, it is
unclear why we should judge the shallower explanations of classical genetics preferable to the deeper accounts provided by molecular genetics.
Moreover, even granting that the classical explanation is more unified,

Waters believes that Kitcher is unduly pessimistic about molecular genetics, with the help of auxiliary assumptions, eventually giving an equally
good explanation. That molecular genetics, apart from such assumptions,
reflects the real diversity in a wide range of mechanisms should not be
held against it.

KITCHER'S DIAGNOSIS OF THE FAILURE OF REDUCTIONISM

Quite apart from the difficulty of explaining classical genetics in molecular
terms, Kitcher believes that the fundamental reason for the failure of reductionism lies with the falsity of (R1): according to Kitcher, we lack an account
of theories that can be suitably applied to classical and molecular genetics.
For if we adopt the standard view that a theory is a set of statements about
laws and particular facts, then it is difficult to decide which statements
make up the theory of classical genetics. What we find in the classical genetic literature at any given time are not laws but claims about particular
organisms—paradigmatic organisms such as fruit flies, bread molds, and
maize. So either classical genetics is merely a motley collection of statements
about various particular organisms, or we must suppose that although there
are general laws in genetics, these laws are never enunciated by geneticists.
Two questions thus arise: What exactly is the relationship between the collection of statements about particular organisms accepted at a given time and
the theory of classical genetics at that time? And what is the relation among
versions of the theory at different times that makes us reckon them versions of
the same theory?

Adapting the approach of T. S. Kuhn to understanding science through
paradigms, Kitcher rejects the standard view that a theory is nothing more
than a set of statements. The set of statements about (say) inheritance at
some time is only one component of a far more complicated entity Kitcher
calls the *practice* of classical genetics at that time. The practice of classical
genetics is determined by a common language about heredity, a set of statements or beliefs about inheritance held at a time, a set of questions taken to
be appropriate about heredity, and abstract patterns of reasoning instantiated by attempts to answer those questions. The beliefs exemplify the theory of classical genetics insofar as they figure in particular problem-solutions
arising from current practice; by applying certain patterns of reasoning, one
puts forward answers to accepted questions. Versions of the theory we call
classical genetics are versions of the same theory because different practices
are linked by relatively small changes in language, accepted questions, and
patterns of reasoning. Moreover, these versions of the theory share a common task, namely, the goal of characterizing hereditary phenomena. Thus,
teaching students classical genetics is not a matter of giving them some set
of fundamental axioms or laws; rather, it involves explaining a vocabulary,
posing a series of crucial questions, and training students to answer those
questions by using certain patterns of reasoning.

Kitcher illustrates this view of classical genetics by describing the initial family of questions to which the theory is applied. This is the family of pedigree problems that arise when we observe various phenotypic distributions among several generations of organisms. Each such problem is characterized by a set of data, a set of constraints, and a question. The data are statements about the distribution of phenotypes among the organisms of the pedigree. The constraints are general cytological facts and descriptions of the chromosomal constitution of members of the species. The questions are about what distribution of phenotypes we should expect from a cross between two individuals of the pedigree, about the probabilities that some phenotypic characteristic will arise from such a cross, and so on. With such problems in hand, one advances to a solution by offering a genetic hypothesis specifying what alleles are relevant, how they are phenotypically expressed, and how they are transmitted through the pedigree. One then answers the questions about the distribution and probability of phenotypes by using the hypothesis to compute the expected distribution of genotypes among offspring.

According to Kitcher, the history of classical genetics is the history of versions of this kind of practice, applied to pedigree and other families of problems. Differences among successive versions of classical genetic theory arise from differences in their genetic hypotheses and the new constraints placed upon them.

Returning now to the relation between classical genetics and molecular genetics, Kitcher is concerned to address how the explanations of hereditary phenomena offered by molecular biology connect with those offered by classical genetics. He describes three major contributions of molecular genetics to our understanding of hereditary phenomena. The first is an account of gene replication, to which classical genetics was committed without any accompanying explanation. As we saw earlier in our brief review, the Watson-Crick model of DNA explains how replication is possible. A second major advance was a molecular-genetic account of mutation. By understanding genes as segments of DNA, we can recognize possible mechanisms whereby mutant alleles are produced by the addition, deletion, or substitution of nucleotides during replication, leading to deviant proteins and, ultimately, to deviant phenotypes. Finally, Kitcher cites the specific explanation of how a substitution error in transcription forms the mutant allele responsible for the deviant hemoglobin in cases of sickle-cell anemia. The resulting change in one of the four amino-acid chains making up the hemoglobin molecule alters its ability to bind with and give up oxygen. These three examples, understood in light of Kitcher's view of theories described above, reflect three different sorts of relations among successive theories: presupposition, conceptual refinement, and explanatory extension. Each of these relations departs from the standard account of intertheoretic reduction in significant ways. Let us consider them in turn.

PRESUPPOSITIONS: THE EXAMPLE OF GENE REPLICATION

Kitcher insists that the assertion that genes replicate is not a central law in classical genetics and compares it unfavorably to bona fide laws such as the Boyle-Charles law used in thermodynamics. His reasons for denying law-like status to gene replication depend on his adopting a functional characterization of laws, since he indicates in a footnote that he is prepared to concede that "genes can replicate" is a true universal generalization. His point is that, even though it is a true universal generalization, "genes can replicate" is far too weak to explain anything interesting. Thus, the generalization fails to qualify as a law because it cannot perform one of the distinctive functions of laws within science, that is, to explain things in conjunction with other laws. (For a discussion of some of the problems with adding functional characteristics to the simple regularity theory of laws, see "Dretske's Criticisms of Ayer's Theory" in the commentary on chapter 7.)

According to Kitcher, gene replication does not appear as part of any classical-genetic explanation; rather, it is presupposed by such explanations. Before the advent of molecular genetics the presupposition of gene replication was problematic because it was hard to understand how such complicated molecules could duplicate themselves so exactly, especially when they had suffered damaging mutations. In answering the question, How is gene replication possible? molecular genetics thus gave a theoretical demonstration of a formerly problematic presupposition of classical genetics.

Kitcher's analysis of this process runs as follows. Theory T is said to presuppose p if instantiations of the problem-solving pattern of T contain statements implying p. Now, suppose that T-theorists recognize an argument from accepted premises to the conclusion that p is impossible: in such a case, p is a problematic presupposition. If a new theory T' were to produce an argument showing that p is possible and showing what is wrong with the existing reasoning that threatens p, then we can say that T' gives a theoretical derivation of the possibility of an antecedently problematic presupposition of T. Kitcher argues that molecular biology provides just such a derivation of the possibility of gene replication. But he notes that this account of the relation between T and T' differs from the standard concept of reduction in two ways. First, it requires no commitment to the view that T and T' can be formulated as sets of fundamental laws (and their consequences). Second, it requires no commitment to the notion that all general claims about genes in T are equally in need of derivation from T'; rather, the derivation is an isolated case in which T' addresses some particular thesis that is problematic but that underlies all T-theoretic explanations. Thus, where the reductionist looks for a global relation between T and all fundamental laws of T', Kitcher's account of the relation between classical and molecular genetics focuses on particular derivations of problematic presuppositions in classical genetic theory.

Conceptual Refinement: The Example of Mutation

The referent of the term *mutant allele* in classical genetics was originally fixed through the description "chromosomal segment producing a heritable deviant phenotype." This description needed amending because it covered, not only the targeted internal changes in the structure of genes, but also cases of gene transposition in unequal crossing-over at recombination. Molecular genetics gives a precise account of the relevant internal changes in mutation (insertion, deletion, substitution), and in thus refixing the referent of *mutant allele* makes it possible to distinguish cases of mutation from the untargeted cases of recombination. In this way, the molecular-genetic account of mutation involves an intertheoretic relation that Kitcher calls the *conceptual refinement* of some theory T by a successor theory T'. Conceptual refinement is accomplished when T' specifies the entities within the extensions of predicates of T in a way that changes our means of fixing the referents of these predicates.[31] After the conceptual refinement wrought by molecular genetics, the term *mutant allele* no longer includes the cases of recombination that had confused earlier researchers.

Explanatory Extension: The Example of Sickle-Cell Anemia

The relation between classical and molecular genetics, as illustrated by the sickle-cell anemia example, might appear to approximate standard reductionist models. Statements ascribing particular phenotypes (ultimately, blockages of blood flow caused by sickle-shaped red blood cells packed with deviant hemoglobin) to particular genotypes (the mutant allele resulting from a nucleotide substitution) would, it seems, be derivable from the molecular characterization of the relevant allele. Encouraged by this apparent success of the standard reductionist model, we might suppose that some general pattern of reasoning in molecular genetics (about the way in which alleles produce proteins) would enable us to generate conclusions about the classical association between each genotype and its corresponding phenotype.

Kitcher proposes that we understand the relation between the patterns of reasoning found in different theories by means of the notion of an explanatory extension of one theory by another. A theory T' is said to provide an explanatory extension of T only if one of the premises of some problem-solving pattern of T can be generated as the conclusion of some new problem-solving pattern of T'. Relevant premises occurring in explanations by T would themselves be explained by explanatory extension arguments of T'. But, Kitcher argues, we cannot infer from this, as the reductionist would have it, that the old T-explanations can always be improved by replacing its relevant premises by the T'-derivations. For—in a way similar to that encountered in the failure of (R3)—what is relevant for the classical genetic explanations of some claim S may not be relevant for explaining some claim S' appearing in the explanatory derivation of S.

Despite its similarity to the derivation condition in the standard reductionist model, Kitcher denies that his relation of explanatory extension is simply reduction under another guise. True, it is a global relation among theories (unlike the replication case above), and it is formulated in a way that frees it from the requirement to accommodate (R1) and (R2). Indeed, in the hemoglobin case it appears that we can, with the assistance of certain limiting boundary conditions (as in the derivation of Galileo's and Kepler's laws from Newton's theory), derive the classical genetic statements about normal and mutant hemoglobin genes from molecular accounts of the genetic structure. But the simplicity of the sickle-cell anemia case makes it a highly atypical example of explanatory extension. Red blood cells are the simplest cells in the human body: lacking a nucleus, they are just membranes containing hemoglobin. The abnormal behavior of sickle cells is a direct consequence of their unusual shape, and their unusual shape is a direct consequence of the deviant hemoglobin molecules they contain. Similarly, the deviant hemoglobin molecules are the result of a single, point mutation on a single allele that has the production of the deviant hemoglobin as its sole effect. Normally, mutations operate in a far more complicated manner, affecting many different phenotypic characteristics, such that derivations of genotype-phenotype associations from molecular genetics alone are not possible. Any appearance of reductionistic promise from explanatory extension is an artifact of the unusual simplicity of the hemoglobin case.

8.5 | Summary

Philosophers of science have made many attempts to characterize the relation between earlier and later scientific theories by using the concept of reduction. Ernest Nagel's influential account characterizes the reduction of one theory T by another T' as a species of explanation. Given that theories can be explicitly formulated as sets of meaningful statements, and given a deductive model of explanation, T reduces to T' on this account if the fundamental laws of T can be derived from those of T'. Two important features of reduction are entailed by this basic formulation: first, the claims of T and T' must be logically consistent, and second, terms shared by the vocabularies of T and T' must have a common meaning. These tenets, fundamental to Nagel's correspondence analysis of reduction, are called the consistency condition and the condition of meaning invariance.

The laws of a reduced theory often contain terms that are not shared by the reducing theory. This leads Nagel to distinguish homogeneous reductions from inhomogeneous (or heterogeneous) reductions. In inhomogeneous (as opposed to homogeneous) reductions, the descriptive terms of the reduced theory T are not a proper subset of the vocabulary of the reducing theory T'. Nagel makes clear that in such inhomogeneous reductions the explanatory derivation can be effected only by supplementing the reducing premises of T' with some additional statements connecting the unique vocabulary of T with the terms of T'. These additional premises are called *bridge laws*. Since the distinctive terms of a reduced theory T have meanings independent of their relation to the new reducing theory T', bridge laws cannot be construed as expressing any analytic or semantic connection between the two vocabularies. Instead, Nagel argues, bridge laws are empirical hypotheses that may take one of two forms. Either the bridge laws state—in terms of T'—conditions under which the properties expressed by the distinctive terms of T are manifest or they assert that apparently distinct objects or processes are in fact identical.

Paul Feyerabend rejects Nagel's correspondence analysis and, along with it, much else of what Feyerabend regards as misguided and pernicious in the logical empiricist account of scientific methodology. Two cornerstones of this methodology are the consistency condition and the condition of meaning invariance, both of which are crucial to Nagel's model of intertheoretic reduction. According to the consistency condition, some proposed theory about a domain of phenomena is to be judged acceptable only if it is consistent with theories already accepted in that domain. According to the condition of meaning invariance, meanings of scientific terms must not alter as science progresses and theories change. Feyerabend argues that these conditions are unreasonable constraints on science that are often violated by actual scientific practice.

Why are these conditions unreasonable? Feyerabend argues that the consistency condition militates against the acceptance of some alternative theory T^* on quite arbitrary grounds—on its temporal relation to T, which just so happens to be the earlier of the two, not its relation to the facts—and so it promotes an unreasonable methodology that no genuine empiricist should tolerate. If we try to judge whether T^* represents an advance over T on the basis of "the facts," we are guilty of assuming the relative autonomy of facts—that the facts exist and are available to us regardless of whether we have already considered alternatives to T. Feyerabend insists that all facts are theory dependent. Thus, the invention of alternative theories in disagreement with an existing theory T—which the consistency condition forbids—is crucial to a properly empiricist methodology. Finally, Feyerabend argues, application of the consistency condition contributes to a false sense of the "success" of existing theories, perpetuating T by a kind of methodological conformism. In the case of the meaning-invariance condition, Feyerabend claims that a methodology allowing mutually inconsistent (but empirically adequate) theories would welcome the use of mutually irreducible theories whose terms are unconnected by anything like factually correct bridge laws. Welcoming such theories brings with it no demand that meanings remain invariant.

Not only are the consistency and meaning-invariance conditions unreasonable restrictions on scientific change, but they are also, according to Feyerabend, violated by actual scientific practice. Against the consistency condition, Feyerabend argues that the relevant (terrestrial and celestial) consequences of Newtonian theory are inconsistent with Galileo's and Kepler's laws: strictly speaking, the Newtonian theory is a replacement for, not a reduction of, the laws of Galileo and Kepler. Against the condition of meaning invariance, Feyerabend argues that, in claiming that the mass of an object depends on its velocity, relativity theory means by *mass* something different than what classical mechanics means by this term. (According to classical mechanics, mass is independent of velocity.) As before, the old theory is not reduced to the new theory in Nagel's sense but is replaced by it. This result applies not only to theoretical terms, but to observational terms as well since observational terms, no less than theoretical terms, are theory laden.

Nagel rejects Feyerabend's claim that the consistency and meaning-invariance conditions are falsified by actual scientific practice. There is no fundamental threat to consistency in granting that reducing theories entail close approximations to the laws of reduced theories, since new theories often specify special conditions under which those approximations hold. Moreover, in actual practice derivations of laws commonly invoke simplifications and approximations of various sorts. Regarding Feyerabend's denial of meaning invariance, Nagel argues that it renders unintelligible any claims to the effect that T and T' are consistent or inconsistent—these being logical relations holding between statements whose terms have shared meanings. Furthermore, in the absence of shared meanings, we cannot intelligibly express the fact that the domains of phenomena to which T and T' apply partially overlap: if any description of the domains is theory laden, there are no means by which to express the claim that T' covers all the phenomena covered by T.

In his oft-cited article, Jerry Fodor takes aim at the reductivist thesis that the laws of the special sciences must be reducible to those of (fundamental) physics. Though inspired by a commitment to physicalism, which Fodor shares, reductivism is a much stronger claim than the assertion that all events are physical events. That assertion, token physicalism, is necessary for reductivism but not sufficient. Fodor argues that reductivism requires in addition to token physicalism the truth of type physicalism, the doctrine that the kinds, types, and properties employed in the laws of the special sciences (S-kinds, for short) are identical to those found in physics (P-kinds). He then rejects type physicalism by appealing to the multiple realizability argument.

The multiple realizability argument rests on the empirical claim that each S-kind (at the level of psychology, biology, and economics, for example) can be realized in a myriad of different ways at the level of microphysics, corresponding to an indefinite variety of P-kinds. Thus there is no identity of any S-kind with any single P-kind, nor can any single S-kind be identified with any finite disjunction of P-kinds. At best each S-kind is correlated with an open-ended disjunction of heterogeneous P-kinds. Because this disjunction does not qualify as a natural kind at the level of microphysics, this means that there are no bridge laws linking predicates at the two levels. Thus no S-law can be deduced from any set of P-laws and reductivism fails.

Fodor's view is that S-laws differ from P-laws by being liable to exceptions. S-laws hold *ceteris paribus*, all other things being equal. Nonetheless they are genuine laws couched in terms of genuine natural kinds. It is a contingent but important fact about the world that are many such kinds at the macro level that allow us to form reliable generalizations without our having to descend to the micro level in each and every case.

Most of the standard examples of intertheoretic reduction have been drawn from the physical sciences. To what extent are recent developments in biology—a special science—instances of reduction? In particular, is the relation between classical genetics and molecular biology sufficiently similar to standard examples in the physical sciences to justify the claim that classical genetics has been reduced to molecular biology?

Philip Kitcher denies any reductionist account of the relation between these two biological theories. Given a familiar model of reduction as a derivational explanation by T' of the fundamental laws of T with the help of suitable bridge laws, the reductionist position would require acceptance of three theses:

R1 Classical genetics contains laws about gene transmission that are conclusions of reductive derivations from molecular genetics.

R2 The vocabulary of classical genetics can be linked by suitable bridge laws to the vocabulary of molecular genetics.

R3 A derivation of laws about gene transmission from principles of molecular biology explains why these laws hold (to the extent that they do).

Kitcher rejects all three of these theses, arguing that there are no laws in classical genetics, that no bridge laws connect terms in classical genetics with terms in molecular genetics, and that no laws (or even approximate laws) of classical genetics can be derived from molecular biology.

Concerning (R1), Kitcher claims that general laws are largely absent from the classical genetic literature. Where they do occur, they play no crucial role in the actual practice and articulation of classical genetics.

Regarding (R2), putative bridge laws of the form "x is a gene if and only if Mx" (where M is a predicate from molecular genetics) are never formulated by molecular biologists, and for good reason: no adequate predicates from molecular genetics are to be found. In particular, no general structural statement can specify which segments of DNA count as genes.

In connection with (R3), Kitcher argues that while molecular biologists can provide the "gory details" of how gene transmission occurs at the molecular level, these details do not explain anything that we did not already understand perfectly well using the general principles of classical genetics and cell biology. For example, our explanatory understanding of why genes segregate at meiosis depends on the process of chromosome separation and migration. This is an instance of what Kitcher calls the *PS-* (or pair-separation) *process*. The explanatory power of pair separation depends on its being a general kind of process that occurs at the level of cells, not at the level of molecules. The many different things going on at the molecular level during particular instances of pair separation are irrelevant to the unified explanation that it gives of gene segregation at the cellular level. One of the fundamental errors of reductionism is to think that, just because systems are composed of smaller components, an explanation at the level of the components must be better than an explanation at a higher level. Often, especially in biology, the reverse is true.

Kitcher traces the failure of the standard model of reduction in genetics to our lack of an adequate account of scientific theories and the relations among them. He proposes that we think of classical genetics as a practice rather than as a set of statements. The practice of classical genetics at a given time includes some common language about heredity, a set of beliefs about inheritance, a set of appropriate questions about heredity, and patterns of reasoning that scientists use to answer those questions. Teaching students classical genetics is not teaching them laws or axioms but training them in this practice. Typically, research involves problems made up of a set of data, a set of constraints, and a question. To solve those problems, a genetic hypothesis will be proposed and then tested against its predictions.

How, then, is molecular genetics related to the practice of classical genetics as described by Kitcher? From three important contributions of molecular to classical genetics, Kitcher isolates three sorts of relations that can hold between the theories: presupposition, conceptual refinement, and explanatory extension.

First, classical genetics will sometimes be committed to, but be unable to account for the possibility of, some general proposition *p*. A new theory (such as molecular genetics) can sometimes provide the premises from which the possibility of that presupposition can be derived. Unlike the reductionist account, this requires neither that *p* be a law of classical genetics, nor that all general claims of the classical theory be equally in need of derivation.

Second, a new theory like molecular genetics will sometimes provide a new means (in its vocabulary) of fixing the referent of terms in the older theory. Kitcher calls this relation *conceptual refinement*.

Third, a new theory such as molecular genetics can sometimes provide an explanatory extension of the older classical genetics, by using some pattern of reasoning to derive general statements of the older theory. But from such an explanatory extension we cannot infer, as the reductionist requires, that classical-genetic explanations can be improved by replacing their premises with the new derivations offered by molecular genetics because what is relevant to a classical-genetic explanation of some claim S may not be relevant to explaining some claim S' figuring in the explanatory derivation of S. In short, a range of intertheoretic relations hold between classical and molecular genetics, none of which are, or contribute to, an instance of reduction as standardly conceived.

■ | Notes

1. The classic account of intertheoretic reduction is offered in chapter 11, "The Reduction of Theories," of Ernest Nagel's *The Structure of Science* (New York: Harcourt, Brace and World, 1961). This chapter, in which Nagel is most explicit about the formal conditions of reduction, is a revised version of his earlier "The Meaning of Reduction in the Natural Sciences," *Science and Civilization*, ed. R. C. Stauffer, (Madison: University of Wisconsin Press, 1949).

2. An argument commits the fallacy of equivocation if it uses a word in different senses, either in two or more premises, or in at least one premise and the conclusion. Take the following argument, for example: "Sugar is a vital ingredient in all human cells. Cotton candy contains sugar. Therefore, cotton candy contains something which is a vital ingredient in all human cells." The equivocation on the word *sugar* renders the argument invalid. In the first premise it means *glucose*; in the second premise it means *sucrose*. Strictly speaking, in defining the fallacy of equivocation we should also stipulate that the word in question be essential to the derivation of the conclusion: merely adding an ambiguous word to the premises and conclusion of an argument that is already valid will not produce an invalid argument.

3. The exceptions: from "P" we can validly infer "P or Q," and from "All dogs are mammals" we can validly infer "All brown dogs are mammals". Nagel explains the irrelevance of such exceptions in chapter 11 of *The Structure of Science* (1961).

4. For more on the role of observation terms in reduction, see the section "The Apparent Violation of the Condition of Meaning Invariance" later in this commentary. The theory-observation distinction is explored at length in chapter 9.

5. Bridge laws are sometimes called *bridge principles*, *correspondence rules*, or *coordinative definitions*.

6. If the stated T' conditions are only necessary (but not sufficient) for the occurrence of the relevant property, then the laws of T in which the property figures could not be derived from the claims of T' conjoined with bridge laws. But Nagel's third assumption of the correspondence analysis of inhomogeneous reduction is that reductions are deductive explanations of the reduced laws.

7. In his paper, Feyerabend uses T to stand for the reducing (explaining) theory, and T' for the theory that is reduced (explained). This is exactly the opposite of Nagel's notation: for Nagel it is T' that reduces (explains) T. To avoid confusion, we have adopted Nagel's notation throughout this commentary, except when quoting Feyerabend on page 1015.

8. In forming a proper assessment of Feyerabend's paper, it is important to remember that when it was published (in 1963) logical empiricism was still a dominant

force in philosophy of science. Kuhn's book, *The Structure of Scientific Revolutions* (Chicago: University of Chicago Press, 1962) had only just appeared, and many of Kuhn's arguments (especially those about observation and meaning) are substantially indebted to Feyerabend's work. Historically, Feyerabend's criticisms of logical empiricism played a major role in its demise. The chorus of disapproval that has greeted Feyerabend's later writings—books such as *Against Method* (London: New Left Books, 1975) and *Science in a Free Society* (London: New Left Books, 1978)—because of their gleeful embrace of relativism and their anarchistic rejection of the authority of science ("anything goes"), should not obscure the significance of his earlier work as a critic of logical empiricism.

9. For a detailed account of attempts to explain Brownian motion in the nineteenth century, see Mary Jo Nye, *Molecular Reality* (London: Macdonald, 1972).

10. The simplest of these predictions was that the kinetic energy of the pollen grains should equal the kinetic energy of the water molecules surrounding them. But the pollen grains were found to be moving much more slowly than predicted. Eventually, Einstein was able to clear up the mystery by pointing out that when we follow the zig-zag path of a pollen grain under the microscope, what we observe is never the instantaneous velocity of the pollen grain (which decays in less than a millionth of a second) but a much smaller, average velocity.

11. For a criticism of Feyerabend's interpretation of this history, see Ronald Laymon, "Feyerabend, Brownian Motion, and the Hiddenness of Refuting Facts," *Philosophy of Science* 44 (1978): 225–47.

12. The second law of classical thermodynamics says that the entropy of an isolated system never decreases. It claims to be a true, universal law of nature. As Maxwell and Boltzmann realized, the kinetic theory implies that there should be random, local fluctuations in entropy even in a fluid that, as a whole, is in thermal equilibrium. Most of these spontaneous variations in density and temperature will be very small and last for only a fraction of a second; the larger the variation, the more improbable its occurrence. Thus, from the standpoint of the kinetic theory, the second law of thermodynamics is not a universal, lawlike truth but a statistical one. Interestingly, although Maxwell and Boltzmann were the two foremost advocates of the kinetic theory in the nineteenth century and the pioneers of statistical mechanics, neither of them thought that the tiny random fluctuations predicted by the theory would ever be observed. Apparently, the significance of Brownian motion never occurred to them.

13. The relevant facts included numerical predictions of how far the particles would move in a given time, how fast they would rotate, and how the number of particles per unit volume of fluid in a vertical column would decrease with height.

14. See the final paragraphs of the section "The Apparent Violation of the Condition of Meaning Invariance," later in this commentary, for a discussion of Nagel's response to this criticism of synonymy and meaning invariance.

15. Kepler's third law asserts that for all the planets, the square of the period divided by the cube of the mean distance from the sun is a constant. Newton's theory entails that this ratio is not constant because it depends on the planet's mass.

16. In arguing that mass in relativity theory is frame-relative, Feyerabend is implicitly adopting the definition of a particle's "relativistic mass" as $m_r = m_0 \gamma = m_0 / \sqrt{(1 - v^2/c^2)}$, where m_0 is the particle's so-called "rest mass," namely, the value of

its "relativistic mass" in the inertial frame in which its velocity is zero. These definitions of "relativistic mass" and "rest mass" were common in physics textbooks in the first half of the twentieth century and still appear in popular works, but most modern physicists have abandoned them in favor of a single concept of mass, designated by the letter m (without a subscript). In this approach, mass is defined by the equation $m^2 = (E/c^2)^2 - (\mathbf{p}/c)^2$, where E is the energy, and $\mathbf{p}$ the momentum vector. Momentum and velocity are related by the equation $\mathbf{p} = E\mathbf{v}/c^2$. These equations hold both for single particles and for composite systems consisting of several particles. In modern treatments of relativity theory, the mass, m, is an invariant; it is the same in all inertial frames. Though energy and momentum vary from frame to frame, mass does not. Thus, modern treatments of relativity reject Feyerabend's claim that mass is frame-relative; but they agree that mass is non-additive. For a defense of the invariance of mass in relativity theory, see Carl G. Adler, "Does Mass Really Depend on Velocity, Dad?" *American Journal of Physics* 55 (1987) 739–43; Lev B. Okun, "The Concept of Mass," *Physics Today* 42 (1989) 31–6; Edwin F. Taylor and John Archibald Wheeler, *Spacetime Physics: Introduction to Special Relativity*, 2d ed. (New York: W. H. Freeman and Company, 1992); Marc Lange, "The Most Famous Equation," *Journal of Philosophy* 98 (2001) 219–38; and Marc Lange, *An Introduction to the Philosophy of Physics: Locality, Fields, Energy, and Mass* (Oxford: Blackwell, 2002), ch. 8.

17. From p. 59 of "Explanation, Reduction, and Empiricism," in *Scientific Explanation, Space, and Time*, ed., H. Feigl and G. Maxwell, vol. 3, *Minnesota Studies in the Philosophy of Science* (Minneapolis: University of Minnesota Press, 1962), 28–97.

18. Length contraction is the well-confirmed phenomenon, predicted by Einstein's special theory of relativity, that moving rods shrink in the direction of their motion when their length is measured by an observer who is not traveling with them. For example, suppose that I have two rods, each exactly one meter long as they lie on the ground at my feet. If one of the rods is now set in motion and I measure its length as it flies past me, it will be less than one meter. Moving rods contract by a factor of $1/\sqrt{(1-v^2/c^2)}$, where v is the velocity of the rod relative to me, the observer, and c is the velocity of light. A person travelling with the rod will make the same judgment about the rod lying at my feet. According to that person, *my* rod has shrunk by the same fraction of its length. The effect becomes noticeable only when the velocity of the rod, v, approaches the velocity of light, c.

19. According to Hempel, the unity of science advocated by logical positivists had three aspects: unity of language, laws, and method. See Carl G. Hempel, "Logical Positivism and the Social Sciences," in P. Achinstein and S. Barker, eds., *The Legacy of Logical Positivism* (Baltimore: The Johns Hopkins University Press, 1969), 163–94. The unity of language would be achieved by defining all scientific terms, however recondite and specialized, in a common vocabulary whose meaning is already well understood. For some this was tantamount to physicalism since the preferred language was that of physics; for others the common language could be that used to describe the common-sense objects of everyday life. The unity of laws thesis comes closest to a statement of physicalism. For example: "all laws of nature . . . are logical consequences of the physical laws, i.e., of those laws which are needed for the explanation of inorganic processes." See Carnap's reply to Feigl in *The Philosophy of*

Rudolf Carnap, ed. Paul A. Schilpp (La Salle, Ill.: Open Court, 1963), 883. The unity of method meant that all branches of science use (or should use) the same concepts of explanation, confirmation, and reduction. It was tantamount to the vision of a general philosophy of science applicable to all the sciences. See also Paul Oppenheim and Hilary Putnam, "Unity of Science as a Working Hypothesis," in H. Feigl, M. Scriven, and G. Maxwell, eds., *Minnesota Studies in the Philosophy of Science*, vol. II (Minneapolis: University of Minnesota Press, 1958), 3–36. Oppenheim and Putnam defend as justified and credible the thesis that all science will eventually be reduced to microphysics.

20. For the very limited extent to which quantum mechanics can be said to have reduced chemistry to physics, see Eric Scerri, "Reduction and Emergence in Chemistry—Two Recent Approaches," *Philosophy of Science* 74 (2007): 920–31.

21. The term *natural kind* is being used loosely here since something's counting as money depends on the beliefs and attitudes of human beings. Thus it is not fully objective in the way that physical and chemical kinds are.

22. See, for example, Fodor's reply to Jaegwon Kim in Jerry Fodor, "Special Sciences: Still Autonomous after All These Years," *Noûs* 31 (1997): 149–63. Fodor stresses the difference between closed disjunctive kinds (such as Kim's example of jade, which is a mixture of two distinct chemical kinds, jadeite and nephrite) and the open multiple realizability of kinds in the special sciences that are "nomologically homogeneous under their functional descriptions *despite the physical heterogeneity of their realizers*" (p. 153).

23. For a comprehensive range of views on provisos, see the papers by Earman and Roberts, Lange, Woodward, Cartwright, and others in the special issue of *Erkenntnis* 57 (2002).

24. Fodor, "Special Sciences: Still Autonomous after All These Years," p. 160. Fodor regards scientists as justified in preferring higher-level laws (such as "pain leads to avoidance behavior" in psychology) over vastly more complicated and incomplete disjunctive generalizations at the micro level because of the epistemic norm that one should choose the strongest claim compatible with one's evidence, all else being equal.

25. Hereafter in this section we will use the terms *reduction* and *reductionism* as shorthand for "reduction of classical to molecular genetics" and for "classical genetics reduces to molecular genetics," respectively, dropping the explicit reference to genetics.

26. Kitcher uses the symbol "↔" to stand for the connective "entails and is entailed by." Thus, (*) is to be read as "*x* is a gene" entails that *x* has property *M*, and "*x* has property *M*" entails that *x* is a gene.

27. Viral genes are composed of RNA.

28. Kitcher is here relying on a view of theoretical explanation as unification that he has defended in "Explanatory Unification," *Philosophy of Science* 48 (1981): 507–31, reprinted in chapter 6. See also Philip Kitcher, "Explanatory Unification and the Causal Structure of the World," in *Scientific Explanation*, ed. P. Kitcher and W. C. Salmon, vol. 13, *Minnesota Studies in the Philosophy of Science* (Minneapolis: University of Minnesota Press, 1989), 410–505; and Philip Kitcher, "Darwin's

1044 | Ch. 8 · Intertheoretic Reduction

Achievement," chap. 2, *The Advancement of Science* (Oxford: Oxford University Press, 1993).

29. C. Kenneth Waters, "Why the Antireductionist Consensus Won't Survive the Case of Classical Mendelian Genetics," in *PSA 1990*, vol. 1, ed. A. Fine, M. Forbes, and L. Wessels (East Lansing, Mich.: Philosophy of Science Association, 1990), 125–39.

30. One might add here that for any wide range of phenomena there is likely to be some abstract description under which they all fall and that such a "natural kind" will undoubtedly appear heterogeneous from many perspectives. This articulation of a putative natural kind would be vindicated to the extent that some general theory encompassing all such instances uses the natural kind in its explanations. Waters's first point is that cytological theory accomplishes nothing of that sort.

31. Kitcher does not discuss the relation of conceptual refinement to the standard model of reduction. Could we view statistical mechanics as offering a conceptual refinement of the term *temperature* in thermodynamics? Presumably not, since Kitcher's account of conceptual refinement seems to presuppose that the predicate whose means of reference-fixing is changed in going from T to T' is present in the vocabulary of both T and T'. This condition is satisfied in the case of *mutant allele*, but not in the case of *temperature*.

9 | Empiricism and Scientific Realism

INTRODUCTION

One of the most lively and persistent debates in philosophy of science is that between empiricists and realists concerning the status of scientific theories. Are scientific theories to be understood as offering (or at least intending to offer) a true account of the world? And if we accept scientific theories as true, should we believe that the entities they appear to postulate really exist?

Empiricists emphasize that the warrant for all scientific claims ultimately rests on experience, on what can be observed, tested, and measured. They are thus suspicious of theoretical assertions that cannot in some fairly direct way be cashed out in terms of experience. This suspicion has, in the hands of some, been reinforced by arguments from underdetermination: given the same body of observational evidence, there are in principle an indefinitely large number of alternative theories that are compatible with that evidence and which entail it. This being so, what could warrant our selecting just one of those theories and asserting that it is true or even probable? Moreover, reflecting on the frequency with which past scientific theories have later been proved false would seem to recommend a cautious skepticism or agnosticism about the unobservable entities postulated by present theories.

Realists concede that many theories accepted in the past have turned out to be false. Nevertheless, they insist that there is a clear sense in which science progresses and improves over time. Present-day theories in physics, chemistry, and biology are objectively better than theories of the eighteenth and nineteenth centuries. Even if present-day theories are not true in every detail, they are closer to the truth, and this makes them more reliable for making predictions and guiding our actions. Although theories are, of course, tested by experience, their claims are not merely about experience, nor are they in some way reducible to it. The fundamental point of theorizing is to go beyond experience to discover deep, hidden truths about the underlying causes of events, regardless of whether these causes can be directly observed. And when such theories are well confirmed by the observable evidence they predict, we have the best reason possible for taking their theoretical claims at face value and for believing that the entities they postulate are real. In this way science has extended our knowledge into the smallest parts of atoms and the farthest reaches of the universe.

Many empiricist philosophers of science are antirealists about the purely theoretical claims of science and the unobservable entities (such as electrons and quarks) to which its theoretical terms allegedly refer. In "The Ontological Status of Theoretical Entities," Grover Maxwell argues that this antirealist view of science cannot be sustained. In particular, Maxwell argues that what counts as observational is itself a theoretical question and hence that the traditional observational-theoretical distinction at work in empiricist accounts of science has no ontological significance whatsoever: no entities are unobservable in principle.

In "Arguments Concerning Scientific Realism"—a piece taken from his 1980 book *The Scientific Image*—Bas van Fraassen argues against Maxwell's position on the way to formulating his own constructive empiricism. According to van Fraassen's version of antirealism, we can accept scientific theories and yet remain agnostic about their truth, requiring only that theories be empirically adequate (i.e., true about observables). From the standpoint of constructive empiricism, van Fraassen offers his diagnosis of why a number of important arguments for realism must be regarded as failures.

Alan Musgrave criticizes van Fraassen's approach to scientific theories in his "Realism versus Constructive Empiricism." One of the most popular arguments for scientific realism, which, according to Musgrave, can survive van Fraassen's attack, is the "miracle" or "ultimate" argument—that if we do not accept our scientific theories as (at least approximately) true and their terms as in fact refering to unobservables, then we can only reckon the success of our theories to be a miracle. In "A Confutation of Convergent Realism," Larry Laudan attacks this inference-to-the-best-explanation defense of realism, arguing from the history of science that the connection between success, reference, and approximate truth is much weaker than realists suppose. In a form of argument that has become known as the "pessimistic induction," Laudan claims that many past theories that were among the most successful in their day involved terms that failed to refer since these theories relied on entities (such as phlogiston) and processes (such as action at a distance) that do not exist according to our present-day science. Since realists acknowledge that reference is necessary for truth, they must agree that many past successful theories were false. When realists respond by insisting that science progresses nonetheless because later (false) theories are closer to the truth than the earlier (false) theories they have replaced and that success is a reliable indicator of approximate truth, Laudan objects that the notion of approximate truth, or verisimilitude, is too imprecise to bear the weight that realists place on it.

The pessimistic induction is one of the chief weapons in the antirealists' arsenal. Because of its importance, many realists have tried to refute the argument by attacking one or other of its premises; but some philosophers of science, notably Marc Lange and Peter Lewis, have argued that the argument is fallacious and that therefore there is no case that realists need to answer. In "On the Pessimistic Induction and Two Fallacies," Juha Saatsi examines these two attempts to undermine Laudan's argument and judges that neither succeeds. He concludes that the pessimistic induction remains a significant challenge for realism.

How then might one try to justify scientific realism? One might narrow one's sights, dropping the troublesome notion that theories are true or approximately true. In "Experimentation and Scientific Realism," Ian Hacking urges that we sever the view that theories are true from a realist commitment to the existence of unobservables. According to Hacking's entity realism, our justification for believing that unobservables exist rests on our ability to manipulate them in scientific experiments. David Resnik replies to this proposal in "Hacking's Experimental Realism," where he argues that Hacking's entity realism can be defended only by relying on some form of traditional success-of-science argument. In addition, Resnik argues that the only initially plausible way of trying to wed realism about entities to antirealism about theories requires a theory of meaning that brings realism about theories along with it after all.

Like Resnik, Martin Carrier is convinced that Hacking's entity realism is unsuccessful; and though he is critical of certain aspects of Laudan's pessimistic induction, he agrees that the customary versions of realism are refuted by examples of false but successful theories from the history of science. Despite this, Carrier thinks that there is something right about the miracle argument. Borrowing the notion of a "natural classification" from Pierre Duhem, Carrier argues that the best explanation of the *strong* success of past theories is due to their correct identification of relevant natural kinds and that commitment to these kinds is retained in future theories even though they often employ radically different concepts.

Arthur Fine suggests that we hunt for some middle ground between scientific realism and antirealism. After criticizing arguments for realism in "The Natural Ontological Attitude," he recommends a broadly minimalist attitude toward scientific theories, the natural ontological attitude (NOA), according to which the confirmed results of science are accepted in just the same way we naturally accept the evidence of our senses about garden-variety objects. According to Fine, this is a nonrealist proposal, a core position to which realists and antirealists add unnecessary and indefensible philosophical accounts of truth. In "NOA's Ark—Fine for Realism," Alan Musgrave responds to Fine by arguing that on any plausible construal of NOA, it is not a nonrealist philosophy of science at all but a thoroughly realist one. Musgrave then offers an alternative construal of Fine's account that is genuinely neither realist nor antirealist, but, Musgrave argues, this alternative account is devoid of philosophical content.

Grover Maxwell

The Ontological Status
of Theoretical Entities

That anyone today should seriously contend that the entities referred to by
scientific theories are only convenient fictions, or that talk about such entities is translatable without remainder into talk about sense contents or
everyday physical objects, or that such talk should be regarded as belonging
to a mere calculating device and, thus, without cognitive content—such
contentions strike me as so incongruous with the scientific and rational
attitude and practice that I feel this paper *should* turn out to be a demoli-
tion of straw men. But the instrumentalist views of outstanding physicists
such as Bohr and Heisenberg are too well known to be cited,* and in a
recent book of great competence, Professor Ernest Nagel concludes that
"the opposition between [the realist and the instrumentalist] views [of theo-
ries] is a conflict over preferred modes of speech" and "the question as to
which of them is the 'correct position' has only terminological interest."[1]
The phoenix, it seems, will not be laid to rest.

The literature on the subject is, of course, voluminous, and a comprehensive treatment of the problem is far beyond the scope of one essay. I
shall limit myself to a small number of constructive arguments (for a radi-

From Herbert Feigl and Grover Maxwell, eds., *Scientific Explanation, Space, and
Time*, vol. 3, *Minnesota Studies in the Philosophy of Science* (Minneapolis: University of Minnesota Press, 1962), 3–15.

* Niels Bohr and Werner Heisenberg, along with Max Born and other pioneers of
quantum physics in the 1920s, were proponents of the so-called Copenhagen interpretation of quantum mechanics, which remains the orthodox textbook account to
this day. Its elements include Bohr's doctrine of complementarity, the insistence
that the wave function contains a complete description of reality, and a profound
reluctance to say anything about the properties of quantum systems when they are
not being observed or measured. As Maxwell says, Bohr and Heisenberg were
instrumentalists (though opinions are sharply divided over what, exactly, Bohr's
position was), and their views had a considerable influence on the development of
logical positivism and logical empiricism. For a good account of Heisenberg's version of the Copenhagen interpretation, see Werner Heisenberg, *Physics and Philosophy* (New York: Harper and Row, 1958).

cally realistic interpretation of theories) and to a critical examination of some of the more crucial assumptions (sometimes tacit, sometimes explicit) that seem to have generated most of the problems in this area.[2]

■ | The Problem

Although this essay is not comprehensive, it aspires to be fairly self-contained. Let me, therefore, give a pseudohistorical introduction to the problem with a piece of science fiction (or fictional science).

In the days before the advent of microscopes, there lived a Pasteur-like scientist whom, following the usual custom, I shall call Jones. Reflecting on the fact that certain diseases seemed to be transmitted from one person to another by means of bodily contact or by contact with articles handled previously by an afflicted person, Jones began to speculate about the mechanism of the transmission. As a "heuristic crutch," he recalled that there is an obvious *observable* mechanism for transmission of certain afflictions (such as body lice), and he postulated that all, or most, infectious diseases were spread in a similar manner but that in most cases the corresponding "bugs" were too small to be seen and, possibly, that some of them lived inside the bodies of their hosts. Jones proceeded to develop his theory and to examine its testable consequences. Some of these seemed to be of great importance for preventing the spread of disease.

After years of struggle with incredulous recalcitrance, Jones managed to get some of his preventative measures adopted. Contact with or proximity to diseased persons was avoided when possible, and articles which they handled were "disinfected" (a word coined by Jones) either by means of high temperatures or by treating them with certain toxic preparations which Jones termed "disinfectants." The results were spectacular: within ten years the death rate had declined 40 per cent. Jones and his theory received their well-deserved recognition.

However, the "crobes" (the theoretical term coined by Jones to refer to the disease-producing organisms) aroused considerable anxiety among many of the philosophers and philosophically inclined scientists of the day. The expression of this anxiety usually began something like this: "In order to account for the facts, Jones must assume that his crobes are too small to be seen. Thus the very postulates of his theory preclude their being observed; they are *unobservable in principle*." (Recall that no one had envisaged such a thing as a microscope.) This common prefatory remark was then followed by a number of different "analyses" and "interpretations" of Jones' theory. According to one of these, the tiny organisms were merely convenient fictions—*façons de parler*—extremely useful as heuristic devices for facilitating (in the "context of discovery") the thinking of scientists but not to be taken seriously in the sphere of cognitive knowledge (in the "context of justification"). A closely related view was that Jones' theory was merely an instrument, useful for organizing observation statements and (thus) for producing desired results, and that, therefore, it made no more sense to ask what was the nature of the entities to which it referred than it did to ask what was the nature of the entities to which a hammer or any other tool referred.[3] "Yes," a philosopher might have said, "Jones' theoretical expressions are just meaningless sounds or marks on paper which, when correlated with observation sentences by appropriate syntactical rules, enable us to predict successfully and otherwise organize data in a convenient fashion." These philosophers called themselves "instrumentalists."

According to another view (which, however, soon became unfashionable), although expressions containing Jones' theoretical terms were genuine sentences, they were translatable without remainder into a set (perhaps infinite) of observation sentences. For example, 'There are crobes of disease X on this article' was said to translate into something like this: 'If a person handles this article without taking certain precautions, he will (probably) contract disease X; and if this article is first raised to a high temperature, then if a person handles it at any time afterward, before it comes into contact with another person with disease X, he will (probably) not contract disease X; and. . . .'

Now virtually all who held any of the views so far noted granted, even insisted, that theories played a useful and legitimate role in the scientific enterprise. Their concern was the elimination of "pseudo problems" which might arise, say, when one began wondering about the "reality of supraempirical entities," etc. However, there was also a school of thought, founded by a psychologist named Pelter, which differed in an interesting manner from such positions as these. Its members held that while Jones' crobes might very well exist and enjoy "full-blown reality," they should not be the concern of medical research at all. They insisted that if Jones had employed the correct methodology, he would have discovered, even sooner and with much less effort, all of the observation laws relating to disease contraction, transmission, etc. without introducing superfluous links (the crobes) into the causal chain.

Now, lest any reader find himself waxing impatient, let me hasten to emphasize that this crude parody is not intended to convince anyone, or even to cast serious doubt upon sophisticated varieties of any of the reductionistic positions caricatured (some of them not too severely, I would contend) above. I am well aware that there are theoretical entities and theoretical entities, some of whose conceptual and theoretical statuses differ in important respects from Jones' crobes. (I shall discuss some of these later.) Allow me, then, to bring the Jonesean prelude to our examination of observability to a hasty conclusion.

Now Jones had the good fortune to live to see the invention of the compound microscope. His crobes were "observed" in great detail, and it became possible to identify the specific kind of *microbe* (for so they began to be called) which was responsible for each different disease. Some philosophers freely admitted error and were converted to realist positions concerning theories. Others resorted to subjective idealism or to a thoroughgoing phenomenalism, of which there were two principal varieties. According to one, the one "legitimate" observation language had for its descriptive terms only those which referred to sense data. The other maintained the stronger thesis that *all* "factual" statements were *translatable* without remainder into the sense-datum language. In either case, any two non-sense data (e.g., a theoretical entity and what would ordinarily be called an "observable physical object") had virtually the same status. Others contrived means of modifying their views much less drastically. One group maintained that Jones' crobes actually never had been unobservable in principle, for, they said, the theory did not imply the impossibility of finding a means (e.g., the microscope) of observing them. A more radical contention was that the crobes were not observed at all; it was argued that what was seen by means of the microscope was just a shadow or an image rather than a corporeal organism.

■ | The Observational-Theoretical Dichotomy

Let us turn from these fictional philosophical positions and consider some of the actual ones to which they roughly correspond. Taking the last one first, it is interesting to note the following passage from Bergmann: "But it is only fair to point out that if this . . . methodological and terminological analysis [for the thesis that there are no atoms] . . . is strictly adhered to, even stars and microscopic objects are not physical things in a literal sense, but merely by courtesy of language and pictorial imagination. This might seem awkward. But when I look through a microscope, all I see is a patch of color which creeps through the field like a shadow over a wall. And a shadow, though real, is certainly not a physical thing."[4]

I should like to point out that it is also the case that if this analysis is strictly adhered to, we cannot observe physical things through opera glasses, or even through ordinary spectacles, and one begins to wonder about the status of what we see through an ordinary windowpane. And what about distortions due to temperature gradients—however small and, thus, always present—in the ambient air? It really does "seem awkward" to say that when people who wear glasses describe what they see they are talking about shadows, while those who employ unaided vision talk about physical things—or that when we look through a windowpane, we can only *infer* that it is raining, while if we raise the window, we may "observe directly" that it is. The point I am making is that there is, in principle, a continuous series beginning with looking through a vacuum and containing these as members: looking through a windowpane, looking through glasses, looking through binoculars, looking through a low-power microscope, looking through a high-power microscope, etc., in the order given. The important consequence is that, so far, we are left without criteria which would enable us to draw a nonarbitrary line between "observation" and "theory." Certainly, we will often find it convenient to draw such a to-some-extent-arbitrary line; but its position will vary widely from context to context. (For example, if we are determining the resolving characteristics of a certain microscope, we would certainly draw the line beyond ordinary spectacles, probably beyond simple magnifying glasses, and possibly beyond another microscope with a lower power of resolution.) But what ontological ice does a mere methodologically convenient observational-theoretical dichotomy cut? Does an entity attain physical thing-hood and/or "real existence" in one context only to lose it in another? Or, we may ask, recalling the continuity from observable to unobservable, is what is seen through spectacles a "little bit less real" or does it "exist to a slightly less extent" than what is observed by unaided vision?[5]

However, it might be argued that things seen through spectacles and binoculars look like ordinary physical objects, while those seen through microscopes and telescopes look like shadows and patches of light. I can only reply that this does not seem to me to be the case, particularly when looking at the moon, or even Saturn, through a telescope or when looking at a small, though "directly observable," physical object through a low-power microscope. Thus, again, a continuity appears.

"But," it might be objected, "theory tells us that what we see by means of a microscope is a real image, which is certainly distinct from the object on the stage." Now first of all, it should be remarked that it seems odd that one who is espousing an austere empiricism which requires a sharp observational-language/theoretical-language distinction (and one in which the former language has a privileged status) should need a theory in order to tell him what is observable. But, letting this pass, what is to prevent us from saying that we still observe the object on the stage, even though a "real image" may be involved? Otherwise, we shall be strongly tempted by phenomenalistic demons, and at this point we are considering a physical-object observation language rather than a sense-datum one. (Compare the traditional puzzles: Do I see one physical object or two when I punch my eyeball? Does one object split into two? Or do I see one object and one image? Etc.)

Another argument for the continuous transition from the observable to the unobservable (theoretical) may be adduced from theoretical considerations themselves. For example, contemporary valency theory tells us that there is a virtually continuous transition from very small molecules (such as those of hydrogen) through "medium-sized" ones (such as those of the fatty acids, polypeptides, proteins, and viruses) to extremely large ones (such as crystals of the salts, diamonds, and lumps of polymeric plastic). The molecules in the last-mentioned group are macro, "directly observable" physical objects but are, nevertheless, genuine, single molecules; on the other hand, those in the first-mentioned group have the same perplexing properties as subatomic particles (de Broglie waves, Heisenberg indeterminacy, etc.). Are we to say that a large protein molecule (e.g., a virus) which can be "seen" only with an electron microscope is a little less real or exists to somewhat less an extent than does a molecule of a polymer which can be seen with an optical microscope? And does a hydrogen molecule partake of only an infinitesimal portion of existence or reality? Although there certainly *is* a continuous transition from observability to unobservability, any talk of such a continuity from full-blown existence to nonexistence is, clearly, nonsense.

Let us now consider the next to last modified position which was adopted by our fictional philosophers. According to them, it is only those entities which are *in principle* impossible to observe that present special problems. What kind of impossibility is meant here? Without going into a detailed discussion of the various types of impossibility, about which there is abundant literature with which the reader is no doubt familiar, I shall assume what usually seems to be granted by most philosophers who talk of entities which are unobservable in principle—i.e., that the theory(s) itself (coupled with a physiological theory of perception, I would add) entails that such entities are unobservable.

We should immediately note that if this analysis of the notion of unobservability (and, hence, of observability) is accepted, then its use as a means of delimiting the observation language seems to be precluded for those philosophers who regard theoretical expressions as elements of a calculating device—as meaningless strings of symbols. For suppose they wished to determine whether or not 'electron' was a theoretical term. First, they must see whether the theory entails the sentence 'Electrons are unobservable.' So far, so good, for their calculating devices are said to be able to select genuine sentences, provided they contain no theoretical terms. But what about the selected "sentence" itself? Suppose that 'electron' is an observation term. It follows that the expression is a genuine sentence and asserts that electrons are unobservable. But this entails that 'electron' is *not* an observation term. Thus if 'electron' is an observation term, then it is *not* an observation term. Therefore it is not an observation term. But then it follows that 'Electrons are unobservable' is not a genuine sentence and does not assert that electrons are unobservable, since it is a meaningless string of marks and does not assert anything whatever. Of course, it could be stipulated that when a theory "selects" a meaningless expression of the form 'Xs are unobservable,' then 'X' is to be taken as a theoretical term. But this seems rather arbitrary.

But, assuming that well-formed theoretical expressions are genuine sentences, what shall we say about unobservability in principle? I shall begin by putting my head on the block and argue that the present-day status of, say, electrons is in many ways similar to that of Jones' crobes before microscopes were invented. I am well aware of the numerous theoretical arguments for the impossibility of observing electrons. But suppose new entities are discovered which interact with electrons in such a mild manner that if an electron is, say, in an eigenstate of position, then, in certain circumstances, the interaction does not disturb it. Suppose also that a drug is discovered which vastly alters the human perceptual apparatus—perhaps even activates latent capacities so that a new sense modality emerges. Finally, suppose that in our altered state we are able to perceive (not necessarily visually) by means of these new entities in a manner roughly analogous to that by which we now see by means of photons. To make this a little more plausible, suppose that the energy eigenstates of the electrons in some of the compounds present in the relevant perceptual organ are such that even the weak interaction with the new entities alters them and also that the cross sections, relative to the new entities, of the electrons and other particles of the gases of the air are so small that the chance of any interaction here is negligible. Then we might be able to "observe directly" the position and possibly the approximate diameter and other properties of some electrons. It would follow, of course, that quantum theory would have to be altered in some respects, since the new entities do not conform to all its principles. But however improbable this may be, it does not, I maintain, involve any logical or conceptual absurdity. Furthermore, the modification necessary for the inclusion of the new entities would not necessarily change the meaning of the term 'electron.'[6]

Consider a somewhat less fantastic example, and one which does not involve any change in physical theory. Suppose a human mutant is born who is able to "observe" ultraviolet radiation, or even X rays, in the same way we "observe" visible light.

Now I think that it is extremely improbable that we will ever observe electrons directly (i.e., that it will ever be reasonable to assert that we have so observed them). But this is neither here nor there; it is not the purpose of this essay to predict the future development of scientific theories, and, hence, it is not its business to decide what actually is observable or what will become observable (in the more or less intuitive sense of 'observable' with which we are now working). After all, we are operating, here, under the assumption that it is theory, and thus science itself, which tells us what is or is not, in this sense, observable (the 'in principle' seems to have become superfluous). And this is the heart of the matter; for it follows that, at least for this sense of 'observable,' there are no a priori or philosophical criteria for separating the observable from the unobservable. By trying to show that we can talk about the *possibility* of observing electrons without committing logical or conceptual blunders, I have been trying to support the thesis that any (nonlogical) term is a *possible* candidate for an observation term.

There is another line which may be taken in regard to delimitation of the observation language. According to it, the proper term with which to work is not *'observable'* but, rather *'observed'*. There immediately comes to mind the tradition beginning with Locke and Hume (No idea without a preceding impression!), running through Logical Atomism and the Principle of Acquaintance, and ending (perhaps) in contemporary positivism.*

---

* Maxwell is referring to doctrines defended by Bertrand Russell and Ludwig Wittgenstein in the early decades of the twentieth century, although both later abandoned them. Wittgenstein's logical atomism holds that all meaningful propositions

Since the numerous facets of this tradition have been extensively examined and criticized in the literature, I shall limit myself here to a few summary remarks.

Again, let us consider at this point only observation languages which contain ordinary physical-object terms (along with observation predicates, etc., of course). Now, according to this view, all descriptive terms of the observation language must refer to that which has been observed. How is this to be interpreted? Not too narrowly, presumably, otherwise each language user would have a different observation language. The name of my Aunt Mamie, of California, whom I have never seen, would not be in my observation language, nor would 'snow' be an observation term for many Floridians. One could, of course, set off the observation language by means of this awkward restriction, but then, obviously, not being the referent of an observation term would have no bearing on the ontological status of Aunt Mamie or that of snow.

Perhaps it is intended that the referents of observation terms must be members of a *kind* some of whose members have been observed or instances of a *property* some of whose instances have been observed. But there are familiar difficulties here. For example, given any entity, we can always find a kind whose only member is the entity in question; and surely expressions such as 'men over 14 feet tall' should be counted as observational even though no instances of the "property" of being a man over 14 feet tall have been observed. It would seem that this approach must soon fall back upon some notion of simples or determinables vs. determinates. But is it thereby saved? If it is held that only those terms which refer to observed simples or observed determinates are observation terms, we need only remind ourselves of such instances as Hume's notorious missing shade of blue.* And if it is contended that in order to be an observation term an expression must at least refer to an observed determinable, then we can always find such a determinable which is broad enough in scope to embrace any entity whatever. But even if these difficulties can be circumvented, we see (as we knew

---

of any degree of complexity are truth-functional compounds of atomic propositions. Atomic propositions have no truth-functional structure and refer directly to nonlinguistic atomic facts. Russell's principle of acquaintance asserts that every proposition we can understand must be composed entirely of elements with which we are acquainted, where acquaintance is the immediate apprehension of an object. By identifying atomic facts with sense data, some of the logical positivists argued that acquaintance with them is necessary both for understanding what propositions mean and for having any empirical knowledge.

* Hume allowed that we can form the idea of a particular shade of blue even though we have never seen it by filling in the gap (so to speak) between the shades of blue that we have perceived. This apparently contradicts Hume's empiricist principle that every simple idea is copied from a preceding simple impression. See David Hume, *An Enquiry Concerning Human Understanding*, Section II.

all along) that this approach leads inevitably into phenomenalism, which is a view with which we have not been concerning ourselves.

Now it is not the purpose of this essay to give a detailed critique of phenomenalism. For the most part, I simply assume that it is untenable, at least in any of its translatability varieties.[7] However, if there are any unreconstructed phenomenalists among the readers, my purpose, insofar as they are concerned, will have been largely achieved if they will grant what I suppose most of them would stoutly maintain anyway, i.e., that theoretical entities are no worse off than so-called observable physical objects.

Nevertheless, a few considerations concerning phenomenalism and related matters may cast some light upon the observational-theoretical dichotomy and, perhaps, upon the nature of the "observation language." As a preface, allow me some overdue remarks on the latter. Although I have contended that the line between the observable and the unobservable is diffuse, that it shifts from one scientific problem to another, and that it is constantly being pushed toward the "unobservable" end of the spectrum as we develop better means of observation—better instruments—it would, nevertheless, be fatuous to minimize the importance of the observation base, for it is absolutely necessary as a confirmation base for statements which do refer to entities which are unobservable at a given time. But we should take as its basis and its unit not the "observational term" but, rather, the quickly decidable sentence. (I am indebted to Feyerabend, loc. cit., for this terminology.) A quickly decidable sentence (in the technical sense employed here) may be defined as a singular, nonanalytic sentence such that a reliable, reasonably sophisticated language user can very quickly decide[8] whether to assert it or deny it when he is reporting on an occurrent situation. 'Observation term' may now be defined as a 'descriptive (non-logical) term which may occur in a quickly decidable sentence,' and 'observation sentence' as a 'sentence whose only descriptive terms are observation terms.'

Returning to phenomenalism, let me emphasize that I am not among those philosophers who hold that there are no such things as sense contents (even sense data), nor do I believe that they play no important role in our perception of "reality." But the fact remains that the referents of most (not all) of the statements of the linguistic framework used in everyday life and in science are *not* sense contents but, rather, physical objects and other publicly observable entities. Except for pains, odors, "inner states," etc., *we do not usually observe sense contents*; and although there is good reason to believe that they play an indispensable role in observation, *we are usually not aware of them when we* (visually or tactilely) *observe physical objects*. For example, when I observe a distorted, obliquely reflected image in a mirror, I may seem to be seeing a baby elephant standing on its head; later I discover it is an image of Uncle Charles taking a nap with his mouth open and his hand in a peculiar position. Or, passing my neighbor's home at a high rate of speed, I observe that he is washing a car. If asked to report these observations I could quickly and easily report a baby elephant and a washing of a car; I probably would not, without subsequent observations, be able to report what colors, shapes, etc. (i.e., what sense data) were involved.

Two questions naturally arise at this point. How is it that we can (sometimes) quickly decide the truth or falsity of a pertinent observation sentence? And what role do sense contents play in the appropriate tokening of such sentences? The heart of the matter is that these are primarily scientific-theoretical questions rather than "purely logical," "purely conceptual," or "purely epistemological." If theoretical physics, psychology, neurophysiology, etc., were sufficiently advanced, we could give satisfactory answers to these questions, using, in all likelihood, the physical-thing language as our observation language and *treating sensations, sense contents, sense data, and "inner states" as theoretical* (yes, theoretical!) *entities.*[9]

It is interesting and important to note that, even before we give completely satisfactory answers to the two questions considered above, we can, with due effort and reflection, train ourselves to "observe directly" what were once theoretical entities—the sense contents (color sensations, etc.)—involved in our perception of physical things. As has been pointed out before, we can also come to observe other kinds of entities which were once theoretical. Those which most readily come to mind involve the use of instruments as aids to observation. Indeed, using our painfully acquired theoretical knowledge of the world, we come to see that we "directly observe" many kinds of so-called theoretical things. After listening to a dull speech while sitting on a hard bench, we begin to become poignantly aware of the presence of a considerably strong gravitational field, and as Professor Feyerabend is fond of pointing out, if we were carrying a heavy suitcase in a changing gravitational field, we could observe the changes of the $G_{\mu\nu}$ of the metric tensor.

I conclude that our drawing of the observational-theoretical line at any given point is an accident and a function of our physiological makeup, our current state of knowledge, and the instruments we happen to have available and, therefore, that it has no ontological significance whatever. . . .

■ | Notes

1. E. Nagel, *The Structure of Science* (New York: Harcourt, Brace, and World, 1961), Ch. 6.

2. For the genesis and part of the content of some of the ideas expressed herein, I am indebted to a number of sources; some of the more influential are H. Feigl, "Existential Hypotheses," *Philosophy of Science*, 17: 35–62 (1950); P. K. Feyerabend, "An Attempt at a Realistic Interpretation of Experience," *Proceedings of the Aristotelian Society*, 58: 144–170 (1958); N. R. Hanson, *Patterns of Discovery* (Cambridge: Cambridge University Press, 1958); E. Nagel, *loc. cit.*; Karl Popper, *The Logic of Scientific Discovery* (London: Hutchinson, 1959); M. Scriven, "Definitions, Explanations, and Theories," in *Minnesota Studies in the Philosophy of Science*, Vol. II,

H. Feigl, M. Scriven, and G. Maxwell, eds. (Minneapolis: University of Minnesota Press, 1958); Wilfrid Sellars, "Empiricism and the Philosophy of Mind," in *Minnesota Studies in the Philosophy of Science*, Vol. I, H. Feigl and M. Scriven, eds. (Minneapolis: University of Minnesota Press, 1956); and "The Language of Theories," in *Current Issues in the Philosophy of Science*, H. Feigl and G. Maxwell, eds. (New York: Holt, Rinehart, and Winston, 1961).

3. I have borrowed the hammer analogy from E. Nagel, "Science and [Feigl's] Semantic Realism," *Philosophy of Science*, 17: 174–181 (1950), but it should be pointed out that Professor Nagel makes it clear that he does not necessarily subscribe to the view which he is explaining.

4. G. Bergmann, "Outline of an Empiricist Philosophy of Physics," *American Journal of Physics*, 11: 248–258; 335–342 (1943), reprinted in *Readings in the Philosophy of Science*, H. Feigl and M. Brodbeck, eds. (New York: Appleton-Century-Crofts, 1953), pp. 262–287.

5. I am not attributing to Professor Bergmann the absurd views suggested by these questions. He seems to take a sense-datum language as his observation language (the base of what he called "the empirical hierarchy"), and, in some ways, such a position is more difficult to refute than one which purports to take an "observable physical-object" view. However, I believe that demolishing the straw men with which I am now dealing amounts to desirable preliminary "therapy." Some non-realist interpretations of theories which embody the presupposition that the observable-theoretical distinction is sharp and ontologically crucial seem to me to entail positions which correspond to such straw men rather closely.

6. For arguments that it is possible to alter a theory without altering the meanings of its terms, see my "Meaning Postulates in Scientific Theories," in *Current Issues in the Philosophy of Science*, Feigl and Maxwell, eds.

7. The reader is no doubt familiar with the abundant literature concerned with this issue. See, for example, Sellars' "Empiricism and the Philosophy of Mind," which also contains references to other pertinent works.

8. We may say "noninferentially" decide, provided this is interpreted liberally enough to avoid starting the entire controversy about observability all over again.

9. Cf. Sellars, "Empiricism and the Philosophy of Mind." As Professor Sellars points out, this is the crux of the "other-minds" problem. Sensations and inner states (relative to an intersubjective observation language, I would add) are theoretical entities (and they "really exist") and *not* merely actual and/or possible behavior. Surely it is the unwillingness to countenance theoretical entities—the hope that every sentence is translatable not only into some observation language but into the physical-thing language—which is responsible for the "logical behaviorism" of the neo-Wittgensteinians.

Bas C. van Fraassen

Arguments Concerning
Scientific Realism

*The rigour of science requires that we distinguish well the undraped figure of nature itself
from the gay-coloured vesture with which we clothe it at our pleasure.*
—Heinrich Hertz, quoted by Ludwig Boltzmann,
letter to *Nature*, 28 February 1895

In our century, the first dominant philosophy of science was developed as part of logical positivism. Even today, such an expression as 'the received view of theories' refers to the views developed by the logical positivists, although their heyday preceded the Second World War.

In this chapter I shall examine, and criticize, the main arguments that have been offered for scientific realism. These arguments occurred frequently as part of a critique of logical positivism. But it is surely fair to discuss them in isolation, for even if scientific realism is most easily understood as a reaction against positivism, it should be able to stand alone. The alternative view which I advocate—for lack of a traditional name I shall call it *constructive empiricism*—is equally at odds with positivist doctrine.

1 | Scientific Realism and Constructive Empiricism

In philosophy of science, the term 'scientific realism' denotes a precise position on the question of how a scientific theory is to be understood, and what scientific activity really is. I shall attempt to define this position, and to canvass its possible alternatives. Then I shall indicate, roughly and briefly, the specific alternative which I shall advocate. . . .

FROM Bas C. van Fraassen, *The Scientific Image* (Oxford: Clarendon Press, 1980), 6–21, 23–25, 31–40.

1.1 Statement of Scientific Realism

What exactly is scientific realism? A naïve statement of the position would be this: the picture which science gives us of the world is a true one, faithful in its details, and the entities postulated in science really exist: the advances of science are discoveries, not inventions. That statement is too naïve; it attributes to the scientific realist the belief that today's theories are correct. It would mean that the philosophical position of an earlier scientific realist such as C. S. Peirce had been refuted by empirical findings. I do not suppose that scientific realists wish to be committed, as such, even to the claim that science will arrive in due time at theories true in all respects—for the growth of science might be an endless self-correction; or worse, Armageddon might occur too soon.

But the naïve statement has the right flavour. It answers two main questions: it characterizes a scientific theory as a story about what there really is, and scientific activity as an enterprise of discovery, as opposed to invention. The two questions of what a scientific theory is, and what a scientific theory does, must be answered by any philosophy of science. The task we have at this point is to find a statement of scientific realism that shares these features with the naïve statement, but does not saddle the realists with unacceptably strong consequences. It is especially important to make the statement as weak as possible if we wish to argue against it, so as not to charge at windmills.

As clues I shall cite some passages most of which will also be examined below in the contexts of the authors' arguments. A statement of Wilfrid Sellars is this:

> to have good reason for holding a theory is *ipso facto* to have good reason for holding that the entities postulated by the theory exist.[1]

This address a question of epistemology, but also throws some indirect light on what it is, in Sellars's opinion, to hold a theory. Brian Ellis, who calls himself a scientific entity realist rather than a scientific realist, appears to agree with that statement of Sellars, but gives the following formulation of a stronger view:

> I understand scientific realism to be the view that the theoretical statements of science are, or purport to be, true generalized descriptions of reality.[2]

This formulation has two advantages: It focuses on the understanding of the theories without reference to reasons for belief; and it avoids the suggestion that to be a realist you must believe current scientific theories to be true. But it gains the latter advantage by use of the word 'purport', which may generate its own puzzles.

Hilary Putnam, in a passage which I shall cite again in Section 7, gives a formulation which he says he learned from Michael Dummett:

> A realist (with respect to a given theory or discourse) holds that (1) the sentences of that theory are true or false; and (2) that what makes them true or false is something external—that is to say, it is not (in general) our sense data, actual or potential, or the structure of our minds, or our language, etc.[3]

He follows this soon afterwards with a further formulation which he credits to Richard Boyd:

> That terms in mature scientific theories typically refer (this formulation is due to Richard Boyd), that the theories accepted in a mature science are typically approximately true, that the same term can refer to the same thing even when it occurs in different theories—these statements are viewed by the scientific realist . . . as part of any adequate scientific description of science and its relations to its objects.[4]

None of these were intended as definitions. But they show I think that truth must play an important role in the formulation of the basic realist position. They also show that the formulation must incorporate an answer to the question what it is to *accept* or *hold* a theory. I shall now propose such a formulation, which seems to me to make sense of the above remarks, and also renders intelligible the reasoning by realists which I shall examine below—without burdening them with more than the minimum required for this.

Science aims to give us, in its theories, a literally true story of what the world is like; and acceptance of a scientific theory involves the belief that it is true. This is the correct statement of scientific realism.

Let me defend this formulation by showing that it is quite minimal, and can be agreed to by anyone who considers himself a scientific realist. The naïve statement said that science tells a true story; the correct statement says only that it is the aim of science to do so. The aim of science is of course not to be identified with individual scientists' motives. The aim of the game of chess is to checkmate your opponent; but the motive for playing may be fame, gold, and glory. What the aim is determines what counts as success in the enterprise as such; and this aim may be pursued for any number of reasons. Also, in calling something *the* aim, I do not deny that there are other subsidiary aims which may or may not be means to that end: everyone will readily agree that simplicity, informativeness, predictive power, explanation are (also) virtues. Perhaps my formulation can even be accepted by any philosopher who considers the most important aim of science to be something which only *requires* the finding of true theories— given that I wish to give the weakest formulation of the doctrine that is generally acceptable.

I have added 'literally' to rule out as realist such positions as imply that science is true if 'properly understood' but literally false or meaningless. For that would be consistent with conventionalism, logical positivism, and instrumentalism. I will say more about this below; and also in Section 7 where I shall consider Dummett's views further.

The second part of the statement touches on epistemology. But it only equates acceptance of a theory with belief in its truth.[5] It does not imply that anyone is ever rationally warranted in forming such a belief. We have to make room for the epistemological position, today the subject of considerable debate, that a rational person never assigns personal probability 1 to any proposition except a tautology. It would, I think, be rare for a scientific realist to take this stand in epistemology, but it is certainly possible.[6]

To understand qualified acceptance we must first understand acceptance *tout court*. If acceptance of a theory involves the belief that it is true, then tentative acceptance involves the tentative adoption of the belief that it is true. If belief comes in degrees, so does acceptance, and we may then speak of a degree of acceptance involving a certain degree of belief that the theory is true. This must of course be distinguished from belief that the theory is approximately true, which seems to mean belief that some member of a class centring on the mentioned theory is (exactly) true. In this way the proposed formulation of realism can be used regardless of one's epistemological persuasion.

1.2 ALTERNATIVES TO REALISM

Scientific realism is the position that scientific theory construction aims to give us a literally true story of what the world is like, and that acceptance of a scientific theory involves the belief that it is true. Accordingly, anti-realism is a position according to which the aim of science can well be served without giving such a literally true story, and acceptance of a theory may properly involve something less (or other) than belief that it is true.

What does a scientist do then, according to these different positions? According to the realist, when someone proposes a theory, he is asserting it to be true. But according to the anti-realist, the proposer does not assert the theory; *he displays it*, and claims certain virtues for it. These virtues may fall short of truth: empirical adequacy, perhaps; comprehensiveness, acceptability for various purposes. This will have to be spelt out, for the details here are not determined by the denial of realism. For now we must concentrate on the key notions that allow the generic division.

The idea of a literally true account has two aspects: the language is to be literally construed; and so construed, the account is true. This divides the anti-realists into two sorts. The first sort holds that science is or aims to be true, properly (but not literally) construed. The second holds that the language of science should be literally construed, but its theories need not be true to be good. The anti-realism I shall advocate belongs to the second sort.

It is not so easy to say what is meant by a literal construal. The idea comes perhaps from theology, where fundamentalists construe the Bible literally, and liberals have a variety of allegorical, metaphorical, and analogical interpretations, which 'demythologize'. The problem of explicating 'literal construal' belongs to the philosophy of language. In Section 7 below, where I briefly examine some of Michael Dummett's views, I shall emphasize that 'literal' does not mean 'truth-valued'. The term 'literal' is well enough understood for general philosophical use, but if we try to explicate it we find ourselves in the midst of the problem of giving an adequate account of natural language. It would be bad tactics to link an inquiry into science to a commitment to some solution to that problem. The following remarks, and those in Section 7, should fix the usage of 'literal' sufficiently for present purposes.

The decision to rule out all but literal construals of the language of science, rules out those forms of anti-realism known as *positivism* and *instrumentalism*. First, on a literal construal, the apparent statements of science really are statements, *capable* of being true or false. Secondly, although a literal construal can elaborate, it cannot change logical relationships. (It is possible to elaborate, for instance, by identifying what the terms designate. The 'reduction' of the language of phenomenological thermodynamics to that of statistical mechanics is like that: bodies of gas are identified as aggregates of molecules, temperature as mean kinetic energy, and so on.) On the positivists' interpretation of science, theoretical terms have meaning only through their connection with the observable. Hence they hold that two theories may in fact *say the same thing* although in form they contradict each other. (Perhaps the one says that all matter consists of atoms, while the other postulates instead a universal continuous medium; they will say the same thing nevertheless if they agree in their observable consequences, according to the positivists.) But two theories which contradict each other in such a way can 'really' be saying the same thing only if they are not literally construed. Most specifically, if a theory says that something exists, then a literal construal may elaborate on what that something is, but will not remove the implication of existence.

There have been many critiques of positivist interpretations of science, and there is no need to repeat them. . . .

1.3 Constructive Empiricism

To insist on a literal construal of the language of science is to rule out the construal of a theory as a metaphor or simile, or as intelligible only after it is 'demythologized' or subjected to some other sort of 'translation' that does not preserve logical form. If the theory's statements include 'There are electrons', then the theory says that there are electrons. If in addition they include 'Electrons are not planets', then the theory says, in part, that there are entities other than planets.

But this does not settle very much. It is often not at all obvious whether a theoretical term refers to a concrete entity or a mathematical entity. Perhaps one tenable interpretation of classical physics is that there are no concrete entities which are forces—that 'there are forces such that . . .' can always be understood as a mathematical statement asserting the existence of certain functions. That is debatable.

Not every philosophical position concerning science which insists on a literal construal of the language of science is a realist position. For this insistence relates not at all to our epistemic attitudes toward theories, nor to the aim we pursue in constructing theories, but only to the correct understanding of *what a theory says*. (The fundamentalist theist, the agnostic, and the atheist presumably agree with each other (though not with liberal theologians) in their understanding of the statement that God, or gods, or angels exist.) After deciding that the language of science must be literally understood, we can still say that there is no need to believe good theories to be true, nor to believe *ipso facto* that the entities they postulate are real.

Science aims to give us theories which are empirically adequate; and acceptance of a theory involves as belief only that it is empirically adequate. This is the statement of the anti-realist position I advocate; I shall call it *constructive empiricism.*

This formulation is subject to the same qualifying remarks as that of scientific realism in Section 1.1 above. In addition it requires an explication of 'empirically adequate'. For now, I shall leave that with the preliminary explication that a theory is empirically adequate exactly if what it says about the observable things and events in this world, is true—exactly if it 'saves the phenomena'. A little more precisely: such a theory has at least one model that all the actual phenomena fit inside. I must emphasize that this refers to *all* the phenomena; these are not exhausted by those actually observed, nor even by those observed at some time, whether past, present, or future. . . .

The distinction I have drawn between realism and anti-realism, in so far as it pertains to acceptance, concerns only how much belief is involved therein. Acceptance of theories (whether full, tentative, to a degree, etc.) is a phenomenon of scientific activity which clearly involves more than belief. One main reason for this is that we are never confronted with a complete theory. So if a scientist accepts a theory, he thereby involves himself in a certain sort of research programme. That programme could well be different from the one acceptance of another theory would have given him, even if those two (very incomplete) theories are equivalent to each other with respect to everything that is observable—in so far as they go.

Thus acceptance involves not only belief but a certain commitment. Even for those of us who are not working scientists, the acceptance involves a commitment to confront any future phenomena by means of the conceptual resources of this theory. It determines the terms in which we shall seek explanations. If the acceptance is at all strong, it is exhibited in the person's assumption of the role of explainer, in his willingness to answer questions *ex cathedra*. Even if you do not accept a theory, you can engage in discourse in a context in which language use is guided by that theory—but acceptance produces such contexts. There are similarities in all of this to ideological commitment. A commitment is of course not true or false: The confidence exhibited is that it will be *vindicated*.

This is a preliminary sketch of the *pragmatic* dimension of theory acceptance. Unlike the epistemic dimension, it does not figure overtly in the disagreement between realist and anti-realist. But because the amount of belief involved in acceptance is typically less according to anti-realists, they will tend to make more of the pragmatic aspects. It is as well to note here the important difference. Belief that a theory is true, or that it is empirically adequate, does not imply, and is not implied by, belief that full acceptance of the theory will be vindicated. To see this, you need only consider here a person who has quite definite beliefs about the future of the human race, or about the scientific community and the influences thereon and practical limitations we have. It might well be, for instance, that a theory which is empirically adequate will not combine easily with some other theories which we have accepted in fact, or that Armageddon will occur before we succeed. Whether belief that a theory is true, or that it is empirically adequate, can be equated with belief that acceptance of it would, under ideal research conditions, be vindicated in the long run, is another question. It seems to me an irrelevant question within philosophy of science, because an affirmative answer would not obliterate the distinction we have already established by the preceding remarks. (The question may also assume that counterfactual statements are objectively true or false, which I would deny.)

Although it seems to me that realists and anti-realists need not disagree about the pragmatic aspects of theory acceptance, I have mentioned it here because I think that typically they do. We shall find ourselves returning time and again, for example, to requests for explanation to which realists typically attach an objective validity which anti-realists cannot grant.

2 | The Theory/Observation 'Dichotomy'

For good reasons, logical positivism dominated the philosophy of science for thirty years. In 1960, the first volume of *Minnesota Studies in the Philosophy of Science* published Rudolf Carnap's 'The Methodological Status of Theoretical Concepts', which is, in many ways, the culmination of the positivist programme. It interprets science by relating it to an observation language (a postulated part of natural language which is devoid of theoretical terms). Two years later this article was followed in the same series by Grover Maxwell's 'The Ontological Status of Theoretical Entities', in title and theme a direct counter to Carnap's. This is the *locus classicus* for the new realists' contention that the theory/observation distinction cannot be drawn.

I shall examine some of Maxwell's points directly, but first a general remark about the issue. Such expressions as 'theoretical entity' and 'observable-theoretical dichotomy' are, on the face of it, examples of category mistakes. Terms or concepts are theoretical (introduced or adapted for the purposes of theory construction); entities are observable or unobservable. This may seem a little point, but it separates the discussion into two issues. Can we divide our language into a theoretical and non-theoretical part? On the other hand, can we classify objects and events into observable and unobservable ones?

Maxwell answers both questions in the negative, while not distinguishing them too carefully. On the first, where he can draw on well-known supportive essays by Wilfrid Sellars and Paul Feyerabend, I am in total agreement. All our language is thoroughly theory-infected. If we could cleanse our language of theory-laden terms, beginning with the recently introduced ones like 'VHF receiver', continuing through 'mass' and 'impulse' to 'element' and so on into the prehistory of language formation, we would end up with nothing useful. The way we talk, and scientists talk, is guided by the pictures provided by previously accepted theories. This is true also, as Duhem already emphasized, of experimental reports. Hygienic reconstructions of language such as the positivists envisaged are simply not on. . . .

But does this mean that we must be scientific realists? We surely have more tolerance of ambiguity than that. The fact that we let our language be guided by a given picture, at some point, does not show how much we believe about that picture. When we speak of the sun coming up in the morning and setting at night, we are guided by a picture now explicitly disavowed. When Milton wrote *Paradise Lost* he deliberately let the old geocentric astronomy guide his poem, although various remarks in passing clearly reveal his interest in the new astronomical discoveries and speculations of his time. These are extreme examples, but show that no immediate conclusions can be drawn from the theory-ladenness of our language.

However, Maxwell's main arguments are directed against the observable-unobservable distinction. Let us first be clear on what this distinction was supposed to be. The term 'observable' classifies putative entities (entities which may or may not exist). A flying horse is observable—that is why we are so sure that there aren't any—and the number seventeen is not. There is supposed to be a correlate classification of human acts: an unaided act of perception, for instance, is an observation. A calculation of the mass of a particle from the deflection of its trajectory in a known force field, is not an observation of that mass.

It is also important here not to confuse *observing* (an entity, such as a thing, event, or process) and *observing that* (something or other is the case). Suppose one of the Stone Age people recently found in the Philippines is shown a tennis ball or a car crash.* From his behaviour, we see that he has noticed them; for example, he picks up the ball and throws it. But he has

---

* The Tasaday people were discovered living as hunters and gatherers in the rain forest of Mindanao in 1971. Contrary to the media hype at the time, they are not a genuine Stone Age tribe nor have they been totally isolated. Though living separate

not seen *that* it is a tennis ball, or *that* some event is a car crash, for he does not even have those concepts. He cannot get that information through perception; he would first have to learn a great deal. To say that he does not see the same things and events as we do, however, is just silly; it is a pun which trades on the ambiguity between seeing and seeing that. (The truth-conditions for our statement '*x* observes *that* A' must be such that what concepts *x* has, presumably related to the language *x* speaks if he is human, enter as a variable into the correct truth definition, in some way. To say that *x* observed the tennis ball, therefore, does not imply at all that *x* observed that it was a tennis ball; that would require some conceptual awareness of the game of tennis.)

The arguments Maxwell gives about observability are of two sorts: one directed against the possibility of drawing such distinctions, the other against the importance that could attach to distinctions that can be drawn.

The first argument is from the continuum of cases that lie between direct observation and inference:

> There is, in principle, a continuous series beginning with looking through a vacuum and containing these as members: looking through a windowpane, looking through glasses, looking through binoculars, looking through a low-power microscope, looking through a high-power microscope, etc., in the order given. The important consequence is that, so far, we are left without criteria which would enable us to draw a non-arbitrary line between 'observation' and 'theory'.[7]

This continuous series of supposed acts of observation does not correspond directly to a continuum in what is supposed observable. For if something can be seen through a window, it can also be seen with the window raised. Similarly, the moons of Jupiter can be seen through a telescope; but they can also be seen without a telescope if you are close enough. That something is observable does not automatically imply that the conditions are right for observing it now. The principle is:

> X is observable if there are circumstances which are such that, if X is present to us under those circumstances, then we observe it.

This is not meant as a definition, but only as a rough guide to the avoidance of fallacies.

We may still be able to find a continuum in what is supposed detectable: perhaps some things can only be detected with the aid of an optical microscope, at least; perhaps some require an electron microscope, and so

---

from surrounding groups, they trade with them for food, tools, and clothing and speak a dialect of the same language.

on. Maxwell's problem is: where shall we draw the line between what is observable and what is only detectable in some more roundabout way?

Granted that we cannot answer this question without arbitrariness, what follows? That 'observable' is a *vague predicate*. There are many puzzles about vague predicates, and many sophisms designed to show that, in the presence of vagueness, no distinction can be drawn at all. In Sextus Empiricus, we find the argument that incest is not immoral, for touching your mother's big toe with your little finger is not immoral, and all the rest differs only by degree. But predicates in natural language are almost all vague, and there is no problem in their use; only in formulating the logic that governs them.[8] A vague predicate is usable provided it has clear cases and clear counter-cases. Seeing with the unaided eye is a clear case of observation. Is Maxwell then perhaps challenging us to present a clear counter-case? Perhaps so, for he says 'I have been trying to support the thesis that any (non-logical) term is a *possible* candidate for an observation term.'

A look through a telescope at the moons of Jupiter seems to me a clear case of observation, since astronauts will no doubt be able to see them as well from close up. But the purported observation of micro-particles in a cloud chamber seems to me a clearly different case—if our theory about what happens there is right. The theory says that if a charged particle traverses a chamber filled with saturated vapour, some atoms in the neighbourhood of its path are ionized. If this vapour is decompressed, and hence becomes super-saturated, it condenses in droplets on the ions, thus marking the path of the particle. The resulting silver-grey line is similar (physically as well as in appearance) to the vapour trail left in the sky when a jet passes. Suppose I point to such a trail and say: 'Look, there is a jet!'; might you not say: 'I see the vapour trail, but where is the jet?' Then I would answer: 'Look just a bit ahead of the trail . . . there! Do you see it?' Now, in the case of the cloud chamber this response is not possible. So while the particle is detected by means of the cloud chamber, and the detection is based on observation, it is clearly not a case of the particle's being observed.

As second argument, Maxwell directs our attention to the 'can' in 'what is observable is what can be observed.' An object might of course be temporarily unobservable—in a rather different sense: it cannot be observed in the circumstances in which it actually is at the moment, but could be observed if the circumstances were more favourable. In just the same way, I might be temporarily invulnerable or invisible. So we should concentrate on 'observable' *tout court*, or on (as he prefers to say) 'unobservable in principle'. This Maxwell explains as meaning that the relevant scientific theory *entails* that the entities cannot be observed in any circumstances. But this never happens, he says, because the different circumstances could be ones in which we have different sense organs—electron-microscope eyes, for instance.

This strikes me as a trick, a change in the subject of discussion. I have a mortar and pestle made of copper and weighing about a kilo. Should I call it breakable because a giant could break it? Should I call the Empire State Building portable? Is there no distinction between a portable and a console record player? The human organism is, from the point of view of physics, a certain kind of measuring apparatus. As such it has certain inherent limitations—which will be described in detail in the final physics and biology. It is these limitations to which the 'able' in 'observable' refers—our limitations, *qua* human beings.

As I mentioned, however, Maxwell's article also contains a different sort of argument: even if there is a feasible observable/unobservable distinction, this distinction has no importance. The point at issue for the realist is, after all, the reality of the entities postulated in science. Suppose that these entities could be classified into observables and others; what relevance should that have to the question of their existence?

Logically, none. For the term 'observable' classifies putative entities, and has logically nothing to do with existence. But Maxwell must have more in mind when he says: 'I conclude that the drawing of the observational-theoretical line at any given point is an accident and a function of our physiological make-up, . . . and, therefore, that it has no ontological significance whatever.'⁹ No ontological significance if the question is only whether 'observable' and 'exists' imply each other—for they do not; but significance for the question of scientific realism?

Recall that I defined scientific realism in terms of the aim of science, and epistemic attitudes. The question is what aim scientific activity has, and how much we shall believe when we accept a scientific theory. What is the proper form of acceptance: belief that the theory, as a whole, is true; or something else? To this question, what is observable by us seems eminently relevant. Indeed, we may attempt an answer at this point: to accept a theory is (for us) to believe that it is empirically adequate—that what the theory says *about what is observable* (by us) is true.

It will be objected at once that, on this proposal, what the anti-realist decides to believe about the world will depend in part on what he believes to be his, or rather the epistemic community's, accessible range of evidence. At present, we count the human race as the epistemic community to which we belong; but this race may mutate, or that community may be increased by adding other animals (terrestrial or extra-terrestrial) through relevant ideological or moral decisions ('to count them as persons'). Hence the anti-realist would, on my proposal, have to accept conditions of the form

If the epistemic community changes in fashion Y, then my beliefs about the world will change in manner Z.

To see this as an objection to anti-realism is to voice the requirement that our epistemic policies should give the same results independent of our beliefs about the range of evidence accessible to us. That requirement seems to me in no way rationally compelling; it could be honoured, I should think, only through a thoroughgoing skepticism or through a commitment to wholesale leaps of faith. But we cannot settle the major questions of epistemology *en passant* in philosophy of science; so I shall just conclude that it is, on the face of it, not irrational to commit oneself only to a search for theories that are empirically adequate, ones whose models fit the observable phenomena, while recognizing that what counts as an observable phenomenon is a function of what the epistemic community is (that *observable* is *observable-to-us*).

The notion of empirical adequacy* in this answer will have to be spelt out very carefully if it is not to bite the dust among hackneyed objections. . . . But the point stands: even if observability has nothing to do with existence (is, indeed, too anthropocentric for that), it may still have much to do with the proper epistemic attitude to science.

3 | Inference to the Best Explanation

A view advanced in different ways by Wilfrid Sellars, J. J. C. Smart, and Gilbert Harman is that the canons of rational inference require scientific realism. If we are to follow the same patterns of inference with respect to this issue as we do in science itself, we shall find ourselves irrational unless we assert the truth of the scientific theories we accept. Thus Sellars says: 'As I see it, to have good reason for holding a theory is *ipso facto* to have good reason for holding that the entities postulated by the theory exist.'[10]

The main rule of inference invoked in arguments of this sort is the rule of *inference to the best explanation*. The idea is perhaps to be credited to C. S. Peirce,[11] but the main recent attempts to explain this rule and its uses have been made by Gilbert Harman.[12] I shall only present a simplified version. Let us suppose that we have evidence E, and are considering several hypotheses, say H and H'. The rule then says that we should infer H rather than H' exactly if H is a better explanation of E than H' is. (Various qualifications are necessary to avoid inconsistency: we should always try to move to the best over-all explanation of all available evidence.)

It is argued that we follow this rule in all 'ordinary' cases; and that if we follow it consistently everywhere, we shall be led to scientific realism, in the way Sellars's dictum suggests. And surely there are many telling 'ordinary' cases: I hear scratching in the wall, the patter of little feet at midnight, my cheese disappears—and I infer that a mouse has come to live with me. Not merely that these apparent signs of mousely presence will continue, not merely that all the observable phenomena will be as if there is a mouse; but that there really is a mouse.

---

* Van Fraassen offers an analysis of empirical adequacy in chapter 3 of *The Scientific Image* (1980), the book from which the present reading is excerpted.

Will this pattern of inference also lead us to belief in unobservable entities? Is the scientific realist simply someone who consistently follows the rules of inference that we all follow in more mundane contexts? . . .

First of all, what is meant by saying that we all *follow* a certain rule of inference? One meaning might be that we deliberately and consciously 'apply' the rule, like a student doing a logic exercise. That meaning is much too literalistic and restrictive; surely all of mankind follows the rules of logic much of the time, while only a fraction can even formulate them. A second meaning is that we act in accordance with the rules in a sense that does not require conscious deliberation. That is not so easy to make precise, since each logical rule is a rule of permission (*modus ponens* allows you to infer B from A and (if A *then* B), but does not forbid you to infer (B *or* A) instead). However, we might say that a person behaved in accordance with a set of rules in that sense if every conclusion he drew could be reached from his premises via those rules. But this meaning is much too loose; in this sense we always behave in accordance with the rule that any conclusion may be inferred from any premise. So it seems that to be following a rule, I must be willing to believe all conclusions it allows, while definitely unwilling to believe conclusions at variance with the ones it allows—or else, change my willingness to believe the premises in question.

Therefore the statement that we all follow a certain rule in certain cases, is a *psychological hypothesis* about what we are willing and unwilling to do. It is an empirical hypothesis, to be confronted with data, and with rival hypotheses. Here is a rival hypothesis: we are always willing to believe that the theory which best explains the evidence, is empirically adequate (that all the observable phenomena are as the theory says they are).

In this way I can certainly account for the many instances in which a scientist appears to argue for the acceptance of a theory or hypothesis, on the basis of its explanatory success. (A number of such instances are related by Thagard.[13]) For, remember: I equate the acceptance of a scientific theory with the belief that it is empirically adequate. We have therefore two rival hypotheses concerning these instances of scientific inference, and the one is apt in a realist account, the other in an anti-realist account.

Cases like the mouse in the wainscoting cannot provide telling evidence between those rival hypotheses. For the mouse *is* an observable thing; therefore 'there is a mouse in the wainscoting' and 'All observable phenomena are as if there is a mouse in the wainscoting' are totally equivalent; each implies the other (given what we know about mice).

It will be countered that it is less interesting to know whether people do follow a rule of inference than whether they ought to follow it. Granted; but the premiss that we all follow the rule of inference to the best explanation when it comes to mice and other mundane matters—that premiss is shown wanting. It is not warranted by the evidence, because that evidence is not telling *for* the premiss *as against* the alternative hypothesis I proposed, which is a relevant one in this context. . . .

4 | Limits of the Demand for Explanation

In this section and the next . . . I shall examine arguments for realism that point to explanatory power as a criterion for theory choice. That this is indeed a criterion I do not deny. But these arguments for realism succeed only if the demand for explanation is supreme—if the task of science is unfinished, *ipso facto*, as long as any pervasive regularity is left unexplained. I shall object to this line of argument, as found in the writings of Smart, Reichenbach, Salmon, and Sellars, by arguing that such an unlimited demand for explanation leads to a demand for hidden variables, which runs contrary to at least one major school of thought in twentieth-century physics. I do not think that even these philosophers themselves wish to saddle realism with logical links to such consequences: but realist yearnings were born among the mistaken ideals of traditional metaphysics.

In his book *Between Science and Philosophy*, Smart gives two main arguments for realism. One is that only realism can respect the important distinction between *correct* and *merely useful* theories. He calls 'instrumentalist' any view that locates the importance of theories in their use, which requires only empirical adequacy, and not truth. But how can the instrumentalist explain the usefulness of his theories?

> Consider a man (in the sixteenth century) who is a realist about the Copernican hypothesis but instrumentalist about the Ptolemaic one. He can explain the instrumental usefulness of the Ptolemaic system of epicycles because he can prove that the Ptolemaic system can produce almost the same predictions about the apparent motions of the planets as does the Copernican hypothesis. Hence the assumption of the realist truth of the Copernican hypothesis explains the instrumental usefulness of the Ptolemaic one. Such an explanation of the instrumental usefulness of certain theories would not be possible if *all* theories were regarded as merely instrumental.[14]

What exactly is meant by 'such an explanation' in the last sentence? If no theory is assumed to be true, then no theory has its usefulness explained as following from the truth of another one—granted. But would we have less of an explanation of the usefulness of the Ptolemaic hypothesis if we began instead with the premiss that the Copernican gives implicitly a very accurate description of the motions of the planets as observed from earth? This would not assume the truth of Copernicus's heliocentric hypothesis, but would still entail that Ptolemy's simpler description was also a close approximation of those motions.

However, Smart would no doubt retort that such a response pushes the question only one step back: what explains the accuracy of predictions based on Copernicus's theory? If I say, the empirical adequacy of that theory, I have merely given a verbal explanation. For of course Smart does not mean to limit his question to actual predictions—it really concerns all actual and possible predictions and retrodictions. To put it quite concretely: what explains the fact that all observable planetary phenomena fit Copernicus's theory (if they do)? From the medieval debates, we recall the nominalist response that the basic regularities are merely brute regularities, and have no explanation. So here the anti-realist must similarly say: that the observable phenomena exhibit these regularities, because of which they fit the theory, is merely a brute fact, and may or may not have an explanation in terms of unobservable facts 'behind the phenomena'—it really does not matter to the goodness of the theory, nor to our understanding of the world.

Smart's main line of argument is addressed to exactly this point. In the same chapter he argues as follows. Suppose that we have a theory T which postulates micro-structure directly, and macro-structure indirectly. The statistical and approximate laws about macroscopic phenomena are only partially spelt out perhaps, and in any case derive from the precise (deterministic or statistical) laws about the basic entities. We now consider theory T', which is part of T, and says only what T says about the macroscopic phenomena. (How T' should be characterized I shall leave open, for that does not affect the argument here.) Then he continues:

> I would suggest that the realist could (say) . . . that the success of T' is explained by the fact that the original theory T is true of the things that it is ostensibly about; in other words by the fact that there really are electrons or whatever is postulated by the theory T. If there were no such things, and if T were not true in a realist way, would not the success of T' be quite inexplicable? One would have to suppose that there were innumerable lucky accidents about the behaviour mentioned in the observational vocabulary, so that they behaved miraculously *as if* they were brought about by the nonexistent things ostensibly talked about in the theoretical vocabulary.[15]

In other passages, Smart speaks similarly of 'cosmic coincidences'. The regularities in the observable phenomena must be explained in terms of deeper structure, for otherwise we are left with a belief in lucky accidents and coincidences on a cosmic scale.

I submit that if the demand for explanation implicit in these passages were precisely formulated, it would at once lead to absurdity. For if the mere fact of postulating regularities, without explanation, makes T' a poor theory, T will do no better. If, on the other hand, there is some precise limitation on what sorts of regularities can be postulated as basic, the context of the argument provides no reason to think that T' must automatically fare worse than T.

In any case, it seems to me that it is illegitimate to equate being a lucky accident, or a coincidence, with having no explanation. It was by coincidence that I met my friend in the market—but I can explain why I was there, and he can explain why he came, so together we can explain how this meeting happened. We call it a coincidence, not because the occurrence was inexplicable, but because we did not severally go to the market in order to meet.[16] There cannot be a requirement upon science to provide a theoretical elimination of coincidences, or accidental correlations in general, for that does not even make sense. There is nothing here to motivate the demand for explanation, only a restatement in persuasive terms. . . . *

6 | Limits to Explanation: A Thought Experiment

Wilfrid Sellars was one of the leaders of the return to realism in philosophy of science and has, in his writings of the past three decades, developed a systematic and coherent scientific realism. I have discussed a number of his views and arguments elsewhere; but will here concentrate on some aspects that are closely related to the arguments of Smart, Reichenbach, and Salmon just examined.[17] Let me begin by setting the stage in the way Sellars does.

There is a certain over-simplified picture of science, the 'levels picture', which pervades positivist writings and which Sellars successfully demolished.[18] In that picture, singular observable facts ('this crow is black') are scientifically explained by general observable regularities ('all crows are black') which in turn are explained by highly theoretical hypotheses not restricted in what they say to the observable. The three levels are commonly called those of *fact*, of *empirical law*, and of *theory*. But, as Sellars points out, theories do not explain, or even entail such empirical laws—they only show why observable things obey these so-called laws to the extent they do.[19] Indeed, perhaps we have no such empirical laws at all: all crows are black—except albinos; water boils at 100°C—provided atmospheric pressure is normal; a falling body accelerates—provided it is not intercepted, or attached to an aeroplane by a static line; and so forth. On the level of the observable we are liable to find only putative laws heavily subject to unwritten *ceteris paribus* qualifications.

This is, so far, only a methodological point. We do not really expect theories to 'save' our common everyday generalizations, for we ourselves have no confidence in their strict universality. But a theory which says that the microstructure of things is subject to *some* exact, universal regularities, must imply the same for those things themselves. This, at least, is my reaction to the points so far. Sellars, however, sees an inherent inferiority in the description of the observable alone, an incompleteness which requires (*sub specie* the aims of science) an introduction of an unobservable reality behind the phenomena. This is brought out by an interesting 'thought-experiment'.

---

* We have omitted the following section in which van Frassen criticizes Reichenbach's principle of the common cause and efforts by Salmon and others to argue that, because the principle is correct, one must postulate unobservable events and processes in order to explain correlations that would otherwise remain inexplicable.

Imagine that at some early stage of chemistry it had been found that different samples of gold dissolve in *aqua regia* at different rates, although 'as far as can be observationally determined, the specimens and circumstances are identical.'[20] Imagine further that the response of chemistry to this problem was to postulate two distinct micro-structures for the different samples of gold. Observationally unpredictable variation in the rate of dissolution is explained by saying that the samples are mixtures (not compounds) of these two (observationally identical) substances, each of which has a fixed rate of dissolution.

In this case we have explanation through laws which have no observational counterparts that can play the same role. Indeed, no explanation seems possible unless we agree to find our physical variables outside the observable. But science aims to explain, must try to explain, and so must require a belief in this unobservable micro-structure. So Sellars contends.

There are at least three questions before us. Did this postulation of micro-structure really have no new consequences for the observable phenomena? Is there really such a demand upon science that it must explain— even if the means of explanation bring no gain in empirical predictions? And thirdly, could a *different* rationale exist for the use of a micro-structure picture in the development of a scientific theory in a case like this?

First, it seems to me that these hypothetical chemists did postulate new observable regularities as well. Suppose the two substances are A and B, with dissolving rates x and $x+y$ and that every gold sample is a mixture of these substances. Then it follows that every gold sample dissolves at a rate no lower than x and no higher than $x+y$; *and* that between these two any value may be found—to within the limits of accuracy of gold mixing. None of this is implied by the data that different samples of gold have dissolved at various rates between x and $x+y$. So Sellars's first contention is false.

We may assume, for the sake of Sellars's example, that there is still no way of predicting dissolving rates any further. Is there then a categorical demand upon science to explain this variation which does not depend on other observable factors? . . . A precise version of such a demand (Reichenbach's principle of the common cause) could result automatically in a demand for hidden variables, providing a 'classical' underpinning for indeterministic theories. Sellars recognized very well that a demand for hidden variables would run counter to the main opinions current in quantum physics. Accordingly he mentions '. . . the familiar point that the irreducibly and lawfully statistical ensembles of quantum-mechanical theory are mathematically inconsistent with the assumption of hidden variables.'[21] Thus, he restricts the demand for explanation, in effect, to just those cases where it is *consistent* to add hidden variables to the theory. And consistency is surely a logical stopping-point.

This restriction unfortunately does not prevent the disaster. For while there are a number of proofs that hidden variables cannot be supplied so as to turn quantum mechanics into a classical sort of deterministic theory, those proofs are based on requirements much stronger than consistency. To give an example, one such assumption is that two distinct physical variables cannot have the same statistical distributions in measurement on all possible states.[22] Thus it is assumed that, if we cannot point to some possible difference in empirical predictions, then there is no real difference at all. If such requirements were lifted, and consistency alone were the criterion, hidden variables could indeed be introduced. I think we must conclude that science, in contrast to scientific realism, does not place an overriding value on explanation in the absence of any gain for empirical results.

Thirdly, then, let us consider how an anti-realist could make sense of those hypothetical chemists' procedure. After pointing to the new empirical implications which I mentioned two paragraphs ago, he would point to methodological reasons. By imagining a certain sort of micro-structure for gold and other metals, say, we might arrive at a theory governing many observationally disparate substances; and this might then have implications for new, wider empirical regularities when such substances interact. This would only be a hope, of course; no hypothesis is guaranteed to be fruitful—but the point is that the true demand on science is not for explanation *as such*, but for imaginative pictures which have a hope of suggesting new statements of observable regularities and of correcting old ones. This point is exactly the same as that for the principle of the common cause.

7 | Demons and the Ultimate Argument

Hillary Putnam, in the course of his discussions of realism in logic and mathematics, advanced several arguments for scientific realism as well. . . .

In . . . 'What is Mathematical Truth?', Putnam gives . . . what I shall call the *Ultimate Argument*. He begins with a formulation of realism which he says he learned from Michael Dummett:

> A realist (with respect to a given theory or discourse) holds that (1) the sentences of that theory are true or false; and (2) that what makes them true or false is something external—that is to say, it is not (in general) our sense data, actual or potential, or the structure of our minds, or our language, etc.[23]

This formulation is quite different from the one I have given even if we instantiate it to the case in which that theory or discourse is science or scientific discourse. Because the wide discussion of Dummett's views has given some currency to his usage of these terms, and because Putnam begins his discussion in this way, we need to look carefully at this formulation.

In my view, Dummett's usage is quite idiosyncratic. Putnam's statement, though very brief, is essentially accurate. In his 'Realism', Dummett begins by describing various sorts of realism in the traditional fashion, as disputes over whether there really exist entities of a particular type. But he says that in some cases he wishes to discuss, such as the reality of the past and intuitionism in mathematics, the central issues seem to him to be about other questions. For this reason he proposes a new usage: he will take such disputes

> as relating, not to a class of entities or a class of terms, but to a class of *statements*. . . . Realism I characterize as the belief that statements of the disputed class possess an objective truth-value, independently of our means of knowing it: they are true or false in virtue of a reality existing independently of us. The anti-realist opposes to this the view that statements of the disputed class are to be understood only by reference to the sort of thing which we count as evidence for a statement of that class.[24]

Dummett himself notes at once that nominalists are realists in this sense.[25] If, for example, you say that abstract entities do not exist, and sets are abstract entities, hence sets do not exist, then you will certainly accord a truth-value to all statements of set theory. It might be objected that if you take this position then you have a decision procedure for determining the truth-values of these statements (*false* for existentially quantified ones, *true* for universal ones, apply truth tables for the rest). Does that not mean that, on your view, the truth-values are not independent of our knowledge? Not at all; for you clearly believe that if we had not existed, and *a fortiori* had had no knowledge, the state of affairs with respect to abstract entities would be the same.

Has Dummett perhaps only laid down a necessary condition for realism, in his definition, for the sake of generality? I do not think so. In discussions of quantum mechanics we come across the view that the particles of microphysics are real, and obey the principles of the theory, but at any time t when 'particle x has exact momentum p' is true then 'particle x has position q' is neither true nor false. In any traditional sense, this is a realist position with respect to quantum mechanics.

We note also that Dummett has, at least in this passage, taken no care to exclude non-literal construals of the theory, as long as they are truth-valued. The two are not the same; when Strawson construed 'The king of France in 1905 is bald' as neither true nor false, he was not giving a non-literal construal of our language. On the other hand, people tend to fall back on non-literal construals typically in order to be able to say, 'properly construed, the theory is true.'

Perhaps Dummett is right in his assertion that what is really at stake, in realist disputes of various sorts, is questions about language—or, if not really at stake, at least the only serious philosophical problems in those neighbourhoods. Certainly the arguments in which he engages are profound, serious, and worthy of our attention. But it seems to me that his terminology ill accords with the traditional one. Certainly I wish to define scientific realism so that it need not imply that all statements in the theoretical language are true or false (only that they are all capable of being true or false, that is, there are conditions for each under which it has a truth-value); to imply nevertheless that the aim is that the theories should be true. And the contrary position of constructive empiricism is not anti-realist in Dummett's sense, since it also assumes scientific statements to have truth-conditions entirely independent of human activity or knowledge. But then, I do not conceive the dispute as being about language at all.

In any case Putnam himself does not stick with this weak formulation of Dummett's. A little later in the paper he directs himself to scientific realism *per se*, and formulates it in terms borrowed, he says, from Richard Boyd. The new formulation comes in the course of a new argument for scientific realism, which I shall call the Ultimate Argument:

> the positive argument for realism is that it is the only philosophy that doesn't make the success of science a miracle. That terms in mature scientific theories typically refer (this formulation is due to Richard Boyd), that the theories accepted in a mature science are typically approximately true, that the same term can refer to the same thing even when it occurs in different theories— these statements are viewed by the scientific realist not as necessary truths but as part of the only scientific explanation of the success of science, and hence as part of any adequate scientific description of science and its relations to its objects.[26]

Science, apparently, is required to explain its own success. There is this regularity in the world, that scientific predictions are regularly fulfilled; and this regularity, too, needs an explanation. Once *that* is supplied we may perhaps hope to have reached the *terminus de jure*?

The explanation provided is a very traditional one—*adequatio ad rem*, the 'adequacy' of the theory to its objects, a kind of mirroring of the structure of things by the structure of ideas—Aquinas would have felt quite at home with it.

Well, let us accept for now this demand for a scientific explanation of the success of science. Let us also resist construing it as merely a restatement of Smart's 'cosmic coincidence' argument, and view it instead as the question why we have successful scientific theories at all. Will this realist explanation with the Scholastic look be a scientifically acceptable answer? I would like to point out that science is a biological phenomenon, an activity by one kind of organism which facilitates its interaction with the environment. And this makes me think that a very different kind of scientific explanation is required.

I can best make the point by contrasting two accounts of the mouse who runs from its enemy, the cat. St. Augustine already remarked on this phenomenon, and provided an intentional explanation: the mouse *perceives that* the cat is its enemy, hence the mouse runs. What is postulated here is the 'adequacy' of the mouse's thought to the order of nature: the relation of enmity is correctly reflected in his mind. But the Darwinist says: Do not ask why the *mouse* runs from its enemy. Species which did not cope with their natural enemies no longer exist. That is why there are only ones who do.

In just the same way, I claim that the success of current scientific theories is no miracle. It is not even surprising to the scientific (Darwinist) mind. For any scientific theory is born into a life of fierce competition, a jungle red in tooth and claw. Only the successful theories survive—the ones which *in fact* latched on to actual regularities in nature.[27]

■ | Notes

1. *Science, Perception and Reality* (New York: Humanities Press, 1962); cf. the footnote on p. 97. See also my review of his *Studies in Philosophy and Its History*, in *Annals of Science*, 34 (January 1977), 73–74.

2. Brian Ellis, *Rational Belief Systems* (Oxford: Blackwell, 1979), p. 28.

3. Hilary Putnam, *Mathematics, Matter and Method* (Cambridge: Cambridge University Press, 1975), Vol. 1, pp. 69f.

4. Putnam, op. cit., p. 73, n. 29. The argument is reportedly developed at greater length in Boyd's forthcoming book *Realism and Scientific Epistemology* (Cambridge University Press).

5. Hartry Field has suggested that 'acceptance of a scientific theory involves the belief that it is true' be replaced by 'any reason to think that any part of a theory is not, or might not be, true, is reason not to accept it.' The drawback of this alternative is that it leaves open what epistemic attitude acceptance of a theory does involve. This question must also be answered, and as long as we are talking about full acceptance—as opposed to tentative or partial or otherwise qualified acceptance—I cannot see how a realist could do other than equate that attitude with full belief. (That theories believed to be false are used for practical problems, for example, classical mechanics for orbiting satellites, is of course a commonplace.) For if the aim is truth, and acceptance requires belief that the aim is served . . . I should also mention the statement of realism at the beginning of Richard Boyd, 'Realism, Underdetermination, and a Causal Theory of Evidence,' *Noûs*, 7 (1973), 1–12. Except for some doubts about his use of the terms *explanation* and *causal relation* I intend my statement of realism to be entirely in accordance with his. Finally, see C. A. Hooker, 'Systematic Realism', *Synthese*, 26 (1974), 409–97; esp. pp. 409 and 426.

6. More typical of realism, it seems to me, is the sort of epistemology found in Clark Glymour's book, *Theory and Evidence* (Princeton: Princeton University Press, 1980), except of course that there it is fully and carefully developed in one specific fashion. (See esp. his chapter 'Why I Am Not a Bayesian' for the present issue.) But I see no reason why a realist, as such, could not be a Bayesian of the type of Richard Jeffrey, even if the Bayesian position has in the past been linked with anti-realist and even instrumentalist views in philosophy of science.

7. G. Maxwell, 'The Ontological Status of Theoretical Entities', *Minnesota Studies in Philosophy of Science*, III (1962), p. 7. [An excerpt from Maxwell's paper is the first reading in this chapter.]

8. There is a great deal of recent work on the logic of vague predicates; especially important, to my mind, is that of Kit Fine ('Vagueness, Truth, and Logic', *Synthese*, 30 (1975), 265–300) and Hans Kamp. The latter is currently working on a new theory of vagueness that does justice to the 'vagueness of vagueness' and the context-dependence of standards of applicability for predicates.

9. Op. cit., p. 15 [p. 1058]. . . . At this point . . . I may be suspected of relying on modal distinctions which I criticize elsewhere. After all, I am making a distinction between human limitations and accidental factors. A certain apple was dropped into the sea in a bag of refuse, which sank; relative to that information it is necessary that no one ever observed the apple's core. That information, however, concerns an accident of history, and so it is not human limitations that rule out observation of the apple core. But unless I assert that some facts about humans are essential, or physically necessary, and others accidental, how can I make sense of this distinction? This question raises the difficulty of a philosophical retrenchment for modal language. This I believe to be possible through an ascent to pragmatics. In the present case, the answer would be, to speak very roughly, that the scientific theories we accept are a determining factor for the set of features of the human organism counted among the limitations to which we refer in using the term 'observable'. . . .

10. See n. 1 above.

11. Cf. P. Thagard, doctoral dissertation, University of Toronto, 1977, and 'The Best Explanation: Criteria for Theory Choice', *Journal of Philosophy*, 75 (1978), 76–92.

12. 'The Inference to the Best Explanation', *Philosophical Review*, 74 (1965), 88–95 and 'Knowledge, Inference, and Explanation', *American Philosophical Quarterly*, 5 (1968), 164–73. Harman's views were further developed in subsequent publications (*Noûs*, 1967; *Journal of Philosophy*, 1968; in M. Swain (ed.), *Induction*, 1970; in H.-N. Castañeda (ed.), *Action, Thought, and Reality*, 1975; and in his book *Thought*, Ch. 10). I shall not consider these further developments here.

13. See n. 11 above.

14. J. J. C. Smart, *Between Science and Philosophy* (New York: Random House, 1968), p. 151.

15. Ibid., pp. 150f.

16. This point is clearly made by Aristotle, *Physics*, II, Chs. 4–6 (see esp. 196^a 1–20; 196^b 20–197^a 12).

17. See my 'Wilfrid Sellars on Scientific Realism', *Dialogue*, 14 (1975), 606–16; W. Sellars, 'Is Scientific Realism Tenable?', pp. 307–34 in F. Suppe and P. Asquith (eds.), *PSA 1976* (East Lansing, Mich.: Philosophy of Science Association, 1977), vol. II, 307–34; and my 'On the Radical Incompleteness of the Manifest Image', ibid., 335–43; and see n. 1 above.

18. W. Sellars, 'The Language of Theories', in his *Science, Perception, and Reality* (London: Routledge and Kegan Paul, 1963).

19. Op. cit., p. 121.

20. Ibid., p. 121.

21. Ibid., p. 123.

22. See my 'Semantic Analysis of Quantum Logic', in C. A. Hooker (ed.), *Contemporary Research in the Foundations and Philosophy of Quantum Theory* (Dordrecht: Reidel, 1973), Part III, Sects. 5 and 6.

23. See n. 3 above

24. Michael Dummett, *Truth and Other Enigmas* (Cambridge, Mass.: Harvard University Press, 1978), p. 146 (see also pp. 358–61).

25. Dummett adds to the cited passage that he realizes that his characterization does not include all the disputes he had mentioned, and specifically excepts nominalism about abstract entities. However, he includes scientific realism as an example (op. cit., pp. 146f.).

26. See n. 4 above.

27. Of course, we can ask specifically why the *mouse* is one of the surviving species, how *it* survives, and answer this, on the basis of whatever scientific theory we accept, in terms of its brain and environment. The analogous question for theories would be why, say, Balmer's formula for the line spectrum of hydrogen survives as a successful hypothesis. In that case too we explain, on the basis of the physics we accept now, why the spacing of those lines satisfies the formula. Both the question and the answer are very different from the global question of the success of science, and the global answer of realism. The realist may now make the *further* objection that the antirealist cannot answer the question about the mouse specifically, nor the one about Balmer's formula, in this fashion, since the answer is in part an assertion that the scientific theory, used as basis of the explanation, is true. This is a quite different argument, which I . . . take up in Ch. 4, Sect. 4, and Ch. 5 [of *The Scientific Image*].

In his most recent publications and lectures Hilary Putnam has drawn a distinction between two doctrines, metaphysical realism and internal realism. He denies the former and identifies his preceding scientific realism as the latter. While I have at present no commitment to either side of the metaphysical dispute, I am very much in sympathy with the critique of Platonism in philosophy of mathematics, which forms part of Putnam's arguments. Our disagreement about scientific (internal) realism would remain, of course, whenever we came down to earth after deciding to agree or disagree about metaphysical realism, or even about whether this distinction makes sense at all.

ALAN MUSGRAVE

Realism versus
Constructive Empiricism

The demise of logical positivism has been followed by a rising tide of scientific realism. Bas van Fraassen is to be congratulated for swimming against that tide. But we must also ask whether he manages to make much headway. I shall argue that he does not. My first section explores van Fraassen's rather attenuated antirealism and the distinction between truth and empirical adequacy on which it depends. My second section argues that van Fraassen succeeds no better than his predecessors in answering a major objection to antirealism. My third section examines the link between realism and explanation and van Fraassen's attempt to sever that link.

I | Truth, Empirical Adequacy, Empirical Equivalence

Scientific realism is an old issue, and over the years both realism and antirealism have taken various forms. Van Fraassen defines realism thus: *"Science aims to give us, in its theories, a literally true story of what the world is like; and acceptance of a scientific theory involves the belief that it is true."* He says that this is a minimal formulation which "can be agreed to by anyone who considers himself a scientific realist."[1] Later, however, van Fraassen extends this minimal formulation by adding to it a realist 'demand for explanation.' As we will see, his antirealism stems in large part from criticisms of this demand. As we will also see, his version of the demand is an absurdly strong one.

What is the nature of van Fraassen's antirealism? The most radical opponents of realism (the instrumentalists) deny that scientific theories have truth-values at all. Van Fraassen's antirealism is not of this radical kind. He

FROM Paul M. Churchland and Clifford A. Hooker, eds., *Images of Science* (Chicago: University of Chicago Press, 1985), 197–221. In this article, the numbers in square brackets refer to pages in this volume.

1084 | Ch. 9 Empiricism and Scientific Realism

accepts a "literal construal of the language of science" whereby "the apparent statements of science really are statements, *capable of* being true or false" (p. 10) [1064]. In the same vein, he rejects positivist interpretations of scientific language, whereby the 'real meaning' of theoretical assertions is somehow cashed out in terms of the observable:

> Most specifically, if a theory says that something exists, then a literal construal may elaborate on what that something is, but will not remove the implication of existence. . . . If the theory's statements include "There are electrons," then the theory says that there are electrons. If in addition they include "Electrons are not planets," then the theory says, in part, that there are entities other than planets. (p. 11) [1064]

Thus, contrary to the positivists, two theories may say exactly the same thing about the observable yet remain distinct and perhaps incompatible theories.

All this puts van Fraassen firmly in the realist camp as far as the *interpretation* of scientific theories is concerned.[2] His antirealism proceeds entirely on the *epistemological or methodological* level. (The same can be said of the antirealism espoused by Larry Laudan in *Progress and Its Problems*.) He thinks that, although scientific theories are capable of literal truth, they "need not be true to be good" (p. 10) [1063]. Accordingly, it is not the aim of science to provide true theories, and to accept a theory is not to believe it to be true. What matters in science is that theories are correct so far as the observations and experiments go. Hence, constructive empiricism: "*Science aims to give us theories which are empirically adequate; and acceptance of a theory involves as belief only that it is empirically adequate*" (p. 12) [1065]. A theory is empirically adequate "exactly if what it says about the observable things and events in this world, is true—exactly if it 'saves the phenomena'" (p. 12) [1065].

The distinction between truth and empirical adequacy, and hence between realism and constructive empiricism, is a subtle one. For theories about the observable, truth and empirical adequacy coincide (p. 21) [1072]. For theories about the unobservable, truth entails empirical adequacy but not vice versa: such a theory may be empirically adequate yet false. Accordingly, to believe that a theory about the unobservable is true is more risky than to believe that it is empirically adequate. Not that the latter is without risk: empirical adequacy "goes far beyond what we can know at any given time" since it requires that the theory save all the phenomena in its field, past, present, and future, not merely all actually observed phenomena (p. 69). Now, the chief difficulty for realism has always been skeptical arguments to the effect that we can never know a scientific theory to be true nor ever be rationally warranted in accepting, however tentatively, a theory as true. This is as much a difficulty for constructive empiricism. The same skeptical arguments might be used to show that we can never know a scientific theory to be empirically adequate nor ever be rationally warranted in accepting, however tentatively, a theory as empirically adequate. Van Fraassen insists, however, that the positions are different:

> There does remain the fact that . . . in accepting any theory as empirically adequate, I am sticking my neck out. There is no argument there for belief in the truth of the accepted theories, since it is not an epistemological principle that one might as well hang for a sheep as for a lamb. (p. 72)

Epistemological or not, the principle that one might as well hang for a sheep as for a lamb is a pretty sensible one. Given two criminal acts A and B whose risks of detection and subsequent penalties are the same, but where A yields a greater gain than B, the sensible criminal will do A. But are the risks and penalties of realism and constructive empiricism the same? And does realism bring with it gains that constructive empiricism does not? Van Fraassen addresses these questions; to evaluate the cogency of his position we must address them too.

Suppose the realist tentatively accepts a theory as true, while the constructive empiricist tentatively accepts it as empirically adequate. The realist does take a greater risk. But he takes no greater risk of being detected in error *on empirical grounds.* So, given strict empiricism (the principle that only evidence should determine theory choice), it seems that we might as well be hung for the realist sheep as for the constructive empiricist lamb.

The trouble is, van Fraassen argues, that realism and strict empiricism do not mix and that realism must pay the penalty of rejecting strict empiricism. He makes the point by considering the case of empirically equivalent yet incompatible theories. This is not, of course, the humdrum case where the *available* evidence fails to discriminate between two incompatible theories. This case need not trouble the realist, who may always hope to show that the two theories are not empirically equivalent and then press for an experimental decision between them. Rather, it is the esoteric case where such hopes are unfounded, where two incompatible theories say exactly the same things about *all* matters observational. The constructive empiricist could accept both theories (believe both to be empirically adequate); the realist cannot on pain of contradiction believe both to be true. But how is the realist to choose between them? In the nature of the case, empirical evidence cannot guide his choice, which must therefore be made on nonevidential grounds. Realism runs counter to strict empiricism and allows nonevidential or 'metaphysical' considerations to intrude into matters of theory choice.

How might the realist respond to this? Presumably, as a realist, he will have no truck with the positivist idea that empirically equivalent theories are really the same theory and not incompatible after all. Nor, as a realist, will he have any truck with the related idea (perhaps it is the same idea in new dress) that there are no 'verification-transcendent truth-conditions' and therefore no truth of the matter for the two theories to disagree about.

These ideas, after all, seem to entail that Berkeley's immaterialism is really the same theory as the commonsense belief in independently existing material objects or that there is no truth of the matter for Berkeley and commonsense to disagree about. And these conclusions are anathema to the commonsense realist, let alone the scientific one.[3]

Taking a cue from this example, one might wonder whether the problem is philosophical or metaphysical rather than scientific, in which case metaphysical considerations would not be an intrusion after all. How often have empirically equivalent but incompatible theories occurred in real science? Van Fraassen gives one example, and it is a notorious one. Newton hypothesized that the center of gravity of the solar system is at rest in absolute space. He also pointed out that the appearances would be no different if that center were moving through absolute space at any constant velocity v. So all of the theories $TN(v)$—Newton's theories of mechanics and gravitation plus the postulate that the center of gravity of the solar system has constant absolute velocity v for any v—were claimed by Newton to be empirically equivalent (p. 46).

Van Fraassen's account can be disputed: Newton only claimed the empirical equivalence of the theories $TN(v)$ *so far as appearances within the solar system are concerned*. Hypothesize that some other star is at rest in absolute space, for example, and the empirical equivalence vanishes: if the solar system has any nonzero velocity, then it will approach or recede from that star and, given sufficient time, the effects of this will become apparent.

Here I resort to a realist ploy whose efficacy van Frassen considers. This is to say that, where equivalent theories occur, by *extending* these theories (that is, embedding them in wider theories) their equivalence will disappear (that is, the wider theories will not be empirically equivalent). In the trivial example just cited, the wider theories are formed simply by adding the statement that some star is at absolute rest to each of the existing theories. The example is trivial because we can, by the Newtonian principle of relativity, construct empirically equivalent theories to each of these wider theories (including the only empirically adequate one): simply consider theories attributing an absolute velocity v to the center of mass of the extended system consisting of the solar system and the star. The process can be continued (assuming the number of masses is finite) until they are all taken into account. And then, again by the Newtonian principle of relativity, we will have an infinite family of empirically equivalent theories each of which consists of Newton's laws plus the hypothesis that the center of mass *of the universe* has velocity v for any value of v.

Van Fraassen considers a more interesting extension or embedding of his Newtonian example, the attempt to combine it with Maxwell's electromagnetism where forces depend upon velocities and not upon accelerations as in Newton. This feature made it possible to devise experiments to detect absolute velocities. The null results of such experiments were an important factor in leading scientists to abandon the Newtonian doctrines of absolute space and time in favor of relativistic ones. And this was to abandon *all* of the empirically equivalent Newtonian theories. Van Fraassen asks us to imagine, however, that null results had not been obtained, that, on the contrary, an absolute velocity for the center of mass of the solar system had been measured. Here it might seem that one of the empirically equivalent Newtonian theories had been confirmed and the rest refuted, and hence that they were not empirically equivalent after all. Van Frassen finds this reasoning spurious (p. 49). But I find his reasoning, if not spurious, at least hard to follow. Operating within this piece of science fiction (or, rather, history-of-science fiction), he says that we could make compensating adjustments in electromagnetic theory so as to retain whichever of the empirically equivalent Newtonian theories we like. In other words, had the history of science been different, we could construct a new family of empirically equivalent combinations of mechanics and electromagnetism.

But, first, van Fraassen has done nothing to impugn the fact that his empirically equivalent Newtonian theories, when combined with Maxwell's electromagnetism, ceased to be empirically equivalent. Second, could the Newtonian readily have accepted that electromagnetic forces depend upon absolute velocities rather than absolute accelerations? Van Fraassen concedes that, had his piece of history-of-science fiction occurred, it would have "upset even Newton's deepest convictions about the relativity of motion" (p. 48). But did not these convictions follow from Newton's laws of mechanics and the doctrine of absolute space? Last, and perhaps most important, all of this is a piece of history-of-science fiction: the historical facts are that in this notorious real example of empirical equivalence, the only good example known to me, the actual development of science removed the problem.

Van Fraassen has a further retort to the idea that empirical equivalence can be removed by extension or embedding. He can say that it is the empirical adequacy of the extended theories which counts and that one should accept the victor only as empirically adequate, never as true. And, to offset the scarcity of empirically equivalent theories in real science, he can point to the fact that we can artificially concoct empirically equivalent alternatives to any theory by resorting to notorious logical tricks. The simplest such trick is to conjoin any theory with "The Absolute is lazy" to form a new theory empirically equivalent with the original.

The standard response to such tricks is to eliminate the concocted theory on the ground of its reduced simplicity or unity. Van Fraassen does not object to the appeal to simplicity, but he insists that simplicity is a *pragmatic virtue* of a theory which has nothing to do with that theory's truth or likelihood of being true. The realist, for whom accepting a theory is believing it true, must forge a link between simplicity and truth if he is to appeal to the former. And the link can be forged only by a metaphysical principle:

> Simplicity . . . is obviously a criterion in theory choice, or at least a term in theory appraisal. For that reason, some . . . suggest that simple theories are

more likely to be true. But it is surely absurd to think that the world is more likely to be simple than complicated (unless one has certain metaphysical or theological views not usually accepted as legitimate factors in scientific inference). The point is that the virtue, or patchwork of virtues, indicated by the term is a factor in theory appraisal, but does not . . . make a theory more likely to be true (or empirically adequate). (p. 90)

So the argument seems to be this: the realist can solve the problem of empirical equivalence only by appealing to simplicity; but he can appeal to simplicity only if he assumes a metaphysical principle ("Nature is simple" or some such); realism therefore involves an illegitimate intrusion of metaphysics into science and the abandonment of strict empiricism.

Is the constructive empiricist in any better position? Presumably he, too, will prefer a respectable theory to an artificially concocted empirically equivalent alternative. He, too, will appeal to simplicity and abandon strict empiricism. But he, apparently, can do this in good conscience, cheerfully admitting that pragmatic virtues such as simplicity have nothing to do with the real aim of science, empirical adequacy. Indeed, how could simplicity have anything to do with that aim? To say that the simpler of two empirically equivalent theories is more likely to be empirically adequate is to contradict oneself.

Returning to the realist, there are several ways he might respond to van Fraassen's argument. The first is simply to admit that there is nothing to choose between empirically equivalent theories. This is hardly satisfactory in view of the ubiquity of the logical tricks. The second is to spice scientific realism with a dash of pragmatism, admitting that there is nothing to choose on realist grounds between empirically equivalent theories but preferring some on the pragmatic ground of simplicity. Despite van Fraassen's argument, I see no reason why the realist cannot appeal to pragmatic virtues just as the constructive empiricist does. The third response is to say that simplicity is not, after all, a merely pragmatic virtue. Realists and constructive empiricists alike value empirical strength; they value it for different reasons, but both connect it with the central aim of science. Is it not a sufficient reason to eliminate concocted alternatives to existing theories that they are not empirically stronger than the theories from which they are concocted?

The realist has a problem here, however. Insofar as simplicity and strength go together (and they do not always), simplicity is not merely a pragmatic virtue. But insofar as simplicity and strength go together, simplicity and truth cannot: the stronger theory is, in some intuitive sense at least, less likely to be true. And here lies the problem for any realist seeking to forge a link between simplicity and truth. Yet the problem may not be completely intractable. "Nature is simple" is a metaphysical principle and a hopelessly vague one to boot. But scientists have made various attempts to say more precisely what it means and to construct theories which conform to it.[4] This transforms it into a metaphysical principle which can, at first remove so to speak, be empirically assessed: roughly speaking, it is acceptable metaphysics if theories constructed under its aegis are empirically successful, while theories which violate it are not.[5] In our postpositivistic age, we should not regard the intrusion of this kind of metaphysical principle into science as illegitimate. If vague appeals to simplicity can be transformed into precise principles of theory construction and if such principles are acceptable (in the sense roughly defined), then the virtue they indicate is not merely pragmatic. It may not be absurd to think that Nature is simple (in some carefully specified sense or sense), if we can point to the empirical success of science in vindication of our belief.

I do not know whether this third response, which I have merely sketched, will work in the end. Perhaps it could be shown (though it would be a far from trivial task) that, for any precise and acceptable sense of the term *simple*, one could concoct empirically equivalent and equally simple theories. I would not see this as the demise of scientific realism, for (and this is the second response again) I cannot see why the realist is barred from invoking a pragmatic virtue to deal with the problem of empirical equivalence just as the constructive empiricist does.

II | Theory and Observation

Antirealists need to draw a dichotomy between theory and observation. Van Fraassen is no exception: after all, he could not even distinguish truth from empirical adequacy without it. An old and powerful objection to antirealist views is that no such dichotomy exists. How does van Fraassen deal with this objection?

He first agrees that no such dichotomy can be drawn in scientific language, agreeing with the realist that "All our language is thoroughly theory-infected" (p. 14) [p. 1067] and pointing out against the positivist that highly theoretical assertions can be made using only so-called 'observational vocabulary' (pp. 54–55). (Here I was reminded of how Popper formulated "There exists an omnipotent, omnipresent, and omniscient personal spirit" in a physicalistic observation language.)[6]

Van Fraassen does insist, however, that some objects and/or events are observable and some not. He concedes the familiar realist point that there is a continuous spectrum between 'directly observing' an object and 'indirectly detecting' it using instruments. This only shows that *observable* is a vague predicate. But a vague predicate is perfectly usable provided it has clear cases and clear noncases—and this one has:

> A look through a telescope at the moons of Jupiter seems to me a clear case of observation, since astronauts will no doubt be able to see them as well from close up. But the purported observation of micro-particles in a cloud chamber seems to me a clearly different case—if our theory about what happens there is

right. . . . [W]hile the particle is detected by means of the cloud chamber, and the detection is based on observation, it is clearly not a case of the particle's being observed. (pp. 16–17) [1069]

What if we had microscopic or electron-microscopic eyes? (Actually, we do, only they are not built into our heads!) Could we not then observe things which at present we can only detect, showing that they were not unobservable in principle? Lockean speculations like this merely change the subject:

> The human organism is, from the point of view of physics, a certain kind of measuring apparatus. As such it has certain inherent limitations—which will be described in detail in the final physics and biology. It is these limitations to which the "able" in "observable" refers—our limitations, *qua* human beings. (p. 17) [1070]

But current physics and biology tell us that what is observable by humans varies (some of us are color-blind) and depends on our particular evolutionary history (other organisms can observe things we cannot). So, even if we can draw a rough and species-specific distinction between what is observable by humans and what is not, should any philosophical significance be attached to it?

Van Fraassen agrees with the realists against the idealists that it has no *ontological* significance: things that humans do not happen to be able to observe may nonetheless exist (p. 18) [1070]. (Actually, I will argue later, there are problems about van Fraassen's making this concession.) But he wants to give the distinction an *epistemological* significance: humans should never believe to be true a theory about what they cannot observe; instead, they should believe such theories only to be empirically adequate, to tell the truth about what they can observe (p. 18) [1070].

Can a distinction which is admitted to be rough-and-ready, species-specific, and of no ontological significance really bear such an epistemological burden? Van Fraassen gives an example of a so-called inference to the best explanation:

> I hear scratching in the wall, the patter of little feet at midnight, my cheese disappears—and I infer that a mouse has come to live with me. Not merely that these apparent signs of mousely presence will continue, not merely that all the observable phenomena will be as if there is a mouse; but that there really is a mouse. (pp. 19–20) [1071][7]

Will not the same style of argument lead us to the conclusion that there really are electrons (or whatever)? Van Fraassen thinks not. He accepts "inference to the best explanation" but puts his own gloss upon it: such inferences should (and do) only lead us to accept the best explanation as empirically adequate (p. 20) [1072]. If the best explanation is a theory about the observable, then empirical adequacy and truth coincide and we can (and do) conclude that there really is a mouse (or whatever). But if the best explanation is a theory about the unobservable, empirical adequacy and truth do not coincide and we cannot (and do not) conclude that there really are electrons (or whatever).

There is an empirical claim here (about what scientists actually do infer) and also a methodological claim (about what they ought to infer). I find the methodological claim quite unreasonable. On any plausible theory of evidential support, one would have to admit that there could be far better evidence for an explanation couched in terms of unobservables than for an explanation couched in terms of observables. Is the evidence for the existence of electrons better or worse than the evidence for the existence of the yeti or of the mouse in van Fraassen's wainscoting? It is a curious sort of empiricism which sets aside the weight of *available* evidence on the ground that a casual observer might one day see his mouse or yeti, while the scientist can never see (but can only detect) his electrons.

Van Fraassen's factual claim (that scientists do infer only the empirical adequacy of theories about the unobservable but never their truth) is even harder to swallow. Admittedly, I have not done a sociological survey to settle the matter. And, even if such a survey were to reveal, as I believe it would, that realism is the instinctive philosophy of working scientists, this would not of course settle the methodological question. But to indicate how difficult it is to avoid realist ways of thinking and talking, let us see how van Fraassen thinks and talks. He talks of *detecting* an electron in a cloud chamber. Can one say truly that one has detected an object without also believing it to be true that the object really exists? Later he describes how Millikan *measured* the charge of the electron (pp. 75–77). Did not Millikan think it true, and does not anyone who accepts Millikan's results think it true, that electrons exist and carry a certain charge? Can one say truly that one has measured some feature of an object without also believing that the object really exists?

I shall quote at some length what I *think* is van Fraassen's answer to very obvious questions like these:

> The working scientist is totally immersed in the scientific world-picture. And not only he—to varying degrees, so are we all. . . . But immersion in the theoretical world-picture does not preclude "bracketing" its ontological implications. . . . To someone immersed in that world-picture, the distinction between *electron* and *flying horse* is as clear as between *racehorse* and *flying horse*: the first corresponds to something in the actual world, and the other does not. While immersed in the theory, this objectivity of *electron* is not and cannot be qualified. *But this is so whether or not one is committed to the truth of the theory.* It is so not only for someone who believes, full stop, that the theory is true, but also for . . . someone who . . . holds commitment to the truth of the theory in abeyance. For to say that someone is immersed in theory . . . is not to describe his epistemic commitment . . . it is possible even after total immersion in the

world of science . . . to limit one's epistemic commitment while remaining a functioning member of the scientific community. (pp. 80–83)

This is, I fear, nothing but a sleight-of-hand and an endorsement of philosophical schizophrenia. The sleight-of-hand converts belief in the reality of electrons (belief in the objectivity of *electron*, belief that the term *electron* corresponds to something in the actual world) into belief in, belief full stop in, and finally commitment to something called "the theory of electrons." But there have been several theories about electrons, and no scientist believes them all to be true. As for the most up-to-date theory about electrons, sensible scientists would do well not to believe it to be wholly true either, for details of it are quite likely to be further refined. All this is quite consistent with a pretty firm belief in the reality of electrons, with a refusal to "bracket" this particular ontological implication of science. The philosophical schizophrenia stems from talk of immersion (even total immersion) in the "scientific world-picture" or the "world of science." These metaphors are meant to suggest, if I understand them rightly, that scientists should believe in electrons or whatever while immersed in their scientific work, but should become agnostic about everything they cannot observe once they leave their laboratories. I suppose that split-minded scientists like this are possible, but I wonder whether they are desirable.

Finally in this section, I want to argue that van Fraassen's treatment of the observable/unobservable distinction verges on the incoherent. He insists that what is observable by humans is a "function of facts about us *qua* organisms in the world," so that it is for *science* to tell us what is observable and what is not (pp. 57–58).[8] Now, suppose some theory T does distinguish "the observable which it postulates from the whole it postulates" (p. 59). T might even be van Fraassen's "final physics and biology," if such a theory is possible. T will say, among other things, that A is observable by humans, while B is not. Of course, if we are to use T to delineate the observable, we must *accept* it. But van Fraassen cannot have us accept it as true, since it concerns in part the unobservable B. The constructive empiricist can accept T only as empirically adequate, that is, believe to be true only what T says about the observable. But "B is not observable by humans" cannot, on pain of contradiction, be a statement about something observable by humans. And, in general, the consistent constructive empiricist cannot believe it to be true that *anything* is unobservable by humans. And, if this is so, the consistent constructive empiricist cannot draw a workable observable/unobservable dichotomy at all.

It might be objected that

1 B is not observable by humans

is logically equivalent with

2 Everything observable by humans is distinct from B

[and] since (2) is a statement about the observable, so is the logically equiva-lent (1). But even accepting that there is a sense in which (2) is "about" the observable, it is *also* about the unobservable B and therefore cannot be accepted as true by the constructive empiricist.

Nor does it help if we say that "observable by humans" is an "observa-tional predicate," that we humans can tell from observation that a thing is observable by us. For one thing, this marks a retreat from van Fraassen's insistence that there is no observable/unobservable dichotomy in scientific *language*. For another thing, "observable by humans" will be a nonstandard observational predicate whose negation is not also observational, a predi-cate akin, for example, to the predicate "is an inscription of finite length." For we cannot observe that anything has a property without also observing that thing. Anyone who claims to have *observed* that something is unobserv-able contradicts himself. But if "unobservable by humans" is *not* an obser-vational predicate, our conclusion stands. We can grant that "observable by humans" is an observational predicate so that the constructive empiricist can accept as true, on the basis of observation, statements of the form "A is observable by humans." But the consistent constructive empiricist cannot accept as true, on the basis of observation or anything else, a statement of the form "B is not observable by humans." Constructive empiricism requires a dichotomy which it cannot consistently draw.

III | Realism and Explanation

Realism and explanation are doubly linked. Realists think science explains facts about the world, and they think realist philosophy of science explains facts about science. I will consider the latter claim first.

The claim is that only a realist philosophy of science can explain the fact that science has had a great deal of predictive success. If the unobserv-ables postulated by (successful) science really exist and if what (successful) science says about them is true or nearly so, then we can explain predictive success. Otherwise, such success is just a lucky accident. As Putnam famously remarks, realism is "the only philosophy that doesn't make the success of science a miracle" (cited on p. 39) [p. 1079].

Van Fraassen given short shrift to this Ultimate Argument for realism:

> The explanation provided is a very traditional one—*adequatio ad rem*, the "adequacy" of the theory to its objects, a kind of mirroring of the structure of things by the structure of ideas—Aquinas would have felt quite at home with it.
>
> ... Will this realist explanation with the Scholastic look be a scientifically acceptable answer? I would like to point out that science is a biological phe-nomenon, an activity by one kind of organism which facilitates its interaction with the environment. And this makes me think that a very different kind of scientific explanation is required.

I can best make the point by contrasting two accounts of the mouse who runs from its enemy, the cat. St. Augustine . . . provided an intentional explanation: the mouse *perceives that* the cat is an enemy, hence the mouse runs. What is postulated here is the "adequacy" of the mouse's thought to the order of nature: the relation of enmity is correctly reflected in his mind. But the Darwinist says: Do not ask why the *mouse* runs from its enemy. Species which did not cope with their natural enemies no longer exist. That is why there are only ones who do.

In just the same way, I claim that the success of current scientific theories is no miracle. It is not even surprising to the scientific (Darwinist) mind. For any scientific theory is born into a life of fierce competition, a jungle red in tooth and claw. Only the successful theories survive—the ones which *in fact* latched on to actual regularities in nature. (pp. 39–40) [1079–80]

Amusing though this is, it does no more than play cat-and-mouse with the argument. The scientist does ask why the mouse runs from the cat and answers in roughly the terms made fun of here: the mouse perceives the cat, perceives the cat as an enemy, and runs. This does not commit the scientist to ascribing thoughts, adequate or otherwise, to the mouse: his response might be quite instinctive. But with this proviso, there is nothing unscientific or un-Darwinian about this kind of explanation. Of course, the Darwinian question is not "Why does the mouse run away from the cat?" but, rather, "How did this piece of mouse behavior evolve?" The Darwinian answers this question roughly in the terms suggested by van Fraassen: given an environment full of mouse-hunting cats, cat-fleeing mice are more likely to survive, reproduce, and pass their cat-fleeing behavior on to future generations. But the Darwinian explanation is not a substitute for the "intentional" one, for they are addressed to quite different questions. The Darwinian explains what the "intentionalist" postulates: that the mouse's perceiving the cat as an enemy (or, better, the mouse's genetically programmed behavioral response to cats) is adequate to the order of nature.

Just as with cats and mice, so also with scientific success. It is one thing to explain why some theory is successful and quite another to explain why only successful theories survive. Van Fraassen's Darwinian explanation of the latter can be accepted by realist and antirealist alike. But to say that only successful theories are allowed to survive is not to explain why any particular theory is successful.

Not that a realist explanation of this in terms of the theory's *adequatio ad rem* will do as it stands. The Ultimate Argument is actually very old, and a brief look at an old example of it should give us pause. Christopher Clavius (in his *Commentary on the Sphere of Sacrobosco* of 1581) said it was incredible to suppose that Ptolemaic astronomy could correctly predict eclipses even though its eccentrics and epicycles were mere figments. But the eccentrics and epicycles were figments, and it was no miracle at all that a geometrical model expressly devised to yield some phenomenal regularity (periodic eclipses) should be successful in doing so. It is different, however, if a theory devised to accommodate some phenomenal regularities should turn out to predict *new* regularities. The realist has a ready explanation: the entities postulated by the theory really exist, and what the theory says about them is true (or nearly so). The antirealist seems forced to say that figments dreamed up for one purpose have turned out, miraculously, to be well adapted for a quite different purpose. So it was that thoughtful realists such as William Whewell distinguished two kinds of predictive success (predicting known effects and predicting new ones) and argued that the antirealist cannot explain the latter. So it was that a thoughtful antirealist such as Duhem, seeing the force of the argument, came to spice his instrumentalism with a whiff of realism: a theory is able to make successful *novel* predictions because it is not "a purely artificial system" but, rather, "a natural classification [whose] principles express profound and real relations among things."[9]

As this brief historical excursion shows, the only form of the Ultimate Argument which *might* work is that which focuses on *novel* predictive success. Yet this focus is lacking in recent discussions, both from prominent defenders of the argument (such as Putnam) and from prominent critics of it (such as Laudan).[10] Difficulties remain, of course, not least that of making precise the intuitive distinction between known effects and novel predictions. These difficulties notwithstanding, van Fraassen has done nothing to impugn the Ultimate Argument in its refined form.

The Ultimate Argument proceeds on the metalevel: epistemology is to be naturalized, and philosophy of science is to explain facts about science. But there is a more direct argument, which proceeds from the assumption that science should explain facts about the world. The connection between the demand for explanation and the realist demand for true theories is apparently very obvious. An explanation is not adequate unless what does the explaining is true. So, given that theories figure in explanations, adequate explanations require true theories. It is worth noting that Duhem found the argument cogent and confessed that since in his view science did not aim at true theories, it could not explain anything either. Others adopt the curious view that science does aim truly to describe the world but cannot really explain any features of it. Van Fraassen's response to our simple argument is twofold. He attacks the realist demand for explanation; and he argues that explanation, where it can be had, does not require true theories, that explanatory power is a pragmatic virtue for which an empirically adequate theory will do just as well as a true one.

He softens us up with a linguistic point. We still speak of explanations even when we think the explanatory principles false:

> I say that Newton could explain the tides, that he had an explanation of the
> tides, that he did explain the tides. In the same breath I can add that this theory

is, after all, not correct. Hence I would be inconsistent if by the former I meant that Newton had a true theory which explained the tides. (p. 99)

Quite so. We can speak without contradiction of a false explanation because truth is not a *defining* condition of explanation but an *adequacy* condition upon it. (In a similar way, we can say that Bode's Law is false without contradicting ourselves.)* Van Fraassen says that we may agree that a theory is false without at all undermining our previous assertion that it explained many phenomena (p. 98). But could we say that a theory *adequately* explained many phenomena even though it is false? Realists think not. Scientists appear to agree: no modern astronomical text cites the vortex theory of planetary motion as the explanation of why the planets all go around the sun in the same direction, though some may have no other explanation of this fact. If this is wrong, and truth is not required for adequate explanation, then it will take more than a linguistic point to show it—van Fraassen gives us more.

First, he attacks the realist demand for explanation. He claims that in science this demand is severely limited, that explanation is not a "preeminent" or "rock-bottom" scientific virtue:

> If explanation of the facts were required in the way consistency with the facts is, then every theory would have to explain every fact in its domain. Newton would have had to add an explanation of gravity to his celestial mechanics before presenting it at all. (p. 94)

But Newton did present his theory. He "declined to explain," admitting famously that he had "not been able to discover the cause of . . . gravity" (p. 94). Thus:

> Newton's theory of gravitation . . . did not (in the opinion of Newton or his contemporaries) contain an explanation of gravitational phenomena, but only a description. (p. 112)

It is the same in modern physics, where the "unlimited demand for explanation leads to a demand for hidden variables, which runs counter to at

---

* Despite its name, Bode's law is not a law. It is a formula that says that the distances of the planets from the sun are in the following ratio: $(0+4)$, $(3+4)$, $(6+4)$, $(12+4)$, $(24+4)$, and so on. The formula fits the planets from Mercury to Uranus quite well, if we include the asteroids as a "missing planet" between Mars and Jupiter. But Neptune violates the series, since its distance calculated from Bode's law is 388, and its actual distance is about 300. So the formula is false and hence not a law. The interesting question, of course, is whether we would count Bode's law as a genuine law if there were no exceptions to it. Our reluctance to do so would stem largely from the fact that, as yet, no one has given a convincing explanation of why it holds (in the cases that it does). It stands isolated from the rest of science as a cosmic coincidence. See chapter 7 for further discussion of the distinction between laws and accidental generalizations.

least one major school of thought" (p. 23) [1073]. And, quite generally, to demand that regularities be explained and shown to be more than cosmic coincidences is self-defeating. For what of the regularities postulated to do the explaining?

Something is obviously wrong somewhere in all this. On the one hand, Newton explained the tides (p. 99) and, on the other hand, Newton's theory did not explain gravitational phenomena at all (p. 112). What has gone wrong is a tacit conflation of realism with essentialism, of the demand for explanation with the demand for ultimate explanation. This will take a bit of explaining.[11]

Suppose we explain the phenomenal regularity that sticks look bent when half-immersed in water by postulating, among other things, that unobservable light rays refract when passing through media of different densities. This is not, of course, to explain the refraction of light, though we might then try to do so. But at any point in our explanatory endeavors there will be things for which we have no explanations, namely the deepest explanatory principles we have reached at that point. One realist response to this situation is to demand that these principles require explanation in their turn. Another realist response, quite the antithesis of the first, is to demand that our deepest explanatory principles should require no explanation, that they should somehow be ultimate or self-explanatory. This second response is central to the tradition of Aristotelian essentialism, the tradition which holds that the only genuine explanations are ultimate or self-explanatory explanations.

The essentialist tradition is buttressed by rhetorical questions like these. Do we *really* explain why sticks look bent in water by postulating some mysterious and unexplained law of refraction? Did Newton *really* explain the tides by postulating his mysterious and unexplained force of gravity? The intuition is that a *real* explanation does not remove one mystery by postulating another one. And behind this intuition lies another one: the intuition that a real explanation should serve the *pragmatic* function of removing puzzlement, setting the curiosity of the inquirer at rest. It is because nonultimate scientific explanations do not serve this pragmatic function, do not remove puzzlement so much as relocate and enhance it, that they are said not to be real explanations at all.

I use the term *pragmatic* here advisedly and roughly as van Fraassen uses it (sometimes). Whether an explanation removes puzzlement depends very much on the person we are considering. What relieves one man's puzzlement may enhance the next woman's. I dare say that some of the most efficacious puzzlement relievers in the history of thought have been explanations which are not scientific at all and which are, from a scientific point of view, quite inadequate (what about "God moves in mysterious ways," said in explanation of anything whatever which is puzzling?). I dare say that on occasions the incurious have had their puzzlement removed by a scientific explanation—but they should not have. For if it is feelings of puzzlement we want to get rid of, we should turn not to science but to the whiskey bottle!

I think that essentialism and the intuitions which lie behind it are to be rejected. And I think it one of Newton's chief claims to methodological fame that he was among the first to see this. Newton admitted that he could not explain gravity *and* that gravity was a perfectly proper thing to try to explain (since it was not an "essential property" of matter). Yet in the same breath he insisted that his theory of gravity did explain celestial motions and the tides:

> Hitherto we have *explained* the phenomena of the heavens and of our sea by the power of gravity, but we have not yet assigned the cause of this power . . . hitherto I have not been able to discover the cause of . . . gravity from phenomena, and I frame no hypotheses . . . to us it is enough that gravity does really exist, and act according to the laws we have explained, and abundantly serves to *account for* all the motions of the celestial bodies, and of our sea.[12]

Elsewhere Newton contrasted his procedure with that of his opponents, the Cartesian essentialists: he gives precise deductive explanations; they mutter about essences and can explain nothing:

> To tell us that every Species of Things is endow'd with an occult specifick Quality by which it acts and produces manifest Effects, is to tell us nothing: But to derive two or three general Principles of Motion from Phaenomena, and afterwards to tell us how the Properties and Actions of all corporeal Things follow from these manifest Principles, would be a very great step in Philosophy, though the Causes of those Principles were not yet discover'd: And therefore I scruple not to propose the Principles of Motion above mention'd, they being of very general Extent, and leave their Causes to be found out.[13]

Van Fraassen quotes from the first of these famous passages (p. 94), but he misunderstands Newton (as Duhem and others have misunderstood him) in a way that can only stem from a tacit conflation of realism with essentialism. He says that Newton "decline[d] to explain" (p. 94) and that Newton's opinion was that his theory "did not contain an explanation of gravitational phenomena, but only a description" (p. 112). But Newton explicitly claimed to have explained or accounted for gravitational phenomena such as the tides *by* describing precisely how gravity works. The antithesis between explanation and description is quite illusory: we explain one thing *by* describing another. Newton did decline to explain gravity. But to take this as a confession that nothing can be explained by the law of gravity is to father upon Newton a view that was not his.

It is, however, the view which van Fraassen calls the "realist demand for explanation" and attacks. He formulates the demand innocently enough as "every theory should explain every fact in its domain" and then takes "every fact" to include *the theory itself* (p. 94, quoted earlier). Only a theory which was somehow self-explanatory could meet this demand. But one can demand explanation without also demanding ultimate or self-explanatory explanation, as Newton tried to teach us. Van Fraassen's rejection of the latter demand leaves the former quite intact.

The tacit conflation of explanation with ultimate explanation also emerges from the delightful joke which forms the last chapter of van Fraassen's book. There, taking his cue from the remark that "everyone believed in the existence of God until the Boyle lectures proved it" (p. 229), he modifies Aquinas's Five Ways into proofs of scientific realism. What gets proved is actually Aristotelian essentialism, which is perhaps not surprising considering the provenance of the arguments. The First Way gives the flavor of the whole:

> So I *argue*: Everything that is to be explained, is to be explained by something else. That some things are to be explained is evident, for the regularities in natural phenomena are obvious to the senses and surprising to the intellect. So we must either proceed to infinity, or arrive at something which explains, but is not itself, a regularity in the natural phenomena. However, in this we cannot proceed to infinity. (pp. 205–6)

Now, what gets "proved" here is that science can achieve something which explains but is not itself a regularity in the natural phenomena, something, in other words, which is not itself to be explained. What gets proved is the essentialist view that science can achieve ultimate explanation. But the "proof" has a missing premise (required to obtain the statement beginning with "So"): the essentialist principle that A does not really explain B if A also requires explanation and has not received it. Van Fraassen hopes that readers skeptical of ultimate explanations will reject the conclusion, miss the missing premise, and infer by *modus tollens* that science cannot explain anything at all. But I am spoiling a clever joke.

I cannot leave this topic without saying a word about van Fraassen's favorite example of how science has transcended realist demands for explanation, quantum mechanics. Hidden variable explanations of quantum mechanics are said to "run counter to at least one modern school of thought" (p. 23) [1073]. A philosophy of science is not refuted by pointing out that it runs counter to a scientific school of thought, not even a dominant school of thought. But this school can point to *proofs* that hidden variable theories are impossible, and this should give the realist pause. Van Fraassen's comments on these proofs (which he seems to approve of) reveal some interesting things. One proof apparently assumes that "if we cannot point to some possible differences in empirical predictions, then there is no real difference at all" between two theories (p. 34) [1077]. In endorsing this proof, van Fraassen's earlier resolute antipositivism has wavered, for this assumption requires a positivist reinterpretation of scientific language to show that empirically equivalent theories are really the same theory. Not that realists would be too happy with an explanation of quantum mechanics which was demonstrably empirically equivalent with it: such an explanation could have no *independent* evidence in its favor. Now, if we assume the truth and completeness of quantum mechanics (or even its empirical adequacy and completeness), we will be able to prove that no explanation of it could have independent evidence in its favor. And if we assume that quantum mechanics is not only true and complete but also ultimately so, we will be able to prove that no explanation of it (independently confirmable or not) will be adequate. These various assumptions are no part of quantum mechanics; rather, they are philosophical assertions about quantum mechanics. Insofar as the various proofs rest upon assumptions like these (I do not know whether they do), they are not so much proofs as philosophical arguments, and pretty questionable ones to boot. Finally, the issue of determinism, important though it is in other contexts, is something of a red herring in this context. It is true that some hidden-variable theorists wanted a deterministic explanation of quantum mechanics. But there is no *a priori* reason why a deeper explanation of quantum mechanics has to be deterministic.

Van Fraassen says nothing to impugn a modest realist demand for non-ultimate explanation. And realists can defend such a demand by pointing to cases where the attempt to explain, even to explain theories regarded as empirically adequate, has paid off handsomely. Van Fraassen is not impressed with the argument:

> Paid off handsomely, how? Paid off in new theories we have more reason to believe empirically adequate. But in that case even the anti-realist, when asked questions about *methodology* will *ex cathedra* counsel the search for explanation! We might even suggest a loyalty oath for scientists, if realism is so efficacious. (p. 93)

Realists might retort that explanation has a payoff in terms of *understanding* the world—but that is unlikely to impress van Fraassen. And realists who are also empiricists are impaled on the horns of a dilemma here (as the case of the hidden-variable theories suggested). Realists who are also empiricists will want any proposed explanation to yield empirical regularities other than those it was devised to explain; otherwise there could be no independent evidence for the truth of the explanation. If an explanation does yield them, then the constructive empiricist can value it, too, but for its predictive rather than its explanatory power. If an explanation does not yield them, then it should be rejected as mere "metaphysical baggage." Heads constructive empiricism wins, tails realism loses:

> I think we must conclude that science, in contrast to scientific realism, does not place an overriding value on explanation in the absence of any gain for empirical results. . . . the point is that the true demand on science is not for

explanation *as such*, but for imaginative pictures which have a hope of sug-gesting new statements of observable regularities and of correcting old ones. (p. 34) [1077]

This true demand is not, it seems, to be vouchsafed to scientists them-selves. They are to take an oath of loyalty to realism, the desire to understand the world, and the search for explanatory truths. Realism is the constructive empiricist's Noble Lie, propounded *ex cathedra* in case scientists should find the true aim (enhancing the empirical adequacy of "imaginative pic-tures") uninspiring! More seriously, is van Fraassen right to say that "the interpretation of science, and the correct view of its methodology, are two separate topics" (p. 93)? I think it preferable to have an interpretation of sci-ence which harmonizes with methodological pronouncements.

At any rate, this is part of what van Fraassen means when he calls explanation a *pragmatic* virtue. The search for explanation *works* because theories which are good explainers will *ipso facto* be good savers of phe-nomena and the real "name of the game is saving the phenomena" (p. 93). But this is not all that he means. The rest is meant to undercut the simple realist idea that adequate explanations must contain theories that are true (or nearly so). Van Fraassen defends a "pragmatic" analysis of explanation according to which theories do not figure in explanations at all but some-how lie behind or underpin them. And he says that good explanations can be underpinned by empirically adequate theories just as well as true ones. At least, I think that is what is going on.

Charles Morris divided the study of language into syntax, semantics, and pragmatics. The last was meant to deal, among other things, with *con-text dependence*, as when the truth or falsity of "I'm hungry" depends upon the context of utterance, upon who says it and when. In philosophical circles, "pragmatic" (or better "merely pragmatic") has also come to mean "useful but not true." There is no obvious connection between these two philosophi-cal usages. Utterances can express both truths and falsehoods in virtue of contextual factors. And an utterance can be useful but not true whether or not contextual factors enter into it. ("He went that-away" may be useful for diverting the pursuer though it is false; it is also heavily context-dependent. "John Brown took the road to California" may be similarly useful but false, and it is less context-dependent.)

Van Fraassen thinks that explanation is also heavily context-dependent. He begins from Bromberger's puzzle about explanatory asymmetries: the height of the flagpole explains the length of its shadow but not vice versa, though the two deductions may be structurally identical. (Actually, van Fraassen also tells a blue story which is meant to show—I do not think it does show it—that there are "contexts" in which the length of the shadow does explain the height of the flagpole.) The obvious solution to this puzzle is an appeal to causality: explanations exhibit causes, while nonexplanatory deduc-tions do not. But this is to jump out of the contextual frying pan into the contextual fire, for which factor is picked out as "the cause" also varies enormously with the context:

> [T]he salient feature picked out as "the cause" . . . is salient to a given person because of his orientation, his interests, and various other peculiarities in the way he . . . comes to know the problem—contextual factors. (p. 125)

Now, John Stuart Mill, who first drew attention to this kind of thing, insisted that only the entire constellation of factors, amounting to a sufficient condition for the event to be explained, is really entitled to be called the cause of it. And John Mackie, with Mill in spirit though more amenable to ordinary ways of talking, said that a cause is an insufficient but necessary part of an unnecessary but sufficient condition. Both Mill and Mackie can admit that contextual factors may influence *which event a person wants explained*. And this is enough to dispose of many of the usual examples adduced to demonstrate context dependence of the explanations given (car crashes, fires, and such like).[14]

Van Fraassen takes a different course: he accepts the context dependence of explanations and tries to make it more precise. An explanation "is an *answer* . . . to a why-question" (p. 134). Every why-question has a *topic* (if we ask, "Why P?" the topic is P), an implied *contrast-class* (what we actually ask is, "Why P rather than Q, R etc.?"), and an implied relation of *explanatory-relevance* which determines what shall count as a possible answer to the question (pp. 142–43). The topic, contrast class, and relation of explanatory relevance all depend upon the context, in particular upon "a certain body K of accepted background theory and factual information," which in turn depends "on who the questioner and his audience are" (p. 145). This gets complicated. The upshot of it is that

> the discussion of explanation went wrong at the very beginning when explanation was conceived of as a relationship like description: a relation between theory and fact. Really it is a three-term relation, between theory, fact, and context. . . . So to say that a given theory can be used to explain a certain fact, is always elliptic for: there is a proposition which is a telling answer, relative to this theory, to the request for information about certain facts (those counted as relevant for *this* question) that bear on a comparison between this fact which is the case, and certain (contextually specified) alternatives which are not the case. (p. 156)

Is it *really* elliptic for all this? At the outset of his discussion, in order to combat "the increasing sense of unreality" the usual examples bring, van Fraassen sets forth three "workaday examples of scientific explanation" (pp. 101–3). And he manages to set these examples forth *without mentioning contextual factors at all*. There are why-questions all right, but there are no implied contrast classes, relevance relations, or anything else which depends "on who the questioner and his audience are." Contextual complications have little to do with explanation in science, if van Fraassen's own "workaday examples" are anything to go by.

Is it true to say that explanation was ever "conceived of as a relationship like description: a relation between theory and fact"? The orthodox account says that explanations are arguments in which three things figure: theories or general laws, initial conditions specifying the cause of the event being explained, and in the conclusion a statement of the event being explained. Explanations *contain* descriptions, but they are not *like* them.

Van Fraassen (his earlier examples notwithstanding) wants to drop the theories out of explanations and relegate them to the "context," to the "background information" relative to which why-questions are asked and answered. (Here I was reminded of the Wittgensteinians, who assimilate theories to rules of inference and insist that these rules do not figure as premises in the inferences constructed in accordance with them.)[15] Still, one might think, the theories must be true if the explanations proffered in the light of them (or constructed in accordance with them) are to be correct. Van Fraassen thinks not. Empirically adequate and empirically strong theories will do just as well as true ones. His argument seems to be as follows. The fact to be explained is always an observable fact. The facts cited in explanation of it are also always observable facts. So what the theory has to get right, to be a good explainer, are just the observable facts. And empirically adequate theories, by definition, do just that:

> So scientific explanation is not (pure) science but an application of science. It is a use of science to satisfy certain of our desires; and these desires are quite specific in a specific context, but they are always desires for *descriptive information* . . . in each case, a success of explanation is a success of *adequate and informative description* [*of the phenomena*]. And while it is true that we seek for explanation, the value of this search for science is that the search for explanation is *ipso facto* a search for empirically adequate, empirically strong theories. (pp. 156–57)[16]

I am not sure that I have got the argument right here—but, if I have, there is lots wrong with it. Sometimes we explain observable facts by citing other observable facts (and laws). But this is not always the case, though it tends to be the case in the usual philosopher's examples, which bring an "increasing sense of unreality" to the subject. The flagpole is a good example again: both its height and the length of its shadow are presumably observable facts. (I can scarcely bring myself to mention the "explanation" of why some bird is black, which consists in pointing out that it's a raven and they are all black!) But, in van Fraassen's own examples of scientific explanations, there are initial conditions such as: the specific heats of water and copper are 1 and 0.1, respectively; the earth's magnetic field at a certain point has a vertical component of approximately $5/10^5$ Tesla; the energy levels associated with stable electron orbits in hydrogen atoms take the form $E_n = -E_0/n^2$ where E_0 is called the ground state energy. These initial conditions are *generalized* ones because the facts to be explained are *general* ones (with the possible exception of the second). But, setting aside the problem of their generality, they do not look much like observable facts, and their provision does not look much like providing descriptions of the observable phenomena. Nor, in science, is it always observable phenomena that we try to explain: we sometimes try to explain theories.

Van Fraassen not only thinks that explanation is a pragmatic affair in Morris's sense (context dependence), he also thinks that explanatory power is one of the *pragmatic virtues*, concerning which he says in general:

> In so far as they go beyond consistency, empirical adequacy, and empirical strength, they do not concern the relation between the theory and the world, but rather the use and usefulness of the theory; they provide reasons to prefer the theory independently of questions of truth. (p. 88)

If what I have said about scientific explanation is right, then the explanatory power of a scientific theory does depend on whether it tells the truth about the unobservable and therefore does go beyond empirical adequacy and empirical strength. But van Fraassen is obviously right when he says:

> Nor can there be any question of explanatory success as providing evidence for the truth of a theory that goes beyond any evidence we have for its providing an adequate description of the phenomena. (pp. 156–57)

This is obviously right, because empirical adequacy is *defined* as correctness so far as the observable evidence is concerned. Explaining things cannot provide a special sort of *evidence* that theories are true rather than just empirically adequate. Realists, made of sterner stuff than constructive empiricists, still demand that a theory be true for the explanations in which it figures to be adequate. And realism carries this much metaphysical baggage: realists can point to no evidence over and above evidence of empirical adequacy that their sterner requirement has been met.

But there is excess baggage of a different kind in the constructive empiricist position. There is, above all, the philosophical excess baggage of defending an observable/unobservable distinction and giving it crucial epistemological significance. There is the excess baggage of providing an alternative to the obvious realist explanation of science's novel predictive success. And there is the excess baggage of a complex account of the pragmatics of explanation.

I suggested earlier that, in comparing constructive empiricism with scientific realism, we should assess the risks, penalties, and gains associated with each. The risks have been discussed, as have the penalties in the form of philosophical "excess baggage" of various kinds. As to the respective gains (or losses), I can only repeat a hackneyed point. The realist values theoretical science as an attempt to *understand* the world and sees continuity between commonsense and scientific knowledge. The constructive empiricist, browbeaten as much by the positivist emphasis on prediction as by esoteric problems in interpreting quantum theory realistically, jettisons understanding and seeks to drive a wedge between theoretical science and commonsense (taking his excess baggage aboard to do so).

Let me conclude by agreeing with van Fraassen that it would be a pity if scientific realism were to become a philosophical dogma. Bas van Fraassen's book certainly roused me from any dogmatic slumber to which I might have been prone. It has, in fact, given me sleepless nights! His antirealism is more viable than earlier antirealist positions. But, in philosophy of science as well as in science, viability directly depends on weakness. Constructive empiricism is weaker than earlier antirealist views in all kinds of ways, and correspondingly closer to realism. This is why I conclude, undogmatically I hope, that realism emerges a little bloodied but unbowed from its encounter with constructive empiricism.[17]

■ | Notes

1. *The Scientific Image* (Oxford: Clarendon Press, 1980), 8 [1062]. Henceforth, all page numbers in the text refer to this book. [The numbers in square brackets refer to pages of the excerpt from this book included in this chapter.]

2. Here and in what follows I ignore, through lack of space, a central feature of van Fraassen's position, his preference for a semantic approach to scientific theories whereby they emerge as sets of models rather than as sets of (true or false) sentences. I have two excuses for this. First, in much of his own discussion van Fraassen ignores it too, and talks as though theories consisted of true or false sentences. Second, and more important, I think that there is little to choose between the two approaches from a logical point of view. As van Fraassen himself once wrote, "There are natural interrelations between the two approaches: an axiomatic theory may be characterized by the class of interpretations which satisfy it, and an interpretation may be characterized by the set of sentences which it satisfies. . . . These interrelations make implausible any claim of superiority for either approach" ("On the Extension of Beth's Semantics of Physical Theories," *Philosophy of Science* 37 [Sept. 1970]: 325–39; cf. p. 326). I am indebted, both for the general point and for the reference to van Fraassen's endorsement of it, to John Worrall's review article of *The Scientific Image*, ["An Unreal Image,"] *British Journal for the Philosophy of Science* [35 (1984): 65–80].

3. Here I assume that Berkeley's immaterialism is empirically equivalent to commonsense realism. I am not sure that this is so. Immaterialism can be formulated so as to be consistent with all possible experience and so irrefutable by it. But empirical adequacy should require more than more consistency with the evidence; it should require (at least) that the theory in question entail the evidence. It might be argued that commonsense realism entails consequences about the stability and

reobservability of physical objects which Berkeley's immaterialism does not. Berkeley does invoke God's benevolence to "explain" *post factum* the stability of tables and trees. But Berkeley cannot predict it because of his admission that God might always make an exception to his "laws of nature" and work a miracle instead.

4. Einstein was always appealing to simplicity or unity. For an analysis of how these vague appeals were articulated into quite powerful principles of theory construction, see E. Zahar, "Why Did Einstein's Programme Supersede Lorentz's (II)?" *British Journal for the Philosophy of Science* 24 (1973): 223–62.

Philosophers of science still, of course, lack a precise and general account of what simplicity is; perhaps there is none to be had and simplicity is, as van Fraassen says, a patchwork of virtues, some pragmatic and some not. Popper's identification of simplicity and strength works nicely sometimes ("All swans are white" is simpler and stronger than "All non-Australasian swans are white") and badly at other times ("All swans are white and ferocious" is less simple and stronger than "All swans are white").

5. For more detail of how certain metaphysical principles can be rationally assessed in this way, see Watkins, "Confirmable and Influential Metaphysics," *Mind* 67 (July 1958): 344–65, especially pp. 363–65. Watkins does not apply these ideas to principles of simplicity.

6. See K. R. Popper, *Conjectures and Refutations* (London: Routledge and Kegan Paul, 1963), 274–76.

7. Here, incidentally, there is a curious prejudice in favor of vision. True, I have not *seen* the mouse, but have I not *heard* it, and is this not a way of observing it? There is a curious tension in the view that, though we can see (and touch) things, we never hear (or taste or smell) things but only the *noises* they make (or the tastes and smells they emit). Notoriously, one way to resolve the tension is to say that we never really see things, either, but only visions (visual sense-data) caused by them (and the same will have to go, even more implausibly, for touch). That way leads to idealism. Realists resolve the tension by saying that we can hear, taste, and smell things as well as see and touch them. Van Fraassen, for all his talk about hearing "an apparent sign of mousely presence" rather than a mouse, is once again with the realists. He says that "sense-data, I am sure, do not exist" (p. 72). And he has to be with the realists if truth and empirical adequacy are to be the same so far as observable things are concerned. If all observable phenomena were only apparent signs of mousely presence (mousely visions, mousely noises, mousely smells, etc.), then "All observable phenomena are as if there is a mouse in the wainscoting" would not entail "There is a mouse in the wainscoting," contrary to what van Fraassen says on p. 21 [1072].

8. In case anyone is reminded here of the talk of "observables" in quantum mechanics, we should remind ourselves that the so-called "observables" of quantum mechanics are in the present context remotely calculable theoretical quantities. If electrons are not observable, neither is their charge, momentum, or spin.

9. P. Duhem, *The Aim and Structure of Physical Theory* (Princeton, N.J.: Princeton University Press, 1954), 28 (see also pp. 297ff).

10. For example, Laudan presents as historical counterexamples successful theories which did not genuinely refer and were not true or nearly true ("A Confutation of Convergent Realism," *Philosophy of Science* 48, no. 1 [March 1981], 19–49). But

few, arguably none, of the theories cited had any *novel* predictive success. Laudan also saddles the realist with the principle that successful reference alone breeds success. I do not know if any realist has thought this, but no realist should think it. For, as Laudan shows, we can construct referring theories which will be quite unsuccessful: take a successful theory containing the term *t* and negate it. Successful reference is a necessary condition for truth (or near truth) but not a sufficient one. And when it comes to success, it is the truth (or near truth) of what a theory says about its theoretical entities which counts, not whether those entities exist. [Laudan's paper follows next in this chapter.]

11. Further detail can be found in my "Explanation, Description and Scientific Realism," *Scientia* 112 (1977), 727–55.

12. *Principia*, Book 3, General Scholium; Motte's translation, revised by Cajori, vol. 2 (Berkeley and Los Angeles: University of California Press, 1962), 546–47. Italics mine. Also important is the famous passage where Newton says he will treat forces "not physically but mathematically" (*Principia*, Book 1, Definition 8; Cajori, vol. 1, 5–6). Newton is saying that he will describe how gravity works in precise mathematical terms, rather than try to explain it physically. He is often misinterpreted as saying that gravity does not really exist.

13. *Opticks*, Book 3, Query 31; Dover edition (New York, 1952), 401–2. The most extended defense of Newton and attack on Cartesian essentialism is, of course, Roger Cotes's preface to the second edition of the *Principia*. Yet Cotes is sometimes interpreted as defending the essentialist view that gravity is, after all that Newton had said to the contrary, essential to matter (for example, by Popper in *Conjectures and Refutations*, 106). It is unlikely that Newton would have allowed Cotes to defend a view he himself had specifically denied. Cotes himself specifically denied that this was the view he was defending, in a letter to Samuel Clarke, who had questioned him on the point (see Cajori's appendix, n. 6; vol. 2, 634–35).

14. This is shown in John Worrall's review article, referred to in note 2 above. Worrall also shows that we can dispose in the same way of van Fraassen's story about the length of a flagpole's shadow explaining its height.

15. For a discussion of this view, see my "Wittgensteinian Instrumentalism," *Theoria* 46 (1980), pts. 2–3, 65–105.

16. The italics in this quotation are mine. And I felt justified in adding "of the phenomena" to "description" because in the preceding sentence (not quoted here) the phrase "adequate description of the phenomena" occurs.

17. Previously published in a shorter form as "Constructive Empiricism versus Scientific Realism," *Philosophical Quarterly* 32 (July 1982): 262–71. Reproduced by permission. I am grateful to Cliff Hooker for his helpful suggestions about what I might focus attention upon in the paper and to Greg Currie, Bob Durrant, and Martin Fricke for comments on earlier versions.

Larry Laudan

A Confutation of
Convergent Realism

The positive argument for realism is that it is the only philosophy that doesn't make the success of science a miracle.

—H. Putnam (1975)

I | The Problem

It is becoming increasingly common to suggest that epistemological realism is an empirical hypothesis, grounded in, and to be authenticated by its ability to explain the workings of science. A growing number of philosophers (including Boyd, Newton-Smith, Shimony, Putnam, Friedman, and Niiniluoto) have argued that the theses of epistemic realism are open to empirical test. The suggestion that epistemological doctrines have much the same empirical status as the sciences is a welcome one: for, whether it stands up to detailed scrutiny or not, it marks a significant facing-up by the philosophical community to one of the most neglected (and most notorious) problems of philosophy: the status of epistemological claims.

But there are potential hazards as well as advantages associated with the 'scientizing' of epistemology. Specifically, once one concedes that epistemic doctrines are to be tested in the court of experience, it is possible that one's favorite epistemic theories may be refuted rather than confirmed. It is the thesis of this paper that precisely such a fate afflicts a form of realism advocated by those who have been in the vanguard of the move to show that realism is supported by an empirical study of the development of science. Specifically, I shall show that epistemic realism, at least in certain of its extant forms, is neither supported by, nor has it made sense of, much of the available historical evidence.

From *Philosophy of Science* 48 (1981): 19–49.

2 | Convergent Realism

Like other philosophical *-isms*, the term 'realism' covers a variety of sins. Many of these will not be at issue here. For instance, 'semantic realism' (in brief, the claim that all theories have truth values and that some theories—we know not which—are true) is not in dispute. Nor shall I discuss what one might call 'intentional realism' (i.e., the view that theories are generally intended by their proponents to assert the existence of entities corresponding to the terms in those theories). What I shall focus on instead are certain forms of *epistemological* realism. As Hilary Putnam has pointed out, although such realism has become increasingly fashionable, "very little is said about what realism *is*" (1978). The lack of specificity about what realism asserts makes it difficult to evaluate its claims, since many formulations are too vague and sketchy to get a grip on. At the same time, any efforts to formulate the realist position with greater precision lay the critic open to charges of attacking a straw man. In the course of this paper, I shall attribute several theses to the realists. Although there is probably no realist who subscribes to all of them, most of them have been defended by some self-avowed realist or other; taken together, they are perhaps closest to that version of realism advocated by Putnam, Boyd, and Newton-Smith. Although I believe the views I shall be discussing can be legitimately attributed to certain contemporary philosophers (and will frequently cite the textual evidence for such attributions), it is not crucial to my case that such attributions can be made. Nor will I claim to do justice to the complex epistemologies of those whose work I will criticize. My aim, rather, is to explore certain epistemic claims which those who are realists might be tempted (and in some cases have been tempted) to embrace. If my arguments are sound, we will discover that some of the most intuitively tempting versions of realism prove to be chimeras.

The form of realism I shall discuss involves variants of the following claims:

R1 Scientific theories (at least in the 'mature' sciences) are typically approximately true and more recent theories are closer to the truth than older theories in the same domain.

R2 The observational and theoretical terms within the theories of a mature science genuinely refer (roughly, there are substances in the world that correspond to the ontologies presumed by our best theories).

R3 Successive theories in any mature science will be such that they 'preserve' the theoretical relations and the apparent referents of earlier theories (i.e., earlier theories will be 'limiting cases' of later theories).[1]

R4 Acceptable new theories do and should explain why their predecessors were successful insofar as they were successful.

To these semantic, methodological and epistemic theses is conjoined an important meta-philosophical claim about how realism is to be evaluated and assessed. Specifically, it is maintained that:

> **R5** Theses (R1)–(R4) entail that ('mature') scientific theories should be successful; indeed, these theses constitute the best, if not the only, explanation for the success of science. The empirical success of science (in the sense of giving detailed explanations and accurate predictions) accordingly provides striking empirical confirmation for realism.

I shall call the position delineated by (R1) to (R5) *convergent epistemological realism*, or CER for short. Many recent proponents of CER maintain that (R1), (R2), (R3), and (R4) are empirical hypotheses which, via the linkages postulated in (R5), can be tested by an investigation of science itself. They propose two elaborate abductive [inference-to-the-best-explanation] arguments. The structure of the first, which is germane to (R1) and (R2), is something like this:

Argument I

1 If scientific theories are approximately true, they will typically be empirically successful.

2 If the central terms in scientific theories genuinely refer, those theories will generally be empirically successful.

3 Scientific theories are empirically successful.

4 (Probably) Theories are approximately true and their terms genuinely refer.

The argument relevant to (R3) is of slightly different form, specifically:

Argument II

1 If the earlier theories in a 'mature' science are approximately true and if the central terms of those theories genuinely refer, then later more successful theories in the same science will preserve the earlier theories as limiting cases.

2 Scientists seek to preserve earlier theories as limiting cases and generally succeed.

3 (Probably) Earlier theories in a 'mature' science are approximately true and genuinely referential.

Taking the success of present and past theories as givens, proponents of CER claim that *if* CER were true, it would follow that the success and the progressive success of science would be a matter of course. Equally, they allege that if CER were false, the success of science would be 'miraculous' and without explanation.[2] Because (on their view) CER explains the fact that science is successful, the theses of CER are thereby confirmed by the success of science and non-realist epistemologies are discredited by the latter's alleged inability to explain both the success of current theories and the progress which science historically exhibits.

As Putnam and certain others (e.g., Newton-Smith) see it, the fact that statements about reference (R2, R3) or about approximate truth (R1, R3) function in the explanation of a contingent state of affairs, establishes that "the notions of 'truth' and 'reference' have a causal explanatory role in epistemology" (Putnam 1978, p. 21).[3] In one fell swoop, both epistemology and semantics are 'naturalized' and, to top it all off, we get an explanation of the success of science into the bargain!

The central question before us is whether the realist's assertions about the interrelations between truth, reference and success are sound. It will be the burden of this paper to raise doubts about both I and II. Specifically, I shall argue that *four* of the five premises of those abductions are either false or too ambiguous to be acceptable. I shall also seek to show that, even if the premises were true, they would not warrant the conclusions which realists draw from them. . . .

3 | Reference and Success

The specifically referential side of the 'empirical' argument for realism has been developed chiefly by Putnam, who talks explicitly of reference rather more than most realists. On the other hand, reference is usually implicitly smuggled in, since most realists subscribe to the (ultimately referential) thesis that "the world probably contains entities very like those postulated by our most successful theories."

If R2 is to fulfill Putnam's ambition that reference can explain the success of science, and that the success of science establishes the presumptive truth of R2, it seems he must subscribe to claims similar to these:

S1 The theories in the advanced or mature sciences are successful.

S2 A theory whose central terms genuinely refer will be a successful theory.

S3 If a theory is successful, we can reasonably infer that its central terms genuinely refer.

S4 All the central terms in theories in the mature sciences do refer.

There are complex interconnections here. (S2) and (S4) explain (S1), while (S1) and (S3) provide the warrant for (S4). Reference explains success and success warrants a presumption of reference. The arguments are plausible, given the premises. But there is the rub, for with the possible exception of (S1), none of the premises is acceptable.

The first and toughest nut to crack involves getting clearer about the nature of that 'success' which realists are concerned to explain. Although Putnam, Sellars, and Boyd all take the success of certain sciences as a given, they say little about what this success amounts to. So far as I can see, they are working with a largely *pragmatic* notion to be cashed out in terms of a theory's workability or applicability. On this account, we would say that a theory is successful if it makes substantially correct predictions, if it leads to efficacious interventions in the natural order, if it passes a battery of standard tests. One would like to be able to be more specific about what success amounts to, but the lack of a coherent theory of confirmation makes further specificity very difficult.

Moreover, the realist must be wary—at least for these purposes—of adopting too strict a notion of success, for a highly robust and stringent construal of 'success' would defeat the realist's purposes. What he wants to explain, after all, is why science in general has worked so well. If he were to adopt a very demanding characterization of success (such as those advocated by inductive logicians or Popperians) then it would probably turn out that science has been largely 'unsuccessful' (because it does not have high confirmation) and the realist's avowed explanandum would thus be a nonproblem. Accordingly, I shall assume that a theory is 'successful' so long as it has worked well, i.e., so long as it has functioned in a variety of explanatory contexts, has led to confirmed predictions and has been of broad explanatory scope. As I understand the realist's position, his concern is to explain why certain theories have enjoyed this kind of success.

If we construe 'success' in this way, (S1) can be conceded. Whether one's criterion of success is broad explanatory scope, possession of a large number of confirming instances, or conferring manipulative or predictive control, it is clear that science is, by and large, a successful activity.

What about (S2)? I am not certain that any realist would or should endorse it, although it is a perfectly natural construal of the realist's claim that 'reference explains success'. The notion of reference that is involved here is highly complex and unsatisfactory in significant respects. Without endorsing it, I shall use it frequently in the ensuing discussion. The realist sense of reference is a rather liberal one, according to which the terms in a theory may be genuinely referring even if many of the claims the theory makes about the entities to which it refers are false. Provided that there are entities which "approximately fit" a theory's description of them, Putnam's charitable account of reference allows us to say that the terms of a theory genuinely refer.[4] On this account (and these are Putnam's examples), Bohr's 'electron', Newton's 'mass', Mendel's 'gene', and Dalton's 'atom' are all referring terms, while 'phlogiston' and 'aether' are not (Putnam 1978, pp. 20–22).

Are genuinely referential theories (i.e., theories whose central terms genuinely refer) invariably or even generally successful at the empirical level, as (S2) states? There is ample evidence that they are not. The chemical atomic theory in the 18th century was so remarkably unsuccessful that most chemists abandoned it in favor of a more phenomenological, elective affinity chemistry. The Proutian theory that the atoms of heavy elements are composed of hydrogen atoms had, through most of the 19th century, a strikingly unsuccessful career, confronted by a long string of apparent refutations. The Wegenerian theory that the continents are carried by large subterranean objects moving laterally across the earth's surface was, for some thirty years in the recent history of geology, a strikingly unsuccessful theory until, after major modifications, it became the geological orthodoxy of the 1960s and 1970s. Yet all of these theories postulated basic entities which (according to Putnam's 'principle of charity') genuinely exist.

The realist's claim that we should expect referring theories to be empirically successful is simply false. And, with a little reflection, we can see good reasons why it should be. To have a genuinely referring theory is to have a theory which "cuts the world at its joints", a theory which postulates entities of a kind that really exist. But a genuinely referring theory need not be such that all—or even most—of the specific claims it makes about the properties of those entities and their modes of interaction are true. Thus, Dalton's theory makes many claims about atoms which are false; Bohr's early theory of the electron was similarly flawed in important respects. Contra-(S2), genuinely referential theories need not be strikingly successful, since such theories may be 'massively false' (i.e., have far greater falsity content than truth content).

(S2) is so patently false that it is difficult to imagine that the realist need be committed to it. But what else will do? The (Putnamian) realist wants attributions of reference to a theory's terms to function in an explanation of that theory's success. The simplest and crudest way of doing that involves a claim like (S2). A less outrageous way of achieving the same end would involve the weaker,

> S2′ A theory whose terms refer will usually (but not always) be successful.

Isolated instances of referring but unsuccessful theories, sufficient to refute (S2), leave (S2′) unscathed. But, if we were to find a broad range of referring but unsuccessful theories, that would be evidence against (S2′). Such theories can be generated at will. For instance, take any set of terms which one believes to be genuinely referring. In any language rich enough to contain negation, it will be possible to construct indefinitely many unsuccessful theories, all of whose substantive terms are genuinely referring. Now, it is always open to the realist to claim that such 'theories' are not really theories at all, but mere conjunctions of isolated statements—lacking that sort of conceptual integration we associate with 'real' theories. Sadly a parallel argument can be made for genuine theories. Consider, for instance, how many inadequate versions of the atomic theory there were in the 2000 years of atomic 'speculating', before a genuinely successful theory emerged. Consider how many unsuccessful versions there were of the wave theory of light before the 1820s, when a successful wave theory first emerged. Kinetic theories of heat in the seventeenth and eighteenth centuries, developmental theories of embryology before the late nineteenth century sustain a similar story. (S2'), every bit as much as (S2), seems hard to reconcile with the historical record.

As Richard Burian has pointed out to me (in personal communication), a realist might attempt to dispense with both of those theses and simply rest content with (S3) alone. Unlike (S2) and (S2'), (S3) is not open to the objection that referring theories are often unsuccessful, for it makes no claim that referring theories are always or generally successful. But (S3) has difficulties of its own. In the first place, it seems hard to square with the fact that the central terms of many relatively successful theories (e.g., aether theories, phlogistic theories) are evidently non-referring. I shall discuss this tension in detail below. More crucial for our purposes here is that (S3) *is not strong enough* to permit the realist to utilize reference to explain success. Unless genuineness of reference entails that all or most referring theories will be successful, then the fact that a theory's terms refer scarcely provides a convincing explanation of that theory's success. If, as (S3) allows, many (or even most) referring theories can be unsuccessful, how can the fact that a successful theory's terms refer be taken to explain why it is successful? (S3) may or may not be true; but in either case it arguably gives the realist no explanatory access to scientific success.

A more plausible construal of Putnam's claim that reference plays a role in explaining the success of science involves a rather more indirect argument. It might be said (and Putnam does say this much) that we can explain why a theory is successful by assuming that the theory is true or approximately true. Since a theory can only be true or nearly true (in any sense of those terms open to the realist) if its terms genuinely refer, it might be argued that reference gets into the act willy-nilly when we explain a theory's success in terms of its truth(like) status. On this account, reference is piggy-backed on approximate truth. The viability of this indirect approach is treated at length in section 4 below so I shall not discuss it here except to observe that if the only contact point between reference and success is provided through the medium of approximate truth, then the link between reference and success is extremely tenuous.

What about (S3), the realist's claim that success creates a rational presumption of reference? We have already seen that (S3) provides no explanation of the success of science, but does it have independent merits? The question specifically is whether the success of a theory provides a warrant for concluding that its central terms refer. Insofar as this is—as certain realists suggest—an empirical question, it requires us to inquire whether past theories which have been successful are ones whose central terms genuinely referred (according to the realist's own account of reference).

A proper empirical test of this hypothesis would require extensive sifting of the historical record of a kind that is not possible to perform here. What I can do is to mention a range of once successful, but (by present lights) non-referring, theories. A fuller list will come later (see section 5), but for now we shall focus on a whole family of related theories, namely, the subtle fluids and aethers of 18th– and 19th–century physics and chemistry.

Consider specifically the state of aetherial theories in the 1830s and 1840s. The electrical fluid, a substance which was generally assumed to accumulate on the surface rather than permeate the interstices of bodies, had been utilized to explain *inter alia* the attraction of oppositely charged bodies, the behavior of the Leyden jar, the similarities between atmospheric and static electricity and many phenomena of current electricity. Within chemistry and heat theory, the caloric aether had been widely utilized since Boerhaave (by, among others, Lavoisier, Laplace, Black, Rumford, Hutton, and Cavendish) to explain everything from the role of heat in chemical reactions to the conduction and radiation of heat and several standard problems of thermometry. Within the theory of light, the optical aether functioned centrally in explanations of reflection, refraction, interference, double refraction, diffraction, and polarization. (Of more than passing interest, optical aether theories had also made some very startling predictions, e.g., Fresnel's prediction of a bright spot at the center of the shadow of a circular disc; a surprising prediction which, when tested, proved correct. If that does not count as empirical success, nothing does!) There were also gravitational (e.g., LeSage's) and physiological (e.g., Hartley's) aethers which enjoyed some measure of empirical success. It would be difficult to find a family of theories in this period which were as successful as aether theories; compared to them, 19th–century atomism (for instance), a genuinely referring theory (on realist accounts), was a dismal failure. Indeed, on any account of empirical success which I can conceive of, non-referring 19th-century aether theories were more successful than contemporary, referring atomic theories. In this connection, it is worth recalling the remark of the great theoretical physicist, J. C. Maxwell, to the effect that the aether was better confirmed than any other theoretical entity in natural philosophy!

What we are confronted by in 19th-century aether theories, then, is a wide variety of once successful theories, whose central explanatory concept Putnam singles out as a prime example of a non-referring one (Putnam 1978, p. 22). What are (referential) realists to make of this historical case? On the face of it, this case poses two rather different kinds of challenges to realism: (1) it suggests that (S3) is a dubious piece of advice in that *there can be* (and have been) *highly successful theories some central terms of which are non-referring*; and (2) it suggests that *the realist's claim that he can explain why science is successful is false at least insofar as a part of the historical success of science has been success exhibited by theories whose central terms did not refer.*

But perhaps I am being less than fair when I suggest that the realist is committed to the claim that *all* the central terms in a successful theory refer. It is possible that when Putnam, for instance, says that "terms in a mature [or successful] science typically refer" (Putnam 1978, p. 20), he only means to suggest that *some* terms in a successful theory or science genuinely refer. Such a claim is fully consistent with the fact that certain other terms (e.g., 'aether') in certain successful, mature sciences (e.g., 19th-century physics) are nonetheless non-referring. Put differently, the realist might argue that the success of a theory warrants the claim that at least some (but not necessarily all) of its central concepts refer.

Unfortunately, such a weakening of (S3) entails a theory of evidential support which can scarcely give comfort to the realist. After all, part of what separates the realist from the positivist is the former's belief that the evidence for a theory is evidence for *everything* which the theory asserts. Where the stereotypical positivist argues that the evidence selectively confirms only the more 'observable' parts of a theory, the realist generally asserts (in the language of Boyd) that:

> the sort of evidence which ordinarily counts in favor of the acceptance of a scientific law or theory is, ordinarily, evidence for the (at least approximate) truth of the law or theory as an account of the causal relations obtaining between the entities ["observational or theoretical"] quantified over in the law or theory in question. (Boyd 1973, p. 1)[5]

For realists such as Boyd, either all parts of a theory (both observational and non-observational) are confirmed by successful tests or none are. In general, realists have been able to utilize various holistic arguments to insist that it is not merely the lower-level claims of a well-tested theory which are confirmed but its deep-structural assumptions as well. This tactic has been used to good effect by realists in establishing that inductive support 'flows upward' so as to authenticate the most 'theoretical' parts of our theories. Certain latter-day realists (e.g., Glymour) want to break out of this holist web and argue that certain components of theories can be 'directly' tested. This approach runs the very grave risk of undercutting what the realist desires most: a rationale for taking our deepest-structure theories seriously, and a justification for linking reference and success. After all, if the tests to which we subject our theories only test *portions* of those theories, then even highly successful theories may well have central terms which are non-referring and central tenets which, because untested, we have no grounds for believing to be approximately true. Under those circumstances, a theory might be highly successful and yet contain important constituents which were patently false. Such a state of affairs would wreak havoc with the realist's presumption (R1) that success betokens approximate truth. In short, to be less than a holist about theory testing is to put at risk precisely that predilection for deep-structure claims which motivates much of the realist enterprise.

There is, however, a rather more serious obstacle to this weakening of referential realism. It is true that by weakening (S3) to only certain terms in a theory, one would immunize it from certain obvious counterexamples. But such a maneuver has debilitating consequences for other central realist theses. Consider the realist's thesis (R3) about the retentive character of inter-theory relations (discussed below in detail). The realist both recommends as a matter of policy and claims as a matter of fact that successful theories are (and should be) rationally replaced only by theories which preserve reference for the central terms of their successful predecessors. The rationale for the normative version of this retentionist doctrine is that the terms in the earlier theory, *because it was successful, must* have been referential and thus a constraint on any successor to that theory is that reference should be retained for such terms. This makes sense just in case success provides a blanket warrant for presumption of reference. But if (S3) were weakened so as to say merely that it is reasonable to assume that *some* of the terms in a successful theory genuinely refer, then the realist would have no rationale for his retentive theses (variants of R3), which have been a central pillar of realism for several decades.[6]

Something apparently has to give. A version of (S3) strong enough to license (R3) seems incompatible with the fact that many successful theories contain non-referring central terms. But any weakening of (S3) dilutes the force of, and removes the rationale for, the realist's claims about convergence, retention and correspondence in inter-theory relations.[7] If the realist once concedes that some unspecified set of the terms of a successful theory may well not refer, then his proposals for restricting "the class of candidate theories" to those which retain reference for the *prima facie* referring terms in earlier theories is without foundation (Putnam 1975, p. 22).

More generally, we seem forced to say that such linkages as there are between reference and success are rather murkier than Putnam's and Boyd's discussions would lead us to believe. If the realist is going to make his case for CER, it seems that it will have to hinge on approximate truth, (R1), rather than reference, (R2).

4 | Approximate Truth and Success: The 'Downward Path'

Ignoring the referential turn among certain recent realists, most realists continue to argue that, at bottom, epistemic realism is committed to the view that successful scientific theories, even if strictly false, are nonetheless 'approximately true' or 'close to the truth' or 'verisimilar'.[8] The claim generally amounts to this pair:

T1 if a theory is approximately true, then it will be explanatorily successful; and

T2 if a theory is explanatorily successful, then it is probably approximately true.

What the realist would *like* to be able to say, of course, is:

> T1' if a theory is true, then it will be successful.

(T1') is attractive because self-evident. But most realists balk at invoking (T1') because they are (rightly) reluctant to believe that we can reasonably presume of any given scientific theory that it is true. If all the realist could explain was the success of theories which were true *simpliciter* [unconditionally], his explanatory repertoire would be acutely limited. As an attractive move in the direction of broader explanatory scope, (T1) is rather more appealing. After all, presumably many theories which we believe to be false (e.g., Newtonian mechanics, thermodynamics, wave optics) were—and still are—highly successful across a broad range of applications.

Perhaps, the realist evidently conjectures, we can find an *epistemic* account of that pragmatic success by assuming such theories to be 'approximately true'. But we must be wary of this potential sleight of hand. It may be that there is a connection between success and approximate truth; *but if there is such a connection it must be independently argued for.* The acknowledgedly uncontroversial character of (T1') must not be surreptitiously invoked—as it sometimes seems to be—in order to establish (T1). When (T1')'s antecedent is appropriately weakened by speaking of approximate truth, it is by no means clear that (T1) is sound.

Virtually all the proponents of epistemic realism take it as unproblematic that if a theory were approximately true, it would deductively follow that the theory would be a relatively successful predictor and explainer of observable phenomena. Unfortunately, few of the writers of whom I am aware have defined what it means for a statement or theory to be 'approximately true'. Accordingly, it is impossible to say whether the alleged entailment is genuine. This reservation is more than perfunctory. Indeed, on the best known account of what it means for a theory to be approximately true, it does *not* follow that an approximately true theory will be explanatorily successful.

Suppose, for instance, that we were to say in a Popperian vein that a theory, T_1, is approximately true if its truth content is greater than its falsity content, i.e.,

$$Ct_T(T_1) \gg Ct_F(T_1).[9]$$

Where $Ct_T(T_1)$ is the cardinality of the set of true sentences entailed by T_1 and $Ct_F(T_1)$ is the cardinality of the set of false sentences entailed by T_1. When approximate truth is so construed, it does *not* logically follow that an arbitrarily selected class of a theory's entailments (namely, some of its observable consequences) will be true. Indeed, it is entirely conceivable that a theory might be approximately true in the indicated sense and yet be such that *all* of its thus far tested consequences are *false*.[10]

Some realists concede their failure to articulate a coherent notion of approximate truth or verisimilitude, but insist that this failure in no way compromises the viability of (T1). Newton-Smith, for instance, grants that "no one has given a satisfactory analysis of the notion of verisimilitude" (Newton-Smith 1981, p. 197), but insists that the concept can be legitimately invoked "even if one cannot at the time give a philosophically satisfactory analysis of it." He quite rightly points out that many scientific concepts were explanatorily useful long before a philosophically coherent analysis was given for them. But the analogy is unseemly, for what is being challenged is not whether the concept of approximate truth is philosophically rigorous but rather whether it is even clear enough for us to ascertain whether it entails what it purportedly explains. Until someone provides a clearer analysis of approximate truth than is now available, it is not even clear whether truth-likeness would explain success, let alone whether, as Newton-Smith insists, "the concept of verisimilitude is *required* in order to give a satisfactory theoretical explanation of an aspect of the scientific enterprise." If the realist would de-mystify the 'miraculousness' (Putnam) or the 'mysteriousness' (Newton-Smith[11]) of the success of science, he needs more than a promissory note that somehow, someday, someone will show that approximately true theories must be successful theories.[12]

Whether there is some definition of approximate truth which does indeed entail that approximately true theories will be predictively successful (and yet still probably false) is not clear.[13] What can be said is that, promises to the contrary notwithstanding, *none* of the proponents of realism has yet articulated a coherent account of approximate truth which entails that approximately true theories will, across the range where we can test them, be successful predictors. Further difficulties abound. Even if the realist had a semantically adequate characterization of approximate or partial truth, and even if that semantics entailed that most of the consequences of an approximately true theory would be true, he would still be without any criterion that would *epistemically* warrant the ascription of approximate truth to a theory. As it is, the realist seems to be long on intuitions and short on either a semantics or an epistemology of approximate truth.

These should be urgent items on the realists' agenda since, until we have a coherent account of what approximate truth is, central realist theses like (R1), (T1), and (T2) are just so much mumbo-jumbo.

5 | Approximate Truth and Success: The 'Upward Path'

Despite the doubts voiced in section 4, let us grant for the sake of argument that if a theory is approximately true, then it will be successful. Even granting (T1), is there any plausibility to the suggestion of (T2) that explanatory success can be taken as a rational warrant for a judgment of approximate truth? The answer seems to be "no".

To see why, we need to explore briefly one of the connections between 'genuinely referring' and being 'approximately true'. However the latter is understood, I take it that *a realist would never want to say that a theory was approximately true if its central theoretical terms failed to refer.* If there were nothing like genes, then a genetic theory, no matter how well confirmed it was, would not be approximately true. If there were no entities similar to atoms, no atomic theory could be approximately true; if there were no sub-atomic particles, then no quantum theory of chemistry could be approximately true. In short, a necessary condition—especially for a scientific realist—for a theory being close to the truth is that its central explanatory terms genuinely refer. (An *instrumentalist*, of course, could countenance the weaker claim that a theory was approximately true so long as its directly testable consequences were close to the observable values. But as I argued above, the realist must take claims about approximate truth to refer alike to the observable and the deep-structural dimensions of a theory.)

Now, what the history of science offers us is a plethora of theories which were both successful and (so far as we can judge) non-referential with respect to many of their central explanatory concepts. I discussed earlier one specific family of theories which fits this description. Let me add a few more prominent examples to the list:

- the crystalline spheres of ancient and medieval astronomy;
- the humoral theory of medicine;
- the effluvial theory of static electricity;
- 'catastrophist' geology, with its commitment to a universal (Noachian) deluge;
- the phlogiston theory of chemistry;
- the caloric theory of heat;
- the vibratory theory of heat;
- the vital force theories of physiology;
- the electromagnetic aether;
- the optical aether;
- the theory of circular inertia;
- theories of spontaneous generation.

This list, which could be extended *ad nauseam*, involves in every case a theory which was once successful and well confirmed, but which contained central terms which (we now believe) were non-referring. Anyone who imagines that the theories which have been successful in the history of science have also been, with respect to their central concepts, genuinely referring theories has studied only the more 'whiggish' versions of the history of science (i.e., the ones which recount only those past theories which are referentially similar to currently prevailing ones).

It is true that proponents of CER sometimes hedge their bets by suggesting that their analysis applies exclusively to 'the mature sciences' (e.g.,

Putnam and Krajewski). This distinction between mature and immature sciences proves convenient to the realist since he can use it to dismiss any *prima facie* counter-example to the empirical claims of CER on the grounds that the example is drawn from an 'immature' science. But this insulating maneuvre is unsatisfactory in two respects. In the first place, it runs the risk of making CER vacuous since these authors generally define a mature science as one in which correspondence or limiting case relations obtain invariably between any successive theories in the science once it has passed 'the threshold of maturity'. Krajewski grants the tautological character of this view when he notes that "the thesis that there is [correspondence] among successive theories becomes, indeed, analytical" (1977, p. 91). Nonetheless, he believes that there is a version of the maturity thesis which "may be and must be tested by the history of science". That version is that "every branch of science crosses at some period the threshold of maturity". But the testability of this hypothesis is dubious at best. There is no historical observation which could conceivably *refute* it since, even if we discovered that no sciences yet possessed 'corresponding' theories, it could be maintained that eventually every science will become corresponding. It is equally difficult to *confirm* it since, even if we found a science in which corresponding relations existed between the latest theory and its predecessor, we would have no way of knowing whether that relation will continue to apply to subsequent changes of theory in that science. In other words, the much-vaunted empirical testability of realism is seriously compromised by limiting it to the mature sciences.

But there is a second unsavory dimension to the restriction of CER to the 'mature' sciences. The realists' avowed aim, after all, is to explain why science is successful: that is the 'miracle' which they allege the non-realists leave unaccounted for. The fact of the matter is that parts of science, including many 'immature' sciences, have been successful for a very long time; indeed, many of the theories I alluded to above were empirically successful by any criterion I can conceive of (including fertility, intuitively high confirmation, successful prediction, etc.). If the realist restricts himself to explaining only how the 'mature' sciences work (and recall that very few sciences indeed are yet 'mature' as the realist sees it), then he will have completely failed in his ambition to explain why science in general is successful. Moreover, several of the examples I have cited above come from the history of mathematical physics in the [19th] century (e.g., the electromagnetic and optical aethers) and, as Putnam himself concedes, "*physics* surely counts as a 'mature' science if any science does" (1978, p. 21). Since realists would presumably insist that many of the central terms of the theories enumerated above do not genuinely refer, it follows that none of those theories could be approximately true (recalling that the former is a necessary condition for the latter). Accordingly, cases of this kind cast very grave doubts on the plausibility of (T2), i.e., the claim that nothing succeeds like approximate truth.

I daresay that for every highly successful theory in the past of science which we now believe to be a genuinely referring theory, one could find half a dozen once successful theories which we now regard as substantially non-referring. If the proponents of CER are the empiricists they profess to be about matters epistemological, cases of this kind and this frequency should give them pause about the well-foundedness of (T2).

But we need not limit our counter-examples to non-referring theories. There were many theories in the past which (so far as we can tell) were both genuinely referring and empirically successful which we are nonetheless loath to regard as approximately true. Consider, for instance, virtually all those geological theories prior to the 1960s which denied any lateral motion to the continents. Such theories were, by any standard, highly successful (and apparently referential); but would anyone today be prepared to say that their constituent theoretical claims—committed as they were to laterally stable continents—are almost true? Is it not the fact of the matter that structural geology was a successful science between (say) 1920 and 1960, even though geologists were fundamentally mistaken about many—perhaps even most—of the basic mechanisms of tectonic construction? Or what about the chemical theories of the 1920s which assumed that the atomic nucleus was structurally homogenous? Or those chemical and physical theories of the late 19th century which explicitly assumed that matter was neither created nor destroyed? I am aware of no sense of approximate truth (available to the realist) according to which such highly successful, but evidently false, theoretical assumptions could be regarded as 'truthlike'.

More generally, the realist needs a riposte to the *prima facie* plausible claim that there is no necessary connection between increasing the accuracy of our deep-structural characterizations of nature and improvements at the level of phenomenological explanations, predictions and manipulations. It *seems* entirely conceivable intuitively that the theoretical mechanisms of a new theory, T_2, might be closer to the mark than those of a rival T_1 and yet T_1 might be more accurate at the level of testable predictions. In the absence of an argument that greater correspondence at the level of unobservable claims is more likely than not to reveal itself in greater accuracy at the experimental level, one is obliged to say that the realist's hunch that increasing deep-structural fidelity must manifest itself pragmatically in the form of heightened experimental accuracy has yet to be made cogent. (Equally problematic, of course, is the inverse argument to the effect that increasing experimental accuracy betokens greater truthlikeness at the level of theoretical, i.e., deep-structural, commitments.)* . . .

---

* The section that follows in Laudan's original paper—"6. Confusions about Convergence and Retention"—has been omitted. In it, Laudan criticizes abductive arguments for realism of sort II, namely those that conclude that (probably) earlier theories in a mature science are approximately true and their terms refer because later, more successful theories in the science preserve those earlier theories as lim-

7 | The Realists' Ultimate 'Petitio Principii'

It is time to step back a moment from the details of the realists' argument to look at its general strategy. Fundamentally, the realist is utilizing, as we have seen, an abductive inference which proceeds from the success of science to the conclusion that science is approximately true, verisimilar, or referential (or any combination of these). This argument is meant to show the sceptic that theories are not ill-gotten, the positivist that theories are not reducible to their observational consequences, and the pragmatist that classical epistemic categories (e.g., 'truth', 'falsehood') are a relevant part of meta-scientific discourse.

It is little short of remarkable that realists would imagine that their critics would find the argument compelling. As I have shown elsewhere (1978), ever since antiquity critics of epistemic realism have based their scepticism upon a deep-rooted conviction that the fallacy of affirming the consequent is indeed fallacious. When Sextus or Bellarmine or Hume doubted that certain theories which saved the phenomena were warrantable as true, their doubts were based on a belief that the exhibition that a theory had some true consequences left entirely open the truth-status of the theory. Indeed, many non-realists have been non-realists precisely because they believed that false theories, as well as true ones, could have true consequences.

Now enters the new breed of realist (e.g., Putnam, Boyd, and Newton-Smith) who wants to argue that epistemic realism can reasonably be presumed to be true by virtue of the fact that it has true consequences. But this is a monumental case of begging the question. The non-realist refuses to admit that a *scientific* theory can be warrantedly judged to be true simply because it has some true consequences. Such non-realists are not likely to be impressed by the claim that a *philosophical* theory like realism can be warranted as true because it arguably has some true consequences. If non-realists are chary about first-order abductions to avowedly true conclusions, they are not likely to be impressed by second-order abductions, particularly when, as I have tried to show above, the premises and conclusions are so indeterminate.

But, it might be argued, the realist is not out to convert the intransigent sceptic or the determined instrumentalist.[14] He is perhaps seeking, rather, to show that realism can be tested like any other scientific hypothesis, and that realism is at least as well confirmed as some of our best scientific theories. Such an analysis, however plausible initially, will not stand up to scrutiny. I am aware of no realist who is willing to say that a *scientific* theory can be reasonably presumed to be true or even regarded as well confirmed just on the strength of the fact that its thus far tested consequences are true.

---

iting cases. Martin Carrier defends a retention condition as a necessary component of scientific realism in "What is Right with the Miracle Argument: Establishing a Taxonomy of Natural Kinds" (later in this chapter).

Realists have long been in the forefront of those opposed to *ad hoc* and *post hoc* theories. Before a realist accepts a scientific hypothesis, he generally wants to know whether it has explained or predicted more than it was devised to explain; he wants to know whether it has been subjected to a battery of controlled tests; whether it has successfully made novel predictions; whether there is independent evidence for it.

What, then, of realism itself as a 'scientific' hypothesis?[15] Even if we grant (contrary to what I argued in section 4) that realism entails and thus explains the success of science, ought that (hypothetical) success warrant, by the realist's own construal of scientific acceptability, the acceptance of realism? Since realism was devised in order to explain the success of science, it remains purely *ad hoc* with respect to that success. If realism has made some novel predictions or been subjected to carefully controlled tests, one does not learn about it from the literature of contemporary realism. At the risk of apparent inconsistency, the realist repudiates the instrumentalist's view that saving the phenomena is a significant form of evidential support while endorsing realism itself on the transparently instrumentalist grounds that it is confirmed by those very facts it was invented to explain. No proponent of realism has sought to show that realism satisfies those stringent empirical demands which the realist himself minimally insists on when appraising scientific theories. The latter-day realist often calls realism a 'scientific' or 'well-tested' hypothesis, but seems curiously reluctant to subject it to those controls which he otherwise takes to be a *sine qua non* for empirical well-foundedness.

8 | Conclusion

The arguments and cases discussed above seem to warrant the following conclusions:

1 The fact that a theory's central terms refer does not entail that it will be successful; and a theory's success is no warrant for the claim that all or most of its central terms refer.

2 The notion of approximate truth is presently too vague to permit one to judge whether a theory consisting entirely of approximately true laws would be empirically successful; what is clear is that a theory may be empirically successful even if it is not approximately true.

3 Realists have no explanation whatever for the fact that many theories which are not approximately true and whose 'theoretical' terms seemingly do not refer are nonetheless often successful.

4 The convergentist's assertion that scientists in a 'mature' discipline usually preserve, or seek to preserve, the laws and mechanisms of earlier theories in later ones is probably false; his assertion that when

such laws are preserved in a successful successor, we can explain the success of the latter by virtue of the truthlikeness of the preserved laws and mechanisms, suffers from all the defects noted above confronting approximate truth.

5 Even if it could be shown that referring theories and approximately true theories would be successful, the realists' argument that successful theories are approximately true and genuinely referential takes for granted precisely what the non-realist denies (namely, that explanatory success betokens truth).

6 It is not clear that acceptable theories either *do* or *should* explain why their predecessors succeeded or failed. If a theory is better supported than its rivals and predecessors, then it is not epistemically decisive whether it explains why its rivals worked.

7 If a theory has once been falsified, it is unreasonable to expect that a successor should retain either all of its content *or* its confirmed consequences *or* its theoretical mechanisms.

8 Nowhere has the realist established—except by fiat—that non-realist epistemologists lack the resources to explain the success of science.

With these specific conclusions in mind, we can proceed to a more global one: it is not yet established—Putnam, Newton-Smith and Boyd notwithstanding—that realism can explain *any* part of the success of science. What is very clear is that realism *cannot*, even by its own lights, explain the success of those many theories whose central terms have evidently not referred and whose theoretical laws and mechanisms were not approximately true. The inescapable conclusion is that insofar as many realists are concerned with explaining how science works and with assessing the adequacy of their epistemology by that standard, they have thus far failed to explain very much. Their epistemology is confronted by anomalies which seem beyond its resources to grapple with.

It is important to guard against a possible misinterpretation of this essay. *Nothing* I have said here refutes the possibility in principle of a realistic epistemology of science. To conclude as much would be to fall prey to the same inferential prematurity with which many realists have rejected in principle the possibility of explaining science in a non-realist way. My task here is, rather, that of reminding ourselves that there *is* a difference between wanting to believe something and having good reasons for believing it. All of us would like realism to be true; we would like to think that science works because it has got a grip on how things really are. But such claims have yet to be made out. Given the *present* state of the art, it can only be wish fulfilment that gives rise to the claim that realism, and realism alone, explains why science works.[16]

■ | Notes

1. Putnam, evidently following Boyd, sums up (R1) to (R3) in these words:

"1) Terms in a mature science typically *refer*.

2) The laws of a theory belonging to a mature science are typically approximately true . . . I will only consider [new] theories . . . which have this property—[they] contain the [theoretical] laws of [their predecessors] as a limiting case" (1978, pp. 20–21).

2. Putnam insists, for instance, that if the realist is wrong about theories being referential, then "the success of science [is] a miracle" (Putnam 1975, p. 73).

3. Boyd remarks: "scientific realism offers an *explanation* for the legitimacy of ontological commitment to theoretical entities" (Putnam 1978, note 2, p. 20). It allegedly does so by explaining why theories containing theoretical entities work so well: because such entities genuinely exist. [The quotation is from Boyd's (unpublished) "Realism and Scientific Epistemology."]

4. Whether one utilizes Putnam's earlier or later versions of realism is irrelevant for the central arguments of this essay.

5. See also p. 3: "experimental evidence for a theory is evidence for the truth of even its non-observational laws." See also (Sellars 1963, p. 97).

6. A caveat is in order here. *Even if* all the central terms in some theory refer, it is not obvious that every rational successor to that theory must preserve all the referring terms of its predecessor. One can easily imagine circumstances when the new theory is preferable to the old one even though the range of application of the new theory is less broad than the old. When the range is so restricted, it may well be entirely appropriate to drop reference to some of the entities which figured in the earlier theory.

7. For Putnam and Boyd both "it will be a *constraint* on T_2 [i.e., any new theory in a domain] . . . that T_2 must have this property, the property that *from its standpoint* one can assign referents to the terms of T_1 [i.e., an earlier theory in the same domain]" (Putnam 1978, p. 22). For Boyd, see (1973, p. 8): "new theories should, *prima facie*, resemble current theories with respect to their accounts of causal relations among theoretical entities."

8. For just a small sampling of this view, consider the following: "The claim of a realist ontology of science is that the only way of explaining why the models of science function so successfully . . . is that they approximate in some way the structure of the object" (McMullin 1970, pp. 63–64); "the continued success [of confirmed theories] can be *explained* by the hypothesis that they are in fact close to the truth . . ." (Niiniluoto 1980, p. 448); the claim that "the laws of a theory belonging to a mature science are typically approximately *true* . . . [provides] an *explanation* of the behavior of scientists and the success of science" (Putnam 1978, pp. 20–21). Smart, Sellars, and Newton-Smith, among others, share a similar view.

9. Although Popper is generally careful not to assert that actual historical theories exhibit ever increasing truth content (for an exception, see his (1963, p. 220)), other writers have been more bold. Thus, Newton-Smith writes that "the historically generated sequence of theories of a mature science is a sequence in which succeed-

ing theories are increasing in truth content without increasing in falsity content."
[See Newton-Smith 1981, p. 184. Laudan is quoting from a draft of Newton-Smith's
book, the wording of which differs from the published version.]

10. On the more technical side, Niiniluoto has shown that a theory's degree of
corroboration co-varies with its "estimated verisimilitude" (1977, pp. 121–147 and
1980). Roughly speaking, 'estimated truthlikeness' is a measure of how closely
(the content of) a theory corresponds to *what we take to be* the best conceptual
systems that we so far have been able to find (1980, pp. 443ff.). If Niiniluoto's
measures work, it follows from the above-mentioned co-variance that an empirically successful theory will have a high degree of estimated truthlikeness. But
because estimated truthlikeness and genuine verisimilitude are not necessarily
related (the former being parasitic on existing evidence and available conceptual
systems), it is an open question whether—as Niiniluoto asserts—the continued
success of highly confirmed theories can be *explained* by the hypothesis that they
in fact are close to the truth at least in the relevant respects. Unless I am mistaken, this remark of his betrays a confusion between 'true verisimilitude' (to
which we have no epistemic access) and 'estimated verisimilitude' (which is accessible but non-epistemic).

11. Newton-Smith claims that the increasing predictive success of science through
time "would be totally mystifying . . . if it were not for the fact that theories are
capturing more and more truth about the world" (Newton-Smith 1981, p. 196).

12. I must stress again that I am *not* denying that there *may* be a connection
between approximate truth and predictive success. I am only observing that until
the realists show us what that connection is, they should be more reticent than they
are about claiming that realism can explain the success of science.

13. A *non-realist* might argue that a theory is approximately true just in case all its
observable consequences are true or within a specified interval from the true value.
Theories that were "approximately true" in this sense would indeed be demonstrably successful. But the realist's (otherwise commendable) commitment to taking
seriously the theoretical claims of a theory precludes him from utilizing any such
construal of approximate truth, since he wants to say that the theoretical as well as
the observational consequences are approximately true.

14. I owe the suggestion of this realist response to Andrew Lugg.

15. I find Putnam's views on the 'empirical' or 'scientific' character of realism
rather perplexing. At some points, he seems to suggest that realism is both empirical and scientific. Thus, he writes: "If realism is an explanation of this fact [namely,
that science is successful], realism must itself be an over-arching scientific *hypothesis*" (1978, p. 19). Since Putnam clearly maintains the antecedent, he seems committed to the consequent. Elsewhere he refers to certain realist tenets as being "our
highest level empirical generalizations about knowledge" (p. 37). He says moreover
that realism "could be false", and that "facts are relevant to its support (or to criticizing it)" (pp. 78–79). Nonetheless, for reasons he has not made clear, Putnam wants to
deny that realism is either scientific or a hypothesis (p. 79). How realism can consist
of doctrines which 1) explain facts about the world, 2) are empirical generalizations
about knowledge, and 3) can be confirmed or falsified by evidence and yet be neither scientific nor hypothetical is left opaque.

16. I am indebted to all of the following for clarifying my ideas on these issues and for saving me from some serious errors: Peter Achinstein, Richard Burian, Clark Glymour, Adolf Grünbaum, Gary Gutting, Allen Janis, Lorenz Krüger, James Lennox, Andrew Lugg, Peter Machamer, Nancy Maull, Ernan McMullin, Ilkka Niiniluoto, Nicholas Rescher, Ken Schaffner, John Worrall, Stephen Wykstra.

■ | References

Boyd, R. (1973), "Realism, Underdetermination, and a Causal Theory of Evidence," Noûs 7: 1–12.

Krajewski, W. (1977), Correspondence Principle and Growth of Science. Dordrecht: Reidel.

Laudan, L. (1978), "Ex-Huming Hacking," Erkenntnis 13: 417–35.

McMullin, Ernan (1970), "The History and Philosophy of Science: A Taxonomy," Stuewer, R. (ed.), Minnesota Studies in the Philosophy of Science V: 12–67. Minneapolis: University of Minnesota Press.

Newton-Smith, W. (1981), The Rationality of Science. London: Routledge & Kegan Paul.

Niiniluoto, Ilkka (1977), "On the Truthlikeness of Generalizations," Butts, R. and Hintikka, J. (eds.), Basic Problems in Methodology and Linguistics: 121–147. Dordrecht: Reidel.

Niiniluoto, Ilkka (1980), "Scientific Progress," Synthese 45: 427–62.

Popper, K. (1963), Conjectures and Refutations. London: Routledge & Kegan Paul.

Putnam, H. (1975), Mathematics, Matter and Method, Vol. I. Cambridge: Cambridge University Press.

Putnam, H. (1978), Meaning and the Moral Sciences. London: Routledge & Kegan Paul.

Sellars, W. (1963), Science, Perception and Reality. New York: The Humanities Press.

Juha T. Saatsi

On the Pessimistic Induction
and Two Fallacies

1 | Introduction

Probably the best known argument against scientific realism is the argument from Pessimistic Induction (Poincaré 1952; Putnam 1978; Laudan 1981). This argument in some form or another has been part and parcel of the realism debate for quite some time now. It is therefore interesting to come across two recent papers which both claim that the argument in its best known form is actually fallacious (Lange 2002; Lewis 2001). Here I want to re-establish the dignity of the Pessimistic Induction by calling to mind the basic objective of the argument, and hence restore the propriety of the realist program of responding to this argument by undermining one or another of its premises.

I take the *Pessimistic (Meta-) Induction* (PMI) against scientific realism to be in essence the argument employed by Laudan (1981). Laudan appeals to an historical record of successful yet false theories to argue against the connection that realists like to draw between successfulness of a theory and its approximate truth—the connection that a successful theory is deemed probably approximately true. This connection is at the heart of the realist's *No Miracles Argument* (NMA), the intuition being that the best explanation of the success of science is the approximate truth of its theories. PMI was devised—in the hands of Laudan, at least—to deliver a lethal blow to NMA. Hence the exact content of PMI is connected in a subtle way to our understanding of NMA, and the latter must be kept firmly in mind in considering the validity of the former.

Laudan's version of PMI can be succinctly reconstructed as the following reductio (Lewis 2001, 373; Psillos 1996), which we may call [PMI]:

From *Philosophy of Science* 72 (2005): 1088–98.

1 Assume that success of a theory is a reliable test for its truth.

2 So most current successful scientific theories are true.

3 Then most past scientific theories are false, since they differ from current successful theories in significant ways.

4 Many of these past theories were also successful.

5 So successfulness of a theory is not a reliable test for its truth (since this leads to contradiction in (3) and (4)).

A typical realist response to this reductio can take issue with, for example, the implicit premise of step (3) by describing (usually via careful case studies) some theoretical elements solely responsible for the successfulness of past theories in a way that renders these theories continuous with otherwise incompatible current theories, and hence candidates for approximate truth in some suitable, restricted sense (e.g., Psillos 1999). I am personally very optimistic about such a line of response, but the purpose of this paper is not to question the premises of Laudan's argument. Here my sole purpose is to stand up for the dignity of such premise defeating work against two lines of thought that claim to remove the antirealist threat of PMI by denying the validity of the argument to begin with.

2 | Lange's Turnover Fallacy

Lange (2002) presents the *turnover fallacy* as a potential source of invalidity of pessimistic inductions in general (and not just of PMI against the realist). The basic idea of this fallacy can be conveyed by the following example:

Assume there is a board of directors comprised of ten members and that you are introduced as a new member to this board replacing someone else. You are told that the company is in turmoil: there has been a change in the assemblage of the board 240 times in the past ten years, but you don't know who's been sitting in the board for how long. You pessimistically infer, inductively, that someone is going to be replaced again very soon. It could be you or it could be someone else for all you know.

You might be tempted to pessimistically infer that the probability of most of you getting the boot within a year, say, is quite high. But this would be to commit the turnover fallacy! For it could be that nine out of ten members of the board have actually sat in throughout the past ten years and it is only your 'predecessors', as it were, who came and went. Just by knowing *the number* of personnel changes in the board does not allow you to inductively infer anything about the probability for *any one particular* individual to get replaced—all you can infer is the high probability for *someone* to get replaced.

Now consider the case of scientific PMI. Looking at the set of current, well confirmed, successful theories we may want to ask: "How likely is it that most of these theories will turn out to be false and will be replaced by new theories incompatible with them?" Given a very bad *numerical* historical record of successful yet false theories we may be tempted—vaguely remembering the intuition behind the PMI argument—to answer "Very likely." But this would be to commit the turnover fallacy! For it could be that most of the current theories have been stable throughout the historical record tracking period, and all the numerous theory changes involve the 'predecessors', as it were, of only one current theory.

Although this is a point about a type of induction in general, Lange takes it to be telling against Laudan's argument in particular. The alleged lesson is that to validly infer the wanted conclusion—that most current theories are probably false—one needs to use a premise much stronger than (3) above in an argument of slightly different form.

> [A] pessimistic induction of a somewhat different and less familiar form is made impervious to the turnover fallacy by employing a historical premise that is not cumulative: at most past moments, most of the theories receiving wide acceptance at that moment are false (by current lights). (Lange 2002, 284)

This is significant since the usual premise "that most of the theories that have ever been accepted were false is inevitably more plausible than the needed premise: that at most past moments, most of the theories then accepted were false" (2002, 285). A fallacy is committed, Lange proposes, since a typical statement of PMI (such as Laudan's) only refers to the *number* of past false theories as an inductive basis, and yet draws a conclusion about the high likelihood of any one of our present theories to be found false and replaced in the course of future science.

It must be admitted that Lange makes a fine point about pessimistic inductions in general, but nevertheless it seems that this potential fallacy *cannot* be incorporated against the scientific PMI of interest (that is, Laudan's PMI). Here we need to be more careful about the real objective of the PMI argument—what is the conclusion being inferred exactly? To begin with, note that the conclusion (5) above makes no reference to future times: what will be found false or whether any theory shifts will take place. This argument [PMI] is therefore *not* an argument to the *time-dependent* conclusion that most of our current theories will be most likely found false and will be replaced. Rather, it is an argument to the *timeless* conclusion that "successfulness of a theory is not a reliable test for its truth." As a matter of fact, in this conclusion no reference is made even to the probable falsity of any one theory of the current successful science; this conclusion would indeed hold even if the current theories were all likely to be true! And nonetheless the force of the argument is considerable given the key role of the claimed explanatory connection between success and approximate truth in the realist's game plan. It is interesting to notice that in the literature the term 'Pessimistic Induction', originally coined by Putnam in 1978, is invariably tagged on to the antirealist line of thought the canonical formulation of which is taken as [PMI] (i.e., Laudan's reductio argument as presented above). The failure to properly distinguish Laudan's argument from Putnam's rhetoric is behind Lange's undue optimism to be able to sidestep the antirealist worry about the historical facts of science as they are typically told.

This reading of PMI—viz. merely as something to counter NMA—may feel unintuitively *neutral* to some.[1] One may feel that PMI should have some pessimistic force on its own and not just as a reactive opposition to NMA, and we can indeed discern different levels of pessimism which PMI is sometimes taken to be an argument for. For example, witness Psillos' informal summary of Laudan's argument:

> Therefore, by a simple (meta-)induction on scientific theories, our current successful theories are likely to be false (or, at any rate, are more likely to be false than true), and many or most of the theoretical terms featuring in them will turn out to be nonreferential. (1999, 101)

This sentence perhaps typifies a more customary reading of PMI as entailing the probable falsity of any one of our current theories, and indeed this is the reading that Lange explicitly adopts. Is this reading of the argument, referring to the probable falsity of our current theories, now subject to the *turnover fallacy* as Lange suggests?

I believe not.[2] First of all, we need to notice that this new argument is no longer just the reductio presented above.[3] Rather, we now add to the above reductio a statistical argument along the following lines, call it [PMI*]:

1* Of all the successful theories, current and past, most are taken to be false by the current lights.

2* The current theories are essentially no different from the past successful theories with respect to their "observable" properties. (Viz. properties potentially figuring in the realist's explanatory argument.)

3* Success of a current theory is not a reliable indicator of its truth (by the reductio argument above), and there is no other reliable indicator of truth for the current theories.

4* Therefore any current successful theory is probably false by statistical reasoning.

This argument concludes that any one current successful theory, *ceteris paribus*, is probably false for all we know. The ceteris paribus clause effectively amounts to the premises (2*) and (3*) above: NMA is taken to be indiscriminating so that the current observer is taken to have no advantage over the past observers in evaluating the truthlikeness of a successful theory. Furthermore, this clause should be also taken to rule out all kinds of 'relativisations' of NMA to specific scientific domains: scientific methodologies and mechanisms are taken to be homogeneous across the domains and the competing realist and antirealist arguments apply across the board. I take the content of these premises to be implicit in the standard construal of PMI.

The argument [PMI*] does not fall foul of the turnover fallacy. However, one may be tempted to *further* infer from such probable falsity the high probability of finding a theory false and it getting replaced, but such an inference would go beyond the confines of—and indeed beyond the validity of—this version of pessimistic induction. Hence a *timeless* conclusion (4*) is inferred from timeless premises and no fallacy of turnover is being committed; this fallacy requires a reference to a time dependent property (e.g., getting the boot within the next two weeks) in the conclusion but 'being false' is not such property.[4] And a further argument to the conclusion that false theories *will be* replaced in the course of future science, whilst perhaps not unthinkable, is surely not part and parcel of the contemporary NMA vs. PMI debate.

Moreover, the conclusion of [PMI*] is clearly compatible with the kind of possible (asymmetric) state of affairs that Lange puts forward as problematic. Assume that all theory changes have taken place within just one domain of scientific enquiry, say. It seems, *pace* Lange, that we nonetheless have reason, ceteris paribus, to believe that all domains of enquiry are currently ridden with false theories. This is because the only feature of theories appealed to in NMA is their successfulness and not, say, the duration of their reign. Once the connection between success and truth has been demolished by [PMI], all the current successful theories (including those which we inductively have *no* reason to expect to get replaced) are on a par with all the past successful theories in one big domain of theories most of which are false, and the conclusion (4*) can be drawn. Furthermore, whilst the assumed asymmetric state of affairs undoubtedly begs for *some* explanation, it is not clear that we have any reason to think that *the best* explanation is achieved by hypothesising the stable theories to be true. What the realist needs is an argument to the conclusion that the combination of successfulness *and* long lifespan of a theory is best explained via truthlikeness, or something like that. As far as I know, no such version of NMA has yet been developed. On the other hand, our degree of confidence in realism as a *possible* explanation of the asymmetric state of affairs is significantly lowered by Laudan's PMI and the availability of numerous other explanations, together with the ceteris paribus clause.

One may, of course, have grave doubts about the ceteris paribus clause in the above portrayal of [PMI*], and many realists indeed argue that at least some current successful theories are not on a par with the past theories which are employed as the basis of the statistical inference above. But while this may offer a way to counter this version of PMI, it does so by undermining one significant premise of the argument and not by virtue of showing it to harbour the turnover fallacy.

I prefer to follow Laudan and read the argument as the reductio [PMI]. We should notice that Laudan's argument is a somewhat atypical case of induction. Usually induction is described as an inference from the particular to the general, and it typically concerns states of affairs at future times being inferred from states of affairs at past times. But we have seen that [PMI] is not best characterised in such terms. Rather, [PMI] should be viewed as a reductio of an indiscriminating realist image—a challenge to the realist's beloved connection between success and truth. Even if none of our current theories eventually succumbed to some incompatible successors, the anti-realist could nonetheless appeal to [PMI] as an anti-NMA. To do this, all that is required is a pool of theories all of which are successful at some time or another, yet most of which have turned out to be false.

So perhaps it is better to regard this meta-induction as a *statistical argument* against the realist claim that one 'observable' feature of our theories—successfulness—is a *reliable statistical indicator* of another, 'unobservable' feature of our theories: their truth(likeness). This is exactly what Peter Lewis (2001) does and claims that it falls victim to another kind of fallacy.

3 | Lewis's False Positives Fallacy

Lewis presents an altogether different rationale for regarding PMI thus understood as harbouring a fallacy. For Lewis the problem is that "the premise that many false past theories were successful does not warrant the assertion that success is not a reliable test for truth" (2001, 374). More specifically: the *fallacy of false positives* that Lewis has in mind concerns the reliability of successfulness as an indicator of (approximate) truth. The notion of statistical reliability is usually characterised in statistics literature in terms of the rates of false positives and false negatives: a reliable indicator is one for which "the false positive rate and false negative rate are both sufficiently small, where what counts as sufficiently small is determined by the context" (2001, 374–375). An instance of false positive (negative) indication is, of course, one in which the existence (absence) of an indication fails to reflect the existence (absence) of the indicated. The rate of false positives (negatives) is then calculated as the number of such cases per all negative (positive) cases.

With statistical reliability characterised in these terms Lewis then takes successfulness to be a reliable indicator of the (approximate) truth of a theory T (picked at random out of *all* theories at time t) if and only if the rate of false-yet-successful theories is small and the rate of true-but-unsuccessful theories is small. With this notion of statistical reliability at hand Lewis explains why Laudan's reductio formulation of PMI is a non sequitur:

> At a given time in the past, it may well be that false theories vastly outnumber true theories. In that case, even if only a small proportion of false theories are successful, and even if a large proportion of true theories are successful, the successful false theories may outnumber successful true theories. So the fact that successful false theories outnumber successful true theories at some time does nothing to undermine the reliability of success as a test for truth at *that* time, let alone other times. In other words, the realist can interpret Laudan's

historical cases, not as evidence against the reliability of success as a test for truth, but merely as evidence of the scarcity of true theories in the past. (2001, 377)

And to do otherwise is, Lewis proposes, to commit the fallacy of false positives.

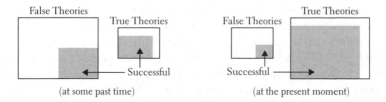

Figure 1. Domains compatible with both statistical reliability and 'bad' historical record. The ratio of successful (gray) and unsuccessful (white) is constant for each domain over time.

The basic intuition behind this argument is made most clear in pictorial terms in Figure 1. We can see immediately that by having a big enough domain of *false and unsuccessful* theories we can satisfy the requirement of statistical reliability even in cases in which, somewhat unintuitively perhaps, the probability of a randomly drawn successful theory to be true is small (less than 0.5, say). At both times pictured the requirement of statistical reliability is satisfied. Furthermore, given that we take *most of our current theories to be successful*, it follows "deductively that most current theories are true, as required by the realist" (2001, 375). This Lewis takes to be a reasonable justification for regarding statistical reliability to be a notion that adequately captures the realist's appeal to the success-versus-truth connection.

So the notion of statistical reliability works for Lewis on the assumption that the statistical reference classes (relative to which statistical reliability is determined) are of the right kind and vary radically as we move from past to current theories: the domain of all theories at some past time t_p must contain a much higher proportion of false and unsuccessful theories than the domain of all current theories. This immediately raises a couple of worries regarding the overall framework in which Lewis casts his realism and allegedly sidesteps the challenge of PMI: (1) How are the crucial reference classes defined in the first place? (2) Has there really been a change in the reference classes such as to enable Lewis's response to PMI to get off the ground?

(1) First of all, it is not at all clear that the notion of reference class of statistical reliability is well-defined in the context of scientific theories. It seems that the relevant domains of *all* true theories and *all* false theories (at some time t) with respect to which the rates of false negatives and false posi-

tives are calculated are not straightforwardly definable in the way a pool of people, say, is readily given in a typical case of medical statistics, for example. Not much has been said in the discussion so far about the putative identity conditions of theories—it just has been surmised that they could in principle be given. But whereas this assumption may be reasonable with respect to both the set of successful theories and the set of true theories, I can make no sense of the idea of delineating a non-arbitrary, well-defined collection of *both false and unsuccessful* theories.

Lewis's realism-friendly scenario which makes Laudan's historical record compatible with success being a reliable statistical indicator depends on there having been a large domain of such false and unsuccessful theories relative to which the rate of false positives is small.[5] But what exactly are the theories which are neither successful nor true? Should we count in only the theory-proposals made by eminent scientists, or perhaps all the proposals actually published in scientific journals, or what? It is easy to imagine a variety of sociological factors, say, yielding scores of unsuccessful and false theories, directly affecting the notion of reliability at stake. But why should we care about *those* theories? It just seems that the debate between NMA and PMI does not involve unsuccessful and false theories (or true yet unsuccessful, for that matter) in anything like the way Lewis projects.

But perhaps a case could be made that the realist should really give us some rough idea of how many false and unsuccessful theories there are per each successful one, given that NMA—being a form of inference to the best explanation—seems to hang on the assumption that this ratio is not high enough to explain away the 'miracle' of successful science by the mere number of trials. But however we decided to delineate the domain of all theories it should not be the case that the realist explanation is held hostage to contingent matters regarding the number of false and unsuccessful theories in the strict manner implied by Lewis's strategy; realism simply cannot depend on the alleged (contingent) fact that most current theories are successful! Rather, it is implicit in the No-Miracles intuition that any feasible fluctuation in the number of false and unsuccessful theories—feasible to science as we know it—is not large enough to overthrow NMA as the *best* explanation around.[6]

(2) So has there been a change in the reference class of the kind that Lewisian realism requires? The idea is that realism only requires that most of our current theories are true which deductively follows, given good statistical reliability of success as an indicator of truth, from the premise that most of our current theories are successful. That is, given any one successful theory—current or past—the best explanation for a Lewisian realist of its successfulness is that *either* it is (approximately) true *or* it is a member of a huge domain of false theories a small portion of which are successful. Regarding past successful science, at least, this is fully amenable to an antirealist reading. To an antirealist like Bas van Fraassen—who persistently denies the force of the No-Miracles Argument—an explanation such as the

above is good enough and fully consonant with his Darwinian selectionist image of science. For van Fraassen, of course, this picture fits the bill with respect to current science just as well; that is, he denies the initial premise of Lewis's that most of our current theories are successful. But the sound-ness of that premise is neither necessary nor sufficient for the realist to make a case against van Fraassen; what is required is NMA as typically understood and the intuition that (approximate) truth is thus connected to successfulness—and for that intuition to have bite is for it to have bite *at all times*, regardless of the number of false and unsuccessful theories present at the time in question.[7]

As a matter of fact, Lewis's unorthodox formulation of the realist posi-tion seems to beg the question against this point to begin with. According to Lewis "convergent realism usually includes the thesis that most of our current theories are true" (2001, 371). But this is certainly an unreason-ably strong thesis for any realist to aspire after: contingent matters regarding the number of false and unsuccessful theories produced by the scientific community depends on factors quite independent from realism and NMA—or so the realist argues—which is why convergence is typically character-ised in terms of increasing level of 'truthlikeness' in a sequence of *successful* theories of cumulative empirical adequacy. Lewis's convergent realist is com-mitted "to the empirical claim that successful theories were rare in the past and are common today" (2001, 377). Such commitment is not generally acknowledged to be part of any contemporary realist position. And it better not be! Keeping in mind how strict a qualification 'successful' can be for the realist and casually glancing through the *Journal of Mathematical Phys-ics*, for example, one is bound to be convinced of the sheer incredibility of this premise upon which realism à la Lewis is erected.

4 | Conclusion

Despite Lange's and Lewis's respective attempts to short circuit the Pessimis-tic Meta-Induction it remains a powerful force to be reckoned with. There is no easy way out for the realist; one or another of the premises must be defeated. To get properly started with this task the realist ought to recognise the variety of forms this intuitively straightforward argument can take when looked at in closer detail. This paper has focused only on how PMI should *not* be understood and much work remains to be done to understand the subtle interplay between PMI and NMA vis-à-vis the notion of success as an indicator of (approximate) truth. To achieve an adequate account of this notion we need to appreciate the timeless character of PMI as a reductio of NMA and not construe the latter in terms of mere statistical reliability.[8]

■ | Notes

I want to thank Angelo Cei, Steven French, and Scott Shalkowski for helpful comments, and Philip Good for sparking the initial interest in the topic.

1. Laudan (1981) does not use the term PMI, but I believe this 'weak' reading of PMI is closest to the use Laudan makes of his pessimistic historical record. This version of the anti-realist's argument is obviously already damaging against the realist, given the respective objectives of the two positions: even if the PMI does not conclude that most current successful theories are probably false, the anti-realist has shown (by undermining NMA) that there is no rationale for taking these theories to be true either, and agnosticism follows. The anti-realist, of course, can be quite happy with this (cf. van Fraassen 1980).

2. This is not to say, of course, that there are no other weaknesses or fallacies that the proposed simplistic classical statistical reasoning may succumb to.

3. The argument is usually presented as a reductio as I have presented it (cf. Laudan 1981; Lewis 2001; Psillos 1999). Lange also refers to Laudan and Psillos in his discussion of the scientific pessimistic induction.

4. Notice that there is a time-dependent part in the above quote from Psillos (1999) invalidly going beyond the confines of PMI. Curiously enough there is no such explicit mistake to be found in Lange's exposition of PMI.

5. Lewis's proposal for testing the history of science for the pessimistic conclusion of PMI in a *valid* way consists of taking "a random sample of theories which are known to be false, and show[ing] that a significant proportion of them are nevertheless successful" (2001, 378). The worry now is that this testing cannot be done since the domain in question is ill-defined.

6. Whether or not this intuition holds is another matter, of course. The point is that Lewis has not only provided a response to PMI but also a particular understanding of realism to go with it. The problems with the former really spring from the inadequacy of the latter.

7. Unless, of course, that 'number' is *so high* as to undermine the credibility of NMA as the best explanation altogether as explained in (1) above. Lewis stresses "the inference that the realist wishes to draw from the success of most *current* theories to their truth" (2001, 378, my italics) but this requires that the realist accounts for some principled difference between the current and the past. And Lewis does not provide such an account.

8. If we insist on casting the debate in such purely statistical terms, then the whole debate threatens to crumble as ultimately nonsensical twiddling of probabilities with undefined and indefinable base rates (cf. Magnus and Callender 2004).

■ | References

Lange, Marc (2002), "Baseball, Pessimistic Inductions and the Turnover Fallacy", *Analysis* 62: 281–285.

Laudan, Larry (1981), "A Confutation of Convergent Realism", *Philosophy of Science* 48: 19–49.

Lewis, Peter (2001), "Why the Pessimistic Induction Is a Fallacy", *Synthese* 129: 371–380.

Magnus, P. D., and Craig Callender (2004), "Realist Ennui and the Base Rate Fallacy", *Philosophy of Science* 71: 320–338.

Poincaré, Henry ([1902] 1952), *Science and Hypothesis*. New York: Dover. Originally published as *La science et l'hypothèse*. Paris: Flammarion.

Psillos, Stathis (1996), "Scientific Realism and the 'Pessimistic Induction'", *Philosophy of Science* 63: S306–S314.

——(1999), *Scientific Realism: How Science Tracks Truth*. London: Routledge.

Putnam, Hilary (1978), *Meaning and the Moral Sciences*. London: Routledge.

van Fraassen, Bas (1980), *The Scientific Image*. Oxford: Oxford University Press.

IAN HACKING

Experimentation and Scientific Realism

Experimental physics provides the strongest evidence for scientific realism. Entities that in principle cannot be observed are regularly manipulated to produce new phenomena and to investigate other aspects of nature. They are tools, instruments not for thinking but for doing.

The philosopher's standard "theoretical entity" is the electron. I shall illustrate how electrons have become experimental entities, or experimenter's entities. In the early stages of our discovery of an entity, we may test hypotheses about it. Then it is merely an hypothetical entity. Much later, if we come to understand some of its causal powers and to use it to build devices that achieve well understood effects in other parts of nature, then it assumes quite a different status.

Discussions about scientific realism or anti-realism usually talk about theories, explanation and prediction. Debates at that level are necessarily inconclusive. Only at the level of experimental practice is scientific realism unavoidable. But this realism is not about theories and truth. The experimentalist need only be a realist about the entities used as tools.

■ | A Plea for Experiments

No field in the philosophy of science is more systematically neglected than experiment. Our grade school teachers may have told us that scientific method is experimental method, but histories of science have become histories of theory. Experiments, the philosophers say, are of value only when they test theory. Experimental work, they imply, has no life of its own. So we lack even a terminology to describe the many varied roles of experiment. Nor has this one-sidedness done theory any good, for radically different

FROM *Philosophical Topics* 13 (1982): 154–72.

types of theory are used to think about the same physical phenomenon (e.g., the magneto-optical effect). The philosophers of theory have not noticed this and so misreport even theoretical inquiry.[1]

Different sciences at different times exhibit different relationships between "theory" and "experiment." One chief role of experiment is the creation of phenomena. Experimenters bring into being phenomena that do not naturally exist in a pure state. These phenomena are the touchstones of physics, the keys to nature and the source of much modern technology. Many are what physicists after the 1870s began to call "effects": the photo-electric effect, the Compton effect, and so forth. A recent high-energy extension of the creation of phenomena is the creation of "events," to use the jargon of the trade. Most of the phenomena, effects and events created by the experimenter are like plutonium: they do not exist in nature except possibly on vanishingly rare occasions.[2]

In this paper I leave aside questions of methodology, history, taxonomy and the purpose of experiment in natural science. I turn to the purely philosophical issue of scientific realism. Call it simply "realism" for short. There are two basic kinds: realism about entities and realism about theories. There is no agreement on the precise definition of either. Realism about theories says we try to form true theories about the world, about the inner constitution of matter and about the outer reaches of space. This realism gets its bite from optimism; we think we can do well in this project, and have already had partial success.

Realism about entities—and I include processes, states, waves, currents, interactions, fields, black holes and the like among entities—asserts the existence of at least some of the entities that are the stock in trade of physics.[3]

The two realisms may seem identical. If you believe a theory, do you not believe in the existence of the entities it speaks about? If you believe in some entities, must you not describe them in some theoretical way that you accept? This seeming identity is illusory. The vast majority of experimental physicists are realists about entities without a commitment to realism about theories. The experimenter is convinced of the existence of plenty of "inferred" and "unobservable" entities. But no one in the lab believes in the literal truth of present theories about those entities. Although various properties are confidently ascribed to electrons, most of these properties can be embedded in plenty of different inconsistent theories about which the experimenter is agnostic. Even people working on adjacent parts of the same large experiment will use different and mutually incompatible accounts of what an electron is. That is because different parts of the experiment will make different uses of electrons, and the models that are useful for making calculations about one use may be completely haywire for another use.

Do I describe a merely sociological fact about experimentalists? It is not surprising, it will be said, that these good practical people are realists. They need that for their own self-esteem. But the self-vindicating realism of experimenters shows nothing about what actually exists in the world. In reply I repeat the distinction between realism about entities and realism about theories and models. Anti-realism about models is perfectly coherent. Many research workers may in fact hope that their theories and even their mathematical models "aim at the truth," but they seldom suppose that any particular model is more than adequate for a purpose. By and large most experimenters seem to be instrumentalists about the models they use. The models are products of the intellect, tools for thinking and calculating. They are essential for writing up grant proposals to obtain further funding. They are rules of thumb used to get things done. Some experimenters are instrumentalists about theories and models, while some are not. That is a sociological fact. But experimenters are realists about the entities that they use in order to investigate other hypotheses or hypothetical entities. That is not a sociological fact. Their enterprise would be incoherent without it. But their enterprise is not incoherent. It persistently creates new phenomena that become regular technology. My task is to show that realism about entities is a necessary condition for the coherence of most experimentation in natural science.

■ | Our Debt to Hilary Putnam

It was once the accepted wisdom that a word like "electron" gets its meaning from its place in a network of sentences that state theoretical laws. Hence arose the infamous problems of incommensurability and theory change. For if a theory is modified, how could a word like "electron" retain its previous meaning? How could different theories about electrons be compared, since the very word "electron" would differ in meaning from theory to theory?

Putnam saves us from such questions by inventing a referential model of meaning. He says that meaning is a vector, refreshingly like a dictionary entry. First comes the syntactic marker (part of speech). Next the semantic marker (general category of thing signified by the word). Then the stereotype (clichés about the natural kind, standard examples of its use and present day associations. The stereotype is subject to change as opinions about the kind are modified). Finally there is the actual reference of the word, the very stuff, or thing, it denotes if it denotes anything. (Evidently dictionaries cannot include this in their entry, but pictorial dictionaries do their best by inserting illustrations whenever possible.)[4]

Putnam thought we can often guess at entities that we do not literally point to. Our initial guesses may be jejune or inept, and not every naming of an invisible thing or stuff pans out. But when it does, and we frame better and better ideas, then Putnam says that although the stereotype changes, we refer to the same kind of thing or stuff all along. We and Dalton alike spoke about the same stuff when we spoke of (inorganic) acids. J. J. Thomson, Lorentz, Bohr and Millikan were, with their different theories and observations, speculating about the same kind of thing, the electron.

There is plenty of unimportant vagueness about when an entity has been successfully "dubbed," as Putnam puts it. "Electron" is the name suggested by G. Johnstone Stoney in 1891 as the name for a natural unit of electricity. He had drawn attention to this unit in 1874. The name was then applied in 1897 by J. J. Thomson to the subatomic particles of negative charge of which cathode rays consist. Was Johnstone Stoney referring to the electron? Putnam's account does not require an unequivocal answer. Standard physics books say that Thomson discovered the electron. For once I might back theory and say Lorentz beat him to it. What Thomson did was to measure the electron. He showed its mass is $\frac{1}{1800}$ that of hydrogen. Hence it is natural to say that Lorentz merely postulated the particle of negative charge, while Thomson, determining its mass, showed that there is some such real stuff beaming off a hot cathode.

The stereotype of the electron has regularly changed, and we have at least two largely incompatible stereotypes, the electron as cloud and the electron as particle. One fundamental enrichment of the idea came in the 1920s. Electrons, it was found, have angular momentum, or "spin." Experimental work by Stern and Gerlach first indicated this, and then Goudsmit and Uhlenbeck provided the theoretical understanding of it in 1925. Whatever we think about Johnstone Stoney, others—Lorentz, Bohr, Thomson and Goudsmit—were all finding out more about the same kind of thing, the electron.

We need not accept the fine points of Putnam's account of reference in order to thank him for providing a new way to talk about meaning. Serious discussions of inferred entities need no longer lock us into pseudo-problems of incommensurability and theory change. Twenty-five years ago the experimenter who believed that electrons exist, without giving much credence to any set of laws about electrons, would have been dismissed as philosophically incoherent. We now realize it was the philosophy that was wrong, not the experimenter. My own relationship to Putnam's account of meaning is like the experimenter's relationship to a theory. I don't literally believe Putnam, but I am happy to employ his account as an alternative to the unpalatable account in fashion some time ago. . . .

▪ | Interfering

Francis Bacon, the first and almost last philosopher of experiments, knew it well: the experimenter sets out "to twist the lion's tail." Experimentation is interference in the course of nature; "nature under constraint and vexed; that is to say, when by art and the hand of man she is forced out of her natural state, and squeezed and moulded."[5] The experimenter is convinced of the reality of entities some of whose causal properties are sufficiently well understood that they can be used to interfere *elsewhere* in nature. One is impressed by entities that one can use to test conjectures about other more hypothetical entities. In my example, one is sure of the electrons that are used to investigate weak neutral currents and neutral bosons. This should not be news, for why else are we (non-sceptics) sure of the reality of even macroscopic objects, but because of what we do with them, what we do to them, and what they do to us?

Interference and intervention are the the stuff of reality. This is true, for example, at the borderline of observability. Too often philosophers imagine that microscopes carry conviction because they help us see better. But that is only part of the story. On the contrary, what counts is what we can do to a specimen under a microscope, and what we can see ourselves doing. We stain the specimen, slice it, inject it, irradiate it, fix it. We examine it using different kinds of microscopes that employ optical systems that rely on almost totally unrelated facts about light. Microscopes carry conviction because of the great array of interactions and interferences that are possible. When we see something that turns out not to be stable under such play, we call it an artefact and say it is not real.[6]

Likewise, as we move down in scale to the truly un-seeable, it is our power to use unobservable entities that make us believe they are there. Yet I blush over these words "see" and "observe." John Dewey would have said that a fascination with seeing-with-the-naked-eye is part of the Spectator Theory of Knowledge that has bedeviled philosophy from earliest times. But I don't think Plato or Locke or anyone before the nineteenth century was as obsessed with the sheer opacity of objects as we have been since. My own obsession with a technology that manipulates objects is, of course, a twentieth-century counterpart to positivism and phenomenology. Their proper rebuttal is not a restriction to a narrower domain of reality, namely to what can be positivistically "seen" (with the eye), but an extension to other modes by which people can extend their consciousness.

■ | Making

Even if experimenters are realists about entities, it does not follow that they are right. Perhaps it is a matter of psychology: the very skills that make for a great experimenter go with a certain cast of mind that objectifies whatever it thinks about. Yet this will not do. The experimenter cheerfully regards neutral bosons as merely hypothetical entities, while electrons are real. What is the difference?

There are an enormous number of ways to make instruments that rely on the causal properties of electrons in order to produce desired effects of unsurpassed precision. I shall illustrate this. The argument—it could be called the experimental argument for realism—is not that we infer the reality of electrons from our success. We do not make the instruments and then infer the reality of the electrons, as when we test a hypothesis, and then believe it because it passed the test. That gets the time-order wrong.

By now we design apparatus relying on a modest number of home truths about electrons to produce some other phenomenon that we wish to investigate.

That may sound as if we believe in the electrons because we predict how our apparatus will behave. That too is misleading. We have a number of general ideas about how to prepare polarized electrons, say. We spend a lot of time building prototypes that don't work. We get rid of innumerable bugs. Often we have to give up and try another approach. Debugging is not a matter of theoretically explaining or predicting what is going wrong. It is partly a matter of getting rid of "noise" in the apparatus. "Noise" often means all the events that are not understood by any theory. The instrument must be able to isolate, physically, the properties of the entities that we wish to use, and damp down all the other effects that might get in our way. *We are completely convinced of the reality of electrons when we regularly set out to build—and often enough succeed in building—new kinds of devices that use various well understood causal properties of electrons to interfere in other more hypothetical parts of nature.*

It is not possible to grasp this without an example. Familiar historical examples have usually become encrusted by false theory-oriented philosophy or history. So I shall take something new. This is a polarizing electron gun whose acronym is PEGGY II. In 1978 it was used in a fundamental experiment that attracted attention even in *The New York Times*. In the next section I describe the point of making PEGGY II. So I have to tell some new physics. You can omit this and read only the engineering section that follows. Yet it must be of interest to know the rather easy-to-understand significance of the main experimental results, namely, (1) parity is not conserved in scattering of polarized electrons from deuterium, and (2) more generally, parity is violated in weak neutral current interactions.[7]

▪ | Methodological Remark

In the following section I retail a little current physics; in the section after that I describe how a machine has been made. It is the latter that matters to my case, not the former. Importantly, even if present quantum electrodynamics turns out to need radical revision, the machine, called PEGGY II, will still work. I am concerned with how it was made to work, and why. I shall sketch far more sheer engineering than is seen in philosophy papers. My reason is that the engineering is incoherent unless electrons are taken for granted. One cannot say this by merely reporting, "Oh, they made an electron gun for shooting polarized electrons." An immense practical knowledge of how to manipulate electrons, of what sorts of things they will do reliably and how they tend to misbehave—that is the kind of knowledge which grounds the experimenter's realism about electrons. You cannot grasp this kind of knowledge in the abstract, for it is practical knowledge. So

I must painfully introduce the reader to some laboratory physics. Luckily it is a lot of fun.

■ | Parity and Weak Neutral Currents

There are four fundamental forces in nature, not necessarily distinct. Gravity and electromagnetism are familiar. Then there are the strong and weak forces, the fulfillment of Newton's program, in the *Optics*, which taught that all nature would be understood by the interaction of particles with various forces that were effective in attraction or repulsion over various different distances (i.e., with different rates of extinction).

Strong forces are 100 times stronger than electromagnetism but act only for a miniscule distance, at most the diameter of a proton. Strong forces act on "hadrons," which include protons, neutrons, and more recent particles, but not electrons or any other members of the class of particles called "leptons."

The weak forces are only $1/10,000$ times as strong as electromagnetism, and act over a distance $1/100$ times smaller than strong forces. But they act on both hadrons and leptons, including electrons. The most familiar example of a weak force may be radioactivity.

The theory that motivates such speculation is quantum electrodynamics. It is incredibly successful, yielding many predictions better than one part in a million, a miracle in experimental physics. It applies over distances ranging from diameters of the earth to $1/100$ the diameter of the proton. This theory supposes that all the forces are "carried" by some sort of particle. Photons do the job in electromagnetism. We hypothesize "gravitons" for gravity.

In the case of interactions involving weak forces, there are charged currents. We postulate that particles called bosons carry these weak forces.[8] For charged currents, the bosons may be positive or negative. In the 1970s there arose the possibility that there could be weak "neutral" currents in which no charge is carried or exchanged. By sheer analogy with the vindicated parts of quantum electrodynamics, neutral bosons were postulated as the carriers in weak interactions.

The most famous discovery of recent high energy physics is the failure of the conservation of parity. Contrary to the expectations of many physicists and philosophers, including Kant,[9] nature makes an absolute distinction between right-handedness and left-handedness. Apparently this happens only in weak interactions.*

---

* Kant did not deny that there is a real, intrinsic difference between right-handed and left-handed objects. Hacking is referring to the belief, held universally prior to 1956, that the laws of nature are indifferent to the left-right distinction: for any process that is physically possible, so too is its mirror image. It is this belief that parity-violation experiments proved false. See Martin Gardner, *The New Ambidextrous Universe*, 3d

What we mean by right- or left-handed in nature has an element of convention. I remarked that electrons have spin. Imagine your right hand wrapped around a spinning particle with the fingers pointing in the direction of spin. Then your thumb is said to point in the direction of the spin vector. If such particles are traveling in a beam, consider the relation between the spin vector and the beam. If all the particles have their spin vector in the same direction as the beam, they have right-handed (linear) polarization, while if the spin vector is opposite to the beam direction, they have left-handed (linear) polarization.

The original discovery of parity violation showed that one kind of product of a particle decay, a so-called *muon neutrino*, exists only in left-handed polarization and never in right-handed polarization.

Parity violations have been found for weak *charged* interactions. What about weak *neutral* currents? The remarkable Weinberg-Salam model for the four kinds of force was proposed independently by Steven Weinberg in 1967 and A[bdus] Salam in 1968. It implies a minute violation of parity in weak neutral interactions. Given that the model is sheer speculation, its success has been amazing, even awe inspiring. So it seemed worthwhile to try out the predicted failure of parity for weak neutral interactions. That would teach us more about those weak forces that act over so minute a distance.

The prediction is: Slightly more left-handed polarized electrons hitting certain targets will scatter, than right-handed electrons. Slightly more! The difference in relative frequency of the two kinds of scattering is one part in 10,000, comparable to a difference in probability between 0.50005 and 0.49995. Suppose one used the standard equipment available at the Stanford Linear Accelerator in the early 1970s, generating 120 pulses per second, each pulse providing one electron event. Then you would have to run the entire SLAC [Stanford Linear Accelerator Center] beam for 27 years in order to detect so small a difference in relative frequency. Considering that one uses the same beam for lots of experiments simultaneously, by letting different experiments use different pulses, and considering that no equipment remains stable for even a month, let alone 27 years, such an experiment is impossible. You need enormously more electrons coming off in each pulse. We need between 1000 and 10,000 more electrons per pulse than was once possible. The first attempt used an instrument now called PEGGY I. It had, in essence, a high-class version of J. J. Thomson's hot cathode. Some lithium was heated and electrons were boiled off. PEGGY II uses quite different principles.

rev. ed. (New York: W. H. Freeman and Company, 1990), ch. 22; James Van Cleve and Robert E. Frederick, eds., *The Philosophy of Right and Left: Incongruent Counterparts and the Nature of Space* (Dordrecht, Netherlands: Kluwer, 1991).

■ | PEGGY II

The basic idea began when C. Y. Prescott noticed (by "chance"!) an article in an optics magazine about a crystalline substance called gallium arsenide. GaAs has a number of curious properties that make it important in laser technology. One of its quirks is that when it is struck by circularly polarized light of the right frequencies, it emits a lot of linearly polarized electrons. There is a good rough and ready quantum understanding of why this happens, and why half the emitted electrons will be polarized, $3/4$ polarized in one direction and $1/4$ polarized in the other.

PEGGY II uses this fact, plus the fact that GaAs emits lots of electrons due to features of its crystal structure. Then comes some engineering. It takes work to liberate an electron from a surface. We know that painting a surface with the right substance helps. In this case, a thin layer of cesium and oxygen is applied to the crystal. Moreover the less air pressure around the crystal, the more electrons will escape for a given amount of work. So the bombardment takes place in a good vacuum at the temperature of liquid nitrogen.

We need the right source of light. A laser with bursts of red light (7100 Ångstroms) is trained on the crystal. The light first goes through an ordinary polarizer, a very old-fashioned prism of calcite, or Iceland spar.[10] This gives linearly polarized light. We want circularly polarized light to hit the crystal. The polarized laser beam now goes through a cunning modern device, called a Pockels cell. It electrically turns linearly polarized photons into circularly polarized ones. Being electric, it acts as a very fast switch. The direction of circular polarization depends on the direction of current in the cell. Hence the direction of polarization can be varied randomly. This is important, for we are trying to detect a minute asymmetry between right- and left-handed polarization. Randomizing helps us guard against any systematic "drift" in the equipment.[11] The randomization is generated by a radioactive decay device, and a computer records the direction of polarization for each pulse.

A circularly polarized pulse hits the GaAs crystal, resulting in a pulse of linearly polarized electrons. A beam of such pulses is maneuvered by magnets into the accelerator for the next bit of the experiment. It passes through a device that checks on a proportion of polarization along the way. The remainder of the experiment requires other devices and detectors of comparable ingenuity, but let us stop at PEGGY II.

■ | Bugs

Short descriptions make it all sound too easy, so let us pause to reflect on debugging. Many of the bugs are never understood. They are eliminated by trial and error. Let us illustrate three different kinds: (1) The essential technical limitations that in the end have to be factored into the analysis of error. (2) Simpler mechanical defects you never think of until they are forced on you. (3) Hunches about what might go wrong.

1. Laser beams are not as constant as science fiction teaches, and there is always an irremediable amount of "jitter" in the beam over any stretch of time.

2. At a more humdrum level the electrons from the GaAs crystal are back-scattered and go back along the same channel as the laser beam used to hit the crystal. Most of them are then deflected magnetically. But some get reflected from the laser apparatus and get back into the system. So you have to eliminate these new ambient electrons. This is done by crude mechanical means, making them focus just off the crystal and so wander away.

3. Good experimenters guard against the absurd. Suppose that dust particles on an experimental surface lie down flat when a polarized pulse hits them, and then stand on their heads when hit by a pulse polarized in the opposite direction. Might that have a systematic effect, given that we are detecting a minute asymmetry? One of the team thought of this in the middle of the night and came down next morning frantically using antidust spray. They kept that up for a month, just in case.[12]

■ | Results

Some 10^{11} events were needed to obtain a result that could be recognized above systematic and statistical error. Although the idea of systematic error presents interesting conceptual problems, it seems to be unknown to philosophers. There were systematic uncertainties in the detection of right- and left-handed polarization, there was some jitter, and there were other problems about the parameters of the two kinds of beam. These errors were analyzed and linearly added to the statistical error. To a student of statistical inference this is real seat-of-the-pants analysis with no rationale whatsoever. Be that as it may, thanks to PEGGY II the number of events was big enough to give a result that convinced the entire physics community.[13] Left-handed polarized electrons were scattered from deuterium slightly more frequently than right-handed electrons. This was the first convincing example of parity-violation in a weak neutral current interaction.*

---

* For competing accounts of why physicists found the SLAC experiment so convincing, see Andrew Pickering, *Constructing Quarks* (Chicago: University of Chicago Press, 1984); and Allan Franklin, *Experiment, Right or Wrong* (Cambridge: Cambridge University Press, 1990).

■ | Comment

The making of PEGGY II was fairly non-theoretical. Nobody worked out in advance the polarizing properties of GaAs—that was found by a chance encounter with an unrelated experimental investigation. Although elementary quantum theory of crystals explains the polarization effect, it does not explain the properties of the actual crystal used. No one has been able to get a real crystal to polarize more than 37 percent of the electrons, although in principle 50 percent should be polarized.

Likewise although we have a general picture of why layers of cesium and oxygen will "produce negative electron affinity," i.e., make it easier for electrons to escape, we have no quantitative understanding of why this increases efficiency to a score of 37 percent.

Nor was there any guarantee that the bits and pieces would fit together. To give an even more current illustration, future experimental work, briefly described later in this paper, makes us want even more electrons per pulse than PEGGY II could give. When the parity experiment was reported in *The New York Times*, a group at Bell Laboratories read the newspaper and saw what was going on. They had been constructing a crystal lattice for totally unrelated purposes. It uses layers of GaAs and a related aluminum compound. The structure of this lattice leads one to expect that virtually all the electrons emitted would be polarized. So we might be able to double the efficiency of PEGGY II. But at present (July 1981) that nice idea has problems. The new lattice should also be coated in work-reducing paint. But the cesium-oxygen compound is applied at high temperature. Then the aluminum tends to ooze into the neighboring layer of GaAs, and the pretty artificial lattice becomes a bit uneven, limiting its fine polarized-electron-emitting properties. So perhaps this will never work.[14] The group are simultaneously reviving a souped up new thermionic cathode to try to get more electrons. Maybe PEGGY II would have shared the same fate, never working, and thermionic devices would have stolen the show.

Note, incidentally, that the Bell people did not need to know a lot of weak neutral current theory to send along their sample lattice. They just read *The New York Times*.

■ | Moral

Once upon a time it made good sense to doubt that there are electrons. Even after Millikan had measured the charge on the electron, doubt made sense. Perhaps Millikan was engaging in "inference to the best explanation." The charges on his carefully selected oil drops were all small integral multiples of a least charge. He inferred that this is the real least charge in nature, and hence it is the charge on the electron, and hence there are electrons, particles of least charge. In Millikan's day most (but not all) physicists did become increasingly convinced by one or more theories about the electron. However it is always admissible, at least for philosophers, to treat inferences to the best explanation in a purely instrumental way, without any commitment to the existence of entities used in the explanation.[15] But it is now seventy years after Millikan, and we no longer have to infer from explanatory success. Prescott et al., don't explain phenomena with electrons. They know a great deal about how to use them.

The group of experimenters do not know what electrons are, exactly. Inevitably they think in terms of particles. There is also a cloud picture of an electron which helps us think of complex wave functions of electrons in a bound state. The angular momentum and spin vector of a cloud make little sense outside a mathematical formalism. A beam of polarized clouds is fantasy so no experimenter uses that model—not because of doubting its truth, but because other models help more with the calculations. Nobody thinks that electrons "really" are just little spinning orbs about which you could, with a small enough hand, wrap the fingers and find the direction of spin along the thumb. There is instead a family of causal properties in terms of which gifted experimenters describe and deploy electrons in order to investigate something else, e.g., weak neutral currents and neutral bosons. We know an enormous amount about the behavior of electrons. We also know what does not matter to electrons. Thus we know that bending a polarized electron beam in magnetic coils does not affect polarization in any significant way. We have hunches, too strong to ignore although too trivial to test independently: e.g., dust might dance under changes of directions of polarization. Those hunches are based on a hard-won sense of the kinds of things electrons are. It does not matter at all to this hunch whether electrons are clouds or particles.

The experimentalist does not believe in electrons because, in the words retrieved from mediaeval science by Duhem, they "save the phenomena." On the contrary, we believe in them because we use them to *create* new phenomena, such as the phenomenon of parity violation in weak neutral current interactions.

■ | When Hypothetical Entities Become Real

Note the complete contrast between electrons and neutral bosons. Nobody can yet manipulate a bunch of neutral bosons, if there are any. Even weak neutral currents are only just emerging from the mists of hypothesis. By 1980 a sufficient range of convincing experiments had made them the object of investigation. When might they lose their hypothetical status and become commonplace reality like electrons? When we use them to investigate something else.

I mentioned the desire to make a better gun than PEGGY II. Why? Because we now "know" that parity is violated in weak neutral interactions.

Perhaps by an even more grotesque statistical analysis than that involved in the parity experiment, we can isolate just the weak interactions. That is, we have a lot of interactions, including say electromagnetic ones. We can censor these in various ways, but we can also statistically pick out a class of weak interactions as precisely those where parity is not conserved. This would possibly give us a road to quite deep investigations of matter and anti-matter. To do the statistics one needs even more electrons per pulse than PEGGY II could hope to generate. If such a project were to succeed, we should be beginning to use weak neutral currents as a manipulable tool for looking at something else. The next step towards a realism about such currents would have been made.

The message is general and could be extracted from almost any branch of physics. Dudley Shapere has recently used "observation" of the sun's hot core to illustrate how physicists employ the concept of observation. They collect neutrinos from the sun in an enormous disused underground mine that has been filled with the old cleaning fluid (i.e., carbon tetrachloride). We would know a lot about the inside of the sun if we knew how many solar neutrinos arrive on the earth. So these are captured in the cleaning fluid; a few will form a new radioactive nucleus. The number that do this can be counted. Although the extent of neutrino manipulation is much less than electron manipulation in the PEGGY II experiment, here we are plainly using neutrinos to investigate something else. Yet not many years ago, neutrinos were about as hypothetical as an entity could get. After 1946 it was realized that when mesons distintegrate, giving off, among other things, highly energized electrons, one needed an extra nonionizing particle to conserve momentum and energy. At that time this postulated "neutrino" was thoroughly hypothetical, but now it is routinely used to examine other things.

■ | Changing Times

Although realisms and anti-realisms are part of the philosophy of science well back into Greek prehistory, our present versions mostly descend from debates about atomism at the end of the nineteenth century. Anti-realism about atoms was partly a matter of physics: the energeticists thought energy was at the bottom of everything, not tiny bits of matter. It also was connected with the positivism of Comte, Mach, Pearson and even J. S. Mill. Mill's young associate Alexander Bain states the point in a characteristic way, apt for 1870:

> Some hypotheses consist of assumptions as to the minute structure and operations of bodies. From the nature of the case these assumptions can never be proved by direct means. Their merit is their suitability to express phenomena. They are Representative Fictions.[16]

"All assertions as to the ultimate structure of the particles of matter," continues Bain, "are and ever must be hypothetical. . . ." The kinetic theory of heat, he says, "serves an important intellectual function." But we cannot hold it to be a true description of the world. It is a Representative Fiction.

Bain was surely right a century ago. Assumptions about the minute structure of matter could not be proved then. The only proof could be indirect, namely that hypotheses seemed to provide some explanation and helped make good predictions. Such inferences need never produce conviction in the philosopher inclined to instrumentalism or some other brand of idealism.

Indeed the situation is quite similar to seventeenth-century epistemology. At that time knowledge was thought of as correct representation. But then one could never get outside the representations to be sure that they corresponded to the world. Every test of a representation is just another representation. "Nothing is so much like an idea as an idea," as Bishop Berkeley had it. To attempt to argue for scientific realism at the level of theory, testing, explanation, predictive success, convergence of theories and so forth is to be locked into a world of representations. No wonder that scientific anti-realism is so permanently in the race. It is a variant on "The Spectator Theory of Knowledge."*

Scientists, as opposed to philosophers, did in general become realists about atoms by 1910. Michael Gardner, in one of the finest studies of real-life scientific realism, details many of the factors that went into that change in climate of opinion.[17] Despite the changing climate, some variety of instrumentalism or fictionalism remained a strong philosophical alternative in 1910 and in 1930. That is what the history of philosophy teaches us. Its most recent lesson is Bas van Fraassen's *The Scientific Image*, whose "constructive empiricism" is another theory-oriented anti-realism. The lesson is: think about practice, not theory.

Anti-realism about atoms was very sensible when Bain wrote a century ago. Anti-realism about *any* sub-microscopic entities was a sound doctrine in those days. Things are different now. The "direct" proof of electrons and the like is our ability to manipulate them using well understood low-level causal properties. I do not of course claim that "reality" is constituted by human manipulability. We can, however, call something real, in the sense in which it matters to scientific realism, only when we understand quite well what its causal properties are. The best evidence for this kind of understanding is

---

* According to the American philosopher John Dewey (1859–1952), the "spectator theory of knowledge" is the mistake, pervasive among Western philosophers from the time of Plato, of thinking that knowledge involves passively representing the world (where these representations—beliefs—are judged by their correspondence with facts) as opposed to actively constructing conceptual frameworks (where these conceptual frameworks are judged by their instrumental value in predicting experience and guiding our actions).

that we can set out, from scratch, to build machines that will work fairly reliably, taking advantage of this or that causal nexus. Hence, engineering, not theorizing, is the proof of scientific realism about entities.[18]

■ | Notes

1. C. W. F. Everitt and Ian Hacking, "Which Comes First, Theory or Experiment?" [Unpublished. See chapter 9, "Experiment," of Ian Hacking, *Representing and Intervening* (Cambridge: Cambridge University Press, 1983).]

2. Ian Hacking, "Spekulation, Berechnung und die Erschaffnung der Phänomenen," in *Versuchungen: Aufsätze zur Philosophie Paul Feyerabends* (P. Duerr, ed.), Frankfurt, 1981, Bd 2, 126–58.

3. Nancy Cartwright makes a similar distinction in a sequence of papers, including "When Explanation Leads to Inference," [*Philosophical Topics* 13 (1982): 111–21]. She approaches realism from the top, distinguishing theoretical laws (which do not state the facts) from phenomenological laws (which do). She believes in some "theoretical" entities and rejects much theory on the basis of a subtle analysis of modeling in physics. I proceed in the opposite direction, from experimental practice. Both approaches share an interest in real-life physics as opposed to philosophical fantasy science. My own approach owes an enormous amount to Cartwright's parallel developments, which have often preceded my own. My use of the two kinds of realism is a case in point.

4. Hilary Putnam, "How Not To Talk About Meaning," "The Meaning of 'Meaning'," and other papers in the *Mind, Language and Reality, Philosophical Papers*, Vol. 2, Cambridge, 1975.

5. Francis Bacon, *The Great Instauration*, in *The Philosophical Works of Francis Bacon* (J. M. Robertson, ed; Ellis and Spedding, Trans.), London, 1905, p. 252.

6. Ian Hacking, "Do We See Through a Microscope?" *Pacific Philosophical Quarterly* 62 (Winter 1981), 305–22.

7. I thank Melissa Franklin, of the Stanford Linear Accelerator, for introducing me to PEGGY II and telling me how it works. She also arranged discussions with members of the PEGGY II group, some of which are mentioned below. The report of experiment E-122 described here is "Parity Non-conservation in Inelastic Electron Scattering," C. Y. Prescott et al., *Physics Letters*. I have relied heavily on the in-house journal, the *SLAC Beam Line*, Report No. 8, October, 1978, "Parity Violation in Polarized Electron Scattering." This was prepared by the in-house science writer Bill Kirk, who is the clearest, most readable popularizer of difficult new experimental physics that I have come across.

8. The odd-sounding bosons are named after the Indian physicist S. N. Bose (1894–1974), also remembered in the name "Bose-Einstein statistics" (which bosons satisfy).

9. But excluding Leibniz, who "knew" there had to be some real, natural difference between right- and left-handedness.

10. Iceland spar is an elegant example of how experimental phenomena persist even while theories about them undergo revolutions. Mariners brought calcite from Iceland to Scandinavia. Erasmus Bartholin experimented with it and wrote about it in 1689. When you look through these beautiful crystals you see double, thanks to the so-called ordinary and extraordinary rays. Calcite is a natural polarizer. It was our entry to polarized light which for 300 years was the chief route to improved theoretical and experimental understanding of light and then electromagnetism. The use of calcite in PEGGY II is a happy reminder of a great tradition.

11. It also turns GaAs, a ¾–¼ left/right hand polarizer, into a 50-50 polarizer.

12. I owe these examples to conversation with Roger Miller of SLAC.

13. The concept of a "convincing experiment" is fundamental. Peter Galison has done important work on this idea, studying European and American experiments on weak neutral currents conducted during the 1970s.

14. I owe this information to Charles Sinclair of SLAC.

15. My attitude to "inference to the best explanation" is one of many learned from Cartwright. See, for example, her paper ["When Explanation Leads to Inference," *loc. cit.*].

16. Alexander Bain, *Logic, Deductive and Inductive*, London and New York, 1870, p. 362.

17. Michael Gardner, "Realism and Instrumentalism in 19th-Century Atomism," *Philosophy of Science* 46 (1979), 1–34.

18. (Added in proof, February, 1983.) As indicated in the text, this is a paper of July, 1981, and hence is out of date. For example, neutral bosons are described as purely hypothetical. Their status has changed since CERN announced on Jan. 23, 1983, that a group there had found W, the weak intermediary boson, in proton-antiproton decay at 540 GeV. These experimental issues are further discussed in my . . . book, *Representing and Intervening* (Cambridge, 1983).

David B. Resnik

Hacking's Experimental Realism

Traditional debates about scientific realism tend to focus on issues concerning scientific representation (broadly speaking) and de-emphasize issues concerning scientific intervention. Questions about the relation between theories and the world, the nature of scientific inference, and the structure of scientific explanations have occupied a central place in the realism debate, while questions about experimentation and technology have not. Ian Hacking's experimental realism attempts to reverse this trend by shifting the defense of realism away from representation to intervention.[1] Experimental realism, according to Hacking, does not require us to believe that our theories are true (or approximately true), nor does its defense depend on inference to the best explanation. For Hacking, the strongest proof for realism is that we can manipulate objects: 'So far as I'm concerned, if you can spray them, then they are real' (ibid., 23).

In this paper I argue (1) that Hacking's argument for experimental realism is, despite his strong denials, another version of the 'success of science' argument; (2) that the experimental realist can only have knowledge about theoretical entities if she assumes that the theories which describe those entities are at least approximately true; and (3) that experimentation is not nearly as theory-free as Hacking maintains. The common thread in these three criticisms is that Hacking does not succeed in shifting the defense of realism away from questions about scientific representation. Experimentation is a form of intervention, but it is intervention strongly guided by representation.

From *Canadian Journal of Philosophy* 24 (1994): 395–412.

■ | I

Since 'realism' is one of the more abused words in the philosophical lexicon, it is important to understand how Hacking uses it. The word 'real,' according to Hacking, is a 'trouser-word' in that its meaning derives from what it is contrasted with: real ducks are not decoys, real diamonds are not made of cubic zirconium, and so forth.* Real, in its scientific usage, is connected to causation: real entities can causally interact with the world; unreal ones cannot.

> Reality has to do with causation and our notions of reality are formed from our ability to change the world. . . . We shall count as real what we can use to intervene in the world to affect something else, or what the world can use to affect us. (ibid., 46)

Given this account of reality, Hacking distinguishes between two types of realism: theory realism and entity realism. Theory realists hold that scientific theories accurately describe the causal structures and properties in the world. They believe that (1) theories are true or false independently of what we believe; (2) truth is determined by the way the world is (i.e., correspondence truth); (3) some theories are approximately true; and (4) theories aim at the truth. Theory anti-realists may deny some or all of these claims. Entity realists, on the other hand, hold only that theoretical entities (or properties, processes, and events) are real; that is, that they are part of the causal structure of the world. Entity realists need not believe that our theories about those entities are true or even close to the truth, according to Hacking. Entity anti-realists claim theoretical entities are useful fictions, logical constructions, or instruments for reasoning (ibid., 27).

Epistemology, not just metaphysics and semantics, also plays a crucial role in Hacking's experimental realism. Hacking, by his own admission, runs together metaphysics and epistemology in his construal of entity realism: the entity realist claims to *know* that some theoretical entities *exist*. Hacking includes an epistemological ingredient in his definition of entity realism because 'the whole issue would be idle' if we believe theoretical

---

* J. L. Austin coined the term *trouser-word* in his *Sense and Sensibilia* (Oxford: Clarendon Press, 1962), p. 70. See section 7, on the meaning of *real*, where Austin argues that *real* is one of those words for which it is the negative uses that "wear the trousers," or, as we might say these days, "call the shots." To say that something is real leather or a real duck is not to say anything positive about leather or ducks but merely to exclude possible ways in which these things can fail to be genuine. Which possibilities are excluded will vary from context to context. Thus, Austin concludes that it is futile for philosophers to search for a property that is common to all things that are or could be called *real*. Hacking agrees. The word *real* is not ambiguous—its general function is the same in all cases—but its application varies from case to case. See Ian Hacking, *Representing and Intervening* (Cambridge: Cambridge University Press, 1983), 33.

entities exist but do not believe that we could have warranted belief in their existence (ibid., 28).

Many realists are both entity realists and theory realists, and quite often entity realism is based on theory realism: if you believe that a theory is true, then you have good reasons for believing that the entities (e.g., objects, causal properties, structures) posited by the theory exist.[2] But Hacking holds that one can be an entity realist without being a theory realist, and that entity realism need not be based on theory realism. Hacking, by his own classification, is an entity realist: 'realism is not about theories and truth. The experimentalist need only be a realist about the entities used as tools.'[3] Since other realists are both entity and theory realists, I will refer to Hacking's realism as theory-free entity realism.

Hacking's argument for theory-free entity realism rests on an analysis of scientific experimentation as well as some controversial metaphysical, epistemological, and semantic views. The experimental part of his argument is straightforward: experimentation largely proceeds without guidance from theories about the phenomena under study: 'much truly fundamental research precedes any relevant theory whatsoever' (*Representing and Intervening*, 158). Experimenters need to rely on some low-level generalizations about their instruments, as well as some theories about phenomena they are not studying, but they do not need to subscribe to full-blown scientific theories about the phenomena they are investigating in order to achieve their results. They need some theories to build instruments, but they do not need theories to use them. Hacking makes similar remarks about observation in science: observation, though guided by theory, need not be completely theory-laden: 'there have been important observations in the history of science, which have included no theoretical assumptions at all' (ibid., 176).

The metaphysical part of his argument turns on an anti-Humean analysis of causation and an instrumentalist account of laws and models. Causal claims, according to Humeans, are based on observed regularities in nature which we formulate as scientific laws. The causes we posit, according to Humeans, are not real; only regularities are real. Hacking reverses this doctrine: the causes are real but the regularities are not. Hacking follows Nancy Cartwright's account of causation and laws in physics and holds that theoretical laws of physics are best viewed as mere instruments ('Experimentation and Scientific Realism'). Theoretical laws are approximations; they are riddled with exceptions and based on idealizing assumptions. Moreover, physicists use a battery of models (some of them mutually inconsistent) to describe the same phenomena. Even though physical laws and models are best viewed as instruments, the entities described by the laws and models are not: the underlying causes are real; the regularities are not. Thus, physicists can make causal claims without basing them on scientific laws or models. This view implies theory anti-realism, provided that one holds that laws and models are the backbone of scientific theories.

Hacking's metaphysics is supported, in part, by a causal theory of reference made popular by Hilary Putnam and Saul Kripke.* In *Representing and Intervening*, Hacking weds himself to Putnam's version of the causal theory of reference and meaning and in another paper he refers to 'Our debt to Hilary Putnam' (ibid., 157). Prior to the development of the causal theory of reference, many philosophers of science held what could be loosely described as a descriptivist theory of meaning.[4] On the descriptivist view, words have a reference and a meaning (or a sense). Words (or phrases) refer to the objects they denote (if these objects exist) and the meanings of words are derived from descriptions we associate with them. Thus, 'water' means 'H_2O' and we can use 'water' to denote (or pick out) objects, i.e. steam, ice, etc. But what happens to meaning when our descriptions change? If we once described water as simply 'clear liquid' and we now describe it as 'compound of two parts hydrogen and one part oxygen,' are we talking about the same thing? A whole host of semantic problems in science resulted from this approach to meaning, the most worrisome of these being Thomas Kuhn's infamous thesis that scientific theories are incommensurable.[5] It is impossible to compare competing theories, according to Kuhn, because successive and competing theories are not even talking about the same things. Two theories, T1 and T2, might employ similar terms, such as 'mass' and 'velocity,' but the meanings of 'mass' and 'velocity' in T1 would not be the same as the meanings of 'mass' and 'velocity' in T2.†

These semantic problems are anathema for scientific realism because they lead to irrationalist conclusions, which, in turn, lead to anti-realist conclusions. If theories are incommensurable, then scientists cannot rationally adjudicate among theories. Even if we had theory-neutral methods and a theory-neutral evidence base, our language would be so theory-dependent that we could not objectively compare theories. We would 'choose' a new theory by learning its language and becoming committed to its assumptions, not by being swayed by the evidence. The anti-realist implication of this view is that reality is also theory-dependent: there is no objective reality, only reality within a given research community using a particular set of theories.[6]

Putnam's causal theory of reference saves philosophers of science from these (and other) difficulties, according to Hacking, by showing how scientists holding competing or successive theories 'may still be talking

* For further discussion of the causal theory of reference and a criticism of Putnam's version of it, see D. H. Mellor, "Necessities and Universals in Natural Laws," in chapter 7 and the section "Mellor's Defense of the Regularity Theory" in the accompanying commentary.

† See Thomas S. Kuhn, "The Nature and Necessity of Scientific Revolutions," in chapter 2 and Ernest Nagel, "Issues in the Logic of Reductive Explanations" and Paul Feyerabend, "How to Be a Good Empiricist—A Plea for Tolerance in Matters Epistemological," in chapter 8.

about the same thing' (*Representing and Intervening*, 75). The meaning of a word, according to Putnam, has four components: the word's syntactic marker, its semantic marker, its stereotype, and its referent.* The syntactic marker classifies the word grammatically, e.g. as a noun, verb, etc. . . . The semantic marker of a word indicates the general category of things it signifies, the stereotype is a description commonly associated with the word, and the referent of a word is that thing (or class of things) it denotes (or refers to) if it denotes at all. Stereotypes change—in science we frequently revise them—but referents, syntactic and semantic markers do not. Thus scientists who hold competing or successive theories can still be talking about the same things even though they describe those things in different ways.

This semantic view has important consequences for realism (in general) because it subverts worries about incommensurability: theories can be compared because different theories can refer to the same things. Scientists may use different descriptions (stereotypes) to talk about those things, but the things talked about exist independently of these descriptions. It also has important consequences for Hacking's version of realism because it shows how we might be anti-realists about theories but not about entities: theories, like stereotypes, come and go; but the entities they describe, i.e., their referents, remain the same. Hacking also hopes that this semantic view has important epistemological consequences, i.e., acceptance of theoretical entities is compatible with skepticism toward theories: 'one can believe in some entities without believing in any particular theory in which they are embedded' (ibid., 29).

■ | II

Now that we have a clearer understanding of the view Hacking attempts to defend, theory-free entity realism, we need to examine his argument for this position. Hacking's argument focuses on our ability to intervene in the world. Intervention is a kind of causal process in which cognizers use the world to *do*, *make*, or *change* things. Intervention (or intervening) contrasts with representation (or representing). Representation is a kind of causal pro-

---

* Hilary Putnam, "The Meaning of 'Meaning,'" in *Language, Mind, and Knowledge*, ed. K. Gunderson, vol. 7, *Minnesota Studies in the Philosophy of Science*, (Minneapolis: University of Minnesota Press, 1975); reprinted in Hilary Putnam, *Mind, Language, and Reality: Philosophical Papers*, vol. 2 (Cambridge: University of Cambridge Press, 1975), 215–71. See also chapter 6 of Ian Hacking's *Representing and Intervening*, for an exceptionally clear summary of Putnam's theory and a criticism of it. Hacking argues that, despite its many virtues for the scientific realist, Putnam's theory places so much emphasis on reference that it cannot do justice to cases (such as phlogiston and caloric) in which scientists agree about what a term means even though it has no referent, and other cases (such as the naming of muons and mesons) in which scientists reassign names on theoretical grounds.

cess in which cognizers use language to represent the world. Representations (the products of the process of representation) are intended to be 'more or less public likenesses' of the world (ibid., 133). Representation is not action in the world; it is talk and thought about the world. Scientists represent the world through scientific theories and concepts.

Traditional arguments for (and against) realism have focused on representation, not intervention, according to Hacking. Their chief concern is to convince us that our theories and concepts do (or do not) accurately represent the world. Scientific realism is a very dubious doctrine if one focuses on representation, according to Hacking. Since we can never get outside of our representations and 'hook-up with the world,' the most plausible doctrine is 'some version of idealism' (ibid., 130). (Once Putnam realized that reference could not rescue realism from idealism, he took this option and became an internal realist, according to Hacking.)

Hacking seeks to reverse this trend by focusing on how we intervene in the world. Ordinary people intervene in the world by building bridges, chopping down trees, driving cars, and so forth. Scientists intervene in the world by experimenting. In experimentation, we take advantage of the world's causal structure in order to produce, control, and observe phenomena. If we use one part of the world to intervene in the world, then we are entitled to believe that our tool for intervention is real. (The structure that we use the tool on may still be regarded as only hypothetical.) Hacking's argument for entity realism now begins to fall in place. Since his writing is a bit enigmatic, highly suggestive, and sometimes vague, it will be useful to quote several passages in which he mounts this argument:

> The best kinds of evidence for the reality of a postulated or inferred entity is that we can begin to measure or otherwise understand its causal powers. The best evidence, in turn, that we can have this kind of understanding is that we can set out, from scratch, to build machines that will work fairly reliably, taking advantage of this causal nexus. Hence, engineering, not theorizing, is the best proof of scientific realism about entities. ('Experimentation and Scientific Realism,' 170) [See original version in this chapter, pages 1153–54.]

> Now, how does one alter the charge on the niobium ball? "Well, at that stage," said my friend, "we spray it with positrons to increase the charge or with electrons to decrease the charge." *So far as I'm concerned, if you can spray them then they are real. (Representing and Intervening, 23)*

> I am told that nobody can yet manipulate a bunch of neutral bosons, if there are any. . . . When might they lose their hypothetical status and become commonplace reality like electrons? When we use them to investigate something else. (ibid., 272)

In a nutshell, Hacking's argument for entity realism goes something like this:

1 We are entitled to believe that a theoretical entity is real if and only if we can use that entity to do things to the world.

2 We can use some theoretical entities, e.g., electrons, to do things to the world, e.g., change the charges on niobium balls.

3 Hence, we are entitled to believe that some theoretical entities, e.g., electrons, are real.

This argument, on the surface, is quite convincing. In experimentation we causally affect and manipulate things: we inject DNA into mouse embryos, we shoot neutrons at atoms, we drop balls off towers. It seems reasonable to believe that the objects we manipulate in experiments are real. When a theoretical entity becomes a tool—when it becomes something we can use and manipulate to achieve certain goals—it ceases to be merely hypothetical and becomes real. Thus, during Mendel's time genes were not real; after genetic engineering, they became real.

Hacking's argument for realism, though widely applicable, is not offered as an argument for an all-embracing realism. In other words, he holds that realism must be defended on a case-by-case basis. He defends what he calls 'realism in particular' not 'realism in general' (ibid., 31). He would have us examine each science's experimental methods, tools, and its theoretical posits. If that science is able to use its posits as tools for experimentation, then the posits are real. If not, then they are merely hypothetical. An implication of this view is that one can be a realist about electrons but not about quarks or weak neutral currents. One can be a realist about DNA but not about species. Indeed, Hacking admits he is an anti-realist when it comes to the objects of astrophysics:

> When we use entities as tools, as instruments of inquiry, we are entitled to regard them as real. But we cannot do that with objects of astrophysics. Astrophysics is almost the only human domain in which we have profound, intricate knowledge, and in which we can be no more than what van Fraassen calls constructive empiricists.[7]

Before criticizing Hacking's argument for realism, we should note that he does not dismiss other arguments for realism altogether: 'I did not say in my [Representing and Intervening] that an experimental argument is the only viable argument for scientific realism about unobserved entities. I said only that it is the most compelling one, and perhaps the only compelling one' (ibid., 560–1). Although Hacking hedges his bets in this and other passages, his disparaging remarks about arguments for realism which focus on representation instead of intervention indicate that he takes a dim view of other arguments for entity realism. Thus, we should read him as attempting to defend the argument for entity realism.

▪ | III

Thus far, most of the critical discussion of Hacking's realism has focused on premise (1) in the argument sketched above. For instance, Dudley Shapere, in opposition to Hacking, argues for realism in astrophysics on the grounds that Hacking's sense of 'use' is ambiguous and vague.[8] 'Use' could be interpreted actively as 'manipulate' or 'control' or more passively as 'employ' or 'exploit.' According to Shapere, only the active sense of 'use' supports anti-realism in astrophysics. But since (according to Shapere) Hacking has no argument for interpreting 'use' in this way, we can employ the passive sense of this slippery word to support realism in astrophysics. For instance, Shapere argues that since we can use gravitational lenses to measure distances, we are entitled to regard gravitational lenses as real even though we cannot do experiments on them (ibid.).

Although I agree with Shapere's critique of Hacking, it is not my aim in this essay to attack this aspect of Hacking's realism. Rather, I will discuss three other objections to his view. My first objection is that his argument is, despite his strong denials, another version of the success of science argument. This objection arises when we consider Hacking's second premise in the argument sketched above, viz., that we can use theoretical entities as tools. My concern is with how we support (or argue for) our claims to use theoretical entities as tools. These claims are far from obvious and need defending. Moreover, since they are causal claims, they demand fairly sophisticated arguments.

The first question to ask then is 'how do you support causal claims?' As David Hume pointed out long ago, we do not deduce causal claims from experience.[9] Hume believed that our causal claims are based on induction: we observe regularities (or effects) in nature and infer causes from these regularities based on the inductive principle 'similar effects have similar causes.' Since Hume regarded induction as an unjustified form of reasoning he held that our causal claims are unjustified as well. Although Hume's doubts about causality still influence all discussions of causation, many philosophers now believe that we can justify causal claims through inductive or possibly abductive reasoning. Today, many scientists defend inductive arguments for causal connections based on statistical reasoning.[10] Many philosophers also hold that one can base causal claims on inference to the best explanation, or abductive reasoning: A causes B if and only if the hypothesis 'A causes B' provides the best explanation of the evidence.

Hacking, as we saw earlier, does not believe that we can base causal claims on observed regularities. Thus, he would not (or should not!) take an inductivist approach to the justification of causal claims. But if he dismisses this approach, the only viable alternative is an abductivist approach. Hacking does not commit himself to any particular methods of scientific reasoning, but his argument for experimental realism bears a strong resemblance to other abductivist arguments for realism, such as the infamous success of science argument. According to J. J. C. Smart's version of this argument, realism is the best explanation of the predictive and explanatory success of particular scientific theories.[11] Smart argues it would be an incredible coincidence—a 'cosmic accident'—if a false theory made accurate predictions, gave good explanations, and so forth. Hence, we should believe that theories which exhibit explanatory and predictive success are true. According to Richard Boyd's version, realism is the best explanation of the success (or instrumental reliability) of scientific methodology.[12] Boyd claims that it would also be an incredible coincidence if successful sciences, which deploy theory-dependent methods, were not making progressive approximation toward the truth. Thus, we should believe that the theories of successful sciences are approximately true and genuinely refer. In both arguments success is an acknowledged fact and realism is supposed to be the best explanation of this fact.

Hacking's argument is also a 'success of science argument' for realism. While Smart and Boyd focus on the success of theoretical science, viz., predictive and explanatory success, Hacking focuses on the success of experimental science. Success in experimentation is equated with the ability to reliably manipulate or control our instruments in order to produce desired effects ('Experimentation and Scientific Realism'). A successful experimenter can also dampen 'noise' and undesired effects. To reiterate an earlier quotation, Hacking says that the best evidence we can have that we use an entity as a tool is that we can 'build machines that will work fairly reliably' (ibid., 170 [1154]). Although Hacking denies that experimenters need to accept any theories about the phenomena they study, he acknowledges that they must rely on some background assumptions and low-level generalizations (*Representing and Intervening*, 149–67). For instance, an experimenter who used electrons to do things, would at least have to assume that electrons are a part of the world's causal nexus. Hacking's argument for premise (2) above now takes shape:

A   Our scientific instruments work fairly reliably in producing desired effects, e.g., changing the charge on niobium balls.

B   The best explanation of this reliability is that these instruments take advantage of a (real) causal nexus in the world. That is, that we are using the instruments to do things in the world.

C   According to our low-level generalizations (theories?) and other assumptions, our instruments take advantage of some theoretical entities in producing their effects.

D   Hence, we can use some theoretical entities, e.g., electrons, to do things in the world, e.g., change the charge on niobium balls.

Hacking's argument, like other success of science arguments, also depends on our recognizing an incredible coincidence that cries out for explanation. The coincidence for the experimenter is that the instruments work reliably in producing desired effects. It would be an incredible coincidence if these instruments worked reliably but were not taking advantage of the world's causal nexus. It would be an incredible coincidence if we managed to use an entity as a tool for inquiry that did not, in fact, exist. Thus, we should believe that the entities we are using exist and that we are taking advantage of the world's causal nexus in order to avoid accepting an incredible coincidence.

Hacking repeatedly asserts that his argument for experimental realism is not just another version of the infamous success of science argument:

> We "infer the best explanation" that the theory is true. The common cause of the phenomena must be the theoretical entities postulated by the theory. As an argument for scientific realism this idea has produced much debate. So it must seem as if my talk of coincidence puts me in the middle of an ongoing feud. Not so! My argument is much more localized. (ibid., 202)

> Once upon a time the best reason for thinking that there are electrons might have been success in explanation. . . . Luckily we no longer have to pretend to infer from explanatory success. (ibid., 271–2)

In these passages Hacking asserts that his argument for realism is not a success of science argument, but he offers no sustained argument to support this assertion. The closest thing I can find to an argument in his work is his claim that experimenters use entities as tools; they do not use them to explain: 'Prescott *et al.* don't explain phenomena with electrons. They know how to use them' (ibid., 272).

The problem with this line of defense is that one cannot *rationally* claim to use a theoretical entity as a tool of inquiry without some evidence, argument, or justification. Our claims about using or manipulating theoretical entities should not be dogmatic assertions: they require some justifications. These justifications can employ either inductive or abductive arguments. If our claims to use theoretical entities as tools are not supported by inductive arguments, as Hacking appears to maintain, then they must be supported by abductive ones. Hence, if one regards an entity as a tool of inquiry, one must also claim that its place in the world's causal structure explains some phenomena.

Putting my reply in a slightly different light, we can distinguish between two additional senses of the word 'use': an epistemic sense and a non-epistemic one. In the epistemic senses of 'use,' one can use an entity only if one claims to know something about it; in the non-epistemic sense of 'use,' one can use an entity without claiming to know anything about it. When we say that we 'use an equation to calculate probabilities' we are using 'use' in the epistemic sense; when we say that we 'use oxygen in respiration,' we are using 'use' in the non-epistemic sense. My argument is that experimentation is, in part, an epistemic activity, and that experimental uses are, in part, epistemic uses. Some activities, such as breathing, can be viewed as non-epistemic: I can breath (or use) oxygen without claiming to know anything about it. We can use many things in this world while remaining completely ignorant about them and having no evidence which might have some bearing on their causal properties. But experimentation is not like this. One cannot use a theoretical entity in an experiment without claiming to know something about that entity and claiming to have some justification for believing that it has specific causal properties.

Thus, using an entity as a tool in experiment also commits us to regarding it as an entity we can use to explain and predict phenomena. But an entity does not explain anything by itself; it explains phenomena within the context of a theory that describes it. Hence, if we believe that we can use a theoretical entity to explain and predict phenomena, we must also believe that our theories which describe the entity are at least approximately true, since theories that are not at least approximately true cannot explain. (More will be made of this point in the next section.)

One reason why Hacking is against inference to the best explanation is that he agrees with the charge made by Bas van Fraassen and others that this form of argument is unscientific because explanation is a hopelessly vague or interest-relative notion.[13] However, some recent theories of explanation attempt to counter this charge by articulating causal accounts of explanation.[14] In an analysis of inference to the best explanation, Peter Lipton proposes an approach to abduction that employs a causal model of explanation, and thus may make this form of inference appear to be less vague or interest-relative.[15] If these approaches to explanation and abduction prove to be fruitful, they may offer plausible arguments for realism, and Hacking's experimental realism can be viewed as a contribution to this trend.

■ | IV

My second and third criticisms of Hacking's view are that he fails to make a plausible case for theory-free, entity realism. His view is on shaky philosophical ground because it does not allow experimental realists to have knowledge about the entities they investigate. Furthermore, it does not accurately portray experimental practice.

Hacking's theory-free, entity realism is a metaphysically consistent position—one *can* believe in theoretical entities without believing in the theories in which they are embedded—but it is not a reasonable position. It is not a reasonable position because it gives the experimenter belief (and perhaps even true belief) in theoretical entities without justified belief. Given that justification is a necessary condition for knowledge, theory-free entity realism cannot yield knowledge about theoretical entities. And if Hacking's realism is a position which does not yield knowledge, then it would be, as he says, 'idle.'

In order to explore this criticism, we need to return to Putnam's semantic theory. This theory, as we saw earlier, provides the semantic backbone for Hacking's metaphysics and epistemology. It enables Hacking to claim that we are justified in believing in a theoretical entity without believing in the theory in which it is embedded because we can continue to successfully refer to that entity even when our theories about that entity undergo radical changes. While this view establishes the cogency of Hacking's metaphysical part of theory-free entity realism, it does not go very far in supporting the epistemological aspect of this view. Theory-free entity realism is on epistemologically solid ground—it can give us knowledge—only if we know that the entities to which we refer are natural kinds; that is, we know that they are fundamental parts of the world's causal structure. However, if we do not know that a theoretical entity is a natural kind, then we cannot claim to continue to successfully refer to that entity when our theories about it change. If a theoretical entity is not a natural kind, then it is an artifact of our theories and classificatory systems. If an entity is a mere artifact, then we cannot have knowledge about that entity.

Some examples will serve to clarify this point. Phlogiston was a theoretical entity used to explain combustion, rusting, metabolism, and other chemical phenomena (see Kuhn). For instance, according to phlogiston theory, during combustion, the burning material emitted smoke and phlogiston— the 'fire substance'—and left behind ashes. This view was supported, in part, by the fact that materials [seem to] lose weight during combustion, and part of this loss was attributed to the loss of phlogiston. The theory also predicted that materials would lose phlogiston (and thus weight) during other chemical processes. However, experiments showed that materials did not lose weight as the theory had predicted but actually gained weight. Since defenders of the phlogiston theory held that mass is conserved in all chemical reactions, they proposed that phlogiston could have no mass or even negative mass. But these ad hoc attempts to save the theory failed, and phlogiston was no longer regarded as a real entity. Phlogiston was not a natural kind; it was merely an artifact of specific chemical theories. Theorists eventually explained the phenomena explained by phlogiston by positing other theoretical entities, such as oxygen and carbon dioxide.

Gregor Mendel posited the existence of hereditary factors (or Mendelian genes) to explain hereditary phenomena. Mendel formulated three laws that governed these entities. The laws were used to explain the phenotypic similarities and differences between parents and offspring and to predict hereditary phenomena. During this century, genetics discovered that Mendel's laws are fraught with exceptions and that the functions performed by Mendelian genes are actually performed by several different kinds of entities, such as cistrons, mutons, and recons, and that the functions performed by these entities are actually performed by DNA.[16] Although geneticists still use the term 'gene,' Mendelian genes can no longer be viewed as distinct, natural kinds. Unlike phlogiston, Mendelian genes are not fictional entities, but they are artifacts of Mendelian theory, nonetheless. Mendelian genes have been, in a sense, split into many different parts by modern genetic theories.

These two examples are neither trivial nor obscure; the history of science contains many episodes where theoretical entities have been shown to be mere artifacts.[17] The question we need to put before Hacking is this: how do we *know* that theoretical entities that we talk about actually exist? How do we know that electrons will not one day be shown to be artifacts of current physical theories? For all we know, electrons could be like Mendelian genes: their effects might really be produced by other entities. Or even worse, they might turn out to be like phlogiston and not exist in any way at all.

Hacking has a simple reply to these questions: we know that electrons (and other theoretical entities) exist because we can refer to them. Our theories about electrons may change, but we can still refer to the same things. But this reply simply begs the question, since we cannot claim to successfully refer to a theoretical entity unless we are already justified in claiming that the entity is a natural kind. Reference, as Hacking himself admits, is not a magical 'sky-hook' which can save us from anti-realism (Hacking, *Representing and Intervening*, 130).

The next issue we need to address, then, is how we can know that a theoretical entity is a natural kind. For if we can know that a theoretical entity is a natural kind, then we can know that we can successfully refer to that entity and that we have knowledge about that entity. Anti-realists, of course, might assert that we can never know that a theoretical entity is a natural kind and that we thus cannot have knowledge about theoretical entities. But some realists, such as Boyd, propose that we can know that an entity is a natural kind provided that we have a well-confirmed, explanatorily successful, approximately true, theory that posits the existence of that entity. We can know that electrons are natural kinds because they are posits of our current, highly successful, theories. This suggestion also leads us back to a conclusion reached in the last section, that the experimental realist is committed to the approximate truth of the theories in which her entities are embedded. Ironically, this realist alternative also brings Hacking back into the traditional representationalist approach to realism.

The trouble with the alternative, of course, is this is precisely that view that Hacking disdains. He wants to maintain that we have knowledge about theoretical entities without believing in any theories in which they are embedded and he wants to dodge the success of science argument for realism. He might try to wriggle his way out of this pigeonhole by claiming that we can know that a theoretical entity is a natural kind if we can use it as a tool for inquiry. But this move does not free him from theoretical bindings. As we saw earlier, we cannot claim to use a theoretical entity as a tool for inquiry unless we also claim that it explains our experimental successes. But how do we explain those successes? By appealing to *theories* that describe the causal processes and structures used in our experiments. Thus, experimenters are justified in believing in electrons because electrons are posited by theories that explain experimental successes.

Hacking, like Houdini, might have one other way to escape this theoretical cage. He could maintain that experimenters need to rely on low-level generalizations about the entities they study, but they do not need to commit themselves to full-blown scientific theories about these entities. Although Hacking does not say what the difference might be between a low-level generalization and a full-blown theory, one commonly accepted way of distinguishing between theories and low-level generalizations is to claim that theories must contain genuine, scientific laws.[18] Scientific laws are unrestricted, universal, generalizations that possess 'nomic necessity' (ibid.).*

There are two problems with this reply. First, it is based on an overly restrictive view of theories. This view is overly restrictive because it implies that the only 'genuine' theories will be found in physics and chemistry. But this position does not accurately reflect scientific practice and belittles theories in the biological, earth, and human sciences. It makes perfect sense to talk about theories of fetal development, plate tectonics, stellar evolution, and animal aggression. A theory need not contain any unrestricted, universal generalizations; there can be theories that apply only to particular things or systems, or which contain only low-level generalizations.

Second, even if we accept this overly restrictive view of theories, it turns out that experimenters actually use full-blown scientific theories about the very entities they study. While one could, for the sake of philosophical argument, claim that experimenters only need rough generalizations, this philosophical fantasy has no basis in experimental practice. Experimenters do not operate without genuine scientific theories and laws about the phenomena they investigate: the gulf between experiment and theory is not nearly as large as Hacking supposes. Indeed, during the scientific revolution there was no sharp distinction between theorists and experimenters: many of the best theorists, such as Galileo, Newton, Harvey, and Boyle, were also excellent experimenters. As scientists have become more specialized, a gap has developed between theorists and experimenters. However, even today these two different communities share some common reference points. A person running experiments with a particle accelerator may not be aware of the latest developments in theoretical physics, but he (or she) is likely to be familiar with most of the commonly accepted background theories in physics, including some theories about the particles he (or she) is studying. Experimenters and theorists both obtain graduate degrees in their chosen fields and are thus both familiar with a common core of background theories.

---

* For more on laws and nomic necessity, see chapter 7.

They also continue to communicate about scientific ideas and instruments as they pursue their different career paths. (Many of these claims about experimentation are empirical points which require further substantiation. Since I do not have the space to provide that support here, I refer the reader to the work of Allan Franklin.[19])

The upshot of this discussion is that Hacking's theory-free entity realism can be attacked on both philosophical and practical grounds. The philosophical problem with this position is that it cannot give us knowledge about theoretical entities; the practical problem with it is that it does not accurately reflect scientific practice.

■ | V

Hacking in right to point out that philosophers of science have placed too much emphasis on scientific representation and have for too long neglected scientific intervention. Given the neglect of experimentation and technology, Hacking's realism is a refreshing and insightful approach. However, Hacking is wrong to think that questions about representation are irrelevant to arguments for realism; the three critiques developed in this essay illustrate this point. Although Hacking eschews theory realism and inference to the best explanation, his experimental realism does not elude either of these traditional realist concerns. My criticism, then, might be properly regarded as a reinterpretation of Hacking's experimental realism, since I do not attack his view per se. I think that experimental realism is a plausible and interesting position, but it is not as non-traditional, non-theoretical, or non-abductive as Hacking assumes it is.[20]

■ | Notes

1. Ian Hacking, *Representing and Intervening* (Cambridge: Cambridge University Press 1983).

2. Wilfrid Sellars, *Science, Perception and Reality* (New York: Humanities Press 1963). According to Sellars, 'to have good reason for holding a theory is *ipso facto* to have good reason for holding that the entities postulated by the theory exist' (97).

3. Hacking, 'Experimentation and Scientific Realism,' in J. Leplin, ed., *Scientific Realism* (Berkeley and Los Angeles: University of California Press 1984), 154 [p. 1140 this volume]. Hacking aligns himself with Nancy Cartwright, who is also an entity realist. She holds that the theoretical laws of physical theories are false, but that we can have good reasons for believing that entities posited by those theories exist. She also believes, like Hacking, that we can have good reasons for believing that theoretical entities exist when we can causally interact with them. See her book *How the Laws of Physics Lie* (Oxford: Oxford University Press 1983).

It should also be noted that by 'theoretical entity' Hacking means nothing more than an entity that cannot be observed with the unaided senses, and that a theory, according to Hacking, is not a mere model of the phenomena or a collection of low-level generalizations or phenomenological laws; a theory contains some high-level generalizations or fundamental laws. Thus Hacking's account of theories also bears a strong resemblance to Cartwright's view.

4. Bertrand Russell, *Introduction to Mathematical Philosophy* (London: George Allen and Unwin 1919), 167–80.

5. Thomas Kuhn, *The Structure of Scientific Revolutions* (Chicago: University of Chicago Press 1962).

6. W. H. Newton-Smith, *The Rationality of Science* (London: Routledge and Kegan Paul 1981).

7. Hacking, 'Extragalactic Reality: The Case of Gravitational Lensing,' *Philosophy of Science* 56 (1989), 578.

8. Dudley Shapere, 'Astronomy and Anti-Realism,' *Philosophy of Science* 60 (1993) 134–50.

9. David Hume, *An Inquiry Concerning Human Understanding* (Indianapolis: Bobbs-Merrill 1955).

10. Richard Johnson and Gouri Bhattacharya, *Statistics: Principles and Methods* (New York: John Wiley and Sons 1985).

11. J. J. C. Smart, *Between Science and Philosophy* (New York: Random House 1968).

12. Richard Boyd, 'The Current Status of Scientific Realism,' in J. Leplin, ed., *Scientific Realism*.

13. Bas van Frassen, *The Scientific Image* (Oxford: Clarendon Press 1980).

14. Wesley Salmon, *Scientific Explanation and the Causal Structure of the World* (Princeton: Princeton University Press 1984).

15. Peter Lipton, *Inference to the Best Explanation* (London: Routledge 1991).

16. Alexander Rosenberg, *The Structure of Biological Science* (Cambridge: Cambridge University Press 1985).

17. Larry Laudan, 'A Confutation of Convergent Realism,' *Philosophy of Science* 48 (1981) 19–49. [Excerpted in this chapter.]

18. Carl Hempel, *Philosophy of Natural Science* (Englewood Cliffs, NJ: Prentice-Hall 1966).

19. Allan Franklin, *The Neglect of Experiment* (Cambridge: Cambridge University Press 1986).

20. I am grateful to Allan Franklin and an anonymous reviewer from the *Canadian Journal of Philosophy* for helpful suggestions and comments.

MARTIN CARRIER

What Is Right with the Miracle Argument: Establishing a Taxonomy of Natural Kinds

It certainly strikes us as one of the most remarkable types of scientific achievement when apparently disparate phenomena are unified theoretically. What appeared to be disparate to the untutored eye turns out to arise from the same underlying mechanism and thus to be identical in kind. When a door slams because the windows are open and it is windy outside, this happens due to the same cause and according to the same mechanism that makes a plane lift off the ground. The *prima facie* conclusion is that science succeeds in going beyond the specious distinctions of the senses. It teaches us what things are truly alike.

My aim in this paper is to examine the viability of this popular view. And the result will be that the view is basically correct. More precisely, I will try to show, first, that in some distinguished cases science arguably manages to induce the right classification or taxonomy among the phenomena, and that, second, this is the only access to reality that science is justifiably able to gain. Accordingly, what I am aiming to do is to support a particular and comparatively weak form of scientific realism.

Scientific realism contends that claims about certain unobservable 'items' which emerge from the theoretical or experimental activity of scientists are literally true; these claims faithfully refer to what is, as it were, going on behind the scenes. There is some quarrel, however, about what these 'items' are legitimately supposed to be. The leading brand of this doctrine is *theory-realism*. According to this position, the successful theories of mature science are approximately true. That is, these theories correctly portray the not-directly-observable processes and mechanisms that make the phenomena occur the way they do.

A more attenuated version of scientific realism is *entity-realism*. On the one hand, entity-realism is an immediate consequence of theory-realism. The truth of a theory implies the existence of its theoretical entities. On the

FROM *Studies in History and Philosophy of Science* 24 (1993): 391–409.

other hand, there is also a more autonomous type of entity-realism which is advanced on the basis of experiment-centered arguments. In this version, entity-realism says that the capacity to manipulate certain unobservable entities, and, in particular, to manipulate them in order to experiment on something else, gives strong evidence for the real existence of these entities. Entity-realism is thus non-committal as to the (near) truth of the accepted scientific account of these entities. It is no theory's entity but the experimenter's entity to which actual existence is attributed. Belief in these entities does not imply belief in any one of the theories involved.

An even more attenuated variant of realism, and the one I set out to support, is a *realism of kinds*. The claim is that science, at least on some rare and distinguished occasions, manages truthfully to forge links among the phenomena. Science sometimes succeeds in collecting phenomena into equivalence classes that reflect truly existing similarity relations among these phenomena. Whereas entity-realism implies a kind-realism, the reverse does not hold. It will be seen that commitment to the experimentally established entities entails commitment to the kind-structures they introduce. By contrast, the assumption that science is sometimes able to unveil true relations of similarity does not imply that the mechanisms employed to establish these relations and the entities they invoke reflect anything in nature.

The argument comes in three steps. The first one is introductory. I sketch the notion of 'natural kinds' and show, using some examples from the history of science, that the structure of natural kinds is theory-dependent and indeed changes in the course of scientific progress. This finding implies that a naive form of a realism of kinds is untenable empirically. In the second step I consider the experimental route to natural kinds. The *prima facie* advantage of this alternative is that it seems to be exempt from the uncertainties and vicissitudes of theoretical reasoning. It turns out, however, that this position is likewise beset with a historical counter-example. Next comes, finally, the constructive step. I take up some earlier results of mine about the actual impact of the so-called 'Miracle Argument' in favor of scientific realism. The argument says that scientific realism is a necessary precondition for any explanation of scientific progress; it is the only choice for divesting progress of its miraculous aspects. I will argue that the Miracle Argument is insufficient for achieving its proper objective, namely, theory-realism, and that its real accomplishment consists in showing that there are instances in which science provides us with a veridical portrait of relations of similarity as they prevail in nature.

1 | Changing Patterns of Natural Kinds

Every theory determines a classification among the objects or events it deals with; it induces a taxonomy into its universe of discourse. The laws of the theory are framed using certain descriptive predicates, and the relevant objects obey the laws in virtue of satisfying these predicates. A law thus serves to bind several objects together. It establishes a link of similarity among them by regarding them as instances of the same law. A so-created equivalence class of like objects is called a *natural kind* (cf. Fodor 1974, 101–102) [958]. Take, for example, the law: All electrons possess the elementary charge *e*. This law generates the natural kind 'electron' by quantifying over electrons. Electrons constitute a natural kind because there is a law that applies to them in virtue of their property of being electrons. Different laws may well pick out the same natural kind. All further laws about electrons (such as: All electrons have a spin-value of ½) obviously specify the same kind, namely, the class of electrons. This implies immediately that a theoretical change need not involve a change in the concomitant kind-structure. Different bodies of laws may select the same objects as being alike.[1]

The concept of natural kinds, as just circumscribed, is entirely non-committal as to the issue of scientific realism. Natural kinds are created by a corpus of laws irrespective of whether these laws are interpreted merely as useful but fictitious unifiers or are thought to refer to a theory-independent reality. Accordingly, natural kinds need not be truly natural, and the attribute 'natural' is, strictly speaking, a misnomer. In order to avoid the realist overtones the expression 'scientific kinds' has been proposed, but this proposal seems not to have gained acceptance. Thus I stick to the usual term, taking its non-committal sense to be understood. The reality of natural kinds constitutes the topic of the present investigation.

The first response to this problem presumably is that we are surely entitled to interpret the taxonomy specified by present-day science realistically. After all, its laws have undergone a large number of severe tests; they are well-confirmed and thus deserve our confidence. And this confidence quite naturally extends to the taxonomy induced by these laws. Most of us would thus be inclined to reason as follows. Since present-day knowledge includes laws about electrons, electrons form a natural kind. For this reason, electrons are in reality of the same kind. Let's examine the tenability of this assessment.

A condition familiar from the discussion of scientific realism in general is the following *retention requirement*. For a theory component to be interpreted realistically, it is necessary that it be retained across scientific change. In particular, if earlier successful theories in the mature sciences are approximately true, then the later even more successful theories in the same discipline will preserve the earlier laws as limiting cases (cf. Putnam 1978, 20–24; Laudan 1981, 234–240). Applied to natural kinds this means that the kind-structure of earlier well-confirmed theories should be reproduced 'in the limit' by the kind-structure of their respective successor theories.

The intuition behind the retention condition is as follows. Once-successful but now-rejected theories in the mature sciences were in their lifetimes well-confirmed according to our present methodological criteria. If such theories turn out to be wrong on all counts, nothing prevents us from the meta-inductive inference that our present most cherished accounts are likewise doomed to complete failure and that nothing will remain from them in the end. If scientific realism is supposed to be a viable position, this meta-inductive move has to be blocked; and it can only be blocked by assuming that these earlier accounts indeed got something right (cf. Putnam 1978, 25). These correct aspects should be retained by their respective successor theories. Realism about a specific item, be it theory, entity or kind, implies a retentionist claim with respect to this item.

This three-fold retention condition constitutes an epistemic requirement. Scientific realism claims not only that there is something real out there, but also that we gain access to reality through science. And we can only maintain that we have managed to lock on to reality if the features accorded this special status will not be discounted through scientific progress. If science is supposed to get hold of reality, it is necessary that the corresponding insights are here to stay. Accordingly, the retention condition does not amount to the assertion that what is not retained can by no means be real. Rather, the claim is that we have no science-based justification for attributing reality to abandoned features.

Scientific realism about theories thus requires preservation of theoretical laws, scientific realism about entities demands preservation of entities, and, finally, scientific realism about kinds necessitates preservation of kinds. On the other hand, it is certainly not requisite that the respective items be carried over unchanged from theory to theory. A successor theory may well have more to say about the corresponding aspect so that revisions cannot be excluded outright. The point rather is that the successor account must not completely overturn the relevant aspects or its predecessor. This is expressed by merely demanding preservation of the relevant item 'in the limit'. Regarding theory-realism this means that the theoretical laws of the earlier account may be reproduced by a 'corrective reduction'. That is, the derivability of the predecessor laws from the successor theory may be restricted to counterfactual initial and boundary conditions (cf. Carrier and Mittelstrass 1991, 42–50). As regards natural kinds, this means that earlier and later taxonomies have to be compatible in the sense that the later taxonomy can be construed as a more fine-grained version of the earlier one. This demands in turn that an earlier kind may comprise several later kinds, to be sure, but that there is no partial overlap or cross-over among the kinds involved.[2]

Now we are in a position to take up the initial question: Is it justified unrestrictedly to interpret the taxonomy of present-day science realistically as it stands? It follows from the discussion that realism implies retention. As a consequence, non-retention favors a non-realist view. Regarding the realist interpretation of kinds, this leads to following empirical requirement. In order that a historical sequence of taxonomic structures be interpreted realistically, the respective kinds must stand in a relation of total inclusion. If, by contrast, cross-over relations among kinds typically occur in the course of theory changes, this tends to discredit a realist view about kinds. I will now argue that the inclusion condition is not satisfied typically.

A claim of this sort is best backed by considering some examples. To begin with, let's cast a brief glance at the transition from classical mechanics to general relativity theory. The basic taxonomic distinction involved in all dynamical theories is the one between force-free and force-induced motion. According to the classical law of inertia, force-free motion is represented by uniform rectilinear motion. Conversely, the motion of a body in physical fields such as the gravitational or electromagnetic field counts as accelerated and thus as force-induced. In contradistinction, it is constitutive of general relativity that it takes gravitation to be part of space–time structure. This means that gravitation-induced motion is construed as inertial motion. More precisely, the motion of a 'test particle' (i.e. a small, non-rotating particle with negligible mass) in the exclusive presence of a gravitational field is considered force-free. This interpretation does not extend to other forces such as the electromagnetic one, however. General relativity maintains that a charged particle in an electromagnetic field moves non-inertially. Accordingly, gravitation-induced motion is now grouped together with classical inertial motion and separated from motion in an electromagnetic field. There is thus non-inclusion, but rather partial overlap, between the respective kind-structures.

Another case in point is the taxonomy induced by Newton's corpuscular optical theory and the modern account. 'Newton's rings', i.e. the colors exhibited by transparent thin plates, are attributed by the current account to the occurrence of interference between the light waves reflected at the upper and lower surfaces of the plate. Newton, by contrast, explained the effect by appeal to his particle model. When light particles enter the plate they produce a longitudinal shock wave in the plate material. This shock wave moves faster through the plate than the light particles and puts the opposite side of the plate in oscillation. If this oscillatory motion happens to be counterdirected to the light particles' motion upon their arrival, the surface throws the particles back; otherwise it lets them pass through. The critical factor is thus the phase of the surface oscillation when the surface is hit by the light particles. And the salient point of Newton's construction is that this phase depends on the passage time of the light through the plate and thus on the distance the light traverses in the plate. Since, furthermore, different colors are refracted differently by the same material, light particles representing different colors have to cover different distances until they reach the opposite surface of the plate. They are thus reflected differently. Consequently, if white light falls at a given angle of incidence on a transparent plate, only one color is reflected and the remainder is let through (cf. Newton 1730, 206–214, 280, 370–371).

The point is that Newton followed the very same approach in order to accommodate the permanent colors of bodies. He assumed the particles of bodies to be transparent and of variable size. Then he identified a body's color with the light reflected by its particles according to the process described. The reflected color is thus dependent on the size of the respective particles (cf. Newton 1730, 248–256).

> The transparent parts of Bodies, according to their several sizes, reflect Rays of one Colour, and transmit those of another, on the same grounds that thin Plates or Bubbles do reflect or transmit those Rays. And this I take to be the ground of all their Colours. [Newton 1730, 251]

Accordingly, the theory forges a link between Newton's rings and the colors of bodies. The two phenomena arise from the same mechanism and thus form a natural kind.

In contradistinction, these two phenomena are interpreted today as growing out of quite different mechanisms. The colors of bodies come about through a process of partial absorption and re-emission of the incident light by the atoms or molecules involved. Newton's rings, on the other hand, are the result of interference. So, what once belonged into the same natural kind is now thought to be of entirely distinct nature. A fundamental taxonomic breach splits the former kind into heterogeneous components. This constitutes another instance of a non-inclusive taxonomic development.

Just one more example. The caloric theory of heat assumed that heat is constituted by a material substance, namely, the matter of heat or caloric. Liquefaction and vaporization were interpreted as combinations with caloric. Liquids and gases are caloric compounds. This implies that vapors are in this respect identical to other compounds. This means that, say, oxygen gas belongs into the same natural kind as, say, a solid metallic oxide. Needless to say, this linkage has been cut by modern theory.

In order not to multiply examples gratuitously let me stop here. I take it to be the upshot of the discussion that the history of science is rife with instances of non-inclusive kind splitting or kind cross-over. This implies that the retentionist claim with respect to kinds is untenable historically, with the consequence that a kind-realism fails—at least in the sweeping and unqualified version given in the first part of this section.

2 | The Experimental Route to Kinds

I: MANIPULATION HERE CREATES KIND-STRUCTURE HERE

The theory-dependence of natural kinds along with the occurrence of profound theoretical changes across history appears to vitiate a theory-based realist interpretation of kinds. An alternative option might be to dispense with theories altogether and to ground the kind-structure in experimental distinctions and experimenters' abilities. It was proposed by Jed Buchwald that it is experimental set-ups and experimental devices that sort effects into

kinds. Buchwald connects the evolving kind-structure of optical phenomena with the available laboratory equipment. The doubly refracting crystal established the distinction between polarized and unpolarized light.
Additional invocation of the Fresnel rhomb produced the sub-kinds linearly and circularly (or elliptically) polarized light. Buchwald's claim is
that 'the apparatus proper often constitutes an embodiment of the relevant kind-structure.'[3]

The idea underlying the experiment-centered approach to kinds seems
to be as follows. The ability to manipulate an object or effect gives evidence
of relations of analogy and disanalogy. If we intervene in an apparently
homogeneous phenomenon and manage to elicit heterogeneous, qualitatively distinct responses, then the components constitute different kinds.
Briefly, what behaves differently in an experiment is different in kind. We
still don't know what it is that manifests itself differently, but we know it's
different.

In the present context the question is whether the experiment-based
kind-structure is trustworthy. It is to be admitted at once that the occurrence
of empirical differences is indeed necessary for sorting effects into distinct
categories. Things that behave alike under all circumstances will hardly
qualify as different in kind. The problem rather lies with the sufficiency part
of the claim. Things that behave differently in an experiment may yet be
alike.

This peculiarity has three possible sources. First, theoretical unification. Different sorts of electromagnetic radiation—such as infrared, visible
light, and ultraviolet—show different behavior. But these sorts of radiation
are still not sufficiently different to be counted as distinct natural kinds.
They rather constitute different instantiations of the same kind, namely,
electromagnetic radiation. The reason is that they merely differ in the values
of the pertinent parameters, namely, frequency and wavelength, whereas
they all obey one and the same corpus of laws.

Second, experiment-induced distinctions may be spurious and insignificant because the experiments may later be discovered to be spoiled by
unnoticed factors and uncontrolled side-effects. Consider eighteenth-century
affinity theory as an example. The theory was created by Newton and dominated important parts of the chemistry of the period. Affinities were conceived as attractive short-range forces between particles, and chemical
reaction and chemical bonding were attributed to their influence. Affinity
forces were thought to be substance-specific, i.e. they should be constant for
a given pair of substances, and to possess a point of saturation. Saturation
means that a particle of a substance A can attract only a limited number of
particles of a substance B. These assumptions suggest the use of substitution
reactions for ordering affinity strengths empirically. When a substance C
replaces a substance A in a compound AB, i.e. if the reaction $AB + C \rightarrow AC + B$
occurs, then the affinity from A to C is stronger than the one from A to B.
If, in addition, D is able to [replace] C in the compound AC, then this

testifies to the fact that D is attracted more strongly by A than C is (cf. Newton 1730, 376–383). Substitution reactions were employed to establish so-called 'affinity series' experimentally. At the top the substance in question was placed, and then followed its reaction partners in the order of decreasing affinity strength. The results sketched thus give rise to the following midget affinity series. A: D, C, B.

The affinities operative in the presence of three substances were called 'simple affinities'. In addition to them, so-called 'double affinities' were assumed. Double affinities were supposed to govern reciprocal substitutions of the general form: $AB + CD \rightarrow AC + BD$. They were considered completely different from simple affinities on the ground that no consistent set of relative affinity strengths could be found that was suitable to accommodate both types of substitutions.

What we have here is a fundamental distinction between two different kinds of chemical reactions. Simple substitutions are determined by the action of simple affinities, reciprocal substitutions are guided by double affinities. The crucial point in the present context is that this distinction between natural kinds was introduced on purely experimental grounds. Reactions of both types appeared to instantiate different behavior; they could not be accommodated by an overall scheme of affinity strengths. That is, one and the same compound was found to behave qualitatively differently according to whether a simple or a double substitution was performed.

This experiment-based distinction later collapsed completely. As Claude Berthollet demonstrated around 1800, the concept of affinity was defective in that the influences of temperature and of the quantities of the relevant substances had been left out of consideration. Due to uncontrolled fluctuations of these factors the experiments on affinity series were unreliable; they were not indicative of any significant trait. Along with the constant affinity forces, the distinction in kind between simple and double affinities was abandoned. Berthollet endeavored to explain both types by one unified scheme.[4]

The moral to be extracted from this episode is that experimentally introduced distinctions can later be recognized as faulty and insignificant. In the case sketched, a kind-distinction was revoked as a result of theoretical progress. What behaved differently experimentally turned out by theoretical reasoning to be of the same kind in the end.

Third, it is sometimes impossible to tell from the facts alone whether various experimental outcomes are really heterogeneous or whether they merely represent different instantiations of the same kind. After all, if an experiment yields gradually varying magnitudes of the same quantity, it does not give rise to a kind-distinction at all. In order that the experimental approach be sufficient to establish natural kinds, we must be able to recognize empirically whether the ensuing experimental results are qualitatively different or whether they merely represent changing magnitudes of one and the same quantity. An example from roughly the same period in the history of chemistry makes it clear, however, that this task is sometimes hard to accomplish.

Consider the distinction between compounds and mixtures that Joseph Louis Proust introduced around 1800. Proust tried to establish empirically the 'law of constant proportions' which involved precisely this distinction. The evidential basis was that some reactions ended up with compounds of definite proportions and some did not. For Proust the former constituted actual compounds and the latter mere mixtures or mixtures of compounds. Proust's proposal was criticized by Berthollet. Berthollet's theory implied that variable proportions are the rule and that constant proportions arise from the influence of particular additional factors. In general, compounds are characterized by variable proportions (cf. Carrier 1986, 374–377).

Both rivaling approaches thus interpreted the same empirical findings differently. In particular, they both led to a different kind-structure. Proust's difference in kind between compounds and mixtures was denied by Berthollet and reinterpreted as gradual change due to gradually varying influences. Berthollet held that the occurrence of definite proportions under some circumstances arose from the presence of ephemeral factors. In fact, there is only one single kind that embraces Proustian compounds and mixtures alike.

The point is that experience alone does not show who is right; it does not show whether the observed differences indicate differences in kind or only differences in magnitude. It is true that Proust finally won the day. But this victory was crucially due to the acceptance of John Dalton's atomic theory. Only in this framework could a theoretical distinction between compounds and mixtures be drawn. The experimental evidence provided too shaky a basis for a non-arbitrary distinction between them.

These examples give sufficient evidence for the conclusion that experiments are insufficient to distinguish among natural kinds. First such experiment-based distinctions may be taken back subsequently as a result of progress in theoretical unification. Second, their introduction may have been flawed right from the beginning due to the unrecognized influence of additional factors. Third, the distinction among natural kinds exclusively on the basis of observable differences may be entirely arbitrary due to the inability always to distinguish unambiguously between differences in quality and changes in quantity.

3 | The Experimental Route to Kinds

II: Manipulation Here Creates Kind-Structure There

Whereas the version of the experimental approach discussed above endeavored to establish a kind-structure with respect to the phenomenon or entity experimented upon, another variant of the same approach leads to the introduction of a kind-structure with respect to the entity used for experimenting on other phenomena. At least this is the by-product of an enterprise whose explicit objective is to support a realist interpretation of entities. In this section I briefly sketch the entity-realism advanced by Nancy Cartwright and Ian Hacking, elaborate its implications [for] a realist interpretation of kinds, and, finally, propose a counter-example.

As just indicated, Cartwright and Hacking attempt to establish realism about entities by focusing on our experimental abilities. That is, their entity-realism is not a consequence of theory-realism but is supposed to stand on its own feet. Theory-realism is inadequate, they argue, since there is no unanimously accepted body of theory about anything that is part of present-day research. Different scientists may hold different and sometimes incompatible models about the same entity; and quite legitimately so, for these different models suit different descriptive purposes best. Every discipline embraces a multitude of distinct accounts of the same theoretical entity. As a result, entity-realism cannot be based on the assumed truth of the relevant theories.

By contrast, entities are rightly attributed reality when we succeed in using them for investigating something else. The successful manipulation of an entity for the sake of intervening in other processes of a more hypothetical nature is the best possible evidence for the existence of this entity. Fictitious entities have no causal powers. When we know how to use an (initially theoretical) entity so as to create new phenomena, then this entity is rightly thought to be real (cf. Cartwright 1983, 87–99; Hacking 1983, 262–265).

The rationale for the existence of theoretical entities thus is that they can be employed to exercise causal influence. Relying on them we can deliberately produce effects. An entity-realism of this sort implies a kind-realism if the entities accepted are the ones that science specifies. For science frequently traces different effects back to the action of one and the same entity. These effects are thus equal in kind in that they are brought about by similar mechanisms or in an otherwise similar way. In this vein, Cartwright holds electrons to be instrumental in the Millikan experiment and likewise to be the cause of cloud-chamber tracks (cf. Cartwright 1983, 92–93, 99). Causal judgments of this sort generate a link between the two phenomena and thus induce a natural-kind structure.[5]

If the argument passed muster, it would indeed be suited to justify a realist view about kinds. In order to examine if it does, I resort to the touchstone of retention: for any item to be interpreted realistically, it is necessary that it is retained across scientific change (see Section 1). As a matter of fact, Cartwright herself makes a claim of roughly this sort. An entity that is experimentally warranted in the way sketched is 'seldom discarded in the progress of science' (Cartwright 1983, 98). I now argue that phlogiston passed the experimental reality test. Nevertheless, it was later abandoned.

The phlogiston theory stands in the tradition of the 'chemistry of principles'. This tradition assumed certain abstract principles or elements that were considered to be bearers of general properties such as hardness, combustibility and volatility. Empirical substances gain their properties by incorporating the property-bearing principles; the principles thus explain the properties of substances found in the laboratory. The theory did not attach one principle to each property; it rather introduced just a few such principles. The challenge was to explain the multitude of empirical properties by recourse to combinations of these few general principles.

In the present context it is the principle of combustibility termed 'phlogiston' by Georg Ernst Stahl, the creator of the phlogiston theory, that deserves particular interest. In the pre-Stahlian tradition, the principle of combustibility was supposed to comprise a number of related sub-principles. It was Stahl's chief objective to dispose of the multitude of distinct but related entities. He tried to demonstrate that there is only one such principle, namely, phlogiston. In particular, all combustion processes and all calcinations (i.e. oxidations of metals in modern terms) are to be interpreted in the same fashion as the release of phlogiston.

This novel claim was backed by the following experiment. If the unified account of combustion and calcination is true, it should be possible to produce a metal from its calx (i.e. the corresponding metallic oxide) by supplying phlogiston that originates from non-metallic combustible substances. Stahl succeeded in verifying this prediction by producing metallic lead out of lead calx (PbO) by heating it with glowing charcoal. Obviously, the calx has accepted the phlogiston released by the charcoal and accordingly turned into the metal.

After this initial empirical conformation of his unified approach, Stahl moved on to bring to bear the causal capacities of his newly established entity. He employed phlogiston so as to bring about one more formerly unknown phenomenon, namely, the synthesis of sulfur from sulfuric acid through the action of phlogiston. There were empirical indications for the view that sulfuric acid was in fact dephlogisticated sulfur, and Stahl managed to confirm this view by his then celebrated sulfur synthesis, published in 1697.

The first step of the experiment consisted in combining sulfuric acid with potash (K_2CO_3) in order to reduce its volatility. The result is 'fixed alkali' (K_2SO_4), interpreted as a compound of sulfuric acid and potash. When this compound is brought to a glow with charcoal—whose combustibility indicates that it contains a large quantum of phlogiston—'liver of sulfur' is formed (which is actually a mixture of various potassium-sulfur compounds such as K_2S_5 or $K_2S_2O_3$). Finally, Stahl precipitated sulfur from the reaction product and thus confirmed that it really contained sulfur (cf. Kopp 1873, 46–47; Partington 1962, 673–674).

What happened in this experiment is roughly and qualitatively represented by the following scheme.

$$K_2SO_4 \quad + \quad C \quad \rightarrow \quad K_2S_5, K_2S_2O_3 \quad + \quad CO$$

potash-sulfuric   charcoal   liver of sulfur   (not identified)
acid compound

The experiment was supposed to embody a phlogiston transfer. The phlogiston released from the charcoal was taken up by the fixed alkali and transformed the sulfuric acid contained in it into sulfur. According to present lights, the core reaction that occurred is $SO_2 + 2C \rightarrow S + 2CO$.

The point of this experiment is the following. What Stahl did was to create a new phenomenon relying on the previously established causal properties of the entity in question. He manipulated phlogiston in order to experiment on another, more hypothetical phenomenon, namely, the composition of sulfur. Stahl employed phlogiston so as to intervene in and actively change other processes. I conclude that phlogiston qualifies as real according to the Cartwright-Hacking criterion. Nonetheless, it was later given up.

The upshot is that this reality-criterion does not single out the right entities as real. As a consequence, the kind-structure induced by such experimentally supported entities cannot be trusted in either. These kind-structures may be real, to be sure, but the Cartwright-Hacking criterion does not license the inference to their reality. We are not justified in attributing reality to kind-structures supported in the experimental fashion.

It is true that Stahl's sulfur synthesis does not strictly entail this conclusion. After all, Cartwright only claims that entities backed in this way are 'seldom discarded' (see above), and one single counter-instance does not trespass this limit. On the other hand, the history of science shows that entities are less frequently abandoned than theories. This stability speaks in favor of entity realism, to be sure, but at the same time it reduces the testability of entity realism by reducing the number of possible counter-examples. And, if among the fairly limited class of discounted entities there is (at least) one that satisfied the Cartwright-Hacking criterion, then this one actual counter-instance is sufficient to justify doubts as to the viability of the experimental approach to entities and, derivatively, to natural kinds.[6]

4 | The Distinguished-Theory Approach to Kinds

I take it to be the upshot of the examples discussed above that neither the theory-based sweeping kind-realism sketched in Section 1 nor the two variants of the experiment-centered approach to natural kinds examined in Sections 2 and 3 succeed in reliably establishing a realism about natural kinds. Are we stuck with this null result? Mercifully not, I believe. Since the experimental approach to kinds has not stood up to empirical scrutiny, we are back with theories. But in light of the results of Section 1, we cannot rest our case on just any theory. Still, we may appeal to a particular subset of distinguished theories: theories, that is, distinguished by their outstanding empirical success. The basis for this distinction is the so-called Miracle Argument.

A sharpened version of the Miracle Argument says that there are two types of empirical success which can only be explained by recourse to realism, namely, novel predictions and consilience of inductions. If a theory succeeds in correctly predicting a novel regularity, formerly unknown and not to be expected against the background knowledge, this capacity is inexplicable (or miraculous for that matter) unless we assume that the theory has got something right. Analogously, if a theory manages to unify regularities that appeared to refer to completely disparate phenomena before, and if it did so naturally, as it were, i.e. without introducing modifications and adjustments for the sake of producing the unification, this ability is simply mysterious unless we presume that the theory has grasped something correctly.[7]

An example of a theoretically anticipated regularity is Einstein's prediction of the correct magnitude of gravitation-induced light-bending within the framework of general relativity theory. The theory managed to foresee what no experimenter had yet seen, and this remarkable achievement strongly suggests that it is veridical in some respect. An example of consilience of inductions is the connection between black-body radiation and the photoelectric effect as forged by Einstein's light quantum hypothesis. Einstein realized that the two phenomena are likewise governed by the quantum relation $E = hv$, and the value of Planck's constant h turned out to be identical in the two cases. It would indeed constitute a strange coincidence if two otherwise unrelated phenomena involved a numerical agreement of this sort. It is much more plausible to assume that the theoretical account that gives rise to this agreement truthfully reflects something real.

The mainstream realist position is theory-based entity-realism. This position involves commitment to the real existence of the entities specified by successful scientific theories, and this existence claim is backed by the supposed (approximate) truth of the relevant theories. It is this supposition that the Miracle Argument is intended to support. The intuition underlying the argument is as follows. There is a particular type of empirical success—that may be termed *strong* success—in which the domain of application of a theory extends itself naturally and without any deliberate adaptation for this purpose to phenomena the theory was not designed to accommodate. The occurrence of strong success is a remarkable and surprising event and is thus in need of an explanation. Realism provides such an explanation by arguing from the (approximate) truth of the theory to the existence of the relevant entities and their modes of interaction. A theory is strongly successful because it is essentially correct. This explanation is admittedly not complete since the empirical success of a basically correct theory may be thwarted by inadequate auxiliary assumptions. Although realism is thus not sufficient for the explanation of strong success it is nonetheless necessary. Without it we are at a loss to account for strong success in any event. Realism thus explains how strong success is possible.

The history of science teaches, however, that the Miracle Argument is in fact unsuitable for supporting the approximate truth of strongly successful theories. This can be gleaned from the sketch given above of Stahl's phlogistic accomplishment. Stahl correctly predicted two novel effects. He produced a metal from its calx through the assistance of a non-metallic substance, and he managed to synthesize sulfur from its purported components (see Section 3). But according to present lights, the theory he employed for this purpose is wrong, and the central entity he invoked is non-existent. So we have come across an instance of strong success that can obviously not be explained along the lines of the Miracle Argument. Two additional counter-instances are provided by Joseph Priestley's prediction of the reductive properties of hydrogen based on a later version of the phlogiston theory and by Dalton's and Joseph Gay-Lussac's prediction of the equality of the thermal expansion rates of all gases based on the caloric theory of heat.[8] In all these instances theories that are wide of the mark descriptively and referentially were nevertheless strongly successful. Accordingly, strongly successful theories may fail the retention test. The conclusion is that the Miracle Argument does not support theory-based entity-realism.

But this is not the end of the story. My claim is that the Miracle Argument is basically all right; it was only applied to the wrong subject matter. What the argument in fact supports is a realism of kinds.[9] The Miracle Argument is right in supposing that there has to be an explanation of strong success. It is wrong in attributing the reason to the truth of theories and to the existence of entities. What explains strong success is that the corresponding theories induced the right relations of similarity among the phenomena in question.

Let's take another look at the phlogiston example. Stahl's experiments worked because combustion and calcination are actually alike; they both constitute oxidation processes. He was only wrong in taking what is actually an oxygen transfer from sulfuric acid or lead oxide to charcoal to be a phlogiston transfer in the opposite direction. Although Stahl's account is now dismissed in its entirety, the classification it induced among the phenomena involved is retained until the present day. Oxidation of metals and nonmetals are still considered as being equal in kind. The same goes analogously with respect to the two further examples mentioned. Priestley's novel prediction was borne out because—speaking in modern terms—it correctly regarded oxidation and reduction as the same process going off in opposite directions. Dalton's and Gay-Lussac's prognosis was successful because it rightly linked the physical constitution of gases with their equal expansion rates (cf. Carrier 1991, 32–33). It is the retention of the kind-structure that provides the sought-for explanation of strong success.

It is to be noted that kind-retention is, in general, restricted to the phenomena connected by strong success. It does not automatically extend to other classifications specified by the same theory. Large parts of the kind-structure introduced by the phlogiston theory were overturned by the subsequent progress of science. In the framework of this theory, sulfur and metals belonged in the same natural kind as wax and oil. All these substances were purported to be alike in that they contain a large proportion of phlogiston. From the contemporary perspective, by contrast, the two groups constitute distinct kinds. The first group includes chemical elements and the second organic compounds.

The overall situation is thus as follows. We are presented with an epistemic argument to the effect that strongly successful theories should correctly reflect an aspect of reality. In order to single out this aspect we apply the retention test. Every theoretical feature that is supposed to reflect something real has to be retained across scientific change. It turns out, then, that theories and entities fail in this test whereas kinds pass it. The conclusion is that kind-structures backed by strong success are real.

5 | Establishing a Natural Taxonomy of Kinds

It might be objected that the vast majority of the examples presented stem from outdated theories that may appear antediluvian to the present-day reader. Some may even doubt whether the phlogiston theory ever qualified as truly scientific. These misgivings are unjustified, however. All the theories I referred to were well-confirmed in their respective lifetimes according to our present methodological criteria and had received high marks from the corresponding scientific communities. They all formed part of 'mature science.' After all, the phlogiston theory scored a strong success at its inception. Moreover, if we want to examine the appropriateness of retentionist claims we simply have to consider possible counter-instances, and such instances, in the nature of the case, are only provided by once successful but now rejected theories. Significant evidence as to retained features can only arise by examining what is left when everything else is gone. Accordingly, theories that appear strange today should not be disallowed as respectable philosophical examples for this reason alone.

It is worth noting, furthermore, that the argument for the reality of distinguished kind-structures is not beset with the following circularity. The judgment about what is real and what is not is based on present-day knowledge which, however, according to the very approach advocated here, is not justifiably reliable in telling what there is. In fact, the argument involves no such circularity; its logic is as follows. Adopting scientific realism implies commitment to the realist interpretation of at least one aspect involved in contemporary scientific theories. This suggests that preceding theories that were once confirmed in roughly the same fashion as their successors are now should likewise be trustworthy ontologically with regard to the aspect in question. From this follows the retention condition: for anything to be counted as real, it is necessary that it is retained across scientific change. This condition is accepted by realists and anti-realists alike. In fact, the anti-realist argument from scientific change proceeds from precisely this condition: because nothing significant is retained, the realist claims are mistaken.[10]

The second step of the argument brings to bear the judgment that strong success is an extraordinary and astonishing phenomenon and deserves an explanation. The Miracle Argument provides us with a possible reason; namely, it attributes strong success to the fact that the relevant parts of the corresponding theory truthfully reflect something real. The Miracle Argument is, however, quite unspecific as to what precisely constitutes this veridical aspect. There are three possible candidates, namely, theoretical mechanisms, theory-introduced entities, and theory-induced kinds. In order to determine which of them fills the bill the retention condition is invoked. Application of this condition to the history of science singles out kinds as the only theoretical features that may legitimately be interpreted realistically.

In short, satisfaction of the retention condition constitutes a necessary precondition for the realist interpretation of a theoretical feature. It is seen, then, that it is *at most* kinds that can be so interpreted. If, in addition, the retained theoretical feature serves as a basis for an explanation of strong success, the Miracle Argument licenses its realist interpretation. Retention together with the capacity to furnish such an explanation is sufficient for attributing reality to it. If a theoretical feature passes both the miracle test and the retention test we are entitled to accept it as part of reality. My contention is that (some) natural kinds indeed pass both tests. From this follows a prediction for the future course of science: phenomena once connected by strong success will continue to be of the same kind in all subsequent theories on the subject.

It may appear problematical to ground such a far-reaching contention on the discussion of no more than three cases. But in fact the evidential basis is not that narrow. It also includes cases in which the occurrence of strong success was due to theoretical aspects we still take to be correct. Consider the prediction of the phases of Venus as implied by Copernicus's theory. In this example of strong success, the relevant aspects of the theory, the entities involved, and the induced kinds are equally retained up to now. We still hold the view that Venus and the Earth revolve around the Sun, we still regard Venus, Earth, and Sun as acceptable entities, and we still stick to the kind-structure induced by the theory; namely, that—in contradistinction to the preceding account—Venus and Earth equally belong in the category 'planet' and the Sun does not. Cases of this type indiscriminately support all three positions at stake. The examples presented in Section 4, by contrast, constitute differential support for kind-realism. They can only be accommodated by kind-realism. And it would be over-demanding to require that the class of differentially supporting instances be large. In that event the alternative explanations would be obviously false and would never have gained the wide acceptance they enjoy.

The realism of kinds advocated here involves a peculiarly non-Aristotelian view of kinds. In the Aristotelian tradition, kinds are individuated through their essential properties. The class of human beings is rightly determined on the basis of the characterization 'rational animal', but not by means of the attribute 'featherless biped'. No such distinction between essential characteristics and accidental features is implied by the present account. Quite the contrary. My claim is that *only* the induced taxonomy, but not the theoretical means used for establishing it, is (in distinguished cases) justifiably to be considered real. It is true, our prime epistemic access to kinds is through theories; kinds are individuated by means of theories. Still, it is only the results, and not the means used for their production, that are arguably reliable ontologically. It is the relations of similarity among phenomena that (sometimes) deserve our confidence. No such case can be made for the theoretical mechanisms employed for specifying these relations and for the entities involved in them. It is clear that the phenomena collected into equivalence classes are equivalent in some respect. But precisely what this respect is cannot reliably be specified. Only the relation of similarity is legitimately to be interpreted realistically.

■ | Notes

I am grateful to Peter McLaughlin, Alexander Rueger and two anonymous referees for their helpful comments.

1. Note that 'natural kind' is used here as a technical term which does not wholly coincide with everyday usage. Biological species, for instance, though they are natural kinds in a rough and ready sense, hardly qualify as natural kinds in the present understanding. Kinds are derived from laws, and it is a matter of dispute in current philosophy of biology whether there are any specifically biological laws. In any event, a regularity of the sort 'All ravens are black' certainly does not count as a law of nature and thus does not give rise to the natural kind 'raven'.

2. This requirement is elaborated in Buchwald 1992, 40–41.

3. Buchwald 1992, 57; for the entire argument cf. *ibid.* 46–58.

4. For this account of affinity theory cf. Carrier 1986, 328–332, 366–371.

5. This feature becomes even more explicit in Cartwright's later work on capacities. Causal claims are made about properties, and an individual object is causally effective because it possesses this property. Aspirins relieve headache because of being aspirins; cf. Cartwright 1989, 141. This specification obviously strongly resembles the definition of natural kinds given above—with the sole difference that she does not refer to laws but to 'causal capacities'.

6. Meehl is certainly right in pointing out that a large number of systematically evaluated case-studies provides a better basis for assessing the merits of metatheo-

retical claims than 'informal, impressionistic (and often biased) reliance on selected case studies' (Meehl 1992, 272). Still, if the set of possible counter-instances—i.e. abandoned entities—is comparatively small, one actual counter-instance serves as a large enough sample.

7. Cf. Putnam 1978, 18–19; Leplin 1984, 203, 205, 217; Musgrave 1988, 232–234, 239; see also the reconstruction in Carrier 1991, 24–28. It is crucial to restrict the Miracle Argument to distinguished types of empirical success. The sweeping version of the argument, which addresses predictive success *simpliciter*, is widely—and rightly—regarded as unsound; cf. Rescher 1987, 65–67; Musgrave 1988, 231; Ben-Menahem 1990, 333–338.

8. Cf. Carrier 191, 29–31 for an analysis of these cases.

9. The view that (in distinguished cases) a theoretically generated classification of phenomena reflects a real order was first advanced by Duhem; cf. Duhem 1906, 24–28.

10. This constitutes the central contention of Laudan 1981.

▪ | References

Ben-Menahem, Y. (1990), 'The Inference to the Best Explanation', *Erkenntnis* 33, 319–344.

Buchwald, J. Z. (1992), 'Kinds and the Wave Theory of Light', *Studies in History and Philosophy of Science* 23, 39–74.

Carrier, M. (1986), 'Die begriffliche Entwicklung der Affinitatstheorie im 18. Jahrhundert. Newtons Traum—und was daraus wurde', *Archive for History of Exact Sciences* 36, 327–389.

Carrier, M. (1991), 'What is Wrong with the Miracle Argument?', *Studies in History and Philosophy of Science* 22, 23–36.

Carrier, M. and Mittelstrass, J. (1991), *Mind, Brain, Behavior: The Mind-Body Problem and the Philosophy of Psychology* (Berlin: de Gruyter).

Cartwright, N. (1983), *How the Laws of Physics Lie* (Oxford: Clarendon Press).

Cartwright, N. (1989), *Nature's Capacities and their Measurement* (Oxford: Clarendon Press).

Duhem, P. (1906), *The Aim and Structure of Physical Theory* (New York: Atheneum, 1974).

Fodor, J. A. (1974), 'Special Sciences (or: The Disunity of Science as a Working Hypothesis)', *Synthese* 28, 97–115.

Hacking, I. (1983), *Representing and Intervening: Introductory Topics in the Philosophy of Natural Science* (Cambridge: Cambridge University Press).

Kopp, H. (1873) *Die Entwicklung der Chemie in der neueren Zeit* (München: Oldenbourg, 1873. Reprinted New York: Johnson; Hildesheim: Olms, 1965).

Laudan, L. (1981), 'A Confutation of Convergent Realism', in Leplin 1984b, 218–249.

Leplin, J. (1984a), 'Truth and Scientific Progress', in Leplin 1984b, 193–217.

Leplin, J. (ed.) (1984b), *Scientific Realism* (Berkeley: University of California Press).

Meehl, P. E. (1992), 'The Miracle Argument for Realism: An Important Lesson to be Learned by Generalizing from Carrier's Counter-Examples,' *Studies in History and Philosophy of Science* 23, 267–282.

Musgrave, A. (1988), 'The Ultimate Argument for Scientific Realism', in R. Nola (ed.), *Relativism and Realism in Science* (Dordrecht: Kluwer Academic Publishers), 229–252.

Newton, I. (1730), *Opticks: or A Treatise of the Reflections, Refractions, Inflections & Colours of Light* (New York: Dover, 1979).

Partington, J. R. (1962), *A History of Chemistry*, Vol. II (London: Macmillan; New York: St. Martin's Press).

Putnam, H. (1978), *Meaning and the Moral Sciences* (London: Routledge & Kegan Paul).

Rescher, N. (1987), *Scientific Realism: A Critical Reappraisal* (Dordrecht: Reidel).

Arthur Fine

The Natural Ontological Attitude

> Let us fix our attention out of ourselves as much as possible; let us chace our imagination
> to the heavens, or to the utmost limits of the universe; we never really advance a step
> beyond ourselves, nor can conceive any kind of existence, but those perceptions, which
> have appear'd in that narrow compass. This is the universe of the imagination, nor have
> we any idea but what is there produced.
> —Hume, *Treatise*, Book I, Part II, Section VI

Realism is dead. Its death was announced by the neopositivists who realized
that they could accept all the results of science, including all the members of
the scientific zoo, and still declare that the questions raised by the existence
claims of realism were mere pseudoquestions. Its death was hastened by the
debates over the interpretation of quantum theory, where Bohr's nonrealist
philosophy was seen to win out over Einstein's passionate realism. Its death
was certified, finally, as the last two generations of physical scientists turned
their backs on realism and have managed, nevertheless, to do science successfully without it. To be sure, some recent philosophical literature . . . has
appeared to pump up the ghostly shell and to give it new life. But I think
these efforts will eventually be seen and understood as the first stage in the
process of mourning, the stage of denial. . . . But I think we shall pass through
this first stage and into that of acceptance, for realism is well and truly dead,
and we have work to get on with, in identifying a suitable successor. To aid
that work I want to do three things in this essay. First, I want to show that the
arguments in favor of realism are not sound, and that they provide no rational support for belief in realism. Then, I want to recount the essential role of
non-realist attitudes for the development of science in this century, and
thereby (I hope) to loosen the grip of the idea that only realism provides a
progressive philosophy of science. Finally, I want to sketch out what seems to
me a viable nonrealist position, one that is slowly gathering support and that
seems a decent philosophy for postrealist times.[1]

From Jarrett Leplin, ed., *Scientific Realism* (Berkeley: University of California
Press, 1984), 83–107.

■ | Arguments for Realism

Recent philosophical argument in support of realism tries to move from the success of the scientific enterprise to the necessity for a realist account of its practice. As I see it, the arguments here fall on two distinct levels. On the ground level, as it were, one attends to particular successes; such as novel, confirmed predictions, striking unifications of disparate-seeming phenomena (or fields), successful piggybacking from one theoretical model to another, and the like. Then, we are challenged to account for such success, and told that the best and, it is slyly suggested, perhaps, the *only* way of doing so is on a realist basis. I do not find the details of these ground-level arguments at all convincing. Larry Laudan has provided a forceful and detailed analysis which shows that not even with a lot of hand waving (to shield the gaps in the argument) and charity (to excuse them) can realism itself be used to explain the very successes to which it invites our attention.[2] But there is a second level of realist argument, the methodological level, that derives from Popper's attack on instrumentalism as inadequate to account for the details of his own, falsificationist methodology. Arguments on this methodological level have been skillfully developed by Richard Boyd,[3] and by one of the earlier Hilary Putnams.[4] These arguments focus on the methods embedded in scientific practice, methods teased out in ways that seem to me accurate and perceptive about ongoing science. We are then challenged to account for why these methods lead to scientific success and told that the best, and (again) perhaps, the only truly adequate way of explaining the matter is on the basis of realism.

I want to examine some of these methodological arguments in detail to display the flaws that seem to be inherent in them. But first I want to point out a deep and, I think, insurmountable problem with this entire strategy of defending realism, as I have laid it out above. To set up the problem, let me review the debates in the early part of this century over the foundations of mathematics, the debates that followed Cantor's introduction of set theory. There were two central worries here, one over the meaningfulness of Cantor's hierarchy of sets insofar as it outstripped the number-theoretic content required by Kronecker (and others); the second worry, certainly deriving in good part from the first, was for the consistency (or not) of the whole business. In this context, Hilbert devised a quite brilliant program to try to show the consistency of a mathematical theory by using only the most stringent and secure means. In particular, if one were concerned over the consistency of set theory, then clearly a set-theoretic proof of consistency would be of no avail. For if set theory were inconsistent, then such a consistency proof would be both possible and of no significance. Thus, Hilbert suggested that finite constructivist means, satisfactory even to Kronecker (or Brouwer) ought to be employed in meta-mathematics. Of course, Hilbert's program was brought to an end in 1931, when Gödel showed the impossibility of such a stringent consistency proof. But Hilbert's idea was, I think, correct even though it proved to be unworkable. Metatheoretic arguments must satisfy more stringent requirements than those placed on the arguments used by the theory in question, for otherwise the significance of reasoning about the theory is simply moot. I think this maxim applies with particular force to the discussion of realism.

Those suspicious of realism, from Osiander* to Poincaré and Duhem to the 'constructive empiricism' of van Fraassen,[5] have been worried about the significance of the explanatory apparatus in scientific investigations. While they appreciate the systematization and coherence brought about by scientific explanation, they question whether acceptable explanations need to be true and, hence, whether the entities mentioned in explanatory principles need to exist.[6] Suppose they are right. Suppose, that is, that the usual explanation-inferring devices in scientific practice do not lead to principles that are reliably true (or nearly so), nor to entities whose existence (or near-existence) is reliable. In that case, the usual abductive methods that lead us to good explanations (even to 'the best explanation') cannot be counted on to yield results even approximately true. But the strategy that leads to realism, as I have indicated, is just such an ordinary sort of abductive inference. Hence, if the nonrealist were correct in his doubts, then such an inference to realism as the best explanation (or the like), while possible, would be of no significance—exactly as in the case of a consistency proof using the methods of an inconsistent system. It seems, then, that Hilbert's maxim applies to the debate over realism: to argue for realism one must employ methods more stringent than those in ordinary scientific practice. In particular, one must not beg the question as to the significance of explanatory hypotheses by assuming that they carry truth as well as explanatory efficacy.

There is a second way of seeing the same result. Notice that the issue over realism is precisely the issue as to whether we should believe in the reality of those individuals, properties, relations, processes, and so forth, used in well-supported explanatory hypotheses. Now what *is* the hypothesis of realism, as it arises as an explanation of scientific practice? It is just the hypothesis that our accepted scientific theories are approximately true, where "being approximately true" is taken to denote an extratheoretical relation between theories and the world. Thus, to address doubts over the reality of relations posited by explanatory hypotheses, the realist proceeds to introduce a further explanatory hypothesis (realism), itself positing such a relation (approximate truth). Surely anyone serious about the issue of realism,

* Andreas Osiander was a Lutheran theologian who contributed a short, anonymous foreword to the first edition of Copernicus's *De Revolutionibus* (1543), arguing that Copernicus's theory should be regarded merely as a means of "saving the appearances," not as a realistic hypothesis. Copernicus, himself a realist, died before he could see the final version of his book, and Osiander supervised its publication. Osiander's imposture was detected and publicized by Kepler, but few astronomers had been deceived by it.

and with an open mind about it, would have to behave inconsistently if he were to accept the realist move as satisfactory.

Thus, both at the ground level and at the level of methodology, no support accrues to realism by showing that realism is a good hypothesis for explaining scientific practice. If we are open-minded about realism to begin with, then such a demonstration (even if successful) merely begs the question that we have left open ("need we take good explanatory hypotheses as true?"). Thus, Hilbert's maxim applies, and we must employ patterns of argument more stringent than the usual abductive ones. What might they be? Well, the obvious candidates are patterns of induction leading to empirical generalizations. But, to frame empirical generalizations, we must first have some observable connections between observables. For realism, this must connect theories with the world by way of approximate truth. But no such connections are observable and, hence, suitable as the basis for an inductive inference. I do not want to labor the points at issue here. They amount to the well-known idea that realism commits one to an unverifiable correspondence with the world. So far as I am aware, no recent defender of realism has tried to make a case based on a Hilbert strategy of using suitably stringent grounds and, given the problems over correspondence, it is probably just as well.

The strategy of arguments to realism as a good explanatory hypothesis, then, *cannot* (logically speaking) be effective for an open-minded nonbeliever. But what of the believer? Might he not, at least, show a kind of internal coherence about realism as an overriding philosophy of science, and should that not be of some solace, at least for the realist?[7] Recall, however, the analogue with consistency proofs for inconsistent systems. That sort of harmony should be of no solace to anyone. But for realism, I fear, the verdict is even harsher. For, so far as I can see, the arguments in question just do not work, and the reason for that has to do with the same question-begging procedures that I have already identified. Let me look closely at some methodological arguments in order to display the problems.

A typical realist argument on the methodological level deals with what I shall call the problem of the "small handful." It goes like this. At any time, in a given scientific area, only a small handful of alternative theories (or hypotheses) are in the field. Only such a small handful are seriously considered as competitors, or as possible successors to some theory requiring revision. Moreover, in general, this handful displays a sort of family resemblance in that none of these live options will be too far from the previously accepted theories in the field, each preserving the well-confirmed features of the earlier theories and deviating only in those aspects less confirmed. Why? Why does this narrowing down of our choices to such a small handful of cousins of our previously accepted theories work to produce good successor theories?

The realist answers this as follows. Suppose that the already existing theories are themselves approximately true descriptions of the domain under consideration. Then surely it is reasonable to restrict one's search for successor theories to those whose ontologies and laws resemble what we already have, especially where what we already have is well confirmed. And if these earlier theories were approximately true, then so will be such conservative successors. Hence, such successors will be good predictive instruments; that is, they will be successful in their own right.

The small-handful problem raises three distinct questions: (1) why only a small handful out of the (theoretically) infinite number of possibilities? (2) why the conservative family resemblance between members of the handful? and (3) why does the strategy of narrowing the choices in this way work so well? The realist response does not seem to address the first issue at all, for even if we restrict ourselves just to successor theories resembling their progenitors, as suggested, there would still, theoretically, always be more than a small handful of these. To answer the second question, as to why conserve the well-confirmed features of ontology and laws, the realist must suppose that such confirmation is a mark of an approximately correct ontology and approximately true laws. But how could the realist possibly justify such an assumption? Surely, there is no valid inference of the form "T is well confirmed; therefore, there exist objects pretty much of the sort required by T and satisfying laws approximating to those of T." Any of the dramatic shifts of ontology in science will show the invalidity of this schema. For example, the loss of the ether from the turn-of-the-century electrodynamic theories demonstrates this at the level of ontology, and the dynamics of the Rutherford-Bohr atom vis-à-vis the classical energy principles for rotating systems demonstrates it at the level of laws. Of course, the realist might respond that there is no question of a strict inference between being well confirmed and being approximately true (in the relevant respects), but there is a probable inference of some sort. But of what sort? Certainly there is no probability relation that rests on inductive evidence here. For there is no independent evidence for the relation of approximate truth itself; at least, the realist has yet to produce any evidence that is independent of the argument under examination. But if the probabilities are not grounded inductively, then how else? Here, I think the realist may well try to fall back on his original strategy, and suggest that being approximately true provides the best explanation for being well confirmed. This move throws us back to the ground-level realist argument, the argument from specific success to an approximately true description of reality, which Laudan has criticized. I should point out, before looking at the third question, that if this last move is the one the realist wants to make, then his success at the methodological level can be no better than his success at the ground level. If he fails there, he fails across the board.

The third question, and the one I think the realist puts most weight on, is why does the small-handful strategy work so well. The instrumentalist, for example, is thought to have no answer here. He must just note that it does work well, and be content with that. The realist, however, can explain why it works by citing the transfer of approximate truth from predecessor theories to the successor theories. But what does this explain? At best, it explains why the successor theories cover the same ground as well as their predecessors, for the conservative strategy under consideration assures that. But note that here the instrumentalist can offer the same account: if we insist on preserving the well-confirmed components of earlier theories in later theories, then, of course the later ones will do well over the well-confirmed ground. The difficulty, however, is not here at all but rather is in how to account for the successes of the later theories in new ground or with respect to novel predictions, or in overcoming the anomalies of the earlier theories. And what can the realist possibly say in this area except that the theorist, in proposing a new theory, has happened to make a good guess? For nothing in the approximate truth of the old theory can guarantee (or even make it likely) that modifying the theory in its less-confirmed parts will produce a progressive shift. The history of science shows well enough how such tinkering succeeds only now and again, and fails for the most part. This history of failures can scarcely be adduced to explain the occasional success. The idea that by extending what is approximately true one is likely to bring new approximate truth is a chimera. It finds support neither in the logic of approximate truth nor in the history of science. The problem for the realist is how to explain the *occasional success* of a strategy that *usually fails*.[8] I think he has no special resources with which to do this. In particular, his usual fallback onto approximate truth provides nothing more than a gentle pillow. He may rest on it comfortably, but it does not really help to move his cause forward.

The problem of the small handful raises three challenges: why small, why narrowly related, and why does it work? The realist has no answer for the first of these, begs the question as to the truth of explanatory hypotheses on the second, and has no resources for addressing the third. For comparison, it may be useful to see how well his archenemy, the instrumentalist, fares on the same turf. The instrumentalist, I think, has a substantial basis for addressing the questions of smallness and narrowness, for he can point out that it is extremely difficult to come up with alternative theories that satisfy the many empirical constraints posed by the instrumental success of theories already in the field. Often it is hard enough to come up with even one such alternative. Moreover, the common apprenticeship of scientists working in the same area certainly has the effect of narrowing down the range of options by channeling thought into the commonly accepted categories. If we add to this the instrumentally justified rule, "If it has worked well in the past, try it again," then we get a rather good account, I think, of why there is usually only a small and narrow handful. As to why this strategy works to produce instrumentally successful science, we have already noted that for the most part it does not. Most of what this strategy produces are failures. It is a quirk of scientific memory that this fact gets obscured, much as do the memories of bad times during a holiday vacation when we recount all our "wonderful" vacation adventures to a friend. Those instrumentalists who incline to a general account of knowledge as a social construction can go further at this juncture, and lean on the sociology of science to explain how the scientific community "creates" its knowledge. I am content just to back off here and note that over the problem of the small handful, the instrumentalist scores at least two out of three, whereas the realist, left to his own devices, has struck out.[9]

I think the source of the realist's failure here is endemic to the methodological level, infecting all of his arguments in this domain. It resides, in the first instance, in his repeating the question-begging move from explanatory efficacy to the truth of the explanatory hypothesis. And in the second instance, it resides in his twofold mishandling of the concept of approximate truth: first, in his trying to project from some body of assumed approximate truths to some further and novel such truths, and second, in his needing genuine access to the relation of correspondence. There are no general connections of this first sort, however, sanctioned by the logic of approximate truth, nor secondly, any such warranted access. However, the realist must pretend that there are, in order to claim explanatory power for his realism. We have seen those two agents infecting the realist way with the problem of the small handful. Let me show them at work in another methodological favorite of the realist, the "problem of conjunctions."

The problem of conjunctions is this. If T and T' are independently well-confirmed, explanatory theories, and if no shared term is ambiguous between the two, then we expect the conjunction of T and T' to be a reliable predictive instrument (provided, of course, that the theories are not mutually inconsistent). Why? challenges the realist, and he answers as follows. If we make the realist assumption that T and T', being well confirmed, are approximately true of the entities (etc.) to which they refer, and if the unambiguity requirement is taken realistically as requiring a domain of common reference, then the conjunction of the two theories will also be approximately true and, hence, it will produce reliable observational predictions. Q.E.D.

But notice our agents at work. First, the realist makes the question-begging move from explanations to their approximate truth, and then he mistreats approximate truth. For nothing in the logic of approximate truth sanctions the inference from "T is approximately true" and "T' is approximately true" to the conclusion that the conjunction "$T \cdot T'$" is approximately true. Rather, in general, the tightness of an approximation dissipates as we pile on further approximations. If T is within ε, in its estimation of some parameter, and T' is also within ε, then the only general thing we can say is that the conjunction will be within 2ε of the parameter. Thus, the logic of approximate truth should lead us to the opposite conclusion here; that is, that the conjunction of two theories is, in general, *less* reliable than either (over their common domain). But this is neither what we expect nor what we find. Thus, it seems quite implausible that our actual expectations about the reliability of conjunctions rest on the realist's stock of approximate truths.

Of course, the realist could try to retrench here and pose an additional requirement of some sort of uniformity on the character of the approximations, as between T and T'.[10] It is difficult to see how the realist could do this successfully without making reference to the distance between the approximations and "the truth." For what kind of internalist requirement could possibly insure the narrowing of this distance? But the realist is in no position to impose such requirements, since neither he nor anyone else has the requisite access to "the truth." Thus, whatever uniformity-of-approximation condition the realist might impose, we could still demand to be shown that this leads closer to the truth, not farther away. The realist will have no demonstration, except to point out to us that it all works (sometimes!). But that was the original puzzle.[11] Actually, I think the puzzle is not very difficult. For surely; if we do not entangle ourselves with issues over approximation, there is no deep mystery as to why two compatible and successful theories lead us to expect their conjunction to be successful. For in forming the conjunction, we just add the reliable predictions of one onto the reliable predictions of the other, having antecedently ruled out the possibility of conflict.

There is more to be said about this topic. In particular, we need to address the question as to why we expect the logical gears of the two theories to mesh. However, I think that a discussion of the realist position here would only bring up the same methodological and logical problems that we have already uncovered at the center of the realist argument.

Indeed, this schema of knots in the realist argument applies across the board and vitiates every single argument at the methodological level. Thus my conclusion here is harsh, indeed. The methodological arguments for realism fail, even though, were they successful, they would still not support the case. For the general strategy they are supposed to implement is just not stringent enough to provide rational support for realism. In the next two sections, I will try to show that this situation is just as well, for realism has not always been a progressive factor in the development of science and, anyway, there is a position other than realism that is more attractive.

■ | Realism and Progress

If we examine the two twentieth-century giants among physical theories, relativity and the quantum theory, we find a living refutation of the realist's claim that only his view of science explains its progress, and we find some curious twists and contrasts over realism as well. The theories of relativity are almost singlehandedly the work of Albert Einstein. Einstein's early positivism and his methodological debt to Mach (and Hume) leap right out of the pages of the 1905 paper on special relativity.[12] The same positivist strain is evident in the 1916 general relativity paper as well, where Einstein (in Section 3 of that paper) tries to justify his requirement of general covariance by means of a suspicious-looking verificationist argument which, he says, "takes away from space and time the last remnants of physical objectivity."[13] A study of his tortured path to general relativity[14] shows the repeated use of this Machist line, always used to deny that some concept has a real referent. Whatever other, competing strains there were in Einstein's philosophical orientation (and there certainly were others), it would be hard to deny the importance of this instrumentalist/positivist attitude in liberating Einstein from various realist commitments. Indeed, on another occasion, I would argue in detail that without the "freedom from reality" provided by his early reverence for Mach, a central tumbler necessary to unlock the secret of special relativity would never have fallen into place.[15] A few years after his work on general relativity, however, roughly around 1920, Einstein underwent a philosophical conversion, turning away from his positivist youth (he was forty-one in 1920) and becoming deeply committed to realism.[16] His subsequent battle with the quantum theory, for example, was fought much more over the issue of realism than it was over the issue of causality or determinism (as it is usually portrayed). In particular, following his conversion, Einstein wanted to claim genuine reality for the central theoretical entities of the general theory, the four-dimensional space-time manifold and associated tensor fields. This is a serious business for if we grant his claim, then not only do space and time cease to be real but so do virtually all of the usual dynamical quantities.[17] Thus motion, as we understand it, itself ceases to be real. The current generation of philosophers of space and time (led by Howard Stein and John Earman) have followed Einstein's lead here. But, interestingly, not only do these ideas boggle the mind of the average man in the street (like you and me), they boggle most contemporary scientific minds as well.[18] That is, I believe the majority opinion among working, knowledgeable scientists is that general relativity provides a magnificent organizing tool for treating certain gravitational problems in astrophysics and cosmology. But few, I believe, give credence to the kind of realist existence and nonexistence claims that I have been mentioning. For relativistic physics, then, it appears that a nonrealist attitude was important in its development, that the founder nevertheless espoused a realist attitude to the finished product, but that most who actually use it think of the theory as a powerful instrument, rather than as expressing a "big truth."

With quantum theory, this sequence gets a twist. Heisenberg's seminal paper of 1925 is prefaced by the following abstract, announcing, in effect, his philosophical stance: "In this paper an attempt will be made to obtain bases for a quantum-theoretical mechanics based exclusively on relations between quantities observable in principle."[19] In the body of the paper, Heisenberg not only rejects any reference to unobservables; he also moves away from the very idea that one should try to form any picture of a reality underlying his mechanics. To be sure, Schrödinger, the second father of

quantum theory, seems originally to have had a vague picture of an underlying wavelike reality for his own equation. But he was quick to see the
difficulties here and, just as quickly, although reluctantly, abandoned the
attempt to interpolate any reference to reality.[20] These instrumentalist moves,
away from a realist construal of the emerging quantum theory, were given
particular force by Bohr's so-called "philosophy of complementarity"; and
this nonrealist position was consolidated at the time of the famous Solvay
conference, in October of 1927, and is firmly in place today. Such quantum
nonrealism is part of what every graduate physicist learns and practices. It is
the conceptual backdrop to all the brilliant successes in atomic, nuclear,
and particle physics over the past fifty years. Physicists have learned to think
about their theory in a highly nonrealist way, and doing just that has brought
about the most marvelous predictive success in the history of science.

The war between Einstein, the realist, and Bohr, the nonrealist, over the
interpretation of quantum theory was not, I believe, just a sideshow in physics, nor an idle intellectual exercise. It was an important endeavor undertaken
by Bohr on behalf of the enterprise of physics as a progressive science. For
Bohr believed (and this fear was shared by Heisenberg, Sommerfield, Pauli,
and Born—and all the major players) that Einstein's realism, if taken seriously, would block the consolidation and articulation of the new physics and,
thereby, stop the progress of science. They were afraid, in particular, that
Einstein's realism would lead the next generation of the brightest and best
students into scientific dead ends. Alfred Landé, for example, as a graduate
student, was interested in spending some time in Berlin to sound out Einstein's ideas. His supervisor was Sommerfeld, and recalling this period,
Landé writes

> The more pragmatic Sommerfeld . . . warned his students, one of them this
> writer, not to spend too much time on the hopeless task of "explaining" the
> quantum but rather to accept it as fundamental and help work out its
> consequences.[21]

The task of "explaining" the quantum, of course, is the realist program
for identifying a reality underlying the formulas of the theory and thereby
explaining the predictive success of the formulas as approximately true
descriptions of this reality. It is this program that I have criticized in the first
part of this paper, and this same program that the builders of quantum
theory saw as a scientific dead end. Einstein knew perfectly well that the
issue was joined right here. In the summer of 1935, he wrote to Schrödinger,

> The real problem is that physics is a kind of metaphysics; physics describes
> 'reality'. But we do not know what 'reality' is. We know it only through physical
> description. . . . But the Talmudic philosopher sniffs at 'reality', as at a fright-
> ening creature of the naive mind.[22]

By avoiding the bogey of an underlying reality, the "Talmudic" origina- tors of quantum theory seem to have set subsequent generations on pre- cisely the right path. Those inspired by realist ambitions have produced no predictively successful physics. Neither Einstein's conception of a unified field nor the ideas of the de Broglie group about pilot waves, nor the Bohm- inspired interest in hidden variables has made for scientific progress. To be sure, several philosophers of physics, including another Hilary Putnam, and myself, have fought a battle over the last decade to show that the quantum theory is at least consistent with some kind of underlying reality. I believe that Hilary has abandoned the cause, perhaps in part on account of the recent Bell-inequality problem over correlation experiments, a problem that van Fraassen calls "the charybdis of realism."[23] My own recent work in the area suggests that we may still be able to keep realism afloat in this whirl- pool.[24] But the possibility (as I still see it) for a realist account of the quantum domain should not lead us away from appreciating the historical facts of the matter.

One can hardly doubt the importance of a nonrealist attitude for the development and practically infinite success of the quantum theory. His- torical counterfactuals are always tricky, but the sterility of actual realist programs in this area at least suggests that Bohr and company were right in believing that the road to scientific progress here would have been blocked by realism. The founders of quantum theory never turned on the nonrealist attitude that served them so well. Perhaps that is because the central under- lying theoretical device of quantum theory, the densities of a complex- valued and infinite-dimensional wave function, are even harder to take seriously than is the four-dimensional manifold of relativity. But now, there comes a most curious twist. For just as the practitioners of relativity, I have suggested, ignore the *realist* interpretation in favor of a more pragmatic attitude toward the space-time structure, the quantum physicists would appear to make a similar reversal and to forget their non-realist history and allegiance when it comes time to talk about new discoveries.

Thus, anyone in the business will tell you about the exciting period, in the fall of 1974, when the particle group at Brookhaven, led by Samuel Ting, discovered the J particle, just as a Stanford team at the Stanford Lin- ear Accelerator Center (SLAC), under Burton Richter, independently found a new particle they called "ψ". These turned out to be one and the same, the so-called ψ/J particle* (Mass 3,098 MeV, Spin 1, Resonance 67 KeV, Strangeness 0). To explain this new entity, the theoreticians were led to intro- duce a new kind of quark, the so-called charmed quark. The ψ/J particle is then thought to be made up out of a charmed quark and an anticharmed quark, with their respective spins aligned. But if this is correct, then there

---

* Often written "J/ ψ" and referred to as the "gypsy" particle (for "J-psi"). For their discovery, Richter and Ting shared the Nobel Prize for physics in 1976.

ought to be other such pairs anti-aligned, or with variable spin alignments, and these ought to make up quite new observable particles. Such predictions from the charmed-quark model have turned out to be confirmed in various experiments.

In this example, I have been intentionally a bit more descriptive in order to convey the realist feel to the way scientists speak in this area. For I want to ask whether this is a return to realism or whether, instead, it can somehow be reconciled with a fundamentally nonrealist attitude.[25] I believe that the nonrealist option is correct, but I will not defend that answer here, however, because its defense involves the articulation of a compelling and viable form of nonrealism: and that is the task of the third (and final) section of this paper.

■ | Nonrealism

Even if the realist happens to be a talented philosopher, I do not believe that, in his heart, he relies for his realism on the rather sophisticated form of abductive argument that I have examined and rejected in the first section of this paper, and which the history of twentieth-century physics shows to be fallacious. Rather, if his heart is like mine (and I *do* believe in a common nature), then I suggest that a more simple and homely sort of argument is what grips him. It is this, and I will put it in the first person. I certainly trust the evidence of my senses, on the whole, with regard to the existence and features of everyday objects. And I have similar confidence in the system of "check, double-check, triple-check" of scientific investigation, as well as the other safeguards built into the institutions of science. So, if the scientists tell me that there really are molecules, and atoms, and ψ/J particles and, who knows, maybe even quarks, then so be it. I trust them and, thus, must accept that there really are such things, with their attendant properties and relations. Moreover, if the instrumentalist (or some other member of the species "non-realistica") comes along to say that these entities, and their attendants, are just fictions (or the like), then I see no more reason to believe him than to believe that *he is* a fiction, made up (somehow) to do a job on me; which I do not believe. It seems, then, that I had better be a realist. One can summarize this homely and compelling line as follows: it is possible to accept the evidence of one's senses and to accept, *in the same way*, the confirmed results of science only for a realist; hence, I should be one (and so should you!).

What is it to accept the evidence of one's senses and, *in the same way*, to accept confirmed scientific theories? It is to take them into one's life as true, with all that implies concerning adjusting one's behavior, practical and theoretical, to accommodate these truths. Now, of course, there are truths, and truths. Some are more central to us and our lives, some less so.

I might be mistaken about anything, but were I mistaken about where I am right now, that might affect me more than would my perhaps mistaken belief in charmed quarks. Thus, it is compatible with the homely line of argument that some of the scientific beliefs that I hold are less central than some, for example, perceptual beliefs. Of course, were I deeply in the charmed-quark business, giving up that belief might be more difficult than giving up some at the perceptual level. (Thus we get the phenomenon of "seeing what you believe," as is well known to all thoughtful people.) When the homely line asks us, then, to accept the scientific results "in the same way" in which we accept the evidence of our senses, I take it that we are to accept them both as true. I take it that we are being asked not to distinguish between kinds of truth or modes of existence or the like, but only among truths themselves, in terms of centrality, degrees of belief, or such.

Let us suppose this understood. Now, do you think that Bohr, the arch-enemy of realism, could toe the homely line? Could Bohr, fighting for the sake of science (against Einstein's realism) have felt compelled either to give up the results of science, or else to assign to its "truths" some category different from the truths of everyday life? It seems unlikely. And thus, unless we uncharitably think Bohr inconsistent on this basic issue, we might well come to question whether there is any necessary connection moving us from accepting the results of science as true to being a realist.[26]

Let me use the term 'antirealist' to refer to any of the many different specific enemies of realism: the idealist, the instrumentalist, the phenomenalist, the empiricist (constructive or not), the conventionalist, the constructivist, the pragmatist, and so forth. Then, it seems to me that both the realist and the antirealist must toe what I have been calling "the homely line." That is, they must both accept the certified results of science as on par with more homely and familiarly supported claims. That is not to say that one party (or the other) cannot distinguish more from less well-confirmed claims at home or in science; nor that one cannot single out some particular mode of inference (such as inference to the best explanation) and worry over its reliability, both at home and away. It is just that one must maintain parity. Let us say, then, that both realist and antirealist accept the results of scientific investigations as 'true', on par with more homely truths. (I realize that some antirealists would rather use a different word, but no matter.) And call this acceptance of scientific truths the "core position."[27] What distinguishes realists from antirealists, then, is what they add onto this core position.

The antirealist may add onto the core position a particular analysis of the concept of truth, as in the pragmatic and instrumentalist and conventionalist conceptions of truth. Or the antirealist may add on a special analysis of concepts, as in idealism, constructivism, phenomenalism, and in some varieties of empiricism. These addenda will then issue in a special meaning, say, for existence statements. Or the antirealist may add on certain methodological strictures, pointing a wary finger at some particular inferential tool, or constructing his own account for some particular aspects of science (e.g., explanations or laws). Typically, the antirealist will make several such additions to the core.

What then of the realist, what does he add to his core acceptance of the results of science as really true? My colleague, Charles Chastain, suggested what I think is the most graphic way of stating the answer—namely, that what the realist adds on is a desk-thumping, foot-stamping shout of "Really!" So, when the realist and antirealist agree, say, that there really are electrons and that they really carry a unit negative charge and really do have a small mass (of about 9.1×10^{-28} grams), what the realist wants to add is the emphasis that all this is really so. "There really are electrons, really!" This typical realist emphasis serves both a negative and a positive function. Negatively, it is meant to deny the additions that the antirealist would make to that core acceptance which both parties share. The realist wants to deny, for example, the phenomenalistic reduction of concepts or the pragmatic conception of truth. The realist thinks that these addenda take away from the substantiality of the accepted claims to truth or existence. "No," says he, "they *really* exist, and not in just your diminished antirealist sense." Positively, the realist wants to explain the robust sense in which *he* takes these claims to truth or existence, namely, as claims about reality—what is really, really the case. The full-blown version of this involves the conception of truth as correspondence with the world, and the surrogate use of approximate truth as near-correspondence. We have already seen how these ideas of correspondence and approximate truth are supposed to explain what *makes* the truth *true* whereas, in fact, they function as mere trappings, that is, as superficial decorations that may well attract our attention but do not compel rational belief. Like the extra "really," they are an arresting foot-thump and, logically speaking, of no more force.

It seems to me that when we contrast the realist and the antirealist in terms of what they each want to add to the core position, a third alternative emerges—and an attractive one at that. It is the core position itself, *and all by itself.* If I am correct in thinking that, at heart, the grip of realism only extends to the homely connection of everyday truths with scientific truths, and that good sense dictates our acceptance of the one on the same basis as our acceptance of the other, then the homely line makes the core position, all by itself, a compelling one, one that we ought to take to heart. Let us try to do so, and to see whether it constitutes a philosophy, and an attitude toward science, that we can live by.

The core position is neither realist nor antirealist; it mediates between the two. It would be nice to have a name for this position, but it would be a shame to appropriate another "ism" on its behalf, for then it would appear to be just one of the many contenders for ontological allegiance. I think it is not just one of that crowd but rather, as the homely line behind it suggests, it is for commonsense epistemology—the natural ontological attitude. Thus, let me introduce the acronym *NOA* (pronounced as in "Noah"), for *natural ontological attitude*, and, henceforth, refer to the core position under that designation.

To begin showing how NOA makes for an adequate philosophical stance toward science, let us see what it has to say about ontology. When NOA counsels us to accept the results of science as true, I take it that we are to treat truth in the usual referential way, so that a sentence (or statement) is true just in case the entities referred to stand in the referred-to relations. Thus, NOA sanctions ordinary referential semantics and commits us, via truth, to the existence of the individuals, properties, relations, processes, and so forth referred to by the scientific statements that we accept as true. Our belief in their existence will be just as strong (or weak) as our belief in the truth of the bit of science involved, and degrees of belief here, presumably, will be tutored by ordinary relations of confirmation and evidential support, subject to the usual scientific canons. In taking this referential stance, NOA is not committed to the progressivism that seems inherent in realism. For the realist, as an article of faith, sees scientific success, over the long run, as bringing us closer to the truth. His whole explanatory enterprise, using approximate truth, forces his hand in this way. But, a "NOAer" (pronounced as "knower") is not so committed. As a scientist, say, within the context of the tradition in which he works, the NOAer, of course, will believe in the existence of those entities to which his theories refer. But should the tradition change, say in the manner of the conceptual revolutions that Kuhn dubs "paradigm shifts," then nothing in NOA dictates that the change be assimilated as being progressive, that is, as a change where we learn more accurately about *the same things*. NOA is perfectly consistent with the Kuhnian alternative, which construes such changes as wholesale changes of reference. Unlike the realist, adherents to NOA are free to examine the facts in cases of paradigm shift, and to see whether or not a convincing case for stability of reference across paradigms can be made without superimposing on these facts a realist-progressivist superstructure. I have argued elsewhere that if one makes oneself free, as NOA enables one to do, then the facts of the matter will not usually settle the case;[28] and that this is a good reason for thinking that cases of so-called "incommensurability" are, in fact, genuine cases where the question of stability of reference is indeterminate. NOA, I think, is the right philosophical position for such conclusions. It sanctions reference and existence claims, but it does not force the history of science into prefit molds.

So far I have managed to avoid what, for the realist, is the essential point, for what of the "external world"? How can I talk of reference and of existence claims unless I am talking about referring to things right out there in the world? And here, of course, the realist, again, wants to stamp his feet.[29] I think the problem that makes the realist want to stamp his feet, shouting "Really!" (and invoking the external world) has to do with the stance the realist tries to take vis-à-vis the game of science. The realist, as it were, tries to stand outside the arena watching the ongoing game and then tries to judge (from this external point of view) what the point is. It is, he says, *about* some area external to the game. The realist, I think, is fooling himself. For he cannot (really!) stand outside the arena, nor can he survey some area off the playing field and mark it out as what the game is about.

Let me try to address these two points. How are we to arrive at the judgment that, in addition to, say, having a rather small mass, electrons are objects "out there in the external world"? Certainly, we can stand off from the electron game and survey its claims, methods, predictive success, and so forth. But what stance could we take that would enable us to judge what the theory of electrons is *about*, other than agreeing that it is about electrons? It is not like matching a blueprint to a house being built, or a map route to a country road. For we are *in* the world, both physically and conceptually.[30] That is, *we* are among the objects of science, and the concepts and procedures that we use to make judgments of subject matter and correct application are themselves part of that same scientific world. Epistemologically, the situation is very much like the situation with regard to the justification of induction. For the problem of the external world (so-called) is how to satisfy the realist's demand that we justify the existence claims sanctioned by science (and, therefore, by NOA) as claims to the existence of entities "out there." In the case of induction, it is clear that only an inductive justification will do, and it is equally clear that no inductive justification will do at all. So too with the external world, for only ordinary scientific inferences to existence will do, and yet none of them satisfies the demand for showing that the existent is really "out there." I think we ought to follow Hume's prescription on induction, with regard to the external world. There is no possibility for justifying the kind of externality that realism requires, yet it may well be that, in fact, we cannot help yearning for just such a comforting grip on reality. I shall return to this theme at the close of the paper.

If I am right, then the realist is chasing a phantom, and we cannot actually do more, with regard to existence claims, than follow scientific practice, just as NOA suggests. What then of the other challenges raised by realism? Can we find in NOA the resources for understanding scientific practice? In particular (since it was the topic of the first part of this paper), does NOA help us to understand the scientific method, such as the problems of the small handful or of conjunctions? The sticking point with the small handful was to account for why the few and narrow alternatives that we can come up with, result in successful novel predictions, and the like. The background was to keep in mind that most such narrow alternatives are not successful. I think that NOA has only this to say. If you believe that guessing based on some truths is more likely to succeed than guessing pure and simple, then if our earlier theories were in large part true and if our refinements of them conserve the true parts, then guessing on this basis has some relative likelihood of success. I think this is a weak account, but then I think the phenomenon here does not allow for anything much stronger since, for the most part, such guesswork fails. In the same way, NOA can help with the problem of conjunctions (and, more generally, with problems of logical combinations). For if two consistent theories in fact have overlapping domains (a fact, as I have just suggested, that is not so often decidable), and if the theories also have true things to say about members in the overlap, then conjoining the theories just adds to the truths of each and, thus, *may*, in conjunction, yield new truths. Where one finds other successful methodological rules, I think we will find NOA's grip on the truth sufficient to account for the utility of the rules.

Unlike the realist, however, I would not tout NOA's success at making science fairly intelligible as an argument in its favor, vis-à-vis realism or various antirealisms. For NOA's accounts are available to these fellows, too, provided what they add to NOA does not negate its appeal to the truth, as does a verificationist account of truth or the realists' longing for approximate truth. Moreover, as I made plain enough in the first section of this paper, I am sensitive to the possibility that explanatory efficacy can be achieved without the explanatory hypothesis being true. NOA may well make science seem fairly intelligible and even rational, but NOA could be quite the wrong view of science for all that. If we posit as a constraint on philosophizing about science that the scientific enterprise should come out in our philosophy as not too unintelligible or irrational, then, perhaps, we can say that NOA passes a minimal standard for a philosophy of science.

Indeed, perhaps the greatest virtue of NOA is to call attention to just how minimal an adequate philosophy of science can be. (In this respect, NOA might be compared to the minimalist movement in art.) For example, NOA helps us to see that realism differs from various antirealisms in this way: realism adds an outer direction to NOA, that is, the external world and the correspondence relation of approximate truth; antirealisms (typically) add an inner direction, that is, human-oriented reductions of truth, or concepts, or explanations (as in my opening citation from Hume). NOA suggests that the legitimate features of these additions are already contained in the presumed equal status of everyday truths with scientific ones, and in our accepting them both as *truths*. No other additions are legitimate, and none are required.

It will be apparent by now that a distinctive feature of NOA, one that separates it from similar views currently in the air, is NOA's stubborn refusal to amplify the concept of truth, by providing a theory or analysis (or even a metaphorical picture). Rather, NOA recognizes in "truth" a concept already in use and agrees to abide by the standard rules of usage. These rules involve a Davidsonian-Tarskian referential semantics, and they support a thoroughly classical logic of inference. Thus NOA respects the customary "grammar" of 'truth' (and its cognates). Likewise, NOA respects the customary epistemology, which grounds judgments of truth in perceptual judgments and various confirmation relations. As with the use of other concepts, disagreements are bound to arise over what is true (for instance, as to whether inference to the best explanation is always truth-conferring). NOA pretends to no resources for settling these disputes, for NOA takes to heart the great lesson of twentieth-century analytic and Continental philosophy, namely, that there *are* no general methodological or philosophical resources for deciding such things. The mistake common to realism and all the antirealisms alike is their commitment to the existence of such nonexistent resources. If pressed to answer the question of what, then, does it *mean* to say that something is true (or to what does the truth of so-and-so commit one), NOA will reply by pointing out the logical relations engendered by the specific claim and by focusing, then, on the concrete historical circumstances that ground that particular judgment of truth. For, after all, there *is* nothing more to say.[31]

Because of its parsimony, I think the minimalist stance represented by NOA marks a revolutionary approach to understanding science. It is, I would suggest, as profound in its own way as was the revolution in our conception of morality, when we came to see that founding morality on God and His Order was *also* neither legitimate nor necessary. Just as the typical theological moralist of the eighteenth century would feel bereft to read, say, the pages of *Ethics*, so I think the realist must feel similarly when NOA removes that "correspondence to the external world" for which he so longs. I too have regret for that lost paradise, and too often slip into the realist fantasy. I use my understanding of twentieth-century physics to help me firm up my convictions about NOA, and I recall some words of Mach, which I offer as a comfort and as a closing. With reference to realism, Mach writes

> It has arisen in the process of immeasurable time without the intentional assistance of man. It is a product of nature, and preserved by nature. Everything that philosophy has accomplished . . . is, as compared with it, but an insignificant and ephemeral product of art. The fact is, every thinker, every philosopher, the moment he is forced to abandon his one-sided intellectual occupation . . . , immediately returns [to realism].

> Nor is it the purpose of these "introductory remarks" to discredit the standpoint [of realism]. The task which we have set ourselves is simply to show why and for what purpose we hold that standpoint during most of our lives, and why and for what purpose we are . . . obliged to abandon it.

These lines are taken from Mach's *The Analysis of Sensations* (Sec. 14). I recommend that book as effective realism-therapy, a therapy that works best (as Mach suggests) when accompanied by historicophysical investigations (real versions of the breakneck history of my second section, "Realism and Progress"). For a better philosophy, however, I recommend NOA.[32]

■ | Notes

1. In the final section, I call this postrealism "NOA." Among recent views that relate to NOA, I would include Hilary Putnam's "internal realism," Richard Rorty's

"epistemological behaviorism," the "semantic realism" espoused by Paul Horwich, parts of the "Mother Nature" story told by William Lycan, and the defense of common sense worked out by Joseph Pitt (as a way of reconciling W. Sellars's manifest and scientific images). For references, see Hilary Putnam, *Meaning and the Moral Sciences* (London: Routledge and Kegan Paul, 1978); Richard Rorty, *Philosophy and the Mirror of Nature* (Princeton: Princeton University Press, 1979); Paul Horwich, "Three Forms of Realism," *Synthese* 51 (1982): 181–201; William G. Lycan, "Epistemic Value" (preprint, 1982) [*Synthese* 64 (1985): 137–64]; and Joseph C. Pitt, *Pictures, Images and Conceptual Change* (Dordrecht: D. Reidel, 1981). The reader will note that some of the above consider their views a species of realism, whereas others consider their views antirealist. As explained below, NOA marks the divide; hence its "postrealism."

2. Larry Laudan, "A Confutation of Convergent Realism," [*Philosophy of Science* 48 (1981): 19–48; excerpted in this chapter].

3. Richard N. Boyd, "Scientific Realism and Naturalistic Epistemology," in *PSA* (1980), vol. 2, ed. P. D. Asquith and R. N. Giere (E. Lansing: Philosophy of Science Association, 1981), 613–662. See also, Boyd's article ["The Current Status of Scientific Realism," *Erkenntnis* 19 (1983): 45–90], and further references there.

4. Hilary Putnam, "The Meaning of 'Meaning'," in *Language, Mind and Knowledge*, ed. K. Gunderson (Minneapolis: University of Minnesota Press, 1975), 131–193. See also his article ["What is 'Realism'?" in *Meaning and the Moral Sciences* (London: Routledge and Kegan Paul, 1978), 18–33].

5. Bas C. van Fraasen, *The Scientific Image* (Oxford: The Clarendon Press, 1980). See especially pp. 97–101 for a discussion of the truth of explanatory theories. To see that the recent discussion of realism is joined right here, one should contrast van Fraassen with W. H. Newton-Smith, *The Rationality of Science* (London: Routledge and Kegan Paul, 1981), esp. chap. 8.

6. Nancy Cartwright's *How the Laws of Physics Lie* (Oxford: Oxford University Press, 1983) includes some marvelous essays on these issues.

7. Some realists may look for genuine support, and not just solace, in such a coherentist line. They may see in their realism a basis for general epistemology, philosophy of language, and so forth (as does Boyd, "Scientific Realism and Naturalistic Epistemology"). If they find in all this a coherent and comprehensive world view, then they might want to argue for their philosophy as Wilhelm Wien argued (in 1909) for special relativity, "What speaks for it most of all is the inner consistency which makes it possible to lay a foundation having no self-contradictions, one that applies to the totality of physical appearances." Quoted by Gerald Holton, "Einstein's Scientific Program: Formative Years" in *Some Strangeness in the Proportion*, ed. H. Woolf (Reading: Addison-Wesley, 1980), 58. Insofar as the realist moves away from the abductive defense of realism to seek support, instead, from the merits of a comprehensive philosophical system with a realist core, he marks as a failure the bulk of recent defenses of realism. Even so, he will not avoid the critique pursued in the text. For although my argument above has been directed, in particular, against the abductive strategy, it is itself based on a more general maxim, namely, that the form of argument used to support realism must be more stringent than the form of argument embedded in the very scientific practice that realism itself is supposed to ground—on pain of begging the question. Just as the abductive strategy fails because

it violates this maxim, so too would the coherentist strategy, should the realist turn from one to the other. For, as we see from the words of Wien, the same coherentist line that the realist would appropriate for his own support, is part of ordinary scientific practice in framing judgments about competing theories. It is, therefore, not a line of defense available to the realist. Moreover, just as the truth-bearing status of abduction is an issue dividing realists from various nonrealists, so too is the status of coherence-based inference. Turning from abduction to coherence, therefore, still leaves the realist begging the question. Thus, when we bring out into the open the character of arguments *for realism*, we see quite plainly that they do not work.

In support of realism there seem to be only those "reasons of the heart" which, as Pascal says, reason does not know. Indeed, I have long felt that belief in realism involves a profound leap of faith, not at all dissimilar from the faith that animates deep religious convictions. I would welcome engagement with realists on this understanding, just as I enjoy conversation on a similar basis with my religious friends. The dialogue will proceed more fruitfully, I think, when the realists finally stop pretending to a rational support for their faith, which they do not have. Then we can all enjoy their intricate and sometimes beautiful philosophical constructions (of, e.g., knowledge, or reference, etc), even though, as nonbelievers, they may seem to us only wonder-full castles in the air.

8. I hope all readers of this essay will take this idea to heart. For in formulating the question as how to explain why the methods of science lead to instrumental success, the realist has seriously misstated the explanandum. Overwhelmingly, the results of the conscientious pursuit of scientific inquiry are failures: failed theories, failed hypotheses, failed conjectures, inaccurate measurements, incorrect estimations of parameters, fallacious causal inferences, and so forth. If explanations are appropriate here, then what requires explaining is why the very same methods produce an overwhelming background of failures and, occasionally, also a pattern of successes. The realist literature has not yet begun to address this question, much less to offer even a hint of how to answer it.

9. Of course, the realist can appropriate the devices and answers of the instrumentalist, but that would be cheating, and it would, anyway, not provide the desired support of realism per se.

10. Paul Teller has made this suggestion to me in conversation.

11. Ilkka Niiniluoto's "What Shall We Do with Verisimilitude?" *Philosophy of Science* 49 (1982): 181–197, contains interesting formal constructions for "degree of truthlikeness," and related versimilia. As conjectured above, they rely on an unspecified correspondence relation to the truth and on measures of the "distance" from the truth. Moreover, they fail to sanction that projection from some approximate truths to other, novel truths, which lies at the core of realist rationalizations.

12. See Gerald Holton, "Mach, Einstein, and the Search for Reality," in his *Thematic Origins of Scientific Thought* (Cambridge: Harvard University Press, 1973), 219–259. I have tried to work out the precise role of this positivist methodology in my "The Young Einstein and the Old Einstein," in *Essays in Memory of Imré Lakatos*, ed. R. S. Cohen et al. (Dordrecht: D. Reidel, 1976), 145–159.

13. A. Einstein et al., *The Principle of Relativity*, trans. W. Perrett and G. B. Jeffrey (New York: Dover, 1952), 117.

between the domain of chairs and tables and atoms, whose existence he plainly accepted, and that of electrons, where he seems to have thought the question of existence (and of realism, more generally) was no longer well defined. An illuminating and provocative discussion of Bohr's attitude toward realism is given by Paul Teller, "The Projection Postulate and Bohr's Interpretation of Quantum Mechanics," [in *PSA 1980*, vol. 2, ed. P. Asquith and R. Giere] pp. 201–223. Thanks, again, to Paul for helping to keep me honest.

27. In this context, for example, van Fraassen's "constructive empiricism" would prefer the concept of empirical adequacy, reserving "truth" for an (unspecified) literal interpretation and believing in that truth only among observables. It is clear, nevertheless, that constructive empiricism follows the homely line and accepts the core position. Indeed, this seems to be its primary motivating rationale. If we reread constructive empiricism in our terminology, then, we would say that it accepts the core position but adds to it a construal of truth as empirical adequacy. Thus, it is antirealist, just as suggested in the next paragraph below. I might mention here that in this classification Putnam's internal realism also comes out as antirealist. For Putnam also accepts the core position, but he would add to it a Peircean construal of truth as ideal rational acceptance. This is a mistake, which I expect that Putnam will realize and correct in future writings. He is criticized for it, soundly I think, by Paul Horwich ("Three Forms of Realism") whose own "semantic realism" turns out, in my classification, to be neither realist nor antirealist. Indeed, Horwich's views are quite similar to what is called "NOA" below, and could easily be read as sketching the philosophy of language most compatible with NOA. Finally, the "epistemological behaviorism" espoused by Rorty is a form of antirealism that seems to me very similar to Putnam's position, but achieving the core parity between science and common sense by means of an acceptance that is neither ideal nor especially rational, at least in the normative sense. (I beg the reader's indulgence over this summary treatment of complex and important positions. I have been responding to Nancy Cartwright's request to differentiate these recent views from NOA. . . . .)

28. "How to Compare Theories: Reference and Change," *Noûs* 9 (1975): 17–32.

29. In his remarks at the Greensboro conference [on realism, March 1982], my commentator, John King, suggested a compelling reason to prefer NOA over realism; namely, because NOA is less percussive! My thanks to John for this nifty idea, as well as for other comments.

30. "There is, I think, no theory-independent way to reconstruct phrases like 'really true'; the notion of match between the ontology of a theory and its 'real' counterpart in nature now seems to me illusive in principle." T. S. Kuhn, "Postscript," in *The Structure of Scientific Revolutions*, 2d ed. (Chicago: University of Chicago Press, 1970), 206. The same passage is cited for rebuttal by W. H. Newton-Smith, in *The Rationality of Science*. But the "rebuttal" sketched there in chapter 8, sections 4 and 5, not only runs afoul of the objections stated here in my first section, it also fails to provide for the required theory-independence. For Newton-Smith's explication of verisimilitude (p. 204) makes explicit reference to some unspecified background theory. (He offers either current science or the Peircean limit as candidates.) But this is not to rebut Kuhn's challenge (and mine); it is to concede its force.

31. No doubt I am optimistic, for one can always think of more to say. In particular, one could try to fashion a general, descriptive framework for codifying and classify-

14. John Earman and Clark Glymour, "Lost in the Tensors," *Studies in History and Philosophy of Science* 9 (1978): 251–278. The tortuous path detailed by Earman is sketched by B. Hoffmann, *Albert Einstein, Creator and Rebel* (New York: New American Library, 1972), 116–128. A nontechnical and illuminating account is given by John Stachel, "The Genesis of General Relativity," in *Einstein Symposium Berlin*, ed. H. Nelkowski et al. (Berlin: Springer-Verlag, 1980), 428–42.

15. I have in mind the role played by the analysis of simultaneity in Einstein's path to special relativity. Despite the important study by Arthur Miller, *Albert Einstein's Special Theory of Relativity* (Reading: Addison-Wesley, 1981), and an imaginative pioneering work by John Earman (and collaborators) ["On Writing the History of Special Relativity," in *PSA 1982*, vol. 2, ed. P. Asquith and T. Nickles (East Lansing, Mich.: Philosophy of Science Association, 1983), 403–16], . . . I think the role of positivist analysis in the 1905 paper has yet to be properly understood. . . . .

16. Peter Barker, "Einstein's Later Philosophy of Science," in *After Einstein*, ed. P. Barker and C. G. Shugart (Memphis: Memphis State University Press, 1981), 133–146, is a nice telling of this story. [See also Arthur Fine, *The Shaky Game*, 2d ed. (Chicago: University of Chicago Press, 1996), ch. 6.]

17. Roger Jones in "Realism About What?" [*Philosophy of Science* 58 (1991): 185–202] explains very nicely some of the difficulties here.

18. I think the ordinary, deflationist attitude of working scientists is much like that of Steven Weinberg, *Gravitation and Cosmology: Principles and Applications of the General Theory of Relativity* (New York: Wiley, 1972).

19. See B. L. van der Waerden, *Sources of Quantum Mechanics* (New York: Dover, 1967), 261.

20. See Linda Wessels, "Schrödinger's Route to Wave Mechanics," *Studies in History and Philosophy of Science* 10 (1979): 311–340.

21. A Landé, "Albert Einstein and the Quantum Riddle," *American Journal of Physics* 42 (1974): 460.

22. Letter to Schrödinger, June 19, 1935. See my "Einstein's Critique of Quantum Theory: The Roots and Significance of EPR," in *After Einstein* (see n. 16), 147–158, for a fuller discussion of the contents of this letter.

23. Bas van Fraassen, "The Charybdis of Realism: Epistemological Implications of Bell's Inequality," *Synthese* 52 (1982): 25–38.

24. See my "Antinomies of Entanglement: The Puzzling Case of The Tangled Statistics," *Journal of Philosophy* 79 (1982), for part of the discussion and for reference to other recent work.

25. The nonrealism that I attribute to students and practitioners of the quantum theory requires more discussion and distinguishing of cases and kinds than I have room for here. It is certainly not the all-or-nothing affair I make it appear in the text. [See Arthur Fine,] "Is Scientific Realism Compatible with Quantum Physics?" [in A. Fine, *The Shaky Game*, 151–171]. My thanks to Paul Teller and James Cushing, each of whom saw the need for more discussion here.

26. I should be a little more careful about the historical Bohr than I am in the text. For Bohr himself would seem to have wanted to truncate the homely line somewhere

ing such answers. Perhaps there would be something to be learned from such a descriptive, semantical framework. But what I am afraid of is that this enterprise, once launched, would lead to a proliferation of frameworks not so carefully descriptive. These would take on a life of their own, each pretending to ways (better than its rivals) to settle disputes over truth claims, or their import. What we need, however, is less bad philosophy, not more. So here, I believe, silence is indeed golden.

32. My thanks to Charles Chastain, Gerald Dworkin, and Paul Teller for useful preliminary conversations about realism and its rivals, but especially to Charles—for only he, then, (mostly) agreed with me, and surely that deserves special mention. This paper was written by me, but cothought by Micky Forbes. I don't know any longer whose ideas are whose. That means that the responsibility for errors and confusions is at least half Micky's (and she is two-thirds responsible for "NOA"). Finally, I am grateful to the many people who offered comments and criticisms at the [Greensboro] conference, and subsequently. I am also grateful to the National Science Foundation for a grant in support of this research.

Alan Musgrave

NOA's Ark—Fine
for Realism

Arthur Fine says that scientific realism is dead, drowned by floods of criticism. In its place he puts the *natural ontological attitude*, NOA, pronounced 'Noah'. Fine thinks that NOA is a minimalist view which is neither realist nor antirealist. I think that NOA is a thoroughly realist view: in NOA's Ark the realist can sail happily above the floods of criticism.

NOA stems from the 'homely line' that we should accept the results of science as true in the same way that we accept the evidence of the senses as true. Fine writes:

> Let us say, then, that both realist and antirealist accept the results of scientific investigations as 'true', on a par with more homely truths. . . . And call this acceptance of scientific truths the "core position". What distinguishes realists from antirealists, then, is what they add onto this core position. (Fine 1984a: 96 [1203])

NOA is the core position all by itself, California-pure, without additives.

This is mysterious. As usually understood, the realism–antirealism issue centres precisely on the question of truth. As usually understood, realists can accept Fine's core position, but antirealists cannot. Positivists deny the existence of the 'theoretical entities' of science, and think that any theory which asserts the existence of such entities is *false*. Instrumentalists think that scientific theories are tools or rules which are *neither true nor false*. Epistemological antirealists like van Fraassen or Laudan concede that theories have truth-values, even that some of them might be true, but insist that no theory should be *accepted as true*. None of these antirealist positions, as usually understood, is consistent with Fine's core position.

From *Philosophical Quarterly* 39 (1989): 383–98. In this article, the numbers in square brackets refer to pages in this volume.

The mystery unravels, it seems, when Fine says that antirealists often go in for some peculiar theory of truth:

> The antirealist may add onto the core position a particular analysis of the concept of truth, as in the pragmatic and instrumentalist and conventionalist conceptions of truth. . . . These addenda will then issue in a special meaning, say, for existence statements. (Fine 1984a: 97 [1203])

So the positivist is seen as saying, not that theoretical science is all false because there are no 'theoretical entities', but that some theoretical science is true because in its application to science 'true' *means useful*. The instrumentalist is seen as saying something similar. Van Fraassen is seen as saying, not that we should never accept a theory about the 'unobservable' as true, but that we sometimes may because in its application to such theories 'true' *means empirically adequate*. Laudan is seen as saying something similar.

It is certainly possible to see the antirealisms this way. But it is not the way the antirealists see themselves, nor is it the clearest way to see them. Adding to the core position a peculiar truth-theory for science actually demolishes that position. The results of science are not being accepted as true on a par with more homely truths. Homely statements are accepted as true in the homely sense of the term 'true'—bits of science are accepted as true in some esoteric sense of the term 'true'. The latter acceptances are not on a par with the former at all. Only the equivocation on the term 'true' could make us think otherwise.

Fine might object that there is no such equivocation: esoteric antirealist truth-theories are meant to apply both to homely truths and to scientific ones. I doubt that this will work. Some scientific theories are true, meaning useful for 'saving the phenomena'—and some statements of the phenomena are true, meaning useful for saving . . . *what?* Some statements about unobservables are true, meaning they yield nothing but truths about observables— and some statements about observables are true, meaning they yield nothing but truths about . . . *what?* The esoteric truth-theories developed for science seem to be parasitic upon the homely conception of truth being applied to homely statements about the phenomena or about observables.

But even if an esoteric truth-theory can be applied across the board, this does not help. Now realists and antirealists cannot both accept the core position, because there is no one core position for them both to accept. Now we have *several different core positions*, depending upon the meaning to be attached (across the board) to the term 'true'. Confusingly, these different positions are all expressed using the same *words*. Realists and antirealists can both assent to the *words*—but only because they mean quite different things by them.

I wrote a paragraph back about 'the homely conception of truth being applied to homely statements'. It may be objected that there is no such 'homely conception of truth', that the core position leaves it open what sense is to be attached to the term 'true', that antirealists attach esoteric senses to the term, *and so do realists*. This does not help either—nor does it seem to be Fine's own view of the matter.

It does not help because now we do not have different core positions confusingly expressed by the same words, but *no core position at all*. With no sense yet attached to the term 'true', neither the realist nor the antirealist knows what it is to assent to the so-called 'core position'. If each assents to it by tacitly reading 'true' a different way, then we are back to the plethora of core positions.

Nor, it seems, does Fine think that the core position (NOA) leaves it open what sense is to be attached to the term 'true'. On the contrary, a very definite conception of truth is built into it:

> When NOA counsels us to accept the results of science as true, I take it that we are to treat truth in the usual referential way, so that a sentence (or statement) is true just in case the entities referred to stand in the referred-to relations. Thus, NOA sanctions ordinary referential semantics, and commits us, via truth, to the existence of the individuals, properties, relations, processes, and so forth referred to by the scientific statements that we accept as true. (Fine 1984a: 98 [1205])

> NOA recognizes in "truth" a concept already in use and agrees to abide by the standard rules of usage. These rules involve a Davidsonian-Tarskian referential semantics, and they support a thoroughly classical logic of inference. Thus NOA respects the customary "grammar" of 'truth' (and its cognates). (Fine 1984a: 101 [1207])

Now 'Davidsonian-Tarskian referential semantics' is already a philosophical theory or analysis of truth. It is hardly part of the 'standard rules of usage' for the term 'true', though it yields or explains many of those rules. No matter. The key point is that referential semantics yields precisely the notion of truth that *realists* want to apply across the board, both to homely truths and to scientific ones. Antirealists of any ilk could not accept the core position once they realized that this committed them to accepting some scientific theories as true *in the usual referential way*. NOA, the core position all by itself, is already a thoroughly realist position.

Fine thinks otherwise. But he has some difficulty in explaining what the realist adds to the core position, what distinguishes the NOA (pronounced 'knower') from the realist. It transpires that the realist shouts while the NOA speaks quietly, that the realist traffics in certain slogans that the NOA avoids, and that the realist has a certain metaphysical picture that the NOA does not have. Let us consider these one by one, beginning with the shouting. Fine writes:

> What then of the realist, what does he add to his core acceptance of the results of science as really *true*? . . . what the realist adds on is a desk-thumping, foot-stamping shout of "Really!" (Fine 1984a: 97 [1204])

So the NOA is a realist who avoids desk-thumping, foot-stamping, and shouting. Whereas older realists shouted and stamped in opposition to anti-realist conceptions of truth, on NOA's Ark realists content themselves with a 'stubborn refusal to amplify' their referential semantic concept of truth. As NOA's Ark sails into the sunset, it carries only polite and restrained realists. I promise to shout no more, so that I can join this happy ship.

The slogans will take longer to dispose of. The realist is supposed to traffick in slogans, like 'Truth is correspondence with Reality', which the NOA avoids. Now for me the content of this slogan is exhausted by something the NOA does say: 'A sentence (or statement) is true just in case the entities referred to stand in the referred-to relations'. I have always thought (with Tarski himself) that the semantic conception of truth is a version of the common-sense correspondence theory of truth. Earlier correspondence theorists gave partial accounts (like Aristotle's) or trafficked in general slogans (like the one under discussion). Tarski (1944) showed how to give a complete account (for each well-defined language) which explains what the slogans mean. In this way Tarski rehabilitates the common-sense correspondence theory.

Most philosophers (including Arthur Fine) think that the correspondence theory of truth is one thing and Tarski's theory another. What does the correspondence theorist provide (or seek to provide) which Tarski does not? Fine thinks he provides (or seeks to provide) a general *account* of correspondence, a *theory* of the way in which language and reality can 'match up' so that truth results. Such a theory would 'explain what *makes* the truth *true*' (Fine 1984a: 97 [1204]). Armed with an account of *the* relation which all truths bear to reality, the correspondence theorist will know what all truths have in common, their 'essence', what makes them a 'natural kind'. And here Fine is sceptical:

> Of course we are all committed to there being some kind of truth. But need we take that to be something like a "natural" kind? (Fine 1984b: 56)

I share the scepticism. One thing Tarski taught us is that an *essentialist* correspondence theory is out, that the *essence* of truth is a chimera. (If we must talk of 'essences' here, let us talk thus: Tarski gives us the essence of the correspondence theory *without* giving truth an essence.) Truths are many and various, and so are the ways they correspond to facts. Tarskian referential semantics captures 'the' correspondence relation as well as it *can* be captured. Michael Levin puts it well:

> Tarski tells us that all true conjunctions have in common the truth of each conjunct, that each true existential generalisation is such that its matrix is satisfied by at least one sequence, and so on. To be sure, at the level of the basis clauses, the definition goes strongly extensional, but that is the way it ought to go. The truths "Ron Reagan is a man" and "This tulip is red" are not shown to have much in common: each is a matter of a different object . . . satisfying a

different primitive open sentence. But *do* they have more in common than being a man has in common with being red, which is to say, very little? I cannot see that they do, or at least more than Tarski gives them. (Levin 1984: 126)

When you think about it, Tarski's work shows that the idea that truths might all share an essence is quite absurd. It is meaningful *linguistic* items that are true or false in Tarski's sense. Languages are largely conventional human inventions, suited to different human purposes. Why suppose that the bewildering variety of truths languages contain form a *natural kind*? It is surely better to suppose that there is no more to the 'correspondence relation' than Tarski gives us. Nor need it be any part of scientific realism to think that there is more.

Fine disagrees. He thinks realists must add to Tarskian truth a hankering after truth's essence. In this connection he mentions Hilary Putnam. Now with friends like Putnam, realism needs no enemies. Perhaps Putnam did hanker after truth's essence—but he did not find it. Instead, he found his model-theoretical argument against realism, and abandoned the realist cause. (By the way, the model-theoretic argument is invalid—but that would be another paper.) Never mind Putnam. Realists *need* not hanker after truth's essence (PROOF: *I don't*). Realists can think that Tarski gives them as much of a correspondence theory as they need. But perhaps, in view of all the dust philosophers have raised with the word 'correspondence' down the ages (only to complain afterwards that they cannot see), realists would do better to drop the word. Perhaps realists would do better to drop the correspondence slogan too in favour of that of the NOA: 'A sentence (or statement) is true just in case the entities referred to stand in the referred-to relations' [1205].

So much for one philosophical worry about Tarski's theory of truth: whether or to what extent it is a *correspondence* theory of truth. There are many other philosophical worries, and I want to digress into one of them, for it will reveal another possible way to drive a wedge between NOA and scientific realism.

I have been arguing that Tarski's theory gives the scientific realist all he needs from a theory of truth. But since Tarski's theory can be applied across the board, so to speak, will it not give all sorts of *suspect* realists all that they need also? That the theory applies across the board can be seen from the following list of instances of Tarski's *Convention T*:

1 The statement 'There is a full moon tonight' is true if and only if there is a full moon tonight.

2 The statement 'Electrons are negatively charged' is true if and only if electrons are negatively charged.

3 The statement 'Two plus two equals four' is true if and only if two plus two equals four.

4 The statement 'Eating people is wrong' is true if and only if eating people is wrong.

5 The statement 'Ronald Reagan gives me the creeps' is true if and only if Ronald Reagan gives me the creeps.

The worry is this. Suppose that (1) yields common-sense realism about the moon, and (2) yields scientific realism about electrons. Then will not (3) yield *Platonic realism* about natural numbers, (4) *moral realism* about wrongness and rightness, and (5) realism regarding a mysterious entity (*the creeps*) which Ronald Reagan gives to me (and simultaneously, no doubt, to others too)? Since creeps-realism is absurd, and moral realism and Platonism philosophically suspect, so is the Tarskian theory which yields them.

This worry is quite groundless. Tarski's Convention T (often called Tarski's 'disquotational scheme', or a disquotational, deflationary, or redundancy theory of truth) *by itself yields none of these realisms.* We all avoid creeps-realism by saying that 'Ronald Reagan gives me the creeps', though true (which it is), is an idiom which is not to be taken at face value for logico-philosophical purposes. We replace it with a non-idiom (say, 'Ronald Reagan makes me nervous') and avoid ontological commitment to *the creeps.* Similarly, one sceptical of moral realism might refuse to take 'Eating people is wrong' at face value for logico-philosophical purposes—which is just what emotivists, prescriptivists, and the like do. Alternatively, one might go in morals the way Hartry Field (1982) goes in mathematics. Field accepts Tarski's scheme *and* takes 'Two plus two equals four' at face value, but eschews arithmetical Platonism by saying that it is false *because* there are no numbers for the numerals 'two' and 'four' to be names of. Moral realism might be eschewed similarly: take 'Eating people is wrong' at face value, apply Tarski's scheme to it, and say that it is false *because* there is no property for 'wrong' to refer to.

Similarly again, those sceptical of scientific realism can *either* refuse to take 'Electrons are negatively charged' at face value (as instrumentalists do in their different ways), *or* take it at face value and say that it is false because there are no theoretical entities for the term 'electron' to refer to (as positivists do), *or* take it at face value and concede that it might be true but insist that we should not accept it as true (as epistemological antirealists do). Finally, those sceptical of common-sense realism can *either* refuse to take 'There is a full moon tonight' at face value, *or* take it at face value and say that it is false because there is no external object for the word 'moon' to refer to (as Bishop Berkeley and the phenomenalists do). The disjunctions here are not, of course, exclusive ones. Indeed, antirealists typically adopt a mixed strategy, first saying that a certain kind of statement is false taken at face value (or taken literally), then trying to soften that conclusion by telling us what such statements 'really mean', that is, how they are to be taken for logico-philosophical purposes. (Think of phenomenalist

'translations', so called, of external-object statements, or of emotivist 'translations', so called, of moral statements.) The idea that Tarski's Convention
T *by itself* begs all kinds of metaphysical questions is simply mistaken. Tarski
himself said as much long ago.

By the way, calling Convention T a 'disquotational theory of truth' is
highly misleading. Tarski's theory of truth is not exhausted by Convention T,
though most of the philosophical worries about it are. Instances of Convention T will not be 'disquotational' if something other than a quotation-mark
name of a statement appears on the left-hand side, or if the statement talked
about comes from a different language from the one in which we talk. It is
cases where neither of these conditions obtains, cases like my (1)–(5), that are
also responsible for the mistaken view that Convention T is trivial *because
circular.* There are similar reservations, for similar reasons, about calling the
theory a 'deflationary' theory or a 'redundancy' theory.

If Tarski's Convention T does not by itself yield realism about the moon,
electrons, natural numbers, moral properties, the creeps, or anything else,
what is its importance for realism? The answer is obvious: it makes realism
about all of these things *possible.* This is its importance for realism; this is
why it is the theory of truth that realists need. Antirealist theories of truth
identify it with some *internal* feature of our beliefs (their coherence, their
usefulness, their self-evidence, their ultimate undisbelievability, or whatever).[1] Such theories make realism impossible; they leave no room for it. So
realists need Tarski's theory of truth—but they also need more. To be a *realist about Xs* (whatever Xs may be) you must:

a take statements about Xs at face value for logico-philosophical
   purposes;
b apply Tarski's Convention T to those statements;
c accept some of those statements (*appropriate* ones: 'There are no Xs'
   will not do) as true.

Let us end the digression and return to Fine's NOA. He does all of
these three things when it comes to the theoretical statements of science.
He insists that such statements are to be taken at face value (unlike the
instrumentalists). He applies Tarski's scheme to them (unlike those who
traffick in unrealistic truth-theories). And he accepts appropriate theoretical statements as true (unlike the positivists or epistemological anti-realists).
All this places Fine's NOA squarely in the realist camp.

But wait! The digression may have a Fine point after all. It reveals a
possible way to drive a wedge between the NOA and the realist. Suppose,
contrary to what was just said, that the NOA does *not* take at face value any
body of statements about Xs, whatever X might be. Suppose that to do this is
already to *add* to the California-pure core position. The NOA is philosophically neutral between realist and instrumentalist interpretations of theoreti-

cal scientific statements like (2). The NOA is also neutral between realist and phenomenalist interpretations of common-sense statements like (1). Bishop Berkeley counts as a NOA—even a solipsist counts as a NOA. The NOA accepts homely statements and scientific statements as true. The NOA reads 'true' Tarski-style, with all this brings in the way of commitment to the entities referred to in accepted statements which are taken at face value. *But* the NOA leaves it open what those commitments actually are, because he leaves it open which statements are to be taken at face value (that is, realistically) and which are not.

This is a possible position. It is consistent (just) with Fine's accounts of NOA. It attributes to the NOA (pronounced 'knower', remember) *a complete philosophical know-nothing-ism*. The NOA is not committed to electrons, the moon, tables and chairs, physical objects, other people, his self; *anything at all*. I first overlooked this possible interpretation of Fine's NOA. It was suggested to me by discussions with Arthur Fine in Indiana in 1987 (discussions for which I am grateful). I now think it might be the correct interpretation—as our unfinished business will show.

The NOA was said to differ from the realist on three counts. We have dealt with two of these: realists need not shout or stamp; and they can confine themselves to slogans acceptable to the NOA, like 'A sentence (or statement) is true just in case the entities referred to stand in the referred-to relations'. I cite this slogan for the third time because it would seem to give the realist all he needs by way of *metaphysics*. Fine disagrees, which brings us to the third alleged point of difference: the realist has a metaphysical picture which the NOA lacks.

Fine describes the realist metaphysic thus:

> For realism, science is *about* something; something *out there*, 'external' and (largely) independent of us. The traditional conjunction of externality and independence leads to the realist picture of an objective, external world; what I shall call the World. According to realism, science is about *that*. Being about the World is what gives significance to science. (Fine 1986a: 150)

What exactly is wrong with this realist metaphysical picture? Is not the NOA committed to precisely the same picture?

Fine talks about 'the obscurity of the correspondence relation and the inscrutability of realist-style reference'. He elaborates:

> The problem is one of access. The correspondence relation would map true statements (let us say) to states of affairs (let us say). But if we want to compare a statement with its corresponding state of affairs, how do we proceed? How do we get at a state of affairs when that is to be understood, realist-style, as a feature of the World? (Fine 1986a: 151)

1222 | Ch. 9 Empiricism and Scientific Realism

What exactly is the problem here? Somebody says 'There is a full moon tonight', and I look up into the night sky and ascertain that the statement is true. (I use humdrum commonsensical examples, rather than esoteric scientific ones, because if there is a problem, it will be a quite general one, which afflicts the common-sense realist metaphysic as much as the scientific one.) I have access to both terms of the 'correspondence relation': my linguistic competence gives me access to what was said, my eyes give me access to the full moon out there in the world (or if you prefer, the World). Of course, *explaining* linguistic competence or sensory awareness gives rise to a host of scientific and philosophical problems—but to explain them is not to explain them away, to show that they do not exist at all.

Perhaps the worry is that if I were to report pedantically (similarity to (1) is intended) 'The statement "There is a full moon tonight" is true since there is a full moon tonight', then I would still be trapped inside language, would not have got 'at a state of affairs . . . understood, realist-style, as a feature of the *World*'. Perhaps the worry is that instances of Convention T, such as (1), do not relate language to the World, but rather one language (that talked about) to another (that in which we talk).

If this is the worry, it is a queer one. (1) speaks about a bit of language *and* about the World. True, to speak about the way in which language relates to the World, one must *use language*. But this is no deep truth; rather, it is a pallid truism. Sweeney had it right:

> . . . I gotta use words when I talk to you
> But if you understand or if you dont
> That's nothing to me and nothing to you
> We all gotta do what we gotta do . . . *

We are not trapped inside language in the *serious* sense that all we ever talk *about* is language. To think otherwise is to ignore the hard-won distinction between use and mention.'[2]†

I hesitate to attribute this worry to Arthur Fine. So let us see how he continues:

> A similar question comes up if we move to reference and try to establish truth-conditions compositionally, for there again, what the realist needs by way of the referent for a term is some entity in the World. The difficulty is that what-

---

* From T. S. Eliot's unfinished poem, "Sweeney Agonistes: Fragments of an Aristophanic Melodrama," in *Collected Poems* 1909–1962 (New York: Harcourt, Brace and Company, 1963), 123.

† The sentence, "*April* has five letters," is about the word *April*, not the month it names; whereas "April is the cruellest month" is about the month, not its English name. The first sentence mentions the word *April* to talk about the name; the second sentence uses the word *April* to talk about April, the month.

ever we observe, or, more generously, whatever we causally interact with, is
certainly not independent of us. This is the problem of *reciprocity*. Moreover,
whatever information we retrieve from such interaction is, prima facie, information about interacted-with things. This is the problem of *contamination*.
How then, faced with reciprocity and contamination, can one get entities both
independent and objective? Clearly the realist has no direct access to his *World*.
(Fine 1986a: 151)

Return to my humdrum example (once again, any problems here will
afflict humdrum examples as much as esoteric ones). There the term 'moon'
referred to the moon, which is an entity out there in the World if anything
is. Do the 'problems' of reciprocity and contamination show that this is
incorrect? Reciprocity is supposed to show that the moon is not independent of us because we can see it or otherwise causally interact with it. But
implicit in this is a silly account of independence: an entity is independent
of us if we cannot causally interact with it. The only independent entities in
this sense will be Platonic entities, which do not exist in space and time,
and which have no relations causal or otherwise with beings (like us) which
do exist in space and time. No scientific realist should accept an account of
independence which means that the only independent entities are Platonic
entities and the only independent reality the Platonic realm of abstract entities. When a scientific realist says that the moon is (largely) independent of
us, he obviously means that it is nonmental, it exists outside of us, we did
not create it, it existed long before we did, it continues to exist when we are
not looking at it, and so forth.

What of the 'problem' of contamination? It is supposed to show that
when we see that the moon is full, we gain information not about an objective moon out there in the World, but rather about an *interacted-with-moon*.
This hyphenated entity is presumably not the same entity as the moon (or,
if hyphens thrill you, as the moon-in-itself). Fine suggests that, unlike the
moon-in-itself, the interacted-with-moon is not objective, not out there in
the World. Where is it then, subjective and inside our heads? This smacks
of the long-discarded view that we do not see external objects like the moon
at all, but rather moonish-sense-data inside our heads. I doubt that Fine
wants to return to that view. I know that Fine's NOA ought not to attach
that bit of bad philosophy to his core acceptance of the results of science.

Perhaps the thought is that the interacted-with-moon, the moon-as-
observed-by-us, is not 'objective' in the sense that it is somehow partly constituted by the moon-concept which is our invention. After all, in order to
see that there is a full moon, I must possess the moon-concept, and in order
to say that there is a full moon, I must possess the word 'moon'. What I can
see (or say) depends partly upon the concepts (words) that I possess.[3] The
world (the world-in-itself, Fine's World-with-a-capital-'W') is not carved up
according to any conceptual or linguistic scheme. It is we who carve things
up according to such schemes. Having carved, we cannot partake of the

world-in-itself, the world-as-it-it-independently-of-any-conceptual-scheme, the World-with-a-capital-'W'. That the moon is full is not a fact about the world-in-itself, since it trafficks in the moon-concept. The statement 'The moon is full tonight' does not state a truth about the World-with-a-capital-'W', since it trafficks in the word 'moon'. The world we experience and talk about is not the world-as-it-is-independently-of-any-conceptual-scheme. Rather, it is a world partly of our own conceptual or linguistic making, a world-as-conceived-by-us or a world-as-talked-about-by-us. This is conceptual (or linguistic) *idealism*.

It quickly turns into conceptual (or Iinguistic) *relativism*. Our concepts vary and change. There is no *one* world-as-conceived-by-us at all. The world-as-conceived-by-the-Aristotelian differs radically from the world-as-conceived-by-the-Newtonian. The world-of-the-Eskimo is not at all the world-of-the-Kalahari-bushman. This gets really exciting once we cease to be human chauvinists and consider non-human animals too. They experience the world too, but differently from us. The world-of-the-chimpanzee is not at all the world-of-Albert-Einstein, and both are *worlds apart* from the world-of-the-honeybee or the world-of-the-three-spined-stickleback. And so on and so forth—tediously.

Kant is, of course, the philosopher who started the rot here. Kant stopped the rot from spreading, blocked the slide from idealism into relativism, by assuming that humans all have the same immutable set of basic concepts. Contemporary philosophical wisdom has outgrown that assumption. Even if contemporary wisdom is misguided, we might still ask Kant and the Kantians about non-human animals. Do they *experience* the world at all? If so, do they possess all those Kantian categories of the understanding deployment of which is a condition of the possibility of all experience? Do chimps, honeybees, and flatworms structure incoming stimuli the way humans do? One alternative is to say that humans are the only animals that have experiences. This must be deemed implausible by anyone who takes Darwinism seriously. Another alternative is to say that the Kantian categories are only conditions of the possibility of *human* experience and that other creatures can do without them. But if chimps can do without them, why not us?

Of course, all this talk of different worlds-as-experienced (conceived, talked about)-by-Xs *need not be taken seriously*. We can see it just as a fancy way of drawing attention to the great diversity of experience, concepts, and talk *of the world*. On this view, all such entities as the moon-as-experienced-by-us are ersatz entities. (After all, is not 'ersatz' German for 'hyphenated', and 'hyphenated' philosopher's English for 'artificial' or 'unnatural' or 'unreal'?) The moon-as-experienced-by-us is just the moon—and similarly for all other hyphenated entities (*including* the Kantian moon-in-itself). On this view, 'The moon-as-conceived-by-Aristotelians was perfectly spherical' is just philosopher's gobbledy-gook for 'Aristotelians thought the moon is perfectly spherical'.

If we do take talk of different worlds-as-experienced (conceived, talked about)-by-Xs seriously, we become experiential (conceptual, linguistic) idealists. And we come to inhabit a strange world indeed. Consider the moon-as-observed-by-us ($moon_0$ for short) and the moon-in-itself ($moon_1$ for short). Is $moon_0$ identical with $moon_1$? Presumably not: if it were, the distinction would have no point, and we could rest content just with the *moon* (without subscript, unhyphenated). But if $moon_0$ is distinct from $moon_1$, then there is presumably some property which the one lacks and the other possesses. But to know this is to know something about $moon_1$, when our knowledge was supposed to be confined to $moon_0$! Certainly, there could be no empirical evidence that $moon_0$ is different from $moon_1$—it is just a piece of idealist metaphysics. Kantians object that we can know, not through evidence but through transcendental argument (whatever that is), that $moon_1$ not only *exists*, but also lacks various properties that $moon_0$ possesses. For example, $moon_1$ has no position in space and time, these being 'forms of sensibility' in which only $moon_0$ is located. Nor does $moon_1$ cause (or help cause) moon-visions down on earth, causality being a category of the understanding which applies only in the world-of-appearance, not in the world-of-things-in-themselves. No wonder that some of Kant's immediate followers, realizing that $moon_1$ is nowhere, at no time, and does nothing, decided that it was an *idle* metaphysical posit, did away with it altogether, and became fully-fledged idealists.* At this point my Kantian friends (and I still have one or two) tell me that Kant was an 'empirical realist' and only a 'transcendental idealist'. I do not understand these Kantian slogans. I am reminded how fond Berkeley was of presenting himself as a defender of common-sense realism. The Kantian metaphysic, seen as it really is, is a form of *idealism*, as is Berkeley's metaphysic. Modern idealism is just Kantian idealism 'gone linguistic' or 'gone conceptual' and generalized.[4]

We have come a long way from Fine's remark that when we observe or otherwise interact with things, the information we retrieve is information

---

* Kant thought that we possess synthetic a priori knowledge about time, space, geometry, and so on—for example, that we know a priori that all external objects are spatial and are causally related to one another—but that we have this knowledge only insofar as the "objects" in question are objects-as-experienced-by-us not things as they are in themselves, independent of human experience. Because things-in-themselves are not subject to the categories of causation, substance, space, time, and so on, some later thinkers (notably, Hegel) dispensed with things-in-themselves entirely, thus precipitating what Musgrave judges to be the disastrous descent into full-fledged idealism. For the idealist, the entire external world is simply a construction of the human mind (or the mind of God) and cannot exist independently of mind. By retaining things-in-themselves in his system, Kant sought to avoid this extreme sort of idealism. For Kant, although things-in-themselves cannot be said to be spatial, or to exist in time, or to cause our perceptions, they nonetheless exist independently of human minds and are the ultimate source (in some noncausal sense) of our perceptions. See Anthony Quinton, "The Trouble with Kant," *Philosophy* 72 (1997): 5–18.

about interacted-with-things. The remark may have been a perfectly innocent one. I do not know whether Fine is an idealist of this kind (or rather, of these kinds). I do think that Fine's NOA should have nothing to do with these idealisms. NOA stands for *natural* ontological attitude. The ersatz hyphenated entities involved in these idealisms are artificial and unnatural entities. These idealisms are dubious, perhaps in the end unintelligible, philosophical theories which no NOA should attach to his core acceptance of the results of science. Indeed, some pretty mundane and well-entrenched results of science tell us that the moon (not some hyphenated moon, not even the Kantian moon-in-itself, just the moon) is objective and independent of us: it exists outside of our heads, it was not created by us, it existed long before we did, and so forth. The NOA who accepts these bits of science as true has precisely the same metaphysical picture as the realist. Fine rejects the realist's metaphysical picture, not as unproved, but as false. Its falsity follows from what might be called an unnatural idealist attachment to science. But the unphilosophical NOA ought not to be trafficking in the 'problems' of reciprocity and contamination—for that traffic is *philosophy*. The unphilosophical NOA should do no more than accept homely truths and scientific ones. Will that not provide the NOA with the same 'metaphysical picture' as the realist?

An affirmative answer to this question overlooks the possibility of the NOA who knows *nothing* philosophical. That NOA's acceptance of bits of science as true implies nothing whatever about the objectivity and independence of the moon. For that NOA leaves it quite open how the accepted statements are to be 'interpreted', what they mean, what their ontological commitments are. Perhaps homely truths and scientific ones are to be taken at face value for logico-philosophical purposes, and *perhaps they are not*. Perhaps the homely truths are to be given a phenomenalist construal and the scientific truths an instrumentalist one. (Remember, Berkeley counts as a NOA in this minimalist sense.) The completely unphilosophical NOA leaves all this open.

In traditional discussions of scientific realism, common-sense realism regarding tables and chairs (or the moon) is accepted as unproblematic by both sides. Attention is focused on the difficulties of scientific realism regarding 'unobservables' like electrons. But Fine's discussion is not a traditional discussion. Fine's NOA, on this interpretation, begs *no* metaphysical question. That is why both realists and antirealists of any ilk (even a solipsist) can accept NOA.

One special consideration suggests that this is the correct interpretation of Fine's position. In his acclaimed biography of Einstein, Abraham Pais tells how Einstein once turned to him and asked 'if I really believed that the moon exists only if I look at it' (Pais 1982: 5). Einstein's question concerned, of course, that interpretation of quantum mechanics according to which entities do not exist in well-defined states unless they are being observed. What if quantum mechanics, thus interpreted, should turn out to be cor-

worlds. The Aristotelian and the Copernican, watching a sunrise, *see different things.* This is, of course, nonsense. What might be true is that the Aristotelian says of the sunrise 'I see that the sun is still orbiting the earth', while the Copernican says 'I see that the earth is still rotating on its axis'. The profundity 'The limits of my language are the limits of my world' is false. What is true is the triviality that the limits of my language limit what I can say of the world.

4. Permit me a true story. I was once told in all seriousness that when the concept 'person with an IQ two standard deviations above the mean' was invented, *new entities* were brought into being. It turns out (I replied) that there are two ways of making babies, the way we all know and love, namely love, and this new way, psychological theorizing! I was told that the new entities are not babies, indeed, are not persons with IQ's two standard deviations above the mean. I could gain no clear idea *what* they were.

▪ | References

Eddington, A. S. (1928) *The Nature of the Physical World* (Cambridge: Cambridge University Press).

Field, H. (1982) 'Realism and Anti-Realism about Mathematics', *Philosophical Topics*, 13: 45–69.

Fine, A. (1984a) 'The Natural Ontological Attitude', in Leplin (1984): 83–107. [Reprinted in Fine (1986b), ch. 7.]

Fine, A. (1984b) 'And Not Anti-Realism Either', *Noûs*, 18: 51–65. [Reprinted in Fine (1986b), ch. 8.]

Fine, A. (1986a) 'Unnatural Attitudes: Realist and Instrumentalist Attachments to Science', *Mind*, 95: 149–79.

Fine, A. (1986b) *The Shaky Game: Einstein, Realism, and the Quantum Theory* (Chicago: University of Chicago Press).

Leplin, J. (ed.) (1984) *Scientific Realism* (Berkeley: University of California Press).

Levin M. (1984) 'What Kind of Explanation Is Truth?' in Leplin (1984): 124–39.

Pais, A. (1982) *'Subtle Is the Lord . . .': The Science and the Life of Albert Einstein* (Oxford: Oxford University Press).

Tarski, A. (1944) 'The Semantic Conception of Truth', *Philosophy and Phenomenological Research*, IV, 341–75.

9 | COMMENTARY

9.1 | Logical Empiricism

In the first half of the twentieth century, the logical positivists and their successors, the logical empiricists, approached the issue of scientific realism in a distinctive way by reflecting on the role of observation in two different types of scientific activity.[1] On the one hand, scientists seek to discover empirical laws, expressing generalizations about observable phenomena such as the motion of bodies or relations between the pressure and volume of a gas. On the other hand, scientists also formulate full-blown scientific theories that unify, explain, and predict such laws and their observational consequences in terms of forces acting on bodies, molecules of gases and their mean kinetic energies, and so on. Thus, it was natural for the logical empiricists to emphasize a distinction between the observational components of a theory, which refer to objects and properties that are directly observable, and the theoretical components, which apparently refer to objects and properties that are not directly observable.

The observational-theoretical distinction thus has two aspects, ontological and terminological (or linguistic). As an ontological distinction, it is used to mark off those objects, properties, and events that are directly perceivable from those that are not. For example, philosophers often distinguish between observable and unobservable objects. By contrast, the terminological distinction applies not to objects, properties, and events but to the language and vocabulary of scientific theories; it is usually made by calling those terms unique to a theory (such as *molecule, gene, electromagnetic field*, and so on) *theoretical terms*, and calling the other descriptive terms figuring in empirical laws *observational terms*.

The ontological distinction has been widely regarded by empiricist philosophers as having important implications for scientific realism. Ernst Mach, for example, (who was an important influence on logical positivism) argued that the aim of natural science is the economical description of experience. Since we can have no experience of unobservables, Mach concluded that they are at best convenient fictions introduced to simplify our theories. In the current idiom, Mach's view represents a kind of antirealism about atoms, forces, and fields.

Switching now to the terminological distinction, consider an empiricist view of language, according to which the meaning of a term is exhausted by its connection with immediate experience. Given this view and the fact that theoretical terms are not connected directly with experience, the cognitive significance of theoretical terms would seem to be in jeopardy, they would seem to be meaningless. This sort of consideration led philosophers of a broadly empiricist persuasion to accord theoretical terms a second-rate semantic status. The strongest version of such an account granted theoretical terms no essential role in scientific theory, regarding them as purely fictional terms to be eliminated altogether.[2] A somewhat weaker approach embraced by some logical empiricists treated theoretical terms as strictly speaking meaningless but nevertheless instrumentally useful in scientific theories; another approach, advocated by operationalists, insisted that theoretical terms be given suitable content by explicit definition in observational terms. Finally, the weakest version of the empiricist approach granted theoretical terms a role in scientific theories but reckoned them to be only "partially meaningful" via their connection with observational terms. Thus, for example, Norman Campbell, Rudolf Carnap, and Carl Hempel viewed scientific theories as partially interpreted formal systems. On this account, the axioms of a theory from which observational consequences are deduced contain both observational and theoretical expressions, but only observational terms can be interpreted directly and fully; theoretical terms acquire a partial and indirect interpretation by contributing to the derivation of observational claims.

Logical empiricism may thus be viewed as a kind of antirealism: while the empirical claims of a theory in which only observational terms figure are judged to be true because those terms connect directly with real objects, events, and properties accessible to perceptual experience, a commitment to the meaningfulness of theoretical terms and the truth of purely theoretical claims in which they occur does not bring with it any commitment to their connection with unobservable objects, events, and properties.

Logical positivism is dead and logical empiricism is no longer an avowed school of philosophical thought. But despite our historical and philosophical distance from logical positivism and empiricism, their influence can still be felt. An important part of their legacy is the observational-theoretical distinction itself, which continues to play a central role in debates about scientific realism. In this commentary, we shall explore a number of questions about empiricism and scientific realism. Is the observational-theoretical distinction sustainable, and in particular, does it possess the ontological significance that many empiricists have taken it to possess? How exactly should we understand the role of theoretical terms in scientific theories? Does acceptance of a scientific theory bring with it commitment to the truth of theoretical claims? Can we explain the success of science without being realists about theories and the entities they postulate? Does the success of our scientific theories count as evidence for realism?

9.2 | Maxwell on the Ontological Status of Theoretical Entities

In his paper "The Ontological Status of Theoretical Entities," Grover Maxwell expresses impatience with those inclining toward what we have called antirealism about theoretical terms and objects. His purpose is to isolate some influential strands of antirealism emerging from the logical empiricist tradition and to argue that all of them are implausible.

To begin, imagine (with Maxwell) a bit of fictional science in which, before the advent of microscopes, a scientist named Jones speculates about the mechanism by which diseases are transmitted from one person to another. Arguing by analogy with observable mechanisms such as lice, Jones reasons that all infectious diseases are transmitted by some form of bug or other, most of which are unobservable. By dint of hard work, he comes to show that preventative measures aimed at killing such "crobes" could drastically lower the incidence of disease. But philosophically inclined scientists were worried: Jones had assumed the existence of objects that, according to his own theory, are unobservable. How should Jones's theory be understood? Some claimed that the tiny organisms were mere fictions; others claimed that such a theory was at best an instrument, helpful to Jones's thinking but not literally asserting the existence of unobservable entities. Still others claimed that sentences in which Jones's theoretical terms appeared were permissible only to the extent that they could be translated into sentences containing just observational terms. But the day arrives when the microscope is invented, and Jones's crobes are observed in great detail, different crobes being identified as the cause of different diseases. The antirealist philosophers respond in various ways. While many convert to realism, others stay their original course. Of these, (*a*) some espouse idealism or phenomenalism, insisting that all claims about physical objects, of whatever sort, be translatable into claims about immediately perceivable sense data;[3] (*b*) others take the less drastic course of claiming that Jones's crobes never had been unobservable in principle, since the theory did not entail the impossibility of finding a means of observing them; (*c*) still others insist that the crobes have not been observed at all: what is seen by means of the microscope is not a physical object or organism, but something far less substantial.

THE OBSERVATIONAL-THEORETICAL DISTINCTION

Maxwell argues against (*a*), (*b*), and (*c*), as they are manifested in actual positions philosophers have defended. In connection with (*c*), Maxwell cites a passage from the logical empiricist, Gustav Bergmann, who describes a certain empiricist rejection of atoms as entailing that strictly speaking ". . . even stars and microscopic objects are not physical things in a literal sense," for when we look through our apparatus all we see "is a patch of color which creeps through the field like a shadow over the wall" (1052). Maxwell notes that if this line is taken, we cannot properly be said to observe garden-variety-sized physical objects through eyeglasses or windowpanes. Shall we say that we can only infer that it is raining, unless we raise the window and observe the rain "directly"? Consider, too, that modern chemistry tells us that there is a continuous transition from very small molecules (of water, say), through medium-sized ones (like proteins or nucleotides), to large ones (like diamonds or crystals of salt). The last of these are directly observable objects, but for all that they are no less genuine, single molecules. The lesson Maxwell encourages us to draw is twofold. First, we have no obvious criteria by which to draw a nonarbitrary line between the observational and the theoretical; while such line drawing as we can make will often prove convenient, its position will vary widely from context to context. Second, the continuous nature of the transition between observable and theoretical shows that the distinction has no ontological significance whatsoever. For surely an entity does not have real existence in one context and lack it in another, and surely objects seen through eyeglasses are not to be judged less real than (or to have a degree of existence inferior to that enjoyed by) objects observed directly by unaided vision.

Although he does not explicitly present them as such, Maxwell's reflections on (c) might be seen to recommend the following sort of argument: Since it is agreed on all hands that (1) there are observable objects about which we must be realists, and (2) given the continuous transition from observable to theoretical shows that there is no nonarbitrary line to be drawn between them, it follows that (3) we are not warranted in claiming that there is a class of in-principle unobservable objects about which we may be antirealists. Maxwell's claim that the observational-theoretical distinction "has no ontological significance whatever" (1058) will on this reading be understood as asserting that, whatever convenience the distinction may offer, it is not a distinction between things we can judge to be real and things we cannot.

Consider again view (b), according to which Jones's crobes never were unobservable in principle: on this view, only putative entities that are in principle impossible to observe must be rejected. Maxwell challenges this antirealist reliance on the notion of unobservability in principle. The current status of electrons is conceivably similar in many ways to Jones's crobes. For, suppose we discover new entities able to interact with electrons in such a way that, with enhancements to our vision and by means of these entities, we are able to perceive various properties of electrons. However improbable such a scenario may be, it does not involve any logical or conceptual absurdity. Maxwell's point is that there are no a priori, purely philosophical criteria for separating the observable from the unobservable; to this extent, any descriptive term is a possible candidate for an observational term.

Might we improve our ability to draw a serviceable distinction here by focusing—as earlier philosophers have done—not on *observable* but on *observed* instead? On such a proposal, observational terms are taken to be those referring to what has in fact been observed. Construed narrowly, this would yield a different observational language for each language user: *sand* would be a theoretical term for some Eskimos, *snow* for some Floridians. Clearly this is too strong. But suppose we construe it broadly, as claiming that observational terms must be members of a kind, some of whose members or their properties have been observed. This is at once too weak and too strong: too weak because the similarity relations needed for specifying a kind are ubiquitous and easy to find; too strong because for any entity we can always specify a kind of which it is the only member. Maxwell concludes that such a strategy can only result in a retreat to some form of phenomenalism, which is position (*a*).

According to the phenomenalist version of antirealism (*a*), all claims about physical objects must be translatable into claims about immediately perceivable sense data. As Maxwell notes, scarcely any contemporary philosopher defends such a view. This is not to underestimate the relevance of sensory impressions themselves, but the fact remains that most statements of our common linguistic framework refer to public entities, not private inner states or objects. Indeed, this fact highlights the importance of an observational base in science as the ultimate grounds of confirmation. In scientific discourse the basic unit is not the observational term, but what Maxwell calls the quickly decidable sentence, a (nonanalytic) sentence that reliable, competent language users can quickly decide whether to assert or deny when reporting on some situation. How is it that we can and sometimes do quickly decide the truth or falsity of some observation sentence—a sentence whose only descriptive terms are those that may figure in a quickly decidable sentence? Maxwell argues that this is itself a scientific-theoretical question, the answer to which might well regard sense data as postulates of a theory. As before, the point is that such issues are not purely logical or conceptual. And thus, as before, it emerges that drawing the observational-theoretical line at any particular point is a purely contingent feature of our physiology, our current state of knowledge, and the instruments we happen to have available at the time. In light of this, Maxwell concludes that the observational-theoretical distinction is entirely void of ontological significance.

9.3 | Van Fraassen's Constructive Empiricism

Maxwell's is but one reply to the antirealist sentiments of logical empiricism in defense of scientific realism. Of course, if Maxwell's arguments fail, realism might still be tenable; for there may be other arguments to support it. Similarly, it would be an error to suppose that if Maxwell's criticism succeeds, then all varieties of antirealism have been shown to to be false; for there may be other forms of antirealism.

The main arguments that have been offered in defense of scientific realism arise from criticisms of logical empiricism. But Bas van Fraassen defends a view that is at odds with both logical empiricism and scientific realism. In "Arguments Concerning Scientific Realism," drawn from his book *The Scientific Image*, van Fraassen proposes to do two things. His first task is to articulate a distinctive form of antirealism, which he calls *constructive empiricism*. His second task is to reply to the main arguments for scientific realism (including Maxwell's), from the perspective of his own antirealism. However untenable logical empiricism might be, the plausibility of scientific realism must be judged against the remaining antirealist alternatives.

SCIENTIFIC REALISM

In rejecting scientific realism, what exactly is van Fraassen denying about the nature of science and its theories? An intuitive characterization of scientific realism would run as follows: the account of the natural world offered by science is true, and the entities it postulates really exist. Van Fraassen believes that this statement is too naive as it stands, since it entails that the scientific realist believes either that today's theories are correct or that science will eventually offer theories true in all respects. Still, the naive characterization is helpful in isolating those virtues of the scientific enterprise most prized by realists. In particular, realists are committed to the view that scientists seek to offer theories that are true and that the claims of such theories are made true by how the external world is. To accept a theory is thus (at the very least) to believe that its claims are true. Here then is van Fraassen's formulation of scientific realism:

> Science aims to give us, in its theories, a literally true story of what the world is like; and acceptance of a scientific theory involves the belief that it is true. (1062)

Notice several things about this statement of scientific realism. First, it is suitably weak in two respects: it commits the realist to the claim that science aims at telling a true story, not that the story science tells is true, and while highlighting truth as the chief virtue of theories, it does not entail that other virtues (such as simplicity, predictive and explanatory power, and so on) are irrelevant to a theory's success. Second, van Fraassen's reference to the goal of literal truth rules out the views of instrumentalists and other logical empiricists who assert that scientific theories may be taken loosely speaking as true if understood in the proper way but are nevertheless strictly speaking (i.e., taken literally) false or meaningless. Finally, acceptance of a theory requires only belief in its truth, without implying that scientists must be fully justified in such beliefs. Since justification comes in degrees, the realist can judge that a scientist's acceptance of a theory is strong or weak or somewhere in between.

Constructive Empiricism

Given this brief account of scientific realism, we can develop a general account of antirealism, with a view towards understanding van Fraassen's own distinctive version of it. According to the antirealist, the aim of science is not to offer literally true theories, and to accept a theory is not to believe it is true. Thus, while the realist will insist that someone who accepts a scientific theory is asserting its literal truth, the antirealist insists that to accept a theory is to do something less than that; it is to claim for it some virtue other than literal truth. Here antirealists will divide. As we have seen in the case of many in the logical empiricist tradition, an antirealist might claim that science aims at theories that are good because they are true when construed nonliterally.[4] Alternatively, the antirealist might claim that theories are to be construed literally but that they need not be true to be good. Van Fraassen's antirealism is of this latter sort. On his literalist view, if a theory's statements include or entail "There are molecules," for example, then that theory says that there are molecules.

But how are we to understand the notion that one can take the language of science literally but deny scientific realism? Van Fraassen urges us to see that the literal construal of scientific language concerns our face-value interpretation of its meaning, while espousal of scientific realism concerns the extent of our belief. Thus, we are encouraged to see that taking theories literally is not believing they are true or believing that the entities they postulate are real. Instead, we are to adopt what he calls *constructive empiricism*:

> Science aims to give us theories which are empirically adequate; and acceptance of a theory involves as belief only that it is empirically adequate. (1065)

A theory is empirically adequate when it "saves the phenomena"—when what it says about *observable* objects, events, and properties is true. The respect in which van Fraassen's antirealism departs both from logical empiricism and from scientific realism is thus apparent. To accept (hold) a theory is to claim that it accurately describes observable phenomena; this does not entail that talk of theoretical entities is meaningless, nor does it entail that such entities are fictional or unreal. By distinguishing in this way between accepting a theory and believing it to be true, the constructive empiricist recommends a position of agnosticism about the theoretical. As van Fraassen says at the end of his book: "To be an empiricist is to withhold belief in anything that goes beyond the actual, observable phenomena. . . . To develop an empiricist account of science is to depict it as involving a search for truth only about the empirical world, about what is actual and observable."[5]

Van Fraassen emphasizes that, in addition to the epistemic dimension of constructive empiricism—concerning how far one's beliefs extend in accepting a theory—there is a pragmatic dimension as well. Accepting a theory brings with it a commitment to explaining future phenomena using that theory's conceptual resources.

9.4 | Van Fraassen's Reply to Arguments for Scientific Realism

We noted earlier that the most influential arguments for scientific realism have come from critics of logical empiricism. Clearly, their success in establishing scientific realism cannot be judged simply by how well they dispense with the antirealism of the logical empiricists, if other forms of antirealism yet remain. Thus, van Fraassen's second task is to evaluate the main arguments for scientific realism in the light of his constructive empiricism.

AGAINST MAXWELL ON THE OBSERVATIONAL-THEORETICAL DISTINCTION

How strongly do Maxwell's arguments support scientific realism? After distinguishing between terminological and ontological strains of the observational-theoretical dichotomy, van Fraassen poses two questions: Can we divide our language into a theoretical and nontheoretical part? Can we classify objects, events, and properties as being either observable or unobservable? To the first of these, recall that Maxwell answers "No." In particular, Maxwell claims that what counts as the fundamental unit of our observational base is not the term but the quickly decidable sentence and that an account of quickly decidable sentences will itself be a scientific-theoretical question. To this extent, then, even experimental and sensory reports threaten to involve theoretical elements. Van Fraassen agrees that, in the end, all language is to some degree theory-infected, but he denies that this shows anything about scientific realism. It may well be true that our language is influenced by theories, but this does not reveal the extent of our beliefs about those theories. An extreme example best makes the point: to speak of the sun rising scarcely reflects a belief in the truth of Ptolemaic astronomy.

The more fundamental question is the second, concerning the possibility of dividing observable objects, events, and properties from unobservable ones. Maxwell answers this question negatively as well. Can van Fraassen follow Maxwell here? Presumably not. For in claiming that acceptance of a theory T involves believing only that T is empirically adequate, the constructive empiricist presumes we can distinguish between those statements of T that are about observable objects, events, and properties and those that are not; and this entails the intelligibility of judging whether or not they are observable.

Maxwell offers three arguments for realism: two against the possibility of distinguishing observable from unobservable objects and one against attributing any ontological significance to such distinctions as typically drawn. The first argument of the former sort leans heavily on the claim that we have no criteria by which to draw a nonarbitrary line between the observable and the unobservable, since there is a continuous transition from looking through air (with unaided vision), through a window, through binoculars, through a microscope, and so on. As we reconstructed the argument earlier (1235):

1 There are observable objects about which we must be realists.

2 No nonarbitrary line can be drawn between observable and unobservable objects.

3 Therefore, we are not warranted in claiming that there is a class of unobservable objects about which we may be antirealists.

Van Fraassen is prepared to grant Maxwell's premises, but he denies that conclusion (3) follows from them. In particular, premise (2) shows at most that *observable* is a vague predicate from which, even in conjunction with (1), nothing about realism follows. Despite the many sophisms that vague predicates have been used to generate, we are safe in using them as long as there are clear cases where they do apply and clear cases where they do not.[6] At best, then, Maxwell is challenging us to find clear cases of unobservable objects. Van Fraassen claims that there are such clear cases. The purported "observation" of microparticles in a cloud chamber is one example. Here, a charged particle traverses the chamber to produce a visible trail of condensed vapor. In this case, the particle is indeed detected, and the detection is based on observation, but we can scarcely insist that the particle itself is being observed. Thus, even if we must be realists about observable objects, and even if the observable-unobservable distinction is vague, it does not follow that we are never warranted in claiming that there are unobservables about which we may be antirealists.

Maxwell's second argument against the observable-unobservable distinction is directed against the antirealist's use of *unobservable* as meaning "whatever the relevant theory entails cannot be observed." Since there are no conceptual or a priori grounds for supposing that there is any object that must remain unobservable regardless of the possible circumstances, the term *unobservable* marks out no class of objects for the antirealist at all. Van Fraassen claims that this argument is a mere distraction and changes the subject. Humans are measuring instruments subject to limitations that a final physics and biology will describe: it is to such limitations that the *able* in observ*able* and unobserv*able* apply. Just as we should not judge the Empire State Building to be portable because some giant might move it, so it is beside the point to note the logical possibility of detection devices outstripping the natural limitations of humans.

Finally, van Fraasen takes up Maxwell's argument against attributing any ontological significance to whatever of the observable-unobservable distinction might remain. The issue for the realist is the reality of entities postulated by science, and as Maxwell notes, something can exist even though it is unobservable. But van Fraassen emphasizes that even if this is so, it does not follow that the observable-unobservable distinction has no significance for realism. Specifically, scientific realism concerns the epistemic stance we take with respect to the claims of science, and on this issue, what we can observe is eminently relevant to what we should believe. The constructive empiricist claims that to accept a theory is to believe only that it is empirically adequate. And a theory's empirical adequacy depends entirely on what it says about observables. Thus, what a theory says about what is observable is relevant to deciding the case between realism and this form of antirealism.

Van Fraassen responds to four more arguments for scientific realism. In each case, van Fraassen's criticisms involve his constructive empiricist version of antirealism.

ON INFERENCE TO THE BEST EXPLANATION

Some have argued that the canons of rational inference require a commitment to scientific realism. In particular, we encounter in science itself a ubiquitous appeal to a rule of inference called *inference to the best explanation*. Given some evidence E and alternative hypotheses H and K, the rule says that we should infer H rather than K when H is a better explanation of E than is K. The argument under consideration is that following this rule consistently leads to scientific realism: to have good explanatory grounds for believing a claim H is to have good grounds for believing that the entities postulated by H exist. (Van Fraassen's homespun example: from the disappearance of my cheese, the midnight scratchings on the wall, and patter of tiny feet, I infer that there really is a mouse in the wainscoting.)

Does our widespread use of this pattern of inference commit us to believing in unobservable entities? Van Fraassen argues that it does not. Consider what it means to claim that we follow a certain pattern or rule of inference. It cannot mean that we explicitly formulate the rule to ourselves and consciously follow it. That is too strong: most people follow various rules of inference implicitly. Nor can it mean that we merely act in accordance with a rule (without conscious deliberation) in the sense that our conclusions could be reached by following that rule. That is too weak: any conclusion whatever could be reached by following the permissive rule that any conclusion can be inferred from any premise. Rather, the appropriate sense of following a rule must entail our willingness to believe what follows from that rule and our unwillingness to believe anything that conflicts with what follows from it. Thus, the claim that we follow a certain rule is an empirical hypothesis about what we are willing and unwilling to believe. The realist's claim that we always follow the rule of inference to the best explanation can be confronted with rival hypotheses. One of these rival hypotheses is van Fraassen's alternative: we are always willing to believe that the theory that best explains the evidence is empirically adequate. This brings with it no commitment to scientific realism whatever. Moreover, it accounts for the many cases where scientists argue for the acceptance of a theory on the basis of its explanatory power.

Against Smart on the Success of Explanation

There are two kinds of arguments for scientific realism based upon explanatory power as a criterion for theory choice. One is offered by J. J. C. Smart; the other, by Wilfrid Sellars, will be taken up presently.

Smart argues that only scientific realism can account for the distinction between correct and merely useful theories. The realist can explain the usefulness of a theory by appealing to its correctness, or if the theory is not true but empirically adequate, by appealing to its relation to some rival that is correct. For example, consider two astronomers: the first is a realist about the Copernican hypothesis (C) but an antirealist about an empirically adequate version of the Ptolemaic hypothesis (P); the second, a thoroughgoing antirealist, accepts both C and P as useful but does not believe that either theory is true. The first astronomer can explain the usefulness of C easily enough (it is true); he can also explain the instrumental usefulness of P, about which he is an antirealist, in terms of its predictions agreeing with those of the true theory C. But how can the second astronomer, the thoroughgoing antirealist, explain the usefulness of the theories he accepts without believing that either C or P (or any other theory) is true?

Van Fraassen replies that empirical adequacy seems equally able to ground explanations of a theory's usefulness. Why think that an antirealist can offer less of an explanation of the usefulness of P than the realist—as long as the antirealist begins, not with the premise that C is true, but instead with the premise that C is empirically adequate and thus gives accurate descriptions of planetary motion as observed from the earth? The realist defender of Smart's argument will of course respond that this only postpones the crucial question one step: for what explains the descriptive accuracy of C and, in particular, what explains the fact that all predictions of planetary motions fit C? It would seem that the antirealist must claim that the observable regularities simply *do* fit C, as a matter of brute fact, there being no need to posit entities and processes behind the phenomena to explain *why* C is a good theory. Smart believes that unless the observable regularities can be explained in terms of some deeper (unobservable) structure, we are left with having to believe in lucky accidents and sheer coincidences on a grand scale. But van Fraassen argues that these explanatory demands are unreasonable. For any theory will eventually bottom out, having to posit some regularities as basic or primitive, without receiving any further explanation at a deeper level. It is only a realist prejudice to suppose that these regularities must concern entities and processes that are unobservable. Moreover, van Fraassen is unprepared to grant the legitimacy of equating coincidence with having no explanation. He points out that inclination to call, say, your meeting me at the market a coincidence reflects only an absence of any plan to meet there, not an absence of explanations for each of us being there.

Against Sellars's Thought Experiment

It is common to distinguish individual observable facts from empirical laws and empirical laws from theoretical hypotheses. The observable fact that this water is boiling might be said to be explained by the empirical law that all water boils at 100°C, and this law itself might be said to be explained by theoretical hypotheses about latent heat and the energies of bonding between molecules and so on. Wilfrid Sellars has suggested that this picture is oversimplified, for our theories do not really explain such empirical generalizations; at best, they explain why observable things obey these regularities to the extent that they do. Moreover, strictly speaking, there may be no genuine regularities of observable phenomena at all: at most pressures water does not boil at exactly 100°C. But Sellars argues that this lack of strict regularity does not infect the theoretical level or genuine explanation would not be possible. Incompleteness or inferiority in our empirical generalizations requires some genuine universal regularities at the theoretical level of unobservables.

Sellars offers the following thought experiment to illustrate the point. Suppose at some early stage of chemistry we found that samples of gold dissolve in aqua regia at different rates even though they are observationally indistinguishable.[7] We postulate two distinct microstructures, A and B, for gold to explain this seemingly unpredictable variation in rates of dissolution by claiming that our various samples are really mixtures of distinct substances (whose microstructures obey strict regularities of dissolution). Clearly, no observational regularities can explain our experimental results; the explanation we offer would be impossible without an appeal to unobservable microstructures and regularities governing them. Since it is the aim of science to explain, science requires belief in unobservable microstructures even if the explanations they make possible have no further observational consequences.

Van Fraassen poses three challenges to the argument based on this thought experiment. First, is it really true that our microstructural postulate has no new observational consequences? If it is indeed true that A and B have distinct dissolution rates x and $x + y$, then all gold samples should dissolve at a rate somewhere between those limits, and any value between x and $x + y$ can be observed. But none of this is implied by our original data. So, it is false that the microstructural postulate has no new testable consequences at the level of observables.

Second, van Fraassen denies that science must explain in every case, even when the explanation brings with it no gain in empirical predictions. He appeals to the rejection of hidden-variable explanations for statistical phenomena in quantum mechanics. In classical statistical mechanics, probabilistic laws describe and predict phenomena that cannot, in practice, be analyzed at the level of single, individual molecules. Nevertheless, the theory still requires that individual particles obey completely deterministic laws, with each particle having a precisely definable position and momentum at every instant; only our ignorance prevents us from following the trajectory of each particle. In quantum mechanics, however, the fundamental descriptions of all systems are irreducibly probabilistic—no particle has a precisely definable trajectory, and we can predict only probability distributions for position and momentum. If quantum mechanics is true, then more precise information cannot be had, even in principle, since according to quantum mechanics, particles simply do not have fully determinate values of position and momentum at all times. Thus, quantum mechanics is fundamentally inconsistent with classical statistical mechanics, and no attempt to reduce quantum mechanics to a classical theory by postulating hidden variables (of the fully determinate classical kind) can possibly succeed.[8]

Sellars, the realist, recognizes this and hence does not insist on explanation in cases where none can be given. But van Fraassen has a further, more subtle, point. He argues that quantum mechanics accepts a principle that prohibits the adoption of theoretical variables when it is impossible to distinguish among them by any experiment or observation. Without this further principle, logical consistency alone would not rule out the possibility of some kind of hidden-variable theory that would be observationally equivalent to quantum mechanics. In fact, such hidden-variable theories have been introduced by physicists such as David Bohm. They are not classical theories, but they do yield predictions that are identical to those of quantum mechanics. Indeed, the main argument that Bohm and others make for their theories is that they, unlike quantum mechanics, *explain* quantum phenomena. (The issue of choosing among observationally equivalent theories is explored at length in chapter 3.)

Third, van Fraassen offers a methodological rationale, as opposed to the explanatory one that Sellars defends, for deploying a microstructural hypothesis in the development of our imaginary science about the dissolving of gold in aqua regia. Perhaps, similar hypotheses might be postulated in connection with other metals, with the eventual result that new observational regularities are predicted about the behavior of alloys and amalgams. In short, perhaps the goal of science and its theories is not to explain, as the realists would have it, but to develop "imaginative pictures which have a hope of suggesting new statements of observable regularities and of correcting old ones" (1077).

Against Putnam's Ultimate Argument

Hilary Putnam has offered an argument for scientific realism that van Fraassen dubs the "Ultimate Argument." (Some have called it the "miracle" or the "no miracle" argument.) According to Putnam,[9] realism is the only adequate explanation for the success of science; if we do not believe that the claims of our theories are true, that many of its terms in fact refer to unobservable objects, then we can only reckon that the success of our theories is miraculous.

Van Fraassen is prepared to grant that an explanation of the success of science must itself be scientifically adequate. If we demand that science explain regularities in the world, the regular success of our scientific predictions also requires an explanation. Realists contend that our scientific predictions are often successful because they are based on theories that correspond with the world; how could such theories fail to describe and predict successfully? As expected, van Fraassen's explanation of scientific success is not a realist one, but it is recognizably scientific in its own way. He offers a Darwinian explanation for the success of science: just as many species of organisms have struggled for survival, so have many theories, and just as organisms that were not adapted to their environment have not survived, so theories that do not have true observational consequences are cast aside. Thus, the success of science is no more a miracle than the survival of any species: empirically inadequate theories die out.

9.5 | Evaluating Constructive Empiricism: Musgrave and Others

Since the appearance of *The Scientific Image* in 1980, considerable attention has been paid to van Fraassen's criticism of realism and to his constructive empiricist version of antirealism. In addition to many discussions in philosophical journals, a collection of papers entitled *Images of Science* (1985) was devoted to an evaluation of van Fraassen's position. In the article "Realism Versus Constructive Empiricism" drawn from this volume, Alan Musgrave takes up three central issues in van Fraassen's account—the connection between truth and empirical adequacy, the theoretical-observational distinction, and the role of explanation in science.

EMPIRICAL ADEQUACY ONCE MORE

For theories about observables, or for those parts of our theories relating only to observables, truth and empirical adequacy amount to the same thing. They diverge only in the case of unobservables, where an empirically adequate theory could be either true or false, given all we can be said to know from the empirical facts. So the constructive empiricist recommends that we accept such theories but not believe them. Accordingly, the realist's epistemic commitment (in believing that a theory about unobservables is true) looks considerably riskier than that of the constructive empiricist. Thus it is that skeptical arguments against realism, to the effect that we could never be rationally warranted in believing theories about unobservables, get their purchase. In this connection, Musgrave argues that skeptical worries apply equally to constructive empiricism. Since an empirically adequate theory must save not merely all actually observed phenomena but all phenomena, past, present, and future, judging that a theory is empirically adequate goes beyond what we can know at any given time. To this same extent, then, we can never be warranted in accepting any theory as empirically adequate.

Van Fraassen insists that the threats (or potential costs) confronting realism and constructive empiricism are different. They are lower, for the antirealist, and "since it is not an epistemological principle that one might as well hang for a sheep as for a lamb" (1085), the epistemically responsible attitude to take is simply acceptance of a theory, not belief in its truth. Musgrave wants to suggest otherwise: if the risks of detection and subsequent penalty for two criminal acts are the same, the sensible criminal will opt for the one yielding the greatest gains. Since the realist takes no greater risk than the antirealist of being proved wrong on empirical grounds, the realist might as well be hung for the realist sheep (believing our theories to be true) as for the constructive-empiricist lamb (accepting our theories only as empirically adequate).

The details of van Fraassen's assessment of this issue are found in later parts of his book not included in our selection. But his position might be summarized as follows: While the *risks* of realism and constructive empiricism may be comparable, the *penalties* are higher for realism. In particular, the realist must pay the penalty of rejecting empiricism, since the realist must concede that empirical evidence is not always the final arbiter of theory choice. Suppose two incompatible theories say exactly the same things about all matters observational. While the constructive empiricist can accept both theories, proclaiming each to be empirically adequate, the realist cannot believe them both to be true (on pain of contradiction). Since evidence cannot guide the choice between them, the realist is forced to allow nonevidential considerations into the decision. Realists commonly respond to the problem of empirically equivalent theories by arguing that putatively empirically equivalent theories can only really be evaluated by extending them—by embedding them in wider theories, where their equivalence will disappear. (A further discussion of issues involved in evaluating theoretical claims that evidence leaves underdetermined can be found in chapter 3.)

Before moving on to Musgrave's discussion of the theoretical-observational distinction and the role of explanation in science, let us note two further points about van Fraassen's concept of empirical adequacy. First, recall again van Fraassen's case of the mouse in the wainscoting. The realist emphasizes that we infer the existence of the unobserved entity (a mouse, in the example) as the best explanation of the observed phenomena (disappearing cheese, scratching noises in the wall). Van Fraassen claims that the inference succeeds in this example because the mouse is an observable thing after all: thus, he writes, "all observable phenomena are as if there is a mouse in the house" and "there is a mouse in the house" are in the present case "totally equivalent" (1072). This reflects van Fraassen's view, noted above, that in the case of observables the realist and constructive-empiricist positions coincide. That is, the realist's successful inference in the mouse example is a limiting case of van Fraassen's own proposal, since for theories solely about observables, theoretical truth and empirical adequacy amount to the same thing. Jeff Foss disagrees. In "On Accepting van Fraassen's Image of Science,"[10] Foss argues that because van Fraassen equates empirical adequacy with saving the phenomena, an empirically adequate theory about observables might be false nonetheless. For example, it might be the case that "all observable phenomena are as if there is a mouse in the wainscoting" (1072) even though there is no mouse producing them. (Perhaps some other creature is devouring the cheese and scampering in the walls, or, just possibly, phenomenalism is true and there are no material objects at all.) Thus, when van Fraassen talks about "observable phenomena" and "saving the phenomena" he must mean by *phenomena* something other than *observations* or *phenomenal appearances*. To avoid inconsistency in his characterization of empirical adequacy, van Fraassen must be using the term *phenomena* rather idiosyncratically to mean *truths about observable things*. But this leads to further trouble, for, from the realist's perspective, there are many truths about observable things that involve entities, processes, and properties that are not observable in van Fraassen's sense. For example, mice have bacteria in their gut, electrons in their tails, and electromagnetic fields emanating from their brains. Thus, realists would reject van Fraassen's claim that realism and constructive empiricism coincide when applied to theories about observable things.[11]

Second, could we not undermine van Fraassen's commitment to constructive empiricism by applying constructive empiricism to itself? Recall that van Fraassen agrees that scientific realism can make sense of scientific activity and so, in that sense, is empirically adequate. His arguments aim to show, not that realism is false or inadequate, but that some other account of science, constructive empiricism, is also empirically adequate. But as John Hawthorne points out in "What Does van Fraassen's Critique of Scientific Realism Show?"[12] believing in constructive empiricism—as presumably van Frassen does—is a far cry from merely proclaiming its empirical adequacy. According to constructive empiricism, we should believe only those claims whose truth or falsity we can settle by observation. Since the observable data cannot settle the truth or falsity of constructive empiricism, the consistent constructive empiricist should be an agnostic about his own doctrine. Thus, a commitment to constructive empiricism would seem to preclude van Fraassen from believing that constructive empiricism is true.

THE OBSERVATIONAL-THEORETICAL DISTINCTION ONCE MORE

The empiricist antirealist needs to distinguish between theory and observation. Van Fraassen agrees with the scientific realist that the observational-theoretical distinction cannot be defended, if the proposal is that scientific

1248 | Ch. 9 Empiricism and Scientific Realism

terms can be divided into two distinct classes. But van Fraassen does not concede that the continuous transition from direct observation (through the air) to indirect detection (through glass, binoculars, a microscope) on its own counts against the distinction between kinds of objects. The vagueness of the predicate *observable* does not entail the absence of a distinction between clear cases where it applies and clear cases where it does not. Thus, although looking at the moons of Jupiter through a telescope seems to van Fraassen to be a clear case of observation, the purported observation of microparticles in a cloud chamber seems to him clearly different: the microparticles are detected but are not themselves observed. And what if we had electron-microscopic eyes, allowing us to observe what now we can only detect? Van Fraassen, recall, says that this changes the subject: the limitations on what is observable do not extend to the logically possible but only to what a final physics and biology would tell us are the inherent limitations on human organisms.

Musgrave argues that this reply is unsatisfactory. Physics and biology tell us that what is observable by humans varies from person to person and is a function of our peculiar evolutionary history. Such a person-relative and species-specific distinction can scarcely be granted the deep philosophical significance the antirealist needs to distinguish observables from unobservables. Van Fraasen claims that the observable-unobservable distinction itself enjoys no immediate ontological significance, but does have an epistemological significance. As his mouse-in-the-wainscoting discussion is meant to show, epistemic proprieties require that humans believe to be true only what a theory says about what can be observed. But Musgrave doubts that a person-relative, species-specific distinction of no ontological significance can bear this epistemological weight. He contends that the constructive-empiricist position has two elements: a methodological prescription (about what scientists ought to infer) and an empirical claim (about what scientists actually do infer) and that both are too strong. The methodological prescription is too strong, for given any plausible theory of evidential support, it is at least possible that there is better evidence for an explanation in terms of unobservables than there is for one in terms of observables. How reasonable is it to accept as evidence phenomena pointing to the existence of mice but not phenomena pointing to the existence of electrons on the grounds that we may one day see the mouse but not the electron? The empirical claim— that scientists in fact infer only the empirical adequacy of a theory, never its truth—is too strong as well: van Fraassen is prepared to agree that we can detect electrons but not observe them, yet how reasonable is it to suppose that a scientist who affirms that she has detected an object does not believe that the object exists?

Musgrave argues further that, worse than resting on unreasonable assumptions about scientific inference, van Fraassen's observable-unobservable distinction cannot be coherently made out. For suppose we grant his claim that our final physics and biology will tell us what is observable and what is not: a theory or set of theories T will say, among other things, that A is observable by humans and B is not. Shall we accept this verdict of T? We cannot accept it as true, since constructive empiricism requires that we accept as true only what T tells us about the observable, and surely "B is unobservable by humans" is not a statement about what is observable by humans. The consistent constructive empiricist cannot, then, believe that anything is unobservable by humans and hence cannot draw a workable distinction between the observable and the unobservable at all. (The constructive empiricist cannot retreat to a position according to which "observable by humans" is an observational predicate which humans can, by observation, see to apply or not; for the construal of the observational-theoretical distinction as a claim about terms has been judged indefensible.)

Van Fraassen replied to this criticism in the *Images of Science* volume, but Musgrave confessed that he was unable to understand it (or find anyone who could explain it to him). Peter Lipton and Paul Dicken have since done a great job of laying out van Fraassen's subtle reply.[13] Suppose that scientific theory T entails that X is unobservable. If we accept T, then we believe that T is empirically adequate. In other words, we believe that everything it says about observables is true. Now suppose that X is observable. In that case, T would entail something false about an observable entity (namely, that it, X, is unobservable). That would be incompatible with T being empirically adequate. So, if one believes that T is empirically adequate, one must also believe everything that T entails of the form "X is unobservable."[14] The problem with this reply is that it presupposes that we have a way of judging that T is empirically adequate (i.e., that everything that T says about observables is true) without relying on T to tell us where to draw the line between entities that are observable and those that are unobservable.

REALISM AND EXPLANATION ONCE MORE

The realist claims not only that science explains facts about the world, but also that scientific realism itself explains facts about science. In particular, the realist believes that only scientific realism can explain the predictive success of science: without the belief that theoretical statements entailing the existence of unobservables are true and that such unobservables indeed exist, we can only judge the success of science to be a miracle. Musgrave reminds us of van Fraassen's reply to the Ultimate Argument: just as we do not ask why mice run from cats (since animals not fleeing their natural enemies are soon killed), so we do not ask why science is successful (theories that are not empirically adequate are soon cast aside).

Does this reply avoid the Ultimate Argument? Musgrave argues that it does not. On the one hand, the respectable scientist *will* of course ask the respectable question, Why does the mouse run from the cat? The answer is that the mouse perceives the cat, perceives it as an enemy, and runs. There is nothing un-Darwinian about such an explanation: it need not gloss the mouse's perception of the cat as its enemy in terms of intentional states in the mind of the mouse. Indeed the explanation can be readily wedded to what, on the other hand, is a more central Darwinian question: Why did that sort of mouse behavior evolve? The answer is as van Fraassen says—in an environment of mouse-hungry cats, it is cat-fleeing mice that tend to survive and thus perpetuate their cat-fleeing behavior. This explanation is not offered in place of the previous answer, but rather tells a story according to which the facts adumbrated in the previous answer are the result of the selective forces of nature. Musgrave goes on to urge that the case of success in science is similar. It is one thing to explain why some particular theory is successful and another to explain why existing scientific theories in general are successful. The latter explanation is the one that van Fraassen gives, and the realist and antirealist can equally accept it: unsuccessful theories do not survive. But that is not to explain why any particular theory is successful, and to this question the antirealist has given no answer.

Musgrave is prepared to grant that the Ultimate Argument cannot be construed simply as the claim that there can be no antirealist explanation of a theory's success. If, for example, a theory is devised precisely to predict phenomenal regularities we know to obtain—as the deferents and epicycles of Ptolemaic astronomy were designed to yield precisely those retrograde motions and periodic eclipses we do observe—then it is no miracle that such a theory should be successful in predicting what it does. But the case is different when a theory T designed to accommodate some range of phenomenal regularities turns out to predict new, as yet unobserved regularities. The realist explanation is near to hand: the objects postulated by T exist, and what T says about them is true. What the antirealist seems forced to say is that T was formulated for one purpose and turned out miraculously also to be well adapted for some other purpose. Thus, Musgrave emphasizes that the Ultimate Argument for realism is most convincing when applied to *novel* predictive success. (For more on the significance of novel prediction, see chapter 4 and the discussion of the article by Martin Carrier later in this commentary.)

9.6 | Laudan against Convergent Realism: Reference and Truth

Our discussion thus far reveals the extent to which philosophers of science are prepared to evaluate realism and its rivals using criteria from the empirical sciences themselves. Scientists' preference for good explanations is chief among such criteria, and the Ultimate Argument represents its most general application by scientific realists: the undeniable success of science remains inexplicable unless the theories of science enjoy all or most of the virtues attributed to them by realists. These virtues, as we have seen, include the truth of scientific theories and the successful reference of their theoretical terms. Other virtues are frequently cited by realists. Given some commitment to the notion that science and its theories undergo not just change but genuine progress, realists will naturally want later theories to preserve and explain whatever approximate truth is captured by their predecessors. Science, it is said, increasingly converges on the truth about the natural world. This view has been called *convergent realism*.

In "A Confutation of Convergent Realism," Larry Laudan questions whether the connections between truth, reference, and success are as sound as the convergent realist believes them to be. To begin, he articulates five theses that make up what he calls "convergent epistemological realism" (CER):[15]

> **R1** Scientific theories (at least in the 'mature' sciences) are typically approximately true and more recent theories are closer to the truth than older theories in the same domain.
>
> **R2** The observational and theoretical terms within the theories of a mature science genuinely refer. . . .
>
> **R3** Successive theories in any mature science will be such that they 'preserve' the theoretical relations and the apparent referents of earlier theories. . . .
>
> **R4** Acceptable new theories do and should explain why their predecessors were successful insofar as they were successful. . . .
>
> **R5** Theses (R1)—(R4) . . . constitute the best, if not the only, explanation for the success of science. (1109–10)

Thesis (R5) expresses in general form the abductive (inference-to-the-best-explanation) argument for scientific realism. That is, realism explains the success of science, and so the success of science empirically confirms scientific realism. Laudan divides these abductive arguments into two general sorts. Arguments of sort I are based on the putative connection between the success of science and the likelihood that theories are true and genuinely referential. They take something like the following form:

> 1 If scientific theories are approximately true, they will typically be empirically successful.
>
> 2 If the central terms in scientific theories genuinely refer, those theories will generally be empirically successful.
>
> 3 Scientific theories are empirically successful.
>
> 4 (Probably) Theories are approximately true and their terms genuinely refer. (1110)

Abductive arguments for realism of sort II are based on the putative limiting-case relations between earlier and later theories and the likelihood that earlier theories are approximately true and genuinely referential. They run roughly as follows:

1 If the earlier theories in a 'mature' science are approximately true and if the central terms of those theories genuinely refer, then later more successful theories in the same science will preserve the earlier theories as limiting cases.

2 Scientists seek to preserve earlier theories as limiting cases and generally succeed.

3 (Probably) Earlier theories in a 'mature' science are approximately true and genuinely referential. (1110)

Laudan's criticism of arguments of sort II, addressing the realist's broadly retentionist view of scientific progress and theoretical change, uses case studies from the history of science to suggest that the following sorts of relations between some theory T and its successor T' do not generally hold: T' entails T; T' retains the true consequences of T; T' preserves T as a limiting case; T' explains why T succeeded to the extent that it did; and T' retains reference for the central terms of T. (This concern with intertheoretic relations is discussed at length in chapter 8 and its commentary, and since Laudan's misgivings about realism are well represented by his criticism of arguments of sort I, we have omitted his discussion of sort II arguments from the latter third of the paper included in the readings of this chapter.)

Let us focus on arguments of sort I. Laudan claims that its main premises (1 and 2 on p. 1109) cannot be sustained. His account may be divided into two parts: the first concerns the relation between reference and success; the second, the relation of approximate truth to success. We examine them in turn and offer an occasional realist response.

REFERENCE AND THE SUCCESS OF SCIENCE

The convergent realist who relies on (R2)—the thesis that terms in our scientific theories genuinely refer—in explaining the success of science must subscribe to the following claims (1111):

S1 The theories in the advanced or mature sciences are successful.

S2 A theory whose central terms genuinely refer will be a successful theory.

S3 If a theory is successful, we can reasonably infer that its central terms genuinely refer.

S4 All the central terms in theories in the mature sciences do refer.

According to the realist, reference explains success—that is, (S2) and (S4) entail and thus explain (S1)—while success warrants a presumption of reference—(S1) and (S3) provide warrant for believing (S4). Laudan is pre-pared to grant these relations; the central question is thus whether (S1) through (S4) are true.

On most readings, the claim of (S1) that theories of mature sciences are successful can be safely judged to be true. What of (S2)—the claim that a theory whose terms refer will be successful? Laudan contends that there is ample evidence that it is false. The history of science is replete with unsuc-cessful theories about atoms and molecules and genes and continents and so on. Moreover, it is easy to see how this could be the case. A theory's terms genuinely refer when the theory cuts the world at its joints, when it postu-lates entities that actually exist, but doing so scarcely entails or even makes likely the truth of that theory's statements about such entities. Thus, the content of a genuinely referential theory might easily be massively weighted on the side of falsehood.

As a speculative and critical digression, consider how a realist might respond to this assessment of (S2). If we encounter a theory massively weighted on the side of falsehood, how likely are we to believe that it succeeds in carv-ing the world at its joints? More specifically, how probable is it that *no* likeli-hood is conferred on a theory by its terms genuinely referring? That genuine reference should place so little constraint on likelihood of truth seems to presume that referential success could be largely accidental, having little or nothing to do with our judgments about the nature of the referents.

Some theories of reference deny this presumption. They claim that scientists cannot successfully refer out of the blue, willy-nilly; rather, suc-cessful reference (to members of a natural kind, for example) is ultimately grounded in some robust, nonaccidental connection between the use of a scientific term and its referent—a causal relation, say. If the realist adopts this view, and if this view is wedded to the doctrine that the properties of objects are (or are uniquely determined by their contribution to) the causal powers of those objects, then the realist can explain how our ability to refer to objects derives from our knowledge of the properties of those objects. If what grounds the possibility of genuine reference is precisely what grounds the empirical scientist's judgments about the properties objects have, then it would seem unlikely that successful reference could be disconnected altogether from the scientist's judgments about the nature of the objects. We can refer because we are causally connected with the properties objects actually have. While such connections to an object's properties in no way guarantees that all our judgments about that object will be true, the realist will urge nevertheless that successful reference in scientific theories is unlikely to be accompanied by wholesale falsehood. At the very least, then, such considerations suggest that the antirealist cannot be so sanguine about the absence or weakness of connections between reference and likelihood of truth, regardless of considerations about the nature of reference.

What about (S3), the realist's claim that success creates a reasonable presumption of reference? Laudan argues that many historical theories have been successful in generating explanatory accounts of various experimental phenomena, without being genuinely referential. In particular, a whole family of ether theories in eighteenth- and nineteenth-century physics and chemistry were used to account for a wide range of phenomena—from the heat of reactions in chemistry to the refraction, interference, and polarization of light waves in optics. Judging from such historical cases, Laudan concludes that (S3) must be rejected. Nor will it do to weaken (S3), to claim that (only) *some* of the terms of a successful science can be inferred genuinely to refer. In the first place, doing so is not open to the typical realist, who claims not that evidence for a theory is evidence for only part of that theory (the parts about observable things, for example), but rather that evidence for a theory is evidence for everything it claims, for the truth of that theory as an account of everything it is about. If the tests to which we subject our theories test only parts of them, then even highly successful theories may have central terms (at the deepest theoretical level, perhaps) that do not refer and may make claims that, as yet untested, we would have little grounds for judging to be true. Moreover, weakening (S3) in this way threatens the realist's rationale for holding (R3)—the thesis that theory succession is and should be retentive. The underlying motivation for (R3) is the realist's conviction that, because the earlier theory was successful, its terms must have been referential, and that retaining referential terms is a constraint on successor theories. If we weaken (S3), this constraint is undermined.

APPROXIMATE TRUTH AND SUCCESS

The other central element in arguments of sort I concerns the relation of truth to success. Realists typically claim that, while full theories (as, say, conjunctions of statements) may well be strictly false, they are nevertheless close to the truth or enjoy some form of verisimilitude. As Laudan puts it:

> The claim generally amounts to this pair:
> T1 if a theory is approximately true, then it will be explanatorily successful; and
>
> T2 if a theory is explanatorily successful, then it is probably approximately true. (1117)

In discussing the "downward path" (1117–18) from approximate truth to explanatory success in (T1), Laudan's primary misgiving concerns the notion of approximate truth. Realists have offered too little by way of an analysis of this concept to show that, as (T1) asserts, approximate truth entails explanatory success. Surely (T1) is not obviously true. Even if the approximate truth of a theory is taken to mean that the size of the set of true claims entailed by the theory is vastly larger than the set of its false entailments, it may yet be the case that most of the theory's consequences thus far tested are false. Without some coherent account of approximate truth that can be seen to entail explanatory success, we are within our epistemic rights to deny (T1).[16]

Laudan argues that the "upward path" (1119) from explanatory success to (probable) approximate truth expressed in (T2) fares no better. He thinks that the main problem, for the realist, is the assumed connection between genuine reference and approximate truth, for Laudan asserts that:

L   "A realist would never want to say that a theory was approximately true if its central theoretical terms failed to refer" (1119–20).

If there were no genes, then genetic theory—however well confirmed it might be—would not be approximately true. According to (L), a necessary condition, especially for the scientific realist, for a theory being close to the truth is that its central explanatory terms genuinely refer. But the history of science is replete with examples of theories that were both successful and nonreferential with respect to their central explanatory terms: among Laudan's examples are the crystalline spheres of ancient and medieval astronomy, the phlogiston of eighteenth-century chemistry, and the optical and electromagnetic ether of nineteenth-century physics. In all such cases, the theories were successful and well confirmed but contained central terms that we now believe not to refer.

In response to Laudan's contention that many successful theories failed to refer, the realist might emphasize again that (R1) through (R5) apply only to theories of a mature science—theories offered at some point sufficiently far into the life of a discipline that all the relevant background theories are reasonably entrenched and well confirmed. Many of the theories on Laudan's list, being drawn from immature sciences, would then fail as counterexamples to the connection that realists claim to hold between approximate truth and explanatory success. Laudan finds this maneuver unsatisfactory in two respects. First, it threatens to make (R3) and (R4) vacuously true. For if the mature sciences are *defined* as those in which correspondence and limiting-case relations hold between successive theories, then (R3) and (R4) become true by definition. Second, the restriction to mature sciences undermines the realist's aim to explain why science in general is successful. No such general explanation will have been given on the restricted account if indeed there are, as Laudan has argued, successful theories in immature sciences.

Before concluding our discussion of Laudan, it is worth noting that Clyde L. Hardin and Alexander Rosenberg see no reason why realists should grant thesis (L), that "a realist would never want to say that a theory was approximately true if its central theoretical terms failed to refer."[17] Laudan's own example of the gene and genetic theories serves as a useful case here. (See the section "Kitcher on Reduction, Classical Genetics, and

Molecular Biology" in the commentary on chapter 8). Mendel's nineteenth-century theory is at the beginning of a sequence of theories enjoying increasing degrees of success, a sequence many biologists would claim is converging on the truth. Despite being credited with approximate truth, despite the ease with which it can be modified to generate increasingly accurate and complete genetic theories, Mendel's theory used the term *gene* to refer to entities that do not in fact exist.[18] That is not to say that the theory was incorrect in ascribing to the fundamental genetic material most of the functions it did; rather, the functions thus ascribed to genes are now parceled out to different sequences and complex combinations of DNA. Indeed, in much of contemporary molecular genetics the use of the term *gene* has simply been dropped in favor of terms better suited to articulate the diverse units of hereditary phenomena and their functions. The causal role Mendel accorded to genes has been divided up among other entities, entities that function together to give the false impression that a unitary genetic item is responsible for all aspects of genetic transmission and expression. Thus, a realist can explain the plausibility of Mendel's theory, and its approximate truth, by appeal to those very diverse functions of diverse objects that reveal that the term *gene* does not refer.

That is one way the realist might go. Another way is to construe Mendel's theory (perhaps with the aid of theories of plural reference or partial reference) as referring successfully after all to configurations of DNA and their polypeptide products, despite Mendel's ignorance of DNA and its role in protein synthesis. Here, unlike the previous account, the realist can opt to sever successful reference from the detailed beliefs of the scientist, granting the Mendelian theory its undeniable measure of success in the light of newer approximations to the truth.

The realist can either trace out the relations between Mendel's theoretical claims and those of modern molecular biology in a way that explains the approximate truth of the former despite their failure of reference or adopt a theory of reference able to preserve the referential success of the term *gene* as described. There being as yet no final and fixed theory of reference in philosophy, a good deal rests on one's views in this area. Nonetheless, whichever way one chooses to understand the relation of Mendelian to molecular genetics, one can be a realist about the success and approximate truth of the earlier theory.

Realism and Begging the Question

The realist strategy that Laudan has been attacking attempts to link empirical success with likelihood of approximate truth and reference. In his final remarks Laudan argues that, leaving aside the status of its premises, such arguments can—from the point of view of antirealists—only be seen as question begging. Antirealists deny that we are warranted in judging any theory true on the basis of its yielding true observational consequences. To suppose that we are is to commit the fallacy of affirming the consequent, since a false theory can have true consequences.[19] Now, realists have been concerned, not with any scientific theory in particular, but with the truth of realism, with the truth of a claim about scientific theories generally. That is, realists have argued that we can reasonably judge scientific realism to be true because it yields true consequences. (If our scientific theories are approximately true and genuinely referential, then they will enjoy empirical success, and they do enjoy empirical success.) But to suppose that realism can be defended in this way is to beg the question against the antirealist, who already denies that the empirical success of any theory is evidence of its truth.

9.7 | The Pessimistic Induction

Realists and empiricists differ over the following, epistemological question: What do our best scientific theories—the ones that are most successful— justify us in believing? Realists understand success in terms of accurate predictions and unifying explanations. They insist that we should accept our best, most successful, scientific theories as true (or, more cautiously, as probably true, or approximately true, or close to the truth), including all the claims that such theories make about unobservable entities and processes. Just as scientists accept the particular scientific theories that give the best explanation of the relevant data and phenomena, so too should philosophers of science accept the best philosophical explanation of why some theories are so successful, namely, *they are successful because they are true* (or approximately true, or close to the truth, or partially true). More cautiously: It is because the electron theory, say, successfully *refers* to electrons (i.e., electrons really exist) and the theory more or less correctly describes their *lawlike* behavior, that the electron theory is successful. As Hilary Putnam put it: Any other explanation would make the success of science a miracle.

Ever since Henri Poincaré described the past of science as littered with the wrecks of abandoned theories and proclaimed that (to an unsophisticated eye) it might appear that science is "bankrupt,"[20] several philosophers have attacked scientific realism by appealing to the large number of past scientific theories that have turned out to be false. Certainly, when one looks at the history of science, especially the scientific revolutions in physics, astronomy, and chemistry chronicled in Thomas S. Kuhn's *Structure of Scientific Revolutions*, the realist seems to be in a bind. For Kuhn and later, Larry Laudan ("A Confutation of Convergent Realism," this chapter), give many examples of theories that were successful in their day and yet we now recognize as based on radically false assumptions about the world. Ptolemaic astronomy, Newton's gravitational theory, the phlogiston theory of chemistry, the caloric theory of heat, the wave theory of light, Maxwell's electromagnetic theory, and so on, all posited entities and processes that, by our lights, do not exist. There are no epicycles, no instantaneous action at a distance, no phlogiston, no caloric, no mechanical ether, no optical ether, no substance-like electromagnetic ether at rest in absolute space. Hence the theories that contained terms that purport to refer to such entities must have been false. As Juha Saatsi points out, the argument based on the apparently large number of successful but false theories in the history of science—what Saatsi and others call the *Pessimistic (Meta-) Induction* (PMI)—is probably the best known modern philosophical argument against scientific realism.[21]

What Is the Pessimistic Induction?

We expect the PMI to be an inductive argument with a pessimistic conclusion. But what is the conclusion, and how does the argument go? It is important to distinguish between two different things that pessimists might be arguing. A pessimist might aim to draw a conclusion that directly contradicts the epistemic optimism of realists. On this reading of the PMI (suggested by Putnam's original wording), the pessimist thinks that the replacement of many theories that were successful in their day but turned out to be false is good inductive grounds for thinking that our current successful theories are likely to be false, too, and will likewise be replaced in the future. This is the "direct" version of the PMI (or what Peter Lipton has called "the disaster argument"): an induction from past failures to present pessimism and the likelihood of replacement in the future.

There is a second, subtler version of the PMI. On this reading, pessimists are not arguing that our current scientific theories are likely to be false; rather, they aim to undermine a key premise in the realist's *No Miracles Argument* (NMA). Realists typically deploy a version of the NMA to conclude that we have good reason to believe that our best, "successful" current theories are true. Why? Because the best explanation of the success of our current theories is that they are true. Inference to the best explanation warrants acceptance of our currently successful theories as true, not just as empirically adequate or observationally reliable, but true in all respects including, especially, what those theories say about unobservable entities and processes. On the second "meta-level" interpretation of the PMI, pessimists are attacking the realist's argument for optimism, not arguing directly for pessimism. (An analogy would be the difference between criticizing a theist's argument for God's existence and offering a direct argument for atheism.) The premise under attack in this version of the PMI is that success (of a scientific theory) is a reliable indicator of truth. It is this argument that Saatsi identifies as the core antirealist reasoning in Laudan's article, which he reconstructs as a reductio.

1 Success is a reliable indicator of truth. [Assumption to be refuted.]

2 Most current successful theories are true. [From 1.]

3 Most past theories are false. [From 2 plus the assumption that past theories are incompatible with current theories in significant ways.]

4 Many of these past theories were successful. [A historical claim.][22]

5 Therefore, assumption (1) is false: success is *not* a reliable indicator of truth. [Since 1 conjoined with 4 contradicts 3.]

Realists have made several objections to the premises of the PMI. Often, part of the realist response to Laudan is to insist on a much stronger notion of success than mere agreement with known data. By requiring novel prediction, for example, the number of successful but false theories on Laudan's list is considerably reduced. (See the section "Why Theory Realism and Entity Realism Fail" in the discussion of Carrier's kind realism, later in this commentary.) Some realists then argue that strongly successful past theories did refer to real unobservable entities even though past scientists had some false beliefs about them. Others see no problem in allowing that a theory can be approximately true even though one or more of its central terms do not refer. (See the discussion of Hardin and Rosenberg in the section "Approximate Truth and Success" in our earlier comments on Laudan.)

A different line of attack on the PMI questions its validity. Saatsi examines two attempts—one by Marc Lange, the other by Peter Lewis—to show that Laudan's conclusion does not follow from his premises.

THE TURNOVER FALLACY

Marc Lange attacks the PMI, not on the grounds that one or more of its premises are false, but because the form of the argument is fallacious.[23] He thinks it is important to show this because the PMI is also a threat to empiricists. Consider all the theories that, at some time or another in the past, have been considered to be empirically adequate on the basis of evidence that supported them roughly as well as the modern evidence supporting our current best theories. Most of those theories in the past were eventually discovered not to be empirically adequate. They made observable predictions that are false, and the theories were abandoned. Therefore, we should believe that most of the theories we currently accept are not empirically adequate.

According to Lange, the PMI in its various versions commits what he calls the *turnover fallacy*. He gives a baseball analogy to illustrate his criticism. Most baseball managers of major-league teams, whose careers are now ended, left the game having lost more games than they won. Therefore, according to the PMI, we should believe of managers who are *now* in charge of teams that most of them will likewise end their careers with losing records. The fallacy arises because baseball managers with losing records are more likely to be fired than those with winning records. In other words, if we look at the *cumulative* track record we are bound to find more losers than winners simply because the losers are replaced more often. Similarly if we look at the history of science, we should *expect* to find more false theories than true ones. Here is another example, from Saatsi. In the last 10 years, there have been 180 changes in membership on a 10-member board of directors. From this I might pessimistically infer that someone will replace me very soon after I join the board. This again is to reason fallaciously. The evidence of the total number of turnovers is consistent with 9 out of 10 of the members retaining their seats for the full 10 years, with just my "predecessors" (as it were) in the demanding position of treasurer/fund raiser being replaced every few weeks.

The moral of Lange's analysis is clear: It is fallacious to infer the chances that currently successful theories will be replaced in the future by looking at the entire history of science and seeing what percentage of successful theories were rejected as false. Rather, the pertinent question to ask is: At each *time-slice* of the past, what proportion of theories that were successful and accepted *at that time* were false as judged by our current lights? In other words, on the assumption that past science was as good as present science at discovering truth and avoiding error, Lange would presumably accept as reasonable a pessimistic inductive inference of the following form:

At time t_1 in the past, X% of successful theories accepted at t_1 were later rejected as false.

At time t_2 in the past, X% of successful theories accepted at t_2 were later rejected as false.

At time t_3 in the past, X% of successful theories accepted at t_3 were later rejected as false.

. . . .

At time t_n in the past, X% of successful theories accepted at t_n were later rejected as false.

Conclusion: X% of currently accepted successful theories will later be rejected as false.

Is this sufficient to dismiss the antirealist PMI as deployed by Laudan and others? Juha Saatsi says "No." The problem as Saatsi sees it is that, while Lange has made a telling point about the fallacious nature of *some* forms of the PMI, this is irrelevant to the form of Laudan's argument. For Laudan is *not* using the PMI to conclude that most current theories will be rejected as false and replaced at some time in the future. Rather, Laudan is arguing for the meta-level epistemological conclusion that the success of a theory is not a reliable guide to its truth. This conclusion, which attacks the key realist premise, leaves open the question of whether or not most current theories are false. Thus, the relevant issue is not: Are most of our current theories true? Rather, the issue is: Are realists justified in inferring from the success of current theories that most of them are true?

Can a stronger pessimistic conclusion about our current theories be drawn from the presumed unreliability of success as an indicator of truth? Saatsi thinks it can. Although he does not endorse the argument as sound, Saatsi thinks that antirealists can argue as follows, without committing the turnover fallacy.

1 Success is not a reliable indicator of truth. [From Laudan's PMI.]

2 There are no other reliable indicators of truth.

3 Most successful theories are false. [Generalization over the entire history of science.]

4 Current successful scientific theories are a representative sample from successful theories throughout the entire history of science.

5 Therefore, any current successful theory is probably false.

Saatsi characterizes this argument (which he labels PMI*) as "statistical" since it is essentially an inductive argument based on a sampling model. A premise about which Saatsi expresses reservations on the realist's behalf is premise (4), asserting that current scientific theories are not relevantly different (in any way that would affect their chances of truth or falsity) from any other successful theories in the history of science. Of course our current accepted theories differ from past rejected ones in an obvious way—we have no strong specific evidence against them. We do not *know* that our present theories are false as, presumably, we do know that past theories are false. The question regarding current theories is: Are we justified in believing that they are true? Antirealists see the situation as analogous to checking a new book scrupulously for typos and finding none. Should one believe that the book is typo-free? Statistically, one knows that the vast majority of books contain spelling mistakes. If one also knows that these other books were subjected to the same kind of scrutiny as the present volume, then the conclusion we should draw is that the new book also contains typos.

Saatsi denies that PMI* commits the turnover fallacy since it is concerned with the "timeless" properties of truth and falsity, not with predicting that theories will be rejected in the future. Even if all the "turnovers" were to have occurred in just one branch of science, Saatsi argues that PMI* would still support its conclusion about the falsity of most current theories because its premises assert the absence of any scientific indicators of truth. Presumably, in that case, we would also inductively predict that turnovers will be rare in those branches of science in which long-lived theories have been the norm up until now.

The Base-Rate Fallacy

According to Peter Lewis and others, there is a flaw in Laudan's PMI, construed as a meta-level argument. Laudan gives numerous examples from the history of science of theories that were once successful but are now regarded as false. Is this sufficient to show that success is not a reliable indicator of truth? Lewis argues that Laudan's PMI commits the *base-rate fallacy*.[24] This fallacy is familiar from medical contexts in which a positive test result, S, is a sensitive indicator of a disease, T, in just the same way that the realist insists that success (S) is a reliable indicator of truth (T) because, according to the realist, most true theories will be successful. In other words, the realist claims that $P(S/T)$ is high, say 0.9. But it does not follow from this alone that the proportion of true theories among successful theories in the past will be high, any more than it follows that most people who show positive on a diagnostic test will have the disease in question. Two other things are relevant to the value of $P(T/S)$. First: How specific is the test? In other words: How likely is someone to show positive on the test even though they are in fact disease-free? The analogous question for the PMI is: How probable is it that a theory will be successful even though false? The realist has to assume that this is less than a half, say 0.25.

The other relevant factor in determining the value of $P(T/S)$ is the base rate, $P(T)$. How rare is the disease in the population from which patients are being selected for testing? If the disease is very rare, then most of the positive test results will be false positives; they will come from people who are disease free. Analogously, in the scientific case, when true theories are rare and false theories common, most cases of success (at least in the relatively short run) are going to come from false theories.

So, in regarding success as a reliable indicator of truth, the scientific realist will typically assume that $P(S/T) = 0.9$, and $P(S/\sim T) = 0.25$. Realists are optimists. They assume that science improves as time goes on: false theories are weeded out, and a higher proportion of true theories are retained. On the realist picture, therefore, a much higher percentage of accepted theories are true now than in the past. Thus we would expect that, in the past, success was a poor guide to truth, not because true theories are not more likely to be successful, but because in the past there was a much higher fraction of false theories among those that were successful. Now, in the present, with the advantage of several hundred years of scientific progress, a much higher percentage of accepted theories are true. Hence, a much higher proportion of successful theories will turn out to be true. Saatsi's diagram (Figure 1, on page 1135) illustrates how the fraction of false

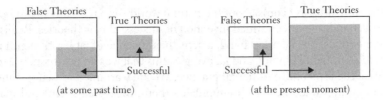

False Theories   True Theories   False Theories   True Theories

Successful

Successful

(at some past time)   (at the present moment)

From Juha T. Saatsi, "On the Pessimistic Induction and Two Fallacies," *Philosophy of Science* 72 (2005): 1088–98.

theories that are successful (0.25) and the fraction of true theories that are successful (0.9) can remain constant over time while the proportion of successful theories that are true increases as true theories begin to outnumber false ones.

There is much here that can be contested. One thing to be clear about is that this is not intended to be a convincing argument for the realist position. Antirealists will reject the assumption that the proportion of true to false theories in science grows over time, and that the majority of current theories are true. The point of the base-rate-fallacy response is to show that, in appealing to numerous cases of successful but false theories in the past of science, the antirealist has not made a compelling case against the realist. In particular, Lewis denies the validity of the inference in Laudan's PMI from "Success is a reliable indicator of truth" to "Most successful theories (whether current or past) are true." This is because Lewis chooses to interpret "Success is a reliable indicator of truth" (by analogy with diagnostic tests in medicine) as "The probability of success given truth is high."[25] The price of identifying reliability as $P(S/T)$ is that no conclusion about $P(T/S)$ can be drawn in the absence of information about the base rate, $P(T)$. Thus Lewis is able to resist Laudan's PMI only by relying on assumptions that no antirealist would be willing to accept. By the same token, realists are unwilling to accept the assumptions about base rates that, according to the same Bayesian analysis, are required for the pessimistic induction of the antirealists. It is for this reason that Magnus and Callender conclude that the realism versus antirealism debate at the "wholesale" level is a standoff.[26]

9.8 | Hacking's Entity Realism

There is a kind of deflationary realism that accepts the existence of unobservable, or theoretical, entities without making any commitment to the truth of scientific theories. Ian Hacking is such a realist. In his paper "Experimentation and Scientific Realism," he claims that most discussions about scientific realism and antirealism, conducted as they are at the level of theories, are necessarily inconclusive. Realism about theories—about the

truth of theories as claims about the natural world—is based only on an optimism attending our continued invention of successful new theories. Realism
about entities, on the other hand, asserts the existence of (at least) the unobservable entities of physics, and Hacking claims that we can justify such assertions at the level of experimental practice. We are warranted in being realists
about entities once we have manipulated them, measured them, and inter-
vened in their causal processes. The relevant considerations for the deflationary realist thus involve not theorizing but experimenting.[27]

One might wonder if the distinction between theory realism and entity
realism can be made out. If one believes a theory, then one surely believes
in the entities it postulates, and if one believes in certain entities, then one
surely believes truths about them. Hacking resists this objection. He contends
that the vast majority of experimental physicists are convinced of the existence of "inferred" and "unobservable" entities but that few believe their theories are literally true. Typically, experimenters will remain agnostic about
the many possible (and mutually incompatible) theoretical frameworks in
which claims about electrons, say, can be cast. This is not simply a sociological report about experimentalists. The point is rather that experimentalists
can be antirealists about scientific theories and models, viewing them—
rather as instrumentalists might—as tools for thinking and calculating while
at the same time believing in the existence of the entities they investigate.
Their enterprise would be incoherent if they did not: realism about entities is
a necessary condition for the coherence of scientific experimentation.

Before presenting an example of scientific experimentation to illustrate
his case, Hacking makes two important points, about reference and about
what he calls *interfering* or *intervening*. The first point (as we shall interpret
it) is in the service of his antirealism about theories; the second point is
central to his views on entity realism.

The realist believes that there are electrons. What is the experimental
realist—who affirms that electrons exist but denies any commitment to theory realism—to make of the notion that *electron* gets its meaning from its
role in theoretical laws? Hacking himself is happy enough to adopt in rough
outline an account offered by Hilary Putnam according to which we must
distinguish two aspects (among others) of the meaning of a natural kind
term: its stereotype, which includes standardly accepted descriptions of the
natural kind, and its actual reference—the thing it denotes or picks out.[28]
Putnam allows that while we may succeed, at an initial baptismal naming or
thereafter, in referring to an object under some stereotype, this does not
entail that stereotypes "determine" referents in any strong way. As we frame
clearer views and discover more about the thing or kind of thing, we can succeed in referring to the same thing or kind of thing even if the stereotype
changes. Now, in the case of electrons, not only has the term *electron* undergone such stereotype changes, but moreover we presently have two incompatible stereotypes—the electron as wave and the electron as particle.
Hacking implies that experimentalists can be confident that they are inves-

tigating the same thing, regardless of their commitments to this or that stereotype.

The notion of interfering or intervening forms the core of Hacking's experimental realism. We have seen that experimental realism recommends a shift away from theoretical representation, toward experimental manipulation. On Hacking's account, the very nature of scientific experimentation is to interfere in the course of nature. Indeed, if we are concerned with reality—with the entities making up the furniture of the natural world—then the fundamental grounds on which we form our beliefs are our causal interactions with such entities. They act on us, we act on them, and when we have understood their causal powers sufficiently well, we use them to interfere elsewhere in nature. Since this is a general fact about observables and unobservables alike, what justifies our belief in the reality of unobservables is no different in kind from what justifies our belief in observable objects. As Hacking puts it elsewhere:

> Reality has to do with causation and our notions of reality are formed from our ability to change the world. . . . We shall count as real what we can use to intervene in the world to affect something else, or what the world can use to affect us.[29]

The implication of Hacking's stress on intervention is clear: we rely on the causal properties of objects—electrons, say—in developing instruments to generate various scientific results, and this very business of experimentation is incoherent unless we believe that such entities exist and have the causal powers they do. Thus, we do not infer the reality of electrons from experimental results; rather, we presuppose that electrons are real when we design and execute our experiments.

Hacking discusses at considerable length an example from experimental physics and then draws some morals from the example. He focuses on PEGGY II, an instrument used to fire polarized electrons at a deuterium target. The aim of the experiment was to observe parity violations in weak neutral interactions. Hacking concludes that constructing PEGGY II and executing the experiment described in his paper were largely nontheoretical affairs. The project began with a core set of beliefs about the properties of elementary particles and atoms, together with the discovery (in an unrelated experiment) that gallium arsenide (GaAs) crystals could serve as a source of polarized electrons. The inventors and designers of PEGGY II did not use a theory about electrons to explain or save any phenomena; rather, they used electrons to produce events that violate parity in weak neutral interactions. The experimenters were realists about electrons because they could intervene with electrons to create new phenomena. Thus, Hacking takes this to be a typical illustration of his thesis that it is experimental practice, not theorizing, that certifies scientific realism about unobservables.

9.9 | Resnik's Evaluation of Entity Realism

Hacking is correct in claiming that the practice of scientific experimentation has been largely ignored in traditional disputes about scientific realism. As we have seen, it is rather questions about the relation between the
world and our representations of the world (theories), about the nature of
scientific inference and explanation that have figured most centrally in the
exchange between realists and antirealists. Hacking's project is to shift the
debate from representation to intervention and to argue that, at the level of
the entities themselves, the strongest argument for their existence is our ability to use them to control and create phenomena. Thus, Hacking is what one
might call a *theory-free entity realist*. In "Hacking's Experimental Realism,"
David B. Resnik argues that this project fails in three respects: experimental
realism is, at the end of the day, just another version of traditional success-of-
science arguments; the experimental realist can claim to have knowledge of
unobservables only if she assumes that the theories describing them are at
least approximately true; and experimentation is not nearly so theory-free as
Hacking claims. We will examine these objections in turn.

According to Hacking, experimental science gives us the best grounds
for believing in unobservable (theoretical) entities. If we use some part of
the natural world to intervene—to control and create phenomena—then we
are entitled to believe that our objects of intervention are real. Resnik
reconstructs Hacking's argument as follows:

1 We are entitled to believe that a theoretical entity is real if and only
   if we can use that entity to do things to the world.

2 We can use some theoretical entities, e.g., electrons, to do things to
   the world, e.g., alter the charges on niobium balls.

3 Hence, we are entitled to believe that some theoretical entities, e.g.,
   electrons, are real. (1162)

While some have criticized Hacking's argument by attacking the first
premise, Resnik is concerned with the grounds an experimental realist
might have for asserting the second premise.

On Hacking's account, the assertion of (2)—that we can use theoretical
entities, or unobservables, to do things to the world—is fundamentally a
causal claim. How are we to support or justify causal claims? Hume argued
that we do not deduce them from experience, but rather base them on
inductive inferences: we observe regularities, or effects, in nature and infer
causes from them according to the principle that similar effects must have
similar causes. While Hume himself regarded such induction as ultimately
unjustified (but the best we can do), many philosophers believe that we can
justify causal claims inductively. Others believe that we can use abductive
reasoning, inferring that A causes B, on the grounds that this best explains

the available evidence. Now, in denying theory realism but accepting entity realism on the basis of causal arguments like the one above, Hacking is clearly committed to denying that we base our causal claims, inductively, on observed lawlike regularities.[30] Thus, while Hacking does not explicitly commit himself to any view about scientific reasoning, it would seem that his only alternative in the present case is abduction.

Resnik's objection is that Hacking's reliance on abductive reasoning places his project firmly in the camp of traditional success-of-science arguments for realism. As we have seen, these arguments conclude that we are justified in being scientific realists on the grounds that there is no other explanation for the success of science: unless the unobservables posited by our theories actually exist, the predictive success of science would be a miraculous coincidence. While these arguments focus largely on the success of theories—on our ability to predict and explain—Hacking's argument focuses on the success of experiments—on our ability to construct and control instruments that use unobservables to create new phenomena. According to Resnik, Hacking is in the position of claiming that it would be an incredible coincidence that our instruments should work as they do, and let us affect the world as we do, if the unobservable entities we take ourselves to be manipulating in our experiments did not in fact exist.

Hacking rejects the traditional success-of-science arguments and denies that his own experimental realist argument is of this sort. Recall that the scientists involved in the PEGGY II experiment did not aim to explain any phenomena with electrons but rather simply to use them, and this is a general feature of experimental science. Clearly, Hacking is right in claiming that the experimentalists believe in the existence of electrons. Indeed, he may well be right in claiming that the experimenter "is *impressed* by entities that one can use to test conjectures," and that "the experimenter is *convinced* of the reality of entities . . . that . . . can be used to interfere elsewhere in nature" (1143; our emphases). But these are psychological facts about the experimenter. What justification Hacking's account provides for belief in those entities remains unclear. Our realist commitments to unobservable entities cannot be secured by dogmatic assertion. The only account available to Hacking, it seems, is the abductive one: such entities explain those very facts about our experiments that most impress us. And this has a further consequence. Entities by themselves explain nothing; rather, they explain only within the context of a theory that describes the phenomena to be explained. Thus, Hacking's theory-free entity realism appears to be jeopardized. This brings us to Resnik's second objection.

Hacking's theory-free entity realism is not incoherent: one can consistently believe in unobservables without believing that the theories about them are true. One can claim, for example, that the explanatory work done by laws of physics can be accomplished only if they are not literally true, while insisting that causal or experimental facts suffice for our beliefs in unobservables. The question is whether this "sufficing" in any way justifies our beliefs in unobservables. Resnik argues that, while Hacking's position is consistent, it is not reasonable, because it grants the experimental scientist belief in unobservables—and perhaps even true belief—without justifying that belief. And since justification is a necessary condition for knowledge, Hacking's experimental realism cannot provide *knowledge* of unobservables. Hacking may have explained why we hold the beliefs that we do, and why we hold them as strongly as we do, but he has not explained why those beliefs should be regarded as knowledge.

Recall again Putnam's view of meaning to which Hacking appeals in defending his experimental realism. As applied by Hacking, this view enables one to believe in unobservable entities belonging to a natural kind without believing in the theory about those entities, since we can successfully refer to a natural kind (electrons, for example) quite independently of theoretically motivated changes in its stereotype. Resnik claims that, while this is cogent, it goes no distance toward providing an epistemological grounding for our belief in electrons. In particular, Putnam's view assists Hacking only if we know that the entities to which we refer are natural kinds (not artifacts or purely theoretical fictions, say), for unless we know that such entities are natural kinds, we cannot claim to know that we continue to refer successfully when our theories change. Thus, phlogiston was once posited as an unobservable, and phlogiston theory enjoyed some measure of success. But after many failures of theory, phlogiston was dismissed as fictitious. Similarly, early work in genetics posited the existence of hereditary units—genes—to explain various hereditary phenomena and formulated laws governing these entities. But these laws were shown to be inadequate, and contemporary molecular genetics has shown that the functions formerly attributed to genes are in fact performed by several different entities. In both cases, putative unobservable entities were shown to be artifacts. How, then, do we know that unobservables exist? How do we know that electrons are not like Mendelian genes or phlogiston? Claiming that we know electrons exist because we continue to refer to them despite changes in our theories will not do; this begs the question, for we cannot claim to refer successfully at all unless we are already justified in believing that the relevant entity is a natural kind.

The standard realist account of how we know that some unobservable entity is a natural kind is familiar: we know that some unobservable entity is a natural kind because we have well-confirmed, explanatorily successful, and approximately true theories positing their existence. We know that electrons are natural kinds because the theories positing them are well-confirmed and highly successful. Hacking, of course, rejects this but offers nothing in its place. Were he to claim that we know electrons are a natural kind because we can use them in scientific experiments, he would still have to explain why the experiments succeed in giving the results they do. The only hope of explaining these facts is by appealing to theories describing the causal goings-on in our experiments: scientists are justified in believing in electrons because theories about electrons explain the success of their experiments.

Hacking might respond to this argument by claiming that scientists can explain their experimental success using only fairly low-level generalizations of various sorts; they do not need full-blown theories—systematic accounts that include genuine scientific laws that are universal and nomically necessary. In reply, Resnik points out that adopting such a narrow view of theories entails that theories are confined to physics and perhaps chemistry, which seems an unacceptable restriction on the use of the term *theory*. Resnik then proceeds to his third and final objection to Hacking's experimental realism.

Even if the restrictive account of theories as full-blown systems of universal scientific laws is granted, it nevertheless remains true that a working knowledge of such theories is essential to the design of experiments and the construction of measuring instruments and apparatus. The PEGGY II experiment, for example, was permeated with theory from beginning to end. Far from being mere recipe-following technicians, experimenters are trained in theory, communicate with theoreticians, and invent new ways of testing theories. Thus, claims Resnik, not only does Hacking's account fail to provide an epistemological grounding for the beliefs of experimenters, it also fails to describe their behavior as working scientists.

9.10 | Carrier's Kind Realism

Given the attractions of realism (reflecting as it does our conviction that science progresses by improving and deepening our knowledge of the world) and mindful of the difficulties attending both full-blown realism about theories and a restricted realism about entities, it is worth considering whether a more attenuated form of realism can be sustained.[31] This is the project undertaken by Martin Carrier. Carrier concurs with Hacking's critics in rejecting entity realism. With some qualifications (noted below) Carrier also agrees with Laudan that the ample historical record of theories, successful in their day but now discarded as false because their central terms do not refer, undermines the miracle argument for what he calls "theory realism." Nonetheless Carrier thinks that the miracle argument has merit when applied to the *kinds* of entities, systems, and processes, postulation of which endure through scientific change. The positive part of Carrier's article is a defense of what he calls "kind realism."

A key assumption in both the negative and positive parts of Carrier's analysis is what he calls the *retention requirement*. Any version of realism worth defending should establish more than the mere *possibility* that, for all we know, scientists occasionally get things right: it should be accompanied by a reasonable degree of epistemic optimism.[32] At the very least, realists want to argue that scientists have *in fact* sometimes (perhaps often) been successful in latching on to objectively real features of the world. But if scientists have genuinely discovered that X is a real feature of the world, then we should expect that X (so to speak) will be retained (or appropriately preserved) in all future scientific developments.[33] In short, realism about X entails retentionism about X. Carrier argues that both theory realism and entity realism fail this test. Later in his article Carrier argues that a suitably attenuated form of kind realism passes the test, especially in those cases in which the more robust forms of realism fail.

WHY THEORY REALISM AND ENTITY REALISM FAIL

Carrier does not endorse Laudan's pessimistic induction in the form in which Laudan originally presented it. Like Laudan's other realist critics, Carrier thinks that when Laudan amasses his list of failed but "successful" theories, his notion of *success* is too weak; for Laudan is prepared to judge as successful any theory accepted by the relevant community of scientists just as long as the theory accommodates most of the facts within its domain with only a few anomalies. That, for realists, is to set the bar far too low. It is no wonder that a theory is able to "fit" phenomena that it was deliberately designed to accommodate. Realists insist on a higher standard of a theory's success: either novel prediction (understood, following Duhem, as the prediction of empirical laws previously unknown) or consilience of inductions (understood, following Whewell, as the unified explanation of laws that played no role in the theory's construction).[34] Adopting this higher standard—*strong* success—significantly reduces the number of false but successful theories on Laudan's list. The list is narrowed still further by insisting that the non-referring term in virtue of which we now regard a theory as false be essentially involved in the derivation of the results that count toward the theory's success. Merely "idle wheels" (or what Kitcher terms "presuppositional posits") that play no such role should not be used as a basis for judging a theory false for the purposes of criticizing realism.[35] Thus, for example, in the case of the (strongly) successful prediction of the Poisson bright spot from Fresnel's wave theory of light, Carrier does not regard Fresnel's commitment to an optical ether as grounds for judging Fresnel's theory as false. That is because the derivation of the bright spot prediction depends solely on the hypothesis (which we still accept as correct) that light is a transverse wave and not at all on beliefs about the presumed medium in which the vibrations occur.[36]

Even with these extra stipulations about what constitutes success, Carrier acknowledges that Laudan is right: there are instances of successful but false theories. Two are described in some detail: Joseph Priestley's prediction of the reductive properties of hydrogen (based on the phlogiston theory of chemistry) and John Dalton's and Joseph Gay-Lussac's (independent) prediction that all gases expand at the same rate when heated (based on the caloric theory of heat). This for Carrier suffices to show that theory realism is false.

Carrier gives two counterexamples to refute the entity realism of Hacking and Cartwright, both from experiments performed by the founder of phlogiston chemistry, Georg Ernst Stahl. Stahl "manipulated" what he took to be phlogiston by using charcoal (assumed to be rich in phlogiston) to recover lead from lead oxide and sulphur from sulphuric acid. As with Carrier's criticism of theory realism, one might wonder whether one (or two) counterexamples, however striking, can suffice to demolish the philosophical positions at issue. Presumably Laudan gave many examples because he recognized that few realists (if any) claim that *all* successful theories are true (or approximately true, or close to the truth, etc.) Similarly Carrier acknowledges that Cartwright writes that entities are "seldom" discarded once they have been experimentally established. Cartwright does not say "never." Thus while Carrier's examples raise doubts about these forms of realism, further episodes (especially from more recent "mature" science) would strengthen his case.[37]

KINDS, NATURAL AND OTHERWISE

Carrier's main focus in his article is a defense of kind realism. Indeed his title refers to "natural kinds," and in a companion article he quotes approvingly Duhem's assertion that the strong predictive success of a theory is a reliable indicator that the theory has achieved a "natural classification" of the phenomena. Duhem writes:

> The highest test, therefore, of our holding a classification as a natural one is to ask it to indicate in advance things which the future alone will reveal. And when the experiment is made and confirms the predictions obtained from the theory, we feel strengthened in our conviction that the relations established by our reason among abstract notions truly correspond to relations among things.[38]

The notion of natural kinds (as opposed to merely artificial kinds) and the associated metaphor of "carving nature at the joints" has been a commonplace in philosophy since at least the time that Plato (in the *Phaedrus*) contrasted the adept butcher with one who hacks at a carcass and breaks its bones. It was certainly a central feature of Aristotle's metaphysics and his philosophy of science in which individual substances belong to the same natural kind if and only if they share the same essence. Carrier wishes to avoid the rich metaphysical overtones of talking about natural kinds. As far as he is concerned, it would be less misleading to refer just to "kinds" and drop the honorific adjective "natural" altogether. Reflecting a general trend toward empiricism and naturalism in modern philosophy, Carrier takes kinds to be defined by science, not by metaphysics or logic. What then are kinds? They are the classes or categories of things (in the broad sense of "thing") that figure in the laws accepted by scientists.[39] Two objects are of the same kind if they obey the same laws. As Carrier notes, if (as many philosophers have argued) there are no laws about biological species, then species (which are an Aristotelian paradigm of a natural kind) are not kinds. Things that do qualify as kinds are a very heterogeneous collection of substances, states of matter, systems, dynamic processes, and properties. They include chemical elements, isotopes, electrons, neutrinos, fermions, bosons, photon emissions, oxidation and reduction reactions, irreversible thermodynamic processes, elastic collisions, the production of virtual particles, mass, charge, magnetic momentum, ideal gases, lasers, diseases, light water reactors, neutron stars, viruses, tropical depressions, inflation rates, and so on.[40]

In section 3 of his paper, Carrier explores at some length whether kinds might not be identified experimentally using laboratory techniques such as those described in a case study of nineteenth-century optics by Jed Buchwald.[41] The reason for considering this route to kinds is that, if it were successful, it would avoid the necessity of having recourse to theories (and the laws they contain) and all the problems of historical change that theories bring with them. Laboratory techniques and experimental setups hold out the promise of a stable, theory-independent way of securing access to kinds. Carrier gives several reasons for doubting that the experimental route to kinds can succeed in the absence of an accompanying theory.[42] First, classes of phenomena that appear distinct (say, by being detected by different instruments) may be fundamentally the same. Carrier gives the example of different segments of the electromagnetic spectrum. Second, the experiments may fail to eliminate or control for factors that distort the results. Carrier describes how attempts to develop Newtonian chemistry in the eighteenth century by measuring affinities were eventually abandoned when Claude Berthollet showed that the differences in reactions (which were thought to result from the affinities of the reactants) were in fact due to differences in concentration, temperature, and pressure. Third, it is sometimes difficult (Carrier says "impossible") to tell from experiments alone whether we are observing a difference in kind or merely a difference of degree within the same kind. Again Carrier gives an apt example from the history of science: the debate between Joseph Proust and Berthollet about whether Proust was right to distinguish compounds from mixtures on the grounds that compounds always contain elements combined in the same definite proportions. Carrier attributes Proust's victory not to experimental evidence but to the growing acceptance of Dalton's atomic theory, which gave a clear theoretical rationale for Proust's distinction. The role of theory in adjudicating disputes about rival experimental ways of classifying phenomena—in this case, types of polarized light—is also illustrated in Buchwald's case study.

Given the failure of attempts to define kinds solely in terms of experimental procedures and techniques, the challenge for Carrier is to characterize a notion of kinds that is robust enough to qualify as realist but that can avoid the problems with theory realism by satisfying his retention requirement. In general, Carrier needs to show what kind of "kinds" endure when theories that were once strongly successful (in either Duhem's or

Whewell's sense) are later overthrown or significantly modified. In particular, Carrier needs to show how his kind realism succeeds in the two cases that he identifies as counterexamples to theory realism.

The key, in Carrier's view, is to focus on the kinds of phenomena that a strongly successful theory regards as relevantly similar in order to generate its correct results. In this way, he claims, we can explain why a false and non-referring theory achieved either its novel prediction or its consilience of inductions. It did so in virtue of what, judged from our perspective, the theory got right even though some of its important theoretical features (assumptions about entities and mechanisms) were radically mistaken. Thus, although phlogiston does not exist—and so it is false that hydrogen and carbon are rich in phlogiston—Priestley correctly regarded hydrogen as relevantly similar to carbon for inducing chemical reactions that we now regard as reduction-oxidation ("redox") processes. From a modern perspective, reduction involves the gain of electrons; oxidation involves the loss of electrons. Hydrogen (like metals and carbon) is electron rich; it acts as a donor of electrons, thereby itself being oxidized (producing water when used to reduce a metal oxide to its parent metal). Oxygen (like the halogens) is electron poor; it receives electrons, thereby itself being reduced (producing water when it combines with hydrogen). Priestley, the phlogistonist, observed that both carbon and hydrogen are rich in phlogiston since they both burn readily in air. Calxes result when metals burn in air, thus losing their phlogiston. Thus, it was reasonable to expect that hydrogen, just like carbon, would prove effective in regenerating metals from their calxes. This line of reasoning maps exactly, category for category, onto modern accounts of redox reactions.

Obviously this sort of reinterpretation of past theories in light of present beliefs has to be handled with care lest it degenerate into special pleading by realists concerned to defend their position at all costs. Obviously, it would be meretricious to use our present theories to decide what has been retained from the past and then, *without further argument*, assume that those features must have been responsible for the past theory's success.[43] In the Priestley case, Carrier has given an independent reason for regarding Priestley's phlogistonist categories (which survive in modern form) as the ones responsible for his theory's predictive success. Carrier's analysis would be further reinforced if there were evidence that Priestley also predicted that some metals (which as a class were presumed to be rich in phlogiston) might be used to reduce the calxes of other metals (metal oxides) to their parent metal.[44]

One constraint that Carrier does impose on the retention of kinds is that later science must never recognize as genuine any case that would belong to more than one category previously regarded as distinct. Refinement (in the sense of subdividing classes) is acceptable, but partial overlap or crossover of kinds is forbidden.[45] The application of this constraint to the Priestley case is not clear, but it is not obviously violated.

Carrier's second example again involves a novel prediction (rather than a consilience of inductions). Dalton (and, separately, Gay-Lussac) correctly predicted that all gases expand at the same rate when heated at constant pressure on the basis of the caloric theory of heat, a theory that we now regard as erroneous. Carrier claims that what the caloric theory got right was the insight that, unlike solids and liquids, gases should behave in a uniform way when heated. This is said to follow from the theory's model of gases as static arrays of particles in which the repulsive forces between the atmospheres of caloric surrounding each particle are totally dominant compared with the much weaker forces between the particles themselves. Thus caloric theorists were led to expect that all gases would behave alike when heated, regardless of their chemical composition.

Carrier's analysis of this second example is unconvincing. For, in what sense is the Gay-Lussac law a *prediction* from the caloric theory? Let us grant that Dalton and Gay-Lussac expected some law or other to characterize all gases, something that might have been anticipated from the much earlier discovery of the Boyle-Mariotte law. The problem lies in deducing the specific form of the Gay-Lussac law from the caloric theory. When Laplace attempted a derivation in the 1820s, he was forced to assume that the attractive forces between the caloric atmosphere of each particle and particles surrounding it were negligible. As John Herepath pointed out at the time, this and other assumptions made by Laplace were ad hoc: they were made solely with an eye on the results (the gas laws) that Laplace wished to obtain. Those results manifestly do not follow from the basic principles of the caloric theory.[46] Of course, this does not refute Carrier's contention that in cases of genuinely novel predictions some classification into kinds responsible for the success is preserved in later theories since it has been argued that, in the caloric theory case, there was no genuine prediction.

What Kind of Realism Is Kind Realism?

As Carrier notes, the kind of realism he defends is "comparatively weak," restricted as it is to the claim science manages to achieve a natural classification, a taxonomy or classification of the phenomena that reflects the way the world is. Carrier says that this is "the only access to reality that science is justifiably able to gain" (1172). In his view, what is right about the miracle argument is that strong success (either novel prediction or consilience of inductions) is best explained by a theory "induc[ing] the right relations of similarity among the phenomena in question" (1185) or, in other words, by the theory latching on to the right kinds where those kinds are the ones responsible for the theory's strong success. While theories are indispensable for specifying kinds, kinds but not theoretical concepts should be interpreted realistically since only kinds are retained (in cases of strong success) as theories change.

What is the difference between concepts and kinds? There is an ambiguity in Carrier's account when it comes to kinds. In one sense, the usual sense, kinds are classes of *things* (such as electrons, elephants, or epileptic seizures); in another sense, the sense in which strong success might justify us in regarding kinds as real, the kinds in question are *concepts*. The primary difference between concepts and kinds in Carrier's account is one of degree: "kinds" in his sense are general concepts. Consider his phlogiston example. Priestley regarded phlogiston as the substance of heat that is released when bodies burn in air and has various other properties. As it turned out, scientists discovered that Priestley's concept of phlogiston is non-referring and that claims such as "Carbon is rich in phlogiston" and "Hydrogen is rich in phlogiston" are false. Nonetheless, Priestley correctly identified a kind, which is specified in part by saying that hydrogen and carbon have a property in common that enables them to reduce metal ores to their parent metals. We still regard *that* claim as correct, though modern chemists understand the property in question in theoretical terms very different from Priestley's. What best explains Priestley's novel prediction is that he correctly assigned hydrogen and carbon to the same general class or category: hydrogen and carbon are relevantly similar with regard to some of their chemical properties. In a similar way, our understanding of the nature of electricity has changed dramatically since the days of Faraday and Maxwell, yet the classification of all charged bodies as being relevantly similar with regard to their lawlike behavior in magnetic and electrical fields is an enduring one. Carrier, following Pierre Duhem, thinks that it is a mistake for scientists to believe that they can ever know what electricity really is. But, according to Carrier, what they can justifiably believe, on the basis of the miracle argument, is that enduring classifications ("kinds") responsible for strong success reflect the structure of reality.[47]

9.11 | Fine's Natural Ontological Attitude (NOA)

A familiar and much-debated argument for realism claims that the best explanation for the success of science is that scientific theories are true (or at least approximately true) and most (if not all) of their terms refer to real entities, properties, and processes. Rejecting traditional defenses of scientific realism at the level of theories, Hacking defends in its place an attenuated realism—a rather more deflationary, theory-free experimental realism solely about entities. Laudan, van Fraassen, and Carrier also, reject traditional arguments for theory realism, van Fraassen going on to defend his own antirealist account of science, constructive empiricism, while Carrier advocates a realism solely about kinds. In "The Natural Ontological Attitude," Arthur Fine agrees that the traditional arguments for theory realism fail. But Fine's own positive account—the natural ontological attitude of his paper's title—falls somewhere between van Fraassen's empiricist antirealism and the different varieties of attenuated realism advocated by Hacking and Carrier. According to Fine, the most compelling view of science and its theories is neither realist nor antirealist. Having examined in some detail responses to arguments for realism, we shall attend somewhat briefly to Fine's criticism of realism, focusing our attention rather more closely on this middle position that Fine calls the natural ontological attitude (NOA).

Against Realism

Fine's criticism of realism is directed along two separate lines: traditional abductive arguments for realism, and the place of realist and nonrealist attitudes in the development of successful twentieth-century scientific theories. What evaluation does Fine offer of success-of-science arguments for realism?

Fine objects to the overall strategy of realist arguments that aim to connect the methods of science with its overall success. He begins with a maxim—gleaned from the history of set theory and the foundations of arithmetic—about the propriety of defending some theory T with argumentative resources taken from T itself. In the case of set theory, the formalist David Hilbert argued that, to the extent that the meaningfulness and consistency of Cantor's set theory was yet in question, only some non–set-theoretic consistency proof could warrant acceptance of Cantor's project. Hilbert thus advocated a more stringent method for establishing the consistency of set theory, a method that did not use arguments from set theory itself. Fine believes that Hilbert's attitude is the correct one: metatheoretic arguments in support of T should meet requirements more stringent than those placed on arguments at work in T itself.

Realists urge us to accept that realism offers the best explanation for why the methods of science lead to empirical success. In an objection reminiscent of Laudan's charge that inference-to-the-best-explanation arguments for realism beg the question, Fine notes that just as anyone who is suspicious of the consistency of set theory would scarcely be moved by set-theoretic arguments for its consistency, so also antirealists already skeptical about the connection between explanation and truth will fail to be moved by best-explanation arguments for realism. More generally, since realists and antirealists disagree about whether the realist's abductive arguments for the existence of unobservables are legitimate, even if realism *is* the best explanation for the success of science, this cannot warrant the conclusion that realism is true. Since it is the status of such best-explanation inferences that divides realist from antirealist, abductive defenses of realism at best beg the question.

Nor is there any pattern of argument more stringent than abductive ones, according to Fine, that would permit the realist to respect Hilbert's maxim. The only obvious candidate acceptable to all parties would be patterns of inductive argument leading to empirical generalizations. But empirical generalizations require observable connections between observables, and no such connection is available to the realist. The connection of relevance for realism is the relation of truth or approximate truth said to hold between theories and the world to which they correspond. To the extent that realism postulates a correspondence relation of truth between theories and the world, what it is committed to is unverifiable.

Fine's argument against the realist strategy rests on the rejection of abductive inferences. In taking this line, it would then seem incumbent on Fine to say what, if not abduction, ever does justify our belief in theories. Of course not all forms of induction will have been rejected. But now Fine and other nonrealists must offer some independent, properly stringent criterion by which to determine which inductive inferences are legitimate, without begging the question against the realist. In short, even if Fine is correct that realist arguments fail (which, of course, is not the same as showing that realism is false), his proposal leaves open whether any coherent nonrealist position can be sustained in place of realism.

In addition to criticizing traditional success-of-science arguments for realism, Fine suggests that the history of science also tells against it. In particular, Fine argues that in two successful twentieth-century physical theories—relativity and quantum mechanics—we can find a refutation of the claim that realism is the only way to explain scientific progress. Both the special and general theories of relativity were the work of Einstein's early period, during which he espoused a broadly positivist methodology. And despite the fact that Einstein thereafter came to have rather more sympathy with realism, practitioners of relativity theory—working scientists— have not followed him in this, adopting instead an instrumental view of general relativity as a tool for solving large-scale gravitational problems. Likewise, Fine argues, the development of modern quantum theory was carried out in a largely nonrealist environment, both Heisenberg and the later Schrödinger expressly avoiding referential commitments to an unobservable reality. Indeed, Bohr is reported to have feared that Einstein's (later) realism, if taken seriously, would thwart the progress of the new quantum physics. By Fine's reckoning, the Bohr-Einstein battle, far from a tempest in a teapot, was in large measure the result of conscious efforts to preserve an increasingly successful theory from stagnation in the hands of realists.

It is, of course, no part of realism that scientific progress depends in any crucial way on *believing* that realism is true. A theory is successful, according to the realist, because it gets the world right, not because of anyone's belief that such a relation does or does not (can or cannot) hold between theories and the world. Nor is it clear how confident Fine or anyone else can be about how differently twentieth-century science would have looked had its founders viewed realism more congenially.[48] Nevertheless, it is undeniable that the two most important and successful theories of the twentieth century—relativity and quantum mechanics—have progressed in a largely positivist and instrumentalist environment.

The Nonrealist "Core": NOA

Despite the nonrealist origins of quantum mechanics, Fine notes that contemporary particle physicists adopt models predicting the existence of new, as-yet-unobserved particles, judging these as discoveries when later confirmed in various experiments. Fine thinks that this talk of "discoveries" reflects not a return to realism but instead a consistent form of nonrealism that is equally at odds with antirealism. Thus, his task in the latter part of his paper is to develop a viable form of nonrealism.

Fine doubts that realists come to adopt their view of science on the basis of abductive reasoning. Rather, he suggests, they engage in a somewhat more homely line of argument of the following sort: "Well, on the whole I trust the evidence of my senses with respect to garden-variety objects. And on the whole I trust the checks and safeguards of working scientists, too. If they tell me that there are atoms or quarks or the like, then I trust them and accept that there are such things. But I can accept the confirmed results of science, in just the same way that I accept the evidence of my senses, only if I am a realist about science. So I should be one."

Now, by Fine's lights, to accept scientific results in the same way that we accept the evidence of our senses is simply to accept them alike as true. But the antirealist could follow the homely line of reasoning sketched above to this extent: nothing about antirealism would incline him either to reject the results of science or to accord the confirmed propositions of science some secondary status distinct from that afforded the garden-variety truths of everyday life. So it is not at all clear that accepting the results of science as true has to bring realism with it.

Thus, Fine urges us to see that both the realist and antirealist alike must in the main toe the homely line above, that is, both must accept the confirmed results of science as quite on a par with other more familiar truths. Fine calls this acceptance of scientific truths the "core position." What distinguishes realists from antirealists is what they add to the core position. Antirealists may add on a pragmatic or instrumental or conventionalist analysis of the notion of truth, or they may add on some broadly epistemological construal of certain concepts, as would idealists, phenomenalists, and other sorts of empiricist, and so on. Realists, according to Fine, simply add on a desk-thumping, foot-stamping emphasis—"Really!" When the realist and antirealist accept the verdict of science that there are electrons, the realist adds the emphatic "There really are electrons, really!" (1204) in part to indicate a rejection of competing, antirealist additions to the core and in part to express the robust sense of truth generally understood to accompany realism—namely, truth as correspondence with the world. (Fine reckons this latter aspect of the realist addition to be merely a superficial decoration of no more rational force than foot-stamping itself.)

From this contrast between what the realist and antirealist add to the core position, Fine urges us to see that an attractive third alternative emerges: the core position itself. Neither a form of realism nor antirealism but at the core of both, this intermediary view seems to be the most compelling and sensible attitude to adopt toward science. Fine calls this position the "natural ontological attitude" (NOA, pronounced *Noah*).

Fine thinks that the adequacy of NOA as a philosophical view of science can be appreciated by seeing what it has to say about ontology and the methods of science—two issues figuring crucially in most assessments of realism and its competitors. Focusing here on the first of these, consider that NOA tells us to accept the results of science as true. To do so, Fine claims, is to treat truth in the normal referential way: statements of science are true only if appropriate entities (individuals, events, properties, etc.) exist and stand in the operative referential relations. In adopting this stance, NOA is not thereby committed to the thesis that long-run scientific success inevitably brings us closer to the truth. A NOAer (pronounced *knower*), who in some particular scientific context or tradition believes in the entities referred to by her theories, can consistently allow that there is nothing progressive about changes in the scientific context or tradition and that Kuhnian paradigm shifts might well permit no stability of reference at all. So, too, can the NOAer resist believing that talk of reference entails belief in some objective external world. According to Fine, those who engage in such talk only pretend that we can step outside the game of reference and science to judge from some higher view what science is about, to mark off a world of external objects that are the referents of our scientific terms. But, Fine insists, we ourselves are in the world, among the objects of science.

Fine suggests that NOA can help to clarify particular connections between scientific method and the realist success-of-science accounts judged earlier to be question begging. But the main focus of Fine's proposal is not NOA's ability to clarify and explain. After all, the core position that NOA represents is at the heart of both realism and antirealism, and if the realist or antirealist additions do not nullify NOA's appeal to the truth of scientific claims, its ability to explain and clarify will to that extent be equally available to its competitors. What Fine wants us to appreciate above all is how minimal an adequate philosophy of science can be. According to NOA, the legitimate aspects of any addition made by realism or antirealism are already implicit in the core belief that the confirmed truths of science are on a par with garden-variety truths, all alike and equally true. NOA brings with it no inflation of the concept of truth, no theory or analysis or picture of truth as correspondence or anything else. In addition to accepting the standard referential semantics and broadly Tarskian conception of truth, the NOAer accepts a standard epistemology by which judgments of truth are based on familiar perceptual and confirmational relations.[49] Unlike either realism or antirealism, NOA makes no pretense to having the resources for settling many disagreements over what is true or even whether best-explanation inferences are truth preserving. And that is as it should be, according to Fine: there are no such resources to be had. NOA's minimalism is thus among its chief virtues. Science does not need what either realism or various versions of antirealism suppose it to need. A global interpretation, a philosophical version of truth to explain the success of science, a grand story purporting to render the significance and goals of science more intelligible—all these things are unnecessary. So the minimalism of NOA reflects the attitude that science can be taken on its own terms, an adequate philosophy of science requiring little more than a homely line of reasoning about the history and practice of science just as we find it. The philosophical theories, the interpretations, the global pictures, the truthmongering "isms," all are idle trappings, unnecessary, unwarranted, perhaps even unintelligible bits of decoration appended to philosophy of science. Nothing else is required, nothing else is legitimate, beyond the natural ontological attitude.

9.12 | Evaluating NOA: Musgrave's Realist Reply to Fine

In "NOA's Ark—Fine for Realism," Alan Musgrave asks, What exactly is there in NOA that runs counter to scientific realism? How nonrealist is Fine's proposal?

Musgrave notes at the outset that, insofar as the debate between realists and antirealists centers largely on the question of truth, NOA seems less a core position that realists and antirealists would then embellish than a position that realists would accept and antirealists reject. The core position would have us accept the claims of science as true. This may be fine for proponents of realism, but not for its naysayers: positivists who deny the existence of theoretical entities judge the theoretical claims of science to be false; instrumentalists reckon scientific theories as neither true nor false; and empiricists like van Fraassen insist that theories should be accepted as empirically adequate, not as true. At times, Fine seems to think that these and other versions of antirealism import a peculiar notion of truth into their philosophies of science. But if that were so, antirealists could scarcely accept the core position's directive to treat the results of science with the same homely conception of truth that we apply to garden-variety beliefs.

Perhaps there is no such homely conception of truth, and the core position itself leaves open what sense realists and antirealists should bring to the word *true*. But Musgrave doubts that this will do justice to Fine's intentions. For without any sense attached to *true*, there is no determinate core position to which realists and antirealists alike could then make additions. Moreover, it looks as if the core position (NOA) attaches a quite definite sense to *true*. Fine says that "we are to treat truth in the usual referential way, so that a sentence (or statement) is true just in case the entities referred to stand in the referred-to relations," adding that "NOA recognizes in "truth" a concept already in use and agrees to abide by the standard rules of usage, [these rules involving] a Davidsonian-Tarskian referential semantics" (1216). In passages like these, Musgrave urges that Fine has already embraced a quite deep philosophical analysis of truth, one that realists would (but antirealists would not) be prepared to apply across the board, to homely and scientific truths alike.

Fine intends NOA to be a nonrealist philosophy of science. What is it that realism adds to the core position that distinguishes the realist from the NOAer (that is, the person who accepts Fine's NOA)? Drawing partly on other work of Fine's, Musgrave isolates three items: first, the realist shouts, emphasizing "really!" whereas the NOAer does not; second, the realist makes a good deal of fuss over certain philosophical slogans that the NOAer avoids; and third, the realist is committed to a metaphysical picture that the NOAer is not. Let us consider these in turn.

The first putative difference is judged by Musgrave to be the shallowest: if the NOAer is a realist who avoids shouting and foot-stamping, the difference is an unphilosophical one, easily dissolved by behavior modification. But, of course, that is too facile: the difference is meant to run more deeply than that. For there is, Fine says, a positive function to the realist's emphasis—namely, an expression of a robust sense of truth, the full-blown version of which involves truth as correspondence to the world. And this brings us to the slogans.

Musgrave's second point is that, according to Fine, realists endorse slogans such as "Truth is correspondence with reality," whereas NOAers will have no truck with correspondence or anything like it. For his own part as a realist, Musgrave wonders what more there can be to the slogan "Truth is correspondence with reality" that is not already expressed by Fine's view that "a sentence (or statement) is true just in case the entities referred to stand in the referred-to relations" (1217). Not all philosophers would agree with Musgrave that Tarski's account of truth is just a codification of the commonsense correspondence theory, though exactly what more there is to correspondence than Tarski's theory plus a standard referential semantics is a point of debate. In any case, it is far from clear that scientific realism requires anything more than Tarski offers. As Musgrave sees it, the realist need not suppose, for example, that in addition to a Tarskian semantics one needs a story about the essence of truth. Indeed, Musgrave agrees with Tarski (and Fine) that there is no property shared by all truths, in virtue of which they constitute a natural kind. If that is correct, then, despite what Fine says, NOA is a realist account of the claims of science.

Musgrave also responds to the concern that, if Tarski's theory supplies everything the realist needs, then it threatens to make us realists about too much. If a Tarskian referential semantics applies across the board (to homely claims, theoretical-scientific claims, and everything in between), then such a liberal application of Tarski's Convention T would seem to recommend realism about numbers, moral claims, and the "creeps," no less than it recommends realism about the homely moon and the scientific electron. All five instances (below) of Convention T should be treated in exactly the same way:

1 The statement 'There is a full moon tonight' is true if and only if there is a full moon tonight.

2 The statement 'Electrons are negatively charged' is true if and only if electrons are negatively charged.

3 The statement 'Two plus two equals four' is true if and only if two plus two equals four.

4 The statement 'Eating people is wrong' is true if and only if eating people is wrong.

5 The statement 'Ronald Reagan gives me the creeps' is true if and only if Ronald Reagan gives me the creeps. (1218–19)

On the Tarskian referential semantics that, according to Musgrave, is part of scientific realism, it seems as though we should be realists about numbers and the "creeps" as well as about some remarkably average family and partial children: the statement "The average American family has 2.3 children" is true if and only if the average American family has 2.3 children.

To avoid such commitments, Musgrave claims that Tarski's theory by itself yields none of these realisms. This is surely right. There is, for example, nothing in Tarski's theory that requires us to take every sentence at face value. If one is charitable about the utterance figuring in (5) above, but is nevertheless keen to avoid a realist commitment to the creeps, one can allow that the relevant sentence is a true but idiomatic expression of a fact about the way Musgrave felt about Reagan—this bringing with it at most a commitment to Reagan's existence and to something Reagan did. Alternatively and less charitably, one can take the sentence in (5) at face value and judge that it is, strictly speaking, false. Similarly, the antirealist about numbers can take "Two plus two equals four" at face value and say that it is strictly false, since there are no numbers answering to the words *two* and *four*. The same options apply to the homely and scientific cases of (1) and (2), respectively.

In short, Tarski's account brings no particular realistic commitments with it. What it does do is make realism possible. Musgrave's concern is to show that Fine's account of NOA renders realism about the claims of science actual. For NOA enjoins us to take the claims of science at face value, to apply the Tarskian referential semantics, and to accept the relevant claims as true. And that is scientific realism.

One might construe NOA as claiming that a face-value reading of sentences is already a departure from the core position, since something has been added to the minimal stance. And here, perhaps, there is room for distinguishing the NOAer from the realist. Where the realist accepts the claims of science (and homely run-of-the-mill claims) as true, reads *true* in the Tarskian way, and takes the claims at face value, the NOAer refrains (on the present reading) from doing the last of these, thus leaving open the issue of commitment to particular entities by virtue of leaving open the issue of what reading (literal and face-value or idiomatic and misleading) to give the claims. As Musgrave notes, this reading renders NOA a far cry from any sort of realism: for it characterizes the NOAer as one who is not committed to anything at all—not to electrons, not to the moon, not to physical objects or to other people.

This interpretation of NOA, as an extreme form of philosophical minimalism, eventually reemerges in the context of the third respect Musgrave notes in which Fine reckons NOA to depart from realism: realism offers a metaphysical picture that NOA rejects. In a later paper entitled "Unnatural Attitudes: Realist and Instrumentalist Attachments to Science"[50] (published two years after the original NOA paper included in our volume), Fine characterizes the realist metaphysic as one according to which science is about something, something out there, external to us—in short, the World. Realists (Musgrave among them) concur with this picture and wonder what it is that Fine and the NOAer reject about it, and what NOA puts in its place. As for what NOA rejects in the realist picture, Fine's misgivings concern what he calls the problem of access: How can we render intelligible the realist's claim of having access to an objective, independent external world? For whatever we observe—or, more properly, whatever we causally interact with—is not independent of us, and whatever information we retrieve from such interaction is information about interacted-with things. Faced with these problems of reciprocity and contamination, Fine contends that the realist can offer no intelligible account of our supposed access to an independent and objective world.

As Musgrave notes, if there is a genuine difficulty for the realist here, it will be perfectly general, infecting both commonsense realism (about the moon, say) and scientific realism (about electrons). So he sticks to commonsense realism: "Somebody says 'There is a full moon tonight,' and I look up into the night sky and ascertain that the statement is true" (1222). This captures the realist notion that sentences express something true if they represent the world to be a certain way when the world is that way. Now when Musgrave claims that there is an independent and objective world about which our sentences can express truths or falsehoods, what he means is not that things in the world are causally insulated, but rather that the existence and nature of these things is not dependent upon any perceptual or other mental goings-on in us. The moon, like an electron, is not something we create by looking at it or thinking of it or speaking about it; indeed, there is some *it* that we see, some object about which we speak and think. By Musgrave's lights, the putative problem of reciprocity is no problem at all. To say that the moon or electrons fail to be independent because we interact with them is to invoke a strange notion of independence that no realist would accept.

What of the problem of contamination? Fine's suggestion is that when we interact with the moon, we gain information not about an objective moon out there in the external world, as it is in itself, but about an interacted-with-moon. Now if this is to show that the interacted-with-moon is not objective, not out there in the external world, then presumably the interacted-with-moon is in some sense subjective and inside our heads. But what sense can be offered and defended here? Musgrave doubts that Fine wants to return to the dusty old view that we do not see external objects (such as moons) but instead merely have commerce with (moonish) sense data.[51] The nonrealist NOAer might urge instead that the interacted-with-moon, the moon-as-seen-by-us, fails to be objective in this sense: insofar as my judging that the moon is full requires, and so depends upon, an application of some moon-concept I happen to possess, this judgment and its truth does not concern any purely external-world moon as it is in itself but rather some world partly of my making, a world-as-conceptualized-by-me.

Some philosophers unsympathetic with realism will speak in this way. Musgrave makes two points about the philosophical view such talk represents.[52] First, this sort of conceptual idealism leads quickly to conceptual relativism. Since concepts vary (from person to person and from culture to culture) and change (from olden days to current enlightened days), there is no single world-as-conceived-by-. . . . Conceived by whom? When? We have, apparently, a plethora of such worlds, all different. Now this point of Musgrave's is not yet an objection: nothing he says indicates what is wrong with relativism, and some nonrealists will find it congenial. But Musgrave offers a second point that is intended to be more damaging—a point concerning not the plethora of such conceived worlds, or the many instances of the moon-as-conceived-by-. . . , but instead the relation of the moon-as-conceived-by-us, say, to the moon-in-itself. What exactly is the relation between them? Are they numerically identical, one and the same moon? If so, there is no point in speaking of a distinction between the moon-as-conceived-by-us and the moon-in-itself: let us simply speak of the moon and be done with it. But if the moon-as-conceived-by-us is not the same thing as the moon-in-itself, then there is some property the one has that the other lacks, by virtue of which they are distinct. And yet to know this is to know something about the moon-in-itself. But this, according to the contamination problem, is impossible, all knowledge being confined to the moon-as-conceived-by-us.[53]

What is the upshot of all this, according to Musgrave? Clearly, the various (and sometimes tedious) routes we have been pursuing most recently in following out Fine's rejection of the realist metaphysic represent something well beyond a natural ontological attitude or a philosophically minimalist one. The accounts of reciprocity and contaminated information, the conceptual idealism, the relativism—all of these are prime examples of philosophical theorizing, precisely the sort of thing Fine had enjoined us *not* to add to the homely core position that is NOA. It rather looks as if Fine rejects the realist's metaphysic only by abandoning the minimal core, by inflating NOA beyond its proper boundaries. By returning to the natural core—judging the claims of science (like those of common sense) to be true, applying the Tarskian semantics, and taking these claims at face value—one has accepted the realist metaphysic.

Musgrave adds one final note, returning us to the extreme philosophical minimalist construal of NOA (broached in our earlier discussion of realist slogans). This philosophical minimalism is an option the NOAer might take. The thoroughly unphilosophical natural ontological attitude would have one accept parts of science as true, without this implying anything at all about the realist's claims of the objectivity and independence of electrons or the moon. By leaving it an open question whether the accepted statements are to be taken at face value or not, it is left open how these statements are to be interpreted and what one is ontologically committed to when one accepts them. This version of NOA—Fine's, perhaps—is not a philosophical view at all. Moreover, since NOA enjoins us to treat the claims of science and the claims of everyday life on a par, pretty much all sentences whatsoever (ordinary, scientific, philosophical) will be "open" in the sense just described. Musgrave concludes that if NOA is to have any content at all, it must at least give the homely, run-of-the-mill claims a realist, face-value hearing. And then, of course, given NOA's insistence that the claims of science are on a par with the run-of-the-mill claims, NOA emerges once more as "fine for realism."

9.13 | Summary

Are scientific theories literally true, and do the entities apparently postulated by them in fact exist? Logical empiricists approached such questions by distinguishing observational terms and observables, on the one hand, from theoretical terms and unobservables on the other. About the observational terms and observables of accepted scientific theories, the logical empiricists espoused a form of realism: claims involving only observational terms were judged to be true and genuinely referential, and observables were said to exist. But they were antirealists about theoretical terms and unobservables. Most logical empiricists were committed to the meaningfulness of theoretical terms and the truth of the theoretical claims in which they figure without also accepting that theoretical terms refer or that unobservables exist.

Grover Maxwell argues that the antirealism of traditional empiricism cannot be sustained. Confronted with scientific advances in which objects previously judged unobservable come to be seen with the help of scientific instruments (such as microscopes), some empiricists had claimed that strictly speaking we see, not the objects, but only images in our instruments.

Maxwell argues that such a view entails that strictly speaking we do not see ordinary objects through binoculars or eyeglasses or even through window-panes. Since we are realists about observables and since there is no nonarbitrary line to be drawn between the observable and the unobservable, we have no grounds for being antirealists about unobservables. Other empiricists restricted their antirealism to what is entailed by accepted theories to be *in principle* unobservable. Maxwell argues that there are no conceptual or a priori limits on what is observable: any object is a candidate for being observable in principle: imagine having electron-microscopic eyes or X-ray vision. Finally, some empiricists insisted that all claims about physical objects be translatable into claims about immediately perceivable sense data. While this view finds few contemporary adherents, it draws our attention to what is widely accepted—that there must be some fundamental observational base on which confirmation in empirical science ultimately rests. Maxwell argues that the question of what things count as observable (and hence qualify as members of the observational base) is ultimately a theoretical question and that the observational-theoretical distinction will be drawn on the basis of our current state of theoretical knowledge, our physiology, and the instruments we happen to have available to us. But this shows that the distinction has no ontological significance whatsoever.

To show that the logical empiricists failed to make their antirealist case is not yet to show that scientific realism is true: other forms of antirealism may be waiting in the wings. Bas van Fraassen defends a kind of antirealism that is at odds both with logical empiricism and with scientific realism. As van Fraassen understands scientific realism, it is the view that science aims to give us literally true theories about the world and that to accept a theory is to accept it as true. Antirealists deny this. Rather than offering a theory as true, the scientist can claim for the theory some other virtue instead, remaining agnostic about its truth. The virtue that van Fraassen claims a proper empiricist will require in accepting a scientific theory is empirical adequacy: a theory is empirically adequate if it "saves the phenomena"—if what it says about observable objects and events is true. According to van Fraassen's constructive empiricism, then, the aim of science is to offer theories that are empirically adequate, and when we accept a theory, we accept it, not as true, but as empirically adequate.

How well can the constructive empiricist defend this distinctive breed of antirealism against the realist arguments of Maxwell and others? Van Fraassen's notion of empirical adequacy entails that the observable-unobservable distinction can be made in an intelligible and defensible way. Maxwell argued that the distinction cannot be made out, since there is no nonarbitrary line to be drawn along the continuum from observable objects (seen through the air or through a window) to unobservable objects (detectable only with the aid of instruments). Van Fraassen argues that this last claim entails only that *observable* is a vague predicate and that we can still use it safely in clear cases where it applies or does not apply. Van Fraassen regards detecting particles in a cloud chamber as a clear case where *observable* does not apply: we observe the vapor trails such particles leave behind, but we do not observe the particles themselves. Van Fraassen also denies Maxwell's claim that the phrase *in principle unobservable* fails to mark off any class of objects whatsoever. Just as we do not judge that the Empire State Building is portable (claiming that there are no conceptual or a priori constraints against the possibility of a giant moving it), so we judge what is unobservable in science on the basis of the natural limitations of humans: given these limitations, some objects *are* unobservable.

Van Fraassen rejects a number of other realist arguments. For example, realists often claim that we are justified in believing in the existence of unobservable objects by an inference to the best explanation: given some class of observable phenomena, our best explanation for what we observe requires us to believe in (posit the existence of) unobservable entities. Van Fraassen agrees that explanation is a legitimate criterion to use in arguing for what should be believed, but he denies that this pattern of inference commits us to realism. Since the claim that we follow such an inferential rule is an empirical one, it can be confronted with a rival hypothesis—van Fraassen's alternative hypothesis is that we are willing to accept that the theory which best explains the evidence is (not true, but) empirically adequate. Van Fraassen also rejects a number of other arguments for realism (from Smart, Sellars, Putnam, and others) intended to show that only a realist construal of theories and unobservables can explain the empirical success of our scientific theories. Perhaps the most influential of such arguments is what van Fraassen calls the Ultimate Argument: if we do not believe that our theories are true and that many of their terms really do refer to unobservable entities, then we can only reckon the success of our theories a miracle. To this "scientific" defense of realism (a defense that takes seriously the notion that even our philosophical theories must meet the demands of adequate explanation), van Fraassen replies with an alternative Darwinian explanation for the success of science. Just as many species of organisms have struggled for survival, so have many past theories; and just as organisms that failed to adapt to their environment no longer exist, so theories that are empirically inadequate no longer exist. There is no miracle in either case. Present-day theories are largely successful because scientists select for empirical adequacy.

Alan Musgrave has offered several criticisms of van Fraassen's constructive empiricism. In going beyond the observable to affirm the existence of unobservable entities, realism seems to be more susceptible than antirealism to skeptical arguments. Musgrave argues that the realist takes no greater risk of being proved wrong on empirical grounds than does the antirealist, given that empirical adequacy requires a theory to save all observable (not simply observed) phenomena. And while van Fraassen believes that the epistemically responsible course to take is mere acceptance of a theory, not belief in its truth, Musgrave claims that, since the risk associated with either option is the same, one should opt for the course offering the chance of the greatest gains. However the details of this issue might be viewed, the observable-unobservable distinction lies at the heart of the debate. Van Fraassen believes that even if the term *observable* is vague, there are clear cases where it does not apply (to particles in the cloud chamber, for example) and, moreover, that the natural limitations of humans make relatively clear what it means to say that something is in principle unobservable. Musgrave notes that such limitations vary from person to person and are a function of our evolutionary history. So he doubts that the distinction can have the philosophical significance the antirealist ascribes to it. Van Fraassen claims that the distinction need only serve the epistemological role of specifying the extent of one's beliefs: believe to be true only what a theory says about what can be observed. Musgrave thinks that this requirement is still too strong. Given any plausible account of evidential support, it is at least possible that there will be better evidence for some claim involving unobservables than for a claim restricted solely to observables. Moreover, he finds it unreasonable to suppose that a scientist can agree that we detect (not observe) a particle in the cloud chamber while yet not believing that the particle actually exists. Finally, Musgrave argues that van Fraassen's Darwinian response to the Ultimate Argument fails. Just as the scientist can ask not only why a particular mouse runs from the cat but also why mice in general have cat-fleeing behavior, so also in the case of the success of science, one can ask not only why some particular theory is successful, but also why existing theories in general are successful. Van Fraassen has answered the second question: existing theories are successful because unsuccessful theories do not survive. But this does not answer the first question about why any particular theory is successful, and Musgrave believes that the antirealist has no answer to this question. In the end, the constructive empiricist seems forced to say that the success of any particular theory about the world is a miracle.

The Ultimate Argument is a popular defense of realism—an inference-to-the-best-explanation argument applied to science itself. According to such abductive arguments, the success of science is inexplicable unless its theories have the virtues typically attributed to them by realists. Chief among these virtues are the approximate truth of scientific theories and the genuine reference of their theoretical terms. Larry Laudan argues that the connection between these virtues and the success of science is much weaker than realists suppose. With regard to reference, realists typically claim that reference explains success while success warrants a presumption of genuine reference. Laudan argues, first, that the history of science supplies numerous cases of unsuccessful theories about objects we believe to exist: reference is no guarantee of success. Likewise, Laudan argues that many historical theories have been successful without being genuinely referential: the success of a theory creates no presumption of genuine reference. With regard to truth, realists typically claim that theories, if strictly speaking false, are nevertheless close to the truth. Laudan argues that realists have not clarified the notion of approximate truth, or verisimilitude, sufficiently for us to evaluate the thesis that approximate truth entails explanatory success. Does the success of our theories warrant belief that they are true or approximately true? Laudan thinks that realists would not want to claim that a theory was approximately true if its theoretical terms failed to refer. This, together with historical cases of successful but nonreferring theories, shows once again that the connection between success and truth is not as close as the realists have assumed. Finally, one might question the fundamental logic of traditional success-of-science arguments. Antirealists have long denied that, on the basis of its yielding true consequences, some theory could reasonably be judged true, for such an inference would commit the fallacy of affirming the consequent. Laudan notes that, while success-of-science arguments are not about any theory in particular but about the truth of realism itself, nevertheless they are of the same form: realism can reasonably be judged true because it yields true consequences (science is successful). And such arguments simply beg the question against the antirealist, who, having rejected the application of this form of argument to particular scientific theories, cannot be expected to accept its application to science as a whole.

Although Larry Laudan has become the acknowledged source of a form of antirealist argument known as the pessimistic induction (PMI), opinions differ about how the PMI should be construed. Juha Saatsi critically examines two reconstructions of the PMI and finds both wanting. The first of these, due to Marc Lange, construes the PMI as an argument for the conclusion that most of our currently accepted theories are false and will eventually be replaced. Lange, with good reason, accuses this form of argument of committing the "turnover fallacy." Saatsi agrees that the turnover fallacy is an invalid form of argument but denies that Laudan's reasoning commits it. Instead, Saatsi argues that Laudan's PMI is most plausibly understood as an attack on the realists' appeal to truth (or probable truth, or approximate truth) as the best explanation of the success enjoyed by the theories we currently accept. As such, Laudan's PMI is not arguing that most of our currently accepted theories are false but rather that the realist's "best explanation" case for the probable or approximate truth of those theories is undermined by the historical track record of similarly promising theories that have turned out to be false in the past. Saatsi then examines Peter Lewis's criticism that the PMI commits the "base-rate fallacy." Saatsi gives several reasons for questioning the applicability of Lewis's Bayesian-style analysis to Laudan's argument.

Ian Hacking proposes that we divorce our commitment to the existence of unobservables from our belief that scientific theories are true. According to his experimental realism, our warrant for believing that electrons exist consists in our ability to manipulate them and to use them to intervene elsewhere in nature. This ability is based on our knowledge of their causal properties and entails no commitment to believing that our theories are true. Hacking adopts Putnam's view of meaning, according to which our theoretical terms can retain the same unobservable objects as referents during periods of scientific progress, even if the theoretical stereotype associated with these terms changes. In addition, Hacking emphasizes that we form beliefs about the existence and nature of objects on the basis of our causal interactions with them. But causal interaction and its role in belief formation is a general fact about observables and unobservables alike; so what grounds our beliefs about unobservables is no different in kind from what grounds our beliefs about observables. Thus, causal interaction, based on our knowledge of the causal properties of unobservables, is a crucial element in Hacking's experimental realism. Scientific experimentation grounds realism about unobservables because it involves causal interaction with them.

Hacking illustrates his view by describing in some detail the PEGGY II experiment—the design, construction, and use of a polarizing electron gun to observe parity violations in the scattering of polarized electrons. The PEGGY II example is meant to show that neither theory nor explanation is central to the practice of intervening. So the point is not to argue for realism by appealing to the explanatory success of theories that posit unobservables. The defense of realism about unobservables rests on our ability to use and manipulate them.

While the attention Hacking has drawn to experimentation is important, David Resnik argues that his attempt to divorce realism about entities from realism about theories does not succeed. Moreover, Resnik objects, if Hacking's experimental realism is to succeed, it must inevitably rely on the same form of reasoning as the traditional success-of-science arguments for scientific realism. Given that causal judgments figure centrally in the notion of experimental intervention with unobservables, how are we to justify causal claims? Since Hacking denies realism about laws and theories, he cannot use causal laws (supported by inductive evidence) to justify his realist beliefs about the causal powers of objects. Since Hacking cannot use inductive inferences to justify causal claims about unobservables, he is left with only abductive arguments to play the justificatory role. While traditional realists focus on the theoretical successes of prediction and explanation, Hacking focuses on experimental success—on our ability to construct apparatus and instruments to create, measure, and control new phenomena. Hacking is thus in the position of claiming, with the traditional realist, that unless we believe in unobservables, we should judge it a coincidence or a miracle that our experiments work as they do.

Moreover, Resnik argues that wedding realism about entities to antirealism about theories, while coherent, is unreasonable. At best, it allows the experimental realist to account for the beliefs we have, without supplying any justification for them. This is because, given Hacking's deployment of Putnam's account of reference, our belief in the existence of electrons, say, can be justified only if we know that electrons are a natural kind. And according to Resnik, this requires an appeal to theory. In other words, electrons can be judged to play the roles they do in experimentation only if we accept as approximately true theories that posit their existence. Thus, despite its initial promise, Hacking's experimental realism does not appear to provide scientific realism with any new justifying arguments.

Martin Carrier argues that even when restricted to the achievement of strong success (such as novel prediction), there have been false theories whose (strong) success crucially involved the postulation of nonexistent entities. Thus he acknowledges the cogency of the historically based arguments (espoused by Laudan and others) against full-blown theoretical realism. Carrier is equally critical of the minimal entity realism espoused by Hacking and Cartwright. Nonetheless he insists that the miracle argument has merit since there must be something about now-discarded theories that explains how they were able to predict novel results. Despite their falsity those theories must, in some respect, have been on the right track. Carrier attributes the success of these theories to their having classified phenomena into kinds that we still recognize as objectively correct. In the terminology of Pierre Duhem, they achieved a *natural classification* of phenomena. Carrier regards this as sufficient grounds for endorsing an attenuated form of realism.

A fundamental sentiment motivating Hacking's entity realism and Carrier's kind realism is a rejection of full-blown realism about theories. Here Arthur Fine concurs: even if the truth of scientific theories is the best explanation for their success, this cannot warrant belief in scientific realism. For since the realist and antirealist disagree over the epistemic value of abductive arguments, such arguments for realism beg the question against the antirealist.

But Fine's belief that no argument for realism is successful does not lead him to espouse any of the traditional forms of antirealism. Indeed, he thinks there is no need to endorse either realism or antirealism. According to Fine, to accept the confirmed results of science is to treat them just as we treat the evidence of our senses: we accept them as true, period. Surely the antirealist need not reject the results of science, nor give them some secondary ranking below the perceptual truths of everyday life. Both realist and antirealist alike must at least adopt this core position; what separates them is what they add to this core. Antirealists add some particular analysis of truth, or some empiricist treatment of certain concepts; realists add an emphatic "Really!" to the claims of science, expressing some more robust notion of truth as correspondence with the world. But Fine insists that an attractive third position between the realist and antirealist philosophies of science is the core position itself—a minimal, commonsense, nonrealist attitude toward science, the natural ontological attitude (NOA).

According to Fine, to accept the results of science as true is simply to treat truth in the normal, referential way, without any further commitment to the long-run progress of science as converging on the truth. Proponents of NOA can consistently believe in the entities referred to by their theories while also allowing that there is no referential stability across scientific traditions or paradigms. Nor does NOA bring with it any inflated notions of truth or any resources for settling particular disagreements over the concept of truth. The commonsense minimalism of NOA brings to philosophy of science all that it needs. Science can be accepted just as the homely core position enjoins us to accept it, without any philosophical trappings, interpretations, or fancy pictures.

Musgrave doubts that on its own, NOA is any sort of nonrealism at all. In the first place, many antirealists deny that we should take the claims of science at face value and accept them as true. To accept the claims of science as being true in just the same homely sense that we judge our ordinary beliefs to be true is the position of the realist, not of the antirealist. In the second place, if there is no common, homely conception of truth about which realists and antirealists can agree, then it is unclear what NOA is recommending when it tells all of us—realists and antirealists alike—to accept the claims of science as true. Indeed, Musgrave suggests that insofar as Fine associates NOA with a standard referential semantics, he has already adopted a philosophical analysis of truth, and it is exactly the analysis that realists would accept.

What is it, then, that realism adds to the core position to distinguish it from NOA? The foot-stamping "Really" is nothing philosophically deep, and Musgrave urges that there is nothing more to a correspondence theory of truth than a broadly Tarskian referential semantics, which the core position endorses. Fine insists that realism brings with it metaphysical commitments that NOA does not. In particular, Fine characterizes realism as embracing the view that science is about an objective, independent external world. The difficulty with such a view, according to Fine, is that we cannot explain how access to such a world is possible: perceptual interaction with objects renders them no longer independent, and whatever information we get from interacting with them will be information about interacted-with things, not objective things after all. Musgrave argues that Fine's notion of independence is nothing the realist would recognize and that Fine's worry about objectivity is in fact a concession to conceptual relativism. Whatever the value of Fine's arguments against the metaphysical commitments of realism, they are clearly philosophical in nature. To this extent, Fine seems able to defend NOA against realism only by adding to the core position just what NOA requires us to omit—philosophical pictures, interpretations, and so on. In the end, Musgrave suggests that if NOA is not to emerge as philosophically empty, we must construe its commitment to the truth of scientific claims as one that is fully congenial to realism.

■ | Notes

1. Logical positivism was the official doctrine of the Vienna Circle or *Wiener Kreis*, a group of philosophers, mathematicians, and scientists led by Moritz Schlick at the University of Vienna in the 1920s and early 1930s. By the mid-1930s it had

evolved into logical empiricism and spread to Britain and America as philosophers of science fled Nazi persecution in Europe. (Schlick was murdered by a deranged student in 1936.) Impressed by the success and rigor of the sciences, and unhappy with the verbose obscurity of most of European academic philosophy, the logical positivists insisted that empirical claims are cognitively meaningful only if they can be conclusively verified by observation. Metaphysics, ethics, and theology were among the traditional branches of philosophy thus reckoned to be cognitively meaningless. By the mid-1930s, leading positivists such as Rudolf Carnap realized that the verifiability criterion of meaning was too strict, since it would rule out as meaningless, not only metaphysics and theology, but also most of science. (Popper had pointed out that no universal generalization—and hence no law of nature—could be conclusively verified.) Similarly, it was realized that many scientific concepts could not be given complete, operational definitions solely in terms of observables, as the positivists had hoped. Thus, the verifiability principle was abandoned and the transition to logical empiricism was begun. Instead of verifiability, the logical empiricists took confirmability as their criterion of cognitive significance, and instead of trying to give complete operational definitions for theoretical concepts, they turned their attention to the analysis of the meaning of theoretical statements in terms of their observational consequences. Logical empiricism was a tremendously influential movement in the philosophy of science, although its popularity began to decline in the 1960s as it confronted a number of tough problems and objections.

Hempel's covering law model of explanation and Nagel's analysis of the logic of reduction are good examples of the logical empiricist approach to understanding science. For many years, logical empiricism was "the Received View" in philosophy of science, and it continues to exert a considerable influence. For a comprehensive account of the rise and fall of logical empiricism, see Frederick Suppe, "The Search for Philosophical Understanding of Scientific Theories," in *The Structure of Scientific Theories*, 2d ed., ed. F. Suppe (Urbana: University of Illinois Press, 1977), 1–241. It was Suppe who popularized Hilary Putnam's phrase "the Received View" to refer to logical positivism and logical empiricism.

2. Various methods of effecting such an elimination have been proposed, including Carnap's reduction sentences, Craigian transcription, and Ramsey sentences. For a critical discussion of some of these eliminative translation projects in logical empiricism, see Israel Scheffler, *The Anatomy of Inquiry* (New York: Knopf, 1963), pt. 2; and Carl G. Hempel, "The Theoretician's Dilemma," reprinted in *Aspects of Scientific Explanation* (New York: Free Press, 1965), 173–226.

3. Historically, traditional idealism is associated most strongly with George Berkeley's view that physical objects are nothing more than collections of sensory ideas. Its phenomenalist cousin, which holds that all statements about physical objects are equivalent in meaning to claims about actual or possible sensory experience, was defended by A. J. Ayer and others.

4. Consider an instrumentalist, for example, who accepts a theory T that includes the sentence "The temperature of a gas is directly proportional to the mean kinetic energy of the molecules composing it." This instrumentalist might take the following position: "Theory T is meaningful and indeed true. But do not suppose that we should take its claims purporting to refer to unobservable molecules or its theoretical terms like *kinetic energy* literally, as referring to or asserting the existence of theoretical objects and properties. To say that the terms are meaningful and the

claims are true is to say that they figure instrumentally, thus-and-so, in T's generation of the following set of observational consequences. . . ."

5. *The Scientific Image*, 202–3. Van Fraassen devotes his entire third chapter to a fuller explication of empirical adequacy as the claim that all actual phenomena fit within a model of the theory. A model of a theory is any structure satisfying all the axioms of the theory.

6. Compare Maxwell's argument (as we reconstructed it) with the following argument, which leans heavily on the claim that there is no nonarbitrary line to be drawn between short and tall people, since there is a continuous transition from persons who are 4'1", 4'2" . . . , to those who are 5'3", 5'4", . . . 6'5", . . . :

  1 There are short people.

  2 No nonarbitrary line can be drawn between short and tall people.

  3 Therefore, we are not warranted in claiming that there are tall people.

7. Aqua regia is a mixture of concentrated nitric acid and hydrochloric acid that can dissolve the "royal" metals, platinum and gold. During World War II the Hungarian chemist George de Hevesy dissolved the Nobel medals of Max von Laue and James Franck in aqua regia to prevent them from falling into Nazi hands. After the war de Hevesy recovered the gold and returned it to the Nobel Foundation. The medals were recast and presented again to von Laue and Franck.

8. Note carefully that we are assuming here that the reduction of quantum mechanics to a classical theory would have to involve the strict logical derivation of quantum mechanics from that theory. For a detailed discussion of other, perhaps more appropriate notions of intertheoretic reduction, see chapter 8.

9. Or at least according to Putnam of the 1970s. Putnam later came to espouse views less sympathetic with realism.

10. Jeff Foss, "On Accepting van Fraassen's Image of Science," *Philosophy of Science* 51 (1984): 79–92. For a reply to Foss's criticism of van Fraassen, see W. Bourgeois, "On Rejecting Foss's Image of van Fraassen," *Philosophy of Science* 54 (1987): 303–8. Foss responds in his "On Saving the Phenomena and Mice: A Reply to Bourgeois Concerning van Fraassen's Image of Science," *Philosophy of Science* 58 (1987): 278–87. In his *Scientific Representation* (New York: Oxford University Press, 2008), van Fraassen characterizes phenomena along the lines recommended by Foss. He writes, "*Phenomena* are observable entities (objects, events, processes, . . .) of any sort, *appearances* are the contents of measurement outcomes" (p. 283).

11. Michael Friedman was the first to make this "incoherence" objection, in his review of *The Scientific Image*. See *Journal of Philosophy* 79 (1982): 274–83. The argument is that, since van Fraassen embraces the theory-ladenness of our observation language, if the constructive empiricist accepts modern physics, then she accepts the characterization of observable macroscopic objects as being composed of elementary particles. This would seem to violate the constructive empiricist injunction against believing as true any statement about unobservables (such as elementary particles). A reply on van Fraassen's behalf by F. A. Muller, "Can a Constructive Empiricist Adopt the Concept of Observability?" *Philosophy of Science* 71

(2004): 80–97, runs as follows. The constructive empiricist accepts the theory that chairs, tables, and so on, are made up of atoms and molecules but she is not obliged to believe it since the proposition in question is not solely about observables. Indeed (according to Muller), the constructive empiricist accepts that unobservable particles exist (since this is logically implied by her acceptance of modern physics); but she does not believe it. For a more recent discussion, see F. A. Muller and B. C. van Fraassen, "How to Talk about Unobservables," *Analysis* 68 (2008): 197–205.

12. John O'Leary-Hawthorne, "What Does van Fraassen's Critique of Scientific Realism Show?" *The Monist* 77 (1994): 128–45. In his later writings, van Fraassen claimed that his aim was to show that constructive empiricism is a rationally permissible position to adopt according to the liberal principle that what is not forbidden is permitted. For further discussion (and criticism), see Marc Alspector-Kelly, "Should the Empiricist Be a Constructive Empiricist?" *Philosophy of Science* 68 (2001): 413–31.

13. Peter Lipton and Paul Dicken, "What Can Bas Believe? Musgrave and van Fraassen on Observability," *Analysis* 66 (2006): 226–33.

14. There is a further small wrinkle about two different ways that "X is observable" could be false: either by X not existing or by X existing but being too small to be seen. So, what T entails (and what a constructive empiricist who accepts T believes) are conditional statements of the form "If X exists, then X is unobservable."

15. One might wonder what is "epistemological" about the five theses of CER given that truth and reference are semantic concepts, and explanation could be regarded as ontic (following Railton and Salmon) rather than epistemic (following Hempel and Kitcher). CER would seem to be epistemological only in the weak sense that evidence is relevant to judging its truth.

16. Laudan claims that without an account of approximate truth, theses like (R1), (T1), and (T2) are "just so much mumbo jumbo" [p. 1119]. But lacking a clear account of some claim does not entail that the claim is meaningless or unintelligible. Many times in the past, scientists and philosophers have been unable to give an adequate account of claims that have later been vindicated as coherent (and sometimes true).

17. See Clyde L. Hardin and Alexander Rosenberg, "In Defense of Convergent Realism," *Philosophy of Science* 49 (1982): 604–15.

18. Mendel himself used the term *factor* to refer to the units of inheritance. The term *gene* was introduced later.

19. The following schema is what Laudan claims antirealists have long disliked about the reasoning of realists:

1 If T is true, then it will have true consequences.

2 T has true consequences.

3 Therefore, T is true.

This form of reasoning is clearly invalid, since the premises could be true and the conclusion false. As Laudan notes, even if the truth of a theory guarantees its having true consequences, the fact that T has true consequences cannot warrant the inference to T's truth, since a false theory could also have true consequences. Having true consequences is necessary but not sufficient for T's being true.

20. See the opening paragraph of Henri Poincaré, *Science and Hypothesis* (1902), Chapter 10, "The Theories of Modern Physics," an address delivered to the 1900 International Congress of Physics. The French phrase Poincaré used was "la faillite de la science": literally, the bankruptcy or insolvency of science. Poincaré went on to argue that this appearance of bankruptcy is merely superficial since the testable laws embedded in theories, expressed in mathematical equations, endure through scientific change. Opinions differ as to whether Poincaré is best regarded as an instrumentalist or a structural realist.

21. The PMI is called a "meta" argument to distinguish it from the way that scientists use first-order evidence to argue for particular scientific theories. Philosophers use the track record of past theories as second-order evidence to either attack (using the PMI) or defend (using the NMA) the reliability of scientific methodology. Thus both the PMI and the NMA are meta-level arguments about science as a whole. Magnus and Callender mark this "first-order versus second-order" distinction by referring to the NMA and the PMI as "wholesale" arguments for realism and anti-realism, respectively, as contrasted with the "retail" arguments that scientists use for particular theories. See P. D. Magnus and Craig Callender, "Realist Ennui and the Base Rate Fallacy," *Philosophy of Science* 71 (2004): 320–38.

22. Premise (4) assumes that Laudan's evidence consists of the fraction of successful theories among past false ones, but his article strongly suggests that his evidence was gathered by looking at the fraction of false theories among past successful ones.

23. Marc Lange, "Baseball, Pessimistic Inductions and the Turnover Fallacy," *Analysis* 62 (2002): 281–85.

24. Peter J. Lewis, "Why the Pessimistic Induction Is a Fallacy," *Synthese* 129 (2001): 371–80. For a similar criticism, see Colin Howson, *Hume's Problem: Induction and the Justification of Belief* (Oxford: Clarendon Press, 2000), ch. 3.

25. Reasons for doubting that the truth of a theory entails its success or makes success likely are given in Timothy D. Lyons, "Explaining the Success of a Scientific Theory," *Philosophy of Science* 70 (2003): 891–901.

26. P. D. Magnus and Craig Callender, "Realist Ennui and the Base Rate Fallacy," *Philosophy of Science* 71 (2004): 320–38.

27. Notice that Hacking's account is broadly anti-Humean. According to Hume, the fundamental causal facts are observed regularities summarized in causal laws. Thus, to speak of causal relations is not to speak of connections between objects in the world, but merely to claim that some sequence of events instantiates a regularity. In denying commitment to the truth of the fundamental laws of science, while affirming the existence of causal connections among entities in the world, Hacking has rejected the Humean account. In this respect his view is similar to Nancy Cartwright's in "Do the Laws of Physics State the Facts?" in chapter 7. Cartwright claims that the fundamental physical laws "do not provide true descriptions of reality" but that nevertheless she has "no quarrel with theoretical entities" [p. 872]. For critical accounts of how Cartwright's "inference to the most probable cause" argument for entity realism differs from Hacking's manipulationist argument, see Steve Clarke, "Defensible Territory for Entity Realism," *British Journal for the Philosophy*

of Science 52 (2001): 701–22; Robert Pierson and Richard Reiner, "Explanatory Warrant for Scientific Realism," *Synthese* 161 (2008): 271–82; and William Seager, "Beyond Theories: Cartwright and Hacking," in *Philosophy of Science: The Key Thinkers*, ed. J. R. Brown (New York: Continuum Press, 2012), 213–35. For other criticisms of Hacking's case for entity realism, see Margaret Morrison, "Theory, Intervention and Realism," *Synthese* 82 (1990): 1–22; and Richard Reiner and Robert Pierson, "Hacking's Experimental Realism: An Untenable Middle Ground," *Philosophy of Science* 62 (1995): 60–69.

28. Putnam's theory of meaning is discussed in the commentary on chapter 7, in the section "Mellor's Defense of the Regularity Theory."

29. Ian Hacking, *Representing and Intervening*, 46. This book is an extended defense of entity realism.

30. See note 27 above.

31. For other attempts to salvage a defensible form of attenuated realism, see John Worrall, "Structural Realism: The Best of Both Worlds?" *Dialectica* 43 (1989): 99–124; and Anjan Chakravartty, *A Metaphysics for Scientific Realism: Knowing the Unobservable* (Cambridge: Cambridge University Press, 2007).

32. For a defense of a minimal conception of realism that primarily responds to skepticism about the possibility of scientific knowledge, see Jarrett Leplin, *A Defense of Scientific Realism* (New York: Oxford University Press, 1997).

33. In the case of theory realism, laws might be preserved as approximations or limiting cases in later theories. Carrier calls this "corrective reduction." Note how the retention requirement, motivated by realism, differs from Laudan's conception of scientific progress as a net increase in problem solutions. Laudan, reacting to Kuhn and others, denies that progress has to be cumulative. Even if some problems (whether empirical or conceptual) are abandoned along the way, research traditions will be progressive (in Laudan's sense) if later theories solve more problems overall than those they replace. Carrier's emphasis on realism precludes him from taking this "instrumentalist" line toward progress through Kuhnian revolutions and radical conceptual change.

34. See, for example, the emphasis on novel prediction in Alan Musgrave, "The Ultimate Argument for Scientific Realism," in Robert Nola, ed., *Relativism and Realism in Science* (Dordrecht: Kluwer, 1988), reprinted in Alan Musgrave, *Essays on Realism and Rationalism* (Amsterdam: Rodopi, 1999); Alan Musgrave, "The 'Miracle Argument' for Scientific Realism," *The Rutherford Journal* 2 (2006–2007); Jarrett Leplin, *A Novel Defense of Scientific Realism* (New York: Oxford University Press, 1997); and Stathis Psillos, *Scientific Realism: How Science Tracks Truth* (New York: Routledge, 1999).

35. Philip Kitcher, *The Advancement of Science* (New York: Oxford University Press, 1993). Stathis Psillos (among other realists) also insists on this restriction. It is part of his "divide and conquer" strategy for defeating Laudan's pessimistic induction. See Stathis Psillos, *Scientific Realism: How Science Tracks Truth* (1999).

36. Stathis Psillos, John Worrall, and Philip Kitcher also argue for this point.

37. Because metatheoretical hypotheses in the philosophy of science are seldom phrased as universal generalizations, Paul Meehl goes further and argues for what

he calls the "strong actuarial thesis" that testing such hypotheses requires random sampling of the history of science, using the same formal analytic methods as those employed in the mature social sciences. See Paul E. Meehl, "The Miracle Argument for Realism: An Important Lesson to Be Learned by Generalizing from Carrier's Counter-Examples," *Studies in the History and Philosophy of Science* 23 (1992): 267–82.

38. Duhem, *Aim and Structure*, p. 28. Quoted in Martin Carrier, "What Is Wrong with the Miracle Argument?" *Studies in the History and Philosophy of Science* 22 (1991): 23–36. Duhem's example is the prediction of the Poisson bright spot from Fresnel's wave theory.

39. In making kinds depend on laws, Carrier is following Jerry Fodor, "Special Sciences (Or: The Disunity of Science as a Working Hypothesis)," *Synthese* 28 (1974): 97–115. See chapter 8.

40. The heterogeneity of so-called natural kinds is emphasized in Ian Hacking, "Natural Kinds: Rosy Dawn, Scholastic Twilight," *Royal Institute of Philosophy Supplements* 82 (2007): 203–39. Hacking's conclusion is that while some natural kinds are more fundamental than others, there is nothing that they share in common. Specifically, Hacking rejects Quine's contention that natural kinds are sets (on the grounds that many natural kinds are not the kinds of thing that can be counted). For an excellent survey of the entire spectrum of views on natural kinds, see Alexander Bird and Emma Tobin, "Natural Kinds," *The Stanford Encyclopedia of Philosophy (Summer 2010 Edition)*, Edward N. Zalta (ed.), http://plato.stanford .edu/archives/sum2010/entries/natural-kinds.

41. Jed Z. Buchwald, "Kinds and the Wave Theory of Light," *Studies in the History and Philosophy of Science* 23 (1992): 39–74.

42. The attempt to define scientific concepts solely in terms of experimental and observational procedures was defended by Percy W. Bridgman in his *Logic of Modern Physics* (New York: Macmillan, 1927). For trenchant criticisms of Bridgman's "operationism," see C. G. Hempel, "A Logical Appraisal of Operationism," *The Scientific Monthly* 79 (1954): 215–20 (reprinted in C. G. Hempel, *Aspects of Scientific Explanation*); and D. A. Gillies, "Operationalism," *Synthese* 25 (1972): 1–24.

43. This objection is raised in Kyle Stanford, *Exceeding Our Grasp: Science, History, and the Problem of Unconceived Alternatives* (Oxford: Oxford University Press, 2006), p. 166.

44. Aluminum, for example, is used to smelt manganese and chromium from oxides that cannot be reduced using carbon.

45. Kuhn endorses the "no-overlap principle" for kinds on the grounds that if items fell into a region of overlap, then we would make incompatible predictions about their behavior. See Thomas S. Kuhn, "Afterwords," in *World Changes: Thomas Kuhn and the Nature of Science*, ed. Paul Horwich (Cambridge, Mass.: MIT Press, 1993), 318. The rationale for the "no cross-cutting of kinds" condition is examined (and criticized) in Alexander Bird, *Thomas Kuhn* (Chesham: Acumen, 2000), 197–202; and Emma Toobin, "Crosscutting Natural Kinds and the Hierarchy Thesis," in *The Semantics and Metaphysics of Natural Kinds*, ed. Helen Beebee and Nigel Sabbarton-Leary (London: Routledge, 2010), 171–91.

46. See the wonderfully clear account in Stathis Psillos, "Is the History of Science the Wasteland of False Theories?" in *Adapting Historical Knowledge Production to the Classroom*, ed. P. V. Kokkotas, K. S. Malamitsa, and A. A. Rizaki (Rotterdam: Sense Publishers, 2011), 17–36. The weakness of Carrier's case can be illustrated further by pointing out that the same caloric model that is alleged to explain the correct result concerning uniform expansion rates would lead one to predict the wrong result for the specific heats on gases at either constant volume (c_v) or constant pressure (c_p). The caloric theorist would presumably expect these specific heats to vary from one gas to another (depending on the particles' "affinity" for caloric). What is found experimentally is that $c_v = (3/2)R$ for *every* monatomic gas, $c_v = (5/2)R$ for *every* diatomic gas, and $c_p = c_v + R$ for *all* gases. The latter equation is often called Mayer's law (after Julius Robert Mayer) and played a major role in establishing the conservation of energy principle and the kinetic theory of heat. Of course, Carrier might claim that the caloric model would lead one to expect there to be other laws common to all gases and that the division of gases into two species (on the basis of their values of c_v) is consistent with the caloric theory's recognition that gases are a state of matter (a natural kind) that shows much more lawlike uniformity than either liquids or solids.

47. On this last point, Carrier and Duhem disagree, despite Duhem's apparent endorsement of the miracle argument concerning natural classifications (which is difficult to square with Duhem's explicit antirealism). See Stathis Psillos, *Scientific Realism: How Science Tracks Truth*, and Karen Merikangas Darling, "Motivational Realism: The Natural Classification for Pierre Duhem," *Philosophy of Science* 70 (2003): 1125–36. Arthur Fine coined the term *motivational realism* to describe Einstein's account of what drives scientific work and gives it meaning. Darling attributes to Duhem the view that while it is psychologically inevitable that scientists will adopt a realist attitude toward natural classifications, such an attitude is not epistemically justified by the miracle argument or by anything else that falls within the legitimate realm of science. What is not clear, either in Duhem or in Darling, is whether this is because Duhem thought that the miracle argument (being nondeductive) has no rational merit or whether it is because the argument has merit but falls outside of the domain of testable science.

48. See James T. Cushing, *Quantum Mechanics: Historical Contingency and the Copenhagen Hegemony* (Chicago: University of Chicago Press, 1994), for a fascinating argument that quantum mechanics would have developed quite differently had realist sentiments prevailed.

49. The broadly Tarskian line, addressed in more detail in Musgrave's article than in Fine's, takes as its starting point the foundational Convention T according to which "snow is white" is true if and only if snow is white. For a good introduction to Tarski's theory of truth, see D. J. O'Connor, *The Correspondence Theory of Truth* (London: Hutchinson and Co., 1975).

50. Arthur Fine, "Unnatural Attitudes: Realist and Instrumentalist Attachments to Science," *Mind* 95 (1986): 149–79.

51. And even if Fine were happy with such a view, there is no obvious reason why a metaphysic of sense data should be thought inconsistent with a realist metaphysic of an objective external world or with correspondence relations holding between the world and linguistic items.

52. Assuming that we take such talk seriously and literally. We might, instead, take such talk as a philosopher's fanciful way of saying something more banal—construing, say, "The sun-as-conceived-by-Copernicus is at rest" and "The sun-as-conceived-by-Ptolemy is in motion" as (respectively) "Copernicus believed that the sun is at rest" and "Ptolemy believed that the sun is in motion." See "The Theory-Ladenness of Observation Argument," in the commentary on chapter 2.

53. One might object to this second point of Musgrave's, on the grounds that it presupposes something the nonrealist need not accept, namely, that that there is a moon-in-itself, in addition to the moon-as-conceived-by-us. Thorough-going relativists will find it natural to claim that there is no moon-in-itself. If there is no moon-in-itself, then Musgrave's question about its relation to the moon-as-conceived-by-us cannot even arise. Perhaps Musgrave would reply that if this route were taken, there could be no claim of contamination at all. It was, after all, Fine's original point that the information we retrieve from interaction with the moon is contaminated—information now about an interacted-with thing, not the thing-in-itself. And this requires the distinction that the nonrealist's objection disallows.

GLOSSARY

ABDUCTION (ABDUCTIVE INFERENCE)   The term *abduction* was coined by C. S. Peirce to describe the inference from a puzzling phenomenon to a theory that, if true, would explain it. N. R. Hanson regarded abduction as a significant kind of logic of discovery that is neither deductive nor inductive in character. Some philosophers treat abduction as distinct from inference to the best explanation: abduction concerns a single theory and reasons for taking it seriously as an interesting candidate for further exploration; inference to the best explanation involves a comparison among several theories and reasons for believing that one of them is true or probable. Others (the majority) use the terms *abductive inference* and *inference to the best explanation* interchangeably. (See *inference to the best explanation.*)

AMBIGUITY OF FALSIFICATION   The doctrine, associated with Pierre Duhem, that testable predictions can be drawn from a scientific theory only with the help of auxiliary hypotheses, theories, and assumptions. Thus, when a prediction turns out to be false, the theory might still be true, since the fault could lie with one or more of the auxiliaries. Some philosophers have espoused the much stronger thesis that no experiment or observation can falsify any theory under any circumstances. (See *underdetermination thesis, holism.*)

ANALYTIC STATEMENTS   Modern philosophers define an analytic statement as one that has its truth (or falsity) completely determined by the meanings of the words and symbols used to express it. (Earlier definitions of analyticity by Immanuel Kant and Gottlob Frege are discussed in the commentary on chapter 3.) All tautologies (logical truths) are analytic. So, too, are statements such as "all squares have four sides" and "all mammals suckle their young." Some philosophers, notably W. V. Quine, have denied that there are any analytic statements (even in logic and mathematics) and have attacked the traditional distinction between statements that are analytic and those that are synthetic. The issue is important because analytic statements (if there are any) would be a priori and thus immune from refutation. (See *synthetic statements.*)

ANTIREALISM   A diverse group of doctrines whose common element is their rejection of realism. In the philosophy of science, antirealism includes instrumentalism, conventionalism, logical positivism, logical empiricism, and Bas van Fraassen's constructive empiricism. Some antirealists (such as instrumentalists) deny that scientific theories that postulate unobservables should be interpreted realistically. Others (such as van Fraassen) concede that such theories should be interpreted realistically, but they deny that we should ever accept as true any theoretical claims about unobservables. (See *constructive empiricism, conventionalism, instrumentalism.*)

A POSTERIORI   An epistemological concept. An a posteriori (or empirical) statement is one that can be known to be true (or false) only through experience: either one's own or someone else's. Apart from the mathematics and logic they contain, the empirical sciences consist of a posteriori statements, knowledge of which can be obtained only through observation and experiment.

A PRIORI   An epistemological concept; the opposite of a posteriori. An a priori statement can be known to be true (or false) independently of experience (although some experience may be necessary to understand what the statement means). The least controversial examples of a priori statements are simple conceptual and logical truths. Examples: everything is identical with itself; all contradictions are false; all vixens are foxes.

AXIOLOGY   Axiology is concerned with values. Typical axiological questions are. What things are worth pursuing? What do judgments of value mean? How can judgments about values be justified? In philosophy of science, Larry Laudan and others use the phrase *axiological level* to refer to views about the aims and goals of science.

BAYESIAN CONFIRMATION THEORY   There are many different ways in which philosophers have applied Bayes's theorem to confirmation theory. The starting place for all of them is Bayes's theorem written in the form: $P(H/E\&B) = P(E/H\&B) \times P(H/B)/P(E/B)$, where E is the evidence for some hypothesis H, and B is a person's background beliefs. Objective Bayesians interpret probabilities as relative frequencies; subjective Bayesians (the majority) interpret probabilities as degrees of belief. According to the subjective Bayesians, $P(H/B)$, the prior probability of the hypothesis given one's background beliefs, will vary from individual to individual. But after acquiring the evidence, E, all rational persons should change their degree of belief in H from $P(H/B)$ to $P(H/E\&B)$. (See *likelihood, posterior probability, prior probability, problem of old evidence.*)

BAYES'S THEOREM   In its simplest form, Bayes's theorem (also called Bayes's rule, law, or equation) asserts that $P(H/E) = P(E/H) \times P(H)/P(E)$. Although versions of Bayes's theorem were derived by the Reverend Thomas Bayes, it was first used systematically by Laplace to calculate $P(H/E)$, the conditional probability of a hypothesis, H, given evidence, E. In the twentieth century, a school of Bayesian statistics and confirmation theory developed in which Bayes's theorem plays a central role. (See *Bayesian confirmation theory, posterior probability, prior probability.*)

BRIDGE LAW   Ernest Nagel and others regard the logical derivability of one theory, T, from another, T', as a necessary condition for the reduction of T to T'. When the reduced theory, T, contains terms that are absent from the reducing theory, T', bridge laws connecting these terms with the vocabulary of T' must be added to T' before T can be deduced from (the augmented version of) T'. The status of bridge laws is controversial. Nagel regards them as empirical hypotheses. A typical example of a bridge law is the statement connecting the temperature of a gas with the mean kinetic energy of the translational motion of its molecules, used to reduce the ideal gas law to the kinetic theory of gases. Bridge laws are sometimes called bridge principles, correspondence rules, or coordinative definitions.

CONDITION OF MEANING INVARIANCE   A necessary condition for intertheoretic reduction according to Ernest Nagel and others. When one theory reduces another, it must leave all the terms of the reduced theory unchanged in meaning. The condition of meaning invariance follows directly from the requirement that the reduced theory be logically derivable from the reducing theory.

CONSISTENCY CONDITION   A necessary condition for intertheoretic reduction according to Ernest Nagel and others. The reducing theory must be logically consistent with the reduced theory. Like the condition of meaning invariance, the consistency condition is a consequence of the requirement that the reduced theory be logically derivable from the reducing theory. The consistency condition has been much criticized by Paul Feyerabend on the grounds that it dogmatically protects old, well-established theories from refutation by new theories that contradict them.

CONSTRUCTIVE EMPIRICISM *Constructive empiricism* is Bas van Fraassen's term for his antirealist philosophy of science. Like logical empiricism, constructive empiricism draws a line between the theoretical and the observable. In van Fraassen's case, observable things and processes are those that can be seen by unaided human vision. Unlike the logical empiricists, van Fraassen is a realist about theories that talk about unobservables. But, van Frassen insists, however successful these theories are in predicting and explaining phenomena, we should never accept them as true but merely as empirically adequate. (See *antirealism, empirical adequacy, logical empiricism, observable.*)

CONTEXT OF DISCOVERY The context of discovery includes anything that is psychologically relevant to the discovery of a theory, hypothesis, theorem, argument, or piece of evidence. (See *context of justification.*)

CONTEXT OF JUSTIFICATION The context of justification includes anything that is logically relevant (either deductively or inductively) to the epistemic justification of a theory, hypothesis, theorem, assertion about an argument, or evidential claim. The contexts of discovery and justification are not exclusive: the factors involved in the discovery of a result might play a role in justifying it, and criteria of justification often guide the search for new theories. (See *context of discovery.*)

CONTEXTUALIST ANALYSIS OF EVIDENCE The *contextualist analysis of evidence* is Helen Longino's term for the thesis, defended by T. S. Kuhn and others, that our judgments of evidential relevance and strength of evidential support for hypotheses depend on our background beliefs and assumptions. For example, what an individual scientist takes to be evidence for a hypothesis depends on the context of that scientist's beliefs. Unlike the standard view of evidence adopted by philosophers of science, on the contextualist analysis, evidence is not a set of statements but the objects, events, or states of affairs that observation statements purport to describe. (See *evidence.*)

CONTINGENCY (CONTINGENT PROPOSITION) A metaphysical concept. Contingent propositions are true in some possible worlds but not in others. True contingent propositions are true in the actual world but false in at least one other possible world. False contingent propositions are false in the actual world but true in at least one other possible world. Examples: sea water contains traces of gold; carnivorous dinosaurs prowled the streets of New York in the nineteenth century. If a proposition is not contingent, then it is necessary. (See *necessity, possible world.*)

CONVENTIONALISM A decision is conventional if it involves choosing from among alternatives that are equally legitimate when judged by objective criteria (such as consistency with observation and evidence): thus, either the decision is entirely arbitrary or it rests on an appeal to factors often presumed to be subjective, such as simplicity, economy, and convenience. Radical conventionalists argue that scientific theories are really definitions (or rules of inference, pictures, conceptual schemes, paradigms) and hence neither true nor false; moderate conventionalists disagree, insisting that once the conventional elements in theories have been isolated, the remaining parts are objectively true or false. Typically, conventionalists appeal to the underdetermination of theories by evidence to bolster their doctrine. Many philosophers of science have been conventionalists in one respect or another. They include Henri Poincaré (on high-level scientific laws as definitions), Pierre Duhem (on the ambiguity of falsification of theories in physics), W. V. Quine (on the decision to retain or abandon any sentence whatever), Hans Reichenbach (on the choice of a geometry to describe physical space), Karl Popper (on the decision to accept basic statements), and T. S. Kuhn (on the decision to switch paradigms). (See *underdetermination thesis*.)

CORRESPONDENCE RULES According to logical empiricists (such as R. B. Braithwaite, Hans Reichenbach, Rudolf Carnap, Carl Hempel, and Ernest Nagel), correspondence rules give a partial interpretation of the theoretical terms of a theory by linking them with observation terms. Thus, the correspondence rules of a theory are mixed sentences that must contain (nonvacuously) at least one theoretical term and at least one observation term. (See *logical empiricism*.)

CORROBORATION *Corroboration* is a term introduced by Karl Popper to characterize the status of theories that have survived severe tests. Since Popper denies that there is any such thing as inductive confirmation, he says that theories that have survived attempts to refute them are trustworthy, not because they have been confirmed or made probable by evidence, but because they have been corroborated. Some critics have charged that either corroboration is confirmation in all but name or, if corroboration refers solely to the past success of the theory, then it can give us no rational justification for trusting that theory in the future.

COUNTERFACTUAL Philosophers of science are particularly interested in counterfactual conditionals, expressed in the subjunctive mood. For example: if my leg were made of gold, it would weigh at least two hundred pounds; if one were to slam together two pieces of uranium-235 the size of elephants, then there would be a nuclear explosion. One key difference between laws of nature and merely true but non-lawlike universal generalizations is that laws do, and nonlawlike generalizations do not, support inferences to true counterfactual conditionals. Thus, even if it were true that all dogs born at sea are cocker spaniels, it would not follow from this that if a golden retriever were to have pups at sea, they would be cocker spaniels.

COVERING LAW MODEL OF EXPLANATION The covering law model is a family of models of explanation, proposed by Carl Hempel and others, in all of which a necessary condition for a scientific explanation is that its premises contain (nonvacuously) at least one law of nature. The main examples of the covering law model are Hempel's deductive-nomological (D-N) and inductive-statistical (I-S) models of explanation. (See *deductive-nomological explanation, inductive-statistical explanation*.)

CRUCIAL EXPERIMENT A crucial experiment is an experiment that would conclusively falsify one of two competing theories or hypotheses, thereby establishing its rival as well confirmed or true. Pierre Duhem argued that no experiment in physics could be crucial in this sense. (See *ambiguity of falsification*.)

DEDUCTION (DEDUCTIVE VALIDITY) An argument is deductively valid if and only if its premises logically imply its conclusion. Hence, if all the premises of a valid argument are true, then the argument's conclusion must also be true. In a valid argument, it is impossible for all the premises to be true and the conclusion false.

DEDUCTIVE-NOMOLOGICAL (D-N) EXPLANATION A popular model of scientific explanation in the twentieth century, the deductive-nomological (D-N) model was given its clearest formulation and defense by Carl Hempel. D-N explanations are deductively valid arguments in which at least one premise is a scientific law. (See *epistemic conception of explanation, irrelevance problem, symmetry objection, thesis of structural identity*.)

DEDUCTIVE-NOMOLOGICAL MODEL OF PROBABILISTIC EXPLANATION (THE D-N-P MODEL) Peter Railton has proposed his deductive-nomological model of probabilistic explanation (the D-N-P model) as an alternative to Carl Hempel's inductive-statistical (I-S) model. Railton denies that explanations are arguments, although he insists that explanations must contain a deductive argument based on a probabilistic law. Probabilistic laws, on this view, are statements, not of frequencies, but of single-case propensities. D-N-P explanations must also contain a theoretical derivation of the relevant probabilistic law based on the indeterministic causal mechanism that brings about the event to be explained. Among the several important differences between Railton's D-N-P model and Hempel's I-S model are these: D-N-P explanations are fully objective (not relative to the state of scientific knowledge at a given time); and complete D-N-P explanations can be given of events no matter how low their probability, but they can be given only when there is genuine indeterminism at work.

DEMARCATION CRITERION A demarcation criterion, if any could be found, would distinguish genuine science from pseudoscience. Past proposals for demarcation criteria such as verifiability (the logical positivists) and falsifiability (Karl Popper) have been abandoned in light of powerful objections. Some philosophers (Imre Lakatos, Paul Thagard) think that a demarcation criterion must include historical considerations; others (Larry Laudan) think that the search for a demarcation criterion is futile.

DETERMINISM According to determinism, there is at any moment exactly one physically possible future; that is, there is exactly one set of future events that is consistent with the past and the laws of nature. A common version of determinism says that every event is causally determined by earlier events and thus, in principle, could have been predicted if we knew enough about those earlier events and the laws of nature. This way of linking determinism with predictability-in-principle has been criticized by John Earman and others on the grounds that deterministic laws do not guarantee predictability.

DISPOSITION A disposition is the power or tendency of a thing to behave in a certain way when placed in a certain kind of situation. For example, soft iron has the disposition to become magnetized when placed in a magnetic field; copper has the disposition to dissolve in hydrochloric acid; tritium (an isotope of hydrogen) has the disposition to undergo radioactive decay. Dispositions have been regarded with suspicion by some empiricists (such as Hume), since they are properties that a thing is supposed to have even when the disposition in question is not being manifested. Thus, the attribution of a disposition to an object goes beyond any set of evidence about how the object actually behaves. Moreover, some empiricists have questioned whether the appeal to dispositions can be genuinely explanatory.

DUHEM-QUINE THESIS It has become common to speak of the Duhem-Quine thesis as if there is one thesis to which both Pierre Duhem and W. V. Quine subscribed. This is historically doubtful. Usually the term refers to the holistic doctrine that no scientific theory can be tested in isolation, since the derivation of an observation sentence from any theory requires a host of auxiliary hypotheses, theories, and assumptions. From this, it is usually concluded that falsification is ambiguous. But some philosophers go further and conclude that the acceptance or rejection of particular scientific theories can never be justified by observational evidence. Thus, the presumed truth of the Duhem-Quine thesis is often appealed to by those who wish to emphasize the role of convention, contextual values, and social forces in the decision to accept or reject scientific theories. (See *holism, underdetermination thesis.*)

EMPIRICAL ADEQUACY The term *empirical adequacy* was introduced by Bas van Fraassen and plays a key role in his constructive empiricism. A scientific theory is empirically adequate when everything that it says about observables is true. As critics of van Fraassen have pointed out, to judge that a theory is empirically adequate goes far beyond claiming that the theory agrees with all the available evidence. (See *constructive empiricism, observable.*)

EMPIRICISM Empiricism includes a wide variety of philosophical doctrines held by philosophers such as Aristotle, Locke, Hume, Mill, and Ayer. What they share is a healthy respect for experience as the ultimate source of all human knowledge (possibly excluding logic and mathematics). Empiricists tend to be suspicious of any claims (such as those made by the theoretical sciences) that cannot be related directly to experience or observation. Some empiricists have tried, unsuccessfully, to reduce theoretical claims to claims about experience. Others (such as the logical empiricists) have tried to analyze the meaning (or cognitive significance) of theoretical assertions by tracing them to observation statements via correspondence rules. This, too, is now generally regarded as a failure. Nonetheless, empiricism in philosophy of science remains vigorous and popular.

ENTAILMENT One statement entails another when it is impossible for the second to be false when the first is true. For example, the premises of a valid argument entail its conclusion.

EPISTEMIC CONCEPTION OF EXPLANATION Wesley Salmon gives Carl Hempel's covering law model as a classic example of the epistemic conception of explanation. On this conception, explanations are a type of argument or inference, and to explain something is to show why that thing was to be expected given the information in the premises of the explanatory argument. It is characteristic of the epistemic conception that explanation and prediction are seen as formally identical and that explanations can be given for events only if their occurrence is certain or highly probable. (See *ontic conception of explanation.*)

EPISTEMIC REGULARITY THEORY OF LAWS A version of the Humean, regularity theory of laws. According to A. J. Ayer and others, laws of nature are true, universal generalizations that have additional features of an epistemic character. These features include our willingness to use the generalization to make predictions and the role that the generalization plays in an organized system of science. (See *regularity theory of laws.*)

EPISTEMOLOGY Epistemology is the theory of knowledge. Epistemologists concern themselves with concepts such as knowledge, belief, epistemic justification, confirmation, and explanation. Epistemic problems include refuting skepticism (the denial that knowledge, either in general or in a particular area, is possible) by explaining why some beliefs (and not others) are justified.

EVIDENCE For most philosophers of science, evidence is an observation statement (or a set of such statements) reporting the outcome of an observation or experiment that either confirms or disconfirms a theory. Positive evidence confirms; negative evidence disconfirms. (See *contextualist analysis of evidence*.)

EXPLANANDUM A Latin word meaning "that which is to be explained." The explanandum is usually taken to be a sentence describing the thing (whether event, law, or theory) for which an explanation is sought. (See *explanans*.)

EXPLANANS A Latin word meaning "that which explains." The explanans is the thing, usually regarded as a set of sentences satisfying certain conditions, that constitutes an explanation of the explanandum. (See *explanandum*.)

EXPLANATIONISM Explanationism is a thesis about confirmation. Explanationists insist that only the explanation of previously known facts has the power to confirm a theory, thus denying that a theory receives any confirmation from novel predictions. (See *historical thesis of evidence, novel prediction, predictionism*.)

EXTENSION The extension of a term (predicate, concept) is the thing or set of things that the term picks out and to which the term is correctly applied. Thus, the extension of *yellow* is the set of yellow things. The extension of a proper name is its bearer. (See *intension*.)

FALSIFICATIONISM A set of methodological doctrines championed by Karl Popper comprising a demarcation criterion and a fallibilist theory of scientific rationality and progress. According to Popper, a theory is scientific if and only if it is falsifiable. No theory can be proven to be true or even probable. Scientific rationality consists in testing theories severely in order to refute them. Progress occurs when falsified theories are replaced by as-yet-unfalsified theories that are more testable than those they replace. A more sophisticated version of falsificationism has been advocated by Imre Lakatos. (See *corroboration, scientific research programme*.)

FUNCTIONAL LAWS Functional laws assert a universal relation between two or more variables in the form of a mathematical equation or inequality. Typical examples are Hooke's law, Snell's law, and Maxwell's field equations.

HIERARCHICAL MODEL OF SCIENTIFIC RATIONALITY A model of scientific rationality, criticized by Larry Laudan, according to which decisions at one level are justified (if at all) by rules and criteria at the next highest level. Inevitably, on this model disagreements about the aims and goals of science at the highest, axiological level cannot be resolved rationally. (See *axiology, reticulational model of scientific rationality*.)

HISTORICAL THESIS OF EVIDENCE The historical thesis of evidence maintains that whether or not a piece of evidence confirms a theory depends on the time that the evidence is known relative to the time at which the theory is proposed. Both explanationism and predictionism entail the historical thesis of evidence. This thesis has been criticized by Peter Achinstein and Laura Snyder, and is rejected by Bayesians as well as by all those (such as Rudolf Carnap) who regard confirmation as a formal or logical relation between propositions. (See *explanationism, predictionism*.)

HOLISM (WHOLISM) A family of doctrines whose common core is the notion that an individual element in a complex whole has the properties it does only insofar as it stands in certain relations to other, similar elements. For example, W. V. Quine is a semantic holist, holding that the meaning of an individual sentence depends on how it is related to other sentences in the same language. Similarly, in epistemology, proponents of coherence theories of justification insist that a particular belief is justified only if it coheres with a large number of similar beliefs. In the philosophy of science, the term *holism* usually refers to Pierre Duhem's doctrine that no individual physical theory (and, more generally, no individual scientific theory) implies any observation statement. Single theories make testable predictions only when they are conjoined with other theories, auxiliary hypotheses, and background assumptions.

HOMOGENEOUS REDUCTION Ernest Nagel calls the reduction of one theory to another homogeneous when the vocabulary of the reduced theory is included in (or can be defined in terms of) the vocabulary of the reducing theory, thus permitting the logical derivation of the one theory from the other. An alleged example of a homogeneous reduction is the derivation of Galileo's law of falling bodies from Newton's gravitational theory. (See *inhomogeneous reduction*.)

HYPOTHESIS Some authors distinguish between hypotheses and theories in the following way: a hypothesis is a scientific claim that is put forward as a plausible conjecture before there is enough evidence to warrant either its acceptance or its rejection; a theory is a well-established scientific claim that is accepted as true or well-confirmed. Hypotheses, if successful, become theories. Other authors (such as the authors of the present book) use the terms *hypothesis* and *theory* interchangeably. (See *theory*.)

HYPOTHETICO-DEDUCTIVE (H-D) MODEL In its simplest and most common form, the hypothetico-deductive (H-D) model of science denies that there is any logic of discovery and affirms we are justified in accepting or rejecting theories only after they have been tested. Testing a theory consists of deriving from it consequences that can then be compared with observations and experimental results. If the results agree with the predictions, the theory is inductively confirmed; if the results disagree, the theory is disconfirmed or refuted.

IDENTITY So-called identical twins are qualitatively identical (having, let us assume, all the same qualitative properties) but not numerically identical (since they are two distinct individuals). When philosophers of science talk about identity, they usually mean numerical identity. For example, flashes of lightning are identical with discharges of atmospheric electricity, genes are identical with segments of DNA, and water is identical with molecules of hydrogen oxide. If an identity statement is true, then the phrases used to express it (such as "a flash of lightning" and "a discharge of atmospheric electricity") have the same extension—they both refer to the same thing. This is not the same as claiming, nor does it entail, that the phrases mean the same thing or that anyone using the terms correctly must be aware that the terms are coextensive. On the rigid designator theory proposed by Saul Kripke and Hilary Putnam, statements of identity are metaphysically necessary but the discovery that they are true can (and, in the empirical sciences, usually does) depend on observation and experiment. (See *extension, necessity, rigid designator*.)

INCOMMENSURABILITY A variety of doctrines fly under the banner of incommensurability. Prominent among them is the claim of Paul Feyerabend and T. S. Kuhn that the meaning of a theoretical term depends on the theory in which it occurs in such a way that it is impossible for rival theories to share meanings. Another variety of incommensurability, defended by Kuhn, is the alleged incommensurability of standards for choosing among theories (what Kuhn calls *values*) associated with rival paradigms.

INDUCTION Induction can be construed broadly or narrowly. Broadly, it is any deductively invalid inference in which the premises make the conclusion probable; narrowly, it is any inference that is an instance of argument forms such as induction by simple enumeration, argument by analogy, statistical syllogism, and induction to a particular. The broad characterization includes as inductive, and the narrow may exclude, argument types such as inference to the best explanation.

INDUCTIVE-STATISTICAL (I-S) EXPLANATION An influential model of explanation, advocated by Carl Hempel, in which the explanans renders the explanandum highly probable but does not deductively entail it. The explanans of an I-S explanation has to include a statistical law, and the form of the argument is a statistical syllogism.

INDUCTIVISM A hopelessly naive model of science according to which scientific knowledge accumulates by using inductive generalization to infer laws and theories from observational and experimental facts. According to inductivism, induction serves as both the logic of discovery and the logic of justification. Inductivism was decisively refuted by Pierre Duhem and Karl Popper. Although everyone agrees that inductivism is false, many philosophers of science insist (contrary to Popper) that the inductive confirmation of theories by the verification of their logical consequences is an essential part of scientific reasoning. (See *hypothetico deductive model*.)

INFERENCE TO THE BEST EXPLANATION A form of argument often relied on by scientific realists in which the fact that a particular theory explains some body of evidence better than any of its rivals is supposed to make the theory probable or reasonable to believe. Usually, inference to the best explanation is regarded as distinct from inductive reasoning. Some of its proponents (such as Gilbert Harman and Peter Lipton) have argued that it underlies most of the reasoning traditionally regarded as inductive. (See *abduction*.)

INHOMOGENEOUS REDUCTION According to Ernest Nagel, inhomogeneous reductions are those in which the vocabulary of the reduced theory is neither included in nor can be defined in terms of the vocabulary of the reducing theory. Nagel argues that the one theory can be logically derived from the other (and thus be reduced by it) only when the reducing theory is supplemented with appropriate bridge laws. An alleged example of an inhomogeneous reduction is the derivation of the ideal gas laws from the kinetic theory of gases. (See *bridge law, homogeneous reduction, reductionism*.)

INSTRUMENTALISM Strictly speaking, instrumentalism is the doctrine that theories are merely instruments, tools for the prediction and covenient summary of data. As such, theories are not statements that are either true or false; they are tools that are more or less useful. But because one has to use the machinery of logic in order to draw predictions from theories, it is difficult to deny that theories have truth values. Thus, instrumentalism has come to be used as a general term for antirealism. Most modern instrumentalists concede that theories have truth values but deny that every aspect of them should be interpreted realistically or that reasons to accept a theory as scientifically valuable are reasons to accept the theory as literally true. In this sense, T. S. Kuhn, who locates the value of scientific theories in their ability to solve puzzles, is an instrumentalist. Theories may have truth values, but their truth or falsity is irrelevant to our understanding of science. (See *antirealism, constructive empiricism, conventionalism, scientific realism*.)

INTENSION A semantic notion; the intension of a general term or concept, A, is its meaning, connotation, or sense. Some authors define the intension of A as all the general terms or concepts B such that "All As are B" is a necessary truth. Others insist on a narrower characterization that restricts the Bs to properties that belong to the explicit definition of A. Still others (such as W. V. Quine) reject the whole notion of intension and meaning as a seductive myth. (See *extension, sense*.)

INTENTIONALITY The intentionality of beliefs and other mental states is the property that enables them to be about or to represent states of affairs. The intentionality of beliefs (hopes, fears, wishes) is usually indicated by a "that" clause followed by a proposition. If Sarah believes (hopes, fears, wishes) that the dodo is not extinct, then the intentional content of her belief (hope, fear, wish) is the nonactual state of affairs of the dodo's not being extinct.

IRRELEVANCE PROBLEM A general term for a variety of alleged counterexamples to Carl Hempel's deductive-nomological (D-N) and inductive-statistical (I-S) models of explanation. These cases satisfy Hempel's conditions for an adequate explanation but in each of them the explanans is explanatorily irrelevant to the explanandum.

LAWS OF NATURE Scientists discover laws of nature but philosophers disagree about what kinds of thing laws of nature are. The main competitors are the regularity theory, the epistemic regularity theory, the necessitarian theory, and the universals theory of laws (all defined elsewhere in this glossary). Laws of nature are not restricted to the things that modern scientists call laws; they include all (true) fundamental equations and inequalities, whether or not they are called laws and regardless of whether they have been or ever will be discovered.

LIKELIHOOD The likelihood of a theory, T, relative to some evidence, E, and background information, B, is $P(E/T\&B)$, the probability of the evidence given the theory and the background information. When T entails E, the likelihood of T (relative to E) is 1.

LOGICAL EMPIRICISM Some authors use the labels *logical empiricism* and *logical positivism* interchangeably. Others (such as the authors of this book) reserve the term *logical empiricism* for the views of Rudolf Carnap, Carl Hempel, Ernest Nagel, Hans Reichenbach, and others after the verifiability principle was abandoned. Among the doctrines of logical empiricism are the theory-observation distinction, the view that theoretical terms are "partially interpreted" (through their role in the derivation of observation statements), Hempel's deductive-nomological and inductive-statistical models of explanation, and Nagel's model of reduction. (See *logical positivism, verifiability principle of meaning*.)

LOGICAL POSITIVISM Logical positivism is the name for the set of doctrines advocated by members of the Vienna Circle from about 1920 to 1936 (when Moritz Schlick, their leader, was assassinated). Prominent members of this group were Rudolf Carnap, Herbert Feigl, Hans Hahn, Otto Neurath, and Friedrich Waismann. Their approach to philosophy relied heavily on the verifiability criterion of meaning and is illustrated in A. J. Ayer's *Language, Truth and Logic*. (See *logical empiricism, verifiability principle of meaning*.)

LOGICISM The thesis, advocated by Gottlob Frege and Bertrand Russell, that mathematics (or, at least, arithmetic) can be reduced to logic by appropriate definitions of terms such as *number* and *is a successor of*. The logicist program foundered, in part because arithmetic requires set theory and the axioms of set theory are not truths of logic.

MEANING INVARIANCE, CONDITION OF Ernest Nagel imposes meaning invariance as a condition for intertheoretic reduction: when one theory reduces another it must leave the meanings of the terms in the reduced theory unchanged.

METAPHYSICS The branch of philosophy that attempts to discover necessary truths about the fundamental constituents of reality (or, more cautiously, about things that, if they exist, would be basic features of the world). Metaphysics includes inquiries into the nature of time, space, matter, causation, minds, freedom, and God and the relations between them. It is widely acknowledged that there is no rigid line separating metaphysics from science (especially physics).

METHODOLOGY In the philosophy of science, *methodology* means the *theory of scientific method*. For example, methodological principles are the rules that are supposed to govern rational choices among scientific theories. More generally, methodology includes criteria for assessing theories, weighing evidence, and evaluating scientific change.

MODUS PONENS The traditional name for any valid argument of the form, "If *P* then *Q*; *P*; therefore, *Q*." (See *deduction*.)

MODUS TOLLENS The traditional name for any valid argument of the form, "If *P* then *Q*; not-*Q*; therefore, not-*P*." (See *deduction*.)

NATURAL KINDS Speaking very generally, natural kinds are classes of things (usually naturally occurring) that form the proper object of scientific theorizing and laws. They include chemical elements, fundamental particles, and (on some accounts) biological species. Decisions about which things belong to natural kinds often depend on our best scientific theories. Oxygen, diamonds, viruses, and neutron stars are natural kinds; phlogiston, valuable gems, pathogens, and things that shine in the night sky are not.

NATURAL ONTOLOGICAL ATTITUDE (NOA) The natural ontological attitude (acronym, NOA, pronounced "Noah") is advocated by Arthur Fine as a solution to the debate between realism and antirealism in the philosophy of science. According to Fine, NOA is the minimal, neutral "core position" that is shared by both realism and antirealism but is itself neither realist nor antirealist. The NOAer (pronounced "knower") accepts the confirmed theories and results of science (as true) in the same way that we accept everyday truths on the evidence of our senses but refuses to adopt any metaphysical theory about the nature of truth (such as truth as correspondence with reality).

NECESSARY CONDITION X is a necessary condition for Y if and only if one cannot have Y without X. Being a reptile is a necessary condition for being an alligator; the presence of oxygen is a necessary condition for the rusting of iron.

NECESSITARIAN THEORY OF LAWS The necessitarian theory of laws insists that there must be a real, objective relation of necessity (*nomic* necessity) between events, objects or properties, if a statement about them can be a law of nature. Necessitarians differ among themselves about whether laws of nature are universal generalizations about particulars or singular statements of fact about universals. But all necessitarians agree on the necessity of laws of nature and reject the regularity account. (See *epistemic regularity theory of laws, regularity theory of laws*.)

NECESSITY   A much-contested metaphysical notion. Propositions are either logically, metaphysically, or physically necessary (where these categories are neither exclusive nor exhaustive). Metaphysically necessary propositions (a category that includes, but is not confined to, all those propositions that are logically necessary) cannot be false; they are true in all possible worlds. Physically necessary propositions are those whose truth is guaranteed by the laws of nature. Propositions that are neither metaphysically nor logically necessary are said to be contingent. (See *contingency*.)

NONINSTANTIAL LAWS   Noninstantial laws are laws that have no instances. Newton's first law (the law of rectilinear inertia) is often cited as an example because, it is claimed, no actual body is ever completely free of a net external force. Nonetheless, Newton's law is true: if there were such a body, it would not accelerate. Likewise, it is a lawlike fact that all pieces of plutonium with a mass greater than 1 million kilograms are excellent conductors of electricity even though the universe never has and never will contain such objects. (See *vacuous laws*.)

NORMAL SCIENCE   A term introduced by T. S. Kuhn to describe science the way it is most of the time, when no scientific revolutions are occurring. During normal science, Kuhn claims that scientists in a given field work under the aegis of a single paradigm that they all accept. (See *paradigm*.)

NOVEL PREDICTION   The prediction of a result by a theory can be novel in many different senses. One straightforward sense, temporal novelty, is that the result predicted was not known by anyone at the time the theory was proposed. Another sense, epistemic novelty, is that the result was not widely known and, in particular, was unknown to the person proposing the theory. Two other senses deemed important by some philosophers of science are design-novelty and use-novelty. A result is design-novel (or heuristically novel) when the theory yielding it was not deliberately constructed so as to yield that result (among others). A result is use-novel when it was not built into the new theory by, for example, using it to fix the value of a parameter. Philosophers of science disagree about whether novelty (in one or more of these senses) is a necessary condition for confirmation or, even if it is not necessary for confirmation, whether novelty enhances the confirming power of a prediction. (See *predictionism*.)

OBSERVABLE   To paraphrase John Stuart Mill, something is observable when it can be observed. Opinions differ as to what this amounts to and what its epistemic significance is. The logical positivists and empiricists defined *observation terms* as those terms whose correct application can be rapidly decided by any person of normal perceptual powers without any special scientific training. Bas van Fraassen defines *observable objects* as those that can be seen by the unaided vision of a normal human being. The intended contrast is between those terms and things that are observable and those that are theoretical. For empiricists, theoretical claims ultimately rest on and are tested by observation reports confined to observables. (See *constructive empiricism, theory-ladenness*.)

ONTIC CONCEPTION OF EXPLANATION Wesley Salmon contrasts the ontic conception of explanation (which he endorses) with the epistemic conception. Unlike the epistemic conception, the ontic conception denies that explanations are arguments (although they may involve arguments) and rejects the requirement that the explanandum be predicted with high probability. On the ontic view, something is an explanation only if it correctly describes the underlying cause or causal mechanism that brought about the explanandum event. A good example of an ontic approach to explanation is Peter Railton's deductive-nomological model of probabilistic explanation (D-N-P model). (See *deductive-nomological model of probabilistic explanation, epistemic conception of explanation.*)

ONTOLOGY Either the part of metaphysics concerned with the nature of existence, or, as in the philosophy of science, the entities (things, processes, properties) postulated by a particular scientific theory or conceptual scheme.

OPERATIONALISM (OPERATIONISM) The doctrine propounded by the physicist Percy W. Bridgman in his book, *The Logic of Modern Physics* (1927). Impressed by the success of Einstein's relativity theory, especially Einstein's analysis of the concept of simultaneity, Bridgman proposed that all scientific concepts and terms are synonymous with the operations and procedures used to measure their values. But as critics such as Carl Hempel have pointed out, very few scientific concepts can be defined completely in terms of a particular set of procedures and, in any case, Bridgman's proposal would lead to the unnecessary proliferation of scientific concepts if each new measuring procedure were to be regarded as defining a new concept.

PARADIGM A multiply ambiguous term that lies at the heart of T. S. Kuhn's philosophy of science. In most of its uses, it means either *exemplar* or *disciplinary matrix*. An exemplar is an important scientific theory or piece of research that serves as a model for further inquiry. Disciplinary matrices contain exemplars as one of their elements. They also include heuristic models, ontological and meta physical assumptions, and methodological principles. In short, paradigms (in the sense of disciplinary matrices) are "super theories" that underlie and guide an entire tradition of scientific research and theorizing.

PHENOMENALISM Proposed in the eighteenth century by Bishop Berkeley and in the twentieth by the logical positivists, phenomenalism is the doctrine that all meaningful talk about physical objects is reducible to talk about actual or possible perceptual experience. (This is the linguistic version of phenomenalism; its ontological version asserts that physical objects are nothing more than sets of actual and possible sense experiences.) John Stuart Mill, Ernst Mach, A. J. Ayer, Bertrand Russell and many other philosophers have been phenomenalists at some point in their careers. It is now generally acknowledged that it is impossible to give the translations required by the linguistic version of phenomenalism.

POSITIVISM An extreme form of empiricism advocated by the French philosopher and sociologist Auguste Comte (1798–1857). Comte denied that it is possible to know anything about unobservables (into which category he placed the underlying causes of phenomena), and he insisted that the sole aim of science is prediction, not explanation. Comte also believed that each branch of science has its own special laws and methods that cannot be reduced to those of other branches. Generally regarded as the founder of sociology, Comte's empiricism and his hostility towards metaphysics were an important influence on logical positivism. (See *logical positivism.*)

POSSIBILITY (LOGICAL, PHYSICAL)   A statement or a state of affairs is logically possible if its denial is not logically necessary. Something is physically possible if, relative to a given set of circumstances, the laws of nature do not forbid it. What is physically possible in one set of circumstances might be impossible in another. Many logically possible things are physically impossible, but everything that is physically possible is logically possible. (See *necessity*.)

POSSIBLE WORLD   The actual world (*this* world) is a possible world. So, too, is any world whose complete description does not contain any logically impossible proposition. (For every proposition, a complete description of a possible world contains either that proposition or its denial.) (See *contingency, possibility*.)

POSTERIOR PROBABILITY   The posterior probability of a theory, *T*, relative to some evidence, *E*, and background information, *B*, is $P(T/E\&B)$, the probability of the theory given the evidence and the background information. (See *Bayesian confirmation theory, prior probability*.)

PREDICATE   In the sentence, "Sodium is monovalent," *sodium* is the subject term and *monovalent* is the predicate term. Similarly, "Socrates is married to Xanthippe" predicates of Socrates the property of being married to Xanthippe. Philosophers disagree about whether every predicate expresses a genuine property. Some authors use the term *predicate* to include not only monadic predicates, but also relations (two-place predicates and higher).

PREDICTIONISM   According to predictionism, when a result is derived from a theory, the result confirms the theory only if it is a novel prediction; a theory is not confirmed by "old evidence," no matter how much of it the theory is able to explain. A less extreme form of predictionism is accommodationism—the view that while the explanation or accommodation of old evidence can have some confirmatory value, the prediction of new evidence always has more. (See *explanationism, novel prediction, problem of old evidence*.)

PRIOR PROBABILITY   The prior probability of a theory is sometimes taken to be the theory's probability when it was first proposed, before any evidence for or against it has been considered. But often the prior probability of a theory means the theory's probability prior to the consideration of a particular piece of evidence. Thus, in the first sense, a theory has only one prior probability, but in the second sense, prior probabilities change as evidence accumulates. (See *Bayesian confirmation theory*.)

PROBLEM OF OLD EVIDENCE   The problem of old evidence, for Bayesian confirmation theorists, is that of explaining why we should not set equal to 1 the probability of any piece of evidence that we already know to be true. The problem is that, on the Bayesian analysis of confirmation, this would prevent that evidence from confirming any theory. (See *Bayesian confirmation theory*.)

REALISM   One can be a realist about many different kinds of thing: numbers, possible worlds, universals, minds, physical objects, quarks, fields, and so on. To call a philosopher a realist requires a specification of what the philosopher is a realist about. Usually, there is an intended contrast with those who deny that the entities in question are real. In the philosophy of science, realists are aligned against instrumentalists, phenomenalists, conventionalists, fictionalists, and others of that ilk. (See *scientific realism*.)

REDUCTIO AD ABSURDUM (REDUCTIO ARGUMENT) A form of valid deductive reasoning in which a claim is refuted by deducing from it a false, contradictory, or "absurd" conclusion. (If the deduction requires additional premises, those statements must be known to be true for the reductio reasoning to be sound.) In mathematics and logic, reductio reasoning is often called the *method of indirect proof.* (See *deduction.*)

REDUCTIONISM Philosophers of science use the term *reductionism* in at least three different senses. Either it refers to the research strategy of trying to understand complex systems by studying their components and the relations among them; or it refers to the claim that a particular scientific theory (such as classical genetics) has been reduced to another theory (such as molecular biology) according to a particular model of intertheoretic reduction (usually the standard model advocated by Ernest Nagel and others); or it refers to the more general claim that scientific knowledge grows primarily by theory reduction (again, as judged by a model of intertheoretic reduction such as Nagel's). The first two senses are quite distinct. The success or failure of reductionism (as a research strategy) has no direct bearing on whether intertheoretic reduction has occurred or is likely to occur: reductionism may be a fruitful research strategy without resulting in theory reduction, and theory reductions may occur without being the result of a reductionistic research strategy. As discussed in chapter 8, several philosophers of biology (notably David Hull and Philip Kitcher) have opposed reductionism in the second sense by arguing that classical genetics cannot be deduced from molecular biology (as Nagel's model of reduction requires), and many philosophers of science (especially Paul Feyerabend and T. S. Kuhn) have criticized the claim that scientific knowledge grows primarily by theory reduction. (See *homogeneous reduction, inhomogeneous reduction.*)

REGULARITY THEORY OF CAUSATION A theory of causation, also called the constant conjunction theory, associated most notably with David Hume. In its simplest form, the theory asserts that an event of some kind *F* causes an event of another kind *G* if and only if *F*-events are, as a matter of empirical fact, always followed by *G*-events. This view denies that there is any kind of causal necessity or power in objects in virtue of which one event or object produces another. The regularity theory of causation may thus be regarded as a conjunction of two claims: (1) causal connections are a species of lawlike connections, and (2) laws of nature merely express regularities of behavior. (See *regularity theory of laws.*)

REGULARITY THEORY OF LAWS A theory of laws of nature, associated with David Hume, which holds that a law of nature is merely a contingently true universal generalization describing how all objects of a certain kind always behave. This view denies that laws express any kind of causal necessity. The regularity theory of laws is part of the regularity theory of causation. An important problem for the regularity theory of laws is its apparent inability to distinguish between genuine laws of nature (e.g., all bodies free from any net external force remain at rest or in uniform motion in a straight line) and merely accidentally true generalizations (e.g., all U.S. presidents are men). (See *counterfactual, epistemic regularity theory of laws, regularity theory of causation, universal generalization.*)

RELATIVISM Moral relativists deny that there are any objective ethical standards for evaluating human conduct independently of particular societies, their beliefs, and institutions. In the philosophy of science, epistemological relativists deny that there are any objective methodological standards for evaluating theories independently of particular scientific research traditions and their associated belief systems. Probably the best known epistemological relativist in the philosophy of science is T. S. Kuhn, who insists that all decisions about theories are relative to particular paradigms.

RELEVANCE CRITERION OF CONFIRMATION   A central feature of the Bayesian approach to confirmation, the relevance criterion holds that a piece of evidence, *E*, confirms a hypothesis, *H*, if and only if *E* raises the probability of *H*. Or, in other words, *E* confirms *H* if and only if the posterior probability of *H* on *E* is greater than the prior probability of *H*. (See *posterior probability, prior probability*.)

RETICULATIONAL MODEL OF SCIENTIFIC RATIONALITY   A model of scientific rationality proposed by Larry Laudan. The distinctive features of the reticulational model are its nonlinear conception of justification and its piecemeal approach to the evaluation of values and rules at every level. (See *hierarchical model of scientific rationality*.)

RIGID DESIGNATOR   An expression is a rigid designator if and only if it has the same reference in every possible world in which it has any reference at all. According to Saul Kripke and Hilary Putnam, proper names (e.g., Marie Curie) and natural-kind terms (e.g., radium) are rigid designators, but definite descriptions (e.g., the discoverer of radium) are not.

SATISFACTION CRITERION OF CONFIRMATION   The satisfaction criterion is a qualitative theory of confirmation proposed by Carl Hempel. The technical details are complicated, but the basic idea is simple: a hypothesis is confirmed by an observation report if the individuals mentioned in the report satisfy the hypothesis. Satisfaction includes, but is not limited to, positive instances. A hypothesis is satisfied by (and hence confirmed by) an observation report if the report entails a description of the world that would be true if the hypothesis were true and if the world consisted only of the individuals that are mentioned, essentially, in the report. Thus, for example, the hypothesis that all ravens are black is confirmed by the report that Adam is a black raven. Now consider the observation report that Beth is not a raven (without saying anything about her color). This report entails that either Beth is not a raven or Beth is black. And that would be a correct description of the world if Beth were the only individual in the world and if it were true that all ravens are black. (This is because the universal generalization "all ravens are black" is logically equivalent to "for all things, either that thing is not a raven or that thing is black.") Thus, according to Hempel's satisfaction criterion, the hypothesis that all ravens are black is confirmed (to some positive degree, however small) not only by the observation of black ravens but also by the observation of nonravens and by the observation of nonblack things (and by the observation of things that are both nonblack and nonravens). Hempel's satisfaction criterion has several controversial features: it requires adopting the regularity theory that laws are nothing more than true universal generalizations, and its application is restricted to hypotheses that can be expressed using the observation terms that appear in observation reports. Many philosophers also find highly counterintuitive Hempel's judgment that the observation of a white shoe, say, can confirm that all ravens are black. (See *regularity theory of laws*.)

SCIENTIFIC REALISM   Scientific realism has several dimensions: metaphysical, epistemological, and methodological. While there is no single, monolithic version of scientific realism that all scientific realists accept, scientific realism is generally taken to be the doctrine that the world studied by science exists and has the properties it does independently of our beliefs, perceptions, and theorizing; that the aim of science is to describe and explain that world, including those many aspects of it that are not directly observable; that, other things being equal, scientific theories are to be interpreted literally; that to accept a theory is to believe that what it says about the world is true, and that by continually replacing current scientific theories with better ones, science makes objective progress and its theories get closer to the truth.

SCIENTIFIC RESEARCH PROGRAMME   A term coined by Imre Lakatos, usually spelt in the English fashion. According to Lakatos, scientific research programmes have three components: a hard core, a protective belt, and a positive heuristic.

SCIENTIFIC REVOLUTION   Scientific revolutions are important scientific advances, such as those associated with Copernicus, Newton, Darwin, and Einstein, in which fundamental concepts and theories were replaced by radical new ones, dramatically changing the course of science. People talked about scientific revolutions long before T. S. Kuhn, but it is Kuhn's views about the nature of those revolutions that has made the term popular. According to Kuhn, scientific revolutions are noncumulative and the new paradigm is incommensurable with the one it replaces. (See *incommensurability, normal science, paradigm.*)

SEMANTICS   The study of the meaning of signs and their relation to what they signify. Semanticists are especially interested in language. They offer theories about how words and sentences mean what they do and what makes sentences true. Meaning, truth, sense, reference, synonymy, and analyticity are all semantic concepts. (See *syntax.*)

SENSE   Following Gottlob Frege, sense is usually contrasted with reference. For example, the terms *Epsom salts* and *magnesium sulfate* differ as to sense but refer to the same chemical compound. Frege thought that all referring expressions have a sense and that it is in virtue of this sense that they refer. More recent philosophers (such as Saul Kripke) disagree, holding that some proper names refer even though they lack a sense. Sense is often referred to as meaning or intension. (See *extension, intension, rigid designator.*)

SENSE DATA   *Sense data*, the plural of *sense datum* (from the Latin *datum*, meaning "given"), are what we are immediately aware of when we perceive something—or think we perceive something—through our senses (e.g., colored shapes, smells, noises). David Hume called them *impressions.* Sense data are supposed to be the raw content of experience, before it has been conceptualized or described in words. Although in the past empiricists regarded sense data as the basis for all human knowledge, the psychological reality of sense data is now widely doubted, even by empiricists.

SOUNDNESS (OF AN ARGUMENT)   An argument is sound if and only if it is valid and all its premises are true. Necessarily, therefore, all sound arguments must have true conclusions. The problem is to figure out which arguments are sound.

STRONG PROGRAMME   The strong programme is a movement in the sociology of science that professes to make the study of science "scientific" by tracing the psychological and sociological causes of scientific beliefs and decisions, especially decisions to accept or reject theories. Particularly controversial among philosophers of science is the strong programme's insistence that all scientific beliefs, whether true or false, rational or irrational, should be explained in the same sort of way in terms of social and cultural factors. The phrase "strong programme" was coined by David Bloor, one of the founders of the Science Studies Unit at Edinburgh University, and many of its proponents are sociologists and historians of science working in the unit.

SUFFICIENT CONDITION If X is a sufficient condition for Y, then anything that is X must also be Y. Having atomic number 88 is sufficient for an atom to be radium; a sufficient condition for the production of alpha particles is the presence of radium.

SYMMETRY OBJECTION A blanket term for a variety of alleged counterexamples to the thesis of structural identity. (See *thesis of structural identity*.)

SYMMETRY THESIS See *thesis of structural identity*.

SYNTAX The syntax of a language is its grammar. In logic, the syntax of a formal system includes rules for constructing well-formed formulas and rules for the derivation of theorems from the axioms of the system. Unlike semantics, syntax is concerned solely with rules for arranging signs, especially words and sentences, independent of what they signify or mean. (See *semantics*.)

SYNTHETIC STATEMENTS Synthetic statements are all the ones that are not analytic. On modern accounts of analyticity, this entails that the truth (or falsity) of synthetic statements is not solely a consequence of the words and symbols used to express them. (See *analytic statements*.)

TAUTOLOGY Any statement that is true solely in virtue of its logical form, e.g., if tigers are herbivorous, then tigers are herbivorous; either xenon combines with phosphorus or it does not. Sometimes the term *tautology* is used more broadly to refer to analytic statements that are true by definition, e.g., no herbivores eat flesh. (See *analytic statements*.)

THEORY Opinions differ as to what a scientific theory is and how it should be analyzed. The traditional view, and still the view of most philosophers of science is that a theory is a set of statements or propositions. For example, Newton's theory of gravity consists of a statement of Newton's law of gravitational attraction together with a few other statements essential to Newton's theory. A newer view, the semantic view advocated by Bas van Fraassen, Ronald Giere, and others, is that a theory is a definition or a set of models. In any case, to call something a theory is not to denigrate it or to imply any unfavorable contrast with facts. A theory becomes a fact (or, at least, is believed to be a fact) when it is well confirmed and established. (See *hypothesis*.)

THEORY-LADENNESS According to some philosophers of science (such as Paul Feyerabend, T. S. Kuhn, and N. R. Hanson), observations are theory laden. What exactly this means is controversial, and whether it is true and what it implies has been hotly debated. One fairly uncontroversial reading of the theory-ladenness thesis is that observation reports, if they are to be evidentially relevant to a theory, must be expressed in the vocabulary of that theory rather than in some theory-neutral observation language. Some philosophers have inferred from this that because observation is theory laden it cannot be appealed to as a neutral ground on which to judge among competing theories. Less tendentiously, it is sometimes claimed, under the rubric of the theory-ladenness-of-observation thesis, that observation and experiment alone cannot rationally compel scientists to accept and reject theories.

THESIS OF STRUCTURAL IDENTITY The thesis of structural identity (sometimes called the symmetry thesis) is the claim that prediction and explanation are formally identical, consisting in the same type of argument. The thesis comprises two claims: every adequate explanation is potentially a prediction, and every adequate prediction is potentially an explanation. Carl Hempel has been the foremost defender of the thesis of structural identity, since it is a corollary of his deductive-nomological (D-N) model of explanation, but he admits that there is some doubt about whether all adequate predictions are potentially explanations.

UNDERDETERMINATION THESIS The thesis, associated with Pierre Duhem and W. V. Quine, that neither the truth nor the falsity of any scientific theory is determined by evidence. From this it is often concluded (by T. S. Kuhn, David Bloor, and others) that the decision to accept or reject scientific theories must depend on sociological factors. Larry Laudan usefully distinguishes between deductive and ampliative versions of the underdetermination thesis, arguing that, while the former is harmlessly true, the latter is perniciously false. (See *ambiguity of falsification, holism, strong programme.*)

UNIVERSAL GENERALIZATION A universal generalization is any statement of the form "All As are B." In first-order predicate logic, this is written as $(x)(Ax \supset Bx)$, where "$\supset$" stands for the material conditional of truth-functional logic, and the formula is read as "for all things x, if x has property A then x has property B."

UNIVERSALS Universals are properties that are can be predicated of any number of particulars. The traditional, Platonist account of universals is that they are real, abstract entities. They exist independently of particulars and are what general terms such as *yellow, umbrella,* and *cucumber* refer to. According to the Platonist, yellow things are yellow only because they are all related to the universal yellow. This view is unpopular with empiricists but has gained new currency with the advent of the universals theory of laws.

UNIVERSALS THEORY OF LAWS Because of the many difficulties afflicting the regularity theory, Fred Dretske, D. M. Armstrong, Michael Tooley, and others have proposed a necessitarian analysis of laws of nature. According to this analysis, laws are not universal generalizations but singular statements of fact about relations between universals—hence the name *universals theory* for this new treatment of laws.

VACUOUS LAWS Vacuous laws are not really laws at all: they are universal generalizations that are true solely because their antecedents are never satisfied. For example, since there are no mountains made of rhodium, it is true that all rhodium mountains are radioactive and also true that none of them are. The simple, Humean, regularity theory of laws insists that laws of nature are nothing more than true universal generalizations. Thus the regularity theorist has the problem of explaining why we do not accept vacuous laws as genuine. (See *noninstantial laws, regularity theory of laws.*)

VALIDITY (OF AN ARGUMENT) See *deduction.*

VALUE-NEUTRALITY THESIS The thesis that scientific decisions about theories should be governed exclusively by cognitive values. Cognitive (or epistemic) values are those values that are linked, directly or indirectly, to the aims of science as a knowledge-seeking enterprise, especially the goal of discovering interesting truths and rejecting error. Cognitive values include predictive power, explanatory scope, and (some would argue) simplicity.

VERIFIABILITY PRINCIPLE (OR CRITERION) OF MEANING   As originally formulated by the logical positivists, the verifiability principle asserts that the meaning of any contingent statement is given by the observation statements needed to verify it conclusively. Observation statements are assumed to be directly verifiable by the experiences they purportedly describe. Unverifiable assertions are declared to be meaningless (or, at least, to lack any cognitive meaning). Criticisms by Karl Popper and others soon led to the abandonment of verifiability as a criterion of meaning in favor of weaker notions such as confirmability and testability. These, too, are controversial insofar as they are intended as explications of meaning.

VERISIMILITUDE   The term used by Karl Popper and others to connote the objective truth content of a theory: the higher a theory's degree of verisimilitude, the closer it is to the truth. Attempts to define verisimilitude in terms of the ratio of the number of a theory's true consequences to the total number of the theory's consequences have run into enormous difficulties.

BIBLIOGRAPHY

Chapter 1 | Science and Pseudoscience

BAMFORD, GREG. "Popper and His Commentators on the Discovery of Neptune: A Close Shave for the Law of Gravitation?" *Studies in History and Philosophy of Science* 27 (1996): 207–32.

COLLINS, HARRY, and TREVOR PINCH. *The Golem: What Everyone Should Know about Science.* Cambridge: Cambridge University Press, 1993.

DENNETT, DANIEL C., and ALVIN PLANTINGA, *Science and Religion: Are They Compatible?* Oxford: Oxford University Press, 2011.

EARMAN, JOHN, and CLARK GLYMOUR. "Relativity and Eclipses: The British Eclipse Expeditions of 1919 and Their Predecessors." *Historical Studies in the Physical Sciences* 11 (1980): 49–85.

GARDNER, MARTIN. *Fads and Fallacies in the Name of Science.* New York: Dover, 1957.

———. *Science: Good, Bad and Bogus.* Buffalo, N.Y.: Prometheus Books, 1981.

———. *The New Age: Notes of a Fringe Watcher.* Buffalo, N.Y.: Prometheus Books, 1991.

———. *On the Wild Side.* Buffalo, N.Y.: Prometheus Books, 1992.

GRIM, PATRICK, ed. *Philosophy of Science and the Occult.* Albany, N.Y.: SUNY Press, 1982.

GROSSER, MORTON. *The Discovery of Neptune.* Cambridge, Mass.: Harvard University Press, 1962.

GROVE, JACK W. *In Defence of Science.* Toronto: University of Toronto Press, 1989.

HALSTEAD, BEVERLY. "Popper: Good Philosophy, Bad Science?" *New Scientist* (17 July 1980): 21517.

KENNEFLICK, DANIEL. "Testing Relativity from the 1919 Eclipse—A Question of Bias," *Physics Today* 62:3 (2009): 37–42.

KITCHER, PHILIP. *Abusing Science: The Case against Creationism.* Cambridge, Mass.: MIT Press, 1982.

LAKATOS, IMRE. "Falsification and the Methodology of Scientific Research Programmes." In *Criticism and the Growth of Science.* Ed. Imre Lakatos and Alan Musgrave. Cambridge: Cambridge University Press, 1970.

LAKATOS, IMRE, and ALAN MUSGRAVE, eds. *Criticism and the Growth of Knowledge.* Cambridge: Cambridge University Press, 1970.

LAUDAN, LARRY. "More on Creationism." *Science, Technology, and Human Values* 8 (Winter 1983): 36–38.

———. "The Demise of the Demarcation Problem." In *Physics, Philosophy, and Psychoanalysis.* Ed. R. S. Cohen and L. Laudan. Dordrecht, Netherlands: D. Reidel, 1983.

LOSEE, JOHN. *Theories on the Scrap Heap: Scientists and Philosophers on the Falsification, Rejection, and Replacement of Theories.* Pittsburgh: University of Pittsburgh Press, 2005.

MAYO, DEBORAH. "Novel Evidence and Severe Tests." *Philosophy of Science* 58 (1991): 523–52.

———. "Ducks, Rabbits, and Normal Science: Recasting the Kuhn's-eye View of Popper's Demarcation of Science." *British Journal for the Philosophy of Science* 47 (1996): 271–90.

PLANTINGA, ALVIN. *Where the Conflict Really Lies: Science, Religion, and Naturalism.* New York: Oxford University Press, 2011.

POPPER, KARL R. *The Logic of Scientific Discovery.* New York: Basic Books, 1959.

———. "Normal Science and Its Dangers." In *Criticism and the Growth of Science.* Ed. Imre Lakatos and Alan Musgrave. Cambridge: Cambridge University Press, 1970.

———. "Autobiography of Karl Popper." In *The Philosophy of Karl Popper.* Vol. 1. Ed. P. A. Schilpp. La Salle, Ill.: Open Court, 1974.

———. "Natural Selection and the Emergence of Mind." *Dialectica* 32 (1978): 339–55.

———. "Letter on Evolution." *New Scientist* (21 August 1980): 611.

———. *Unended Quest: An Intellectual Autobiography.* London: Routledge, 1992.

QUINN, PHILIP L. "The Philosopher of Science as Expert Witness." In *Science and Reality.* Ed. J. T. Cushing, C. F. Delaney, and G. M. Gutting. Notre Dame, Ind.: University of Notre Dame Press, 1984.

———. "Creationism, Methodology, and Politics." In *But Is It Science?* Ed. Michael Ruse. Buffalo, N.Y.: Prometheus Books, 1988.

RUSE, MICHAEL. *Darwinism Defended: A Guide to the Evolution Controversies.* Reading, Pa.: Addison-Wesley, 1982.

———. "The Academic as Expert Witness." *Science, Technology, and HumanValues* 11 (Spring 1986): 6873.

RUSE, MICHAEL, ed. *But Is It Science?* Buffalo, N.Y.: Prometheus Books, 1988.

RUSE, MICHAEL, and ROBERT T. PENNOCK, eds. *But Is It Science?* 2d ed. Amherst, N.Y.: Prometheus Books, 2009.

SCHICK, THEODORE, Jr., and LEWIS VAUGHN. *How to Think about Weird Things: Critical Thinking for a New Age.* 6th ed. New York: McGraw-Hill, 2011.

SOBER, ELLIOTT. *The Nature of Selection.* 2d ed. Chicago: University of Chicago Press, 1993.

———. *Philosophy of Biology.* 2d ed. Boulder, Colo.: Westview Press, 2000.

———. *Did Darwin Write the Origin Backwards? Philosophical Essays on Darwin's Theory.* Amherst, N.Y.: Prometheus Books, 2011.

———. "Evolution without Naturalism." In *Oxford Studies in Philosophy of Religion.* Ed. J. Kvanvig. Vol. 3. Oxford: Oxford University Press, 2011.

THAGARD, PAUL. "Pseudoscience." Chap. 9, *Computational Philosophy of Science.* Cambridge, Mass.: MIT Press, 1988.

THOMPSON, R. PAUL. "Is Sociobiology a Pseudoscience?" In *PSA 1980*. Vol. 1. Ed. P. D. Asquith and R. N. Giere. East Lansing, Mich.: Philosophy of Science Association, 1980.

Chapter 2 | Rationality, Objectivity, and Values in Science

AUDI, ROBERT. "Scientific Objectivity and the Evaluation of Hypotheses." In *The Philosophy of Logical Mechanism*. Ed. M. H. Salmon. Dordrecht, Netherlands: Kluwer Academic Publishers, 1990.

BIRD, ALEXANDER, *Thomas Kuhn*. Princeton, N.J.: Princeton University Press, 2000.

———. "What Is Scientific Progress?" *Noûs* 41 (2007): 64–89.

BROWN, HAROLD I. *Perception, Theory, and Commitment: The New Philosophy of Science*. Chicago: University of Chicago Press, 1977.

———. "Incommensurability." *Inquiry* 26 (1983): 3–29.

———. "Response to Siegel." *Synthese* 56 (1983): 91–105.

CEDARBAUM, DANIEL GOLDMAN. "Paradigms." *Studies in History and Philosophy of Science* 14 (1983): 173–213.

CURD, MARTIN. "Kuhn, Scientific Revolutions, and the Copernican Revolution." *Nature and System* 6 (1984): 1–14.

DOPPELT, GERALD. "Kuhn's Epistemological Relativism: An Interpretation and Defense." *Inquiry* 21 (1978): 33–86.

———. "A Reply to Siegel on Kuhnian Relativism." *Inquiry* 23 (1980): 117–23.

———. "Relativism and the Reticulational Model of Scientific Rationality." *Synthese* 69 (1986): 225–52.

GLYMOUR, CLARK. *Theory and Evidence*. Princeton, N.J.: Princeton University Press, 1980.

GUTTING, GARY, ed. *Paradigms and Revolutions*. Notre Dame, Ind.: University of Notre Dame Press, 1980.

HARDING, SANDRA. *Whose Science? Whose Knowledge? Thinking from Women's Lives*. Ithaca, N.Y.: Cornell University Press, 1991.

HARDING, SANDRA, and JEAN F. O'BARR, eds. *Sex and Scientific Inquiry*. Chicago: University of Chicago Press, 1987.

HORWICH, PAUL, ed. *World Changes: Thomas Kuhn and the Nature of Science*. Cambridge, Mass.: MIT Press, 1993.

HOYNINGEN-HUENE, PAUL. *Reconstructing Scientific Revolutions: Thomas Kuhn's Philosophy of Science*. Chicago: University of Chicago Press, 1993.

KINKAID, HAROLD, JOHN DUPRÉ, and ALISON WYLIE, eds. *Value-Free Science: Ideal or Illusion?* Oxford: Oxford University Press, 2007.

KORDIG, CARL R. *The Justification of Scientific Change*. Dordrecht, Netherlands: D. Reidel, 1971.

———. "The Theory-Ladenness of Observation." *Review of Metaphysics* 24 (1971): 448–84.

KOURANY, JANET A. "The Nonhistorical Basis of Kuhn's Theory of Science." *Nature and System* 1 (1979): 46–59.

KUHN, THOMAS S. *The Essential Tension.* Chicago: University of Chicago Press, 1977.

———. "Rationality and Theory Choice." *Journal of Philosophy* 80 (1983): 563–70.

———. *The Structure of Scientific Revolutions,* 3d ed. Chicago: University of Chicago Press, 1996.

LACEY, HUGH. *Is Science Value-Free? Values and Scientific Understanding.* New York: Routledge, 1999.

LAKATOS, IMRE, and ALAN MUSGRAVE, eds. *Criticism and the Growth of Knowledge.* Cambridge: Cambridge University Press, 1970.

LAUDAN, LARRY. *Science and Values: The Aims of Science and Their Role in Scientific Debate.* Berkeley: University of California Press, 1984.

———. "Relativism, Naturalism, and Reticulation." *Synthese* 71 (1987): 221–34.

———. *Science and Relativism: Some Key Controversies in the Philosophy of Science.* Chicago: University of Chicago Press, 1990.

LLOYD, ELISABETH A. *The Case of the Female Orgasm: Bias in the Science of Evolution.* Cambridge, Mass.: Harvard University Press, 2005.

LONGINO, HELEN E. "Beyond 'Bad Science': Skeptical Reflections on the Value-Freedom of Scientific Inquiry." *Science, Technology, and Human Values* 8 (1983): 7–17.

———. "Can There Be a Feminist Science?" *Hypatia* 2 (1987): 51–64.

———. *Science as Social Knowledge: Values and Objectivity in Scientific Inquiry.* Princeton, N.J.: Princeton University Press, 1990.

———. "In Search of Feminist Epistemology." *The Monist* 77 (1994): 427–85.

———. *The Fate of Knowledge.* Princeton, N.J.: Princeton University Press, 2002.

MACHAMER, PETER, and GEREON WALTERS, eds. *Science, Values, and Objectivity.* Pittsburgh: University of Pittsburgh Press, 2004.

MARTIN, JANE R. "Ideological Critiques and the Philosophy of Science." *Philosophy of Science* 56 (1989): 1–22.

McMULLIN, ERNAN. "Values in Science." In *PSA 1982.* Vol. 2. Ed. P. D. Asquith and T. Nickles. East Lansing, Mich.: Philosophy of Science Association, 1982.

MUSGRAVE, ALAN. "Kuhn's Second Thoughts." Review of *The Structure of Scientific Revolutions,* 2d ed., by Thomas S. Kuhn. *British Journal for the Philosophy of Science* 22 (1971): 287–97. Reprinted in Gary Gutting, ed., *Paradigms and Revolutions.* Notre Dame, Ind.: University of Notre Dame Press, 1980.

NICKLES, THOMAS, ed. *Thomas Kuhn.* Cambridge: Cambridge University Press, 2003.

PINNICK, CASSANDRA. "Feminist Epistemology: Implications for Philosophy of Science." *Philosophy of Science* 61 (1994): 646–57.

RICHARDSON, ROBERT C. "Biology and Ideology: The Interpenetration of Science and Values." *Philosophy of Science* 51 (1984): 396–420.

SALMON, WESLEY C. "Rationality and Objectivity in Science, *or* Tom Kuhn Meets Tom Bayes." In *Scientific Theories*. Ed. C. W. Savage. Vol. 14, *Minnesota Studies in the Philosophy of Science*. Minneapolis: University of Minnesota Press, 1990.

SANKEY, HOWARD. "Kuhn's Ontological Relativism." In *Issues and Images in the Philosophy of Science*. Ed. D. Ginev and R. S. Cohen. Dordrecht, Netherlands: Kluwer, 1997.

———. "Taxonomic Incommensurability." *International Studies in the Philosophy of Science* 12 (1998): 7–16.

SCHEFFLER, ISRAEL. *Science and Subjectivity*. Indianapolis, Ind.: Bobbs-Merrill, 1967.

SHAPERE, DUDLEY. Review of *The Structure of Scientific Revolutions* by Thomas S. Kuhn. *Philosophical Review* 73 (1964): 383–94. Reprinted in Gary Gutting, ed., *Paradigms and Revolutions*. Notre Dame, Ind.: University of Notre Dame Press, 1980.

———. "The Paradigm Concept." *Science* 172 (1971): 706–9.

SHIMONY, ABNER. "Comments on Two Epistemological Theses of Thomas Kuhn." In *Essays in Memory of Imre Lakatos*. Ed. R. S. Cohen et al. Dordrecht, Netherlands: D. Reidel, 1976. Reprinted in Abner Shimony, *Search for a Naturalistic World View*, Vol. 1, *Scientific Method and Epistemology*. Cambridge: Cambridge University Press, 1993.

SIEGEL, HARVEY. "Epistemological Relativism in Its Latest Form." *Inquiry* 23 (1980): 107–17.

———. "Brown on Epistemology and the New Philosophy of Science." *Synthese* 56 (1983): 61–89.

Chapter 3 | The Duhem-Quine Thesis
and Underdetermination

BARKER, STEPHEN F. "Logical Positivism and the Philosophy of Mathematics." In *The Legacy of Logical Positivism*. Ed. P. Achinstein and S. F. Barker. Baltimore, Md.: The Johns Hopkins University Press, 1969.

BLOOR, DAVID. "The Strengths of the Strong Programme." *Philosophy of the Social Sciences* 11 (1981): 199–213.

———. "Durkheim and Mauss Revisited: Classification and the Sociology of Knowledge." *Studies in History and Philosophy of Science* 13 (1982): 267–97.

———. "A Reply to Gerd Buchdahl." *Studies in History and Philosophy of Science* 13 (1982): 305–11.

———. *Knowledge and Social Imagery*, 2d ed. Chicago: University of Chicago Press, 1991.

BROWN, JAMES R., ed. *Scientific Rationality: The Sociological Turn*. Dordrecht, Netherlands: D. Reidel, 1984.

BUCHDAHL, GERD. "Editorial Response to David Bloor." *Studies in History and Philosophy of Science* 13 (1982): 299–304.

DUHEM, PIERRE. *The Aim and Structure of Physical Theory*. Trans. Philip P. Wiener. Princeton, N.J.: Princeton University Press, 1954.

FINE, ARTHUR. "Science Made Up: Constructivist Sociology of Scientific Knowledge." In *The Disunity of Science*. Ed. P. Galison and D. J. Stump. Stanford, Calif.: Stanford University Press, 1996.

FRANKLIN, ALLAN. *No Easy Answers: Science and the Pursuit of Knowledge*. Pittsburgh: University of Pittsburgh Press, 2005.

GLYMOUR, CLARK. *Theory and Evidence*. Princeton, N.J.: Princeton University Press, 1980.

GRICE, H. P., and P. F. Strawson. "In Defense of a Dogma." *Philosophical Review* 65 (1956): 141–58.

HAACK, SUSAN. *Philosophy of Logics*. Cambridge: Cambridge University Press, 1978.

———. *Deviant Logic, Fuzzy Logic*. Chicago: University of Chicago Press, 1996.

HARDING, SANDRA, ed., *Can Theories Be Refuted? Essays on the Duhem-Quine Thesis*. Dordrecht, Netherlands: D. Reidel, 1976.

HEMPEL, CARL G. "Geometry and Empirical Science." *American Mathematical Monthly* 52 (1945): 7–17. Reprinted in *The Philosophy of Carl G.Hempel*. Ed. James H. Fetzer. New York: Oxford University Press, 2001.

———. "On the Nature of Mathematical Truth." *American Mathematical Monthly* 52 (1945): 543–56. Reprinted in *The Philosophy of Carl G.Hempel*. Ed. James H. Fetzer. New York: Oxford University Press, 2001.

HESSE, MARY B. "The Strong Thesis of Sociology of Science." In *Revolutions and Reconstructions in the Philosophy of Science*. Bloomington: Indiana University Press, 1980.

HOEFER, CARL, and ALEXANDER ROSENBERG. "Empirical Equivalence, Underdetermination, and Systems of the World." *Philosophy of Science* 61 (1994): 592–607.

HORWICH, PAUL. *Probability and Evidence*. Cambridge: Cambridge University Press, 1982.

HOWSON, COLIN, and PETER URBACH. *Scientific Reasoning: The Bayesian Approach*, 3d ed. La Salle, Ill.: Open Court, 2006.

KUKLA, ANDRÉ. "Laudan, Leplin, Empirical Equivalence, and Underdetermination." *Analysis* 53 (1993): 1–7.

LAKATOS, IMRE. "History of Science and Its Rational Reconstructions." In *Method and Appraisal in the Physical Sciences*. Ed. C. Howson. Cambridge: Cambridge University Press, 1976.

LAUDAN, LARRY. "Grünbaum on 'The Duhemian Argument'." *Philosophy of Science* 32 (1965): 295–99.

———. "The Pseudo-Science of Science?" *Philosophy of the Social Sciences* 11 (1981): 173–98.

———. "More on Bloor." *Philosophy of the Social Sciences* 12 (1982): 71–74.

———. "The Epistemic, the Cognitive, and the Social." In *Science, Values, and Objectivity*. Ed. P. Machamer and G. Walters. Pittsburgh: University of Pittsburgh Press, 2004.

LAUDAN, LARRY, and JARRETT LEPLIN. "Empirical Equivalence and Underdetermination." *Journal of Philosophy* 88 (1991): 449–72.

LEPLIN, JARRETT, and LARRY LAUDAN. "Determination Undeterred: Reply to Kukla." *Analysis* 53 (1993): 8–16.

MCCARTHY, THOMAS. "Scientific Rationality and the 'Strong Program' in the Sociology of Knowledge." In *Construction and Constraint.* Ed. E. McMullin. Notre Dame, Ind.: University of Notre Dame Press, 1988.

NAGEL, ERNEST. *The Structure of Science.* London: Routledge and Kegan Paul, 1961.

PUTNAM, HILARY. "Is Logical Empirical?" in *Boston Studies in the Philosophy of Science.* Vol. 5. Ed. R. S. Cohen and M. W. Wartofsky. Dordrecht, Netherlands: D. Reidel, 1968. Revised version reprinted as "The Logic of Quantum Mechanics." In *Mathematics, Matter, and Method, Philosophical Papers.* Vol. 1. Cambridge: Cambridge University Press, 1975.

———. "'Two Dogmas' Revisited." In *Contemporary Aspects of Philosophy.* Ed. G. Ryle. Boston, Mass.: Oriel Press, 1976. Reprinted in *Realism and Reason, Philosophical Papers.* Vol. 3. Cambridge: Cambridge University Press, 1983.

———. "There Is at Least One A Priori Truth." *Erkenntnis* 13 (1978): 153–70. Reprinted in *Realism and Reason, Philosophical Papers.* Vol. 3. Cambridge: Cambridge University Press, 1983.

QUINE, WILLARD V. *Philosophy of Logic.* Englewood Cliffs, N.J.: Prentice-Hall, 1970.

———. "On Empirically Equivalent Systems of the World." *Erkenntnis* 9 (1975): 313–28.

———. "Two Dogmas in Retrospect." *Canadian Journal of Philosophy* 21 (1991): 265–74.

QUINN, PHILIP. "What Duhem Really Meant." In *Methodological and Historical Essays in the Natural and Social Sciences.* Ed. R. S. Cohen and M. W. Wartofsky. Dordrecht, Netherlands: D. Reidel, 1973.

ROSENKRANTZ, ROGER D. "Does the Philosophy of Induction Rest on a Mistake?" *Journal of Philosophy* 79 (1982): 78–97.

SLEZAK, PETER. "A Second Look at David Bloor's Knowledge and Social Imagery." *Philosophy of the Social Sciences* 24 (1994): 336–61.

———. "The Social Construction of Social Constructionism." *Inquiry* 37 (1994): 139–57.

SOBER, ELLIOTT. "Quine's Two Dogmas." *Proceedings of the Aristotelian Society* (Supplementary Volume) 74 (2000): 237–80.

Chapter 4 | Induction, Prediction, and Evidence

ACHINSTEIN, PETER. "The Method of Hypothesis: What Is It Supposed to Do, and Can It Do It?" In *Observation, Experiment, and Hypothesis in Modern Physical Science.* Ed. Peter Achinstein and Owen Hannaway. Cambridge, Mass.: MIT Press, 1985.

———. *Particles and Waves.* New York: Oxford University Press, 1991.

———. "Are Empirical Evidence Claims A Priori?" *British Journal for the Philosophy of Science* 46 (1995): 447–73.

———. *The Book of Evidence*. New York: Oxford University Press, 2001.

ACHINSTEIN, PETER, ed. *The Concept of Evidence*. Oxford: Oxford University Press, 1983.

ACHINSTEIN, PETER, and OWEN HANNAWAY, eds. *Observation, Experiment, and Hypothesis in Modern Physical Science*. Cambridge, Mass.: MIT Press, 1985.

BARKER, STEPHEN F., and PETER ACHINSTEIN. "On the New Riddle of Induction." *Philosophical Review* 69 (1960): 511–22.

BARNES, ERIC CHRISTIAN. *The Paradox of Predictivism*. Cambridge: Cambridge University Press, 2008.

BRUSH, STEPHEN G. "Prediction and Theory Evaluation: The Case of Light Bending." *Science* 246 (1989): 1124–29.

———. "Dynamics of Theory Change: The Role of Predictions." In *PSA 1994*. Vol. 2. Ed. D. Hull, M. Forbes, and R. M. Burian. East Lansing, Mich.: Philosophy of Science Association, 1994.

COLLINS, ROBIN. "Against the Epistemic Value of Prediction over Accommodation." *Noûs* 28 (1994): 210–24.

GARDNER, MICHAEL R. "Predicting Novel Facts," *British Journal for the Philosophy of Science* 33 (1982): 1–15.

GINGERICH, OWEN. "Neptune, Velikovsky, and the Name of the Game." *Scientific American* 275 (September 1996): 181–83.

GLYMOUR, CLARK. "Hypothetico-Deductivism Is Hopeless." *Philosophy of Science* 47 (1980): 322–25.

———. *Theory and Evidence*. Princeton, N.J.: Princeton University Press, 1980.

GODFREY-SMITH, PETER. "Goodman's Problem and Scientific Methodology." *Journal of Philosophy* 100 (2003): 573–90.

GOODING, DAVID, TREVOR PINCH, and SIMON SCHAFFER, eds. *The Uses of Experiment*. Cambridge: Cambridge University Press, 1989.

GOODMAN, NELSON. "Positionality and Pictures." *Philosophical Review* 69 (1960): 523–25.

———. *Fact, Fiction, and Forecast*, 4th ed. Cambridge, Mass.: Harvard University Press, 1983.

GRÜNBAUM, ADOLF. "Is the Method of Bold Conjectures and Attempted Refutations Justifiably the Method of Science?" *British Journal for the Philosophy of Science* 27 (1976): 105–36.

HAACK, SUSAN. "The Justification of Induction." *Mind* 85 (1976): 112–19.

HARKER, DAVID. "Accommodation and Prediction: The Case of the Persistent Head." *British Journal for the Philosophy of Science* 57 (2006): 309–21.

HEMPEL, CARL G. "Studies in the Logic of Confirmation," *Mind* 54 (1945): 1–26, 97–121. Reprinted in *Aspects of Scientific Explanation*. New York: Macmillan, 1965.

———. *Philosophy of Natural Science*. Englewood Cliffs, N.J.: Prentice-Hall, 1966.

HORWICH, PAUL. *Probability and Evidence.* Cambridge: Cambridge University Press, 1982.

JACKSON, FRANK. "Grue*," *Journal of Philosophy* 72 (1975): 113–131. Reprinted in D. Stalker, ed., *Grue! The New Riddle of Induction.* Chicago: Open Court, 1994.

JEFFREY, RICHARD C. "Probability and the Art of Judgment," in *Observation, Experiment, and Hypothesis in Modern Physical Science.* Ed. Peter Achinstein and Owen Hannaway. Cambridge, Mass.: MIT Press, 1985.

LAUDAN, LARRY. "Why Was the Logic of Discovery Abandoned?" In *Scientific Discovery, Logic, and Rationality.* Ed. Thomas Nickles. Dordrecht, Netherlands: D. Reidel, 1980.

LEPLIN, JARRETT. "The Bearing of Discovery on Justification." *Canadian Journal of Philosophy* 17 (1987): 805–14.

LIPTON, PETER. "Prediction and Prejudice." *International Studies in the Philosophy of Science* 4 (1990): 51–65. Reprinted in *Inference to the Best Explanation,* 2d ed. New York: Routledge, 2004.

———. *Inference to the Best Explanation,* 2d ed. New York: Routledge, 2004.

MAYO, DEBORAH G. "Novel Evidence and Severe Tests." *Philosophy of Science* 58 (1991): 523–52.

MCINTYRE, LEE C. "Accommodation, Prediction, and Confirmation." *Perspectives on Science* 9 (2001): 308–23.

MELLOR, D. H. "The Warrant of Induction." In *Matters of Metaphysics.* Cambridge: University of Cambridge Press, 1991.

MILLER, DAVID. "Can Science Do Without Induction?" In *Applications of Inductive Logic.* Ed. L. J. Cohen and M. B. Hesse. Oxford: Clarendon Press, 1980.

MUSGRAVE, ALAN. "Logical versus Historical Theories of Confirmation." *British Journal for the Philosophy of Science* 25 (1974): 1–23.

NAGEL, ERNEST. "Carnap's Theory of Induction." In *The Philosophy of Rudolf Carnap.* Ed. P. A. Schilpp. LaSalle, Ill.: Open Court, 1963.

NICKLES, THOMAS. "Justification and Experiment." In *The Uses of Experiment.* Ed. David Gooding, Trevor Pinch, and Simon Schaffer. Cambridge: Cambridge University Press, 1989.

NICKLES, THOMAS, ed. *Scientific Discovery, Logic, and Rationality.* Dordrecht, Netherlands: D. Reidel, 1980.

POPPER, KARL R. *The Logic of Scientific Discovery.* New York: Basic Books, 1959.

———. *Conjectures and Refutations.* New York: Harper and Row, 1963.

REICHENBACH, HANS. *Experience and Prediction.* Chicago: University of Chicago Press, 1938.

SALMON, WESLEY C. *The Foundations of Scientific Inference.* Pittsburgh, Pa.: University of Pittsburgh Press, 1967.

———. "Unfinished Business: The Problem of Induction." *Philosophical Studies* 33 (1978): 1–19.

SCERRI, ERIC R. "Prediction and Accommodation: The Acceptance of Mendeleev's Periodic System." Chap. 5, *The Periodic Table: Its Story and Its Significance.* New York: Oxford University Press, 2007.

SKYRMS, BRIAN. *Choice and Chance: An Introduction to Inductive Logic*, 3d ed. Belmont, Calif.: Wadsworth, 1986.

SNYDER, LAURA J. "Is Evidence Historical?" In *Scientific Methods: Conceptual and Historical Problems*. Ed. P. Achinstein and L. J. Snyder. Malabar, Fla.: Krieger Publishing Company, 1994.

STALKER, DOUGLAS, ed. *Grue! The New Riddle of Induction*. Chicago: Open Court, 1994.

SWINBURNE, RICHARD G., ed. *The Justification of Induction*. Oxford: Oxford University Press, 1974.

VAN CLEVE, JAMES. "Reliability, Justification, and the Problem of Induction." In *Midwest Studies in Philosophy*. Vol. 9. Ed. P. French, T. Uehling, Jr., and H. Wettstein. Minneapolis: University of Minnesota Press, 1984.

WATKINS, JOHN. *Science and Scepticism*. Princeton, N.J.: Princeton University Press, 1984.

WORRALL, JOHN. "Fresnel, Poisson, and the White Spot: The Role of Successful Predictions in the Acceptance of Scientific Theories." In *The Uses of Experiment*. Ed. David Gooding, Trevor Pinch, and Simon Schaffer. Cambridge: Cambridge University Press, 1989.

———. "Theory-Confirmation and History." In *Rationality and Reality: Conversations with Alan Musgrave*. Ed. Colin Cheyne and John Worrall. Dordrecht, Netherlands: Springer, 2006.

WORRALL, JOHN, and ERIC R. SCERRI. "Prediction and the Periodic Table." *Studies in History and Philosophy of Science* 32 (2001): 407–52.

Chapter 5 | Confirmation and Relevance: Bayesian Approaches

BERGER, JAMES O., and DONALD A. BERRY. "Statistical Analysis and the Illusion of Objectivity." *American Scientist* 76 (1988): 159–65.

DORLING, JON. "Bayesian Personalism, the Methodology of Scientific Research Programmes, and Duhem's Problem." *Studies in History and Philosophy of Science* 10 (1979): 177–87.

EARMAN, JOHN. *Bayes or Bust? A Critical Examination of Bayesian Confirmation Theory*. Cambridge, Mass.: MIT Press, 1992.

EARMAN, JOHN, ed. *Testing Scientific Theories*. Vol. 10, *Minnesota Studies in the Philosophy of Science*. Minneapolis: University of Minnesota Press, 1983.

FORSTER, MALCOLM. "Bayes and Bust: Simplicity as a Problem for a Probabilist's Approach to Confirmation." *British Journal for the Philosophy of Science* 46 (1995): 399–424.

GARBER, DANIEL. "Old Evidence and Logical Omniscience in Bayesian Confirmation Theory." In *Testing Scientific Theories*. Ed. John Earman. Vol. 10, *Minnesota Studies in the Philosophy of Science*. Minneapolis: University of Minnesota Press, 1983.

GILLIES, DONALD. "Bayesianism Versus Falsificationism." *Ratio (New Series)* 3 (1990): 82–98.

GLYMOUR, CLARK. *Theory and Evidence.* Princeton, N.J.: Princeton University Press, 1980.

GOOD, I. J. "The White Shoe Is a Red Herring." *British Journal for the Philosophy of Science* 17 (1967): 322.

HEMPEL, CARL G. "Studies in the Logic of Confirmation." *Mind* 54 (1945): 1–26, 97–121. Reprinted in *Aspects of Scientific Explanation.* New York: Macmillan, 1965.

HESSE, MARY B. *The Structure of Scientific Inference.* Berkeley: University of California Press, 1974.

HORWICH, PAUL. *Probability and Evidence.* Cambridge: Cambridge University Press, 1982.

HOWSON, COLIN. "Theories of Probability." *British Journal for the Philosophy of Science* 46 (1995): 1–32.

HOWSON, COLIN, and PETER URBACH. *Scientific Reasoning: The Bayesian Approach,* 3d ed. La Salle, Ill.: Open Court, 2006.

JEFFREY, RICHARD C. "Bayesianism with a Human Face." In *Testing Scientific Theories.* Ed. John Earman. Vol. 10, *Minnesota Studies in the Philosophy of Science.* Minneapolis: University of Minnesota Press, 1983.

———. *The Logic of Decision,* 2d ed. Chicago: University of Chicago Press, 1983.

KYBURG, HENRY E., JR. *Probability and Inductive Logic.* London: Collier-Macmillan, 1970.

———. "Subjective Probability: Criticisms, Reflections, and Problems." *Journal of Philosophical Logic* 7 (1978): 157–80.

MACKIE, J. L. "The Paradox of Confirmation." *British Journal for the Philosophy of Science* 13 (1963): 265–77.

———. "The Relevance Criterion of Confirmation." *British Journal for the Philosophy of Science* 20 (1969): 27–40.

MAYO, DEBORAH G. *Error and the Growth of Experimental Knowledge.* Chicago: University of Chicago Press, 1996.

MAYO, DEBORAH G., and ARIS SPANOS, eds. *Error and Inference: Recent Exchanges on Experimental Reasoning, Reliability, and the Objectivity and Rationality of Science.* New York: Cambridge University Press, 2010.

NAGEL, ERNEST. "Carnap's Theory of Induction." In *The Philosophy of Rudolf Carnap.* Ed. P. A. Schlipp. LaSalle, Ill.: Open Court, 1963.

NELSON, DAVID E. "Confirmation, Explanation, and Logical Strength." *British Journal for the Philosophy of Science* 47 (1996): 399–413.

NIINILUOTO, ILKKA. "Novel Facts and Bayesianism." *British Journal for the Philosophy of Science* 34 (1983): 375–79.

PLANTINGA, ALVIN. "Bayesian Coherentism and Warrant" and "Bayesian Coherentism and Rationality." Chaps. 6 and 7, *Warrant: The Current Debate.* Oxford: University of Oxford Press, 1993.

——. "Epistemic Probability: Some Current Views" and "Epistemic Conditional Probability: The Sober Truth." Chaps. 8 and 9, *Warrant and Proper Function*. Oxford: University of Oxford Press, 1993.

ROSENKRANTZ, ROGER D. *Inference, Method and Decision.* Dordrecht, Netherlands: D. Reidel, 1977.

——. "Does the Philosophy of Induction Rest on a Mistake?" *Journal of Philosophy* 79 (1982): 78–97.

SALMON, WESLEY C. *The Foundations of Scientific Inference.* Pittsburgh, Pa.: University of Pittsburgh Press, 1967.

——. "Confirmation." *Scientific American* 228 (May 1973): 75–83.

——. "Confirmation and Relevance." In *Induction, Probability, and Confirmation.* Vol. 6, *Minnesota Studies in the Philosophy of Science.* Ed. G. Maxwell and R. M. Anderson, Jr. Minneapolis: University of Minnesota Press, 1975.

——. "The Appraisal of Theories: Kuhn Meets Bayes." In *PSA 1990.* Vol. 2. Ed. A. Fine, M. Forbes, and L. Wessels. East Lansing, Mich.: Philosophy of Science Association, 1991.

——. "Reflections of a Bashful Bayesian: A Reply to Peter Lipton." In *Explanation: Theoretical Approaches and Applications.* Ed. Giora Hon and Sam S. Rakover. Dordrecht, Netherlands: Kluwer, 2001.

STEEL, DANIEL. "Bayesianism and the Value of Diverse Evidence." *Philosophy of Science* 63 (1996): 666–74.

SWINBURNE, RICHARD G. "The Paradoxes of Confirmation—A Survey." *American Philosophical Quarterly* 8 (1971): 318–30.

——. *An Introduction to Confirmation Theory.* London: Methuen and Co., 1973.

VINEBERG, SUSAN. "Eliminative Induction and Bayesian Confirmation Theory." *Canadian Journal of Philosophy* 26 (1996): 257–66.

WAYNE, ANDREW. "Bayesianism and Diverse Evidence." *Philosophy of Science* 62 (1995): 111–21.

WORRALL, JOHN. "Kuhn, Bayes, and 'Theory Choice': How Revolutionary Is Kuhn's Account of Theoretical Change?" In *After Popper, Kuhn, and Feyerabend: Recent Issues in Theories of Scientific Method.* Ed. Robert Nola and Howard Sankey. Dordrecht, Netherlands: Kluwer, 2001.

Chapter 6 | Models of Explanation

ACHINSTEIN, PETER. "Can There Be a Model of Explanation?" *Theory and Decision* 13 (1981): 201–27. Reprinted in Peter Achinstein, *Evidence, Explanation, and Realism: Essays in the Philosophy of Science.* New York: Oxford University Press, 2010.

BARNES, ERIC. "Explanatory Unification and the Problem of Asymmetry." *Philosophy of Science* 59 (1992): 558–71.

BAUMGARTNER, MICHAEL. "Interdefining Causation and Intervention." *Dialectica* 63 (2009): 175–94.

COFFA, J. ALBERTO. "Hempel's Ambiguity." *Synthese* 28 (1974): 141–63.

FRIEDMAN, MICHAEL. "Explanation and Scientific Understanding." *Journal of Philosophy* 71 (1974): 5–19.

———. "Theoretical Explanation." In *Reduction, Time and Reality*. Ed. Richard Healey. Cambridge: Cambridge University Press, 1981.

HEMPEL, CARL G. "The Function of General Laws in History." *Journal of Philosophy* 39 (1942): 35–42. Reprinted in *Aspects of Scientific Explanation*. New York: Free Press, 1965.

———. "Deductive-Nomological vs. Statistical Explanation." In *Scientific Explanation, Space, and Time*. Vol. 3, *Minnesota Studies in the Philosophy of Science*. Ed. H. Feigl and G. Maxwell. Minneapolis: University of Minnesota Press, 1962.

———. "Aspects of Scientific Explanation." In *Aspects of Scientific Explanation*. New York: Free Press, 1965.

———. "Maximal Specificity and Lawlikeness in Probabilistic Explanation." *Philosophy of Science* 35 (1968): 116–33.

HEMPEL, CARL G., and PAUL OPPENHEIM. "Studies in the Logic of Explanation." *Philosophy of Science* 15 (1948): 567–79. Reprinted in *Aspects of Scientific Explanation*. New York: Free Press, 1965.

HUMPHREYS, PAUL. "Why Propensities Cannot Be Probabilities." *Philosophical Review* 94 (1985): 557–70.

JONES, TODD. "How the Unification Theory of Explanation Escapes Asymmetry Problems." *Erkenntnis* 43 (1995): 229–40.

KITCHER, PHILIP. "Explanation, Conjunction, and Unification." *Journal of Philosophy* 73 (1976): 207–12.

———. "Explanatory Unification and the Causal Structure of the World." In *Scientific Explanation*. Ed. Philip Kitcher and Wesley C. Salmon. Vol. 13, *Minnesota Studies in the Philosophy of Science*. Minneapolis: University of Minnesota Press, 1989.

KITCHER, PHILIP, and WESLEY C. SALMON, eds. *Scientific Explanation*. Vol. 13, *Minnesota Studies in the Philosophy of Science*. Minneapolis: University of Minnesota Press, 1989.

MEIXNER, J. "Homogeneity and Explanatory Depth." *Philosophy of Science* 46 (1979): 366–81.

MORRISON, MARGARET. *Unifying Scientific Theories*. Cambridge: Cambridge University Press, 2000.

MYRVOLD, WAYNE C. "A Bayesian Account of the Virtue of Unification." *Philosophy of Science* 70 (2003): 399–423.

PITT, JOSEPH C., ed. *Theories of Explanation*. Oxford: Oxford University Press, 1988.

PSILLOS, STATHIS. *Causation and Explanation*. Montreal: McGill-Queen's University Press, 2002.

———. "Causal Explanation and Manipulation." In *Rethinking Explanation*. Ed. Johannes Persson and Petri Ylikoski. Dordrecht, Netherlands: Springer, 2007.

RUBEN, DAVID-HILLEL. *Explaining Explanation*. New York: Routledge, 1990.

RUBEN, DAVID-HILLEL, ed. *Explanation*. Oxford: Oxford University Press, 1993.

SALMON, WESLEY C. *Statistical Explanation and Statistical Relevance*. Pittsburgh, Pa. University of Pittsburgh Press, 1971.

——. *Scientific Explanation and the Causal Structure of the World*. Princeton, N.J.: Princeton University Press, 1984.

——. "Four Decades of Scientific Explanation." In *Scientific Explanation*. Ed. Philip Kitcher and Wesley C. Salmon. Vol. 13, *Minnesota Studies in the Philosophy of Science*. Minneapolis: University of Minnesota Press, 1989. Also published as *Four Decades of Scientific Explanation*. Minneapolis: University of Minnesota Press, 1989.

——. *Causality and Explanation*. New York: Oxford University Press, 1998.

STREVENS, MICHAEL. "Review of Woodward, *Making Things Happen*." *Philosophy and Phenomenological Research* 74 (2007): 233–49.

——. "Comments on Woodward, *Making Things Happen*." *Philosophy and Phenomenological Research* 77 (2008): 171–92.

WOODWARD, JAMES. "Explanation and Invariance in the Special Sciences." *British Journal for the Philosophy of Science* 51 (2000): 197–254.

——. *Making Things Happen. A Theory of Causal Explanation*. New York: Oxford University Press, 2003.

——. "Causation with a Human Face." In *Causation, Physics, and the Constitution of Reality: Russell's Republic Revisited*. Ed. Huw Price and Richard Corry. Oxford: Oxford University Press, 2007.

WOODWARD, JAMES, and CHRISTOPHER HITCHCOCK. "Explanatory Generalizations, Part I: A Counterfactual Account." *Noûs* 37 (2003): 1–24.

Chapter 7 | Laws of Nature

ARMSTRONG, D. M. "Laws of Nature as Relations between Universals, and as Universals." *Philosophical Topics* 13 (1982): 7–24.

——. *What Is a Law of Nature?* Cambridge: Cambridge University Press, 1983.

BEEBEE, HELEN. "The Non-Governing Conception of Laws of Nature." *Philosophy and Phenomenological Research* 61 (2000): 571–94. Reprinted in *Readings on Laws of Nature*. Ed. John W. Carroll. Pittsburgh, Pa.: University of Pittsburgh Press, 2004.

BIRD, ALEXANDER. "On Whether Some Laws Are Necessary." *Analysis* 62 (2002): 257–70.

——. *Nature's Metaphysics: Law and Properties*. Oxford: Oxford University Press, 2007.

BRAITHWAITE, RICHARD B. *Scientific Explanation*. Cambridge: Cambridge University Press, 1953.

CARROLL, JOHN W. *Laws of Nature*. Cambridge: Cambridge University Press, 1994.

CARROLL, JOHN W., ed. *Readings on Laws of Nature*. Pittsburgh, Pa.: University of Pittsburgh Press, 2004.

CARTWRIGHT, NANCY. *How the Laws of Physics Lie*. Oxford: Clarendon Press, 1983.

——. *Nature's Capacities and Their Measurement*. Oxford: Clarendon Press, 1989.

——. *The Dappled World: A Study of the Boundaries of Science*. Cambridge: Cambridge University Press, 1999.

CHALMERS, ALAN. "So the Laws of Physics Needn't Lie." *Australasian Journal of Philosophy* 71 (1993): 196–205.

——. "Cartwright on Fundamental Laws: A Response to Clarke." *Australasian Journal of Philosophy* 74 (1996): 150–52.

CLARKE, STEVE. "The Lies Remain the Same: A Reply to Chalmers." *Australasian Journal of Philosophy* 73 (1995): 152–55.

CREARY, LEWIS. "Causal Explanation and the Reality of Natural Component Forces." *Pacific Philosophical Quarterly* 62 (1981): 148–57.

DONNELLAN, KEITH S. "Kripke and Putnam on Natural Kind Terms." In *Knowledge and Mind*. Ed. C. Ginet and S. Shoemaker. Oxford: Oxford University Press, 1983.

FEYNMAN, RICHARD P. *The Character of Physical Law.* Cambridge, Mass.: MIT Press, 1965.

GIERE, RONALD N. "The Skeptical Perspective: Science without Laws of Nature." In *Laws of Nature: Essays on the Philosophical, Scientific, and Historical Dimensions*. Ed. Friedel Weinert. New York: Walter de Gruyter, 1995.

HEMPEL, CARL G. "Provisos: A Problem Concerning the Inferential Function of Scientific Theories." In *The Limitations of Deductivism*. Ed. A. Grünbaum and W. C. Salmon. Berkeley: University of California Press, 1988.

KLINE, A. DAVID, and CARL A. MATHESON. "How the Laws of Physics Don't Even Fib." In *PSA 1986*. Vol. 1. Ed. A. Fine and P. Machamer. East Lansing, Mich.: Philosophy of Science Association, 1986.

KNEALE, WILLIAM C. "Natural Laws and Contrary-to-Fact Conditionals." *Analysis* 10 (1950): 121–25. Reprinted in Tom L. Beauchamp, ed., *Philosophical Problems of Causation*. Belmont, Calif.: Dickenson, 1974.

KRIPKE, SAUL A. *Naming and Necessity.* Cambridge, Mass.: Harvard University Press, 1972.

LANGE, MARC. *Laws and Lawmakers: Science, Metaphysics, and the Laws of Nature.* New York: Oxford University Press, 2009.

LAYMON, RONALD. "Cartwright and the Lying Laws of Physics." *Journal of Philosophy* 86 (1989): 353–72.

LEWIS, DAVID. "New Work for the Theory of Universals." *Australasian Journal of Philosophy* 61 (1983): 343–77.

MELLOR, D. H. "Natural Kinds." *British Journal for the Philosophy of Science* 28 (1977): 299–312.

MITCHELL, SANDRA D. "Dimensions of Scientific Law." *Philosophy of Science* 67 (2000): 242–65.

MUMFORD, STEPHEN. *Laws in Nature.* London: Routledge, 2004.

MUSGRAVE, ALAN. "Wittgensteinian Instrumentalism." *Theoria* 47 (1981): 65–105.

NAGEL, ERNEST. "The Logical Character of Scientific Laws." Chap. 4, *The Structure of Science*. New York: Harcourt, Brace and World, 1961.

SCHWARTZ, STEPHEN P., ed. *Naming, Necessity, and Natural Kinds.* Ithaca, N.Y.: Cornell University Press, 1977.

SKLAR, LAWRENCE. "Dappled Theories in a Uniform World." *Philosophy of Science* 70 (2003): 424–41.

SUCHTING, W. A. "Regularity and Law." In *Boston Studies in the Philosophy of Science*. Vol. 14. Ed. R. S. Cohen and M. W. Wartofsky. Dordrecht, Netherlands: D. Reidel, 1974.

SWARTZ, NORMAN. *The Concept of Physical Law*. Cambridge: Cambridge University Press, 1985.

———. "A Neo-Humean Perspective: Laws as Regularities." In *Laws of Nature: Essays on the Philosophical, Scientific, and Historical Dimensions*. Ed. Friedel Weinert. New York: Walter de Gruyter, 1995.

SWOYER, CHRISTOPHER. "The Nature of Natural Laws." *Australasian Journal of Philosophy* 60 (1982): 203–23.

TOOLEY, MICHAEL. "The Nature of Laws." *Canadian Journal of Philosophy* 7 (1977): 667–98.

TWEEDALE, MARTIN. "Universals and Laws of Nature." *Philosophical Topics* 13 (1982): 25–44.

URBACH, PETER. "What Is a Law of Nature? A Humean Answer." *British Journal for the Philosophy of Science* 39 (1988): 193–210.

VAN FRAASSEN, BAS C. *Laws and Symmetry*. Oxford: Clarendon Press, 1989.

WEINERT, FRIEDEL, ed. *Laws of Nature: Essays on the Philosophical, Scientific, and Historical Dimensions*. New York: Walter de Gruyter, 1995.

WOODWARD, JAMES. "Realism about Laws." *Erkenntnis* 36 (1992): 181–218.

———. "Law and Explanation in Biology: Invariance is the Kind of Stability that Matters." *Philosophy of Science* 68 (2001): 1–20.

Chapter 8 | Intertheoretic Reduction

BATTERMAN, ROBERT W. *The Devil in the Details: Asymptotic Reasoning in Explanation, Reduction, and Emergence*. New York: Oxford University Press, 2002.

BEDAU, MARK A., and PAUL HUMPHREYS, eds. *Emergence: Contemporary Readings in Philosophy and Science*. Cambridge, Mass.: MIT Press, 2008.

FEYERABEND, PAUL. "Explanation, Reduction, and Empiricism." In *Scientific Explanation, Space, and Time*. Ed. H. Feigl and G. Maxwell. Vol. 3, *Minnesota Studies in the Philosophy of Science*. Minneapolis: University of Minnesota Press, 1962.

———. "Problems of Empiricism." In *Beyond the Edge of Certainty*. Ed. R. G. Colodny. Englewood Cliffs, N.J.: Prentice-Hall, 1965.

———. *Against Method*. London: New Left Books, 1975.

FODOR, JERRY. "Special Sciences: Still Autonomous After All These Years." *Philosophical Perspectives* 11 (1997): 149–63.

HEMPEL, CARL G. "Reduction: Ontological and Linguistic Issues." *Philosophy, Science, and Method: Essays in Honor of Ernest Nagel*. Ed. S. Morgenbesser et al. New York: St. Martin's Press, 1969.

HULL, DAVID. "Reduction in Genetics—Biology or Philosophy?" *Philosophy of Science* 39 (1972): 491–99.

———. "The Reduction of Mendelian to Molecular Genetics." Chap. 1, *Philosophy of Biological Science*. Englewood Cliffs, N.J.: Prentice-Hall, 1974.

KITCHER, PHILIP. "Explanatory Unification and the Causal Structure of the World." In *Scientific Explanation*. Ed. P. Kitcher and W. C. Salmon. Vol. 13, *Minnesota Studies in the Philosophy of Science*. Minneapolis: University of Minnesota Press, 1989.

———. "Darwin's Achievement." Chap. 2, *The Advancement of Science*. Oxford: Oxford University Press, 1993.

KUHN, THOMAS. "The Nature and Necessity of Scientific Revolutions." Chap. 9, *The Structure of Scientific Revolutions*, 3d ed. Chicago: University of Chicago Press, 1996.

LOEWER, BARRY. "Why Is There Anything Except Physics?" *Synthese* 170 (2009): 217–33.

NAGEL, ERNEST. "The Reduction of Theories." Chap. 11, *The Structure of Science*. New York: Harcourt, Brace and World, 1961.

NICKLES, THOMAS. "Two Concepts of Intertheoretic Reduction." *Journal of Philosophy* 70 (1975): 181–201.

ROSENBERG, ALEXANDER. "Whatever Happened to Reductionism, and Why?" and "Reductionism and Explanation in Molecular Biology." Chaps. 2 and 3, *Instrumental Biology: or the Disunity of Science*. Chicago: University of Chicago Press, 1994.

———. *Darwinian Reductionism or How to Stop Worrying and Love Molecular Biology*. Chicago: University of Chicago Press, 2007.

SCERRI, ERIC. "Reduction and Emergence in Chemistry—Two Recent Approaches." *Philosophy of Science* 74 (2007): 920–31.

SCHAFFNER, KENNETH F. "The Watson-Crick Model and Reductionism." *British Journal for the Philosophy of Science* 20 (1969): 325–48.

———. "Reduction: The Cheshire Cat Problem and a Return to Roots." *Synthese* 151 (2006): 377–402.

SKLAR, LAWRENCE. "The Reduction of Thermodynamics to Statistical Mechanics." Chap. 9, *Physics and Chance: Philosophical Issues in the Foundations of Statistical Mechanics*. Cambridge: University of Cambridge Press, 1993.

SOBER, ELLIOTT. "The Multiple Realizability Argument Against Reductionism." *Philosophy of Science* 66 (1999): 542–64. Reprinted in *Conceptual Issues in Evolutionary Biology*. 3d ed. Cambridge, Mass.: MIT Press, 2006.

TOBIN, EMMA. "What Makes the Special Sciences So Special? Exploring Scientific Methodology in the Special Sciences." In *Noesis: Essays in the History and Philosophy of Science, Philosophy of Language, Epistemology, and Political Philosophy*. Ed. Stephen Rainey and Barbara Gabriella Renzi. Newcastle: Cambridge Scholars Press, 2005.

WATERS, C. KENNETH. "Why the Antireductionist Consensus Won't Survive the Case of Classical Mendelian Genetics." In *PSA 1990*. Vol. 1. Ed. A. Fine, M. Forbes, and L. Wessels. East Lansing, Mich.: Philosophy of Science Association, 1990.

———. "Genes Made Molecular." *Philosophy of Science* 61 (1994): 163–85.

WEISKOPF, DANIEL. "The Functional Unity of Special Science Kinds." *British Journal for the Philosophy of Science* 62 (2011): 233–58.

WIMSATT, WILLIAM. C. "Reductionism and Its Heuristics: Making Methodological Reductionism Honest." *Synthese* 151 (2006): 445–75.

Chapter 9 | Empiricism and Scientific Realism

ACHINSTEIN, PETER. "Is There a Valid Experimental Argument for Scientific Realism?" *Journal of Philosophy* 99 (2002): 470–95.

ACHINSTEIN, PETER, and STEPHEN F. BARKER, eds. *The Legacy of Logical Positivism*. Baltimore, Md.: The Johns Hopkins University Press, 1969.

ALSPECTOR-KELLY, MARC. "Should the Empiricist Be a Constructive Empiricist?" *Philosophy of Science* 68 (2001): 413–31.

ARONSON, JERROLD A. "Verisimilitude and Type Hierarchies." *Philosophical Topics* 18 (1990): 5–28.

CARRIER, MARTIN. "What Is Wrong with the Miracle Argument?" *Studies in History and Philosophy of Science* 22 (1991): 23–36.

CARTWRIGHT, NANCY. *How the Laws of Physics Lie.* Oxford: Clarendon Press, 1983.

CHEYNE, COLIN, and JOHN WORRALL, eds. *Rationality and Reality: Conversations with Alan Musgrave.* Dordrecht, Netherlands: Springer, 2006.

CHURCHLAND, PAUL M., and CLIFFORD A. HOOKER, eds. *Images of Science: Essays on Realism and Empiricism.* Chicago: University of Chicago Press, 1985.

CLARKE, STEVE. "Defensible Territory for Entity Realism." *British Journal for the Philosophy of Science* 52 (2001) 701–22.

CLENDINNEN, F. JOHN. "Realism and the Underdetermination of Theory." *Synthese* 81 (1989): 63–90.

CUSHING, JAMES T. *Quantum Mechanics: Historical Contingency and the Copenhagen Hegemony.* Chicago: University of Chicago Press, 1994.

FINE, ARTHUR. "And Not Anti-Realism Either." *Noûs* 18 (1984): 51–65. Reprinted in *The Shaky Game: Einstein, Realism, and the Quantum Theory.* 2d ed. Chicago: University of Chicago Press, 1996.

———. "Unnatural Attitudes: Realist and Instrumentalist Attachments to Science." *Mind* 95 (1986): 149–79.

———. "Piecemeal Realism." *Philosophical Studies* 61 (1991): 79–96.

———. *The Shaky Game: Einstein, Realism, and the Quantum Theory.* 2d ed. Chicago: University of Chicago Press, 1996.

———. "The Scientific Image Twenty Years Later." *Philosophical Studies* 106 (2001): 107–122.

FOSS, JEFF. "On Accepting van Fraassen's Image of Science." *Philosophy of Science* 51 (1984): 79–92.

FROST-ARNOLD, GREG. "The No-Miracles Argument for Realism: Inference to an Unacceptable Explanation." *Philosophy of Science* 77 (2010): 35–58.

GHINS, MICHEL. "Putnam's No-Miracle Argument: A Critique." In *Recent Themes in the Philosophy of Science.* Ed. S. P. Clarke and T. D. Lyons. Dordrecht, Netherlands: Kluwer, 2001.

HACKING, IAN. "Do We See through a Microscope?" *Pacific Philosophical Quarterly* 62 (1981): 305–22. Reprinted in *Images of Science: Essays on Realism and Empiricism.* Ed. Paul M. Churchland and Clifford A. Hooker. Chicago: University of Chicago Press, 1985.

———. *Representing and Intervening.* Cambridge: Cambridge University Press, 1983.

HARDIN, CLYDE L., and ALEXANDER ROSENBERG. "In Defense of Convergent Realism." *Philosophy of Science* 49 (1982): 604–15.

HARKER, DAVID. "Two Arguments for Scientific Realism Unified." *Studies in History and Philosophy of Science* 41 (2010): 192–202.

HAUSMAN, DANIEL M. "Constructive Empiricism Contested." *Pacific Philosophical Quarterly* 63 (1982): 21–28.

HEMPEL, CARL G. "The Theoretician's Dilemma." *Aspects of Scientific Explanation.* New York: Free Press, 1965.

HITCHCOCK, CHRISTOPHER R. "Causal Explanation and Scientific Realism." *Erkenntnis* 37 (1992): 151–78.

KUKLA, ANDRÉ. "Scientific Realism, Scientific Practice, and the Natural Ontological Attitude." *British Journal for the Philosophy of Science* 45 (1994): 955–75.

———. "The Two Antirealisms of Bas van Fraassen." *Studies in History and Philosophy of Science* 26 (1995): 431–54.

———. "The Theory-Observation Distinction." *Philosophical Review* 105 (1996): 173–230.

LANGE, MARC. "Baseball, Pessimistic Inductions and the Turnover Fallacy." *Analysis* 62 (2002): 281–85.

LAUDAN, LARRY. *Science and Hypothesis.* Dordrecht, Netherlands: D. Reidel, 1981.

LEPLIN, JARRETT, ed. *Scientific Realism.* Berkeley: University of California Press, 1984.

———. *A Novel Defense of Scientific Realism.* New York: Oxford University Press, 1997.

LEWIS, PETER J. "Why the Pessimistic Induction Is a Fallacy." *Synthese* 129 (2001): 371–80.

LIPTON, PETER. "Tracking Track Records." *Proceedings of the Aristotelian Society,* Supplementary Volume 74 (2000): 179–205.

LUGG, ANDREW. "Pierre Duhem's Conception of Natural Classification." *Synthese* 83 (1990): 409–20.

MAGNUS, P. D., and CRAIG CALLENDER. "Realist Ennui and the Base Rate Fallacy." *Philosophy of Science* 71 (2004): 320–38.

MCMULLIN, ERNAN. "Comment: Duhem's Middle Way." *Synthese* 83 (1990): 421–30.

———. "Comment: Selective Anti-Realism." *Philosophical Studies* 61 (1991): 97–108.

Monton, Bradley, ed. *Images of Empricism: Essays on Science and Stances, with a Reply from Bas C. van Fraassen*. New York: Oxford University Press, 2007.

Morrison, Margaret. "Theory, Intervention and Realism." *Synthese* 82 (1990): 1–22.

O'Connor, D. J. *The Correspondence Theory of Truth*. London: Hutchinson and Co., 1975.

Oddie, Graham. "Truthlikeness." In *The Encyclopedia of Philosophy Supplement*. Ed. D. M. Borchert. New York: Simon and Schuster Macmillan, 1996.

Pierson, Robert, and Richard Reiner. "Explanatory Warrant for Scientific Realism." *Synthese* 161 (2008) 271–82.

Psillos, Stathis. *Scientific Realism: How Science Tracks Truth*. New York: Routledge, 1999.

———. "Thinking about the Ultimate Argument for Realism." In *Rationality and Reality: Conversations with Alan Musgrave*. Ed. Colin Cheyne and John Worrall. Dordrecht, Netherlands: Springer, 2006. Reprinted in *Knowing the Structure of Nature*.

———. *Knowing the Structure of Nature: Essays on Realism and Explanation*. Houndmills: Palgrave Macmillan, 2009).

Putnam, Hilary. "The Meaning of Meaning." In *Language, Mind, and Knowledge*. Vol. 7, *Minnesota Studies in the Philosophy of Science*. Ed. Keith Gunderson. Minneapolis: University of Minnesota Press, 1975. Reprinted in *Mind, Language, and Reality: Philosophical Papers*. Vol. 2. Cambridge: Cambridge University Press, 1975.

Reiner, Richard, and Robert Pierson. "Hacking's Experimental Realism: An Untenable Middle Ground." *Philosophy of Science* 62 (1995): 60–69.

Salmon, Wesley C. *Scientific Explanation and the Causal Structure of the World*. Princeton, N.J.: Princeton University Press, 1984.

Scheffler, Israel. *The Anatomy of Inquiry*. New York: Alfred A. Knopf, 1963.

Seager, William. "Ground Truth and Virtual Reality: Hacking vs. van Fraassen." *Philosophy of Science* 62 (1995) 459–78.

———. "Beyond Theories: Cartwright and Hacking." In *Philosophy of Science: The Key Thinkers*. Ed. James Robert Brown. New York: Continuum Press, 2012.

Sellars, Wilfrid. *Science, Perception and Reality*. New York: Humanities Press, 1962.

———. "Is Scientific Realism Tenable?" In *PSA 1976*. Vol. 2. Ed. F. Suppe and P. D. Asquith. East Lansing, Mich.: Philosophy of Science Association, 1977.

Smart, J. J. C. *Between Science and Philosophy*. New York: Random House, 1968.

Stanford, Kyle. *Exceeding Our Grasp: Science, History, and the Problem of Unconceived Alternatives*. New York: Oxford University Press, 2006.

Suppe, Frederick. "The Search for Philosophical Understanding of Scientific Theories." In *The Structure of Scientific Theories*, 2d ed. Ed. F. Suppe. Urbana: University of Illinois Press, 1977.

Teller, Paul. "Whither Constructive Empiricism?" *Philosophical Studies* 106 (2001): 123–50.

TOBIN, EMMA. "Crosscutting Natural Kinds and the Hierarchy Thesis." In *The Semantics and Metaphysics of Natural Kinds*. Ed. Helen Beebee and Nigel Sabbarton-Leary. London: Routledge, 2010.

VAN FRAASSEN, BAS C. "Wilfrid Sellars on Scientific Realism." *Dialogue* 14 (1975): 606–16.

———. *The Scientific Image*. Oxford: Clarendon Press, 1980.

———. "Empiricism in Philosophy of Science." In *Images of Science: Essays on Realism and Empiricism*. Ed. Paul M. Churchland and Clifford A. Hooker. Chicago: University of Chicago Press, 1985.

———. *Laws and Symmetry*. Oxford: Clarendon Press, 1989.

———. "Constructive Empiricism Now." *Philosophical Studies* 106 (2001): 151–70.

———. *Scientific Representation: Paradoxes of Perspective*. Oxford: Clarendon Press, 2008.

PERMISSIONS ACKNOWLEDGMENTS

Donald Gillies: "The Duhem Thesis and the Quine Thesis," from Donald Gillies, *Philosophy of Science in the Twentieth Century* (Oxford: Blackwell Publishers, 1993), pp. 98-116. Reprinted by permission of Blackwell Publishers Ltd.

Nelson Goodman: Reprinted by permission of the publisher from "The New Riddle of Induction" in *Fact, Fiction, and Forecast* by Nelson Goodman, pp. 72-81, Cambridge, Mass.: Harvard University Press, Copyright © 1979, 1983 by Nelson Goodman.

Ian Hacking: "Experimentation and Scientific Realism" from *Philosophical Topics* 13 (1982): 154-172. Copyright © 1982 by the Board of Trustees of the University of Arkansas. Reprinted with the permission of The Permissions Company, Inc., on behalf of the University of Arkansas Press, www.uapress.com.

Carl G. Hempel: "Criteria of Confirmation and Acceptability," *Philosophy of Natural Science*, 1st Edition, © 1967. Reprinted by permission of Pearson Education, Inc., Upper Saddle River, NJ.

"Inductive-Statistical Explanation." Reprinted with the permission of Free Press, a Division of Simon & Schuster, Inc., from *Aspects of Scientific Explanation and Other Issues* by Carl G. Hempel, pp. 381-383, 394-403. Copyright © 1965 by The Free Press. All rights reserved.

"The Thesis of Structural Identity." Reprinted with the permission of Free Press, a Division of Simon & Schuster, Inc., from *Aspects of Scientific Explanation and Other Issues* by Carl G. Hempel, pp. 366-376. Copyright © 1965 by The Free Press. All rights reserved.

"Two Basic Types of Scientific Explanation." Excerpts from "Explanation in Science and in History," by Carl G. Hempel, from *Frontiers of Science and Philosophy*, edited by Robert G. Colodny, © 1962. Reprinted by permission of the University of Pittsburgh Press.

Paul Horwich: "Wittgensteinian Bayesianism," *Midwest Studies in Philosophy*, Vol. 18 (1993), eds. P. A. French, T. E. Uehling, Jr., and H. K. Wettstein. © 1993 by the University of Notre Dame Press. Reprinted with permission.

Colin Howson and Peter Urbach: "The Duhem Problem." Reprinted by permission of Open Court Publishing Company, a division of Carus Publishing Company, Chicago, IL, from *Scientific Reasoning: The Bayesian Approach* by Colin Howson and Peter Urbach, copyright © 2006 by Carus Publishing Company.

Philip Kitcher: "1953 and All That: A Tale of Two Sciences," from *Philosophical Review* 93 (1984): 335-373. Copyright © 1984 Cornell University. Reprinted by permission of the author and the publisher.

"Explanatory Unification," from *Philosophy of Science* 48 (1981): pp. 507-531. © 1981 by the Philosophy of Science Association. All rights reserved. Reprinted with permission of the University of Chicago Press.

Thomas S. Kuhn: "Logic of Discovery or Psychology of Research?," from Imre Lakatos and Alan Musgrave, eds., *Criticism and the Growth of Knowledge*, pp. 4-10. Copyright © 1970 Cambridge University Press. Reprinted with the permission of Cambridge University Press.

"The Nature and Necessity of Scientific Revolutions," from Thomas S. Kuhn, *The Structure of Scientific Revolutions, 2nd Edition* (Chicago: University of Chicago Press, 1970), pp. 92-110. © 1962 by the University of Chicago. All rights reserved. Reprinted with permission of the University of Chicago Press.

"Objectivity, Value Judgment, and Theory Choice," from Thomas S. Kuhn, *The Essential Tension: Selected Studies in Scientific Tradition and Change* (Chicago: University of Chicago Press, 1977), pp. 320-339. © 1977 by the University of Chicago. All rights reserved. Reprinted with permission of the University of Chicago Press.

Imre Lakatos: "Science and Pseudoscience," from John Worrall and Gregory Currie, eds., *The Methodology of Scientific Research Programmes: Philosophical Papers, Vol. 1* (Cambridge: Cambridge University Press, 1977), pp. 1-7. Copyright © 1978 Imre Lakatos Memorial Appeal Fund and the Estate of Imre Lakatos. Reprinted with the permission of Cambridge University Press.

Larry Laudan: "A Confutation of Convergent Realism," from *Philosophy of Science* 48 (1981): pp. 19-49. © 1981 by the Philosophy of Science Association. All rights reserved. Reprinted with permission of the University of Chicago Press.

"Demystifying Underdetermination," from C. Wade Savage, ed., *Scientific Theories, Vol. XIV, Minnesota Studies in the Philosophy of Science* (Minneapolis: University of Minnesota Press, 1990), pp. 267-297. © 1990 by the Regents of the University of Minnesota. Reprinted with permission of the publisher.

"Commentary: Science at the Bar—Causes for Concern," *Science, Technology, and Human Values* 7, No. 41 (Fall 1982), pp. 16–19, copyright © 1982 by SAGE Publications. Reprinted by permission of SAGE Publications.

"Kuhn's Critique of Methodology," *Science and Values: The Aims of Science and Their Role in Scientific Debate*, pp. 87-102. Copyright © 1986, The Regents of the University of California. Reprinted by permission of the publisher.

Peter Lipton: "Induction," from *Inference to the Best Explanation* (London: Routledge, 1991), pp. 6-22. © 1991 Peter Lipton. Reprinted by permission of the Estate of Peter Lipton.

Helen E. Longino: "Values and Objectivity" from *Science as Social Knowledge: Values and Objectivity in Scientific Inquiry*, pp. 62-82. © 1990 Princeton University Press. Reprinted by permission of Princeton University Press.

Grover Maxwell: "The Ontological Status of Theoretical Entities," from Herbert Feigl and Grover Maxwell, eds., *Scientific Explanation, Space, and Time, Vol. 3, Minnesota Studies in the Philosophy of Science* (Minneapolis: University of Minnesota Press, 1962), pp. 3-15. © 1962 by the University of Minnesota. Reprinted with permission of the publisher.

Deborah G. Mayo: "Thomas Kuhn Meets Thomas Bayes, Introductions by Wesley Salmon," from *Error and the Growth of Experimental Knowledge*, pp. 112-127. © 1996 by The University of Chicago. Reprinted with permission of the University of Chicago Press.

Ernan McMullin: "Rationality and Paradigm Change in Science," from Paul Horwich, ed., *World Changes: Thomas Kuhn and the Nature of Science* (Cambridge,

Mass.: MIT Press, 1993), 55-78. © 1993, Massachusetts Institute of Technology. Reprinted with permission.

D. H. Mellor: "Necessities and Universals in Natural Laws," from D. H. Mellor, ed. *Science, Belief, and Behaviour: Essays in Honor of R. B. Braithwaite* (Cambridge: Cambridge University Press, 1980), pp. 105-125. © 1980, Cambridge University Press. Reprinted by permission of the author.

Alan Musgrave: "NOA's Ark – Fine for Realism," from *Philosophical Quarterly* 29 (1989): pp. 383-398. © The Management Committee of the Philosophical Quarterly. Reprinted with permission of Blackwell Publishers Ltd.

"Realism versus Constructive Empiricism," from Paul Churchland and Clifford A. Hooker, eds., *Images of Science* (Chicago: University of Chicago Press, 1985), pp. 197-221. © 1985 by the University of Chicago. All rights reserved. Reprinted with permission of the University of Chicago Press.

Ernest Nagel: "Issues in the Logic of Reductive Explanations," *Telelogy Revisited* (New York: Columbia University Press, 1974), pp. 95-113. Reprinted by permission of the Estate of Ernest Nagel.

Kathleen Okruhlik: "Gender and the Biological Sciences," from *Biology and Society; Canadian Journal of Philosophy*, supplementary vol. 20 (1994): pp. 21-42. © 1994 University of Calgary Press. Reprinted with permission.

Karl Popper: "Science: Conjectures and Refutations," from Karl Popper, *Conjectures and Refutations* (London: Routledge and Kegan Paul, 1963), pp. 33-39. Reprinted with the permission of University of Klagenfurt/Karl Popper Library.

"The Problem of Induction," from Karl Popper, *The Logic of Scientific Discovery* (New York: Basic Books, 1959), pp. 27-34. Reprinted with the permission of University of Klagenfurt/Karl Popper Library.

Willard V. Quine: "Two Dogmas of Empiricism" reprinted by permission of the publisher from *From a Logical Point of View: Nine Logico-Philosophical Essays* by Willard Van Orman Quine, pp. 20-46, Cambridge, Mass.: Harvard University Press, Copyright © 1953, 1961, 1980 by the President and Fellows of Harvard College. Copyright © renewed 1981, 1989 by Willard Van Orman Quine.

Peter Railton: "A Deductive-Nomological Model of Probabilistic Explanation," from *Philosophy of Science* 45 (1978): pp. 206-226. © 1978 by the Philosophy of Science Association. All rights reserved. Reprinted with permission of the University of Chicago Press.

David B. Resnik: "Hacking's Experimental Realism," from *Canadian Journal of Philosophy* 24 (1994): pp. 395-412. © 1994 University of Calgary Press. Reprinted with permission.

Michael Ruse: "Creation Science Is Not Science," *Science Technology, and Human Values* 7, No. 40 (Summer 1982), pp. 72-78, copyright © 1982 by SAGE Publications. Reprinted by permission of SAGE Publications.

Juha T. Saatsi: "On the Pessimistic Induction and Two Fallacies," from *Philosophy of Science* 72 (2005): pp. 1088-1098. © 2005 by the Philosophy of Science Association. All rights reserved. Reprinted with permission of the University of Chicago Press.

Name Index

SUBJECT INDEX

Note: Entries marked with an asterisk (*) are defined in the Glossary